Essentials of Geomorphology

Essentials of Geomorphology is an introductory textbook covering the latest research on landforms, both on Earth as well as on planets and moons. This easy-to-read, non-quantitative textbook homes in on the knowledge of leading scientists from around the globe, and presents the practicality, applications, and necessity of geomorphology. Replete with beautiful color figures and photographs, it contains in-depth discussions on fluvial and glacial geomorphology while also covering topics such as planetary geomorphology, bio-geomorphology, Earth history and climate change, and periglacial systems. Descriptive, but also process-driven, it is intended for readers interested in physical landscapes, regardless of their previous background or level of training in geography or geology. To this end, it only includes the basic mathematics needed to understand the concepts presented.

Randall Schaetzl is a soil geomorphologist whose work spans the disciplines of geography, soils, geology, and Earth science. Many know him by virtue of his best-selling textbook *Soils: Genesis and Geomorphology* (2nd Edition, 2015, Cambridge University Press). His work on soils, loess, and glacial geology has earned him the title of University Distinguished Professor at Michigan State University. Both the Geological Society of America and the Soil Science Society of America have rewarded him with the distinction of Fellow status. That said, he derives his greatest joy from teaching, and especially, from being in the field with students.

Richard Marston is a University Distinguished Professor Emeritus in the Department of Geography and Geospatial Sciences at Kansas State University. He is a geomorphologist specializing in human impacts on landform stability, including the effects of mining, wildfires, deforestation and reforestation, grazing, agriculture, river regulations, and military maneuvers. He has been appointed a Fellow in the American Association of Geographers (2017), American Association for the Advancement of Science (2006), Geological Society of America (1997), The Explorers Club (1995), and Royal Geographical Society (1991). He has been awarded Distinguished Career Awards by the Geomorphology and Mountain Geography specialty groups of the American Association of Geographers. Kansas State University has established the Marston Earth Systems Science (MESS) Lab for physical geography research.

"A clear and engaging introduction to geomorphology that will appeal to both beginners and experienced Earth science readers."

Faran Ali, University of Northern British Columbia

"*Essentials of Geomorphology* represents a remarkable achievement. Written by a team of the world's leading specialists, the volume provides a balanced treatment of theory, application, and illustration for all major geomorphological subfields and process regimes. The result is both comprehensive and cohesive, approachable and erudite. The volume will be an essential resource for students, researchers, and professionals for years to come."

Mike Daniels, University of Denver

"*Essentials of Geomorphology* is a readable geomorphology textbook that does not sacrifice thoroughness. The authors skillfully construct conceptual bridges between key foundational concepts and advanced ideas that make the content applicable to a range of learners, with differing levels of expertise and disciplinary backgrounds. Engaging, high quality photographs of landscapes and landforms are consistently paired in the same figure with equally high-quality diagrams that help reveal and explain the processes behind the scenery and help train the 'geomorphologist's eye.' These are just a few of *Essentials'* many outstanding qualities that are sure to make it the 'go to' choice for anyone teaching geomorphology."

Lisa Davis, University of Alabama

"This is the perfect textbook to introduce geomorphology. It clearly, simply, and thoroughly illustrates and explains landform development. There are many photos and diagrams that show the landscapes and landforms being discussed, and the graphics clearly illustrate how those same landforms develop over time. It is accessible and readable, and it provides simple explanations for complex topics. What a great textbook!"

Donald A. Friend, Minnesota State University

"*Essentials of Geomorphology* is an accessible and visually rich volume that offers a comprehensive introduction to Earth surface processes. Authored by 22 leading geomorphologists, it supports diverse pedagogic approaches with clarity and rigor. Schaetzl and Marston have developed a compelling resource for survey or topically focused courses. Its 800+ color graphics enhance understanding, making it an invaluable resource for educators, students, and general readers alike."

Francis A. Galgano, Villanova University

"*Essentials of Geomorphology* is accessible, beautifully illustrated, and thorough – a first-choice textbook for geomorphology classes and personal reference. Schaetzl and Marston, both highly respected geomorphologists, and their outstanding, international team of contributing authors are clearly superb teachers, too. I applaud their engaging, direct writing style and the abundance of excellent illustrations."

Carol Harden, University of Tennessee

"The breadth and depth of the book distinguishes it from others, and it is the most comprehensive geomorphology text of the past 25 years. Topics span the discipline, and include chapters on Earth's internal influences on landforms as well as planetary geomorphology. Indeed, in addition to including stand-alone chapters on specific topics within geomorphology (e.g., coastal, fluvial, etc. ...) the book contains five chapters on structural and lithologic drivers to landscape evolution. This attention to integrating 'core' Earth science concepts within a modern process-driven framework assures that the text is equally useful to geomorphology education in both Geology and Physical Geography programs. The book is ideal for upper-division undergraduate courses in geomorphology, as well as focused graduate-level courses. The text provides a fresh perspective on a range of topics within geomorphology. It does not get bogged down in well trodden examples, but rather draws from an international range of case studies, including updates to classics. The assembled collection of expert authors ensures that each chapter represents the state-of-the-science. And the extensive editorial experience of Schaetzl and Marston provides consistency in presentation style and writing, essential to effective instruction."

Paul Hudson, Leiden University

"This pioneering textbook presents a sophisticated synthesis of traditional and contemporary geomorphological paradigms, emphasizing the dynamic processes that have sculpted our planet and beyond. It offers a thorough analysis of climate variability throughout Earth's geological history, underscores the critical role of biotic agents in shaping the surface, and investigates extraterrestrial landforms, providing profound insights into planetary geomorphology."

Daehyun Kim, Seoul National University

"*Essentials of Geomorphology* is an outstanding textbook, providing a truly comprehensive overview of the field of geomorphology, and the structure and presentation of the chapters make it accessible for students from an introductory level to professionals in the field. What stands out from this textbook compared to others is the inclusion of chapters dedicated to Earth's climatic history, feedbacks between landforms and plants and animals, numerous chapters on landforms in different geologic landscapes, and planetary geomorphology. The detailed figures clearly illustrate complex geomorphic processes, and the exhaustive glossary of key concepts and terminology at the end of the textbook makes this a one-stop resource for understanding geomorphic processes and how landscapes develop and change through time. The textbook is impressive in that it includes contributions from over 20 world-renowned experts in the field of geomorphology, yet each chapter is excellently written in the same clear narrative, making it engaging and easy to read from chapter to chapter. I am excited to adopt this textbook for my undergraduate and graduate courses on geomorphology."

Kory Konsoer, Louisiana State University

"I didn't think it was possible to write a geomorphology book that is both comprehensive and readable to the non-expert, but Schaetzl and Marston, along with many co-authors, have done just that. I see this as the standard text and reference book for many years."

Jeffrey A. Lee, Texas Tech University

"Kudos to Schaetzl and Marston for providing the geomorphic community with a textbook that is exceptionally comprehensive in scope yet rich in detail. Each chapter has a nicely balanced writing style that makes the book intellectually attractive for a mid-level undergraduate class, yet it can also serve as an important research reference for graduate students."

Frank Magilligan, Dartmouth College

"Authors Schaetzl and Marston along with a strong cast of contributors have created a richly illustrated and approachable geomorphology text that is an important contribution to the discipline. This book will find a useful home in introductory courses for both non-science and science majors interested in the Earth's surface and its processes. Intermediate and even advanced students of the discipline will appreciate the complete descriptive coverage provided across 23 chapters. I recommend this as an addition to your shelves."

Doug Miller, Penn State University

"*Essentials of Geomorphology* is an engaging text inviting professionals, students, and citizen scientists into a critical field with deep relevancy and applications for understanding our surface world. With contributions from leading Earth scientists in the discipline, this textbook is written in an easy style with straightforward illustrations and thoughtful dialogue."

Amalie Jo Orme, California State University, Northridge

"In *Essentials of Geomorphology*, Schaetzl and Marston bring together experts in geomorphology, including their own expertise, to provide a comprehensive overview for how our planet's landforms have formed and evolved. The book is nicely illustrated with an approachable writing style for the beginning student and the level of detail required for the advanced student. This book serves as a great resource for students studying our diverse landscapes."

Mark Sweeney, University of South Dakota

"*Essentials of Geomorphology* by Schaetzl and Marston is a modern and up-to-date introductory textbook with thoughtful organization, including chapters on the history and development of the fields of geomorphology, biogeomorphology, and planetary geomorphology that are seldom included in similar textbooks. The figures and diagrams are clear and add to the understanding of the text, as do the review questions for each chapter. The contributing authors are well respected in the field and provide an extra level of detail to each chapter."

Gregory Vandeberg, University of North Dakota

Essentials of
Geomorphology

Randall Schaetzl
Michigan State University

Richard Marston
Kansas State University

Chapter contributions by

Devon M. Burr, formerly University of Tennessee
David R. Butler, Texas State University
Patrick M. Colgan, Grand Valley State University
Jo De Waele, University of Bologna
Ronald I. Dorn, Arizona State University
Douglas J. Faulkner, University of Wisconsin-Eau Claire
Warren D. Huff, University of Cincinnati
L. Allan James, University of South Carolina
Ara Jeong, Kyung Hee University
J. Steven Kite, West Virginia University
Phillip H. Larson, Minnesota State University
Richard Marston, Kansas State University
Lewis A. Owen, North Carolina State University
Kevin J. Patrick, Indiana University of Pennsylvania
Gregory A. Pope, Montclair State University
Bruce L. Rhoads, University of Illinois at Urbana-Champaign
Randall Schaetzl, Michigan State University
Thomas Stevens, Uppsala University
Markus Stoffel, University of Geneva
Ethan J. Theuerkauf, Michigan State University
Heather A. Viles, University of Oxford
Stephen A. Wolfe, Natural Resources Canada

CAMBRIDGE
UNIVERSITY PRESS

CAMBRIDGE
UNIVERSITY PRESS

Shaftesbury Road, Cambridge CB2 8EA, United Kingdom

One Liberty Plaza, 20th Floor, New York, NY 10006, USA

477 Williamstown Road, Port Melbourne, VIC 3207, Australia

314–321, 3rd Floor, Plot 3, Splendor Forum, Jasola District Centre, New Delhi – 110025, India

103 Penang Road, #05–06/07, Visioncrest Commercial, Singapore 238467

Cambridge University Press is part of Cambridge University Press & Assessment,
a department of the University of Cambridge.

We share the University's mission to contribute to society through the pursuit of education,
learning and research at the highest international levels of excellence.

www.cambridge.org
Information on this title: www.cambridge.org/highereducation/isbn/9781108748728

DOI: 10.1017/9781108781459

First published 2026

Printed in the United Kingdom by CPI Group Ltd, Croydon CR0 4YY

Cover image: Cavan Images / Per-Andre Hoffmann / Getty Images. The Chocolate Hills of the Bohol
province of the Philippines span an area of more than 50 square kilometers, varying in height from 30 to
50 meters. These karst features (fenglin and fengcong) form due to the gradual dissolution in a tropical
climate of rubbly, poorly consolidated and porous, marine limestones, dominated by corals.

A catalogue record for this publication is available from the British Library

Library of Congress Cataloging-in-Publication Data
Names: Schaetzl, Randall J., 1957- author | Marston, Richard A. author
Title: Essentials of geomorphology / Randall Schaetzl, Richard Marston.
Description: Cambridge, United Kingdom ; New York, NY : Cambridge University Press, 2026. |
Includes bibliographical references and index.
Identifiers: LCCN 2025007795 | ISBN 9781108748728 paperback | ISBN 9781108781459 ebook
Subjects: LCSH: Geomorphology
Classification: LCC GB401.5 .S33 2025 | DDC 551.41–dc23/eng/20250417
LC record available at https://lccn.loc.gov/2025007795

ISBN 978-1-108-74872-8 Paperback

Additional resources for this publication at www.cambridge.org/geomorphology

Cambridge University Press & Assessment has no responsibility for the persistence or
accuracy of URLs for external or third-party internet websites referred to in this publication
and does not guarantee that any content on such websites is, or will remain, accurate or appropriate.

For EU product safety concerns, contact us at Calle de José Abascal, 56, 1°, 28003 Madrid,
Spain, or email eugpsr@cambridge.org

Contents

Contributors

Devon M. Burr
Formerly The Earth and Planetary Science Department
University of Tennessee
Knoxville, Tennessee 37996, USA

David R. Butler
Department of Geography
Texas State University
San Marcos, Texas 78666, USA

Patrick M. Colgan
Department of Geology
Grand Valley State University
Padnos Hall of Science, 1 Campus Drive
Allendale, Michigan 49401, USA

Jo De Waele
Department of Biological, Geological and Environmental
  Sciences
University of Bologna
Via Zamboni 67 – 40126
Bologna, Italy

Ronald I. Dorn
School of Geographical Sciences & Urban Planning
Arizona State University
PO Box 85287–5302
Tempe, Arizona 85287, USA

Douglas J. Faulkner
Department of Geography and Anthropology
University of Wisconsin-Eau Claire
105 Garfield Avenue
Eau Claire, Wisconsin 54702, USA

Warren D. Huff
Department of Geosciences
University of Cincinnati
Cincinnati, Ohio 45221, USA

L. Allan James (deceased)
Department of Geography
University of South Carolina
Calcott Hall
Columbia, South Carolina 29208, USA

Ara Jeong
Department of Geography
Kyung Hee University
26, Kyungheedae-ro, Dongdaemun-gu
Seoul, 02447, Republic of Korea

J. Steven Kite
Department of Geology and Geography
West Virginia University
330 Brooks Hall
Morgantown, West Virginia 26506, USA

Phillip H. Larson
Earth Science Programs
Department of Anthropology and Geography
Minnesota State University
206 Morris Hall
Mankato, Minnesota 56001, USA

Richard Marston
448 Seville Avenue
Newport Beach, California 92661, USA

Lewis A. Owen
Department of Marine, Earth, and Atmospheric
  Sciences
North Carolina State University
Raleigh, North Carolina 27695, USA

Kevin J. Patrick
Department of Geography, Geology, Environment, and
  Planning
Indiana University of Pennsylvania
Indiana, Pennsylvania 15705, USA

Gregory A. Pope
Department of Earth and Environmental Studies, Center for
  Environmental and Life Sciences
Montclair State University
1 Normal Avenue
Montclair, New Jersey 07043, USA

Bruce L. Rhoads
Department of Geography and Geographic Information
  Science
University of Illinois at Urbana-Champaign
1301 West Green Street
Urbana, Illinois 61801, USA

Randall Schaetzl
Department of Geography, Environment, and Spatial
  Sciences
Michigan State University
673 Auditorium Road
East Lansing, Michigan 48824, USA

Thomas Stevens
Department of Earth Sciences
Uppsala University
Geocentrum, Villavägen 16
Uppsala, 75236, Sweden

Markus Stoffel
Climate Change Impacts and Risks in the Anthropocene
    (C-CIA)
Institute for Environmental Sciences, University of Geneva
Geneva, Switzerland

Ethan J. Theuerkauf
Department of Geography, Environment, and Spatial
    Sciences
Michigan State University
673 Auditorium Road
East Lansing, Michigan 48824 USA

Heather Viles
School of Geography and the Environment
University of Oxford
South Parks Road
Oxford, OX1 3QY, UK

Stephen Wolfe
Natural Resources Canada
Geological Survey of Canada
601 Booth Street
Ottawa, Ontario, K1A 0E8, Canada

Preface

This geomorphology textbook is intended for readers interested in physical landscapes, irrespective of their level of geologic or geographic training. It is a basic, yet comprehensive overview of the field of geomorphology. Inside, we discuss all the major types of landforms, on Earth as well as on planets and moons – both rocky and icy – of our Solar System, and examine the processes that formed them. Because we believe that illustrations are important in a highly visual subject like geomorphology, we have developed an impressive list of photos and illustrations to support the reader's journey. Our global view of the field of geomorphology is written in a clear, accessible way. We hope that the book appeals to students who love the outdoors and relish seeing the physical landscape through better lenses, and in sharper focus.

Our goals in writing this book are many, all equally important. We hope to generate the same excitement about landforms that we, the authors, developed when we first started learning about geomorphology. The field of geomorphology is incredibly interesting and exciting, and we hope our zeal for physical landscapes shows through in the pages that follow. We strive to draw as many young minds as possible into the ever-expanding and personally rewarding field of geomorphology, as well as to help more seasoned minds to understand the discipline a little better. By illustrating the practicality and applications – and yes, even the necessity – of geomorphology to modern society, we want to show that landforms are everywhere, and that they are important.

KEY FEATURES OF THIS TEXTBOOK

- Combines the experience of 22 different professional geomorphologists, bringing together a "global collection" of geomorphic knowledge.
- Contains in-depth discussions (as pairs of chapters) on fluvial and glacial geomorphology, while also covering topics such as planetary geomorphology, biogeomorphology, Earth history and climate change, and periglacial systems.
- Written for non-specialists, explaining topics and concepts in non-quantitative ways that introductory readers can understand. As Bill Bryson wrote in his 2003 book *A Short History of Nearly Everything*, a readable college text is "that rarest thing." The authors of *Essentials of Geomorphology* hope that Bryson would deem our book not only "readable" but also accessible and intelligible; that certainly was our goal.
- Includes more than 800 color graphics and photographs of Earth's landforms and even some on other planets to help readers to visualize landforms, thus facilitating learning.
- Highlights and clearly defines key terms when they first appear in the text. Then, these same terms are assembled and defined again, in a rich glossary with almost 1,500 entries.
- Each chapter ends with a selection of Review Questions and a list of Further Readings. The latter covers in greater depth some of the topics that we could only touch upon in this textbook and will be of interest to those who wish to learn more.
- Online resources (at www.cambridge.org/geomorphology) include the book's figures and photos as PowerPoint slides.

So then, who should use this book? Naturalists who love and cherish the physical landscape will find this book to be enlightening. Students in introductory Geomorphology and Earth Science classes will find this book ideally suited to their needs. More advanced geomorphology students, practicing Earth scientists, and other professionals will find the book to be an invaluable reference. Bright, inquisitive land-lookers who relish the outdoors and love seeing the landscape through new lenses will find the book the perfect gateway to geomorphology.

HOW TO USE THIS BOOK

Most geomorphology instructors will choose to use only a select group of chapters in their course. Few will have the time to use the entire book. The appeal of *Essentials of Geomorphology* is that its breadth accommodates a variety of pedagogic approaches. For example, instructors in tropical climates may choose to not use the glacial and periglacial chapters, whereas those in the mid-latitudes may emphasize these topics. In this way, *Essentials of Geomorphology* can support a wide variety of geomorphology courses and approaches, from the most "geology-intensive" course to a general survey course. You choose!

The 23 chapters in this book cover all the major topics in the field:

- geomorphology's guiding principles, theories, and history;
- methods used to establish the age of landforms and their sediments;
- climate and climate change throughout Earth's history;
- the resisting framework – rocks, minerals, and the biosphere;
- tectonic processes – how mountains are built;
- heat – landforms on volcanic and plutonic rocks, plus geysers and associated hydrothermal features;
- landforms dominated by bedrock and bedrock structure – those built on horizontal and weakly dipping rocks, those developed on folded rocks, and those impacted by faulting;
- weathering and the myriad forms it takes in geomorphology;
- landforms impacted by groundwater – caves and karst;
- soils, slopes, and paleosols;
- mass movements and slope processes;
- the work and landscapes of running water;
- the constantly changing world of ocean and lake coasts;
- the important, lasting legacy of Quaternary glaciers and ice sheets;
- geomorphology's cold side – periglacial processes and landforms;
- wind-driven sediments and landforms;
- biogeomorphology – plants and animals as geomorphic agents; and
- the geomorphology of planetary bodies with solid surfaces.

We invite you to discover anew the exciting field of geomorphology. Thank you for reading on!

Acknowledgements

Editors/authors Schaetzl and Marston have many, many people and organizations to thank for helping us produce this book. At the top of the list are the editors and production personnel at Cambridge University Press. Matt Lloyd, Ilaria Tassistro, Rachel Norridge, and the rest of the Cambridge team never wavered in their support of *Essentials of Geomorphology*, even when the Covid pandemic slowed our progress. Their highly professional manner and approach to this book has seamlessly carried it through the production process. Lastly, we thank David Hemsley of Aardwolf Books for his careful copyediting of the manuscript and Julie M. Jackson for her detailed proofread.

We owe a huge debt of appreciation to HaJin Kim, our incredible and dedicated graphic artist. HaJin's talent, vision, organization, and attention-to-detail skills are on display in every page of this book. We cannot thank her enough for all the hard work that she has put into this project.

Five undergraduate Geology and Geography students at Michigan State University served as informal reviewers for almost every chapter. After each chapter was "finished," it went to this team for their evaluation and review. In every instance, they found errors and injected constructive comments that improved the chapters – immensely. By the time that the chapters had gotten to this stage, most of the authors were understandably weary of edits and revisions. But every single author, after having seen the comments made by this team, commented on how useful the student reviews had been in making their chapter just a little bit better. We thank (in alphabetical order) Abby Wolff, Chloe Mikelsons, John (JR) Nosal, Kate Henry, and Nick Catanzaro. We note that JR Nosal also took on the glossary. His Herculean efforts with the glossary really helped to improve it in countless ways – small and large.

Other colleagues also helped to review many of the completed chapters. For this, we thank (in alphabetical order) Kevin Kincare, Fritz Nelson, and Vasya Tolmanov. We sent several figures to colleagues for their review and want to thank them here as well: John Attig, Bill Blewett, B. Brandon Curry, Sebastien Huot, Henry Loope, Fritz Nelson, and Vasya Tolmanov.

Heidi Schaetzl reviewed every chapter, placed the text in its proper format, and created the initial list of glossary words and definitions. This Herculean effort saved a great deal of time at the end of the process and was completed on time and with tremendous attention to detail.

Essentials of Geomorphology is filled with myriad images of landforms and landscapes; many are from the authors' private collections, but many are not. And so, we thank the many people who have kindly granted us permission to use their images in this book – mentioned individually in each image caption. Nonetheless, we particularly want to thank Jürg Alean and Michael Hambrey for sharing images from their amazing website (https://Swisseduc.ch) of glacial imagery, Jani Radebaugh and her Geophotos website, Martin Mergili and his website of rich and varied imagery of landforms from around the world (www.mergili.at/worldimages), and Marli Miller, whose Geologic Time Pics website (https://geologypics.com) is a treasure-trove of high-quality geology images. Additionally, we want to particularly thank the following people, who went the extra mile to help us obtain imagery from various organizations and/or their websites: Alexa Whipple of the Methow Beaver Project, Steven Reidel of Washington State University, and Ted Stout of the National Park Service, at Craters of the Moon National Monument and Preserve.

Structural (hardware and software) support was provided by the Department of Geography, Environment, and Spatial Sciences at Michigan State University.

Lastly, and most importantly, we thank the authors of *Essentials of Geomorphology*, for their belief in this project and their support of its editors. By coming onto this project, they acknowledged that writing for introductory learners is a critical part of Earth Science education, and that such efforts will contribute to the long-term success of the field of geomorphology. The authors endured far more editing and figure production iterations than they had "signed up for," and yet never did we hear a complaint. To think that such a world-renowned group of scientists would go so far and put aside so many other tasks just to write an introductory chapter for a textbook says so much about their commitment to the field of geomorphology and to education. We profoundly thank them all.

This book is dedicated to the memory of Dr. L. Allan James (1949–2022), who co-authored two chapters in this volume before his untimely passing. AJ was a world-class fluvial geomorphologist, perhaps best known for his award-winning publications about legacy fluvial sediment. Much of his career was spent as an educator in the Department of Geography at the University of South Carolina (USC), where he was honored with the title of Distinguished Professor Emeritus of Geography in 2018.

We acknowledge and appreciate the tremendous scientific, collegial, and mentoring impacts that AJ has had on the geomorphology community. At USC, his courses in physical geography, geomorphology, hydrology, and watershed science were important training grounds for undergraduate and graduate students across the university. He especially loved bringing students into the field, where he could introduce them to methods of data collection and foster a sense of inquisitiveness and appreciation for the study of the environment. Most importantly, AJ was a wonderful human being, with a great sense of humor and always had a positive attitude. His smile and laugh will be greatly missed, but his memory and scholarship will live on.

Randall Schaetzl
Richard Marston

1 Introduction to Geomorphology

Richard Marston and Randall Schaetzl

Geomorphology is the study of **landforms** – their evolution, shape (morphology), and composition. The word comes from the Greek (*geo*, Earth, *morphos*, referring to form, and *ology*, a branch of knowledge). Landforms come in all types, shapes, sizes, compositions, and ages. There is a landform for everyone, and no two are exactly alike. Understanding Earth's landforms – how they are formed, altered, destroyed, and/or buried by various geologic processes – is at the core of geomorphology. This textbook will teach you the language and concepts that will help you to understand the workings of many of Earth's physical systems. Our goal is to equip you with the vocabulary and toolkit for understanding why Earth's physical landscapes look the way they do. This knowledge will help us all to better manage our fragile natural resources.

The fundamental building blocks of geomorphology are landforms and the sediments that comprised them. Decades of study have taught us that Earth's landforms have formed through various types of surficial and geologic processes, driven by the actions of water, wind, ice, biota, and gravity. These "drivers," and the landforms that result from them, are constantly changing, always evolving. Studying the evolutionary stories of Earth's landscapes are at the very heart of the discipline of geomorphology.

At the core of geomorphology are the depositional and **tectonic processes** that are constantly forming and raising up the land surface, balanced out by the erosional processes that wear it down. The landforms geomorphologists study are caught in the middle of this endless recycling struggle, existing somewhere in between their initial formation and destruction. It is this very cycle that explains why the summit of Mt. Everest is composed of uplifted limestone that had initially formed at the bottom of an ancient ocean, and why this limestone has been glacially carved into jagged peaks. Change at that scale is awesome in the truest sense of the word.

Asking questions about landforms is central to the mindset of every geomorphologist. How did this landform form? Why in this location, formed in this sediment, and in this particular way? Where did the sediment come from and how did it get here? How rapidly did the landform develop, and when did this all take place? Did its formation occur all at once or in a series of depositional and/or erosional events?

What did this area look like during the time of formation? What was the climate, and how has it changed? How has the landform been modified since its initial formation? How is it connected to a larger regional story, or to the broader landscape? Answering essential questions like these is what drives, and what fulfills, those who study geomorphology.

Landforms are intricately interconnected with almost every aspect of our daily lives, and as a result, geomorphology is relevant to everyone, everywhere, and at all times. We use sediments from landforms to build roads, to make concrete and pottery, and to construct earthen dams. We ski down glaciated mountains, sunbathe on broad ocean beaches, play golf on glacial moraines, fish in lazy rivers, and explore deep caves. And of course, the soils that feed us have developed in sediments with rich geomorphic histories. Geomorphology lies at the heart of every soil map, and soils sustain our lives. We rely on our knowledge of geomorphology and geology to keep us safe from hazards such as landslides, earthquakes, tsunamis, cave collapses, volcanic eruptions, dust storms, and floods. The water that sustains us is found through our knowledge of the saturated subsurface that we so often take for granted. Most of our drinking water comes from wells set within rock and sediment, or pulled from rivers and lakes that have developed in concert with the local geomorphology. And so, the better we understand the surface and the sediments below it, the better we will be able to manage our hydrological resources. Geomorphology even has extraterrestrial implications! Geomorphic knowledge developed here on Earth helps planetary geologists interpret the geologic past of the many planets and moons within our galaxy. In every part of our everyday lives, and even beyond Earth itself, geomorphology is present, and it is essential.

Landscapes may seem static and unchanging at first glance. But because geomorphologists view landscape development across both short and long timescales, they can see and interpret how landscapes change. Short-term changes are easy to see. In a matter of days, a flood inundates and deposits sediment across large areas. A windstorm uproots thousands of trees, disturbing soils and over the next decade, facilitating a downslope cascade of sediment. Sand dunes, stable for thousands of years, suddenly activate and move swiftly across the landscape, even burying homes and villages. Rockfalls catastrophically wreak havoc with

Figure 1.0 Moraine Lake, in Banff National Park, Alberta Canada, is a classic example of a lake formed in glaciated mountains. Source: Vicki Jauron / Getty Images.

everything below. And then, over longer timescales, landscapes are also constantly changing. Even mountains wear away, their eroded remains washing into the ocean basins. These systems are all unquestionably geomorphology in action – whether we can actually watch it all happen, or not. Geomorphologists work across this range of timescales, expressly *because* landforms change at varying rates.

Geomorphology is both a basic and an applied science. It is a basic science in that landforms and landscapes deserve scientific attention for their own sake. By engaging in geomorphic study, we can improve our understanding of Earth and our place on it. The sheer beauty and wonder of the physical landscape draws many to it, whether in national or state parks, government-managed recreation areas, or private land that we use simply for getaways. Unquestionably, landforms have inherent value simply in their beauty. But geomorphology is also an applied science in that it helps us to understand human impacts on landforms and landscapes, and the impact of landforms on society. Geomorphologists are instrumental in helping to locate important geological resources such as **rock**, **sand**, gravel, **clay**, and **peat**. For example, all societies need sand and gravel, otherwise known as **aggregate**, for their economies to efficiently function. The average American uses, indirectly and directly, five *tons* of aggregate each year. Aggregate is needed for roads, sidewalks, building materials, and myriad other applications! No one is better suited to finding aggregate than a geomorphologist. As mentioned above, geomorphology is also vital to finding and properly managing our water resources – wells, surface water, and wastewater. But perhaps nowhere is geomorphology more applied than in its links to natural hazards (**Fig. 1.1**). Geomorphologists help societal managers keep us safe by studying the hazards (and resources) formed by geomorphic processes, for example, coastal erosion, shifting river channels, cave collapses, landslides, massive floods,

sandstorms and dust storms, thawing permafrost, among many others. Geomorphology is a part of our everyday lives – far more than most people imagine.

Recently, humans have become a driving geomorphic force. Many geologists consider today's world to be so uniquely modified (both biologically and physically) by human-caused agents that it deserves its own distinction in the geologic timescale – the **Anthropocene** (see Chapter 4). Although its inclusion as an official segment of geologic time is still being debated, the name refers to the current geologic epoch in which the effects of human activity have been so dramatic that they are overwhelming many "natural" phenomena. Ravaging the planet for resources, draining aquifers, leveling rainforests, driving thousands of species into extinction, and causing unprecedented desertification and soil erosion, not to mention the human-caused changes to the atmosphere that are dramatically changing our climate – there is no doubt that humans are changing the planet in ways and at rates that have never been seen before. Permafrost is thawing, ocean levels are rising, and islands are being flooded, even as large-magnitude weather events are becoming more common. Few specialists are as prepared as geomorphologists are, to assess the impacts of climate change on human societies and systems.

The long history of geomorphology is rich in adventure and discovery of the great outdoors. Geomorphology's blend of lab science and field studies continues to attract scholars even today. The community of geomorphologists is growing ever more diverse and international, working with an increasing variety of field, lab, and analytical tools. Yet, much remains to be done, as our understanding of many landforms and regions is still in its infancy. Armed with the plethora of new, detailed topographic data that are emerging today, and with rapidly increasing options for dating landforms and sediments, the geomorphic community now has

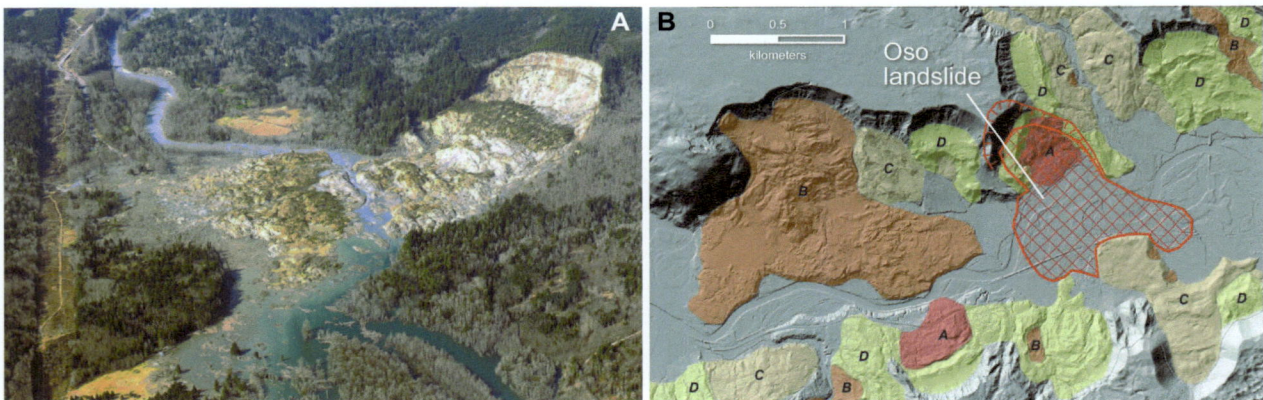

Figure 1.1 The Oso, Washington (USA) landslide is an example of geomorphology in action. **A.** This massive March 22, 2014 landslide resulted in the loss of 47 lives and 49 homes, completely burying a local neighborhood known as "Steelhead Haven." Local residents described the event as a "fast-moving wall of mud." The event was triggered by 200% of normal rainfall over the previous few weeks. **B.** Scientists at the US Geological Survey had predicted that this area could expect to see such events, based on their geomorphic studies of the area. Topographic information for this region clearly shows areas that were affected by similar landslides in the past. Colored areas in this figure indicate older landslide deposits, from youngest (A) to oldest (D). Improved knowledge of geology and geomorphology could have saved many lives. Source: Photo and graphic from the US Geological Survey; in public domain.

the ability to understand and help manage Earth's surface better than ever before. This book might be just the start of your geomorphic journey.

1.1 LANDFORMS AND LANDSCAPES

Landforms are natural features, formed in sediment or rock, that can be observed, described, mapped, and classified. Most landforms possess three fundamental characteristics. They (1) have a distinctive shape and dimensions, (2) occur on a typical position in the landscape, and (3) have formed from distinctive and characteristic materials. Due to their critical importance to the field of geomorphology, landforms are the focus of this book.

Although *Essentials of Geomorphology* contains discussions of hundreds of different types of landforms, for now let's examine just one – an **alluvial fan** (**Fig. 1.2**). As the name implies, the landform is fan shaped. Alluvial fans are typically positioned where steep-flowing, mountain streams deposit sediment from uplands onto the plain below. When the stream emerges from its confined mountain canyon, the channel widens and much of the water infiltrates into the ground, such that the stream loses water and power, leading to sediment deposition and the formation of the fan. Most alluvial fans have lengths of 2–8 km, but can range from a few hundred meters to tens of kilometers long. They are steeper near the apex (where the stream emerges from the mountain canyon), and flatten farther out, resulting in a slightly concave-upward shape from their apex to their distal (far) end. Geomorphologists refer to this as a **longitudinal profile**. Stream channels on the fan shift location frequently, often with each flow event, slowly building up the fan surface over time. Because alluvial fans become larger over time, it is difficult to assign a single age to any particular fan. Rather, we attempt to date the various segments of the fan and in so doing, determine how dynamic its formation has been.

Over time, parts of the fan may be destroyed by river incision, provided the water–sediment mix tips in the favor of more water and less sediment. Streams that traverse the

fan may then pick up sediment by cutting into the fan. Other floods may transport large amounts of sediment *onto* the fan, causing it to enlarge and thicken. Geomorphologists have ways of explaining and predicting these various outcomes.

Alluvial fans are common on desert landscapes, where they can be a vital source of **groundwater**, which is one reason that agriculture, for example, citrus groves in southern California, have encroached onto so many fans. Many of the western suburbs of Las Vegas, Nevada, and the cities of the San Gabriel Valley in southern California (USA) are built on the outer margins of alluvial fans. However, the same flash floods and debris flows that formed alluvial fans pose potential future hazards for these urban areas. This example points out that the better we understand landforms, the better we can manage land use on them.

An alluvial fan is a landform. **Landscapes** are organized assemblages of genetically interconnected and interrelated landforms, like alluvial fans. To refer back to our example (**Fig. 1.2**), an alluvial fan is but one landform in a desert landscape that might be comprised of a variety of interconnected landforms such as mountain ranges, fans, sand dunes, and desert basins (**Fig. 1.3**). Although each of these landforms has a different geologic history, they are all linked via the processes that have formed them. Water flowing out of (and forming deep valleys in) the mountains helps to form the alluvial fans, which grade into other landforms such as the flat, wet, and salty lake plains at their distal margins (see Chapter 10).

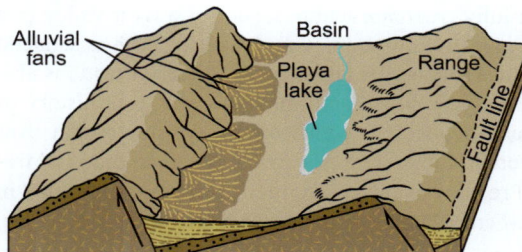

Figure 1.3 Alluvial fans are but one landform in an arid region landscape.

1.1.1 Categorizing Landforms

Landforms come in all shapes and sizes. The largest ones can extend across large regions of Earth's surface, such as mountain ranges or island arcs. These features tend to be formed mainly by **endogenic** processes driven by sources of energy within the Earth, particularly heat left over from when the planet initially formed, and from the decay of radioactive elements. Examples of endogenic processes include crustal uplift, faulting, and deformation, and volcanism. Endogenic processes often are associated with **tectonism**, which refers to deformation of Earth's lithosphere, or crust (**Fig. 1.4A**). These types of landforms also tend to form over long periods of time, or temporal scales (**Fig. 1.4B**) – often millions of years. Local-scale landforms, which form quickly – like ripples in a sand dune – have short temporal scales, on the order of a few hours.

Figure 1.2 An alluvial fan in Death Valley National Park, California, USA. Source: R. Dorn.

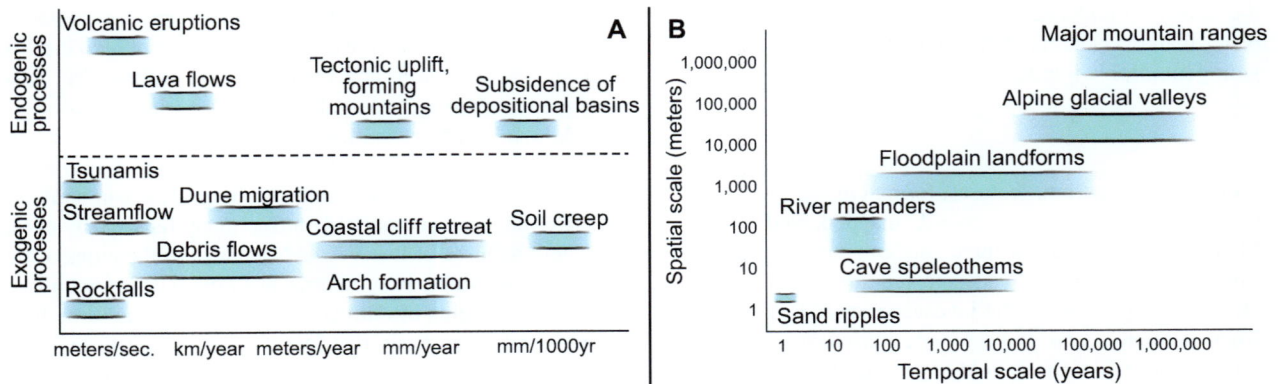

Figure 1.4 A selection of landforms and the timescales over which they may form. **A.** Landforms owe their origins to processes that are along a continuum of forces – from endogenic (from within the crust) to exogenic (at or above the land surface). These forces operate across a wide variety of temporal scales, from the very slow to the extremely rapid. **B.** Landforms vary widely in size, as do the timescales over which they form, as the examples here show. Source: Adapted from *Landscapes and Geomorphology: A Very Short Introduction* by Andrew Goudie and Heather Viles (2010). Reproduced with permission of Oxford University Press through PLSclear.

Landforms are usually shaped by erosion and deposition. They rely on processes driven (ultimately) by solar energy and/or on gravity – think of heavy rain, landslides, flowing **glaciers**, and the frozen/thawing processes of today's Arctic. Collectively these are termed **exogenic** processes because they rely on sources of energy that do not originate within the Earth.

Landforms can also be classified along a continuum ranging from fully *erosional* features, such as a river valley or a glacial cirque, to fully *depositional* features, such as a glacial moraine or a sand dune. Other types of categorizations also exist (**Table 1.1**). Most landforms have a hybrid history of erosion and deposition, which often vary in strength over time. Geomorphic processes also vary in magnitude and frequency. Frequent events occur often, but are usually small in impact, whereas large events can produce substantial change, but occur rarely. On the other hand, moderate events often have the right combination of magnitude and frequency to strongly influence the shape of steady-state landforms.

1.2 DRIVING FORCES, THE RESISTING FRAMEWORK, AND TIME

Landforms result from the interplay, over time, of the **resisting framework** and various **driving forces**. The resisting framework includes mainly (1) rock type (principally, their hardness and resistance to **weathering** and **erosion**), (2) stratigraphy (the layered sequencing of the various rock types), (3) the underlying geologic structure (the geometric arrangement of rock layers, as affected by uplift/subsidence, folding and faulting, intrusion, and emplacement), and (4) biota – the ability of plants and animals to impact erosion rates (**Figs. 1.5, 1.6, 1.7**). Driving forces act on these rocks and sediments to create landforms. They are a mix, or balance, of endogenic and exogenic processes, as well as (possibly) forces that originate outside Earth's system (**extraterrestrial processes** such as meteorite and asteroid impacts). Driving forces can either build up (**diastrophism**) or tear down (**denudation**) the landscape. It is the interplay

Table 1.1 End members of some general landform classifications

Classification	Explanation
Depositional (aggradational) vs erosional	Formed by deposition of sediment (depositional) vs those landforms formed by erosion of preexisting sediment or rock (erosional)
Subaerial vs subaqueous	Occurring on dry land (subaerial) vs those landforms occurring underwater (subaqueous)
Stable vs unstable	Exhibiting little change over time due to erosion (stable) vs those landforms that are actively eroding or becoming buried by sediment (unstable)
Bedrock vs unconsolidated	Formed on bedrock vs those landforms that are developed on soil or other loose (unconsolidated) sediment
Relict vs present-day	Formed in the past and have been little-changed since vs those that are currently forming due to erosion or deposition (timescale dependent)

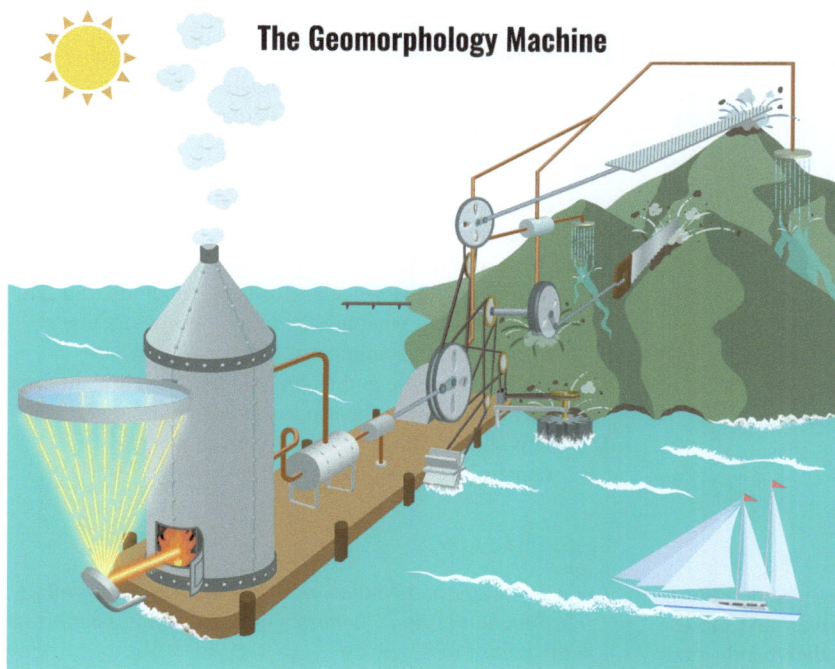

Figure 1.5 The geomorphology we see today represents the long-term actions and balance between the resisting framework and the driving forces, as conceptualized here by the "geomorphology machine," first drawn by geomorphologist Arthur Bloom (1969). The energy for the machine is mainly exogenic – coming from the Sun. The Sun's energy powers the hydrological cycle, which then provides the energy for weather (pressure differences, winds, storms, etc.), thereby forming waves and currents, as well as precipitation. Gravity is also important as an energy source, pulling water, ice, and earth material downward. In this figure, the land represents the resisting framework, whereas the remaining parts of the "machine" are the driving forces that act to wear it down. The driving forces (the geomorphic agents of gravity, water, ice, wind, waves, and currents) abrade, excavate, incise, and then transport any eroded materials from the land to the ocean basins. Source: After Bloom (1969).

Figure 1.6 Illustrations of the role of underlying geology on landform development. A. The interesting landforms ("Goblins") in this landscape at Goblin Valley State Park in Utah, USA, are the result of differential erosion. Erosion is more pronounced, and goblin necks are thinner, on the softer rock layers (see Chapter 8). Source: R. Schaetzl. B. The Grand Canyon of the Colorado River is an excellent example of how rock structure and stratigraphy, coupled with rock resistance, influence landform development. Some rock layers (limestone and sandstone) form cliffs and spires, whereas the weaker shale beds erode to form broad, flat benches (see Chapter 8). Source: R. Schaetzl. C. Rocks in this broad valley, near Vercors, France, have been folded downward into a geologic structure called a **syncline** (see Chapter 9), illustrating the effect of geologic structure on landform development. Source: R. Marston.

of these two opposing suites of processes that create the landforms we see today. Part of a geomorphologist's role is to better understand how this interplay functions, and the rates at which these processes operate.

Although we often cannot see their evolution in real time, landforms are not static and unchanging features. They continually change over time, in response to the interplay between the resisting framework and the various driving

Figure 1.7 The Chaco River near Farmington, New Mexico, USA, illustrates the interplay between the endogenic forces that operate below the crust and the exogenic forces that operate on and above the surface. The river (the exogenic force) is cutting its valley down into and through the small mountain range, which owes its existence to geologic uplift (an endogenic force). Source: L. Maher.

forces. Most landforms respond very slowly to long-term processes of erosion, which in turn are driven and moderated by climate and vegetation. On the other hand, landforms can also change seemingly overnight in response to disturbances, many of which occur suddenly, like wildfires, but also human-caused activities such as dam construction, urbanization, and coastal development.

The international team of geomorphologists who have collaborated to write this book invite you to explore the world of geomorphology with us. Landforms are everywhere, and never before have geomorphologists had such a wide array of techniques to help us understand them. And of course, every day we have a richer array of literature to assist us in this task. With regard to our knowledge base, we still stand on the shoulders of giants in the field, and every day the pioneering work they have done becomes, in hindsight, more impressive. Together, we can do even more.

REVIEW QUESTIONS

1.1 What is geomorphology and what are the main questions that geomorphologists try to answer?

1.2 How do geomorphologists define the term "landscape"?

1.3 Provide an example of a landform that develops over a short timescale and one that develops over a longer timescale.

1.4 What are the basic characteristics of landforms? As an example, list the characteristics of an alluvial fan.

1.5 Besides time, what are the two other factors that influence the formation of landforms?

1.6 Define exogenic and endogenic factors. How do these factors fit within the scope of geomorphology?

1.7 How do landforms create opportunities and constraints for human activities?

FURTHER READING

Anderson, R. S. and Anderson, S. P. 2010. *Geomorphology: The Mechanics and Chemistry of Landscapes*. Cambridge University Press.

Bierman, P. R. and Montgomery, D. R. 2014. *Key Concepts in Geomorphology*. W.H. Freeman.

Bloom, A. L. 1998. *Geomorphology: A Systematic Analysis of Late Cenozoic Landforms*. 3rd ed. Prentice-Hall.

Easterbrook, D. J. 1999. *Surface Processes and Landforms*. Prentice-Hall.

Murck, B. W. and Skinner, B. J. 1999. *Geology Today: Understanding our Planet*. Wiley.

Rice, R. J. 1988. *Fundamentals of Geomorphology*. 2nd ed. Longman Scientific and Technical.

Ritter, D. F., Kochel, R. C., and Miller, J. R. 2011. *Process Geomorphology*. 5th ed. Waveland.

Smith, G.A. and Pun, A. 2010. *How Does Earth Work?* 2nd ed. Pearson.

ANOTHER VIEW OF THE GRAND CAÑON.

2 History and Foundations of Geomorphology

Randall Schaetzl and Bruce Rhoads

As defined in Chapter 1, geomorphology is the study of landforms – plain and simple. Whether they are formed on **bedrock** or on loose sediment, by erosion or deposition of sediment, and whatever their age, landforms are the building blocks of Earth's physical landscapes. In essence, *landscapes* are organized and interconnected *assemblages* of landforms. These interconnections may be **temporal**, **genetic**, or **spatial**. With regard to temporal connections, some landforms on a landscape may have all formed at roughly the same time. They may share a similar origin (genetic connections). On many landscapes, however, the landforms may have formed at different times and in different ways. Because geomorphic processes operate over all parts of the landscape, spatial connectivity between the various parts of these landscapes is common – or even expected! Consider the example where the upper part of a hillslope – formed mainly by erosion – merges seamlessly with the lower part of the same hillslope formed in the sediment transported off it (**Fig. 2.1**). The upper and lower parts of this hillslope are connected spatially. Connections like the one in this example are almost always present in landscapes, and geomorphologists strive to better understand them.

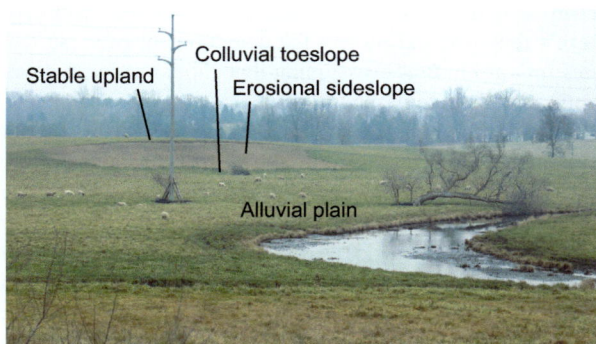

Figure 2.1 Even this low-relief landscape in southern Michigan exhibits interconnectivity among slopes, sediment, and landforms. Source: R. Schaetzl.

The purpose of this chapter is to document and illustrate the basic characteristics of geomorphology as a field of science. Because geomorphology has a colorful history, we also explore some important aspects of how modern geomorphology evolved from its earliest roots. Who first began to think about and explore the geomorphology of landscapes, and what were their contributions? How has the science evolved since then, and what of the future?

Because so many landscapes have been modified by human actions, it is essential that geomorphology also account for anthropogenic impacts. And so, lastly, we will strive to document the importance and utility of the field in the modern world. What can geomorphology contribute to society, and what has it already taught us about the varied ways that surficial processes interact with human activity? Let's begin our brief overview of the field of geomorphology.

2.1 THE SCIENCE OF GEOMORPHOLOGY

Earth's physical landscapes are dynamic and always changing. Geomorphic processes are constantly at work to mold and reshape the land surface. Geomorphologists strive to understand how these processes operate and interact to form landscapes, both today and in the past.

A fundamental principle of geomorphology is that landscapes and landforms are formed by movements of mass. Transfer of mass (sediment) is, in turn, performed by a suite of surficial processes. Often, that movement/transfer is directed downhill, under the influence of gravity, toward a river, a lowland, or a water body. Nonetheless, some processes can move sediment to higher elevations, for example, volcanism, tectonics, or wind. Humans are also capable of moving sediment against gravity, particularly through the use of machinery.

Landscapes are the result of processes that deposit sediment on them or remove sediment from them, coupled with processes that lift up the crust or move it downward. Many of the former (erosional and depositional processes) are driven by flowing water, waves and currents, wind, and/or glaciers. Processes driven by forces within the Earth, such as tectonism, volcanism, and isostasy, are important because they create vertical movements of the land surface. The overall lowering and flattening of a landscape is called **denudation**. Other processes may add **relief** (the elevation difference between high and low points on a landscape). The intricate balance (or lack of balance) between uplift,

Figure 2.0 This nineteenth-century engraving depicts the Grand Canyon, in the southwestern United States, as early geologists saw it. Source: duncan1890 / Getty Images.

additions and removals of mass, and denudation is often a focus of geomorphic study.

Rates of geomorphic change vary widely. Some processes that shape landscapes occur quickly and dramatically, such as landslides, volcanic eruptions, or floods. Many geomorphic processes, though, operate so slowly that changes in landforms are not readily visible across human lifetimes.

To better understand how different surficial processes interact to form landscapes, geomorphologists often link field observations of processes with theoretical models. By doing so, they can better understand how complex interactions among processes change landforms across different scales of space and time. As we shall see below, geomorphic modeling has been around almost since the inception of the field.

Geomorphology is also an applied science, in that it can help (1) identify important resources such as groundwater, surface water, sand, gravel, rock, metals, and clay, and (2) determine how various types of human activities may disrupt geomorphic processes that sustain environmental systems. Geomorphologists are also adept at identifying when and where natural hazards may occur and at helping to mitigate the effects of these hazards.

Soils form the skin of many landforms, and so all manner of soil-based applications rely on geomorphic knowledge and data. The old adage that "Everything we have is either mined or grown" is not only true (think about it!), but also, most of the "mining" and "growing" done on Earth is guided by geomorphology. We mine gravel from glacial and fluvial sediments, pump water from aquifers, grow cotton on coastal plain sediments, and use Earth's deltas to bring foodstuffs of all kinds to market. Geomorphology is the backbone of so much of what we do and have. It is intimately intertwined with our lives.

2.2 AN HISTORICAL VIEW OF THE FIELD OF GEOMORPHOLOGY

Before we discuss the tools and the methods of contemporary geomorphologists, let's examine how the science got to where it is today. Humankind has always been interested in the varied forms of Earth's landscapes. Perhaps it is for this reason that the roots of geomorphology can be found in the science of geology. Geology began to develop in its modern form with the work of Scottish naturalist James Hutton (1726–1797), often called the Father of Geology. Hutton first proposed that the Earth is not static, but changes over time. He also thought that the geologic processes of the past were largely the same processes that we can observe today – a radical proposal at a time when the Biblical flood was thought to have formed many of Earth's landscapes.

Hutton's ideas have since been proven to be both accurate and useful. Thus, we know today that evidence of past processes is preserved in geologic materials, like rocks. As a result, we can use our understanding of modern processes to interpret what caused past changes in the landscape, and estimate how rapidly these changes may have taken place.

Because many modern geologic processes operate quite slowly compared to human lifespans, the time required for geologic change can be very long – often many millions of years.

Hutton's way of thinking about the Earth provided the basis for what has become known as the **principle of uniformitarianism**, which, simply stated, is that "the present is the key to the past." Prior to Hutton's work, many people had ascribed the features of Earth's surface to extraordinary occurrences in the past, and which defy modern understanding. These theories, grouped under the term **catastrophism**, asserted that short-lived, violent events of the past were mainly responsible for shaping large parts of the land surface and the rocks below. Proponents of catastrophism attributed Earth's features to divine creation, to the great Biblical flood, or to catastrophes like earthquakes and volcanoes. Alternatively, advocates of uniformitarianism held that surficial processes in the past resembled those that can be observed today. Uniformitarianism helped scientists of the day explain geologic features by studying contemporary processes and understanding their long-term consequences, rather than by invoking some past catastrophe.

Hutton's views inspired his close friend John Playfair (1747–1819) to produce what has become known as **Playfair's Law** – a statement that captures nicely the integrated relations between erosion by rivers and the form of their valleys. In essence, Playfair's Law states that each stream cuts its own valley, and that stream valleys are proportional in size to the streams that cut them.

Hutton's ideas represented a major sea change for the field of geology and laid the foundation for the development of the science of geomorphology. Because he studied rocks near the surface and related their structure to the form of landscapes, his work was very much like that of a geomorphologist. Earth scientists, including geomorphologists, began to extrapolate into the past, to infer the origins of Earth's features, and project the development of these features forward in time – assuming that similar processes will continue to operate in the future.

In the second half of the nineteenth century, several major expeditions were conducted by the United States to explore newly acquired lands in the American West. In 1869, John Wesley Powell (1834–1902) and his fellow explorers bravely floated down the Grand Canyon of the Colorado River, not knowing what lay ahead. During the journey, Powell was struck by how the river had incised into the surrounding rock and by the amount of sediment it carried. Considering the processes associated with this type of erosion, Powell deduced that rivers cannot erode below some lower limit. Thus, he paved the way for the concept of **base level** – the level (elevation) below which a land surface cannot be eroded. Sea level is the ultimate base level, but each river has its own local base level – the elevation at which it enters the next larger trunk stream, or lake.

Other geologists also began to add to the body of geomorphic theory. Clarence Dutton (1841–1912), developed important ideas about **isostasy** – the vertical adjustments of the land surface in response to changes in the loading of mass onto it (by water, glacial ice, or sediment). Grove Karl Gilbert (1843–1918), who explored the Henry Mountains in Utah, introduced the notion of a **graded river** – the idea that rivers undergo adjustments in form to efficiently transport the amount of sediment supplied to them by the surrounding watershed. Gilbert's ideas combined geological concepts with those from physics – a development that has led to him being called the Father of Modern Geomorphology. Later in his career, Gilbert also introduced rigorous experimental methods to geomorphology, by using flumes to study the movement of sediment in rivers. In sum, his contributions added to the theoretical advances made by Powell.

Work by Powell, Dutton, Gilbert, and several others not mentioned here set the stage for the scientific exploration and analysis of landform genesis. The term geomorphology was formally introduced immediately prior to the turn of the twentieth century to describe the emerging science of landform study. A theme that became prominent at this time is that landforms may evolve sequentially and predictably over time. This theme would be expanded on by William Morris Davis, who developed it into the overarching geomorphic theory of the early twentieth century.

2.2.1 William Morris Davis' Cycle of Erosion

Any discussion of the history of geomorphic theory must include the **cycle of erosion**, developed and advocated by William Morris Davis, a geologist and geographer at Harvard (**Fig. 2.2**). Davis' model is a simple, yet elegant, visualization of how landscapes evolve. The cycle of erosion, often called the geographic cycle, reigned supreme in geomorphology for several decades from the early to middle twentieth century.

Figure 2.2 William Morris Davis (1850–1934). Source: Universal History Archive / Contributor / Getty Images

Davis was influenced by ideas about sequential development over time, which were prominent throughout the Earth and biological sciences in the late 1800s – when he first formulated his ideas. For example, think of the evolutionary theories of the English naturalist Charles Darwin. Davis travelled widely and was a keen observer. Based primarily on observations in humid-temperate environments in the United States, he conceptualized that landscapes progress through long but predictable cycles. According to Davis, the cycle begins with an uplift event that raises an old, low-relief landscape to heights well above its ultimate base level (sea level). The uplifted landscape then evolves through sequential stages, over long periods of time, until it becomes a nearly flat plain just above sea level. Influenced by ideas about sequential biological development, Davis named the stages that landscapes undergo, as they move through the cycle, in developmental terms – youth, maturity, and old age (**Fig. 2.3**).

The model begins when tectonic forces (assumedly rapidly) uplift a low, poorly drained, nearly flat surface (**Fig. 2.3**). Uplift increases the potential energy of the streams, allowing them to quickly and efficiently incise, forming deep valleys (**Fig. 2.3**). This stage (**youth**) is characterized by broad, flat uplands (remnants of the former, uplifted landscape) with deep, widely spaced canyons, carved by powerful rivers, as they attempt to reach base level (**Figs. 2.3, 2.4, 2.5A**). The Colorado Plateau region of the western United States is a classic example of a youthful landscape, having been uplifted in the past, and is now being deeply incised by rivers such as the Colorado and Green.

As the downstream segments of the major rivers transition into lower and lower gradients, that is, as they approach their base level, their capacity to incise decreases. As a result, these sections of the rivers begin to meander laterally, depositing sediment delivered from steep upstream sections of the rivers, and developing floodplains (**Fig. 2.3**). The upper (headwater) reaches of the streams remain steep and, as a result, the rivers develop a concave-upward longitudinal profile. At this time, rivers also "smooth out" their longitudinal profiles, as incision rates decrease. Erosion along hillslopes works to eliminate the remnants of the initial landscape. During this **maturity** stage, most of the landscape is sloping, valleys are broad, the main rivers have well-developed floodplains, and little or nothing of the originally uplifted landscape remains on the uplands (**Figs. 2.3, 2.4, 2.5B**). At this stage, **local relief** (the difference in elevation between the local uplands and the nearby lowlands) is greatest. Davis saw this type of landscape at numerous locations worldwide – one of steep slopes, high hills, low mountains, and deep, broad river valleys.

Given enough time and tectonic stability, continued erosion will decrease the relief of the landscape, producing a nearly flat, **old age** landscape as the uplands get worn down, as the lowlands – near base level – gradually fill with sediment. Because hillslopes are gentle and erosion rates small, most of the old age landscape becomes mantled with a thick cover of sediment. Davis famously referred to this nearly flat landscape of low relief as a **peneplain** ("almost a plain") (**Figs. 2.3, 2.5C**). One of the main criticisms of the cycle of erosion is that few such peneplains exist on Earth – to which Davis rebutted that the crust was too dynamic, and that only rarely were landscapes afforded enough time to erode down to this low-relief surface. Davis acknowledged that few areas complete the cycle; rather, the cycle is restarted by a new uplift event before a peneplain can develop.

Davis also noted that most old age landscapes retained a few isolated, relict uplands, typically formed on resistant bedrock, or located far from the major streams. These features he called **monadnocks**. This type of long-term evolution of the fluvial system, which can take millions of years, is the essence of the cycle of erosion.

The concept of rapid uplift followed by long-term landscape stability has since been criticized by many geomorphologists. That said, Davis also recognized that this aspect of the model was a simplification. Nonetheless, his qualitative model organized – incorrectly in some instances

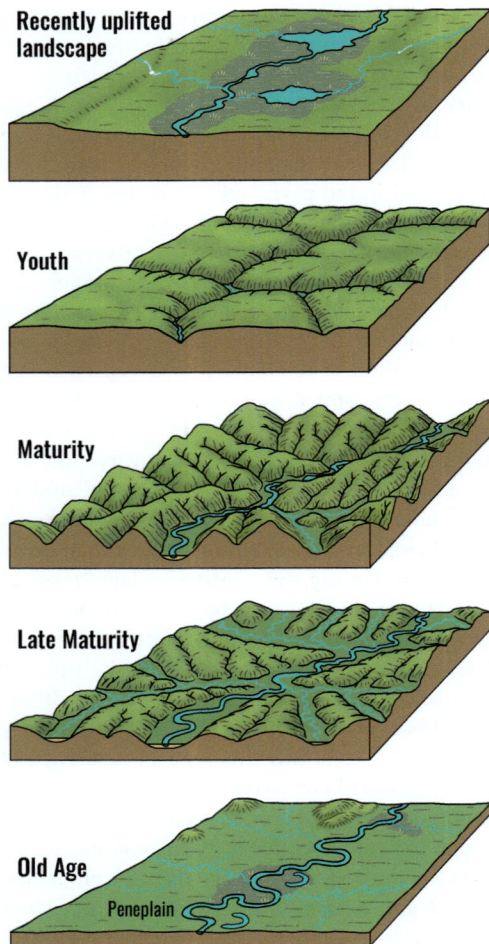

Figure 2.4 Changes in slope and elevation throughout the cycle of erosion.

Figure 2.3 General model of the initial uplift of a landscape and its progressive denudation by running water, as envisioned in Davis' cycle of erosion.

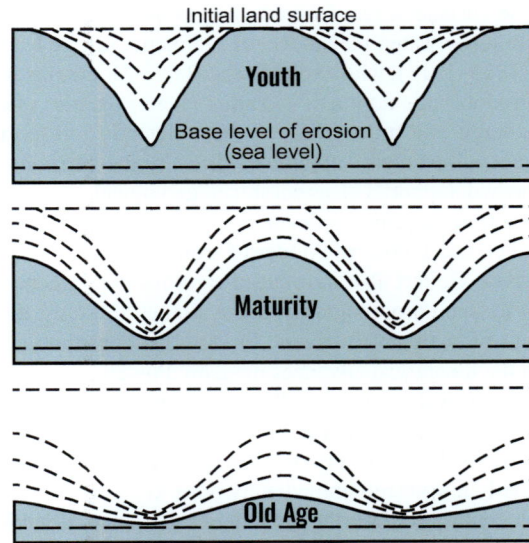

– relations among hillslopes, valleys, rivers, and overall landscape form, as erosion progresses on an uplifted, tectonically stable portion of a land mass. Although Davis' model is no longer used to guide scientific advances in geomorphology, it remains a highly useful teaching tool. The model illustrates an idealized form of progressive erosion that some landscapes might undergo if and when the assumptions of the model hold. Its visual character is also easy to comprehend for students with different geologic backgrounds.

Although the cycle of erosion is an oversimplification of how landscapes evolve, it served to draw many people into geomorphology because of its elegance and simplicity. It made landscape evolution and development easy to comprehend and inspired many in the field to think about landscapes in different ways. For example, in Davis' model, slopes are assumed to go through a progression from the steep slopes of the youthful landscapes to the low-angle slopes of old age. Davis coined the term **downwasting** for the way in which slope gradients become lower (gentler) over time (**Fig. 2.6**). This assumption drove many geomorphologists to study how slopes change over time. It is now generally assumed that slope gradients do not decline over time if the erosional debris at their bases is continually removed by rivers, as is commonly the case.

Davis' model was based on the notion that landform development and appearance is a function of the underlying geology (mainly rock type, structure, and erodibility), process (the various agents of weathering, erosion, and sediment transfer), and time. If nothing else, Davis drew attention to these three aspects of geomorphic development, all of which are viable avenues of contemporary geomorphic research.

Davis also incorporated into his model the concept of **grade**, first introduced by G. K. Gilbert. In doing so, he supported the idea that rivers can (over time) *adjust*, so that they

Figure 2.5 Images of landscapes that represent what Davis might have called (**A**) youth, (**B**) maturity, and (**C**) old age. **A.** The Grand Canyon, incising into the uplifted Colorado Plateau, in Arizona, USA. Source: L. Maher. **B.** The Adirondack Mountains of New York State, USA, where most of the land surface is in slope. Source: R. Schaetzl. **C.** The Bayanaul region in Kazakhstan, a low-relief landscape that Davis might have considered a peneplain. Source: Ekamaloff, Public domain, via Wikimedia Commons.

Figure 2.6 Idealized depictions of two different scenarios for hillslope erosion and landscape denudation. Backwasting was generally supported by King's model of landscape evolution. Downwasting was generally supported by Davis' cycle of erosion. Source: From Davis (1932); © Geological Society of America, used with permission.

have sufficient energy to transport the sediment supplied to them. In short, grade represents a state of balance between erosion and deposition. Davis envisioned that streams reach grade during maturity, when downcutting slows. At this time, the stream forms and widens its floodplain. The floodplain is used to store sediment that the stream cannot currently transport downstream. The concept of grade has since been refined and revised. The shortcomings of Davis' view are that he restricted the attainment of a graded condition to a particular stage of landscape development, and that he associated the achievement of grade with adjustments of channel slope. Today, it is widely accepted that the development of a graded condition can occur at any time within a river system, and that adjustments to attain grade can involve many different aspects of river form (besides just channel slope).

2.2.2 Alternative Models of Landscape Evolution

Although the cycle of erosion reigned supreme throughout the early twentieth century, prominent geomorphologists in parts of the world other than the United States found it lacking. Born in Austria, Walther Penck (1888–1923) was one of the first to challenge Davis' ideas. Penck developed a model of landform

evolution based on observations in South America and Europe. A particularly noteworthy aspect of his model is that uplift does not occur abruptly and then cease, as Davis proposed, but instead can occur at any time. That is, uplift can occur even as landscapes are undergoing erosion and denudation. Penck also believed that peneplains could only develop where uplift rates are less than erosion rates over long periods of time. If uplift rates equal erosion rates, relief will remain constant over time. On the other hand, if uplift rates exceed erosion rates, the elevation and relief of the landscape will increase.

Changes in the rate of uplift in Penck's model are also important for understanding the *form* or shape of hillslopes. When uplift rates are accelerating, they outpace erosion, and slopes develop convex-upward forms. When uplift rates are constant, erosion rates equal uplift rates, and slopes may be straight (linear) in cross-section. Finally, when uplift decelerates over time, erosion rates will outpace uplift and slopes become concave upward, like the inside of a bowl.

Another prominent geomorphologist to challenge Davis' ideas was Englishman Lester King (1907–1989), whose experience came mainly from dryland areas. Based on his work in Africa, he proposed an alternative to the cycle of

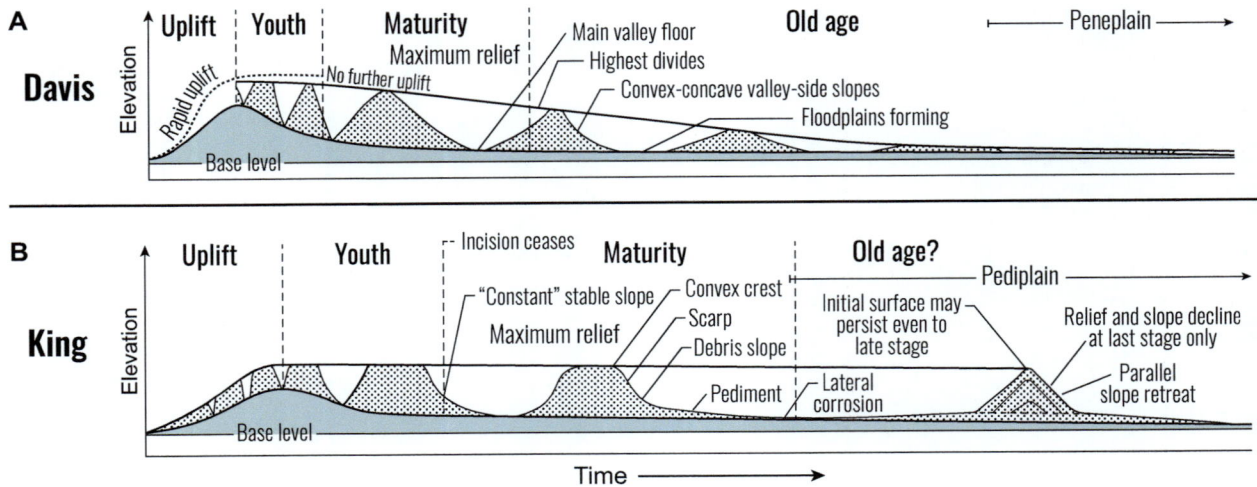

Figure 2.7 Schematic diagrams showing changes in elevation, relief, and slope profiles over time (for the stages of youth, maturity, and old age), following an initial uplift event. The figure contrasts Davis' and King's views on landscape evolution. Source: Oldroyd (2013). Used with permission of Elsevier Science & Technology Journals, from *Treatise on Geomorphology*, Shroder, John F. (2013); permission conveyed through Copyright Clearance Center, Inc.

Figure 2.8 In this scene from Canyonlands National Park (Utah, USA), slopes maintain their relative steepness as they retreat by erosion – a concept called backwasting. Surfaces are primarily eroded backward. "Downward erosion" is not widespread. Source: R. Schaetzl.

erosion and is credited for introducing the term **backwasting**, or parallel retreat, for slope development in dryland areas. Backwasting is widespread in desert landscapes today, especially those developed on bedrock, and King envisioned most slope development as being dominated by this process (**Figs. 2.7, 2.8**). He suggested that backwasting led to the production of widespread, gently sloping surfaces known as **pediments**, which form at the base of the steep, backwasting slopes (see Chapter 10). Pediments are erosion surfaces cut on bedrock and may be covered by a thin veneer of sediment slowly being transported downslope (**Fig. 10.44**). In landscapes dominated by parallel retreat, pediments coalesce over vast areas to form **pediplains**. Thus, King's alternative view of dryland landscape evolution championed the notion of **pediplanation**, as opposed to Davis' peneplanation concept for humid climates. In short, Davis envisioned hillslopes, and the landscape as a whole, as flattening over

time, but King felt that steep hillslopes produced by rapid, initial uplift remain steep, in essence shaving off layers of rock (by backwasting) as they retreat. In the final analysis, King's model was not markedly different than Davis' cycle of erosion, as both saw landscape evolution involving initial uplift followed by a long period of stability, with progressive erosion occurring until uplift resets the cycle.

2.2.3 The Rise of Modern Geomorphology, Focusing on Quantification and Process

By the middle of the twentieth century, geomorphologists were becoming increasingly dissatisfied with qualitative, descriptive models of landscape evolution that neglected details of geomorphic processes. They wanted to know *how* the surface had been formed and sculpted, not just what it looks like. Davis, Penck, and King had focused on the general *form* of landscapes. They had paid little, if any, attention to the *processes* that had produced those forms. Eventually, the lack of "process" in all these models became viewed as a weakness, motivating geomorphologists to engage in detailed studies of geomorphic processes. To do this well, they also needed better – and more quantitative – data.

Initial signs of discontent can be found in work by hydrologist Robert E. Horton (1875–1945), who in 1945 proposed models of hillslope erosion, the initiation of stream channels on sloping landscapes, and the development of fluvial (river) drainage networks. His work drew upon basic physical principles governing runoff and erosion. Horton noted that when rainfall rates exceed rates at which the soil can absorb the water, water will flow across the surface as overland flow, or runoff (see Chapter 15). If the erosive forces associated with this runoff exceeded the resistance of the soil to these forces, channels will then be cut. Thus, Horton's work linked the processes of runoff to the formation of stream channels and stream networks.

In a similar "process vein," in the 1930s, arose the work of British geologist and geomorphologist Ralph Bagnold (1896–1990). Bagnold had spent time exploring the barren, wind-swept, desert landscapes of northern Africa. During his travels, he became interested in the physics underlying the transport of sand by wind, and how sand transport can form dunes. To understand the physical processes associated with blown sand, he conducted many experiments, using wind tunnels. In doing so, he linked sand transport to dune formation.

The studies by Horton and Bagnold in many ways represented a return to the type of work initiated in the late 1800s by Grove Karl Gilbert, who combined his knowledge of geology with the laws of physics to tackle geomorphic problems – by investigating the processes that shape landforms. Gilbert's approach, however, had long been overshadowed by the cycle of erosion of William Morris Davis, which, for the most part, lacked a process component.

The trend toward continued process-based study was further amplified by geomorphologist Arthur Strahler (1918–2002), who, inspired by the work of Gilbert, Horton, and Bagnold, argued in 1952 that the understanding of geomorphic processes must be grounded in basic physical and chemical principles. This approach connected geomorphology to a vast arsenal of concepts from other disciplines. This perspective, now known as **process geomorphology**, transformed the field.

Today, the qualitative, descriptive models of Davis, Penck, and King are no longer used to guide geomorphic research. Instead, geomorphologists draw upon theory from the basic sciences to understand how surficial processes shape landscapes, and the rates at which these processes operate. Modern geomorphologists use observations and instruments to collect data on processes in the field and laboratory. They analyze data using statistical or mathematical methods (**Fig. 2.9**) and compare their findings to computer-based simulations of landform change, as guided by theory. Through this stepwise cascade of data collection, data analysis, and theoretical evaluation, geomorphologists are continually refining their understanding of how landforms develop and change. In short, contemporary research in geomorphology is very much a "high-tech" endeavor that takes advantage of the latest developments in technology and science. And yet, it all starts (and often ends!) in the field.

2.2.4 Modern Geomorphology's Increased Emphasis on Landscape Dynamics

Geomorphologists have always known that landscapes are dynamic and ever-changing. Millennia of geologic activity, tectonism, weathering, and denudation have formed the complex landscapes we see today. Geomorphologists strive to understand not only how surficial processes work and where they work most effectively, but how they may have changed over time. Because landscapes often preserve evidence of past processes, proper interpretations can help us learn much about what occurred in the past. It is not easy and is often quite challenging. But therein lies the fun, and eventually, the reward, of this science.

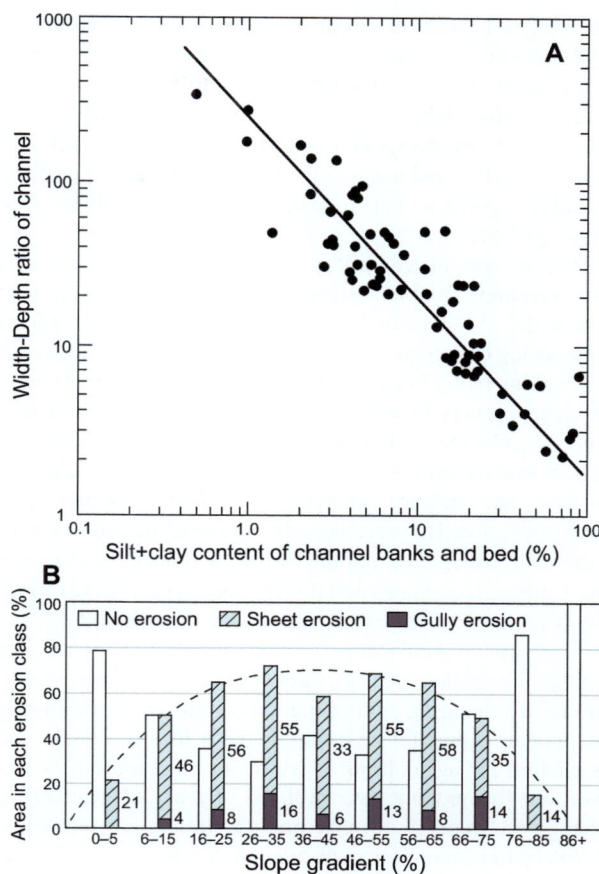

Figure 2.9 Examples of quantitative research undertaken by geomorphologists, in the mid-twentieth century. **A.** A scatterplot of the relationship between the width/depth ratio of river channels to the percentage of silt and clay in the channel bed and banks. Source: After Schumm (1960), via the US Geological Survey; in public domain. **B.** Bar graphs showing how the different forms of hillslope erosion vary in relation to their steepness (gradient). Source: From Horton (1945); © Geological Society of America, used with permission.

Studies of landscape dynamics – how they change over time – are particularly applicable to the modern world. The climate is changing rapidly today, at least in part because of human impacts, and as it changes, so are the processes that shape landforms. Permafrost is thawing, leading to massive thaw slumps and water-driven erosion processes that are new to many of Earth's cold landscapes (see Chapter 20). Dune fields, once active but since stabilized, are beginning to move again, as droughts become more prevalent. As sea level rises, and violent storms become more common, coastal erosion is chewing away at beaches. Increases in the frequency of large precipitation events are causing rivers to flood more frequently, forcing them to adjust their channels to accommodate these changes. Humans also are impacting geomorphic processes directly, by changing vegetation cover on land surfaces and by deliberately moving sediment. Agricultural and urban areas continue to expand, altering volumes of runoff and amounts of sediment entering streams and rivers. The construction of dams, mainly for hydropower, continues to increase around the world, starving coastal areas of sediment needed to maintain beaches and sensitive delta environments. For these reasons and many more, contemporary geomorphologists are actively trying to better understand all types of geomorphic systems in today's world, so as to help mediate their possible harmful consequences.

Under natural conditions, many geomorphic systems are seemingly "stable," especially when viewed over short timescales. As short-lived human observers, we have a hard time seeing change in most geomorphic systems. So, let's start by working with a landscape like that – one that is unchanging, or changing very little over time, even over short timescales. Such a landscape is often described as being in a condition of **steady-state equilibrium** (**Fig. 2.10A**). Short-term changes in form can (and do) occur in the landscape, as individual events affect it. But over periods of several decades or centuries, the average form of the landscape remains constant, as the landscape oscillates about some long-term average form. Short-term, event-related changes counteract each other.

But not all landscapes remain unchanged, even over human timescales. Instead, they may be adjusting to slowly or rapidly changing environmental conditions. Such adjustments can occur, for example, when a forested landscape is cleared for agriculture. If we adopt a geologic perspective, one that extends over millions of years, all landscapes change or "evolve" because neither climate nor tectonism is likely to remain constant over such long spans of time. Changes in landscape form/shape around a slowly changing average condition is called **dynamic equilibrium** (**Fig. 2.10B**). An example is the progressive decrease in relief related to long-term denudation. Blips and bumps occur along the way, perhaps lasting several decades or even centuries, but the overall trend continues for millions of years.

Some disruptions to, or changes in, geomorphic systems can occur so suddenly, and are so large, that the system cannot readjust back to its former state. These types of large changes are often external, driven by rare, extraordinary events that produce long-lasting effects, such as earthquakes, major floods, volcanic eruptions, rapid changes in climate, or human alteration of landscapes. Nonetheless, in some cases, sudden, irreversible landform and landscape changes can result from *internal* dynamics. An example might be the gradual weaking of bedrock on a hillslope by persistent weathering, causing the slope to fail, producing a catastrophic landslide. Such changes can occur independently of any external factors.

Whether internally or externally caused, abrupt changes in landscape systems often involve the crossing of a **geomorphic threshold** – the point or value in the system that, once exceeded, triggers such large changes that the system shifts to a new, more stable configuration. Perhaps the best way to understand geomorphic thresholds is to look at some examples.

Prior to the eruption of Mt. St. Helens in the state of Washington in 1980, the Toutle River had a meandering pattern. After the eruption, the river quickly transformed into a braided pattern because of the large amounts of ash and debris that had entered the fluvial system. The eruption, an external factor, pushed the river system over a geomorphic threshold, changing it from meandering to braided. For an example of a threshold *within* a system, let's look at alluvial fans. As rivers flow from steep mountains to flat adjacent lowlands, sediment accumulates at the transition zone from the mountains to the lowlands, forming alluvial fans (**Fig. 1.2**). As sediment accumulates on the fan surface, its slope gradually – almost imperceptibly – steepens (**Fig. 2.11**). These changes in slope steepness might be considered a type of dynamic equilibrium (**Fig. 2.10B**). Eventually, however, because of continued deposition and steepening, the surface slope on the fan exceeds some critical angle, causing the river on it to flow so fast that it begins to incise. As incision takes place, the zone of deposition shifts downslope, permanently changing the form of the alluvial fan. In other words, an internal threshold has been crossed, triggering rapid adjustment of the river and resetting the trajectory of dynamic equilibrium. The implication of this example for geomorphic systems is that small changes, when they accumulate, may eventually cause a system to cross a geomorphic threshold, thereby triggering a large, persistent response. If a series of events moves the system close to a threshold, the event that causes it to transcend the threshold may not be an exceptional one, but a rather ordinary one.

In geomorphology, we can sometimes identify systems that are about to exceed a threshold. In that case, we may be able to take action to prevent the change or mediate the damage caused when it does happen. Because of human actions, increasing numbers of landscapes are experiencing rapid change, crossing thresholds with increasing frequency. In short, the modern world is a much more dynamic place than it was prior to the Industrial Revolution, and geomorphologists stand ready to help understand it – armed with appropriate theories, excellent data, and sophisticated instruments. In the chapters that follow, you will learn how geomorphologists perform this important work, but first let's examine some of the tools used by practicing geomorphologists.

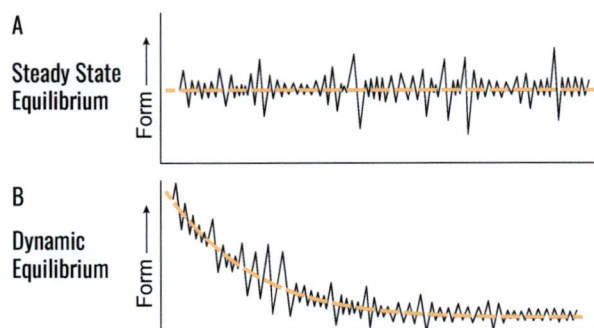

Figure 2.10 Schematic illustrations of two types of equilibrium in geomorphic systems: **A.** steady-state equilibrium; **B.** dynamic equilibrium. Source: From Chorley and Beckinsale (1980) courtesy of the Geological Society of America, which is based on earlier illustrations from Schümm and Lichty, 1965; Chorley and Kennedy, 1971; and Schümm, 1977.

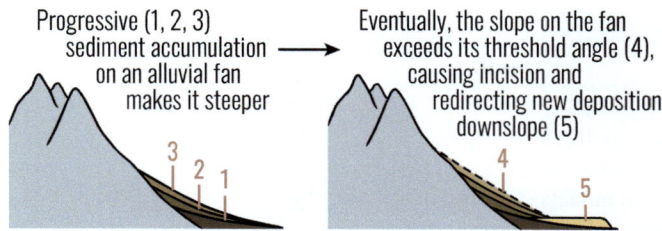

Progressive (1, 2, 3) sediment accumulation on an alluvial fan makes it steeper → Eventually, the slope on the fan exceeds its threshold angle (4), causing incision and redirecting new deposition downslope (5)

Figure 2.11 Diagram showing how the slope on a slowly aggrading alluvial fan eventually comes to exceed a critical steepness – a threshold value. Crossing this threshold leads to instability, incision, and a new mode and location for further deposition – farther down-fan.

2.3 TOOLS OF THE GEOMORPHOLOGIST

You may ask, "How is geomorphology done?" First and foremost to geomorphology is basic field observation. Our eyes are often our best "tools." Geomorphologists inspect landscapes at all scales. Basic "field observation" will always be essential to geomorphology.

Geomorphologists also combine inference with observation. Early geomorphologists were exceptional at applying these two simple skills. They saw things, and put them together in ways that others did not, to help them understand what they saw. Being able to infer how landscapes develop, based on observation, requires in-depth knowledge of geomorphic processes, along with insight into how this knowledge relates to what is being observed. Proficient geomorphologists are adept at "seeing" the landscape and linking landscape components together, to arrive at a defensible model of how the landscape developed. If you are an avid landscape "observer," or as the famous geographer Francis D. Hole often said, a "land looker," then you are a geomorphologist!

Although today's geomorphologists still rely on visual observations, they also generate landscape data using high-tech tools, instruments, and techniques. A vast amount of data and a plentiful toolkit are now available to the modern geomorphologist (**Fig. 2.12**, **Table 2.1**).

The reliance on both basic observation and high-tech data means that geomorphology has a place for people with varying skills and abilities – those who prefer working with spatial data on computers or smartphones, those who excel in the controlled lab environment, and "field scientists" who are good at "seeing" things outside. Geomorphology utilizes the skills of map lovers and GIS (geographic information science) analysts, as well as geographers, geologists, chemists, climate scientists, physicists, ecologists, and soil scientists. Everyone can contribute; this diversity makes the discipline stronger (**Table 2.1**).

A common form of representing landscape data is in the form of various kinds of digital maps – maps of topography, hydrology, sediment, land cover, etc. Maps (spatial data) have always been critical to geomorphic research, but geomorphologists now have at their disposal high-resolution spatial data that can display aspects of the physical landscape in rich detail and with unparalleled accuracy and areal coverage (**Table 2.1**, **Figs. 2.12, 2.13**). Spatial data that previously were variable in coverage and quality are now available in high quality for most areas. Such datasets can be easily downloaded, making them readily accessible in the field, lab, or office (**Fig. 2.13**). Such data are a necessary first step to any geomorphic work, as they provide context and background information to help

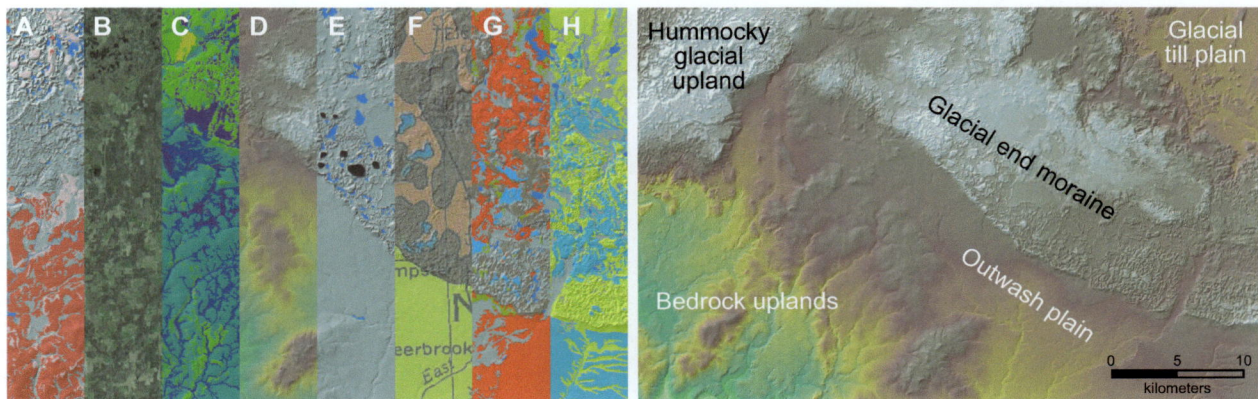

Figure 2.12 A collage of pieces of eight different maps, each exported from a geographic information system, displaying a different aspect of the same landscape in east-central Wisconsin, USA. Many more such exports are possible, because of the rapidly expanding array of databases worldwide. The data are displayed using either discrete colors or a gradational color ramp. **A.** Loess thickness shown in shades of red (gray = no loess). **B.** A true-color aerial photograph. **C.** Natural soil wetness, based on an index that was derived specifically for this purpose (purple = wettest, yellow = driest). **D.** Topography overlain on a hillshade elevation model. **E.** A hillshade elevation model, showing ice-walled lake plains in black. **F.** A small-scale map of glacial sediments and landforms, compiled prior to the advent of GIS. **G.** Soil parent materials. **H.** Texture of the uppermost mineral soil horizon. The map on the right depicts the topography and elevation of the landscape (similar to D), with the landform regions labelled.

Table 2.1 Some examples of tools and techniques geomorphologists commonly utilize

Tool category	Examples
Maps and spatial data	Topographic maps Air photos and satellite imagery Detailed topographic data, e.g., LiDAR Digital elevation models (DEMs) Soil, vegetation and geological maps Land-use maps
Field equipment	Excavation and coring equipment Observational equipment, e.g., cameras, drones Locational equipment, e.g., global positioning systems, total stations Sediment sampling equipment Instruments for measuring flows of water, air, or ice, and sediment
Modeling, mathematical, coding, and statistical tools	Computer modeling and simulation software Statistical techniques Coding and programming software, e.g., R, Python
Laboratory equipment tools to analyze:	Sediment chemistry and mineralogy, e.g., X-ray diffraction or scanning electron microscopy Pollen and macrofossils Grain size and shape, e.g., laser diffraction Moisture content
Dating techniques and tools (see Chapter 3)	Radiocarbon Luminescence Dendrochronology Rock coatings and rock weathering Isotopic techniques Cosmogenic nuclides (and several others)
Experimental techniques	Rainfall simulations Wind experiments, such as in wind tunnels Water flow experiments, such as in flumes and stream tables

Figure 2.13 An illustration of the increased topographic detail available to geomorphologists, over just the past few decades, for an area in Michigan, USA (as an example). **A.** A standard, 7.5-minute series, topographic map – the gold standard for landscape study for most of the twentieth century. **B.** A hillshade, 10-meter digital elevation model, exported from a GIS. **C.** The increased detail afforded by LiDAR data. Note how some features, not previously discernable, such as the sand dunes, are now easy to pick out, using LiDAR data.

characterize the landscape and provide opportunities for hypothesis generation. Topographic maps – the bedrock of the mid-twentieth-century geomorphologist – have been largely replaced by digital terrain data, called digital elevation models (DEMs). The recent introduction of LiDAR terrain data, with spatial resolutions as small as 0.1 m, has been a boon to geomorphic interpretations (**Fig. 2.13C**). Aerial imagery, from a variety of sources and spanning a wide range of ages, are now freely available on public platforms such as Google Earth and from various government websites. Digital soil maps contain detailed spatial information on the upper two meters of the land surface, including

Figure 2.14 Geomorphology in action – in the field and lab. Source: R. Schaetzl and B. Rhoads.

"Doing Geomorphology" - The CUP Model

Characterize the **landforms**, so as to...

Understand the processes that govern their formation, distribution and behavior, as well as the sediments that comprise them. This information enables us to...

Predict the future occurrences and behavior of the landforms, thereby assisting in management and use.

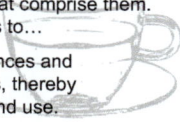

Figure 2.15 The CUP model of geomorphology.

texture, wetness, depth to bedrock, and organic matter content (**Fig. 2.12**), allowing surficial geology to be characterized in detail. The variety of spatial data and tools available to geomorphologists has never been more diverse, and the data quality has never been better.

Armed with ample spatial data, many geomorphologists' next step is to go to the field and collect information, often to answer a research question. Typically, some sort of excavation (by auger, shovel, coring device, or backhoe) is required (**Fig. 2.14**) to remove and characterize the materials in which landforms have developed. That said, most of us have been more than happy to examine exposures in road cuts and gravel pits as well! Any exposure of sediment and rock is often helpful. The next step in geomorphic study is often some type of sampling, assisted by the requisite equipment. Sediment samples brought back to the lab are variously analyzed for grain size, mineralogy, color, moisture content, chemical signatures, etc., all designed to fully characterize them (**Fig. 2.14**). *Characterization* is usually a necessary first step in many geomorphic research projects – we must establish what is "out there" before we can answer questions about the landscape (**Fig. 2.15**). And yet, data derived from lab work or pulled from digital datasets can be used to do more than just characterize the landform; they can serve as inputs to mathematical or simulation models. Such modeling and analyses help geomorphologists to better *understand* the landscape and the surficial systems that produced it. Characterization helps geomorphologists to understand what is "out there" – the first two steps in the *CUP model* of geomorphic science (**Fig. 2.15**). To summarize, geomorphologists often aim to characterize, describe, and analyze the landforms under study, by taking key samples and then by deriving information about them in the lab.

In many cases, the measurement of geomorphic *processes*, such as the flow of water that moves sediment in a river or the speed of wind that blows sand across dunes on a beach, is also necessary to address research questions about landform dynamics (**Fig. 2.15**). The goal is to determine how various surficial processes have actively shaped and changed landforms over time and space. Measurement of processes in the field can be both exciting and challenging (such as trying to capture how sediment moves, or channel banks erode, in a flooded river!). Nonetheless, contemporary instrumentation now provides unprecedented opportunities to collect field data on geomorphic processes.

We also note that a key part of most geomorphic work involves *dating* the feature or sediment under study (see Chapter 3). How old is it? When did it form? Is it stable or is its form changing? How rapidly does it change? Indeed, never before have geomorphologists had a wider array of dating techniques at their disposal (**Table 2.1**). In many cases, we have the capability to establish when (1) the *sediment* was deposited, as well as when (2) the *surface* of that sediment (the landform) last stabilized, that is, when the landform itself was formed. Knowing how old the sediment is, or the age of the landform, helps us establish the rates of various formative processes. It is a common goal to determine when and at what rates a landform may have formed, and how rapidly it might change in the future. Humankind often needs to know this, as all of our daily activities are done on some kind of landform; wouldn't it be good to know how stable it is?

This brings us to the last part of the CUP model of geomorphology – *prediction* (**Fig. 2.15**). Armed with information about what the landform or landscape is like, and understanding how (and when) it formed (the processes that shaped it), geomorphologists may then be able to assist in the goal of better land management. Can we *predict* what this landscape will be like in the future, how resilient it is to change? Can we predict where future geologic or hydrologic resources might be hiding in plain view? Those goals and questions are always in the background of most geomorphic studies, with the objective of assisting land managers of all kinds.

It is easy to understand why this is an exciting time to be a geomorphologist! Opportunities now exist to view, measure, date, and model landscapes in ways that were previously unimaginable. As a result, geomorphologists are shedding light on longstanding problems and answering questions about changes in the Earth's surface that we did not have the data or tools to address only a few decades ago. And we are even taking our work to other planets (see Chapter 23)!

REVIEW QUESTIONS

2.1   What is the difference between a landform and a landscape?

2.2   What methods do geomorphologists use to collect and analyze data?

2.3   What types of datasets are used in geomorphic research and applications?

2.4   Describe the general characteristics of the youth, maturity, and old age stages of Davis' cycle of erosion.

2.5   Provide an example of a system in steady-state equilibrium and one in dynamic equilibrium. Avoid using examples from the chapter.

2.6   What types of external factors or causes might trigger a geomorphic system to cross a threshold? What happens to systems that cross a threshold?

2.7   Compare and contrast the landscape evolution models of William Morris Davis and Lester King.

2.8   Describe the difference between backwasting and downwasting. How have ideas of grade changed over time, and what is the current theory?

2.9   Describe the main components of the CUP model of geomorphology, and how it is operationalized.

3 Dating Methods in Geomorphology

Randall Schaetzl

How old is the Grand Canyon? When did the glaciers last retreat from this area? How long does it take to form an inch of topsoil? When did the earthquake occur that formed these rock scarps? These are the questions that geomorphologists ponder. This chapter will outline the tools and approaches we use to answer such questions.

Establishing how old a landform might be, that is, when it formed, has always occupied the mindset of geomorphologists. If we know how OLD a landform is, then we can begin to understand how it is evolving, how fast it might be changing, and how it formed in the first place. Fortunately, various dating principles and techniques now exist to address these issues. These techniques require the ability to measure change in a system or a landform over time, with the (usual) goal of establishing the age of a sediment package or a landform. The geomorphologist's "dating toolkit" is rich indeed, and new methods are seemingly always being developed!

3.1 BASIC PRINCIPLES AND CONCEPTS OF DATES AND DATING

Let's first discuss some basic terminology. Many physical scientists use the term **time**$_{zero}$ to describe the start of something. Geomorphologists might use the term to indicate when a landform was formed, or when a sediment was deposited; time continues to pass since this time$_{zero}$.

A **date** is a specific point in time, for example, AD 1066 (the Battle of Hastings) or July 20, 1969 (the first Moon landing). Dates can also be written in terms of "years ago." For example, we may be able to estimate that sediments on a floodplain were deposited about 750 years ago, or that a glacial moraine formed about 13,560 years ago. Note the modifier "about"; many times, a date cannot be precisely determined. We can only make a good estimate. And so, some degree of uncertainty is almost always built into our discussion of dates. That said, if we KNOW the exact date of an event, like the May 31, 1889 Johnstown (Pennsylvania) flood, then we can say with certainty that it happened on that date, or describe it as having happened 137 years ago.

An **age** is an interval of time, often measured back from the present, for example, 11,700 years ago, or the Bronze Age. Because most geomorphic features were not formed recently, the terms **ka**, **Ma**, and **Ga** are commonly used for ages – shorthand for "thousands of years ago," "millions of years ago," and "billions of years ago," respectively. The ages 11,700 and 11.7 ka are identical. For most ages, it is assumed that they mean "years ago," so one does not have to always write "8.35 ka ago" (the "ago" is assumed).

A key factor in any dating exercise centers on the accuracy and precision of the dates obtained (Fig. 3.1). **Precision** describes the statistical uncertainty that is associated with a date, that is, how reproducible is the estimate? Once a date has been derived (precision), the researcher must also consider the **accuracy** of the date in relation to the landscape, event, or depositional environment that is being dated, that is, how close is the estimate to the true age of the material being dated? Although we often do not know a great deal about the accuracy of a date, we should always keep its possible accuracy – or inaccuracy – in mind when we report and discuss it.

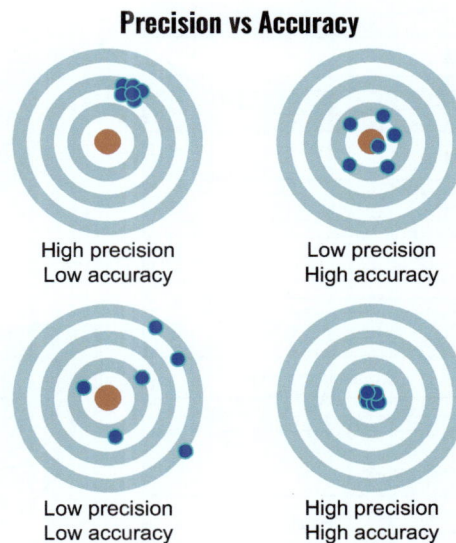

Precision vs Accuracy

High precision
Low accuracy

Low precision
High accuracy

Low precision
Low accuracy

High precision
High accuracy

Figure 3.1 Diagrams illustrating the concepts of precision and accuracy.

Figure 3.0 This geomorphologist is taking a sample of a sand wedge, for the purpose of establishing when the sand inside it was emplaced there by strong winds. Source: R. Schaetzl.

3.2 GENERAL TYPES OF DATING IN GEOMORPHOLOGY

Landforms, sediments, and their surfaces can be dated in a number of different ways. However, each method falls into three basic groups of dating techniques: **relative dating**, **correlated age dating**, and **numerical dating**. We start with the last method first.

In numerical (sometimes called absolute) dating, a numerical estimate of the age of the feature is determined or estimated, using some type of natural chronometer built into the system. For example, by using a numerical dating technique we may be able to estimate that a volcanic ash deposit is ≈11,300 years old, or that a fault scarp formed by an earthquake happened in 1817 AD. Numerical ages allow for quantification of absolute differences in age among sediments or units. For example, sediment A is ≈2,500 years older than sediment B. The latter part of this chapter is focused on some of the most commonly used methods of numerical dating in geomorphology.

Correlated dating methods, sometimes referred to as calibrated methods, are also able to provide information about time intervals between events. But because they depend on numerical techniques to establish some sort of quantitative relationship, correlated methods cannot stand alone. They require numerical dates on similar features or landforms, as something to correlate with. For example, rinds of decayed rock material (weathering rinds) tend to thicken over time, just as lichens on rocks in alpine settings tend to increase in diameter over time. To establish the actual duration of exposure for these rocks, the rates at which the rocks develop weathering rinds, or the growth rates of lichens on rocks, must be determined by *correlating* them to numerical dates from other rocks, in similar climatic settings.

Before the advent of numerical dating, sediments and rocks could only be dated by relative means. Early geomorphologists, for example, could be sure that landform X was older than landform Y by its appearance, or some form of untold intuition. They were performing relative dating, which indicates if a sediment, surface, or landform is *relatively* older or younger than another – no numerical age estimates are determined. Relative dating cannot tell us exactly how long ago a sediment was deposited or how long ago a geomorphic surface formed, only that it followed or preceded the deposition of another sediment, or the formation of another surface. However sometimes landforms, surfaces, and sediments can be placed in their proper *sequence* of formation, in other words, which came first, second, third? But even though we may be able to establish that sediment A is older than sediment B, which is older than sediment C, we may not be able to tell *how much* older A is than B, or B than C. Relative assessment of age is nonetheless very useful, especially when coupled with available numerical dates of sediments above or below the units of interest. Theoretically, relative dating has no temporal limit.

3.3 GEOLOGIC DATING PRINCIPLES

Several geologic principles are available to assist us with dating applications (**Fig. 3.2**). Nicolaus Steno (1638–1686), a Danish geologist and priest, developed one of the earliest of

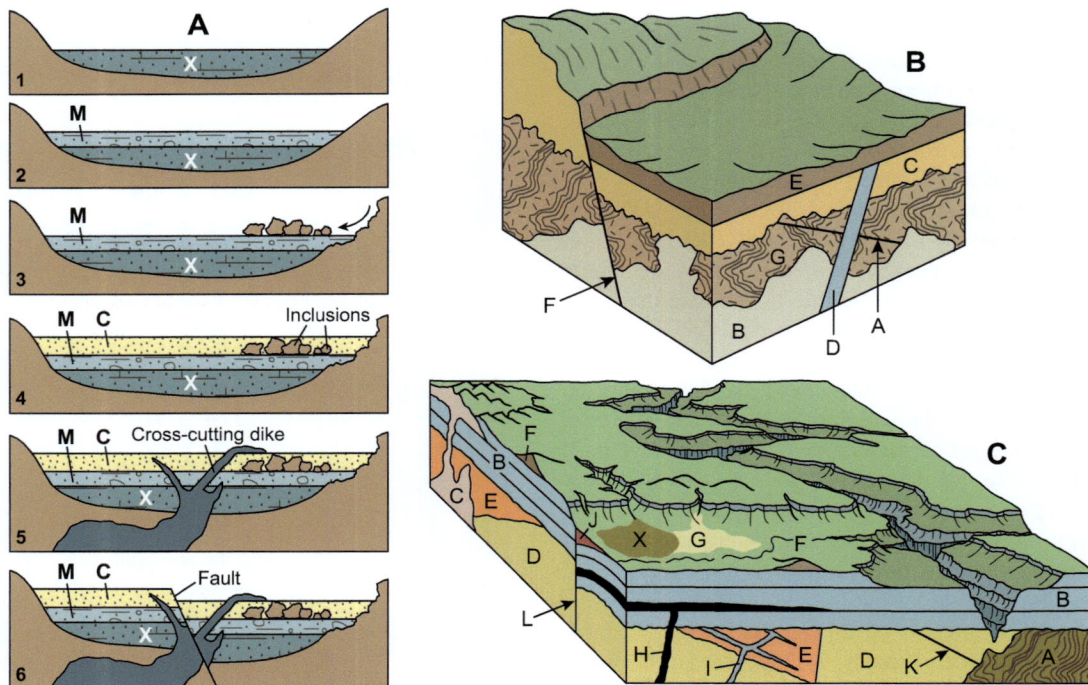

Figure 3.2 Diagrams that illustrate some of the basic geomorphic principles that can help determine the relative ages of rocks, sediments, and geologic events.

these principles – the **principle of original horizontality** – which states that layers of sediment are generally deposited in a horizontal position. From the principle of original horizontality, we assume that layers of sediment in **Fig. 3.2A**, which are flat-lying, have not been disturbed and maintain their original horizontality, as is true of many sedimentary rocks (see Chapter 5). Steno also recognized the **principle of superposition** – that in an undeformed sequence of sedimentary rocks or other similar sediments, each bed is older than the one above, and younger than the one below. In other words, deeper layers are relatively older. Layer X in **Fig. 3.2A**-2 is clearly older than layer M, using this principle. A third principle that can help further with relative dating is the **principle of inclusions**. Specifically, inclusions within a host rock (or deposit), such as fragments of basalt within a mud flow, or logs embedded in a glacial moraine, must be *older* than the formation that contains/encloses them. The rocks that originated in a landslide in **Fig. 3.2A**-4 must be older than the sediment (layer C) that eventually buries ("includes") them. Finally, the **principle of cross-cutting relationships** states that when a rock unit (**Fig. 3.2A**-5) or a fault (**Fig. 3.2A**-6) cuts across a body of rock or sediment, it must be younger than the host rock. This principle is most commonly applied for faults, which are always younger than the rocks they cut.

Let's use the block diagram in **Fig. 3.2B** to expand on some of these points. Using the principles discussed above, we can ascertain the relative ages of the various deep-seated rocks. Rock B is younger than rock G because of the **fault** at A, which cross-cuts G, but not the younger rock B. Rock B must then have formed as magma intruded into the pre-existing rock G. The deposition of the rocks at C follows in time, and then the rocks at B, G, and C got intruded by the dike D. The rocks that form E then follow, and lastly, the entire sequence is faulted (F). Many of the principles discussed above are used to help determine the sequence of geologic events here.

This type of relative dating logic can be applied to even more complex landscapes, as summarized in **Table 3.1**. In **Fig. 3.2C**, sedimentary rocks overlie a complex series of older rocks, using the principle of superposition. With regard to the older rocks, the rocks at A must be a younger intrusion into D, because it is not cross-cut by the fault at K, as are the rocks at D. We can assume that rock E is younger than rock D (superposition), and that the igneous intrusion at C is younger than the rocks at B, E, and D (cross-cutting relationships). Likewise, the igneous intrusion at I is also younger than the rocks at E and D. Of the two intrusions (I and C), C must be younger, because it cuts across the rocks above (B), but I does not. The sedimentary rocks at B lie

Table 3.1 Relative age information for the block diagram in Fig. 3.2C

	Reason/principle
Youngest rock or geologic event	
Rocks at X	Fan deposits post-date the rocks at B and the fault at L, and are younger than the deposits at G (superposition)
Rocks at G	Fan deposits post-date the rocks at B and the fault at L
Rocks at F	Must be older than the rocks at B (superposition)
Intrusion at C	Must be younger than the host rock (B), due to cross-cutting relationships. Its age relationship to the intrusion at H cannot be determined
Intrusion at H	Must be younger than the fault at L, because the intrusion does not cross it. Based on superposition, this intrusion must be older than at least the lowermost layer in B
Fault at L	Cross-cuts the rocks at B, so it must be younger than those rocks, and anything else that is older than B
Rocks at B	Overlie the rocks at E, D, and A, and hence, are older
Rocks at I	Igneous intrusion must be younger than the rocks (E and D) it cross-cuts, but older than the rocks above (B), which it does not cross-cut
Rocks at E	Overlie the rocks at D; relationship to the rocks at A cannot be established
Rocks at A	Not cross-cut by the fault at K; must have intruded into the rocks at D
Fault at K	Cross-cuts the rocks at D, but not the rocks at A
Rocks at D	Lowest in the stratigraphic sequence
Oldest rock or geologic event	

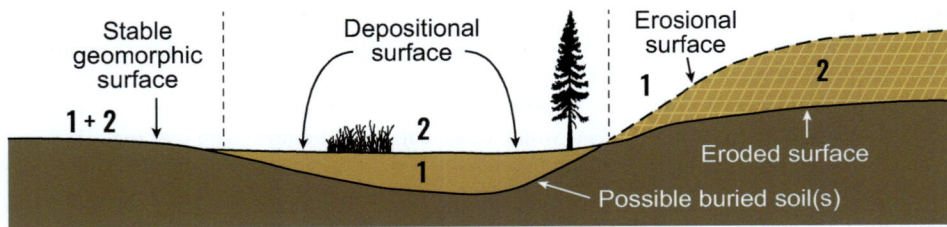

Figure 3.3 Diagram through an imaginary landscape, illustrating the concepts of stable, erosional, and depositional surfaces.

above the deeper rocks, so they must be younger than the rocks at A, D, and E, but older than the igneous intrusion at C. The rocks at B also cut across the older intrusion at I. The intrusion at H must be younger than at least some of the lower rocks at B. After they were already in place, the rocks at B and D were faulted (L). The intrusion at H is clearly older than the upper rocks at B, but younger than the fault at L. Subsequent to all this geologic activity, the sediments at J, F, X, and G were deposited on top of the rocks at B (superposition). Although we cannot ascertain their ages relative to each other, we can determine that the fan at X formed later than the fan at G, however, because of its overlapping relationship.

So, in summary, we often can place some, but maybe not all, of the age relationships in a landscape in chronological order using the basic geologic principles of relative dating discussed above. Of course, much of this type of dating requires detailed subsurface information, which is not always available.

3.4 DATING GEOMORPHIC SURFACES

Before we delve further into dating techniques, let's discuss the concept of **geomorphic surfaces**, because often it is the *surface* of a landform that we are dating, and care most about.

All landforms have surfaces. These surfaces mark the intersection of the landform and the surrounding environment; they define the limits of the feature in space. Geomorphologists refer to them as geomorphic surfaces if they can be defined and mapped, and are generally associated with a landform (not with several). Most geomorphic surfaces are **subaerial surfaces**: *sub* (under), *aerial* (the air, or atmosphere). In other words, the top and sides of a landform's subaerial surfaces are simply where "dry land" meets the atmosphere. The other two types of surfaces are **buried surfaces** and **subaqueous surfaces** (under the water) (**Fig. 3.3**).

Most geomorphic surfaces were formed by a unifying set of processes (erosional, constructional, or both), which occurred (or are occurring) over a defined interval of time – the period we are hoping to determine when we date the surface. Geomorphic surfaces may be flat, rolling, or steeply sloping. They may face any compass direction and have any elevation. They may be composed of any kind of sediment. All of these terms are descriptive, but perhaps one of the most *genetically descriptive* terms for a geomorphic surface is whether it is stable or (currently) unstable. Stable

geomorphic surfaces are being neither eroded nor buried at the moment, nor recently (**Fig. 3.3**). The presence of a soil cover is a good indication that a surface has been stable for some time. That said, some landforms, particularly those in deserts, can be quite stable but may have thin soils, or no soil at all.

We can sometimes determine the age of the surface in numerical terms, but we should always at least be able to discern its relative age vs adjoining surfaces, again using the principles discussed above. For example, let's examine **Fig. 3.4** – a landscape with both some active sand dunes and some older, stabilized sand dunes on it, as well as an actively migrating river cutting into it. Consider surface A to be the oldest and currently stable surface. The river is cutting into surface A and as blocks of sediment slump into the river, new surfaces at B are episodically being formed (a cutbank, see Chapter 16). Surface B is an erosional surface that is younger than surface A.

This brings us to another useful principle, the **principle of ascendancy and descendancy**, which states that an erosion surface is younger than the youngest *deposit or surface* that it cuts across, or truncates. This principle is widely applied in geomorphology. A corollary to the ascendancy–descendancy principle states that all surfaces are older than any valleys cut into them, and any deposits that lie within those inset valleys. This principle is especially useful in riverine settings, where younger materials are often inset within older sediments.

Buried geomorphic surfaces are usually fairly easy to relate to nearby surfaces. Buried surfaces (and their sediments) are always older than the sediments that bury them, and by extension, older than the surfaces formed on (or *in*, if they are erosional surfaces) those burying sediments. Thus, the buried surface at F in **Fig. 3.4** must be older than any of the dune sand that buries it, and older than the surfaces

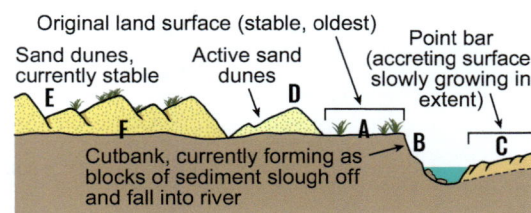

Figure 3.4 The concept of geomorphic surfaces and relative ages, illustrated in a cross-section through a fictional landscape.

that comprise the dunes. And so, we can now surmise that, with respect to relative surface age, A = F > B. Although we cannot conclude anything about surface B relative to the surface at D, it is nonetheless clear that the actively forming surface at B must be older than the long-stable surfaces on the dunes at E. Likewise, the surfaces and sediments associated with the stabilized dunes are older than the surfaces on the active dunes. Other information, such as soils data, could have been used to determine whether the dunes here are active or stable.

The last point to make is that some surfaces are **time-transgressive**, meaning that their age varies, usually predictably, across the surface. If the sand dune at D were to stop moving tomorrow, due to rains that stabilize it and plants that then quickly colonize it, the dune surface would be all the same age, across all parts of the dune. The dune stabilized all at the same time. But the constructional, point bar surface at C is slowly but steadily growing outward, as the river migrates to the left, cutting into and episodically creating new surfaces at B. Thus, the age of the surface at C gets progressively and predictably younger to the left – this is the definition of a time-transgressive surface – one whose age changes systematically across its extent.

3.5 WHAT DO WE DATE, EXACTLY?

In geomorphology, we are usually concerned with the formation, composition, and ages of landforms, as well as events that formed them. But what exactly are we dating? Do we date when the landform finally came to be, generally, as it appears now, in its current state? Or do we date when it started forming? And what of post-formative changes to the landform? And of course, as discussed above, are we dating the surface or the sediment? In other words, we might try to estimate the age of a glacial moraine, but the moraine may have formed over a period of hundreds of years. Normally, our preferred "date" estimate for this moraine would be for the *end* of this formative period – when it finally started to "look like" it does now (absent some minor erosion or other trivial changes since its time$_{zero}$). Geomorphologists refer to this endpoint in the development of a landform as the time when it – and the surfaces that comprise it – finally *stabilized*. And we will use that point in time too, when we discuss the ages of landforms.

So, when we date a landform, do we estimate when the sediment was deposited or when it finally stabilized into the form it currently displays? Depending on the technique we employ, we can date either, or both – when the sediment was deposited, and/or when the surface of the landform finally stabilized. Establishing the numerical age when a surface finally became stable is called **surface exposure dating** – an estimate of the length of time that a particular surface of a landform has been subaerially present and stable.

For some landforms, the age of the sediment, that is, when the sediment was formed, or deposited, is the same as the age of the landform and its surface. For example, the surfaces on the stabilized dunes in **Fig. 3.4** are the same age as

the eolian sand within them. Or think of a 15 cm thick flood deposit on a river bottom – the constructional surface formed by the flood deposit is the same age as the sediment below. For these types of settings, we have a choice to date the sediment or the surface – their ages are essentially the same! But for many other landforms and landscapes, there is a temporal disconnect, as we will examine in **Fig. 3.5**, which illustrates some additional "dating" concepts for another fictional landscape. Here, the original landscape, formed in glacial sediment, is shown at time window Time$_1$. By some later point, Time$_2$, a river has incised the landscape and formed a new surface – called an **erosion surface**. And as often happens, when erosion surfaces are formed, the sediment removed from them accumulates nearby, to form another surface – a **constructional surface**, also known as a **depositional surface**, on the bottom of the river valley (**Figs. 3.3**, **3.5**). Both the erosional and the constructional surfaces formed (more or less) simultaneously, and are of similar age, and both are younger than the original, unaltered, and stable surface upslope (ascendency/descendency). But how do the ages of the

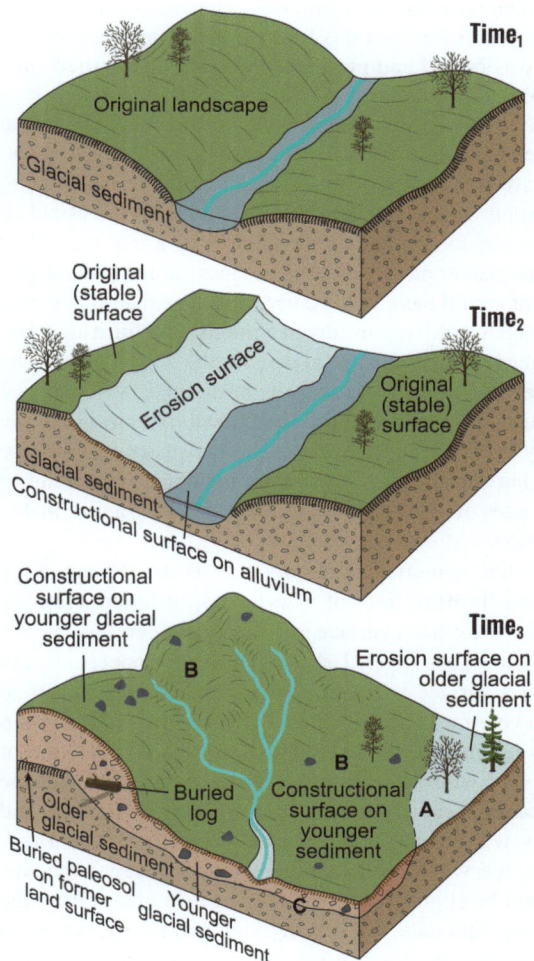

Figure 3.5 An imaginary landscape at three periods in its past, illustrating various concepts associated with surface exposure dating.

surfaces compare to the ages of the sediment that underlies them? The erosional surface is clearly *younger* than the sediment beneath it, whereas the depositional surface is the *same age* as the sediment beneath it, just like the dunes and the flood deposit examples above. As you may now understand, it is possible to use geologic principles and surface exposure dating techniques to establish when many different types of surfaces formed.

3.6 "LIMITING" DATES

The landscape in the third time window (Time$_3$) of **Fig. 3.5** adds additional dimensions to our discussion. Let's assume that the landscape shown in Time$_2$ was subsequently overridden by glacial ice, completely reorganizing much of it. Using the principle of superposition it is clear that the upper sediment was deposited by a later ice advance.

We initially observe that, in one area of the landscape the second advance of the ice did not deposit any sediment. In these locations, it simply eroded the preexisting landscape and formed a new erosion surface (A) in the older sediment (**Fig. 3.5**). Across the remainder of the landscape, a layer of sediment was deposited and new, constructional surfaces formed. Across much of this former landscape – surface C – its soil was buried and preserved; we call such buried soils **paleosols**.

How could we determine, with more specificity, when this glacial event occurred? First, organic materials that had previously laid on top of the paleosol could be dated using radiocarbon dating (see below). This date would establish when the organic materials there, such as a tree root, died when the glacier advanced over it. Because the second glacial event would have to have been *younger* than this wood, such an age would be considered a **maximum-limiting date**, that is, it provides a maximum date for the glacial event. The glacial event *could be no older* than the date we obtain on the buried soil. We might also try to establish the age of the glacially eroded surface at A, or the constructional surface at B, using surface exposure dating methods. This assumes that the ages of these two surfaces are similar – a reasonable assumption.

Many dating methods can only give us an estimate of the **minimum-limiting date** of a surface or sediment. That is, they can tell us that a surface is *at least as old* as the date we obtain, and likely older. Let's explore this concept. Rocks lying on a surface weather from the outside-in, and in so doing develop weathered "rinds." Weathering rind data from rocks lying on surface B could give us a good indication of its age. But with these ages come the assumption that the rocks started weathering as soon as the surface formed, and that they were lying on the surface as soon as it stabilized. Such an age would be a minimum-limiting age – the surface could be slightly older than the age we obtain from the weathering rind data, but not any younger. Lastly, we could try to date the sediment itself, because its deposition here would be equivalent to the glacial event. The most common way to date glacial sediment is to find within it some sort of buried organic material, like wood, picked up by the glacier and deposited with the other sediment. Such a date would give us a maximum-limiting date for the glacial event, because the tree was there *before* the ice picked it up, and certainly before it got buried in the sediment. That is, the sediment is younger than the buried log.

Let's discuss one last example of dating and landform relationships. In northern Arizona, near the Utah border, several impressive landforms have developed on former lava flows. At some time in the past, lava flowed across this landscape, cooling into dense, resistant layers of basalt. Over time, the softer rocks beneath the basalt weathered and eroded, leaving the lava flows to stand up as mesas that rise almost 300 meters above the modern landscape (**Fig. 3.6**). This type of landscape development is referred to as **topographic inversion** – what was once a lowland is now an upland (or vice-versa). How much time was necessary for this inversion to happen? To find out, we need a date on the lava flow. Dating basalt *is* possible by numeric dating means, but for now, let's assume that all we can do is date the sediment immediately beneath the basalt. Such a date would be a maximum-limiting date for the lava flow; the lava flow must have occurred *after* that date. Put another way, the lava flow could be no older than the date on the underlying sediment. Alternatively, the age of the oldest tree (if one even existed!), or some wind-blown sand, on top of the mesa would provide a minimum-limiting date for the lava flow; the lava flow must be older than that date.

3.7 TOOLS AND METHODS USED IN SURFACE EXPOSURE DATING

Geomorphologists have at their disposal a vast suite of surface exposure dating (SED) "tools," used to estimate the ages of stable, subaerial surfaces (**Table 3.2**). These tools assume that as a surface gets older, some aspect of it changes, in a predictable and irreversible manner. These **post-depositional modifications** (PDMs) to the surface usually

0   0.5   1
kilometers

Figure 3.6 A satellite image of East Mesa, formed on an ancient lava flow, in northern Arizona, near Hurricane, Utah, USA. Source: NASA.

Table 3.2 The major surface exposure dating methods and tools[1]

Method	Assumptions of the method	Application of the method	Approximate maximum timespan utility (years)
Rock-based post-depositional modifications			
Rock weathering	Rocks lying on a surface become more weathered with time	Rocks are struck with a hammer and then categorized by ease with which each breaks apart	10,000
Sound velocity in rocks	Speed with which P waves move through rock is correlated to degree of weathering	Compressional (P) waves are made to travel through clasts; the speed with which the waves move through the rock is measured	>100,000
Rock surface hardness	An impact from a small, specially crafted, device onto a rock surface will cause a pin to rebound farther if the rock is harder and less weathered	Rocks are repeatedly impacted with a Schmidt hammer	100,000
Rock angularity	Sharp edges on rocks lying on a surface become more rounded with time, as the rocks weather	Rock angularity is measured by comparing them to standard charts	10,000
Hornblende etching	Sand-sized grains composed of hornblende become progressively more etched along their edges, as they weather	Hornblende grains are examined under a microscope and their degree of etching is placed into various categories	100,000
Rock weathering rinds	Rocks lying on a surface weather from the outside-in and develop weathering rinds	Rocks are broken and their weathering rind thickness are measured	100,000
Pitting of rock surfaces	Rocks lying on a surface develop pits due to weathering	Weathering pits are identified on rocks and measured	100,000
Veins on rock surfaces	As rocks lying on a surface weather, less weatherable areas (typically quartz) stand up as veins or ridges; assumes that veins do not break off over time	Veins due to weathering are identified on rocks and measured	10,000
Surface boulder frequency	Over time, the number of surface boulders present on surfaces decreases, as they weather away or become buried	Boulder densities are determined in the field	10,000
Lichenometry	Lichens growing on rocks get larger with time	Lichens of a particular species, or of several species, on rocks are identified and the diameters of their thalli are measured	4,500
Rock varnish (varnish microlamination dating)	Rocks lying on dryland surfaces develop a dark, Mn-rich varnish, which gets darker and thicker over time, and covers larger proportions of the rock	Varnish thickness and layering are measured in the lab, under a microscope; varnish color and coverage are determined in the field	100,000

Table continues...

Method	Assumptions of the method	Application of the method	Approximate maximum timespan utility (years)
Soil-based post-depositional modifications			
Soil thickness	The thickness of soil profiles and their individual horizons increase with time	Soils are examined in pits and exposures, and their horizon thicknesses are measured in the field	100,000
Soil contents in various horizons	Contents of "mobile" materials in soils that accumulate in B horizons (or which are depleted from E horizons), e.g., clay, carbonates, salts, gypsum, change predictably with time	Soils are excavated and sampled, and samples are analyzed in the lab	100,000
Soil color	Soils become redder with age (in certain climates)	Soil color is evaluated in the field or lab	10,000
Soil clay mineralogy	Changes in clay mineralogy of soils follows a predictable pattern (based on climate)	Soils are excavated and sampled; samples are analyzed in the lab	10,000
Soil sand and silt mineralogy	Ratios of resistant/weatherable minerals in the sand or silt fraction increase with time and are correlated to age	In the lab, contents of various mineral types are determined on samples of sand and silt grains	200,000
Desert pavement	Rocks in desert pavements become more effectively interlaced (fewer gaps) with time, and thickness of underlying fine sediment increases	Measurements are made of desert pavement characteristics in the field	10,000

[1] These methods all assume that the changes examined here all are predictable and correlate to the length of surface exposure, or age.

fall into one of two families: (a) changes to *rocks* lying on the surface (we assume that the rocks first came to be on the surface when it first became stable), and (b) developmental changes to *soils* that have developed since the surface last stabilized. Application of PDM data to real-world landscapes has repeatedly confirmed the efficacy of these tools, and in some cases, even developed a semi-quantitative association between the data and actual surface ages.

3.7.1 Rock-Based Post-Depositional Modifications

A time-tested family of SED methods centers on the degree of alteration of rocks that are lying on stable surfaces (**Table 3.2**). We assume that the rocks on (or shallowly buried beneath) a freshly exposed or formed surface were, at the time, unweathered and uncoated, and that they then undergo slow, predictable, and irreversible changes with time. Over time, we assume that the rocks slowly weather, and in some instances, accumulate coatings. Use of rock weathering information as an SED tool is most useful in deposits that have many large rocks on the surface, so that the samples can consist of rocks of similar size and lithology.

Rock weathering can be associated with the overall "weakening" of the rock. This "weakening" can be measured in several ways. We could determine how difficult it is to break the rock with a hammer, or how fast sound waves travel through the rock (**Table 3.2**). Rocks may develop pits due to weathering, their edges may become more rounded, or veins of more resistant minerals will stand up more with time as other parts of the rock weather and fall away. All of these methods may be useful in determining surface exposure.

Rocks weather from the outside-in, developing a "rind" of more weathered rock (**Fig. 3.7**). Quite commonly, **weathering rind** thickness is well correlated with surface exposure (**Fig. 3.8**). Weathering rinds develop mainly due to the *dissolution* of minerals and materials, or the oxidation of iron-bearing minerals, rather than the accumulation of weathering byproducts. Assessing the degree of rock weathering is usually accomplished by examining rind thicknesses (or other rock SED attributes) on ≈30–50 rocks scattered across a surface. The main concern with using weathering rinds as an SED tool is that, after they form, parts

Figure 3.7 Examples of rocks with thick, distinct weathering rinds. Source: R. Schaetzl.

may spall off. In this case, the thickness of the rind may not be directly proportional to the exposure age, and provide only a minimum-limiting age for the surface.

Relationships between surface exposure duration and rock weathering are most commonly applied, and therefore presumably most dependable, in dry, semi-arid, and alpine regions, where vegetation is less likely to cover them and alter their weathering rates. Also, in such environments, rocks weather more slowly, and thus the limit of the method is extended back farther in time. **Figure 3.8** illustrates how weathering rinds on rocks on glacial moraines of different age increase over time, and that rind thicknesses vary across a geomorphic surface. Variability in rind thicknesses is why sampling a large number of rocks is essential, in order to provide good statistical rigor for the age–rind relationship.

Most studies have shown a clear relationship between rock weathering and time. Analysis of rock-related features

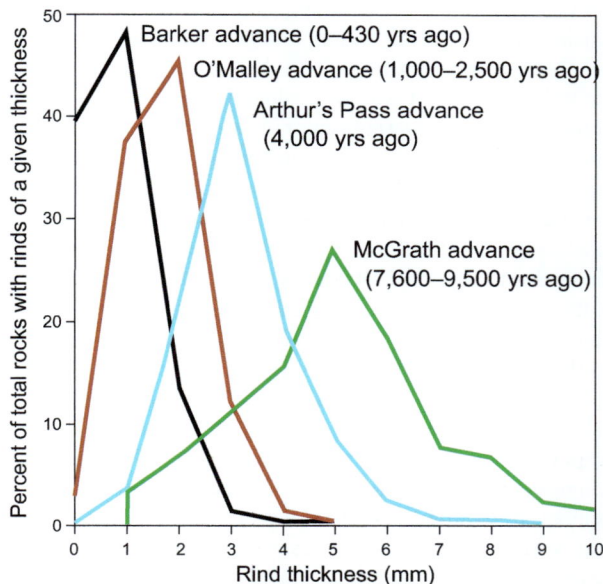

Figure 3.8 Thicknesses of weathering rinds on rocks sampled across four glacial moraines of different age in the Alps of New Zealand. Source: After Chinn, T. J. H. (1981). Reprinted by permission of the publisher (Taylor & Francis Ltd, http://www.tandfonline.com).

has an advantage over soil development as an SED tool because it is much easier to sample a large number of rocks than a large number of soils. Plus, most rock characteristics can often be determined quickly and inexpensively in the field.

Even something as simple as the density of rocks on a surface can change predictably as a surface gets older; rocks weather away, to become regolith, or more often they slowly sink into the sediment below due to soil mixing. How does this work? Certain forms of mixing processes slowly but continually bring sediment to the surface, gradually lowering anything that is ON the surface, especially rocks. Freeze–thaw processes can perform this type of mixing, as can shrinking and swelling in certain types of clay-rich soils, but mainly it is accomplished by soil animals, in which case the process of soil mixing is called **bioturbation** – a process that is more widespread than many realize. The world is replete with soil animals like worms, ants, termites, and small mammals, always **burrowing** and bringing fine sediment to the surface. In the process, anything large (like rocks) that is on the surface gradually sinks (**Fig. 3.9**). After many years of this type of activity, a zone near the surface, generally devoid of large stones, forms. This **biomantle** also thickens with time, but because it is more often a function of the intensity of mixing rather than the length of time mixing has occurred on the surface, biomantle thickness is not an optimal SED tool over long timespans.

A hybrid (rock–soil) type of SED tool involves the formation of **desert pavement** of small rocks, even as larger rocks develop a coating of varnish. A desert pavement is a dense cover of rocks on the surface, usually in a desert, and often containing a layer of fine-grained sediment underneath (see Chapter 21). Over time, the layer of surface rocks becomes increasingly interlocked and closely spaced (**Figs. 3.10, 21.14, 21.15**). Pavement clasts also get increasingly shattered, making their angularity and size, plus any varnish on them, additional SED tools. In addition, as desert pavements form, they develop a layer of porous, fine sediment below, which thickens with time. Although the formation of desert pavement is not normally regarded as a highly precise SED tool, it can help differentiate young from old surfaces, or establish the rank order of a series of surfaces. Under optimal

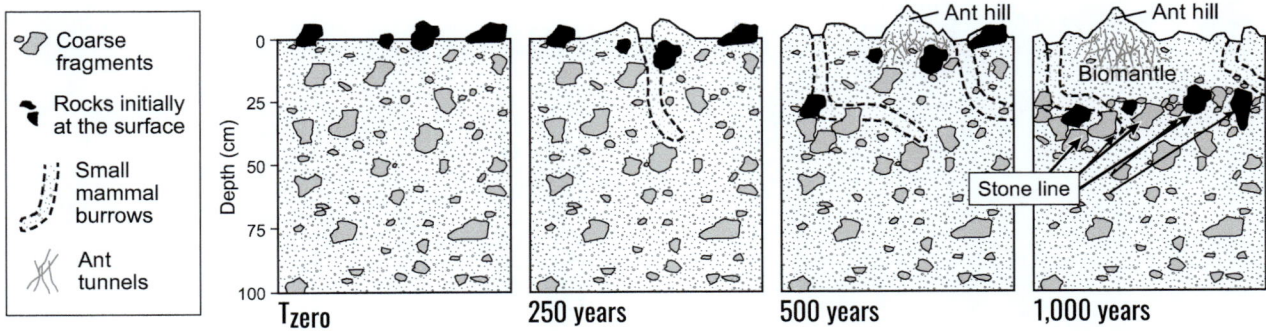

Figure 3.9 Diagram showing how large rocks on the soil surface are lowered over time by the burrowing of small mammals and insects, forming a subsurface stone line, with a stone-poor **biomantle** above.

Figure 3.10 Desert pavement in the Sonoran Desert of southern Arizona, USA, and the stone-free sediment that often lies immediately below. Rock hammer provides scale. Source: R Dorn.

conditions, desert pavements can form in a few thousand years. And they are known to form in cold deserts, hot deserts, and those in between!

Rocks lying on surfaces may acquire various kinds of coatings, such as silica, carbonates, or even coverings of living organisms such as lichens. Particularly common in many warm desert areas is the formation of a dark, almost black, manganese-rich coating, called **rock varnish**. Although it is sometimes referred to as desert varnish, that term is a misnomer because similar coatings can form in a wide variety of terrestrial settings. In deserts, where these coatings have been most commonly studied, the continued accumulation of varnish causes the surfaces of rocks to darken, as it thickens and increases in coverage. Another example of a rock coating are the orange iron films that coat the walls of rock crevices (or the undersides of desert pavement clasts). An interesting but typical example of these types of rock coatings is shown in **Fig. 3.11**. Chuckwalla lizards, common in the deserts of the American Southwest, take shelter in rock crevices. Deep in the crevice where rock fracture walls are in constant contact with alkaline dust, beautifully orange, iron-rich (and manganese-poor) films form. The lizard in **Fig. 3.11** has evolved a protective coloration that reflects these rock coatings, probably to hide more effectively from predators. In wider fractures, however, where water can episodically wash the dust away, the typical black,

manganese-rich varnish forms – just as it does on more exposed rock surfaces (**Fig. 3.11**). This sequence of black and orange rock coatings is common in rocks on desert pavements as well – black on top, orange underneath.

Rock varnish coatings have long been studied by geomorphologists as a way of evaluating surface age. The degree of varnish covering (color, coverage) can be evaluated in a relative sense in the field, and even under the transmitted light microscope. Even more quantitative estimates of varnish can be determined using microlaminations of the varnish, sampled in ultra-thin (5–10 μm thick) cross-sections. How does this work? Due to its sedimentary origin, varnish often contains a layered microstratigraphy comprised of black, orange, and yellow layers (**Fig. 3.12**). These layers differ in chemical composition, for example, in their Mn, Fe, Si, and Al contents, and their formation is affected by climate. As a result, the layering sequence in varnish is replicable in a given desert region (like tree rings) and, if numerically calibrated, can be correlated with regional or global climatic events of known ages. Once this is done, the varnish layering sequence can be used as a correlated dating method for otherwise undated surfaces, sometimes as far back as far as ≈100,000 years and even back to 300,000 years if the rock surfaces have been stable enough. The main applications of this method rest in providing minimum-limiting ages for landforms like alluvial fans, debris flows, and

Figure 3.11 The Chuckwalla lizard in **A** has just emerged from a rock crevice with its protective coloration that mimics coatings found in rock crevices and surrounding rock surfaces. Source: Gene Hanson via Wikimedia Commons. In **B**, this type of rock coating contrasts with a more typical, dark coating, which many call desert varnish. Source: R. Dorn.

Figure 3.12 Rock varnish on a debris flow fan in Death Valley, California. Note how variably dark the rocks are on this fan. Darker colors here imply thicker and more complete coverage of varnish with more microlaminations, which in turn record a longer period of past climate variations, and hence, greater age. Source: (microlaminations) T. Liu; (fan) NASA.

bare rock surfaces in deserts, as well as for human artifacts and petroglyphs. This application, technically referred to as varnish microlamination dating (**Table 3.2**), has corollary applications beyond dating; layers found within rock varnish can be dated and correlated to paleoclimatic events. This technique is limited to warm deserts, where rock varnish is biogeochemically stable.

The rock-based PDMs discussed above are all largely abiotic, even though rock weathering and the development of rock coatings is assisted by microorganisms. A purely biological rock PDM method – one that has been used since Roland Beschel's pioneering work in 1950 – involves the growth of crustose lichens on rock surfaces, or **lichenometry** (**Table 3.2**). The body of a lichen is called a *thallus* (plural, thalli). Lichens that grow on rock surfaces are referred to as epipetric lichens (**Fig. 3.13**). These lichens usually occur as circular forms that get progressively larger over time, thus providing a unique bio-chronometer of surface exposure.

Figure 3.14 An example of Roland Beschel's (1950) original lichen growth curves for rocks on moraines in the Austrian Alps.

be determined by directly monitoring the increase in size of individual thalli over a period of years. Lichen growth curves are unique to each lichen species; a different curve must be developed for each species under study. However, once constructed, growth curves can then be used to estimate the exposure ages of other surfaces for which numerical age control has not been established. Thus, lichenometry is a type of correlated age dating. That said, lichen sizes could be used simply as a relative dating tool, without calculating a growth curve, that is, rocks on older surfaces should have larger average lichens.

Lichenometry is mainly applied in cold, alpine regions, where rocky surfaces are common, lichen growth is slow, and vascular plant competition is low. Although its useful range can extend back beyond 4,000 years, lichenometry is most useful on young (late Holocene) surfaces that are < 1,000 years old. Its main applications have been for such features as glacial moraines, debris avalanches, and other rocky landforms in cold, dry climates.

3.7.2 Soil-Based Post-Depositional Modifications

Soils start forming as soon as a surface has stabilized and become subaerial; this is considered time$_{zero}$ for the soil and the surface. Many soil properties and characteristics change predictably with time, making soils data excellent PDMs (**Table 3.2**). Depending on the environment (climate), some soil properties increase progressively over time, for example, contents of clay, iron, or carbonates. These make for good natural PDMs and "soil chronometers." Other soil properties increase or decrease rapidly to some steady state level, after which they change little with time; the most notable is organic matter content, but pH is also another example. And still other soil properties are steadily and irreversibly depleted from soils, due to weathering and leaching, for example, contents of weatherable minerals, or any of a number of soluble substances. Knowing which properties to use to assess surface age – which is the best soil chronometer to

Figure 3.13 Examples of crustose epipetric lichens on the surfaces of rocks. Source: R. Schaetzl.

Crustose species of epipetric lichens are the slowest growing of all; their slow growth and longevity made them the group of choice for many lichenometry studies.

Lichenometry as an SED tool assumes that lichens begin colonizing rocks shortly after the rocks are subaerially exposed, and that they expand in size predictably over time. In practice, the method involves measurement of lichen sizes on rock surfaces of different, but known, ages, and using these data to create a lichen growth curve (**Fig. 3.14**). (Another simple option is to determine the overall lichen cover on rocks across a surface, regardless of species, as this should increase with time.) Usually, the largest few thalli are used to characterize the age of surface, because we assume that the largest thalli started growing first, shortly after the surface (and the rock) was exposed. Growth curves can also

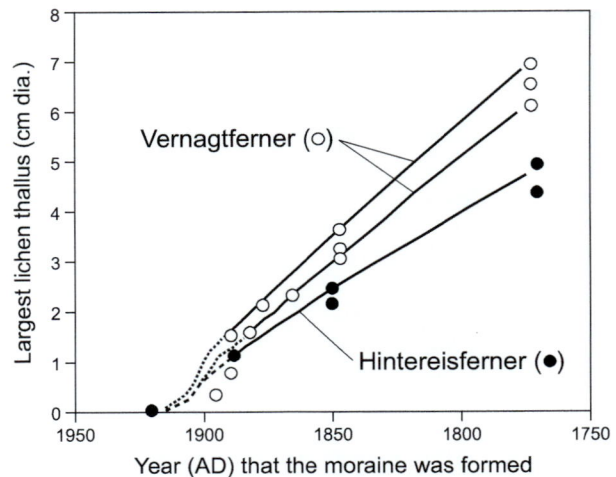

use – requires an understanding of soil formative processes at the site, as well as a ballpark estimate of surface age. For example, some soil properties would not be useful on "old" surfaces, because they would no longer be changing with time, having achieved a steady state condition. Or another example – one would not use carbonate contents as a dating tool for humid climates, in which soil carbonates do not accumulate.

Applications of soils as a relative dating tool involve correlating soil development to surface age. To use this approach, some facet of soil development must be evaluated in at least a semi-quantitative manner – we have to put a number on it. The science of **soil geomorphology** does just that; it uses soil attributes and properties to better understand geomorphic surface ages, soil formation rates, and surface stability over time.

A key tool of the soil geomorphologist is the quantification of soil properties vs time – when displayed in a graph this statistical relationship is referred to as a **soil chronofunction** (*chrono* = time). **Figure 3.15** is a collection of some typical soil chronofunctions. Note that each chronofunction puts soil (surface) age on the *x* axis, as the independent variable. A soil characteristic then is listed on the *y* axis, as the dependent variable. What soil properties are most useful in

this regard? Well, the list of potential soil properties is a long one, for example, soil horizon or soil profile thickness, accumulation (or loss) of mobile constituents from specific soil horizons, or even some integrative index of soil development. One of the better SED tools is mineralogy, both clay mineralogy and sand or silt mineralogy. It all depends on the likely age of the surface and the types of soils present there.

Some of the more common soil-based PDMs involve soil or soil horizon thicknesses, many of which are easily measured in the field. Most soils get thicker with time, as do their individual horizons (**Fig. 3.15A**). A "thinning" soil chronofunction might indicate that the surfaces under study are not stable; instead, they are eroding. We know from published research how long it takes for specific types of soil horizons to develop. Thus, we can simply use the presence of soils with that type of horizon as an indication of a minimum-limiting date for that surface (the surface must be at least *x* years old, where *x* is the time required for that horizon to form). Many soils, as long as they are not influenced by a water table, get redder with age, making color (which can be easily quantified) another, easily calculated, soil PDM tool. Other soils, especially in dry climates, accumulate airborne dust, making contents of the "dust-borne" minerals, and the overall soil texture, a good PDMs there.

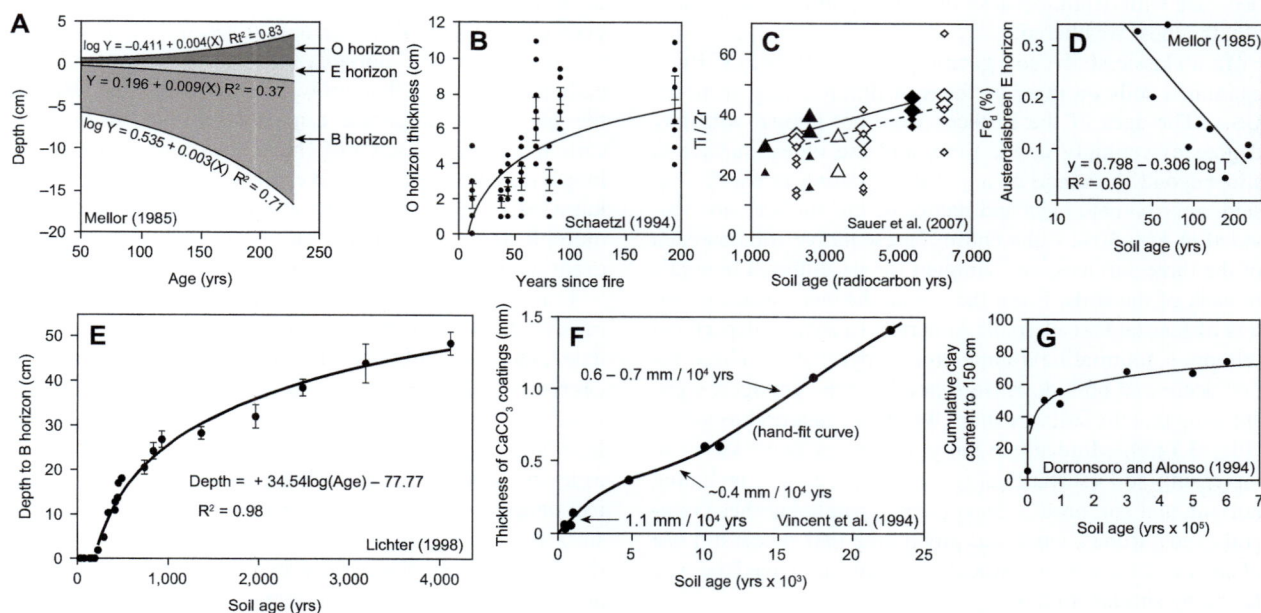

Figure 3.15 Examples of soil chronofunctions. **A.** Soil horizon thicknesses as a function of age. **B.** O horizon thicknesses in the years after a fire. **C.** Titanium/zirconium ratios in soils as a function of soil age. **D.** Extractable iron contents in E horizons as a function of soil age. **E.** Depth to the top of the B horizon as a function of soil age. **F.** Thickness of CaCO₃ coatings on rocks within soils, as a function of soil age. **G.** Cumulative clay contents in the upper 150 cm of soils, as a function of their age. See Schaetzl and Thompson (2015) for information about the original studies. Source: (A, D) After Mellor, A. (1985); reprinted by permission of the publisher (Taylor & Francis Ltd, http://www.tandfonline.com). (B) After Schaetzl, R.J. (1994), reprinted by permission of Springer Nature. (C) Used with permission of Elsevier, from Sauer D., Schellmann G., Stahr K. (2007); permission conveyed through Copyright Clearance Center, Inc. (E) Used with permission of Elsevier, from Lichter, J. (1998); permission conveyed through Copyright Clearance Center, Inc. (F) After Vincent, K. R., Bull, W. B., & Chadwick, O. A. (1994). Copyright © National Association of Geoscience Teachers, reprinted by permission of Taylor & Francis Ltd, https://www.tandfonline.com on behalf of National Association of Geoscience Teachers. (G) After Dorronsoro, C. and Alonso, P. (1994). © American Society of Agronomy, Crop Science Society of America, and Soil Science Society of America.

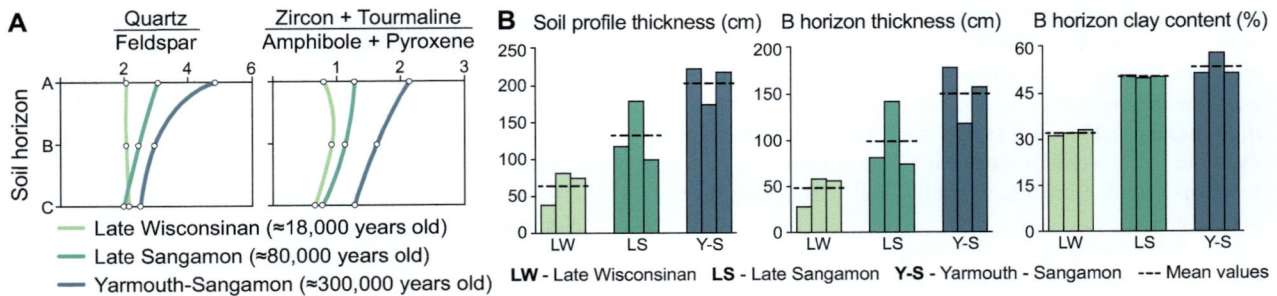

Figure 3.16 Data for soils on surfaces of different age in Iowa, illustrating the utility of soils as an SED tool. Source: After Ruhe (1956).

Perhaps one of the most reliable soil PDMs involves mineral weathering. Unlike some soil properties that can increase and then decrease through time, once minerals are weathered out of a soil, there is no going back. It is a one-way street. In almost all soils, minerals in the sand and silt fractions of soils weather away over time; those most vulnerable to weathering will weather faster, and first, and eventually will be completely lost. Minerals that are more resistant to weathering will mostly remain unaltered, and thus, increase in relative abundance over time. For this reason, the ratio of resistant/weatherable minerals should increase in soils as they get older. In most soils, these ratios should also decrease with depth, because the upper part of a soil is usually the most weathered.

In a classic study, soil geomorphologist Robert V. Ruhe examined soils on three surfaces of different age, in Iowa, USA. The ages of the surfaces were not known with any degree of certainty at the time, and the oldest surface is still beyond the range of any current dating method. That said, approximate ages and names of the surfaces are provided in **Fig. 3.16**. Ruhe chose representative soils on each of the three surfaces, and sampled for six different minerals in each of the soils. From these data, he calculated two ratios of resistant/weatherable minerals: (quartz/feldspar) and (zircon + tourmaline)/(amphibole + pyroxene). These ratios decreased both downward in all but the youngest soils, showing that the soils are more weathered nearer the surface (**Fig. 3.14A**). More importantly, the ratios increased with increasing soil (surface) age. Other soil data – including horizon and soil profile thicknesses – paralleled these mineral ratios. Ruhe's work was among the first successful use of soils as a surface exposure dating tool, and paved the way for many similar studies.

3.8 CORRELATED AGE DATING

As we transition from relative dating to numerical dating, we first visit a hybrid type of dating, one which uses elements of both – correlated age dating. Let's discuss an example. Across the American Southwest, airborne dust, rich in carbonates such as $CaCO_3$, is seemingly everywhere. This dust, constantly falling onto soils, is often later blown away just as quickly it had arrived. But some of this carbonate-rich

dust is washed into the soil, where it accumulates in the subsurface. And because there is never enough rainfall to leach the carbonates out of the soils, they continue to accumulate these **secondary carbonates**. This trait renders carbonate content an excellent soil PDM tool in many desert regions. The tool is so useful that a series of carbonate accumulation *stages* has been developed, which often correlate well to surface age.

At first, secondary carbonates show in the soils as thin white filaments lining pores (stage I), then as coatings on rocks (stages II–III), and by stage IV, the soil pores have become impregnated with it (**Fig. 3.17**). After this thick, impermeable horizon of soil carbonate forms, any new carbonate that washes downward, into the soil, accumulates *on top* of the "plugged" horizon, forming dense layers of white carbonate material that grow upward with time. Because the age of soil carbonate can be determined using radiocarbon dating, soil scientists have determined roughly how long each carbonate stage takes to develop. Armed with this knowledge, other soil scientists can assess the age of geomorphic surfaces in the region by simply digging a pit and examining the soil – a great relative dating model.

Use of soil carbonates in soils is a good example of a relative dating tool. But if some of the surfaces can also be dated, then soil carbonates become even more useful – as a **correlated age dating** tool. In it, we use numerical dating to establish the ages of various surfaces, and then use PDMs from that (dated) surface (in this case, carbonate stages) to estimate the age of other surfaces, for which we lack numerical dates. We do not need to numerically date those other surfaces, we simply correlate PDM data among them to arrive at a date for the surface of unknown age. Correlated age assessment requires that numerical ages have been determined for *some* of the surfaces, and that these surfaces have also been assessed using relative dating techniques. For example, weathering rind data from rocks on a glacial moraine (D) of unknown age might be correlated to similar data for moraines A, B, and C (with known ages) to arrive at a statistical function/correlation. Using this relationship, we can then estimate (within limits) the age of moraine D (**Fig. 3.18**). Thus, correlated age assessment can provide semi-quantitative estimates of the magnitude of age differences among sites, surfaces, or materials.

Figure 3.17 Soils in various stages of carbonate accumulation in the Desert Southwest, USA. Source: B. Buck, C. Monger, and R. Schaetzl.

STAGE I

STAGE II

STAGE III

STAGE IV

STAGE V

STAGE VI

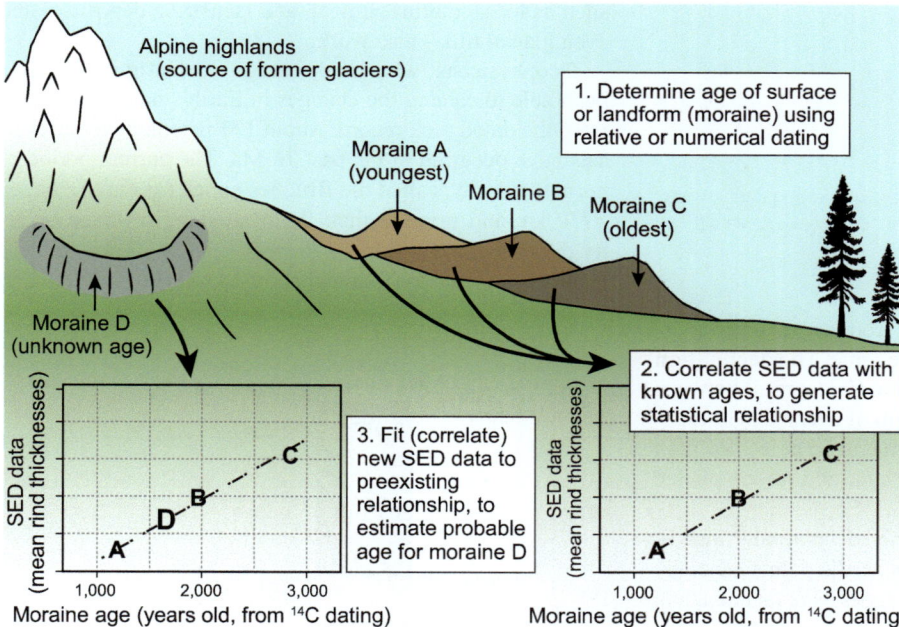

Figure 3.18 Schematic diagram illustrating the concept of correlated age dating for a suite of alpine moraines. Moraine D correlates to an age that is between that of moraines A and B.

Alpine highlands (source of former glaciers)

1. Determine age of surface or landform (moraine) using relative or numerical dating

Moraine A (youngest)

Moraine B

Moraine C (oldest)

Moraine D (unknown age)

2. Correlate SED data with known ages, to generate statistical relationship

3. Fit (correlate) new SED data to preexisting relationship, to estimate probable age for moraine D

SED data (mean rind thicknesses)

1,000   2,000   3,000
Moraine age (years old, from ¹⁴C dating)

SED data (mean rind thicknesses)

1,000   2,000   3,000
Moraine age (years old, from ¹⁴C dating)

3.9 PALEOMAGNETISM

A unique type of dating involves **paleomagnetism** – the magnetic signature of Earth that is preserved in rocks and sediments. Neither a relative dating technique nor a numerical dating technique (or maybe a bit of both),

paleomagnetism is often used in geomorphology to inform us whether landforms are older or younger than a key datum at ≈770 ka.

North is north, right? Well, not exactly. Earth's *true* (Geographic) North Pole is determined by its rotation, and is shown by the latitude–longitude grid. But Earth also has

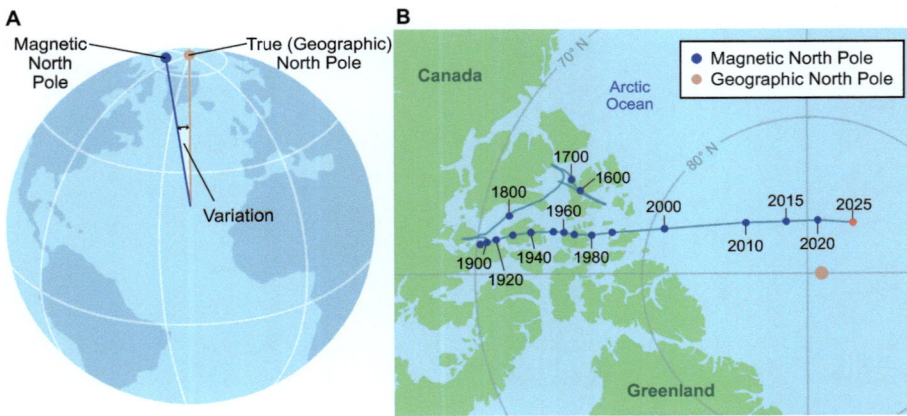

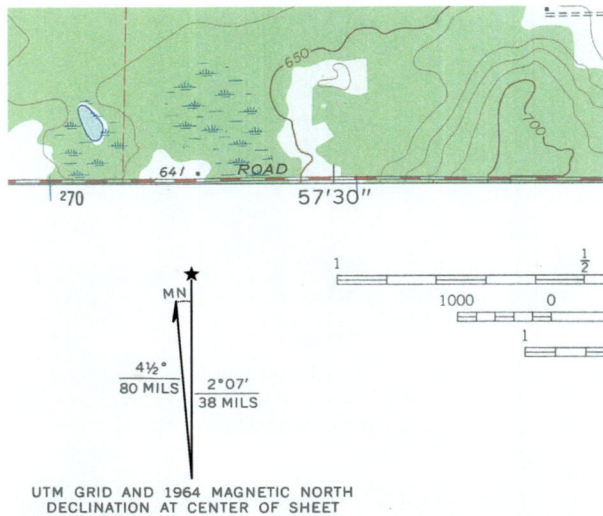

Figure 3.20 Part of the bottom of a standard USGS topographic map, showing the declination at the center of the map sheet. For simplicity, the arrow for *grid north* has been removed.

a North *Magnetic* Pole – to which compasses point (**Fig. 3.19**). These two locations are not the same. The **declination** – the direction and amount of variation (in degrees) between the two poles – is shown at the bottom of every USGS topographic map (**Fig. 3.20**). The north magnetic pole's location is determined by Earth's magnetic field, and to complicate matters, it wanders over time (**Fig. 3.19B**).

Earth's current polarity is considered "normal," that is, the north-seeking end of the compass needle points generally toward the north magnetic pole. If one goes back far enough in geologic time, however, periods of **normal polarity** have alternated with periods of **reversed polarity**, when the north-seeking end of the compass needle would have pointed generally south. Essentially, at times in the geologic past, the magnetic field did a flip-flop, as south became north. The cause of these magnetic reversals is not clearly understood. But because the transitional periods are usually quite short (<10 ka), the long periods of normal or reversed polarity are useful stratigraphic markers for correlation and dating, worldwide.

Rocks that solidify from a melt, as well as many kinds of unconsolidated sediments, can preserve within them the polarity of the Earth's magnetic field, if they contain ferromagnetic minerals (particularly magnetite). This property is called paleomagnetism. Basically, magnetic minerals in these rocks and sediments become aligned to the Earth's magnetic field, as they form. If undisturbed, the rock or sediment can then preserve a record of the Earth's magnetic field for a point in the geologic past – when the rocks formed or when the sediments were deposited. Igneous rocks and volcanic deposits are particularly good at preserving paleomagnetic information, although other fine-grained sediments such as loess, marine, fluvial, and lacustrine deposits – and even glacial tills – also work.

Geophysicists who study **magnetostratigraphy** have been able to catalog the changes in Earth's paleomagnetism from the dated rock record. About 171 paleomagnetic reversals have occurred in the past 76 Ma. The current period of normal polarity, called the Brunhes epoch (or chron), began ≈770 ka, and prior to that, the Matuyama reversed polarity event spanned ≈2,480–770 ka (**Fig. 3.21**). Each of these events contained within them short-lived (100–1,000 years)

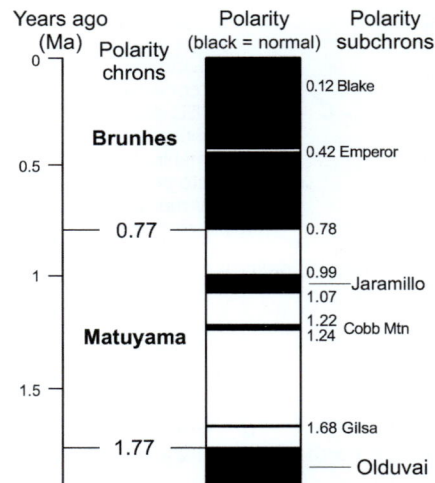

Figure 3.21 Earth's paleomagnetic timescale for the past ≈2 Ma.

geomagnetic excursions, or subchrons. During these excursions, the magnetic pole moved sharply toward the equator and then returned to a more stable position near a pole.

Paleomagnetism in rocks and sediments is useful in geomorphology because it provides several key chronostratigraphy baselines, the most important of which is the ≈770 ka boundary between the Brunhes and Matuyama chrons (**Fig. 3.21**). Most sediments geomorphologists study have normal polarity. But, for example, glacial tills in central Missouri that are unable to be dated by conventional means are now known to be older than 770 ka, because they have reversed polarity. Loess also preserves paleomagnetism fairly well and has proven to be very useful in China, where the oldest loess deposits easily pre-date the Brunhes–Matuyama boundary.

3.10 TOOLS AND METHODS USED IN NUMERICAL DATING

A wide variety of numerical dating techniques exist that have applicability in geomorphology, and the number of such techniques continues to expand as new technologies are developed. Numerical dating tools provide a quantitative estimate of surface or sediment age. Each dating method has its own assumptions, age range limitations, and inherent uncertainties.

3.10.1 Tephrochronology

Tephra (Greek *tephra*, ashes) refers to volcanic materials that have been transported aerially (see Chapter 7). **Tephrochronology** uses tephra (usually ash) deposits to understand the chronologies of stratigraphic sediment sequences. For example, consider a loess deposit that overlies an ash bed dated at 245 ka; the ash bed provides a maximum-limiting date for the beginning of loess deposition.

Volcanic ash is composed primarily of glass shards, enabling it to be dated directly (numerically), using K–Ar, fission track, or neutron-activation dating, or even by luminescence methods. The age of a young ashfall can even be dated if it is mentioned in the historical record, or by radiocarbon dating if it has organic materials incorporated within (or immediately above or below – as limiting dates). On a geologic timescale, ash is deposited in an instant. Ashfall beds are, therefore, **isochronous** horizons – for all intents and purposes an ash bed is the same age everywhere.

Tephrochronology has a unique utility for dating – each ash deposit in the field does not need to be dated directly. After a numerical age is obtained for an ash layer, that age can be transferred to other sites where the same ash bed is present, using either stratigraphic principles and/or by fingerprinting. **Ash fingerprinting** works because each ashfall deposit has a unique geo-signature of shard size and shape, elemental composition, mineralogy, hydration, or weathering. Individual, wide-ranging ash deposits are often so unique that they can be identified anywhere, often just by sight. This is great news, for if we know the age of an ashfall at one location, that age can be transferred to anywhere that

Figure 3.22 A layer of volcanic ash from the eruption of Mt. Mazama (now Crater Lake) ≈7,700 years ago, exposed in a roadcut near Helena, Montana. Source: R. Benson, Bigskywalker.com

we are able to locate the same ash deposit. How useful! For this reason, ash deposits – most of which are already dated – can be utilized as key geostratigraphic marker beds across wide areas, and with minimal cost or lab time. Many ashfall beds, tens of meters in thickness near the source, remain identifiable as geostratigraphic marker beds hundreds of kilometers away (**Fig. 3.22**).

3.10.2 Dendrochronology

Dendrochronology is a technique that utilizes tree ring data for establishing the ages of events. Tree ring ages can be determined with great accuracy, an obvious advantage of the method. In its most basic application, tree ring data can only establish the minimum-limiting age of a geomorphic surface by counting the annual rings of the oldest trees growing there. A surface can be no younger than the oldest tree growing on it. Because most sites have trees that are less than a few hundred years old, dendrochronology usually sacrifices longevity for this high (annual) level of SED (and paleoclimate) accuracy.

Because trees respond to geomorphic stimuli, the tree ring record is a good proxy for the geomorphic record. For example, trees on a mountainside may have been scarred by an avalanche, rock fall, or debris flow. Such scars are usually grown over and covered by callous tissue. A count of the number of growth rings outside of the scar or callous tissue will date the geomorphic event. Trees that are partially buried by flood deposits or dune sand may show a response in their ring history, as the tree undergoes post-burial stress; this ring width variation is called **reaction wood**. For example, a tree made to lean due to an earthflow from upslope will develop wider rings on its downslope side, providing a date for the event. Some tree ring records have even been shown to respond to volcanic eruptions. Because of the utility of using tree rings to date geomorphic events, a new science has evolved – **dendrogeomorphology**. The numerical accuracy of dendrogeomorphology has been widely applied in analyzing erosion rates and alpine hazards such as avalanches on timescales of tens to thousands of years.

Tree rings are also useful as paleoenvironmental proxy data. Commonly, the width of a tree ring is correlated to some aspect of local climate, usually temperature or moisture availability. In cold areas, where temperature limits growth, for example, at the alpine treeline, the width of the tree rings is instead correlated to warm-season temperatures. In dry areas, tree ring widths might respond best to paleoprecipitation. Dendrochronologists frequently use the technique of **crossdating** (**Fig 3.23**) to take the tree ring sequence, and hence the climate record, back much farther than the record contained in living trees. In crossdating, each individual tree ring is assigned its exact year of formation by matching patterns of rings from wood recovered from different locations. Cores are first extracted from living trees to develop a modern chronology. Then, that chronology is extended back in time by obtaining cores from dead, standing trees nearby, and from wooden beams inside old buildings (**Fig. 3.23**). This method is referred to as **chronology building**. After a chronology is built, it may be possible to establish the age of wood buried, for example, in recently

formed glacial moraines or flood deposits. For buried wood, the outermost (youngest) ring provides a maximum-limiting age for the sediments. The burial event can be no older than the youngest ring in the buried wood. Crossdating is also important because a simple annual ring count, without crossdating, can produce errors in the age of geomorphic episodes due to locally absent, or false, rings. By applying crossdating techniques in the Bristlecone Pine (*Pinus longaeva*) forests of California, dendrochronologists have extended the tree ring record back to almost 9,000 years (**Fig. 3.24**).

3.10.3 Radiocarbon

Radiocarbon dating is part of a family of techniques that utilize isotopes – both stable and unstable – to arrive at numerical dates for sediments and surfaces. Unstable isotopes decay over time to daughter products; such isotopes are termed "radioactive." Radioisotopic methods – of which there are many – utilize data from unstable isotopes. Ages can be determined by knowing either the loss of the isotope over time, due to radioactive decay, and/or the accumulation of daughter products. This section focuses on radiocarbon – the most popular radioisotope used for numerical dating.

Compounds containing carbon (C) occur in many forms, continually cycling among several major reservoirs, such as CO_2 in the atmosphere and trapped as bubbles in ice, in biomass (plant and animal matter) and in soils, and even in water. **Radiocarbon dating** determines the ages of C-bearing materials based on their content of the radioisotope ^{14}C (carbon-14). For formerly living organisms, ^{14}C dating indicates when that organism died. Dating non-living materials like seawater or soil carbonate is slightly more complicated and beyond the scope of this discussion. But it, too, can be done!

Of the carbon on Earth, about 98.9% is ^{12}C, with an atomic number of 12. Another 1.1% is ^{13}C. An even smaller

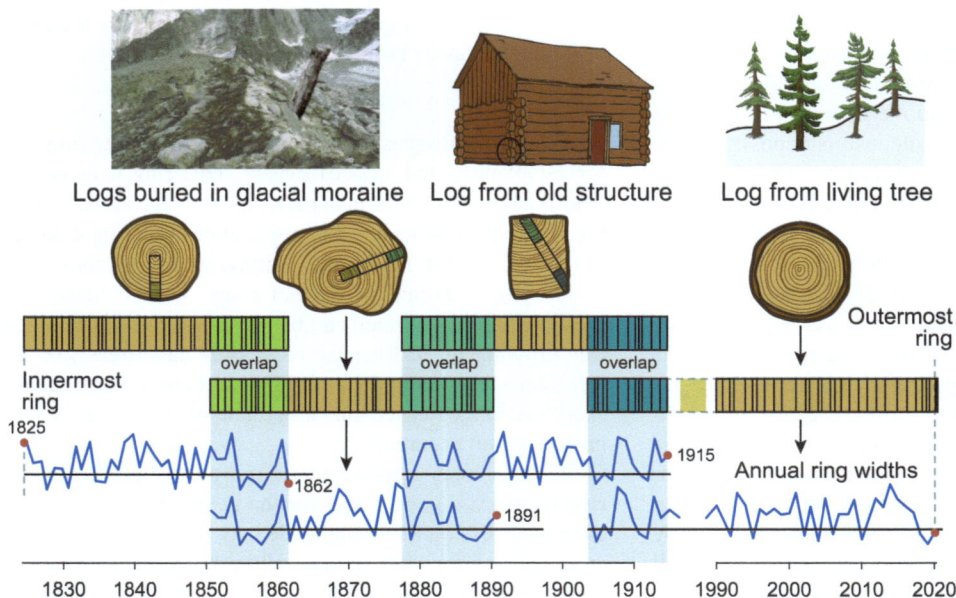

Figure 3.23 A diagram depicting how crossdating of tree rings is applied.

Logs buried in glacial moraine   Log from old structure   Log from living tree

Outermost ring

Innermost ring

overlap   overlap   overlap

1825

1862   1891   1915

Annual ring widths

1830 1840 1850 1860 1870 1880 1890 1900 1910   1990 2000 2010 2020

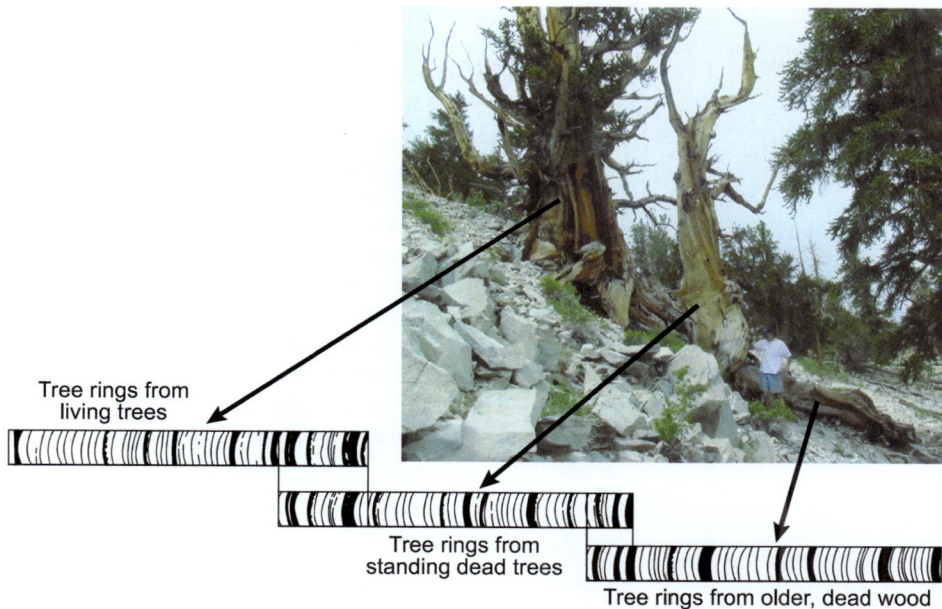

Figure 3.24 The use of tree rings to establish a chronology is illustrated here for a stand of Bristlecone Pine in the White Mountains of California, USA. Source: R. Schaetzl.

Tree rings from living trees

Tree rings from standing dead trees

Tree rings from older, dead wood

proportion is ^{14}C, an unstable isotope called **radiocarbon**. With two extra neutrons in its nucleus, ^{14}C is produced when neutrons emitted from cosmic rays impact nitrogen ($^{14}N_2$) gas in the atmosphere (**Fig. 3.25**). Both ^{12}C and ^{13}C are stable, but ^{14}C is not. Therefore, it decays at a known rate.

In order to fully understand radiocarbon dating, it is important to view ^{14}C in terms of its cycle (**Fig. 3.25**). Carbon-14 produced in the atmosphere is quickly converted to $^{14}CO_2$. Much of this gas becomes dissolved in the oceans, but some is also taken up by plants during photosynthesis and converted to biomass. The amount of radiocarbon in plant biomass quickly equilibrates to the amount in the atmosphere. As long as the plant is alive, this value stays fairly constant. Because plants form the base of the food chain, other organisms that ultimately derive their energy from plant matter have radiocarbon in their biomass in similar proportions to that of plants and the atmosphere. But then, after the plant or animal dies, their uptake of radiocarbon stops, and the store of radiocarbon in the dead plant (seed, wood, leaf, root, cotton) or animal (bone, hair, teeth, ivory, shell) begins to exponentially decay along a known decay curve (**Fig. 3.25**). The half-life of radiocarbon is 5,730 ± 40 years; in 5,730 years, half of the radiocarbon will have decayed to ^{14}N, and after another 5,730 years another half will be gone (leaving only ¼ of the original store), and so on. After about 8–10 half-lives, the amount of residual ^{14}C is so small that it cannot be determined, meaning that samples older than ≈50 ka are not normally datable.

Radiocarbon dates on biologic samples provide an estimate of the time since the death of the organism. The date is determined by measuring the residual ^{14}C content in a sample, knowing the half-life decay constant. The amount of residual ^{14}C in a sample is usually determined via accelerator mass spectrometry, which is why most ^{14}C dates are referred to as AMS dates (to distinguish them from an older form of ^{14}C determination that is only rarely used today). AMS dating can be completed successfully on samples as small as 2–30 mg; one could theoretically date an individual needle from a tree, a small seed, or a fragment of a leaf.

One must always consider the statistical uncertainty (accuracy) of a reported ^{14}C date. Carbon-14 dates are reported with an envelope of error, expressed as a standard deviation, or sigma (σ), for example, 7,560 ± 120 radiocarbon years BP. The radiocarbon date is therefore a range of years within which the true age lies, according to certain probability rules. Always, the most probable age is the number obtained by the measurement – in our example, 7,560 years. If the reported error range is one standard deviation, one sigma, the true age has a 68% probability of falling within that envelope. Some labs also present errors for two standard deviations, a 2-σ range, implying that the true age has a 95% chance of falling within the reported range.

Radiocarbon ages are reported in radiocarbon years *Before Present*, or BP. Because the method was devised in the year 1949 by Willard Libby, the term "BP" customarily refers to radiocarbon years before 1950 AD. This date has stood as the standard, largely because after 1950 the open-air detonation of nuclear devices caused the ^{14}C content of the atmosphere and plants to spike. As a result, radiocarbon ages on post-bomb materials are erroneously enriched in ^{14}C.

Unfortunately, a radiocarbon year is longer than a calendar year. This discrepancy occurs, in part, because ^{14}C production in the atmosphere has varied in the past due to cosmic ray fluctuations. To correct for this problem, and to make radiocarbon ages comparable to our traditional calendar, ^{14}C dates *must be calibrated*. Most radiocarbon laboratories report both uncalibrated (raw) and calibrated ^{14}C dates, although various websites are also available for calibration.

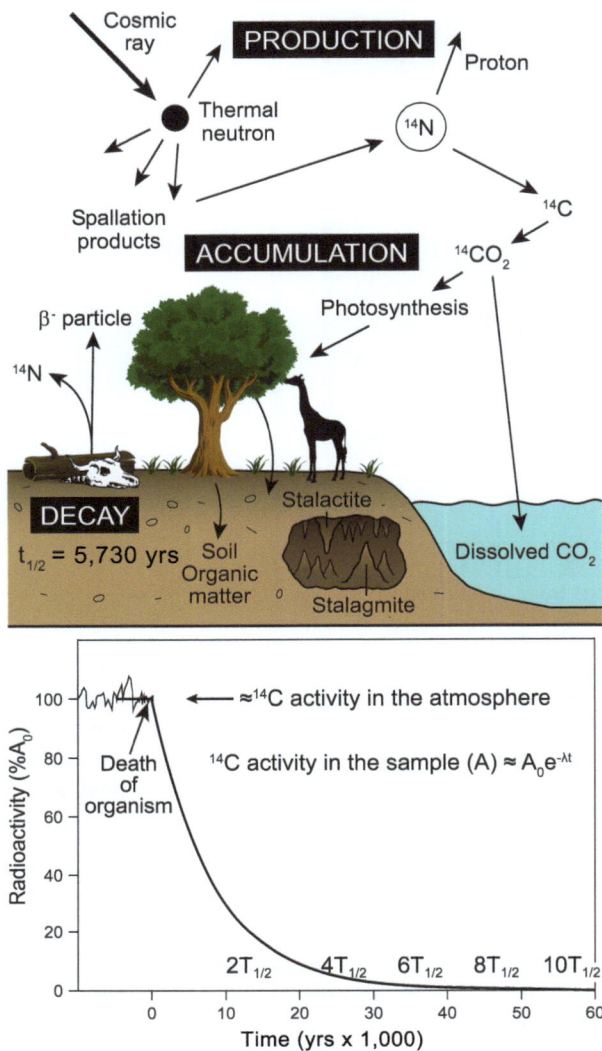

Figure 3.25 The global cycle of radiocarbon, illustrating how ^{14}C atoms, formed in the atmosphere, eventually cycle through the biosphere, hydrosphere, and lithosphere.

For example, a ^{14}C date of 5,000 ± 125 (radiocarbon years) BP would calibrate to 5,748 ± 146 cal. yr BP. Note that the calibrated date remains a "BP' date, but that the "calendar years" modifier is included to inform the user that this is a *calibrated* radiocarbon date. In general, the divergence between ^{14}C ages and the calibration curve is not significant over the last 3,500 years. But before this time the calibration uncertainties get progressively larger. For example, a radiocarbon date of 24,000 ± 1,200 BP calibrates to 28,344 ± 1,185 cal. yr BP.

The parameters used for radiocarbon calibration derive mainly from the tree ring record, which provides detailed information to about 10 ka. Information from varves (carbonate-rich layers of lake sediment which accumulate in annual layers and thus can be dated to the exact year of formation) and sea corals have extended the calibration curve to ≈50 ka.

Theoretically, any material containing C that at some time has been in exchange with atmospheric CO_2 can be dated using radiocarbon. The most common type of materials used in ^{14}C applications are organic materials such as wood (including samples of individual tree rings) and charcoal, although sediments rich in inorganic C, such as cave speleothems, soil carbonate, and soil organic matter, are also datable. Charcoal is particularly attractive because it is readily preserved in sediments. Non-woody plant parts such as leaves, seeds, reeds, and their derivatives (papyrus, paper) are also datable. Animal parts such as bone, tooth, hide, and hair are also good, as are some types of shells. Even bubbles trapped in ancient **ice sheets** can be dated, because they preserve $^{14}CO_2$ from the distant past. Knowing the ages of ice layers in the Greenland and Antarctic ice sheets has helped paleoclimatologists and geomorphologists delve into past climates with great accuracy.

3.10.4 Cosmogenic Isotopes

Cosmogenic isotopes are produced as cosmic radiation, high-energy radiation impacting Earth from space, interacts with atoms. These interactions can occur *in situ*, for example, within rocks or soils, or in the atmosphere, as with ^{14}C. Both of these types of isotopes have multifaceted applications in the dating of surfaces and sediments. Geomorphologists use cosmogenic isotopes to date moraines, debris flow deposits, fluvial terraces, soils, even surface exposures of bedrock. Isotopic dating methods have the advantage of not only providing information about surface exposure duration, but also about erosional or burial histories of surfaces.

As we have seen, ^{14}C produced in the atmosphere quickly joins with oxygen to become $^{14}CO_2$. Another, less chemically reactive, isotope produced during spallation reactions between cosmic rays and ^{16}O and ^{14}N in the atmosphere is ^{10}Be (beryllium-10). Beryllium-10 is particularly attracted to atmospheric aerosols such as cloud droplets or dust (**Fig. 3.26**). It is then delivered to the surface, mainly in snow or rain but also as dustfall, where it is either retained in soils or carried away by water or wind; some gets locked in ice sheets. Beryllium-10 has two distinct advantages for dating: (1) the rate of ^{10}Be "fallout" has been fairly constant over time, and (2) its half-life is very long (1.39 million years). Therefore, the content of ^{10}Be in soils and sediments has become a useful surface exposure dating tool.

After meteoric ^{10}Be (also known as "fallout ^{10}Be") nuclides reach the surface, they may be leached, taken up by plants, eroded and washed away, or remain/accumulate in soils (**Fig. 3.26**). The assumption that most of the fallout ^{10}Be is retained in the soil is important, because then ^{10}Be contents can be used as long-term dosimeters for surface exposure. Fortunately, ^{10}Be is insoluble at pH values >4 and therefore, retained (chemically) for long periods of time in most soils, especially clay-rich soils.

In all but the oldest, most acidic, and sandiest soils, ^{10}Be is found in highest concentrations near the surface, and its concentrations decrease with depth, suggesting that it is retained fairly well in most soils (**Fig. 3.26**). However, in older soils,

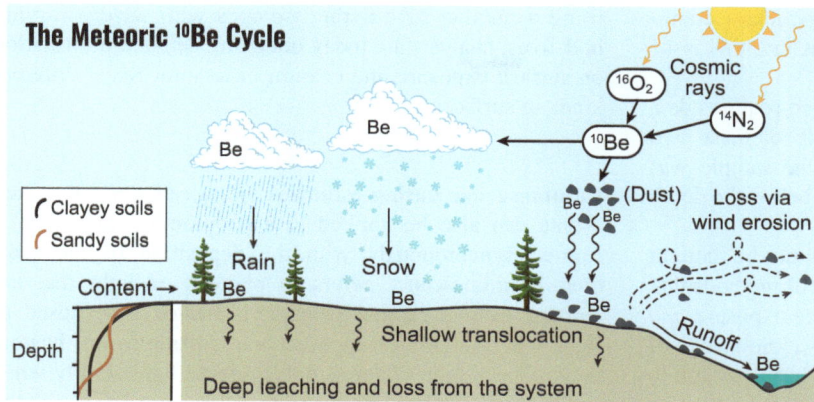

Figure 3.26 The cycling of meteoric ^{10}Be in the atmosphere–land system.

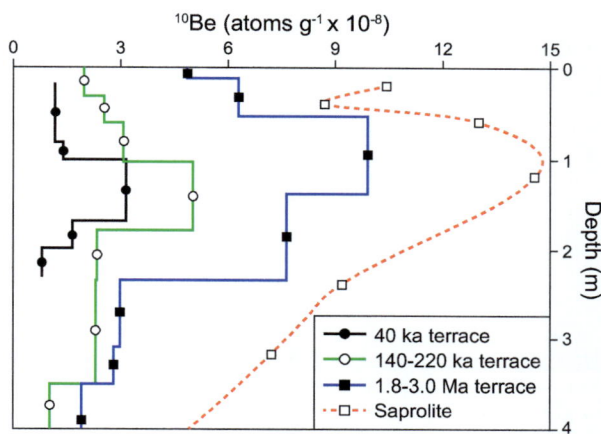

Figure 3.27 ^{10}Be contents with depth in a soil-saprolite sequence in Virginia. Source: Used with permission of Elsevier, from J. A. Graly et al. (2010); permission conveyed through Copyright Clearance Center, Inc.

the highest ^{10}Be concentrations can occur at depth, usually at 50–200 cm (**Fig. 3.27**). Because all surfaces undergo some amount of erosion, and all soils incur some amount of leaching, small amounts of ^{10}Be are inevitably lost from the soil–sediment system over time. Thus, SED data derived from ^{10}Be data must always be considered minimum-limiting exposure dates. And lastly, it is also possible that soils inherit some ^{10}Be in their initial parent materials. But for old soils and surfaces, the amount of inherited ^{10}Be is negligible as compared to the amount that has accumulated via fallout. Therefore, if we assume some degree of surface stability and minimal amounts of ^{10}Be leaching and inheritance (from the parent material), one need only know the rate of ^{10}Be influx and its decay constant to estimate the minimum age of a surface.

A second group of methods utilizes isotopes that form directly in minerals within rocks, again due to exposure of the rocks to cosmic rays (sunlight). The five most widely used nuclides used in these *in situ* applications – ^{3}He, ^{10}Be, ^{21}Ne, ^{26}Al, and ^{36}Cl – enable surface exposure dating of rocks and surfaces over ranges of 100 to 10 million years. *In situ* cosmogenic nuclides have a wide variety of applications.

Examples include erosion rates on bedrock surfaces, fluvial incision rates, denudation rates of entire drainage basins, burial histories of surfaces (whether formed in rock or sediment), rock-to-soil conversion rates, and the ages of fault scarps. The future is indeed bright for this new family of SED methods, nicknamed "cosmo dating."

How are *in situ* nuclides applied in dating applications (**Fig. 3.28**)? Consider the analogy of tanning on human skin, where the degree of tanning is mainly proportional to the *duration* of exposure to sunlight (= surface exposure). But also important are sunscreen (in this case, burial of the sediment, which shields it from the Sun), cloudiness, and latitude. Rocks are similar to skin in many ways; they accumulate nuclides mainly as a function of exposure *duration*, making them excellent SED tools, but they are also affected by latitude, cloudiness, and shielding (burial), for example, by regolith or snow. Most cosmic rays can penetrate rock to ≈50–60 cm; below that depth the rock will accumulate almost no nuclides, that is, it is effectively shielded. In soils, because of their higher porosities, shielding is not complete until a depth of ≈3 m. People tan at different rates, just as *in situ* nuclide production rates vary in different minerals. Tanned and burned skin can peel, just as erosion may result in an underestimation of surface exposure duration in rocks. Some rocks inherit nuclides from a period of previous exposure, just like a tanned person can go back outside for more sun. Most importantly, a tan will gradually wear away, just as most cosmogenic nuclides decay over time. To review, accumulation rates of various cosmogenic nuclides is dependent upon a number of site factors: duration (mainly), but also altitude, latitude, rock mineralogy and density, geometry of the exposed rock, depth of burial or shielding, and cosmic ray flux. Sounds complicated? It is. But the methods are also extremely insightful for landscapes and surfaces that cannot be dated in other ways.

For stable isotopes, like ^{3}He and ^{21}Ne, accumulation rates in rocks are linear, making them useful dating tools on surfaces that have had no prior exposure. During erosion, isotopes are lost, but with correct interpretation, stable isotopes can be quite useful in the determination of surface erosion rates and history. For radioactive nuclides, which decay over time, net accumulation is more complicated – rocks accumulate and lose isotopes concurrently. Nonetheless, this issue can be solved mathematically.

Dating how long a surface has been buried requires data from two different nuclides. Then, the ratio of these two nuclides can be used to determine when the sample was buried, provided that data are from samples below the depth of shielding.

In situ cosmogenic nuclide dating should not be thought of as a single method, but rather a briefcase of methods that can be used as chronometers for both surface exposure *and* for length of burial. But the analyst must have knowledge of (or assume) events that the rock surface has been through to determine its most likely exposure history (**Fig. 3.28**). By using a number of different isotopes with widely varying half-lives, analysts are today obtaining detailed information on surface exposure and erosion history for many different kinds of surfaces.

3.10.5 Luminescence

Luminescence dating estimates the age of *sediment*; such a date can also be applied to a geomorphic *surface* if it formed synchronously with the deposition of the sediment. Luminescence refers to photons of light that are emitted from a sample when it is heated or exposed to light. Thermoluminescence occurs as samples are heated; they release light. Other samples release light simply when excited by light; these methods are called optically stimulated

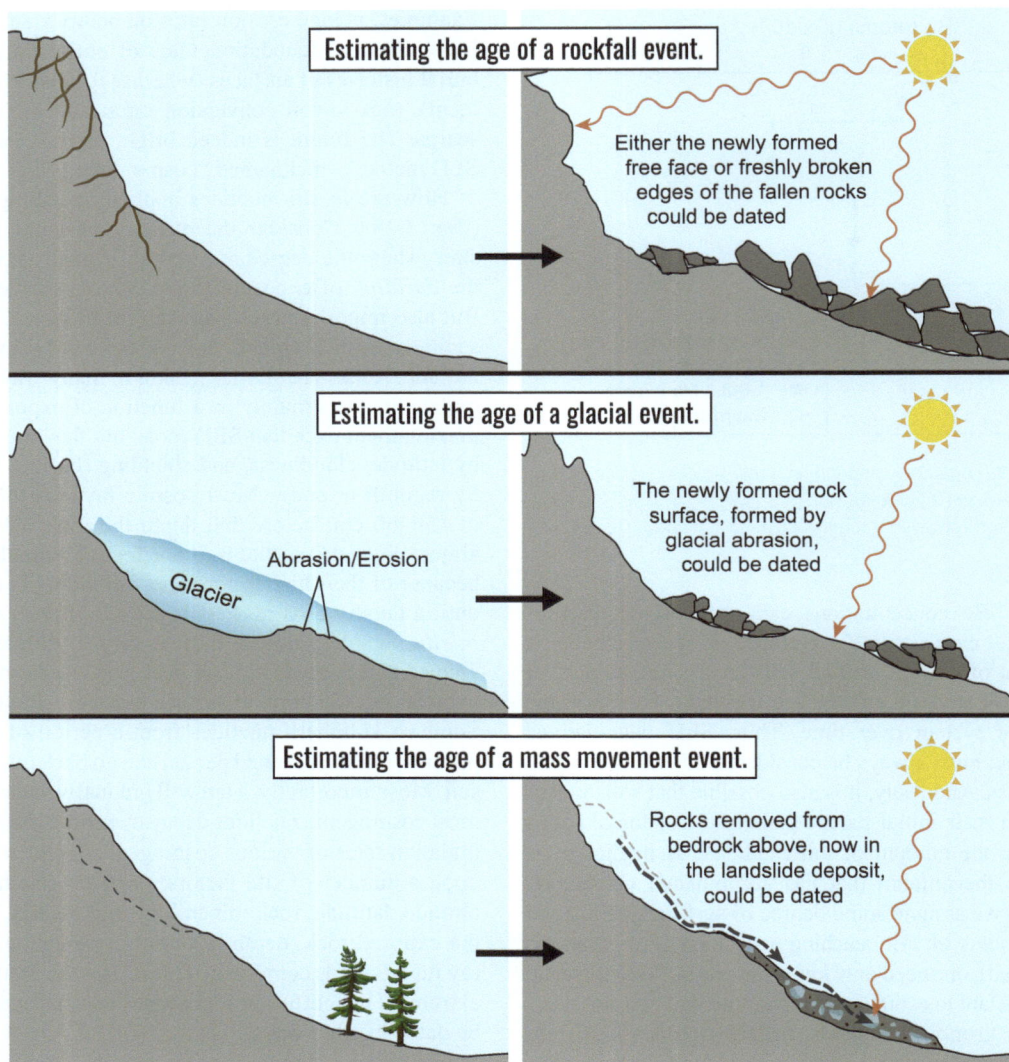

Figure 3.28 Examples of different exposure histories for rocks sampled for age determination by accumulation of *in situ* cosmogenic nuclides. In all three cases, the rocks must have not have inherited any of the isotopes for the method to be appropriate. **A.** The rocks in the rockfall begin accumulating cosmogenic isotopes as soon as they broke off the escarpment above, thereby becoming exposed to sunlight and effectively dating the rockfall event. **B.** The eroded rock surface begins accumulating cosmogenic isotopes as soon as the glacier melts and uncovers it, effectively dating the retreat of the ice. **C.** The rocks deposited in a mass movement event both expose these new rocks for dating but also bury (shield) other rocks from sunlight, creating a useful comparison for the geomorphologist. The burial depth must be at least 3 m for the method to yield accurate results.

luminescence, commonly referred to as OSL. In short, luminescence dating is used to determine the last time sediment was exposed to sunlight, and then (presumably) buried.

Luminescence dating is possible because most silicate minerals (mainly quartz and feldspars) contain crystal lattice defects and imperfections that can "trap" and store free electrons. The primary source of these free electrons, also known as *beta particles*, is the decay of radioactive isotopes in minerals in the surrounding sediment. As these isotopes decay, the electrons they release become caught in the traps and accumulate there, such that the traps become a type of long-term dosimeter. It often takes tens of thousands of years before these traps become filled; but when they do, they cease being an effective dosimeter, like a glass full of water that can hold no more. A short period of exposure to heat or light in the lab releases the beta particles from the traps, emptying them, and the sample luminesces. The more "full" the traps are, the more it luminesces. Emptying the traps in the lab is a way of "re-zeroing" the sample (**Fig. 3.29**). Sediment is re-zeroed naturally by exposure to even short periods of sunlight. Thus, sediment that was exposed to sunlight, such as dune sand, and then buried, are the typical targets for luminescence dating. The method thus estimates when a sample was last exposed to light; since then, its traps have been accumulating beta particles. By measuring the amount of luminescence coming from a sample, the analyst can determine how many beta particles were trapped in the sample, and equate that to how long it has been accumulating them.

The number of beta particle stored in traps is mainly a function of (1) the length of time since it was last re-zeroed (by exposure to sunlight), and (2) the radioactivity of the surrounding sediment. The rate at which electrons are produced by radioactive decay in the ambient sediment is called the **dose rate**. How fast is the water filling the glass? The dose rate is determined by obtaining some of the sediment closely surrounding the sample and analyzing it in the lab.

So then, the dose rate tells us how fast beta particles are produced and accumulated in the sample. The **equivalent dose** then tells us how many electrons or beta particles have actually been stored/trapped in the sample – how full is the glass? Another analogy – the dose rate would be similar to

your weekly earnings from your job, and the equivalent dose would be your bank account balance, assuming that you had no balance when you first started your job!

The simplified equation for sample age is:

Luminescence age (years) = (equivalent dose (De, in Grays) / dose rate (in Grays yr^{-1})

Knowing this, we assume that the amount of stored electrons in a sample is a function of (1) elapsed time since the sediment was last re-zeroed and then deposited (the unknown that we are trying to determine), and (2) the radioactivity of the surrounding sediment, or dose rate (a "known" that we can determine in the lab). Also of importance, although less so, are the (a) sample density and (b) long-term water content of the sediment; dense, wet sediment slows the electrons as they travel through the sediment, meaning that a smaller sphere of surrounding sediment is available to send electrons to the traps. Water contents and sample densities can usually only be estimated. Nonetheless, water content data are important, for every 1% increase in estimated water content, the sample age changes by 1%.

Samples for luminescence dating are typically taken using metal cans or opaque plastic tubes, driven into a freshly exposed sediment face (see **Figure 3.0** at the start of the chapter). In the lab, samples are opened under controlled light conditions. Lab personnel will discard sediment near either end of the tube; sediment farther inside will not have been exposed to sunlight since its initial deposition. Exposure of this sediment to light of the proper wavelengths releases the electrons, empties the traps, and re-zeroes the grains. This release of the stored electrons (in the lab) is detected as a glow, or *luminescence*. In essence, the luminescence obtained in the lab is equivalent to the radiation energy deposited within traps in the grains since its last exposure to sunlight (or heat) in the natural environment.

Luminescence dating is best applied to sandy-silty sediment that was previously fully exposed to sunlight and re-zeroed, either by transportation in the atmosphere or in clear water. Grains can be re-zeroed in as little as a minute or two under full sunlight. In practice, quartz-rich mineral samples such as dune sand, loess, beach sand, or even sandy riverbed sediment, work best; these would likely have been fully zeroed immediately before final deposition and burial. Think of a sand dune – that sand bounced along for decades or centuries, and then finally stopped saltating and became buried. That last date is the time$_{zero}$ for the dune sand and the sand dune surface.

Some types of luminescence dating can inform us of the last time a sample was *heated*; the method is useful for pottery or hearth materials recovered from an archeological site. But the most common type of luminescence dating – OSL – tells the user how long it has been since a sediment was last exposed to *light*. Thus, a luminescence date reflects the length of burial, for example, the age of the sand in a dune. Traditionally, luminescence ages refer to calendar years before present, unlike ^{14}C ages, which refer to radiocarbon

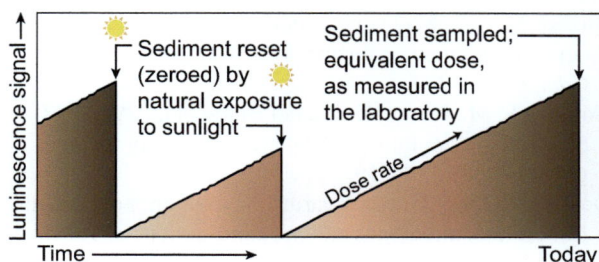

Figure 3.29 Illustration of how, in luminescence dating, the equivalent dose grows over time, and then can become re-zeroed by exposure to light – either by the Sun or in the lab.

years before 1950 AD. Thus, luminescence dates do not need to be calibrated.

As discussed above, dose rates are affected mainly by the radioactivity of the surrounding sediment, as well as moisture content and density. However, latitude, elevation, and depth of burial also affect dose rate, due to the exposure to ionizing radiation from sunlight (cosmic rays) *during burial*; thus, laboratories always request this information. Sunlight will not re-zero shallowly buried sediments, but it can penetrate some distance below the surface, and thus add some beta particles to the traps. This effect will be more of an issue at high elevations and low latitudes, and in sunny climates, where solar radiation is more intense. For these reasons, OSL samples should be taken from at least one meter depth, in order to minimize the exponentially decreasing effects of cosmic radiation penetrating into the soil, that is, the cosmic dose. Samples also must be taken as deep as possible to minimize the effects of soil mixing (**pedoturbation**), which can recycle grains between depth and the surface. While at the surface, grains can become re-zeroed. If they are later mixed into the subsurface, they may be sampled along with the grains that were buried much longer ago, producing an erroneously young age.

Luminescence dating is not without methodological issues. Analysts are always concerned about the **partial bleaching** of samples. Optically stimulated luminescence dating relies on the assumption that all the traps were fully emptied prior to burial, but this does not always happen. In this case, the sample is said to have only been partially bleached, that is, its traps were not completely emptied. Partial bleaching is not uncommon for samples that were deposited in water, for example, glacial outwash or sand from a riverbed. **Anomalous fading** – the loss of electrons from the traps – is like a water glass with a pinhole leak. It is primarily a problem with feldspars, resulting in an age estimate that is too young. Despite these potential pitfalls, luminescence dating has taken off in geomorphology and its applications continue to grow.

3.11 IMPORTANCE OF CROSS-CHECKING DATES

When you get your first dating result, either by your own hand when you compare weathering-rind thicknesses with a calibration curve, or when your hands are shaking as you open the email with your radiocarbon, cosmogenic, or OSL result, the feeling can be intoxicating. You alone in the universe have the power of new knowledge that combines your understanding of the geomorphology with the dating result. It's easy to get carried away and report your new "numbers."

But first, stop and think critically. How sure are you of the date? What are the error bars (uncertainties)? Have you determined if the numerical or correlated ages actually match what you know about the relative sequence of geomorphic events? Geomorphologists with lots of experience in using dating methods always cross-check their new data against one (or even more) different strategies. It's easy to forget some of the assumptions of a single method and how

false assumptions can lead you to incorrect interpretations. That new date may have been waiting for thousands of years to be determined; take a few more days to see if it fits with the wider geomorphic story that is unfolding through your work.

REVIEW QUESTIONS

3.1   What are the major geologic principles used for relative dating, and in what settings are they applicable?

3.2   How does calibrated age dating work, and what data are needed to create calibrated age dates for a surface?

3.3   Explain how fallout ^{10}Be and *in situ* ^{10}Be are used as numerical dating methods, and how they differ. What assumptions are made in each method? Provide an example of an application for each method.

3.4   What does a luminescence date tell the investigator? Is it a method used to date surfaces or sediments, or both? What are the dose rate and equivalent dose?

3.5   Define and provide examples of minimum-limiting and maximum-limiting dates. Explain their significance.

3.6   What is a soil chronosequence, and what information does it provide? What types of data are typically used as dependent variables in a soil chronosequence?

3.7   What is rock varnish and how can it tell us about the age of a surface, in a relative sense and in a numerical sense?

3.8   What kinds of materials are amenable to radiocarbon dating, and what does a radiocarbon date actually tell the investigator? Why do radiocarbon dates need to be calibrated?

3.9   Describe how dendrochronology works and crossdating figures into this method. What is dendrogeomorphology?

3.10 What kinds of dating applications do soils data offer to geomorphologists?

FURTHER READING

Benedict, J. B. 2009. A review of lichenometric dating and its applications to archaeology. *Am. Antiquity* 74:143–172.

Brookes, I. A. 1982. Dating methods of Pleistocene deposits and their problems: VIII. Weathering. *Geosci. Can.* 9:188–199.

Colman, S. M., Pierce, K. L., and Birkeland, P. W. 1987. Suggested terminology for Quaternary dating methods. *Quat. Res.* 28:314–319.

Dorn, R. I. 2009. The rock varnish revolution: new insights from microlaminations and the contributions of Tanzhuo Liu. *Geography Compass* 3:1–20.

Lowe, D. J. 2011. Tephrochronology and its application: a review. *Quat. Geochronol.* 6:107–153.

Mahaney, W. C. (ed.) 1984. *Quaternary Dating Methods*. Elsevier.

Oguchi, C. T. 2013. Weathering rinds: formation processes and weathering rates. In: Shroder, J. F. (ed.) *Treatise on Geomorphology*, Vol. 4. Academic Press. pp. 98–110.

Olsson, I. U. 2009. Radiocarbon dating history: early days, questions, and problems met. *Radiocarbon* 51:1–43.

Preusser, F., Degering, D., Fuchs, M., Hilgers, A., Kadereit, A., Klasen, N., Krbetschek, M., Richter, D., and Spencer, J. Q. G. 2008. Luminescence dating: basics, methods and applications. *Quat. Sci. J.* 57:95–149.

Schaetzl, R. J. and Thompson, M. L.. 2015. *Soils: Genesis and Geomorphology*. 2nd ed. Cambridge University Press.

4 Climate Change Throughout Earth History

Thomas Stevens

Climate and landforms are intimately tied together. Indeed, much of geomorphology is concerned with how landforms, climate, and other surficial processes (like erosion) interact. Landforms are often studied to understand past climates, and vice versa. Thus, a complete understanding of landform genesis requires knowledge of past climates, generally termed **paleoclimate**.

Climate can be viewed as the prevailing weather/atmospheric conditions for a site, but over long timescales. If a geomorphologist was interested in how sand dunes in a modern desert migrate, they might look at climate over the last few decades. However, a geomorphologist interested in the origin and evolution of the entire desert would need to examine climate over tens of thousands, or even millions, of years. Thus, climate is a somewhat slippery concept, especially when one considers that climate is always changing.

The critical importance of climate in landform analysis led to a branch of geomorphology, prominent in the mid-twentieth century, called **climatic geomorphology**. One of its objectives was to understand the linkages between climate zones and landform types. For example, hot desert areas in the mid-latitudes today are closely associated with certain types of sand dunes. Climatic geomorphologists spent considerable time delineating these **morphoclimatic zones**, within which most of the landforms can be correlated to the present climate. Since then, however, we have confirmed that climates are always changing, rendering morphoclimatic zones only partially useful. Indeed, many landforms have a complex history, having been formed under more than one climate. Nonetheless, the linkages between climate and landforms remain important even today; by understanding past climates we can better understand landform genesis and distribution.

Climate is driven by inputs of solar energy, and then in turn by a whole series of other factors such as variations in Earth's orbit, atmospheric composition, and local geography. None of these factors are constant, often leading to complex interactions between climate and geomorphology. For example, global cooling over the past few million years may have driven desert formation in central Asia. In turn, erosion in these deserts may then have sent copious amounts of **dust** (fine-grained silts and clays) into the atmosphere. These dust particles may then further drive cooling by reflecting incoming solar radiation. This example of a climate–landscape **feedback** (in this case, a **positive feedback**) is typical of the kinds of concerns that geomorphologists have to consider when interpreting landform genesis.

This chapter provides a tour through geologic time, focusing on the major changes in Earth's climate. As this tour draws closer to the present, more detail is known about past climates, and the discussion therefore becomes more comprehensive. Most of our focus will be on the **Quaternary Period** – within which we currently live and during which most of Earth's landforms formed. Throughout, every effort is made to provide not only the *record* of past climates, but the *drivers* and *causes* of the climatic shifts that occurred in the past. The chapter then delves into some of the main archives and methods that scientists have used to reconstruct and understand climate changes.

4.1 CLIMATE CHANGE – A TOUR THROUGH TIME

Because most landforms are less than two million years old, and many are far younger, geomorphologists are primarily interested in climate change during the Quaternary Period – from 2.58 million years ago to the present (**Fig. 4.1**). The Quaternary Period is often called the **Ice Age**, because continental glaciers dominated Earth history during this time. This chapter focuses on the Ice Age because of the profound effects that its extreme climate changes have had on the land surface. Across the middle and high latitudes, the massive ice sheets (and related events) of the Quaternary removed or strongly affected most evidence of earlier landforms. Nonetheless, many pre-Quaternary landscapes and landforms still exist, even if heavily modified by more recent processes. Most of these more "ancient" landscapes are in the tropics or subtropics, which escaped most of the effects of the Ice Age.

Geologists separate Earth history into (from longest to shortest) **eons, eras, periods, epochs**, and **stages** (**Fig. 4.1**). These divisions are generally based on changes in the sediment and rock record. The Quaternary Period – the most recent period and the one we are currently residing in – is split into the **Pleistocene** and **Holocene Epochs**, with the latter being the current warm (interglacial) phase of Earth's Ice Age climate. The Quaternary Period forms part of the

Figure 4.0 An artist's rendering of what parts of the world might have looked like during the Pleistocene Epoch, when glaciers were widespread across much of the northern hemisphere. Source: Mauricio Antón.

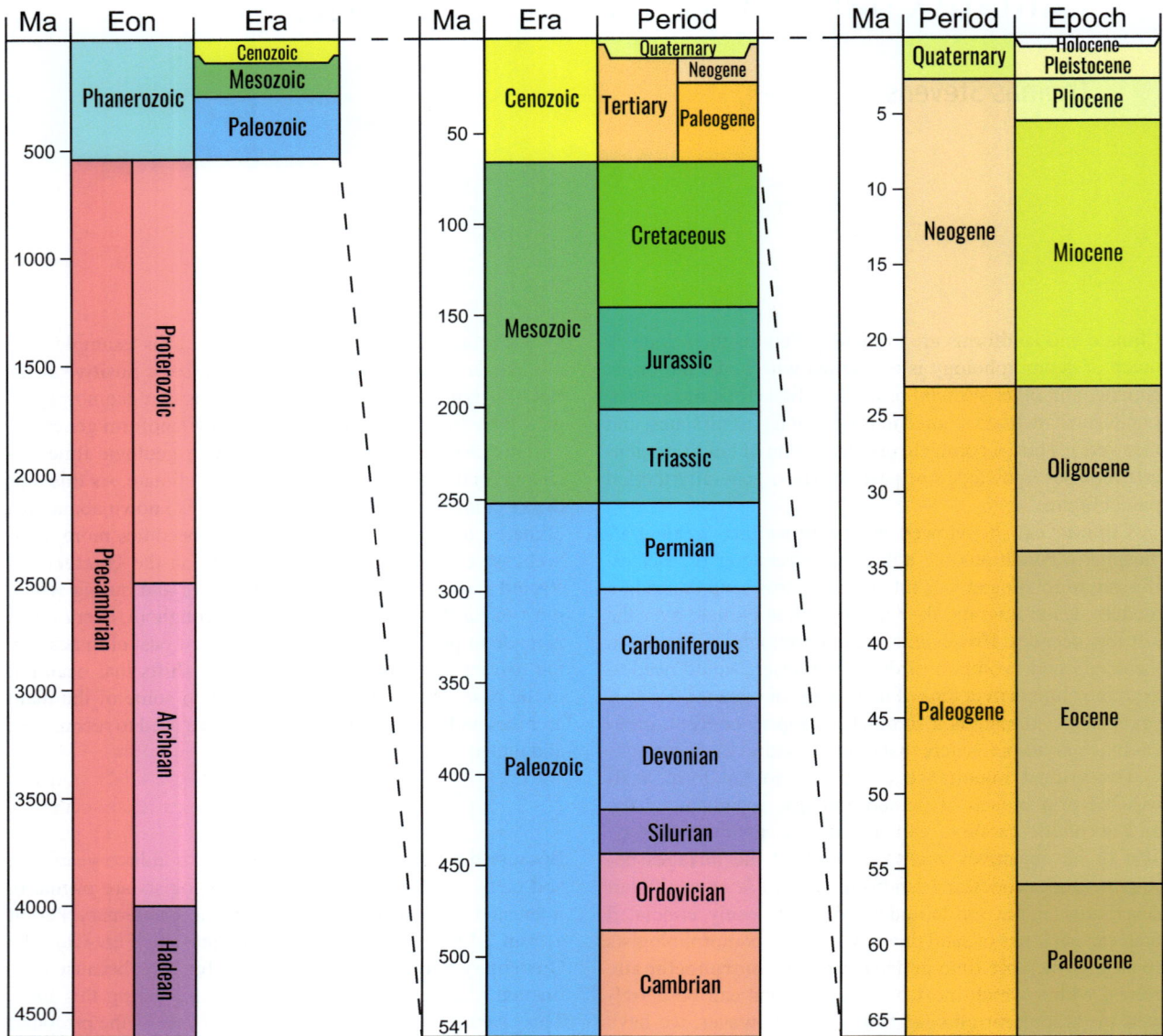

Figure 4.1 Earth's geological timescale. In the United States, the Carboniferous Period is typically split into the Mississippian (≈359–332 Ma) and Pennsylvanian (≈323–299 Ma) Periods.

Cenozoic Era, the last 66 Ma, and the Cenozoic in turn is part of the **Phanerozoic Eon**, covering the last 539 Ma (**Fig. 4.1**). Prior to the Phanerozoic Eon, Earth's history is made up of the **Proterozoic**, **Archean**, and **Hadean Eons**, collectively grouped under the informal term "**Precambrian**." Note that "Ma" (*mega annum*) is the abbreviation for millions of years ago, just as "Ga" (*giga annum*) and "ka" (*kilo annum*) are abbreviations for billions and thousands of years ago, respectively. (The "k" in ka is not capitalized.)

Breaks in the geologic timetable are often made at times of major events that have had a profound effect on life and the rock record, such as the asteroid impact 66 Ma that caused a sudden, mass extinction of almost three-quarters of Earth's plant and animal species. Other breaks occur at times

of significant shifts in climate or environment, for example, when large ice sheets formed in the northern hemisphere, ushering in the Quaternary Period.

4.2 EARTH'S DEEP, HIDDEN HISTORY: THE PRECAMBRIAN (≈4,600–539 MA)

The Precambrian (which itself is not a true eon or era) is the first major division of geologic time. It starts with the formation of our planet, ≈4.6 Ga, and ends when complex life becomes more abundant and diverse in the rock record (**Fig. 4.1**). Little is known about this mysterious and vast interval of time, even though it spans >85% of Earth's history. Most Precambrian landforms have either been eroded, buried, or subducted into the mantle and melted. No Hadean rock remains on the surface today, and only a few rocks of Archean age have been found on the surface. Indeed, during the Hadean Eon, Earth was covered by magma oceans and experienced regular extraterrestrial impacts. Only during the subsequent Archean Eon came the first "solid land" and the resultant birth of plate tectonics.

Little is known about the climate of the Archean and the earlier parts of the Proterozoic. However, this was certainly a period of climatic extremes, with a very hot, CO_2-dominated atmosphere. Eventually, the atmosphere became increasingly oxygen-rich, due largely to the evolution of photosynthesis. In turn, this process caused CO_2 decline and drove global cooling that may have resulted in repeated, large-scale global glaciations – periods referred to as **Snowball Earth** climates. Conditions broadly similar to today may not have existed on Earth until the earliest parts of the Paleozoic Era.

Geomorphic processes during the Precambrian would have been vastly different than those operating today. With no plants and soil, erosion would have been rapid and widespread. Although few geomorphic traces remain from the Precambrian, the isolated remnants that have persisted reflect this erosive environment. In the far northwestern corner of Scotland, an ancient landscape underlain by metamorphic rocks of Archean–Proterozoic age called the Lewisian gneiss is exposed and partly buried by 1.2–1.0 Ga old sandstones (**Fig. 4.2**). These sandstones used to completely bury the ancient Lewisian rock surface, until more recent (probably Quaternary) erosion removed some of them. The now-exhumed surface of the Lewisian gneiss formed before 1.2 Ga and is considered to be a **peneplain** – a low-relief plain formed by long-term erosion (see Chapter 2). It is likely that this landscape would have looked similar to how it appears today (minus the vegetation), 1.2 billion years ago! This landscape offers a glimpse into a wildly different world – one that lacked land plants and where soils were thin and largely barren. Nonetheless, climate and uplift-driven erosion processes drove geomorphic evolution even then.

A similar example comes from southern Scandinavia, Finland, and Russia, where much of the land surface likely formed at the end of the Proterozoic. This "sub-Cambrian peneplain" (formed just before the Cambrian Period) still defines many of the topographic features in southern Finland and Sweden today. Thus, many of southern Sweden's landscapes actually date to events prior to 540 Ma. These landforms here have survived repeated coverage by thick, Quaternary ice sheets. Nonetheless, ancient land surfaces are relatively scarce globally, because tectonic and erosional processes are constantly uplifting, deforming, burying, and/or eroding them.

4.3 THE AGE OF VISIBLE LIFE: THE PHANEROZOIC EON (539 MA TO PRESENT)

Increased diversity and complexity of life defines the onset of the Phanerozoic Eon (Greek *phanerós*, "visible" and *zōḗ*, "life"). With the advent of complex and more abundant life, details in the climate record increase. However, the geologic record, especially early on, is still very patchy.

Climate change during this period can be broadly categorized into periods of warm (**greenhouse**) and cold (**icehouse**) climates, each often lasting tens of millions of years (**Fig. 4.3**). During greenhouse climates, Earth's climate is warm and large ice sheets are absent. The late **Cretaceous** to **Paleogene**, late **Permian** to early **Triassic**, and late **Devonian Periods**, appear to have been the warmest of these greenhouse climates (**Fig. 4.4**). Greenhouse climates would have promoted deep, tropical-style chemical weathering over vast areas, and polar regions would have been ice-free. The higher sea levels during periods of greenhouse climate promoted the widespread deposition of sediments on continental shelves, forming extensive areas of sedimentary rock. Indeed, many of Earth's sedimentary rocks that are currently exposed at the surface date from intervals of greenhouse

Figure 4.2 An ancient landscape in northwestern Scotland, formed over one billion years ago. Source: Outcrop of Lewisian Gneiss by Robert Bone, CC BY-SA 2.0, via Wikimedia Commons

Figure 4.3 The Cretaceous chalk cliffs of southeastern England, a remnant of a past greenhouse climate. Chalk is a sedimentary rock formed on the sea floor from the remains of microscopic marine organisms. The chalk deposits of England formed under a warmer climate, with higher sea levels than today. They are currently exposed due to the lower global sea levels in today's icehouse climate. Source: Suicasmo, CC BY-SA 4.0, via Wikimedia Commons

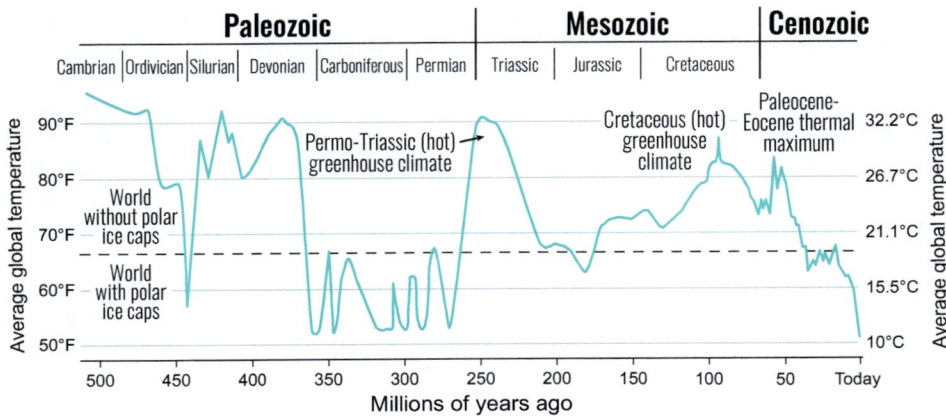

Figure 4.4 Global temperatures over the last 500 million years. Greenhouse climates occur when Earth temperatures are much higher than at present. During icehouse climates, polar ice caps are present. Source: NOAA graphic.

climate. The vast late Cretaceous chalk deposits of northern Europe are stark evidence of this warm, wet world (**Fig. 4.3**).

Conversely, icehouse climates are defined by the presence of continental glaciers and ice sheets, which lead to lower sea levels. Icehouse climates are less frequent, with only five, fully agreed-upon, icehouse climates in Earth history – in the Phanerozoic, the Ordovician–Silurian, the Carboniferous–Permian, and most recently, the Neogene–Quaternary Ice Age, also called the **Late Cenozoic Ice Age** (**Fig. 4.4**). It is important to note that icehouse climates are defined by when ice sheets existed on Earth. Icehouse climates, which last millions to tens of millions of years, should not be confused with much shorter glacial–interglacial cycles, when ice sheets repeatedly grow and shrink *within* these icehouse climates. Overall, icehouse climates are characterized by increased physical erosion (often by glaciers), glacial and periglacial weathering, and highly unstable land surfaces.

Relying mainly on the rock record, paleoclimatologists have been able to reconstruct Earth's temperatures over the past 500 Ma (**Fig. 4.4**). These data illustrate the extreme swings between icehouse and greenhouse climates, spanning from the Cambrian to the Quaternary Period. No one fully understands the causes of these swings, but they likely involve changes in atmospheric CO_2 levels (driven by volcanism and rock weathering) and plate tectonics. This record also shows that the climate that we are currently living in is quite cold, compared to most of Earth's past climates!

Few direct landform traces can be found from the time prior to the Cenozoic. However, climate over this interval continues to indirectly affect landforms today, via the properties of the rocks that formed during this time. For example, many of the chalk deposits of northern Europe formed during the Cretaceous Period, a time of greenhouse climate (**Figs. 4.3, 4.4**). A ridge of this chalk that runs from northwestern France to southeastern England used to form a permanent land-bridge between Britain and continental Europe. However, the ridge has been breached and its remnants are now below sea level, forming what we know as the English Channel (**Fig. 4.5A**). This breach occurred sometime during the last few hundred thousand years, when Quaternary ice sheets dammed up vast amounts of meltwater in huge glacial lakes. These lakes were banked up against the ridge. At some point, the ridge was overtopped and the catastrophic flood that followed eroded the ridge, cutting the island of Britain off from the rest of Europe (**Fig. 4.5B**). This event illustrates the interplay between icehouse Quaternary climate and landforms, as controlled by the properties of

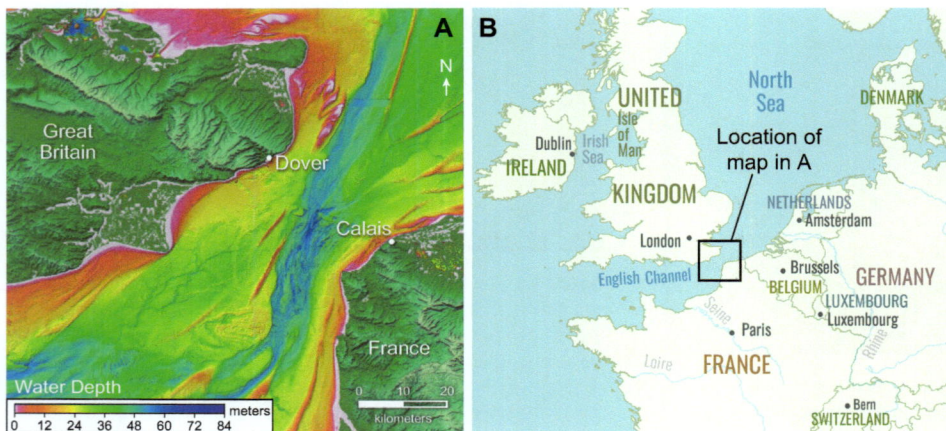

Figure 4.5 Landforms under the sea. **A.** A map of the topography and bathymetry of the English Channel and the coastal areas of southeastern England and northwestern France. Source: After Gupta, S., Collier, J., Garcia-Moreno, D. et al. (2017) with permission from Springer Nature. **B.** This map shows a wider, regional context. Note the breached area of higher land that used to connect England and France and the deep valley of the "Channel River" that is currently drowned by high sea levels.

Cretaceous-age rocks, which as we know formed under a greenhouse climate. Had this ridge been built on harder rock, the breach might not have occurred. It is interesting to consider just how drastically this one geologic event has shaped the history of our modern world – defining national borders, regional history, and trade/transportation routes, and even influencing the outcomes of world wars!

4.4 THE DESCENT FROM GREENHOUSE TO ICEHOUSE CONDITIONS: THE CENOZOIC ERA (66.5 MA TO PRESENT)

A mass extinction, likely caused by a massive asteroid impact and volcanism, defines the start of the Cenozoic Era, which continues to this day. The extinction event heralded major changes in the flora and fauna of the planet. With the exception of some sea turtles and crocodilians, no animals weighing more than 25 kg are thought to have survived. Prior to the asteroid impact, the Cretaceous Period had been very warm (**Fig. 4.4**), and dinosaurs were generally flourishing, although some herbivore taxa were in decline. To a greater or lesser extent, this warmth extended into the Paleocene and Eocene Epochs of the Cenozoic Era (**Fig. 4.4**). However, following the peak warmth during the Eocene, the story of the Cenozoic is one of gradual, but punctuated, descent into the full icehouse climates of the Quaternary Period (**Fig. 4.6**).

During the Paleocene and early **Eocene Epochs** of the Cenozoic Era, Earth's climate was unusually warm (**Fig. 4.6**). With this warmth came significant climatic instability, with abrupt episodes of extreme global warming. The most dramatic of these events occurred at the end of the Paleocene, an event called the **Paleocene–Eocene Thermal Maximum**, widely referenced as a potential analogue for future warming on Earth (**Fig. 4.6**). The cause (or causes) of the Paleocene–Eocene Thermal Maximum is still unclear, but it certainly involved substantial increases in atmospheric CO_2, resulting in 5–8 °C of global warming. During this event, palm trees and crocodiles lived above the Arctic Circle. Furthermore, this event was likely associated with considerable numbers of extinctions.

By the middle Eocene, Earth's climate had begun to cool (**Fig. 4.6**). This cooling was not continuous and was punctuated by short, abrupt, warming and cooling events, during which polar ice caps appeared and disappeared. However, the overall trajectory during the mid to late Cenozoic was toward increased ice volumes on the continents. After several episodes of ice sheet growth and retreat, the Antarctic ice sheet finally became a permanent fixture at the South Pole (**Fig. 4.6**). Cooling accelerated at about 14 Ma, during the middle **Miocene**, which caused extinctions and the rapid expansion of the Antarctic ice sheet. Ice sheets may also have expanded in the northern hemisphere. Northern hemisphere glaciation progressed even further in the late Miocene and **Pliocene**, with the growth of ice sheets on Greenland. This trend continued through the end of the Pliocene, when another substantial drop in global temperatures facilitated further increases in global ice volume (**Fig. 4.6**), especially in the northern hemisphere. As the ice sheets grew, global sea levels dropped by many tens of meters, and weathering processes in many mid- and high-latitude regions became driven more by physical, rather than chemical, processes. Importantly, Earth's climate became significantly more variable at this time, with repeated phases of expansion and retreat of ice sheets over Eurasia and North America, about every 40,000 years. This transformation – to a planet dominated by cycles of northern hemisphere ice sheet growth and decay – marked the start of the Quaternary Period.

Paleoclimatologists still debate what exactly precipitated the change from greenhouse to icehouse climates during the Late Cenozoic. One of the main theories suggests that Cenozoic uplift of major mountain belts like the Himalayas led to increased chemical weathering of silicate rocks, which drew down atmospheric CO_2 contents (see **Fig. 11.29**). In this scenario, a major geologic event (mountain building in the Tibetan Plateau) would have driven a fundamental shift in the Earth's climate, from greenhouse conditions to the icehouse climates of today. However, other paleoclimatologists argue that changes in ocean circulation or sea floor spreading may have been more important.

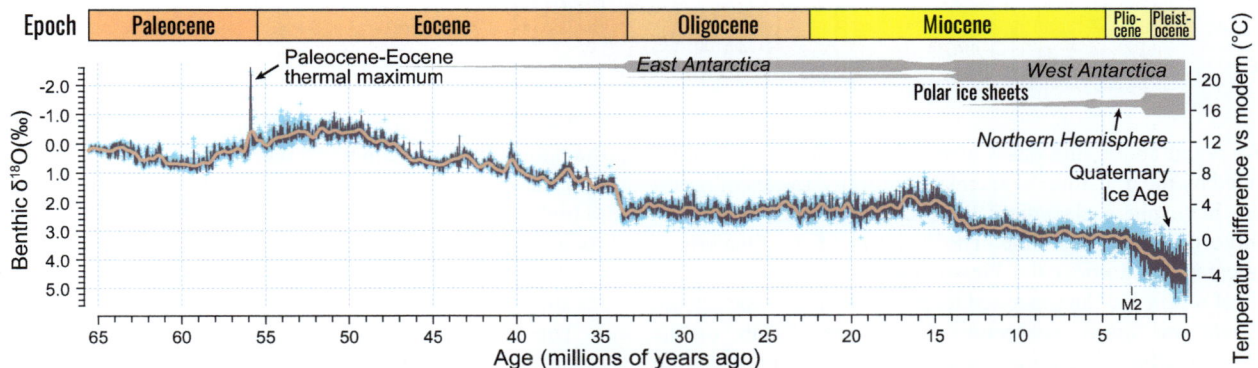

Figure 4.6 Temperature changes and ice sheet volumes throughout the Cenozoic Era. Source: Thomas Westerhold et al., An astronomically dated record of Earth's climate and its predictability over the last 66 million years. *Science* 369, 1383–1387 (2020). Used with permission of American Association for the Advancement of Science.

The huge swings in climate that characterize the Quaternary Period have had massive impacts on Earth's geomorphology. Glacial landforms are prevalent across many parts of the globe (see Chapter 19). However, in regions that escaped glaciation, the legacy of the warm, wetter conditions that prevailed during the preceding Neogene and Paleogene Periods still endures. This legacy is primarily expressed in deep weathering profiles, typical of warm, wet, tropical climates. Many of these landscapes are covered in residuum that is rich in **oxide clays**, attributed to long-term weathering and related volcanism under the warm Cenozoic climate.

A good example of this geomorphic legacy are the granite **tors** (isolated, rocky pinnacles and uplands) of unglaciated southwestern Britain (**Figs. 4.7, 11.0**). Here, the warm Cenozoic climate drove chemical weathering deep into the granitic bedrock, building up a large volume of weathered material. After the climatic cooling during the Quaternary, remnants of the weathered rock were unearthed by **periglacial** processes, exposing in localized areas masses of hard, underlying bedrock. Today, the resulting granite tors are a testament to this very different paleoclimate.

Figure 4.7 Haytor, in Dartmoor, southwestern England, is a granite tor resulting from prolonged Cenozoic weathering under a warm climate, followed by erosion during the cold Quaternary climate that followed. See also Fig.11.24. Source: Nilfanion, CC BY-SA 3.0, via Wikimedia Commons.

4.5 EARTH'S LATEST ICEHOUSE: THE QUATERNARY PERIOD (2.58 MA TO PRESENT)

As noted above, the onset of the Quaternary Period is defined by the transition to extensive, repeated cycles of northern hemisphere glaciation; these cycles became larger and more dramatic over time. The majority of the Quaternary is categorized under the **Pleistocene Epoch**, with only the most recent 11.7 ka separated from it as the Holocene Epoch. Could the warm, comparatively ice-free climate of the Holocene have been a major facilitator of the proliferation

and development of human civilizations? This important question is still being debated.

The geomorphology of most of the middle and high latitudes is dominated by the effects of Quaternary climate change. The gradual shift to colder conditions during the Cenozoic led to changes from warm-climate chemical weathering processes to cold-climate physical weathering processes. As a result, geomorphic impacts derived from Quaternary glaciations include (1) erosion and sediment transport by ice sheets, (2) changes in global and local sea levels, often leading to (3) episodic incision and aggradation of rivers that drained to the sea, and (4) increased windiness and expansion of deserts. The Quaternary, with its large swings in climate, led to very geomorphically dynamic conditions.

During the Quaternary, ice sheets repeatedly expanded and contracted over Europe, North America, and Siberia. At their maximum extent, during **glacial** phases, ice sheets 3–4 km thick (or more) advanced as far south as New York City and northern Germany and Poland (**Fig. 4.8**). During **interglacials** like today, these same ice sheets have retreated greatly, or melted completely away. Compared to interglacials, glacial phases were characterized by overall lower temperatures and reduced seasonality. Globally, glacials were also windier and drier, with lower sea levels associated with the greatly enhanced ice sheets.

Northern Hemisphere Ice Coverage

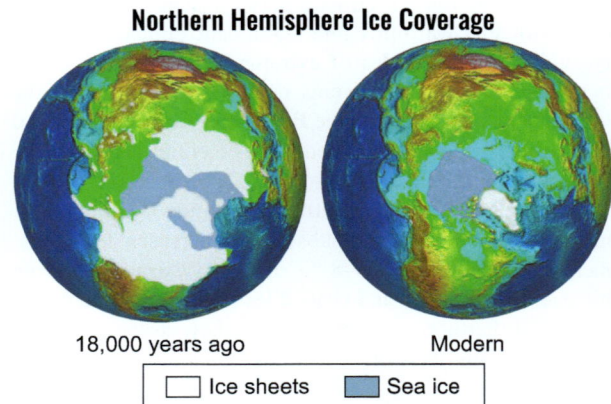

18,000 years ago   Modern

☐ Ice sheets   ■ Sea ice

Figure 4.8 Northern hemisphere ice cover, 18,000 years ago (during the last glacial maximum) and at present. Source: NOAA graphic.

Late Cenozoic cooling eventually accelerated into the dramatic climatic swings of the Pleistocene Epoch (**Fig. 4.9**), leading especially to great temperature drops in the mid to high latitudes, and around the perimeter of the ice sheets. As a result, across vast areas of Siberia an extensive periglacial **permafrost** zone developed near the margins of the ice sheets.

The colder intervals of the Quaternary usually ended much more quickly than they had begun (**Fig. 4.10**).

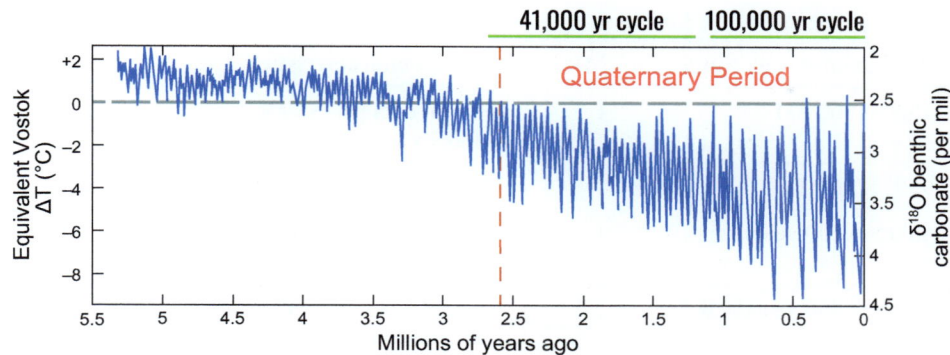

Figure 4.9 The temperature record (difference from present) for the past 5.5 million years, derived mainly from benthic foraminifera data (see below). Based on a figure by Robert A. Rohde, svg by Jo, CC BY-SA 3.0, via Wikimedia Commons, with data from Lisiecki and Raymo (2005).

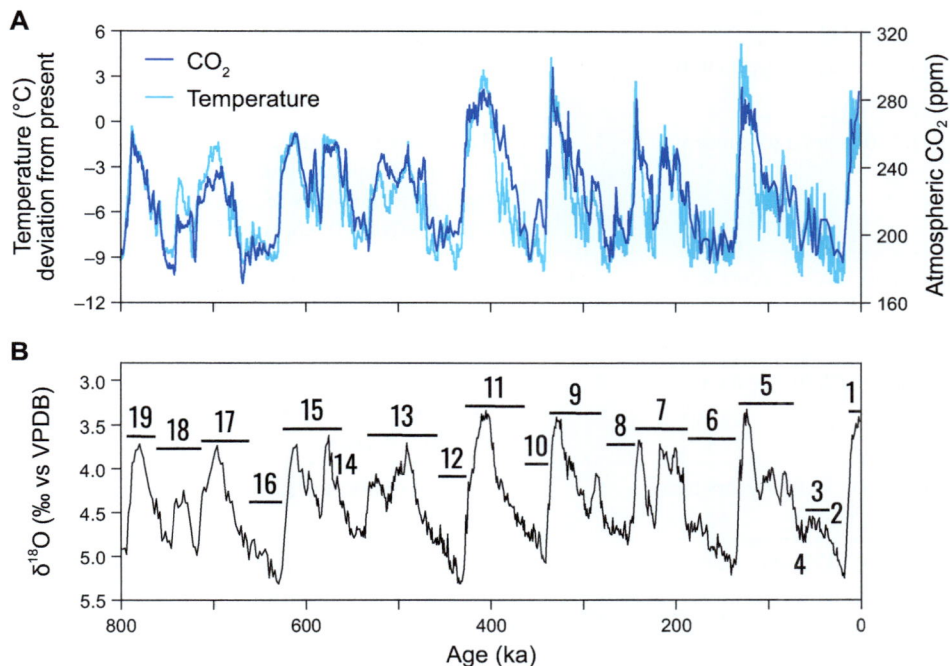

Figure 4.10 Global climate over the past 800,000 years. **A.** Temperatures of, and CO_2 contents in, Earth's atmosphere, as recorded in the EPICA Dome C ice core (Antarctica). Source: NOAA graphic. **B.** The marine $\delta^{18}O$ record (ice volume/temperature) as it correlates to the marine oxygen isotope stages. Source: Based on a figure by Robert A. Rohde, svg by Jo, CC BY-SA 3.0, via Wikimedia Commons, with data from Lisiecki and Raymo (2005).

Although the accumulation of ice in an ice sheet took tens of thousands of years, their collapse (and the shift to warmer interglacial conditions like today) took just a few thousand years. These comparatively short periods of ice sheet retreat and collapse are referred to as **terminations**. Terminations are associated with rapid temperature rises, increases in global sea level, drainage reorganization, and catastrophic flooding from glacial meltwater. The oscillations between glacial and interglacial periods – a hallmark characteristic of the Pleistocene Epoch – have occurred many times, with predictable, regular pacing (**Figs. 4.9, 4.10**). In addition to rapid terminations, Quaternary glacial phases are also characterized by multiple, rapid changes in climate, occurring over tens to hundreds of years.

To facilitate effective communication, scientists have developed names (and a numbering scheme) for many of the glacial and interglacial periods, as well as other climatic events and cycles during the Quaternary (**Table 4.1**). Most of the names derive from locations where a key finding (related to some aspect of paleoclimate) was first studied. Geomorphologists understand best the most recent glaciation – the Wisconsinan in North America, *aka* the Weichselian (in western Europe). At this time, ice covered much of northern Europe, almost all of Canada, and parts of the northern United States. Maximum ice extent occurred globally at around 26–19 ka; a time referred to as the **Last Glacial Maximum**, or LGM. The previous interglacial (the Sangamonian or Eemian; **Table 4.1**) was an especially warm period during the Pleistocene, and the preceding glacial phase (the Illinoian or Saalian) produced some of the largest ice sheets in all of the Pleistocene – a testament to the increasing climatic variability (larger cold-to-warm "swings") that has developed over the last million years (**Fig. 4.9**).

Table 4.1 Major climate events and cycles in the Quaternary, generally from youngest to oldest

Event name(s)	Approximate age range	MIS[1]	Impact/characteristics
Little Ice Age	1500–1800 AD	1	Recent cool period, glaciers advanced globally
Medieval Warm Period	950–1250 AD	1	Short, warm period in the Late Holocene
Climatic Optimum; Holocene Thermal Maximum	8,000–6,000 years ago	1	Peak postglacial warmth; warmest part of the Holocene
8.2 kiloyear event	8,200–7,900 years ago	1	Abrupt, cool interval during the early Holocene
Younger Dryas	12,900–11,700 years ago	1	Brief return to glacial conditions during a period of overall warming and worldwide ice retreat
Bølling–Allerød	14,700–12,900 years ago	1	Abrupt, warm interval during the wider period of deglaciation
Holocene Epoch; Flandrian (in Europe)	11,700 years ago–present	1	The current interglacial period and the only geological Epoch defined by a climate change (rapid warming), as recorded in a Greenland ice core (NGRIP core).
Wisconsinan (Weichselian in western Europe). Also called "Last Glacial"	71,000–11,700 years ago	2–4	The last major glacial phase of the Pleistocene Epoch; contains the LGM
Last glacial interstadial[2] (Farmdalian in the central USA)	57,000–29,000 years ago	3	A relatively warm period during the Wisconsinan/Weichselian; glaciers retreated in many locations
Sangamonian (Eemian in Europe)	129,000–71,000 years ago	5	The last major interglacial; even warmer than today
Illinoian (Saalian in Europe)	191,000–129,000 years ago	6	Penultimate glacial phase; ice sheets grew as large or larger than during MIS 2–4

1. The marine oxygen isotope stage (MIS) nomenclature used in the table is described in the text (Section 4.11.1).
2. **Stadials** and **interstadials** are smaller divisions of the larger/longer "glacial" and "interglacial" events. During stadials, ice advances in some areas but perhaps not worldwide, and only for comparatively short periods of time. Interstadials represent short events of glacial retreat. With the exception of MIS 3, most stadials and interstadials are too small to be given a marine oxygen isotope stage number.

4.6 THE PACEMAKER OF ICE AGE CLIMATES: EARTH'S ORBITAL GEOMETRY

What caused the longer-term oscillations between cold glacial and warm interglacial phases during the Quaternary, as seen in **Figs. 4.9** and **4.10**? The peaks (interglacials) and troughs (glacials) in **Fig. 4.9** have a semi-regular period of 40–100 ka. It is now recognized that these periods coincide with changes in the amount and distribution of incoming solar radiation, or **insolation**, which are governed by slow but continual changes in Earth's orbit (**Fig. 4.11**). Three main aspects of Earth's orbit are responsible for these changes:

1. **Eccentricity**, or changes in the *shape* of the orbit, from more circular to more elliptical.

2. **Obliquity**, or the *degree of tilt* of the planet with respect to the orbital plane.

3. **Precession**, or the changing *direction of the planet's tilt* relative to the Sun, at different points in the orbit.

These changes in Earth's orbit follow predictable cycles. The cycles are named for the Serbian mathematician Milutin Milanković, who hypothesized in the 1920s that orbital variability was responsible for changes in the amount and distribution of solar radiation reaching Earth, and that these changes would have had an impact on climate. Needless to say, Milanković was ahead of his time. Paleoclimatologists today recognize that these orbital variations drove the swings in climate between glacial and interglacial stages during the Quaternary, and probably also during other episodes of

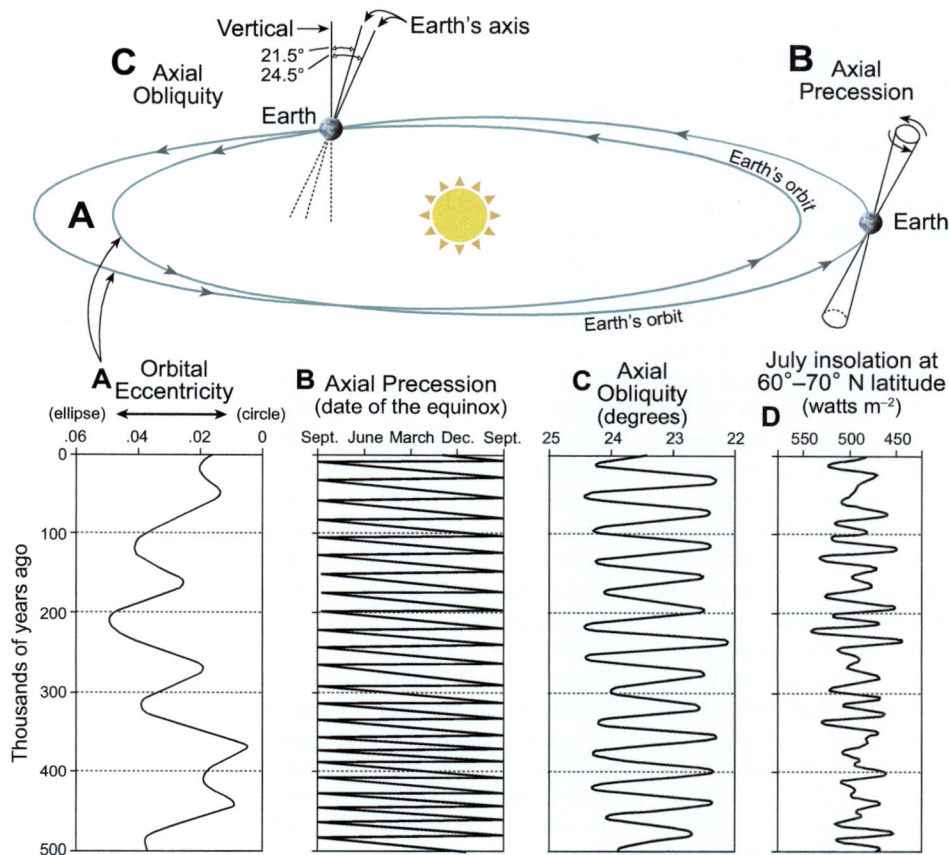

Figure 4.11 Milanković cycles and changes in Earth's orbital parameters. Changes in the shape of Earth's orbit around the Sun drive changes in the amount and distribution of incoming solar radiation (insolation) at the surface (**D**), here given for 60–70° N latitude. These changes involve three parameters: (**A**) eccentricity, (**B**) obliquity, and (**C**) precession.

icehouse climate in Earth's history. In other words, the many Quaternary climate shifts were mainly driven by these three orbital **forcings** – variables that control the climate.

Changes in eccentricity (**Fig. 4.11A**) occur over 100 ka and 400 ka cycles, as Earth's orbit changes from more circular to more elliptical, and back. In a highly elliptical orbit, Earth's distance from the Sun varies considerably over a year; these differences are smaller when the orbit is more circular. Changes in eccentricity affect the amount of incoming solar radiation (insolation) by only a few percent. More importantly, in combination with the direction of the planet's tilt (precession), eccentricity changes the **seasonality** of climate – referring to the overall differences between winter and summer. During periods of low seasonality, for example, winters are warmer than normal and summers are comparatively cooler.

Axial precession (**Fig. 4.11B**) varies over a period of ≈23 ka and changes the date that each hemisphere points most directly toward the Sun. For example, the northern hemisphere is currently maximally inclined toward the Sun in late June (its summer **solstice**). However, at this time of year, Earth is *farthest* from it (its **aphelion**). This orbital configuration means that during winter in the northern hemisphere, Earth is actually closest to the Sun (our **perihelion**). Thus, because perihelion currently occurs in January, northern hemisphere

winters are relatively warm, and summers are relatively cold; in other words seasonality is relatively low. Conversely, at ≈11.5 ka, during northern hemisphere summers, Earth was *nearest* the Sun. Thus, seasonality in the northern hemisphere was relatively high, with warmer summers and colder winters. The importance of this is explained below.

The impacts of Milanković cycles on climate are complex. However, the changes they impart to climatic *seasonality* are thought to be the most important aspect, as they control the long-term, glacial–interglacial changes by driving the accumulation (or melting) of snow and ice. Why is this so? Somewhat counterintuitively, cold winters and high seasonalities do not lead to ice sheet growth. Instead, ice sheets grow when winters are warmer and summers are cooler, that is, during periods of low seasonality. Think of it this way: if summers are very warm, then more melting of ice and snow occurs and less snow persists on the land. Inter-annual accumulation of snow is needed for the growth of ice sheets. Conversely, if winters are very cold, less snow falls because cold air holds less moisture. Warm winters in the high latitudes still mean snowy conditions, and because of the greater amount of moisture in the atmosphere, more snow falls than during colder winters. Thus, seasonality is a key factor in the growth and decay of Quaternary ice sheets. Furthermore, because the largest ice sheets of the Quaternary were located in the northern hemisphere (where plenty of high- and mid-latitude land is available for ice growth), seasonality in the northern hemisphere tends to control the swings in climate from cold glacial to warm interglacial phases. What happens in the southern hemisphere is less important.

The tilt of the planet's axis relative to the orbital plane (obliquity) also impacts seasonality. The angle of tilt varies from around 22° to 24.5° over the course of a 41 ka cycle (**Fig. 4.11C**). Indeed, it is this tilt that causes Earth to have winter and summer seasons. As Earth revolves around the Sun over the course of a year, the orientation of Earth's tilt (relative to the Sun) changes. The hemisphere that is tilted toward the Sun experiences summer, and the one that is tilted away is in winter. But, the greater the *angle* of tilt, the greater the degree of seasonality – particularly in high latitudes. If Earth had no axial tilt, we would have no seasons at all!

Although orbital changes explain glacial–interglacial cycles, one of the great mysteries of Earth's paleoclimate remains: Why did the time interval between glacial phases change from around 41 ka to 100 ka? This change began at ≈1.0 to 0.7 Ma (**Fig. 4.9**). Associated with this change were increases in the amplitudes of the glacial–interglacial cycles; the interglacials became slightly warmer, but most of all, the glacial phases became much colder and ice sheets much larger. Prior to a million years ago, peak glacial conditions on Earth occurred around every 41 ka, strongly suggesting that obliquity primarily controlled the glacial–interglacial cycles. The shift to 100 ka dominance after 1 Ma, which hints at overall eccentricity control, is somewhat of a paradox. In absolute amounts of solar radiation received on Earth, the changes caused by the 100 ka eccentricity cycle are relatively small. Paleoclimatologists lack a widely agreed-upon answer to this paradox. But what is clear is that the northern hemisphere ice sheets became much larger after this point, and the swings in sea level became more extreme (with > 100 m differences in global sea level between glacials and interglacials).

4.7 ACCELERATORS AND BRAKES: HOW CLIMATIC FEEDBACKS MODIFY ORBITAL CLIMATE FORCINGS

One thing is now clear – Milanković cycles, or the variations in Earth's orbit, drove the major climatic swings from glacial to interglacial climates during the Quaternary. However, the magnitude of these "swings" is much larger than what would be expected from orbital changes alone. The mismatch between Milanković forcings and climatic responses is often explained by invoking climatic feedbacks, which enhance (**positive feedback**) or reduce (**negative feedback**) an initial change in climate that may have been initiated by Milanković forcings. A great number of feedbacks appear to operate over glacial–interglacial cycles; three are discussed here (**Fig. 4.12**). In the first, an initial cooling occurs, perhaps driven by orbital forcings. The cooling is then enhanced by changes in the reflectivity of the surface – its **albedo** (**Fig. 4.12A**). As the climate cools, ice sheets start to grow. Ice reflects more insolation

than the land that once occupied the same area. Increased amounts of reflected insolation cool the climate even more. In turn, ice sheets grow even faster, further increasing Earth's albedo, leading to more cooling – an excellent example of a positive feedback in the climate system. In a similar but opposite manner, the loss of glacial ice during glacial retreat helps to decrease Earth's overall albedo, and that in turn helps to increasingly warm the climate, and melt the ice.

Other feedbacks also exist. For example, during a period of climate warming, such as at the end of a glacial phase, increased evaporation (from warmer oceans) leads to increased amounts of water vapor in the atmosphere (**Fig. 4.12B**). Water vapor is a potent greenhouse gas. Higher amounts of water vapor would accelerate warming, leading to more evaporation, more absorption of insolation, and so on. Both of the above examples involve positive feedbacks: an initial change in the system is amplified, causing even more change (but in the same direction). It is through these positive feedbacks that the changes initiated by Milanković forcings became amplified into the huge climatic swings of the Quaternary.

However, this amplification does not go on forever. If it had, runaway cooling or warming would have occurred, leading to an uninhabitable planet. Something must exist to put the brakes on these positive feedbacks. Indeed, a separate set of negative feedbacks also exist (**Fig. 4.12C**). In one such example, an initial climate cooling is reduced through the combined effects of water vapor, clouds, and albedo. Cooling reduces evaporation, which in turn leads to less water vapor in the atmosphere, and fewer low clouds. Low clouds reflect insolation, so that with fewer clouds, more insolation is absorbed by the land and water, leading to warming. In this example, an initial cooling leads to changes that put the brakes on this cooling – a negative feedback.

Many other feedbacks exist and interact with each other in the climate system. Many of these interactions, like those that involve greenhouse gases, aerosols, vegetation, albedo, ocean circulation, clouds, and sea level, are still only incompletely understood.

4.8 RAPID CLIMATE CHANGES DURING THE QUATERNARY

Feedbacks also drive much shorter (decades to centuries) shifts in climate, which cannot be explained by Milanković forcings. As seen in **Fig. 4.13**, which focuses on the period between 50 and 30 ka, many abrupt climate changes occurred repeatedly during the last glacial period. These short-term climate events in the Quaternary may derive from a variety of causes. Some, called **Heinrich events**, involve huge releases of ice from ice sheets, leading to swarms of icebergs in the North Atlantic. Heinrich events are associated with dramatic fluctuations in global climate, with abrupt cooling occurring in many places in the northern hemisphere. The abrupt cooling phases at ≈31, 38, and 45 ka are associated with Heinrich events (**Fig. 4.13**). Three or four additional Heinrich events probably occurred during the last glacial period, each generally lasting <1,000

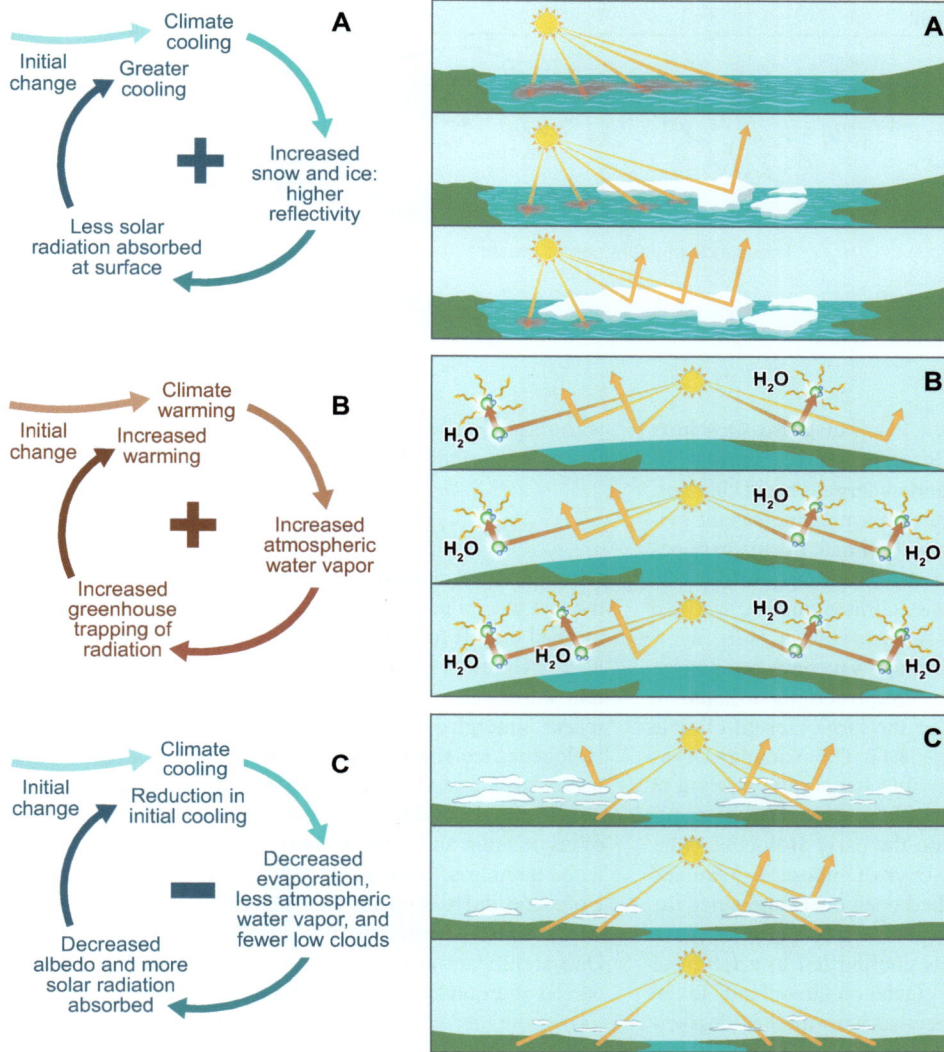

Figure 4.12 Examples of climate feedbacks. **A.** and **B.** Examples of positive feedbacks, where an initial change in the climate–Earth system becomes amplified. **C.** A negative feedback, where an initial change is counteracted by a climate response.

years. The drivers of Heinrich events are unclear, but they likely relate to a variety of feedbacks, including the internal dynamics of ice sheets, ocean circulation patterns, and solar forcings.

Not all of the cooling phases shown in **Fig. 4.13** are Heinrich events. Even more frequent (but equally abrupt) climate oscillations are **Dansgaard–Oeschger (D–O) cycles**, which seem to occur with a periodicity of ≈1.5 ka (**Table 4.1**). These events are defined by rapid, short-term warming occurring over years to decades, followed by cooling occurring over hundreds to thousands of years. Dansgaard–Oeschger cycles are most strongly expressed during glacial phases, but also occurred to a lesser extent during the Holocene. As with Heinrich Events, D–O cycles are strongest in the North Atlantic area. This suggests a link between North Atlantic Ocean circulation patterns and the large ice sheets in the northern hemisphere. Temperatures over Greenland during D–O events reveal rapid (decade-scale) warming of around 5–8 °C, followed by cooling

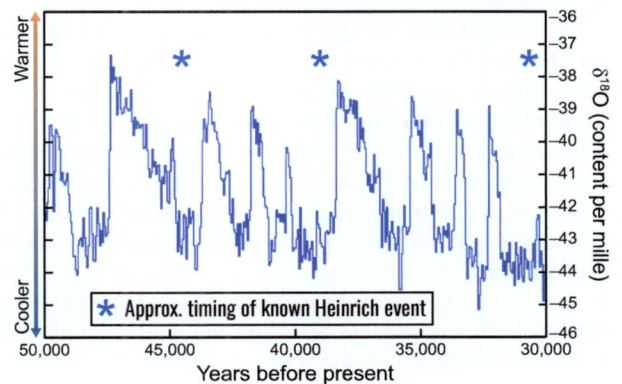

Figure 4.13 Temperature and abrupt Dansgaard–Oeschger cycles, as recovered from the Greenland NGRIP ice core. Also shown are the timing of Heinrich events that coincide with some of the D–O cooling phases. These temperature data, derived from oxygen isotopes ($\delta^{18}O$) over the period from 50–30 ka (the last glacial phase), show repeated D–O cycles of abrupt warming, followed by slow cooling. Source: Adapted from Leland_McInnes at en.wikipedia, CC BY-SA 3.0, via Wikimedia Commons

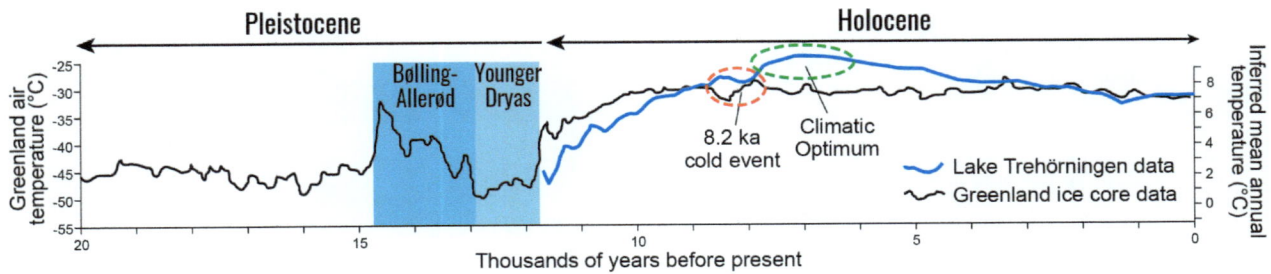

Figure 4.14 Post-glacial temperature changes, as indicated by proxy data from a Greenland ice core and a lake in Sweden. Source: Daniel E. Platt, et al., CC BY-SA 4.0, via Wikimedia Commons

lasting a few hundred years (**Fig. 4.13**). Dansgaard–Oeschger events indicate that, under certain circumstances, Earth's climate system becomes unstable and undergoes rapid changes. Such behavior is of great concern, because increasing evidence suggests that similarly abrupt climate events may become more likely under our currently warming world.

At the close of the Pleistocene, climate quickly warmed as Earth entered the **Bølling–Allerød** phase (**Fig. 4.14**; **Table 4.1**). However, this warming was only temporary. Less than 2,000 years later, an abrupt cooling ushered in the **Younger Dryas** – the last of the large, abrupt climate events of the Pleistocene (**Table 4.1**). The Younger Dryas was characterized by a dramatic, but temporary, cooling at around 12.9–11.7 ka over the northern hemisphere, terminated by an abrupt warming into the early Holocene (**Fig. 4.14**; **Table 4.1**). The Younger Dryas temporarily reversed the overall warming trend that had been occurring since the end of the last glaciation. Cooling over Greenland may have been as much as 10 °C and this cooling led to a return to conditions like those of the last glacial on Greenland. Taken together, the abrupt events associated with the last deglaciation changed the global climate, led to rapid rises in sea level, and may even have altered the course of human history by affecting the movement of agriculture across Eurasia.

What caused the Younger Dryas? Many scientists believe that the main factor was the catastrophic drainage of glacial Lake Agassiz. This huge lake formed along the margin of the **Laurentide ice sheet** in North America during the last glacial period (**Fig. 4.15**). The land surface in eastern Canada, where most of glacial Lake Agassiz existed, slopes toward the center of the former ice sheet, forcing meltwater to pond in front of it. At its peak, glacial Lake Agassiz was larger than any lake on Earth, covering an area four times that of the modern Great Lakes of North America. Eventually, retreat of the ice margin opened up low outlets in eastern Canada, through which the lake catastrophically and episodically drained, sending massive influxes of cold water into the North Atlantic. The cold water shut down ocean circulation patterns and likely caused abrupt climate shifts. The largest of these lake discharge episodes coincides with the **Younger Dryas**, while an abrupt cooling at 8.2 ka (**Fig. 4.14**) is also believed to be related to final drainage of Lake Agassiz. Although the postglacial period was one of overall

warming, climate feedbacks caused it to be punctuated by a few, brief periods of cooling.

4.9 THE HOLOCENE (11.7 KA TO PRESENT)

After the rapid changes in climate that occurred at the end of the last glacial period, the onset of the Holocene interglacial marked a shift to more stable and warmer conditions (**Fig. 4.15**). Although the **Holocene Epoch** is defined as the current warm climatic interval, in actuality, it is simply the most recent interglacial within the Quaternary. During the Holocene, ice sheets retreated and permafrost thawed, sea levels rose, and the monsoons strengthened. The Holocene also marks the rise of agriculture, fostering the development of civilization and urban centers.

Nonetheless, short-term changes in climate during the Holocene did still occur. Dansgaard–Oeschger type cycles appear to have continued, but at greatly reduced amplitudes. One significant abrupt cooling event in the Holocene occurred at around 8.2 ka (**Fig. 4.14**; **Table 4.1**); it lasted only

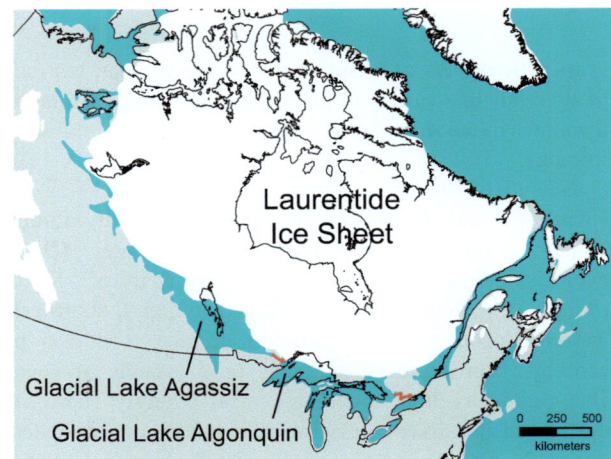

Figure 4.15 The maximum extent of glacial Lake Agassiz during the Pleistocene, at ≈12 ka. Red arrows show the direction of flow through the major outlets of these proglacial lakes, which opened up as the ice margin periodically but steadily retreated to the north. The circuitous drainage of this lake into the North Atlantic Ocean had a major impact on global climate.

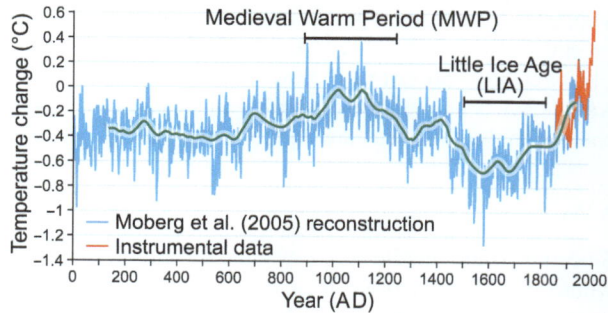

Figure 4.16 Northern hemisphere temperatures over the last 2,000 years, based on proxy and instrumental data. After Moberg et al. (2005) with permission from Springer Nature.

Figure 4.17 The ruins of a Norse church in Greenland attest to the flourishing culture that once existed there, when the climate was favorable during the Medieval Warm Period. Source: J. Barrett.

200–400 years – dwarfed by the changes that had happened at the end of the last glacial (Fig. 4.14). However, this event cooled the climate by a few degrees in many regions of the world, changed precipitation patterns, and may even have led to changes in early human civilizations. For example, in ancient Mesopotamia, one of the main areas where agriculture first developed and spread to the rest of the world, this cooling coincided with significant changes in animal husbandry and early farming techniques.

The gradual warming of the early to middle Holocene peaked at about 8–7 ka, at the so-called **Climatic Optimum** (**Table 4.1**). The Climatic Optimum, also known as the **Holocene Thermal Maximum**, was particularly pronounced in the higher and middle latitudes, with less warming occurring in the tropics. During this time, the monsoons in Asia were strongest and many of the remaining ice sheets, like Greenland, had not been as small in over 100,000 years. Nonetheless, this peak warmth did not last. Temperatures gradually declined in the late Holocene, accompanied by the slow expansion of ice sheets and reductions in monsoon strength. This pattern is typical of how many interglacials develop over time – with peak warmth early on, followed by gradual cooling.

It should come as no surprise that the overall cooling during the late Holocene was punctuated by numerous short-term cycles of warming and cooling. The latest two cycles, the **Medieval Warm Period** (MWP) and the **Little Ice Age** (LIA) (**Table 4.1**), are arguably the most recent in the sequence of Holocene D–O style cycles (**Fig. 4.16**). Because these events occurred during a period of recorded human history, we know far more about them than we do about previous climatic swings. Occurring between ≈900 and 1250 AD, the MWP primarily affected the North Atlantic region. Generally, temperatures during the MWP did not achieve the levels they had reached in the Climatic Optimum, but in the North Atlantic region they were sufficiently high to permit Norse settlers to colonize Greenland (**Fig. 4.17**) and parts of North America in the late tenth century AD. Erik the Red, the Viking who first landed on, and settled, Greenland, felt it necessary to be liberal with the truth in order to draw more settlers to his new colony. Thus, he named the island "Greenland," despite the fact that it was mostly covered with ice. It worked; Viking settlements flourished on Greenland for 400–500 years.

Following the mild MWP, the cool-down of the Little Ice Age was a shock to much of Europe (**Fig. 4.16**). Mainly occurring from ≈1500–1800 AD, the cool-down dramatically impacted human history. Viking settlements in Greenland disappeared around this time, and although climate is unlikely to have been the only factor, it was certainly important in making farming far more difficult there. In Europe, the Little Ice Age is associated with crop failures and "frost fairs" on frozen rivers like the Thames in London (**Fig. 4.18A**), a river which at present does not freeze. Solid ice was reported to have developed for many kilometers off the coasts of England, France, and the Low Countries, disrupting shipping and closing off many harbors in the North Sea. Glaciers began to advance again, all over Europe (**Fig. 4.18B**). It may also have driven considerable global migration, including to North America.

Shortly after (in 1816) the LIA was "The Year Without a Summer." Although probably unrelated to the LIA, 1816 was one of the coldest years in the last few hundred years, particularly in northwestern Europe and eastern North America. The cause of this cold event was likely the eruption of Mt. Tambora in Indonesia. When it erupted, clouds of volcanic ash and sulfuric acid droplets circled the northern hemisphere, causing substantial cooling. As crops failed, famines developed in China, Europe, and North America. Incredibly, accounts in Massachusetts (USA) documented that severe frosts occurred in every month of that year. Snow even fell in June. The Year without a Summer illustrates how isolated events like volcanic eruptions can add to, or counteract, regular oscillations in climate, all of which can have profound consequences for human history.

Numerous causes have been proposed for the swings in Holocene climate. Many have been linked to volcanism, while others have suggested that changes in solar activity may be key, as recorded in sunspot numbers. Sunspot numbers vary across regular to semi-regular cycles, most notably throughout an 11-year cycle. The LIA coincides with the

Maunder Minimum, a period when sunspot numbers were exceptionally low (**Fig. 4.19**). Could this have triggered the LIA? Scientists are not sure. However, additional feedbacks would have been required to amplify this change and cause cooling on the scale seen during the LIA.

4.10 THE ANTHROPOCENE

Since about the middle of the twentieth century, Earth has witnessed a sharp increase in the rate and amount of climatic warming, ushering in a reversal of the gradual Holocene cooling that had been occurring since the Climatic Optimum (**Fig. 4.16**). Today's warming exceeds any of the previous Holocene warm phases. Climatologists nearly universally agree that this warming is mostly due to the effects of large increases in atmospheric CO_2 – a consequence of human activity (**Fig. 4.20**). Indeed, recent evidence suggests that current atmospheric CO_2 levels (>422 ppm) are higher than at any point in the Quaternary, suggesting that today's warming climate is being driven mainly by anthropogenic forcings. Contemporary CO_2 levels are similar to those seen in the Pliocene, when global sea level was tens of meters higher than today, and when crocodiles thrived in northern Canada! Under the currently warming climate, glaciers are melting worldwide, and catastrophic events such as landslides and ice-dammed lake outburst floods are becoming more common (**Figs. 18.30, 18.31**).

Has the human impact on the planet has been so strong and widespread that it might warrant a distinction as new geological epoch? Think of the combined effects of anthropogenic warming, deforestation, farming, species extinction, soil loss, and resource exploitation. These dramatic changes have led some to suggest that Earth has now entered a new period in geological history – the **Anthropocene**. The Anthropocene is an unofficial geological epoch that would follow the Holocene. It is defined by the many significant human impacts on the planet – not only in climate, but also with respect to ecosystems, soils, surficial processes, and atmospheric composition. The Anthropocene concept highlights that the human impact on the planet has become so pervasive and fundamental that it will persist into future geologic records.

Recall that many of the breaks in the geologic timescale are defined by major changes in the rock record (**Fig. 4.1**). These changes were usually caused by extinction events, climate shifts, landscape transformations, or changes in atmospheric

Figure 4.18 Effects of the Little Ice Age in Europe. **A.** Frost fairs on the River Thames in London. Today, the River Thames does not freeze over. Nonetheless, in 1683–1684 (as shown here), the river was completely frozen over for two months, with the ice reaching a thickness of ≈30 cm in London. Source: Thomas Wyke, Public domain, via Wikimedia Commons. **B.** Images of the Rhone Glacier (Switzerland), showing how expansive it was during the Little Ice Age, and how it has been steadily shrinking in area since. Source: B1: unknown source; B2: Chronicle / Alamy Stock Photo; B3: Juerg Alean, Eglisau, SwA.

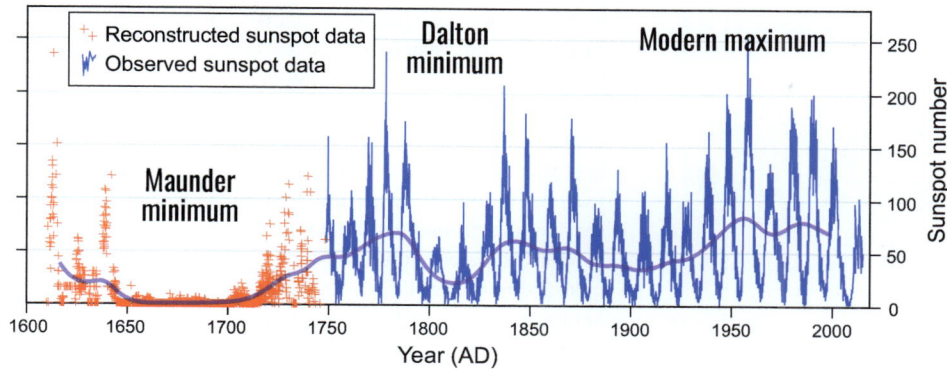

Figure 4.19 Sunspot activity (indicating solar activity) since 1600. Source: Based on a figure prepared by Robert A. Rohde and is part of the Global Warming Art project, via Wikimedia Commons.

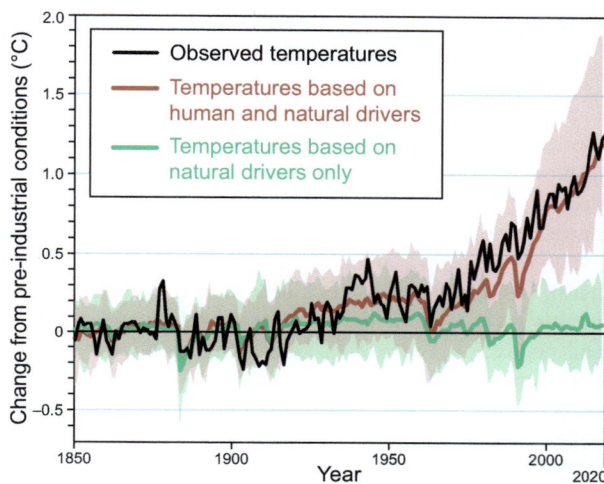

Figure 4.20 Global temperature change from 1850 to present, relative to 1850. The black line shows observed changes. The green and red lines show modelled climate data, based on natural forcings only (green) and natural forcings plus anthropogenic forcings combined (red). Shaded areas show the uncertainties of the climate model simulations. Source: Adapted from figure SPM.1 Panel (b) in IPCC, 2021: Summary for Policymakers. In: Climate Change 2021: The Physical Science Basis. Contribution of Working Group I to the Sixth Assessment Report of the Intergovernmental Panel on Climate Change [Masson-Delmotte, V., et al. (eds.)]. Cambridge University Press, pp. 3–32, doi: 10.1017/9781009157896.001.

history, but the current extinction event is on course to become the sixth. Additionally, some species, notably farm animals like cattle and chickens, will be vastly overrepresented in the geologic record for the Anthropocene. For example, humans now consume 65 billion chickens annually, leading some to suggest that we are living in the "Geologic Age of the Chicken." Will future paleontologists conclude from the global distribution of bones that chickens were the most widespread, dominant life form during the Anthropocene? Ecosystems and landscapes have also been transformed by direct human intervention via agriculture, river regulation, mining, urban development, and the widespread use of chemicals and plastics.

Indeed, the scope and depth of human impact is so large that geoscientists cannot fully agree about when to define the beginning of the Anthropocene, or even what defines its starting point. There are numerous possibilities. Some argue that the best starting point is the advent of agriculture in Europe and Asia, which led to CO_2 release from deforestation and enhanced methane production via rice cultivation in East Asia. Because methane is a potent greenhouse gas, increases in methane likely led to major changes in climate, starting thousands of years ago. Others argue that the dramatic increase in growth rates in human activity in the mid-twentieth century should mark the start of the Anthropocene. A precise marker of geologic change might be the peak in radionuclide fallout caused by the first atomic bomb test on July 16, 1945 – the Trinity test in New Mexico, USA. Regardless, the inception and nature of global change during the Anthropocene is debatable. But the overall concept is a useful one, as it reinforces the fact that humans have become (and will remain) major agents of planetary change.

composition. In the Anthropocene, all of these causes remain in play. In essence, the Anthropocene concept posits that the human species has now become a geologic force of global import – changing patterns and rates of weathering, erosion, extinction, and climate at an exceptional scale and breadth – and with no self-directed end in sight.

Even if emissions of greenhouse gases stopped today, research suggests that we have now broken the glacial–interglacial cyclicity that are driven by Milanković forcings. We are also in the middle of a human-driven mass extinction event that will certainly be reflected in the fossil record. Only five mass extinctions have been recorded in Earth's

4.11 THE RECORDS OF CLIMATE CHANGE: ARCHIVES AND PROXIES

We know about past climates from a variety of data sources, and the list continues to grow. Humans have been recording climate with instruments, that is, the instrumental record, for only the past few hundred years. To understand climates

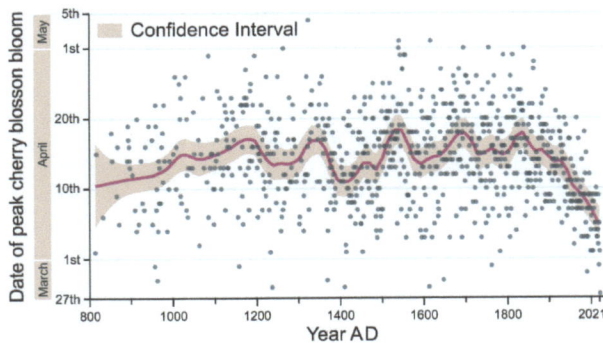

Figure 4.21 Dates of the peak cherry blossom time at Kyoto, Japan, from 800 to 2021 AD. Note how the trees are blooming much earlier in recent years, due to anthropogenic warming. Slightly later blooming during the LIA (≈1500–1800 AD) is also evident in the figure. Source: Graphic courtesy Y. Aono.

farther in the past, for which no instrumental data exist, **proxy data**, or substitute records, are needed. For example, useful proxy paleoclimate data have recently emerged within the science of **phenology**, which is the study of changes in the seasonal timing of biological events and life cycles. Phenologists may examine the date of the first cherry blossoms each year, the arrival of certain species of migratory birds in spring, or the first date of lake freeze-over, all of which are influenced by weather (**Fig. 4.21**). Long-term variations in the timing of these phenomena can be used to examine changes in climate. By observing the timing of these phenomena and recording the dates of their occurrence through time, it is possible to reconstruct how recent climates have changed. For example, the usual date of full-blooming in Washington, DC, whose cherry blossom festival is a relative newcomer to the scene, is now happening (on average) five days earlier than its first recorded date in 1921. Fortunately, people have been recording natural events for longer periods of time, all across the globe. As you see, phenological data can be used as a proxy for climate change in areas where instrumental records do not exist today, or for older periods that lack instrumental data entirely.

Phenology still requires people to write things down. Thus, for the vast extent of geologic time, humans with writing abilities simply did not exist to create such a record. Therefore, we must look for most proxies of past climate changes in rock, sediment, ice, or landform archives. The best geomorphic or sediment archives have physical, chemical, or biological properties that become changed in some way by the climate or the environment. By analyzing these properties, paleoclimatologists can estimate how climate has changed in the past. Of course, this process requires that the approximate age of this change is known.

In practice, proxies used to examine past climates are affected by a variety of different environmental parameters, not just one convenient climatic variable. Thus, records from *multiple* proxies and archives are often needed to compile a truly comprehensive understanding of past climates.

4.11.1 Deep Sea Sediments

Far away from continental areas, the deep ocean basins represent some of the best places to preserve sediment for millions of years. Far from tectonic plate boundaries, the deep water allows deposits to settle and be preserved, by protecting them from disturbances like waves and currents, making them ideal archives of climate change. The deep ocean basins are, therefore, a dependable and enduring **depositional environment**. In contrast, shifts between erosive and depositional processes *on land* or in shallower ocean environments affect the sediment record – so much so that it reflects more than just climatic drivers. In such locations, the sediment record is more fragmented and variable, making it difficult to interpret.

An advantage of utilizing climate archives from the deep ocean basins lies in the fact that they accumulate the shells of single-celled organisms (**foraminifera**) that live in ocean water (**Fig. 4.22**). Foraminifera (*forams* for short) grow shells of calcite ($CaCO_3$). Over time, the shells of many generations of forams accumulate in thick sequences of deposits on the ocean floor, forming the backbone of much of our understanding of Quaternary climates.

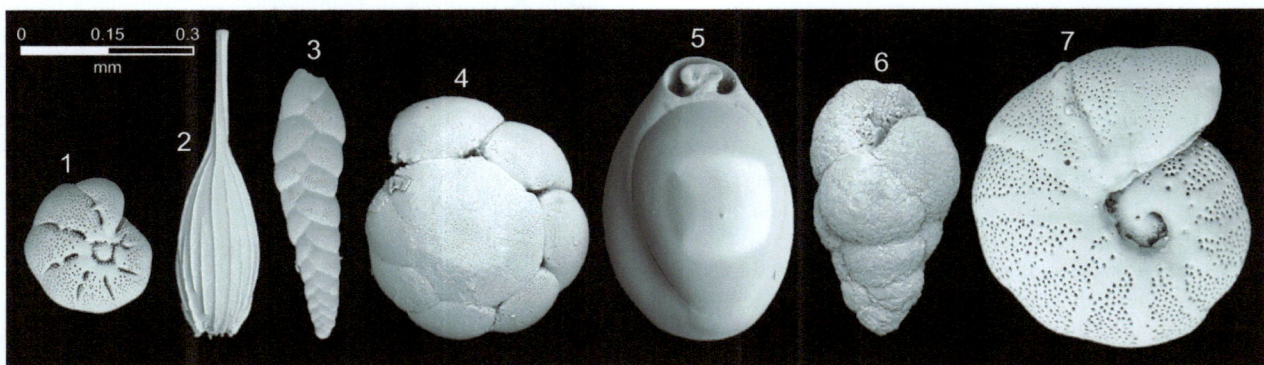

Figure 4.22 Scanning electron microscope images of benthic (living on, or close, to the sea floor) foraminifera: 1. *Elphidium excavatum.* 2. *Lagena sulcata.* 3. *Bolivinellina pseudopunctata.* 4. *Ammonia* sp. 5. *Pyrgo williamsoni.* 6. *Eggerelloides scaber.* 7. *Melonis barleeanus.* Source: I. Polovodova Asteman, reproduced with permission from Encyclopedia of Marine Geosciences.

Important to this discussion is the fact that the chemistry of foram shells reflects the composition of the sea water in which they were formed. Because ocean chemistry is in part a function of climate, forams give us key insights into the climate at the time of their formation. One of the most important elements in this respect is oxygen, which, together with hydrogen, makes up water molecules (H_2O). Like all elements, oxygen comes in various forms, called **isotopes**. Isotopes of an element have the same number of protons (and therefore, the same atomic number) but different numbers of neutrons (and therefore different atomic masses) in their nuclei. Isotopes of an element have the same *chemical* properties but different *physical* properties. Chemical properties refer to how atoms of one element interact with atoms of another. Physical properties include such things as density, changes in phase (solid, liquid, gas), and boiling or melting points. Variable physical properties imply that environmental and biological processes may affect each isotope differently.

Oxygen atoms occur in three natural forms, ^{16}O, ^{17}O, and ^{18}O. They each contain eight protons, but have eight, nine, or 10 neutrons in their nucleus, respectively. Of these, ^{16}O is by far the most abundant, accounting for more than 99.7% of all oxygen atoms, whilst ^{17}O and ^{18}O make up the rest. Advances in measurement techniques in the 1940s facilitated the accurate determination of the proportions of these different isotopes in various substances, such as foram shells. This pioneering work led to suggestions that compositional changes in sea water could be reflected in the isotopic composition of the foram shells.

In the 1950s and 1960s, Italian-American scientist Cesare Emiliani obtained core samples of deep marine sediments, using drilling rigs on research vessels (**Fig. 4.23**). He hypothesized that the large swings in the climate of the Quaternary were recorded in changes in the oxygen isotope composition of the foram shells in these cores. In essence, he hoped that data from forams could serve as a new and important type of paleoclimate proxy. Furthermore, he thought that the variations in oxygen isotope composition could be used to develop a timescale for the glacial–interglacial cycles of the Quaternary. He was right. Indeed, the findings from the deep sea record have revolutionized our understanding of past climates.

Emiliani was able to demonstrate that Earth's climate had gone through major swings in the past, paced in part by the orbital variations, as proposed more than 50 years earlier by Milutin Milanković. This record – known as the **oxygen isotope record** – survives today as the gold standard for understanding Quaternary climates (**Figs. 4.9, 4.10**).

Paleoclimatologists initially debated what specific kinds of proxy data could be gleaned from foram shells. They first determined that *water temperature* affected the ratio of ^{18}O to ^{16}O in the shells. But later, in the 1960s and 1970s, British scientist Nicholas Shackleton argued successfully that, during icehouse climates like the Quaternary, it was not so much temperature that drove fluctuations in oxygen isotope ratios in forams, but rather changes in the *volume* of the ice sheets, as they waxed and waned. This important revelation is explored below, but remains the fundamental premise that underpins our understanding of the deep sea climate record.

Numerous paleoclimate proxies can be recovered from ocean cores. One of the most important is the ratio of ^{18}O to ^{16}O in foram shells, as initially suggested by Emiliani. This proxy, called $\delta^{18}O$ (delta ^{18}O), is based on the ratio of ^{18}O to ^{16}O, relative to a standard value. When these data are obtained from **benthic foraminifera** (forams that live on the sea floor; **Fig. 4.22**), they only partially reflect the temperature of the deep ocean (which does not change much anyway). Instead, $\delta^{18}O$ data from benthic forams serve as an excellent proxy for past *global ice volumes*. Why is this so? We know that water molecules containing the lighter isotope (^{16}O) evaporate more easily, such that clouds and precipitation are enriched in ^{16}O relative to ocean water (**Fig. 4.24**). During periods of ice sheet growth (like during Quaternary glacial phases), some of this evaporated water ends up as snow and eventually becomes locked up in expanding ice sheets. As more water is locked up on land in the ice sheets, the oceans preferentially lose more and more ^{16}O, relative to ^{18}O, increasing the $\delta^{18}O$ in sea water. When ice sheets reach their maximum size, like they did during the last glacial maximum, the $\delta^{18}O$ of deep ocean water achieves its highest value. Conversely, during periods when the ice sheets melt, the meltwater, enriched in ^{16}O, drives decreases in the $\delta^{18}O$ of sea water (and foram shells). Thus, during glacial phases, the $\delta^{18}O$ of sea water (and foram shells) is high, while sea water and shells have lower $\delta^{18}O$ during interglacials. This important relationship

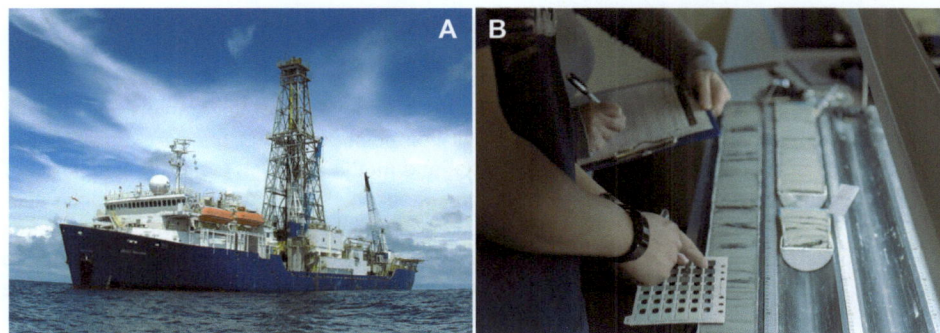

Figure 4.23 Ocean drilling to understand climate change. **A.** The drilling ship *Joides Resolution*, part of the International Ocean Discovery Program. Source: Arito Sakaguchi & IODP/TAMU, Attribution, via Wikimedia Commons. **B.** Sampling a core of oceanic sediment. Source: Marum, CC BY 4.0, via Wikimedia Commons.

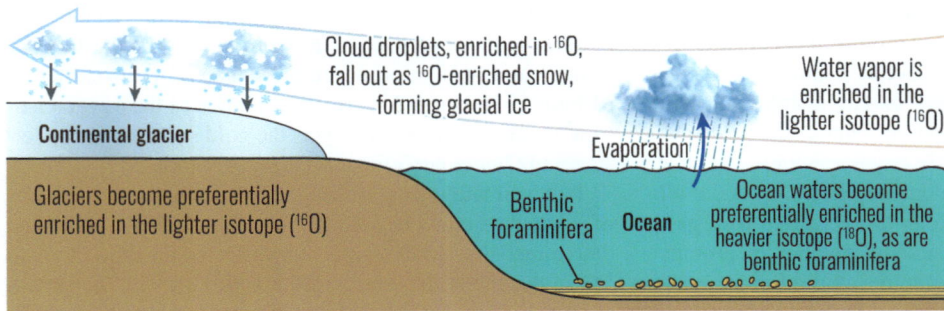

Figure 4.24 The cycle of oxygen isotopes during a glacial period, as it pertains to the climate record in deep sea (benthic) forams.

makes the benthic foram record an excellent proxy for past global ice volumes. The climate record shown in **Fig. 4.10B** is based on this technique, demonstrating the incredible number and range of swings in climate that have occurred in the past.

Benthic forams are used as proxies for Quaternary climate because the deep ocean does not experience local climatic effects; instead, it reflects global trends. Reconstructions using benthic $\delta^{18}O$ data are so well studied and reliable that they now define *marine oxygen isotope stratigraphy*. Under this scheme, peaks and troughs in $\delta^{18}O$ through time are designated **marine oxygen isotope stages** (MIS) (**Table 4.1**) – a nomenclature used worldwide. Each stage is numbered and assigned an age, based on correlations with shifts in insolation driven by Milanković forcings. One of the most important findings from the benthic $\delta^{18}O$ MIS record is the large number of stages that have occurred during the Quaternary, each representing a shift between a cold, glacial climate and a warm interglacial climate (**Fig. 4.10B**). In the MIS numbering system, interglacials are given odd numbers and glacial phases are given even numbers. The Holocene is MIS 1, with the last glacial represented by stages 2, 3, and 4 (**Table 4.1**). The last interglacial is represented by MIS 5, and the second-to-last glacial phase correlates to MIS 6. MIS 3 represents a comparatively short period of climate warming in some parts of the world, set within a broader glacial interval (MIS 4–2). Currently, over 100 marine oxygen isotope stages have been identified, going back over 6 Ma. The chronology is likely to be extended in the future, as more and deeper cores are recovered.

Marine oxygen isotope stages define the temporal framework for Earth's Quaternary climate and ice sheet history. They are the framework around which other climate records are referenced and compared. As such, it is difficult to overstate the importance that proxy climate data from ocean sediments have had in deciphering Quaternary paleoenvironments worldwide.

Benthic forams also provide additional climate proxies, because they preferentially incorporate certain isotopes of carbon (not just oxygen!) in their shells. Thus, the shells are also a good proxy for the carbon isotope ratio of sea water. Furthermore, ratios of elements like magnesium (Mg) and calcium (Ca) can be used as proxies for water temperature changes, because forams increasingly utilize Mg over Ca

in their shells as water temperatures rise. Even the various foram species themselves can be useful in understanding past climates (**Fig. 4.22**), as different species have different temperature requirements.

Marine cores also provide proxy paleoenvironmental data beyond even what forams can supply. All ocean cores contain sediments that were washed or blown into the ocean basins from land masses. The mass, mineralogy, chemistry, and other properties of this sediment are all valuable paleoclimate proxies.

4.11.2 Ice Cores

Although deep ocean sediments can provide extremely long-term records of global climate, they lack temporal detail and reveal little about past atmospheric compositional changes. Here, the climate records derived from ice cores can have their moment in the spotlight (**Fig. 4.25**). Glaciers form as snow gets compressed and compacted (see Chapter 18). Glacial ice accumulates almost continuously in the coldest parts of the planet. Some glaciers even preserve annual bands of ice, allowing for an extraordinarily detailed picture of past climates and atmospheres. The Greenland and Antarctic ice sheets have accumulated ice, layer by layer, for tens of thousands of years, and even millions of years in the case of Antarctica. In order to ensure the most complete record, ice cores can be retrieved from the centers of these ice sheets, where accumulations are thickest.

Of importance here is the fact that ice cores contain a record of past atmospheric composition, including greenhouse gases. As glacial ice forms, it traps small pockets of air (**Fig. 4.26**). The trapped air is so well preserved and isolated from its surroundings that these air bubbles become miniature time capsules. Air from past atmospheres can be stored within glacial ice for thousands to millions of years. Ice cores thus provide the rare opportunity for direct examination of the composition of Earth's past atmospheres.

Many paleoclimatologists are very interested in the greenhouse gas composition of Earth's past atmospheres, particularly CO_2 and methane (CH_4). Continuous records of atmospheric greenhouse gas concentrations have been generated from ice cores. Records of this kind from Antarctica span over nearly a million years. Pockets of even older ice have been found that pre-date the Quaternary. These ice core records have revealed two striking things. First, past

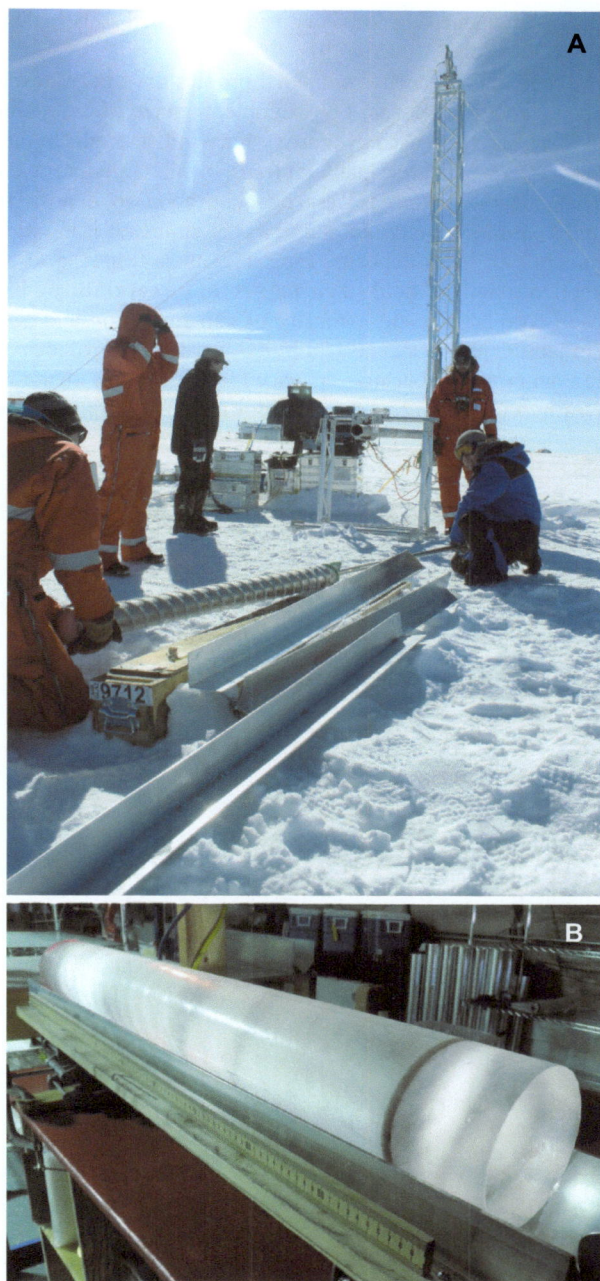

Figure 4.25 Ice core drilling. **A.** Coring the summit of the Greenland ice sheet (the EastGRIP core). Source: Helle Astrid Kjær, CC BY 4.0, via Wikimedia Commons. **B.** An ice core from the West Antarctic ice sheet. Note the layering and the dark layer of volcanic ash deposited on the ice ≈21 ka. Source: Heidi Roop, University of Minnesota.

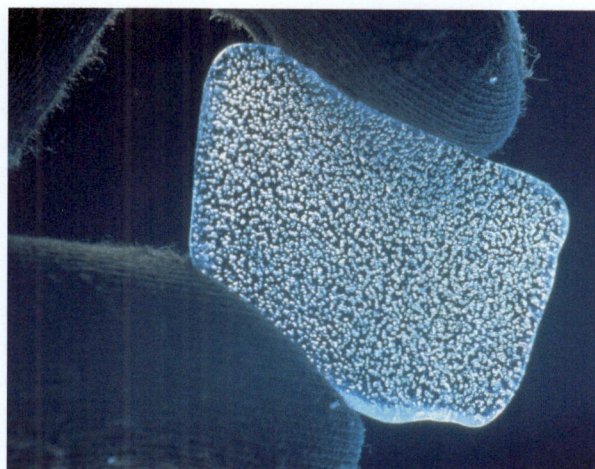

Figure 4.26 A fragment of the Antarctic ice sheet, showing numerous small air bubbles – time capsules of a former atmosphere. Source: CSIRO, CC BY 3.0, via Wikimedia Commons.

anthropogenic CO_2 emissions have interrupted the preexisting, natural CO_2 cycle.

In addition to records of past atmospheres, ice cores can act as paleoclimate proxies in other ways as well. Of importance here is the fact that ice chemistry reflects the temperatures that existed when it formed. The temperature records shown in **Figs. 4.10A** and **4.14** come from the oxygen and hydrogen isotopic compositions of ice in Antarctic and Greenland ice cores, respectively. Furthermore, ice can record the presence of far travelled impurities such as volcanic ash (**Fig. 4.25B**) and dust. These clastic (mineral) impurities in the cores provide insight into past volcanism and wind directions. Even more important is the fact that **aerosols** (atmospheric solids like volcanic ash, mineral dust, pollen, etc.) can directly impact climate by (among other things) affecting how much sunlight is absorbed. Thus, records of past changes in aerosol contents from ice cores, as well as greenhouse gases, highlight some of the possible *causes* of changes in past climates.

Notably, the dust abundance in ice cores matches nicely with past temperatures. Ice core studies in Greenland have shown that both dust concentrations and temperatures changed abruptly in the later part of the Quaternary. These abrupt warming (D–O) events appear coupled to reductions in dustiness, as well as shifts in oceanic circulation patterns. Their remarkable impact has been made clear in the detailed records preserved in Greenland ice.

4.11.3 Loess

Cores of marine sediment and glacial ice may be excellent proxies for global paleoclimates, but they do not necessarily record climate changes occurring on the land masses. To understand these changes better, archives of climate change on land are necessary. But a confounding issue exists – the land surface is subject to much more changeable conditions than are the deep ocean basins, or the crests of massive ice caps. Terrestrial sediment is more susceptible to erosion,

temperatures on Earth are strongly coupled to greenhouse gas concentrations (**Fig. 4.10A**). For example, glacial–interglacial cycles during the Quaternary are coupled to shifts of up to 100 ppm in concentrations of CO_2, with higher levels during interglacial periods like the Holocene. Secondly, current levels of CO_2 in the atmosphere are greater than at any point in the Quaternary, highlighting the fact that making long records difficult to find. One exception to this is **loess** – wind-blown, dominantly silt-sized sediment (**Fig. 4.27**). Loess deposits are estimated to cover 10% of the world's continents (**Fig. 21.33**), and the properties of loess make it an excellent climate archive. Loess can also be dated with reasonable precision, using luminescence dating techniques (see Chapter 3).

In areas of thick loess, such as in China or eastern Europe, loess tends to accumulate during cool, dry, glacial periods. Then, during the warmer, wetter interglacials, when loess accumulation rates are lower, soils form in the loess. Studies have also shown that these warmer, wetter phases are also characterized by less windy, dusty conditions, as was also seen in many ice cores. Over time, some areas accumulate

stacked sequences of loess deposits and buried soils (paleosols) (**Figs. 4.27, 4.28**).

Numerous paleoenvironmental proxies have been developed for loess deposits. For example, the size of the grains in loess has been used as a proxy for wind strength. In theory, faster winds are capable of transporting larger particles. Thus, coarser-textured loess deposits point to deposition under stronger winds. In addition, the nature of iron oxide minerals in loess deposits, especially their response to an applied magnetic field (their **magnetic susceptibility**), has been correlated to the intensity of soil formation and precipitation. Magnetic susceptibility works as a paleoenvironmental proxy because enhanced rainfall drives chemical weathering, which can cause the formation of very fine iron

Figure 4.27 Thick loess deposits. **A.** Millions of years of loess deposits are exposed near Luochuan, on the central Chinese Loess Plateau. The faint, darker bands in the loess are ancient soils (paleosols), formed in warmer climates and later buried by subsequent additions of loess. **B.** Loess from the last glacial is shown here at the Irig loess section in Serbia, where it buries a paleosol that formed during the last interglacial phase (MIS 5). Source: T. Stevens.

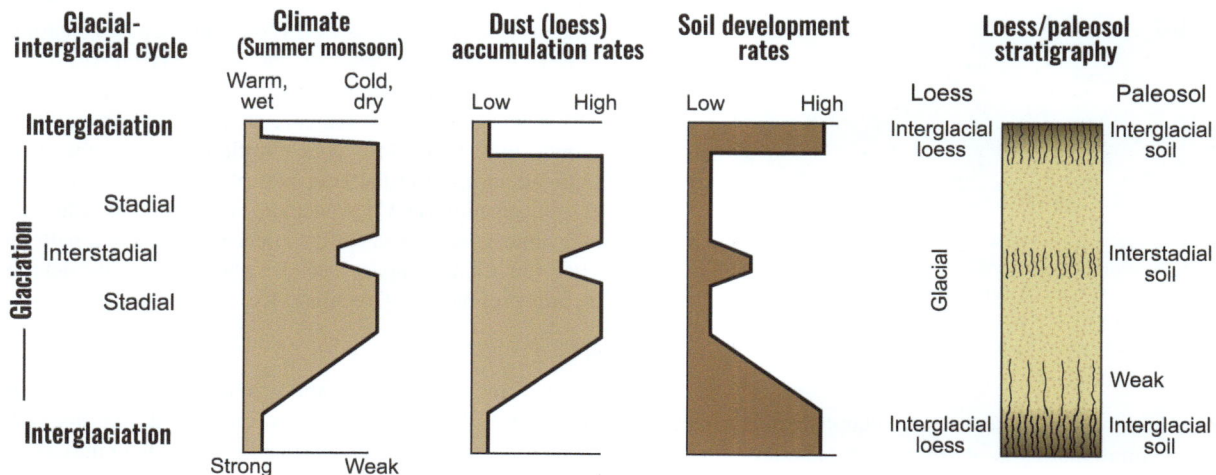

Figure 4.28 Climate variation, as it correlates with loess deposition and soil development rates on the Chinese Loess Plateau. Low rates of dust accumulation and high rates of soil development occur during interstadials and interglacials – times of strong summer monsoon rainfall. In contrast, faster dust accumulation rates and slow or no soil development occur during stadial and glacial phases, when the summer monsoon is weaker. The resultant loess stratigraphy is shown on the rightmost diagram. Source: Used with permission of Elsevier, from S. C. Porter (2001); permission conveyed through Copyright Clearance Center, Inc.

oxide particles in soils. These particles respond strongly to an applied magnetic field. Other climate proxies from loess include its elemental chemistry, clay mineralogy, and color, all of which have been tied to the intensity of weathering and soil formation – which can then be correlated to climate. Isotopes of oxygen and carbon have also been analyzed in loess deposits, potentially revealing past temperature and precipitation changes.

Some of the most-studied loess deposits in the world are found in China, in an area called the Chinese Loess Plateau (Fig. 4.29). This area, larger even than Germany or California, is draped in hundreds of meters of wind-blown loess, representing millions of years of climate change. The oldest deposits may be up to 40 million years old, but the thickest and most widespread are of Quaternary age (Fig. 4.27A). The

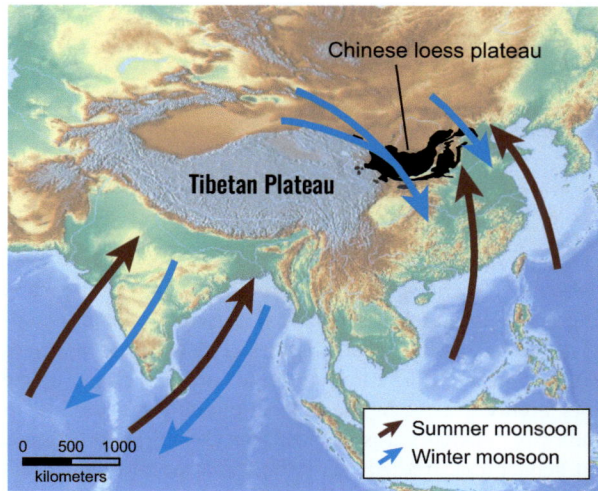

Figure 4.29 Location of the Chinese Loess Plateau relative to the influence of the Asian Monsoon. Red arrows denote warm, wet summer monsoon winds, whereas blue arrows show the wind direction during the cold, dry, winter monsoon.

Chinese Loess Plateau lies at the edge of the monsoon rainfall belt in East Asia. Because of this location, the Chinese loess deposits have been used to reconstruct past changes in the East Asian Monsoon, which is a major component of the global climate system. Monsoon climates involve the seasonal alternation of winds and precipitation (Fig. 4.29). Monsoon areas in East Asia have rainy summers, as air masses move from the Indian and Pacific Oceans, onto the land mass. Monsoonal winters are then characterized by dry, cold winds blowing off the continental interior.

The East Asian Monsoon directly affects the loess depositional system on the Chinese Loess Plateau. Loess is transported from dry areas of the Asian interior during the dry, windy, winter monsoon. Very little loess is transported during the wet, summer monsoon period; instead, soil formation occurs in loess deposits already present. Thus, loess deposits grow fastest during periods when the winter monsoon dominates, and soils form in loess deposits when the climate is warmer and wetter, under a stronger summer monsoon.

Using loess records of winter and summer monsoon changes, researchers have shown that the East Asian Monsoon has been active for millions of years, and that large changes in the strength of the monsoon occur in tandem with shifts between glacial and interglacial climates (Fig. 4.30). Monsoon variability is recorded in numerous proxies in the Chinese loess, particularly in its many buried soils, but also in its magnetic susceptibility record. **Magnetic susceptibility** (MS) values in the loess increase during warm, interglacial periods, and decrease during cold glacial stages. The loess MS data from the Chinese Loess Plateau correlate well with the benthic marine oxygen isotope record (Fig. 4.30). These observations show that the major climate swings of the Quaternary drove not only ice advance and retreat in the high and mid latitudes, but also the low- and middle-latitude climate systems like the East Asian monsoon.

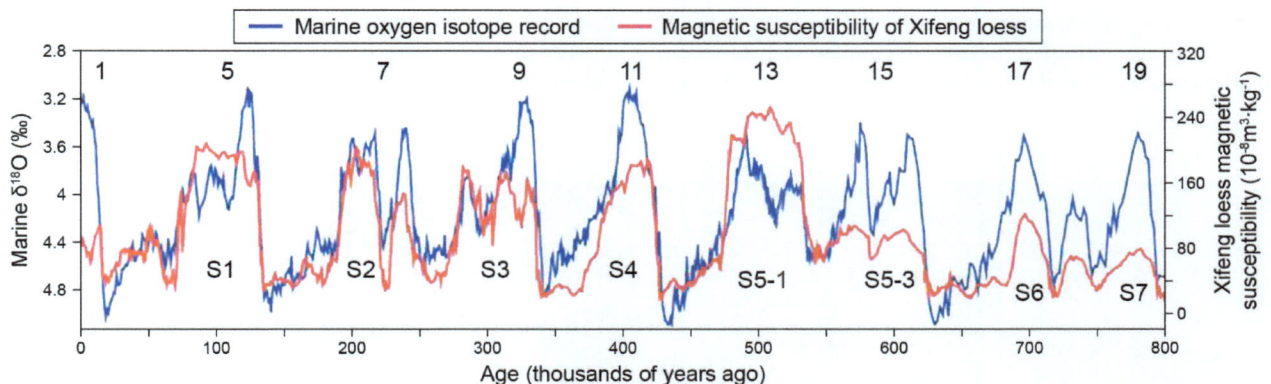

Figure 4.30 The magnetic susceptibility record of loess on the Chinese Loess Plateau, using data from the Xifeng site, as compared to the benthic $\delta^{18}O$ marine record, over the last 800 ka. These records show how the East Asian Monsoon (derived from the loess record) fluctuates in tandem with changes in global ice volume (derived from the marine record). Numbers near the top of the figure correspond to warm, marine oxygen isotope stages (interglacials), whereas S1, S2, etc., are paleosols in the loess sequence. Source: Used with permission of Elsevier, from B. A. Maher (2016); permission conveyed through Copyright Clearance Center, Inc.

4.11.4 Pollen and Macrofossils

One of the most useful paleoenvironmental archives, and one which is particularly useful for recently glaciated locations, are **peat** bogs. Peatlands are waterlogged sites where organic remains of all sorts can accumulate under anoxic conditions, which prevent or slow the decay of organic materials. The peat deposits thicken over time within these settings, as new organic materials are added, layer after layer. Eventually, thick sequences of peat accumulate in waterlogged lowlands. These deposits can preserve a nearly complete record of past vegetation and climate, from the initiation of the water body until today (**Fig. 4.31**).

Both peat bogs, and even clastic lake sediments in areas that do not develop peat, can preserve a range of past climate proxies, particularly amounts and assemblages of **pollen grains**. These grains can be preserved quite well in acidic peat deposits, as the waterlogged conditions inhibit their decay. Pollen grains produced from plants growing near to the bog (or lake) fall into the water and settle to the bottom, where they are preserved in relatively undisturbed sequences. The grains serve as a proxy for the paleovegetation at the site, which in turn is an excellent indicator of the paleoclimate.

As with many climate proxies, paleoclimatologists apply the principle that "the present is the key to the past," and assume that the ecological tolerances of plants today are broadly the same as in the past. Thus, when pollen grains from cold-tolerant tundra plants are found in a mid-latitude peat bog, we can infer that a colder climate once existed there. And because peat can be easily dated using radiocarbon dating (see Chapter 3), we can connect the past ecology and climate to a specific date in time. Pollen data can therefore provide a good picture of the regional climate and plant assemblages for the site from which they were obtained. The study of past climates using pollen data is called **palynology**.

Pollen data are used to reconstruct plant communities of *regional extent*, because pollen grains can travel many kilometers by wind. Pollen is considered a plant microfossil, as it can only be seen under a microscope. Paleoecologists also use a variety of plant **macrofossils** (visible to the naked eye) to reconstruct past environments. Examples include seeds, leaf fragments, fragments of bark and flowers, and other small parts of the plant. Because they are larger, macrofossils do not travel far from the host plant. Thus, they provide information about more *localized* plant communities.

Using a combination of approaches, paleoclimatologists can make detailed reconstructions of temperature and precipitation. For example, analysis of different types of pollen recovered from a Holocene peat bog in the Great Lakes region of the United States indicate that arboreal (tree) pollen increased and then decreased there, at various times in the past (**Fig. 4.32**). As pollen from tree species such as pine (*Pinus*) increase, non-arboreal species such as grasses and herbs decreased proportionally (**Fig. 4.32**). At this site, more trees indicate a wetter climate. How is this inferred? Again, the present is key to the past; in this region today, drought is the biggest stress or limiting factor on tree growth, so it is reasonable to assume that this was also the case over the Holocene. Thus, lower pollen counts from tree species indicate periods of increased moisture stress when grassland vegetation expanded, at the expense of trees. Many more such interpretations are possible from pollen and macrofossil data obtained from peat and lake cores.

4.12 LANDFORMS AND OTHER ARCHIVES OF CLIMATE CHANGE

In the above sections we have covered just a few of the many different climate archives and proxies used by paleoclimatologists to reconstruct past climates. Paleoclimate analyses require different (and often multiple) proxies. Some of the other archives used in the paleoclimatologist's toolkit include cave **speleothems** (deposits of $CaCO_3$ in stalagmites and stalactites), tree rings, shallow marine sediments and coastal deposits, fluvial sediments, sand dunes of all kinds, and paleosols.

Furthermore, because landforms themselves are fundamentally influenced by climate and climate changes, they can also be used as past climate archives. Periglacial landforms, glacial moraines, paleolake shorelines, sand dunes, and fluvial terraces all provide geomorphic evidence of climate change. One example provided here is a permafrost feature formed in organic-rich, frozen, loess-like sediments called **yedoma** (**Fig. 4.33**). Ice wedges and frost mounds

Figure 4.31 A peat core collected from the Great Dismal Swamp, VA, USA. Source: K. Koefke, US Geological Survey, Florence Bascom Geoscience Center.

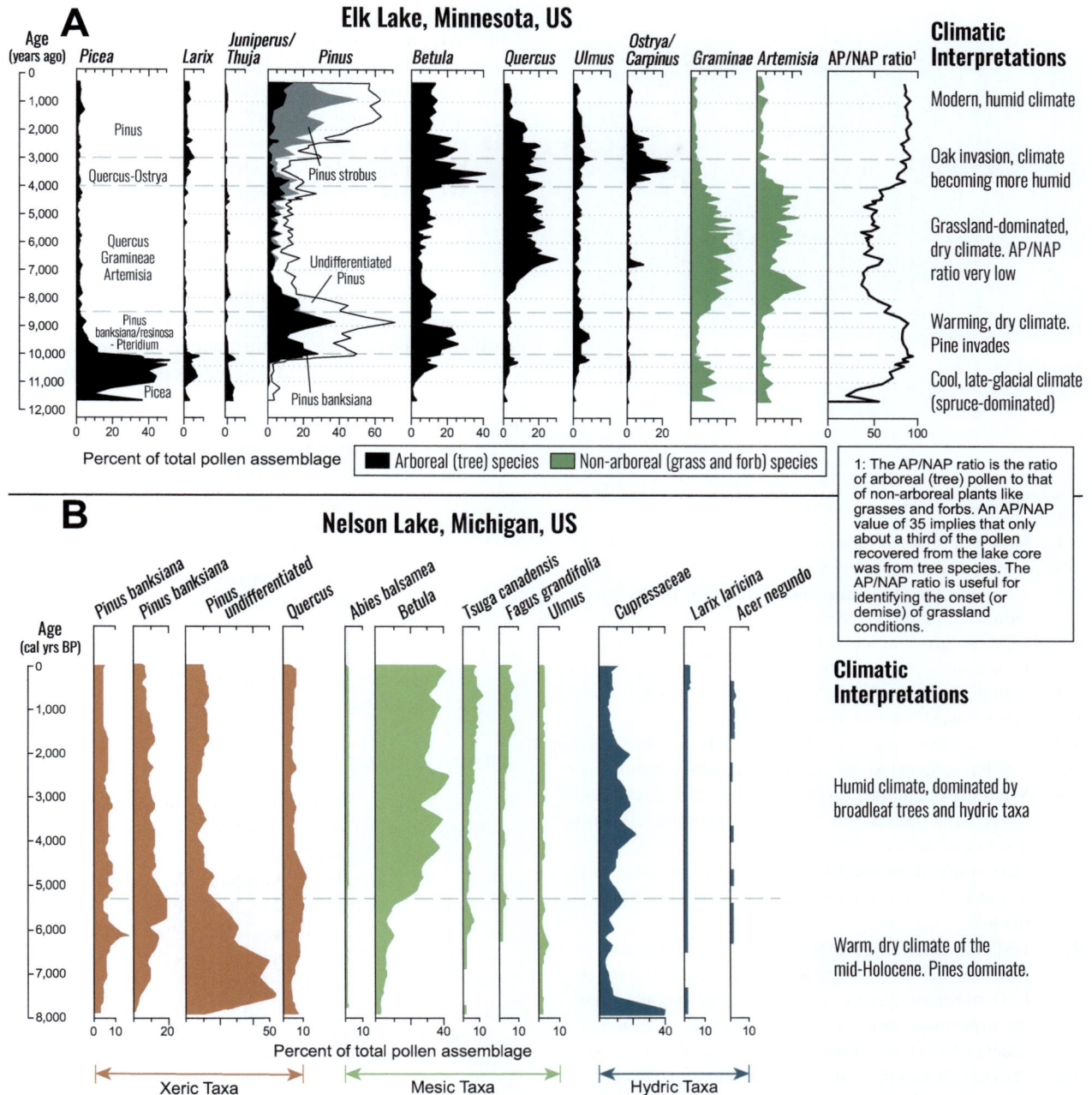

Figure 4.32 Pollen diagrams from two lakes in the northern United States. Both show how vertical changes in the pollen spectra can help identify past climate shifts. Source: (**A**) After Whitlock et al. (1993), © Geological Society of America; (**B**) Used with permission of Elsevier, from P. A. Delcourt, P. L. Nester, et al. (2002); permission conveyed through Copyright Clearance Center, Inc.

in these deposits can reveal periods when active annual freezing and thawing occurred in upper levels of these deposits. Yedoma deposits preserve this evidence perfectly, as long as they remain frozen.

Yedoma deposits are widespread in Arctic Siberia and Alaska (USA), but are thawing rapidly today. As they thaw, yedoma deposits often reveal exceptionally well-preserved remains of extinct animals such as Woolly Mammoths and other large, Pleistocene animals (**Fig. 4.0**). The yedoma permafrost acts like a deep freezer that preserves the

carcasses (hair, skin, blood, bone, and even the contents of their stomachs) for thousands of years, providing a snapshot of the animals that lived on the vast, frozen, Ice Age tundra.

Although landforms can be useful paleoclimate proxies, they cannot provide the detailed, *continuous* records of past climate change that form the backbone of our understanding of past climate. What they can do, though, is provide a means by which we can verify these more complete records. In doing so, geomorphology gives us insight into the effects on the landscape of the huge range of climate changes that have occurred in the past.

Figure 4.33 Yedoma outcrop on the Viluy River, Siberia, Russia. The upper brown part is composed of loess and yedoma, overlying gray sandstone. The scalloped darker parts within the loess are ice wedges. Source: A. Galanin, Russian Academy of Science.

REVIEW QUESTIONS

4.1 How might some modern landscapes reveal past climates, even before the Cenozoic?

4.2 How was Earth fundamentally different in the Precambrian than it is today?

4.3 Explain how and why climate has changed during the Cenozoic.

4.4 Define "icehouse" and "greenhouse" worlds. Which type of world are we currently in?

4.5 Describe the main glacial and interglacial changes that have occurred during the Quaternary Period. What are the names given to the last few glacial and interglacial phases, and the significant characteristics associated with each?

4.6 How might drainage from massive glacial lakes have influenced past climates? Provide an example from the text.

4.7 List and describe the three main types of orbital forcings that are thought to have controlled changes in Quaternary climate. Which of these is considered the most important for the development of large ice sheets on northern hemisphere continents, and why?

4.8 Provide an example of a positive and a negative feedback within Earth's climate system that may have modified the various Milanković forcings.

4.9 Describe the main, abrupt, climate change events in the North Atlantic during the last glacial period. What drove these events?

4.10 Define and discuss the Holocene and the Anthropocene. When did each begin, and what defines their beginnings? What are the key arguments for recognizing the Anthropocene as an official epoch in the geologic timescale?

4.11 What is a climate proxy and what kinds of characteristics make for a good, accurate proxy? List the main paleoclimate proxies discussed in this chapter.

4.12 What is the science of phenology, and what types of data do phenologists use?

4.13 How do oxygen isotopes in benthic foraminifera help us understand past changes in ice volume? What type of data in foram shells correlates to the climate record, and what specifically are the correlations?

4.14 What are the possible causes of the Little Ice Age and the "Year Without a Summer" cold climate periods during the Holocene?

4.15 Describe two key climate archives and their strengths and weaknesses in reconstructing past environmental change.

4.16 What can ice cores tell us about past climates? What key paleoenvironmental proxy do they contain that the marine record does not?

4.17 Explain what loess is, and how thick loess deposits record changes between glacial and interglacial climates of the Quaternary. How do East Asian loess deposits and the paleosols within them serve as paleoclimate records of the East Asian Monsoon?

4.18 What can the pollen and macrofossil data from cores in a peat bog tell us about past environments? What are the strengths of each of these two proxies?

FURTHER READING

Alley, R.B. 2015. *The Two-Mile Time Machine: Ice Cores, Abrupt Climate Change, and Our Future*. Updated ed. Princeton University Press.

Alley, R. B., Marotzke, J., Nordhaus, W. D., Overpeck, J. T., Peteet, D. M., Pielke Jr, R. A., Pierrehumbert, R. T., Rhines, P. B., Stocker, T. F., Telley, L. D., and Wallace, J. M. 2003. Abrupt climate change. *Science* 299(5615):2005–2010.

Bell, M. and Walker, M. J. C. 2014. *Late Quaternary Environmental Change: Physical and Human Perspectives*. 2nd ed. Routledge.

Bender, M. L. 2013. *Paleoclimate*. Princeton Primers in Climate. Princeton University Press.

Bradley, R. 2013. *Paleoclimatology: Reconstructing Climates of the Quaternary*. 3rd ed. Elsevier.

Chiotis, E. 2018. *Climate Changes in the Holocene: Impacts and Human Adaptation*. Taylor and Francis.

Ellis, E. C. 2018. *Anthropocene: A Very Short Introduction*. Oxford University Press.

Lowe, J. J. and Walker, M. 2014. *Reconstructing Quaternary Environments*. 3rd ed. Routledge.

5 Geomorphology's Resisting Framework: Minerals, Rocks, and Biota

Gregory A. Pope

G. K. Gilbert is considered one of the founders of modern geomorphology (see Chapter 2). In his 1877 report on the geology of the Henry Mountains of Utah, he wrote that (p. 109):

Certain rocks, of which the hard are most conspicuous, oppose a stubborn resistance to erosive agencies; certain others, of which the soft are most conspicuous, oppose a feeble resistance. Erosion is most rapid where the resistance is least ... .

Gilbert's observations parallel a rule of thumb in geomorphology – in a bedrock-controlled landscape, areas with harder rocks (which are better at resisting the forces of denudation, or wearing down of the landscape) tend to stand up in bold relief as uplands, ridges, hills, and mountains. Conversely, those parts of the landscape underlain by softer rocks will eventually be worn down to become valleys and lowlands. These outcomes occur due to the **differential weathering** and **erosion** of the rocks (Fig. 5.1).

Rock hardness and resistance are a function of its **lithology** (the kind of rock, and its physical and chemical characteristics), as conditioned by climate and other factors, adding caveats to this rule of thumb. For example, hard rocks can be "made softer" as they develop cracks, or joints, and soft rocks can develop more resistance to erosion by a cover of vegetation or a surface crust. In this chapter we provide some general associations between lithology and hardness, as they affect erosional and denudational processes. Let's start examining this "resisting framework" by focusing on the major rocks that comprise Earth's upper crust.

5.1 GEOMORPHIC FORCES AND PROCESSES

Two opposing groups of processes or forces are responsible for the landforms we see today (Fig. 5.2). **Endogenic processes** (from within the Earth) include the tectonic forces of doming, folding, and faulting (see Chapters 8–10), as well as volcanism (see Chapters 6 and 7). These forces build up terrain to higher elevations. Many geologists refer to the large-scale deformation of Earth's crust as **diastrophism** (Fig. 5.2). In diastrophism, the rocks of the crust are lifted

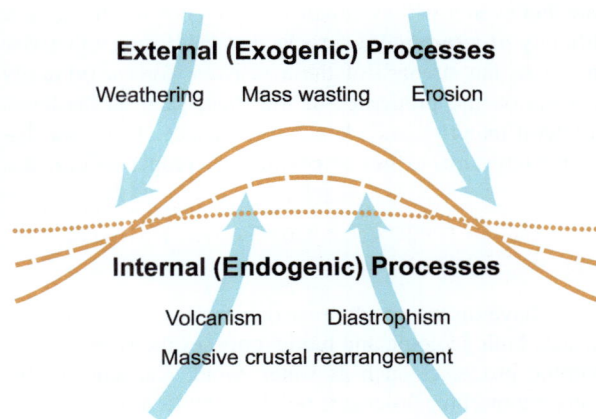

Figure 5.1 The impact of bedrock type and resistance on geomorphology is shown here in the inclined strata of the Morrison Formation (Brushy Basin Member [Jurassic]) in the Salt Valley Anticline in Arches National Park (Utah, USA). The harder sandstone layers, deposited in ancient streams and floodplains, stand up as light-colored, "ledgey" ridges. Alternating with the sandstone are thicker, softer beds of gray and tan siltstone which have eroded to form lowlands between the sandstone beds. Source: R. Schaetzl.

External (Exogenic) Processes

Weathering   Mass wasting   Erosion

Internal (Endogenic) Processes

Volcanism   Diastrophism

Massive crustal rearrangement

Figure 5.2 Diagram of the opposing forces that operate on and near the surface, leading to the various landforms that are present today.

Figure 5.0 The Siamese Twins, near Pikes Peak, Colorado, USA, formed due to the weathering of sandstone bedrock. Source: Ronda Kimbrow Photography / Getty Images.

up and (usually) bent and broken. Conversely, **exogenic processes** (exterior to Earth's surface) generally wear down the crust. The geomorphic agents that wear down the crust include **weathering** and **mass wasting**, agents of sediment transport (water, ice, wind, and gravity), and biota. So then, although endogenic processes are responsible for forming positive relief (uplands), the persistence of this relief over geologic time is largely governed by rock resistance to the exogenic forces.

Various types of mechanical, chemical, and biological processes all work to shape the land surface, giving us the geomorphology we see today (**Fig. 1.4**). These processes include gravity (more of a force than a process), glacial ice, flowing water, and/or wind, among others. But the end results are the same – to dislodge rock material and transport it. Examples include the collapse of a cliff face due to gravity, abrasion of bedrock under the flow of a glacier, the transport of sediment in rivers, and the excavation of rotten rock by ants. In all these examples, rock and sediment are dislodged and moved, forming a cascade of materials from hard bedrock to the ultimate depositional setting – the ocean basins.

Weathering is an important rock "preparatory agent" in this sediment cascade. **Weathering** is not a transport mechanism per se, but rather an *in situ* ("in place") process, fracturing or disaggregating rock into smaller pieces, or dissolving it chemically (see Chapter 11). Thus, weathering makes it possible for other geomorphic processes to remove and transport rock, as fragments or as dissolved substances. Chemical weathering, the attack on minerals by chemical agents such as acids, although not a physical process, is still responsible for the liberation of molecules and atoms, which can then be carried away in solution. In short, weathering is a necessary *precursor* to the transport of sediment. Rocks that best resist weathering will persist as ridges and mountains. Rocks that succumb to weathering will be more easily removed from the landscape, forming valleys.

A mountain is seemingly immovable over short spans of time, but even it will eventually be worn down. The relative difficulty of wearing down rocks and landforms, otherwise known as the strength of these resistive forces, primarily depends on the properties of the rock, such as hardness, structural integrity, and chemical resistance. Thus, our discussion now turns to the composition and properties of rocks.

5.2 ROCKS AND THEIR COMPONENTS: THE "MATERIAL SYSTEM"

As we have discussed, the resisting framework at the land surface, both geologic and biotic, opposes the work of geomorphic processes such as water, wind, ice, and gravity. Every geomorphic process is regulated by its ability to *overcome* the resisting framework, so it is essential for geomorphologists to understand basic geology.

Rocks are either an aggregation of minerals, or aggregations of particles of other rocks. They are categorized by the minerals they include, their chemical composition, and the manner in which they formed. **Minerals** are inorganic,

naturally occurring, crystalline solids with specific physical properties and chemical compositions. Rocks and minerals also comprise unconsolidated surface materials like gravel, sand, and clay. Unconsolidated material at the surface, such as soil, sediment, or decomposed rock, is referred to as **regolith**. These terms are not limited to Earth; the rocky planets and moons have rock and regolith as well (see Chapter 23). The resisting framework of geomorphology, however, hinges primarily on rock, which is usually much harder than regolith.

Rocks composed of well-cemented or interlocking, resistant minerals are difficult to weather and erode. That said, rock resistance is also a function of many other variables – the nature of their cementing agents, the **rock fabric** (how the minerals within them are arranged), and the presence of fractures and deformations, all at scales ranging from molecular and upward.

The resisting framework is not limited to rock. Regolith also exhibits varying resistance to erosion. Nonetheless, landscapes underlain by thick regolith are usually more vulnerable to erosion than are bedrock-controlled landscapes. Organisms, although often capable of breaking up rock and moving sediment, can also help stabilize it and, thus, add additional resistance. For example, both living and dead biomass impedes erosion by anchoring unconsolidated materials (think "roots"), as do microorganisms that co-exist with mineral material at micro-scales.

5.2.1 Primary Minerals

Before we discuss the major rock types, we turn to minerals – the building blocks of rocks. Thousands of mineral species have been identified, some of them well-known as gems, are culturally significant, or important in industrial applications. In this chapter, we focus on only a few of the common rock-forming minerals (**Figs. 5.3, 5.4, Table 5.1**). **Primary minerals** (like **quartz**, **feldspars**, and **biotite**) form *with* the rock,

Major rock types	Rock-forming minerals								
	Plagioclase feldspar	Alkali feldspar	Quartz	Pyroxenes	Amphiboles	Olivine	Micas	Clays	Calcite/Dolomite
Basalt/Gabbro	●			●	•	●	•		
Granite/Rhyolite	●	●	●		●		•		
Andesite/Diorite	●	•	•	•	●		•		
Sandstone	•	•	●				•	•	
Shale			•				•	●	
Carbonate rocks			•					●	●

Figure 5.3 The relative abundance of the major rock-forming minerals in Earth's most abundant rock types. Basalt and gabbro comprise most of the rock in the oceanic crust. Granite and andesite are abundant in the continental crust. Sandstones, shales, and carbonate rocks (like limestone) comprise much of the sedimentary cover of the continents. Source: After Hobart King, Geology.com.

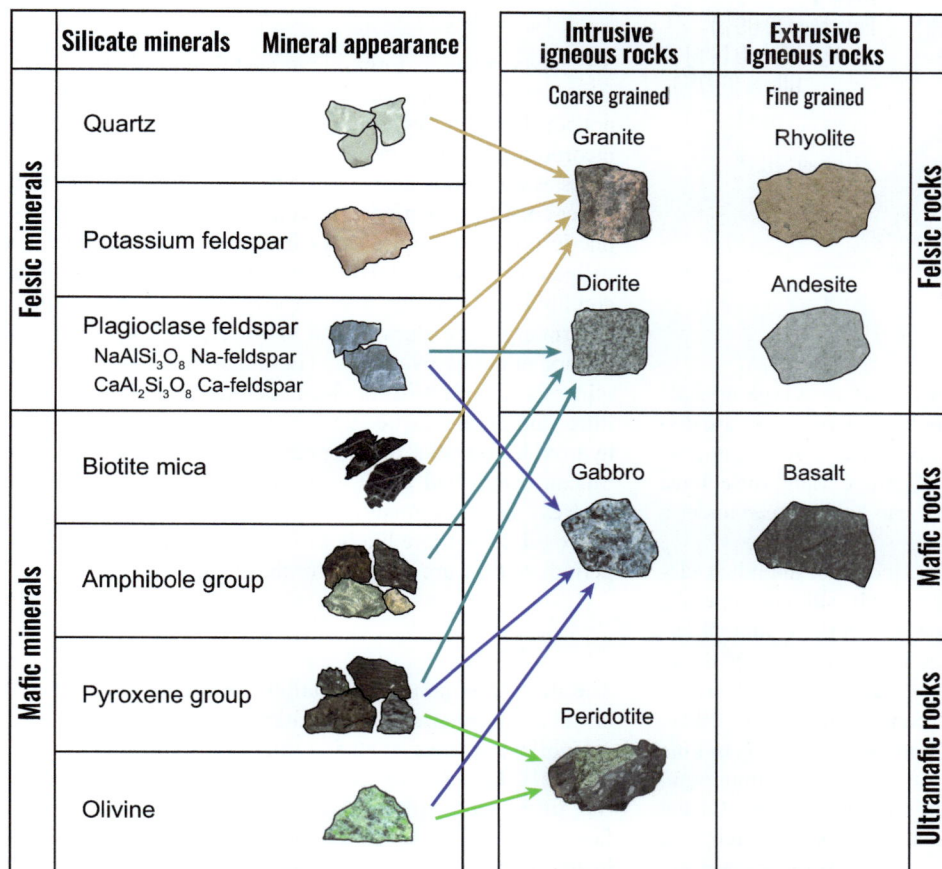

Figure 5.4 The most common primary minerals, and examples of the most common rocks in which they occur (as shown by arrows). Rocks that comprise the continental crust tend to be composed of felsic minerals, whereas oceanic crust rocks tend to be composed of mafic minerals. The less common ultramafic rocks derive from deep within the crust.

Table 5.1 Earth's most important and common minerals

Mineral group	Mineral species/series	Chemical formula	Examples in geomorphology
	Quartz	SiO_2	White sand beaches
Feldspars	Orthoclase	$KAlSi_3O_8$	"Pink" granite and rhyolite from crustal rifts and hotspots
	Albite	$NaAlSi_3O_8$	Granite of the continental crust
	Anorthite	$CaAl_2Si_2O_8$	Magma and metamorphic rocks
Micas	Muscovite	$KAl_2(AlSi_3O_{10})(OH)_2$	Sandstone, siltstone, and shale
	Biotite	$K(Fe,Mg)_3(AlSi_3O_{10})(OH)_2$	Granite, diorite, gabbro, schist, gneiss
Amphiboles		$((Na,Ca,Fe,Mg)_7(Si,Al)_8O_{22})(OH)_2$	Metamorphic and igneous rocks
Pyroxenes		$(Ca,Na,Mg,Fe)_2(Si,Al)_2O_6$	Metamorphic and igneous rocks
Olivines		$(Mg,Fe)_2SiO_4$	Volcanic sediment, ultramafic rocks
Carbonates	Calcite	$CaCO_3$	Limestone and shells of marine organisms

out of molten material (magma or lava), or by precipitation from fluids, for example, at hot springs or in salty lake bottoms. The environment in which most primary minerals and rocks form is usually associated with high temperatures and pressures – both radically different from conditions at the land surface. Thus, primary minerals are out of equilibrium at typical surface temperatures and pressures. Weathering occurs as these minerals adjust to the new equilibrium (see Chapter 11). Minerals that have been altered (by weathering) into different chemical or crystalline forms are called **secondary minerals**.

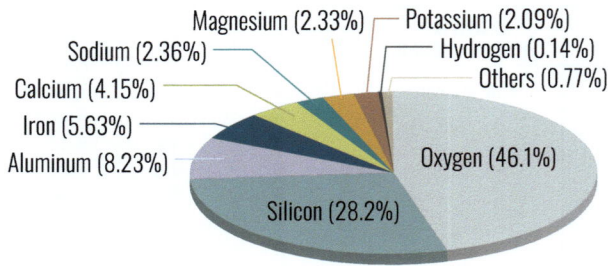

Figure 5.5 The abundance of elements, by mass, in Earth's crust.

All minerals have a crystalline structure. Large crystals are visible to the naked eye. Many of the larger ones are displayed in museums, such as the hexagonal prisms of quartz, the cubes of pyrite, or polyhedra of garnet. Many rocks have minerals that are so small that they can only be seen under a microscope – but they are there!

Oxygen and silicon are, by far, the most abundant elements in the crust (**Fig. 5.5**). As a result, **silicate** minerals, which contain SiO_2 groups somewhere in their mineral lattice, dominate the rocks of the continental crust. Of these, feldspars are the most common. Except for quartz (pure SiO_2), most silicate minerals also contain cations (positively charged molecules) of other elements in their crystalline structure. Common accessory cations include aluminum, iron, magnesium, calcium, potassium, and sodium. Of the silicate minerals, quartz is the most resistant to weathering, that is, most stable at the surface. Thus, most beaches are composed of quartz sand. Most of the other minerals have weathered away.

Primary minerals (and the rocks that contain them) can be divided into felsic and mafic groups. **Felsic rocks** are dominated by the mineral feldspar and various minerals rich in silicon, such as quartz (**Fig. 5.4**). Felsic rocks and minerals tend to be lighter colored and have low specific gravities, that is, they are "light." **Mafic rocks**, containing minerals with large amounts of magnesium and iron (ferric) in their crystalline structures, tend to be dark and comparatively "heavy." They usually contain little quartz (**Fig. 5.4**).

5.2.2 Secondary Minerals

Secondary minerals form due to chemical or physical alterations of preexisting (usually primary) minerals. They are not considered rock-forming minerals, although they may be incorporated (within pores and fractures) into rocks, sometimes as cementing agents. Because they form at the surface, under conditions of moderate temperature and regular atmospheric pressures, and usually in the presence of abundant oxygen, secondary minerals are very stable and in chemical equilibrium. Thus, they weather very slowly, or not at all.

Primary minerals will tend to weather to clay-sized secondary minerals over time. The major types of secondary minerals include silicate clays such as kaolinite, smectite, illite, and others, carbonates such as calcite, and oxide and hydroxide minerals like hematite, goethite, and boehmite. Silicate clays tend to form under cool climatic conditions, whereas in hot climates, clays composed of oxides and hydroxides are more likely to form. In both instances, long periods of time are required for these transitions to occur.

5.3 THE MAJOR ROCK TYPES

The most common classification system for rocks, based on their environment of formation, places them into three categories: igneous, metamorphic, and sedimentary (**Figs. 5.6, 5.7**). **Igneous rocks** crystallize out of molten material, often deep within the crust but sometimes at the surface (**Figs. 5.6A, 5.7A**). Primary minerals form and grow in the melt as it cools and solidifies. As a result, igneous rocks (like granite) are a mass of closely interlocking minerals (**Fig. 5.7A**). **Metamorphic rocks** form as preexisting rocks are "changed" (Greek *meta*, "changed, altered," and *morph*, "shape, form") by high pressures and/or temperatures, deep in the crust (**Figs. 5.6B, 5.7B**). However, the heat and pressure are not so great that the preexisting rocks completely melt. Partial melting, wherein metamorphism affects only those minerals with lower melting points, is typical. Metamorphism may generate some new primary minerals in the rock, even as other minerals are minimally altered, or even unaltered. Typical of many metamorphic

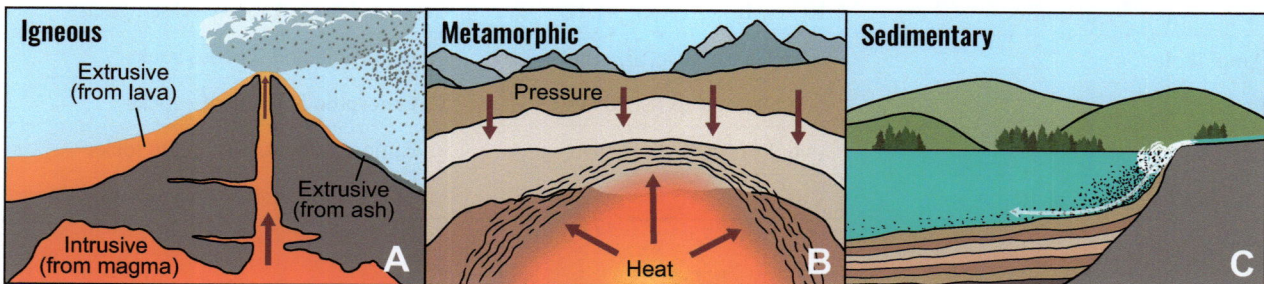

Figure 5.6 Illustration of the formation environments for the three major rock types. **A.** Igneous rocks form as molten rock materials cool. **B.** Metamorphic rocks form deep in the crust under intense heat and pressure – but not enough to completely melt the preexisting rock. **C.** Sedimentary rocks form as sediments accumulate, become compacted, and then cemented/lithified (lithification).

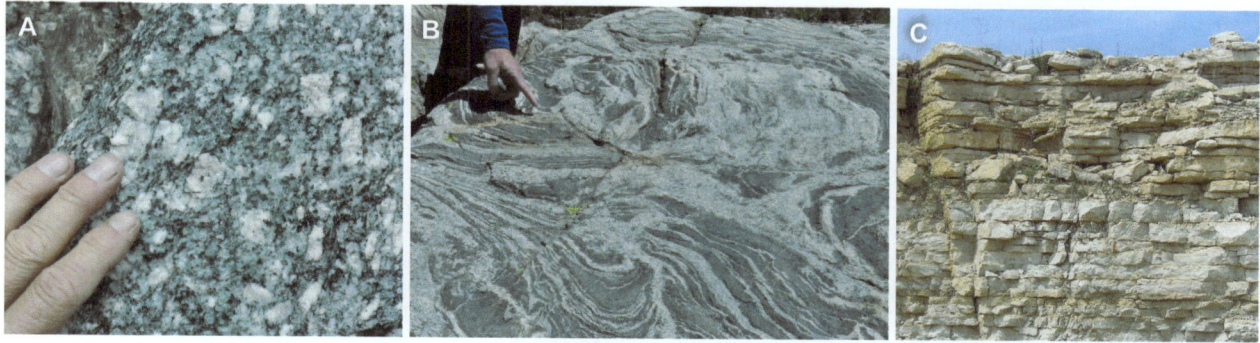

Figure 5.7 Examples of the three major rock types. **A.** Coarse-grained granite (igneous), showing the various primary mineral crystals. **B.** Highly foliated gneiss (metamorphic). **C.** Limestone (sedimentary), showing its layers, or strata. Source: R. Schaetzl.

rocks are areas of **foliation** (Latin *folium*, "leaf"), where the minerals align in layers, or more commonly, in swirls or curving "bands," suggestive of partial melting (**Fig. 5.7B**). **Sedimentary rocks** form out of particles (sediments) of previously weathered rocks; these weathered sediments are then transported to some sort of depositional (sedimentary) basin (**Fig. 5.6C**). The depositional area may be an ocean or a lake basin, but it can sometimes be on dry land as well. The sediments first need to be compacted, cemented, and then **lithified** (turned to rock) to become a sedimentary rock. The pressures necessary for this to occur usually are present only when the stacks of sediments become very thick, or at the bottom of an ocean. Sedimentary rocks typically exhibit distinctive layers, called **strata** (singular: *stratum*) (**Fig. 5.7C**).

All rocks and earthen materials exist in a constant cycle of formation, weathering, erosion, transportation, deposition, and re-formation into new rocks. For this reason, very few examples of truly old rocks remain on Earth. Over time, they have been "re-made" into new rocks, within what Earth scientists call the **rock cycle** (**Fig. 5.8**). One component of the rock cycle involves rocks that undergo weathering and erosion to form particles that are transported to a basin; there, they may become a sedimentary rock. Similarly, if any rock is subjected to extreme pressures and/or heat, it can metamorphose and become a metamorphic rock. Finally, any of the rocks may, if heated to the melting point, melt to become **magma** (molten rock, underground). As

magma cools it will form an igneous rock. (Molten material is referred to as **lava** if it extrudes onto the surface.) The rock cycle illustrates that rocks form from preexisting rocks, and that the rocks of today will eventually change to become rocks of the future.

Another way to classify rocks is whether they are crystalline, detrital, or precipitates. **Crystalline rocks** are composed of interlocking crystals. They include both igneous and metamorphic rocks, which are dominated by crystals of primary minerals. **Detrital rocks** are composed of particles of preexisting rocks – small, intact pieces of rock that have been weathered out of other rocks. Detrital rocks are therefore synonymous with some sedimentary rocks. Detrital particles can be mixtures of more than one mineral type or consist mainly of a single mineral (such as quartz in a "quartz sandstone"). Detrital particles can range from clay-sized (microscopic) to boulder-sized. **Precipitates** are a category of rock, fitting neither crystalline nor detrital categories. They form via chemical precipitation of minerals out of an aqueous solution, such as a salty lake or the ocean.

5.3.1 Igneous Rocks

Igneous rocks form as minerals crystallize out of molten material, or what is often called a "melt." Mineral crystals form and grow in size in this molten material, as it cools. As a rule, the longer it takes for the melt to cool, the larger the crystals can become. Molten materials cool very slowly at locations deep within the crust, faster when on land (as

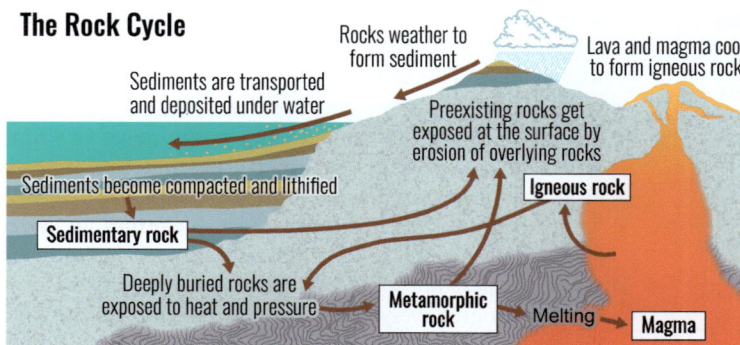

The Rock Cycle

Figure 5.8 The rock cycle. Although not all the possible interactions among the three rock types is shown, the general nature of the cycle is accurate.

lavas), and almost instantly when they are blown into the air (as tephra) or extruded into water (as happens under the ocean). Geologists use this information to define two broad end-member groups of igneous rocks: (1) **intrusive igneous rocks**, which cool relatively slowly from magmas within the crust, and (2) **extrusive igneous rocks**, which form at the surface, as lavas quickly cool. Extrusive igneous rocks are almost always associated with **volcanism** (see Chapter 7). Where molten material makes direct contact with water, it cools so fast that mineral crystals have little time to grow. Minerals in such rocks are either absent or indiscernible without the use of a microscope. Truly instantaneous cooling of lavas will form amorphous silicate glass, common examples being **obsidian**, **pumice**, and **volcanic ash**. Conversely, magmas form deep within the crust, taking years to cool. This long timespan allows crystals of primary minerals to grow to large sizes that are visible to the naked eye (**Fig. 5.7A**). Slow cooling can be enhanced when the heat source remains relatively close, or where hot, hydrothermal solutions are present.

Besides crystal size, we also classify igneous rocks by their mineralogy and chemistry. Recall that most minerals can be grouped into "felsic" or "mafic" end-member groups; igneous rock compositions also follow this spectrum, regardless of whether they formed extrusively or intrusively (**Figs. 5.4, 5.9**). The spectrum of igneous rocks ranges from **felsic**, to intermediate, to **mafic**, to **ultramafic**. Iron contents increase along this gradient. Thus, mafic and ultramafic rocks (rich in iron-bearing minerals) tend to be darker colored and more dense than felsic types. A common, readily

recognized example of a felsic igneous rock is granite (**Fig. 5.7A**). **Granite**, strictly defined, has 20–60% quartz, with most of the remaining minerals being feldspar, and with a small percentage of dark accessory minerals like biotite or hornblende. **Diorite** is a darker version of granite, with an intermediate composition (less quartz and more plagioclase feldspar) (**Fig. 5.4**). Rocks at the mafic and ultramafic end of the igneous spectrum contain almost no quartz and <50% plagioclase feldspar, allowing mafic minerals to dominate those rocks. Examples include **gabbro** (mafic), as well as the ultramafic rocks **peridotite** and the diamond-bearing **kimberlite**.

Extrusive igneous rocks follow a similar compositional spectrum, but have much smaller mineral crystals (**Figs. 5.4, 5.9**). The lightest-colored of these are **rhyolites**, similar in composition to granite. **Andesite** is the compositional equivalent to diorite, while the dark gray or black **basalt** is compositionally equivalent to gabbro (**Figs. 5.4, 5.9**). Overall, mafic rocks are darker and denser than felsic rocks and tend to be more resistant to erosion.

The composition of the tectonic plates that comprise Earth's crust can be generally classified along the same felsic–mafic spectrum. Rocks comprising the oceanic plates are mainly mafic in composition, whereas most rocks in the continental plates tend to be felsic (see Chapter 6). As a result of the contrasting densities, the continental crust tends to "float" higher up on the more dense rocks below, contributing to mountain ranges and uplands (see Chapter 6).

You may ask why doesn't molten rock crystallize as a mass of all one type of mineral, each with similar chemistry and crystallography? In other words, why are there collections of *different* minerals, all scattered about, in igneous rocks? The answer has to do with the temperatures at which the different minerals crystallize in the melt, which are in turn influenced by melt chemistry. We are assuming that the melt has ample amounts of oxygen and silicon atoms, as they are so abundant in the crust (**Fig. 5.5**). In general, mafic minerals have the highest crystallization point temperatures, meaning that as the melt cools, they crystallize out first, at high temperatures. Felsic minerals, of which quartz is the best example, crystallize out at lower temperatures; they also melt at the lowest temperatures. This generalized relationship is known as **Bowen's Reaction Series**, named for geochemist Norman Bowen who discovered it in the 1920s (**Fig. 5.10**). Bowen recognized that, as molten rock cools, groups of minerals crystallize in a discrete *sequence*; mafic minerals crystallize first and quartz crystallizes last. Notice that, at higher temperatures, Bowen's Series has two "arms." On the left is a **discontinuous series**; each of the minerals along this arm has a limited temperature range of formation. As the temperature of the melt drops, different minerals, each with a unique crystal structure, will form. The right "arm" of Bowen's Series is considered a **continuous series**. Here, only plagioclase minerals crystallize out of the melt. As the temperature cools, calcium-rich plagioclase feldspars become progressively more sodium-rich, until by the end of the series, only sodium-rich plagioclase feldspars will form.

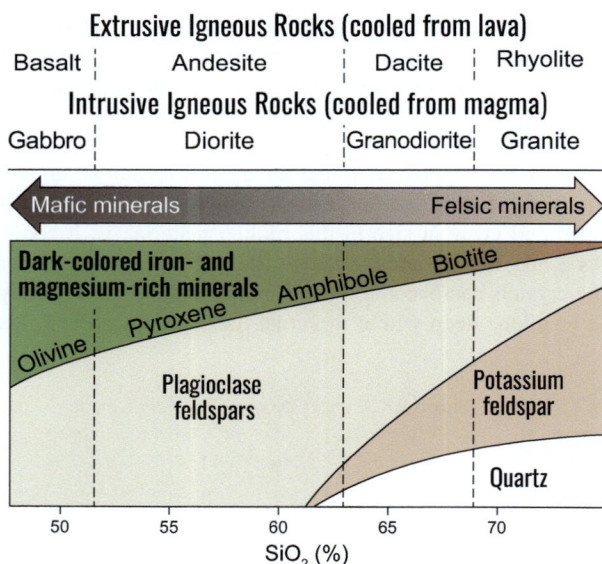

Figure 5.9 The main types of igneous rocks, classified by mineralogy and mode of formation (extrusive vs intrusive). The lighter-colored felsic rocks are higher in silica and potassium (alkali) feldspars. The darker-colored mafic or ultramafic rocks are high in iron- and magnesium-rich minerals like olivine and pyroxene. Source: US Geological Survey graphic.

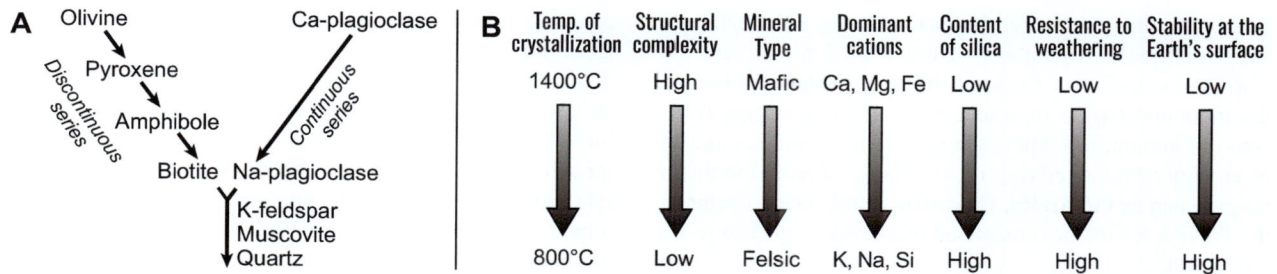

Figure 5.10 Bowen's Reaction Series, shown diagrammatically in **A.** Part **B** illustrates the range of properties that occur in minerals as a result of their various formation temperatures and crystalline structures.

Bowen's Series explains why the composition of the original melt determines the proportion of mafic to felsic minerals that form; melts with high contents of iron, magnesium, and calcium will end up with a small proportion of minerals on the "cool" end of the series, like mica and quartz. Quartz (pure SiO_2) is the last mineral to form in a melt, crystallizing out after all of the other accessory elements (Al, Fe, Ca, Na, Mg, and K) have been taken up in other minerals. Generally, the last minerals to crystallize from a melt (quartz and K-feldspar) tend to be most resistant to weathering (**Fig. 5.10B**). Bowen's Series is a simplification of what geochemists now recognize as the complex dynamics that occur as magma cools. Other factors, such as the proportion of water in the melt, changes in pressure as the melt migrates to different levels in the crust, and the addition or loss of other components in the melt, all work to alter the crystallization dynamics of the cooling magma.

Another key point to make is that the composition of the melt changes as minerals crystallize out of it. For example, olivine is always one of the first minerals to crystallize (**Fig. 5.10A**). But because it has a relatively high specific gravity (density), olivine minerals can settle out of a melt by gravity. Thus, igneous rocks may have a lower, deeper layer in the crust that is olivine-rich, while upper layers, depleted in iron and magnesium by the crystallization of olivine, have a different mineralogical mix, likely with more felsic minerals.

From the perspective of geomorphology, igneous rocks are often some of the more resistant to weathering and erosion, especially extrusive igneous rocks like basalt (**Fig. 5.11A**). Granite is an exception, as its mica minerals weather quickly in humid climates, causing the rocks to crumble into a loose mass of minerals called **grus** (**Fig.**

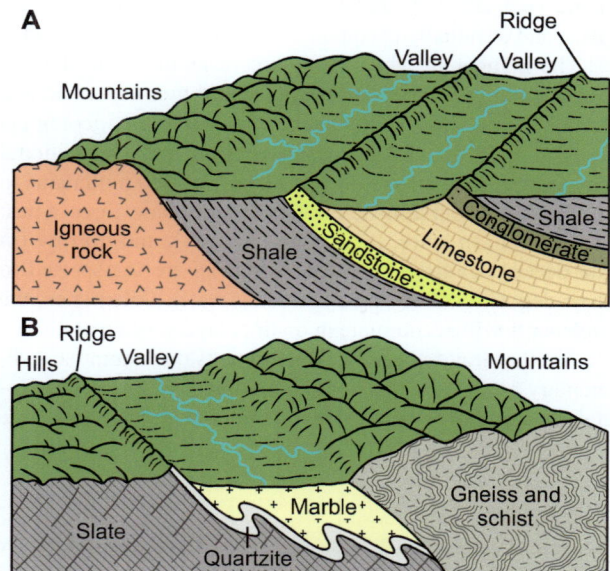

Figure 5.11 Idealized landscapes, showing the various resistances of different rock types in a humid climate. Resistant rocks stand up as uplands or ridges. **A.** Igneous and sedimentary rocks. **B.** Metamorphic rocks. Source: © A. N. Strahler (1992). Used by permission.

Figure 5.12 Granite (**A**) often weathers, grain-by-grain (**B**), into a type of regolith called grus (**C**). Note the large sizes of the mineral crystals in granite. Source: R. Schaetzl.

5.12). Granites dominate, for example, the Sierra Nevada and Adirondack Mountains in the United States and the Cornish Peninsula of England. In most of these instances, the mountain ranges represent deep intrusive masses (**plutons**) of granite, brought to the surface by millions of years of erosion of the overlying rocks. Many volcanic mountain ranges such as the Andes, Cascades, or the island chains in the Pacific, are mainly composed of extrusive igneous rocks like basalt.

5.3.2 Sedimentary Rocks

Sedimentary rocks form as sediments are transported and deposited in some type of basin, where they are compacted, cemented, and lithified. Sedimentary rocks reflect both the texture and the mineralogy of the previously weathered rocks, i.e., of the sediment. Sedimentary deposits may be ephemeral, to be later eroded and transported to a new location, or permanent, depending on long-term changes to the environment. Typically, though, once deposited, sediments remain in place and eventually become lithified into sedimentary rocks.

The manner in which sediments accumulate in depositional settings impacts the type of sedimentary rock that forms there. Thus, sedimentary rocks reveal a geological story of their former depositional environment. Depositional environments occur everywhere, from uplifted mountains to the deep sea, but can be broadly divided into three types: **terrestrial** (on land), **lacustrine** (in lakes), and **marine** (in the ocean) (**Table 5.2**). Most sediments are transported to their depositional setting by running water. Oftentimes, the sediments will accumulate in distinct layers, or **strata**.

Another way to classify the sediments that comprise sedimentary rocks is by their origin: detrital or chemical (precipitated). **Detrital sediments**, sometimes termed "**clastic**," are composed of particles or **clasts** of former rocks and minerals. Examples include sand, gravel, and clay. We further classify detrital sedimentary rocks based on their dominant particle size, from the smallest (clay-sized) up through gravel and cobbles (**Table 5.3**). Along an increasing particle size gradient, these include **shale** (clay), **mudstone** (silt and clay), **siltstone** (silt), **sandstone** (sand), and **conglomerate**. Conglomerate is a mixture (a "conglomeration") of detrital particles of various sizes, and must include at least some gravel. Sedimentary rocks can also be classified by the mineralogy of their detrital particles, by the nature of the cementing agents between the particles, and how well sorted the particles are (**Table 5.3**). Because of its resistance to weathering, quartz is not only a very abundant mineral but also the most "survivable" mineral in sedimentary environments. Most sandstones are therefore dominated by sand-sized quartz grains.

Some sandstones have an **eolian** origin, that is, they are former sand dunes. Many sandstones of the southwestern United States (such as the Navajo Sandstone of the Colorado Plateau and the Aztec Sandstone of Nevada, USA) formed from ancient sand dunes (**Fig. 21.22**). These rocks are comprised mainly of quartz grains, all similar in size and shape, and nicely rounded due to eolian abrasion during transport (**Fig. 5.13**).

Sedimentary rocks composed primarily of **chemical sediments** have usually precipitated out of an aqueous (watery) solution, primarily in oceans or lakes. Such rocks are called **evaporites**. Examples of such precipitates include salts, gypsum, and lime ($CaCO_3$). Carbonates are the most common such deposit. Precipitated calcium carbonate forms **limestone** ($CaCO_3$) or **dolomite** ($Ca,Mg(CO_3)_2$), its magnesium-rich equivalent. Dissolved silica may also precipitate into a rock called **chert**. Chemical precipitates

Table 5.2 The various major depositional environments that can host sediments

Sediment class/type	Depositional environment	Mechanism of transport
Glacial	Mountains or ice sheets	Flowing glacial ice
Colluvial	At the bases of slopes	Gravity, sometimes assisted by water
Alluvial, Fluvial	Rivers or streams	Flowing water
Deltaic	Rivers flowing into a standing body of water like a lake or ocean	Flowing water
Lacustrine	Lakes	Quiet water
Lagoonal	Coastal lagoons	Quiet water
Tidal	Ocean shores, estuaries	Tidal currents
Littoral (beach and nearshore)	Ocean and lake shores	Waves and near-shore currents
Reef	Living coral in a warm, shallow ocean	Biogenic (coral)
Oceanic	Deep ocean basins	Ocean currents, turbidity currents
Eolian	Sand dune fields, loess plateaus	Wind

Table 5.3 The main types of clastic sedimentary rocks

Dominant grain size (dia.) (name)	Rock type	Other distinguishing characteristics
>2 mm (gravels and pebbles)	Conglomerate	Rounded gravels
>2 mm (gravels and pebbles)	Breccia	Fractured and angular gravels
0.05–2 mm (sand)	Sandstone	Mostly quartz sand
0.05–2 mm (sand)	Arkose sandstone	>25% of the sand is feldspar
0.002–0.05 mm (silt)	Siltstone, mudstone	Sometimes as lithified loess (eolian silt)
<0.002 mm (clay)	Shale	Soft and easily weathered

Figure 5.13 Navajo Sandstone is a widespread and durable rock in the western United States. **A**. This rock formed from well-sorted and cross-bedded quartz sand, originally deposited as sand dunes during the Triassic and Jurassic Periods in Zion National Park, Utah, USA. Source: P. Dwyer. **B**. Near Canyonlands National Park in Utah. Note how this exposure even preserves the original dune shape. Source: R. Schaetzl.

are also important as accessory components of sedimentary rocks, where they usually form the cement between particles. Silica, iron oxides, and calcite are common cementing agents in many detrital sedimentary rocks.

Chemical precipitation occurs when an aqueous solution becomes supersaturated due to additions of chemicals or loss of water. This situation can sometimes be exacerbated by changing conditions, such as cooling. Precipitation of halite salt (NaCl) in the Michigan Geologic Basin, a shallow arm of the sea during the Silurian Period, provides a good example of this process (**Fig. 5.14**). Shallow coral reefs ringed the deeper water in the center of the basin. The warm waters in the basin provided ample opportunity for evaporation, even as additional salt-rich seawater flowed in from outside, through gaps in the reefs. This setup led to supersaturation in the center of the basin, forcing the excess salt to precipitate on the sea floor as **rock salt** – a sedimentary rock. Eventually, over several million years, more than a trillion tons of salt precipitated onto the seafloor of the Michigan Geologic Basin. And as the chemistry of the sea water changed over time, the composition of salts also changed; gypsum and potassium salts are other common precipitates in the sedimentary rocks here.

Sedimentologists also classify sediments based on their mineralogy, degree of sorting, and shape. Together, these characteristics reflect a sediment's **maturity**, referring to

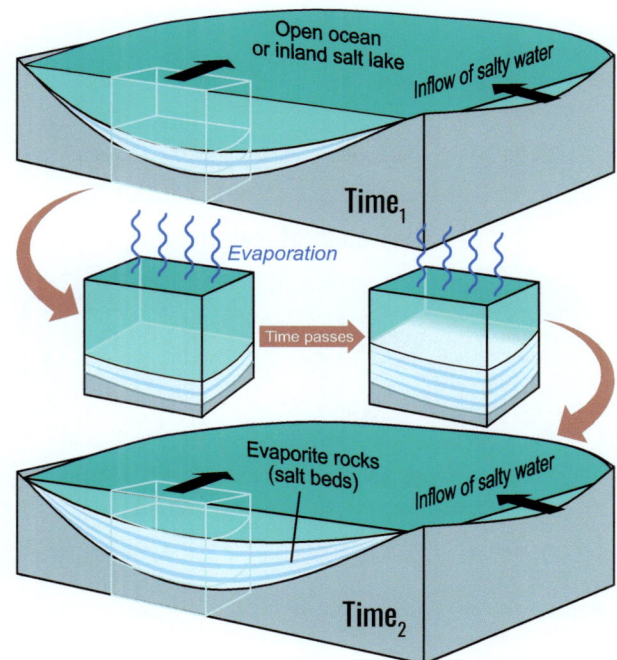

Figure 5.14 Model showing how salts came to be precipitated onto the sea floor of the Michigan Geologic Basin. The model is widely applicable to the formation of other kinds of evaporates.

Grain size

Transport of sediment

Closer to source   Farther from source

Sorting

Very poorly sorted   Poorly sorted   Moderately-well sorted

Well sorted   Very well sorted

Angularity

Angular   Subangular   Subrounded   Rounded

Mineralogical maturity

🐚 Lithic clast (fragment of a former rock)   ⌇ Clay flakes

◯ Quartz grain   ♥ Silt grain   ⌇ Feldspar grain

Less mature   More mature

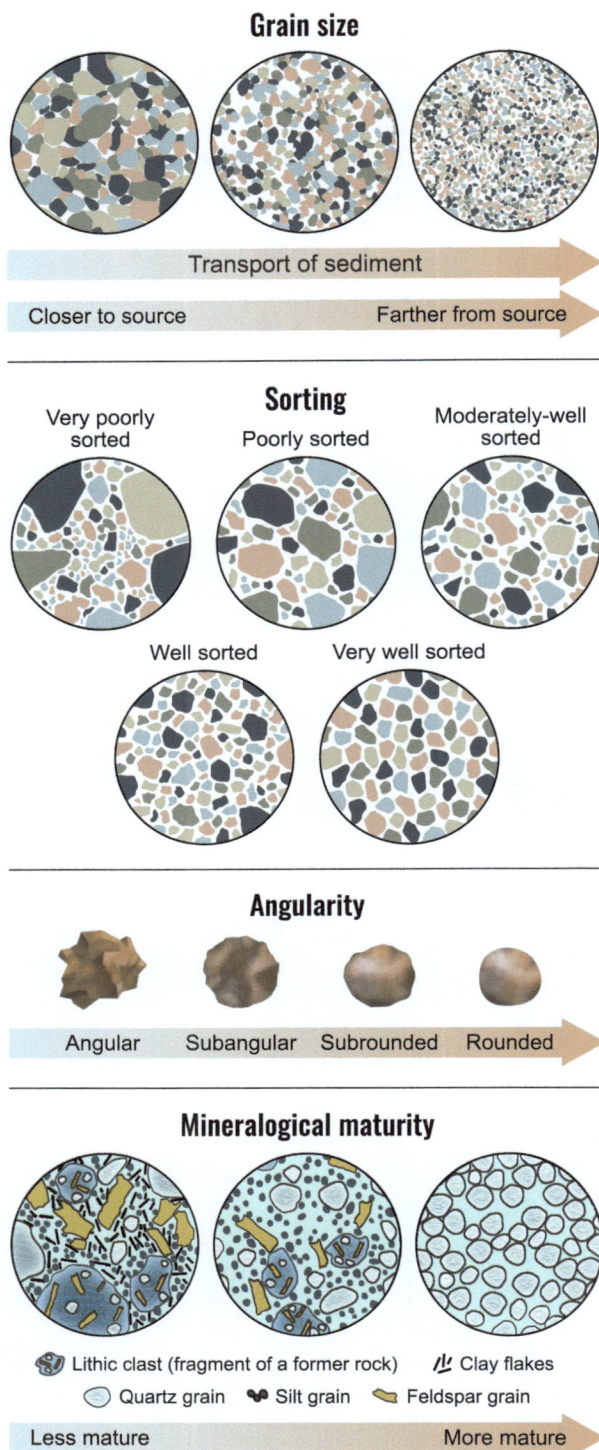

Figure 5.15 Potential effects of transportation processes on grain size, roundness, sorting, and mineralogical maturity. Source: After Marshak (2019). Used by permission of W. W. Norton & Company, Inc.

how weathered and "evolved" (old) the sediments might be (**Fig. 5.15**). Mature sediments have been, as the saying goes, "through the mill" a few more times than less mature sediments. Thus, grains in mature sediment tend to be

rounder, and contain fewer grains from minerals that are more easily weathered. A highly "mature" sand would contain many rounded quartz grains, as quartz is a highly persistent mineral. Less mature sediments contain other minerals that have not yet had a chance to weather completely. **Arkose** is a type of sandstone that includes many feldspar grains, and sometimes other minerals like mica (**Table 5.3**); it would be classified as having formed from a less mature sediment than a quartz sandstone. Lastly, mature sediments trend toward a single grain size, because sediments become better **sorted** while in transport. Sorting refers to the degree of uniformity in grain size. Well-sorted sediments have grains that are mostly within a narrow size range (**Fig. 5.15**). Sediments closer to their source, such as in alluvial fans, tend to be less well-sorted and less mature. Such sediments have not been transported very far, and thus have had fewer opportunities to becomes sorted as to size. Wind and water are both effective sorting agents, such that deposits farther "downwind" or "downstream" tend to be better sorted. Arkose and conglomerate are examples of poorly sorted sedimentary rocks. **Greywacke** is a type of sandstone, sometimes described as "dirty sandstone," with (often, angular) sand or rock fragments set within a clay matrix. This type of sediment would be considered both poorly sorted and "immature."

Finally, some sedimentary rocks form from detrital organic deposits – the remains of living organisms. One such rock, a cemented layer of fossil shells and coral, is a type of limestone called **coquina**. When plant materials in terrestrial wetlands and swamps become compressed and lithified, they can become **lignite** or **coal**. Coal can even be metamorphosed into a much harder rock called **anthracite**. Similarly, organic-rich marine muds and silts can become **oil shale**.

Sedimentary rocks comprise the bulk of the rock-based landforms on Earth (**Fig. 8.2B**). Thus, how well each of these rock types resist the forces of denudation impacts landscapes worldwide. Sedimentary rocks vary widely in their resistance to denudation, depending not only on their degree of cementation and jointing, but also on the climate where they occur. For example, limestone and dolomite are reasonably hard rocks in dry climates, often forming uplands and prominent escarpments and even in some instances, free faces of bare bedrock. However, in humid climates, acidic rainwater chemically wears away at these carbonate rocks, causing them to be some of the weaker rocks on the landscape (see Chapters 11 and 12). Most of the valleys in the Appalachian Mountains of the United States are floored with limestone, while sandstones hold up most of the ridges (**Fig. 5.16**). That said, many of the more prominent ridges and escarpments in the Great Lakes region of the United States are formed on dolomite, including the bedrock ledge over which Niagara Falls drops. In general, dolomite is harder than limestone in a humid climate, primarily since magnesium carbonate is less soluble than calcium carbonate.

Sandstones vary widely in their durability, largely depending on their cementation. Additionally, in some sandstones the sphericity of the grains makes for less efficient packing, leading to many more voids in the rock, weakening it.

Figure 5.16 Morrison Cove, near Woodbury, Pennsylvania, USA, is a broad limestone valley. The nearby uplands are formed on sandstone. Source: K. Patrick.

Almost universally, shale and evaporite rocks are weak, so they typically underlie valleys. Evaporite rocks such as limestone and rock salt are particularly susceptible to dissolution by water and its dissolved acids, rapidly forming caves. As these caves enlarge and collapse, they form sinkholes and lowlands on the surface above (see Chapter 12). Thus, they influence the surface geomorphology even when deeply buried. Shale is almost everywhere a valley-former. Several of the Great Lakes of North America owe their origins to the shale bedrock that underlies them, which was preferentially scoured out by Pleistocene glaciers.

5.3.3 Metamorphic Rocks

Metamorphic rocks form from preexisting (parent) rocks, due to changes induced by heat and pressure. These conditions dominate the deep parts of the crust, where most metamorphic rocks form. The heat sources may derive from rising magma or geothermal fluids, and pressures develop due to deep burial or subduction of tectonic plates. The composition of a metamorphic rock is largely determined by its predecessor rock, called a **protolith**. Recall that any rock, even a metamorphic rock, can undergo metamorphism.

Metamorphic rocks are good indicators of former temperature and pressure conditions in the deep crust. They range from "low-grade" varieties (formed under low temperature and/or pressure conditions), through "intermediate" and "high-grade" varieties (**Table 5.4**). Below low-grade conditions, no metamorphism takes place. Under conditions of high-grade metamorphism, minerals and rocks begin to melt. Beyond that, complete melting takes place, recycling the parent rock back into magma and eventually forming new igneous rocks (**Fig. 5.8**).

Metamorphism can be localized, as when magma makes contact with adjacent rock, causing partial melting or **contact metamorphism** (**Fig. 5.17**). **Hydrothermal metamorphism** is another, more localized type of metamorphism. It occurs as hot, volatile solutions percolate into cracks and fissures in the protolith, allowing for metamorphism near the cracks. Higher grades of metamorphism, occurring across

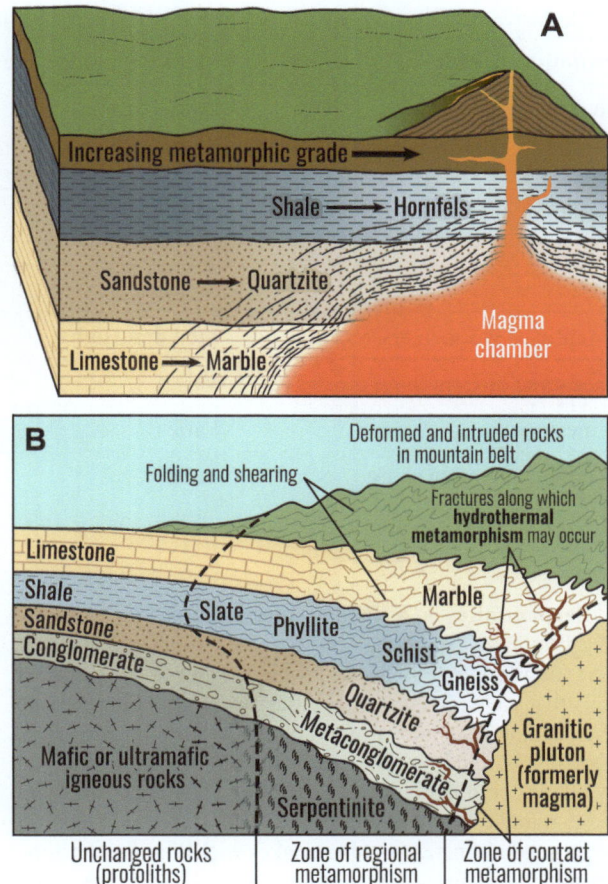

Figure 5.17 Grade and type of metamorphism. **A.** In contact metamorphism, heat from nearby magma can partially melt or metamorphose parent rock (protolith) near the surface, shallower than ≈5 km and usually no more than 1 km from the heat source. These rocks display increasing metamorphic grade nearer to the heat source. **B.** Regional metamorphism occurs across hundreds of kilometers and to depths of tens of kilometers. The degree of metamorphism increases nearer to the source of the heat but also nearer to areas of intense pressure, such as in a developing mountain range.

large regions, happen during mountain building events and at locations of plate subduction. Typically, these kinds of events form vast swaths of new metamorphic rocks.

Metamorphic rocks can also be classified by their outward appearance (**Table 5.4**). Many appear **foliated**, with visible banding, plates, or sheets, sometimes in intricate folds, for example, **gneiss** and **schist** (**Fig. 5.7B**). Foliation develops due to compression and/or shearing during metamorphism. **Non-foliated** metamorphic rocks lack sheets or bands, appearing instead to be more "granular," somewhat like granite. Common examples include **quartzite** and **marble** (**Table 5.4**). Although they may be locally non-foliated, these metamorphic rocks may still be warped into larger-scale folds indicative of physical stress. **Hornfels** is a non-foliated metamorphic rock formed by contact metamorphism, without pressure stress, typically in volcanic areas. Hornfels is typically harder than its protolith.

Table 5.4 The main types of metamorphic rocks based on protolith and degree of metamorphism

Protolith (parent rock)	Metamorphic grade	Metamorphic rock	Other
Shale	Low–High	Slate Phyllite Schist	Metamorphic grades: slate (low), phyllite (medium), schist (high)
Peridotite or Dunite	Low	Serpentinite	Contain abundant ultramafic minerals
Basalt	Medium	Amphibolite	
Granitic or sedimentary rocks	Medium–High	Gneiss	Pronounced like "nice"
Sandstone or Conglomerate	Varies	Quartzite	Very durable/hard
Limestone	Varies	Marble	
Sedimentary, often shale	Varies	Hornfels	Forms only by contact metamorphism

Serpentinite is a unique metamorphic rock formed of ultramafic minerals in oceanic subduction zones. Serpentinite is relatively weak, weathering to regoliths and soils that have high contents of heavy metals such as nickel (Ni), cobalt (Co), and chromium (Cr), which can even reach toxic levels, and low contents of calcium (Ca) and potassium (K). This combination makes serpentine residuum inhospitable to many types of plants (**Fig. 5.18**).

In general, metamorphism makes rocks harder and more resistant to erosion than their igneous or sedimentary protoliths (**Fig. 5.11B**). Among the metamorphic rocks, quartzite is one of the hardest – almost always standing up as ridges or uplands. Another very hard metamorphic rock is **slate** – the metamorphic product of shale. Metamorphic rocks such as gneiss, phyllite, and schist vary considerably in their ability to withstand denudation, depending on their mineralogy, foliation, and degree of metamorphism. Marble (the metamorphic equivalent of limestone) is very hard in dry climates, but is quite susceptible to chemical attack in wetter areas, where it may undergo dissolution and form caves (see Chapter 12).

5.4 UNCONSOLIDATED MATERIALS

Unconsolidated sediment refers to rock fragments and mineral crystals that are not lithified. Examples include incoherent accumulations of loose volcanic ash or cinders, completely disaggregated rock formed by weathering, soil materials, and miscellaneous sediments ranging from mud through sand, gravel, and boulder sizes.

Usually, unconsolidated sediments are much weaker than rocks, and thus they erode to form valleys and low areas on the landscape. However, in some settings, unconsolidated materials can be resistant to erosion, and therefore behave more like bedrock. For example, large boulders may be immobile except under events of high magnitude, such as during large floods or landslides. Examples include cobble bars on rivers (**Fig. 5.19**) or boulder beaches, which were probably initially emplaced during high-magnitude storms. Slope-related deposits, for example, **talus** at the base of a cliff (see Chapters 8 and 14) are also sometimes very resistant (**Fig. 8.7B**). Except for gravity-influenced mass movement, the talus is too large to move, and water simply "sieves" through it. Similar to talus are rock fields or **felsenmeer**, areas of dense cobbles and boulders on flat or gradually sloping surfaces, relics of a former periglacial climate. Rocks in felsenmeer deposits were moving slowly downslope when the landscape was under the influence of permafrost, but have moved little since the permafrost has thawed (**Fig. 20.23**).

Figure 5.18 The serpentine "barrens" of Mt. Eddy, in the Klamath Mountains of California, USA. Note the lack of trees on the weathered serpentinite, which crops out here. Source: US Forest Service.

5.5 PROPERTIES THAT IMPACT ROCK DURABILITY

Rocks resist erosion and denudation mainly because of factors inherent to the rock itself, but as conditioned by climate and

Figure 5.19 Armoring of surfaces by large, unconsolidated sediments is illustrated here by the cobbly banks and mid-channel bars within the Lake Fork of the Gunnison River in Colorado, USA. These sediments are relatively immobile except in extreme flooding conditions. Source: US Geological Survey.

Figure 5.20 In this example from the Paria Wilderness, near Buckskin Gulch, southeastern Utah, USA, veins of cemented quartz stand out against the weaker sandstone matrix. Source: S. Wagon.

vegetation. Rock composition and (for sedimentary rocks) cementation are the most important lithological determinants of a rock's resistance to denudation. However, because rocks undergo changes induced by weathering, particularly near the Earth's surface, their "strength" changes over time.

Think of rock strength (resistance) as being a function of its "weakest link." For instance, quartz grains can survive tens of millions of years in the surface environment, and within sandstone can persist for hundreds of millions of years. A quartz sandstone, however, can deteriorate rapidly if it loses the cement that holds the quartz grains together. The cement is its weakest link (Fig. 5.20). A silica-cemented sandstone would be more durable than one weakly cemented by calcite, which is far more susceptible to dissolution. Likewise, in igneous and metamorphic rocks, quartz crystals can be resistant, but the rock begins to disintegrate when other silicate minerals between them, such as biotite, break down. Secondary cements, emplaced after a rock has formed, can be more resistant. Silica-saturated solutions can penetrate rock fractures and joints, only to precipitate as crystallized quartz. These types of quartz veins persist in positive relief as the less resistant rock matrix weathers and erodes away (Fig. 5.20).

Rock heterogeneity also impacts its overall durability. For example, if a portion of a rock is comparatively weak, that part will weather first and compromise the integrity of the entire rock, regardless of how strong the rest of the rock may be. Properties in play here include rock structure (such as folding deformation, fracturing, and faulting), stratigraphy or foliation, porosity, and fabric (of grains or crystals).

5.5.1 Rock Structure

Earth's crust is not static. Both **exogenic** and **endogenic** forces continually act to compress, extend (via tension), and shear (break) rocks. Rocks develop various structures as the result of these forces (see Chapters 8 and 9). And as should

now be evident, landforms on bedrock-controlled landscapes mimic the general lithology and structure of the rock. Rock **structure** and hardness control the shape of vast parts of the land surface, and as a consequence, the patterns of the streams on it. Thus, a distinctive assemblage of landforms and stream patterns tends to develop on different kinds of geology (Fig. 5.21).

5.5.2 Rock Hardness

How hard is that rock, anyway? And does hardness matter? Geoscientists and engineers use different terms, such as rock mass strength and rock resistance, in addition to hardness, to describe the overall durability of rocks. These physical parameters can be measured in various ways to get a better idea of the hardness of the rock. The simplest method is to strike a rock with a hammer. A field scientist can interpret both the feel and sound of the response. A hard rock resists a hammer blow, does not deform, and the hammer blow resounds with a sharp "ping." A softer rock will absorb the impact, may deform or break easily, and produces a sound more like a "thud." This method is, of course, difficult to quantify. A similar but more quantitative approach uses a device called a rebound hammer, which contains a small, spring-loaded piston that strikes the rock with the same force every time. The user then measures the amount of bounce-back, as a surrogate for rock hardness. Rebound hammers were invented to test the hardness of concrete, for quality control, but are now also used by geomorphologists.

Geomorphologists also use other methods to assess rock durability and resistance. For example, they may measure the compressive (or tensile) strength of rocks. Here, a laboratory device exerts stress until the rock breaks at some (measured) failure threshold. In the field, water and oil drillers can get a general sense of rock hardness by observing the difficulty that the drill experiences. For example, during a recent tunneling project in New York City, geologists discovered an unexpected mass of garnet gneiss, because it slowed the drilling. Geotechnical engineers can also measure shallow

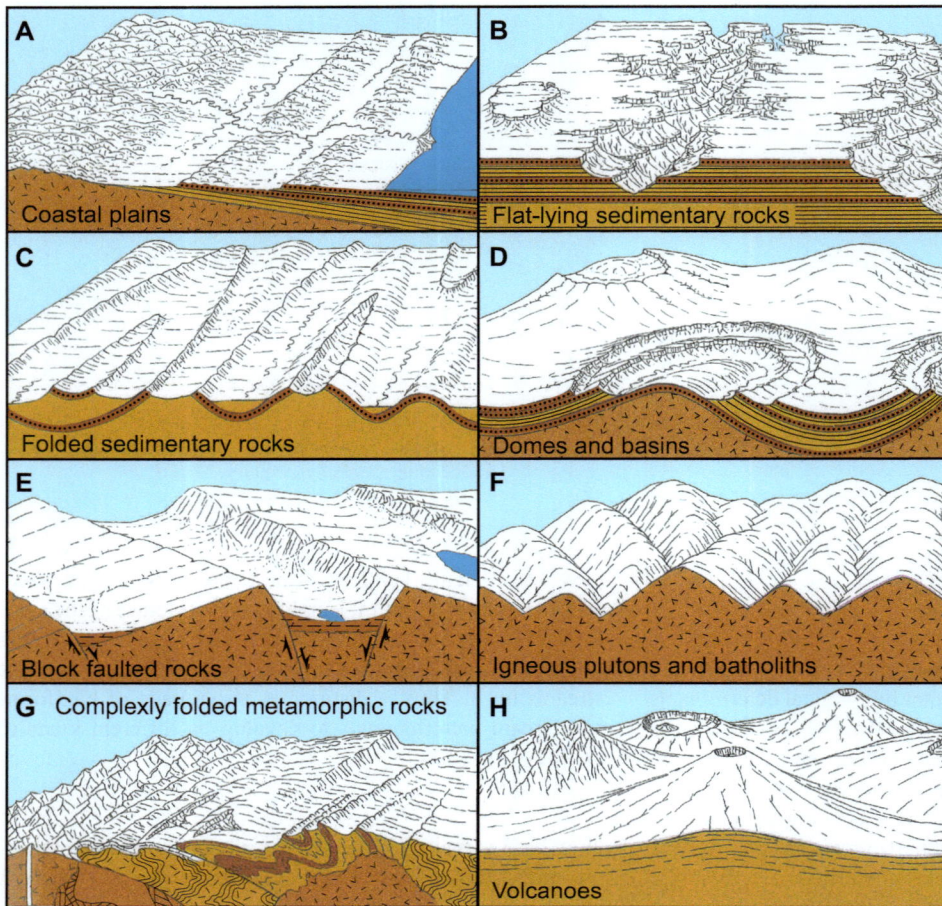

Figure 5.21 Idealized landscapes, showing how rock structure and lithology can affect surface topography and stream patterns. Source: © A. N. Strahler (1992). Used by permission.

rock integrity by inducing small explosions on the surface, and then measuring the response by using a seismic wave detection array.

Some rocks can withstand a certain amount of elastic deformation, but beyond some point they will fracture into cracks and **joints** (**Fig. 5.22**). Joints may appear as cracks on the surface, but extend as two-dimensional planes of weakness into the rock. Like joints, fractures can range from microcracks in a mineral crystal or sedimentary grain, to major faults that extend for many kilometers. Rocks that undergo stress from multiple directions can, over time, develop an array of intersecting joints, which if mapped can help explain the stress history. Fractures in rocks are also conduits for water and roots. Water in joints can freeze and force them apart, as can roots. Whenever fractures are compromised in these ways, they can weaken the rock. Conversely, secondary mineral precipitates from groundwater may actually penetrate joints and cement such fractures, rendering the rock more resistant, much like mortar solidifying a brick wall (**Fig. 5.20**). Mineral-filled fractures are known as **veins**.

Faults are similar to fractures. During faulting, parts of the rock displace vertically and/or horizontally, along either side of the fault (see Chapter 10). Small faults can be simple

fracture planes in a rock, but larger faults are likely to exist as **fault zones** – wide corridors of variously fractured rock. Faults and fault zones are notoriously weak areas in rock, which erosion can (and does) exploit. Examples include the placement of stream courses (**Fig. 5.23**), drainage networks (see Chapter 15), and cave systems (**Fig. 5.24**), among many others.

Like all rocks, sedimentary rocks are capable of deformation by **folding**. In this case, tectonic forces distort and bend the horizontal beds, but do not break the beds as faulting would. The resulting folded rocks often display a telltale relief that is recognizable in many settings. Folding often weakens the rock and disrupts the original fabric. The weaker rock is then subject to preferential weathering and erosion (**Figs. 5.23, 5.24**).

5.5.3 Rock Anisotropy

Sedimentary rocks that lack clear bedding and are of homogeneous composition and grain size are said to be **massive**. These rocks are rare. Instead, most sedimentary rocks exhibit clear **strata**, or layers. Sedimentary rocks with well-formed strata can then pass these structures along as foliations, if they were to become metamorphosed. Strata in sedimentary rocks, or foliation heterogeneities in

Figure 5.22 The joints in this granite exposure at Acadia National Park, Maine, USA have been widened by weathering and erosion. In addition to tectonic forces from past mountain building events, stresses from both glaciation and the release of overburden rock mass have contributed to the fracture pattern. Source: G. Pope.

metamorphic rocks, are areas of weakness. Such rocks are referred to as being **anisotropic**, that is, having unequal resistance to stresses from different directions. **Schist** is a good example of an anisotropic metamorphic rock, with a compressive strength that is greater perpendicular to the crystal grain orientation (and foliation) than it is in the parallel direction. Overall, anisotropy makes most rocks weaker.

5.5.4 Rock Hydrology and Porosity

Additions of water make most rocks weaker. Water content in rocks is largely a function of their **porosity**, that is, the amount of void space in the rock. **Macroporosity** refers to large void spaces, usually between fractures and bedding planes. However, most of the porosity in sedimentary rocks and in unconsolidated sediments and soils consists of **microporosity**. These smaller pores occur between mineral grains or within grain microfractures.

How does water in rocks make them weak? Generally, weakness is driven by pore pressure. Because water is not very elastic, in confined spaces such as tiny pores in the rock, it exerts a small amount of stress on the rock surfaces, especially at depth (where it exists under pressure). In addition, water acts as a lubricant. This aspect of water is especially relevant in clayey rocks or sediments, for example, shales, where the water between the layers facilitates movement and sliding.

Water also weakens rocks by making them more susceptible to weathering. Water is involved in weathering in a variety of ways, from being a key part of chemical and physical weathering processes, to an agent whereby

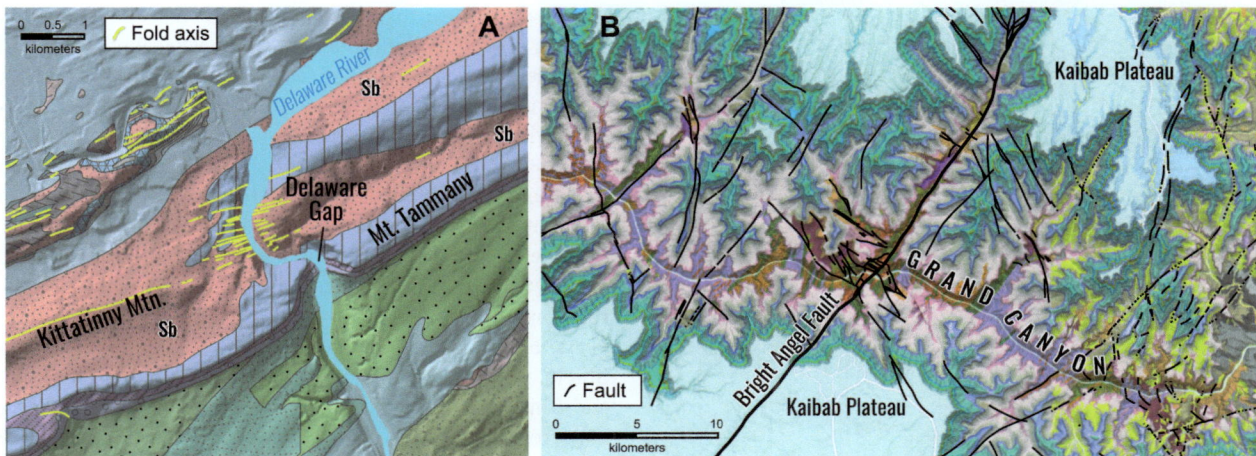

Figure 5.23 The effects of faulting and folding on rock integrity and landscape denudation. **A.** The Delaware Water Gap, where the Delaware River passes through, not around, Kittatinny Mountain (a ridge of folded sedimentary rock) in the Appalachian Mountains, between Pennsylvania and New Jersey, USA. The ridge is composed of the very resistant Shawangunk conglomerate (lavender color on the map, on Mt. Tammany) and the somewhat less resistant but still competent Bloomsberg sandstone (Sb, pink color on the map, on Kittatinny Mountain). Note that the fold axes are concentrated at the gap, suggesting that the deformation of the rock has weakened it by producing fractures, providing an easier path for the river through the ridge. Source: After Jackl (2013) and Epstein (1966). **B.** A geological map of the Grand Canyon region, Arizona, USA, where the Bright Angel Fault traverses the Canyon. Note the large and unusually straight tributary that follows the fault. Many of the side canyons in this area also follow similar (but smaller), fault lines. Source: After R. Clark, Arizona Geological Survey, based on work by G. Billingsley and colleagues at the US Geological Survey. Note: the various colors in these maps represent different rock types.

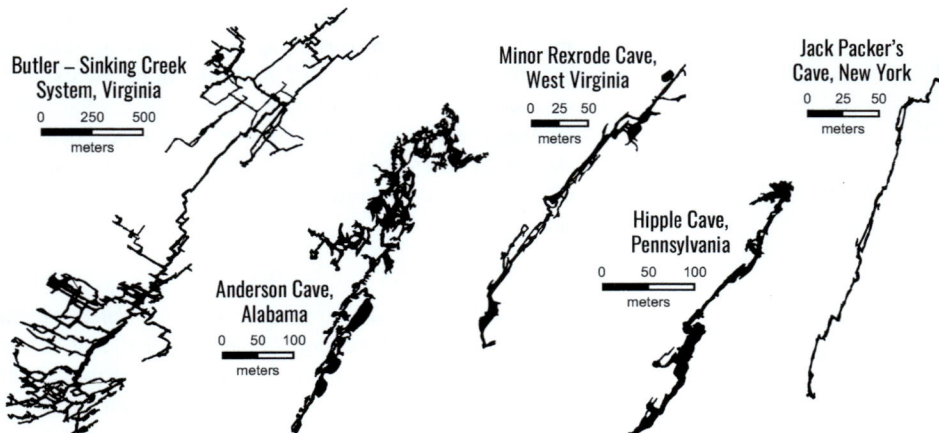

Figure 5.24 Maps of cave systems in limestone bedrock of the Valley and Ridge Province of the Appalachian Mountains, in the eastern United States. All these caves show linearity parallel to the regional folded rock structure, as well as near-perpendicular intersections of secondary passages, probably following joints and fault zones. Source: After Palmer (2009). Used with permission of the National Speleological Society (www.caves.org)

weathering byproducts are removed from the rock (see Chapter 11).

5.5.5 Erosion of Overlying Sediment

When rocks are uncovered by erosion of overlying materials, many of the confining pressures that held the rock in place are relieved. In a sense, the exposed rock can "relax" and expand outward and upward. Undercutting of rock by rivers or waves also enables gravity-induced stresses in the overhanging rock. The outward stresses (from exposure) and downward stresses (from undercutting) often cause joints to develop in the rock, weakening it (**Fig. 5.25**).

Human activity can lead to stress release in rocks as well. Excavations for mines, tunnels, roads, and building foundations remove overburden or confining pressures, allowing the surrounding rock to expand and crack. The presence of water and faults in the rock can further exacerbate the problem. Dangerous and deadly "rock bursts" (sudden, explosive stress failures in mines, tunnels, and at excavations) are widely known in human history. Engineers attempt to minimize this hazard by inserting reinforcing bolts into the rock, and erecting shielding plates or cable mesh nets to catch any shattered or fallen rock masses. Fortunately, these safety measures have eliminated many such rock bursts.

5.6 ROCK RESISTANCE AND GEOMORPHOLOGY

By now it is clear that hard, competent rocks resist erosion to form positive relief (uplands), whereas weaker rocks tend to erode into negative relief (valleys, lowlands) (**Fig. 5.26**). Additionally, areas with dense fold, fracture, and joint patterns will preferentially wear down and become lowlands, or become exploited by rivers (**Fig. 5.23**). The inverse is also true: isolated summits or domes in regions of uniform terrain typically lack dense joint networks (**Fig. 5.27**).

Where ancient, igneous and metamorphic rocks form the cores of the continents, these **continental shields**, or **cratons**, have been tectonically stable for more than a billion years. Examples include the Brazilian Shield, the Canadian Shield, and the Fennoscandian Shield of Norway, Sweden, and Finland. Hundreds of millions of years of weathering and erosion have led to a subdued, undulating landscape on most of Earth's cratons, sometimes above a deep cover of weathered regolith (**Fig. 4.2**). But in a few areas where highly resistant rocks crop out, **inselbergs** rise above the gradually undulating plains, attesting to the greater resistance of the bedrock on that area (**Fig. 5.28**).

Hard caprocks, typical in sedimentary strata, protect underlying softer rocks, forming landforms like buttes and mesas (see Chapter 8, especially **Fig. 8.12**). Resistant caprock types include iron- or silica-cemented sandstones, as

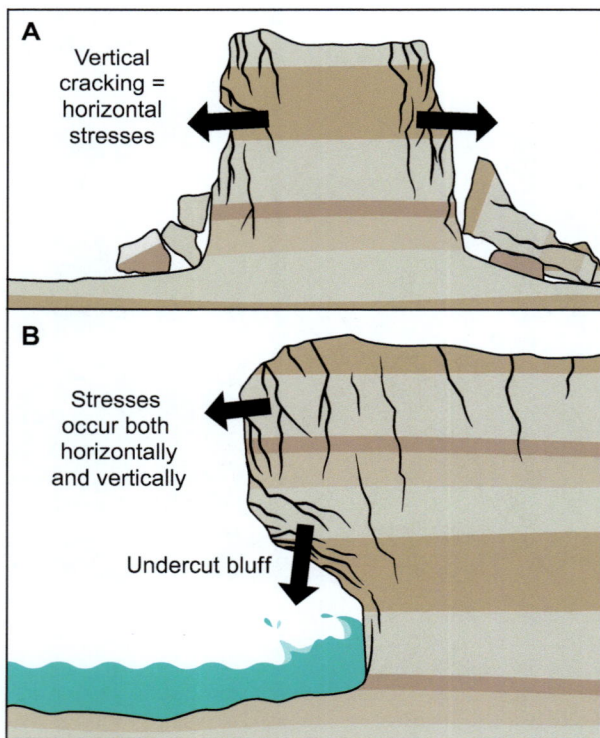

Figure 5.25 Examples of how release of pressure from the sides (**A**) or base (**B**) of a sedimentary rock mass can induce fracturing, weakening the rock and inducing failure.

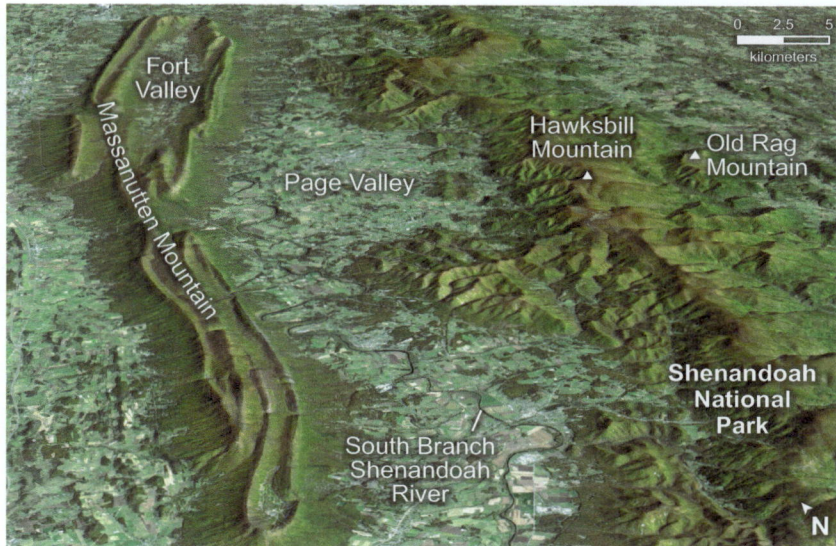

Figure 5.26 Examples of rock resistance on the resultant topography. In Virginia, USA, upturned ridges of sandstone underlie Massanutten Mountain, and igneous and metamorphic rocks comprise the irregular range of Shenandoah National Park. Granite underlies the upland at Old Rag, whereas a metamorphic rock (metabasalt) forms most of the highest crest of Shenandoah National Park. Valleys within and between these uplands are underlain by weaker shales and limestones. Source: Landsat 8 image from the NASA Earth Observatory, draped over digital elevation model.

Figure 5.27 The isolated monolith of arkose sandstone that forms Uluru (also known as Ayers Rock) in the Northern Territory of Australia is an anticlinal fold. Both the hardness of the rock type and its relative lack of joints, as compared to surrounding rocks, probably accounts for its resistance. Source: DYJ, CC BY-SA 3.0, via Wikimedia Commons.

Figure 5.28 The tropical landscape of Namibia, with a relict inselberg in the distance. Source: Zairon, CC BY-SA 4.0, via Wikimedia Commons.

well as volcanic rocks such as basalt (**Fig. 3.6**). Hard rock, formed from lava flows, can lead to a situation called **inversion of topography**. This phenomenon occurs, for example, when lava fills a stream valley. If the basalt (hardened lava) is more resistant than the surrounding rock, erosion will lower the surrounding terrain, such that the basalt from the lava flow will eventually become an upland (**Fig. 3.5**). Inversion

of topography can also be driven by a resistant armoring of gravel or boulders in a stream valley, as talus boulders fill in a steep gully, or when a soil **crust**, cemented by silica, iron, or calcite, holds up an ancient soil surface.

5.7 EFFECTS OF ROCK RESISTANCE ON RIVERS

Shallow bedrock has an important influence on the morphology of streams and their drainage networks, leading to varied (but predictable) patterns of river channels (**Fig. 16.4**). Rock structure and lithology also influence the longitudinal profiles of stream channels. A river's **longitudinal profile** shows its elevation along the river course. Where rivers flow on unconsolidated sediments, they are able to develop and maintain fairly smooth, concave-upward, longitudinal profiles (**Figs. 16.29A, 16.30A**). However, in bedrock-floored channels, rock hardness and structure have strong influence on the channel slope. For example, where rivers flow across outcrops of hard bedrock, rapids and even waterfalls may form, because the resistant rocks (or basalt layers from former lava flows) are more difficult for the river to "smooth out" (**Figs. 16.29B, 16.30A**).

Waterfalls can also form where a hard rock layer overlies weaker rock (**Figs. 5.29, 16.30**). The harder rock forms the ledge over which the water plummets, because the softer rocks below have been undercut by the falling water. Undercutting is essential to maintain the vertical escarpment that forms the waterfall, or else it will quickly develop into a **rapids**. The falling water also helps to break up any large fragments that fall from the caprock. As it breaks up, the rim of the waterfall, that is, the rock lip that forms the falls, retreats upstream (**Fig. 5.29**; see Chapter 16).

Viewed more broadly, some landscapes have **fall lines**, where a string of waterfalls or rapids on a number of different rivers all align with a resistant rock layer. A fall line usually develops where rapids or waterfalls have retreated upstream to it, but this retreat slows or even gets "stuck" at the fall

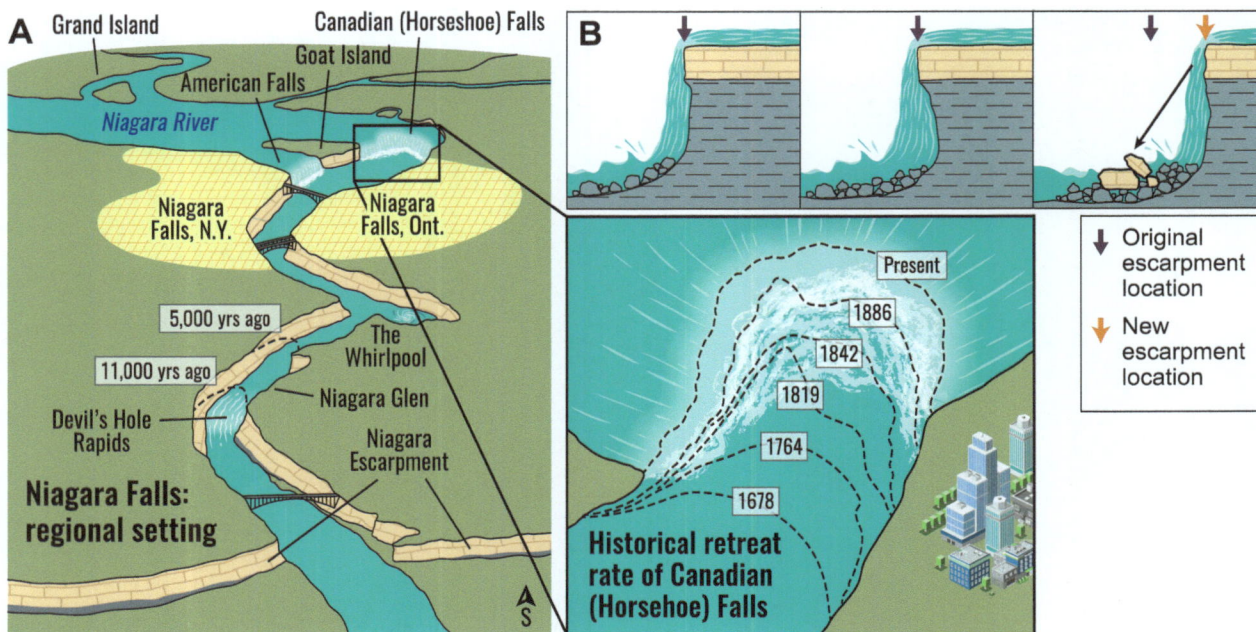

Figure 5.29 Waterfalls are excellent indications of variable rock hardness within stream valleys. **A.** A diagrammatic, oblique view of the Niagara River gorge and Niagara Falls, between New York (USA) and Ontario (Canada). Niagara Falls is held up by a thick, resistant layer of dolomite, which overlies a layer of shale. The falls have been retreating upstream at almost 2 m/yr (or > 11 km total) since it formed, that is, when the last ice sheet retreated from the area. Downstream of the falls, the river flows in a distinct gorge. Source: After the May 28, 2015 *Hudson Valley Geologist*. **B.** Diagrammatic sketch of the typical rock stratigraphy for a waterfall on sedimentary rock, showing how a waterfall retreats upstream over time as its caprock is undercut and periodically collapses.

line because the rocks there are so hard to erode. The best-known example of a fall line occurs on the Piedmont of the southeastern United States, running from Georgia to New England (**Fig. 5.30**). Other examples exist in western Australia, Brazil, and India, all on passive continental margins. In the USA, the Fall Line occurs where softer rocks and sediments of the Coastal Plain meet the harder rocks of the Appalachian Piedmont. Here, rivers flowing east "fall" off their bedrock channels onto the softer Coastal Plain sediments. Any waterfalls or rapids that might have existed on the Coastal Plain have migrated upstream, only to become stalled at the contact with the Piedmont.

In the past, the Fall Line presented a significant barrier to boats travelling upstream, many of which could not get past the waterfalls of the Fall Line. Thus, settlements often

developed at this ship-going terminus. Later, settlers realized the hydropower energy potential afforded by these waterfalls, fueling the development of a variety of industries in the eighteenth and nineteenth centuries. Today, the Fall Line shows as a long line of major cities along the Eastern Seaboard, all of which trace their beginnings back to a time when ships could not traverse the falls.

On regional scales, broadly similar rock types often result in similar, sometimes repeatable, stream patterns, much like the Fall Line example. For example, in drainage basins that become "tilted" due to regional uplift, the rivers will develop steeper gradients and begin to incise, while also forming stepped surfaces at lithologic contacts. Landscapes dominated by large basalt plateaus, for example, the Siberian Traps in Russia, the Deccan Traps of India (**Fig. 7.15**), and

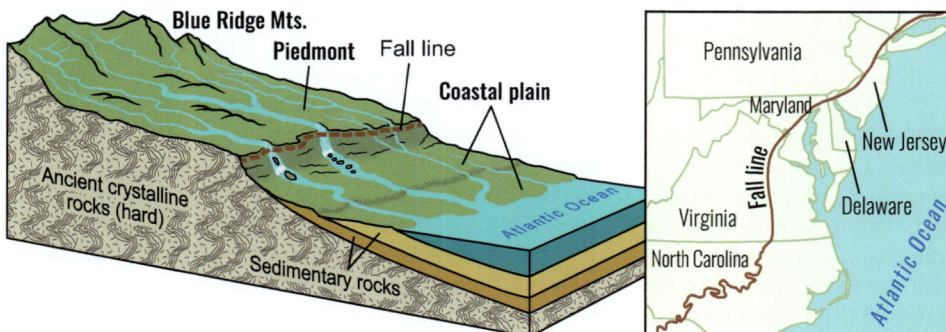

Figure 5.30 The Fall Line on the Atlantic Seaboard of the eastern United States, formed at the boundary between the resistant metamorphic and igneous rocks of the Piedmont and the softer sediments of the Coastal Plain. Here, waterfalls and rapids are quite common in eastward-flowing rivers. Source: After Encyclopedia Britannica, and Meng and Harsh (1988).

the Columbia Plateau in the northwestern United States (**Figs. 7.12, 7.16**) are examples of this, with individual lava flows forming steps within the stream longitudinal profiles.

5.8 THE BIOSPHERE'S ROLE

Geomorphology's resisting framework is not the sole responsibility of rocks. Biota are also in play, helping to protect or bind together Earth materials (see Chapter 22).

5.8.1 Rock Coatings

Organisms can function as effective agents of weathering (see Chapter 11). Examples include the chemical dissolution of rocks by organic acids, and the physical prying apart of rock by roots and burrowing animals. Our focus, however, is on how biota can *enhance* the resisting framework of rocks and sediments.

Some biochemical reactions, when they occur on rock surfaces, can form coatings, specifically rock varnish (**Fig. 5.31**). Although rock varnish is readily recognized in deserts (see Chapter 10, and **Figs. 3.11** and **3.12**), it can form anywhere, on any rock type. Rock varnish can develop in association with organisms such as lichens, fungi, bacteria, or their remains, as well as chelating or oxalating substances derived from these organisms. Most varnishes also include inorganic materials such as dust or clay minerals, as well as byproducts of chemical weathering such as silica, iron, and magnesium. Together, the organisms and their byproducts help to bind the clay, silica, iron, and magnesium to the rock surface, forming a thin (micrometers thick) protective coating. Some of the thicker coatings or rinds can produce indurated crusts that (at least temporarily) act like armor.

5.8.2 Vegetation as a Resisting Agent

Generally, vegetation protects the land surface and the rocks below from erosion. It is easy to understand this role – simply observe locations where vegetation is lacking. Arid lands have

Figure 5.32 Mounds (0.5–2.0 m dia.) of Sicilian Milkvetch (*Astragulus siculus*), a low shrub growing on freshly formed land surfaces on Mt. Etna, Sicily. Where Milkvetch grows on the thin, cindery soils of the volcano, it helps stabilize the slopes from erosion. Source: G. Pope.

such minimal protective vegetation cover that even the infrequent rain events there can be erosive. Rates of denudation in such areas can be high. Denuded landscapes occur elsewhere as well, such as on mined lands or overgrazed grasslands, and in areas of burned or cut forests. The lack of vegetation – even if temporary – contributes to increased erosion and denudation. Initially, erosion impacts only the soil cover, but continued and extreme erosion incises into regolith and even bedrock.

Vegetation inhibits surface erosion by (1) intercepting and slowing the impact of precipitation, (2) slowing and lessening the amount of runoff, and (3) binding unconsolidated sediment (with roots) (**Fig. 5.32**). Vegetation helps to lessen the erosional impact of precipitation by intercepting much of it before it hits the surface. Aboveground vegetation slows the speed and intensity of falling raindrops and runoff waters. How well vegetation can slow the impact of runoff is affected by a number of factors; chief among them are vegetation density and type.

Figure 5.31 Petroglyphs have been carved into this sandstone at Desolation Canyon on the Green River, Utah, USA, illustrating that the rock has a coating of varnish. Darker orange/brown varnish is generally a thicker coating, developed over a longer time. Source: K. Cann of the US Geological Survey.

Figure 5.33 An eroded trail in sandy soil in southern Utah (USA) contrasts with the undisturbed soil, which hosts a biological soil crust. The crust, a mixture of soil particles and microorganisms, gives the surface a darker and rougher appearance. It helps the surface to resist erosion of rain and wind but cannot withstand the effects of foot traffic. Source: R. Schaetzl.

Biological soil crusts, or cryptobiotic soils, are increasingly being examined for their importance in slowing erosion on bare soils. These crusts represent a specific type of symbiotic organism suite, consisting of cyanobacteria, fungi, and lichen, typically found in dryland areas (Fig. 5.33). Without these crusts, loose, bare soils are more easily eroded. Once disturbed, these crusts can take many decades to reform.

REVIEW QUESTIONS

5.1   What are the main exogenic and endogenic forces that shape the surface of the Earth, and thereby produce landforms?

5.2   List and describe some examples of harder and softer rocks. What factors contribute to their durability or lack thereof?

5.3   What are the differences between primary and secondary minerals? Give some examples of each.

5.4   Explain how igneous, sedimentary, and metamorphic rocks form. Provide examples of each type.

5.5   What can the crystal sizes in an igneous rock tell you about its environment of formation?

5.6   Explain the rock cycle and how it accounts for the absence of "really old" rocks on Earth.

5.7   What is Bowen's Reaction Series?

5.8   Define sedimentological and mineralogical maturity. How does a sediment become "mature"? Also, explain sedimentological sorting.

5.9   What methods can be used to estimate or measure the hardness of a rock?

5.10   How and why do joints weaken a rock, and how can jointing also act to harden a rock?

5.11   How does variability in rock hardness lead to the formation of rapids and waterfalls?

5.12   How does variability in rock resistance contribute to positive or negative topographic relief or the orientation of rivers?

5.13   In what ways does vegetation present resistance to erosion forces?

FURTHER READING

Gerrard, A. J. 1988. *Rocks and Landforms*. Unwin Hyman.

Klein, C. and Philpotts, A. R. 2012. *Earth Materials: Introduction to Mineralogy and Petrology*. Cambridge University Press.

Marshak, S. 2019. *Essentials of Geology*. W.H. Norton.

Migoń, P. 2006. *Granite Landscapes of the World*. Oxford University Press.

Prothero, D. R. and Schwab, F. 1999. *Sedimentary Geology: An Introduction to Sedimentary Rocks and Stratigraphy*. W.H. Freeman and Co.

Short, N. M. and Blair Jr., R.W. (eds.) 1986. *Geomorphology from Space: A Global Overview of Regional Landforms*. NASA Goddard Space Flight Center.

Young, R. W., Wray, R. A. L., and Young, A. R. M. 2009. *Sandstone Landforms*. Cambridge University Press.

6 Tectonics and Mountain Building

Lewis A. Owen and Warren D. Huff

Mountains are among the most prominent and inspiring landforms on Earth. Earth's internal (tectonic, or **endogenic**) and external (surface, or **exogenic**) processes have conspired to produce a wealth of mountainous landscapes that span almost every region of our planet. No strict definition of a mountain exists, other than they rise abruptly and prominently above the surrounding land, usually in the form of peaks and ridges. Thus, mountains have considerable **local relief**. Some mountains may rise only a few hundred meters above sea level (asl), such as the highest mountain in the United Kingdom, Ben Nevis (1,099 m asl [above sea level]). Nonetheless, it is one of the most formidable mountains in the Scottish Highlands (**Fig. 6.1A**). Other mountains are far more prominent. Mount Everest, the highest point on Earth at 8,849 m asl (**Fig. 6.1B**), is undoubtedly the most famous of all mountains.

Most mountains occur within elongated ranges or chains that may stretch from tens to many thousands of kilometers in length, and many tens to hundreds of kilometers across (**Fig. 6.2**). However, some mountains such as Mt. Kilimanjaro, the highest mountain in Africa (5,885 m asl), rise as prominent, isolated peaks above the neighboring plains. Mountains arouse a passion in people, whether to escape to their beauty and solitude, to explore and hike them, to provide religious significance, or to inspire scientific investigation.

Award-winning author John McPhee, in his book *Basin and Range*, expressed that if there was a single sentence to sum up all of geology, it would be that "the summit of Mt. Everest is marine limestone." McPhee was eloquently highlighting the mighty forces that must have been at play in order to move limestone that formed in a shallow sea all the way up to the highest point on Earth (**Fig. 6.1B**). It is our role as geomorphologists to explain how this has happened. The mechanism here is simple: **tectonic processes** have done this. The forces that have acted here to uplift this ancient seafloor into such high mountains are also known as **orogenic forces**. Thus, **orogeny** refers to a mountain-building event. The resulting mountains, are, therefore, often referred to as **orogenic belts**.

This chapter focuses on mountain building and orogenic forces. We also consider how individual isolated mountains, such as Mt. Kilimanjaro, have formed, by exploring the forces and processes that have produced some of Earth's highest topography.

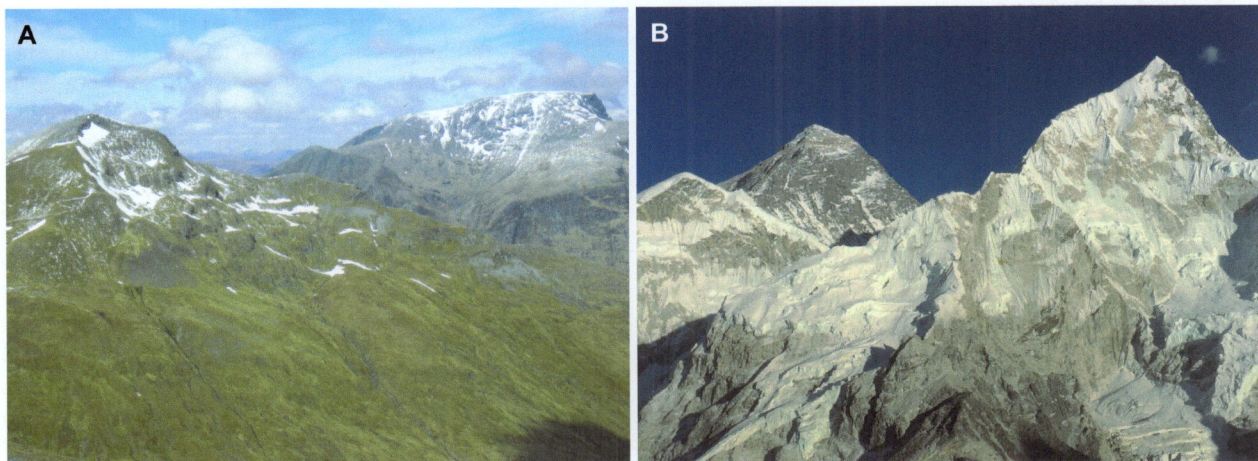

Figure 6.1 Earth's contrasting mountains. **A.** The highest mountain in the United Kingdom, Ben Nevis, along with Sgùrr a' Mhàim in the foreground, form prominent peaks in the Scottish Highlands part of the Caledonian orogenic belt. **B.** Mount Everest, along with Lhotse (8,516 m asl) in the foreground. The pale buff band capping Mt. Everest is limestone, formed in a shallow sea. Source: L. Owen.

Figure 6.0 The Trango Towers, in the Karakoram Mountains of northern Pakistan, reach elevations exceeding 6,200 meters. Source: L. Owen.

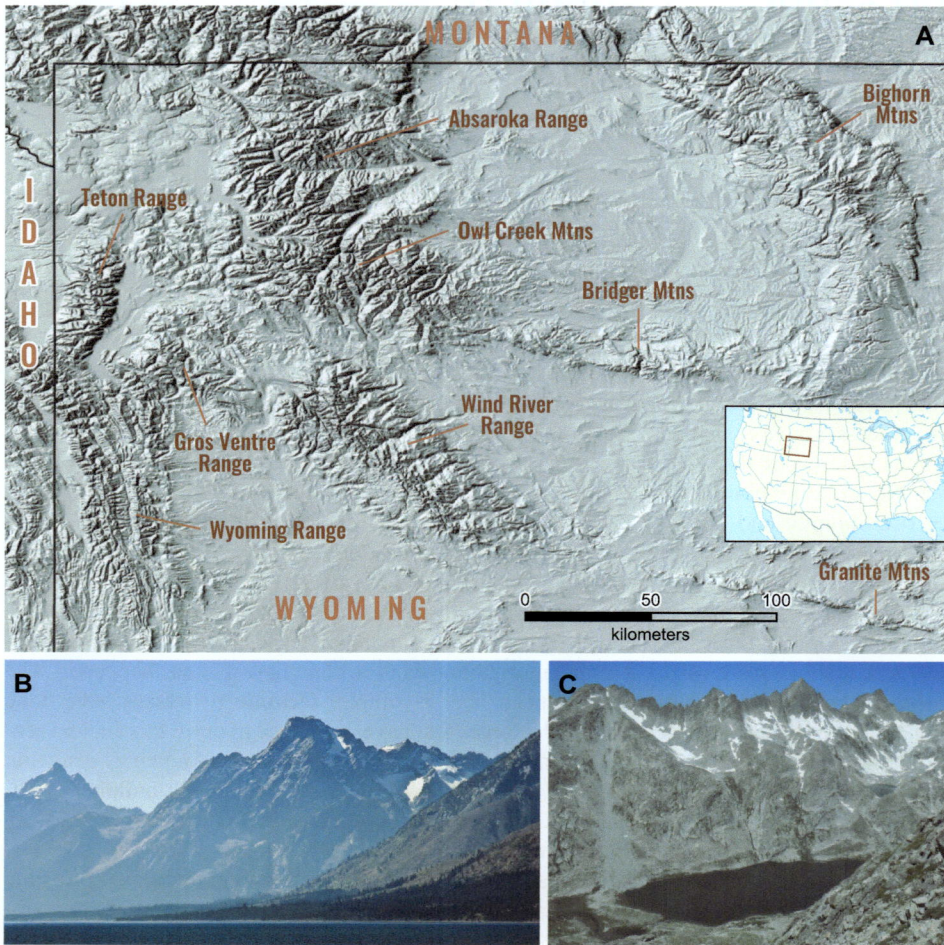

Figure 6.2 The mountain ranges of western Wyoming, USA. Often, mountain ranges are separated by broad plains, across which orogenic activity has been less active. Wyoming's topography, shown in **A**, is typical of this scenario. **B.** Typical view of the Grand Teton Mountains, near Jackson, Wyoming. Source: Acroterion, CC BY-SA 3.0, via Wikimedia Commons. **C.** Wyoming's heavily glaciated Wind River Mountains. Source: R. Schaetzl.

6.1 THE GEOGRAPHIC EXTENT AND AGES OF GLOBAL MOUNTAIN BELTS

Although inspiring and commonly politically significant, as they often form political boundaries, high mountain ranges make up only a small fraction of the continents (**Fig. 6.3**). For example, if we assumed that every land area above 1,000 m asl is a mountain, then only about 10% of Earth's land surface would be considered mountainous. Only about 3% of Earth's land surface lies above 5,000 m asl, and only 14 peaks rise above 8,000 m asl (**Fig. 6.3**). However, the concept of mountainous terrain involves more than just elevation. Mountains are areas of high **relief** *and* high **elevation**. For example, much of the expansive plateaus of the continental interiors, such as the Colorado Plateau and the Great Basin of the western United States, are well above 1,000 m asl. But because these regions are high plateaus of generally low relief, they are not considered mountainous. Thus, if we exclude high plateaus, the percentage of the land surface that is considered to be mountainous is much less than the 10% cited above. Nonetheless, the extent and grandeur of mountain belts are impressive and provide a

framework to consider the linkages between global tectonics and geomorphology.

The Alpine (Alps)–Himalayan–Tibetan mountains are the largest and longest mountain system on the continents. They stretch from the Baetic Mountains in southern Spain through the European Alps, the Turkish–Iranian Plateau, the Zagros Mountains of the Middle East, the Himalayas, and the Tibetan Plateau of southern Asia, to the Sumatra Arc of Indonesia. All told, this mountain chain is ≈7,000 km long and > 2,000 km wide at its widest part (**Fig. 6.4**). The Tien Shan of southern Asia and Gobi Altai Mountains of Mongolia are also part of this system. Several major zones of crustal plate collision have helped form this mountain system: (1) the European Alps (African–European plate collision), (2) the Turkish–Iranian Plateau (Arabian–Asian plate collision), and (3) the Himalayan–Tibetan orogen (Indian–Asian plate collision). This massive mountain system involves the pervasive deformation of rocks, including thick sequences of shallow-water sandstones, limestones, and shales deposited on continental crust, and oceanic deposits characterized by deep-water sediments, commonly mixed with volcanogenic sediments. In short, the Alpine–Himalayan–Tibetan system

A

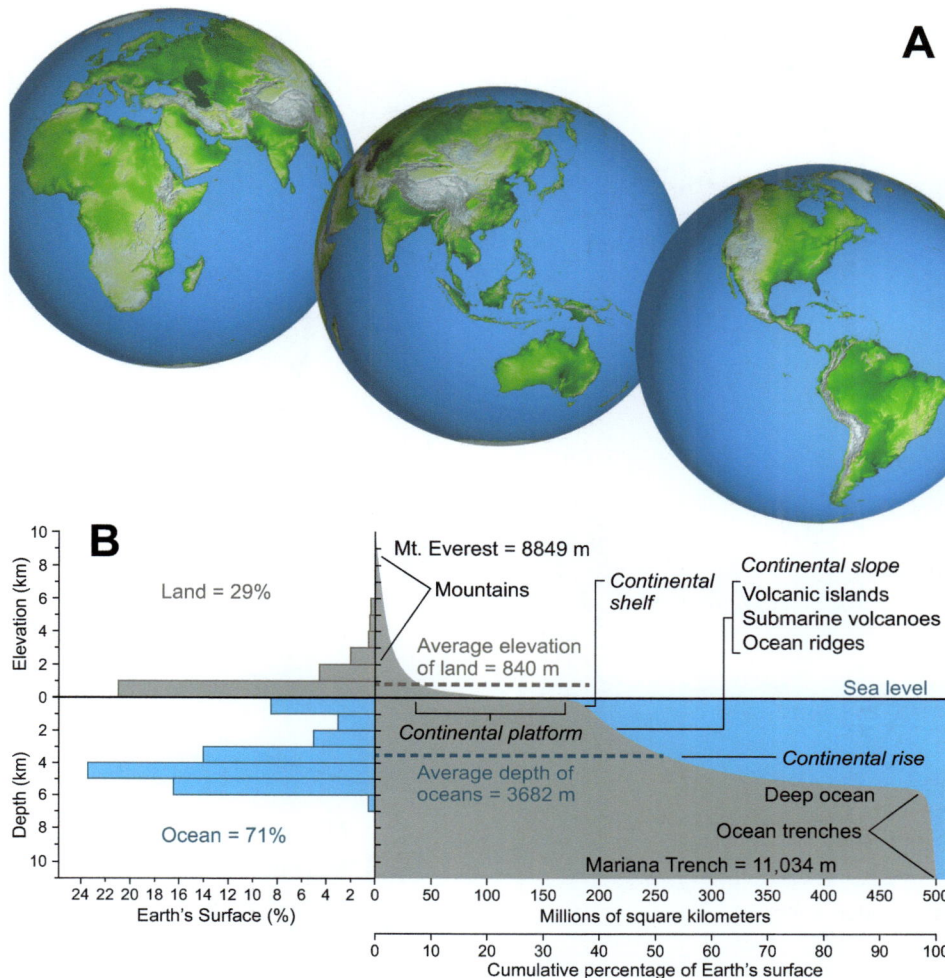

Figure 6.3 Earth's topography. **A.** Earth topography as shown in a digital elevation model. Source: NASA/JPL/NIMA. **B.** A histogram and cumulative frequency curves for Earth's land surface and ocean basins. These graphs highlight the small percentage of Earth's surface that constitutes terrestrial mountain systems. However, these graphs do not adequately show the extensive mid-oceanic ridge systems that form the longest mountain system on Earth. Source: After Wyllie (1976), with permission from John Wiley & Sons Limited.

is geologically very complex, and no one model completely explains its tectonic evolution. The Alpine sector in southern Europe, for example, involved the formation and collision of island arcs and a microcontinent, as the African plate moved north into Eurasia, starting ≈120 million years ago. Since then, the rocks have undergone continued convergence and crustal shortening, resulting in giant folds bounded by **thrust faults** (see Chapter 9).

Other significant systems of mountains lie within the Circum-Pacific orogenic belt, also known as the **Pacific Ring of Fire** (see Chapter 7). This belt includes the Antarctic Peninsula, the Andes Mountains of South America, the Western Cordillera of North America, and the volcanic island arcs of the Aleutian Islands (Alaska), through to Japan and the Philippines, ending in New Guinea (**Fig. 6.4**; see also Chapter 7). As the name suggests, it circles the margin of the Pacific Ocean and owes its origin to oceanic–oceanic and continental–oceanic collision zones at tectonic plate boundaries (**Fig. 7.1**).

The longest mountain system on Earth is nearly inaccessible (and is often ignored) because it lies on the ocean floor. Earth's **mid-oceanic ridge** systems extend for >40,000 km

and rise several km above the seafloor, being among some of the world's most impressive geomorphic and tectonic features (**Figs. 6.4, 6.5**). The main ridges (from west to east) include the Mid-Atlantic Ridge, the Southwest Indian Ridge, the Central Indian Ridge, the Southeast India Ridge, the Pacific-Antarctic Ridge, the East Pacific Rise, and the Juan de Fuca Ridge. Because the ocean basins have an average depth of about 5 km, mid-oceanic ridges seldom reach the water surface. Only at Iceland, where the Mid-Atlantic ridge traverses the Icelandic hotspot, do we get to see a mid-oceanic ridge above sea level. And even here, due to successive subaerial and subglacial eruptions, as well as uplift and rifting related to the Icelandic hotspot, the ridge at Iceland has a very different form than its continuation beneath the ocean.

A comparison of topographic and tectonic maps of the world indicates that the major mountain systems occur along, or are parallel to, lithospheric plate boundaries (**Fig. 6.4**; see Chapter 7). Mountain belts that are not associated with active (**leading edge**) lithospheric plate boundaries are usually related to boundaries that were once active in the past, or occur at areas of continental crust extension, such as those of the Basin and Range Province in the western USA

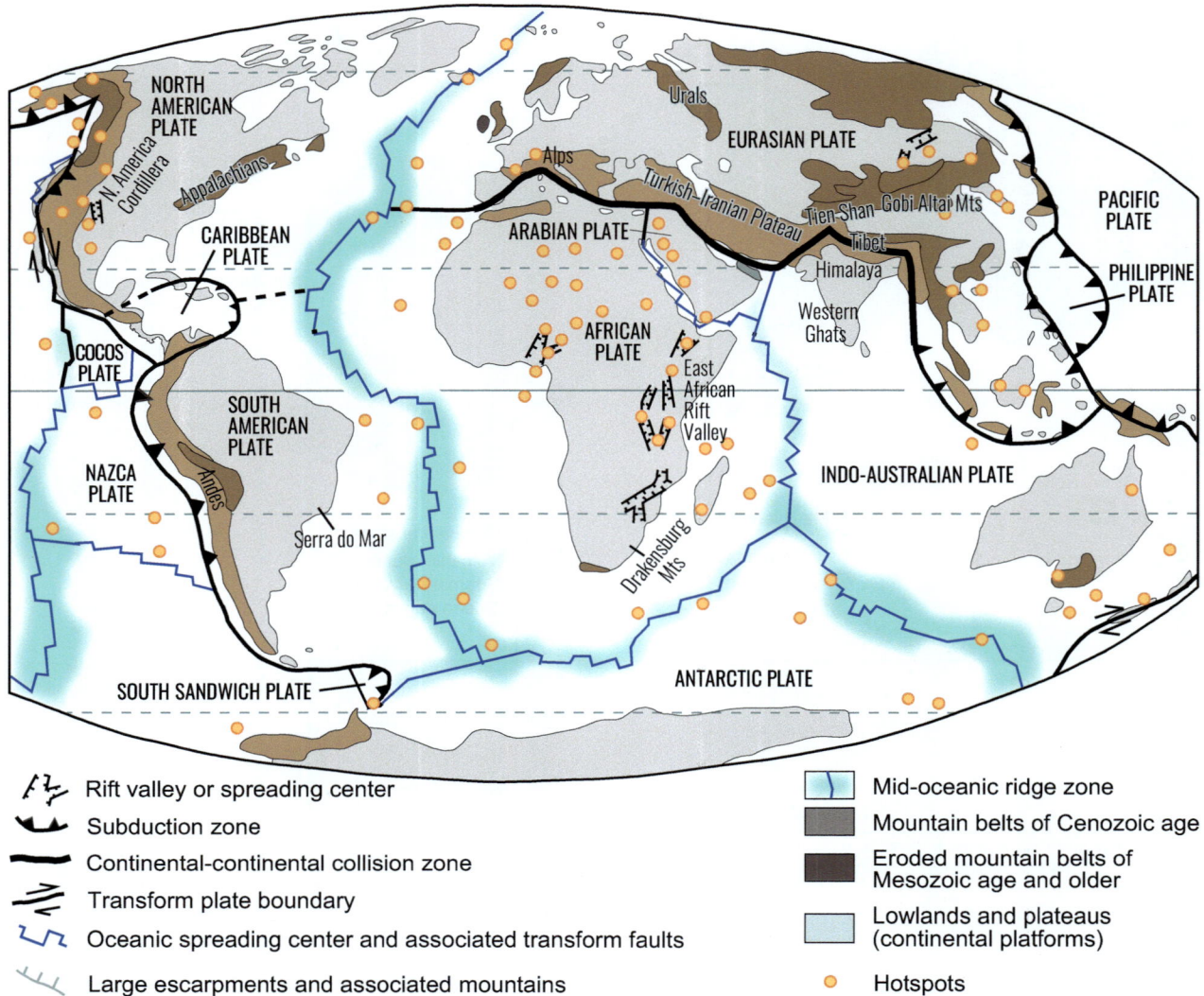

Figure 6.4 The relationships among mountain systems, tectonic plate boundaries, and the various types of global tectonic settings.
Source: Reproduced from Owen (2004), reprinted by permission of the publisher (Taylor & Francis Ltd, www.tandfonline.com).

and northern Mexico (**Fig. 10.23A**), or in the East African Rift valley. Other impressive mountain ranges, such as the Western Ghats of India and the Drakensburg Mountains of South Africa, have formed along **passive plate margins**. In these cases, the mountains have other origins, as explained later in this chapter.

Many of Earth's "minor" mountain systems are associated with **transform plate boundaries** and major strike-slip faults (see Chapter 10). At transform boundaries, crustal plates are sliding past each other, with little or no upward motion or subduction. However, compression may result in mountain development along curved stretches of transform plate boundaries. Mountain systems formed at these types of boundaries include the Alps of New Zealand, which

rise to 3,724 m asl at Mt. Cook along the Alpine Fault, and the Transverse Ranges of Southern California, which rise from a few hundred meters asl to San Gorgonio Mountain at 3,506 m asl. The Transverse Ranges are associated with the famous San Andreas Fault, which runs along the Pacific–North American transform plate boundary from the Gulf of California and through southern California, exiting into the Pacific Ocean at San Francisco (**Fig. 10.17**). This process will be expanded upon in a subsequent section.

Mountains can also form by regional warping and rifting of the crust, as well as via volcanism at hotspots (see Chapter 7). Among the largest hotspot volcanoes are the Hawai'ian Islands, which rise to 4,205 m asl at Mauna Kea from the ocean depths, over the Hawai'ian hotspot (**Fig. 7.13**). The

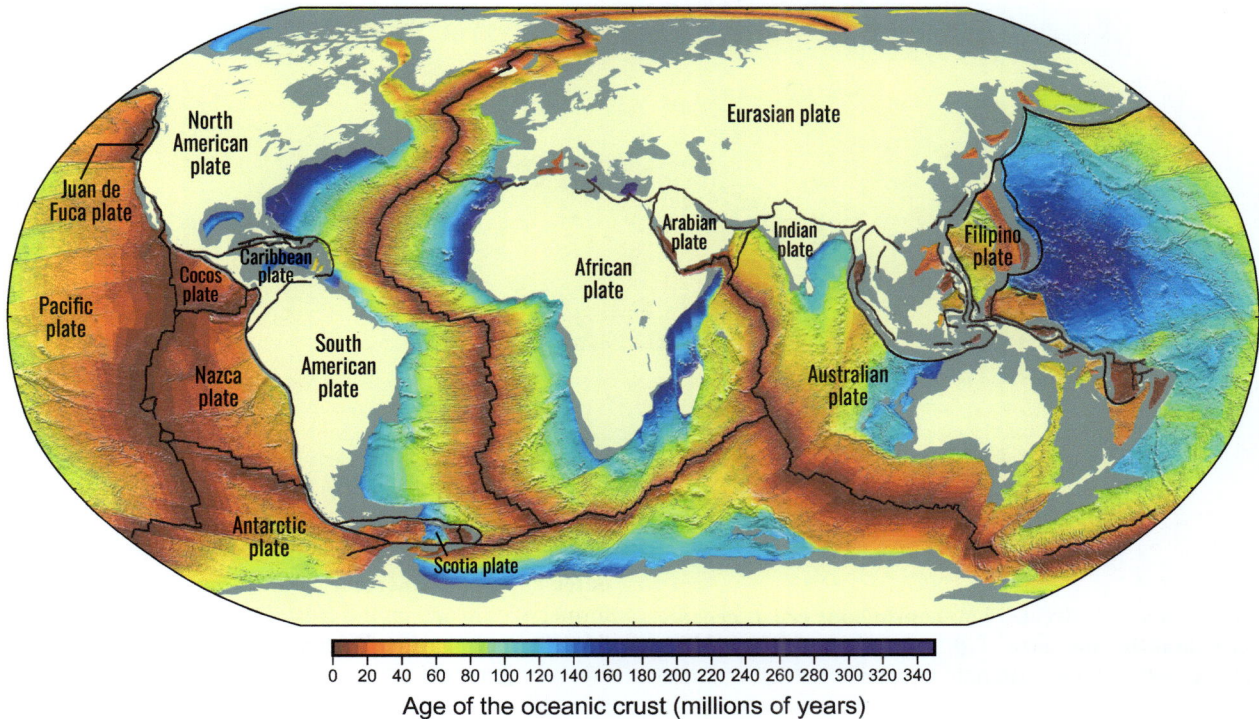

Figure 6.5 Map showing the age of the sea floor crust, major tectonic plate names and boundaries, and the locations of the major mid-oceanic ridges. The sea floor is youngest at and near the ridges, where it is formed, and gets progressively older farther away. Source: Reproduced from Seton, M., Müller, R. D., Zahirovic et al. (2020), © American Geophysical Union.

Grand Teton range in Wyoming (USA) is an example of mountains formed by uplift associated with the Yellowstone hotspot (**Figs. 6.2, 7.12**).

Most of Earth's mountains began to form and primarily evolved during the Cenozoic Era (the last 65 million years). Geologists refer to these types of mountain systems as "young" and "active," especially if they are presently experiencing uplift, earthquakes, and/or active volcanism. Mountain ranges along much older and less active plate margins are referred to as "ancient." These older and lower mountain systems, such as the Appalachian Mountains of the eastern United States, may be regionally extensive and significant, but bear little relationship to present-day lithospheric plate boundaries or zones of regional extension (**Fig. 6.6A**). Although ancient mountain systems formed many hundreds of millions of years ago, they are continuing to evolve due to **unloading**. In unloading, rock is removed by erosion and the crust adjusts itself by uplifting, helping to sustain the elevation of the mountains. This process is known as **isostatic adjustment** (see Chapter 9).

The Appalachians are part of a more extensive mountain system known as the Appalachian–Caledonian orogenic belt (**Fig. 6.6**). This belt stretches over 6,000 km and includes the Cordillera Oriental in Mexico, the Venezuelan Andes, the Ouachita Mountains of south-central USA, the Appalachian Mountains of the eastern USA and Canada, the Innuitian mountains of Arctic Canada and Greenland, and the Caledonides of eastern Greenland, Svalbard, Ireland, Britain, and Scandinavia. The belt even extends southward, into the West African fold belt. Three major orogenies between 570 and 250 million years ago helped to build these mountains, which subsequently split apart when the Atlantic Ocean began to form, about 200 million years ago. Continued erosion has carved these mountain ranges into some of the world's most spectacular landscapes. The Appalachian–Caledonian mountain system is the most studied of all ancient orogenic belts. It was broken apart tectonically ≈150 million years ago, with the Appalachians confined to North America and the Caledonides forming the mountains of the Scottish Highlands and Norway (**Fig. 6.1A**). The Appalachian Mountains are a classic **fold and thrust belt**, having formed due to continental–continental collision that formed the supercontinent Pangaea between 470 and 270 million years ago (see Chapter 9). Other examples of ancient mountain belts include the Ural Mountains of Russia, the Apennines of Italy, the Black Hills of South Dakota and Wyoming (USA), the Guiana Highlands of Brazil, the Makhonjwa and Waterberg Mountains of South Africa, and the Hamersley Range of Western Australia.

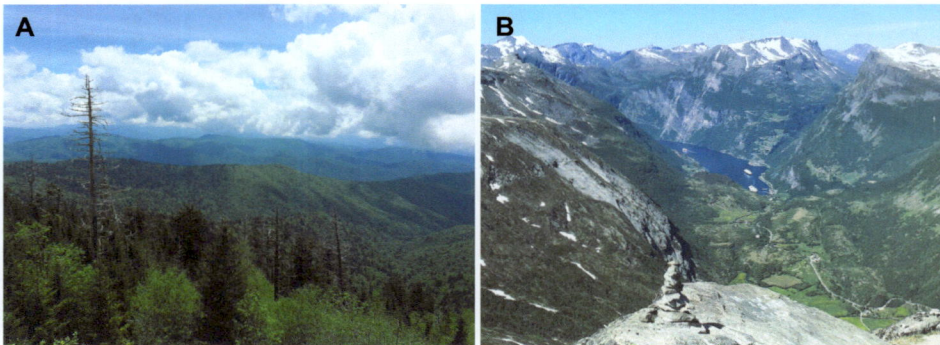

Figure 6.6 Views of mountains associated with the Appalachian–Caledonian orogenic belt. **A.** View toward Mt. Mitchell, North Carolina, the highest peak (2,037 m asl) in the eastern USA, from Clingmans Dome in Tennessee, the highest peak in the Great Smoky Mountains (2,025 m asl). **B.** View into Geiranger Fjord in Norway, illustrating the topography of the Scandinavian (Sacnades) Mountains. This World Heritage Site is famous for its beautiful, deep, glacially carved fjords and rugged mountains. Source: L. Owen.

6.2 EARTH'S INTERNAL STRUCTURE AND COMPOSITION

Earth's interior is divided into **continental crust**, **oceanic crust**, **mantle**, and **core** (**Fig. 6.7**). The mechanical (aka *rheological* – how materials respond to deformation) subdivisions of Earth are the **lithosphere**, **asthenosphere**, **outer core**, and **inner core**. All of Earth's landforms are formed within and on the crust. Oceanic crust is generally thin, dense, and primarily composed of basalt. Continental crust is generally thick, less dense and composed of

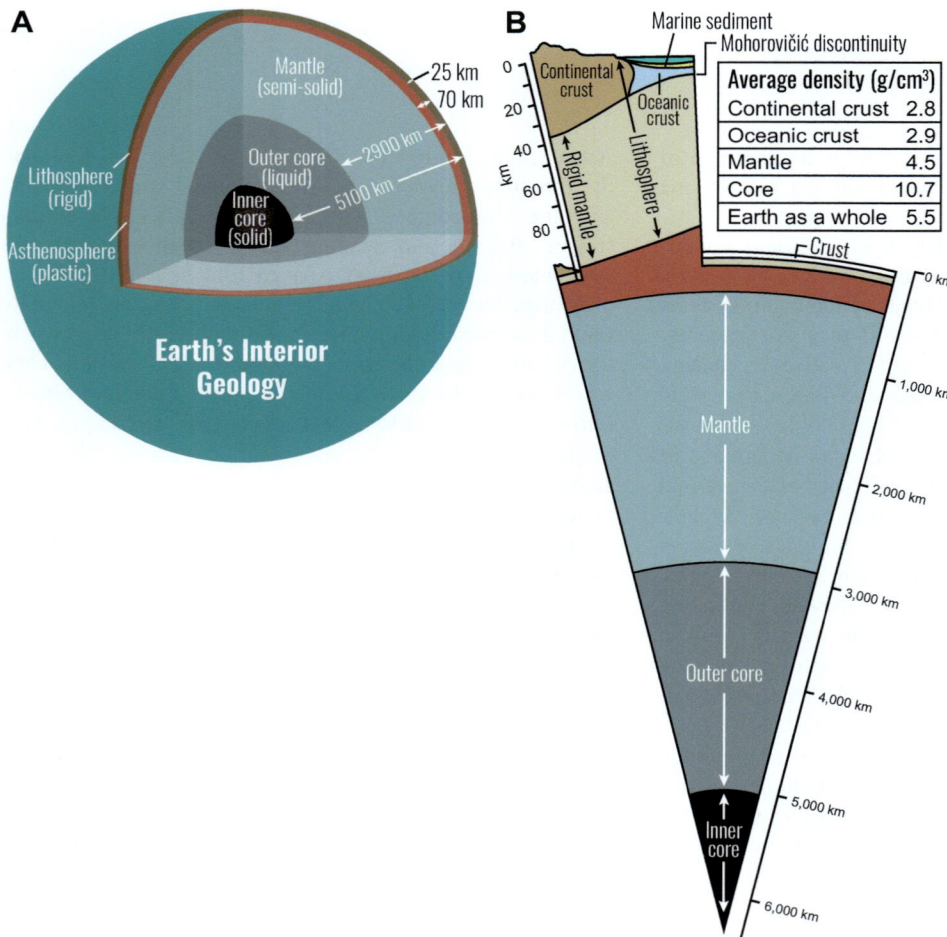

Figure 6.7 Diagrams of the generalized interior structure of Earth.

Average density (g/cm³)	
Continental crust	2.8
Oceanic crust	2.9
Mantle	4.5
Core	10.7
Earth as a whole	5.5

granitic and felsic rocks. Between the core and the crust is the mantle. The mantle–crust boundary, known as the **Mohorovičić discontinuity** (aka the Moho), has an average depth of ≈35 km beneath the continents but only 5–10 km beneath the ocean basins. Below the roots of the largest mountain chains, the Moho can be as deep as 70 km.

The mantle is solid. However, due to its high temperatures and great pressures, mantle material flows – but at extremely slow strain rates. Although we have not yet been able to drill into the mantle, we can survey the boundaries and character of the mantle using **seismology** and geophysics. Additionally, studies have found evidence of relatively unaltered mantle material, where pieces of oceanic crust known as **ophiolites** have been thrust (obducted) onto continents.

The mantle likely has the consistency of warm asphalt – flowing very slowly. It is composed of dense, ultramafic magnesium- and iron-rich minerals such as olivine, which lead us to believe that mantle material is dark green with scattered flecks of black minerals. Velocities of seismic waves in the upper part of the mantle increase with depth to ≈100 km and then decelerate slightly to a depth of ≈300 km.

The upper part of the mantle, known as the low-velocity zone, corresponds to the asthenosphere (**Fig. 6.7**). The asthenosphere is a region of partial melting, and as a result, the upper mantle exhibits its most fluid-like behavior here. Rocks here "flow" when forces are applied to them. The partial melting of these rocks derives from heat generated by the decay of radioactive elements.

Earth's core is iron-rich, metallic, and divided into two regions, the outer and inner core. The outer core is the liquid region below the mantle, and the inner core at the very center of our planet is a solid sphere (due to incredibly high pressures) that is responsible for our planet's magnetic field, which protects us from solar radiation.

All Earth's landforms occur on the coherent, semi-rigid layer called the **lithosphere**. The lithosphere is generally composed of two regions: the uppermost section of crust (oceanic or continental), and a thicker, lower region of lithospheric mantle (the uppermost rigid part of the mantle). The lithospheric mantle and the crust above it move together as a combined unit. The thickness of the lithosphere varies considerably and ranges up to 80 km thick. Generally where the mountains are highest, the lithosphere is thickest. Mass added or removed from the crust causes the surface to sink or rise, respectively, and move toward an equilibrium state called **isostasy**.

6.3 PLATE TECTONICS AND PLATE BOUNDARIES

The lithosphere is comprised of seven major and at least a dozen minor lithospheric plates, which move with respect to each other at velocities averaging ≈70–100 mm/yr (**Fig. 6.8**). Both continental and oceanic crust comprise Earth's major lithospheric plates, whereas many of the minor lithospheric plates are primarily composed of oceanic crust.

Three main types of active plate boundaries occur between the lithospheric plates. At **convergent plate boundaries**, lithospheric plates are moving together and colliding. If both plates are oceanic, one plate is taken below (subducted) beneath the other. If one plate is comprised of continental crust and the other is oceanic crust, the thinner and denser oceanic plate will be subducted beneath the continental plate. If both plates are comprised of continental crust, one will reluctantly be subducted under the other, and impressive mountain building results. At **divergent plate boundaries**, tectonic plates are pulling apart and as a result, molten material wells up from below to form new oceanic crust. This process begins as mantle materials rise in convective currents and spread apart. Rift valleys form (eventually) at divergent plate boundaries, becoming new, narrow oceans, as is occurring at the Red Sea today. Lastly, at transform plate boundaries, tectonic plates are sliding past each other laterally, and neither is subducted.

Passive plate margins occur where oceanic and continental crusts meet, but there is no differential movement. The

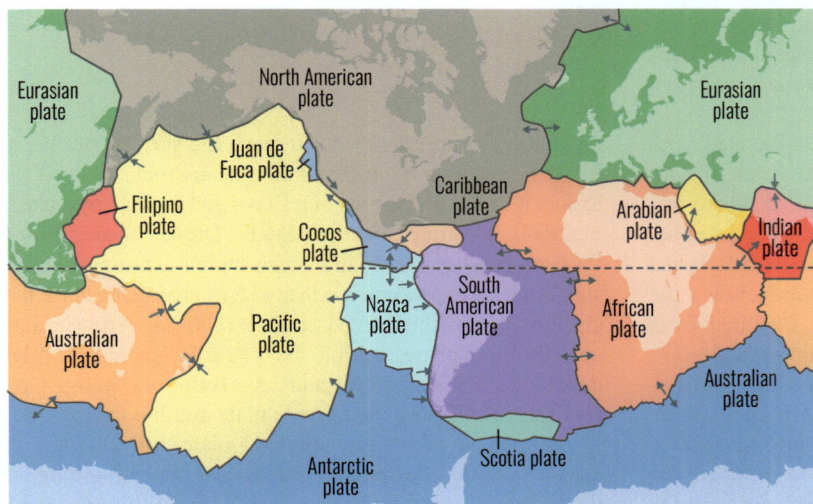

Figure 6.8 A map of Earth's lithospheric plates. Source: US Geological Survey.

oceanic plate is attached to the continental plate, and both are moving in the same direction, as one unit, attached to each other. The eastern coast of North America, where continental crust of the Northern American Plate is joined with the oceanic crust of the western Atlantic Ocean, is a good example of this type of a **passive plate margin**. Typically, along such passive margins, tectonic activity is minimal and often, a wide continental shelf lies offshore (**Fig. 6.3B**).

The driving mechanism for lithospheric plate motion is Earth's internal heat, which produces convection cells in the mantle. Within these cells, upwelling of magma drives plates apart, forming spreading centers, which are the origin of the mid-oceanic rift system (**Figs. 6.4, 6.5, 7.7**). Oceanic plates eventually get subducted back into the mantle beneath oceanic or continental crust, at a convergent plate boundary (**Fig. 7.6**). Ocean spreading is one part of the machine that drives the relative motions of lithospheric plates, which ultimately lead to mountain building.

Young, geologically active mountains are almost always associated with plate margins or boundaries. Five types of boundaries are possible: (1) continental–continental collision zones, (2) oceanic–oceanic and continental–oceanic collision zones, where the subduction of one plate is occurring, (3) oceanic spreading centers, which develop into mid-oceanic ridges, (4) transform plate boundaries, and (5) passive plate margins, where active tectonism is not currently happening, but erosion plays an important role in shaping the margin and creating mountains and escarpments. In addition, young mountains may be associated with **hotspots** in the mantle, where volcanism is active (see Chapter 7), as well as areas of continental crust extension, such as the Basin and Range province of the western United States. In the list above, (1) and (2) are examples of convergent plate boundaries (**Fig. 6.9**), (3) is an example of a divergent plate boundary, and (4) is at a conservative plate boundary, where subduction is not occurring. Details regarding these various types of plate boundaries are discussed below.

6.3.1 Convergent Plate Boundaries

The two largest, young mountain systems that are *not* a part of the mid-oceanic ridge system are the Alpine–Himalayan–Tibetan and the Circum-Pacific mountain systems – mountain belts of Cenozoic age (**Fig. 6.4**). Both are at convergent plate margins.

The Circum-Pacific mountain systems, associated with the Pacific Ring of Fire, incorporate several types of convergent plate margins (**Fig. 6.9A, B, C**). At areas where two oceanic plates converge, one of the plates gets subducted beneath the other, such as east of the Mariana Islands in the South Pacific, forming a trench near the subduction zone (**Fig. 6.9A**). The subducting oceanic plate, hydrated from sitting below the ocean, dehydrates as it plunges and melts in the asthenosphere. This influx of water hydrates the surrounding mantle material, causing it to partially melt. Some of this hot, buoyant magma then rises and pushes through the adjacent oceanic lithosphere, making its way towards the surface, where it can form undersea volcanoes. Some of the

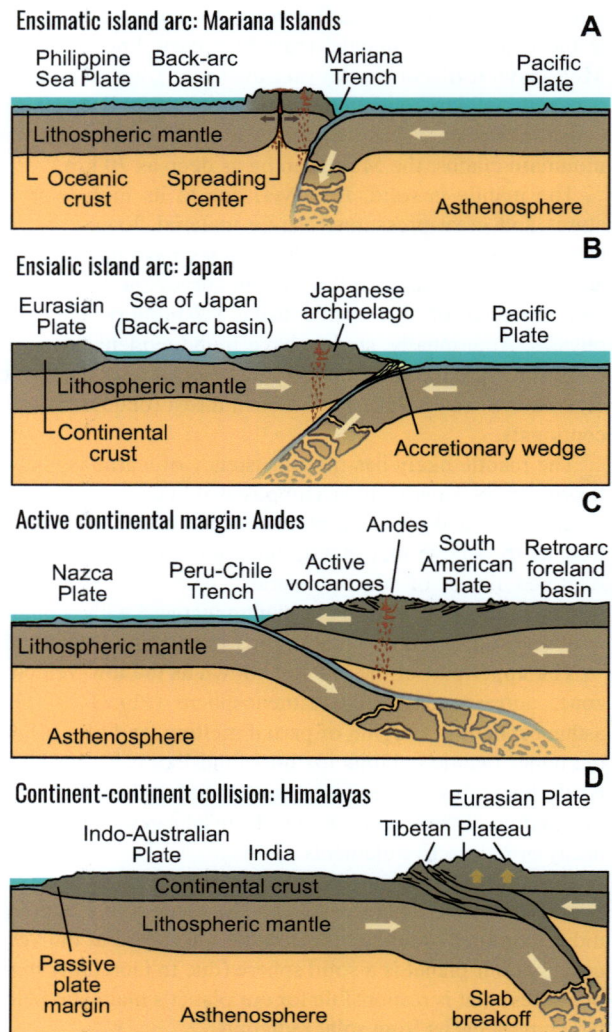

Figure 6.9 Examples of the various types of convergent plate margins and their associated terrains. The convergence of (**A**) two oceanic plates to form the Mariana Trench and its island arc, (**B**) oceanic and continental plates to form the volcanic islands of Japan, with an oceanic back-arc basin, (**C**) oceanic and continental plates to form the volcanic Andes Mountains, and (**D**) two continental plates to form the Himalayas and the Tibetan Plateau. Source: After Frisch et al. (2011) with permission from Springer Nature.

magma wells up too slowly to reach the surface and begins to cool, crystallizing at depth as igneous rock. Magma that makes it to the surface becomes lava and various pyroclastic materials, forming volcanic islands. These volcanic islands form in **arcs** parallel to the ocean trench and margin of the subducting plate, which are themselves arcuate because they are moving on the curved surface of Earth. The (trailing) oceanic lithosphere, behind the volcanic island arc, might develop into a spreading center. **Accretionary wedges** may also form along oceanic–oceanic plate margins (**Fig. 6.9B**).

At sites of oceanic–continental plate convergence, the denser oceanic lithosphere is subducted beneath the lighter, thicker, continental crust (**Fig. 6.9B, C**). In this scenario, a deep ocean trench again forms at the surface expression of the plate boundary, although in some cases it fills with sediment eroding off the continental crust. Additionally, a series of faults may develop within an accretionary wedge of sediment. The hydrated oceanic crust again dehydrates as the lithospheric slab descends into the asthenosphere, causing the surrounding mantle to partially melt. The rising molten material can form volcanoes on the continental crust above. These continental volcanic arcs, like volcanic island arcs, parallel the subducting plate margin. The Andes of South America and the Cascades of Oregon and Washington (USA) states are good examples of this type of geology. In some areas, such as near Japan, the crust begins to spread behind the island arcs to form **back-arc basins** (Fig. 6.9B).

The Circum-Pacific orogenic belt is broadly divided into eastern and western sectors. The western sector is the result of the convergence of oceanic plates, including the Pacific, Philippine, and Indian-Australian plates, and the eastern margin of the Eurasian plate (Fig. 6.8). This sector, however, is discontinuous and includes volcanic island arcs and arc-collision zones. Its associated mountains are not very geographically extensive but are impressive in terms of their relative relief. Taiwan, the Philippines, New Guinea, and the Vanuatu arc in the southeastern Pacific are the best examples of these types of tectonic collisions. The eastern sector includes the Andes and North American Cordillera, a mountain chain that stretches almost continuously for more than 20,000 km. Mountain ranges and individual mountains in this sector evolve mainly as a consequence of the convergence of the oceanic and continental plates. This sector includes the collision of the Pacific, Juan de Fuca, Cocos, and Nazca oceanic plates with the North and South American continental plates. Presently, the margin is being consumed beneath Alaska, the US Pacific Northwest and southwestern Canada, Central and South America, the Scotia Arc, and the Antarctica Peninsula. Most of the mountain building that produced the present landscapes has occurred during the last 200 million years.

The Alpine–Himalayan–Tibetan chain is a classic example of mountains that have developed along a continental–continental collision zone (Fig. 6.9D). Before continental–continental collision occurs, however, the ocean between the continents has to close and the oceanic crust between them must be removed by subduction. As a result, continental convergence settings will have geologic evidence for subduction-related features and may record a lengthy, complex geologic history, extending over many hundreds of millions of years. As a result, these types of mountain belts contain thick deposits of both shallow-water and deep-water marine sediments; hence, the marine limestone at the top of Mt. Everest (Fig. 6.1B). Eventually, the ocean crust gets completely subducted, leading to volcanism and the production of volcanic islands or continental arcs. Rocks from the islands, or continental arcs, are incorporated into the mountain-building event as the sea begins to close. Ocean crust may even be **obducted** onto the top of the continental crust to form an ophiolite. As the continental plates continue to collide, the

crustal "shortening" is taken up within a foreland fold-and-thrust belt (see Chapter 9); many rocks get folded and faulted. Continued deformation and metamorphism occur together, with the intrusion of massive granitic bodies (**batholiths**) into the core zone of the evolving mountains. Sediments eroded off the continent are then deposited and partially deformed along the front of the mountains. Over time, the continental crust continues to thicken as it is continually deformed, and the mountains develop deep roots. The continental crust under the Tibetan Plateau is, for example, >80 km thick – twice the global average. Eventually, the mountain belt begins to collapse under its own mass, and block faulting, the development of fault-bounded basins, and the intrusion of scattered dikes and intrusive bodies occurs, as magma cools within the surrounding crust. This sequence of events, although somewhat simplified, typifies many such settings.

Most orogenic belts at continental–continental plate boundaries grow outwards from a central core, implying that their various segments may form at different times. Crustal collision at this type of plate margin produces a complex geology, but there are many similarities in their evolution and geologic characteristics. In the mature stages of an active young orogen, and still recognizable in ancient ones, are broad geologic and topographic belts, which display these characteristics (Fig. 6.10):

1. An outer foredeep or **foreland basin** in front of the mountains, where sediments eroded from the mountains are deposited in the ocean or on land.
2. A foreland fold and thrust belt along the margins of the mountains. Here, tectonic forces are working to actively extend the mountains outwards as they evolve. Earthquakes are common in this area.
3. A crystalline core complex, composed of sedimentary rocks and their basement, volcanic and igneous rocks, and associated sediments, along with fragments of metamorphosed ocean crust (ophiolites), gneissic terranes with abundant ultramafic bodies, and granitic batholiths.
4. High-angle fault zones representing the gravitational collapse of the overthickened mountain mass.

The Himalayan–Tibetan orogen began in the early Paleozoic Era, when microcontinents and island arcs collided with the southern margin of Eurasia. These collisions were followed by the collision of the Indian and Eurasian continental plates ≈50 million years ago. Since then, the Indian plate has been moving northward at ≈50 mm/year, resulting in between 1,400 and 2,000 km of total crustal shortening. This shortening is expressed at the surface as discrete thrust belts with relatively narrow zones of contraction or regional detachment structures/faults. This crustal collision formed the Tibetan Plateau and its adjacent mountains. The timing of the Tibetan Plateau uplift is uncertain, but by ≈14 million years ago, the Plateau had largely been formed. The continued northward movement of the Indian plate into Eurasia makes the Himalayas one of the most seismically active regions on Earth.

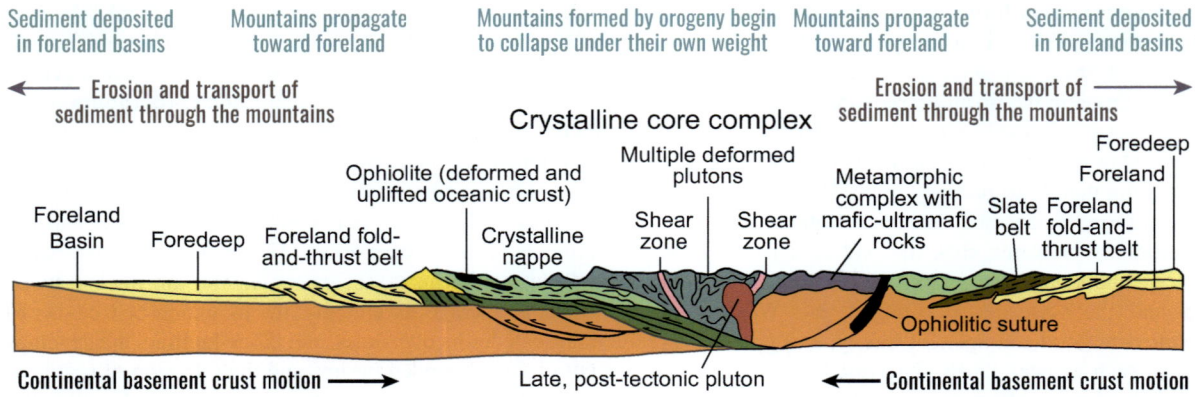

Figure 6.10 Schematic diagram of a composite orogenic belt, showing the many major structures and tectonic components. Source: From Owen (2004), reprinted by permission of the publisher (Taylor & Francis Ltd, www.tandfonline.com).

Many ancient mountain belts are, today, only remnants of a former continental–continental plate collision. Arguably the most well studied is the Appalachian–Caledonian mountain system (**Fig. 6.6**). The Appalachians are a classic fold and thrust belt, resulting from the continental–continental collision that formed the supercontinent Pangaea. They rise from the broad, low-elevation coastal plain to an eroded landscape known as the **Appalachian Piedmont**. Beyond

that, are the higher-elevation Blue Ridge, and Valley and Ridge Provinces, which define the spine of the range, and the lower-relief landscape known as the Appalachian/Allegheny Plateau on the western edge (**Fig. 6.11**). The topography of the Valley and Ridge reflects the underlying geology, with ridges formed by erosion-resistant rock (such as quartzite and sandstone) and valleys formed on more erodible shales and limestones. The Blue Ridge is composed of

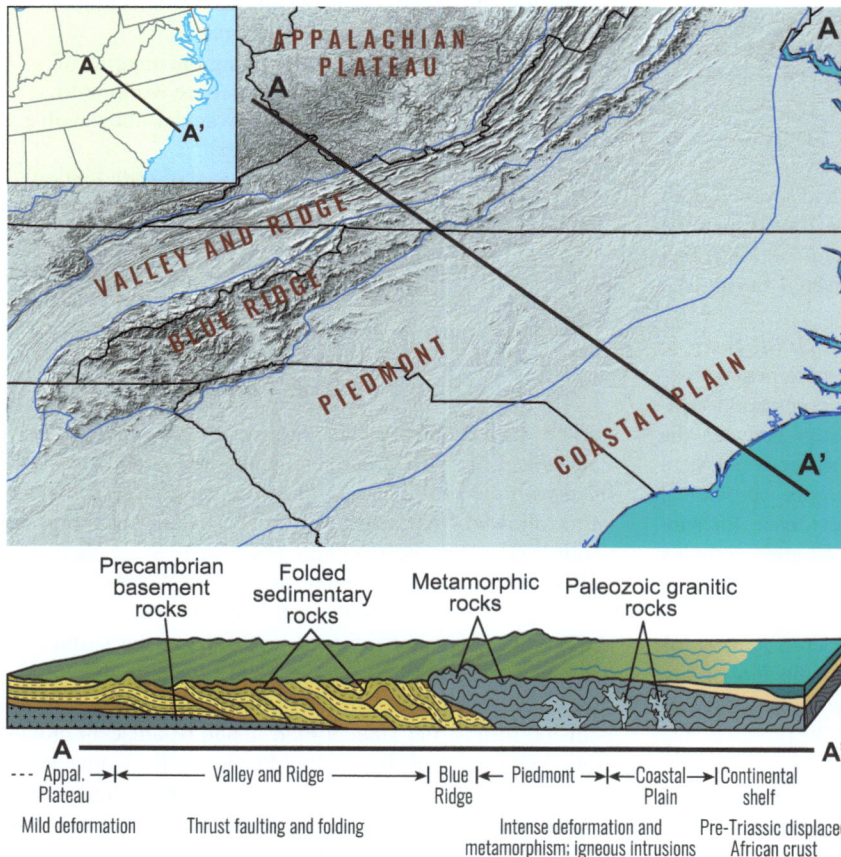

Figure 6.11 Digital elevation model showing the topography and extent of the Appalachian Mountains and related landscapes in the southeastern United States, with a schematic diagram below, illustrating the geology across a typical transect across these landscapes.

highly metamorphosed Precambrian and Cambrian crystalline rocks that were thrust into place over sedimentary rocks some 300 Ma. The Piedmont is a lower-relief, hilly region on eroded, metamorphosed, Precambrian and Paleozoic sedimentary and volcanic rocks, intruded by granites.

6.3.2 Divergent Plate Boundaries: Mid-Oceanic Ridge Systems

Divergent plate boundaries occur where oceanic lithospheric plates are pulling away from each other (**Fig. 7.7**). Rates of spreading range from a few mm/yr, such as in the Gulf of Aden, to as much as 6 cm/year for the East Pacific Rise. The area of divergence fills with magma, which cools as it rises from below, eventually crystallizing to form oceanic crust (see Chapter 7). Areas at divergent oceanic plate boundaries will eventually form mid-oceanic ridges. The ridges are elevated above the surrounding ocean floor because they consist of rock that is hotter and less dense than the adjacent oceanic crust, and also because hot mantle material is rising beneath the ridges to fill the gap created by the spreading plates, further increasing their elevation. As the mantle material rises, it decompresses and partially melts, forming magma chambers within the oceanic crust. The crystallized magma and lava help produce new oceanic crust that is then pulled away from the spreading center (see Chapter 7). As this new crust moves apart, it cools, contracts, and subsides. Most of these ridges occur within ocean basins, and so they are commonly referred to as mid-oceanic ridges. Nonetheless, they can also occur within back-arc basins (**Fig. 6.9B**).

Most of Earth's mid-oceanic ridges are between 1,000 and 4,000 km wide (**Figs. 6.5, 6.12**). They typically rise 2–3 km above the surrounding ocean floor, and their crests have an average depth of 2,500 m below sea level – comparable in height and width to continental mountain ranges, but much longer overall (**Fig. 6.12**). Mid-oceanic ridges with slow spreading rates have a well-defined (1.5–3 km deep), symmetrical, rift valley, bounded by faults (**Fig. 6.13A**). In contrast, the fastest spreading ridges have subdued topography, more reminiscent of Hawai'ian volcanoes, with

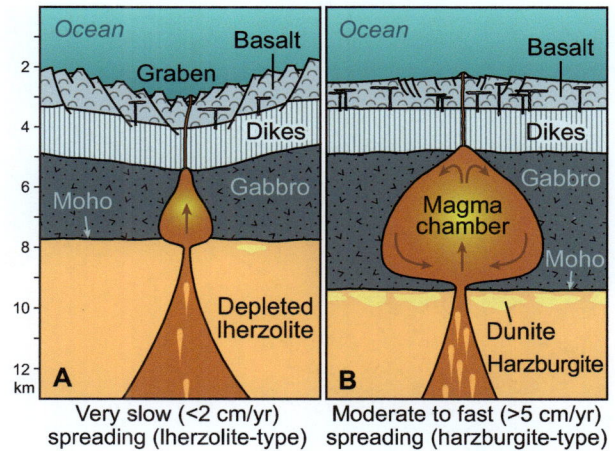

Figure 6.13 Schematic representation of cross-sections through mid-oceanic ridge systems that are spreading at different rates, and the transform faults along them. Source: After Frisch, W., Meschede, M., Blakey, R. (2011), with permission from Springer Nature.

a small summit ridge or graben, and well-developed axial valleys that may drop to depths below that of the surrounding ocean floor (**Fig. 6.13B**). **Hydrothermal activity** (hot water and steam) is commonly associated with these systems, producing springs of extremely hot water that may form columnar structures known as hydrothermal vents or chimneys.

6.3.3 Transform Plate Boundaries

Transform plate boundaries, also known as conservative plate boundaries, occur where lithospheric plates slide past each other, along **transform faults** (see Chapter 10). Here, lithospheric crust is neither destroyed nor formed. Mid-oceanic ridges are broken into segments by transform faults that displace them by tens or even hundreds of kilometers (**Fig. 6.12**). Thus, most transform faults occur in the deep ocean, in oceanic crust. These faults are sub-vertical and may produce fault scarps that exceed 500 m

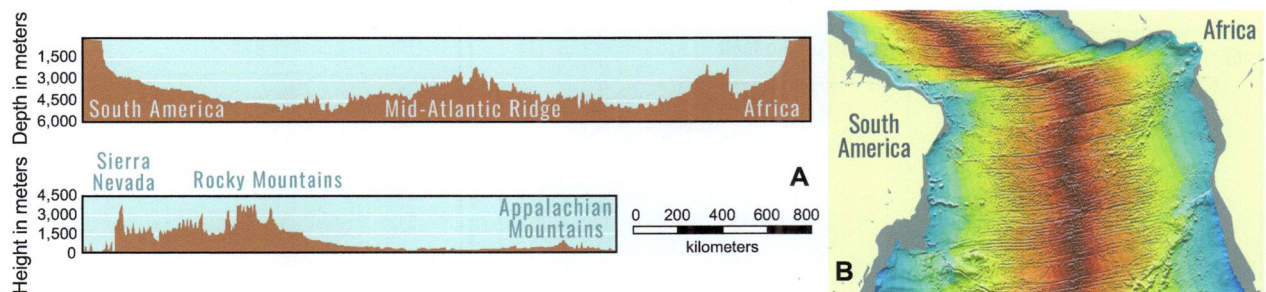

Figure 6.12 The topography of the Mid-Atlantic ridge, formed by the divergence of the North American and South American plates from the Eurasian and African plates. **A.** The seafloor topographic profile between South America and Africa (top), compared to the topographic relief across the USA, from the Pacific to the Atlantic coast, showing the major mountain ranges (bottom). **B.** Seafloor topography across the Atlantic Ocean; note that numerous transform faults offset the ridge. See **Fig. 6.5** for color symbology, which depicts the age of the various segments of the seafloor.

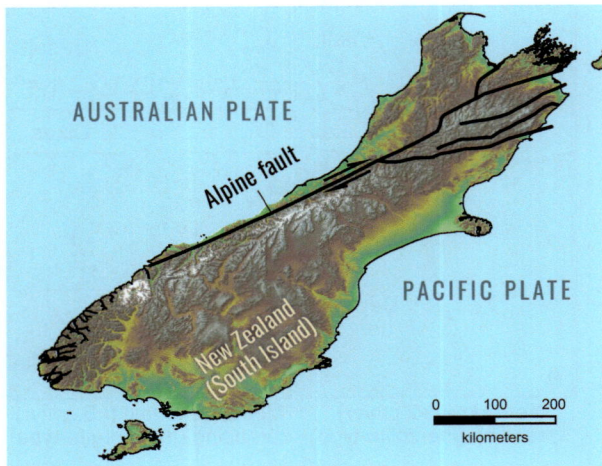

Figure 6.14 Topography of the South Island of New Zealand, showing the trace of the strike-slip Alpine Fault and its relation to the Southern Alps. Source: Used with permission of Elsevier, from Frankel and Owen (2013); permission conveyed through Copyright Clearance Center, Inc.

in height. The faults are a consequence of accommodating the movement of lithospheric plates on the curved surface of the Earth.

Within continental plates, motion along transform plate boundaries typically occurs along **strike-slip faults**. The Southern Alps of New Zealand is probably the most impressive mountain range to form along this type of boundary (**Fig. 6.14**). Here, the Australian and Pacific lithospheric plates meet at the Alpine Fault. Rising to an impressive 3,724 m asl at Mt. Cook, the Southern Alps of New Zealand are rapidly uplifting. Strike-slip faults at transform plate boundaries range in length from a few tens of meters to >1,000 kilometers. Bends in the surface expression of these faults (fault lines) produce areas of crustal extension (**transtension**), where basins may develop. Areas of compression (**transpression**) within these same faults produce thrust faulting and uplift, resulting in uplands and, potentially, mountains (**Fig. 6.15**). Mountainous terrain may also be formed along restraining bends within strike-slip faults, such as at the Transverse Ranges in Southern California, which follow the San Andreas Fault (**Figs. 6.16, 10.17, 10.19**), and the Gobi Altai Mountains in Mongolia, along the Gobi-Tein Shan and North Gobi faults. Similar transtensional and transpressive landforms are present along the transform faults on the ocean floor. These faults may split at their ends into horsetail splays and become thrust faults, also producing mountainous topography (**Fig. 6.15**).

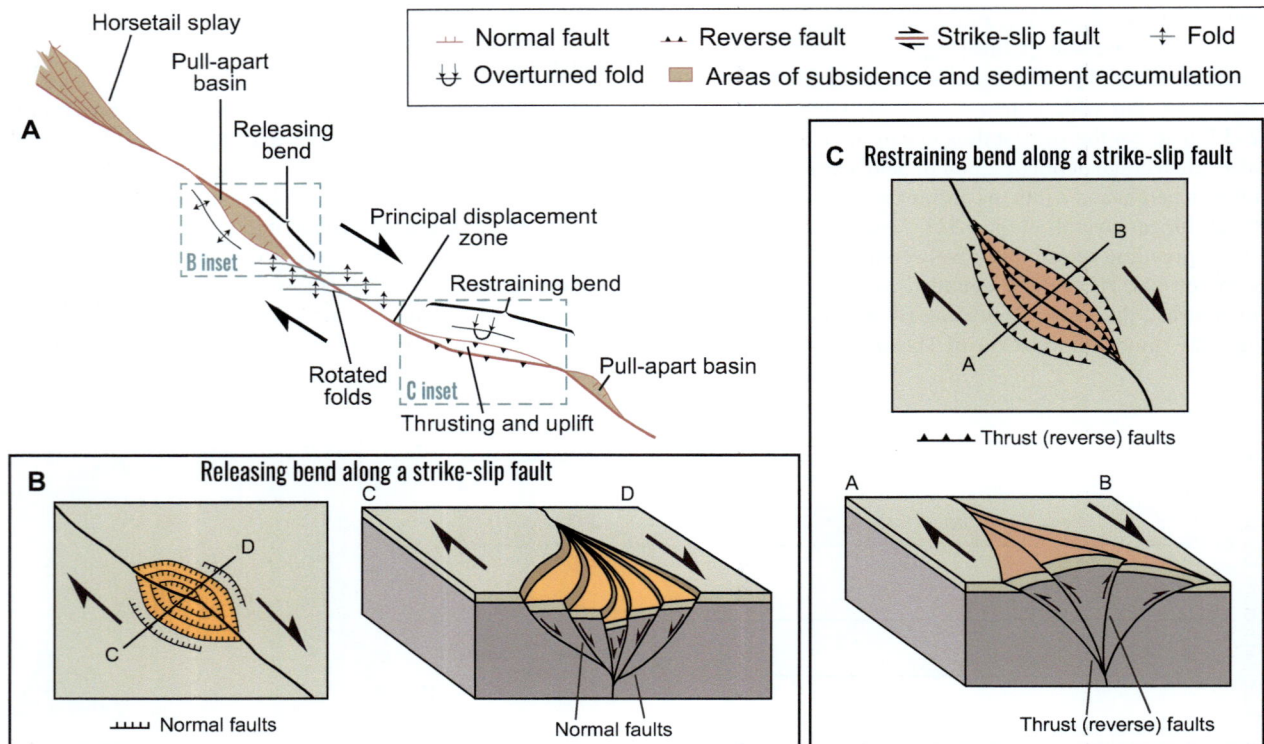

Figure 6.15 Motion along transform plate boundaries. **A.** A schematic view, looking down on the surface, of the geometry of strike-slip faults within transform settings, and their associated landforms. The insets summarize the landforms associated with strike-slip faults. Source: Used with permission of Elsevier, from Frankel and Owen (2013); permission conveyed through Copyright Clearance Center, Inc. **B.** Releasing bends result in pull-apart basins, in which crustal extension is accommodated as a suite of nested normal faults, resulting in a lowland. **C.** Restraining bends must accommodate crustal compression, resulting in uplift of the surface, and potentially, mountainous terrain.

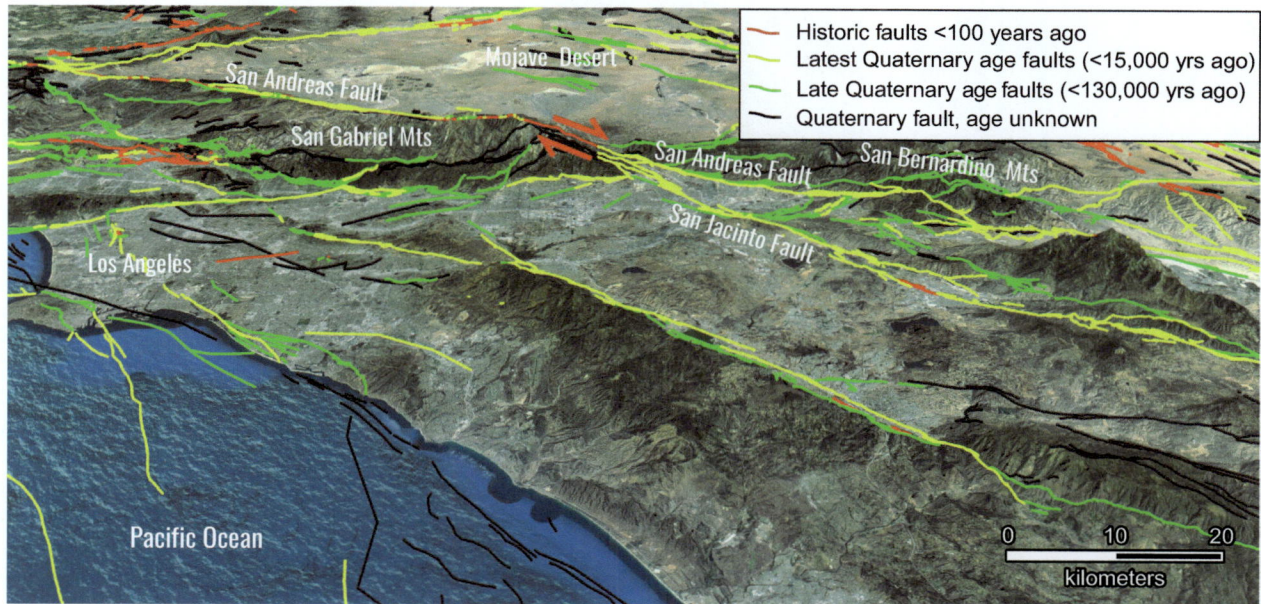

Figure 6.16 The Transverse Ranges in Southern California are an example of mountains formed within a restraining bend of a strike-slip fault. Here, the San Andreas Fault has a double bend, producing the San Gabriel and San Bernardino Mountains. This satellite image shows Quaternary-age faults, including the San Andreas (superimposed on the image). Source: Fault data are from the US Geological Survey Quaternary Fault and Fold Database.

6.4 CONTINENTAL CRUSTAL EXTENSION

Mountains also develop where regions of continental crust begin to thin, due to prolonged crustal extension. This thinning (extension) may be related to processes such as gravitational collapse, delamination, or regional warping related to hotspots. The best example of crustal extension occurs in a vast physiographic region covering much of the western United States and northwestern Mexico, called the Basin and Range Province (**Figs. 6.17, 6.18, 10.23A**). The Basin and Range is characterized by alternating up-faulted mountains and broad basins which have been down-faulted. The region is bounded on the west by the Sierra Nevada Mountains, which include Mt. Whitney (4,421 m asl), the highest peak in the contiguous 48 states. Death Valley, with the Basin and

Range, contains the lowest point on land in North America (Badwater), at 86 m below sea level (**Fig. 6.17C**). Notable valleys include Owens, Panamint, and Saline, and important mountain ranges here include the White and Sandia mountains, as well as the Grand Tetons (**Fig. 6.2B**).

The Basin and Range province is the result of the extension and thinning of the lithosphere that began ≈17 million years ago. This thinning produced **listric faults** (**Fig. 10.22C**). These types of faults dive into the crust as shallow angles, bounding the ranges and defining the margins of the basins. The average crustal thickness in this area is 30–35 km, as compared to 25–70 km elsewhere. Similarly, the base of the lithosphere here is only 60–70 km deep, whereas it is 40–280 km deep elsewhere. Geologists argue that the region experienced about 100% lateral extension (varying from ≈60

Figure 6.17 Typical Basin and Range landscapes. **A.** Aerial view of several parallel, fault-block mountain ranges and the intervening basins between the Last Chance Range and Eureka Valley (in Death Valley National Park, California, USA) and the Sierra Nevada Mountains. The Eureka Sand Dunes can be seen in the lower left part of the photo. Source: M. Miller. **B.** Extensive valley floors and bounding mountain ranges of the Mina Valley in Nevada, USA. Source: L. Owen. **C.** Badwater, the lowest point in North America (Death Valley, California) in the middle ground, showing the low angle normal fault that bounds the Black Mountains to the right. Source: L. Owen.

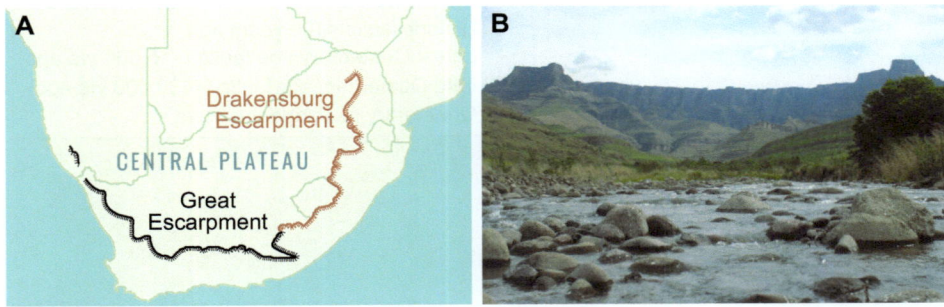

Figure 6.18 The Drakensburg Escarpment. **A.** A map of South Africa's Great Escarpment and the Drakensburg Escarpment. **B.** View of the Drakensberg Amphitheatre from the upper Tugela River, KwaZulu-Natal, South Africa, with the escarpment in the background. Source: Bothar, Public domain, via Wikimedia Commons.

to 300 km). Details on exactly how crustal extension can develop into these types of block-faulted mountains are discussed in Chapter 10.

6.5 PASSIVE PLATE MARGINS

Impressive mountain ranges can also be present along passive margins, which occur where oceanic and continental crusts meet, but there is no differential movement between the crustal plates. The oceanic plate is attached to the continental plate, and both plates move together, in the same direction. Thus, tectonism is not presently or directly responsible for mountains at such locations.

Mountain ranges at passive margins include, for example, the Western Ghats, which rise to 2,695 m asl and stretch some 1,600 km along the western coast of India, the Serra do Mar along the southeastern coastline of Brazil, which stretches ≈1,500 km and rises to 2,255 m asl, and the >1,000 km-long Drakensburg Mountains in South Africa, which rise to 3,482 m asl (**Fig. 6.18**). Large **escarpments** that descend to the coast are present on the coastal sides of these mountains. The formation of mountains on passive late margins has been greatly debated, but likely relates to continental breakup, coastal erosion, and progressive inland migration of the escarpments. These processes lead to crustal **unloading** inland, coupled with loading in the ocean and along coastal regions associated with deposition of sediment from the eroded mountains. The differential loading and uploading results in warping of the crust over a distance of several hundred kilometers, a process known as **flexural isostatic uplift**, which can form mountains (**Fig. 6.19**).

6.6 MOUNTAIN BUILDING AT HOTSPOTS

"Plumes" of magma are argued to form in the mantle. Here, upwelling of molten materials carry heat toward the surface from deeper within the Earth. Once these plumes reach the surface, they cause extensive eruptions of basaltic lava in the form of oceanic plateaus and continental flood basalts. **Hotspots** are the surface expressions of these relatively stationary mantle plumes. Their presence is inferred by anomalous volcanism at a location that is usually far from a plate boundary (**Fig. 6.4**). Hotspot trails are the evidence left behind when a plate continues to move across a subsurface plume (**Figs. 7.12, 7.13**). Dozens of hotspots occur around the world; the Hawai'ian hotspot is the most well-known. The Hawai'ian hotspot has been active for at least 70 Ma, producing a volcanic chain that extends ≈6,000 km across the floor of the Pacific Ocean (**Fig. 7.13**). Many volcanoes formed at hotspots within the ocean basins are far more massive than those at subduction zones (**Fig. 6.20**).

Other hotspots include the Azores and the Canary hotspots in the eastern Atlantic Ocean, and the East Australia hotspot. A particularly noteworthy hotspot track extends from India to the island of Réunion in the Indian Ocean. About 66–68 Ma, while present-day India was above the hotspot, enormous volumes of basaltic lava erupted to produce the impressive highlands of the Deccan Traps (**Fig. 7.15**).

Hotspots also develop beneath continental plates. For example, the Yellowstone hotspot has been active for at least 15 million years, producing a chain of calderas and volcanic features along the Snake River Plain that extends for ≈650 km (**Fig. 7.12**).

Hotspots may lead to uplift over a broad, circular area (doming) and then to continental extension, accounting for

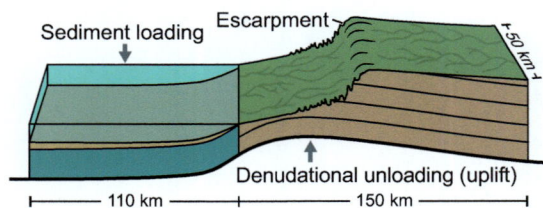

Figure 6.19 Conceptual model for the development of escarpments and mountains along passive plate margins. As the escarpment erodes inland, sediment is deposited in the coastal region, leading to unloading and loading, respectively, and flexural isostasy. Source: After Tucker, G. E. and R. L. Slingerland (1994), with permission from John Wiley and Sons.

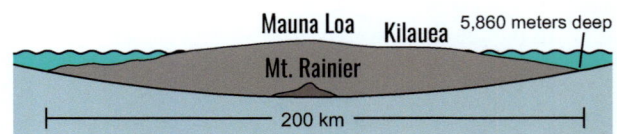

Figure 6.20 Comparison of the profiles of Mauna Loa, a hotspot volcano on the Island of Hawai'i, and Mt. Rainer in the Cascades of Oregon, a more typically sized stratovolcano. Source: Used with permission of Elsevier, from W.D. Huff, L.A. Owen (2013); permission conveyed through Copyright Clearance Center, Inc.

much of the rifting and volcanism along the East African Rift Valley and into Afar and the Red Sea. Such processes help produce isolated volcanoes such as Mt. Kilimanjaro in Kenya (**Fig. 6.21**).

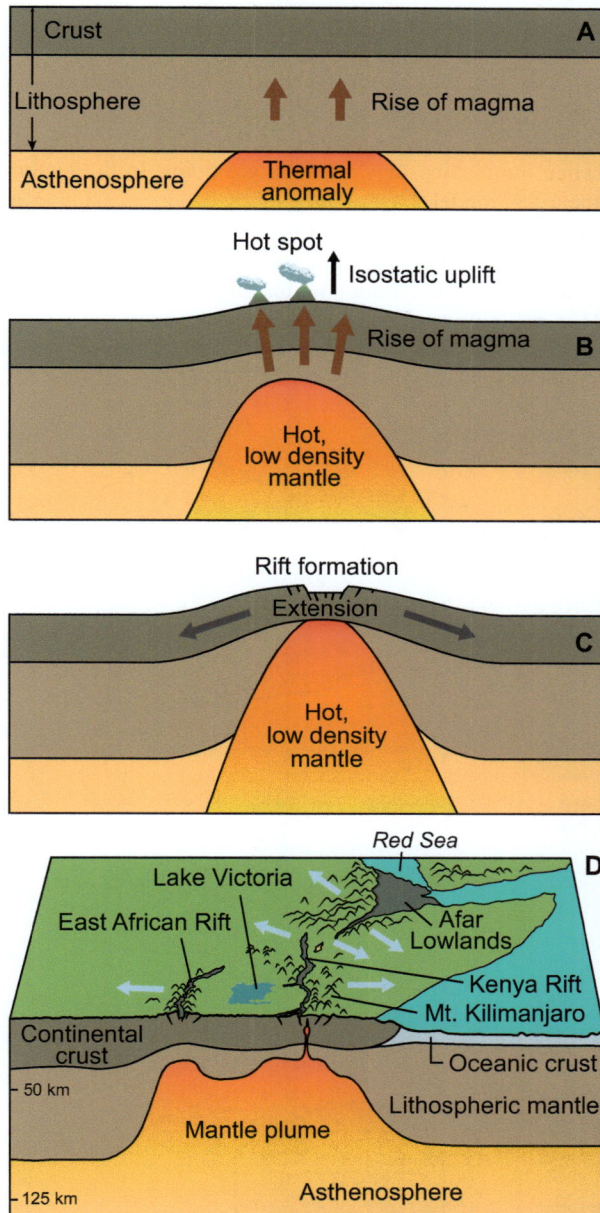

Figure 6.21 Schematic diagrams illustrating how hotspots below continental crust can produce doming, crustal thinning, and eventually, mountains. **A.–C.** Development of domal uplift, volcanism, and rifting above a developing hotspot. Source: After Summerfield (1991), reprinted by permission of the publisher (Taylor & Francis Ltd, www.tandfonline.com). **D**. Diagrammatic sketch of the East African graben system and Rift Valley. The cross-section illustrates the thinning of the lithosphere, associated with warping and volcanism. Source: After Frisch, W., Meschede, M., and R.C. Blakey (2011) with permission from Springer Nature.

6.7 OROGENIC PROCESSES

In addition to the tectonic settings related to divergent, convergent, and transform plate boundaries, other processes can also interact to form mountains. Some of these are described below.

6.7.1 Delamination

Delamination refers to the breakup, sinking, and eventual loss of a portion of the lower continental crust and mantle lithosphere, effectively separating it from the upper continental crust (**Fig. 6.22**). This process occurs in crustal convergence zones, especially where continental–continental collisions occur, such as in Tibet, as well as in areas of crustal extension. In delamination, the lower part of the continental crust is colder, and hence, denser than the asthenosphere below it. The negative buoyancy of the lower lithosphere then drives the process. Thick crustal roots beneath what used to be mountains get pulled down into the warmer mantle below. The ensuing collapse of the mountain belt forms granitic plutons such as **batholiths** and **stocks** (see Chapter 7), as the plate fragment is melted

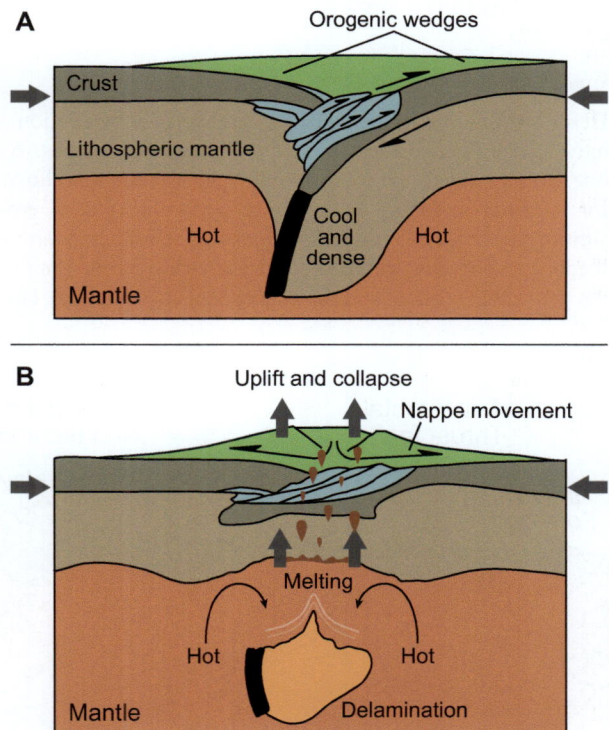

Figure 6.22 Model showing how delamination can cause orogenic collapse and orogenic root collapse. **A**. Note the cold and dense part of the continental crust that is deep within the aesthenosphere. **B**. Delamination and descent of the crustal root leads to uplift and collapse in the upper part of the collision zone. Source: After Fossen (2010), with permission from Cambridge University Press & Assessment.

by the intense heat; the roots of the mountains literally melt in the warmer mantle material below. Delamination is postulated to have been the formative process for the Sierra Nevada Range of California, as well as the Basin and Range Province and the Colorado Plateau of the western United States.

6.7.2 Crustal Extrusion and Channel Flow

The crustal extrusion/channel flow model theorizes that **ductile** (moldable) middle and lower crust material can flow and extrude laterally through a mountain, thereby "replenishing" the range as it erodes on its flanks. The model is derived from studies in the Himalayas, where geologists observed crustal extrusion/channel flow between two major faults, the South Tibetan fault system and the Main Central Thrust (**Fig. 6.23**). This extruding zone has since ground its way to the surface, and it is argued that channel flow is occurring even today along the eroding Himalayan front. This view was enhanced by geomorphologists that suggested that feedbacks between erosion and extrusive flow are closely linked. In short, uplift and erosion rates are highest in the zone of crustal extrusion, where erosion allows the crust to be more readily extruded. Thus, extrusion creates enhanced relief that in turn helps produce more precipitation, leading to increased erosion, forming a positive feedback loop.

6.7.3 Denudational Unloading

Denudation – the lowering of the land surface by erosion – may also influence tectonism in mountainous regions. As one would expect, the loss of weight by denudation allows for **isostatic uplift** of the crust. In particular, glacial erosion of the cores of mountain ranges might initiate positive feedback, whereby tectonic uplift increases the heights of the peaks that remain, and thereby enhances glaciation. This

process may be particularly applicable to the Himalayas, which are very high and have significant numbers of large glaciers.

6.7.4 Tectonic Aneurysm

The **tectonic aneurysm** model has become popular as a means of explaining regions of anomalously high topography, such as the Nanga Parbat and Namche Barwa syntaxis at the east and west ends of the Himalayan arc, and the St Elias syntaxis in Alaska (**Fig. 6.24**). A geologic **syntaxis** is a convergence of geologic structures and hence, mountains, towards a single point. In the tectonic aneurysm model, focused erosion by deeply incising rivers and glaciers, plus the mass of the mountain, helps weaken the crust. The weakened crust enhances the flow of crustal material under the mountain, leading to accelerated uplift. This uplift drives further incision, enhancing the effect – another example of how positive feedback can help form mountains.

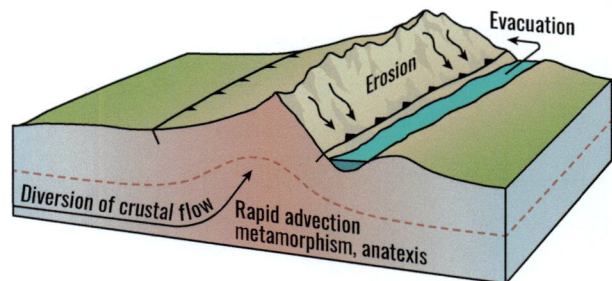

Figure 6.24 The development of a tectonic aneurysm, showing the dynamic interactions of focused erosion, topographic stresses, uplift, rapid exhumation, and thermal weakening of the lithosphere, causing deformation. Source: After Zeitler et al. (2001), © Geological Society of America, Inc., used with permission.

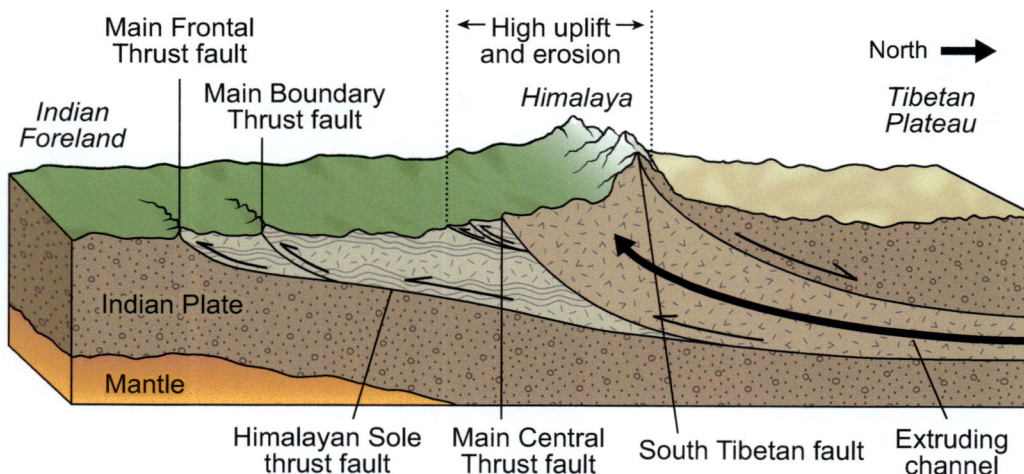

Figure 6.23 Schematic section across the Himalayas and southern Tibet, illustrating channel flow and extrusion. The greatest amount of uplift, the steepest hillslopes and channels, and the highest precipitation totals, are broadly coincident with the region of extrusive channel flow.

6.8 RATES OF MOUNTAIN BUILDING

Contemporary studies and research on the evolution of mountain systems involve most branches of geology. Studies of mountainous topography are becoming ever more multidisciplinary. Nonetheless, they all make use of direct observations and measurements, particularly estimates of the rates of vertical and horizontal displacement of the crust, and rates of denudation.

Direct observation of mountain building is challenging because tectonic processes are quite slow. However, occasionally, such as during an earthquake, the uplift of a mountain range can be directly observed and measured. Measuring uplift can be undertaken using GPS (global positioning systems) instruments, as they continuously measure elevations and lateral movements of the crust (**Figs. 6.25, 6.26**). More commonly, the denudation of a mountain during a massive storm, as sediment and rock are carried away in response to a flood, or when a landslide happens, provide data on rates of erosion and hence, loss of mass.

6.8.1 Estimating Rates of Crustal Displacement

Rates of vertical and horizontal displacement can also be estimated and quantified in the field. To do this, geomorphologists identify landforms that have been offset by faults, measure the amount of movement along the fault, and then (if possible) determine when that movement occurred. Displacement could be lateral, as at the San Andreas Fault, or vertical. To measure the movement of the latter, geomorphologists might utilize landforms that can be related to a base level such as the ocean, a lake, or a riverbed. They might also measure the heights of dated,

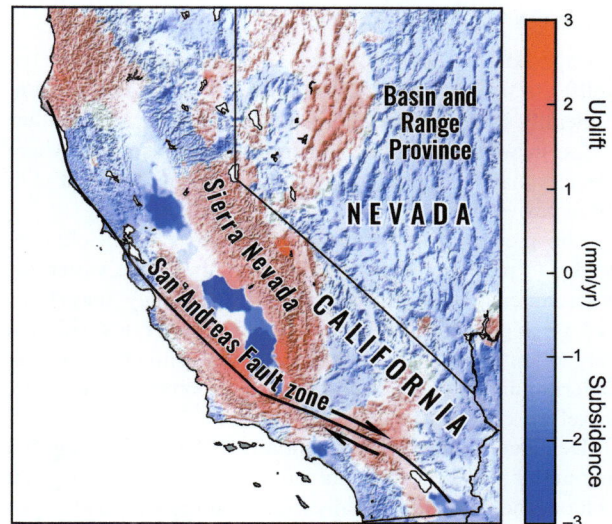

Figure 6.26 Rates of surface uplift and subsidence across California and Nevada, USA, based on GPS studies. The highest uplift rates are in the Sierra Nevada Mountains. The curved area of uplift parallel to the southern California coast follows the San Andreas Fault, resulting in the topography of the Transverse Ranges and Coastal Ranges. The Basin and Range Province is dominantly an area of subsidence, due to crustal extension and thinning. Areas of deep subsidence (dark blue) in California's Central Valley are due to decades of groundwater withdrawal for agriculture, and are not directly related to tectonics. Source: After Hammond et al. (2016), with permission from John Wiley and Sons.

uplifted, marine terraces, raised lake shorelines, and fluvial terraces, to estimate the amounts and rates of vertical displacement (uplift).

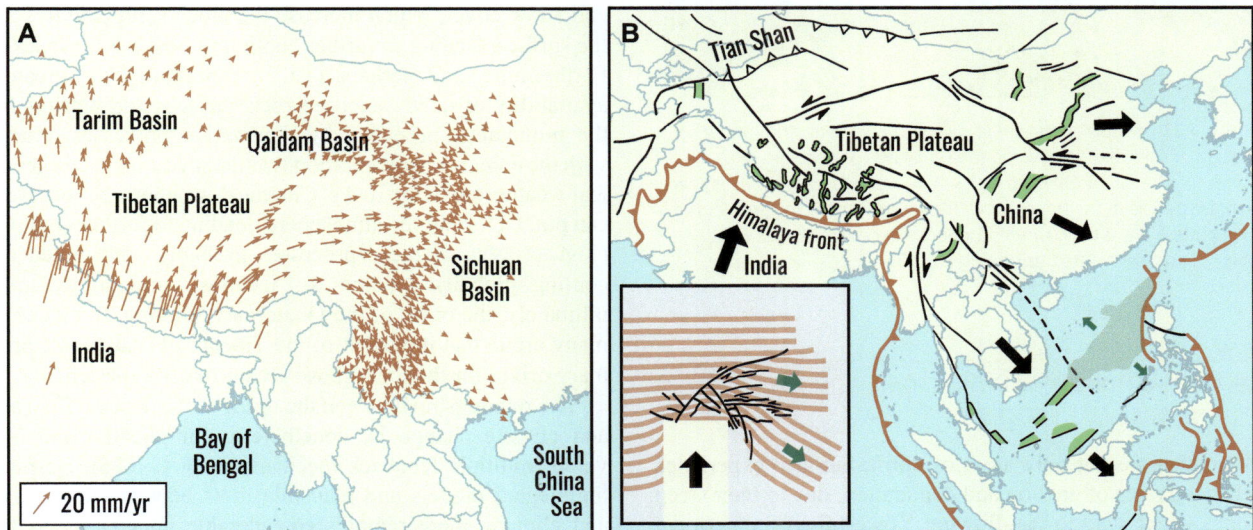

Figure 6.25 The movement of crustal plates can be determined in great detail using modern instrumentation. **A.** Arrows from global positioning systems indicate the motion of the land surface in southern Asia. **B.** Due to the continued northward migration of the Indian Plate, these data help to establish the directions of crustal block motion in southeastern Asia towards the Pacific region. Crustal extension accompanies this motion, as revealed by graben structures and metamorphic domes, shown in green. The box shows an analog experiment with plasticine where the pushing indenter and the escaping wedges duplicate the geologic structures in southeastern Asia.

6.8.2 Estimating Rates of Denudation

Juxtaposed to geomorphologists determining rates of uplift are their attempts at estimating rates of **denudation**. Generally speaking, rates of denudation are greatest in softer rocks, in wetter climates, and in higher landscapes, where glaciers may be active and where rivers flow fast and with more power.

The greatest mountain systems traverse many climatic belts and may include along their length a wide variety of environmental and geomorphic settings. They may have had a rich glacial history as well. Denudation rates will then vary dramatically across these different climatic regimes. Determining rates of denudation for mountain ranges is, therefore, challenging.

Denudation rates are essentially calculated using the amount of sediment exiting a mountain system over a given unit of time. Sediment usually exits mountainous landscapes in rivers. Thus, most denudation rate estimates are based on fluvial data. Modern rivers can provide contemporary denudation rates, whereas studies of former rivers (in their terrace deposits or bedrock surfaces) can take this method back into the geologic past (**Fig. 6.27**). Lastly, we can examine the rates at which rocks within the mountains themselves are eroding, using cosmogenic nuclides (see Chapter 3).

Figure 6.27 A bedrock (strath) terrace that represents the former riverbed of the Chandra River in the Himalayas of northern India. Higher terraces are present up to 15 m above the river, slightly higher than the people in the upper right. The terraces mark the progressive downcutting (≈3 mm/yr) of the Chandra River over the past 6,000 years, having been dated using ^{10}Be cosmogenic nuclides. Source: L. Owen.

6.9 INTERACTIONS BETWEEN TECTONICS, CLIMATE, AND SURFICIAL PROCESSES

Mountains are not only areas of uplift, but also experience enhanced rates of erosion and denudation, due to their steep slopes which lead to fast-flowing, powerful rivers, periodic glaciation, and the potential for mass movement events. Mountain ranges also commonly have an altitudinal precipitation gradient (more precipitation at higher elevations) and drier climates (termed "rain shadows") on their leeward slopes. Together, these environmental factors make mountainous areas some of the most geomorphically active and complex places on Earth.

Dramatic climatic changes, most notably the cooling of Earth's climate, have taken place throughout the Cenozoic, particularly during the Quaternary Period (see Chapter 4). These climate changes have caused the magnitude and frequency of surficial processes in mountainous regions to vary widely over time. Geologists have postulated that the uplift of large mountain ranges, particularly the Himalayas, may have *contributed to* these climate changes by affecting global atmospheric circulation patterns, initiating and enhancing monsoons, and altering biogeochemical cycles. How does this work?

Uplift of the Tibetan Plateau may have driven changes in global circulation, forcing cold, polar air into critical locations in the northern hemisphere, particularly North America and northwestern Europe, where large ice sheets were able to develop. The high elevations of the Himalayas would also have led to the formation of the South Asian **monsoon** – a rainy season that occurs there every summer, coupled with the winter dry season (see Chapter 4). Monsoon rains are enhanced as moisture-laden air masses are drawn onto the high plateau, cooling as they rise. Cooler air cannot hold as much moisture, and thus, torrential rains can result. The South Asian monsoon was likely present in the region before the formation of the Himalayan Plateau, but uplift intensified it. Uplift of the Tibetan Plateau also enhanced its **rain shadow**, creating aridity across the northern Plateau and Central Asia. Increased aridity is usually accompanied by higher annual and daily temperature ranges, which enhance weathering rates. All of these changes would have fostered greater denudation rates across the Plateau.

Conversely, the climate of the Tibetan Plateau would have cooled as it rose. Cooling would have enhanced glaciation and snow cover, which increase the albedo (reflectivity) of the surface, leading to further cooling (a positive feedback). Furthermore, and perhaps just as importantly, the greater availability of newly exposed rock surfaces and detritus in the mountains, produced by denudation processes, along with the enhanced precipitation, would have enhanced chemical weathering in the rocks. Chemical weathering processes can pull CO_2 from the atmosphere, leading to further cooling (**Fig. 11.29**). All these processes probably contributed to cooling globally, leading to "icehouse" conditions and, ultimately, the onset of the Quaternary Ice Age. In essence, many argue that the uplift of the Tibetan Plateau was a primary driver for the Ice Age we are currently experiencing.

The relationship between the uplift of the Tibetan Plateau and climate change is, nonetheless, complicated and involves multiple feedback mechanisms (**Fig. 6.28**). Despite the elegant models and plausible links between this uplift and climate, there remains considerable uncertainty as to whether the proposed mechanisms can explain progressive cooling throughout the Cenozoic and the onset of the Ice Age. One issue that plagues the models involves the timing of uplift of the Tibetan Plateau and other plateaus, such as in the American West. Geologists are still struggling to

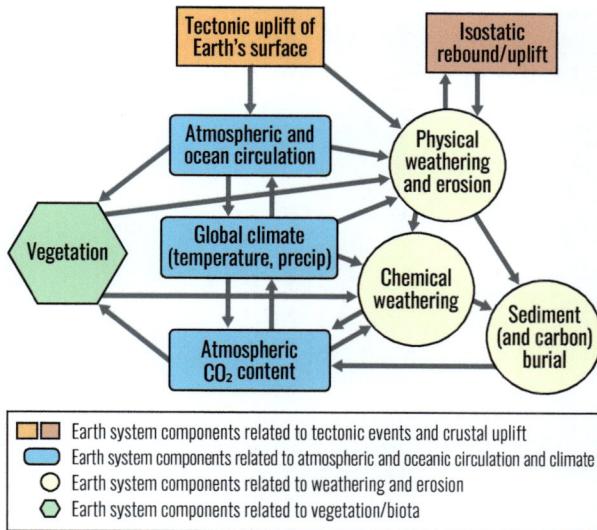

Figure 6.28 A model of the many potential interactions and feedbacks between tectonic uplift and Earth's climatic and environmental systems.

thoroughly understand when these uplift events occurred; the timing of the glaciation on Earth is better understood (see Chapter 4). Plus, the complex feedbacks involved are not fully understood, and the rates of weathering and drawdown of atmospheric CO_2 are not well established. Furthermore, many other tectonic events that were occurring elsewhere on Earth, such as the opening and closing of oceanic gateways and passages, and volcanism during the Cenozoic, could have helped drive recent climate changes as well.

6.10 HAZARDS IN MOUNTAINOUS TERRAIN

As people live, work, and play in mountains, they become exposed to the processes that have shaped those landscapes. Exogenic processes that constitute natural hazards

in mountains include landslides, snow and ice avalanches, floods, and dust storms. Endogenic hazards are most common in geologically young, active mountain belts. These include earthquakes and volcanic eruptions that commonly occur at plate boundaries (**Fig. 6.29**). Mountains might be thought of as uplifting – being thrust upward – each time there is an earthquake. This uplift may be centimeters to meters at a time, and is often associated with faulting. Tectonic activity and its resultant earthquakes are more common in young mountain belts.

Earthquakes develop due to built-up stresses in the crust. The stress is released by displacement along a fault, as the rocks break and slide past each other. If the fault reaches the surface, the displacement can often be seen as a **fault scarp** (**Fig. 6.30**; see Chapter 10). Fault scarps form primarily at locations where rocks have been suddenly uplifted. They can provide important information on the vertical motion associated with a faulting event.

Earthquakes can also contribute to the denudation of mountains by triggering landslides and other mass movements (**Fig. 6.31**). The 2008 moment magnitude (M_w: the conventional scale for measuring earthquake magnitude) 7.9 Wenchuan earthquake on the margin of Tibet, and the 2015 M_w 7.8 Gorkha earthquake in the Himalayas of Nepal, triggered ≈200,000 and ≈50,000 landslides, respectively. The effects of earthquakes can be devastating to the people living in the mountains and adjacent regions, as dwellings collapse or are buried. And of course, landslides can kill people and destroy their possessions.

The highest magnitude earthquakes are generated on the largest faults, which in turn create the greatest amount of displacement. The largest faults are associated with subduction zones, such as those along the Andean and Alaskan margins of the Circum-Pacific mountain systems. Indeed, these regions have experienced the greatest earthquakes on record. Yet, similar large faults bound other mountain ranges, such as the Main Frontal Thrust along the edge of

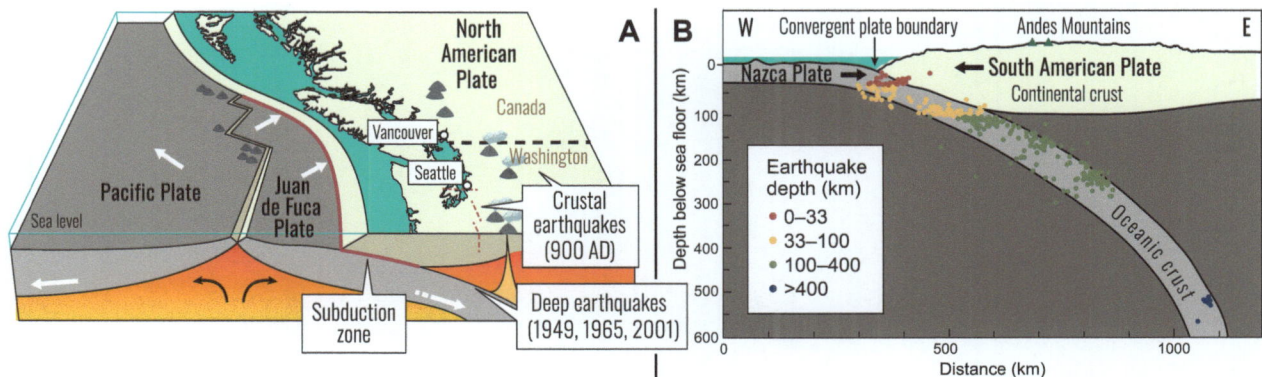

Figure 6.29 Subduction and earthquakes. **A.** The relationship between subducting Juan de Fuca Plate (beneath the North American Plate) and historical earthquakes (dates provided) and volcanoes. The "deep earthquakes" arrow points to the magnitude (M_w) 6.8 (1949), M_w 6.5 (1965), and Nisqually M_w 6.8 (2001) earthquakes, which caused >$100 million, $50 million, and $500 million in damage, respectively. Recent growth and development in the Puget Sound area increase the likelihood of costly losses from future earthquakes. Source: After the Washington Division of Geology and Earth Resources (2008). **B.** Schematic representation of the subduction of the Nazca plate under the South American plate, which has helped form the Andes Mountains. The diagram illustrates the distribution, by depth and location, of earthquakes, as they relate to the subducting plate. Source: Modified from the original by Anne E. Egger as published by Visionlearning.

Figure 6.30 Fault scarps (at arrows). **A.** Fault scarp produced during the 1915 M_w 6.8 Pleasant Valley earthquake in Pleasant Valley, Nevada, USA. **B.** Fault scarp produced by the 2010 M_w 7.2 El Mayor–Cucapah earthquake in Baja California, near the Mexico–US border. **C.** The impressive Sierra Nevada Frontal fault that bounds the eastern side of the Sierra Nevada Mountains, which formed due to the combined impact of tens to perhaps hundreds of earthquakes. Note the road in the lower right, for scale. Source: L. Owen.

Figure 6.31 Earthquake-triggered landslides produced during the October 8, 2005, M_w 7.6 Kashmir earthquake in northern Pakistan. **A.** The Hattian Bala landslide buried three villages, killing ≈700 people. **B**. Scars left behind after extensive rock avalanches, near the town of Muzaffarabad. The mountains rose 1–5 m during the earthquake, ≈70,000 lives were lost, and ≈1,000,000 individuals were left homeless. Source: L. Owen.

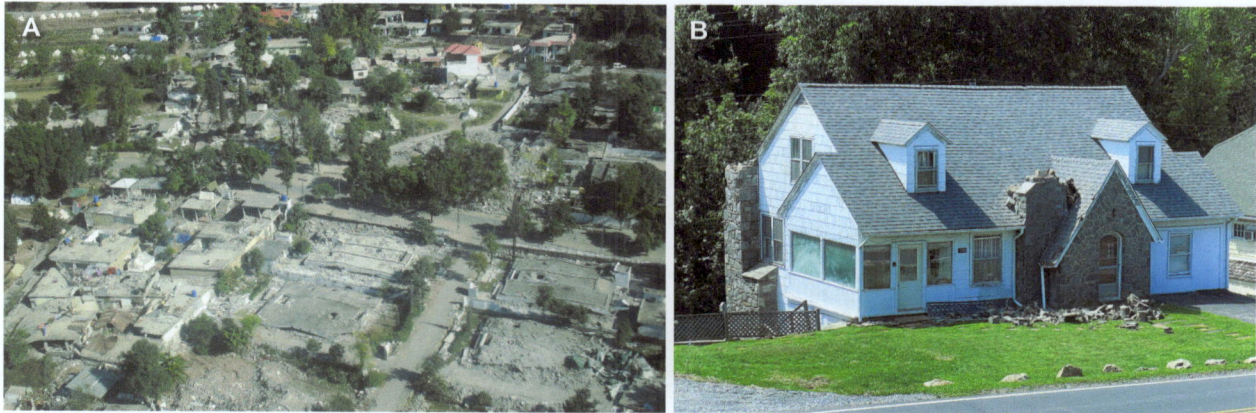

Figure 6.32 Damage to buildings due to earthquakes in both a young mountain and an ancient mountain belt. **A.** Buildings flattened during the October 8, 2005, M_w 7.6 Kashmir earthquake in Northern Pakistan. **B.** House condemned due to earthquake damage during the August 9, 2010, M_w 5.1 earthquake in Sparta, in the Blue Ridge Mountains of North Carolina, USA. Note that the chimney has fallen. The house foundations, basements, and walls were all cracked and unrepairable. Source: L. Owen.

the Himalayas. Geologic studies of sediments offset by this fault show that earthquakes have occurred in the past here and will, of course, occur again (**Fig. 6.32A**). Ancient mountain belts also experience earthquakes, but these are generally of smaller magnitude. Nonetheless, occasionally, larger earthquakes even occur within ancient mountain belts, such as the August 9, 2010, M_w 5.1 Sparta earthquake in North Carolina, USA (**Fig. 6.32B**).

REVIEW QUESTIONS

6.1   Where are Earth's major mountain ranges located in relation to tectonic plate boundaries?

6.2   What are the three major types of tectonic plate boundaries, and what geologic processes occur at each?

6.3   What are the main tectonic settings associated with Earth's young, continental-scale mountain belts?

6.4   How do unloading and isostatic adjustment contribute to the formation of mountainous topography?

6.5   What are the various layers that comprise the interior of our planet? For each, describe its physical characteristics.

6.6   Where are volcanic mountain belts most common, and what kind of geologic setting is typical for their formation?

6.7   How do mountains form within continents, far from plate boundaries (intracontinental settings)?

6.8   What are the differences between crustal transpression and transtension?

6.9   How did the Himalayas and the uplands of Tibet form?

6.10   What is the longest mountain range on Earth and how did it form?

6.11   Describe the main characteristics of a mountain belt/ orogen.

6.12   How do erosion and denudation contribute to mountain building?

6.13   What is a "hotspot," and what types of mountains form at hotspots?

6.14   What is a tectonic aneurysm, and how can it lead to the formation of mountains?

6.15   How do tectonics, mountain building, and erosion influence climate?

6.16   What are some of the main natural hazards associated with mountainous terrain?

FURTHER READING

Frisch, W., Meschede, M., and Blakey, R. C. 2011. *Plate Tectonics: Continental Drift and Mountain Building.* Springer.

Lamb, S., 2006. *Devil in the Mountain: A Search for the Origin of the Andes.* Princeton University Press.

McPhee, J. 1996. *Annals of the Former World.* Douglas and McIntyre, Ltd.

Searle, M. 2013. *Colliding Continents: A Geological Exploration of the Himalaya, Karakoram, and Tibet.* Oxford University Press.

7 Landforms on Volcanic and Plutonic Rocks

Warren D. Huff and Lewis A. Owen

Volcanoes are exciting – yet dangerous – features, capable of reworking entire landscapes overnight. With more than a dozen volcanic eruptions occurring at different locations on Earth at any one time, volcanoes are a very real part of our world. Each of these volcanoes, and the many extinct ones around the world, has a different eruptive history. Some burst to life in explosive eruptions, like the 1991 eruption of Mt. Pinatubo in the Philippines. Others bubble up rivers of lava in what is known as an effusive eruption, like the 2018 activity on Hawai'i's Kilauea volcano.

This chapter will examine the landforms associated with **endogenic** (Earth's internal) processes, particularly those associated with heat and which lead to eruptions of molten rock, hot water, and steam. The latter are referred to as **hydrothermal processes**. Therefore, we will not only discuss volcanic and plutonic landforms, but also the intriguing features associated with hydrothermal processes, such as geysers and hot springs.

7.1 VOLCANIC MATERIALS AND LANDFORMS

Volcanoes exist on every continent, including Antarctica, some 1,500 of which are still considered potentially active. About 75% of active volcanoes occur around the Pacific Rim, in an area called the **Pacific Ring of Fire** (Fig. 7.1), threatening about 50 million people, mostly in Indonesia,

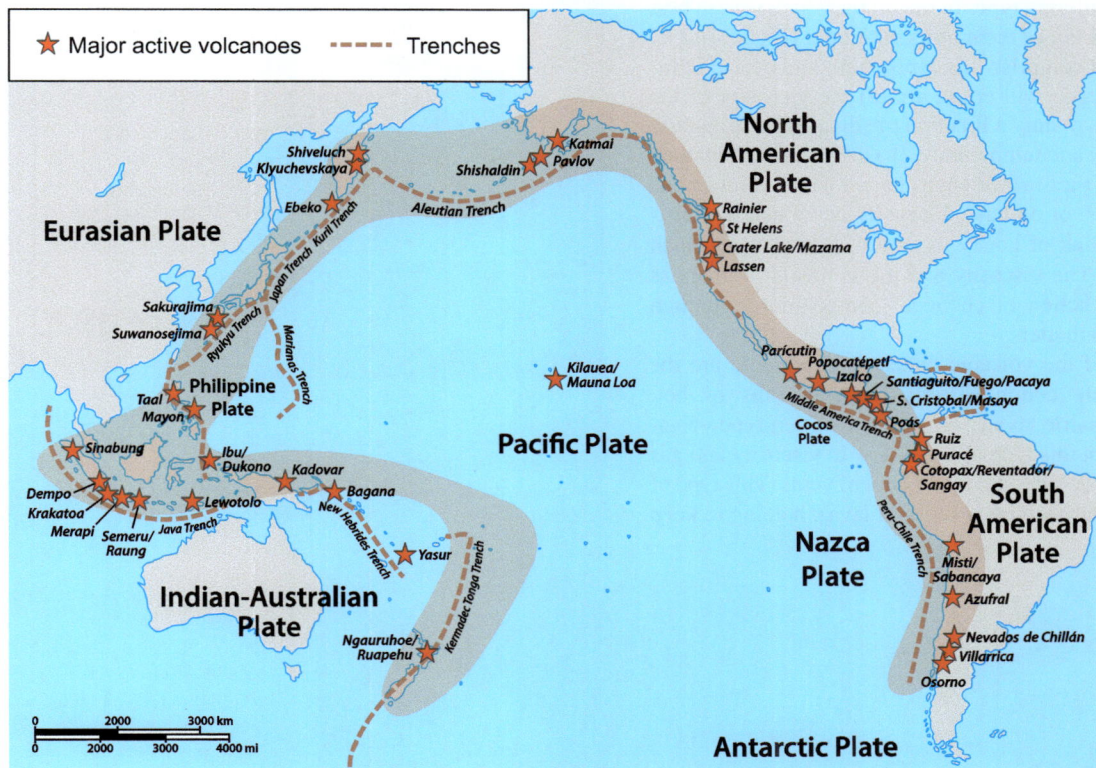

Figure 7.1 The Pacific Ring of Fire contains a series of volcanoes, all near the margins of the Pacific Ocean. Most of these volcanoes are associated with plate boundaries and subduction zones, which are typically manifested as deep oceanic trenches. Source: Based on Rainer Lesniewski / Alamy Stock Vector.

Figure 7.0 Mount Vesuvius, near modern-day Naples, Italy, is one of the most active and destructive volcanoes on Earth. Source: Petr Svarc / Getty Images.

Japan, Russia, Ecuador, Chile, Philippines, New Guinea, New Zealand, Nicaragua, and the United States. About 150 of Earth's active volcanoes are found within the United States. Some 80% of our planet's surface is in some way tied to volcanic processes, or **volcanism**. Throughout Earth history, volcanic and magmatic processes have continued to recycle geologic materials as a major component of the geologic/rock cycle (**Fig. 5.8**).

Volcanic landforms are produced by the crystallization of **lava** – molten rock that pours onto the surface, plus **volcanic ash** and other materials explosively ejected and blown out of volcanic **vents (pyroclastic debris)**. Volcanic materials extruded onto the surface will form the extrusive igneous rocks that are the signature geology of many types of volcanic terrains (see Chapter 5). The explosive force of volcanoes helps to form mountains as well as craters. Lava flows and ashfalls also produce dramatic and unique landscapes.

Volcanic and **magmatic** (associated with underground molten rock, or magma) processes are a key part of Earth's geomorphic system. Magma that cools and crystallizes in the subsurface forms **plutons**. Volcanic landforms change continuously through the interplay between constructive processes such as eruptions, and the denudational processes that erode them and wear them down. As a result, many plutons are now exposed at the surface as volcanic landforms.

Volcanic activity comes in a wide variety of "styles" and includes many different types of materials, which we can best examine by thinking of them as a continuum (**Fig. 7.2**). The main factors that comprise this continuum, and hence, influence how a volcano will erupt, are (a) the viscosity of the magma, which is mainly a function of silica content, amount of trapped gases, and temperature, (b) the volume of magma erupted, and (c) the surface environment, such as under the ocean or glaciers, or on land. **Viscosity** is defined as resistance to flow. Think of honey vs water – water has a much lower viscosity. The viscosity and gas contents of magmas are usually a function of chemical composition, although temperature also matters.

At one end of the volcanic materials continuum are the gentle, seemingly continual, **effusive eruptions** of hot, non-viscous, **basaltic lava** (**Fig. 7.3**). Hawai'ian-type volcanoes tend to erupt this type of lava – low in viscosity and gas content, flowing or bubbling readily out of vents. This type of lava is often hotter than other types of lavas, helping to keep its viscosity low. Basaltic lavas are rich in iron and magnesium and relatively low in silica. Hence, they are sometimes called **mafic** (Ma – magnesium, Fi – iron [Fe]) lavas. Gases present in these lavas can readily escape, and thus, little pressure builds up. Lava flows of this type rarely cause loss of life, because they flow predictably and comparatively slowly.

Rhyolitic lavas are at the other end of the continuum. They are thick and sticky (viscous), such that trapped gases cannot as easily escape (**Fig. 7.2**). Gases escape violently when the pressures build up and overcome the mass of the rock above, sometimes leading to an **explosive eruption**. Explosive volcanoes are what most people envision when the term "volcano" is mentioned. The eruption of Mt. St. Helens in Washington State on May 18, 1980 is a classic example of this type of eruption (**Fig. 7.4**). Another descriptive name for these lavas is **felsic** (Fe = iron, Si = silicon). The comparatively high silica contents of rhyolitic lavas render them highly viscous, and so, as the magma moves upward and into the vent of the volcano, the pressure drops, and the gases attempt to escape. However, the high-viscosity magma resists expansion. Eventually, enough gas bubbles

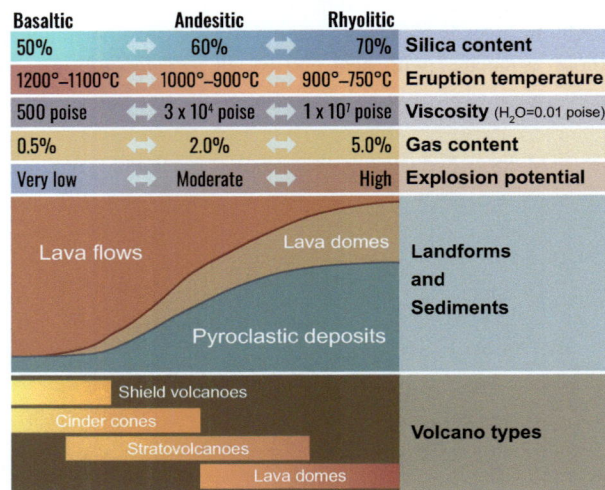

Basaltic	Andesitic	Rhyolitic	
50% ⟷	60% ⟷	70%	**Silica content**
1200°–1100°C ⟷	1000°–900°C ⟷	900°–750°C	**Eruption temperature**
500 poise ⟷	3×10^4 poise ⟷	1×10^7 poise	**Viscosity** (H_2O=0.01 poise)
0.5% ⟷	2.0% ⟷	5.0%	**Gas content**
Very low ⟷	Moderate ⟷	High	**Explosion potential**
Lava flows		Lava domes	**Landforms and Sediments**
		Pyroclastic deposits	
	Shield volcanoes		**Volcano types**
Cinder cones			
	Stratovolcanoes		
		Lava domes	

Figure 7.2 The continuum of magma types, from basaltic through andesitic through rhyolitic. Each type of magma has a range of characteristics with respect to silica content, temperature, viscosity, and gas content, which greatly influence its potential for explosive eruptions, and lead to distinctive types of volcanic features.

Figure 7.3 Effusive eruptions, commonly known as Hawai'ian-style eruptions, of basaltic lava from the 2018 eruption of Kilauea on the island of Hawai'i. Source: US Geological Survey.

Figure 7.4 The eruption and pyroclastic flow that formed during the May 18, 1980 eruption of Mt. St. Helens in Washington, USA. Source: US Geological Survey.

expand to blow the magma into fragments that are ejected into the atmosphere.

Material explosively ejected from a volcano is called **pyroclastic debris** or **tephra**. Pyroclastic debris may be juvenile (formed of magma involved in the eruption) or accidental (derived from preexisting rock, blown out during the eruption). The various types of pyroclastic debris are categorized largely by size. The largest fragments, >64 mm in diameter, are **bombs** and **blocks**. They form as large blobs of lava and rock fragments are ejected, cooling as they fall. Blocks are defined as having rough and jagged edges, whereas bombs are more elongated and fluidal/spiral, probably because they were ejected in a more molten state, twisting and turning as they flew through the air. Slightly smaller are **lapilli**, typically in the small gravel, or pea, size range. Most people can easily conceptualize **volcanic ash**, a common type of pyroclastic debris. Volcanic ash is unlike ash from wood fires; instead, it is composed of tiny shards

of volcanic glass and ground-up rock, often mixed with toxic, sulfur-rich gases. By definition, ash shards are sand size or smaller, and they can form into a low-density, light-colored, porous rock called **tuff**. In summary, depending on the explosivity of the eruption, pyroclastic debris can range in size from tiny particles of ash to house-sized boulders.

Explosive volcanic eruptions can be both dangerous and deadly. They can blast out clouds of hot tephra from a central vent or from vents on the side of the volcano. Fiery ash clouds, called **pyroclastic flow**s, are a major hazard associated with these types of eruptions. They race down mountainsides, destroying almost everything in their path. Less violent ashfalls can nonetheless cover vast parts of the landscape, suffocating plants, animals, and people. Hot ashfalls can cause fires that burn down homes and forests. When hot volcanic materials mix with water from streams or melted snow and ice, mudflows called **lahars** can form and race down river valleys as raging floods of soupy, gray sludge.

Magmas may cool above or beneath the surface, forming **igneous rocks**. Magmas that cool belowground form one end of the igneous rock continuum – with large mineral crystals, easily visible to the naked eye (**Fig. 5.7A**). Such rocks are referred to as **phaneritic**. The slow cooling associated with belowground magma allows ample time for the large mineral crystals to grow. The classic example of a phaneritic rock is **granite**. Rocks formed from lavas are characterized by fine-grained textures because their rapid cooling at or near the surface did not allow enough time for large mineral crystals to grow. Igneous rocks with this fine-grained texture are called **aphanitic**. The most common extrusive volcanic rock, an aphanitic one, is **basalt**. Lavas that cool extremely rapidly, typical of those quenched in water, have a glassy texture; they cool so fast that mineral crystals cannot form, forming volcanic glass instead. Glasses do not have an orderly atomic structure; in the strict sense, therefore, they lack minerals. Volcanic glass is called **obsidian**.

Figure 7.5 The two main, end-member, lava types. **A.** Pāhoehoe lava formed during the 2017 eruption of Kilauea. Source: Ekrem Canli, CC BY-SA 4.0, via Wikimedia Commons. **B.** 'A'ā lava photographed on January 29, 2008 in the Royal Gardens subdivision on the island of Hawai'i. Source: US Geological Survey.

More than 90% of all volcanic rock on Earth is basalt. Names given to the various forms of hardened lava (which *is* basalt, in this case), based on their appearance, come from the language of the indigenous Hawai'ian people. The most common terrestrial lava texture is **pāhoehoe**, with a rope-like pattern, formed as the liquid lava continues to move underneath, while the thin "skin" above bends and deforms (**Fig. 7.5A**). Lava that flows more slowly develops a thicker skin, which may later crack and break to form sharp edges. This lava, with its surface made of sharp, angular, and jagged blocks of basalt, is referred to as **'a'ā** (**Fig. 7.5B**). The cooled surface layer breaks up into these small, jagged fragments as molten lava below continues to flow. In this case, the hard surface layer is too thick to bend/deform (as in pāhoehoe), and so it breaks randomly.

7.2 VOLCANISM AND PLATE TECTONICS

Volcanic eruptions tend to occur only in certain places; they do not occur randomly. Driven by endogenic processes, volcanism is largely associated with **tectonic plate** boundaries, but also with (a) anomalously hot areas/plumes (hotspots) of Earth's interior, and (b) areas where Earth's crust is extending and thinning. To better understand volcanism, we must first turn to a discussion of plate tectonics.

Plate tectonic processes involve the movement of the rigid outer layer of the Earth. These **lithospheric plates** comprise Earth's outer crust. Earth's lithospheric plates essentially "float" on a hotter, softer layer, the **asthenosphere** (**Fig. 7.6**). Lithospheric plates are of two main types: (a) thinner and denser oceanic plates and (b) thicker but less dense continental plates (**Fig. 7.6**). As the plates move, they spread apart, collide, or slide past each other; most volcanism occurs at or near these boundaries.

A force that has long been thought to be a driving mechanism behind plate motion is **mantle convection**. Convection currents circulate throughout the mantle, caused by differences in temperature at depth. Over millions of years, hot material from deep within the Earth rises slowly through the mantle. As it rises, the hot material cools and begins to sink.

These convection currents act like a conveyor belt, moving the lithospheric plates above.

7.2.1 Volcanism at Divergent Plate Boundaries

Divergent plate boundaries occur where lithospheric plates are moving apart or spreading (**Fig. 7.6**). On continental crust, a **rift** or **rift zone** forms where plates diverge, leading first to crustal thinning, and then the development, on the surface, of an elongated **rift valley** bounded by normal faults (**Fig. 7.7**). On land, divergent boundaries, also called **spreading centers**, initially produce broad rift valleys, where volcanism can occasionally be active, as new crustal material is being formed from magma below. The East Africa Rift Valley is an example. Rift valleys and rift zones that form within continental plates often eventually open up far enough, and become low enough, to become ocean basins. The Red Sea represents an old and well-expressed rift zone, already invaded by the Indian Ocean.

Oceanic crust forms at spreading centers (**Figs. 7.6, 7.7**). Eventually, a broad ocean will develop, as has happened in the Atlantic. At the center of this oceanic plate is its **mid-oceanic ridge**. These ridges connect to form a single global mid-oceanic ridge system that is part of every ocean – making the mid-oceanic ridge system the longest mountain range in the world (**Fig. 6.4**). More volcanoes are located here than on land, although they are rarely seen and thus, not included in the count of active volcanoes. Most mid-oceanic ridge systems were once a rift valley on a continent (see Chapter 6). **Transform plate boundaries** form at mid-oceanic ridges, developing where plates slide past each other. These boundaries are only rarely the site undersea of volcanic activity.

In the deep ocean bottoms, basaltic eruptions at the mid-oceanic ridges produce new seafloor and oceanic crust. These underwater eruptions – developed at spreading centers – are the most volcanically active features on Earth. Here, effusive, lower energy, basaltic lavas flow out of fissures and harden into blob-shaped **pillow lavas**. Their rounded forms develop as lava is suddenly quenched when it comes into contact with sea water. The outer rim of the

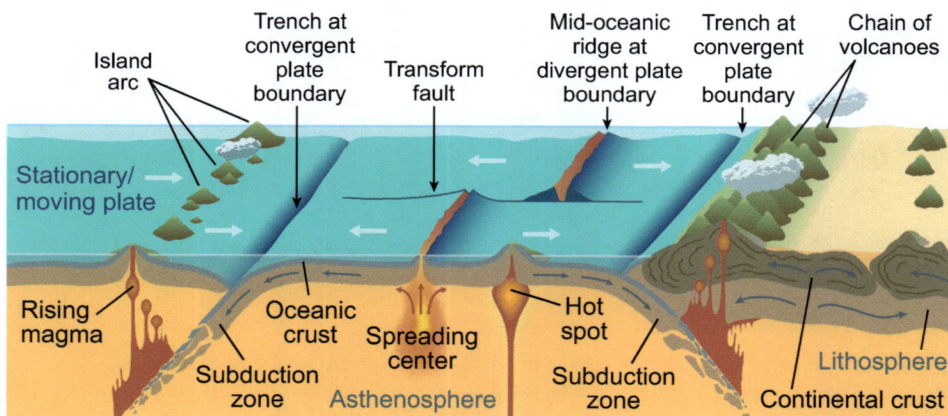

Figure 7.6 Schematic cross section through Earth's crust, showing the relationships among volcanic processes and lithospheric plates (see also Chapter 6). Source: Based on van Andel and Murphy (2024).

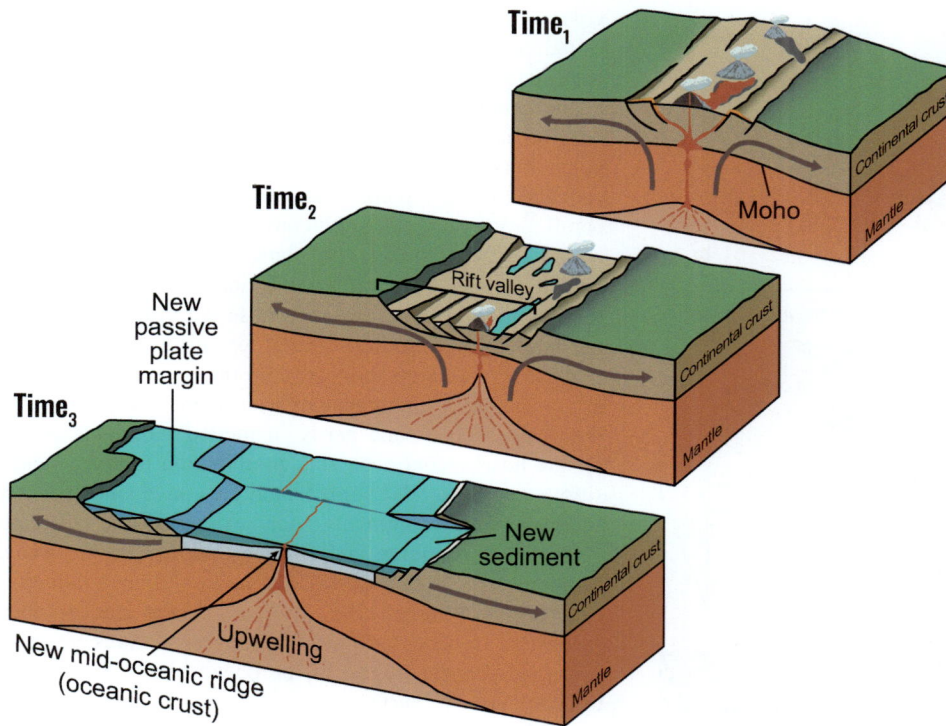

Figure 7.7 Rifting of continental crust, leading to the formation of a rift valley and, finally, an open ocean. Source: After Marshak © 2019. Used by permission of W. W. Norton & Company, Inc.

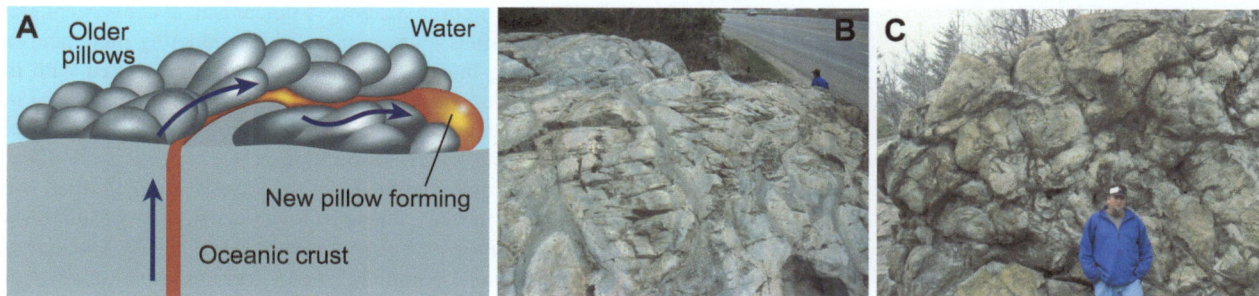

Figure 7.8 Pillow lavas. **A.** Schematic representation of the formation of pillow lavas, as magma erupts underwater. **B** and **C.** Pillow lavas formed in a Precambrian ocean, now exposed in northern Michigan, USA, near Marquette. Note the darker outer ring of the pillow lava basalts. Source: R. Schaetzl.

lava then instantly "freezes" into obsidian (volcanic glass). Lava inside crystallizes more slowly, forming a pillow that is mainly composed of finely crystalline basalt (**Fig. 7.8**).

Magmas erupted in deep-water environments are subject to physical constraints very different from those of subaerial (on dry land) eruptions. Subaqueous (underwater) eruptions are subject to high hydrostatic pressure, bulk modulus, thermal conductivity, heat capacity, and the density of water mass, which are generally orders of magnitude greater than subaerial eruptions experience. Thus, explosive eruptions underwater are rare, and explosive expansion of gas bubbles in lavas is limited to shallower waters.

7.2.2 Volcanism at Convergent Plate Boundaries

Convergent boundaries occur where two plates collide. Three types of convergent plate boundaries are recognized:

oceanic–oceanic, oceanic–continental, and continental–continental (**Figs. 7.6, 7.9**). If both plates are continental, they will smash together and eventually form a high and geologically complex mountain range. If one (or both) of the two converging plates is of the oceanic type, it will be subducted at a subduction zone (**Fig. 7.9**) and eventually melt. Let's discuss each of these three different scenarios of plate collision, below.

Where continental crust converges with continental crust, both crusts are too light to subduct. Instead, what occurs is uplift and large-scale **diastrophism** of both plates, forming large and geologically complex mountain ranges (**Fig. 7.9A**; see Chapter 6). The Himalayas of southern Asia are the most spectacular example. Volcanoes are rare in these settings, as magma does not readily extrude through the thick continental crust. However, the magmas at depth will crystallize

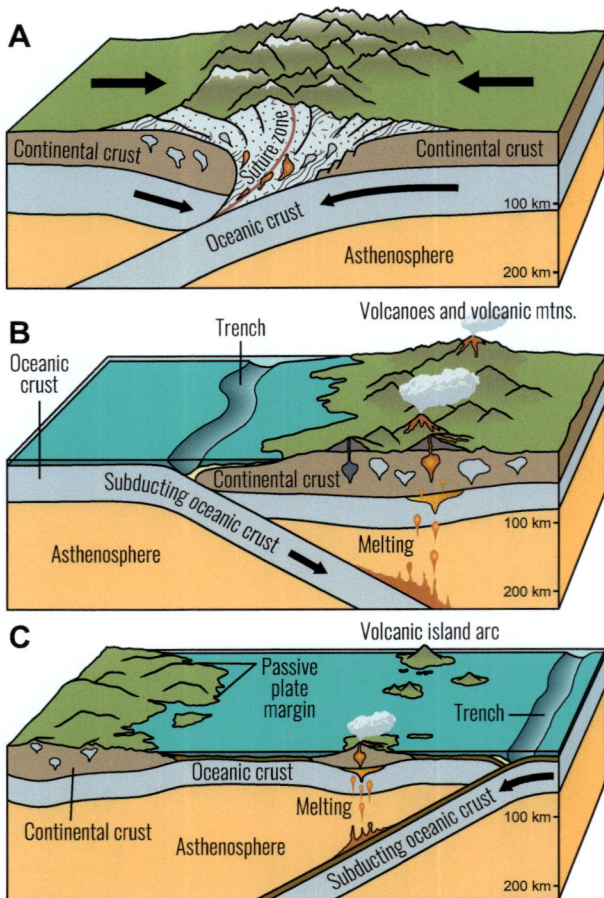

Figure 7.9 Schematic diagrams showing the configuration of crustal plates at locations where convergence occurs. **A**. Convergence of two continental crustal plates. **B**. Convergence of an oceanic crustal plate with a continental crustal plate. **C**. Convergence of two oceanic crustal plates.

there to form bodies of igneous rock called **plutons**, named for Pluto, the Roman god of the underworld (see below). Over time, plutons become exposed by uplift and erosion to yield some of the world's most spectacular landscapes, such as the Sierra Nevada of California, and the Karakoram of Pakistan (**Fig. 6.0**).

Wherever thin, dense oceanic crust enters a tectonic collision zone, it is subducted, leading to volcanism nearby. That volcanism will be expressed as a chain of volcanic mountains on a nearby continent, as occurs at the Cascade Range in the USA, if the oceanic plate has collided with a continental plate (**Fig. 7.9B**). The Pacific Ring of Fire includes a variety of subduction zones and volcanic mountains that generally surround the margin of the Pacific oceanic plate (**Fig. 7.1**). Volcanoes at convergent plate boundaries within the Pacific Ring of Fire are especially notable along the margins of the Pacific, Nazca, and Cocos Plates (see Chapter 6).

Where two oceanic plates are converging, subduction of one of them will lead to the formation of underwater

volcanism, possible forming a chain of volcanic islands called an **island arc** (**Fig. 7.9C**). Examples of island arcs include the Japanese and Philippine Archipelagos, the Kuril Islands, and the Aleutian Islands of Alaska. As the name implies, island arcs typically occur as a curving chain of volcanic islands. The arcs are convex toward the ocean and concave toward the continent, with a deep trench running parallel to the arc along the convex (ocean) side. The typical rock type of subduction island arcs is **andesite**, named after the Andes Mountains. Andesite is very different than the basalt of oceanic crust, because it formed by the partial melting and mixing of basaltic crust and oceanic sediments, as both are subducted into the trench. Thus, island arc volcanoes are composed partly of melted basalt (from the oceanic crust) and partly of melted sediments eroded off the nearby continent; this combination produces andesitic lavas (**Fig. 7.2**). Some island arcs have since become accreted onto continents, for example, the Cascade Range of Washington and Oregon (USA), and the Andes Mountains of South America. Common to every subduction zone is a deep trench (**Fig. 7.9B, C**). Island arcs and their associated trenches are major structural features, together with mid-oceanic ridges, of the ocean basins. In essence, trenches are the surface expression of subduction zones, marking where oceanic crust subducts into the mantle. Arcs and trenches can be hundreds of kilometers long.

Crustal melting, leading to volcanism, at convergent plate boundaries has many causes. The subducting plate heats as it is drawn into the mantle. Also, water is mixed with the sediments of the subducting plate. Water subducted into the mantle with these sediments rises into the overlying mantle material and lowers its melting point. Magma created from melting of the mantle above the subducting plate then rises toward the surface, due to its lower density and greater buoyancy, relative to the surrounding mantle. Along this journey it precipitates key minerals and integrates other minerals and rock fragments from surrounding materials, all of which can significantly alter the magma's chemical signature. Volcanic eruptions at the surface are then formed as the magma pours out, onto the surface as lava and pyroclastic materials.

7.2.3 Volcanism at Hotspots

Although ≈95% of the world's volcanoes occur at plate boundaries, volcanism can also occur at **hotspots**. Geologists argue that, below hotspots, plumes of hot material well up from deep in the mantle, much like plumes rise buoyantly in a lava lamp (**Fig. 7.10**). When a plume reaches the shallow mantle, it partially melts the surrounding rock, allowing the melted material to upwell to the surface, erupting as a volcano or lava flow. Longstanding wisdom within the geological community holds that the sources of hotspots is as deep as the core–mantle boundary (≈3,000 km), and that they are nearly stationary with respect to the plate movement above. As the plate continues to move away from the hotspot below, any volcanoes (on the plate above) are dragged along with it. Then, as these volcanoes cool and subside, they can produce lines of volcanic islands, atolls, or seamounts along

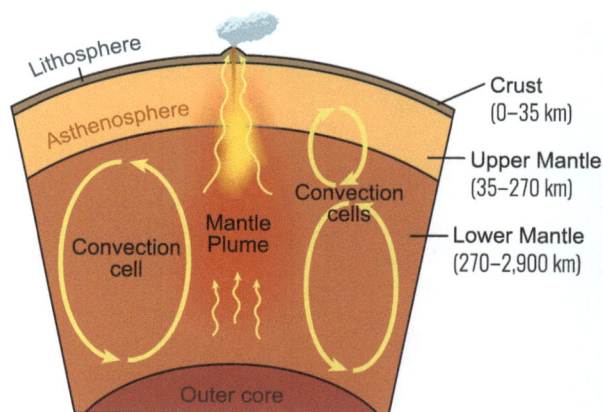

Figure 7.10 A conceptual model of a hotspot plume, originating in the mantle.

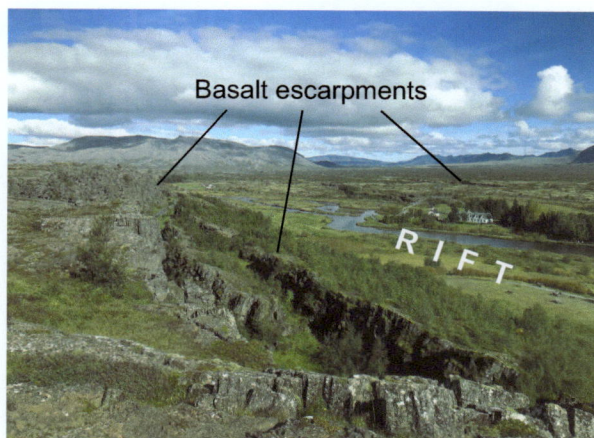

Figure 7.11 The rift valley at Thingvellir National Park, Iceland, where the North American and Eurasian plates are diverging. The basalt escarpments formed as lava was extruded from the rift. The ridge system is exposed above sea level because Iceland is also astride a hotspot. Source: L. Owen.

these "hotspot tracks." The youngest, most active volcanoes are situated directly over the plume – their point of origin. Progressively older (extinct) volcanoes populate the plate in a downstream direction. Hotspots can therefore be used as a reference frame from which to derive plate motion (direction and speed). Likewise, the geochemistry of the lava within hotspot volcanoes has been used to provide information about the composition of the lower mantle.

Hotspot volcanoes occur somewhat randomly around the globe. Their relationship (or lack of one) to the lithospheric plate boundaries is still being debated. More than 100 hotspots are known to have been active during the last ≈10 Ma (**Fig. 6.4**). Many of them occur on mid-oceanic ridges, such

as at the Icelandic hotspot, which also overlies a divergent plate boundary (**Fig. 7.11**). Whether this is a coincidence or not is debatable. Hotspots also develop on continents. The Yellowstone hotspot, arguably the most well-known hotspot on a continent, has been active for at least 15 Ma. As the crust has slowly moved across the hotspot, its heat has produced a chain of volcanic features along the Snake River Plain that extends for ≈650 km westward, from northwestern Wyoming to the Idaho–Oregon border (**Fig. 7.12**).

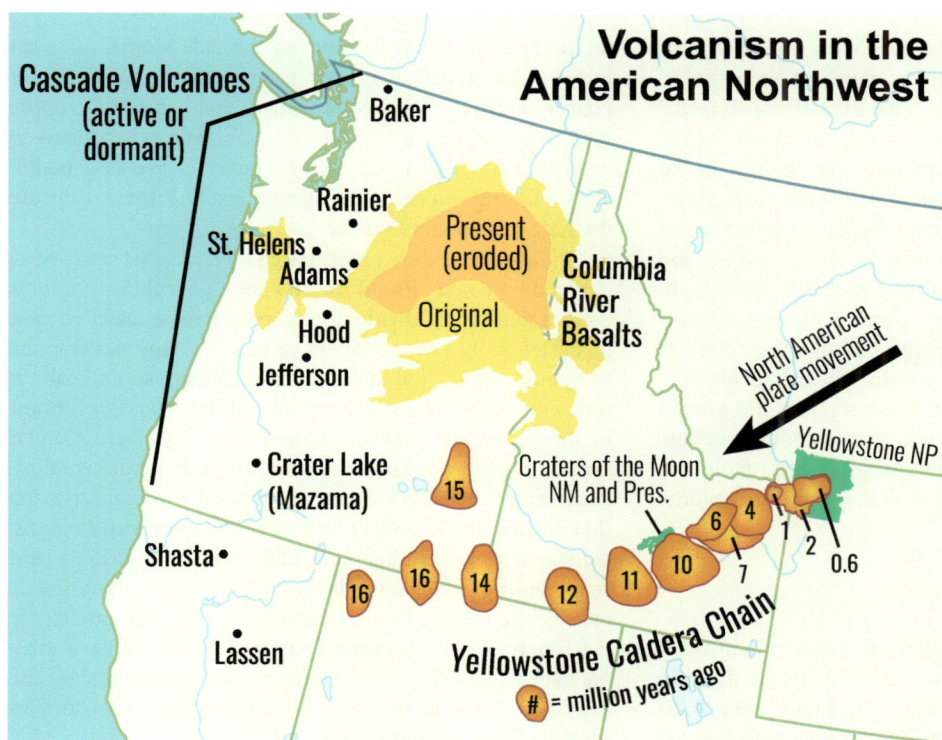

Figure 7.12 Volcanism in the Pacific Northwest, USA, focusing on that which is associated with the Yellowstone hotspot. Also shown are the Columbia Plateau flood basalts (original and present extent), and the largest volcanoes within the Cascade Range.

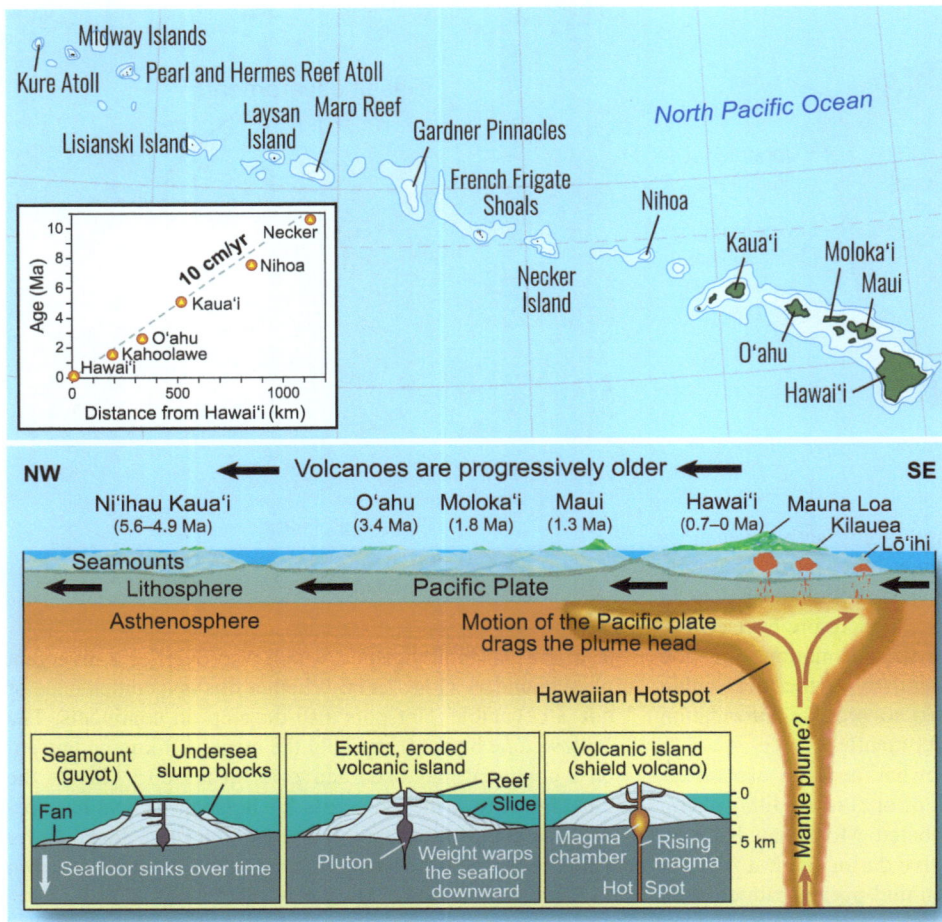

Figure 7.13 Evolution of the Hawai'ian Islands, as it pertains to current hotspot theory.

The Hawai'ian Islands, coupled with the Emperor Seamount chain, represent perhaps the best example of islands along a hotspot track. The Hawai'ian Islands' chain of volcanic features extends >6,000 km across the Pacific basin, where the hotspot has been active for at least 70 Ma (**Fig. 7.13**). Contemporary volcanic activity occurs over the hotspot, on and near the easternmost island of Hawai'i. Nonetheless, volcanism here is slowing, and is being replaced by activity on nearby Lō'ihi Seamount, offshore and just east of the main island. Although the hotspot volcanism theory successfully explains the relationship between age and distance within volcanic chains (as at Hawai'i), it does not explain the temporal and spatial pattern of volcanism within other chains. Today, debate among geologists about whether hotspots are "fixed" in position has led some to suggest that the hotspot plume theory be reconsidered.

7.3 VOLCANISM ASSOCIATED WITH FLOOD BASALTS

Not all volcanic eruptions are violent. **Fissure eruptions** occur when magma flows up through cracks in the surface and pours out effusively (**Fig. 7.14**). Fissure eruptions occur at spreading centers, where plate movement has led to fractures in the crust, and on the side slopes of volcanoes. These types of eruptions are sometimes characterized by a **curtain of fire**, a wall-like structure of erupting lava, spewing upwards. Fissure eruptions can produce very extensive flows of lava, even though the lava is generally slow moving. Nonetheless, eruptions like these typically dwindle after hours or days.

Widespread fissure eruptions can lead to vast areas being covered with lava. **Flood basalts** are high-volume eruptions of lava that cover broad regions, producing expansive areas of horizontally bedded layers of basalt. Many develop due to mantle convection at hotspots, occurring sporadically in time and space. The great continental flood basalt eruptions of the geological past are perhaps the largest eruptions of lava on Earth, with known volumes of individual lava flows exceeding 2,000 km³. Some flood basalts are so large that they form entire igneous provinces. The term **trap** (Swedish *trappa*, steps of a staircase) is often used to describe a landscape built of layer-upon-layer of lava flows. Flood basalts can also be formed in a submarine environment; others, like the Siberian Traps and the Deccan Traps of India, formed on land (**Fig. 7.15**). The Ontong Java and North Atlantic Province flood basalts are notable submarine examples, forming large submarine plateaus.

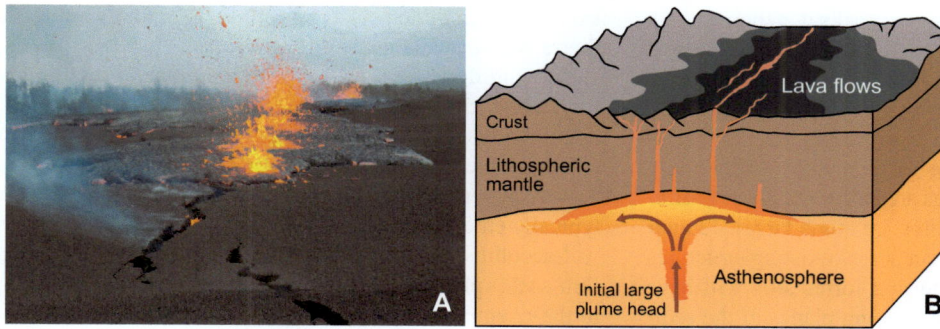

Figure 7.14 Fissure eruptions. **A.** A fissure eruption on the surface of Kilauea in Hawai'i. Source: Hawai'i Volcanoes Observatory, US Geological Survey. **B.** A schematic diagram of fissure eruption dynamics.

Figure 7.15 Map of the extent of the Deccan Traps of India – an area of former flood basalts.

Figure 7.16 An exposure of the flood basalts on the Columbia River Plateau, along the Snake River in Idaho, USA. These basalts form a plateau of 164,000 km² between the Cascades and the Rocky Mountains. Source: S. Reidel.

An additional characteristic of flood basalts is the rapidity with which these large volumes of basaltic magma erupt. The Deccan Traps, which erupted ≈65 Ma, have an estimated volume of ≈106 km³ that erupted in less than a million years; probably >80% of that erupted in <500,000 years (**Fig. 7.15**). The Columbia River basalts in Washington State are another example of an extensive area of flood basalt. They formed between 17.5 and 17.6 Ma, during which ≈100 km³ of magma were erupted (**Figs. 7.12, 7.16**).

Flood basalts also have wide paleoenvironmental implications. For example, continental flood basalt eruptions, which de-gas to the atmosphere during the eruption, have been associated with mass extinctions. The link between eruptions and extinctions is thought to be linked to CO_2 and sulfur-rich gases, which are released as the lava degasses. The SO_2 is converted to sulfuric acid aerosols that linger in the atmosphere, possibly impacting the solar energy flux to the land surface, leading to global-scale cooling. Huge volumes of sulfur-rich magma, erupted over a short period of time (≈2 Ma), can increase this type of "natural pollution" to deadly levels. Because of their age and altered state, the amounts and signatures of the minor volatile elements in ancient lavas are difficult to determine. Ascertaining whether eruptions have degassed large amounts of sulfur-rich gases and CO_2 is only possible for more recent lavas, such as some in the Columbia River province (**Fig. 7.16**).

7.4 LANDFORMS DEVELOPED ON PLUTONS

When magma cools underground, as it usually does, it forms a variety of igneous intrusions, which collectively form a family of subsurface landforms known as **plutons**. In ascending order of silica content, they are composed of rocks called gabbro, diorite, granite, and pegmatite (see Chapter 5). Rocks like these are by far the most common rock types in the crust (**Fig. 8.2A**). Over time, erosion of the overlying rock brings many plutonic rocks to the surface, where they form identifiable landforms.

In most cases, as magma melts its way through the crust, toward the surface, it usually intercepts overlying sedimentary rock. At most locations on Earth, the uppermost rock is a sedimentary rock (see Chapter 8; **Fig. 8.2A**). The intruding magma tends to seek out the most accessible conduits through this sedimentary rock cover – commonly between bedding planes. The magma fills and then widens these preexisting cracks, melts some of the surrounding rock (called **country rock**), pushes the country rock aside (where the rock is hot enough and under enough pressure to deform without breaking), and even breaks it. When magma forces itself into cracks, it often breaks off pieces of rock and then envelops them. The resulting fragments, or inclusions, are called **xenoliths**.

Plutons are named based on their shape, size, and relationships with the surrounding country rock (**Fig. 7.17**). Large, irregularly shaped plutons are called **stocks** or **batholiths**, depending on size; batholiths are much larger. Tabular plutons are called **dikes** if they cut across existing structures and **sills** if they do not. Magma that forms sills will have flowed along and parallel to the bedding planes of the sedimentary rock, that is, conformally. **Laccoliths** are like sills, except they have caused the overlying rocks to noticeably bulge upward, and **lopoliths** are a type of "inverted" laccolith. **Pipes** are cylindrical conduits that often feed a volcanic vent at the surface. Other plutons, less commonly encountered, include **chonoliths** (like a stock but with a flatter and definable base) and **phacoliths** (a lens-shaped intrusive body that has formed inside of folded sedimentary rocks).

Batholiths are tremendously large bodies of intrusive igneous rock – the largest of all plutons. They form as large storehouses – motherships, if you will – of magma cooling beneath the surface, forming a rock body with an aerial coverage of at least 100 km^2, and extending to depths sometimes exceeding 15–25 km. Batholiths are commonly composed of coarse-grained, felsic rocks, for example, granite or granodiorite. Often associated with subduction zones, they form the cores of many large mountain ranges (**Fig. 7.18**). The shape or outline of a batholith is typically highly irregular, with sidewalls that incline steeply against the host rock. Most batholiths intrude across mountain folds and are elongated along the dominant axis of the range; faulting and contact metamorphism of the enveloping rock near the batholith is typical.

Batholiths become exposed at the surface due to uplift and/or erosion of the overlying rocks. Today, landscapes formed on batholiths are expressed as seemingly endless expanses of granitic peaks and rolling hills, for example, the Sierra Nevada Mountains of California (**Fig. 7.18**).

Moving down in size, but with a similar formation history, are stocks – plutons with far less surface area and volume than batholiths. Stocks sometimes represent a neck of material emerging from the top or side of a batholith (**Fig. 7.18B**). When exposed by erosion of overlying rocks, stocks typically emerge as areas of hilly terrain or small mountains.

Laccoliths, lopoliths, and phacoliths are meso-size (km-scale) plutons, defined by their shape and relationship to the country rock (**Fig. 7.17**). Of these, laccoliths are the most common. A laccolith is a concordant, broadly mushroom-shaped plutonic body of rock with a flat floor (aligning with a sedimentary rock layer) and a domed roof, usually with a feeder pipe below. The term means "lake of rock" (lacc = lake, lith = rock). Laccolith formation is straightforward. As magma intrudes into the overlying (usually sedimentary) rock, the intrusion pushes the overlying layers upward, while also spreading out laterally (**Fig. 7.19**). Given that most of the uppermost rocks in the crust are sedimentary, laccolith formation is commonplace. The intruding magma will preferentially flow between weak zones in the sedimentary strata and the deformable rock layers above will yield to the magma, bulging upward. With time, laccoliths usually erode to form small areas of hills and mountains, with cuestas and hogbacks surrounding a granitic core (**Fig. 7.19**). The formation of laccoliths usually takes centuries to complete. The state of Colorado (USA) is the first place where laccoliths were recognized and named. The region known as the "Laccolith triangle" extends from Carbondale on the north to Crawford on the southwest to Crested Butte on the southeast, and is one of the best places for examining laccoliths in all states of erosion.

A large, lenticular, igneous intrusion with a depressed central region is called a lopolith (**Fig. 7.17**). Think of it as an upside-down laccolith – a body of hardened magma with a roughly flat top (usually concordant with sedimentary rock

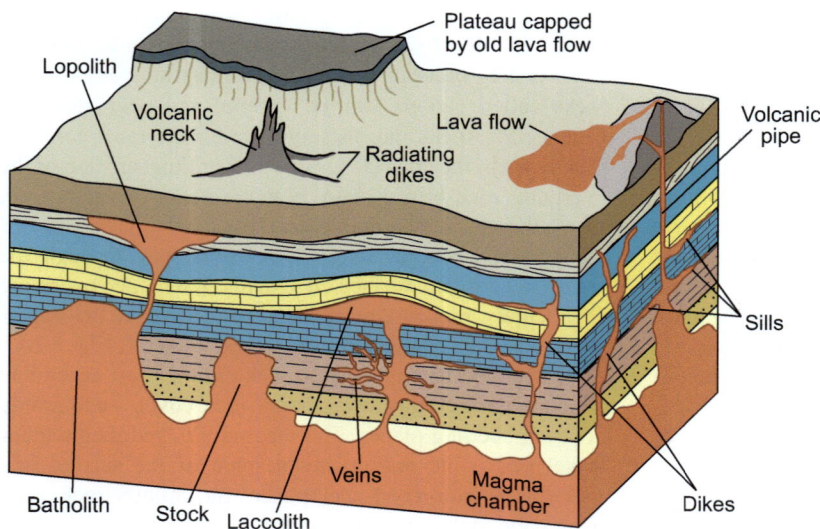

Figure 7.17 Block diagram of sedimentary rock overlying a volcanically active area, showing the various types of plutons and volcanic features that might form.

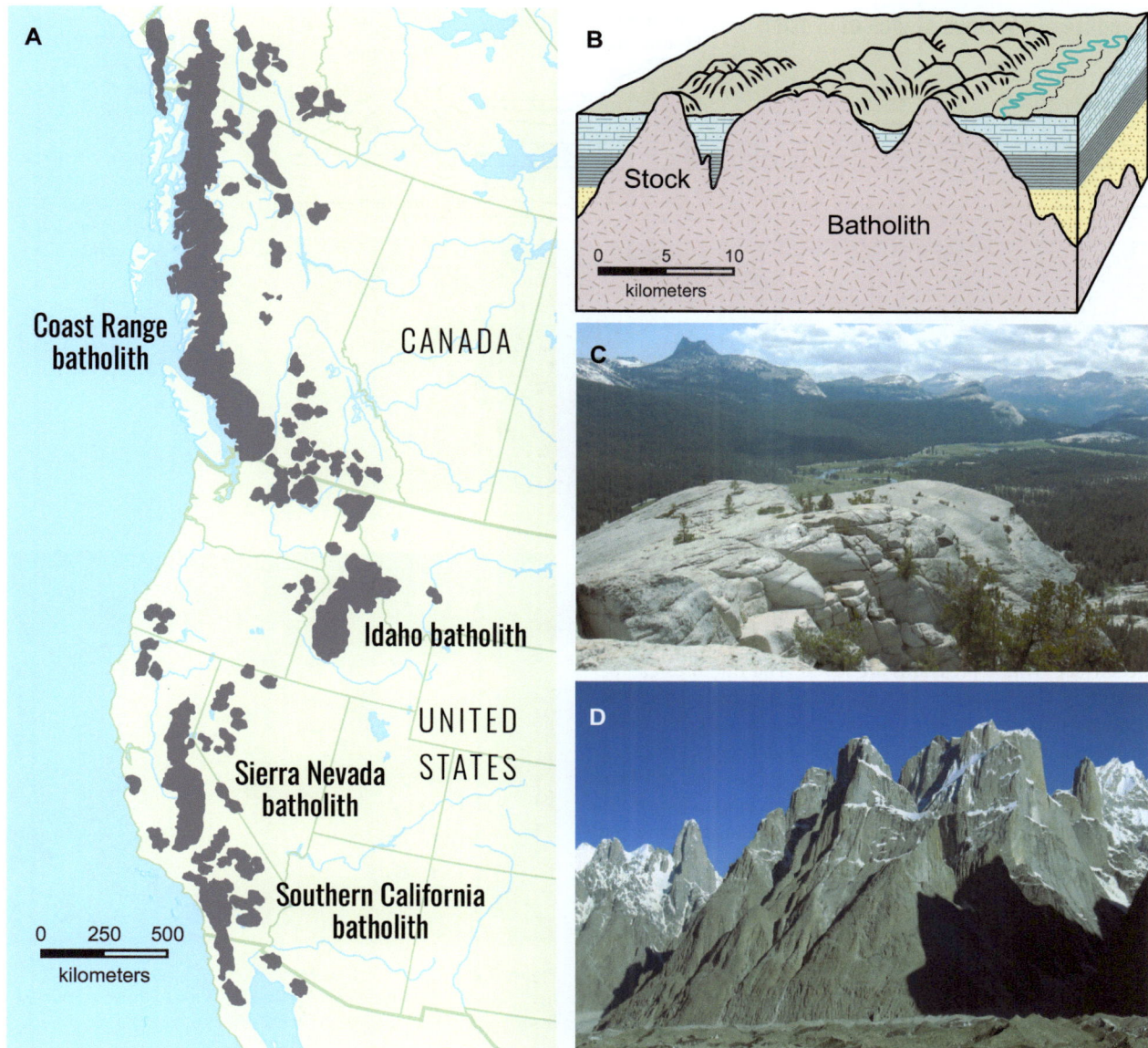

Figure 7.18 Batholiths. **A.** Map of the extent of some of the major batholiths in western North America. **B.** Schematic diagram illustrating the typical underground geology of batholiths and stocks, which become subaerially exposed only after erosion of overlying rocks. **C.** The Sierra Nevada Mountains, within Yosemite National Park, California, consist of a massive granite batholith that has been uplifted by faulting, and subsequently heavily glaciated. The view in the foreground shows an ice-polished granite surface. Source: R. Schaetzl. **D.** The Karakoram Mountains, which span the Pakistan–China border region, are one of the most heavily glaciated and highest mountain ranges in the world, and include K2, the second-highest peak in the world. Source: L. Owen.

strata) and a shallow, convex base that may have a feeder dike or pipe below. The term was originally proposed to describe the Duluth Complex of rocks near Lake Superior, Minnesota (USA). Igneous intrusions also assumed to be lopoliths include the Great Dyke in Zimbabwe, the Freetown Complex in Sierra Leone, and the Stillwater intrusion in Montana.

Continuing along the same theme, a **phacolith** is another concordant, lens-shaped pluton, but the magma, in this case, has been intruded into *folded* sedimentary beds. Phacoliths usually have their greatest thickness along the axes of synclines or anticlines. In rare cases, the magma body may extend as a sill from the crest of an anticline through the trough of an adjacent syncline, such that in cross-section it is S-shaped. Phacoliths form in this manner because, in intensely folded terrain, the hinges of folds are areas of reduced pressure and thus, preferred sites for magma emplacement.

Increasing in number but decreasing in size are a family of smaller plutonic features: dikes (or dykes), sills, and veins. A **dike** is a tabular intrusive rock body that cuts *across* strata or other structural features of the surrounding rock (**Fig. 7.17**). Many dikes are relatively thin, and often nearly vertical. Key to the definition of a dike is that it cuts

Laccoliths in various stages of erosion

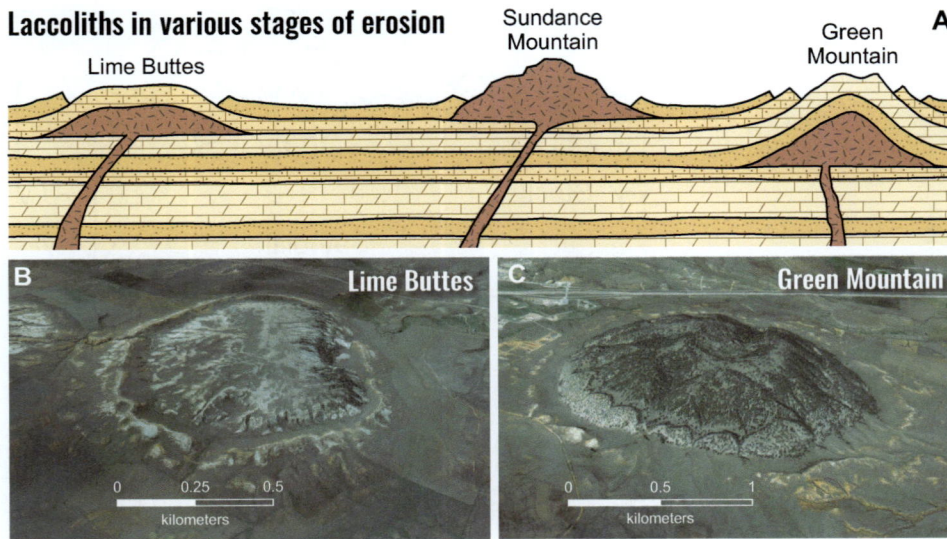

Figure 7.19 Laccoliths are especially common in the western USA. **A.** The general geology associated with three small laccoliths in western South Dakota, USA. **B.** Oblique images of Lime Buttes and Green Mountain, both prototypical laccoliths. Both show the hogbacks and flatirons that have developed on the sedimentary rocks that have been uplifted and deformed along their margins (see Chapter 8). Source: NASA.

unconformably through, that is, *across* the layers of, older rocks or sediments. Dikes could be envisioned as conduits or narrow pathways of (formerly) rising magma, forming as magma follows fractures in country rock, on its way to the surface, where pressures are lower. The fractures can be caused by the intrusion of pressurized magma, or the rise of magma can be simply exploiting existing fractures. Thus, host rocks for dikes must be brittle enough that they fracture as magma intrudes. If magma reaches the surface and lava pours out, such a portal would be called a **vent**. Very often, dikes occur as swarms of several to hundreds, emplaced more or less contemporaneously.

The interior of a large volcano can be crisscrossed with dozens to hundreds of dikes, acting as feeders to the main vent. Hardened lava in the vent will come to form a tubular, roughly vertical body of igneous rock. This rock, typically basalt, is usually much harder than the surrounding rocks, which may have formed with large components of "softer" pyroclastic debris. As a result, after the volcano has gone extinct and started to erode, the neck and some of the larger dikes will often persist as either radial or concentric features. The former are known as **radiating dikes**. Shiprock, in New Mexico, USA, is the classic example of an eroded volcanic neck with several large radiating dikes (**Fig. 7.20**). It is easy to see that the dikes near Shiprock stand up much better to weathering than the surrounding country rock.

Alternatively, a **sill** is a tabular sheet of basalt that has formed by the intrusion of magma between layers of sedimentary rock, beds of volcanic lava or tuff, or even along the direction of foliation in metamorphic rock (**Fig. 7.17**). Thus, the term implies that the intrusion was *concordant* with the surrounding rock, that is, a sill does not cut across preexisting rocks. Alternatively, dikes – discordant intrusive sheets – *do* cut across older rocks. The magma that forms sills follows weakened zones within the country rock, and is usually originally emplaced horizontally. Sills can be fed by

Figure 7.20 Shiprock, in New Mexico, USA, is an eroded volcanic neck with several radiating dikes (one of which is in the foreground). Source: el ui, CC BY 3.0, via Wikimedia Commons.

dikes, and vice versa. Tectonic processes may subsequently rotate horizontal sills into near vertical orientation.

Sills may be confused with solidified lava flows. However, intruded sills can be distinguished because they show partial melting and incorporation of the fragments of the surrounding country rock as **xenoliths**. Additionally, in sills, evidence of heating and **contact metamorphism** might be observed on both the upper and lower contact surfaces of the surrounding country rock. Lava flows only show evidence of heating on the lower side of the flow. In addition, lava flows typically will show evidence of vesicles (or bubbles) where gases have escaped. Because sills generally form at shallow depths below the surface, the pressure of overlying rock limits the escape of gases. One of the world's most famous sills is the Whin Sill, which runs close to the border of England and Scotland. It forms the foundation for parts of Hadrian's Wall that marked the outer reaches of the Roman Empire.

The smallest of the intrusive magmatic features are **veins**. Veins can be as small as millimeters in width. Like sills and dikes, they form as magma is intruded into country rock. However, the magma did not have enough time (or volume) to become large enough to form sills or dikes. Usually, where dikes exist, veins are nearby, and connect to them.

7.5 VOLCANIC LANDFORMS

Volcanoes are familiar to most people. We think of explosive eruptions of ash and cinders, or red-hot lava flowing down the slope of a mountain. Those types of volcanoes are clearly **active**. According to the Global Volcanism Program (GVP), volcanoes considered to be active must have erupted since the retreat of the last continental glaciers. The GVP lists about 40–50 volcanoes that are currently erupting, 531 that have erupted since 1800, and 1,559 that are known (or thought) to have erupted in the Holocene. A **dormant volcano** is one that has not erupted in the past 10 ka but is expected to erupt again. Most of the world's volcanoes are **extinct** – and not expected to erupt again. But be careful, there have been several eruptions from "extinct" volcanoes!

7.5.1 Lava Plains and Plateaus

Plutons become surface landforms only after erosion of the overlying rock. Alternatively, most volcanic features form at and above the surface; these are the "volcanic landforms" that come to mind for most people.

The least volatile type of eruption does not form a "volcano" at all. Fissure eruptions form broad lava plateaus, where extensive lava flows (called flood basalts or traps) and volcanic ash overwhelm and completely bury preexisting terrain, as exemplified by the Columbia Plateau (**Figs. 7.12, 7.16**). This type of volcanism is commonly associated with hotspots or rift zones on continental plates. Lava associated with these types of eruptions flows long distances from their sources. The resulting topography is generally undulating and not dominated by volcanic cones, although the upper surface of flood basalts can have sharply incised canyons and valleys. The erupted material, typically basaltic and of low viscosity, can be as thick as tens to even hundreds of meters. The Laki fissure eruption in southern Iceland – the largest such eruption in historical times – is a classic, modern example of a fissure eruption (**Fig. 7.21**). The eruption began on June 8, 1783 and continued until early February 1784.

7.5.2 Volcanoes

Volcanoes come in a variety of sizes, shapes, and geologic histories. They are formed of different rock types and vary markedly in the rates at which they erode (**Fig. 7.22**).

Shield volcanoes are the largest volcano in volume, diameter, and height, sometimes tens to hundreds of kilometers across, or more (**Figs. 7.22, 7.23**). Shield volcanoes get their name from their morphology; viewed from above, they look like a warrior's shield – broad, slightly raised in the center, and with long, gently sloping sides. Large in terms of

Figure 7.21 The extensive plain of basalt produced by the Laki fissure eruption in Iceland in 1783–1784. Source: L. Owen.

area, shield volcanoes have much gentler slopes than other volcanoes.

Basaltic lava, typical of shield volcanoes, has low viscosities, implying that it can flow quickly and far, rapidly covering large areas. Explosive eruptions are rare. The lava from each eruption eventually hardens, just as the next eruption covers the previous one. Over time, the multiple eruptions build up multiple layers of basalt, one on top of the other.

Shield volcanoes have such low, broad slopes because of the low viscosity of the lava, but also because of where the lava has erupted. Unlike many volcanoes, where lava exits only through a vent in the center of the feature, shield volcanoes often have **flank eruptions**, where lava flows out of fissures on their side slopes (**Fig. 7.22**). Lava that flows out of the main vent must first fill and then overflow a deep central crater – a **summit crater**. A few shield volcanoes have persistent lava lakes in these craters. Among them are Mt. Erebus (Antarctica) and Nyiragongo in the Congo. Flank eruptions from shield volcanoes assist lava in flowing far downslope, helping to form their broad dome-like shape.

The Hawai'ian Islands are a chain of shield volcanoes (**Fig. 7.13**). Perhaps the most active shield volcano on Earth is Kilauea on the island of Hawai'i. Its low-viscosity lavas mainly issue from the Halema'uma'u summit crater. Nonetheless, many recent eruptions have come from fissure vents as well, mainly on the southeastern and eastern flanks of the volcano.

Shield volcanoes obtain their unique shape because of their free-flowing, low-viscosity, basaltic lavas, which retain little gas and are erupted at high temperatures (**Fig. 7.2**). Thus, eruptions on shield volcanoes are non-explosive. Freely flowing, the lava can cover large areas rapidly, bringing homes and cities far downslope into the sphere of influence of an eruption. Lava fountains are a common occurrence within summit craters, as the lava bubbles and spouts almost continuously, sometimes even forming lava curtains (**Fig. 7.24**).

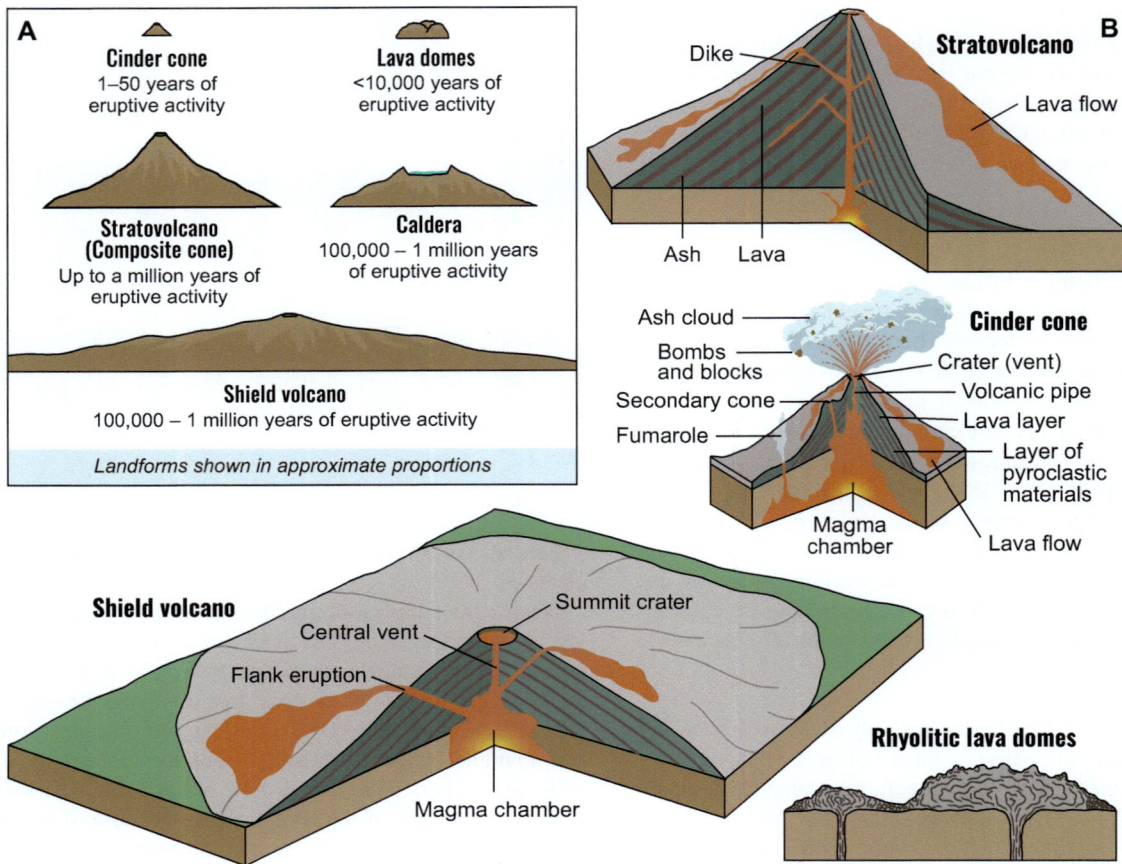

Figure 7.22 The major types of volcanoes and volcanic features. **A.** The size and morphology of the major types of volcanoes. **B.** Internal structure of the four major types of volcanoes; relative sizes are not proportional, as they are in A.

Figure 7.23 Profile view of Mauna Loa, a shield volcano on the main island of Hawai'i. Source: US Geological Survey.

Shield volcanoes become so large because they can erupt regularly over long periods, each time depositing a new layer of lava. Another phenomenon that allows shield volcanoes to grow so large are **pyroducts**, or **lava tubes**. They form as lava flows away from the site of an eruption, cooling and forming a surface crust. Depending on the conditions, that outer shell/crust may insulate the molten lava below so that it stays molten for some time, flowing downslope, far from

the eruption site. Eventually, the vacated subterranean space becomes a type of elongated, natural cavern. The pyroducts at Lava Beds National Monument in northern California (USA) are large enough to walk in.

Many of the shield volcanoes on the Hawai'ian Islands, which formed formerly above a hotspot, remain as subaerial features today. After these volcanoes migrate off the hotspot, they eventually erode. Eroded, submarine (below sea level) shield volcanoes are called **seamounts**. Seamounts are common at the northern end of the Hawai'i islands chain, in a section known as the Emperor Seamounts (**Fig. 7.13**).

Basaltic lava also erupts along mid-oceanic ridges during seafloor spreading. Therefore, many basaltic, volcanic islands (including Iceland) occur along, for example, the mid-Atlantic Ridge (**Fig. 7.21**).

On the opposite end of the "size spectrum" from shield volcanoes are **cinder cones**, also referred to as **scoria** or **pyroclastic cones** (**Fig. 7.22**). These short-lived volcanic features are only tens to hundreds of meters in height and monogenetic, meaning they are often formed by a single eruption. Cinder cones are steep-sided, conical hills composed mainly of loose pyroclastic fragments, for example, volcanic clinkers, ash, and cinders. As cinders and pyroclastic debris pile up to form the cone, lava can also sometimes

Figure 7.24 Three-hundred-meter-high lava fountains feed lava flows from the Mauna Ulu vent area, on the eastern rift zone of Kilauea, on December 30, 1969. Shortly after the photo was taken, the wind shifted, and the vehicles, still parked there, were pelted with falling pumice. Source: US Geological Survey.

Figure 7.25 A typical Strombolian-type eruption of Mt. Pacaya, one of Guatemala's most active volcanoes. Source: R. Cosar.

stream down the slopes in fiery rivers. All these deposits typically harden into a dark, porous rock called **scoria**. Most cinder cones have a bowl-shaped summit crater – the remnant of its central vent.

Many cinder cones occur as parasitic cones on larger, polygenetic volcanoes. Others occur in groups within volcanic fields. Because cinder cones erupt with material that has a dominantly basaltic composition, they do not "explode." Their eruptions are instead governed by the speed of the rising magma, which in turn is primarily determined by viscosity and gas content. These conditions produce a type of eruption that looks like a fireworks display, termed a **Strombolian-type eruption**, after the volcano Stromboli in Sicily. Strombolian eruptions propel huge blobs of lava, cinders, and hot rocks upward and outward from a central vent (**Fig. 7.25**). Strombolian eruptions are driven by gas inside the volcano, which coalesces into bubbles. As the bubbles grow large enough to rise through the magma column and reach the top, they burst apart because of the reduced pressures, hurling magma into the air as bombs, blocks, and all

manner of other pyroclastic debris. Once airborne, the lava breaks into small fragments that solidify and fall as either cinders, clinkers, or scoria. These fragments are then deposited around the vent to form a cone that is often usually symmetrical, with slopes between 30–40°, and nearly circular in plan view. During a Strombolian eruption, gas bubbles can be popping every few minutes. Typically, Strombolian-style eruptions at most scoria cones are completed by additional, Hawai'ian-style lava fountains.

Parícutin in Mexico is one of the most famous cinder cones. Its initial (1943) eruption was observed by a farmer as a pillow of smoke; within days a large cone had formed, which over the coming years reached heights >500 m. Scoria from the eruption covered two villages. Two excellent places to see cinder cones in the United States are Craters of the Moon National Monument and Preserve in Idaho, and Sunset Crater Volcano National Monument in Arizona (**Fig. 7.26**).

Stratovolcanoes comprise the largest percentage (≈60%) of Earth's volcanoes. Most are characterized by eruptions of viscous, rhyolitic (silica-rich) lavas that harden into andesite and dacite (**Figs. 7.22, 7.27**). Due to their viscosity, these lavas are resistant to flow, thereby allowing gas pressures

Figure 7.26 Classic examples of small cinder cones. **A.** Crescent Butte at Craters of the Moon National Monument and Preserve. Crescent Butte is 60–90 m high, and ≈15,000 years old. Source: Courtesy of the US National Park Service. **B.** Sunset Crater at Sunset Crater Volcano National Monument. Sunset Crater is ≈340 m high and formed from eruptions in the late eleventh century. Both cones are considered to be extinct. Source: Courtesy of the US National Park Service. **C.** S P Crater and a lava flow in northern Arizona. This cone is thought to be ≈55,000 years old. Source: Arizona Geological Survey.

to build up to high levels; explosive eruptions can result. Indeed, eruptions at stratovolcanoes are the most powerful, multifaceted, and dangerous type of volcanic eruptions, involving lava flows, pyroclastic flows, ash, and bombs. The 1980 eruption of Mt. St. Helens in Washington State (USA) was a classic stratovolcanic eruption, sending ash kilometers into the air, blasting apart the side of the mountain, and sending torrents of debris down its side slopes.

Stratovolcanoes are usually composed of layers of lava and pyroclastic material, giving them their other common name – **composite cones**. Most active stratovolcanoes occur near the Pacific Ring of Fire, where they are associated with subduction zones (**Fig. 7.1**). Stratovolcanoes are large, often isolated, mountain peaks, sometimes with a well-defined vent or summit crater. They are steeply sloping and much larger than cinder cones (**Fig. 7.22A**). Indeed, most of the world's most famous and well-known volcanoes are stratovolcanoes, for example, Mt. Kilimanjaro in Tanzania, Mt. Fuji in Japan, Mt. Rainier and Mt. St. Helens in Washington State, Mt. Shasta in California, and Tambora in Indonesia. The Cascade Range in the western United States is essentially a string of stratovolcanoes, many of which are extinct, and the surrounding uplands (**Fig. 7.12**).

Figure 7.27 Redoubt Volcano, an active stratovolcano in the Kenai Peninsula of Alaska. Historical eruptions included those in 1902, 1966, 1989, and 2009. Source: US Geological Survey.

Lava domes are hemispherical to irregularly shaped mounds of volcanic rock, formed by one or more eruptions of rhyolitic lava above a vent. Too viscous to flow very far, the lava accumulates as irregularly shaped, often overlapping, domes (**Figs. 7.22, 7.28**). The extrusion of lava that forms these domes can occur above the central vent of a volcano, on the lower slopes of the volcano (either in isolation or in groups), or along fault lines. Like lava flows, the eruptions that produce lava domes typically do not have enough embedded gas or pressure to erupt explosively, although they may sometimes be preceded or followed by explosive activity. Eruptions that produce lava domes can last from a few days to several decades. In most cases, the impacts of these typically small eruptions are short-lived. Most eruptions at andesitic volcanoes are small, and many are brief.

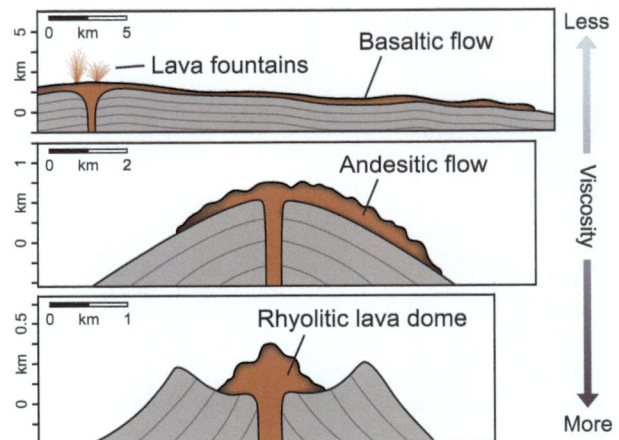

Figure 7.28 A comparison of the types of large-scale volcanic landforms produced by lavas of different viscosities. See **Fig. 7.2** for more details on the characteristics of the different types of lavas. Source: After Marshak © 2019. Used by permission of W. W. Norton & Company, Inc.

Many examples of volcanic domes can be found in the Cascade Range of the northwestern United States. Lassen Peak in California is part of a larger volcanic center, and has erupted between 1914 and 1917. Lassen Peak is part of the Mono Lake-Inyo Craters volcanic chain, which consists of several lava domes and thick, viscous, rhyolitic lava flows; four of these domes erupted only ≈600 years ago. Mount St. Helens, Mt. Shasta, and Glacier Peak are among the most active stratovolcanoes in the Cascades. Before the 1980–1986 eruptions of Mt. St. Helens, its symmetrical summit cone was capped by a lava dome that had formed between the 1500s and late 1700s. These examples highlight the often-close relationship between stratovolcanoes and lava domes.

Although lava domes are built by non-explosive eruptions, they can by themselves generate lethal pyroclastic flows. The sides of a dome can collapse to form an avalanche of hot lava fragments and gas. Thousands of people were forced from their homes during the eruptions of lava domes at Unzen Volcano in Japan in 1991, and at Soufriere Hills in Montserrat, starting in 1995.

7.5.3 Calderas and Summit Craters

All of the major types of volcanoes typically have one feature in common – a central vent, often expressed as a circular depression near the summit. By definition, if this depression is <1.6 km in diameter, it is referred to as a **summit crater**, as in most cinder cones and shield volcanoes. Larger features of similar shape, regardless of origin, are **calderas**.

Both features can form by collapse, when the vent and magma chamber below become emptied by the eruption (**Fig. 7.29**). Structural support for the roof of the magma chamber may be lost, allowing the ground surface to collapse. Some calderas form due to a violent explosion or eruption that blasts away the topmost portion of the volcano – a formative mechanism for stratovolcanoes only.

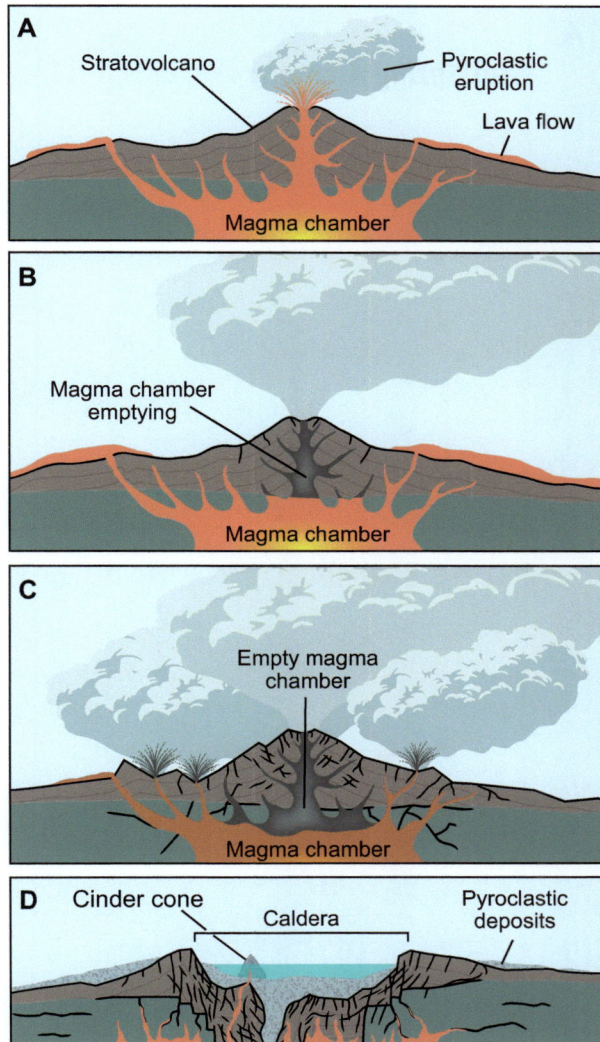

Figure 7.29 Schematic representation of the formation of a large caldera (here, mainly by collapse).

Figure 7.30 Crater Lake, Oregon (USA) – perhaps the world's best known caldera, and Wizard Island, a small cinder cone that formed at a later point in time. Source: K. Patrick

7.6 FEATURES THAT ERUPT HOT WATER AND STEAM

Beneath many volcanoes is a **magma chamber** that in most cases extends into and through the crust. Depending on the setting, magma may rise directly from the mantle or be staged in one or more chambers, before erupting (**Fig. 7.31**). The uppermost part of these chambers often hosts an active hydrothermal system where groundwater mingles with magmatic volatiles, heated by deeper magma. Identifying the extent and vigor of this hydrothermal activity is important because (a) much of the unrest at volcanoes occurs in hydrothermal systems, such that understanding the interactions between hydrothermal and magmatic systems is important for forecasting eruptions, (b) pressures at depth can cause sudden and potentially deadly explosions from the hydrothermal system itself, such as occurred on Montserrat in 2009 (**Fig. 7.32**), and (c) (geo)hydrothermal systems are potential energy resources. These types of **phreatic eruptions**

The world's most well-known caldera is Crater Lake, in the Cascade Range of Oregon (**Fig. 7.30**). Crater Lake formed when a former stratovolcano, named Mt. Mazama, exploded ≈7,600 years ago. Ash from this explosion is found throughout the Pacific Northwest of the United States. A subsequent eruption formed a small cinder cone within the lake, called Wizard Island.

Related to calderas are **maars**, shallow volcanic craters with steep sides formed in tephra deposits. Maars form by one or more explosions that develop as hot magma comes into contact with shallow groundwater. The result is a violent steam explosion that fractures the overlying rocks and throws them into the air, along with steam, water, ash, and molten material. The ejecta usually travel nearly vertically and return as tephra that surrounds the crater. Most maars are hundreds to thousands of meters across, and up to 200 m deep. As with calderas, maars often fill with water after they form.

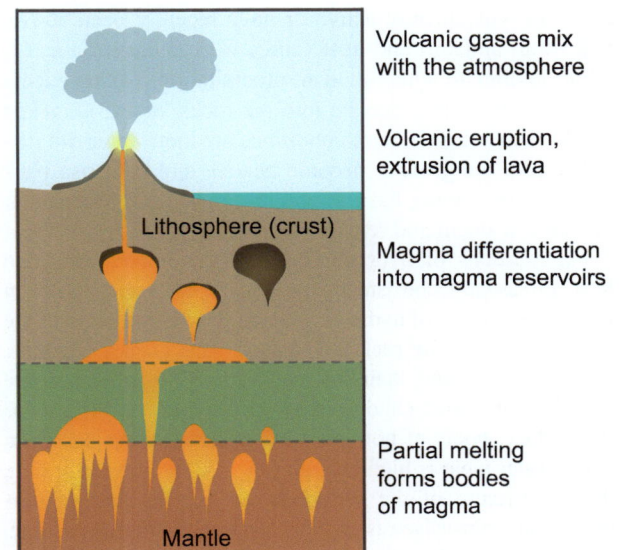

Figure 7.31 Upwelling of magma from deep magma chambers.

Figure 7.32 An image from the international space station of the 2009 steam (hydrothermal) eruption of Montserrat, an island volcano in the Lesser Antilles island chain in the Caribbean Sea. Source: Image taken by the NASA Expedition 21 crew, Public domain, via Wikimedia Commons.

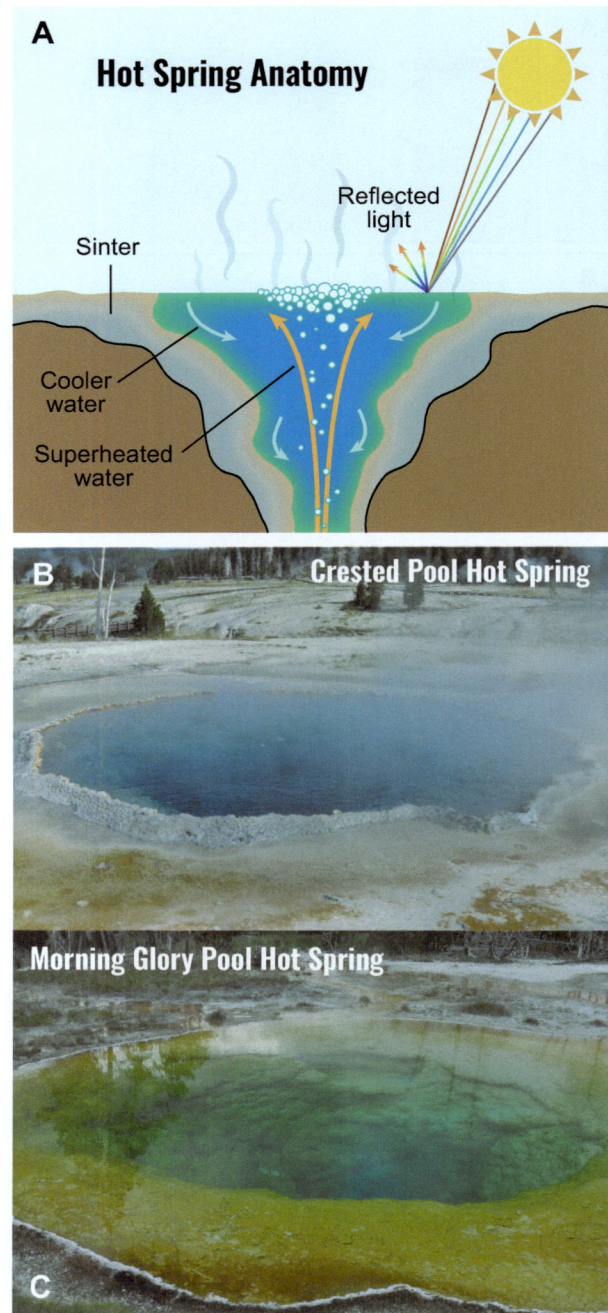

Figure 7.33 Hot springs – classic types of hydrothermal features. **A.** The underground architecture of a hot spring. Source: Adapted from US National Park Service. **B.** and **C.** Two of the many classic hot springs from Yellowstone National Park. Source: R. Schaetzl.

are steam-driven events, formed as magma or hot rock heats groundwater, causing it to boil and flash to steam. The phreatic eruption that ensues is an explosion of steam, water, ash, blocks, and bombs.

Because of abundant heat in the near-surface environment, areas of volcanism – especially near hotspots – are also sometimes home to less explosive **hydrothermal activity**, associated with hot water and steam. Examples of hydrothermal features include geysers, fumaroles, mud pots, and hot springs. These features are generally found in regions of relatively young volcanism. Hydrothermal activity is especially prominent at Yellowstone National Park, in Wyoming, USA, as well as in parts of Iceland and New Zealand. All of these locations have the three prerequisites necessary for hydrothermal activity: (1) a subterranean heat source – these areas are volcanically active or have recently been active, and so an underground heat source is near the surface; (2) a humid climate with abundant precipitation, so that excess surface water can percolate into the rocks; and (3) cracked and broken rocks. If these conditions are met, water can percolate into the ground, become heated, and then rise back to the surface along fissures and cracks to form spectacular displays of steam and related "explosions."

At locations of hydrothermal activity, ample water and heat are usually consistently present. What varies, to form the different types of hydrothermal features, is the depth of the water table, and the rock architecture – its cracks, conduits, and reservoirs, and their interconnectivity. In places where a fairly wide, open conduit exists in the rock, all the way to the surface, pools of hot water called **hot springs** can form if the water table is high enough (**Fig. 7.33**). In a hot spring, the architecture of the rock between a wide surface portal and the subsurface is an open and unobstructed passage. Because this architecture is simple and uncomplicated, hot springs are the most common type of hydrothermal feature.

Hot water – sometimes well exceeding the boiling point – may bubble up and discharge out of the hot spring, spilling over the lip, or it may simply just boil and steam away, depending on factors such as the rate at which water circulates through the system of underground channels, the amount of heat at depth, and the extent of dilution of the heated water by cool water near the surface. Many hot springs simply produce steam, but some slowly or even continuously bubble. Eruptions are rare.

Hot springs often display a kaleidoscope of colors. Many hot springs are an intense, deep blue – formed as sunlight passes through the deep, clean water. Of the light entering the water, blue (with the shortest wavelength) is scattered most by the water, and therefore, the hot spring waters appear blue (**Fig. 7.33A, B**). In some hot springs, colonies of bacteria and blue-green algae form layers of different kinds on the rock surface; each has its preference for the specific temperature range within which they thrive. Grand Prismatic Spring in Yellowstone National Park (Wyoming, USA) is one of the best examples of the kinds and ranges of color that can develop in a hot spring (**Fig. 7.34**). It is also one of the world's largest.

Fumaroles, another interesting type of hydrothermal feature, continually emit mixtures of steam and other gases (**Fig. 7.35**). Many fumaroles have a similar rock architecture to that of hot springs but simply have a deeper water table. In essence, they are **steam vents**; the water table is too deep for liquid water to make it up to the surface. Therefore, only the ever-present steam, the foul smells of H_2S gases, and various types of bubbling/hissing sounds come from the vent portals. Hydrogen sulfide, one of the typical gases issuing from fumaroles, forms because the heat and water combine to oxidize sulfuric acid and native sulfur compounds in the rocks. The acids lead to intense chemical weathering, which helps to maintain the open architecture of the bedrock. Precipitated minerals from

Figure 7.34 Grand Prismatic Spring – a large and especially beautiful hot spring in Yellowstone National Park. Source: Clément Bardot, CC BY-SA 4.0, via Wikimedia Commons.

these sulfur-rich waters add to the bright colors in many thermal areas.

Mud pots are hydrothermal features that are a bit of a hybrid between a hot spring and a fumarole (**Fig. 7.36**). They slowly gurgle and bubble, hiss, and plip-plop as bubbles rise within a pool of thick, muddy soup, forming spectacular shows as the mud bubbles burst and spurt mud into the air. Consider mud pots to be like hot springs filled with a thick cauldron of mud instead of water.

Figure 7.35 Fumaroles constantly emitting steam in Norris Geyser Basin in Yellowstone National Park. To appreciate a fumarole, though, one must hear and smell it! **A.** and **B.** Fumaroles in the Norris Geyser Basin. **C.** Red Spouter Fumarole. Source: R. Schaetzl.

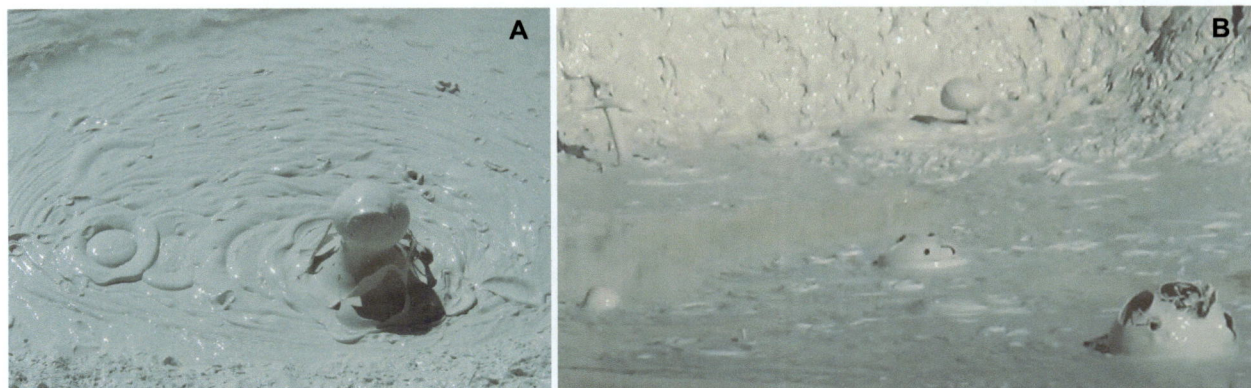

Figure 7.36 Mud pots in Yellowstone National Park. Source: (A) D. Monniaux, CC BY-SA 3.0, via Wikimedia Commons; (B) R. Schaetzl.

Why a mud pot and not a hot spring? At mud pots, locally, water is in limited supply and the rocks have been weathered to silt and clay by the high acidity levels, forming the mud. Mud pots are especially common where the local bedrock is rich in volcanic ash, which weathers more easily to mud than do the more competent volcanic rocks. The mud, therefore, is a byproduct of the heat and intense acidity in the underground hydrothermal system. The acidity comes from H_2S gases exiting the rocks. These gases are partially consumed by heat-loving thermophile bacteria, who convert much of it to sulfuric acid, which then weathers rock to mud. Mud pots can dry up during long periods of dry weather, and so, at any one time, they vary in consistency from slightly gray water to gray mud to hard-baked, cracked dirt.

The mud in mud pots is generally whitish-gray, but sometimes it takes on reddish or pink spots from iron compounds (**Fig. 7.36B**). When the slurry is particularly colorful, the feature may be referred to as a **paint pot**. If the bubbling mud is very viscous, large, intact blobs of it are ejected, forming a little mound at the opening; these features are informally called **mud volcanoes**.

Geysers, the most well-known and exciting of all hydrothermal features, are simply a type of hot spring with a specific type of underground geometry and rock structure; this structure permits geysers to emit episodic, spasmodic bursts of hot water and steam (**Figs. 7.37, 7.38**). The word geyser comes from the Icelandic word *geysa*, which means "to rush forth furiously." The classic geysers produce explosive eruptions of steam and hot water that occur at fairly regular intervals, from minutes to days to years apart.

Like the other types of hydrothermal features, geysers occur above a heat source, and where fractured rock can trap abundant water. Typically, geysers tend to cluster in volcanic basins where all three of these prerequisites occur. The basins are areas of more intense hydrothermal activity, and because of their low-lying topography, water freely collects in them. At Yellowstone National Park, almost all of the active geysers occur in five, possibly six, different geyser basins. Old

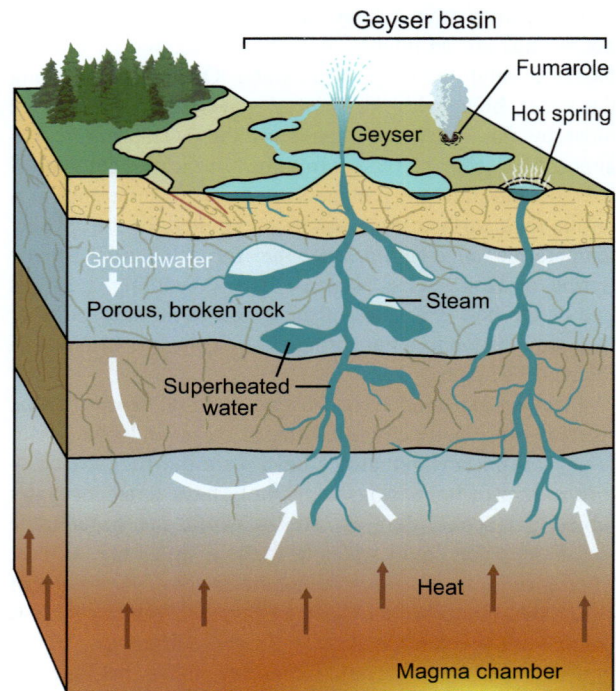

Figure 7.37 Characterization of the hydrogeology of a geyser basin.

Faithful Geyser, perhaps the world's most famous geyser, is only one of over 400 named geysers in its geyser basin – the Upper Geyser Basin.

Recall that, in places where there is an unobstructed passage down into the bedrock, hot springs may form. Geysers form where such passages are narrower and more constricted; only a small portal exists at the surface for water and steam to escape (**Fig. 7.38**). But exactly how does the bizarre plumbing system of geysers operate? In geyser basins, water accumulates in subterranean chambers (a type of "pocket") in the fractured rock, and seeps down, toward hotter layers below. Here, the water becomes superheated,

Figure 7.38 Many geysers of various types exist in Yellowstone National Park. Each has its own rhythm and style. **A.** Castle Geyser. **B.** Beehive Geyser. **C.** Lone Star Geyser. Note how each of these **cone-type geysers** has a conspicuous cone of sinter surrounding their vent. Source: R. Schaetzl.

reaching temperatures > 100 °C (which can only occur under pressure). The pressure to form this superheated water is provided by the weight of the overlying water, as well as the intricate labyrinth of rock crevices and cracks, which holds the water in and resists releasing the pressure that builds up as the water gets increasingly hotter and starts to boil. Key to all this: the underground labyrinth of chambers and conduits beneath a geyser needs to be *pressure-tight*. This is accomplished by deposits of **sinter**, a silica-rich precipitate formed from the water of thousands of previous eruptions. Although the vents of many geysers are surrounded by a low cone of sinter deposits, called **geyserite**, sinter deposits continue well into the subsurface. Over time, more and more sinter is deposited in the underground channels of the geyser, making them increasingly pressure-tight. The formation of sinter from the hot waters of a geyser is determined mainly by the amount of silica in solution, which varies considerably at the different localities and may have varied still more in the past. In many geyser basins, one-fifth to one-third of the mineral matter held in solution consists of silica, the remaining constituents being soluble salts carried off by surface drainage.

As water and bubbles continue to collect in the rock chambers belowground, the increased pressures continue to raise the boiling point, causing more and more of the water to become superheated. For example, water temperatures at ≈90 m below the surface can exceed 200 °C. When the water finally does reach its pressure-altered boiling point, some of it (nearest the surface) vaporizes, which causes the uppermost water in the geyser to bubble out at the surface. This period of bubbling and minor splashing is usually a good precursor (commonly called an "indicator") to an eruption. The bubbling and spillage from the geyser vent release some of the pressure on the trapped, superheated water below. Quickly then, pressures fall, and the volume of the water

held belowground expands until it finally bubbles up and flashes into steam – the eruption has started. As the geyser erupts, columns of water surge and pulsate out of the surface vent, as belowground chambers empty, one after another. Eruptions can last from a few seconds to almost an hour. What a sight to see!

Not all geysers erupt with high flashes of steam like Old Faithful; many just wildly bubble and splash. These "bubblers" are referred to as **fountain geysers (Fig. 7.39)**. There are far more of these than the prototypical cone geysers (like Old Faithful), and their eruptions usually last longer. Regardless, after the eruption, the pressure is relieved, and the eruption ceases.

Lastly, much of the water at the surface trickles back into the fractured rock, and the process begins anew – for both geyser types. Then, because the chambers below a geyser take approximately the same amount of time to refill after each eruption, many geysers erupt on a fairly regular schedule, which may range from a few minutes to months. Others go dormant for years, and then trigger back into action, often following a small earthquake which presumably alters the underground geology.

With all the acidic, high-temperature water and easily weathered volcanic rocks in hydrothermal areas, much of the groundwater is rich in dissolved substances. As this water exits the surface, it cools and many of the dissolved substances are forced to precipitate. Therefore, **evaporite deposits** are common in and near hydrothermal areas; they form as water, rich in dissolved substances, evaporates.

The main evaporite deposits are sinter (mainly silica) and **travertine** (mainly carbonates). The word travertine is derived from Tibur, which was the former name of Tivoli, Italy, where travertine has been quarried since Roman times. Travertine is a unique type of limestone; it is often deposited at mineral springs, such as hot springs, and occurs in a

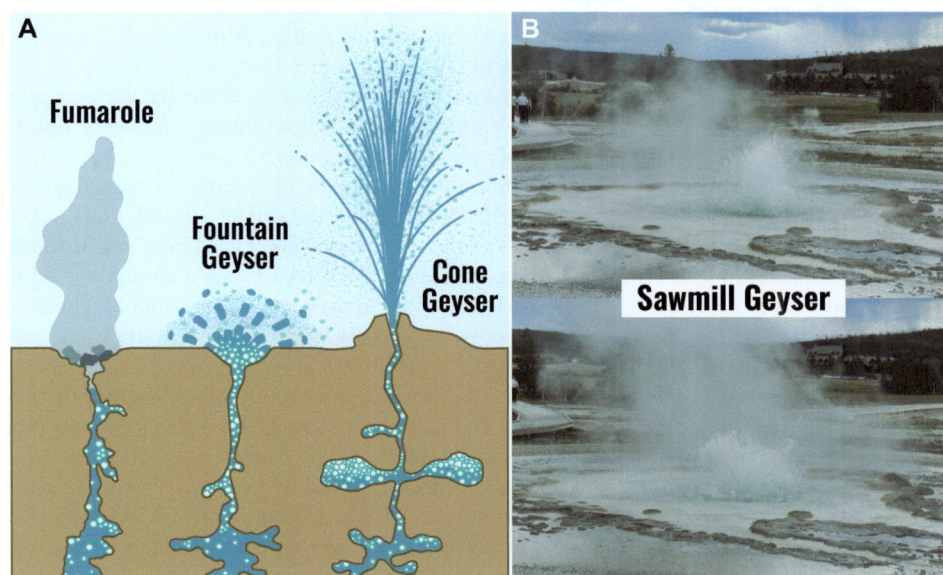

Figure 7.39 A. Schematic diagram showing the main geologic and hydrothermal differences among the two types of geysers, vis-à-vis a fumarole. **B.** Photos of Sawmill Geyser, a fountain geyser in Yellowstone National Park. Source: R. Schaetzl.

Figure 7.40 Mammoth Hot Springs – excellent examples of travertine deposits at Yellowstone National Park. Source: Don Graham from Redlands, CC BY-SA 2.0, via Wikimedia Commons.

variety of colors such as white, tan, and beige, as well as red and coral (**Fig. 7.40**). Travertine deposition requires very specific conditions, involving rapid precipitation of calcium carbonate ($CaCO_3$). Deposition often occurs at the beginning point of a hot spring or in a limestone cave. Features such as stalactites, stalagmites, and other speleothems may form in a limestone cave, but when formed in and near geothermal springs, travertine often precipitates in shelf-like forms (**Fig. 7.40**), often connected to sinter structures. Colonial microorganisms often preserve the exterior surface of travertine by covering it and providing it with additional porosity.

7.7 VOLCANIC HAZARDS

Volcanoes can produce a variety of hazards, many of which can be deadly. Volcanic hazards vary, depending on the chemical composition and gas content of the lava (as well as on other factors). Although sometimes spectacular, volcanic hazards lead to fewer fatalities than earthquakes, floods, and severe storms. Many deaths associated with volcanoes are not from the eruption per se, but due to indirect consequences, such as famine from crop damage, or secondary hazards such as lahars (see below). Thus, we can think of volcanic hazards as being composite. The *primary* volcanic hazard may be associated with the eruption, for example, lava and pyroclastic flows, phreatic explosions, jökulhlaups, earthquakes, noxious and poisonous gases, and the fallout of tephra. *Secondary* hazards include ground deformation, lahars, landslides, and tsunamis (for eruptions on the ocean floor).

The 1883 submarine volcanic eruption of Krakatoa, in Indonesia, was one of the most violent eruptions in the last 3,000 years. The eruption blew apart the entire northern and lower areas of the volcano, creating an explosion that was heard nearly 5,000 km away. Its volcanic ash was transported through the atmosphere for years, with the effects of the ash felt in North and South America, Europe, Asia, southern Africa, and Australia. Nonetheless, the worst consequence of this eruption was the **tsunami** that inundated coastal areas of the islands of Sumatra and Java, resulting in 36,000 fatalities.

Pyroclastic flows are mixtures of rock, lava, ash, and volcanic gas that move downslope at speeds greater than 80 km/h, often at temperatures of 200–700 °C (**Fig. 7.4**). Consider them to be fast-moving currents of dense, hot gases and tephra. Hot pyroclastic flows sometimes glow red in the dark; these flows are called **nuée ardente** (French, "burning cloud"). Driven by gravity, they start by moving down valleys, but can quickly overwhelm the valley and flow across large swaths of the volcano. A related type of hazard – a **pyroclastic surge** – has much higher proportions of gas to rock. Pyroclastic surges can flow right over ridges and hills rather than being primarily confined to valleys like many pyroclastic flows.

Pyroclastic flows are mainly associated with volcanoes that are undergoing explosive eruptions. Most travel several km down the slopes of the volcano. They pose lethal hazards from incineration, asphyxiation, burial, and blunt-force impact, because of their high speeds and hot temperatures. Their speed of flow makes them difficult or impossible to escape; evacuation of likely hazardous areas must take place before eruptions occur. A pyroclastic flow will usually destroy almost everything in its path, and buries everything under a mixture of rock, ash, and debris incorporated along the way. Landforms produced by pyroclastic flows and their associated ashfalls include extensive soft layers of sediment that are later prone to landsliding, although if deposited hot, ash can fuse to become a rock called **tuff** that is more resistant to later deformation.

The eruptions at El Chicón in Mexico (1982) and Mt. Pinatubo in the Philippines (1991) provide notable examples of destructive pyroclastic flows. A pyroclastic flow produced by the 1980 eruption of Mt. St. Helens travelled downslope at speeds of up to 530 km/h, with temperatures up to 700 °C, causing total destruction in its path, before flowing into and filling Spirit Lake (**Fig. 7.4**). This event, in turn, caused lahars (see below), adding to the destruction.

Fallout of tephra and other ballistic projectiles are also dangerous volcanic hazards. They endanger life and property by (1) the force of impact of falling fragments, but this occurs only close to an eruption, (2) the loss of agricultural lands by burial in raw tephra, (3) producing suspensions of fine-grained particles (mainly ash) in air and water, which clog filters and vents of motors and machines, and are harmful if inhaled, and (4) their association with noxious gases, acids, salts, and (close to the vent) heat. Fallout of tephra can collapse roofs, break power and communication lines, and damage or kill vegetation.

Ash clouds can travel hundreds of kilometers from an eruption site and cover the land and sea surface in a blanket of ash. Even thin ashfalls can damage infrastructure due to their weight. When dispersed widely over a drainage basin, ash and tephra can change rainfall/runoff relationships. The low permeability of fine ash deposits can lead to increased runoff, accelerated erosion, stream-channel changes, and hazardous floods. In contrast, thick, coarse-grained deposits close to the source can increase the infiltration capacity of the soils and essentially eliminate surface runoff.

Volcanic ash is comprised of tiny pieces of rock and volcanic glass. Thus, it poses a serious health hazard; inhalation of ash can lead to major respiratory problems. Air traffic may be disrupted by the large ash clouds such as occurred during the recent eruptions in Iceland and Indonesia.

Many of the above hazards can be mitigated with proper planning. This includes clearing tephra from roofs as it accumulates, designing roofs with steep slopes, strengthening roofs and walls, and designing filters for machinery. Wearing respirators or wet clothes over the mouth and nose is suggested, because tephra can contain harmful gases adsorbed on the ash particles, as acid aerosols and salt particles.

Volcanic gases such as sulfur dioxide (SO_2), carbon dioxide (CO_2), hydrogen sulfide (H_2S), and many others are dissolved in magma and released as it reaches the surface. The release of gases often leads to eruptions, although some volcanoes produce large amounts of gas even when not erupting. When in sufficiently high concentrations, these gases can pose a variety of dangers to surrounding populations. Sulfur dioxide can irritate the eyes, skin, and respiratory system, and H_2S is toxic. Carbon dioxide is denser than air, and large releases of CO_2 can be lethal if breathable air is displaced, as happened in 1986 at the Lake Nyos disaster in Cameroon. Here, CO_2 that had accumulated in the deep waters of a volcanic lake rapidly rose to the surface, possibly triggered by a volcanic landslide. As the CO_2 gas flowed down nearby valleys, it displaced the surrounding air and led to the deaths of 1,746 people and 3,500 livestock. The eruption of the Icelandic volcano Laki in 1783 is another example of volcanic activity; this one had global consequences. The eruption lasted for nearly a year and released 15 km³ of lava over just eight months. The lava released huge amounts of SO_2 and fluorine gas, which affected the climate, causing famine and weather anomalies across the northern hemisphere. In Iceland, some 9,350 people and thousands of livestock died as a result of drought, famine, and gas poisoning. This type of volcanic smog forms by a complex series of chemical reactions, involving volcanic gases, air, and sunlight.

Deaths and injuries from lava flows are rare because lava typically flows at a walking-running pace, or slower. Lava flows, however, may cause fires, and engulf everything in their path. Recent lava flows from Kilauea, on Hawai'i, have buried roads, burned forests, and destroyed houses. Several people have died while walking on lava that had recently flowed into the ocean, as it collapsed. Such collapses are not uncommon. Other hazards associated with the entry of lava into the ocean include waves of scalding water or plumes of steam that rain down hydrochloric acid and tiny volcanic glass particles.

A **jökulhlaup**, an Icelandic term that means "glacial run," is a type of flash flood often triggered by the sudden emergence of a heat source beneath a glacier. Although the processes that lead to a volcanically induced jökulhlaup are complex, the basic concept is that magma rising to the surface, or even superheated groundwater mobilized by magma, melt large amounts of glacial ice rapidly, producing a sudden, massive outburst of floodwater. Initially, the heat forms subglacial lakes, which are then capable of catastrophic breakouts at a later date. Iceland, with its many glaciers and abundant volcanic activity, is the site of frequent jökulhlaups. An exceptionally large jökulhlaup triggered by the 1755–1756 eruption of Katla produced more water than the typical discharge of the Nile, Mississippi, Amazon, and Yangtze Rivers combined. Although this jökulhlaup was clearly a hazard of colossal proportion, even the more typical jökulhlaups pose serious danger to local populations.

Lahars are volcanic mudflows. Consider them as sediment-laden deluges composed of water and volcanic debris (commonly, mostly ash) that flow rapidly (50–100 km/h) down the slopes of a volcano, usually following the local drainage systems. They form because many volcanoes, especially large stratovolcanoes, are covered with glaciers and/or thick snowpacks; this snow melts rapidly during an eruption, forming the lahar. The sediment-to-water ratio in lahars can range from 20–90% by weight; if this proportion increases during flow as the water picks up ash and sediment, the shear stress, viscosity, and velocity of the flow also change, and the impact of the lahar increases.

Although largely confined to valleys, lahars are a greater threat to life and property in valley communities than any other volcanic phenomenon. Damage occurs due to impacts from large boulders or logs carried in the mud, by burial in mud, and by damage from the mud itself. Lahars can wipe out mature forests and any structures in their paths, including houses, bridges, and roads (**Fig. 7.41**). The most significant long-term environmental impacts of lahars include the (1) complete infilling of valley bottoms with thick accumulations of unconsolidated deposits of gravel, sand, and silt, (2) diversion of drainageways, and (3) destruction of wildlife habitats associated with preexisting rivers and wetlands. Given enough warning, people can quickly climb or drive to safety and avoid the lahar by evacuating valley floor areas.

Figure 7.41 Lahar damage from the 1980 eruption of Mt. St. Helens. Lahars that swept down this river valley carried the bridge on Highway 504 more than 500 m downstream and partially buried it. Source: US Geological Survey.

Landslides are masses of rock, sediment, and soil that travel rapidly downslope. In contrast with a lahar, a landslide involves failure of the slope of the volcano itself (see Chapter 14). Landslides, sometimes called **debris flows**, can be caused by earthquakes or periods of high rainfall, but volcanic activities such as magma intrusions and volcanic gas explosions can also be triggers. Large landslides caused by eruptions can create dams that block rivers and bury roads, bridges, and homes. Underwater and coastal landslides can also trigger tsunamis.

Volcanic earthquakes are caused by a sudden change in subsurface rock architecture. These earthquakes differ from the deep, tectonic events that occur along faults. Thus, they are considerably less devastating and may continue at a detectable level for weeks or months (usually before an eruption). Fortunately, they can be monitored to provide important insights into volcanic processes before eruptions, aiding scientific study and hazard mitigation.

Although volcanic earthquakes are usually isolated events, separated in time, **volcanic tremors** show rather spasmodic or harmonic behavior. They involve long-duration, more or less continuous, volcanic vibrations. Tremors are associated with activities such as the flow of underground magma, oscillations in magma reservoirs, and eruptions of volcanic gases.

REVIEW QUESTIONS

7.1   What are the differences between magma, lava, and pyroclastic debris?

7.2   What is the difference between an effusive eruption and an explosive eruption? How does the viscosity of the erupted material affect this?

7.3   What are the two end-member types of lava? How do they differ chemically and how do they impact the style of eruption?

7.4   What are the primary factors that affect the viscosity of lava? And how does viscosity affect the ability of lava to flow and deform?

7.5   What are the three major types of plate boundaries and what kinds of volcanism and landforms are typically found at each?

7.6   What are pillow lavas and how do they form?

7.7   How do the four major types of volcanoes erupt, and why are their eruptions so different? How do they differ in form, size, explosivity, and character of the lava?

7.8   What are hotspots and what kinds of volcanic features are associated with them? Name and describe at least two major hotspots on Earth.

7.9   What is a "trap" and what kinds of natural hazards are often associated with outpourings of lava at traps?

7.10   What are the main types of plutonic features, and how does each one form?

7.11   What do calderas look like, where are they found, and how do they form? How does a caldera differ from a summit crater?

7.12   What is a hydrothermal feature? What are the various kinds of hydrothermal features and how do they differ, geologically?

7.13   Explain the process that a geyser goes through, leading up to an eruption.

7.14   What are the main hazards associated with volcanoes?

FURTHER READING

Carson, R. 2000. *Mount St. Helens: The Eruption and Recovery of a Volcano*. Sasquatch Books.

Olson, S. 2016. *Eruption: The Untold Story of Mount St. Helens*. Norton and Co.

Oppenheimer, S. 2011. *Eruptions that Shook the World*. Cambridge University Press.

Sigurdsson, H. (ed.). 2015. *The Encyclopedia of Volcanoes*. 2nd ed. Academic Press.

Winchester, S. 2003. *Krakatoa: The Day the World Exploded: August 27, 1883*. Harper Perennial.

Witz, A. and Kanipe, J. 2014. *Island on Fire: The Extraordinary Story of a Forgotten Volcano That Changed the World*. Pegasus Books.

8 Landforms on Horizontal and Weakly Dipping Rocks

Randall Schaetzl

Arches, hoodoos, buttes, mesas … these are the picturesque landforms that most tourists and landscape-lovers know about, and which are the focus of many parks and recreation areas. All of these landforms are **bedrock-controlled**, with rock at or immediately beneath the surface. This chapter introduces a wide array of bedrock-controlled landforms. Most have formed on sedimentary rock, the most common rock in Earth's upper crust. Thus, much of the focus in this chapter will be on landforms developed on flat-lying bedrock strata (layers) that have experienced minimal tectonic disturbance throughout their history. Chapters 9 and 10 focus on bedrock-controlled landforms formed on much more tectonically active landscapes.

8.1 BEDROCK-CONTROLLED SLOPES

Many of the examples in this chapter are taken from arid or semi-arid climates, where erosion on rocks with minimal vegetation cover has produced distinctive, bedrock-controlled landforms. Nonetheless, most of the landforms in this chapter can (and do) also form in humid climates. Here, they are less spectacular, more rounded and subdued, and covered with soil and vegetation. Perhaps that is why geomorphology students everywhere flock to the drylands of the world to see classic landforms like arches, buttes, balanced rocks, and hogbacks.

Earth's landscapes are composed of many bedrock-controlled landforms, but some are less well formed and more difficult to discern. Many are overlain by thick **regolith** or overlain by dense vegetation, but they are there. Differential weathering and erosion has formed them, allowing harder rock layers to stand up as ridges and uplands, and wearing down the softer rocks into valleys and basins. Recall that **denudation** is the gradual wearing down, or lowering, of the land surface. Denudation works hand-in-hand with **erosion**, whereby sediment is moved from one location to another, usually lower, location. The main agents of erosion are water, wind, ice, and to a lesser extent biota, all operating in concert with gravity.

As denudation proceeds across a landscape, the loose, unconsolidated regolith that overlies the bedrock is eroded and removed, bringing the underlying bedrock closer to the surface. Uplands and areas of steep slope are stripped first. Eventually, after most of the regolith is stripped away, more and more of the surface begins to take on the pattern of the underlying bedrock, as influenced by the rock's composition, structure, and form. The landscape becomes increasingly bedrock-controlled, wherein the overall topography is largely determined by the resistance of the rocks to erosion. Areas above hard rocks stand up as uplands, and weaker rocks underlie valleys and lowlands.

Bedrock-controlled landscapes have many **weathering-limited slopes** and surfaces, that is, where the agents of erosion and denudation outpace the agents of weathering (see also Chapter 14 and **Fig. 14.2**). Weathering-limited slopes typically have only a thin cover of regolith, and are more common in dry climates where weathering is slow and the scattered vegetation is unable to hold all the sediment in place (**Fig. 8.1**). In humid climates and on more gentle

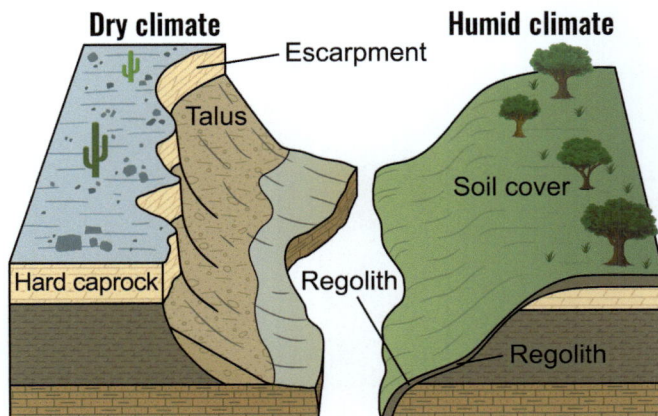

Figure 8.1 Bedrock-controlled slopes in arid vs humid climate settings. Weathering is slow and regolith is thin in arid landscapes. Weathering debris sometimes accumulates as talus below the eroding, steep slopes. In humid climates, weathering proceeds faster and the regolith is held in place by vegetation, limiting transport. Thus, a thick blanket of regolith covers all slopes. Nonetheless, in both climate types, the harder rock will stand up as steeper slopes, or even, as in the arid climate example, a free face (escarpment).

Figure 8.0 Monument Valley, in the southwestern United States, has many breathtaking examples of classical mesa-and-butte landforms. Source: Jeremy Edwards / iStock / Getty Images Plus.

slopes, weathering proceeds faster and erosion is slower due to a denser cover of vegetation. Here, most slopes have thick regolith and are **transport-limited**. Generally, though, in all climates, steeper slopes are more likely to have thinner soils and are more likely to be weathering-limited.

8.2 ROCKS OF THE CRUST

When viewed by volume, Earth's crust is composed mainly of crystalline (igneous and metamorphic) rocks (**Fig. 8.2A**). This situation is just as would be expected – crystalline rocks form under heat and pressure, both of which are in ample supply deep beneath the surface. Thus, most of the deep crust is composed of rocks like granite – an intrusive **igneous rock** – and different kinds of highly metamorphosed rocks. Lying above these deep-seated crystalline rocks is, at most places, a veneer of **sedimentary rocks** (**Fig. 8.2B**). Sedimentary rocks form as sediments which are deposited either on land or in a body of water such as an ocean or lake (see Chapter 5). That is, they form naturally *on top of* other rocks, in the uppermost portions of the crust. As a result, the first rocks that one usually encounters in the crust, for example, in a drill rig or an outcrop, are sedimentary (**Fig. 8.2B**). For this reason, many of the bedrock-controlled landforms on Earth are formed on sedimentary rock – the focus of this chapter.

8.3 SEDIMENTARY ROCK LITHOLOGY AND STRUCTURE

The complex evolution of bedrock-controlled landscapes over time is primarily determined by the rock's lithology (physical characteristics, including color, texture, grain size, and composition) and structure (three-dimensional characteristics and orientation), as conditioned by climate and vegetation cover. As shown in **Fig. 8.2C**, the most common sedimentary rock lithologies are **shale**, **sandstone**, and **limestone**. Of these, shale is easily the weakest in most settings, and so it almost always forms valleys or lowlands when exposed at the surface. In glacial regions, such as the upper Midwest of the USA, areas of shale bedrock have been preferentially scoured out and now form large parts of the modern Great Lakes' Basins. Sandstone and limestone can either be comparatively soft or resistant, depending on their degree of lithification, cementing agents, jointing, etc. Limestone is typically much weaker in humid climates, where it is attacked by carbonic and other acids (see Chapter 12). But in arid climates, limestones, along with many sandstones, are often very hard rocks.

Structure in sedimentary rocks is usually expressed as the **strike** and **dip** of the rock layers (**Fig. 8.3**). These two measures are used to describe the orientation and attitude of strata in space. They work particularly well for sedimentary rocks,

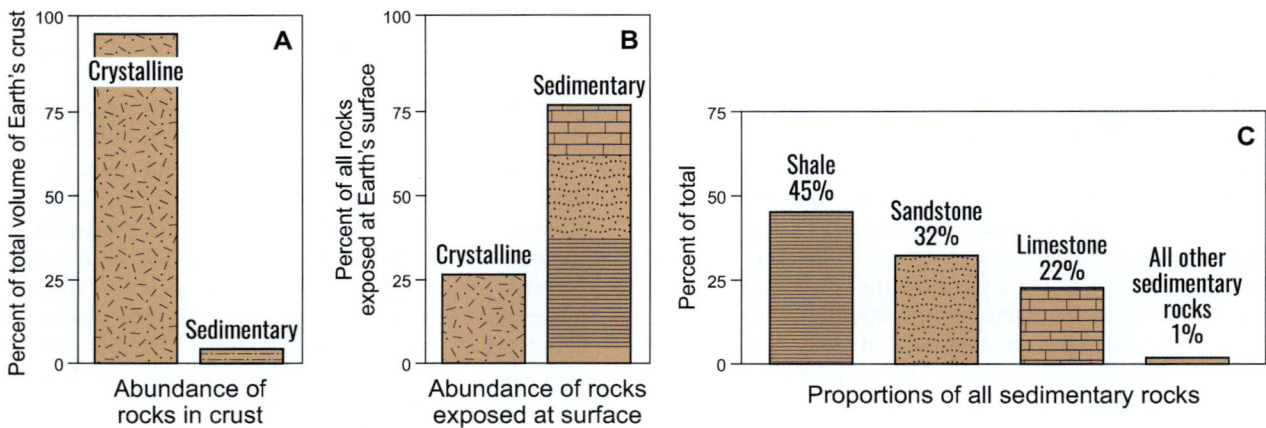

Figure 8.2 Composition of Earth's crust. **A.** By volume of rock types. **B.** By abundance of rocks exposed at the surface, that is, the local bedrock. **C.** By frequency of occurrence of the various sedimentary rock types.

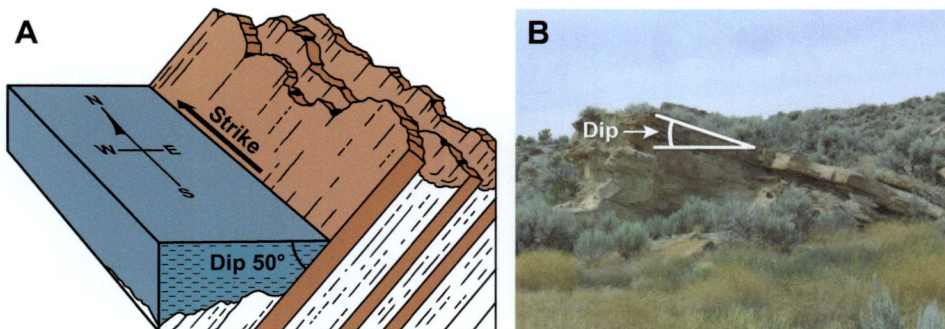

Figure 8.3 The concepts of rock strata strike and dip, as shown (**A**) graphically, and (**B**) in an outcrop. Source: (A) © A. N. Strahler (1992). Used by permission; (B) R. Schaetzl.

because of their natural planes of weakness that are often quite obvious. The acute angle formed between a rock layer and an imaginary horizontal plane is its dip. In **Fig. 8.3A**, a water surface is used to mimic such a horizontal plane. Flat-lying rocks have zero dip, whereas those standing vertically have a dip of 90°. Strike is defined as the compass direction of a line that perpendicularly intersects the rock layer and a horizontal plane. In **Fig. 8.3A**, the strike of the uppermost rock layer is north–south, with a dip of 50°.

8.4 LANDFORMS ON WEAK SEDIMENTARY ROCKS

We begin this discussion with landforms that have formed on rocks with *lithologies* that are very susceptible to weathering, for example, shales and mudstones. In such rocks, *structure* is often of little concern, because the rocks are so weak that they weather and erode rapidly, irrespective of structure. In humid climates, shale is quickly eroded to form lowlands. Few recognizable landforms will result; the geomorphologist would only see a lowland, and the main landforms would have been cut by running water. Vegetation may cover much of the surface, and under it a thick cover of regolith, hiding the weak bedrock.

Lithology again trumps structure on these rocks where they occur in arid or semi-arid climates. Here, vegetation cover is often minimal and exposures of bare rock are commonplace. In such landscapes, where the shale and its regolith have low permeabilities, water is prone to run off. Heavy but infrequent precipitation – typical of dry climates – on exposed shale bedrock can produce copious amounts of runoff and drive high rates of erosion, thwarting any plant growth. Thus, exposures of shale typically have thin or non-existent regolith and soils, resulting in a nearly barren landscape with little vegetation. Denudation rates on shale landscapes can be extremely high. Erosion outpaces soil formation, forming transport-limited slopes with a maze of rills and gullies on bare bedrock, a type of landscape called **badlands topography**. Intricate dissection of shale and mudstone bedrock characterizes badlands (**Fig. 8.4**). All the rocks exposed are weak and highly erodible.

Badlands landscapes are self-sustaining. Exposed surfaces will continue to retreat, but *en masse*. Therefore, the landscape will not appear to change in overall appearance until denudation takes it to the point where shale is no longer the uppermost rock.

Perhaps the best and most well-known area of badlands topography is in the semi-arid climate of Badlands National Park, South Dakota, USA. Here, weakly cemented rocks, rich in silt and clay (not all are shales, some are siltstones), with interbedded deposits of volcanic ash, are exposed along a retreating **escarpment** (**Fig. 8.4**). Here, a perfect combination of conditions promotes the formation of badlands topography – thick deposits of soft, erodible sediment, along a retreating escarpment in a semi-arid climate with its minimal vegetation cover. Because of the relatively uniform geology across most badlands landscapes, dendritic drainage patterns typically develop in the eroding sediments (see Chapter 16). Badlands are some of the most picturesque landscapes, with a seemingly different palette of color and topography around every turn. They are favorites for tourism across the globe. Paleontologists also value badlands because fossils are usually well preserved in the shale bedrock, and easy to extract. Key examples, in addition to Badlands National Park, include Theodore Roosevelt National Park in North Dakota, USA, the Cheltenham Badlands in Ontario, Canada, the Badlands near Zabriskie Point in Death Valley National Park, California, USA, and Makoshika State Park in Montana, USA. Many of these sites have also contributed to important fossil discoveries.

8.5 LANDFORMS ON FLAT-LYING SEDIMENTARY ROCK OF VARYING HARDNESS

In arid and semi-arid climates, many slopes are weathering-limited. Here, bare rock is often exposed at the surface, especially in areas where erosion is rapid, as along steep slopes with thin or non-existent soils. Along such steep outcrops, the effects of rock lithology and structure on the geomorphology can be readily seen.

Figure 8.4 The classic badlands landscape of Badlands National Park, South Dakota, USA. Source: R. Schaetzl.

Let's begin with rocks in the simplest situation – sedimentary rocks that have little or no dip, that is, they are nearly flat-lying. As these types of landscapes evolve, erosion becomes most pronounced along river valleys. The rivers are able to incise their valleys because slope processes deliver little sediment to their channels (**Fig. 14.2**) Thus, they cut down and often flow in deep, narrow valleys. Many of these rivers may be **exotic streams**, if they originate in a wetter and more humid location and flow today through an arid region. The Nile River of eastern Africa and the Colorado River of the American Southwest are classic exotic streams. Escarpments often form along the sides of stream valleys, as the rivers incise into the bedrock. An **escarpment** is a steeply sloping face, usually of rock, formed by erosion into preexisting rock or sediment; some might call it a cliff, a wall, or a bluff (**Fig. 8.5**). As they weather and rock fragments fall off, escarpments retreat backward, into the parent rock or sediment (**Fig. 8.6**). This process continues more-or-less unabated as long as the debris (**talus**) that accumulates at the base of the escarpment can be removed. If these materials are not removed, they eventually bury the escarpment, and erosion is slowed or stopped (**Fig. 8.5**). The larger the fragments, the more difficult they are to remove.

At Badlands National Park, most of the topography is formed in thick beds of siltstone and shale within a retreating escarpment known locally as "The Wall" (**Fig. 8.6**). Flowing near the base of The Wall, the White River is able to carry away the sediment eroded from it. The sediment here is fine-textured and easily washed downslope and into the river. Consequently, little sediment accumulates at the base of this retreating escarpment, sustaining its rapid retreat.

Rocks and materials weathered and eroded off retreating escarpments, called talus (Old French *talu*, "slope"), typically accumulate at the base of an escarpment (**Fig. 8.7**). Talus can be pronounced either as *tay*-lus or *tah*-lus. Talus can be of many sizes, from sand to large boulders (**Fig. 8.7**). The angle of the slope that is formed by the accumulating talus – its **angle of repose** – is remarkably consistent, with a range of values between 35° and 45° (**Fig. 8.8**). Coarser particles and larger blocks, however, tend to develop steeper angles of repose (**Fig. 8.8**).

Talus is not a landform. Rather, it is the material that has accumulated below an eroding escarpment, or **free face**. The term "free face" implies that the slope is so steep that anything not physically attached to it is "free" to fall off. Material removed from a free face is delivered to the base of

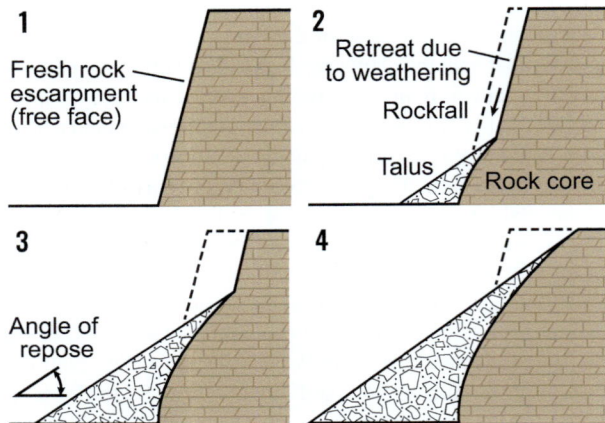

Figure 8.5 Idealized retreat of a bedrock escarpment over time. This process is best developed in arid landscapes. **1.** A fresh, bedrock escarpment. **2.** Rock debris (talus) falling off the retreating escarpment accumulates at its base. **3.** Talus continues to accumulate at the angle of repose. If processes that remove the talus proceed faster than those that produce it, the escarpment will continue to retreat. But if the talus accumulates too rapidly or the escarpment is not tall enough, it will become buried in its own talus, as shown in **4**. Then, the retreat of the escarpment will be slowed or stopped. Source: After Young (1972).

Figure 8.6 Two views from the top of The Wall – a retreating escarpment – at Badlands National Park, South Dakota, USA. Park personnel report that the rate of retreat is ≈1–3 cm yr⁻¹. Source: R. Schaetzl.

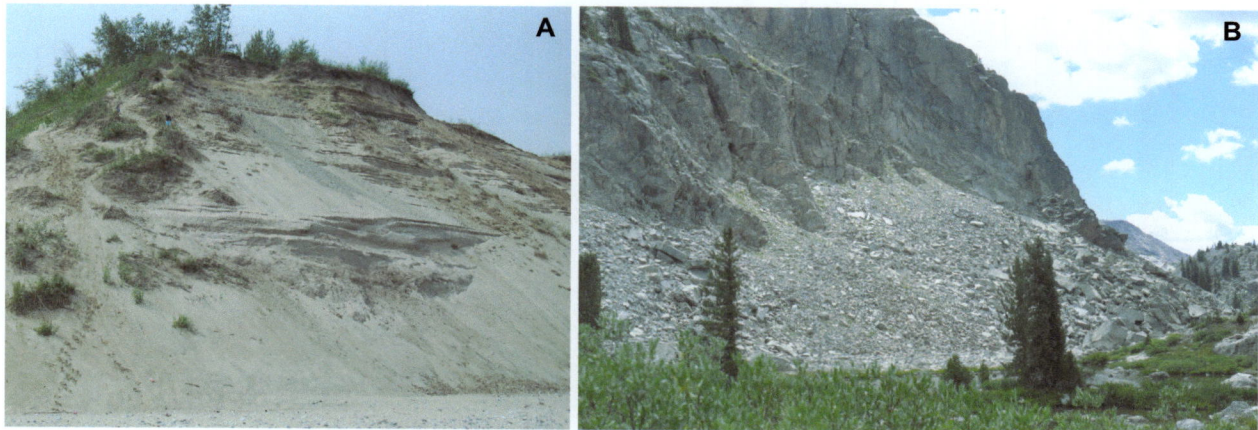

Figure 8.7 Talus in different settings, and of different sizes, has angles of repose that generally fall within the 35°–45° range. **A.** Sandy talus on the shore of Lake Superior, Michigan, USA. **B.** Bouldery talus in the Wind River Mountains, Wyoming, USA. Source: R. Schaetzl.

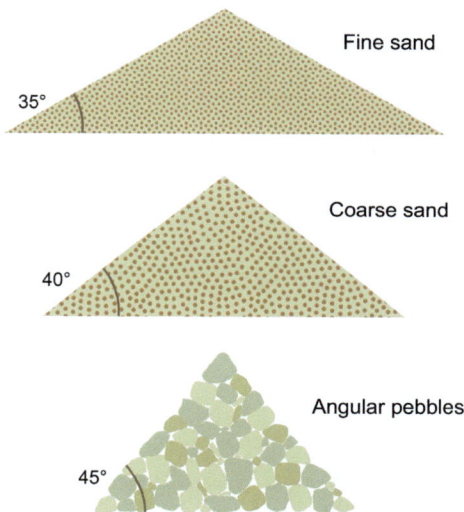

Figure 8.8 Examples of how the angle of repose in sediment varies as a function of sediment size.

the slope as talus by falling, rolling, and/or bouncing (**Fig. 14.18**). Many free faces weather as water gets into cracks in the rock and freezes, prying rock off the free face and allowing it to fall (**Figs. 14.16, 14.17**). For this reason, some of the best examples of talus are in alpine areas, where freeze–thaw processes are frequent and strong.

Rockfalls are especially common in escarpments held up by a resistant stratum, or layer (**Fig. 8.9**). Such layers are called **caprocks**. As shown in **Fig. 8.9A**, escarpments beneath caprocks often retreat as weaker rock undercuts the resistant caprock above it. Fragments of rock will break off the undercut caprock and fall onto the talus below, continuing the retreat of the escarpment.

Accumulating talus can form distinct landforms. Where the free face is long and continuous across an escarpment,

a broad belt of talus can accumulate as a landform called a **talus slope** (**Fig. 8.10**). Talus slopes are common in glaciated valleys, where a previous glacier has eroded so deeply that long, nearly vertical, bedrock faces form the valley-side walls. Sometimes, the eroding free face has valleys cut into it. Rockfalls on these slopes will preferentially deposit talus within these valleys and at their base, forming **talus cones** (**Figs. 8.10, 8.11**). Materials weathered out of uplands can also be delivered to the base of the slope by running water, sometimes forming alluvial fans (see Chapter 9, and **Fig. 8.10**), or by other forms of mass movements, such as small rock- and debris-rich avalanches or debris flows (see Chapter 14, and **Fig. 8.10**), or even **rock glaciers**.

The type of escarpment retreat shown in **Fig. 8.9** is especially apparent in dry climates, where the rocks are often exposed because of thin (or no) soil and limited vegetation cover. Here, rivers can incise and cut up bedrock uplands or **plateaus**, which are usually capped by a resistant caprock (**Fig. 8.12**). In many areas of dry climate, the caprock is a **competent** (generally hard, or hard to weather) sandstone or limestone. The Colorado Plateau in the southwestern United States is the classic example of a broad plateau held up by a resistant caprock – usually the Kaibab Limestone. Much of the underlying geology is then exposed as large rivers cut down, into the underlying strata. The most prominent and well-known of these is the Colorado River and its major tributaries.

Over time, the escarpments at the edges of plateaus retreat, as the lowlands between them become wider (and possibly deeper). As a result, the plateaus get "cut up" into isolated upland segments known as **mesas** and **buttes** (**Fig. 8.12**). The distinctions made between plateaus, mesas and buttes are based on areal extent, given that they all formed in generally the same way – through escarpment retreat around their edges. All are usually held up by a resistant caprock.

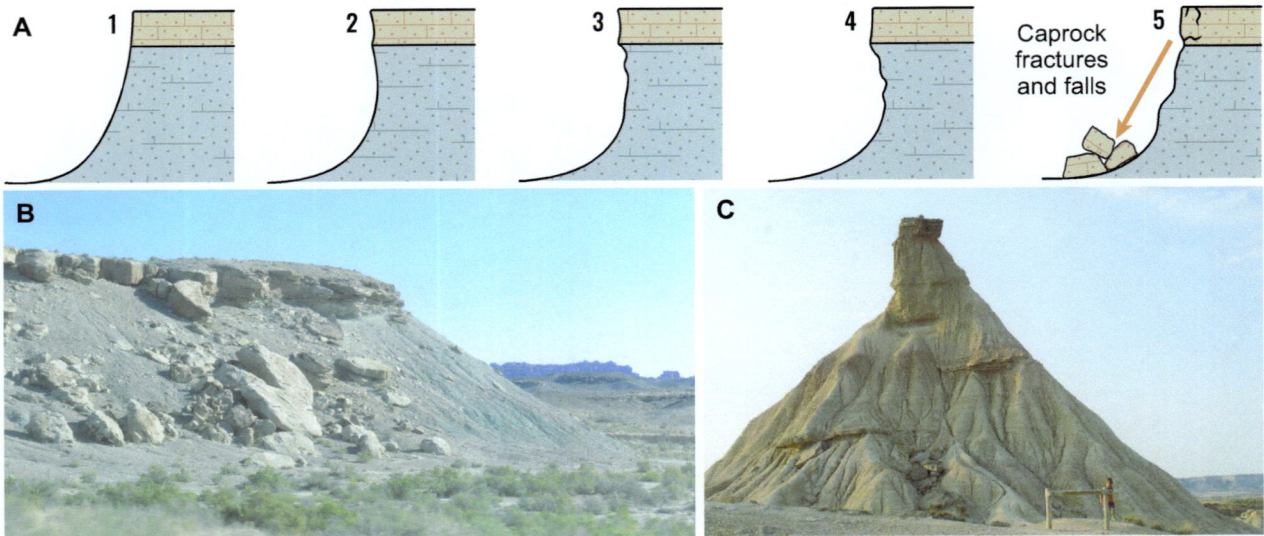

Figure 8.9 Development and retreat of escarpments with hard caprocks. **A.** A schematic illustration of an eroding escarpment with a prominent caprock. Over time, undercutting of the softer rock leaves the caprock in an overhanging condition, causing (5) collapse and eventual retreat of the escarpment. Of course, if significant amounts of rubble accumulate at the base of the slope, as shown in **Fig. 8.5**, the retreat of the escarpment is slowed. **B.** This escarpment in Utah, USA, is held up by a resistant limestone caprock, underlain by a thick layer of weak shale. Note how the limestone retreats mainly via breakage of large blocks – as rapidly as the undercutting shale will allow. Source: R. Schaetzl. **C.** This small butte has lost almost all of its caprock. Source: Machulin, CC BY-SA 3.0, via Wikimedia Commons.

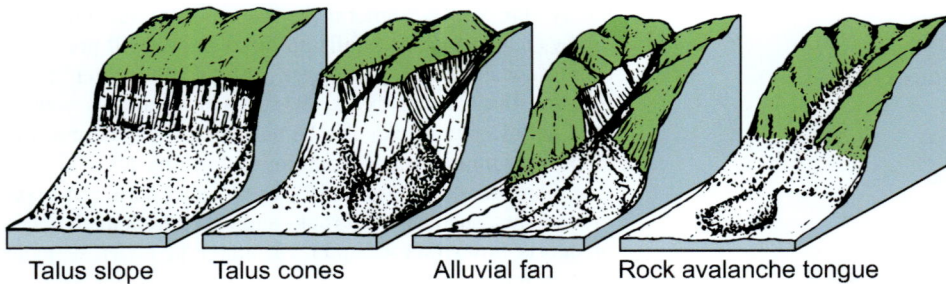

Talus slope   Talus cones   Alluvial fan   Rock avalanche tongue

Figure 8.10 Some of the different types of landforms that can develop at the base of an eroding free face or upland. Source: After Selby (1985), with permission from Oxford University Press.

Figure 8.11 A talus cone in the Wind River Mountains of Wyoming, USA. The growth of vegetation on the cone suggests that additions of talus have recently slowed. The majority of the talus probably accumulated during the Little Ice Age, when frost action would have been much more prominent (see Chapter 4). Source: R. Schaetzl.

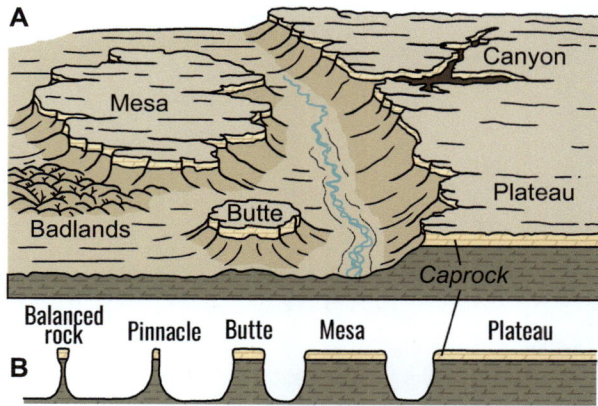

Figure 8.12 The distinctive suite of bedrock-controlled landforms that forms in arid and semi-arid landscapes on horizontally bedded sedimentary rock. Note the pronounced caprock on most of these landforms. **A.** Landscape view. **B.** Side view.

The type of landscape that forms as this process unfolds is often referred to as a **mesa-and-butte landscape**. Perhaps the most photogenic buttes and mesas in the world exist in Monument Valley, part of the Navajo Nation's Tribal Park in northeastern Arizona, USA (**Fig. 8.0**). Here, a thick layer of hard sandstone has capped an eroding layer of shale (**Fig. 8.13**). Thinner layers atop the hard sandstone are eroding as well, but they will only retreat as fast as the sandstone beneath it does. In this way, the sandstone caprock controls the retreat of the strata above and below it.

Additional terms for the slopes and landscapes that form on horizontally bedded rock, under this type of erosion/denudation scenario, include **stepped structural surfaces** and **cliff-and-bench topography**. Both terms are excellent, because the steps (or benches) form directly in concert with the lithology and structure of the rocks (**Fig. 8.14**). In a series of stepped structural surfaces, the hardest rock layers will limit the rate of escarpment retreat, with most of them forming vertical and free faces. The weaker strata will usually be worn down and covered with talus from above, forming a distinctly flat or gently sloping "step"

or platform surface (**Fig. 8.15**). Some of the most well-known "platforms" of the Grand Canyon, for example, the Tonto Platform and the Esplanade, have formed on beds of weak shale, held up by, and resting upon, resistant strata below. Equally impressive stepped structural surfaces are present in Dead Horse Point State Park, near Moab, Utah, USA, exposed here by deep incision of the Colorado River (**Fig. 8.16**).

Over time, buttes and mesas continue to erode and get progressively smaller until they eventually become a tall, narrow, bedrock feature known by many names, such as **pinnacle**, **pillar**, or **tower** (**Figs. 8.12B, 8.17**). As soon as the pinnacle loses its caprock, it begins to erode much more rapidly. But even when the caprock remains intact, erosion continually works to wear away at the feature, especially on the neck. At some point, the neck of the pinnacle gets so narrow that it will topple. This process will take longer if it can preserve its caprock. Sometimes, however, while the caprock remains, a unique type of landform can develop, where the caprock on the pinnacle appears seemingly "balanced" on the neck below (**Fig. 8.17**). These **balanced rocks** occur where

Figure 8.14 Vista overlooking the Grand Canyon, in the southwestern United States, from the north rim. The flatness of the Colorado Plateau, held up by the Kaibab Limestone, is very apparent from this view, as are the stepped structural surfaces that form the sidewalls of the Canyon. Source: R. Schaetzl.

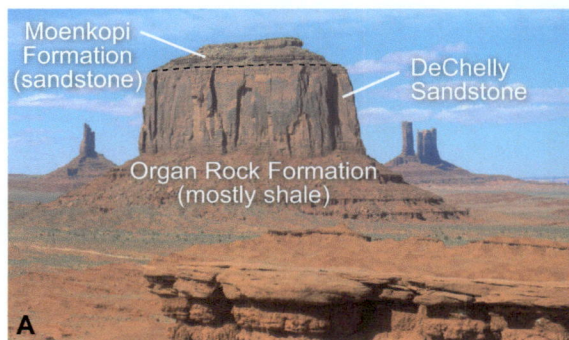

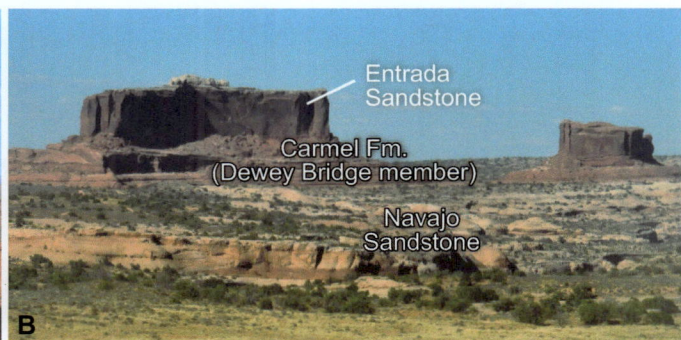

Figure 8.13 Buttes. **A.** A classic view of the buttes in Monument Valley, Arizona. Source: W. Ranney. **B.** The famous Monitor (on the right) and Merrimac Buttes, near Moab Utah, USA. Although a sandstone, the Carmel Formation is much weaker than the overlying, massive Entrada Sandstone, because it contains softer beds of gypsum. Source: R. Schaetzl.

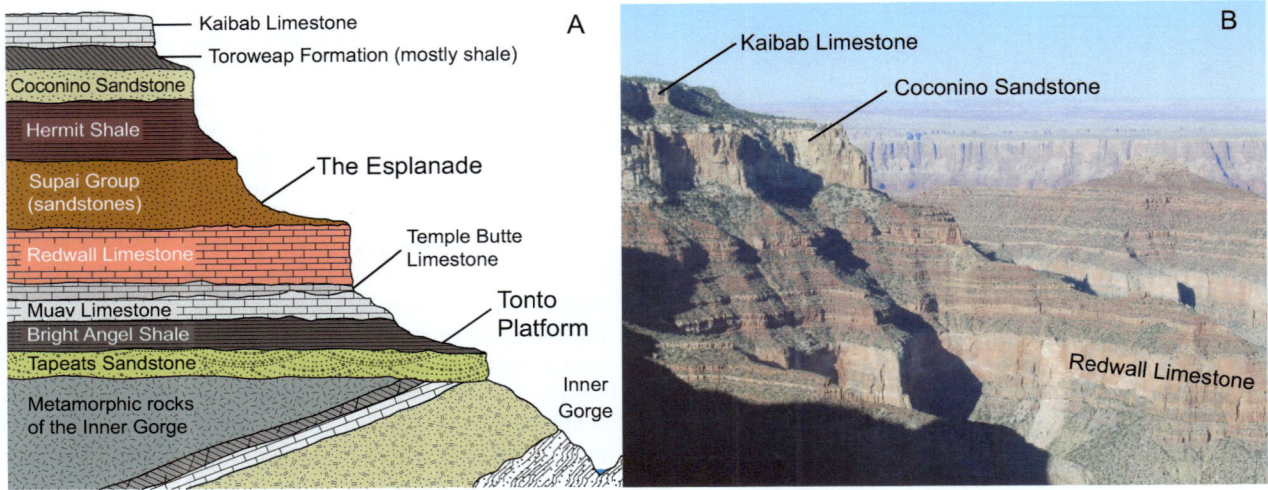

Figure 8.15 The rocks and "rock steps" of the Grand Canyon. **A.** A geologic cross-section through the Grand Canyon, illustrating how the various rock strata form escarpments vs platforms and steps. **B.** This photo illustrates many of the same strata in the upper part of the canyon. Source: R. Schaetzl.

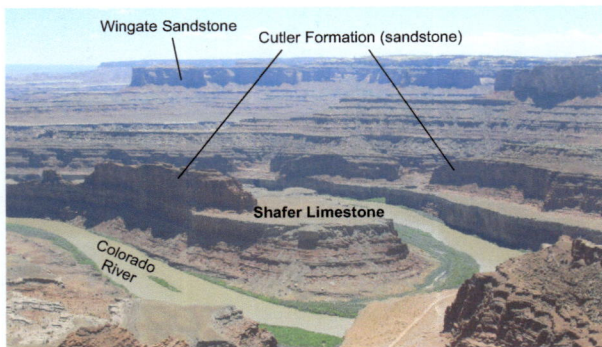

Figure 8.16 The distinctive stepped structural surfaces near Dead Horse Point State Park, Utah, USA. Hard sandstone and limestone strata form the vertical escarpments that flank the valley. Source: R. Schaetzl.

flat-lying sedimentary strata, such as shales, limestones, and sandstones, occur in weathering-limited environments. They are fairly common on the Colorado Plateau (**Fig. 8.17**), where flat-lying sedimentary rocks with different resistances to weathering and erosion are widespread. Shale and poorly cemented siltstones or sandstones erode much faster than the well-cemented sandstones that often cap these balanced rocks.

Balanced rocks are sometimes thought to form by wind-driven abrasion by sand, focused near the base of a pinnacle (**Fig. 8.18A**). Although this process may happen in some extremely sandy and windy areas, like the western deserts of Egypt (**Fig. 8.18B**), abrasion is not normally an important process in the formation of balanced rocks – or the erosion of their necks. How do we know this? Should you ever visit these locations, try a simple test for whether a surface was

Figure 8.17 Balanced rocks. **A.** The Goblet of Venus, perhaps the world's most famous balanced rock, which once stood in Natural Bridges National Monument, Utah, USA. Sadly, in 1948, vandals used a chain to tear down this unique geological feature. Source: Utah State Historical Society. **B.** A large balanced rock in Arches National Park, Utah. Source: R. Schaetzl. **C.** The myriad "goblins" in Goblin Valley State Park, Utah, are classic examples of balanced rocks. Source: R. Schaetzl.

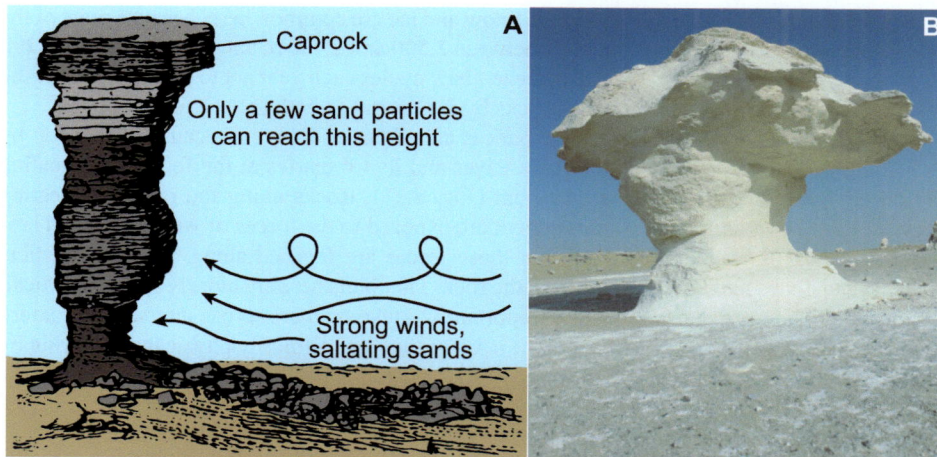

Caprock

Only a few sand particles
can reach this height

A

B

Strong winds,
saltating sands

Figure 8.18 Balanced rocks. **A.** A schematic representation of how a balanced rock might form by erosion from saltating sands. Such erosion would be focused on the lowest part of the neck. **B.** A typical "mushroom rock" near the Bahariya Oasis in Egypt's White Desert, where abrasion and erosion by sand is especially strong. Source: Nearodney, Public domain, via Wikimedia Commons.

eroded and abraded by **saltating** sands (see Chapter 21). Rub your hand against the surface of the rock. If the rock in the neck was eroded by "sand blasting," its surface will be smooth. Sand abrasion is like sandpaper; it smooths the rock. **Ventifacts** – rocks eroded by this type of sandblasting – typically have smooth surfaces (**Fig. 21.17**). However, if you try this test on the neck of most pinnacles, you would likely need a bandage for the scrape on your hand. The necks of pinnacles are rough, because they were produced by **weathering**, not sandblasting. Erosion of the necks of many balanced rocks involves various forms of weathering, accompanied by rainsplash and water flowing down the neck, picking up mineral grains made loose by weathering and transporting them away. Sandblasting by wind-driven sand is an effective agent of erosion in only a few areas (**Fig. 8.18**).

Many factors – most of them geological – must come together to *form* and *maintain* a balanced rock. Because it requires a type of geological "perfect storm" to form a balanced rock, these features are not very common. But when one also considers how *fragile* they appear to be, how *top-heavy* they are, and how they often are so precariously *balanced*, it makes one wonder why there are even any at all! One hypothesis that helps to explain why so many balanced rocks exist – and persist – is shown in **Fig. 8.19**. In this model, weathering and erosion are, of course, focused on the lower parts of the neck. But should these agents preferentially wear away one side of the neck, the induced weakness there will cause the balanced rock to lean in that direction. This "lean" increases the pressure on the rocks on that side of the neck, strengthening them and facilitating increased erosion on the *other* side. This to-and-fro may continue for

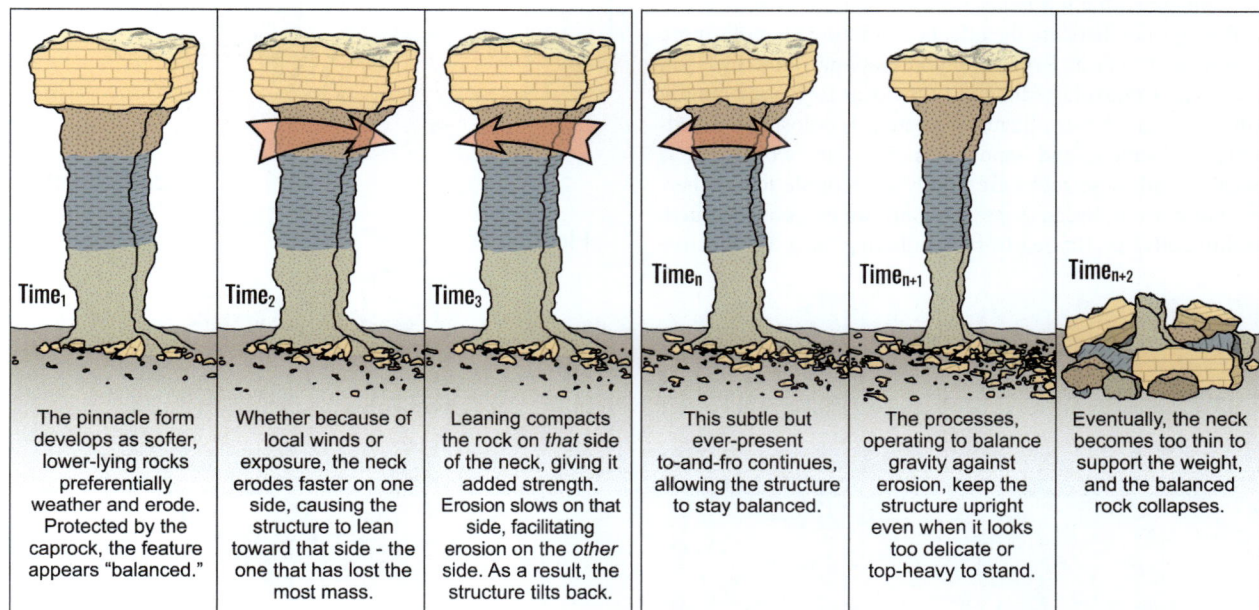

Time₁ — The pinnacle form develops as softer, lower-lying rocks preferentially weather and erode. Protected by the caprock, the feature appears "balanced."

Time₂ — Whether because of local winds or exposure, the neck erodes faster on one side, causing the structure to lean toward that side - the one that has lost the most mass.

Time₃ — Leaning compacts the rock on *that* side of the neck, giving it added strength. Erosion slows on that side, facilitating erosion on the *other* side. As a result, the structure tilts back.

Timeₙ — This subtle but ever-present to-and-fro continues, allowing the structure to stay balanced.

Timeₙ₊₁ — The processes, operating to balance gravity against erosion, keep the structure upright even when it looks too delicate or top-heavy to stand.

Timeₙ₊₂ — Eventually, the neck becomes too thin to support the weight, and the balanced rock collapses.

Figure 8.19 A model (one possible explanation) of how balanced rocks stay balanced.

some time, always keeping the rock in balance, until the neck becomes too thin and weak to support the caprock. What an amazing geomorphic system!

8.6 EFFECTS OF JOINTING

The discussion above has focused on landforms developed on generally flat-lying sedimentary rocks. The assumption is that these rocks, whether weak strata as in Badlands National Park or strong and competent sandstones and limestones elsewhere, are generally similar in erosion resistance across/within each rock layer. That is, each bed of Navajo Sandstone is generally of similar resistance across an outcrop or even across a span of several kilometers. As a result, their escarpments retreat predictably and more-or-less uniformly.

But what if the rock layer had weaker and stronger areas *within* itself? This variability could occur because of changes in lithology, cementation, or mineralogy across the extent of the layer. Such lateral changes are not at all uncommon in sedimentary rocks, because of **facies** variations that developed when they first formed, or variations that may have developed later by such processes as **diastrophism** – deformation involving bending/folding stresses. Diastrophism and bending can lead to areas of excessive jointing in rocks. **Joints** are cracks in rock along which no movement has occurred. If movement occurs along a joint we refer to it as a fault (see Chapter 10).

Jointing in sedimentary rocks occurs as the rock is bent or slightly folded, because of crustal upheaval or downwarping, that is, diastrophism. Any force that "flexes" a rock layer can lead to cracks and jointing, and the joints will later become focused areas of weathering and erosion. Jointing weakens rocks, because the rock may wear away (internally) along joints, as well as at its edges (escarpments). Areas between the joints are often left behind as erosional residuals.

Perhaps nowhere are the effects of jointing on rocks more pronounced than at Bryce Canyon National Park in Utah, USA. The Cenozoic sedimentary rocks at Bryce are mainly soft and weakly consolidated limestones, dolostones, mudstones, siltstones, and sandstones. The key word here is "weak" – all these rocks are highly susceptible to weathering and erosion, because they are only weakly consolidated. Additionally, at Bryce, frost weathering is a very active

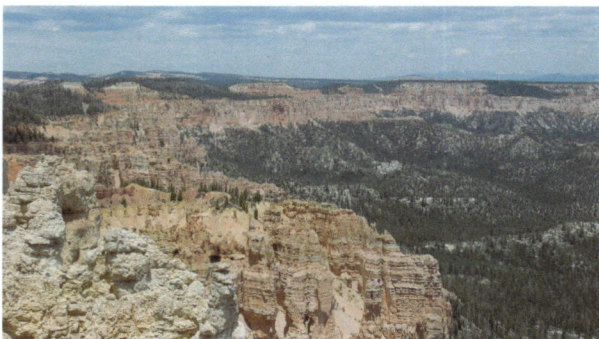

process. Snow and ice are common here, because most of the park is between 2,500 and 2,700 m above sea level. Over 200 daily freeze–thaw cycles each year act to rapidly weather the weak rocks by ice crystal growth (see Chapter 11).

The park is called Bryce Canyon because of incision by the Paria River and its tributaries at the foot of a retreating escarpment (**Fig. 8.20**). Rocks along the rapidly retreating escarpment are exposed to the forces of weathering and erosion, but these forces are focused along the many vertical joints in the rock. The joints here are closely spaced, vertical, and – importantly – take on a crisscross intersecting pattern (**Fig. 8.21**). As weathering, primarily physical weathering by

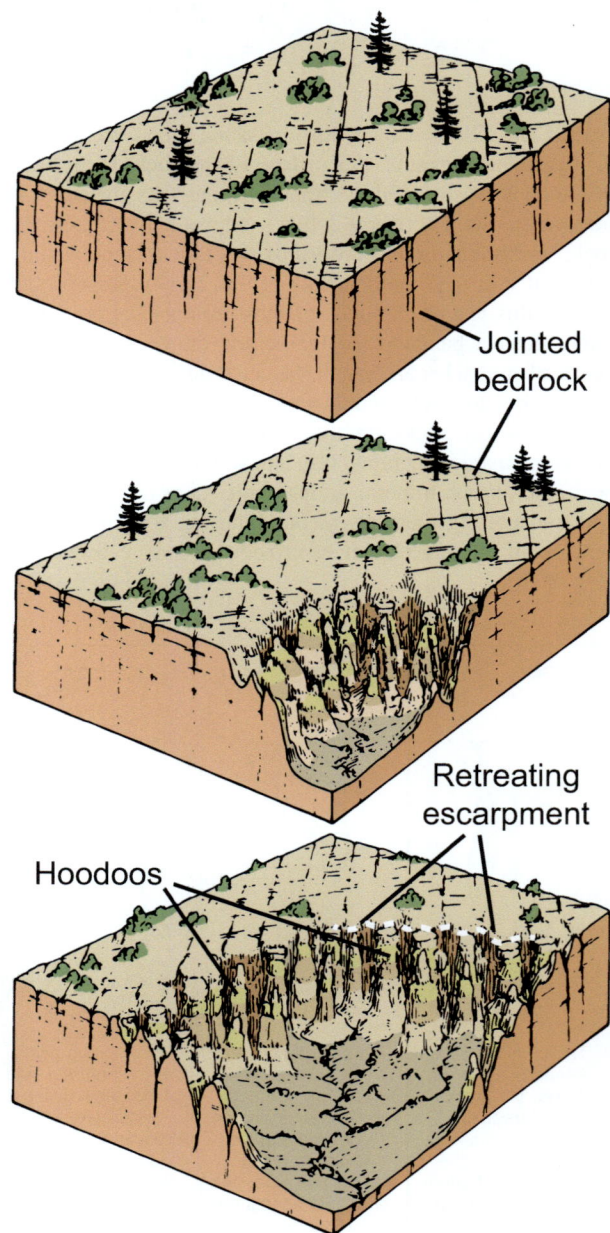

Figure 8.21 The jointing pattern at Bryce Canyon National Park, as depicted over time, and how it impacts the retreat of the escarpment and the resultant hoodoos.

Figure 8.20 The retreating escarpment at Bryce Canyon National Park, Utah. Source: R. Schaetzl.

ice crystal growth within joints and cracks, works on and opens these joints, particles are dislodged from the rocks by rain and melting snow. The rain and melting snow that remains in the joints provides fuel for another round of freezing that might arrive the next night. Tall spires, a type of pinnacle with a narrow, sometimes pointed top, called **hoodoos** result from this unique rock and climate intersection (**Figs. 8.21, 8.22**).

Hoodoos are a unique, picturesque, and ever-changing type of landform, with lifespans of years to decades. Weathering and erosion quickly transform hoodoos into narrower and narrower spires, until they eventually topple. But never fear – retreat of the nearby escarpment exposes more jointed rock, driving the formation of more hoodoos. The cycle continues (**Fig. 8.21**). Key to the continued existence of the hoodoos is the balance between erosion of the rock (formation of the hoodoos) and the removal of the debris that is constantly being created (**Fig. 8.22C**). As long as the sediment is removed, more hoodoos can and will form. If the sediment builds up, hoodoos become buried in **talus**.

Lastly, we examine a situation where competent rock has developed *parallel* (but not intersecting) joint patterns. As with intersecting joints, this type of joint pattern is typical of rock that has undergone slight folding or bending (see Chapter 10; **Fig. 8.23A**). Weathering along the vertically oriented, parallel joints exposes vertically oriented "slabs" of rock – called **fins** – that over time become more widely spaced, as weathering and erosion widen the openings between them (**Fig. 8.24B**).

Fins are a common occurrence at Arches National Park in Utah, USA, where they have formed in competent Entrada Sandstone (**Fig. 8.24**). In the semi-arid climate of Arches National Park, like many of the landscapes we have been discussing, regolith cover is thin and bare rock is exposed in many places. This setup facilitates the weathering of the bare rock along joints. Without a soil cover, water can more easily get into the joints, and the lack of soil also facilitates wider diurnal temperature swings within the rock itself. Thus, any water in the joints has a higher likelihood of freezing on cold nights, slowly wearing away the rock and opening up

Figure 8.22 The hoodoos at Bryce Canyon National Park. **A.** and **B.** Overlooks showing the hoodoos and in **B**, illustrating how even the thinnest layer of more durable rock can temporarily form a caprock. **C.** Sediment eroded from hoodoos accumulates as talus and is eventually transported away by streams; shown here is a dry stream bed at the foot of the main escarpment in the park. Source: R. Schaetzl.

Figure 8.23 The formation of arches. Parallel jointing in competent sedimentary rocks (**A**) leads to the development of fins (**B**), which may then weather into arches.

Figure 8.24 Fins at Arches National Park, Utah. **A.** As the parallel joints get increasingly weathered and eroded, the spacing between the fins widens. **B.** Eventually, in some locations, the fins stand alone as isolated landforms. Shown here is one of the most prominent fin groups in the park, known as "Wall Street." Source: R. Schaetzl.

joints by frost action. Plant roots, seeking this water, also help weather and erode the rock within and near joints (**Fig. 8.24A**).

Over time, the gaps between fins gets increasingly wider, due to weathering and erosion. Blowing sand also impacts the bottom few meters of the fins, preferentially eroding and thinning them at the base, while having little effect on the upper parts of the fins. The lower parts of the fins are also wetter, because rainwater frequently trickles down to the base, and because this part of the fins is more often shaded. The result: wet–dry cycles, freeze–thaw activity, frost weathering are maximized along the *bases* of the fins. Plants, seeking out the extra water near the bases of the fins (**Fig. 8.24A**), also continue to pry apart the sandstone grain-by-grain with their roots – a type of **biophysical weathering** (**Fig. 11.10**). Seeps of water may develop at preferred sites along the base of the fins. Over time, the fins thin out, especially along their bases. Some of the ever-thinning fins develop indentations called **apses** or alcoves (**Figs. 8.25, 11.23**). With continued erosion, the apse may weather completely through the fin, forming a majestic landform called a natural **arch** (**Fig. 8.26**). To qualify as an arch, the hole in

the base of the fin must be large enough for a person to crawl through – highly quantitative indeed!

Instead of developing gaps near their bases, some fins will widen and open up at their crest. These openings, called **windows**, are also common in fins as they evolve. Some exceptionally wide windows in fins are shown in **Fig. 8.24B**.

Arches are some of the most iconic of landforms, with an attraction that is hard to describe. They are not common because their formation requires the right kind of circumstances – rock (hard, competent sandstone is probably best), with parallel but mainly vertical jointing, in a semi-arid landscape to help maintain bare rock exposures, and a climate cold enough to have abundant freeze–thaw events. The elevation at Arches National Park provides excellent conditions for freeze–thaw activity, at between 1,245 and 1,720 m above sea level. Snow is present in the winter, and diurnal freeze–thaw events can occur at most times of the year. It's the perfect environment for the formation of magnificent sandstone arches.

8.7 LANDFORMS ON GENTLY DIPPING SEDIMENTARY ROCK

Bedrock-controlled landscapes on sedimentary rock are common. As indicated in **Fig. 8.2**, on >75% of Earth's land surface, the uppermost rock is sedimentary. Until now, this discussion has focused on landforms formed on sedimentary rocks with minimal dip – those which are nearly flat-lying. However, on many landscapes, the sedimentary rock layers are dipping, because of crustal downwarping or upwarping, typical of rocks that flank broad domes or basins. In some instances, these rocks may even be tightly folded and contorted (see Chapter 9). Regardless, similar denudational processes occur; escarpments form (often by incision along river valleys) and continue to retreat by weathering and erosion. The differences often lie in the direction of retreat. On horizontal strata, escarpments are generally free to retreat in any direction, always eroding *into* uplands such as plateaus, mesas, and buttes. When the rock layers are inclined and dipping, however, the escarpments always

Figure 8.25 This outcrop of Entrada Sandstone near Arches National Park is being preferentially weathered and eroded at its base, forming apses, or incipient arches. Plants seek out these areas, further enhancing the effect. Source: R. Schaetzl.

Figure 8.26 Some of the classic and prototypical arches in and near Arches National Park, Utah. **A.** Delicate Arch – the park's most famous landmark. **B.** Corona Arch, just outside the park. **C.** Double Arch (left). Note the incipient arches (apses) forming in the nearby sandstone. **D.** Landscape Arch – the world's longest arch, at 93 m. In 1991, a 22 m-long slab of rock fell from underneath the thinnest section of the arch, but as of this writing, the arch remains intact. Source: R. Schaetzl.

retreat down-dip – toward lower elevations. This property is shown in two examples – around a **basin** and a **dome** – in **Fig. 8.27**. A dome is an area of uplifted or up-folded rock, whereas in basins the rocks have been down-folded, "sagging" in the center. In both cases, erosion and formation of valleys and escarpments starts at the highest parts of the landscape and proceeds downward, toward lower areas. Likewise, in both cases, the escarpments generally will parallel the strike of the rock strata. Therefore, in a basin, the escarpments that tend to form around the basin will retreat downward, into the basin center. Escarpments in this situation will face outward, away from the center of the basin (**Fig. 8.27B**). The opposite setup develops on domes, where escarpments face inward and retreat away from the center of the dome (**Fig. 8.27A**). This type of retreat is sometimes enhanced because the sedimentary rock layers are thinner on the tops of domes. In both cases, though, the escarpments retreat toward lower elevations.

The landforms developed on slightly dipping sedimentary rock are referred to as **cuestas** (**Fig. 8.28**). Cuestas have two components – an escarpment, sometimes a steep slope in arid climates but more rounded and covered with regolith in humid climates, and a **dip slope**, the long slope that

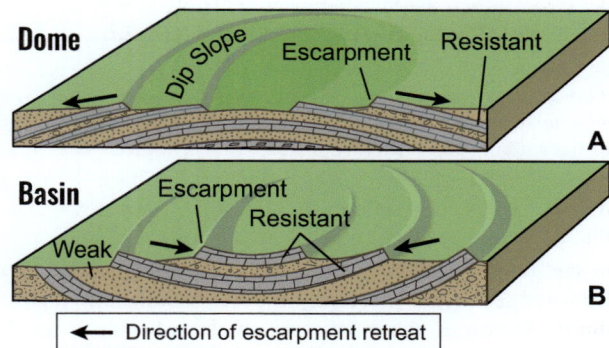

Figure 8.27 Schematic illustration of escarpments on sedimentary rocks that ring a (**A**) dome and (**B**) basin. Source: After *Physical Elements of Geography*, Trewartha, et al., 1967. © McGraw-Hill.

generally follows the dip of the underlying rock (**Figs. 8.27, 8.28A**). Usually, at the far end of the dip slope of one cuesta is the escarpment of the next. One after another, a series of cuestas comprises a landscape type known as a **cuesta-form plain**, or belted plain. Much of the Great Plains, the southern US Coastal Plain, and central Europe are formed

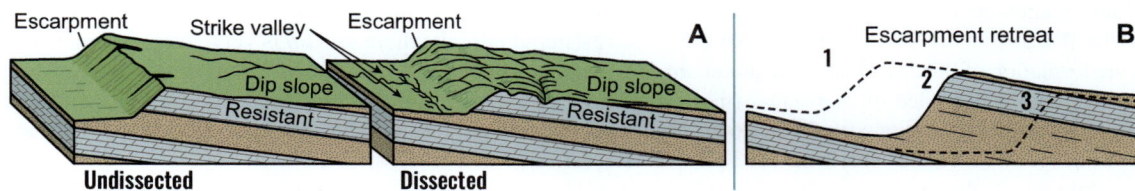

Figure 8.28 Cuestas: landforms formed on gently dipping sedimentary rocks. **A.** Cuesta structure and components: escarpments, strike valleys, and dip slopes. Source: After *Physical Elements of Geography*, Trewartha, et al., 1967. © McGraw-Hill. **B.** Schematic illustration of the retreat of a cuesta escarpment.

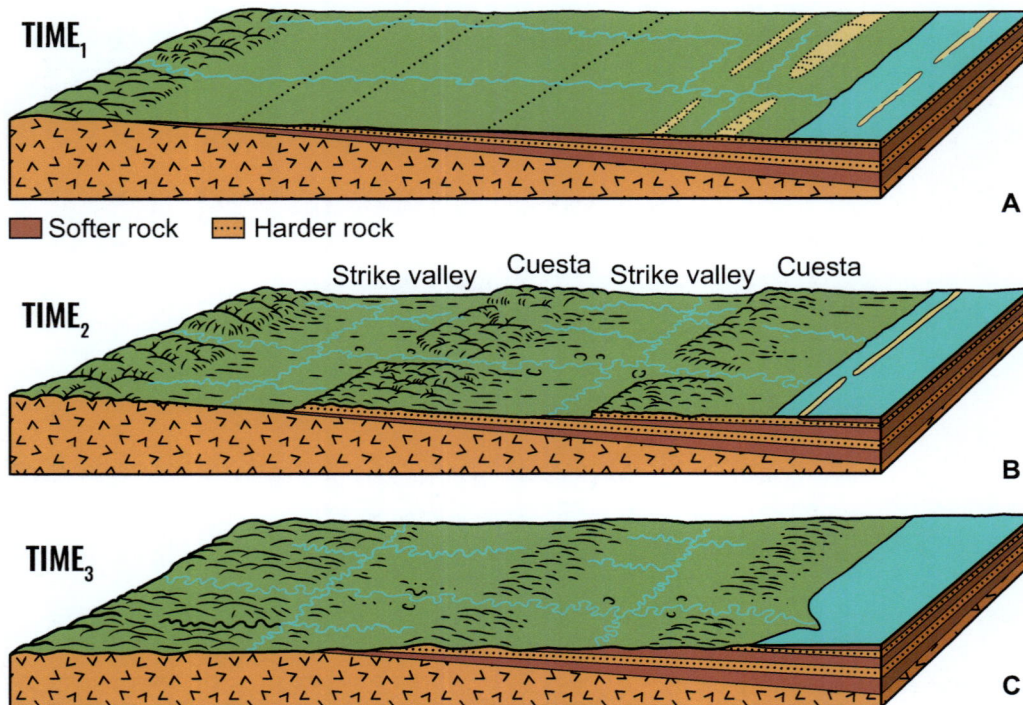

Figure 8.29 The development of a cuestaform plain. Sediments deposited beneath the sea lithify into alternating strata of soft and resistant sedimentary rock. **A.** By Time$_1$ the sea has retreated, exposing the strata to erosion. **B.** By Time$_2$, the softer rocks have been eroded to form strike valleys, and cuestas have formed on the harder rocks. This landscape is a classic cuestaform plain. **C.** As the cuestaform plain undergoes continued denudation and approaches base level (sea level) in Time$_3$, erosion processes become less effective and valleys begin to fill, producing a muted version of the cuestaform plain that preceded it.

on cuestaform plains. These landscapes began as freshly deposited sediments or recently exposed sedimentary rocks. Shortly thereafter, the most resistant strata began to stand up as ridges, beginning the development of the cuestaform plain (**Fig. 8.29**).

Escarpment retreat in cuestaform plains, as elsewhere, depends on the removal of weathering and erosion byproducts. This process is accomplished most efficiently by streams that flow parallel to, and often directly in front of, the escarpment. The lowlands that are aligned with the strike of the rock and which parallel the escarpments are called **strike valleys**. Streams in strike valleys operate to remove the sediment delivered to them by erosion of the escarpment and nearby areas. Over time, cuesta escarpments slowly retreat down-dip, toward lower elevations, as the landscape is lowered by erosion and **denudation**.

Broad geologic domes often develop ringing cuestas. A classic case is the Black Hills of South Dakota, USA (**Fig. 8.30**). During the formation of this broad dome, sedimentary rocks were uplifted and later eroded off the center (highest part) of the dome, exposing the granite core. These sedimentary rocks stand up today as cuestas that ring the uplifted rocks of the central part of the dome. In such a case, strike valley streams flow around the dome, developing an **annular drainage pattern** (**Fig. 16.4**). In most cases, an annular

pattern indicates that the streams are flowing around either a dome or a basin, trapped between cuestas.

Because cuestas form on gently dipping strata, the rate of escarpment retreat can be slow, especially in humid climates and where the dip is low. Here, the escarpments can become rugged and difficult to discern on the ground due to a thick cover of regolith. However, if the dip of the rock is steeper, and in drier climates, more rapid retreat is likely, and the landforms take on a different character. **Hogbacks** are cuesta-equivalent ridges developed on bedrock, but on more steeply dipping rocks (**Fig. 8.31**). These striking landforms are common in dry climates (**Fig. 8.32**). Most hogbacks could be considered **asymmetrical hogbacks**, forming where bedrock dip is steep, with steep escarpments but with more gently inclined dip slopes. **Symmetrical hogbacks** have generally the same slope gradient on either side of the ridge. They develop on resistant bedrock where the dip is very steep. Symmetrical hogbacks developed on nearly vertical or vertical bedding may have bedrock cliffs (free faces) on both sides. As on cuestaform plains, valleys between hogback ridges are referred to as strike valleys.

Hogbacks are common on the flanks of uplifted domes, occupying the areas closer to the dome center, where rocks dip more steeply (**Fig. 8.30**). Cuestas often ring the dome

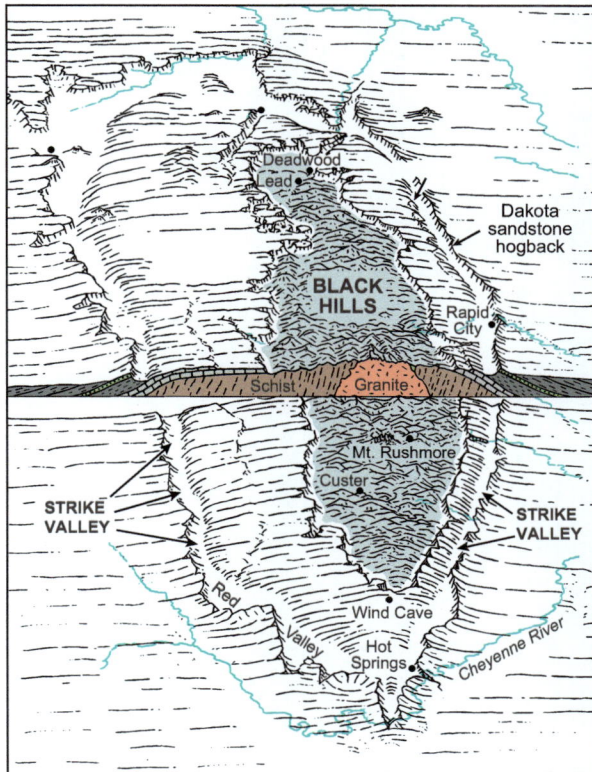

Figure 8.30 Geology and structure of the rocks in the Black Hills region of South Dakota, USA. Source: © A. N. Strahler (1992). Used by permission.

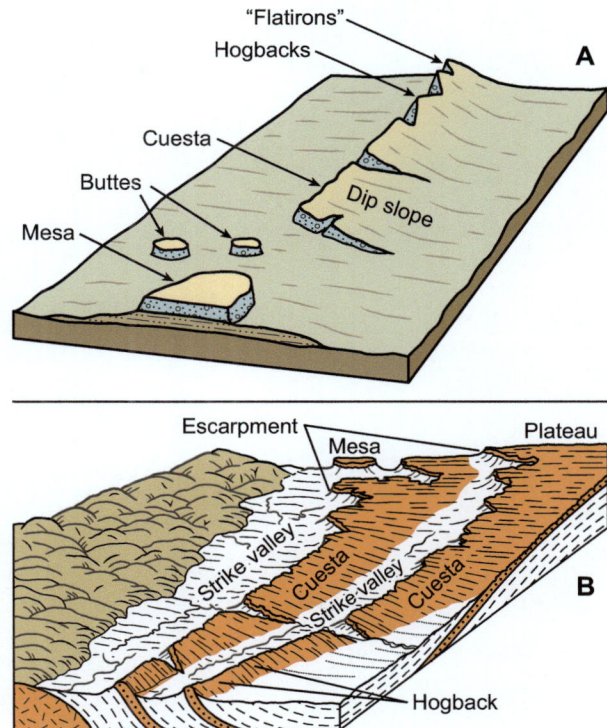

Figure 8.31 Schematic diagrams of the effects of dip on sedimentary rock, and the resultant landforms. **A.** The main upland landforms that result from increasing dip of the rock strata. **B.** Diagram showing how strike valleys form in the lowlands between the upland landforms. Source: © A. N. Strahler (1992). Used by permission.

farther out, where the sedimentary rocks have been less steeply uplifted. Erosion of a hogback ridge usually leads to distinctive gaps in the escarpment, such that the uplifted strata between the gaps take on a more jagged or pointed appearance (**Figs. 8.31, 8.32**). These types of hogbacks are called **flatirons**. With their distinctive triangular facets, they resemble the steel flatirons used by pioneers to iron and press clothes. Lastly, should a rock layer become so steeply inclined as to be vertically dipping, or nearly so, the landform name given to it is a **razorback**.

Escarpments on sedimentary rocks always retreat downdip, and as they do, various landforms evolve and become apparent on the landscape. The names given to them depend on the degree of that dip – cuestas, hogbacks, flatirons, etc. (**Fig. 8.31A**). However, the retreat of escarpments is not always ideal; it seldom results in a smooth front like those shown in **Figs. 8.27** and **8.28**. Rather, the escarpment retreats faster along areas where the caprock is more fractured, thinner, or where it is inherently weaker due to lithology. For example, a sandstone caprock may contain more shale in places, and at those locations the escarpment will retreat faster.

The variable rates of erosion on the escarpment lead to irregular and jagged escarpments on cuesta fronts. Over time, the escarpment may become increasingly scalloped. Often, as this process proceeds, isolated pieces of the cuesta

(uplands) are left behind, out in front of the rest of the retreating escarpment; these uplands are referred to as **outliers** (**Fig. 8.33**). As the name suggests, these isolated uplands lie ahead of, or outside of, the main body of the cuesta (or hogback). Outliers are present because something about their caprock protected them from the more rapid retreat of the caprock elsewhere, or because they lie midway between valleys incised into the cuesta and hence, farthest from the most "erosive" areas.

Another feature of the bedrock geology, similar in some ways to an outlier, is an **inlier**. An inlier represents an upland of older rock, isolated among (and below) younger rocks (**Fig. 8.33**). Inliers of this older rock protrude up and through the dip slopes of cuestas. Inliers form in one of two ways, or sometimes both ways, simultaneously. The first and most common reason for inliers to form on landscapes of sedimentary rock involves the surface of the rock strata underlying the cuesta. If this buried surface is uneven, with isolated knobs or up-arched areas, erosion will thin the overlying rock and these "higher" areas will be exhumed first, as inliers (**Fig. 8.33**). The second method of inlier formation involves deeply entrenched streams or other areas of focused erosion, which carve into the cuesta dip slopes, exposing the underlying rock.

Figure 8.32 Hogbacks and flatirons at various locations in the American Southwest. Source: (A, B, and C) R. Schaetzl; (D) G. Bevenger.

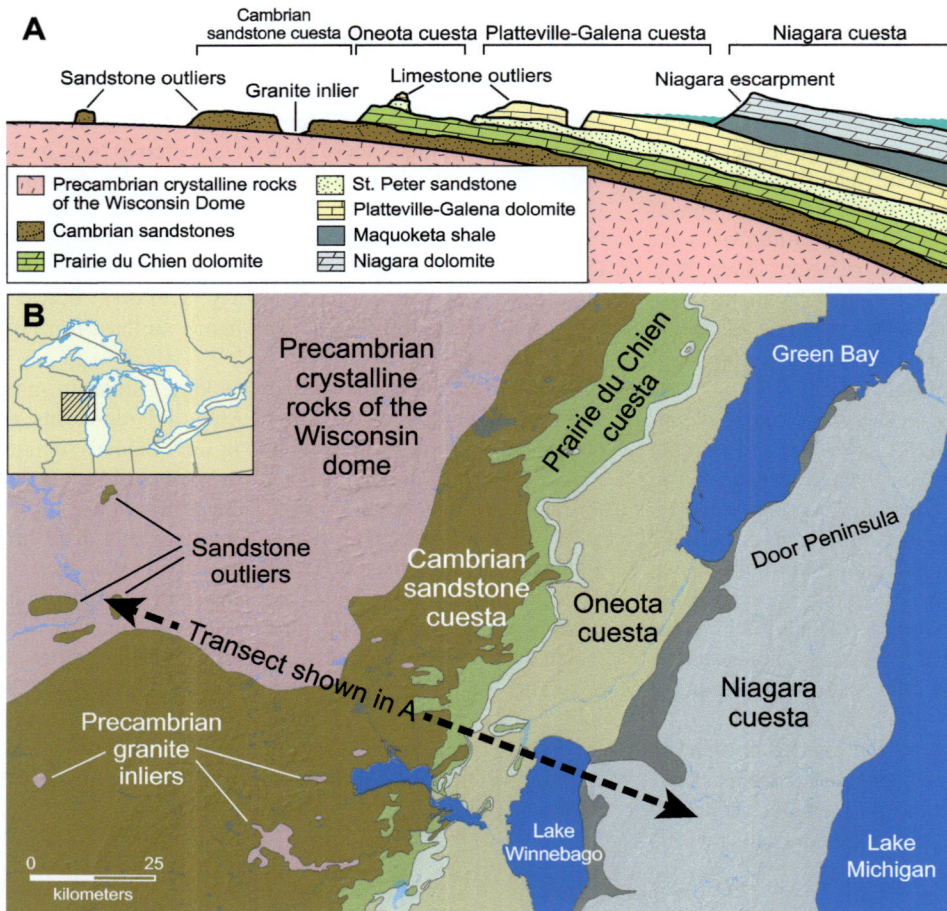

Figure 8.33 Geology of the glaciated cuestaform plain of eastern Wisconsin, USA, shown in cross-section (**A**) and in map view (**B**).

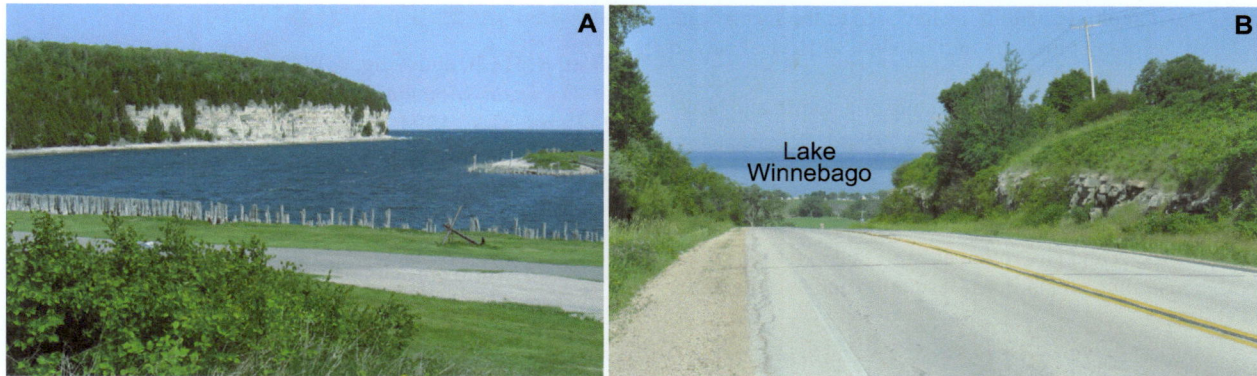

Figure 8.34 The Niagara cuesta and escarpment. **A.** The bright white Niagara Dolomite that underlies the Niagara cuesta is very apparent here, in Michigan's Upper Peninsula. Erosion by waves has kept the escarpment steep by continually removing debris from its base. **B.** A view off the top of the Niagara cuesta, looking west and down into Lake Winnebago. Note the bedrock that crops out. Source: R. Schaetzl.

The cuestaform plains of eastern Wisconsin, USA provide good examples of some of these classic cuesta features (**Fig. 8.33**). Here, sedimentary rocks are retreating off a broad geologic dome, composed mainly of crystalline rocks like granite, into the Lake Michigan Basin. Four prominent cuestas have formed here. The first (bottommost) cuesta is formed on a resistant bed of Cambrian sandstone (with a weaker bed of sandstone underneath, and one above). Outliers of this cuesta occur as isolated sandstone uplands on the dome proper. In some areas, this cuesta has developed inlier "windows" into the underlying Precambrian rocks. The remaining three cuestas (above) are all formed on hard dolomite bedrock (**Fig. 8.33A**). The highest and most impressive of these is the Niagara cuesta. Its escarpment is very prominent (**Fig. 8.34**), not only because the dolomite is so hard but because it is underlain by a thick sequence of very weak shale. The famous Niagara Falls is formed on these same rocks, just a few hundred kilometers to the east. The strike valley formed on these shales, in front of and paralleling the Niagara escarpment, is home to Green Bay and the largest (but extremely shallow) lake in the state – Lake Winnebago (**Fig. 8.34B**).

The other two cuestas in this cuestaform plain, sandwiched in between, are less prominent but nonetheless still well-formed features. Formed on slightly "softer" dolomites and limestones, these cuestas have been variously worn away and buried by glacial deposits, such that more of the landscape has a "glacial" look than a "bedrock-controlled" look. Glacial erosion has also made the escarpments on these two cuestas more jagged, with more outliers. Nonetheless, across this cuestaform plain we still can see evidence of major river systems flowing within strike valleys, lakes nestled into these valleys, and prominent escarpments in a handful of areas. The forces of erosion and denudation have operated on this landscape for so long that today it is mainly a plain, despite the alternating layers of hard and soft bedrock.

Cuestas and their associated landforms represent features that form as the sedimentary rock cover of the continents erodes and undergoes denudation. Cuestas and cuestaform plains are common features of many landscapes worldwide, but often go unnoticed!

REVIEW QUESTIONS

8.1 Why are most bedrock-controlled landforms formed on *sedimentary* rock?

8.2 Describe and explain the concepts of dip and strike for sedimentary rocks.

8.3 What is the difference between erosion and denudation?

8.4 What are the main differences between slopes formed on bedrock in arid vs humid climates? Which is transport-limited and which is weathering limited, and why? What is meant by these terms?

8.5 Generally speaking, what are some of the harder rocks on Earth, and what are some of the weaker ones?

8.6 Describe the geology of Earth's crust in general terms. What rocks make up most of the volume of the crust, and which ones comprise most of the upper crust?

8.7 What geologic and environmental conditions are necessary for badlands topography to form?

8.8 What is talus, how does it form, and what kinds of landforms develop on talus?

8.9 Balanced rocks can form by two different mechanisms. Outline and explain each.

8.10 How do geologic arches form? What types of bedrock are conducive to arch formation, and what preconditions are necessary for arches to develop?

8.11 What processes lead to the retreat of escarpments?

8.12 What are the differences and similarities among pinnacles, buttes, mesas, and plateaus? What types of rocks and geologic conditions lead to their formation?

8.13 What geologic conditions have led to the formation of the many pinnacles and hoodoos at Bryce Canyon National Park?

8.14 What are the main components of a cuesta and how do cuestas differ on domes vs basins?

8.15 How does differential weathering play a foundational role in the formation of the major features discussed in this chapter? What are the main characteristics of the rocks and environment itself that determine the constraints on landform formation and evolution?

FURTHER READING

Dott, R. H., Jr. and Attig, J. W. 2004. *Roadside Geology of Wisconsin*. Mountain Press.

LaBerge, G. L. 1994. *Geology of the Lake Superior Region*. Geoscience Press.

Martin, L. M. 1965. *The Physical Geography of Wisconsin*. University of Wisconsin Press.

Williams, F., Chronic, L, and Chronic, H. 2014. *Roadside Geology of Utah*. Mountain Press.

9 Landforms Developed on Folded Rocks

Kevin J. Patrick, J. Steven Kite, and Randall Schaetzl

From the Blue Ridge overlook in Shenandoah National Park, Virginia, USA, one can see the broad Shenandoah Valley, split by Massanutten Mountain, with more ridges and valleys in the distance (**Fig. 9.1**). This view of the Appalachian ridges and valleys provides a classic example of an eroded **fold and thrust belt**, where parallel ridges of hard, resistant rocks are separated by valleys underlain by comparatively softer rocks. Fold and thrust belt topography develops on folded bedrock structures called anticlines and synclines (**Fig. 9.2**). But this type of geologic structure is not without a long back-story. Most of the folded rocks underlying these mountains were originally deposited as flat-lying sediments, hundreds of millions of years ago. The folding occurred much later, driven by compressive forces associated with continental collision. Millions of years of subsequent erosion on these rocks were then required to give us the landscapes we see today.

In this chapter, we discuss the interplay between the folding and bending that contorts sedimentary rocks into anticlines and synclines, and the erosional processes that then form the resultant topography like the great mountain belts of the Appalachians. These beautiful landscapes, serene and idyllic as they appear today, belie their violent and storied past!

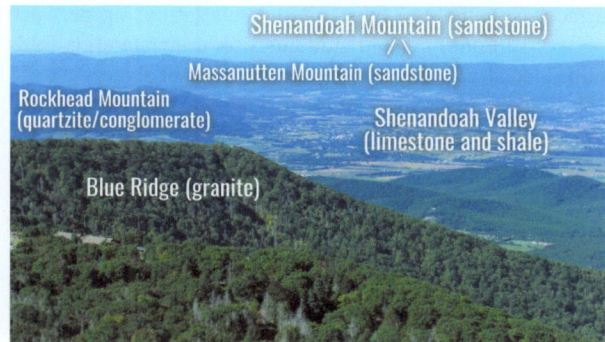

Figure 9.1 The view (looking west) from Stony Man Mountain in Shenandoah National Park (Virginia, USA), shows the geomorphic rhythm of ridges and valleys in the Appalachian Mountains. Resistant rocks like sandstone, quartzite, conglomerate, and granite form the ridges. Less resistant limestone and shale bedrock units underlie the valleys. Source: K. Patrick.

The key to understanding fold and thrust belt landscapes lies in the fact that folding brings rock types of differing structure, lithology, and resistance to the surface. This point is important because rocks of different lithologies erode at different rates, leading to uplands and lowlands that mimic the

Figure 9.2 The folded and tilted strata (rock layers) along Route 48 near Old Fields, West Virginia, USA, show a down-folded ("U-shaped") syncline and up-folded anticlines at the margins. The highest point on this particular ridge is above the syncline, indicating that the link between folding and topography is not always straightforward. Source: J. S. Kite.

Figure 9.0 The Raplee Ridge in southeastern Utah, USA, illustrates the variety of landforms that can develop on folded sedimentary rocks. Source: Cavan Images / Getty Images.

rock type as much as, or more than, the rock structure. This concept is referred to as **differential erosion**. Differential erosion in bedrock landscapes is predictable – thick layers of resistant rock types form ridges, whereas less resistant rocks form valleys (see Chapters 5 and 8). But before we discuss differential erosion in detail, let's first examine folding, and the geologic processes that drive it.

9.1 TERMINOLOGY AND GEOMETRY OF FOLDED ROCKS

Terms involved in fold geology and fold-dominated topography refer to the structure of the bedrock, not the land surface. Nonetheless, there often is a relationship between the two. Terminology and guidelines for understanding folds and folded strata are consistent, regardless of scale.

Just as the folds of a rug pushed across a floor may stand upright, tilt, flop over, or (possibly) tear, rocks in geologic folds can do likewise. Their shapes provide clues into their geologic past.

Anticlines are rock structures that have been folded *upwards*, like an arch, whereas **synclines** are rock structures that have been folded *downwards* (**Figs. 9.2, 9.3**). The sides of a fold are called **fold limbs**. Synclines usually share limbs with neighboring anticlines, just as anticlines share limbs with neighboring synclines. The two limbs of a rock layer within a fold converge at a **fold axis**, sometimes called a **hinge line**. The plane that marks the center of the fold is called an **axial plane**.

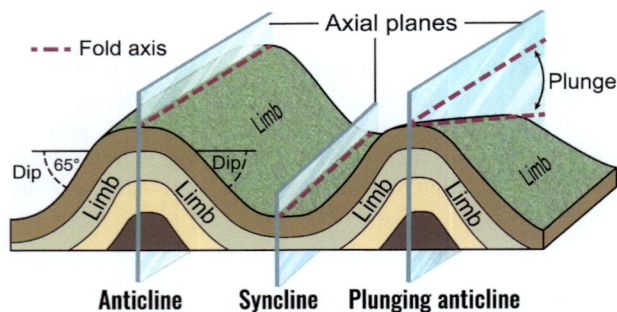

Figure 9.3 Schematic diagram of folded rocks, showing some of the terms used to describe fold geometry. Geoscientists measure a fold's plunge at its axis using the angle (0° = horizontal and 90° = vertical), as well as the compass direction of the rock's inclination (the direction a ball would roll if placed on top of a layer of rock). Source: Reprinted from Murck, B. W. and B. J. Skinner (1999), with permission from Wiley & Sons.

Locally, vertical forces within the crust can cause sedimentary layers to buckle, forming a structure called a **monocline** – a fold with just one dipping limb (**Fig. 9.4**). Although some monoclines are simply the outermost fold (developed on the margins of fold belts), they are more common in regions where extensional (pull-apart) forces dominate the geology. Monoclines are formed as crustal extension causes part of the rock column to "fall down" or bend downward. Forces that form monoclines may also cause the rocks

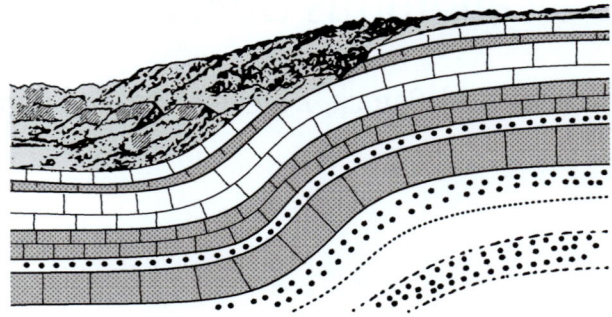

Figure 9.4 Reproduction of an 1875 drawing of a monocline in the Grand Canyon area (Arizona, USA), as reported by geologist John Wesley Powell. Powell's classic monocline illustration shows that strata on one side of the fold have moved downward relative to strata on the other side. Source: After Powell (1875), via the US Geographical and Geological Survey of the Rocky Mountain Region. In public domain.

to break and fault, just as they do during compression. A monocline may further develop into a fault zone if the vertical stresses are sufficient to cause the bedrock to rupture and be offset from its original position.

Rock layers along the crests of anticlines (or the troughs of synclines) need not be parallel to the ground surface; they can also **plunge** downward (**Fig. 9.3**). Non-plunging anticlines and synclines have horizontal fold axes, while plunging folds have fold axes that are inclined. Anticlines and synclines with conspicuous plunge, that is, **plunging folds**, are common in fold and thrust belts. For example, the altitude of the axis of the Wills Mountain Anticline near Cumberland, Maryland, USA, varies significantly because of changes in plunge throughout its 250 km length.

Folds in bedrock tend to become larger and more complex under greater compressive forces. Such folds can have wavelengths ranging from less than 1 meter to more than 100 kilometers. Structural complexity is added to these folded rocks when smaller folds develop within larger folds. A very large anticline or syncline whose overall structure contains smaller folds is known as an **anticlinorium** or a **synclinorium** (**Fig. 9.5**).

Many anticlines and synclines are narrow and/or complexly folded features, with "tight," elongated folds (**Figs. 9.2, 9.3**).

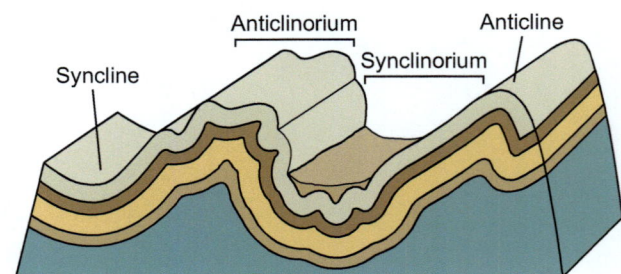

Figure 9.5 Complexly folded rocks can develop into an anticlinorium or a synclinorium. Each of these structures contains smaller, inset folds, like the ruffles on a bent potato chip.

On the other end of the spectrum, some deformed rocks have very broad, almost circular or ovoid, deformation patterns (**Fig. 9.6**). Gently up-warped areas are called **domes**, whereas down-warped areas are called **basins** (see Chapter 8). Think of a dome as a large, broad anticline, shaped something like a blister (**Fig. 8.27**). Unlike anticlines and synclines, which form due to crustal compression, domes often form due to the rising of a large body of magma below, pushing up broad swaths of overlying rock. A large, elongated dome is called an arch. Nonetheless, rocks can also be forced upward by compressive tectonic forces. Geologic forces can also allow rock layers to sag downward, forming broad, down-folded, bowl-shaped basins.

A good rule of thumb is that bedrock exposed at the surface always gets *younger* as one moves toward the center of an eroded basin, or toward the axis of an eroded syncline. Conversely, the outcropping bedrock always gets *older* as one moves toward the center of an eroded dome, or toward the axis of an eroded anticline (**Fig. 9.6**). These guidelines hold in all instances because erosion exposes deeper, older rocks in the centers of anticlines. Conversely, shallower, younger rocks are preserved in the centers of synclines, where they have not yet been eroded away.

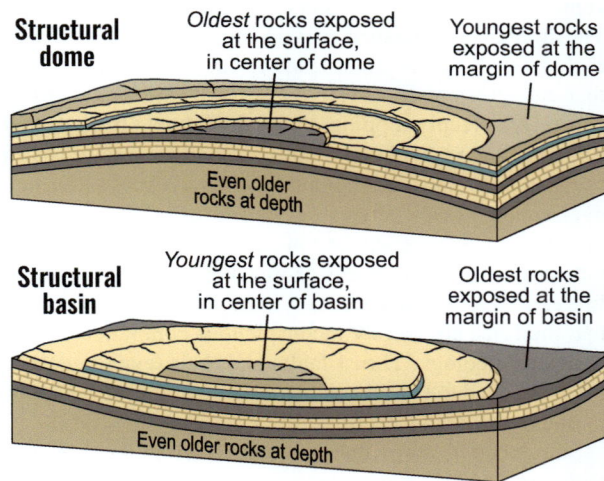

Figure 9.6 Structural domes and basins typically form ovate patterns in the rocks exposed at the surface. As in other folded rock structures, the *oldest* rocks at the surface are in the center of domes, whereas the *youngest* rocks crop out in the centers of basins.

Even more terminology exists to describe the geometry of folded rocks. A fold with a vertical axial plane is called a **symmetrical fold**. These folds have a limb on one side that closely mirrors the geometry of the other limb (left-hand side of **Fig. 9.3**). **Asymmetrical folds** have axial planes that are slightly tilted, where the rocks on one limb dip at a steeper angle than the other. Rocks with a severely tilted axial plane can even develop into an **overturned fold**, where the original tops of the beds in the overturned limb are upside down.

Although some folds may not plunge significantly along much of their length, all folds plunge to some extent (**Figs. 9.3, 9.7**). Plunge may end abruptly where a fold tapers out,

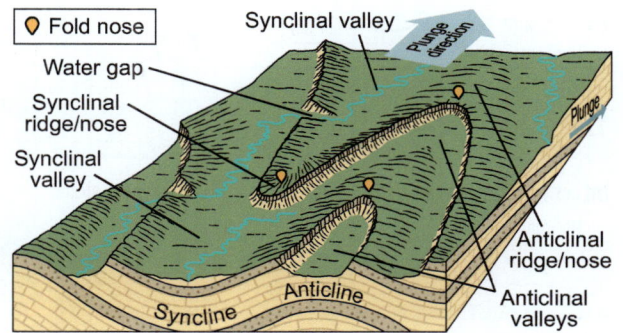

Figure 9.7 Plunging anticlines and synclines eroded into ridges and valleys that "zigzag" across the landscape. Source: © A. N. Strahler (1992). Used by permission.

such that the geology flattens out or transitions into other structures. Nonetheless, some folds rise and fall, plunging repeatedly over many kilometers, much like a porpoise jumping in and out of water. Plunge geometry has a large impact on fold belt topography. Where folds plunge markedly across the landscape, ridges and valleys commonly zigzag across the landscape, changing orientation at plunging fold axes (**Fig. 9.7**). Such zigzag patterns are reflected in distinct **fold noses**, which indicate changes in the strike direction of the bedrock. Anticline noses point in the direction of plunge, whereas syncline noses point in the opposite direction. The land surface along **anticlinal ridges** tends to gradually converge with the adjacent valley bottom, whereas noses of **synclinal ridges** end abruptly in steep escarpments (**Fig. 9.7**).

9.2 GEOLOGY OF FOLD AND THRUST BELT MOUNTAINS

The complex history of folding and mountain building in the central Appalachian Mountains – a focus of this chapter – of the eastern United States began 335–260 million years ago. At that time, the tectonic plate of northwestern Africa collided with eastern North America, helping to form the **Pangaea** supercontinent. The sedimentary rocks of these two ancient continents, originally deposited in horizontal layers, were caught in this collision and compressed into massive folds. Some layers were also broken by thrust faults (**Fig. 9.8**). Because folding and faulting processes are best developed at the margins of the colliding continents, the folded rocks broadly parallel the continental boundaries, occurring in distinct fold and thrust *belts*. The term **fold belt** may be applied to fold and thrust belts where the thrust faults at depth are relatively inconspicuous at the surface. Once folded and faulted, rocks more-or-less stay that way. But eventually the forces of erosion wearing the rocks down to a plain of lower relief. But deep below the surface, the marks of folding and faulting persist in the rocks that remain.

The Appalachian Mountains of the eastern United States – the poster child for fold and thrust belt topography – figure prominently in this chapter. The region's geologic and tectonic history has also played many important roles in North

Early stage
Horizontal strata

Thrust fault

Initial folding

Initial force from
continental collision

Intermediate stage
Horizontal strata

Nearly horizontal
strata

Highly deformed strata

Force from continued
continental collision

Late stage
Nearly horizontal strata
(Appalachian Plateaus)

Highly deformed strata
(Valleys and ridges)

Force from late stages of
continental collision

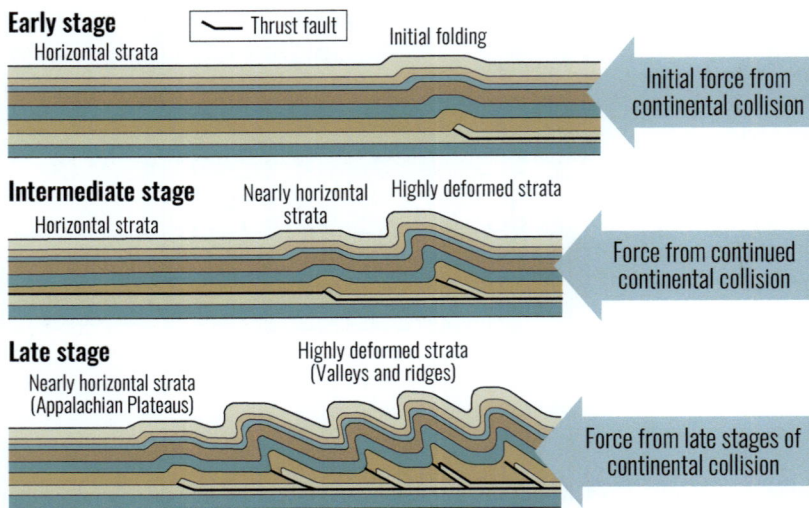

Figure 9.8 Schematic diagram showing how tectonic collision formed the Appalachian fold and thrust belt, where sedimentary rocks were folded and faulted over tens of millions of years. Thrust faults within the weaker rocks, like shale or rock salt, are linked to anticlines and synclines in stronger rocks above, like sandstone or limestone. Eventually, many of these folded rocks will be eroded away, to form a more modest, low-relief landscape with distinct ridges and valleys.

American history and culture. By studying the geomorphology of the Appalachians, we can gain insight into other mountain ranges with similar geology, including the Alps, Apennines, and Pyrenees in Europe, the Atlas Mountains in northwestern Africa, the Ouachitas and parts of the Rockies in North America, and many others.

Tectonic collision compresses the rocks of the continental plates, squeezing them especially hard near the collision zone (**Fig. 9.8**). Like two highway vehicles involved in a collision, where some parts of the vehicles are crumpled while others are broken, some rock layers in a plate collision are folded while others are faulted. Everyday experience suggests that rock is unlikely to bend or fold, yet incredibly strong tectonic forces, acting over almost unfathomable lengths of time, can crumple bedrock as surely as a muscular superhero can smash an automobile. Folding is especially likely when the rocks have been heated to high temperatures, and/or under great pressure, as happens at depth. The high temperatures make the rock more ductile and deformable.

Some weak rock types, like shale, deform more easily, typically by shearing, faulting, and breaking. Stronger, more competent bedrock types like limestone and sandstone are less likely to fault; they often compress into folds, somewhat like what would happen if you were to push one side of a rug across a hardwood floor. In this example, the sliding of the rug across the floor corresponds to the faulting of weak strata, while the folds in the rug mimic the strong, compressed rocks.

9.2.1 Faulting Within Fold and Thrust Belts

The folding *and* faulting of thrust belts adds to their geologic complexity. Thrust faults are the driving mechanisms beneath fold and thrust belts. They develop as parts of the crust are compressed against another part, causing slippage of rocks against each other. These slip zones, called **fault planes**, typically occur within weak strata, such as shale or rock salt.

Geophysical studies have revealed massive, low-angle thrust faults at great depths beneath fold and thrust belt mountains (**Figs. 9.8, 9.9, 10.6, 10.10D, 10.15**). During thrust faulting, thick sheets of bedrock get displaced horizontally, sometimes by tens or even hundreds of kilometers. The immense resistance to shoving thick sheets of bedrock great distances often leads to secondary thrust ramps that cut across rocks nearer the surface (**Fig. 9.9**). **Thrust ramps** are steeply dipping, upward offshoots from deeper, horizontal thrust faults. Faulting along closely spaced thrust planes can also completely detach rocks, forming overlapping **thrust sheets**. Within thrust sheets, overlapping rock layers are stacked in exactly the same sequence, rather than reflected in mirror-image sequences (typical of folded rocks). Some mountain belts are so heavily faulted that the folds are merely inconspicuous components of the structural geology. Thrust faults and their related geologic folds have led to the formation of many of the world's famous alpine landscapes.

Thrust faults form due to intense deforming pressures within rocks. Thus, they commonly form in the weaker rock layers, which are more susceptible to deformation. They are more common nearer to the collision zone and deeper in the crust. Thus, folded strata without significant faulting tend to be more common nearer the surface and farther from the collision zone. Under extreme compression, thrust faults, deep within folded rocks, can completely offset one fold limb from another (**Fig. 9.9**). Because these relationships are predictable, rock structures can help geologists better understand fold histories.

Glacier National Park in western Montana (USA) shows good examples of thrust faulting, which is often accompanied, to some degree, by folding (**Fig. 9.10**). The crust here was shortened by thrust faulting, as rocks slid across and on top of one another. Most of the long-distance tectonic compression – possibly as much as 100 kilometers – was along nearly horizontal thrust faults, ≈50–80 million years ago.

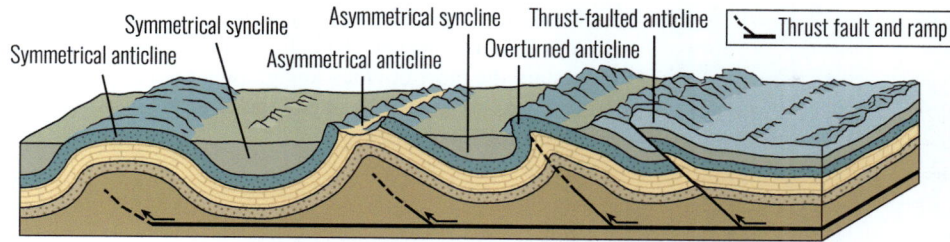

Figure 9.9 A depiction of how fold geometry reflects the extent and intensity of strain the rocks have experienced. Progressing from left to right, the rocks in this figure have experienced increasing deformation and tilting of their axial planes, leading to increased fold asymmetry. Major anticlines are linked to thrust ramps. The nearer a thrust ramp comes to the surface, the more asymmetrical the associated fold is likely to be.

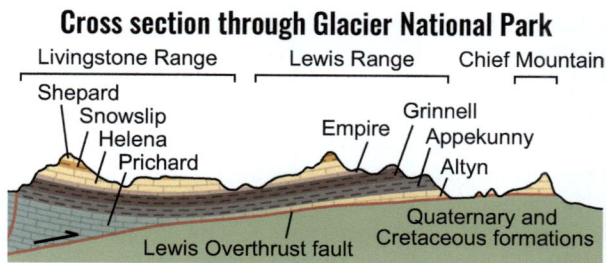

Figure 9.10 A simplified cross-section of the geology at Glacier National Park, where the sedimentary bedrock of a broad syncline was moved across the Lewis Overthrust fault. Source: Adapted from Chakraborty, T., Taral, S., More, S., Bera, S. (2020), with permission from Springer Nature.

9.2.2 The Folded and Eroded Geology of the Appalachian Mountains

Fold and thrust belt topography epitomizes the Appalachian Valleys and Ridges physiographic province (**Fig. 9.11**). The metamorphic and igneous rocks in the neighboring Blue Ridge and Piedmont have also been folded and faulted, but have less overall variation in their resistance to erosion, so the topography there is less compelling. Likewise, the rocks in the Appalachian Plateaus and Coastal Plain are only minimally folded. Thus, our emphasis is on the Appalachian Valleys and Ridges. As is obvious in **Fig. 9.11**, the topography of the southern and northern Appalachians is strikingly different. Let's examine why.

The few kilometers of bedrock eroded off the folded rocks in central Pennsylvania since the last major mountain building event may represent a lot of erosion from a human perspective. However, this amount of erosion is less than what has occurred across most of the southern Appalachians. Imagine the folded rocks shown in **Figs. 9.8** and **9.9**, eroded down to successively lower levels. A little erosion will expose relatively young strata deformed by intense folding, but relatively few major faults; this situation epitomizes much of modern-day central Pennsylvania (**Fig. 9.11A**). The zigzag pattern exists because both limbs of most folds remain connected and are exposed at the surface in mirror image patterns across the landscape. Now, imagine greater/deeper erosion of similarly folded rocks, which will expose even deeper, folded strata that have been offset and truncated by thrust fault ramps. This situation is common in the more linear ridge-valley topography of the southern Appalachians (**Fig. 9.11B**), where many, if not most, major folds have been thrust faulted long ago and their limbs are now totally detached from one another. Instead of a zigzag of mirrored fold limbs, the heavily thrust-faulted topography is dominated by isolated valleys and ridges where the geology is repeated over and over.

The topographic contrast between the northern and southern Appalachian fold and thrust belt is further enhanced

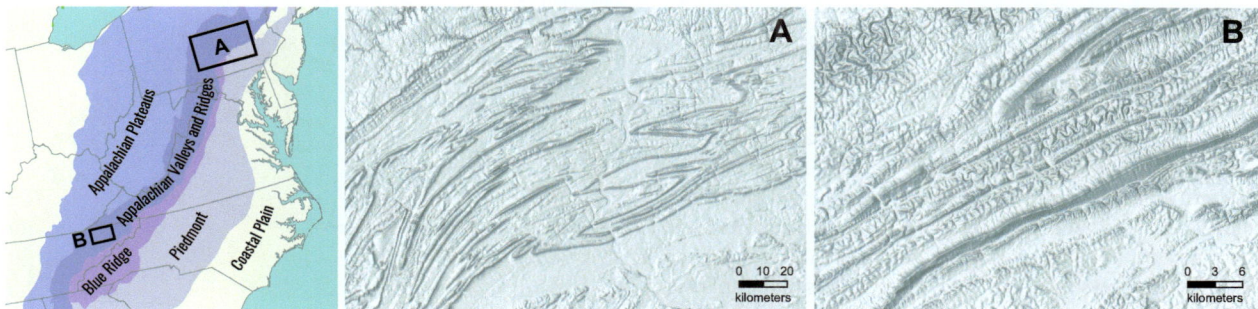

Figure 9.11 Northern vs southern landscapes in the Appalachian Valleys and Ridges. **A.** In the north, a zigzag, fold-dominated topography dominates, where many of the folds have prominent plunge. **B.** In the south, a more linear, thrust-faulted topography occurs.

Figure 9.12 Typical view of the Appalachian Mountains of Pennsylvania, USA, east of State College. The highest ridge on the horizon is held up by very resistant Tuscarora Sandstone and Quartzite. Less resistant rocks of the Juniata Formation (shales and siltstones) underlie a narrow valley that is partly obscured by a lower ridge held up by resistant Bald Eagle Sandstone. The valley in the foreground has developed on easily eroded Ordovician shales and limestones. Source: K. Patrick.

because of different geology. The younger, shallower bedrock units in the northern Appalachians tend to include more resistant sandstones than in the older, deeper units found farther south (mainly weak shales, limestones, and dolomites). Sandstones develop into prominent ridges in central Pennsylvania (**Fig. 9.12**), whereas the less resistant rocks, more common in the southern Appalachians, develop into broad valleys.

Thrust faults – a common partner to folding processes – are more prominent in the local geology when the tectonics and local rocks are weaker and more conducive to faulting. These faults become more important as a geomorphic factor when the bedrock has been so deeply eroded that most of the shallow folds are now gone, exposing the deep thrust ramps. This latter situation is typical of the southern Appalachians, which have a generally linear valley and ridge geomorphology associated with thrust sheets and tight folds (**Fig. 9.11B**). This geomorphology differs markedly from the zigzag patterns of plunging folds 500 km farther north, in Pennsylvania (**Fig. 9.11A**). Some of the topographic differences may have been inherited from the style of tectonics, but geologic processes since the last major mountain building event have differed significantly between these regions.

In effect, long-term weathering and erosion have been larger players in the southern Appalachians, carrying away more rock than in the folded mountains of Pennsylvania (in the north). This long-term erosion has stripped away the minimally faulted folds that likely existed at the surface in the southern Appalachians. As a result, the southern Appalachian landscape tells us what the northern Appalachians may look like many millions of years from now, after differential erosion has progressed down to deeper levels, where thrust faults are more common.

A similar topographic contrast can be seen in east to west transects across the Appalachian Valleys and Ridges, where deeply eroded landscapes toward the east (nearer old

tectonic collision zones) are typically dominated by wide valleys of limestone and shale. In contrast, zigzag sandstone ridges are the most obvious topographic features in the less eroded and somewhat younger rocks to the west (farther from old collision zones).

9.3 THE GEOMORPHIC EVOLUTION OF FOLD AND THRUST BELTS

In "mature" fold and thrust belts like in the Appalachians, the geologic structure doesn't entirely determine whether an anticline or a syncline develops into a ridge or valley. Rather, the surface topography is largely determined by the differential erodibility of the rocks (**Fig. 9.12**). As mentioned above, in bedrock landscapes, differential erosion causes thick layers of resistant rock types to stand up as ridges (**Figs. 9.13, 9.14**), whereas adjacent, less resistant, rock layers are preferentially eroded into valleys. All of these rocks are hard, but it is their *variable* resistance to erosion that determines the eventual topography (see Chapter 5).

It is important to also note that the erodibility (weakness vs strength) of rocks depends not only on lithology but also on climate. For example, limestone is physically strong, but chemically weak because the carbonate minerals that make up the rock dissolve in rain and surface waters, which tend to be weakly acidic (see Chapter 12). Thus, limestone is a common ridge-former in dry climates (**Fig. 9.15**), but a valley- and cave-former in a humid climate like the eastern United States.

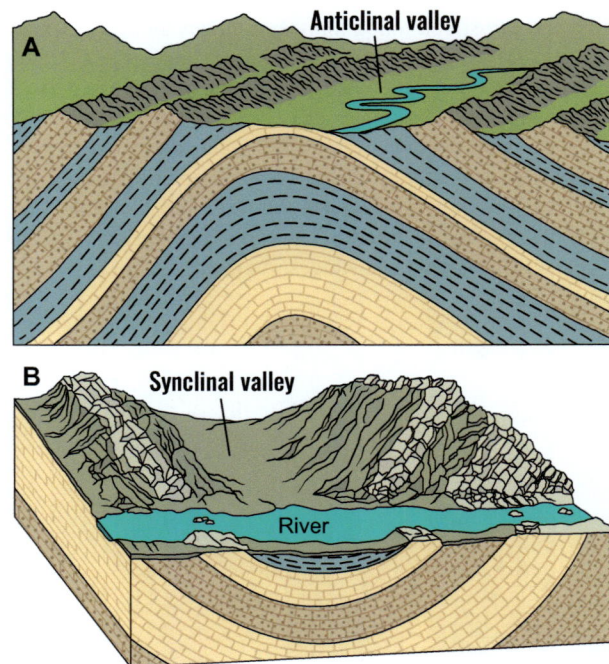

Figure 9.13 Schematic diagrams of (**A**) an anticline and (**B**) a syncline, showing how either can evolve and erode into a valley, depending on rock resistance.

In a humid climate like the eastern United States, most ridges in fold belts have formed on hard sandstone, conglomerate, or quartzite (**Fig. 9.14**). Other rock types, such as chert, can be ridge-formers locally. Nonetheless, fresh, unweathered bedrock at the top of any Appalachian ridge will almost always be composed of relatively insoluble, well-cemented, hard minerals such as quartz. Similarly, most of the intervening valleys in the same region will be floored with soft rocks like shale and siltstone, or soluble rocks like limestone and dolomite (**Fig. 9.12**).

Figure 9.14 An image of Kittatinny Mountain on the New Jersey (USA) side of the Delaware Water Gap. In fold and thrust belts like this one, the ridges are underlain by dipping, resistant sandstone and conglomerate. The slope on the far right is underlain by weaker siltstones and shales, but is now mantled with regolith, colluvium, and talus. Source: K. Patrick.

Figure 9.15 The Madison Limestone is a major cliff-forming bedrock unit in Spearfish Canyon in the Black Hills of South Dakota (USA). This region receives about half the average rainfall of the Appalachians, where limestone typically underlies valleys. Source: US Geological Survey.

Because of differential erosion, thick-bedded, resistant rocks typically form prominent ledges, bedrock cliffs, or bouldery talus (**Fig. 9.14**). Conversely, softer, more easily eroded rocks often underlie valleys – locations typically covered in regolith, colluvium, broken rock, and deep soils. Much of this material has been weathered and moved downslope from more resistant rock outcrops in upslope locations.

To fully grasp the effects of differential erosion in fold belts, we can compare the topography in geologically old vs. young settings. In young and active mountain belts like the Zagros Mountains of Iran, the topography closely matches the geologic structures, that is, anticlines tend to form topographic highs and synclines tend to form valleys (**Fig. 9.16**). Broad-scale erosion has not yet dominated the topography like it does in older, less active terranes like the Appalachians.

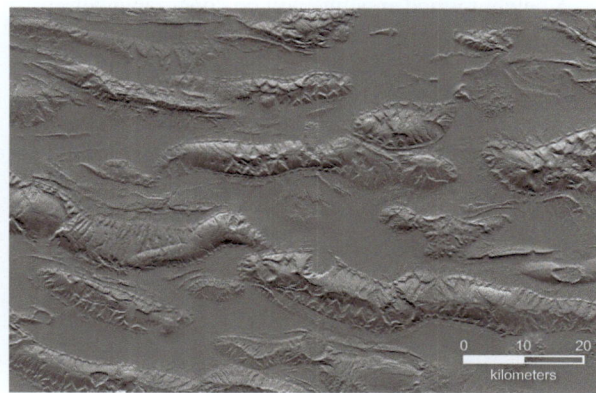

Figure 9.16 Topography of the young, tectonically active Zagros Mountains of Southern Iran, as indicated on this hillshade image from the NASA Space Shuttle. Here, rugged anticlinal ridges rise > 1,000 m above low-relief, synclinal valleys filled with sediment. The Zagros fold belt is located on an active margin of the Arabian Plate, which is colliding with the Eurasian Plate at a rate of about 1 cm/yr. Source: NASA.

Over time, as tectonic forces decrease and the rock structures stabilize, erosional processes increasingly dominate fold belt topography. More and more of the landforms display erosional origins. Erosional dominance stems from a loss of competition from the processes of folding, rather than from an increase in rates of erosion. In fact, rates of erosion are greater in tectonically active areas, but when tectonic activity diminishes, erosion becomes dominant, until the uplands are worn away. The underlying geologic structure is typically easier to identify on ridges than in valleys, because many valley bottoms are buried by thick coverings of **regolith**, weathered off the surrounding rocks. These sediments hide the underlying rock lithology and structure.

Because of the great age of the landscape, most of the topography of the Appalachian fold and thrust belt has clear erosional origins. Nonetheless, although the Appalachians are much older than the Zagros Mountains, they are not tectonically "dead." Like many regions far from plate boundaries, eastern North America continues to experience some amount of uplift and tilting, long after the continental collision events that folded and faulted its bedrock. Nonetheless, in old fold belts like the Appalachians, erosion of the land surface continues to expose ever-deeper rocks. Differential erosion wears away preferentially at weaker rocks, while harder rocks increasingly stand up as ridges. This process causes the topography along a fold axis to "invert," as bedrock units of different resistance gradually become exposed by erosion. What starts out as a topographically high area may erode to a valley (**Fig. 9.13A**). For example, a resistant sandstone ridge may be eroded away, exposing softer shale or limestone below. There are four possible outcomes to this type of geomorphic evolution, and all are caused initially by folding, followed by differential erosion. These geomorphic "rules" are nearly universal in fold belt topography:

1. A newly formed anticline will be, topographically, an anticlinal ridge, simply by nature of its geologic structure (**Figs. 9.3, 9.7, 9.16, 9.17**).
2. After long-term erosion, rocks at the top of the anticlinal ridge will be removed, exposing non-resistant rocks beneath. The non-resistant rocks erode quickly, deepening and transforming the eroding ridge into an **anticlinal valley** (**Figs. 9.7, 9.13A, 9.18**). This evolution is fairly common, especially where shale or limestone underlies the ridge. Nonetheless, as erosion

Figure 9.17 The Wills Mountain Anticline is here expressed as an anticlinal ridge, near Cumberland, Maryland, USA. The asymmetrical structure of the fold is apparent where Wills Creek cuts through the ridge at The Narrows Water Gap. Shadows, vegetation patterns, and bright Tuscarora Sandstone cliffs indicate the fold's asymmetry. Source: Jstuby at English Wikipedia, Public domain, via Wikimedia Commons.

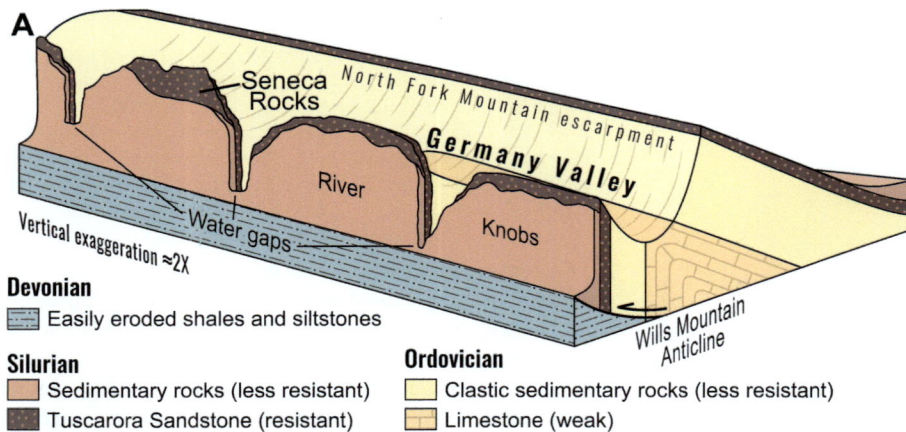

Devonian
▦ Easily eroded shales and siltstones

Silurian
▦ Sedimentary rocks (less resistant)
▦ Tuscarora Sandstone (resistant)

Ordovician
▦ Clastic sedimentary rocks (less resistant)
▦ Limestone (weak)

Figure 9.18 Germany Valley. **A.** A simplified geological diagram of the Wills Mountain Anticline, where it has been eroded to form Germany Valley (an anticlinal valley) in West Virginia, USA. The resistant Tuscarora Sandstone along the axis of the Wills Mountain Anticline has been eroded to expose the weaker Ordovician shales and limestones in the valley. The sandstone now stands up as distinct ridges on the limbs of the anticline, including Seneca Rocks, a famous rock-climbing locality. The asymmetry of Germany Valley reflects the asymmetry of the rock layers in the anticline. The asymmetrical form of the ridge (North Fork Mountain) stems from the 30° dip of the bedrock on the anticline's southeast limb, whereas the symmetrical hogback form of Seneca Rocks and other River Knobs stems from the near-vertical dip of the bedrock on the northwest limb. Source: Reprinted with permission from Lebold & Wilkinson, *Roadside Geology of West Virginia*, © 2018 by Mountain Press Publishing Co. **B.** Germany Valley, bounded by ridges held up by Tuscarora Sandstone. Source: J. S. Kite.

continues, deeper, resistant rock layers may become exposed at the surface, transforming the structure back to an anticlinal ridge (**Fig. 9.19**).

3. By virtue of their geologic structure, most synclines begin as synclinal valleys (**Figs. 9.13B, 9.20B**).
4. As erosion wears away at the margins of the synclines, they, too, may eventually develop into synclinal ridges, wherever resistant rocks core the features along a fold axis (**Figs. 9.2, 9.7, 9.20A**).

Thus, in landscapes of folded rock, the topography is as much as (or more of) a function of rock lithology as it is rock structure. The folded structures are primarily responsible for the *repetition* of ridges and valleys across the fold belt.

Figure 9.19 Sheep Mountain, Wyoming, USA, is a striking example of an anticlinal ridge with ringing cuestas and hogbacks. In this photo, the anticline plunges toward the viewer. The Madison Limestone forms the prominent ridge along the anticlinal axis. The cuesta and hogback ridges on either side of the mountain have developed on younger rocks in the fold limbs. This example illustrates how the direction and angle of bedrock structures can often be interpreted from the topography, especially by the patterns formed by outcrops of resistant strata. Source: L. J. Maher, Jr.

Differential erosion causes fold and thrust belt topography to continually evolve. Millions of years from now, the anticlinal ridges along the Wills Mountain Anticline (**Fig. 9.17**) will be breached by erosion and transformed into anticlinal valleys. Similarly, the modern-day Germany Valley, an anticlinal valley (**Fig. 9.18**), would have been an anticlinal ridge (held up by Tuscarora Sandstone) in the distant past. These examples illustrate that the topography we see today is but a *snapshot in time*. Fold and thrust belt landscapes continually evolve.

9.4 LANDFORMS DEVELOPED ON FOLD LIMBS

Because of the repetitive geologic patterns associated with fold belts, a ridge-forming or a valley-forming bedrock unit on one fold limb typically will also crop out on the opposite limb (**Fig. 9.19**). In a symmetrical fold, the topography on one side of the fold axis is often an almost mirror image of the topography on the other side. Nonetheless, because most folds are asymmetrical, the exact shapes of ridges and valleys on opposite limbs of a fold tend to differ. Thrust faults that extend to the surface also impact the topography; they significantly offset or completely displace fold limbs, disrupting the rhythm of ridges and valleys.

Our previous discussion focused on ridges and valleys that develop along fold *axes* (**Fig. 9.3**). However, most ridges in eroded fold belts develop on fold limbs (sides), rather than along the fold axes (**Fig. 9.21**). Think of it this way – a single ridge may form on an anticline axis, but two (or many more!) ridges may also develop on the limbs, or sides, of the same structure. The side of a ridge that follows the direction of bedrock dip is called its **dip slope** (**Figs. 8.3, 8.28, 9.21A, 9.22**). Then, on the opposite side of the ridge, where the topographic slope is in the opposite direction as bedrock dip, one typically finds a steep, bedrock escarpment (**Figs. 8.28, 9.21B, 9.22**). Dip slopes typically look different enough from escarpment slopes that non-geologists can readily recognize them too!

Figure 9.20 Synclines. **A.** The Sideling Hill road cut on Interstate 68, west of Hancock, Maryland, USA, is a classic cross-section through a synclinal ridge. Source: K. Patrick. **B.** Narcea Valley is a remarkable example of a synclinal valley in Asturias, Spain. Resistant quartzite beds have been folded into a broad syncline. Source: G. Gutierrez-Alonso.

Figure 9.21 Landforms on the limbs of folded rocks. **A.** The dip slope on Tussey Mountain in Bedford County, Pennsylvania, USA, shows bedding planes of resistant Tuscarora Sandstone beneath the ridge. Note that the dip of the bedding planes parallel the land surface. Source: K. Patrick. **B.** The bedrock escarpment on North Fork Mountain, overlooking Germany Valley (see Fig. 9.18). The cliff of Tuscarora Sandstone that forms the prominent escarpment contrasts with its gentle dip slope to the right. Below the escarpment, slopes are even gentler, because they are developed on less resistant (but still dipping) bedrock. Source: Pat & Chuck Blackley / Alamy Stock Photo.

Figure 9.22 Cretaceous Lakota Sandstone outcrops at the top of a homoclinal ridge overlooking older Spearfish Shale in Red Valley, just west of Rapid City, South Dakota, USA. The ridge and the Red Valley encircle the Black Hills, which lie in a broad structural dome (**Fig. 8.30**). Source: K. Patrick.

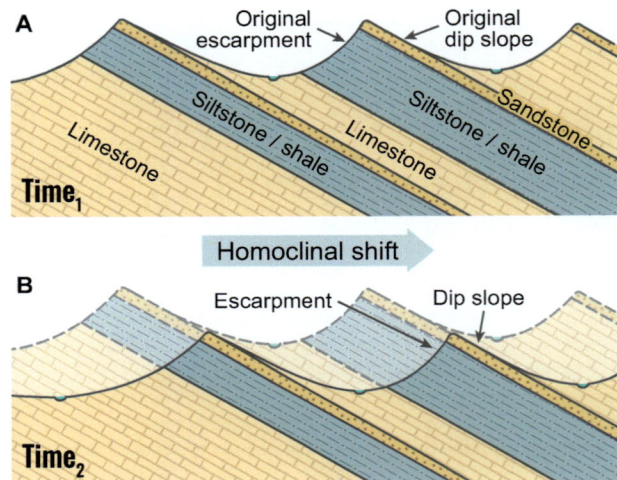

Figure 9.23 Homoclinal shifting of erosional topography formed on dipping strata. **A.** Time$_1$ shows how the topography correlates to bedrock type. **B.** Time$_2$ illustrates how the position of ridges, valleys, rivers, and their component slopes have shifted in the direction of dip (to the right, in this case), due to homoclinal shifting. This process occurs even as the correspondence between topography and geology remains the same.

A **homocline** is a geologic structure in which dip angle and direction are fairly consistent. **Homoclinal ridges** form on resistant dipping beds on fold limbs (**Figs. 9.22, 9.23**). They tend to have a steep escarpment on one side and a gentler dip slope on the other. Cuestas (**Fig. 8.28**) and hogbacks (**Figs. 8.31** and **8.32**) are two distinct types of homoclinal ridges discussed in Chapter 8.

The summits and side slopes of fold limbs are not static. Not only does erosion lower them but, over time, the ridges and streams on them also shift position laterally. This process is known as **homoclinal shifting**. Homoclinal shifting always occurs in the down-dip direction, that is, in the direction of bedrock dip, as the landscape is lowered by erosion (**Fig. 9.23**). Although the general *shape* of a ridge or valley formed on a fold limb may change little over millions of years, the location of the ridge or valley will change significantly due to homoclinal shifting. Fold limbs with low bedrock dip angles migrate farther and more rapidly than those with steep dip.

Bedrock structure involves more than just the dip of the beds. Rock layers also have **strike**, as shown diagrammatically in **Fig. 8.3**. A **strike ridge** is an elongated ridge, typically along a fold limb, that parallels the strike of the bedrock (**Fig. 9.19**). Between strike ridges lie **strike valleys** (**Fig. 9.24**). Many strike valleys are filled with thick accumulations of regolith, obscuring the underlying bedrock geology.

Figure 9.24 The 2.5 km-wide Strike Valley of the Waterpocket Monocline in Capital Reef National Park, Utah, USA, is aptly named, because it follows the *strike* of the bedrock. Most of the valley bottom is underlain by weak sedimentary rocks, such as shale, although a low sandstone-dominated hogback does dominate the middle of the valley. Hence, the small strike ridges within the larger strike valley. Source: R. Langstraat.

9.5 GAPS THROUGH RIDGES

No ridge can continue indefinitely. Thus, ridges may end in noses (**Fig. 9.7**) or at distinct **gaps** – short, narrow, stream-carved valleys that cut through the ridge. Geomorphologists sometimes apply the term **breached fold** to a partially eroded anticlinal or synclinal ridge that contains gaps. Generally, two types of gaps are recognized – **water gaps** (**Figs. 9.17, 9.25, 9.26A**) contain active through-flowing rivers or streams, whereas **wind gaps** (**Fig. 9.26B**) do not. Contrary to the name, wind gaps have not been formed by wind erosion. Rather, most wind gaps once held streams, but have since lost their stream flow and become dry. Most of the thousands of Appalachian water gaps are relatively young features – some perhaps only a few million years old.

Water gaps have long been star attractions in fold belts. Many years before becoming president, Thomas Jefferson wrote about Potomac Gap at Harper's Ferry, in modern-day West Virginia, USA: "The passage of the Patowmac [*sic*]

Figure 9.25 Water gaps. **A.** Two water gaps cut by the Potomac River through the quartzite bedrock of the Blue Ridge Mountains, near the junction of West Virginia, Maryland, and Virginia, USA. Source: Jstuby at English Wikipedia, Public domain, via Wikimedia Commons. **B.** Ground-level view of the water gaps near Harper's Ferry, West Virginia. Source: K. Patrick.

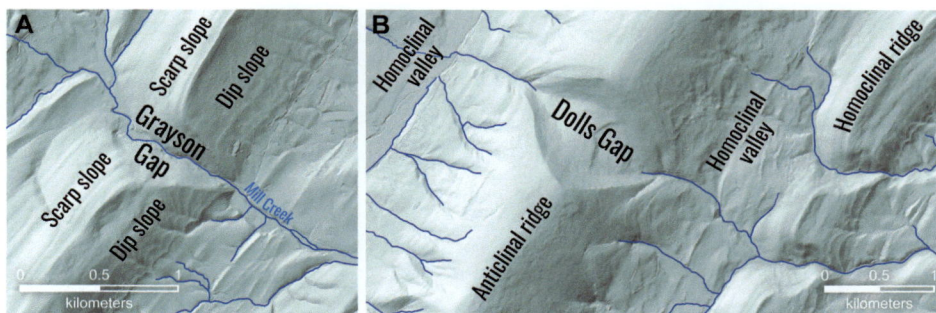

Figure 9.26 Hillshade model of (**A**) Grayson Gap (a water gap) and (**B**) Dolls Gap (a wind gap), both in West Virginia, USA. Mill Creek has cut a 200 m deep valley through the sandstone of Knobly Mountain, forming Grayson Gap. Streams on the eastern dip slope of Knobly Mountain have formed small, V-shaped valleys, separated by small **flatirons** (see Chapter 8). Dolls Gap only partly breaches the crest of the Wills Mountain Anticline. Thus, the gap has no through-flowing stream, unlike the water gaps on the right side of the map. Although most wind gaps form after water gaps are abandoned, Dolls Gap may have formed where there is a structural weakness in the rock. It could even become a water gap in the future.

through the Blue Ridge is perhaps one of the most stupendous scenes in Nature … worth a voyage across the Atlantic." Jefferson clearly appreciated the scenery formed by this gap, and how it had played key roles in history and culture.

Although it may seem counterintuitive for a river to carve water gaps through resistant bedrock ridges, several factors explain why this happens. In some cases, local structural weaknesses, such as faults and "kinks" in the fold geology, lead to fracture zones, weakening the rock and making it easier for a river to cut through (**Fig. 5.23A**). Alternatively, the route of the river may have been **superposed**, that is, inherited from an earlier time when the river was flowing over more erodible strata above (**Fig. 16.39**). At that time, the ridge was buried and did not even yet exist! Then, after a long period of erosion, the river encounters the folded, resistant strata. But by then, the river had become "locked" into that location. And so, it stays in place, forming a gap in what was becoming a ridge. Thus, superposition happens when a preexisting drainage system gets "lowered" onto a landscape by erosion and incision, so that it now crosses geologic structures to which it bears no relation (see Chapter 16). In short, rivers sometimes stay "in place," even as the landscape erodes around them, exposing buried geologic structures.

As geologic structures become "uncovered" by superposition, smaller rivers may change their course to accommodate the resistant rocks that are becoming exposed; they "go around" harder rocks and ridges. But the larger rivers are able to cut right through these resistant rocks and ridges, forming water gaps.

During superposition, the buried roots of ridges "rise" to the surface due to erosion of overlying sediment, but also because of the release of overlying weight, like an unloaded boat rises in the water. This process is known as **isostatic rebound**. As rebound occurs, most, but not all, of the drainage networks rearrange themselves, according to the geology.

9.6 STREAM PATTERNS AND PIRACY IN FOLD BELTS

The geomorphology of rivers and streams in fold belts reveals several interesting characteristics. Although water gaps present unique scenery, most large rivers in fold belts flow in deep valleys *between*, and *parallel to*, the prominent ridges. These valleys have usually formed in less resistant bedrock, for example, shale or limestone, following the strike of the rocks.

Nonetheless, the "uniqueness" of rivers and river patterns in fold belts lies not in the major streams, but in the many small, ephemeral creeks and "hollows" that drain into the larger streams. These small streams, typically too small to appear on topographic maps, are dry for much of the year. They cut across the strike of the bedrock, flowing more-or-less straight down the dip slope of the ridges, into the adjacent valleys, where they join the larger, perennial streams. This type of **trellis drainage pattern** is so common in fold belt topography that it epitomizes such landscapes (**Figs. 9.27**, **16.4**).

Stream piracy is another common and interesting aspect of the hydrology of fold belts. That said, streams can be "pirated" on any landscape where drainage basins "compete" for runoff waters, ranging from tiny headwater streams to major river systems.

Piracy occurs when a stream is diverted from its own channel, into the channel of a neighboring stream. For example, let's assume that stream A slowly lengthens by headward erosion, as its channel extends upslope, farther into its headwater reaches. Piracy occurs when the lengthening channel of stream A intersects the channel of a second stream (stream B), "capturing" the water that had been flowing in B. Water flow is then diverted from stream B into stream A because the more aggressively eroding stream – the one that does the pirating – almost always has a steeper local gradient. Thus, stream A gets enlarged, enabling it to erode and grow upstream even faster. More captures are then possible, and

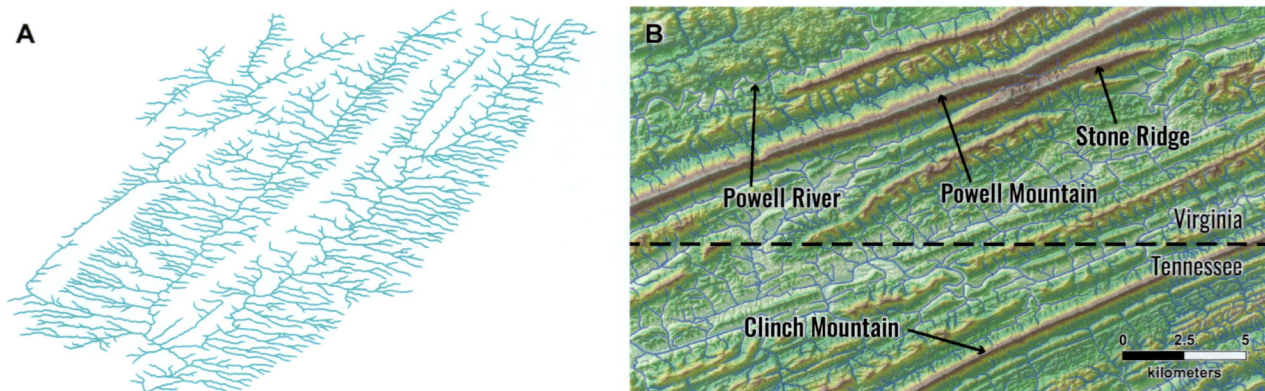

Figure 9.27 Trellis drainage patterns, typical of fold belts. **A.** A classic trellis pattern from the Appalachian Mountains of Pennsylvania, USA. Source: After Mejía, A. I., and J. D. Niemann (2008), Identification and characterization of dendritic, parallel, pinnate, rectangular, and trellis networks based on deviations from planform self-similarity, *J. Geophys. Res.*, 113, F02015, doi:10.1029/2007JF000781. **B.** Topography and trellis drainage patterns in the valleys and ridges of southern Virginia and northern Tennessee, USA.

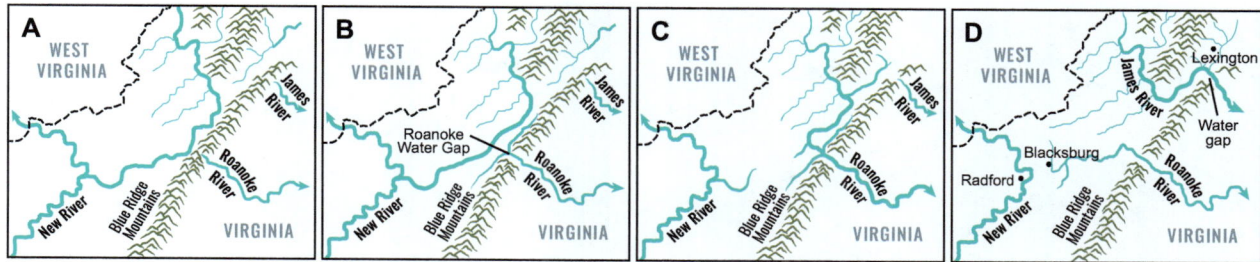

Figure 9.28 The upper James River drainage in western Virginia, USA, illustrates how stream piracy favors rivers with the shortest, steepest path to the ocean. Long ago (**A**), most Appalachian valleys and ridges in this area drained into the west-flowing New River. Headward erosion by the steeper and more powerful Roanoke River (**B**) breached the Blue Ridge and captured some of the New River drainage, by virtue of the Roanoke's much shorter path to the sea. The additional flow into the Roanoke River system helped it to erode faster and deepen its route through the Blue Ridge, forming a conspicuous water gap. The Roanoke's tributaries continued to extend farther headward into nearby valleys and ridges (**C**). Next, when the aggressive James River breached the Blue Ridge (**D**), it provided an even shorter, steeper path to the Atlantic for the uppermost Roanoke River. Thus, the pirating stream later got pirated! Together these piracy events captured over 10% of the New River's former drainage.

even likely. After a stream is pirated, all the water in its tributaries, that is, upstream from the point of intersection, is redirected into the pirating stream. Then, because the flow in stream B is greatly diminished or totally eliminated by the capture, the remnants of its old channel will dry up. Former water gaps it had carved will be left "high and dry" as wind gaps. Wind and water gaps are commonly the geomorphic legacies of stream piracy in fold belts.

For a stream channel to be pirated, its channel must be at a higher elevation, so that its water can be diverted into the lower-elevation pirating stream. Prior to a piracy event, the stream undergoing more vigorous headward erosion ("stream A" in the example above) typically has both a lower elevation *and* a steeper local gradient. The steeper gradient facilitates more rapid headward erosion. The steeper stream gradient may have arisen by virtue of a path over weaker rocks, a shorter flow path, or a deeper channel. In short, piracy is driven by the local gradient of the rivers in play. As long as the bedrock is similar, rivers with steeper gradients tend to erode headward more rapidly, enhancing their opportunities for piracy.

Interestingly, the headward erosion rates of some Appalachian rivers, as affected by recent tectonic activity, local geology, climate, and other factors, are typically greater than that of other rivers that also drain the same mountain range. Particularly, rivers in the Appalachians that drain (generally westerly) to the Gulf of Mexico via the Ohio and Mississippi Rivers (2,700–3,650 km downstream) have much lower overall gradients than east-draining streams that ultimately flow into the Atlantic Ocean (only 550–650 km downstream). Hence, the low-gradient streams that drain to the Gulf of Mexico are often the victims of stream piracy.

Let's illustrate this point by examining a piracy event from the James River in Virginia, USA (**Fig. 9.28**). Although many other pirating rivers in the Appalachians show ample evidence of headward erosion, the James River is a local heavyweight, eroding at rates two to five times faster than

most rivers in the region. Studies of the distribution of Appalachian fish species indicate that the uppermost James River Basin developed through two steps (**Fig. 9.28**). First, part of the New River (a west-draining river, flowing to the Gulf of Mexico) was captured by the east-draining Roanoke River. Next, the east-draining James River captured part of the Roanoke River. Left to nature, this regional drainage reorganization will continue over time. In fact, the James River has yet another capture victim in sight. Its headwaters are actively capturing Potomac River headwaters where the two rivers share a drainage divide.

9.7 NATURAL RESOURCES AND HAZARDS IN FOLDED MOUNTAINS

Bedrock structure and stratigraphy play fundamental roles in the land uses in fold belts. Surface mining is one of the most important "human-caused" geomorphic processes in the world today. In fold belts, the potential locations of mines for mineral resources like limestone, gravel, and coal are controlled primarily by topography and geology. Most mines occur where these factors combine to bring a resource near to the surface (**Fig. 9.29**). Although mineral rights ownership, transportation costs, and environmental concerns influence where mines are located, these factors are secondary to the geology.

Because of the many steep slopes, slope failures such as landslides and slumps (see Chapter 14) are common natural hazards in fold belts. Indeed, the hazard risk to people and property due to slope failure over much of the Appalachians is second only to flooding. Geomorphologists have determined that landslide susceptibility tends to be closely linked to specific rock types and structures. Formations composed of interbedded sandstone and shale tend to be most prone to landslides, particularly where this bedrock combination occurs on steep slopes.

Within a given rock type, the dip of the rock layers is important in determining landslide susceptibility. Dip slopes

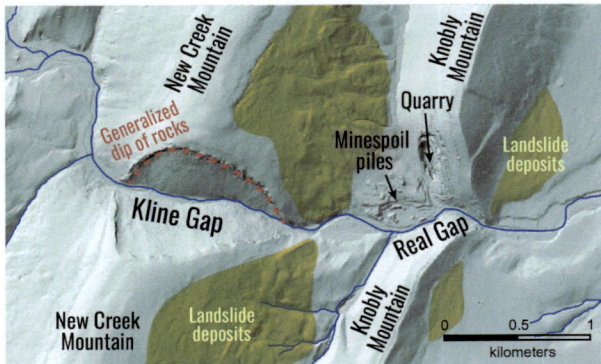

Figure 9.29 The importance of dip of the rock units on slope stability is apparent at Kline Gap, a prominent water gap through the Wills Mountain Anticline in Grant County, West Virginia, USA. Cliffs of Tuscarora Sandstone (in the gap) reflect the anticline's asymmetry. The hummocky topography on the southeast flank of New Creek Mountain is developed on dozens of landslide deposits on the unstable dip slope. In contrast, the topography on the same bedrock units on the (geomorphically stable) northwest side of the mountain has had very few landslides. Instead, we see many small, southwest–northeast ridges, aligned with the strike of bedrock. Knobly Mountain's southeast dip slope shows several landslides, while the escarpment on the west face reveals bedrock ribs, a limestone quarry aligned along the strike of the bedrock, and nearby mine waste piles, but no landslides.

Figure 9.30 February 2019: debris flow on a dip slope dominated by shale and sandstone in the Appalachian Mountains near Bluefield, West Virginia, USA. The house with the red roof was carried 30 meters downhill, coming to rest at the edge of a major railroad line. Tan-colored, unpaved roads near the top of the ridge do not appear on satellite imagery pre-dating 2019, suggesting a human-induced trigger for this failure, which remained active for many months. Source: L. Lackey, InFlight Aerial Imaging Services.

are notorious bad actors wherever the topographic slope is nearly the same as (or steeper than) the dip of the geologic beds (**Fig. 9.30**). Escarpments where the rock units dip at low angles, or in the opposite direction to the slope, may produce rock falls or modest debris flows, but tend to host only comparatively small landslides. Slopes underlain by horizontal or vertical beds usually have relatively few (but perhaps large) landslides.

Slope failures can be triggered or caused by any activity that alters surface topography, drainage, or soil moisture, such as infrastructure development, mining, deforestation, irrigation, or storm-water management. Improper site design or human error often cause some of the largest failures. Thus, human actions are becoming a more important contributor to landslide susceptibility, especially in fold belts. Nonetheless, most landslides are the unintended result of disturbance by individuals lacking sufficient knowledge to safely attempt a project on unstable slopes (**Fig. 9.30**). In short – knowing the local geology can help greatly to keep everyone safe and to avoid unintended slope failures.

REVIEW QUESTIONS

9.1 Explain the geologic structure of anticlines and synclines. How does a fold axis relate to these rock structures?

9.2 Restate the "rule of thumb" that allows you to determine the age sequence of the underlying bedrock of domes and basins. Provide an example for both a dome and a basin.

9.3 With which type of tectonic plate boundary are fold and thrust belts associated?

9.4 What forces and causes produce a thrust fault?

9.5 With regard to folded rocks, what is plunge?

9.6 What types of quartz-rich rocks are most likely to outcrop as ridges in fold belt mountain systems?

9.7 What types of sedimentary rocks are likely to occur as valleys in humid-climate fold belt landscapes?

9.8 What determines whether the axis of an anticline or syncline forms a ridge or a valley?

9.9 What landform will breached anticlines eventually become?

9.10 How do zigzag mountains form in folded rock terranes?

9.11 What is stream piracy and how are streams pirated by other streams?

9.12 What are wind gaps and water gaps? Can one evolve into the other? If so, how?

9.13 What type of drainage pattern is typically associated with fold and thrust belt mountains? How does this pattern form? What does it look like from above?

FURTHER READING

Blakely, R. C. and Ranney, W. D. 2018. *Ancient Landscapes of Western North America*. Springer.

DiPietro, J. A. 2018. *Geology and Landscape Evolution: General Principles Applied to the United States*. Elsevier.

Harris, A. G., Tuttle, E., and Tuttle, S. D. 2004. *Geology of National Parks*. 5th ed. Kendall/Hunt.

Lebold, J. G. and Wilkinson, C. 2018. *Roadside Geology of West Virginia*. Mountain Press.

Lillie, R. J. 2005. *Parks and Plates: The Geology of Our National Parks, Monuments and Seashores*. W.W. Norton.

Schultz, C. H. (ed.). 1999. *The Geology of Pennsylvania*. Pennsylvania Geological Survey and Pittsburgh Geological Society.

Spencer, E. W. 2017. *Guide to the Geology and Natural History of the Blue Ridge Mountains*. University of Virginia Press.

10 Landforms on Faulted Rocks

Ronald I. Dorn, Phillip H. Larson, and Ara Jeong

Faulting within rocks and sediment creates some of the most dramatic landscapes (Fig. 10.1). A favorite trip for many visitors to the Western United States follows the route from San Francisco to Las Vegas, or from San Francisco to Phoenix, where much of the mountainous scenery along these routes has been formed by faulting. Because of this faulting, the high alpine landscapes contrast greatly with the desert landforms in the lowlands below. After reading this chapter, you will be able to take this excursion and understand the stunning geomorphology of these faulted landscapes, and others.

A good way to understand how and why rocks break and fault is to think of the **stress**, or the force applied to solid materials. Look at your hands as you wash them. Just as in rocks, stresses involve pushing together (compression), pulling apart (extension), and slipping side-by-side (shearing). The stress you apply to your hands while washing may not rupture skin or break bones. But **faulting** can and does break rocks and moves the broken pieces along a **fault plane** – the surface along which the rocks will move.

To continue the analogy, extreme stress to your hands can break skin or bones, just as such stresses can break rocks. In actuality, the solid earth – rocks – responds to stresses by both bending (folding) and breaking (faulting), like a stick bends or snaps when force is applied (see Chapter 9). Geologists consider a **fault** to be a planar fracture or discontinuity in a rock across which there has been significant displacement. When the processes of folding and faulting cause large-scale deformation of the Earth's crust, they are summarily referred to as **diastrophism**. Diastrophic movement is often associated with mountain building, and when it is, the term **orogeny** is used (see Chapter 6).

Faulting moves the parts of the land surface up, down, and/or sideways. These changes then govern the effectiveness and pathways of subsequent erosional processes. Geomorphologists take a long-term view of both of these processes by studying the cumulative effects of faulting *and* erosion on the land surface, and on timescales of centuries to millions of years. And that is the focus here.

10.1 ORDER OF RELIEF PERSPECTIVE IN GEOMORPHOLOGY

Stresses that build within the Earth's crust – that lead to folding and faulting – often produce changes in the relief (also called **topographic relief**) of the land surface. Stresses and forces thrust parts of Earth's crust upward, and/or allow it to drop down, impacting the relief of the landscape, that is, the elevation difference between the highest and lowest places in an area.

Most people think of mountain ranges when they think of high-relief landscapes, for example, the Himalayas of southern Asia (see Chapter 6). Others picture deep canyons, like the Grand Canyon in the southwestern United States (Fig. 10.2). Thoughts of low-relief areas conjure up images of flat and open spaces like the continental grasslands of the Great Plains in the United States, the Steppes of Central Asia, or the glacially scoured Canadian Shield.

Figure 10.1 Faulting can lead to dramatic, high-relief landscapes. **A.** Mount Whitney, the highest point in the continental United States, was separated vertically by faulting from the area in the foreground, and is now a popular tourist destination. **B.** The Mormon Point "turtleback" in Death Valley National Park, California, and the down-faulted valley next to it. **C.** The Sierra Estrella fault block, Arizona (USA), once >4,500 m high during the peak of normal faulting in the Miocene, has since been eroded to an elevation of about 1,300 m. Source: R. Dorn.

Figure 10.0 Rocks appear to have moved across the monotonously flat and usually dry floor of Racetrack Playa, in Death Valley National Park, California, USA. Source: Photography by Deb Snelson / Getty Images.

Figure 10.2 Contrasts in topographic relief. Examples of areas of high relief include (**A**) the Grand Canyon of the southwestern United States and (**B**) the Wind River Mountains of Wyoming, USA. **C.** The incredibly flat, low-relief steppes of southern Russia display almost no relief. Source: R. Schaetzl.

Figure 10.3 The 1st order of relief on Earth involves the topographic highs of continents and the topographic lows of ocean basins. Source: NOAA, National Centers for Environmental Information.

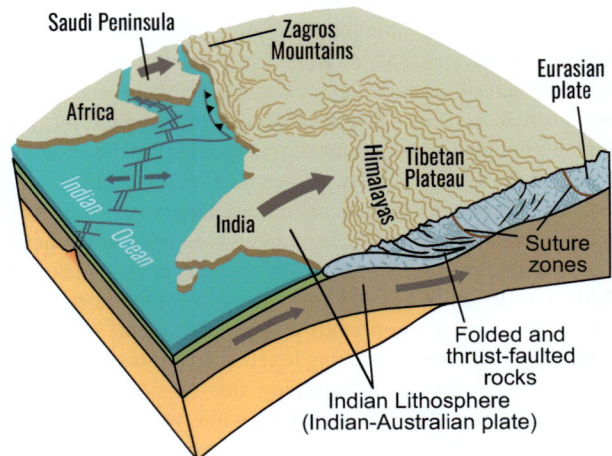

Figure 10.4 The Himalayas and Tibetan Plateau exemplify the 2nd order of Earth's relief features. These mountains were produced by a series of compressive thrust and reverse faults associated with the collision of the Asian and Indian tectonic plates (see Chapter 6).

Geomorphologists often categorize Earth's surface as having three orders of relief. The **1st order of relief** refers to the overall differences in elevation between continents and the ocean basins (**Fig. 10.3**). Although faulting is involved in the growth of continents and the opening and closing of ocean basins, via plate tectonics, the faults themselves cannot be seen at this scale.

Earth's **2nd order of relief** encompasses mountain ranges and valley systems – features that are also usually the product of large-scale plate tectonics (see Chapter 6). This order of relief is associated with the core concept of diastrophism. A good example would be the Himalayas of southern Asia (**Fig. 10.4**). The relief associated with the extensional crustal tectonics of the East African Rift Valley, with its deep (600–900 m) valleys, is yet another example (**Fig. 6.21D**).

The **3rd order of relief** consists of landforms that are often formed by faulting and folding of rocks, followed by erosional and depositional processes. For example, individual mountains and river valleys are 3rd order relief

features. Denali, the highest mountain in North America, owes its relief to the Denali Fault that runs from British Columbia through Central Alaska (**Fig. 10.5**). Although the fault is mostly a strike-slip fault, compression along the fault zone has led to considerable uplift. The combination of compression and shearing is sometimes called **transpression**, where *trans*- is for transform (side-by-side motion, as in a strike-slip fault) and -*pression* is for compression.

The Canadian Rockies illustrate linkages that can occur between the 2nd order of relief (formed by tectonic compression) and the individual faults that lead to the 3rd order of relief (as individual mountains). **Figure 10.6** illustrates how the collision of the Farallon Plate against Canada's west coast shortened the crust in this region. This shortening was accomplished by a series of low-angle thrust faults and higher-angle reverse faults between the west coast of Canada and Calgary, Alberta. The mountains in this region are, therefore, a direct result of these faulting processes.

The remainder of this chapter focuses on geomorphic processes and landforms associated with the 3rd order of relief, in areas dominated by faulting.

Figure 10.5 At over 6,000 m elevation, Denali's relief is a product of faulting in combination with erosional processes. This photograph of the Cantwell Glacier was taken just after a major earthquake in 2002, which produced a fault scarp that appears as a dark zigzag feature in this photo. Source: US Geological Survey.

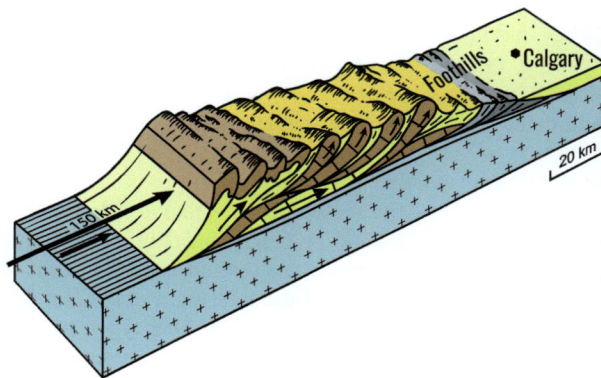

Figure 10.6 This geological cross-section illustrates that the Canadian Rockies in parts of Alberta and British Columbia are a product of reverse and thrust faults acting together to shorten the crust and build a series of mountains with intervening valleys. Larger-scale tectonic forces (2nd order of relief) caused the crust to be compressed, and that compression was expressed as the thrust faults and folds that drove mountain growth (3rd order of relief).

10.2 STRESS, EARTHQUAKES, AND FAULTING

When rock material is stressed by compression (pushing together), extension (pulling apart), or shearing (side-by-side motion), it tends to react in one of two ways. It may behave as a **brittle** material and snap or break, forming a **fault**. In other circumstances, rocks behave as a **ductile** material and bend, resulting in folding. Sticks, like rocks, can behave similarly – they can break or bend!

Imagine working with clay as you form a ceramic pot. Such material is easily manipulated. It is easy to make it thinner or thicker. The clay stretches as you pull on it, just as ductile or bendable rocks behave. However, after you fire the pot in a kiln, the clay becomes brittle. Then, when you exert stress on this brittle material, that is, the pot, it breaks, just like rocks break along a fault.

Whether a rock behaves like a brittle or ductile material depends on its lithology, its temperature and pressure conditions, the rate of deformation (**strain**), and also the type of stress (**Fig. 10.7**). Generally, rocks fold (but not break) under conditions of high temperature and pressure, typical of great depths, and under slower rates of deformation (see Chapter 9). And of course, various types of rocks respond to these conditions differently. Faulting of rocks is generally favored under conditions of lower temperatures *and* pressures, more typical of near-surface conditions, as well as in situations of rapid deformation. Rock type also matters; sedimentary rocks tend to fold more easily, whereas metamorphic and igneous rocks tend to facilitate faulting.

When rocks break (fault), shock waves (also called **seismic waves**) emanate from an initial slippage point – a **focus point**. Observers on the surface may then feel shaking and trembling as an **earthquake** – vibrations (seismic waves) in the crust caused by the rupture and sudden movement of rocks that have been strained beyond their elastic limits. Earthquakes occur in association with faulting events. Focus points almost always occur at depth. The location directly above the focus point, on the surface, is called the earthquake's **epicenter** (**Fig. 10.8A**).

When rocks break, the sudden movement sometimes, but not always, results in the rupture and displacement of rock

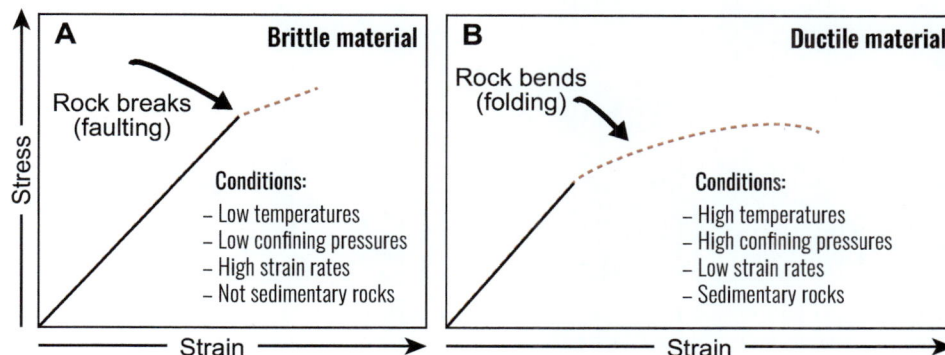

Figure 10.7 Conditions of rock stress and strain (deformation) that lead to faulting (brittle failure/breakage) vs folding (ductile behavior/bending). The change in line type identifies when rock either breaks (**A**) or folds (**B**).

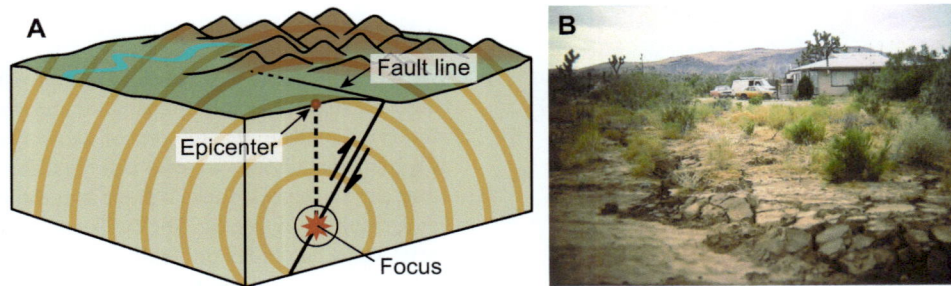

Figure 10.8 Earthquakes are associated with faulting. **A.** Diagram illustrating the relationships between the epicenter of an earthquake, its focus, and the resultant fault line (or trace) that may be formed at the surface. **B.** Photograph taken near the epicenter of the 1992 Landers, California (USA) earthquake (magnitude 7.3). This area experienced mainly strike-slip (side-by-side) motion, but also some vertical displacement, resulting in the small fault scarp seen here. Source: R. Dorn.

and sediment at the surface, forming a fault. Sometimes, when the focus point is shallow or when rocks are strongly faulted, the fault itself is visible at the surface, as a **fault line** or **fault trace** – the trace of the fault on the surface (**Fig. 10.8A**). Some fault lines can be traced along the surface for many kilometers.

If rocks are pushed upward or dropped downward along a fault line, a **fault scarp** may form, marking the location of the deformation (**Fig. 10.8B**). A fault scarp (short for "escarpment") is the near-vertical face left behind by the upwardly or downwardly displaced rock. The height of the scarp is a good indication of the vertical component of movement along the fault. But because faulting is often episodic, the height of the scarp may reflect multiple offsets or faulting events. Scarps are probably the most common landform produced by faulting. Most fault scarps are only a few meters in height.

Faulting associated with earthquakes, even large ones, may leave a geomorphic expression of the event – the fault scarp – for decades or centuries. Only very recently formed scarps have a sharp edge and steep face (**Fig. 10.9**). Soon after, they are subjected to the forces of erosion, becoming more rounded, and their overall slope angle declines. They also may develop ravines and channels across them. Debris

that accumulates at the base of fault scarps eventually buries them. However, repeated earthquakes along the same fault can "freshen up" previous scarps by initiating more vertical motion.

10.3 TYPES OF FAULTS AND FAULTING

Brittle failure of rocks causes them to break, or fault – the heart of this chapter. Faults are rarely isolated features. Instead, there may be one or several main faults, and several additional, associated faults nearby. Many of these faults will parallel the main fault plane/line. A **fault zone** is the term for these mostly parallel faults.

Faults are classified based on the relative movement of the rocks on either side of the fault plane (**Fig. 10.10**). Side-by-side (also called **transform**) motion in rocks produces **strike-slip** faulting. Think of the rocks in a strike-slip fault as simply sliding past each other. The other types of faults involve some component of vertical motion. When faulting occurs due to extension (stretching) of the crust, one or more blocks are dropped *down*; this process is referred to as **normal faulting**. Normal faulting dominates rift zones like East Africa. Crustal compression can generate either high-angle **reverse faulting** or lower-angle **thrust faulting**,

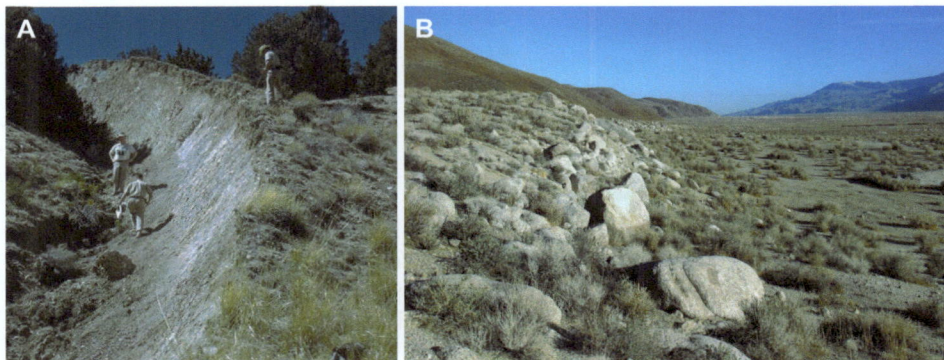

Figure 10.9 Historic fault scarps. **A.** Vertical displacement along the newly formed, Fairview Peak fault scarp, formed in 1954, near Fallon, Nevada (USA). Source: Steinbrugge Collection, Earthquake Engineering Research Center, UC-Berkley. **B.** An eroded fault scarp at the base of the Sierra Nevada Mountains, formed during the Owens Valley Earthquake of 1872. The 2–4 m vertical movement seen here is from a single earthquake. Source: M. Miller.

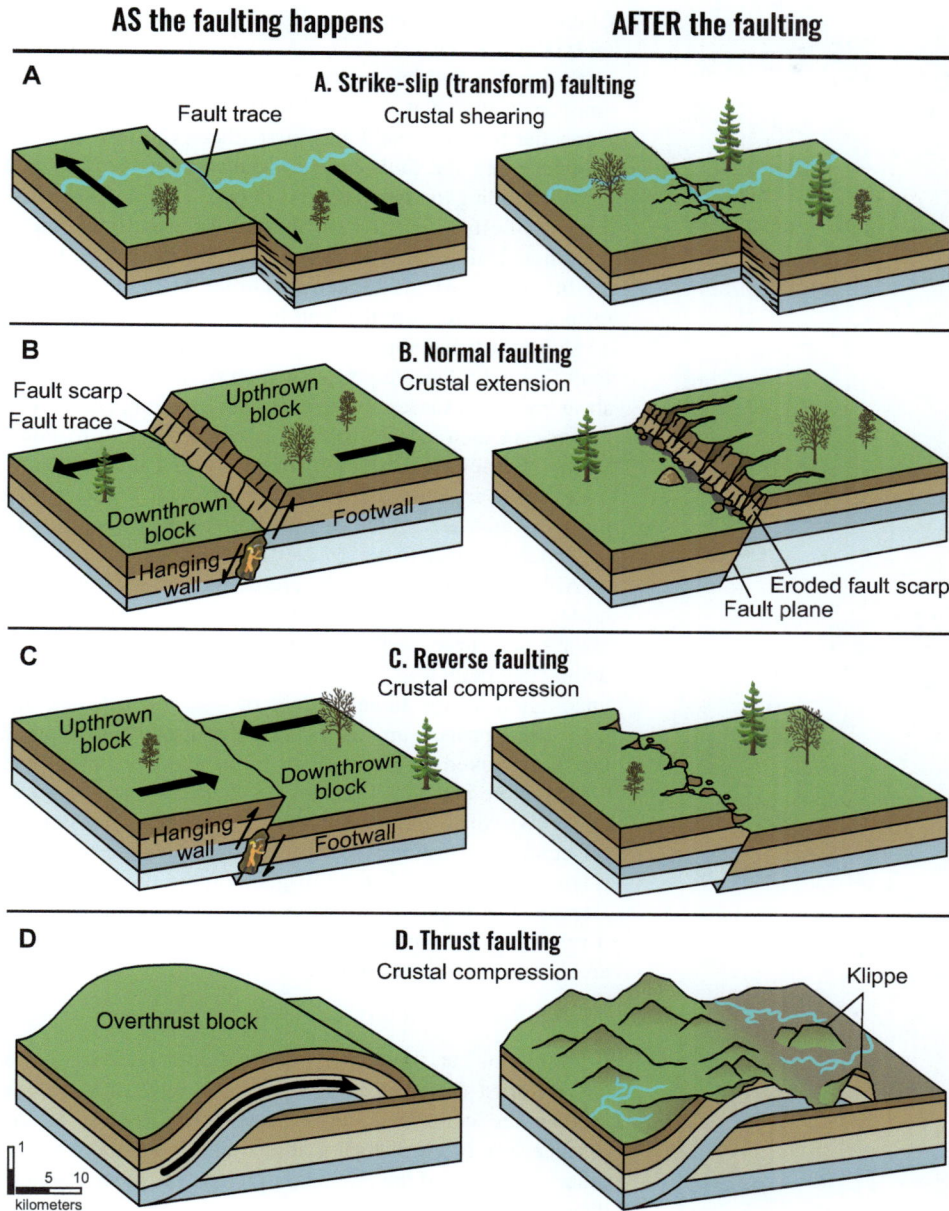

Figure 10.10 The major categories of faulting, resulting from the three different types of applied stresses. **A.** Horizontal or transform stress leads to strike-slip faulting. **B.** Extensional stress leads to normal faulting. Compressional stress leads to (**C**) high-angle reverse faulting and/or (**D**) low-angle thrust faulting. Notice that with thrust faulting, one block is thrust over (overthrust) another block.

depending on how the stress is applied. Compressional stresses are commonly associated with tectonic subduction zones and orogenic belts. The Himalayas, Alps, and Andes Mountains largely formed by compressional tectonic stresses (**Figs. 6.8, 6.22**). Shearing effects often dominate areas near transform plate boundaries.

10.3.1 Strike-Slip Faulting

Perhaps the most easily visualized type of faulting results from horizontal (shearing) motion; no upward or downward motion need occur. This type of *transform* motion leads to strike-slip faulting, in which the movement is parallel to the fault line, or strike (**Fig. 10.10A**). The fault plane in most strike-slip faults is nearly vertical, plunging deeply into the crust below. Horizontal movement along strike-slip faults separates and *offsets* rocks (**Fig. 10.11**). Strike-slip faults impact many regions of Earth's surface, such as the Anatolian Fault in Turkey, the Talas-Fergana Fault in Central Asia, and the San Andreas Fault in California, USA.

Geomorphologists utilize adjectives to communicate to others the direction of movement along strike-slip faults. Examine the fault diagram in **Fig. 10.11A**. Now, place yourself on one side of the fault and picture someone else standing on the opposite side, as the faulting occurs. That person will move to your right. Now, try to visualize what that other person sees happening to you; it's the same thing; to them, *you* are moving to *their* right! Thus, the fault in **Fig. 10.11A** is a right-lateral strike-slip fault. All strike-slip faults are

A

Before faulting

During faulting

After faulting

B

Moves to the right

Fault trace (Fault line)

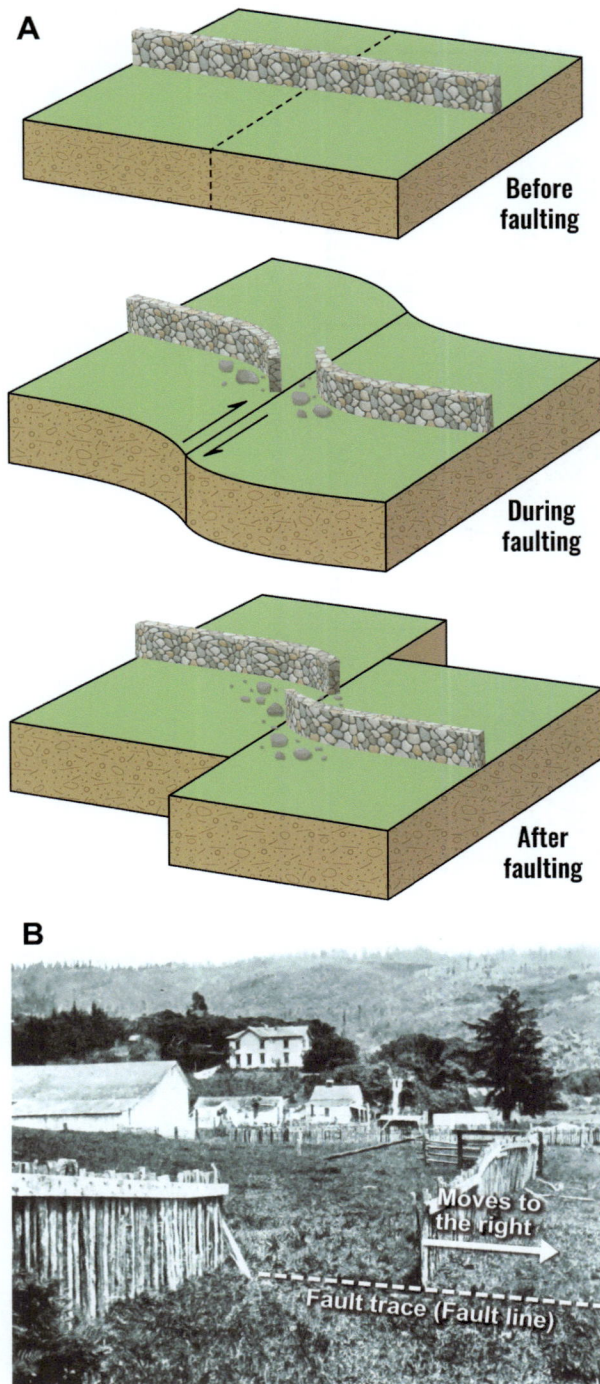

Figure 10.11 A. Diagram showing the motion and offsets associated with strike-slip faulting. **B.** G. K. Gilbert's famous photograph of an offset fence, taken near Bolinas, California (USA), associated with the 1906 San Francisco earthquake along the San Andreas Fault. No matter your perspective, the movement appears to have been to the right, making this a right-lateral fault. Source: US Geological Survey.

either **right-lateral** or **left-lateral** faults. The San Andreas Fault shown in **Fig. 10.11B** is a right-lateral strike-slip fault. The Piqiang Fault of the Taklamakan Desert in Western

China is an excellent example of a left-lateral strike-slip fault (**Fig. 10.12**).

Strike-slip faulting produces a variety of characteristic landforms, as shown in **Fig. 10.13** for an idealized landscape along the San Andreas Fault in southern California, USA. Motion along the fault can physically separate rivers that cross it, leading to **offset streams** (**Figs. 10.13, 10.14A**). The amount of offset for such streams can be used to determine the amount of recent movement along the fault line. Rocks are often crushed and broken within the fault zone, leading to springs and seeps at the fault line.

During a strike-slip faulting event, rocks in and near the fault line get intensively broken and crushed, rendering this shear zone area susceptible to enhanced erosion (**Fig. 10.13**). Rivers can then exploit this eroded area, leading to linear valleys that follow these fault zones. Along and within these valleys, lower areas can become filled with water, as **sag ponds**, which are produced by localized down-warping of the ground surface (**Figs. 10.13, 10.14B**).

Although strike-slip faults are defined as having mostly horizontal motion, some amount of vertical motion is almost always present as well, leading to other features that help define the fault line (**Fig. 10.8B**). For example, it is not uncommon to find **linear ridges** of less crushed rock that have been pushed up, and **shutter ridges** where a small hill was moved in front of an offset river channel (**Fig. 10.13**). Note in **Fig. 10.13** how the shutter ridge of slightly up-thrown rock formed by the faulting has blocked the offset stream, forcing it to change course. The term shutter ridge comes from this type of geomorphic "consequence." Because the ridge is a barrier to anything crossing the fault, it reminded nineteenth-century geologists of the *shutter* of an old-fashioned box camera.

10.3.2 Reverse and Thrust Faulting

Compression can shorten the crust and often leads to reverse and/or thrust faults, as rocks are pushed upward to accommodate the compressional stresses (**Fig. 10.10C, D**). In essence, one block of rock is forced to ride up and over another. The angle of the fault plane (relative to the horizontal) determines the name of the fault. Fault planes with angles < 30°, that is, shallow angles of motion, are referred to as thrust (or overthrust) faults (**Figs. 10.10D, 10.15**), whereas movement along steeper (closer to vertical) angles is classified as reverse faulting (**Figs. 10.10C, 10.16**). Most thrust faults have fault planes with very shallow angles, often less than 10°. Reverse faulting can create an overhanging fault scarp, that usually then collapses under its own weight, leading to a rubble and debris pile at the base of the fault scarp (**Fig. 10.10C**). Because of collapse and/or burial, seldom is an intact fault scarp observed, and even when it is, it will be an ephemeral feature, lasting a few years at most.

Some terms for faulted rocks derive from miners. As they encountered tilted or inclined fault planes in the mine walls, miners would refer to the block of rock under their feet as the **footwall**. The block of rock above the miners' heads, that is, the roof, was termed the **hanging wall**. These terms persist

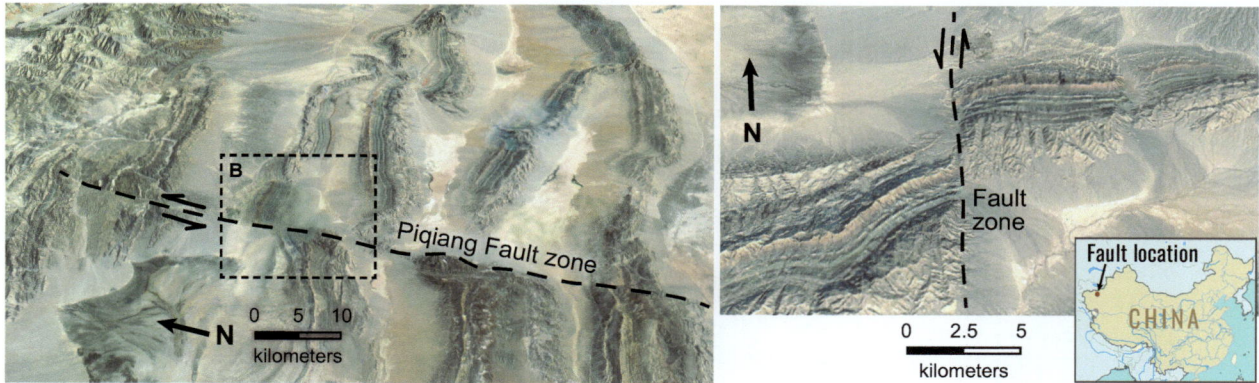

Figure 10.12 The Piqiang Fault in the Taklamakan Desert of Western China, showing several kilometers of left-lateral motion. A careful inspection will allow you to match up the rocks across the fault. Source: NASA.

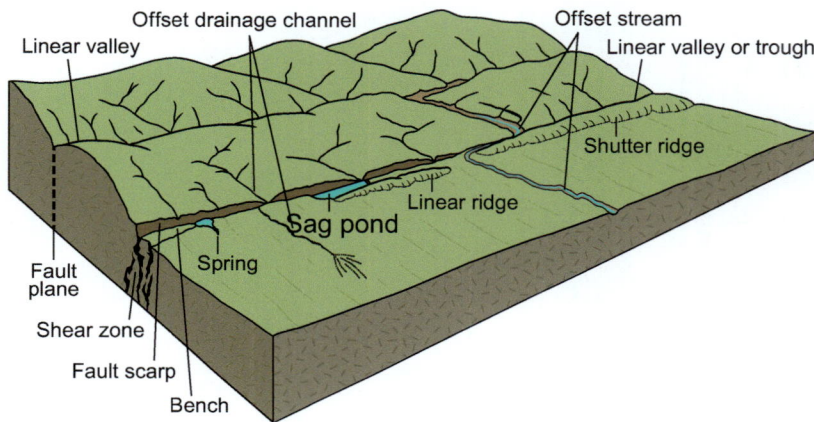

Figure 10.13 Landforms commonly associated with strike-slip faulting, as shown on this idealized diagram of the area near the San Andreas Fault in southern California (USA). Source: After Wesson et al. (1975), via the US Geological Society; in public domain.

Figure 10.14 Landscapes affected by strike-slip faulting on the San Andreas Fault. **A**. Offset streams caused by right-lateral motion on the San Andreas Fault, which runs from right-to-left across the photo (the white line is a dirt road). Source: NASA. **B**. Small vertical movements can occur along a strike-slip fault. When they do, sag ponds fed mostly by groundwater can form, as illustrated here along the southern boundary of Carrizo Plain National Monument in California, USA. The fault runs north along the east (left) flank of the distant Elkhorn Hills. Source: US Geological Survey.

today in describing the different fault blocks in both reverse and normal faults (**Fig. 10.10B, C**). In reverse faults, the hanging wall always moves upward, relative to the footwall; normal faults are the opposite.

Compression that causes thrust and reverse faulting is often driven by tectonic activity, for example, in the Himalayas and Andes Mountains (**Fig. 10.4**) and the Canadian Rockies (**Fig. 10.6**). Although thrust and reverse

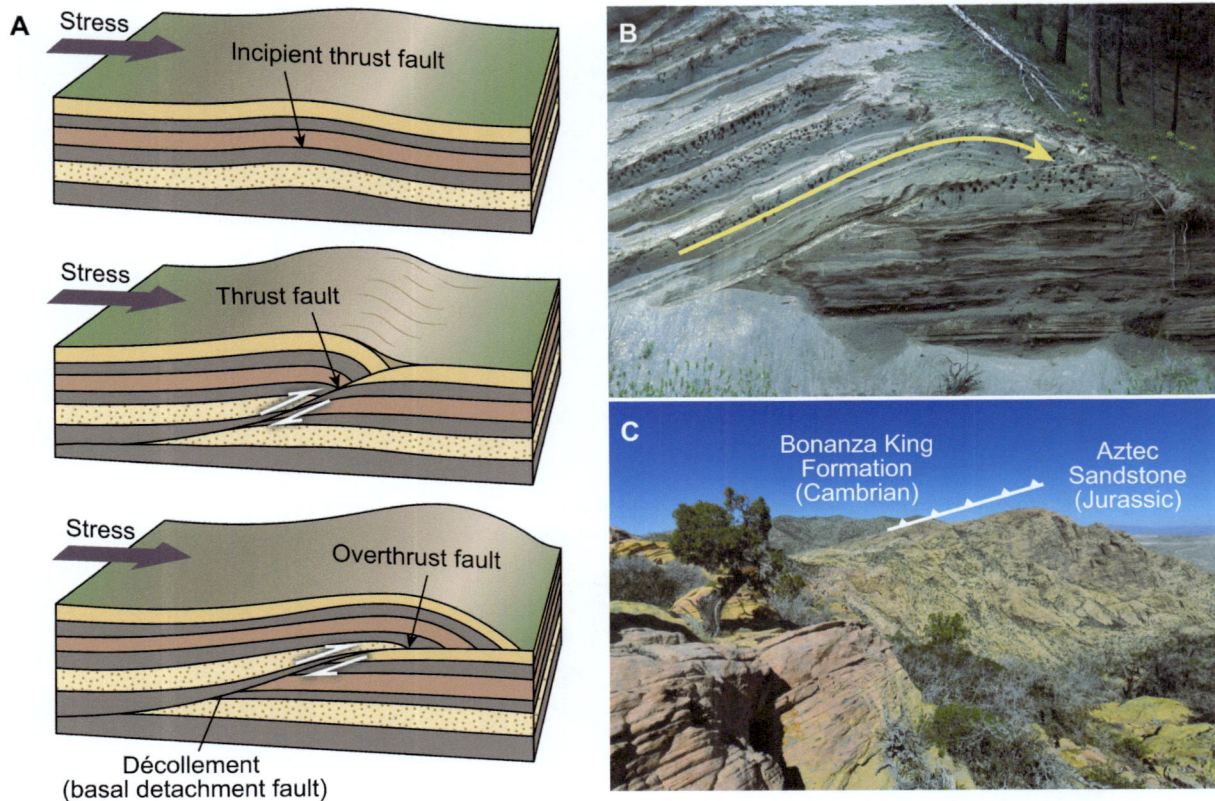

Figure 10.15 Thrust, or overthrust, faults. **A.** Schematic diagram of a thrust fault in sedimentary rocks. **B.** A small thrust fault and bend-fold in eastern Washington State, USA. Most thrust faults are much larger and more expansive phenomena. Source: M. Miller. **C.** The famous Keystone Thrust Fault, west of Las Vegas, Nevada, USA. Here, gray, Cambrian-aged limestones were thrust over red and white, Jurassic-aged sandstone. The overall lateral displacement of rocks by this 66 Ma thrusting event was between 11 and 22 km. Note the low angle of the fault plane. Source: H. Fossen (2010), with permission from Cambridge University Press & Assessment.

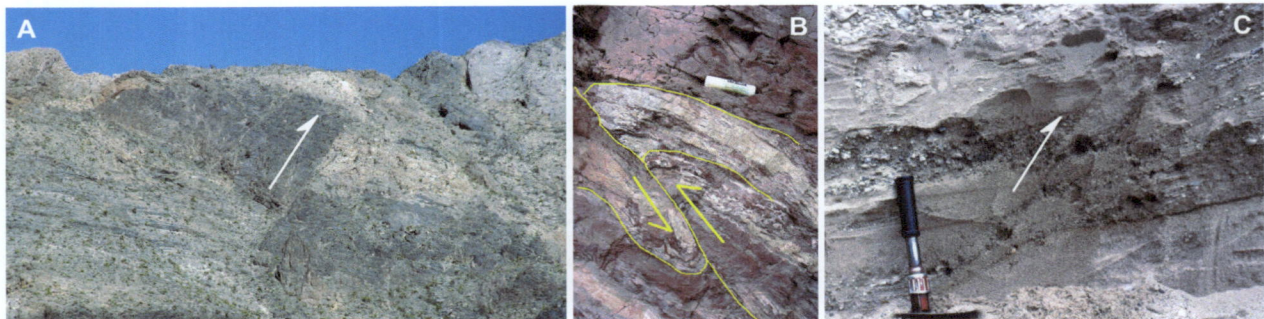

Figure 10.16 Reverse faults. **A.** A small reverse fault in the Providence Mountains of the Mojave National Scenic Preserve of California, USA. Source: Garry Hayes, via Geotripperimages.com. **B.** A small reverse fault in Proterozoic-aged rocks along the Grinnell Glacier trail in Glacier National Park, Montana, USA. Source: Garry Hayes, via Geotripperimages.com. **C.** Even loose sediment can be reverse faulted, as in this sandy glacial outwash sediment, which was probably frozen at the time of displacement. Source: R. Schaetzl.

faulting can and does occur in areas *not* undergoing collision of tectonic plates, there must nonetheless be compression for this type of faulting to occur. Such compression can occur within a restraining bend of a strike-slip fault (**Fig. 6.15**). This situation occurs in southern California, USA, along the San Andreas Fault, the world's most famous strike-slip fault (**Fig. 10.17**). Although most of the stress along the fault is

accommodated by shearing, a large bend in the fault line has resulted in compression in this part of the fault zone, leading to reverse faulting and uplift, forming the Transverse Ranges.

Thrust faults sometimes begin and propagate along zones of weakness within the rock column, for example, within mudstones, shales, or salt beds. Such rocks will

Figure 10.17 Satellite view of the landscape around the San Andreas Fault, looking northwards. The "big bend" in the fault has resulted in localized compression (see Figs. 6.15 and 6.16 for more information). Note that the area experiencing the compression (between the large white arrows) has been uplifted into low mountain ranges oriented transverse to the rest of the landscape. Mountain building and valley subsidence are in close proximity in this area. Source: NASA

break or slide more easily than the harder, more competent rocks above and below. The displacement can then sometimes extend for many tens of kilometers, such that rocks can overlie others of completely unrelated geology. For example, at the Keystone Thrust, Cambrian limestones now overlie Jurassic-aged sandstones that are ≈350 million years younger (Fig. 10.15C).

Lower-angle thrust faulting can set the stage for the development of **klippe** – where a mountain formed in older (but resistant) rocks occurs *on top of* younger, more erodible rocks (Fig. 10.10D). This situation develops as erosion of the older (thrust) rocks leads to their physical separation. Chief Mountain in Montana (USA) is a classic example of a klippe (Fig. 10.18). Here, the Lewis overthrust has moved Precambrian rocks on top of Cretaceous rocks. Erosion gradually separated the Precambrian rocks from the Lewis Range, leading to the isolation of Chief Mountain and smaller mountains nearby. Erosion of the younger, "overthrown" rocks can also form inlier "windows" into the less disturbed rocks below.

In areas rife with faulting, like within the Los Angeles (California) area, faults of many different kinds can occur, and even overlap. The San Andreas Fault is only one of many faults in this area, some of which are strike-slip faults,

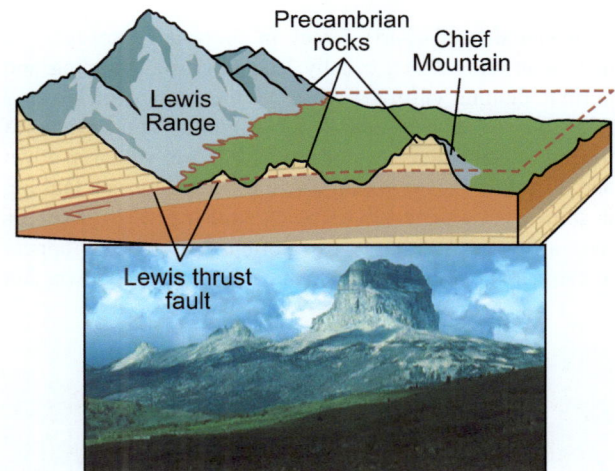

Figure 10.18 Illustration showing how the topography in the Chief Mountain region of Montana has formed due to thrust faulting, followed by erosion of the surrounding rock material.

whereas others are reverse faults (Fig. 10.19). Some of the faults are exposed at the surface, like the Sierra Madre Fault at the base of the San Gabriel Mountains, whereas others are hidden, buried, and thus, unmapped. The many high-angle

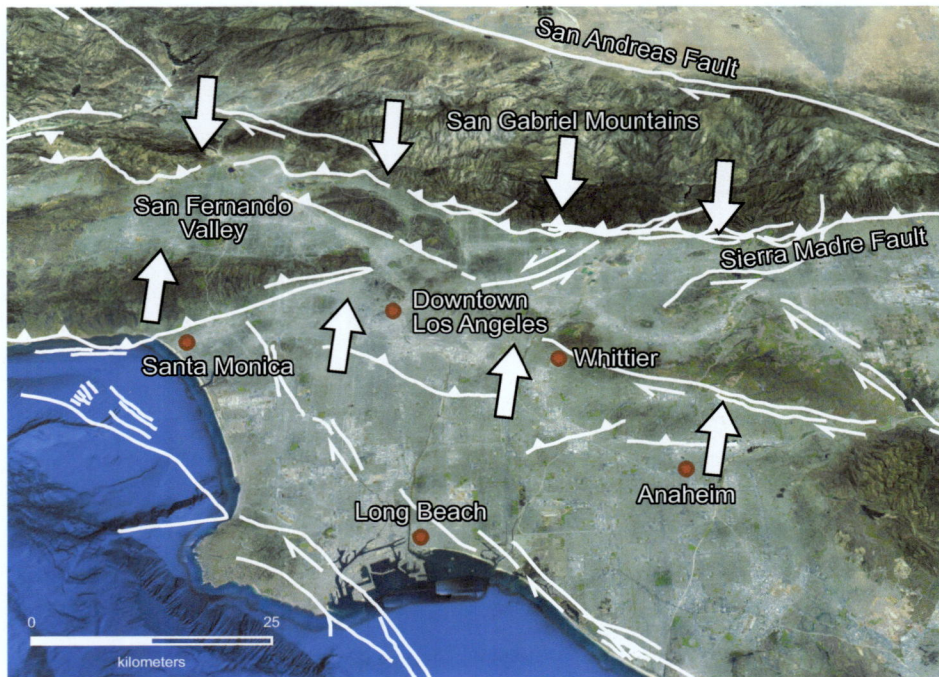

Figure 10.19 Faults are seemingly everywhere within the Los Angeles metropolitan region in California (USA). The San Andreas strike-slip fault is the most famous one, but compression associated with the "big bend" in the fault has produced a number of associated faults, most of which are thrust faults that bound mountains near the city, such as the San Gabriel and Santa Monica Mountains. Because of all this faulting, the Los Angeles Basin is undergoing hazardous tectonic compression. Source: NASA.

reverse faults in the Los Angeles area contrast with the lower-angle thrust faults of the Rocky Mountains of Canada (**Fig. 10.6**).

10.3.3 Normal Faulting

Extensional (pull-apart) forces in the crust, especially if they span wide areas, usually lead to crustal thinning and normal faulting (**Fig. 10.10B**). As the crust stretches and thins, blocks of rock separate along normal faults. Some of the blocks may then rise up but most of them will drop down and/or rotate (**Figs. 10.20, 10.21**). Thus, normal faulting is sometimes referred to as **block faulting** when it occurs over a region. A side view of normal faulting may appear to show that one block moves up and one moves down. But in actuality, the up-thrown block seldom moves up at all. Instead, it only appears as such when viewed *relative* to the downthrown block (**Figs. 10.10B, 10.20**).

Examples of locations with widespread normal faulting include the East African Rift Valley, the Central Apennines of Italy, parts of the Korean Peninsula, parts of China such as the Hetao Graben, and the Dead Sea in the Middle East. These areas contain large masses of down-dropped rock; depending on the climate, they may hold water as lakes, for example, the Dead Sea. Down-dropped blocks of crust are called **grabens** (German *graben*, "to dig") (**Fig. 10.21**). Blocks of crust that are either rising or not falling, that is, which have moved *relatively* upward, are called **horsts** (German *horst*, "mass, heap"; **Fig. 10.22A**).

Figure 10.20 Examples of normal faults, shown in outcrop. Note the people for scale in **A** and **B**. Source: (A and B) Garry Hayes, via geotripperimages.com; (C) M. Miller.

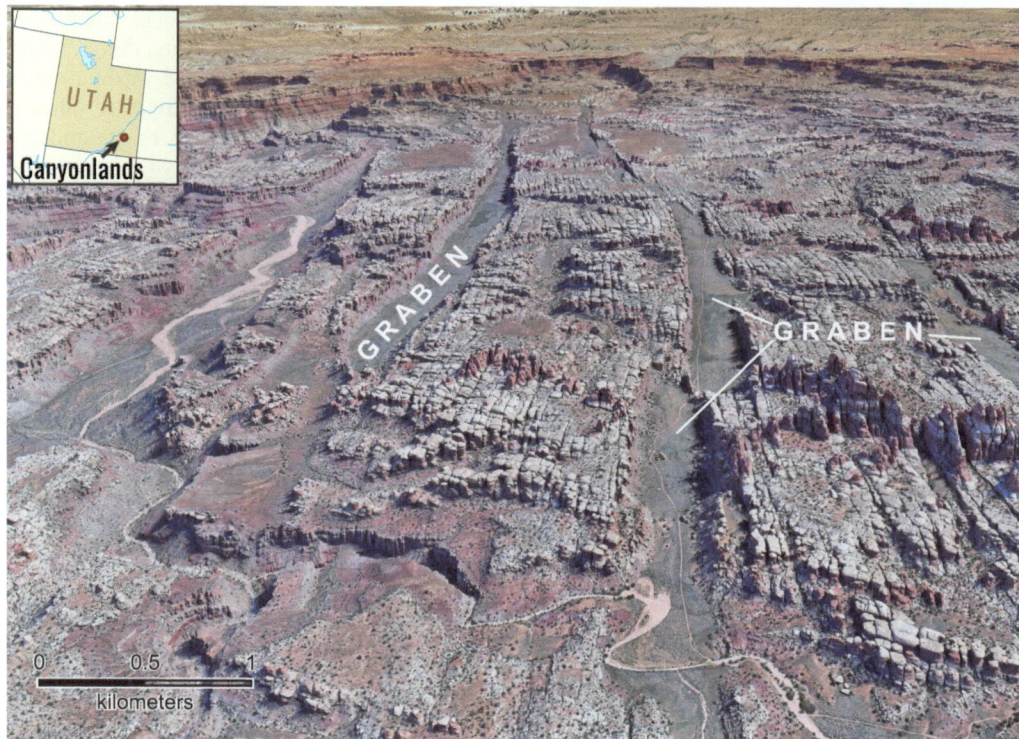

Figure 10.21 A satellite image (view to the north–northwest) of the Needles District in Canyonlands National Park, Utah, USA. The heavily jointed sandstones here have been slowly rifting apart, aided not so much by crustal extension as to the presence of soft rock salt beds at depth, which facilitate their slow lateral "sliding." Clearly defined grabens have formed in some areas. Source: NASA.

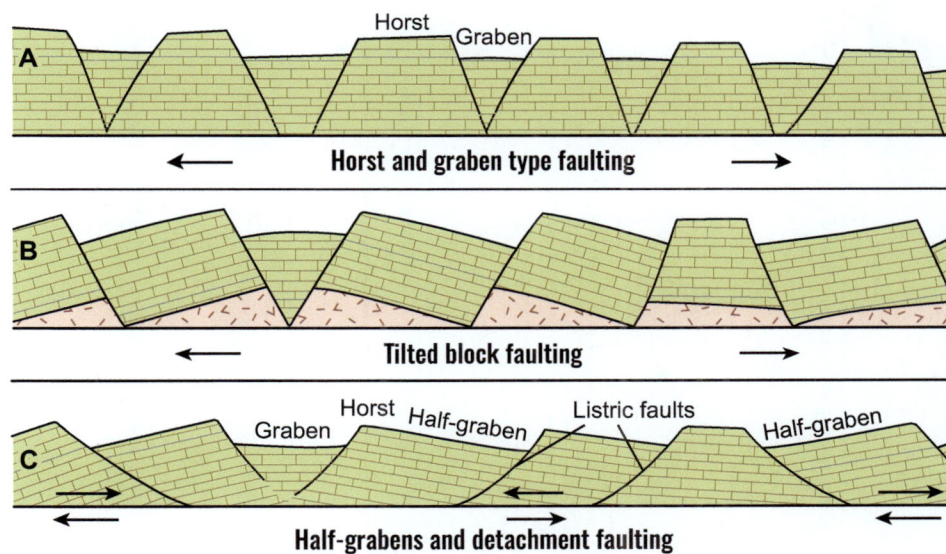

Figure 10.22 The various types of crustal configurations formed by normal faulting. Down-dropped blocks are called grabens or half-grabens and an up-thrown mass is called a horst.

When recently formed, fault scarps are often visible and quite apparent on the edges of horsts and grabens (**Fig. 10.9**). Normal faults usually have fairly steep fault scarps (**Fig. 10.10B**). One can usually distinguish a normal fault in that the fault scarp (even belowground) is never overhanging (>90°), as it would be in a reverse fault (**Fig. 10.10C**).

10.3.4 Block Faulting

Normal faulting can lead to repetitive range–basin (horst–graben) configurations on the landscape. How does this happen? The process begins as stresses induced by crustal stretching/thinning lead to a series of parallel (normal) faults that develop perpendicularly to the direction of stretching. As a result, suites of (parallel) horsts and grabens will form along normal faults, generally within three different configurations (**Fig. 10.21**). Horst and graben forms have normal faults (or zones of normal faulting) on the sides of the blocks. Typically, the down-dropped blocks fall vertically, or nearly so. More commonly, however, some amount of *rotation* occurs during normal faulting (**Fig. 10.22B, C**). **Tilted block faulting** is one such example of this. Rocks that move in tilted block faulting act like books on a shelf that have toppled over, falling back, against each other. In other instances, the rocks start with a steeper-angled normal fault near the surface, which then curves to a flatter angle and merges into a low-angle **detachment fault** deeper in the crust. This situation forms **half-grabens**, where only *one* side of the graben is on a fault plane (**Fig. 10.22C**). The curved faults that merge into the (deeper) detachment fault are called **listric faults**. Listric faults are often thought of as extensions to normal faults, found deeper within the crust and with a curved fault plane.

Rocks undergoing extensional stresses will also often endure some amount of shearing, that is, side-by-side movement. Think again of the hand-washing example; different types of stresses can occur together. Shearing may form block faulted landscapes with horsts and grabens that are laterally offset – another way of releasing the stress that is placed on rock in such settings.

10.4 LANDFORMS OF THE BASIN AND RANGE PROVINCE

Worldwide, the best-known region of block faulting is the Basin and Range Province of Western North America, centered on (and best expressed in) the State of Nevada (**Fig. 10.23**). The Basin and Range Province is named for the many basins (grabens) and ranges (horsts) that give this area its distinctive geomorphology. Nowhere on Earth are landforms associated with block faulting better expressed, and over such a vast area, than here, where the crust has experienced widespread extensional stresses.

How did the Basin and Range landscape form? From at least 80 to 30 million years ago, convergence of tectonic plates off the west coast of North America produced regional crustal compression. This compression is still going on in areas farther north, as the Juan de Fuca Plate subducts under the North American Plate (**Figs. 6.29A, 10.23**). The

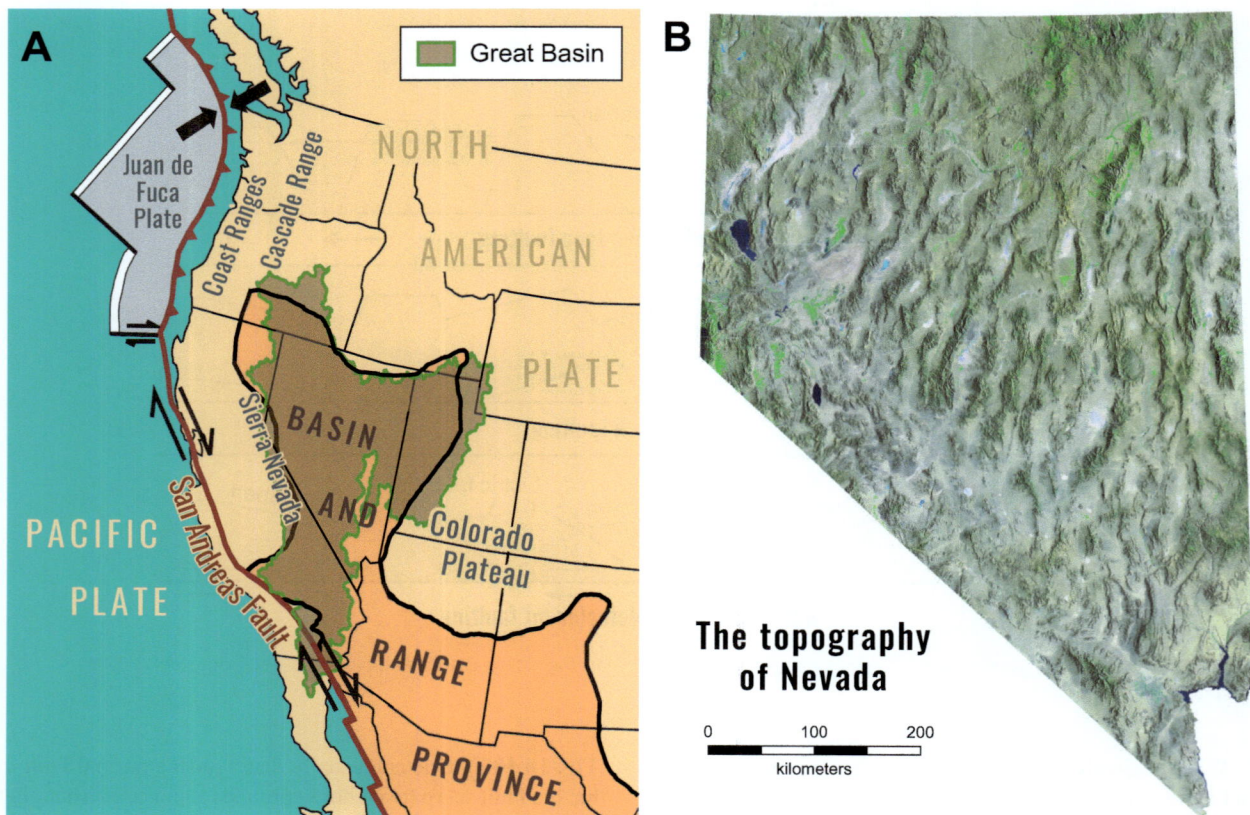

Figure 10.23 The Basin and Range Province. **A.** Map of the Basin and Range Province, outlined in black, as it compares with the Great Basin. The Basin and Range is an area of crustal extension and widespread block faulting, whereas the Great Basin represents an area of internal drainage within which surface waters cannot drain to the sea. Source: Adapted from a US National Park Service map. **B.** The topography of Nevada, dominated by fault block mountains. Source: US Geological Survey.

San Andreas Fault – a strike-slip fault – originated about 30–25 million years ago along the western margin of the North American tectonic plate. Over time, the fault extended both to the north and south. Later, east of the fault, where the North American Plate was no longer being compressed, extensional forces took over. As a result, blocks of crust cracked apart and dropped down – opening up the Basin and Range Province.

Crustal extension in the Basin and Range has effectively doubled the crustal extent, leading to crustal thinning. Chapter 6 covered isostasy, where loading and unloading of mass on the crust results in compensation, such that the land surface may sink or rise, respectively. For example, isostatic uplift occurred when the massive Quaternary ice sheets retreated from North America (**Fig. 19.42**). Isostasy has also been a player in the development of the Basin and Range. Facilitated by the crustal extension, the deeper, more ductile (typically crystalline or granitic) rocks were able to dome up in response to crustal thinning, leading to the development of a **metamorphic core complex** (**Fig. 10.24**). How does such a feature form? First, half-grabens with listric faults develop nearer the surface, in response to the crustal thinning (**Fig. 10.22C**). These faults curve at depth and join into a larger, mainly horizontal, detachment fault. In a metamorphic core complex, so much shearing occurs along the detachment fault that the adjoining rocks are heated and begin to be metamorphosed, hence the name metamorphic core complex. What starts as a bit of crustal extension and thinning ultimately ends with the core complex rising up (isostatically) and emerging at the surface, surrounded by half-grabens, with their listric faults merging into detachment faults at depth (**Figs. 10.24, 10.25**).

The cities of Phoenix and Tucson, Arizona, along with Las Vegas, Nevada, all within the Basin and Range Province, are surrounded by metamorphic core complexes. Visitors to Major League Baseball spring training games in Arizona have views of small mountain ranges that are the half-graben blocks sliding off the detachment faults of the giant metamorphic core complexes below.

With high mountains on several sides, the Basin and Range occupies a vast area of **internal drainage**, lending the name **Great Basin** to most of this landscape. Internal drainage implies that rivers within this area do not drain to the sea. The Great Basin is a subset of the Basin and Range Province, characterized (today) by a dry climate and small, steep river systems that flow off the ranges, ending in closed basins (**Fig. 10.23**). Areas of internal drainage like these are called **endorheic basins** (Greek *endo*, within, and *rheic*, an ocean of the past). Most of the world's endorheic basins are formed by faulting, as blocks of bedrock are down-faulted to form deep grabens and half-grabens, the best example being the Dead Sea in the Middle East. In older literature, the term "bolson" (Spanish *bolsón*, "large purse") was used to describe basins of interior drainage. We avoid this term here, because its use has changed to include other meanings. Water flowing into endorheic basins must evaporate or percolate into the ground; it has no other option. Even the rivers that flow out of one endorheic basin in the Great Basin dead-end into another, lower basin.

Although the Basin and Range is usually portrayed simplistically as a region of horsts and down-dropped grabens, this view does not do justice to its complex geology. Even **Fig. 10.22** is a simplification of the many, complex forms

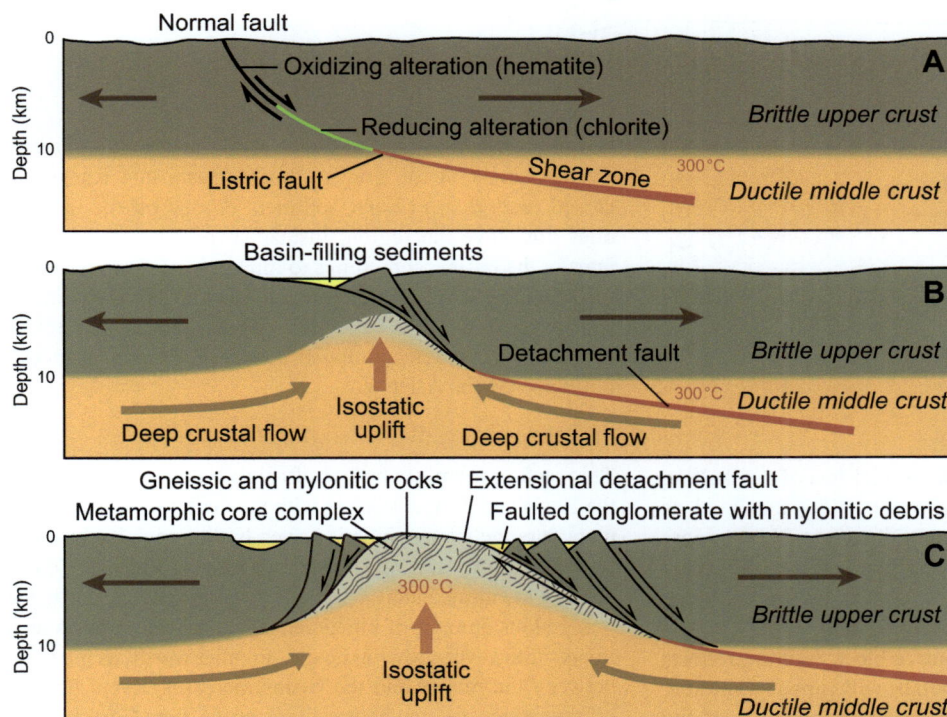

Figure 10.24 Schematic diagram showing how crustal thinning and faulting can lead to the formation of a metamorphic core complex. **A.** Crustal extension initiates normal faulting. **B.** The thinning crust initiates isostatic uplift of deeper rocks, causing the brittle rocks at the surface to develop even more normal faults, which merge into low-angle detachment faults at depth. Friction along the fault squeezes and shears the rocks, beginning the metamorphism process. **C.** Continued crustal thinning and erosion at the surface enables the core of the uplifted area, called a metamorphic core complex, to reach the surface, with half-grabens and detachment faults on its flanks.

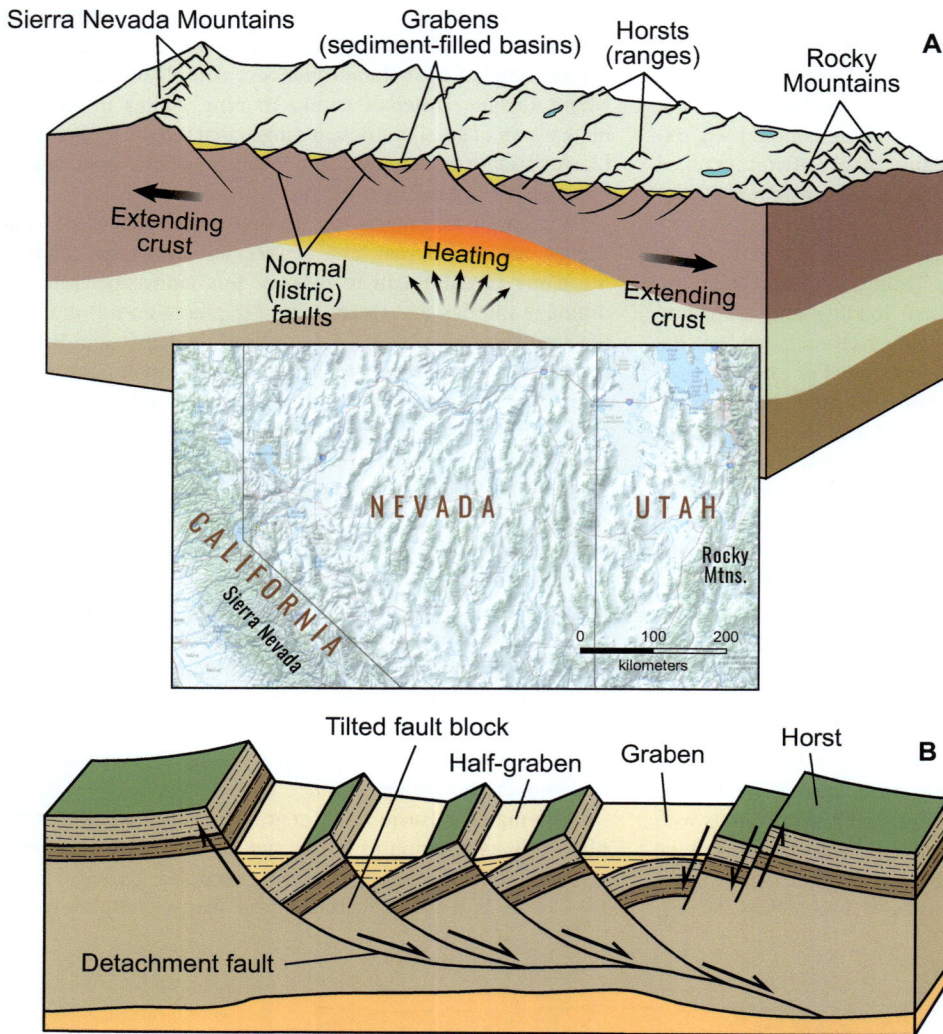

Figure 10.25 **A.** The topography of the Basin and Range is dominated by block-faulted horsts, grabens, and half-grabens, whose formation was driven by widespread crustal thinning, coupled with uplift associated with isostatic adjustment. **B.** The faulted structures across the Basin and Range illustrated in greater detail.

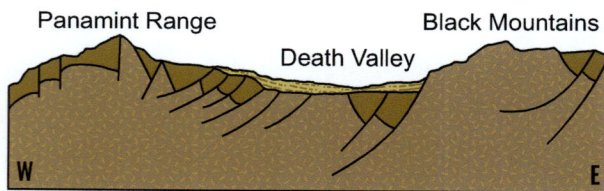

Figure 10.26 Generalized geologic cross-section of the Death Valley (California, USA) area, illustrating horsts and half-grabens that comprise its geologic structure. Source: US Geological Survey.

that can result from extensional faulting. Most of the faulting in the Basin and Range is actually of the half-graben kind, as shown in **Fig. 10.22C**. **Figure 10.26** illustrates this in more detail for Death Valley – perhaps North America's most famous graben, and one whose bottom is below sea level. Note the listric faults that dominate this area.

The topography in areas of active block faulting reflects a complex interplay between horsts with upward motion, down-dropped blocks that underlie basins, erosion of the ranges, and infilling of the basins with the eroded sediment. Because most of the grabens in the Basin and Range are closed (endorheic) basins, sediment eroded off the uplands does not leave the region. Instead, it continually accumulates in the basins, sometimes to thicknesses of thousands of meters. Over 2,700 m of sediment have accumulated in the lowest point of Death Valley (**Fig. 10.26**), and over 4,300 m of sediment have accumulated underneath much of metropolitan Phoenix, Arizona.

10.4.1 Evolution of the Range Front

This section examines how landscape evolution progresses in the Basin and Range, starting with the evolution of landforms along a range front, which is driven by the interplay between faulting (uplift) and erosion. Fault scarps formed by block faulting are immediately attacked by running water, due to the tremendous amount of kinetic energy that water attains as it flows off the ever-growing ranges. In other words, as the relief between the ranges and the basins increases, rivers flowing off the ranges become better able to erode them (**Fig. 10.27**).

Figure 10.27 A satellite image of the Inyo Mountains, which tower 800 m above Deep Springs Valley, eastern California, USA. This is a classic example of a block fault landscape, driven by normal faulting. The fault scarp here is growing by basin down-dropping at the very fast rate of ≈150 meters/million years. Source: NASA.

Initially, faulting forms a steep and fairly straight fault scarp (**Fig. 10.28A**). As faulting continues, the scarp episodically grows in height. Concurrently, erosion reduces the angle of the older (upper) parts of the fault scarp, a process called **slope decline**. Rivers draining the range encounter steep slopes as they cross the scarp, facilitating rapid erosion and deepening of their channels. As these valleys deepen, they also grow longer, by headward erosion, up and into the range (**Fig. 10.28B**). Over time, rivers crossing the fault scarp carve it into a series of V-shaped valleys, interspersed with planar remnants (spurs) of the fault scarp, known as **triangular facets** due to their shape (**Figs. 10.28C, 10.29**). Triangular facets typically start near the **angle of repose** for the weathered rock of which they are composed. Their bases align with the fault line below. The ongoing interplay between faulting and erosion may develop additional channels inside of the older and more eroded triangular facets, forming smaller facets, inset within the older, larger ones (**Fig. 10.29**). As faulting slows or stops, the fault scarp continues to erode, the inclination (slope angle) of the facets decreases (slope decline) (**Fig. 10.28D**), overall relief decreases, and the range front erodes backward from the now-inactive fault scarp (**Fig. 10.28E**).

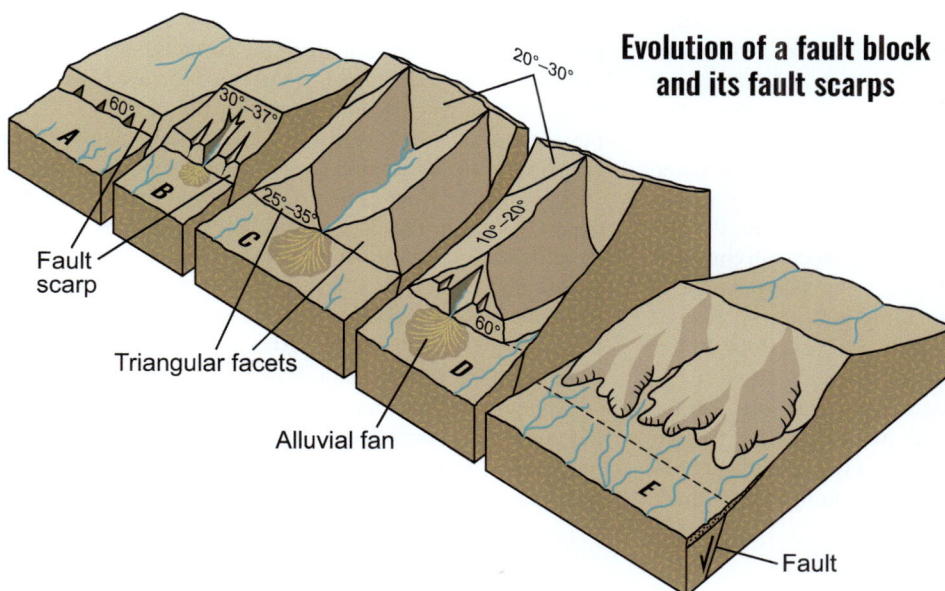

Evolution of a fault block and its fault scarps

Figure 10.28 A block diagram showing the growth and erosion of fault scarps, leading to the formation, and eventual destruction, of triangular facets at the range front. Source: After Wallace (1978), via the US Geological Society; in public domain.

Figure 10.29 The Spanish Fork segment of the Wasatch Fault in Utah, USA, exemplifies how ongoing uplift and fluvial incision can form triangular facets along the fault scarp. Notice how smaller triangular facets are inset into larger ones. Source: R. Dorn.

The hardness of the rocks along the range front, their rate of uplift, and the climate of the region all influence how the range front evolves. For example, faulting in the Deep Springs Valley in eastern California (**Fig. 10.27**) began only recently (≈3 million years ago) by geologic standards. Its rapid rate of uplift has outpaced fluvial incision processes, and hence, its ≈800 m-high fault scarp has not had time to develop triangular facets. In essence, down-faulting of the valley and fault scarp growth "won out" over processes of fluvial erosion. Contrast this with the Spanish Fork segment of the Wasatch Front in Utah, which has multiple triangular facets and river channels that break the scarp into distinct segments (**Fig. 10.29**). The slower rate of faulting at this location, the greater erodibility of the rocks, and the wetter climate have all contributed to the formation of triangular facets here.

As **Fig. 10.28** shows, rivers can cut deep, narrow, often V-shaped valleys into the range front at the junction of the range and the basin below, as well as upslope from the fault scarp. Geomorphologists have given the descriptive name **wineglass valley** (or **canyon**) to these features (**Figs. 10.30, 10.31**). The name comes from the profile of the valley as seen from the basin (looking upstream); they have the basic shape of a wineglass with a narrow glass stem, almost like a Y-shape. The gradual steepening of the sidewalls of a wineglass canyon towards its bottom may indicate that the rate of faulting is accelerating, facilitating channel downcutting and thus, forming an ever-narrowing valley/gorge. Wineglass canyons typify rivers debouching from rapidly uplifting ranges or rapidly down-dropping valleys (or a combination of the two).

The narrowness of a wineglass canyon forces water debouching from the range to flow rapidly and with considerable turbulence. This setup increases the **competence** of the stream (see Chapter 15), enabling it to transport even large rocks and boulders out and onto the basin floor. Rivers flowing in wider valleys, typical of ranges formed in less consolidated materials, will tend to "spread out" within wider channels and thus encounter more resistance to flow, slowing their velocities. In this case, many of the larger sediments upstream will first have to be **comminuted** (made smaller) by abrasion within the channel before they can be transported out of the range.

10.4.2 Evolution of the Basin

As ranges erode, the adjacent endorheic basins fill with sediment, delivered there by rivers and mass wasting events. Sediment accumulation in the basins is usually a one-way street, from the range front to the basin floor. Trending away from the fault scarps are **piedmonts** – gently sloping surfaces that lead from the mountain front to the center of the basin. Piedmonts have both depositional and erosional components. The most common depositional landform on the piedmont is an alluvial fan. Erosional components of piedmonts include pediments. Lastly, the distant centers of the basins are often occupied by playas – flat surfaces marking former lakes. The remainder of this chapter will track sediment as it moves from ranges, out and onto the basin floor, and discuss all of the associated landforms.

Endorheic basins in the Basin and Range are dominated by sediment accumulation via two main types of processes – mass wasting and fluvial transport (**Figs. 10.30, 10.31, 10.32**). Rivers flowing off up-thrown blocks have steep gradients, and thus can erode rapidly. They usually flow in response to strong, isolated storms, resulting in torrents of muddy water rushing downstream. Erosion rates can thus be exceedingly rapid, even in the dry climate of the Great Basin. Most sediment eroded off the ranges is transported to the basins during these high-magnitude, low-frequency runoff events. Debris flows and mud flows are also not uncommon, mainly occurring where steeply sloping river channels debouch along the steep mountain front (**Fig. 10.33**). Debris flows often produce a landform that is a hybrid between a talus cone and an alluvial fan – a **debris flow fan** (**Fig. 10.32**). Sediment moved rapidly in debris flows can even dominate some fluvial features (alluvial fans) at the range front. Many of these mass wasting events can be extremely hazardous to structures built near the mountain front (**Fig. 10.33**).

Although sediment moved catastrophically by mass wasting processes is impressive, deposits of this kind are largely restricted to areas near the range front. The majority of Basin and Range landforms are developed in sediment transported

Figure 10.30 Wineglass canyons. **A.** and **B.** The wineglass valley at Coffin Canyon, Death Valley, California, shown in an oblique aerial photograph and from the ground. The mouth of Coffin Canyon displays a classic wineglass shape. Source: R. Dorn. **C.** Oblique view of a wineglass canyon in the Black Mountains of Death Valley. Note the sheer face of the fault scarp. Source: M. Miller.

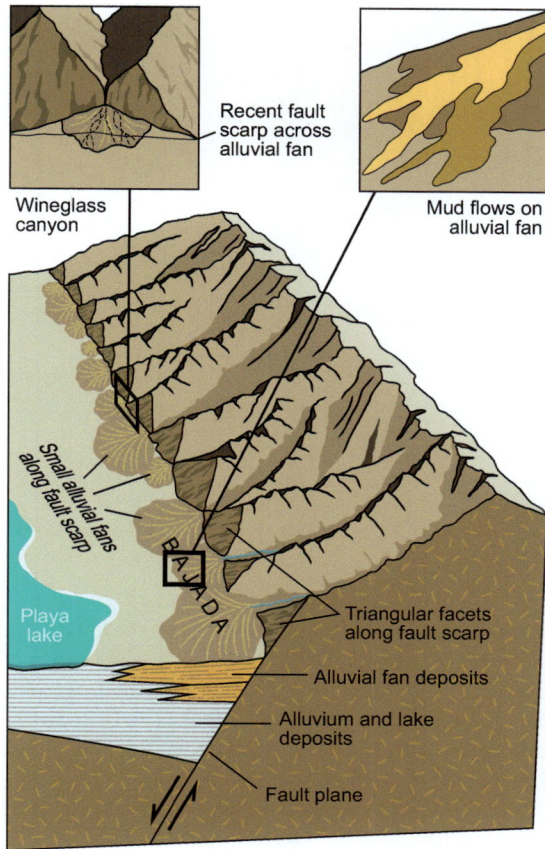

Figure 10.31 Schematic illustration of the landforms that develop along a range front in a landscape dominated by block faulting, modeled after the Black Mountains context seen in **Fig. 10.35**.

Figure 10.32 The sediment that has accumulated at the base of this mountain front in Death Valley is not an alluvial fan or a talus cone (**Figs. 8.10** and **8.11**). Instead, it has formed from a mixture of debris flows and **sieve flows** – open-network deposits of rock, lacking fine sediment like silt and clay. Some geomorphologists call this type of feature a **colluvial fan**. Source: R. Dorn.

The alluvium transported and deposited by streams varies greatly in type along the piedmont. Alluvium deposited near the mountain front is often very coarse, with abundant gravels, but gets progressively finer-textured away from the mountain front (**Fig. 10.34A**). This gravelly, coarse-textured alluvium is often called **fanglomerate**. Most of the alluvium eventually gets deposited in alluvial fans that have their upslope end at the mouth of a wineglass canyon (**Figs. 10.31, 10.34B**). Streams that exit the range must then flow across the gravelly surfaces of these fans, and as a result, much of their water will infiltrate. Streams like this are called **losing streams**, because they lose water downstream, as it percolates into the ground (see Chapter 15). Within these losing streams, only the finest sediment, which can be carried in smaller channels and at lower flow velocities, is able to be transported off the fan and into the center of the basin. For this reason, alluvial fans can be tapped as groundwater reservoirs. Sediment is so quickly deposited by these losing streams that much of the fan surface is one dry river channel after another (**Fig. 10.34B**).

Despite being the prototypical landform of the Basin and Range Province, alluvial fans are not restricted to these types

farther out into the basin, by running water. Although most streams in the Basin and Range Province are usually dry, they can become torrents of water, mud, and debris after a long winter rain or a thunderstorm passes through the range. These types of river systems are called **ephemeral**; they only flow during and immediately after rainfall or snowmelt events.

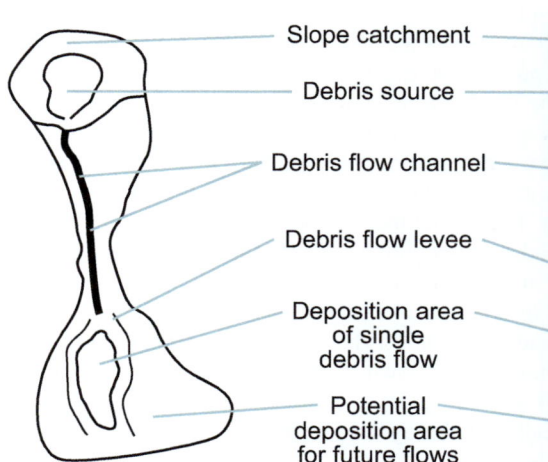

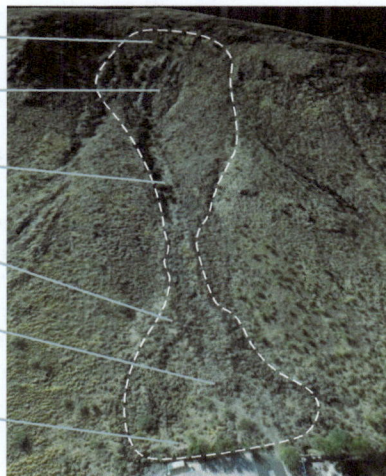

Slope catchment
Debris source
Debris flow channel
Debris flow levee
Deposition area of single debris flow
Potential deposition area for future flows

Figure 10.33 A debris flow system on a steep range front in Phoenix, Arizona, USA. Extreme rain events transported sediment downstream as mud and debris slurries. As the sediment moved through and overflowed the debris flow channel, it deposited levees on either side, eventually stopping on the upper part of an alluvial fan. Houses built on these deposits are thus at risk. Source: Image on the right is from NASA.

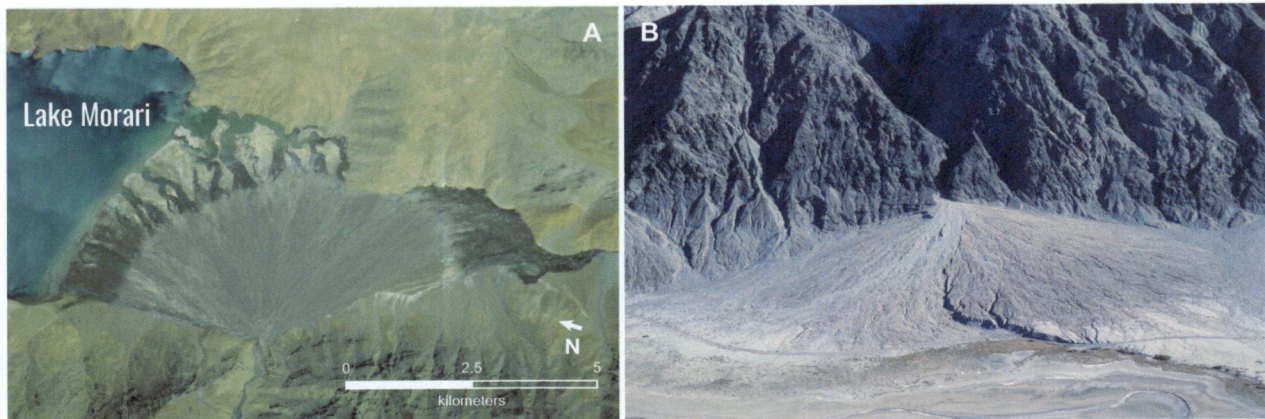

Figure 10.34 Alluvial fans. **A.** A satellite image of a large alluvial fan that has blocked the southern outlet of a valley, forming Lake Morari in Tibet. The apex of the fan is > 40 m above the lake surface. Beyond the margins of the fan are deltaic plains, building out into the lake. Source: NASA. **B.** A classic-shaped alluvial fan in Death Valley, California, USA. Darker areas represent older gravelly deposits with rock varnish coatings. The variation in varnish intensity across the fan surface enables geomorphologists to determine how the locus of deposition has shifted over time. Source: M. Miller.

of landscapes. They can occur wherever water debouches from a faster-flowing river onto a flatter plain. They even occur on Mars (**Fig. 23.28**)!

Each fan's apex begins where a stream channel debouches from the range front (**Fig. 10.35**). The fan shape is developed and maintained as the streams and their distributary channels swing back and forth across the fan surface, as they attempt to find the steepest route off the fan. Nonetheless, the stream channel always maintains the same position at the top of the fan – its **hydrographic apex**. Below the hydrographic apex, the channel shifts from side to side across the

Figure 10.35 Massive alluvial fans debouch from the Black Mountains into Death Valley, California (USA). Although debris flows do occur, most of the sediment transport in this area is by ephemeral floods. Within this short distance, particle sizes on the alluvial fans decrease from boulders near the mountain front, to sand and silt at the fan margin. The white material is salt that precipitates when the floodwaters that reach the floor of the valley evaporate, forming a playa. Faulting of the Black Mountains along the eastern margin of Death Valley (**Fig. 10.26**) is associated with ongoing down-dropping of the basin, leading to smaller alluvial fans, which have in most places not coalesced into bajadas. Instead, they are slowly being buried by playa deposits at their toe. Source: R. Dorn.

fan, along the steepest slope. Put another way, the location of the head of the stream on the alluvial fan is "locked in place" at the canyon mouth. River flow, focused at the hydrographic apex, then can branch out across the fan surface. Because the most (and coarsest) sediment is deposited at the head of the fan, with finer sediment deposited farther out, the fan grows fastest in its upstream reaches. Mature fans will therefore eventually develop a slightly convex-upward surface slope.

Alluvial fans must have space in the basin to "grow" and thicken – this is called their **accommodation space**. That is, sediment being "dumped" into the basin is competing for a limited amount of space. Rivers that produce a lot of sediment can form large fans that dwarf the fans of smaller rivers nearby. The real "losers" in the battle for space end up being the playas, which slowly shift position – out and away from the mountain front that produces the most sediment. Where multiple rivers exit from a range front, but are not widely spread out along it, their alluvial fans will eventually merge and overlap into a broad, sloping surface called a **bajada** (**Figs. 10.31, 10.36**).

As alluvial fans grow and develop into bajadas, sediment continues to accumulate in the basin. Fluvial systems at this point are **aggradational** – filling their channels and building the fan surface upward. But perturbations such as climate change might alter this dynamic, driving the fluvial system toward incision. Incision may also be driven by tectonics, as uplift makes the slope across the fan (or bajada) steeper. The fluvial incision process typically begins at the fan head (or topographic apex), where rivers are flowing fastest and have the most discharge, and progresses downslope, toward the toe of the fan. This process is called **fan-head incision**. At the same time, deposition may be occurring farther down the fan, away from the mountain front (**Fig. 10.37A, B**).

Recall that the hydrographic apex represents the location where the distributary channels on the fan first begin

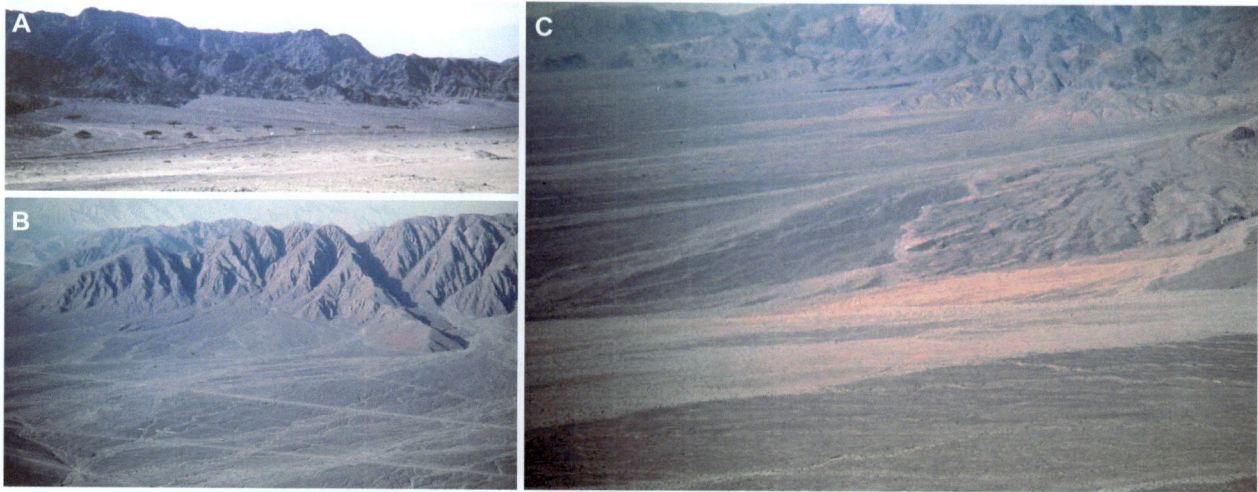

Figure 10.36 Bajadas. **A.** Individual alluvial fans emerging from uplands on the eastern side of the Sinai Peninsula, Egypt have merged into a bajada. **B.** Alluvial fans form the foundation material for the famous Nazca lines of Peru (seen in this image as lighter areas), made by removing desert pavement clasts, exposing the lighter-colored soil below. The bajada here has its topographic apex near the range front, but its hydrographic apex is down-fan. **C.** Fans at the base of the Panamint Range in Death Valley, California, USA, have coalesced into an extensive bajada. Source: R. Dorn.

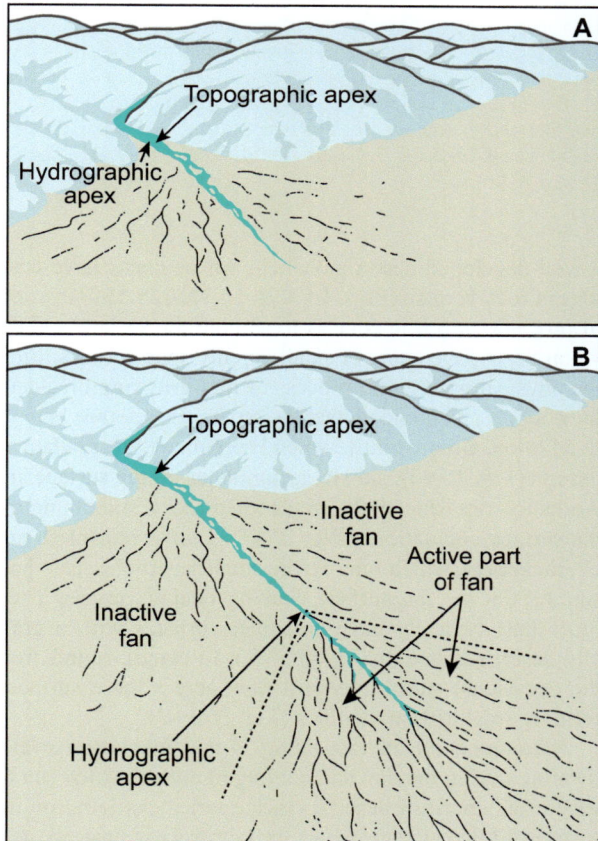

Figure 10.37 Schematic diagram illustrating the change in location of the hydrographic apex on an alluvial fan undergoing incision, from (**A**) an up-valley location to (**B**) a location down-fan. **C.** This situation is shown in the Warm Springs Canyon fan in Death Valley. Source: R. Dorn.

to spread, usually at the top of the fan, and typically right at the mountain front (**Figs. 10.37A, 10.38C**). For some fans, the hydrographic apex also coincides with the **topographic apex** – the highest point on the fan. But for incised fans, their hydrographic apex can be much farther out, well down the fan (**Figs. 10.36B, 10.37, 10.38B**). Some geomorphologists call this "telescoping"; they envision a "mariner's telescope" of the sort that folds in on itself for storage but can get

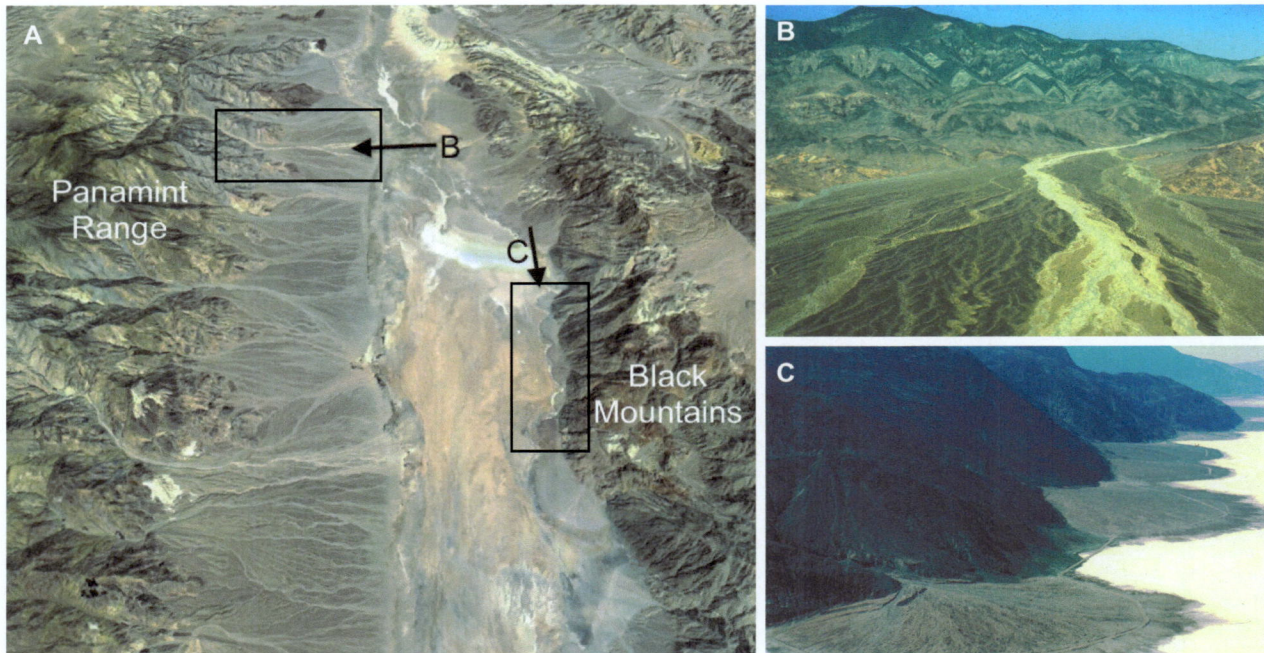

Figure 10.38 Landscapes of Death Valley, California, USA. **A.** A northward-looking view of the southern Death Valley. The left (west) side of the valley, bordering on the Panamint Range, has developed large alluvial fans and bajadas, with their hydrographic apexes shifted far down-fan. Alternatively, the right (east) side fans have grown more slowly out of the Black Mountains, and as a result, have retained their hydrographic apexes at the heads of the fans. That is, their topographic and hydrographic apexes coincide at the base of the mountains. Arrows and inset boxes identify the locations where the next two aerial photographs were taken. Source: NASA. **B.** Trail Canyon fan on the west side of Death Valley, where faulting has tilted the range, causing the hydrographic apex to shift down-fan, forcing large sections of the fan surface to become inactive. Rocks on these older, inactive fan surfaces have thus had time to develop a coating of dark-colored rock varnish. Source: R. Dorn. **C.** Alluvial fans along the Black Mountains, which delineate the eastern margin of the basin. Source: R. Dorn.

stretched outward to see something. Geomorphologists picture fans "telescoping" gradually down the fan as the hydrographic apex shifts down-fan. This is typical – as alluvial fans evolve, their hydrologic apexes move further into the basin. Thus, opportunities for fluvial incision into the old depositional surfaces (nearer the range front) become more numerous.

In addition to incision, fan surfaces change in other ways. Relatively young (< 5,000 years) fans often have rough, gravelly surfaces with many small channels, reflecting the fluvial depositional system with its myriad bars and dry channels (**Fig. 10.39A**). However, some surfaces on the upper parts of fans are left behind and made inactive when the hydrographic apex moves down-fan (**Figs. 10.37, 10.38B**). After that, deposition on these isolated surfaces ceases. In the Basin and Range Province, the transition from bar and channel topography (**Fig. 10.39A**) to a smoother surface with desert pavement (**Fig. 10.39B**) takes from 5,000 to 15,000 years for most alluvial fans. The process is faster on fans with smaller particle sizes, for example, gravel and sand vs rocks and boulders, and on those where dust deposition is more active.

After thousands of years, any exposed and stable fan surface becomes smoother and is sometimes covered with

a well-developed desert pavement whose clasts have coatings of rock varnish (**Figs. 10.38B, 10.39B, 21.15**). A variety of processes work in concert to smooth the surface and develop these pavements of small, evenly spaced rocks. First, the physical weathering process of **dirt cracking** breaks up the cobbles and boulders that lay on the surface (see below). In addition, airborne dust, which is seemingly everywhere in deserts (**Fig. 10.40**), slowly accumulates on the surface and works its way beneath the fragmented rocks, smoothing out the surface undulations (**Fig. 21.15**). Lastly, rainsplash and overland flow, which remove the smaller particles, also help smooth out the fan surface, leaving behind gravels. Thus, older fans typically have smoother surfaces with a gravelly desert pavement (**Fig. 10.39B**). In places, runoff from the occasional storm may still flow across these surfaces, forming small rills and gullies (**Fig. 10.39C**).

A common misunderstanding is that **insolation cycles** (heating during the day and cooling at night), which can be very pronounced in deserts, crack apart rocks lying on the surface. Little evidence exists for this process. Instead, geomorphologists are now pointing to the importance of **dirt cracking** as an important surficial process. In dirt cracking, dust enters existing rock fractures and breaks apart the rock, because the dust contains clays that expand when wetted.

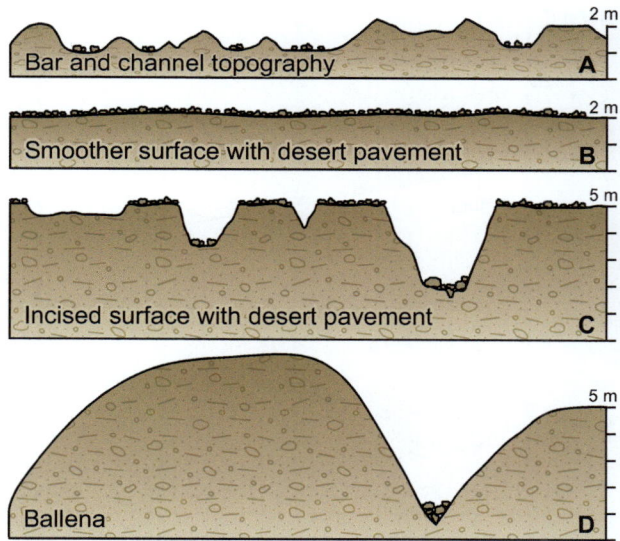

Figure 10.39 Schematic diagram showing typical changes in surface topography of alluvial fans over time. **A.** The bar and channel form derives from the topography of the ephemeral channels. **B.** As stream channels shift to other locations, parts of the fan surface get isolated. On these surfaces, desert pavements develop and gradually smooth the surface to the point where it is difficult to see the original bar and channel form. **C.** Incision of the fan surface then forms rills and gullies between relict patches of the smooth pavement. **D.** Old fan surfaces (abandoned > 2 Ma) slowly erode to a rounded, cross-profile form known as ballenas.

Figure 10.40 A haboob dust storm approaches Phoenix, Arizona, USA in 2011. The wall of dust was ≈400 m high. Source: Alan Stark from Goodyear, AZ, United States, CC BY-SA 2.0, via Wikimedia Commons.

The wetting of this dust also solubilizes $CaCO_3$ – which is also seemingly everywhere in deserts. The solution thus formed can then move deeper into crevices. Later, as the $CaCO_3$ precipitates out of the solution upon drying, more pressure is exerted on the rock, wedging the fractures further apart. This process is yet another example of a positive feedback in geomorphology – as dust accumulates and opens cracks farther, they can accumulate even more dust and $CaCO_3$, leading to more dust/carbonate wedging, etc., until eventually the rock fractures.

Finally, with continued channel incision, the original surface erodes even more deeply, leaving behind rounded ridge crests called **ballenas** (pronounced *ba yay' nuhs*) (**Figs. 10.39D, 10.41**). Ballenas are characterized by rounded crests with little or nothing remaining of the original depositional surface.

Until recently, geomorphologists have had a difficult time discerning how long it takes to go from abandonment of a fan surface to the development of ballenas. Estimates have ranged from a few hundred thousand years (mid-Pleistocene) to 3–5 million years (Pliocene). The first precisely dated ballenas were recently reported for the McDowell Mountains of Arizona, USA (**Fig. 10.42**). Ballena formation started here about 2.5 Ma, when the closed basin was breached and developed a through-flowing river system. This change in

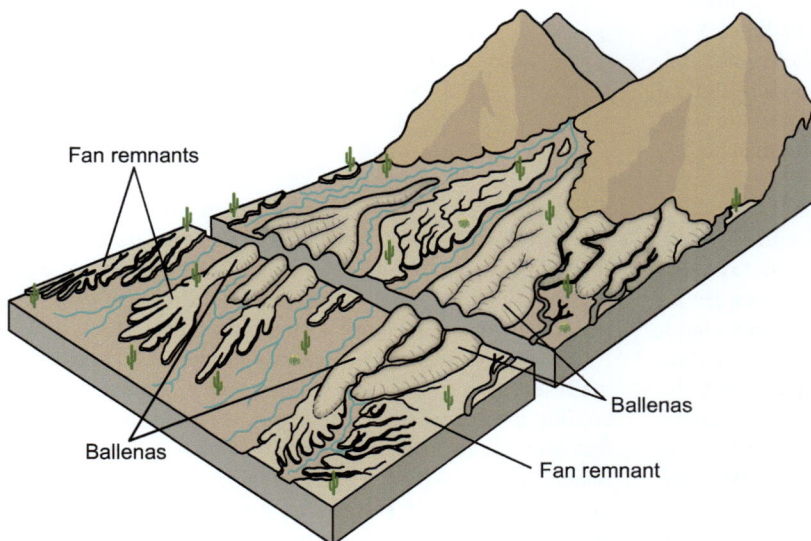

Figure 10.41 Schematic diagram showing isolated groups of ballenas on a heavily dissected alluvial fan. Erosional fan remnants – precursors to ballenas – have flat surfaces and typically retain a desert pavement. Source: After Peterson (1981); in public domain.

Figure 10.42 Heavily eroded alluvial fans in the McDowell Mountains, which now have developed the rounded forms of ballenas, Central Arizona, USA. Source: R. Dorn.

hydrology led to a large drop in base level, initiating fan-head incision. Parts of the fans then became isolated from further deposition, initiating ballena formation.

10.4.3 Pediments and Inselbergs

In Basin and Range landscapes, tectonic activity (faulting) eventually slows and then ceases, allowing erosion of the uplands and basin infilling to become the dominant surficial processes (**Fig. 10.43**). Today, some 10 to 15 million years after faulting has ended in the Basin and Range, many ranges in Arizona have been eroded to the point where they appear as low, isolated bedrock hills – **inselbergs** (literally, "island rock"). Inselbergs are the erosional remnants of the original fault block (**Fig. 10.43C**).

The piedmont (mountain front) that slopes down and away from an inselberg may look like a bajada to a casual bystander, but many of these gentle slopes have not formed due to accumulations of alluvium. They are, in fact, **pediments**, underlain shallowly by bedrock. Recall that the *piedmont* surfaces that extend out from the ranges can be depositional *or* erosional in form – *pediments* are the erosional ones. The pediment surface in front of an inselberg is a gently inclined plain, formed as water flows across the same bedrock that comprises the inselberg. Pediments are commonly overspread with a carpet of gravelly material, only centimeters to a few meters thick. The thin alluvial layer often leads the unwary observer to diagnose the slope as a constructional feature, such as a bajada. But in actuality, pediments are erosional surfaces.

Death Valley is an excellent location to study the geomorphology in an area of active faulting (**Figs. 10.26, 10.38**). In contrast, central Arizona (USA) represents an area that has had a similar geologic history, but where faulting ceased or slowed to a crawl, starting ≈10 Ma. Although both landscapes may appear similar to the untrained eye, important differences exist. In Death Valley, broad bajadas underlie the mountain-front piedmont surfaces, reflecting massive accumulations of sediment (**Fig. 10.36C**). However, in central Arizona, the piedmont surfaces are often erosional pediments, mantled only with a thin (< 1–5 m) veneer of sand and gravel (**Figs. 10.44, 10.45**). The gently inclined

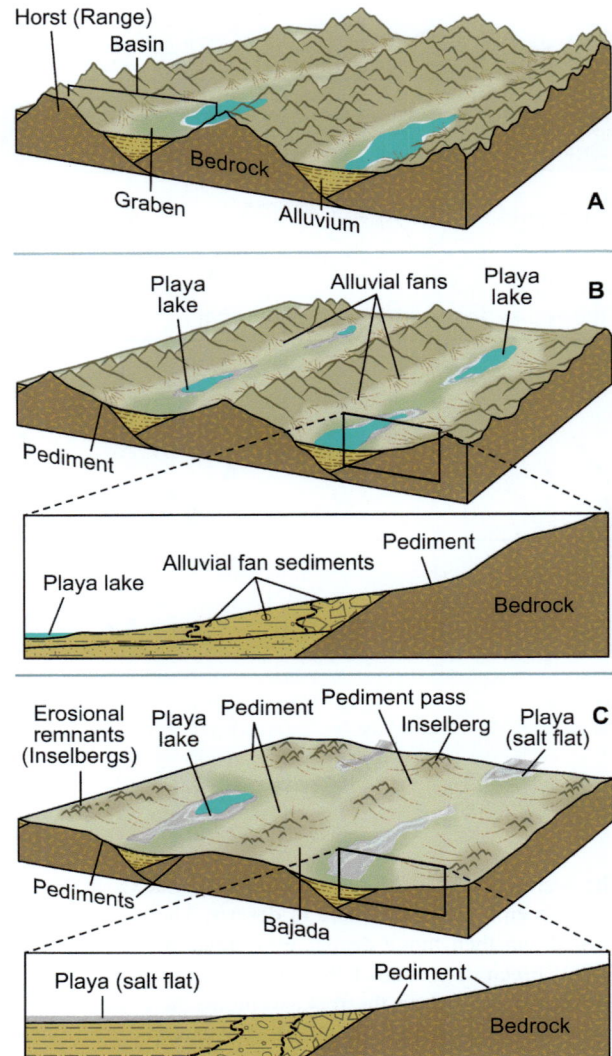

Figure 10.43 Schematic illustration of the evolution of a fault block landscape in an arid climate, with endorheic basins. The landscape of active faulting in **A** transitions to the erosional landscape shown in **B**, which depicts the same landscape at the end of the period of active faulting, and then in **C**, which depicts the same landscape ≈10–15 million years after faulting has ceased. Source: Adapted from Fig. 23.19 of E.H. Christiansen and W.K. Hamblin. 2015. *Dynamic Earth*. Jones & Bartlett Learning.

pediments have formed as the mountain front has retreated. They are essentially bedrock-floored slopes of transportation for sediments eroded off the range, helping to facilitate the erosion of the mountain front. Sediment moves across the pediments mainly by slopewash within small rills.

Recall that ranges are constantly eroding. Old ones eventually erode into inselbergs. Sediment produced from the eroding inselbergs will be transported across the pediments that span the foot of the range (**Fig. 10.43C**). Think of pediments as the shaved (eroded) remnants of ancient block faults. Their bedrock surfaces terminate at their lower end at an ancient range-front fault line. Eventually, the *upper* end

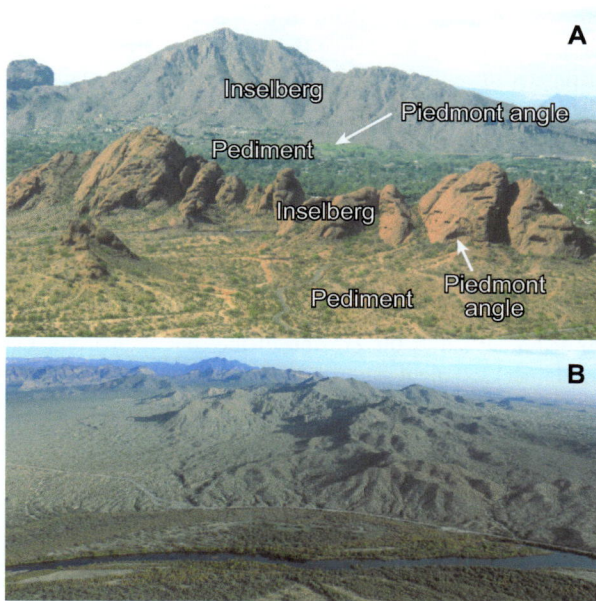

Figure 10.44 Pediments in the Sonoran Desert of central Arizona form in bedrock that can be composed of granitic, metamorphic, and/or sedimentary rock, or even welded tuff. **A.** The pediment and inselberg in the background are both granitic, whereas the one in the foreground is sedimentary. **B.** The granitic Usery Mountains in Mesa, Arizona, are surrounded by broad, low-sloping pediments. Granitic residuum (grus) eroded from mountain fronts is being transported by ephemeral streams to the Salt River in the foreground. Source: R. Dorn.

of pediments erode through to the other side of the range, meeting the pediment there to form a **pediment pass** (Fig. 10.43C). At the *distal* margin of the pediment, sediment transportation ends in playas if the basin is still endorheic, or into trunk drainages if the basin is no longer hydrologically closed. This change from endorheic to **exorheic** (outwardly draining) river systems has occurred in the southern portion of the Basin and Range Province, for the Salt, Gila, Rio Grande, and Colorado Rivers.

This terminology can get confusing. Both alluvial fans/bajadas and pediments are components of desert piedmonts. The bedrock ramp (pediment) and the sediment ramp (bajada) have only one similarity – they appear as gentle slopes in front of a steep mountain range, perhaps explaining why the word piedmont is used as a descriptive term for any gently sloping surface in front of a mountain range. The Appalachian Mountains also have a piedmont to their east – albeit a very large one (Fig. 6.11). Geomorphologists used to believe that pediments are relict features that developed during a past, wetter climate. Research has shown, however, that pediments are actively forming features.

On their upper end, most pediments intersect the mountain front with an abrupt change in slope, where the gentle slope of the pediment changes to the steeper slope of the range front. This change in slope is quantified in a term called the **piedmont angle** (Fig. 10.45A). This abrupt change in slope often has nothing to do with the faulting that formed the range

long ago. Recall that ranges (horsts) that are actively being uplifted, or which were recently uplifted, have fairly straight range fronts that *do* align with the fault zone (Fig. 10.31). Old, eroded ranges and inselbergs, on the other hand, have meandering range fronts that do not align with the fault zone because the mountain front has been eroding back, away from it, for millions of years. Thus, the abrupt face of an inselberg front does not typically align with the former fault scarp.

As a result of these processes, pediments are, on first principle, underlain by the same rock as the eroding range behind them, because they represent a surface that has been scalped/eroded from that same rock. To reiterate this point, notice the pediments in Fig. 10.45B and C, formed on eroded

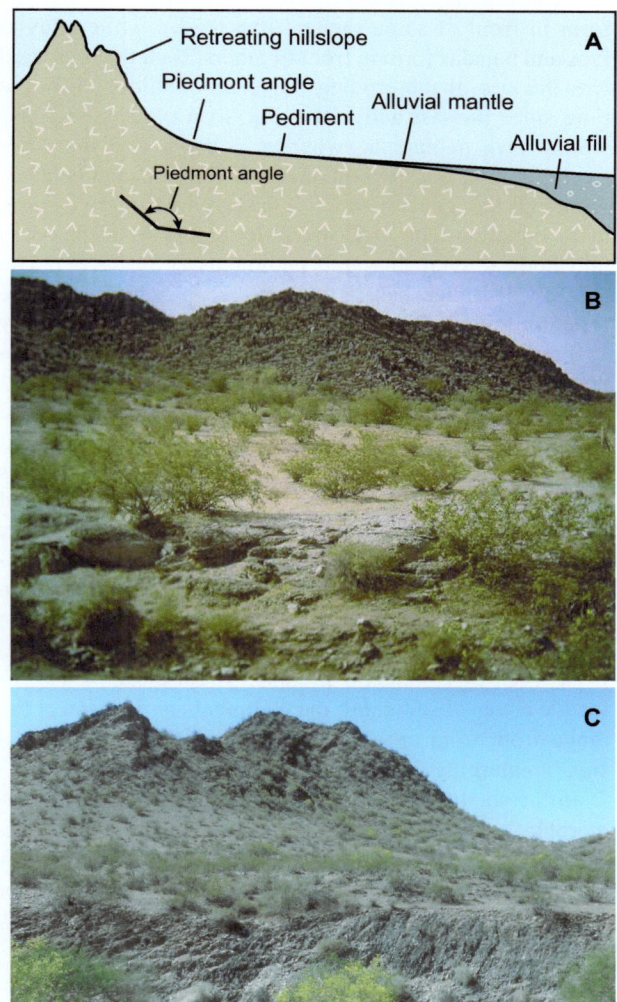

Figure 10.45 Pediments. **A.** An idealized cross-section through a pediment that has formed at the foot of a range that was up-faulted millions of years ago and where faulting is no longer active. Source: After *Desert Landforms* by J. A. Mabbutt (1977), reprinted courtesy of The MIT Press. **B.** A pediment formed on granite, in front of the Maricopa Mountains, Central Arizona, USA. The pediment continues off photo, to the right. Source: R. Dorn. **C.** A range and pediment, both formed on schist, in the Phoenix Mountains of Central Arizona, USA. Source: R. Dorn.

granite bedrock. In both cases, the mountain front is formed in the same type of rock, as is the pediment. (In **Fig. 10.45B** the range appears darker because of rock varnish.) Notice also that there is no fault line to explain the sudden break in slope. Instead, the break in slope simply represents the edge of the retreating bedrock (mountain) front.

Much of Central Arizona, in the southern Basin and Range, is often described as a "pediment and inselberg" landscape. Pediments here are the most studied in the world, with inselbergs that were once as large and extensive as those in the Death Valley region today. But in the ≈15 million years since, erosion (but with no additional uplift) has worn down the ranges and, as a result, the amount of sediment coming off them has diminished.

Geomorphologists have long wondered why pediments form in front of some ranges (like these), while alluvial fans and bajadas form in front of others. If all that mattered was the *size* of the eroding range, as conditioned by the time since the last active faulting, then older landscapes would have pediments, whereas younger and more actively faulting ones would have alluvial fans and bajadas. Although this generalization is largely true, there is more to the equation. The McDowell Mountains in Arizona contain both alluvial fans *and* pediments (**Fig. 10.46**). The McDowell Range had its "Death Valley stage" of active relief building in the late Oligocene and early Miocene. By ≈10 million years ago, faulting had ceased or slowed to a crawl, and the range has since been greatly eroded down.

But importantly, these mountains provide insight into pediment formation because they have, in places, different lithologies. The northern end of the range, on granite, has pediments, while the southern end of the range has formed on metamorphic rocks and has alluvial fans. Pediments seem to form best where both the amount of sediment coming off the range, and its particle size, are small. Erosion of granite often leads to pediments, because it decays to sandy grus that is readily transported away from the range front, even by ephemeral streams. Swept clean of its protective cover of sediment, the pediment cannot protect the range front from erosion, and so it continues to erode – slowly enlarging the pediment over time (**Fig. 10.45B**).

In the southern McDowell Mountains, erosion of metamorphic rocks produces sediment of much larger size. This larger sediment can only be moved by steep mountain streams during large flood events. But even they cannot carry it all away and so much of the sediment gets stored in alluvial fans. Then, once these boulders and cobbles are deposited in alluvial fans, they remain at the surface until they break apart into small enough pieces for small desert streams to transport. The thicker cover of sediment protects the rock below and slows the erosion of the range front that is upslope.

So, it comes down to a *balance*. More (and coarser) sediment supplied to the piedmont forms slopes underlain by alluvial fans, because the sediment cannot be readily transported away from the mountain front. Here, the thick cover of sediment slows the erosion of the mountain front. Less (and finer) sediment supplied to the piedmont is more readily transported away, resulting in bare bedrock that is exposed to erosion, leading to pediment extension/growth into the eroding mountain front. Only a thin veneer of alluvium remains on the pediment proper, and alluvial fans are largely absent.

10.4.4 Playas in the Centers of Closed Basins

The Great Basin is characterized by large numbers of closed (endorheic) basins, where water flowing towards the center of the basin cannot flow out (**Fig. 10.23A**). Endorheic basins in dry climate regions typically have playas (dried lake beds) or **playa lakes** at their lowest points. Playas represent the end member of the depositional sequence – from the mountain front, across the piedmont, and into the basin center (**Fig. 10.43C**). Often, bajadas grade imperceptibly into playas, without a distinct break in slope (**Fig. 10.35**). The finest (smallest) sediment accumulates in playas; it has nowhere else to go, unless it leaves as eolian dust. Playas often will retain water, as a playa lake, after rainfall events. In most instances, this lake dries up shortly after it fills, due to percolation and evaporation.

Playas exist in two forms – mud playas and salt playas. Playa muds (primarily silt and clay) and salts are both initially derived from weathering of the rocks in the ranges, and also from dust that falls on the landscape. Salts of various kinds form as a byproduct of weathering, but in most landscapes the salts – being soluble – are flushed out of the system and into the ocean by through-flowing rivers. In endorheic basins, however, there is no exit plan for the salts (or the muds), and so they accumulate.

The formation of **mud playas** requires that the groundwater in the basin can percolate into the subsurface, through

Metamorphic alluvial fan   Granitic pediment

Figure 10.46 The eastern side of the McDowell Mountains, Arizona, USA, has both alluvial fans and pediments. Source: R. Dorn.

Figure 10.47 Racetrack Playa, a mud playa in Death Valley National Park, California, USA. The mud comes from the suspended load of streams that reach the center of the basin. Racetrack Playa is a very special place to geomorphologists because, for years, they had been unable to explain why rocks appear to "race" across its mud surface. Dozens of hypotheses have been advanced over the years, but none were ever fully accepted or confirmed. Recently, though, scientists actually *observed* the process, helping to end the confusion. Remote cameras filmed rocks "racing" across the playa, embedded in thin sheets of ice that were floating on water. The ice forms as a result of a rainfall event, followed by a night-time freeze. As the ice breaks apart during the morning heat, the sheets of floating ice, driven by light winds, push the rocks across the wet, muddy playa surface. After the ice melts, no record remains of the mysterious geomorphic agent that pushed the rocks around! Source: R. Dorn.

alluvium and the bedrock below. As it flows, the water carries any dissolved salts it contains to a deeper aquifer, leaving only silts and clays at the playa surface (**Fig. 10.47**).

Salt playas occur where dissolved solids in the lake water are unable to percolate into a lower aquifer (**Fig. 10.48**). Other salt playas occur where the basin center is so far from any upland that clastic sediments like silt and clay cannot reach it. As a result, the water in the ephemeral, shallow, playa lake is clean but salty (**Fig. 10.49A**). Later, the water evaporates and the salts precipitate on the playa surface, often as a thick salt crust (**Fig. 10.48**). Names for salt playas vary geographically. **Sabkha** is a common synonym in the Arab-speaking world. Playa and **salar** are commonly used in the Spanish-speaking world. Alkali flats and salt flats are synonyms in the English-speaking world.

The types of salts that precipitate as water evaporates from playa lakes tend to be grouped based on their solubility. Carbonates, being the least soluble, precipitate first. Sulfates tend to precipitate second, and chlorides last, as they are the most soluble. Thus, within a salt playa, the salts often change in composition from the outer ring toward the center (**Fig. 10.49B**). Sodium carbonate (Na_2CO_3, also known as soda ash) would be a typical carbonate mineral found toward the outer ring of salts, along with dolomite ($CaMg(CO_3)_2$). Gypsum ($CaSO_4 \cdot 2H_2O$) is commonly found in the middle ring of salts, whereas halite (NaCl) and polyhalite ($K_2Ca_2 Mg(SO_4)_4 \cdot 2H_2O$) tend to precipitate in the basin center. Dunes (discussed below) may also occur on the margins of playas. Interestingly, the sands in some playa-marginal dunes can include gypsum and other salts.

In the 1800s, salt deposits of various kinds were extensively mined from salt playas around the Great Basin. Particularly famous were borax salts, mainly sodium borate and sodium tetraborate ($Na_2H_4B_4O_9 \cdot nH_2O$), which were used in soaps. Large teams of mules were used to haul the salts out of the basins and over the surrounding ranges by wagon, leading to the name for a famous (and still produced) soap called 20 Mule Team Borax (**Fig. 10.50**). Salt mining is still practiced in playas in the region, today.

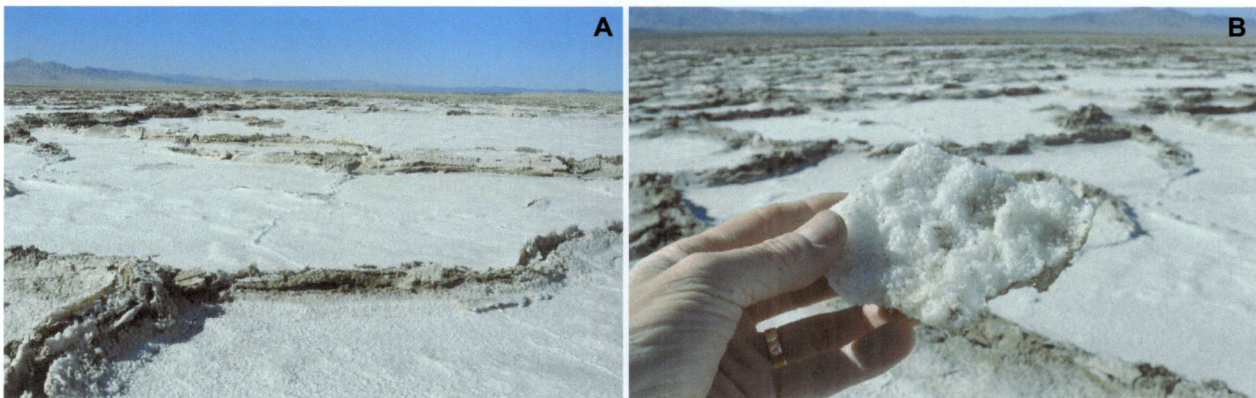

Figure 10.48 Images of a salt playa and its salts, in southeastern California, USA. Source: R. Schaetzl.

Figure 10.49 Playas in the Great Basin. **A.** The evaporating water in a playa is sometimes so salty (with sodium salts) that swimmers can float, as shown in this scene from Death Valley, California, USA. Source: R. Dorn. **B.** In the Deep Springs Valley of eastern California, carbonates tend to precipitate first, along the playa margin, sulfates such as gypsum precipitate next, and finally, chlorides such as halite precipitate in the basin center. Source: Dcrjsr, CC BY 3.0, via Wikimedia Commons.

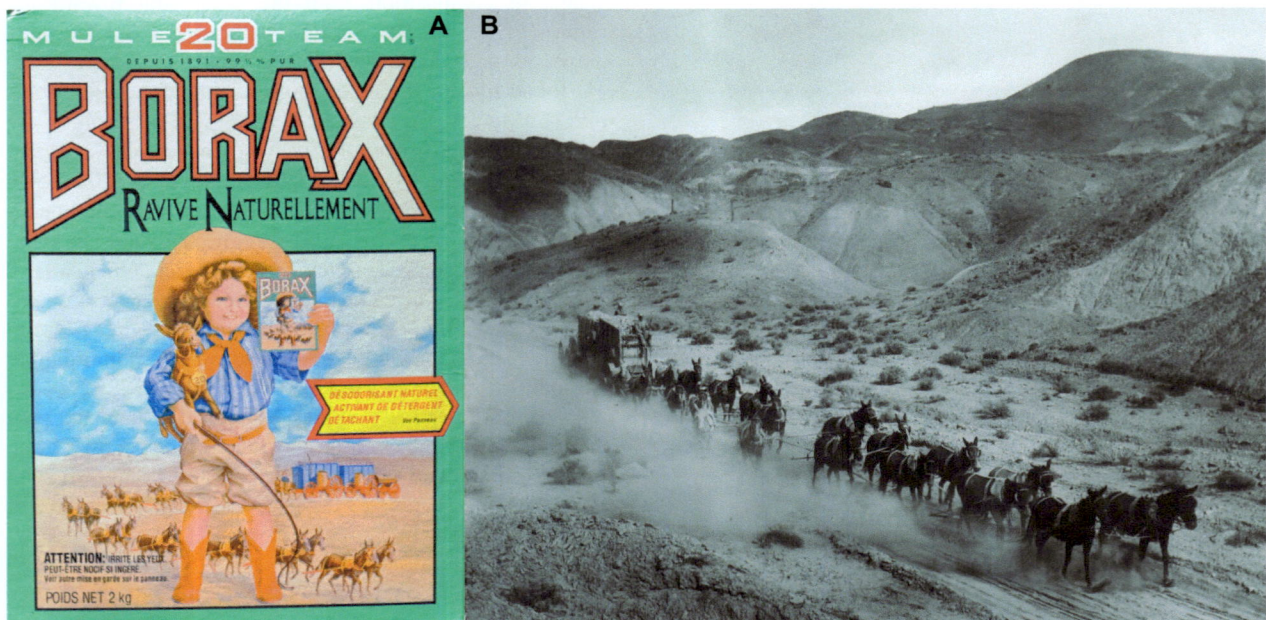

Figure 10.50 **A.** The famous 20 Mule Team Borax soap, as shown in this image of a vintage soap box. Source: Rick Rudnicki / Alamy Stock Photo. **B.** A mule team hauling a wagon filled with borax salts out of Death Valley, ca. 1900. Source: California Historical Society Collection, 1860–1960.

10.5 PALEOLAKES IN CLOSED BASINS

Today, the Basin and Range Province has an arid climate. As a result, playa lakes that form there are typically short-lived. But this has not always been the case. The cooler, wetter climate of the Quaternary Period altered the balance between evaporation and accumulation of water in the American West. As a result, vast lakes formed in many of the endorheic basins of the Great Basin. The two largest lakes were Lake Bonneville in northwestern Utah and Lake Lahontan in northwestern Nevada (**Fig. 10.51**). Lakes that formed in former cool, moist climates were sometimes referred to as **pluvial lakes** (Latin *pluvia*, "rain") in past decades, but are more often called **paleolakes** today. As the glacial period started to end, the Great Basin became warmer and drier, causing the levels of many of these paleolakes to fall, leaving behind shorelines as evidence of their higher levels (**Fig. 10.52A**). The Great Salt Lake in Utah is but a small relic of Lake Bonneville, the largest of the paleolakes in the Great Basin (**Fig. 10.51B**). Continued climatic drying has caused most of the paleolakes of the Great basin to completely dry up and become playas.

Paleolakes and playas are found in climatically dry areas throughout the world, where faulting has formed endorheic basins. Some of the largest salt playas are found in the Andes Mountains. The lowest subaerial elevation on Earth, the Dead Sea, similarly experienced wetter episodes during the last glacial period. Peak climatic wetness ≈24,000 years ago enlarged the Dead Sea's precursor (Lake Lisan) to an elevation of about 170 m below sea level, as evidenced by well-formed shorelines (**Fig. 10.52B**).

Another interesting geomorphic feature that can form in pluvial lakes is **tufa towers**. These deposits of secondary carbonate originally formed underwater, as carbonate-rich spring waters entered the highly alkaline (soda) lakes. Recent

drops in lake level have exposed many of these towers, as at Mono Lake, in California (USA) (**Fig. 10.53**). Because tufa forms subaqueously, it took centuries of water drawdown to form these well-recognized tufa towers.

Blowing sand is a common occurrence in areas of dry climate, and the Great Basin is no exception, with its many dry playas and minimal vegetation cover. Regardless, sand dunes occur in closed basins worldwide, where wind deflates sand from the dry, barren margins of playa lakes and alluvial fans. Water levels in playa lakes fluctuate widely, hindering plant growth that might otherwise stabilize the sands at the surface. Alkaline conditions also hinder plant growth. Winds can then easily transport the barren sands. Playa-sourced

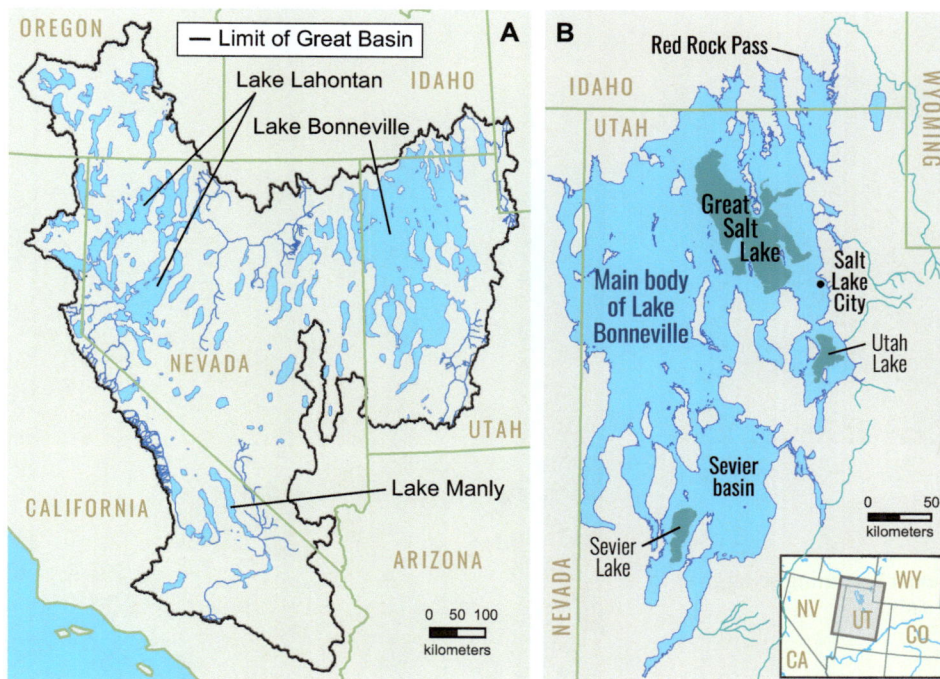

Figure 10.51 A. Paleolakes were widespread across the Great Basin during the last glacial period, centering on ≈18 ka. **B.** Map of the extent of Lake Bonneville, the largest paleolake in the Great Basin. Relicts of the lake remain mainly as three smaller lakes – Sevier Lake, Utah Lake, and the Great Salt Lake. Lake Bonneville had its outlet at Red Rock Pass.

Figure 10.52 Paleolake shorelines can provide important information about former lake dynamics and paleoclimate. **A.** Shorelines of Lake Lahontan, abandoned ≈14 ka, at the north end of modern-day Pyramid Lake, Nevada, USA. **B.** Shorelines of paleolake Lisan on the West Bank, with the Dead Sea in the background. Source: R. Dorn.

Figure 10.53 Tufa deposits in Mono Lake, California, USA. Source: M. Miller.

Figure 10.55 A scene from San Francisco after the 1906 earthquake that devastated the city. Source: US National Archives.

Figure 10.54 A satellite image of the source-bordering Khar-us Nuur sand dunes, Mongolia. Source: NASA.

dunes can range widely in size, even becoming so large as to be considered a dune field (**Fig. 10.54**).

10.6 NATURAL HAZARDS ASSOCIATED WITH FAULTING

Areas of active faulting are usually associated with earthquakes, and can lead to a variety of other natural hazards. The largest earthquake ever recorded along the San Andreas Fault in California occurred in 1906 near San Francisco, devastating that city (**Fig. 10.55**). The sudden shaking motion that accompanies large earthquakes can do a tremendous amount of damage to buildings, especially when the different parts of the building move in different directions, as bricks often do.

Arid regions like the Basin and Range are also notorious for hazards associated with intense rainstorms, such as localized debris and mud flows. Even Death Valley, known for its intense aridity and the hottest temperatures on Earth, experiences debris flows from periodic, heavy rains. For example, on August 5, 2022, Death Valley received 37 mm of rainfall in three hours – just 2 mm shy of its all-time record daily

Figure 10.56 Scenes from the floods of August 5, 2022 in Death Valley National Park. These kinds of hazards, although uncommon, can and do happen in desert landscapes. Source: Gabriele Notariâ / National Park Service photos.

rainfall (**Fig. 10.56**). Death Valley receives < 4 mm of rain in a typical August.

Another, more long-term, geomorphic hazard in the Basin and Range involves groundwater withdrawals, which leads to uneven compaction/settling. Underground irregularities, such as changes in particle size or a subsurface bedrock block, can produce fissures at the surface due to differential rates of ground subsidence (**Fig. 10.57**).

Figure 10.57 Queen Creek, a town on the fringe of metropolitan Phoenix, USA, is a hotspot for earth fissures. Groundwater withdrawal here has led to irregular amounts of ground subsidence and fissure formation. Source: B. Gootee, Arizona Geological Survey.

REVIEW QUESTIONS

10.1 What are the main types of stresses on rock, and what type of faulting is associated with each?

10.2 What are the main orders of relief for Earth's landforms? Which order of relief is represented by fault scarps (or fault escarpments)?

10.3 Generally, what determines whether stress applied to rocks leads to bending or breaking?

10.4 Define and discuss the terms fault trace and fault scarp. What does each represent, geologically?

10.5 What are the main differences between thrust and reverse faulting?

10.6 What landforms would you expect to see in association with strike-slip faulting?

10.7 What processes are associated with block faulting, and what types of landforms initially result from it?

10.8 What is the difference between the Basin and Range Province and the Great Basin?

10.9 Generally, how did the Basin and Range Province form? What are some common configurations of block faulting found there?

10.10 What is an endorheic basin, and how do most endorheic basins form?

10.11 Describe the evolution of a range front formed by block faulting. What types of landforms develop here over time? Use the piedmont concept in your answer.

10.12 What two main processes are involved in the filling of endorheic basins with sediment?

10.13 What distinguishes an alluvial fan from a bajada?

10.14 What are the main characteristics of pediments, and how do they differ from bajadas?

10.15 Explain the difference between the hydrographic and topographic apex of an alluvial fan.

10.16 What generally happens to an alluvial fan over time, as it evolves from an active feature to a relict feature?

10.17 What are ballenas and how do they form?

10.18 As basins and ranges evolve, both pediments and piedmonts form. What exactly are each of these features and how do they differ?

10.19 How does the activity-level of faulting (whether the faulting is ongoing or whether it has been "turned off" for millions of years) influence whether you would expect to see a pediment or an alluvial fan along the base of a mountain range?

10.20 What is a playa, how does it form, and where would you normally find one in the basin-range sequence?

10.21 Why do mud playas form in some endorheic basins and salt playas form in others?

10.22 What sorts of geomorphic evidence indicate that much larger lakes once occupied closed basins in many parts of the world?

10.23 What are some of the main natural hazards of the Basin and Range?

FURTHER READING

Burbank, D. W. and Anderson, R. S. 2012. *Tectonic Geomorphology*, 2nd ed. Wiley.

Hunt, C. B. 1975. *Death Valley: Geology, Ecology, Archaeology*. University of California Press.

Jeong, A., Cheung, S. Y., Walker, I. J. and Dorn, R. I. 2018. Urban geomorphology of an arid city: Case study of Phoenix, Arizona. In: Thornbush, M. J. and Allen, C. D. (eds.) *Urban Geomorphology: Landforms and Processes in Cities*. Elsevier. pp. 177–204.

Mabbutt, J. A. 1977. *Desert Landforms*. Massachusetts Institute of Technology, Cambridge University Press.

Peterson, F. F. 1981. Landforms of the Basin and Range Province. *Nevada Agric. Exp. Station Tech. Bull.* 28:52pp.

Wallace, R. E. (ed.) 1990. *The San Andreas Fault System, California*. US Geological Survey Professional Paper 1515.

11 Weathering

Heather A. Viles and Randall Schaetzl

Weathering is central to geomorphology; without it, landforms would not exist. Weathering sculpts rocks and landscapes at all scales, from producing tiny pits on rock surfaces to forming large valleys. It is everywhere.

However, weathering does not work alone. Instead, it operates alongside other surficial processes to produce the landscapes we see around us. Weathering is often defined as the *in situ* (meaning "in position") breakdown of rocks and minerals. It is distinct from erosion, which involves the removal and transport of material, usually downslope. Often, weathering *preconditions* rocks for erosion by making them weaker and less coherent. Together, weathering and erosion operate to form landforms via denudation – the overall lowering of the land surface.

Although weathering is often a slow and subtle process, it punches above its weight in terms of geomorphic importance. It occurs in all terrestrial environments on Earth, including on bare rock surfaces, within and below soils, along coasts and far inland, on mountaintops, plateaus, and in low-lying valley bottoms. It occurs in all climatic zones – from the driest deserts to the wettest parts of the humid tropics. By weakening rocks and producing regolith (unconsolidated sediment), weathering paves the way for other geomorphic processes such as erosion by wind and water, as well as mass movements.

Weathering occurs within the Earth's **critical zone** – the permeable, near-surface layer, where atmosphere, biota, and the lithosphere meet and interact (**Fig. 11.1**). The critical zone extends from the atmosphere down to where groundwater ends, which can be up to 100 meters below the surface. The study of weathering requires understanding a wide range of interconnected processes that operate within the critical zone.

Weathering of rocks and minerals at the land surface can occur in a variety of ways, through a range of chemical and physical weathering processes, often mediated by organisms of all kinds. Weathering can be conceptualized as a dynamic interaction between force (determined by environmental conditions) and resistance (controlled by rock and mineral characteristics). For just that reason, Chapter 5 – on rocks and minerals – is titled the "Resisting Framework." In short, the rocks that exist at a site set the table for the weathering processes that follow.

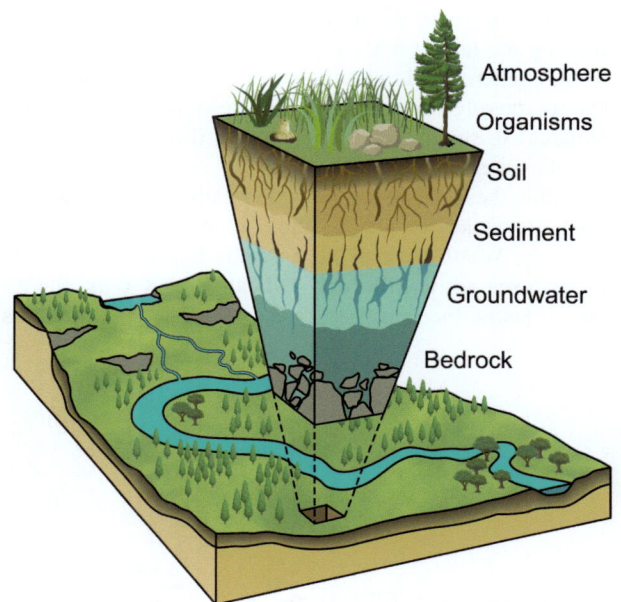

Figure 11.1 Schematic representation of the many components of Earth's critical zone.

11.1 SETTING THE SCENE: WEATHERING AND ITS GEOMORPHIC CONTEXT

Weathering is the way that rocks, formed deep within the crust or at the ocean floor, adapt and change to the environment of the surface. Igneous and metamorphic rocks form in hotter, higher-pressure conditions with far less oxygen (and usually less water) than is found at the surface. Most sedimentary rocks are produced under higher-pressure conditions than at the surface, where conditions are generally cooler and wetter, with abundant oxygen and hosts of organisms. Thus, newly formed rocks and minerals must eventually interact with the surficial environment – either as bare, exposed surfaces or while covered in **regolith**. In these new "surficial" conditions, rocks become unstable, causing them to weather. These very "changed" conditions of the land surface ultimately drive all weathering processes.

Geomorphologists use the term regolith to refer to all the unconsolidated material from the surface down to intact

Figure 11.0 Rough Tor, near Bodmin Moor in Cornwall, England, shows the effects of rock structure and lithology on patterns of weathering. Source: James Osmond / Getty Images.

bedrock below (**Fig. 11.2**). Put another way, regolith is everything between fresh rock and fresh air. Soils form in the upper part of the regolith, and in some settings, there is a layer of **saprolite** beneath the soil, within the upper part of the weathering rock. Saprolite is weathered rock, but not so much that it has lost its original rock fabric, that is, it looks like "soft rock" or "rotten rock." Not all of the rock in the saprolite layer is completely weathered; rock fragments here commonly occur as rounded **corestones** of unweathered rock (**Fig. 11.3**). The lower part of the saprolite is often considered to be **saprock** – rock that has started to weather, but less than a third of its primary minerals have been altered. At the base of the saprock is the **weathering front**, marking the lowest detectable limit of weathering activity. Fresh, unaltered bedrock occurs below the weathering front. The weathering front should not be seen as a sharp, clearly demarcated, simple boundary. Instead, it is a dynamic and complex interface.

Within the bedrock and near the weathering front, weathering may take its first foothold along joints and other openings. Weathering at these depths is often referred to as **deep-seated weathering**, indicating that it is operating far below the land surface. Deep-seated weathering is

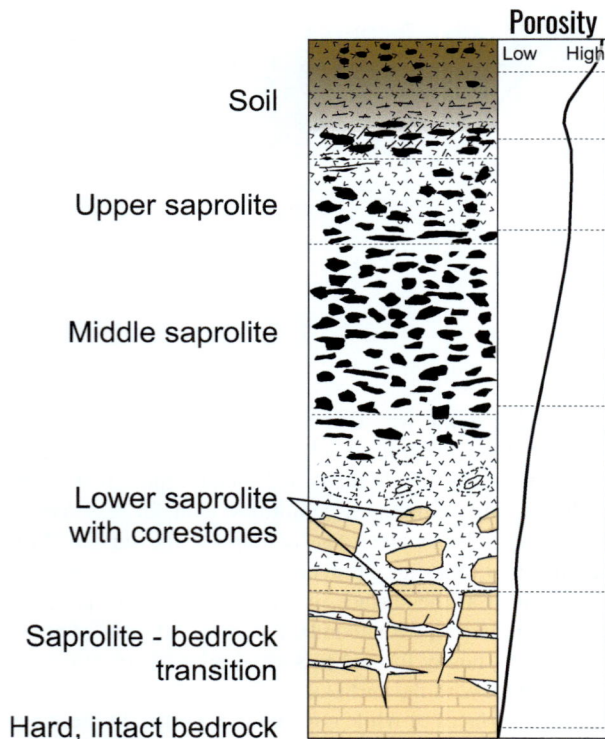

Figure 11.3 Corestones formed from weathering of basalt, set within regolith, at Golden Gate National Park, South Africa. Source: A. Goudie.

Figure 11.2 Schematic illustration of a weathering profile forming in regolith, from the soil down to the bedrock, with corestones at depth. Source: Used with permission of The Geological Society of London, from The development of crystalline basement aquifers in a tropical environment, The Geological Society of London, Vol. 20, 1967; permission conveyed through Copyright Clearance Center, Inc. Acworth (1987)

dominated by chemical processes within the mantle of regolith, and will continue to operate as long as ample water exists to remove the byproducts of chemical weathering reactions. Geomorphologists use the term **weathering profile** to refer to all the various zones and layers within the regolith where complex associations of physical, chemical, and biological weathering processes occur (**Fig. 11.2**). The weathering profile includes everything between the land surface and the top of the fresh bedrock – all of it is currently weathering!

Across landscapes, weathering profiles, regolith, and soils vary with topography. Some areas are dominated by *in situ* accumulations of weathering byproducts, whereas others are dominated by sediment transported there by processes such as wind or water (**Fig. 11.4A**). The term **residuum** (or **residual regolith**) is used to refer to unconsolidated or partly weathered mineral material that has accumulated due to weathering *in situ*. Residuum is regolith that has *not* been significantly transported or moved since it formed. Residuum is but one form of regolith. Other forms of regolith include materials transported to the site; these materials may be called **transported regolith** (**Fig. 11.4A**).

Because it generally involves a suite of rather slow processes, especially in drier and colder climates, weathering is often the rate-limiting factor in landscape evolution, resulting in weathering-limited slopes (**Figs. 8.1, 11.4B**). Weathering-limited slopes evolve without a significant cover of regolith, because rates of erosion and transport of any loose material on top of the surface exceed the rate at which weathering can produce it. Thus, erosion removes the regolith so rapidly that fresh, bare rock surfaces are constantly being exposed, resulting in slopes with a considerable amount of exposed bedrock. Alternatively, transport-limited slopes are exemplified by thick accumulations of weathered debris (regolith) and soil, as weathering occurs relatively rapidly in comparison with removal of material (**Fig. 11.4B**). Most transport-limited slopes are able to support a cover of vegetation, which

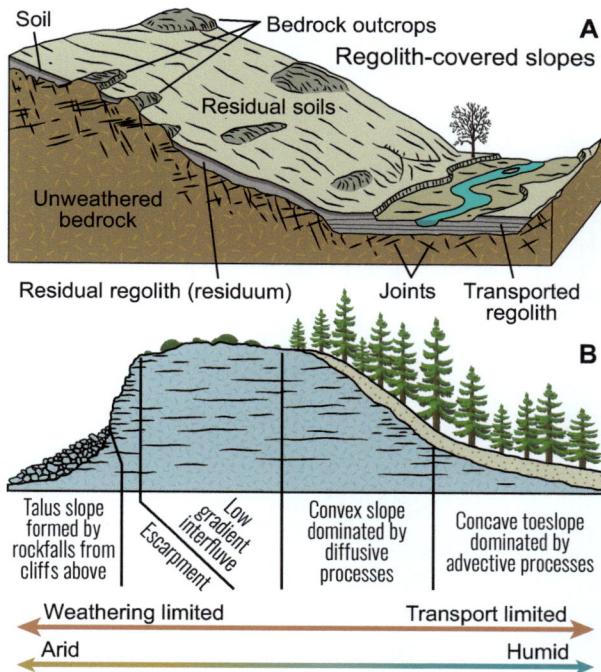

Figure 11.4 Soils, regolith, and weathering profiles across landscapes. **A.** A simplified representation of the variable nature of regolith across a valley-side slope. © A. N. Strahler (1992). Used by permission. **B.** A conceptual model of bedrock below weathering-limited and transport-limited slopes. Diffusive processes include creep, rainsplash, and biogenic transport of sediment. Advective processes include those in which sediment is moved by runoff.

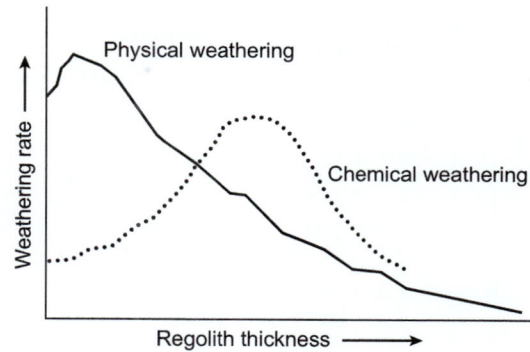

Figure 11.5 Variation in physical and chemical weathering rates as a function of regolith thickness.

downward water movement is retarded and even chemical reactions are inhibited. In short, weathering rates at the base of the weathering front are slow when regolith thicknesses are great.

11.2 PHYSICAL WEATHERING

Most geomorphologists distinguish two, or sometimes three, broad categories of weathering: (1) physical weathering, (2) chemical weathering, and (3) biotic weathering. **Physical weathering** (also called **mechanical weathering**) processes lead to the disintegration of rocks and minerals into smaller pieces. This usually preserves the chemical make-up of the rock. In essence, physical weathering involves the propagation of fractures and cracks, until rocks fall apart into smaller pieces, with considerably more surface area (**Fig. 11.6**). It is on these surfaces that subsequent weathering processes can operate. Thus, physical weathering is often a necessary precursor for chemical weathering.

Rocks can break up – physically – in a variety of ways, into fragments of many different shapes, depending on the type of rock and the fracture patterns within it. Many igneous rocks, especially coarse-grained igneous rocks like granite, break apart along mineral crystal boundaries. This type of **grain-by-grain disintegration** produces a gravelly residuum called grus, which is dominated by grains of minerals such as quartz and feldspar (**Fig. 5.4**). Many sedimentary rocks fracture along bedding planes, producing rectangular-shaped blocks; this type of breakdown is referred to as **block-by-block disintegration**. Loss of large sheets that break off from rock surfaces is called **exfoliation**. The terms **sheeting** or **spalling** are equivalent terms, but used for smaller, thinner sheets that break off of rocks. Lastly, some rocks break apart along seemingly random planes of weakness, perhaps best described as **shattering** (**Fig. 11.7**).

Perhaps the simplest physical weathering process involves the heating and cooling of a rock surface by **insolation** (sunlight) or fire. Heat expands the rock. Repeated expansion and contraction of the surface layers – **thermal** or **insolation weathering** – can effectively break down a rock. Environments that experience high diurnal cycling of air

helps hold the regolith in place. The vegetation and soil cover retain moisture, which then can help to enhance weathering of the rocks below, producing a positive feedback loop that tends to maintain a regolith cover on the slope. Alternatively, on weathering-limited slopes, vegetation has a difficult time colonizing the bare bedrock, and as a result, water runs off readily, regolith is removed, and the rock surfaces are often dry, causing weathering processes to slow down – another case of feedback in the geomorphic system.

Weathering also includes a temporal dimension. As **Fig. 11.5** demonstrates, the balance between physical and chemical weathering processes changes over time, often driven by changes in regolith thickness. Physical weathering rates are strong when regolith is thin, as may occur on young landscapes or steep slopes. On these types of (often, bare rock) surfaces, physical weathering processes like freeze–thaw and wet–dry cycles are most frequent. On such surfaces, plants can root directly in the bedrock, prying it apart (a physical weathering process). Conversely, chemical weathering rates increase as regolith gets thicker, because the regolith can retain both water and various organisms, which are necessary for many chemical weathering processes (see below). The thick regolith cover has the opposite effect on physical weathering processes, slowing them by inhibiting wet–dry and freeze–thaw cycles at the weathering front (**Fig. 11.5**). Eventually, the cover of regolith becomes so thick that

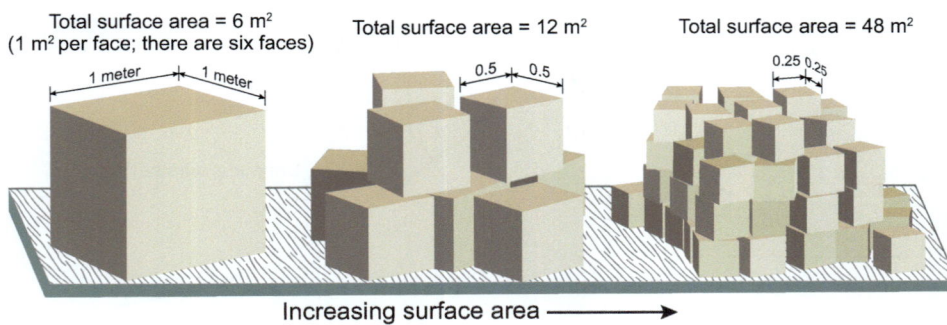

Total surface area = 6 m²
(1 m² per face; there are six faces)

1 meter ← → 1 meter

Total surface area = 12 m²

0.5 ← → 0.5

Total surface area = 48 m²

0.25 ← → 0.25

Increasing surface area ———————→

Figure 11.6 Illustration of the relationship between decreasing block size and increasing surface area. Physical weathering produces smaller blocks, and as a result produces more overall surface area on which chemical weathering processes can operate. In all three examples, the total volume of the blocks is the same..

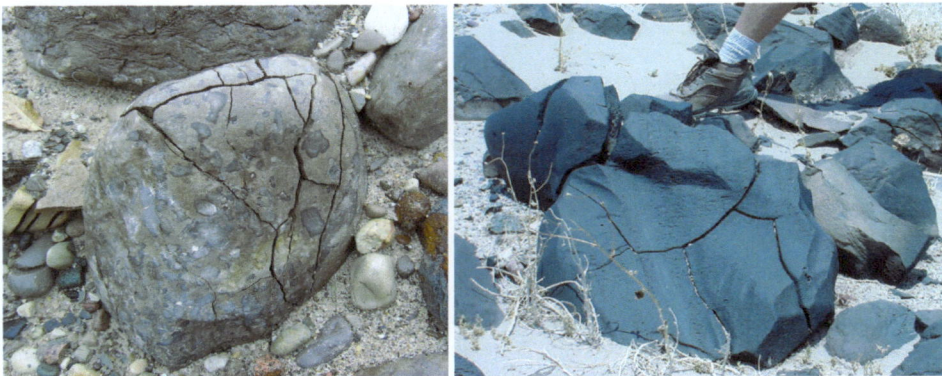

Figure 11.7 Rocks shattered along seemingly random planes of weakness, probably by freeze–thaw activity. Source: R. Schaetzl.

temperatures are likely to be particularly vulnerable to thermal weathering. Repeated diurnal cycling usually produces weathering by the gradual build-up of stresses in the rock, a process called **thermal stress fatigue**, rendering the rock more vulnerable to other weathering processes.

The albedo (reflectivity) and thermal conductivity of rocks are crucial in determining their sensitivity to thermal weathering, as they control the surface temperature and the rock's internal thermal gradient. Igneous and metamorphic rocks containing a range of different minerals with varying **coefficients of thermal expansion** can be particularly prone to thermal weathering. That is, heating and cooling cycles cause stresses because of differential expansion and contraction of adjacent mineral grains. Marble is particularly prone to thermal weathering, because its interlocking calcite crystals respond anisotropically to heating, that is, different crystal axes have very different coefficients of thermal expansion. Thermal weathering can also result from other agents of heating, such as wildfires. The intense and episodic heating by fires weathers rock by thermal shock. Thermal weathering can result in sheeting, **granular disintegration**, or cracking, and can even split or spall apart large boulders (**Fig. 22.15**).

Frost and **salt weathering** are two other, related forms of physical weathering. Both result from phase changes between crystalline and liquid states, which result in formation and expansion of ice (or salt) crystals within rocks. Both require water to operate. They are likely to be important in environments that experience frequent and intense cycles of temperature and moisture, and are particularly effective in weak, porous rocks such as limestone or sandstone.

Frost weathering involves the growth of ice crystals and ice lenses in rocks. Growth of ice crystals in rocks can generate significant pressures, opening existing cracks in rocks even wider (**Fig. 11.7**). Frost weathering works in two ways. First, when water freezes, it expands by ≈9%, producing pressures which can, in some circumstances, break rock apart. Second, and of more general importance, is **segregated ice**, wherein water that starts freezing in pores attracts unfrozen water to it, resulting in ice lenses that can, over time, grow to large sizes (see Chapter 20).

The frequency, intensity, and duration of freezing cycles are all important controls on the efficacy of frost weathering. As long as ample moisture is available, rocks that undergo more freeze–thaw cycles will be more intensively fractured. High alpine areas are particularly good places to witness the effects of frost weathering, as many of these locations cross the 0 °C threshold dozens of times annually. Usually, they freeze at night but warm up during the day, facilitating a new suite of ice crystal growth every night. Finally, it is important to note that extremely cold areas, such as polar landscapes, may not experience as much frost weathering as one might initially expect. These areas stay below freezing for long periods of time each winter, and as a result, ice crystal formation (or, more correctly, re-formation) does not occur. The moisture content of rocks is also an important control of frost weathering, with rocks in wetter climates experiencing more rapid breakdown (other things being equal). In the past, frost weathering was likely enhanced in areas affected by the melting of ice sheets and glaciers, associated with the amelioration of climatic conditions. Large swaths of land near the ice sheets were locked in a periglacial climate, with its frequent pulses of freeze–thaw activity.

Salt weathering is most common in dry (desert) landscapes because it is driven by the repeated crystallization and growth of salts within porous rocks. Three main causes of stress are developed during salt weathering. Pressure is exerted (1) as salts crystallize from **solution**, (2) as some salts hydrate, that is, incorporate water into their crystal lattice, and thereby expand, and (3) as existing salt crystals expand when heated. The first of these processes is generally the most important. Typically, this process occurs as salty liquids, present within rocks, develop salt crystals upon drying. The crystals pry apart fractures to break up the rock. Rain can then rehydrate the salt crystals, and the process repeats as the rocks again dry out.

Salt weathering can produce granular disintegration, cracking, and exfoliation in susceptible rocks (**Fig. 11.8**), and is particularly prevalent in hot, arid areas, where high rates of evaporation encourage salts to crystallize out of solution. Additionally, in many desert areas, salts are more commonly a part of the geomorphic system. Here, they can deflate from salt flats (playas) and get transported by wind to sites downwind. Later, rains will dissolve these salts and allow the salty liquids to penetrate fractures in rock. Such salt sources are rare in humid climates. Common salts contributing to salt weathering are $NaCl$ (halite) and Na_2SO_4 (thenardite).

Some rocks are prone to breakdown by **wetting and drying**, especially those that contain minerals which experience high rates of swelling when wetted. Particularly vulnerable in this regard are rocks that contain smectite clay minerals, such as some shales. Wetting and drying can lead to a type of rock breakup called **slaking**, as well as granular disintegration, cracking, and flaking. In slaking, very thin sheets appear to peel off the rock surface. It is often difficult to separate the effects of wetting and drying from other processes such as salt and frost weathering, as similar agents and end results are involved – a geomorphic condition called **equifinality**.

Finally, many rocks can be weathered mechanically as a result of unloading. Unloading refers to the release of weight from rocks, usually bedrock, as weathering and erosion strip away regolith over long periods of time. The loss of overpressure allows the rock to expand upward, developing a series of cracks and joints that form parallel to the land surface. The likelihood and intensity of unloading is largely controlled by the topographic setting. Where a large load of overlying rock or ice is removed (perhaps through tectonic processes or glaciation), confining stresses acting on the rock are also lost, and large-scale exfoliation may result. Exfoliation commonly affects brittle, crystalline rocks like granite, as well as some sandstones. It is typically expressed as multiple sheets of rock that get thicker with depth (**Fig. 11.9**). Over time, extensive exfoliation forms rounded bedrock landforms known as **exfoliation domes**.

Recent research has demonstrated the importance of the **subcritical cracking** process to physical weathering. Also called dirt cracking, subcritical cracking results from a suite of chemo-physical mechanisms. The process is common in deserts. Understanding fracture mechanics and how and why

Figure 11.8 Exfoliation of granite boulders and outcrops, near Gobabeb, Namibia. Salt weathering is likely to have been a major causal factor in the development of these features in this hyperarid climate, which encourages evaporation and the crystallization of salts from both ground and surface waters. Source: H. Viles.

Figure 11.9 Exfoliation, a dominant weathering process in Yosemite National Park, California, USA. **A.** and **B.** Exfoliation sheets are very apparent in these granite outcrops. **C.** Half Dome; its left side was sheared off by a Pleistocene glacier, but the crest is the prototypical exfoliation dome. Source: R. Schaetzl.

Figure 11.10 Biotic weathering is clearly at play in these images. **A.** Tree roots have pried apart large boulders with ease. Source: R. Schaetzl. **B.** The desert tortoise (*Gopherus agassizii*) is a prolific and powerful burrower, seeking shelter underground to escape the desert heat. Source: Lake Mead NRA Public Affairs, CC BY-SA 2.0, via Wikimedia Commons.

they form are important goals for an improved understanding of mechanical (physical) weathering.

In a variety of ways, biota also function as agents of physical weathering. Plant roots can pry apart even the largest of rocks (**Fig. 11.10A**). Animals that dig and burrow in the regolith inadvertently break up pebbles and gravels, and some, like the wombat of Australia and the desert tortoise of the United States (**Fig. 11.10B**), can easily burrow into hard bedrock. A full section on biotic weathering follows (below).

11.3 CHEMICAL WEATHERING

Chemical weathering processes are broadly grouped under the term **decomposition**, wherein the chemical and mineralogical makeup of rocks and minerals is changed. Decomposition produces a different suite of secondary minerals, the loss of some minerals (as chemical elements and ions) in solution, and the release of unweathered minerals through granular disintegration. Chemical weathering of rocks results from the interactions of oxygen, water, CO_2, and other chemical agents such as acids. Biota are an important source of various organic acids, which play key roles in many types of chemical weathering. Some rocks and minerals, such as limestone and calcite, are particularly prone to specific forms of chemical weathering (see Chapter 12). Chemical weathering processes often result in distinctive surface textures on exposed rock surfaces, or on minerals and rocks within soil profiles, such as rounded surfaces, pitting, and open joints.

Rocks and minerals are prone to chemical weathering because they formed in an environment (deep within the crust or at the seafloor) that is chemically very different from what exists at the land surface. Thus, most minerals in rocks are in disequilibrium at the land surface and slowly weather into other forms, mainly secondary clay minerals and various types of ions. The clay minerals may then accumulate in soils or get washed into the sea, to become shales. The ions are variously soluble, and thus are available for water to carry them into the groundwater and, eventually, to the sea, or to be biocycled by plants. Along the way they may

participate in myriad biochemical and mineral synthesis reactions.

Like most chemical reactions, chemical weathering processes generally are facilitated by warmer temperatures. Thus, chemical weathering is particularly important in warm, tropical and subtropical climates. Many chemical reactions also either involve water or require water to flush away the reaction byproducts. If these reactants are not removed, most chemical reactions will slow and then cease. Therefore, chemical weathering is important in regoliths in humid climates, through which abundant amounts of water can flow. In summary, look for strong evidence of chemical weathering, and exceptionally deep weathering profiles, in hot, humid climates.

Chemical weathering within rocks and regolith has some similarities to a drip (or filter) coffee machine (**Fig. 11.11**). Water that falls onto a rock or regolith percolates through it, interacting with a suite of primary minerals along the way. These minerals are reconstituted (or remain unaffected by interaction with water), but regardless, some amount of dissolved materials (ions, in the case of chemical weathering)

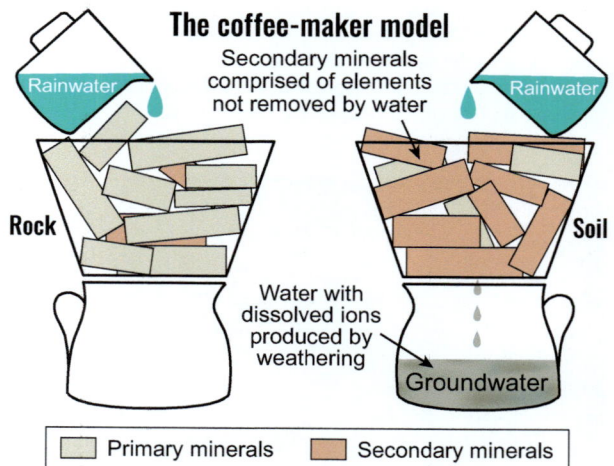

Figure 11.11 The "coffee maker" model of chemical weathering of rocks and minerals.

are removed and leave the system through groundwater. New, secondary minerals are then formed and left behind.

Bare rock surfaces or surfaces with thin regolith are often too dry to foster chemical weathering processes (**Fig. 11.5**). Chemical weathering is only important on bare rock surfaces for carbonate and/or evaporate rocks such as limestone, gypsum, and halite, where **solution weathering** is the dominant process. Solution processes can slowly dissolve the rock to produce a range of weathering products, including soluble compounds (such as calcium and carbonate ions released in dissolved forms).

Chemical weathering affects mainly primary minerals, that is, those formed from molten materials in igneous rocks or precipitated out of solution. During chemical weathering, primary minerals less affected by weathering reactions (such as quartz, which is very resistant to chemical weathering), and new, secondary minerals produced by weathering reactions (such as phyllosilicate clays and various oxide compounds) tend to accumulate over time (**Fig. 11.11**). The various primary minerals show very different susceptibilities to chemical weathering. As a result, weathering profiles develop over time, with more resistant minerals becoming dominant, as the more susceptible minerals are weathered to other forms. Typically, within a weathering profile, the most-weathered materials will be nearest the surface and the regolith will become less weathered with depth.

Common mechanisms of chemical weathering include **hydration**, congruent and incongruent **dissolution**, oxidation–reduction processes, **hydrolysis**, and **chelation**. Hydration involves the incorporation of water molecules into a mineral lattice, producing a new mineral form and causing the mineral to expand and, potentially, fail due to this stress. For example, calcium sulfate (anhydrite) hydrates to produce the mineral gypsum:

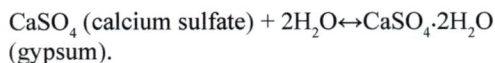

$$CaSO_4 \text{ (calcium sulfate)} + 2H_2O \leftrightarrow CaSO_4 \cdot 2H_2O$$
(gypsum).

This reaction causes swelling and exerts physical pressure, resulting in physical breakdown.

Solution weathering (also called dissolution) occurs where susceptible minerals are partially or fully dissolved in water. Rock salt, composed of the mineral halite ($NaCl$), is particularly prone to dissolution. Solution weathering that affects limestone is often called **carbonation** (see Chapter 12). Solution processes, like many other processes of chemical weathering, can be enhanced under acidic conditions.

Some minerals are prone to weathering by **redox** processes (short for reduction–oxidation reactions). Particularly vulnerable to redox reactions are minerals rich in iron and manganese, as well as sulfate minerals like pyrite. For example, in the presence of oxygen, ferrous iron (Fe^{2+}) in ferrous oxide can oxidize to ferric iron (Fe^{3+}) in the mineral hematite:

$$4FeO + O_2 \rightarrow 2Fe_2O_3.$$

In turn, the Fe hematite can be reduced and change form if conditions become reducing, that is, where oxygen is lacking.

Reducing conditions are not uncommon in the lower parts of regolith. Rocks at depth can repeatedly undergo oxidation–reduction cycles where water table depths fluctuate or where regoliths wet and dry. These redox cycles are a common and effective form of chemical weathering, particularly those with large amounts of Fe-bearing minerals. In some cases, oxidation can also produce further agents of weathering. For example, oxidation of pyrite (FeS_2) produces sulfuric acid, which can then lead to even further dissolution of susceptible minerals:

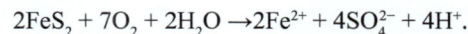

$$2FeS_2 + 7O_2 + 2H_2O \rightarrow 2Fe^{2+} + 4SO_4^{2-} + 4H^+.$$

As the name implies, hydrolysis (hydro, "water" + lysis, "split") involves the breaking down of minerals through reactions that "split" water molecules, separating them into H^+ and OH^- ions. Hydrolysis can be differentiated from hydration, as hydration involves the water being taken up by the mineral as a complete molecule, whereas in hydrolysis both the mineral and the water split up and recombine into new forms. Hydrolysis is an important process for the weathering of some silicate minerals such as feldspar. For example, K-feldspars undergoing hydrolysis release K^+ and OH^- ions and produce silicic acid as follows:

$$KAlSi_3O_8 + H_2O \rightarrow HAlSi_3O_8 + K^+ + OH^-.$$

Chelation (Greek *chēla*, "crab's claw") is a process whereby a chemical complex or ligand – usually an organic one – removes a metal ion directly from a mineral. The chelated metal is then soluble, allowing it to go into solution. Thus, the mineral loses some of its constituent ions, and is more susceptible to breakdown. Chelation often affects iron minerals, but can also affect other minerals that contain metals such as zinc, copper, manganese, calcium, and magnesium.

11.4 BIOLOGICAL WEATHERING

Weathering is almost everywhere influenced by biota. Plants, animals, and microbes can be agents of both physical and chemical weathering. Even in the harshest of hot desert environments, a host of microbial species are active on rock surfaces. Within soils, animals such as ants and earthworms, plant roots, and microbial communities all play roles in weathering. On bare rock surfaces, whole communities of lower plants (lichens and mosses) and microbes can form biofilms or biocrusts, which drastically change the chemical and hydrological conditions of the rock surface. Lastly, the geomicrobiology of regolith plays other roles in the geomorphic system, especially in the consumption of nutrients (pollutants, as well as naturally present ones) and by forming acids that contribute to even more weathering.

Biological contributions to weathering can be direct or indirect. Indirectly, plants and microbes can influence the thermal and water regimes on rock surfaces, which can then affect processes such as thermal weathering and frost weathering. Microorganisms in soils also are important for their contributions of CO_2 and organic acids, both of which factor prominently in some chemical weathering processes,

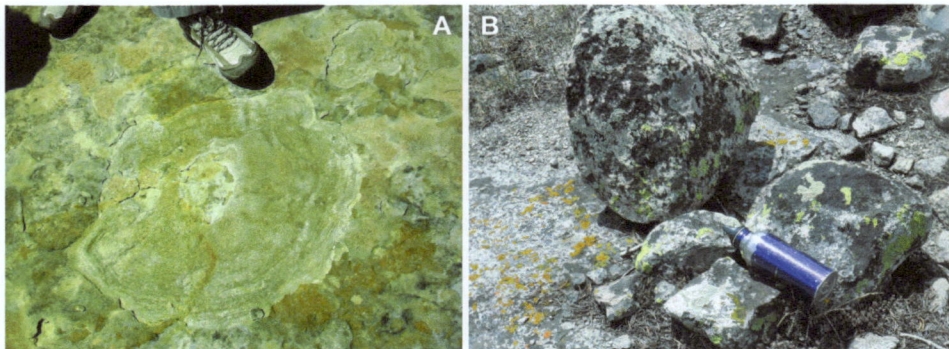

Figure 11.12 Lichens on rock surfaces. **A.** A mixture of biochemical and biophysical processes operating beneath a large lichen thallus (of the genus *Lecidea*) in the center of the image has produced a ≈1 cm deep depression. The lichen has, in effect, caused extensive weathering of the underlying sandstone at Golden Gate National Park, South Africa. Source: H. Viles. **B.** A variety of different lichen species are attacking these rocks, causing them to develop distinctive surface textures and pits. Source: R. Schaetzl.

particularly carbonation. Directly, plants can be important agents of **biophysical weathering** through the growth of roots, which contribute to the enlargement of joints in rocks (**Fig. 11.10A**). Soil-dwelling animals (**infauna**) such as ants, termites, and earthworms are also effective agents of weathering, mainly because they break down rocks and minerals into smaller particles (biophysical weathering). Some lichen species growing on bare rocks are known to be able to physically pluck small grains from rock surfaces by expanding and contracting as they wet and dry. Many different types of **biochemical weathering** are also driven by plants and animals. Many lichens, biofilms, and biocrusts interact chemically with the underlying rocks, producing acids and chelating agents which have a range of chemical weathering impacts (**Fig. 11.12**). Biological weathering processes often lead to distinctive surface textures on bare rock surfaces, or on minerals and rocks within or at the base of soil profiles, including pitting and opened joints.

11.5 WEATHERING SYSTEMS

Although it is convenient to differentiate between the various, individual physical and chemical weathering processes, in reality, overlap and interactions exist. Hydration, for example, produces chemical changes in some minerals as a result of the uptake of water molecules into their crystal lattices. However, in some cases it can also produce volume changes in other ways, which can produce physical breakdown. For example, sodium sulfate in the form of thenardite (Na_2SO_4) hydrates, forming the mineral mirabilite ($Na_2SO_4 \cdot 10H_2O$), producing damaging effects in susceptible rocks. The new salt species (mirabilite) formed through the addition of 10 water molecules, leads to a volume expansion of 320%. If this process occurs within near-surface pores in a porous rock, the pressure induced can drive physical weathering, expressed as flaking and exfoliation.

Furthermore, in any environment and on any given rock type, a multitude of individual weathering processes are likely to operate, because organisms, water, thermal cycling,

and a range of acidic compounds are usually present. Such processes can sometimes act synergistically, for example, chemical weathering processes can produce minerals that are more susceptible to physical breakdown via wetting and drying. It is also clear that some key weathering phenomena (such as granular disintegration, cracking, flaking, exfoliation, pitting, etc.) can be produced by a range of different processes. So, it is perhaps folly to try to link a specific weathering outcome to an individual process. Increasingly, geomorphologists studying weathering conceive of these interactions as part of a broader "weathering system."

So, what are some general statements about weathering systems? Initially, climate acts as a first order control, influencing the availability of water, daily and seasonal temperature regimes, and affecting the nature of biological communities. Geology and topography act as the other first order controls, adding complexity to any clear links between climate and particular weathering regimes. Lastly, weathering systems are often very complex and can exhibit non-linear behavior over time. In essence, a change in climatic conditions can lead to complex responses in the weathering system.

11.6 SMALL-SCALE LANDFORMS PRODUCED BY WEATHERING

Weathering is a dominant process in the production of a wide variety of small-scale landforms. These landforms are not only of interest in their own right, but the degree to which they are expressed can assist in establishing the age of the rock surface (surface exposure dating – see Chapter 3).

11.6.1 Surface Alteration Features

Many chemical and biochemical weathering processes produce new (secondary) minerals that can accumulate on, or close to, rock surfaces. These **alteration features** have been variously referred to as weathering rinds, varnishes, crusts, and case-hardened layers (**Figs. 3.7, 3.11, 3.12**). Characteristically on the order of a few millimeters to a few centimeters in thickness, these layers may form *on top of* rock, or alternatively, the rock may be weathered and weakened because the new layer is forming *within* the rock. Similar features can also develop on rocks buried within regolith.

Water that penetrates rocks on and within regolith can dissolve minerals. The soluble products can then move back up, towards the surface, driven by capillarity and evaporation, or continue to percolate downward, possibly to the groundwater. Weathering rinds, crusts, and case-hardened layers can then form on the rock surface, as cycles of water percolation and evaporation cause preferential accumulation and precipitation of mobile elements in the near-surface zone. Alternatively, or additionally, weathering rinds, crusts, and case-hardened surfaces can be produced by a suite of biochemical weathering processes associated with microbial biofilms or **biocrusts** (**Fig. 5.33**).

Weathering rinds, crusts, and case-hardened surfaces can help protect the underlying rock from other, ongoing weathering processes. However, they can also deteriorate the rock over time because of differences that develop in characteristics such as porosity and thermal expansion coefficients between the altered zone and the underlying rock, leading to cracking, blistering, and flaking.

An unusual example of small-scale surface sculpting, driven by weathering, involves the **polygonal cracking** features that develop on some sandstones and granites. These features are also sometimes called "elephant skin weathering" or *Schildkrötenmuster* (German, "tortoise pattern"). They appear as convex polygonal plates, separated by shallow cracks, sometimes also associated with a weathering rind or crust (**Fig. 11.13**). Well-documented examples are found near Fontainebleau, France, where each polygon is about 10–20 cm in diameter, as well as near Boulder, Colorado and in Zion Canyon National Park in Utah, USA. Geomorphologists are still debating the exact origin of these features.

Figure 11.13 Polygonal weathering features developed on sandstone, Golden Gate National Park, South Africa. Source: H. Viles.

Figure 11.14 The light-colored limestone in this valley around Lake Mead, Arizona, USA, is covered with rock varnish. Note the dark color of the varnish, which has been removed from the rock below the former waterline during former periods of higher lake levels. Source: H. Viles.

A distinctive type of weathering rind, rock varnish is a thin (< 1 mm) rock coating, formed of clays (deposited on the surface in the form of dust), cemented onto the rock surface by hydroxides and oxides of manganese and iron (**Figs. 3.12, 11.14**). Because rock varnish is formed in arid climates, it is sometimes called desert varnish. Much debate exists about the processes involved, although bacteria appear to play a role. The varnish can easily be weathered itself, as environmental conditions change, for example, by lichens or by a change in climate. Because rock varnish gets progressively darker and thicker over time in many desert regions, it has been used as a relative dating tool (see Chapter 3).

11.6.2 Sculpted Features

Many weathering processes can produce sculpted rock surfaces by preferentially removing material from a rock surface. In this suite of processes – **differential weathering** – the net effect is to increase the local relief on a surface. **Weathering pits** or **basins** are small, closed depressions found on horizontal or gently sloping rock surfaces; they are common on granites, sandstones, and limestones (**Fig. 11.15**). The pits can range in size from < 1 cm to > 10 m in diameter. Local names for such pits include vasques, pia, gnamma, and Opferkessel. Much debate persists regarding which weathering processes are mainly responsible for these types of pits. Solution, hydration, frost weathering, salt weathering, and biochemical weathering associated with lichens have all been variously invoked to explain their formation. Nonetheless, it is likely that different processes are operative in different places, and that more than one process may be involved in any one location.

Various other weathering processes can also produce small-scale features, such as **grooves**, **flutes**, and **runnels**, on rock surfaces. These features are elongated or sinuous excavations on bare rock surfaces (or on buried

Figure 11.15 Weathering pits developed on sandstones, southeastern Utah, USA. Source: H. Viles.

rock surfaces). On karst surfaces, such grooves are known as various types of **karren** (**Fig. 12.14**). Nonetheless, they can also be found on other rocks such as granites and sandstones, where they can be in the order of 1 m wide and several meters long.

Another distinctive, small-scale weathering feature produced by the differential removal of material is **alveoli** or **honeycomb weathering** – a form of cavernous weathering similar to, but smaller than, **tafoni** (see below). Alveoli are centimeter-scale, closely spaced, cavities on a rock surface (**Fig. 11.16**). They are commonly found in sandstone,

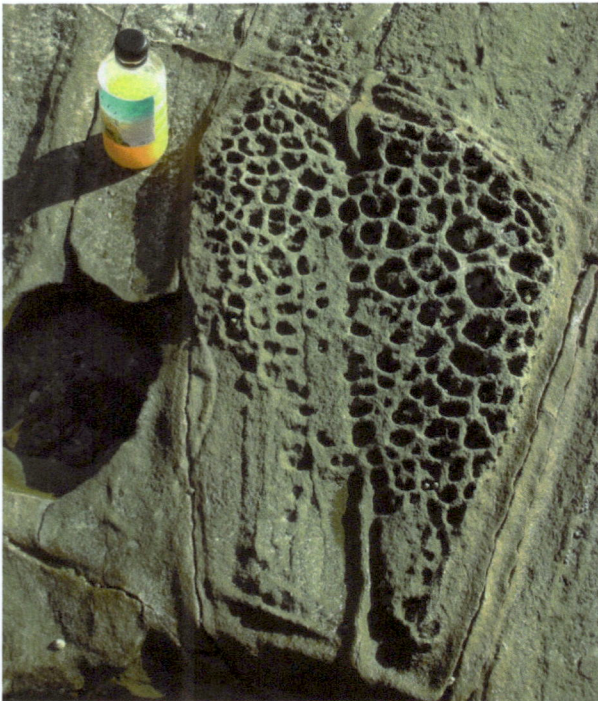

Figure 11.16 Alveoli carved into jointed sandstone on the south coast of Victoria, Australia. Source: H. Viles.

granite, and limestone surfaces in a wide range of environments, including arid (even in the Dry Valleys of Antarctica) and coastal settings. As with weathering pits, debate continues about the processes responsible for their formation, although salt weathering is commonly invoked, and case hardening may also play a role. Explanations for their formation need to consider why weathering becomes concentrated in regular, rounded zones, and why the intervening spurs remain in place.

11.7 DEPOSITIONAL FEATURES ASSOCIATED WITH WEATHERING

Weathering processes generate soluble byproducts and clastic debris, any of which can then accumulate to produce distinctive landforms. Good examples occur in karst terrain, as speleothems and **tufa** deposits. Solution of $CaCO_3$ (covered more fully in Chapter 12) is a reversible process, such that it can be redeposited when environmental conditions change. Many limestone rock faces exhibit these *secondary* calcite deposits in the form of **tufa curtains**, sometimes called aussen stalactiten. These features form where water saturated with $CaCO_3$ flows out of a rock opening, allowing CO_2 to be removed, leading to reprecipitation of secondary $CaCO_3$. Similar tufa deposits can form around springs or in lakes, where they are usually called **spring mounds** (**Fig. 11.17**).

Figure 11.17 A large, lens-shaped tufa deposit, formed as spring waters flow out of the limestone in the Napier Range, Kimberley, Western Australia. Source: H. Viles.

11.8 LANDFORMS PRODUCED BY WEATHERING

At the scale of meters to hundreds of meters, weathering also contributes to the production of a wide range of landforms, usually in conjunction with a host of other processes. Some of the features produced are highly distinctive and can be of conservation value in their own right, contributing to geoheritage and geotourism. A great majority of these features are **residual** – implying that their morphology develops as other material is removed by weathering and erosion.

11.8.1 Alteration Features

Analogous to the weathering rinds and varnish found at smaller scales, crusts are large-scale accumulations of indurated/cemented material within (or even on top of) soils (Fig. 5.33). Weathering is an important factor in the production of all crusts, as are inputs from dust, rain, plant residues, and dissolved substances in groundwater. Individual types of crust features can be identified based on their dominant mineralogy, such as **calcrete** (dominated by $CaCO_3$), **ferricrete** (dominated by iron oxides), **gypcrete** (dominated by gypsum), and **silcrete** (dominated by silica). All of these subsurface crusts can be variously cemented; some are completely indurated and resemble bedrock. Calcrete is most commonly formed in deserts where calcium-rich dust falls on the soil surface and is washed into the subsoil during infrequent rains. Over time, carbonates can accumulate at the typical depth of infiltration to form dense, subsurface crusts known as calcrete, or in the soil science literature, a **petrocalcic horizon** (Figs. 3.17, 11.18B). Duricrusts enriched in silica minerals are most common in climates with wet winters and dry summers, typical of the Mediterranean region, southern California, and parts of Australia. Iron-rich crusts, often referred to as **laterite**, are widespread in the wet–dry tropics, where iron-rich, red tropical soils are widespread. Iron in such soils can become mobile in the wet season but then as the soils dry out, the iron precipitates to form dense crusts that are so hard (after drying) that they are sometimes even mined as building materials (Fig. 11.18A).

All crusts that form in soils can, over time, continue to develop, becoming sometimes tens of meters (or more) thick. Many crusts take hundreds of thousands to millions of years to form. They are widespread, especially in the tropics, and can have many impacts on landscape evolution. Many landforms worldwide are held up by such crusts – the crusts form a hard cap on the surface that limits erosion (Fig. 11.18B).

11.8.2 Sculpted Features

Slopes formed on bedrock, including cliffs and steep escarpments, commonly exhibit a range of large features resulting from differential weathering and erosion. One example are tafoni (singular *tafone*) – cavernous weathering features that often form at the base of outcrops, but also can occur higher up. Tafoni can also form within large boulders (Fig. 11.19A). They occur on many rock types and in many different environmental settings, but are commonly found on granite and sandstone outcrops in arid and coastal areas. Tafoni are usually elliptical in shape, with concave inner walls, an archlike entrance, and an overhanging visor (Fig. 11.19).

Figure 11.18 Crusts developed within soils. **A.** Laterite, an iron-rich crust, is being mined here in Nigeria, for building materials (bricks). Source: J. Olson. **B.** The calcrete in this exposure near Las Vegas, Nevada, USA is so thick and indurated that it holds up this entire mesa. No, that's not bedrock – it's a soil crust! Source: B. Sekretarev.

Figure 11.19 Tafoni. **A.** A tafone in a granite boulder in the Anti-Atlas Mountains, Morocco. Source: H. Viles. **B.** Tafoni in sandstone near Sunset Bay, Oregon, USA. The tafoni here likely have been influenced by salt crystallization from sea water, brought in from the Pacific Ocean. Source: M. Miller.

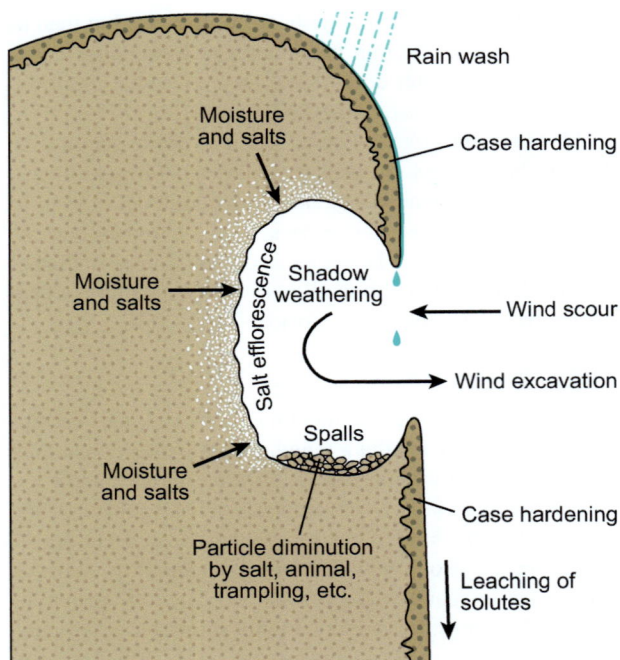

Figure 11.20 Diagram showing the many possible processes that might contribute to the formation of tafoni.

Figure 11.21 Wave Rock, ≈14 m high and 100 m long, is a famous example of a flared rock. Source: Kaliumfredrik at the English-language Wikipedia, CC BY-SA 3.0, via Wikimedia Commons.

Some geomorphologists refer to the process by which tafoni form as honeycomb weathering. The formation of tafoni is often ascribed to salt weathering, but a range of processes have been invoked. Indeed, tafoni appear to develop as intense granular disintegration and flaking occur locally, in rock interiors, in combination with case hardening and preservation of the arch and visor (**Fig. 11.20**). Tafoni take tens of thousands of years to form in most environments. Where they occur in groups with case-hardened rims standing between, they produce complex cliff-side features, sometimes called **niches** or **heads**.

At the base of outcrops on massive rocks, **foot caves** or **rock shelters** are often produced by weathering processes. Foot caves may be a form of tafoni, but they can also be produced by solution processes, for example, in towers in tropical karst. The caves result from the local intensification of weathering at the intersection of the rock outcrop and the rock surface at its base. Weathering intensification might happen, for example, where water becomes ponded or concentrated on an impermeable surface below, or where biotic weathering is enhanced by plants growing at the base of the outcrop.

Unlike the more extensive cave networks found in many karst terrains, foot caves are usually quite small (a few meters to tens of meters in diameter) and do not extend far back, into the rock. A similar feature, a **flared slope**, is a shallow basal concavity found on many granite outcrops. A famous example of a flared rock occurs at Wave Rock in Western Australia (**Fig. 11.21**). Features like these are (again) probably caused by the accentuation of subsurface chemical weathering at the base of the slope. Here, water becomes concentrated and thus, chemical weathering is enhanced (as is the capacity to remove weathering byproducts).

11.8.3 Residual Features

Some of the most awe-inspiring and distinctive weathering landforms are *residual* features that have been formed as a result of weathering processes interacting with a range of other processes. Where weathering and erosion have removed much material, residual landforms remain behind, as evidence of these processes. These types of features are often good examples of equifinality in geomorphology, in which any of several processes can produce the same type of landform. A similar statement could be made for many of the small-scale sculpted features that form by weathering of rock surfaces.

Perhaps the simplest residual weathering features are **rock pinnacles**, pillars, and towers which form in bedrock. Here, a combination of weathering and erosion exploit joints and other weaknesses, leaving only tall, isolated bedrock remnants behind, often capped by harder and more resistant rock. A good example is the sandstone pinnacles near Zhangjiajie, China (**Fig. 11.22**). These types of residual pinnacles, pillars, and towers often form in clusters, with their dimensions partially controlled by spacing of joints and other planes of weakness. Individual towers (or groups of towers) may be associated with the retreat and/or undercutting of an escarpment, as exemplified by the Finger of God ("vingerklip") in southern Namibia, which eventually collapsed, or the hoodoos at Bryce Canyon National Park in Utah, USA (**Figs. 8.21, 8.22**). In coastal settings, such features are often called **stacks**. Good examples include the Needles on the Isle of Wight, England, the Twelve Apostles, near Victoria, Australia, and other examples shown in **Fig. 17.33**. Weathering *and* erosion are required to produce these features, with the processes being most effective at the bases of the rock and along joints. As a result, the landform is considered a *residual* feature.

Figure 11.22 Pinnacles, tens of meters in height and carved into sandstone, near Zhangjiajie, China. Source: H. Viles.

Figure 11.23 An apse developed in sandstone, Zion National Park, Utah, USA. Over time, the apse will likely develop into an arch. Source: H. Viles.

Arches are complex, weathering-related features that form when weathering and wind erosion are followed by collapse to produce a gap through a competent, vertical slab of rock (see Chapter 8). They are particularly common in well-jointed sandstone, such as at Arches National Park in Utah, USA. Arches often develop from the widening of **apses** (also known as **alcoves**), as large slabs on a sandstone outcrop become detached along curved fractures, forming a recess near the base of the rock (**Figs. 8.25, 11.23**). Weathering caused by seepage of water into the sandstone then enlarges the recess. With continued weathering and erosion, the apse may break through to the other side and form an arch. Many of the largest and most impressive natural arches are found in the sandstones at Arches National Park, although the highest concentration of arches is thought to be found in the Tassili National Park in Algeria. Once formed,

natural arches are surprisingly stable. Similar features are also found along rocky coasts in sandstone, limestone, and granite.

Another, more complex, residual landform that forms due to long-term weathering is called a **tor**. Other, local, names for these features include **kopjes**, or **castle koppies** (in Africa). Tors are large, free-standing, residual masses of rock. They form commonly in upland areas on granitic rocks, but also in sandstone, schist, and dolerite. Tors are found on all continents, but the most well-known examples come from Dartmoor in southwestern England, where more than 150 tors occur within a 500 km² area. They vary markedly in appearance and shape, but their basic outline is controlled by the orientation, density, and curvature of fractures and joints within the rock (**Figs. 11.0, 11.24**). Most tors have three main fracture sets that form an orthogonal pattern which together

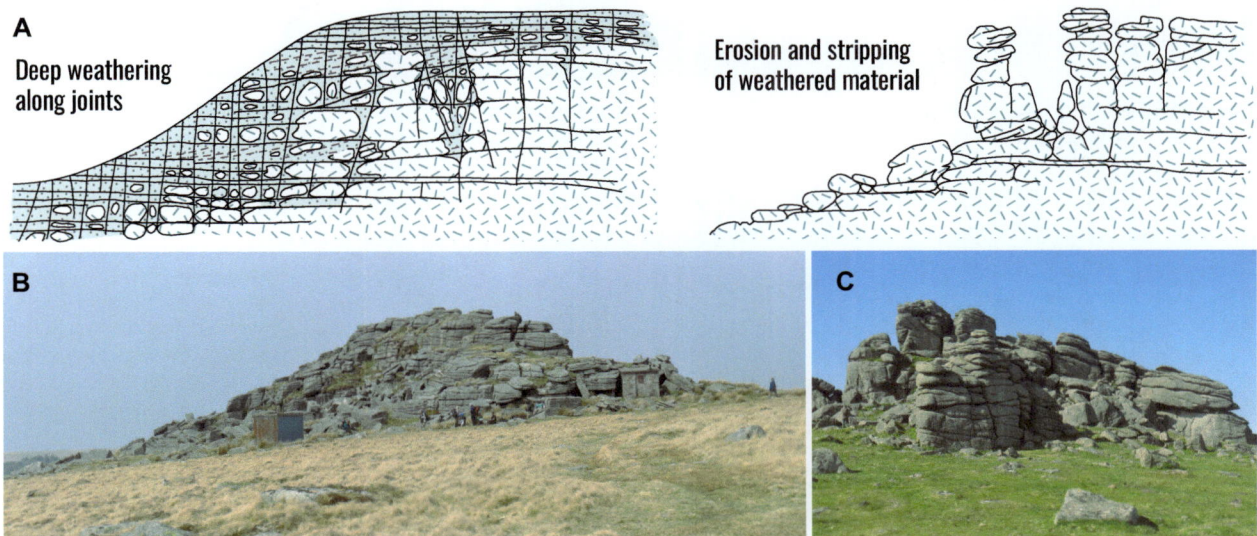

Figure 11.24 Tors. **A.** Theoretical illustrations of how tors might form in jointed rock. Note the displacement of some of the rocks. After Linton (1955), via Selby (1985), with permission from Oxford University Press. **B** and **C.** Tors in the Dartmoor region of the United Kingdom (B: Oke Tor. C: Hound Tor). Source: Nilfanion, CC BY-SA 3.0, via Wikimedia Commons.

guide erosion processes to produce a blocky, or castellated, morphology. Tors range from around 1 m in height and a few meters in diameter, to tens of meters in height and diameter. The process of **spheroidal weathering** is often important in the production of tors – where weathering produces concentric layers which then erode off like the skin of an onion, forming rounded boulders.

Several theories have been proposed to explain the formation of tors (**Fig. 11.24A**). Some geomorphologists have proposed that tors evolve along a two-stage process – initially driven by chemical weathering within deep regolith, but focused along joints. This episode is then followed by erosion, which preferentially removes the weathered rock along joints. Alternatively, a one-stage model proposed by Palmer and Neilson (1962) invokes solifluction and frost weathering under periglacial conditions as the main processes in tor formation. In this case, tors are relict features from a former climate. Third, Lester King proposed a scarp retreat theory to account for tors in southern Africa. He assumed that these tors were simply residual uplands left by erosion of less resistant granites. Arguments and debates still rumble on about the development of tors in any one area, but it is clear that many different combinations of process histories can produce them. And so, they are yet another good example of equifinality in geomorphology – different processes can form the same feature.

Like tors, **inselbergs** and **bornhardts** are large, residual, bedrock landforms that owe much of their origin to weathering and erosion. Both inselbergs and bornhardts are usually much larger than tors. The term inselberg (German, *insel*, "island," and *berg*, "mountain") was coined to describe isolated rock hills which rise from surrounding plains, as often seen in the savannas of eastern and southern Africa (**Fig. 11.25**). The name is also used for residual bedrock uplands/landforms of the Basin and Range Province of the western USA (see Chapter 10). Bornhardts are monolithic, dome-shaped hills, found in a range of topographic contexts. A world-renowned example of an inselberg is Uluru in Australia which is made of steeply dipping sandstone bedrock (**Fig. 5.27**).

Although inselbergs and bornhardts look similar, they have slightly different origins. Bornhardts are usually found in igneous rocks like granite and rhyolite. Their dome shape is thought to have developed as a result of differential weathering and erosion, often involving chemical weathering in deep regolith, followed by erosion of the weathering products. Inselbergs can be more varied in shape, from domed to castellated to bouldery.

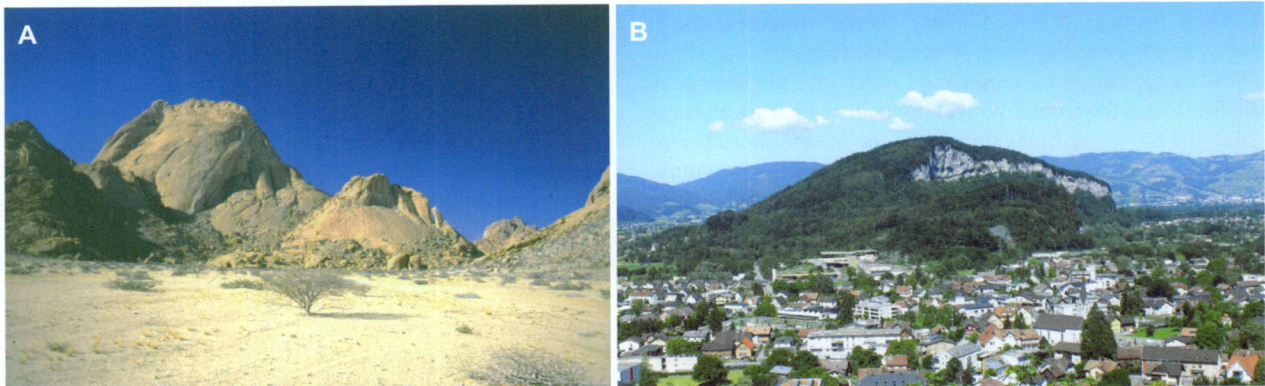

Figure 11.25 Inselbergs. **A.** Spitzkoppe, a granite inselberg in Namibia. Its upper surfaces are peppered with weathering pits. Source: H. Viles. **B.** Kummaberg, an inselberg near Koblach, Austria. Source: Asurnipal, CC BY-SA 4.0, via Wikimedia Commons.

Figure 11.26 Ruiniform landscapes. **A.** An aerial view of the sandstone ruiniform landscape of the Bungle Bungles, eastern Kimberley, Western Australia. Source: H. Viles. **B.** Walls alongside a narrow corridor in a sandstone rock city in Pennsylvania, USA. Source: R. Schaetzl.

One of the most impressive weathering features are **ruiniform landscapes**, also known as **rock cities** (**Fig. 11.26**). These curious features occur where widening of (usually vertical) joints by erosion across a large area of bedrock produces complex networks of slots and corridors, interspersed with pinnacles and towers. When formed on sloping surfaces or summits, the relict rocks can even slide slightly downslope, widening the fissures between them. Rock cities are essentially bedrock labyrinths. Excellent examples occur in quartzite in Roraima in Venezuela (where the towers are called "tepuis"), in the limestone Nahanni karst in northern Canada, and in the sandstones of Canyonlands National Park, Utah, USA, at Bohemian Paradise, Czech Republic, and the Stołowe Mountains of Poland. Large tracts of ruiniform landscapes also occur in the northern parts of Australia, such as the sandstone Bungle Bungles in the Kimberley region of Western Australia. In several of these examples, it is hypothesized that solution of quartz plays an important role in weathering and forming the openings.

Although no general theory of ruiniform landscape development has yet been accepted, geomorphologists point to the interplay of a range of different weathering and erosional processes in the formation of rock cities. Two fundamental preconditions are that the rocks (1) need to be mechanically strong enough to support tall, steep slopes, and (2) have a series of intersecting joints. Weathering processes such as frost and salt weathering are thought to be involved in the widening of the joints, as well as solution processes. Large-scale loss of rock in the form of rockfalls and other mass movements is also a critical process in many rock cities. For the ruiniform landscapes in Bohemia, Vaclav Cílek and colleagues (2007) have suggested a four-phase process of formation (**Fig. 11.27A**). Phase 1 begins with preparatory weathering of the rock as fluids circulate along joints. A second phase of formation, ascribed to cold conditions during the Pleistocene, involves removal of rock debris from the jointed areas. The third, mature phase, also during a cold paleoclimate, involves a range of additional weathering processes. In contrast, Jiri Bruthans and colleagues (2013) have proposed another theory. In this model, subsurface erosion from deeply circulating water in joints leads to *collapse* of the overlying rock along cavities (**Fig. 11.27B**). Undoubtedly, many different sets of processes operating over varying histories can produce very similar ruiniform relief – yet another case of equifinality!

At the landscape scale, weathering can also have an important impact on geomorphology. For example, physical weathering along rock faces and escarpments contributes to the production of coarse, angular debris which can accumulate at the base of the slope as **scree** or **talus** (**Figs.**

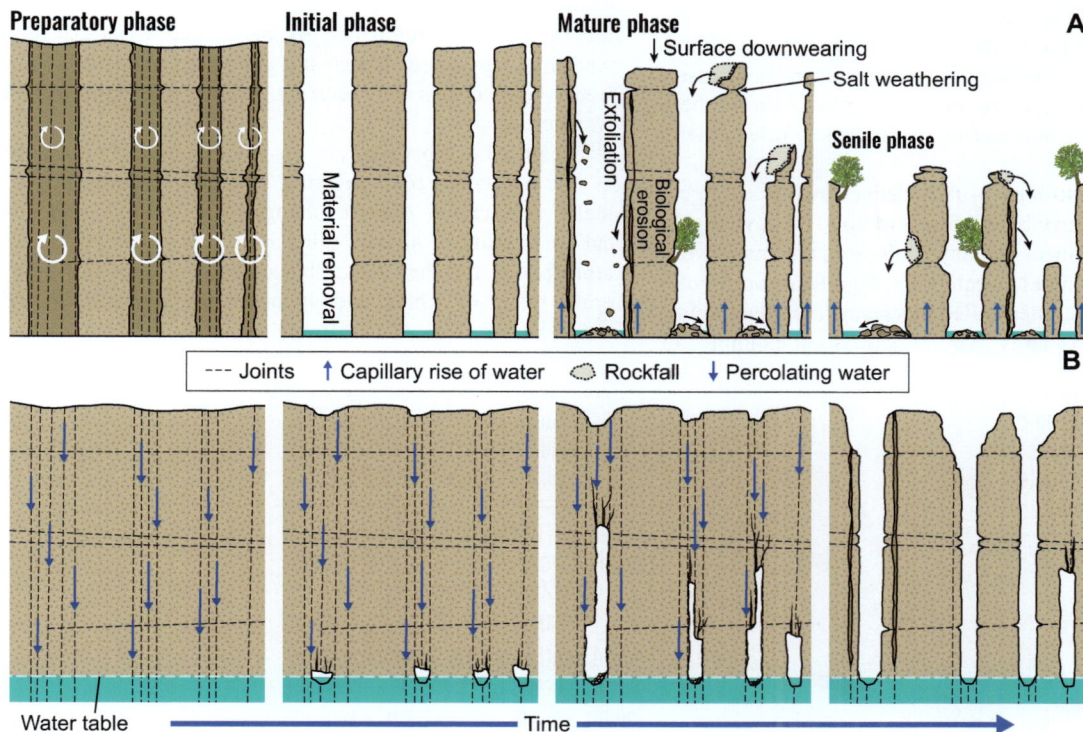

Figure 11.27 Schematic diagrams of two evolutionary models of the development of ruiniform landscapes in Bohemia. **A.** Sandstone weathering is aided by circulation of fluids deep in rock joints and crevasses, followed by removal of residuum by erosion. **B.** Erosion is focused on joints and fractures, along with upward propagation of voids developed at depth. The subsurface voids collapse, opening up wide fissures in the rock. Source: Used with permission of Elsevier, from Migoń, P., Duszyński, F. and Goudie, A. (2017); permission conveyed through Copyright Clearance Center, Inc.

Figure 11.28 Stone runs on Weddell Island, part of the Falkland Islands. Source: Kelperwitch, CC BY 3.0, via Wikimedia Commons.

8.7, 8.11, 14.14). Talus accumulates as a result of rockfalls, which can be triggered by events such as rain, snowstorms, or earthquakes (see Chapter 14). However, in most cases, long-term weathering is an important precursor to these rockfalls – with frost weathering playing a key role in many mountainous areas.

More widely, **boulders** (usually defined as rocks > 25 cm in diameter, with much bigger forms of up to 30 m wide) are an important component of many landscapes and may owe at least part of their nature to weathering. Significant accumulations of boulders are often called block streams, **blockfields**, felsenmeer, or **stone runs** (**Fig. 20.23**). Good examples of

stone runs can be found on the Falkland Islands in the southern Atlantic Ocean. Here, accumulations of quartzite boulders occupy valley floors (**Fig. 11.28**). Geomorphologists still debate whether these types of rocky surfaces are the product of frost weathering under periglacial conditions, or whether they formed through deep-seated chemical weathering under a warmer paleoclimate.

11.9 HUMAN IMPACTS ON WEATHERING

Humans influence weathering by changing the environmental conditions which cause weathering. Major human impacts on weathering are driven by air pollution and, increasingly, through anthropogenic climate change. Air pollution, especially when produced by combustion of coal and petroleum products, leads to enhanced acidity in rainfall. The sulfuric and nitric acids in "acid rain" can dramatically enhance chemical weathering. For example, within the Bohemian Paradise Geopark in the Czech Republic, the weathering of sandstone outcrops and rock arches has been accelerated by decades of air pollution, driven by the burning of coal. Additionally, the burning of fossil fuels produces carbon-rich particulate material which can coat rock surfaces.

Land-use changes can also have large effects on weathering. For example, because all irrigation water contains some amount of salt, irrigated agriculture in many desert areas has accelerated salt weathering in soils and on rocks. Climate change is also undoubtedly having an impact on weathering in many environments. As periglacial areas are now warming, permafrost is thawing and more frequent diurnal freeze–thaw cycles are occurring, leading to enhanced frost weathering.

Weathering processes are also of potential importance as one of many geoengineering *solutions* to anthropogenically accelerated climate change. For example, chemical weathering of silicate minerals removes CO_2 from the atmosphere and then locks that carbon in secondary minerals, many of which end up becoming sediment in the ocean basins (**Fig. 11.29**). In theory, as the climate warms,

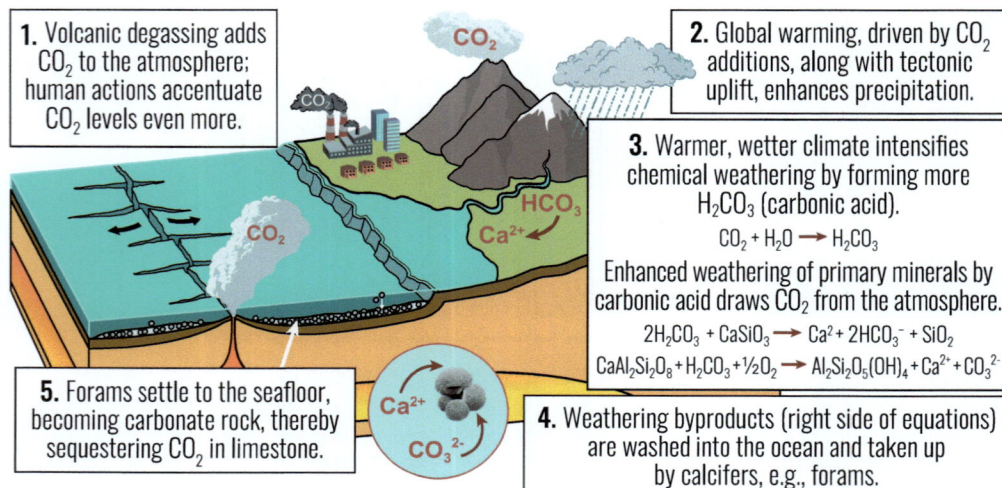

1. Volcanic degassing adds CO_2 to the atmosphere; human actions accentuate CO_2 levels even more.

2. Global warming, driven by CO_2 additions, along with tectonic uplift, enhances precipitation.

3. Warmer, wetter climate intensifies chemical weathering by forming more H_2CO_3 (carbonic acid).

$$CO_2 + H_2O \longrightarrow H_2CO_3$$

Enhanced weathering of primary minerals by carbonic acid draws CO_2 from the atmosphere.

$$2H_2CO_3 + CaSiO_3 \longrightarrow Ca^{2+} + 2HCO_3^- + SiO_2$$
$$CaAl_2Si_2O_8 + H_2CO_3 + \tfrac{1}{2}O_2 \longrightarrow Al_2Si_2O_5(OH)_4 + Ca^{2+} + CO_3^{2-}$$

4. Weathering byproducts (right side of equations) are washed into the ocean and taken up by calcifers, e.g., forams.

5. Forams settle to the seafloor, becoming carbonate rock, thereby sequestering CO_2 in limestone.

Figure 11.29 Simplified model of how carbon circulates within the major Earth systems, and how weathering processes are involved in this cycle.

the rate of these reactions will increase and thus, more carbon will be removed from rocks and sent to the ocean basins. Once in ocean-bottom reservoirs, this carbon, tied up in carbon-rich minerals, is effectively locked up and removed from the atmosphere for long periods of time. More details about this weathering system are shown in **Fig. 11.29**. Scientists have proposed that this effect could be manipulated through, for example, spraying finely ground basalt over large areas of land in order to enhance the drawdown of atmospheric CO_2, and thereby mitigating global warming. Unfortunately, because this process operates over millions of years, it is likely too weak and slow to offset contemporary global warming over human lifespans. However, on geologic timescales, this process has global implications.

11.10 HAZARDS ASSOCIATED WITH WEATHERING

Despite its often slow, imperceptible rates of operation, weathering can sometimes pose a hazard to human societies. Rock weathering can facilitate mass movements such as rockfalls and landslides, which can have devastating consequences. Mass movements are often triggered by events such as severe rainstorms or earthquakes, but weathering often acts as an important conditioning factor, making rock slopes more prone to failure in the event of such storms or tectonic processes (see Chapter 14).

Weathering is also hazardous when it affects buildings and structures, causing deterioration and, in some cases, collapse. Some of the world's most valuable cultural heritage buildings and structures are threatened by salt weathering, including the UNESCO world heritage sites of Kiva, Bukhara, and Samarkand in Uzbekistan. All of these sites have been impacted by irrigation-induced rises in groundwater, which have brought soluble salts into contact with vulnerable stonework. Salt weathering is also increasingly a problem in coastal cities such as Venice, Italy, where high rates of sea level rise are coupled with local subsidence as a result of groundwater extraction.

Weathering is also problematic in that it reduces the value of the cultural heritage, sometimes leading to the erasure of important carvings, sculptures, and other important historical artifacts. Air pollution is causing enhanced weathering in cities, changing the fabric of ancient and modern buildings and structures. For example, many of the buildings in the historic district of Budapest, Hungary, were constructed from a soft, porous limestone which has been badly affected by enhanced weathering under acidified, particulate-rich environmental conditions caused by the decades-long burning of coal (**Fig. 11.30**).

REVIEW QUESTIONS

11.1   What are the three major types of weathering and how does each generally operate?

11.2   What is the difference between weathering and erosion?

11.3   What is meant by Earth's critical zone and how does weathering fit into studies of the critical zone?

11.4   Define and differentiate among the terms regolith, residuum, and saprolite.

11.5   Explain what may make some rocks and minerals more susceptible to weathering than others. How does the thickness/thinness of regolith influence the rates of physical and chemical weathering?

11.6   What are some of the ways that rocks can break up, physically?

11.7   Explain how frost weathering and salt weathering operate to break rocks apart. In what types of environments are each most common?

11.8   What is unloading, what drives this process, and what is the main type of landform formed by unloading?

11.9   What are the major outputs of chemical weathering?

11.10 What are case-hardened layers and how do they form?

11.11 What are some typical small-scale weathering features and how do they form?

11.12 What are the main types of larger-scale weathering landforms? Compare and contrast them.

11.13 Why is it difficult to explain the genesis of larger-scale weathering landforms such as weathering pits and tafoni?

11.14 What are weathering-limited and transport-limited slopes and how does the nature and importance of weathering differ between them?

11.15 What are some of the more common types of crusts formed in soils, and for each, what is the cementing agent?

11.16 What is meant by equifinality in geomorphology? Give at least one example.

11.17 What is a tor and how might it have formed? How do tors differ from inselbergs? From bornhardts?

11.18 Describe some of the main theories of ruiniform landscape development and explain why it is so difficult to explain what role weathering plays in their formation.

Figure 11.30 In Budapest, Hungary, blackened and weathered limestone facades (right) are being cleaned and restored (left), as a means of enhancing tourism. Source: R. Schaetzl.

11.19 What are some ways that human activity is affecting weathering?

11.20 Describe how weathering influences Earth's long-term carbon cycle, and how weathering relates to global climate change in a geologic context.

11.21 What are some common weathering hazards and why are they important?

FURTHER READING

Anderson, R. S. and Anderson, S. P. 2010. *Geomorphology: The Mechanics and Chemistry of Landscape*. Cambridge University Press.

Bland, W. and Rolls, D. 1998. *Weathering: An Introduction to the Scientific Principles*. Arnold.

Goudie, A. S. (ed.). 2004. *Encyclopedia of Geomorphology*. (2 vols.). Routledge.

Goudie, A. S. and Viles, H. A. 1997 *Salt Weathering Hazards*. Wiley.

Goudie, A. S. and Viles, H. A. 2010. Weathering hazards. In: Alcantara-Ayala, I. and Goudie, A. S. (eds.) *Geomorphological Hazards and Disaster Prevention*. Cambridge University Press, pp. 145–159.

Migon, P. 2006. *Granite Landscapes of the World*. Oxford University Press.

Migoń, P., Duszyński, F., and Goudie, A. 2017. Rock cities and ruiniform relief: Forms–processes–terminology. *Earth-Sci. Rev.* 171:78–104.

Young, R. W., Wray, R. A. L., and Young, A. R. M. 2009. *Sandstone Landforms*. Cambridge University Press.

12 Karst and Landforms of Dissolution

Jo De Waele

Water, in all its forms, is the most important agent responsible for shaping the landscape. Some water is at the surface in rivers and lakes (surface water), but much of it eventually penetrates underground. Groundwater, present in the pore spaces of soil, regolith, and bedrock, plays a fundamental role in our lives, and (a focus of this chapter) in the dissolution of bedrock, which is perhaps the most important geomorphic effect of groundwater. Because all rocks are at least partially soluble, parts (or all) of them will dissolve and go into solution when exposed to water and its associated acids – the essence of dissolution (**Fig. 12.1**).

12.1 WHAT IS KARST?

Dissolution of particular types of rocks gives rise to a terrain so distinctive that it is given a special name – **karst**. Humankind has lived on or near karst landscapes all around

the world, since prehistoric times. For example, early humans used caves as shelters, as well as for rituals and burials. And as we shall see later in this chapter, karst landscapes are some of the most dynamic – and vulnerable – on Earth.

Karst is the Germanized form of "carso" or "kras," a term used to describe the region between Trieste, in northern Italy and Slovenia. These words derive from the pre-Indo-European word "karra," meaning "stone" or "barren stony ground." The Balkan region of southeastern Europe has long been regarded as the "classical karst" landscape, and it is here that much of the pioneering karst investigations were developed in the nineteenth century (**Fig. 12.2**).

Karst is a terrain with distinctive geomorphology and hydrology, formed through a combination of high rock solubility and well-developed secondary (fracture) porosity. By definition, karst can occur in any kind of rock or geologic context, as long as the landforms are driven by dissolution. Our understanding of karst formation may even apply to extraterrestrial bodies like Mars, where dissolution may have shaped areas of both its surface and subsurface.

Karst landscapes form where soluble bedrock is close to the surface, giving rise to classical karst features such as

Figure 12.1 These meter-long soda straw stalactites in Valdemino Cave in northern Italy illustrate that the rock above is undergoing dissolution, and that the dissolved products are now being precipitated in the **cave**. The inset shows the tip of a soda straw, with its hanging water droplet. Note the transparent calcite crystals growing along the edge of the drop. Source: V. Balestra.

Figure 12.2 A 1689 copper engraving of Adelsberger Grotte by Valvasor, which today is Postojna Cave in Slovenia, one of the most famous show caves in Europe. The oldest writings on the cave walls date to 1213, confirming that visits to the cave have been occurring for centuries. Source: Artwork in the public domain.

Figure 12.0 Buso del Vallon Cave in the Lessini Mountains of northern Italy has an impressively deep entrance. Source: S. Sedran S-Team.

sinking streams, caves, closed depressions, and fluted rock outcrops (karren). Most rocks on Earth are variously soluble in water, but only a few are so soluble that they produce karst. Extensive karst landscapes mainly develop on **carbonate rocks** such as limestone, dolostone, and marble. Many people think of karst as being restricted to limestone landscapes. Nonetheless, over long periods of time, even poorly soluble rocks such as quartz sandstone can develop karst morphologies, such as in Venezuela, Australia, and Brazil. Karst features can develop rapidly in highly soluble evaporite rocks such as gypsum, anhydrite, and halite, but these rocks occupy less than 5% of Earth's land surface. Most of the world's karst landscapes are therefore developed on limestones and dolostones, which occupy ≈15% of the ice-free land surface of Earth (**Fig. 12.3**).

Karst landforms are worthy of study not only because of their dramatic form and unique origins, but also because of their abundance. Overall, about a quarter of Earth's surface is underlain by rocks that show evidence of dissolution (**Fig. 12.3**). Karst is particularly abundant in the northern hemisphere, where soluble rocks crop out on over ≈20% of the land surface. South America, Africa, and Australia contain smaller and more scattered outcrops of karst (except for the Nullarbor, in southern Australia, the most extensive karst area in the world).

Not all carbonate rocks are susceptible to dissolution. For example, thinly bedded alternations of limestones and shales or marls hinder the formation of karst. Densely fractured limestones also do not allow for the underground drainage necessary to form a karstic flow network. Important to karst formation are favorable combinations of rock lithology and structure, as well as ample water to dissolve the rock, which is why karst is not active in areas of permafrost.

12.2 CARBONATE ROCK DISSOLUTION

Dissolution is the primary process that forms karst. During dissolution, rocks transform into ions that are transported away as solutes, forming cavities in the rock and beginning the formation of karst. Dissolution of halite (NaCl) and gypsum ($CaSO_4 \cdot 2H_2O$) does not need any acid exposure, because they dissolve by simple dissociation in water. Carbonate rocks require acids for dissolution. Groundwater, surface water, and rainwater all contain small amounts of dissolved CO_2, which forms weak carbonic acid:

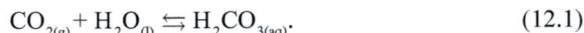

$$CO_{2(g)} + H_2O_{(l)} \leftrightarrows H_2CO_{3(aq)}. \tag{12.1}$$

Carbonic acid will readily dissociate into its cations (H^+) and anions (bicarbonate, HCO_3^-):

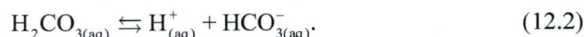

$$H_2CO_{3(aq)} \leftrightarrows H^+_{(aq)} + HCO^-_{3(aq)}. \tag{12.2}$$

Calcite ($CaCO_3$), the main mineral in limestone, dissociates according to the reaction

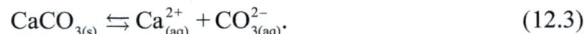

$$CaCO_{3(s)} \leftrightarrows Ca^{2+}_{(aq)} + CO^{2-}_{3(aq)}. \tag{12.3}$$

The H^+ of equation (12.2) combines with the carbonate anion (CO_3^{2-}) of equation (12.3) to form another bicarbonate anion:

$$CO^{2-}_{3(aq)} + H^+_{(aq)} \leftrightarrows HCO^-_{3(aq)}. \tag{12.4}$$

The overall process of dissolution of calcite by carbonic acid is summarized by the reaction

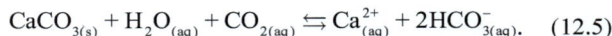

$$CaCO_{3(s)} + H_2O_{(aq)} + CO_{2(aq)} \leftrightarrows Ca^{2+}_{(aq)} + 2HCO^-_{3(aq)}. \tag{12.5}$$

and for dolomite

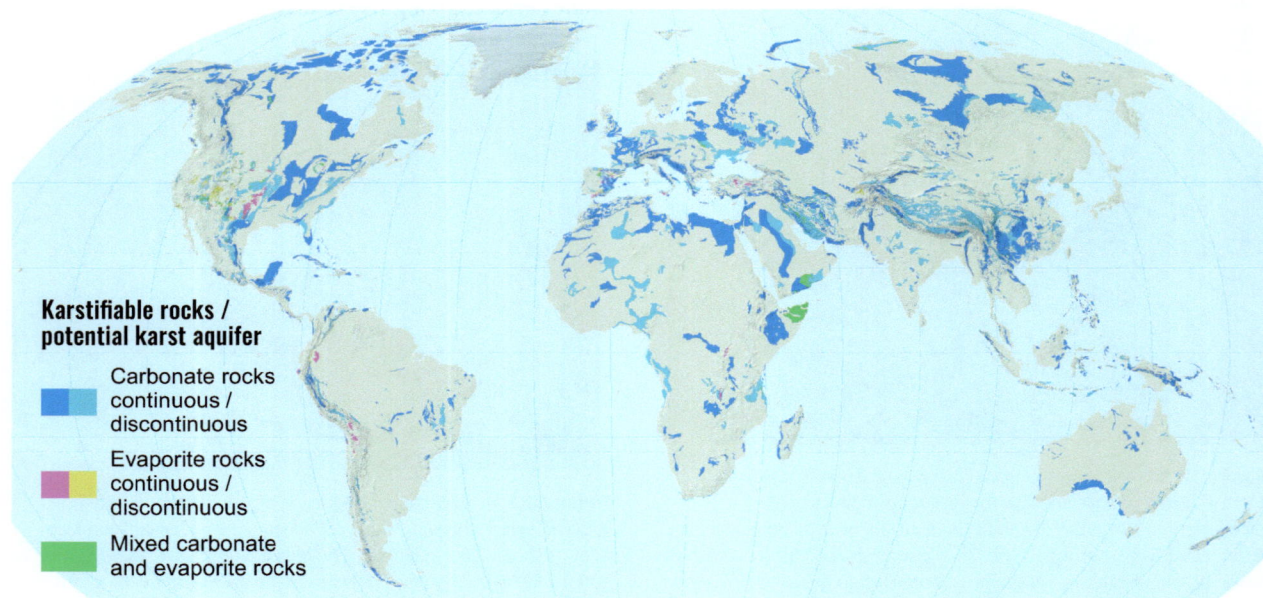

Karstifiable rocks / potential karst aquifer

- Carbonate rocks continuous / discontinuous
- Evaporite rocks continuous / discontinuous
- Mixed carbonate and evaporite rocks

Figure 12.3 Map of areas with potential karst **aquifers**. Source: Courtesy of WHYMAP.org – World Karst Aquifer Map (2017).

$$CaMg(CO_3)_{2(s)} + 2H_2O_{(aq)} + 2CO_{2(aq)}$$
$$\leftrightarrows Ca^{2+}_{(aq)} + Mg^{2+}_{(aq)} + 4HCO^-_{3(aq)}. \qquad (12.6)$$

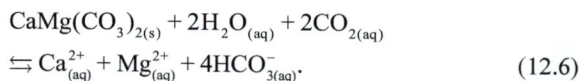

The upshot of these reactions is that the solubility of $CaCO_3$ is mainly controlled by the amount of CO_2 that can be made available to water that is in contact with carbonate rock, because CO_2 is necessary to form the carbonic acid.

Importantly, the two (four for dolomite) bicarbonate (HCO^-_3) ions come from different sources. One (two for dolomite) comes from the dissociation of carbonic acid (equation 12.4), whereas the other(s) form by the reactions shown in equations (12.5) and (12.6). The reaction shown in equation (12.5), also known as carbonation (see Chapter 11), is essential to understand the dissolution of carbonate rocks (**Fig. 12.4**). The reaction moving to the right illustrates carbonate weathering (dissolution), whereas the one moving to the left represents the precipitation of carbonate, which can occur in the oceans, in sedimentary rocks (as carbonate cement or calcareous tufa), or in caves (forming speleothems). Carbonate dissolution is an atmospheric carbon sink, because it consumes molecules of CO_2 that had previously resided in the atmosphere, whereas carbonate precipitation releases this greenhouse gas. The karst process is thus a fundamental part of the global carbon cycle.

The carbonate–water–CO_2 system is in reality more complicated, involving a series of mutually interdependent, reversible reactions. In general, dissolution of carbonate rock consumes CO_2 (equations 12.5 and 12.6), which causes more CO_2 to dissolve in the water, and thus more carbonate to dissolve whenever the solvent (water) remains in contact with CO_2-containing air. Thus, dissolution of carbonate rock primarily depends on the amount of CO_2 actually dissolved in the water, which in turn depends on the partial pressure of CO_2 (P_{CO_2}) in the air that is in contact with the water, as well as temperature. Importantly, the partial pressure of CO_2 is higher in soil air (normally between 1% and 2%, but up to 10% in tropical soils) than it is in caves (between 0.05 and

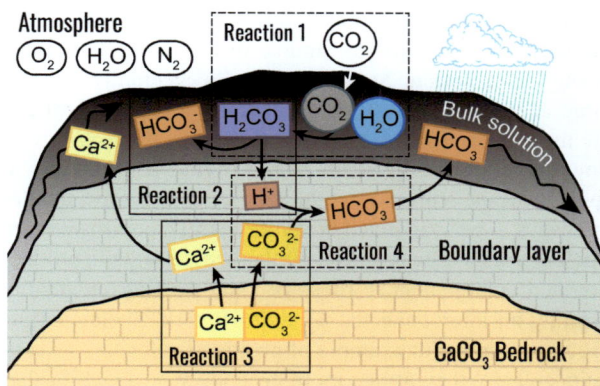

Figure 12.5 Typical CO_2 partial pressure values (concentrations) in the atmosphere, in soil air, and in caves.

1%), or in the atmosphere (0.042%) (**Fig. 12.5**). The higher CO_2 concentrations in soil air are largely due to the respiration of plant roots and to the decomposition of soil organic matter.

Simply stated, the karst system is as follows. Rainwater, which acquires low contents of dissolved CO_2 from the atmosphere, percolates through the soil, acquiring more CO_2. Through this process, the percolating water accumulates carbonic acid, which may then react with and dissolve soluble carbonate rock.

Carbon dioxide dissolves more readily into cold water than warm water, at any given partial pressure. Cold water is thus more aggressive toward carbonate rocks. Recall, however, that much of the CO_2 used for dissolution comes from soil air, and CO_2 production is greater in soils in warmer climates. Additionally, reaction rates of the different steps in carbonate dissolution also increase with temperature, whereas water viscosity decreases. As a result of these complex temperature effects, karst is usually better developed in warmer climates, as long as ample water is available.

Most dissolution processes occur within ≈10 meters of the surface. Deeply buried rocks dissolve more slowly because they are farther from the main source of CO_2 (soil air). The uppermost limestone below the soil is thus much more dissolved, contrasting with the less dissolved, lower-permeability rock below (usually non-weathered limestone). This highly permeable, weathered, and "karstified" upper part of the rock, normally 10–30 meters thick, is called **epikarst** (**Fig. 12.6**). Water moves through the large openings and fractures in epikarst, into the less permeable rock below.

As mentioned above, water is most able to dissolve rock when it contains high concentrations of CO_2, a situation that develops as water percolates through the soil. The ability of water and acids to dissolve rock is referred to as **solutional aggressiveness**. Percolating water loses some of its aggressiveness at depth, as it moves through the regolith (farther from the soil) and approaches the bedrock. However, this loss of aggressiveness is in part counterbalanced by a

Figure 12.4 Schematic illustration of the main steps involved in the dissolution of limestone in CO_2-bearing waters. This figure shows a limestone upland in contact with a static boundary layer, and water in flowing bulk solution, which is in contact with the atmosphere.

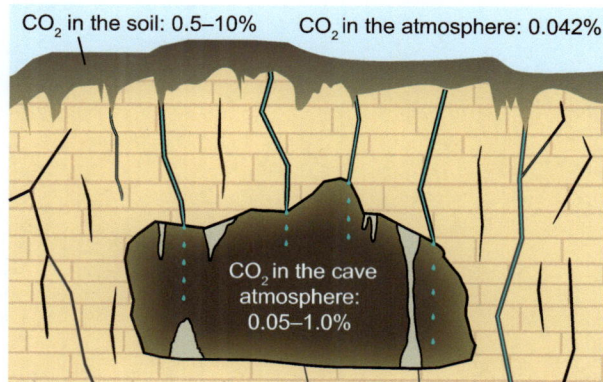

Figure 12.6 Epikarst in a limestone quarry along the borders of Popovo Polje in Bosnia and Herzegovina. Source: A. Palmer.

process known as **mixing corrosion**. This phenomenon occurs because the solubility of $CaCO_3$ in CO_2-containing waters is a non-linear function (**Fig. 12.7**). Mixing corrosion, sometimes called the "mixing effect," can be stated as follows: the mixing of two $CaCO_3$-saturated solutions with different CO_2 contents will give rise to a new solution that is undersaturated with respect to $CaCO_3$.

Let's look at **Fig. 12.7** for further elaboration on this important process. Each solid line represents water in equilibrium between the amount of dissolved CO_2 it is holding (which in turn indicates how much H_2CO_3 it can produce) and how much $CaCO_3$ it can dissolve. These lines represent **saturation curves**, because water falling along these lines is saturated with $CaCO_3$. Note the positive relationship displayed by the solid lines in the figure – more CO_2 in the

water implies that more $CaCO_3$ can be dissolved. By comparing the blue, red, and green lines, one can also see that colder water is capable of both dissolving and holding more $CaCO_3$ in solution than warm water, as discussed above. Water resting below the line (at its respective temperature) is aggressive because it is **undersaturated**. This water is capable of dissolving more $CaCO_3$ (if available). Water above the line – supersaturated water – does not stay that way for long. Instead, it quickly precipitates some of its $CaCO_3$ as secondary carbonates, which can then be manifested as cave deposits (speleothems). Key to this graph is the realization that any two waters, when mixed, become undersaturated. For example, **Fig. 12.7** shows that the mixing of equal amounts of solution A (with low amounts of dissolved calcite), and B (with more dissolved calcite) at 20 °C (along the red line) will produce a new solution (D), which will fall *under* the saturation curve, that is, the red line. This situation occurs even though both solutions were initially saturated with dissolved $CaCO_3$. Thus, the mixed solution (D) will be undersaturated. More $CaCO_3$ will thus be dissolved, consuming CO_2 (in these typical closed system conditions) following the pink arrow until saturation is reached at point C. Although important, the mixing effect is rather small – generally around 1–2% in natural waters.

The amount of $CaCO_3$ dissolved in water depends not only on the concentration of CO_2 in the water, but also on the amount of time available for water–rock interactions to occur. In very slowly percolating water, solutional byproducts are not carried away rapidly enough to allow dissolution to continue. Dissolution occurs faster in waters that are moving more rapidly, allowing the reactions to continue. However, if the water is flowing too rapidly, less time exists for reactants to actually dissolve the limestone. Rapidly flowing waters can thus remain undersaturated with respect to $CaCO_3$, even over long distances. Turbulence in these faster flowing waters can, however, facilitate the erosion and transport of clastic, sedimentary particles. In such instances, physical erosion dominates over chemical (dissolution) processes.

To summarize the karst process, water percolating through narrow fissures as laminar flow enlarges the cracks through dissolution. Some cracks will eventually become large enough to sustain turbulent flow, which can enlarge the fissures even faster by physical erosion. This change of flow conditions corresponds to the **breakthrough time** in speleogenesis, representing the time it takes for the smaller pathways to be abandoned in favor of larger, dominant ones. **Speleogenesis** refers to the natural processes of void production and enlargement in rocks of karst terrains, eventually forming an interconnected conduit network. Speleogenesis provides a splendid example of positive feedback – as conduits enlarge, they attract more groundwater, which further enlarges them, and so on.

Karst development processes are clearly controlled by climate. Dissolution is severely hindered in both arid and cold climates, because of the lack of liquid water and the low amounts of biogenic CO_2. Halite (NaCl salt) karst, however, can survive there, even though dissolution occurs only during

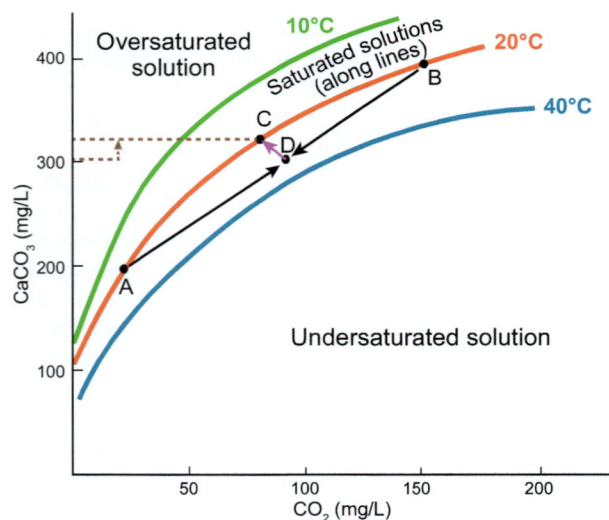

Figure 12.7 Graph of mixing corrosion, or the mixing effect, for waters of varying $CaCO_3$ and CO_2 concentrations, and at different temperatures. In waters at 20 °C, the mixing of solutions A and B results in a new solution (D). Once formed, solution D, because it is undersaturated, will dissolve more $CaCO_3$ and consume some CO_2 (dashed lines) before reaching chemical equilibrium at point C.

intermittent and rare precipitation events. Examples include the Atacama Desert in Chile, the Zagros Mountains in Iran, and on Mt. Sedom, on the shores of the Dead Sea in Israel.

Carbonate karst is actively forming today in both temperate and tropical areas. Limestone dissolution rates mainly depend on precipitation totals (climate), the amount of surface runoff, and the thickness of the soil and the vegetation cover. Temperature influences not only CO_2 solubility (**Fig. 12.7**), but also the reaction rates of the process. Temperature also mediates the amount of water available belowground, by affecting evaporation (from surface waters) and transpiration (from plants). Nonetheless, evaporation amounts appear to be less important than the quantity of water available for dissolution, which correlates well with runoff totals. Landscapes where more water runs off generally also have higher precipitation totals and less evapotranspiration, implying that more water is also available for dissolution. On landscapes with similar amounts of runoff, carbonate dissolution is more efficient at the cooler sites. Although karst is affected by climate, the differences are not significant enough to facilitate a climatic classification of karst.

12.3 HYDROGEOLOGY OF KARST LANDSCAPES

Groundwater flow in karst aquifers occurs primarily through conduits and enlarged fractures. Karst aquifers are extremely dynamic and vulnerable to contamination because of the rapid transfer of water from the surface to the subsurface, and from recharge to discharge areas. Pollutants that enter the aquifer from surface sites make their way into the aquifers quickly, often with little or no filtration. For these reasons, land uses in karst areas must be monitored and regulated more than in most other landscapes.

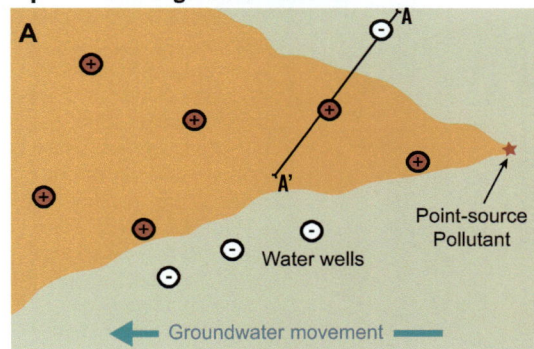

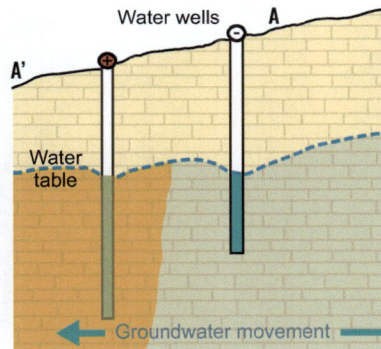

Rocks in karst terrain are characterized by highly irregular and complex flow through a maze of different conduits. The more an aquifer is **karstified**, that is, developed fractures, fissures, and large openings by dissolution, the less it can be described using traditional groundwater flow models. Well-developed karst aquifers can actually have triple porosity, that is, a primary matrix porosity, a secondary porosity encompassing more or less enlarged fractures, and karst conduits. Water flow velocities and hydraulic regimes in such rock will vary widely, from very slow in matrix porosity, faster in fractures, to very fast (even similar to surface streams) in large karst conduits and caves. Therefore, the water table concept, as used in more homogeneous rocks and regolith, does not work as well in karst aquifers (**Fig. 12.8**). As a result, the **water table** in karstified rocks is very irregular and dynamic, with the conduit system showing rapid responses to hydraulic changes across time and space. Springs in very well-developed karst aquifers exhibit rapid changes in flow rate and geochemistry in response to recharge events, showing changes in discharge of up to three orders of magnitude (from 1 to 1,000 liters/second) over only a few hours.

The heterogeneous behavior of karst aquifers makes them difficult to study and model. Besides the classical geological, geomorphological, and structural analyses, a combination of methods must be undertaken to fully understand water flow directions and rates. The best way to understand

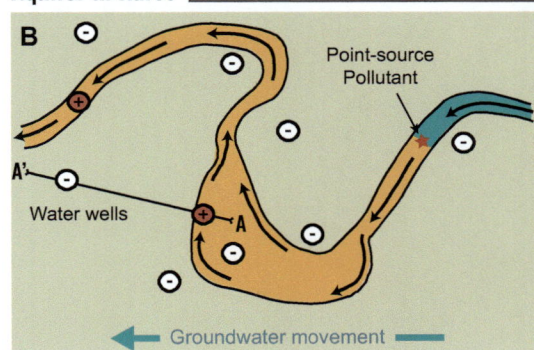

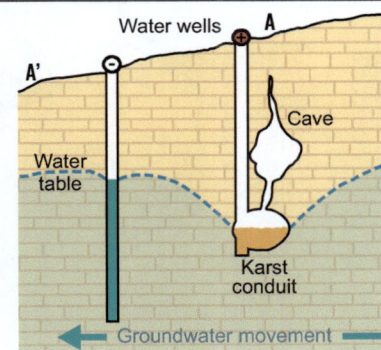

Figure 12.8 Groundwater flow in homogeneous porous media (**A**) vs heterogeneous karst aquifers (**B**), showing how a point-source pollutant (red star) behaves in each system. Pollutants in porous aquifers spread radially downstream, contaminating water wells in that direction, following a clear "plume" (The cross-section does not show the surface contamination, but only the contaminant water in the aquifer.) Contrast that to karst aquifers with large conduits and passageways, where pollutants are transported rapidly by and through the conduit system. Wells downstream, but far from such passageways, may not be contaminated (initially).

Figure 12.9 Injection of the yellowish-green tracer "fluorescein" in an underground (cave) river during a karst hydrogeology study. Dyes like this one are the standard way to trace the flow of water within karst aquifers. Because of the heterogeneity of karst aquifers, it is not unusual for nearby wells and springs to be unaffected by dye injections, while outlets farther away show prompt responses. Source: V. Crobu.

the complex behavior of a karst aquifer might be to include hydro-chemical monitoring of springs, chemical analyses of waters, flow rate measurements, and multi-tracer tests. Such tests would need to be undertaken under different hydrological conditions (**Fig. 12.9**).

12.4 KARST AT THE LAND SURFACE

Karst landscapes are unique in that they display a suite of landforms that derive, at least in part, from dissolution of bedrock. Dissolution can occur by descending meteoric waters (epigenic karst, **Fig. 12.10A, D**), rising fluids (hypogene karst, **Fig. 12.10B**), or in fresh–salt water mixing conditions (e.g. coasts, **Fig. 12.10C**) (see details in Section 12.7.5). Solution is often not the only active process; it can be subordinate to other processes such as fluvial or glacial erosion in some areas. True karst, or **holokarst**, is a landscape where dissolution is the primary, active, land-forming process. Holokarst landscapes are what most geomorphologists envision when they think of karst. In **fluviokarst** and **glaciokarst** terrains, erosion by running water or glaciers dominates over dissolution, respectively.

Dissolution of rock is most active at (and close to) the surface, where waters are still low in dissolved minerals but high in CO_2 – so-called epikarst (**Fig. 12.6**). Dissolution also occurs at and near the water table, due to the mixing effect. Karst surface features are unique, making these landscapes among the most beautiful on Earth, attracting large numbers of tourists (**Fig. 12.11**).

Typical of karst landscapes is the presence of bare rock surfaces; recall that the word "kras" derives from this type of landscape. Unlike most rocks, where weathering produces regolith that can later become soil, dissolution of almost pure limestones can completely remove the rock in solution. This situation leaves almost no weathering residue and only a thin soil cover. This non-soluble residue is often composed of oxy-hydroxides of iron and aluminum, as well as any clay minerals that may have been present in the limestone, forming a residual red sediment/soil known as **terra rossa** (**Fig. 12.12**). Terra rossa regolith gives many soils in karst landscapes their distinct red hue.

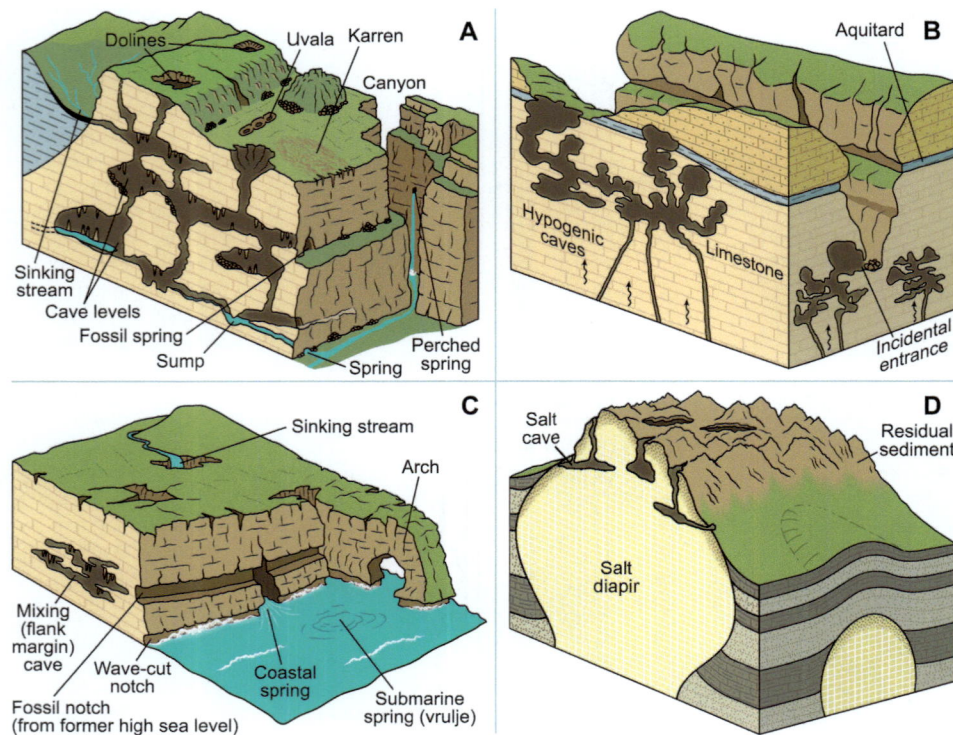

Figure 12.10 Schematic representations of different types of karst landscapes. **A.** An epigenic holokarst landscape, with dolines, sinking streams, springs, and caves. **B.** A hypogene karst system intersected by subsequent fluvial erosion. Note the lack of karst features at the land surface. **C.** A **coastal karst** system, with both epigenic imprints (sinking streams and dolines) but also mixing corrosion features at the interface between the fresh groundwater and the denser, salty water below. **D.** Karst forming in a salt diapir. Source: Original drawings by V. Chiarini.

Figure 12.11 Karst landscapes. **A.** Fengcong karst in Guangxi Province, China. Note the large underground river passage, the broad corrosion plain, and the residual valleys and cones. Source: A. Klimchouk. **B.** Fenglin (tower) karst in Yangshuo area (Guilin), around River Li in China. Source: T. Waltham. **C.** The karst landscape of Guilin has inspired artists since the Tang Dynasty (618–907 AD), as shown in these vases, and continues to recur in Chinese art. Source: A. Klimchouk.

Figure 12.12 Terra rossa. **A.** Agriculture in the Berici Mountains of northern Italy, where the red terra rossa sediment has been derived from the weathering of Upper Cretaceous limestones. Source: U. Sauro. **B.** Terra rossa with rounded bauxite pea-size concretions (pisoliths), derived from Cretaceous limestones, near Otranto in southern Italy. Source: J. De Waele. **C.** Roadcut in northern Italy, showing a small doline with its dark reddish-brown terra rossa infilling. Source: U. Sauro.

In many karst landscapes, runoff or glacial scouring can easily produce bare rock surfaces called **limestone pavements** (Fig. 12.13A). These limestone plains may be disrupted by open fractures, and enlarged in places by dissolution (Fig. 12.13B, C).

An important factor in the development of karst is the focused percolation of acidic water. On highly permeable limestones such as chalk and reef limestones, water can easily infiltrate through myriad pores. Such landscapes may lack surface streams and many of the classical karst dissolution features, because percolating water is not focused within joints or fractures. Many limestones, however, have pronounced jointing, which focuses percolating water, becoming areas of intensive dissolution. Joints get preferentially widened and deepened by dissolution, to produce narrow and deep openings. Many of these initially form beneath a soil cover but eventually become exposed as the soil cover gets eroded. The presence of a soil cover adds CO_2 to the soil air, which greatly increases the effectiveness of the dissolution process when compared to regions where a similar process is occurring on bare rock.

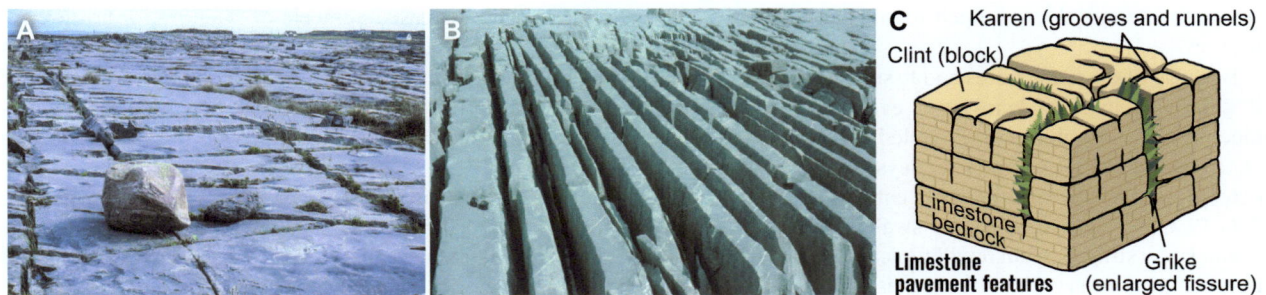

Figure 12.13 Limestone pavements. **A.** A limestone pavement in the Burren Geopark, County Clare, western Ireland, where glacial erosion has flattened the surface, erasing all surficial karst. Nonetheless, solutional forms have started forming since deglaciation. Photo by T. Waltham. **B.** Spectacular clint and grike topography in Muotatal, Switzerland. Source: U. Sauro. **C.** Schematic diagram of classical clint and grike topography, developed on limestone.

Figure 12.14 Karren landscapes. **A.** A typical Mediterranean karren landscape in Mallorca, Spain. Source: U. Sauro. **B.** An amazing karren surface in the French Jura. Source: J.-Y. Bigot. **C.** An alpine karren landscape in Marguareis, Ligurian Alps, northern Italy. Source: B. Vigna.

On a barren limestone surface, wide and intersecting openings lead to a type of topography called **clint** and **grike** (**Fig. 12.13C**). Deep cracks in the limestone, enlarged by dissolution, are referred to as grikes. The small "upland" areas of rock between the openings are known as clints. Clint and grike topography is but one of a family of barren limestone surfaces grouped broadly under the name **karren** ("lapiés" in French). Karren are small-scale, sculptured landforms, ranging in size from a few millimeters to several meters. Karren landscapes display grooves and fissures, typically separated by sharp ridges (**Fig. 12.14**). They can take on a variety of morphologies, such as sharp ridges, pinnacles, U-shaped (straight or meandering) channels, pits, downslope channels and flutes, and flat-floored basins, among others. Karren edges are sharp where they are formed by dissolution on bare rock surfaces, but are more rounded when they develop beneath a soil or vegetation cover (subsoil karren). On sloping surfaces in mountains (alpine karst), karren development depends on a complex suite of chemical and physical factors such as (1) amount of rainfall and snowmelt, (2) concentrations of CO_2, (3) rock texture and mineralogy, (4) degree of fracturing, (5) slope, and (6) climate (**Fig. 12.14C**).

Much of the surface water in karst landscapes enters the ground through open fissures or bedding planes, which allow for further dissolution. For this reason, true karst landscapes are characterized by a lack of valleys and very little surface water, but which may still sporadically flow during times of higher precipitation. Most water flow in karst occurs underground, through open fissures and conduits, and in caves.

Because of the many pores and fissures in the rock, many of the rivers in karst landscapes eventually get diverted underground through openings called **swallow holes** (**Fig. 12.15**). **Blind valleys** (or **karst valleys**) end abruptly against a steep rock face where the river enters the swallow hole (**Fig. 12.15D**). Downstream from swallow holes, the stream channel may still exist, but it will typically be dry. Karst geomorphologists refer to this feature as a **half-blind valley**. Water will flow there only during wet periods, when the swallow hole is unable to take the entire discharge of the river (**Fig. 12.15C, D**). Disappearing streams and blind valleys are classic features of most karst landscapes, illustrating that much of the water there is flowing underground.

Eventually, subterranean rivers rise again and flow on the surface. The location where the water rises to the surface is called a **karst spring** (or **rise**). Rises also give birth to surface streams and short, steep-headed (pocket) valleys, which start abruptly at the foot of sometimes towering cliffs, where the spring is located (**Fig. 12.16**).

12.5 DOLINES: DIAGNOSTIC KARST FEATURES

The prototypical karst landform is the **doline**, which, together with caves, form the analog of rivers in a fluvial landscape. Dolines (also known as **limestone sinks** or **sinkholes**) are isolated, closed, often bowl-shaped depressions that range from one to several hundred meters in diameter. Dolines are wider than they are deep; otherwise they are termed **shafts** (**Fig. 12.0**). Dolines are almost always intimately connected to underground drainage systems. Dissolved and detrital materials that enter dolines (in the epikarst zone) get transported into the subsurface, and may emerge again at karst springs. Dotting the landscape, dolines form an intricate network of depressions, which tend to be randomly distributed but are sometimes aligned along structural weaknesses or lithological contacts (**Fig. 12.17**).

Dolines are the surface expressions of dissolution processes. Although it is a slow process, dissolution is responsible for the enlargement of bedrock fissures. This process is focused in areas with high secondary permeability, such as areas with increased numbers of fractures, or brittle bedrock (**Fig. 12.18**). Percolation of surface waters, and thus dissolution and doline formation, are concentrated in these types of locations where the bedrock is particularly vulnerable. This process is sometimes referred to as **point-source dissolution**.

Areas with dense fractures and fissures, more favorable for downward flow than others, become preferential pathways for point-source dissolution. Water flows towards and through these locations, which then increases dissolution and further enhances their permeability. This is another example of positive feedback in karst terrain – accelerated dissolution over these more permeable areas will lower the

Figure 12.15 Disappearing (sinking) streams and swallow holes. **A.** The sinking stream and swallow hole of Hull Pot, after a heavy rainfall, in the Yorkshire Dales, United Kingdom. The river falls into a fracture-controlled, collapse sinkhole before "disappearing" into a cave. Source: T. Waltham. **B.** The swallow hole of a disappearing stream on the Mitchell Plain, a classic karst region in southern Indiana, USA. Source: R. Schaetzl. **C** and **D.** The Lost River, a classic disappearing stream, shown upstream of the Mitchell Plain and then, as a half-blind valley, downstream from its many, small swallow holes. Source: R. Schaetzl.

Figure 12.16 Karst springs or rises. **A.** The karst spring of Mas d'Azil in the French Pyrenees. Source: Jean-Yves Bigot. **B.** The karst spring at Tytoona Cave, Pennsylvania, USA. Source: R. Schaetzl.

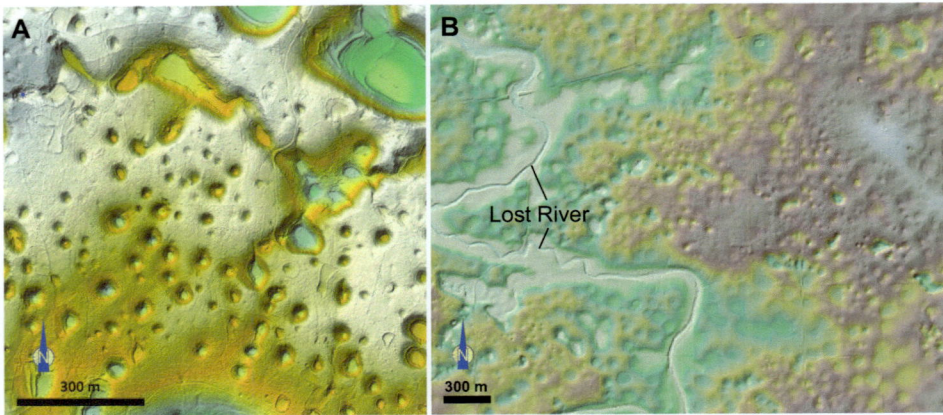

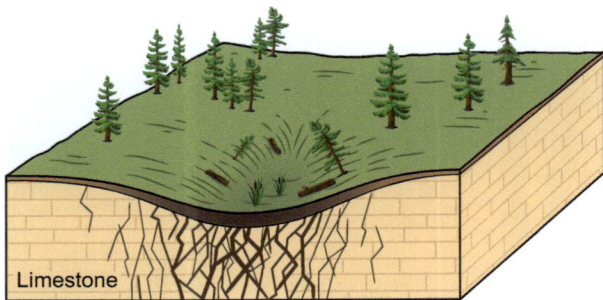

Figure 12.18 Schematic diagram illustrating how a subsidence doline can form above an area of highly fractured bedrock.

land surface, leading to even more water entering the site, enhancing dissolution (**Figs. 12.18, 12.19**). In these kinds of situations, the width of a solution doline is generally determined by fissure spacing; wider fractures facilitate wider dolines, whereas more focused areas of fracturing will tend to produce deeper and narrower dolines.

The subsidence that forms dolines occurs because of sagging, suffosion, and collapse of the underlying bedrock (**Fig. 12.20**). **Sagging** is the ductile deformation (slow bending) of bedrock, caused by the lack of basal support. This process is particularly common in areas shallowly underlain by evaporite deposits, where these very soluble rocks are dissolved allowing the overlying rocks to slowly sag into the lowering subsurface. **Suffosion** is the downward migration of unconsolidated material into and through open fractures, fissures, and pipes in the bedrock. This process mainly operates where fissures are wide. In this case, finer material gets transported downward, while coarser sediments remain behind, closer to the surface. Some dolines form via collapse of the surface. **Collapse** is the brittle deformation (breakage) of bedrock into an underlying void, usually a **cavern**. Collapse causes a rupture in the land surface, and is the most dangerous form of subsidence because it can occur suddenly (**Fig. 12.21**). Collapse can be triggered by weakening of the roof of an underground cavity, or by upward propagation of still-deeper collapses, a process known as **stoping**. Collapse can also be caused as the water table, which may have been holding up the roof of a cave, drops. Falling water table levels may be caused by factors such as climate change, variations in sea level, nearby river entrenchment, or by direct withdrawals for mining, irrigation, or drinking water, etc.

Sinkholes formed by suffosion and sagging are generally saucer- or bowl-shaped, with width/depth ratios > 5. As seen from above, they are circular to elliptical, and only

Figure 12.19 Dolines. **A.** Solution dolines in the Lessini Mountains of northern Italy, illustrating their alignment along a structural weakness in the limestone. Source: U. Sauro. **B.** A doline-dotted landscape in Kentucky. Source: R. Schaetzl. **C.** A deep doline in the high mountain karst of Peru. Source: J.-Y. Bigot.

Processes that form dolines

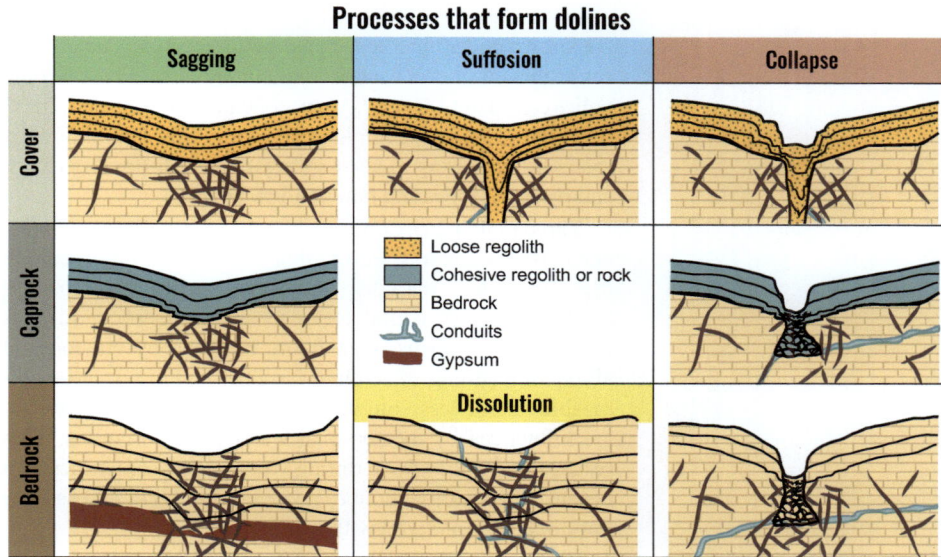

Figure 12.20 Classification of dolines according to the type of process involved and the material that deforms at the surface.

Figure 12.21 The collapse sinkhole that formed directly below the National Corvette Museum in Bowling Green, Kentucky, USA, on February 14, 2014. Because it occurred at night, no one was injured, but eight vintage corvettes were damaged as they fell into the hole. This event provided an opportunity to educate the public on karst, through the development of a permanent exhibit at the museum. Note the exposed limestone in **B**. Source: Courtesy of the Corvette Museum.

rarely are they elongated. Collapse sinkholes are initially steep-sided, often with bedrock walls (**Fig. 12.21B**). They have lower width/depth ratios than other sinkhole types. Collapse sinkholes tend to transform into gentler shapes through degradation of the bedrock walls and infilling of their bottoms. These processes cause the sinkhole to eventually develop a bowl-shape similar to that of suffosion and sagging sinkholes.

Dolines continue to evolve as long as they remain foci of infiltration. However, these conditions can be interrupted during, for example, glacial periods. Sinkholes buried by glacial materials, or eroded or destroyed by the weight of the ice itself, must start all over again when the climate becomes suitable for karstification. Dolines filled with glacial sediments can survive as buried dolines. Some buried dolines can be of economic importance, containing bauxite ores that form as aluminum-rich residuum in the limestone becomes concentrated by dissolution.

Dolines are point-sources for pollution of underground waters. Preventing surface contamination through these landforms is thus essential to the protection of underlying aquifers.

12.6 EVOLUTION OF KARST LANDSCAPES

Where dissolution can operate over long timespans, such as in warm temperate and tropical climates, dolines can continue to enlarge and deepen. Over long periods of time, they eventually intersect each other in complex morphologies called **uvalas**. Uvalas generally have irregular floors with multiple sinking points (**Fig. 12.22**). Think of an uvala as a coalesced set of dolines. Most uvalas have floors that are well above the local water table.

When dissolution proceeds to the point that the surface lowers to the water table, corrosion will be enhanced. This scenario eventually levels out the bedrock surface, forming large

Figure 12.22 Large karst depressions called uvalas. **A.** An uvala with multiple sinking points and agricultural modifications in the Velebit Mountains of Croatia. Source: U. Sauro. **B.** Funtensee uvala in the Berchtesgaden Alps of Germany. The lake filling one of the depressions is an exception, rather than the rule, in uvalas. Source: Thorsten Hartmann, CC BY-SA 3.0, via Wikimedia Commons.

plains bordered by steeply rising slopes. These broad lowlands are known under the Slavic name **polje** (meaning "field"), because they are often used for agriculture (**Fig. 12.23**). Some poljes can span 1,000 km². Many contain thick deposits of terra rossa. Poljes have a complex hydrology, with both inflowing rivers (often arising at karst springs), through-flowing rivers, and rivers disappearing into swallow holes.

Three main types of poljes are recognized. **Border poljes** occur where streams are eroding laterally into the bedrock that forms their margins. **Structural poljes** develop their borders along structural elements in the bedrock, such as fold axes or grabens, with floors flattened by erosion and alluviation (filling of sediment) on the underlying, less permeable and/or less soluble rock. Finally, **base level poljes** are influenced by the seasonal water table variations. As a result, these poljes flood during wet periods, sometimes also because the swallow hole of a stream that flows across their floor becomes plugged with debris.

In the humid tropics, karst processes are highly operative. Not only do the high amounts of rainfall and runoff lead to rapid dissolution, but the high evaporation rates favor deposition of secondary carbonate as well. This process can cause **case hardening**, during which a hard, protective, surface layer is formed on and within the bedrock. Then, as dissolution continues in lowlands, areas of bedrock left behind become increasingly case hardened, making them difficult to erode. As a result, the landscape takes on a unique appearance – with broad lowlands separated by uplands held up by case-hardened limestone. This spectacular landscape, known as **cockpit karst**, is also variously referred to as **cone karst** (**fengcong**), **tower karst** (**fenglin**), and **kegelkarst** (**Fig. 12.24**). Individual uplands in cockpit karst are sometimes called **mogotes** (Cuba). The most impressive examples of cockpit karst are found in tropical China (where it is named **fengcong**), Indonesia, Vietnam, and Jamaica. Where dissolution is even more extreme, the lowlands reach local base level. These landscapes have large plains dotted with residual hills that stand up as tall (sometimes > 100 m) towers with almost vertical walls (fenglin) (**Figs. 12.11, 12.24A**). The intervening plains can be filled with residual sediment, as occurs in Cuba (**Fig. 12.24B**), or drowned by the sea, as at Ha Long Bay, Vietnam, or in New Guinea

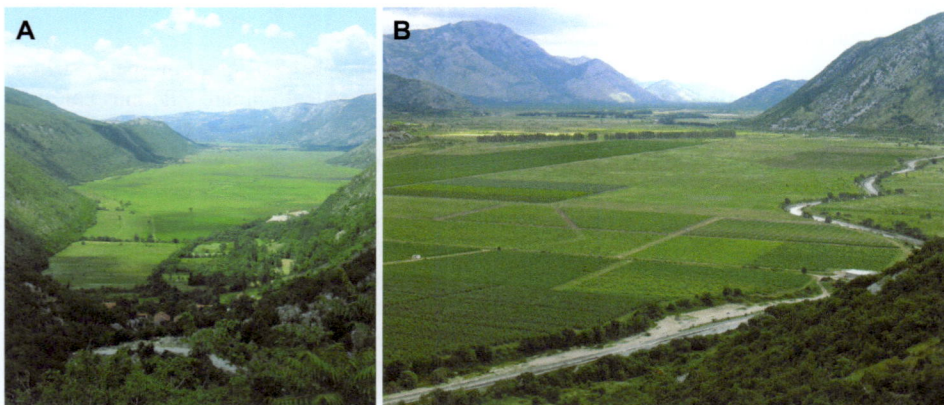

Figure 12.23 Large karst depressions – poljes in Bosnia and Herzegovina. **A.** The spectacular Dabar Polje; note the flat floor and steep side slopes. Source: J. De Waele. **B.** The Trebišnjica River flowing on the alluviated floor of Popovo Polje, in Bosnia and Herzegovina. Source: A. Palmer.

Figure 12.24 Tropical (cockpit) karst landscapes. **A.** Fenglin, China, with its large planation surfaces and residual karst towers. Source: U. Sauro. **B.** A mogote standing above an undulating plain of residual sediment in Cuba. Source: U. Sauro. **C.** A flooded tower karst landscape NW of Waigeo Island, New Guinea. Source: L. Plan.

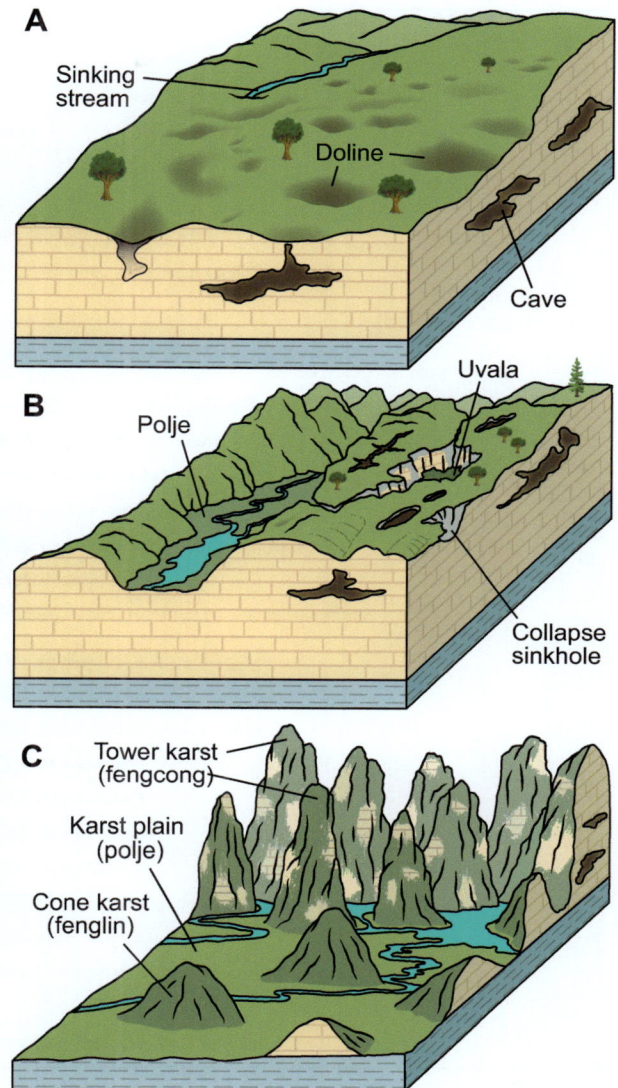

Figure 12.25 Landscape evolution in limestone karst regions under a temperate climate. **A.** Doline-dotted terrain. **B.** Typical, mature holokarst. **C.** The extreme end-member – fenglin and fengcong karst in tropical China. Source: Original drawings by V. Chiarini.

(**Fig. 12.24C**). Sometimes, the towers are shorter, less steep, and even asymmetric because the prevailing direction of the wind-driven rainfall favors directional case hardening on the wetter slopes.

To summarize, the evolution of karst landscapes is driven by dissolution, which lowers the land surface at focused locations. The landscape then progresses from a few scattered dolines to one with an increasing number of sinking streams and collapse features, to larger and more complex forms like uvalas and poljes. Finally, if conditions are right, the landscape evolves to one with large plains dotted with residual hills (**Fig. 12.25**). Along the way, a multitude of other, less apparent, landforms may evolve as well.

As you can see, karst geomorphology is rife with terminology. Most early workers named the various karst features using their native language, many of which were then translated into the languages of nearby countries. The karst vocabulary is also so diverse because the classical European karst areas span a variety of countries and languages (**Fig. 12.26**). Only recently has there been international integration of terms and theories among hydrogeologists, speleologists, and geomorphologists.

12.7 CAVES

Just as dolines are the prototypical landform of karst terrain at the surface, caves are the prototypical underground karst feature. Most of the world's caves are formed in association with karst.

Caves are often defined in a rather anthropocentric way as cavities large enough for people to enter. Geologists may

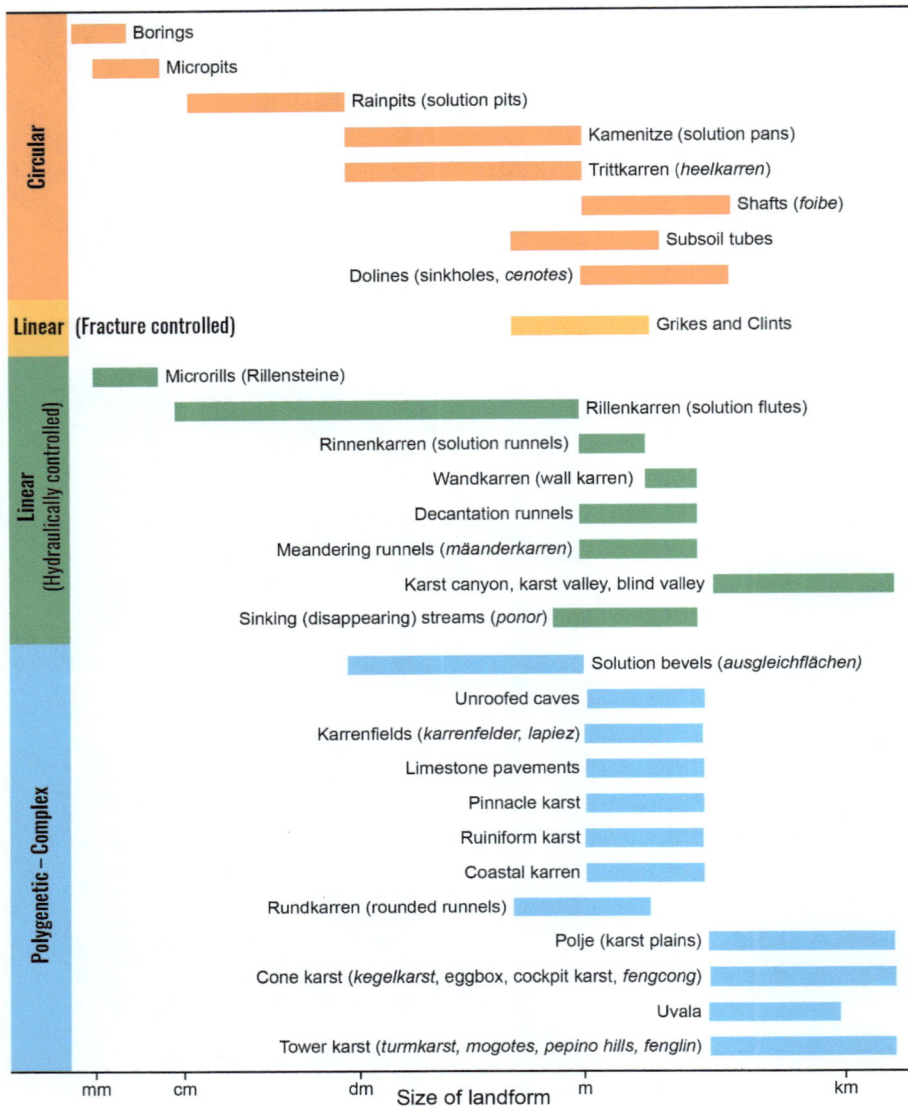

Figure 12.26 A synthesis of the major karst landforms, classified according to their shape and size. Terms shown in italics are often-used synonyms for the same features.

define caves differently – as solutionally enlarged, underground spaces that allow for the turbulent flow of water. In a karst landscape, the presence of caves may not be apparent at the surface. Nonetheless, they often are linked to surface topography by collapse of underground chambers and via narrow conduits in the bedrock. Caves also greatly influence surface drainage patterns, thus causing the unique surface morphology of karst landscapes. For example, many swallow holes connect to caves.

Caves form through a combination of dissolution and erosion of bedrock by underground waters, acting across both the **vadose** (unsaturated) and the **phreatic** (saturated) zones. The process of speleogenesis (cave formation) depends on the structural and lithological properties of the rock, as well as groundwater chemistry and flow paths.

Karst geomorphologists divide caves into two categories, based mainly on their water source and chemistry: (1)

epigenic caves, formed by descending (downward-percolating) fluids that derive their aggressiveness from the surface, and (2) **hypogenic caves**, which form from fluids that are ascending, and therefore acquire their aggressiveness from sources deep underground.

12.7.1 Epigenic Caves

Most epigenic caves are intimately linked to the surface hydrology. They form wherever ample quantities of groundwater percolate downward within soluble bedrock, and where there has been ample time to generate voids large enough to sustain turbulent flow. Rapidly flowing water is necessary for the cave to pass the breakthrough time, after which water flow is rapid enough to promote corrosion and erosion of the cave.

In most cases, epigenic speleogenesis occurs in successive steps. The first step is the formation of a hydrological connection from the surface to the bedrock below. Water that enters the underground system eventually exits at a discharge area, that is, at a karst spring.

As the acidic water moves through the tiny bedrock pathways (often as small as 0.02 mm), they become enlarged by dissolution. Solutional aggressiveness (the ability of a fluid to dissolve rock) can be sustained by mixing corrosion (**Fig. 12.7**) or by the localized production of acids, as when sulfides in the bedrock get oxidized to form sulfuric acid. At this point, the bedrock is a fractured aquifer, maintaining a more or less stable water table, similar to what happens in other fractured aquifers. Under these "slow flow" conditions, with long water–rock contact times, most dissolution takes place at the *upstream* ends of the flow pathways. During flow, the water becomes saturated with $CaCO_3$ – long before it emerges back aboveground. So, in general, shorter flow routes through the bedrock are enlarged faster than longer ones. Think of it this way – a given amount of water can only dissolve a certain quantity of rock, such that when water flows through shorter conduits, the opening will enlarge faster than it would if spread out over longer routes. Eventually some pathways reach a size (5–10 mm) that makes turbulent flow possible, so that clay and silt particles can be entrained. The faster flow through the enlarged openings enhances dissolution because the water can remain undersaturated over longer distances. These larger openings eventually become preferential pathways for most of the water, and adjacent, smaller pathways are progressively abandoned in favor of these "victor tubes." The aquifer now has enlarged fractures and solution conduits, and is considered truly "karstified."

An active epigenic cave system will experience enlargement in both the vadose zone (above the water table) and phreatic (saturated) zone. Vadose cave evolution will mostly be concentrated in upstream areas, where the water is more aggressive. Phreatic enlargement of flow pathways will mainly occur in lower-lying, deeper pathways downstream in the karst system. Think of phreatic conduits like cylinders that enlarge in all directions. Their size is an indication of the water flow – the larger the discharge through them, the greater their diameters will become. When the size of the conduit exceeds that of the greatest water flow (as will eventually happen), vadose conditions will start to prevail. Subsequent enlargement will then occur only on the floor and wetted sides of the conduit, as the roof remains open to the air. At this point, the cave floor may become entrenched, sometimes forming an underground canyon.

12.7.2 Hypogenic Caves

An increasing body of evidence indicates that ≈15% of all caves are formed by (often rising) groundwater, in which the acidity does not derive (only) from surface sources. These hypogenic caves represent some of the longest and most complex caves on Earth. Hypogenic cave formation does not rely on a surface source of acidity, so they often have poor connections with the surface. This is also why most explored caves have epigenic origins. However, our knowledge of

Figure 12.27 Gypsum crystals in the Chandelier Ballroom in Lechuguilla Cave in New Mexico, USA, considered by many to be the most beautiful cave in the world. Source: L. Plan.

hypogenic caves is destined to increase rapidly, as more are discovered and recognized.

The acids that dissolve rock to form hypogenic caves can be of a variety of types, but those formed by sulfuric acid (H_2SO_4) are probably the most spectacular. Examples include Frasassi Cave in Italy and Lechuguilla Cave and Carlsbad Caverns in New Mexico, USA (**Fig. 12.27**). These caves have (or had) H_2S gases, which derived from the microbial reduction of sulfates (gypsum or anhydrite) deep in the bedrock, in the presence of organic carbon (hydrocarbons). When these H_2S-rich gases rise along faults and fissures, they mix with oxygenated waters near the water table (where the groundwater is more aerated). The H_2S then oxidizes to H_2SO_4, which readily dissolves calcite in the carbonate bedrock. The reaction produces gypsum ($CaSO_4 \cdot 2H_2O$) and releases CO_2, which can then become H_2CO_3 further contributing to dissolution.

Other, sometimes spectacular, hypogenic caves are formed by rising thermal (> 4 °C warmer than the mean annual air temperature of the local area) waters. This water gets heated deep within the crust, rises along deeply rooted fractures, and occasionally reappears in caves or at the surface as thermal springs. These waters have variable chemistries, depending on their origin and how they mix with fluids during their ascent. Some are rich in H_2S, and form thermal sulfuric acid cave systems, as described above. Many thermal caves,

Figure 12.28 Speleothems in Pál-völgyi Cave, a thermal cave in Budapest, Hungary. **A.** Raft cones made of accumulations of floating calcite rafts in a former (but now dry) pool. Note the large stalagmites and stalactites in the background. **B.** Folia on an overhanging cave wall. Source: Cs. Egri.

however, are enriched mainly in CO_2. As the rising thermal waters in these caves cool, they become more solutionally aggressive, because colder water is capable of dissolving more CO_2 (**Fig. 12.7**). Then as the water gradually rises more, the decrease in pressure allows some CO_2 to escape, which in turn drives the precipitation of secondary carbonates. As a result, in most thermal caves, net dissolution occurs at depth, whereas deposition is dominant closer to the surface.

Most of the known thermal caves are dewatered (and thus explorable), and are often lined with hydrothermal calcite and interesting speleothems. Calcite rafts consist of floating flakes in thermal lakes, forming because of rapid CO_2 degassing and/or evaporation. **Raft cones** are stalagmite-like deposits growing on lake bottoms beneath drip points,

composed of accumulated calcite rafts (**Fig. 12.28A**). **Folia** are calcite linings resembling bracket fungi, often covering large areas on overhanging walls or ceilings. They form in the oscillation zone of a thermal water body (**Fig. 12.28B**). Good examples occur in Budapest, Hungary, where theories on thermal cave development were first developed.

Some of the geologically oldest known caves are Wind and Jewel Caves in South Dakota, USA. These extremely long (248 and 346 km, respectively) caves have a complex and long history that included an early thermal phase. Both form complex, multi-story mazes. Interestingly, they both also host spectacular calcite spar coatings on walls and roofs, and calcite veins known as **boxwork**, which protrudes from the walls (**Fig. 12.29A, B**). Boxwork is composed of veins of

Figure 12.29 Wind Cave, South Dakota, USA. **A.** and **B.** Wind Cave's famous boxwork. **C.** Wind blows out of the natural entrance to Wind Cave, implying that air pressures in the atmosphere are currently falling, causing air inside the cave to rush out of the small cave opening. Air rushes into the cave when atmospheric pressures are rising. Source: (A) A. Palmer; (B and C) R. Schaetzl.

Figure 12.30 Megacusps on the roof of the Cocci Abyss Cave, in Sicily (Italy), formed by condensation–corrosion of warm air. Note that the lower walls are covered by calcite popcorn (coralloids), which form by evaporation of cold air flowing across the lower parts of the passage. Source: M. Vattano – La Venta Geographic Association.

>242 km long, also has a single, narrow entrance, and thus has enormous airflow as well, regulated by two airtight doors.

In caves where thermal waters are still present, or where warm, moist air still rises from below, warm air masses inside the cave will rise and cool, becoming denser and descending again. This setup leads to the formation of thermal convection cells. Water vapor moving within these cells will condense on the colder, upper part of the cave roof and walls, to form condensation–corrosion forms such as **cupolas** (deep hemispherical concavities) and cuspated walls and roof features (wavy ceilings and walls) (**Fig. 12.30**). Condensation–corrosion is particularly important in air-filled thermal caves, but is also an efficient enlarging process in any caves filled with moist, warm air and which have significant (daily or seasonal) temperature differences.

Another hypogenic cave situation occurs in artesian aquifers, where a soluble rock unit is sandwiched between two non-soluble, but still permeable, beds (**Fig. 12.31A**). In this case, water will flow under pressure from the lower-lying poorly soluble beds, through the soluble rock unit, and into the overlying permeable but less soluble beds. This slow upward (and diffuse) recharge of water will exploit almost all the possible passageways in the soluble rock, dissolving and enlarging them. This process will ultimately form **maze caves** composed of a complex and dense grid of intersecting passages. Maze caves occur in western Ukraine, where water flows from limestone into the much more soluble gypsum beds above (**Fig. 12.31B**). Optymistychna Cave in Ukraine is ≈260 km long, and the nearby Ozernaja Cave in Russia is longer than 140 km.

Caves in coastal karst have little or no direct relationship with surface topography (**Fig. 12.10C**). There, the mixing of two water bodies (fresh groundwater and the denser, lower-lying salt water of the ocean) with contrasting chemistry drives the dissolution process and forms a special kind of hypogenic cave. Caves develop parallel to the coastline and close to sea level, at the elevation where the mixing zone is usually located. In young, poorly cemented limestone, these types of caves, also known as **flank margin caves** (because

secondary calcite, initially formed in cracks and voids within the bedrock of the cave walls. It only becomes a positive feature after the surrounding rock is lost due to weathering. As the name indicates, Wind Cave is characterized by very strong airflows at its small (natural) entrance, which indicate its great volume. Pressures changes in the atmosphere drive large air exchanges through the narrow opening of the cave. Almost continually, air rushes into the cave, or out (**Fig. 12.29C**). Lechuguilla Cave in New Mexico, USA, which is

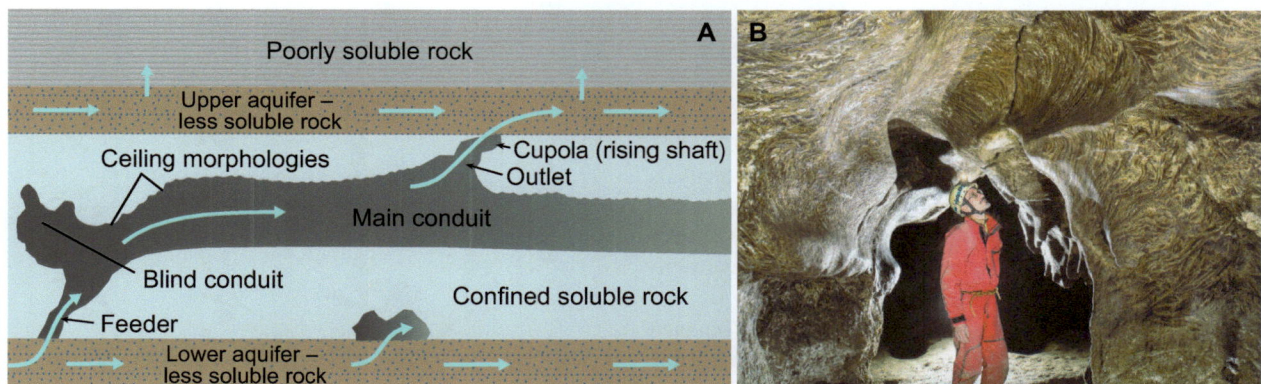

Figure 12.31 Hypogenic caves. **A.** Schematic representation of artesian hypogenic cave development, where the more soluble rock unit (often gypsum, in light gray) is confined between two aquifers composed of less soluble rocks (brown), the lower of which feeds the karst dissolution processes working in the rock above. **B.** Wavy roof morphologies in Krisztaljeva Cave, a maze cave in Ukraine. Source: Cs. Egri.

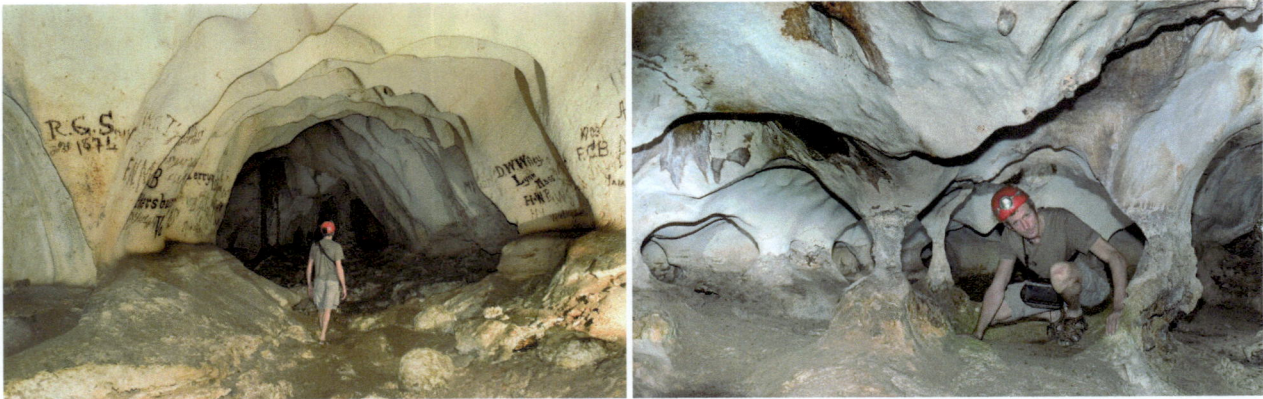

Figure 12.32 Hatchet Bay Cave, on Eleuthera Island in the Bahamas. Typical of flank margin caves, it has only local sediment, no scallops, and a wavy roof and walls. Like most large flank margin caves in the Caribbean region, it is now dry and well above sea level, having been mainly formed during the last interglacial, when sea level was much higher. Source: A. Palmer.

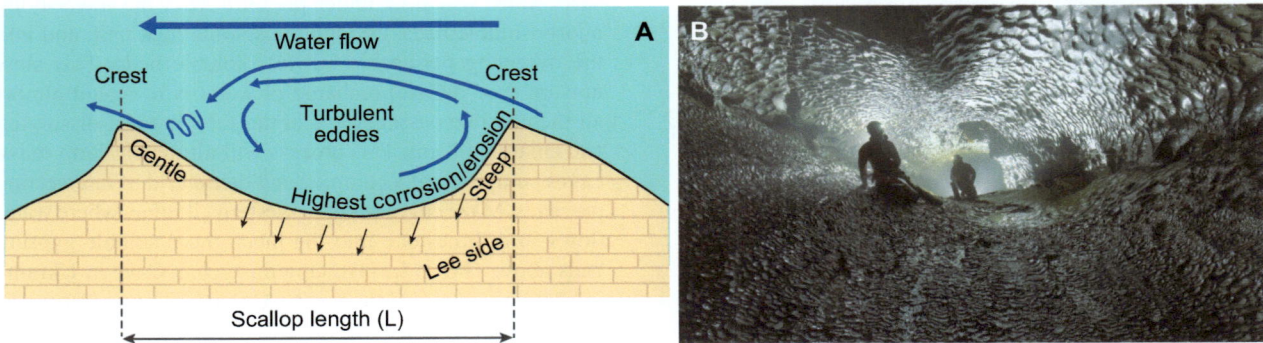

Figure 12.33 Scallops in phreatic caves. **A.** A schematic cross-section of a typical scallop along a cave wall, showing the locus of highest dissolution and erosion. **B.** Scallops carved on the limestone walls of Markov Spodmol Cave, Slovenia. Source: S. Sedran S-Team.

they form at the coastward flank of a freshwater lens) can form in only a few thousand years (**Fig. 12.32**). In general, flank margin cave formation is driven by mixing corrosion, in slowly moving waters, forming wavy sculpturing on the cave walls. Fluvial sediments, deposited by running water within the cave, and **scallops** (**Fig. 12.33**) are generally absent. Instead, small grains of secondary carbonate are the typical cave sediment (**Fig. 12.32**). Sea level changes during the Quaternary have led to a series of karst/cave levels in these types of caves, each reflecting a former high stand.

12.7.3 Analyzing Cave Systems

Determining how a cave has formed requires the analysis of the overall cave pattern, both in plan and profile view. Geomorphologists will also examine the shapes of the cave passages and the details of the wall sculpturing, cave sediment, and secondary deposits.

In phreatic (epigenic) caves, the walls are often sculpted with asymmetrical, shell-like hollows known as scallops, due to erosion and dissolution in turbulently flowing water (**Fig. 12.33**). Their shape and size are indicative of flow direction and velocity, respectively, with the steep sides of the hollows facing upstream. Smaller scallops indicate faster flowing waters.

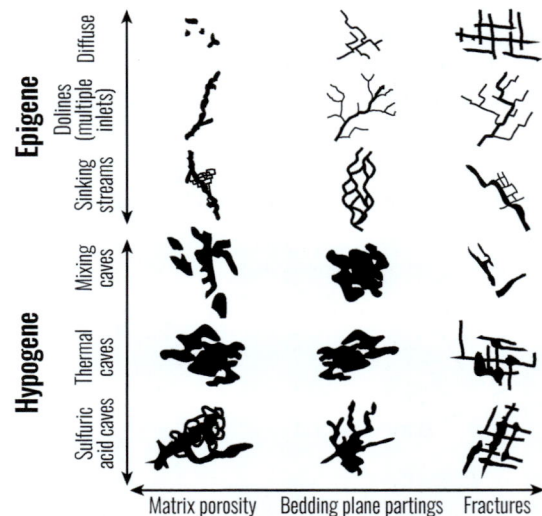

Figure 12.34 Matrix of cave patterns as seen from above (plan view). Such data are often used to determine the specific formation history of the cave. Source: After Palmer (2009). Used with permission of the National Speleological Society (www.caves.org).

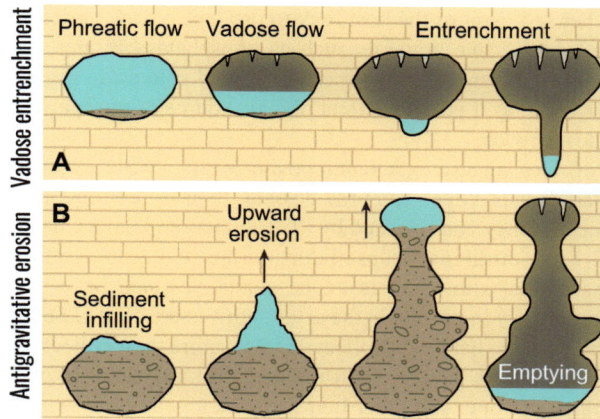

Figure 12.35 Schematic representation of downward and antigravitative (upward) erosion in caves. **A.** The normal evolution of a cave passage, where a canyon is carved into the cave floor by flowing water, leading to a keyhole passage. **B.** In caves with active sediment deposition on the floor, dissolution is driven upward, into the cave ceiling, forming an antigravitative underground canyon.

Cave patterns mainly depend on the type of recharge and the structural characteristics of the bedrock (**Fig. 12.34**). The pathways that are exploited and widened by dissolution could originally have been fractures, pressure-solution seams, bedding planes, and/or intergranular pores. Each of these will result in a different type of cave pattern (**Fig. 12.34**). Recharge of waters into the cave also impacts the pattern of its passages. Water may enter through sinkholes and karren fields at the surface (low amounts of water, dispersed across many points), at swallow holes (high amounts of water entering at only a few points), or as diffuse inflows coming from permeable, but insoluble, rocks above. Recharge can also occur from below the cave, in hypogenic sulfuric, thermal, coastal, or deep mixing zones. The various combinations of these parameters yield widely varying cave patterns.

Yet another way to unravel the history of a cave involves examination of its sediments and their influence on passage shape. Vadose flow – water that is freely flowing through the cave – can form vertical shafts through the bedrock, with water films, or even waterfalls, cascading down the walls. Cave passages can develop "canyons" – elongated, sinuous, deep-but-narrow passages – containing an underground stream that is entrenching into the cave floor (**Fig. 12.35A**). In completely drowned (phreatic) conditions, flowing water will enlarge the cave passages in all directions, forming generally rounded tubes with lenticular, elliptical, or circular cross-sections. Phreatic tubes can both display descending and rising passages, when viewed longitudinally, implying that, at times, water is flowing under pressure. These types of passages can also be produced during floods in caves that normally would be only partially filled with water. In this type of cave, the downstream, deeper sectors are purely phreatic in shape, whereas the upper-but-descending portions show clear signs of vadose entrenchment. When a

phreatic (rounded) tube passes into a vadose entrenching passage, the cross-section will resemble a keyhole (**Fig. 12.35A**).

Water may at times flow only slowly through drowned portions of caves, which can lead to sediment deposition. Such sediment can shield the floor from further erosion and dissolution. The water will thus carve upward into the roof, keeping pace with the slowly accumulating sediments on the floor (**Fig. 12.35B**). Features that indicate this type of process include ceiling half tubes, sinuous channels cutting across the roof (but independent of fractures in the rock), and **antigravitative canyon**s that very much resemble upside-down canyons on the cave roof when emptied during a later, more turbulent flow stage (**Fig. 12.35B**).

12.7.4 Detrital Sediments in Caves

Material that gets transported into, or precipitated in, caves can be very insightful regarding cave evolution. Let's start with detrital cave sediment (pebbles, sand, and clay) that gets transported into the cave by flowing water, gravity, ice, or even wind. Sediment deposited by rapidly flowing waters is typically well-rounded (**Fig. 12.36**). In caves where the

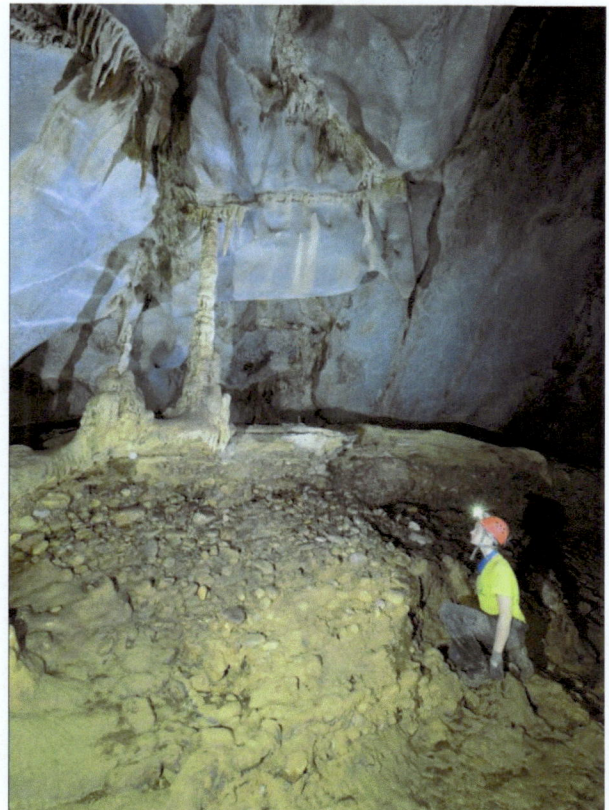

Figure 12.36 An alluvial deposit of large pebbles is covered with a more recent flowstone deposit dated to 71,000 years ago. These gravels were probably deposited during the sea level high stand of 125,000 years ago in this branch of the Puerto Princesa Underground River network in Palawan Cave, in the Philippines. Source: M. Vattano – La Venta Geographic Association.

flow velocity has decreased over time, coarse sediments such as pebbles and sands will be buried by later deposits such as slackwater muds. Cave muds are typically best preserved in higher locations and in sheltered recesses of the cave.

In areas where sediments are washed into caves from the surface, for example, in swallow holes, cave sediment can contain large quantities of pebbles of exotic origin. The geology of these sediments can shed light on the recharge area of the karst system. It can sometimes also be used as evidence of changes in the drainage basin through time. If quartz is present in cave sediments, luminescence or cosmogenic isotope dating can determine how long the sediments have been underground, shielded from cosmic rays (see Chapter 3). Clastic sediments in caves can also retain their original magnetic polarity, allowing the investigators to determine if they were deposited before or after the Brunhes–Matuyama reversal, 0.78 Ma (**Fig. 3.21**).

Many caves also fill with **breakdown** – sediment that has broken off of the roof and walls and fallen to the floor. Breakdown material can be composed of tiny flakes, piles of hand-sized rocks, and sometimes massive blocks and slabs.

Another very typical deposit in caves is **guano**, excrement produced by bats or birds (**Fig. 12.37**). Guano contains phosphoric, nitric, and sulfuric acids that react with other sediment or the host rock to produce a wide variety of secondary cave minerals such as phosphates, nitrates, and sulfates. Guano is sometimes mined and sold as an organic fertilizer, as it is an excellent organic source of nitrogen and phosphorous.

Caves are also excellent repositories for a wide variety of plant fossils, which can provide clues to paleo-environments because they can be dated by ^{14}C (radiocarbon) (see Chapter 3). Cave **middens** (heaps of dung, urine, and general refuse) associated with packrats of the genus *Neotoma* have been particularly important in the dry, southwestern United States. Packrats are habitual "collectors" who store their collections in caves (**Fig. 12.38**). Their middens contain pieces of plant material, bone, and other items, bound together by deposits of sticky, viscous urine. Middens preserve very well in dry desert environments, enabling paleo-environmental scientists to extract both pollen and a variety of plant macrofossils (such as seeds).

Vertical cave entrances can trap large vegetal and animal debris, sometimes becoming excellent fossilization sites (**Fig. 12.39**). Many of the oldest human remains have been found in caves. For example, Sterkfontein and Rising Star Caves in South Africa yielded a >2,000,000-year-old *Australopithecus* skeleton, and 350,000–250,000-year-old *Homo naledi* skeletons, respectively.

Caves have often been used as shelters by both humans and animals, and some vertebrates are especially connected to this habitat, like *Ursus spelaeus*, or the Cave Bear (**Fig. 12.40A**). Many caves in southern France and northern Spain are also known for their extraordinary upper Paleolithic rock art. These drawings are of animals such as bison, mammoth, lion, rhino, horse, and many others, using charcoal and ochres (**Fig. 12.40B**).

12.7.5 Speleothems

Caves are widely known for their attractive and unusual speleothems – secondary mineral deposits formed by flowing, dripping, ponded, or seeping water. Speleothems decorate roofs, walls, and floors of myriad caves. "Show caves" that are filled with lights and boardwalks are used to "show off" cave formations to tourists. Show caves are popular because of their varied and intricate speleothems.

Most speleothems are composed of secondary calcite, formed by the slow deposition of slightly supersaturated water, driven by CO_2 degassing or evaporation of water as it drips off the ceiling (**Figs. 12.1, 12.41**). How does this process work? The CO_2 content of cave air is normally between 0.05% and 1%, a value higher than that normally found in the atmosphere ($\approx$0.042% in 2022), but far below normal CO_2 levels in the soil and in infiltrating water (**Fig. 12.5**), which are in equilibrium with soil CO_2. When seeping waters enter a cave, CO_2 is lost from the solution, and consequently, $CaCO_3$ precipitates, forming speleothems

Figure 12.37 Guano. Fresh (blackish and wet) guano of vampire bats in Lapinha Cave, Bahia, Brazil. Note the group of bats in the cupola above. Source: J. De Waele.

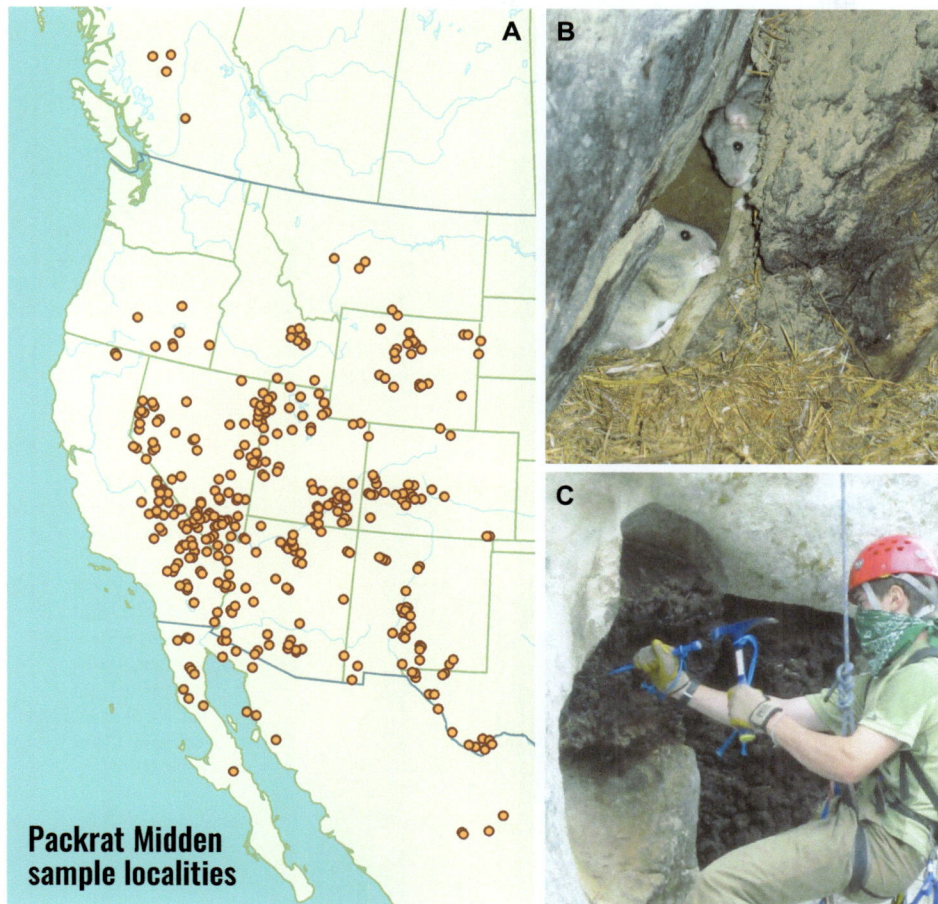

Figure 12.38 Packrat middens. **A.** Locations of packrat middens studied in the United States and Canada, almost all of which have been dated. **B.** A pair of bushy-tailed woodrats (*Neotoma cinerea*) next to both a modern, loose midden and an indurated fossil midden to their right. The Pleistocene-aged midden contained limber pine (*Pinus flexilis*) fossils, a tree that no longer grows near this site in Idaho, USA. **C.** Indurated, fossil woodrat middens like the one in the photo are usually so hard that they require a rock hammer and chisel to sample. In the case of this midden in Idaho, dated at ≈5225 yrs BP, sampling also required the use of technical climbing equipment. Source: J. Betancourt.

Figure 12.39 The skeleton of a Neanderthal man, covered with abundant calcite coralloids, in Lamalunga Cave, Apulia, southern Italy. Source: P. Forti.

Figure 12.40 Indications of cave habitation by paleo-vertebrates. **A.** Two cave bear skulls on the sandy floor of Govještica Cave, Bosnia and Herzegovina. Note the stalagmites that have grown on top of them. Source: V. Chiarini – GSB/USB Bologna expedition. **B.** The exceptional cave art of Chauvet Cave, Ardèche, France. Source: S. Jaillet – Edytem – CNRS – MCC.

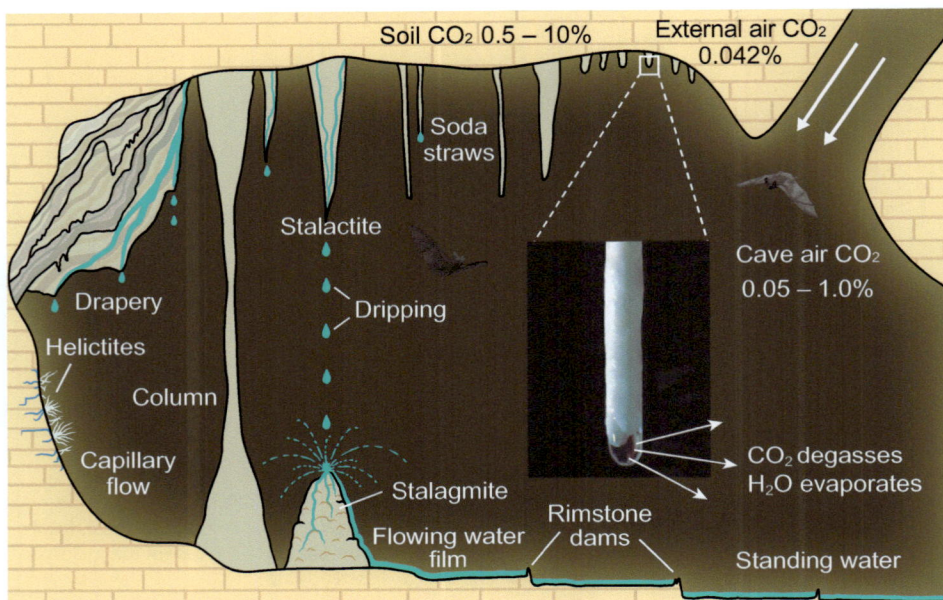

Figure 12.41 Schematic representation of the formation of speleothems from dripping and capillary water, caused by CO_2 degassing and/or H_2O evaporation. Note that CO_2 concentrations in cave air, although often 10× higher than in the atmosphere, are well below that of soil air, leading to CO_2 degassing, which drives calcite deposition.

(**Fig. 12.41**). Evaporation is a less common process, because cave air usually has very high relative humidity. Nonetheless, as even small amounts of water evaporate off dripping water in caves, the process will also contribute to the formation of speleothems. Caves that are more "open" to the atmosphere are likely to contain air with lower relative humidity and CO_2 concentrations, perhaps explaining why such caves often have more and better speleothems.

Deposits of secondary calcite, in the form of speleothems, are given the name **calcareous tufa**, unless they form from thermal waters, in which case they are known as travertine. In some conditions, seeping waters deposit aragonite, not calcite, or even magnesium-rich carbonate minerals such as huntite and hydromagnesite. Although gypsum can also form speleothems, because of its high solubility such speleothems survive best in dry caves; gypsum speleothems are only rarely found in wet caves.

Speleothems display wide variations in shape and size (**Figs. 12.41, 12.42, 12.43**). Generally, these differences are

related to the manner in which water is or is not flowing across the locations where the minerals are deposited – as falling drops from a small spot on the ceiling, as flowing films of water across the cave wall or floor, within standing water bodies, and/or as capillary fluids. Drops of water exiting from small, discrete pores will often deposit calcite in narrow, cylindrical **soda straws** (**Figs. 12.1, 12.41, 12.42**). Soda straws slowly build downward, but stay "hollow" for some time. If water flows along the sides of a speleothem that hangs from the cave roof, conical **stalactites** can form. Stalactites will maintain their distinctly "pointed" form as long as the locus of the dripping water is at the end, or tip. If droplets flow down along an inclined part of the cave roof, calcite will be deposited along a trail, eventually creating a **drapery**-like formation. Thin draperies may be partially transparent and resemble a slice of bacon (**Fig. 12.44**). Many show caves are proud of their drapery deposits, locally named "cave bacon," which often end in a soda straw or a stalactite.

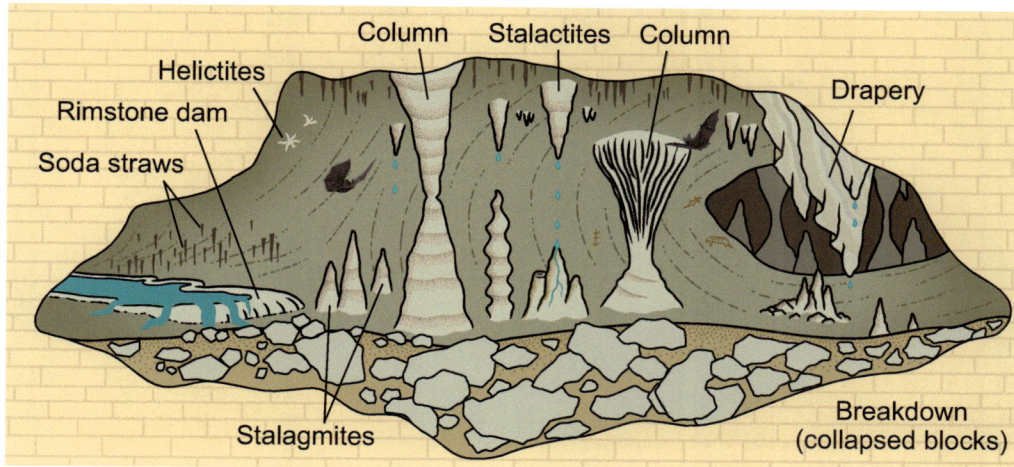

Figure 12.42 A schematic diagram of a cave with a rich display of speleothems.

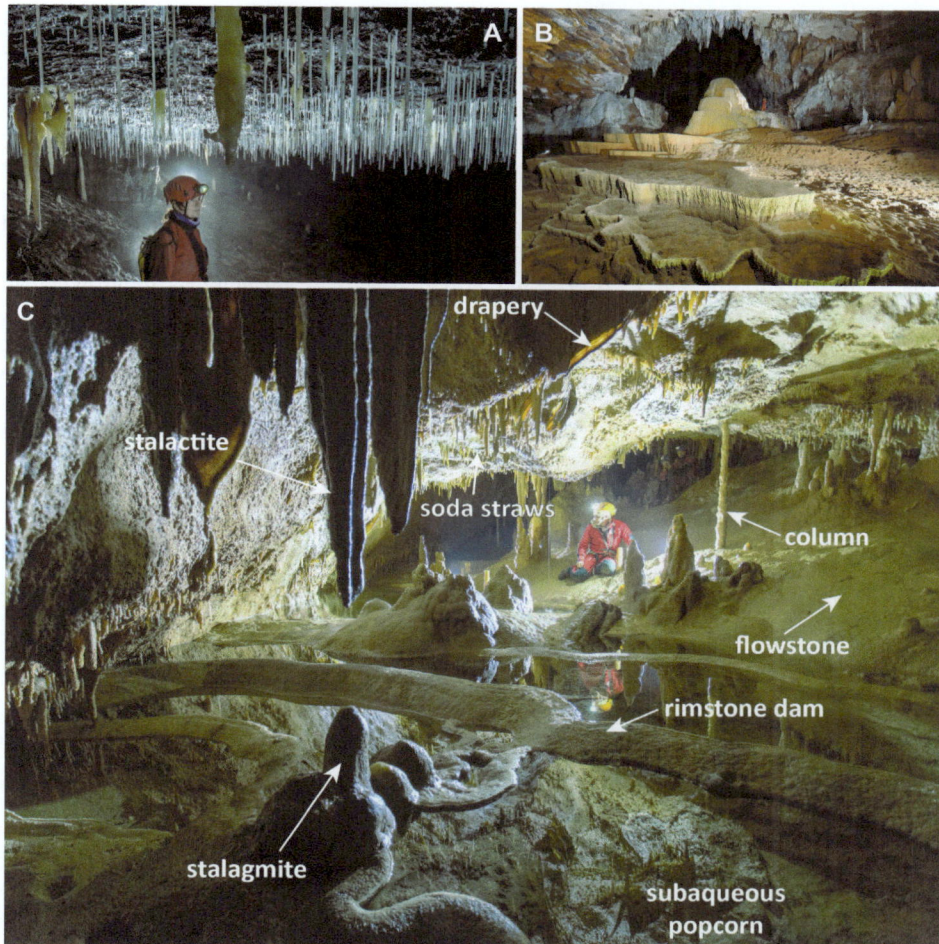

Figure 12.43 Speleothems. **A.** Soda straws and stalactites in Ease Gill Cave system, Yorkshire, UK. Source: M. Burkey. **B.** A huge stalagmite, grading into a flowstone first, and rimstone farther downstream, in São Bernardo Cave, Mato Grosso do Sul, Brazil. Note the person right of the large stalagmite, for scale. Source: M. Renda. **C.** A beautiful display of different speleothems in Martinska Jama, Slovenia. Source: S. Sedran S-Team.

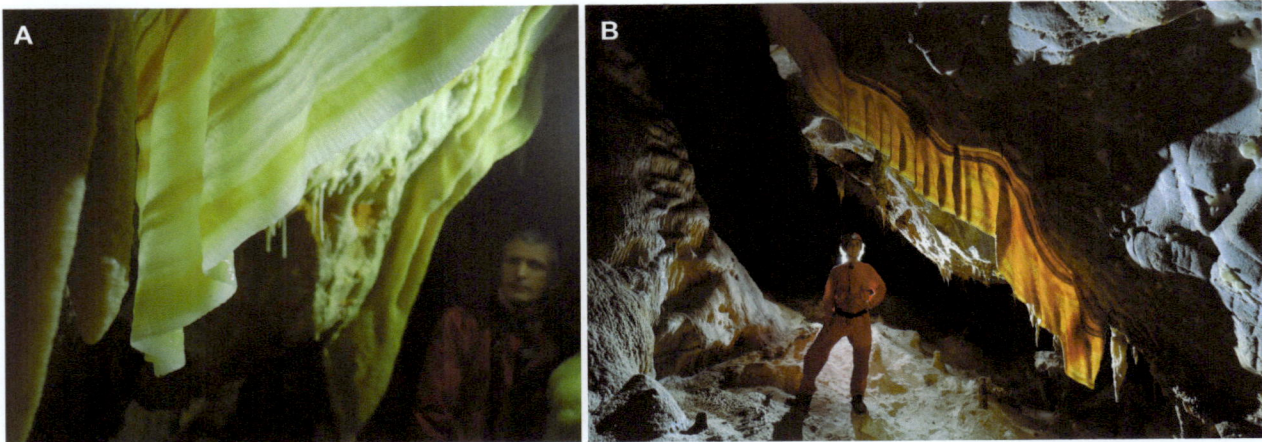

Figure 12.44 Draperies. **A.** A semi-transparent drapery in Valdemino Cave, Italy. Note the colored layers that parallel the roof. Source: B. Vigna. **B.** An orange-reddish colored drapery in Toca da Boa Vista Cave, Bahia, Brazil. Source: M. Renda.

On the floor of the cave, usually directly below stalactites, broad, stubby **stalagmites** may form where drops fall onto the cave floor (**Figs. 12.42, 12.43**). Stalagmites are wider than stalactites, because as the drops of water hit the floor, they splatter, spreading the deposition of secondary calcite over a wide area. The splatter also facilitates additional loss of CO_2 and H_2O, which leads to additional calcite deposition.

When stalactites grow downward far enough to intersect stalagmites, the structures combine to form a **column** that connects the cave roof and floor (**Fig. 12.42**). Water that flows as a sheet over the cave floor can deposit a blanket of calcite called **flowstone**. If enough calcite is deposited, flowstones will resemble a solid, rippled, wavy waterfall of crystalline material. In conditions of low supersaturation of the fluid (and thus less carbonate deposition), calcite will be mainly deposited in areas where turbulence is greater (steeper). Water flowing across the cave floor may then form **rimstone dams** (**Fig. 12.43**). As they grow upward,

rimstone dams can pond water to form pools and lakes on the cave floor. Within the pools, supersaturation leads to the formation of subaqueous **coralloids** (also called "cave popcorn") on the walls and floor. Calcite can also precipitate underwater, around small clastic grains. Water dripping onto pools in the cave creates minor ripples and waves, driving the formation of spherical, free-standing speleothems known as **cave pearls**. If small fissures along the cave walls and roof emit only small amounts of water, capillary forces may overwhelm gravity, enabling speleothems to grow in all directions. Generally, growth is perpendicular to the wall, but it can also follow random orientations – even upward. These formations, called **helictites**, can be monocrystalline or made of aggregates of tiny crystals (**Fig. 12.45**).

Most speleothems are white or light-colored, because they are composed of an intricate mosaic of finely crystalline calcite crystals that reflect all wavelengths of light. Speleothem calcite can also be transparent. Nonetheless, speleothems are usually opaque, with different tones of

Figure 12.45 Helictites. **A.** The extraordinary, white aragonite helictites in the Asperge Cave, in southern France. The growth of these speleothems appears to be in part biomediated by actinobacteria. Source: V. Crobu. **B.** Some of these speleothems are bluish, due to the inclusions of copper. Note that the aragonite needles are white, whereas the calcite speleothems are blue. Source: J.-Y. Bigot.

yellow, orange, and gray. The creamy, yellowish and orange colors are often incorrectly ascribed to (poorly soluble) iron oxide impurities. They are actually due to the presence of minute amounts of humic and fulvic acids, which originate in the overlying soil and are transported in solution into the cave. Humic acids have higher molecular weights than fulvic acids, giving darker (orange) colors, whereas fulvic acids create creamy-yellowish tones. In warm tropical climates, where organic acids are degraded to smaller molecules more readily, one finds mainly lighter-colored speleothems, whereas caves in mid-to-high latitudes often have more vividly colored speleothems. Aragonite speleothems are usually white because aragonite incorporates far fewer organic molecules in its crystalline lattice than does calcite. Nonetheless, the presence of bivalent metal ions, such as copper, can give even them exotic colors (**Fig. 12.45**).

Stalagmites are sometimes referred to as "trees of stone," because they can record past environmental changes, like trees do. Variations in drip rate and air temperature cause calcite (or aragonite) to deposit in different petrographic fabrics and incorporate varying geochemical signals (**Fig. 12.46**). Stable carbon and oxygen isotopes within the stalagmite minerals can be used as proxies for paleo-temperature and paleo-precipitation – and therefore, climate, soil, and vegetation conditions as well. Trace element analyses can provide even more detailed paleo-environmental data. Calcite and aragonite in speleothems can be precisely dated with the uranium/thorium (U/Th) method (up to 500,000 years) or, under ideal conditions, with the uranium/lead (U/Pb) method for up to billions of years. Speleothems mainly grow when ample water is available above ground. As a result, periods of lack of growth (which can be ascertained using dating methods) can often be equated to drier climates in the past, or to conditions of permafrost.

Figure 12.46 Cross-section of a stalagmite from Mračna Pećina Cave in Bosnia and Herzegovina. Note the differences in petrography and color of the calcite layers, which indicate changes in growth conditions, primarily drip rates. Source: V. Chiarini.

12.8 RESOURCES OF KARST LANDSCAPES

About 15% of Earth's land surface is underlain by soluble rock (**Fig. 12.3**), and at least a quarter of the world's population lives on or near karst. As a result, large amounts of resources are exploited from karst (**Fig. 12.47**). Perhaps 15% of the world's drinking water comes from carbonate aquifers. In fact, many important cities such as Vienna (Austria), Rome (Italy), Damascus (Syria), and Taiyuan (China) obtain most of their water from karst springs. Sustainably managing these water supplies will become increasingly important as societal pressures on karst aquifers are exacerbated due to population growth.

Carbonate rocks, including limestone, dolostone, and marble, are the most quarried rocks in the world. These rocks are used as building stone and as base materials for roads, railways, and buildings. Limestone is also used to

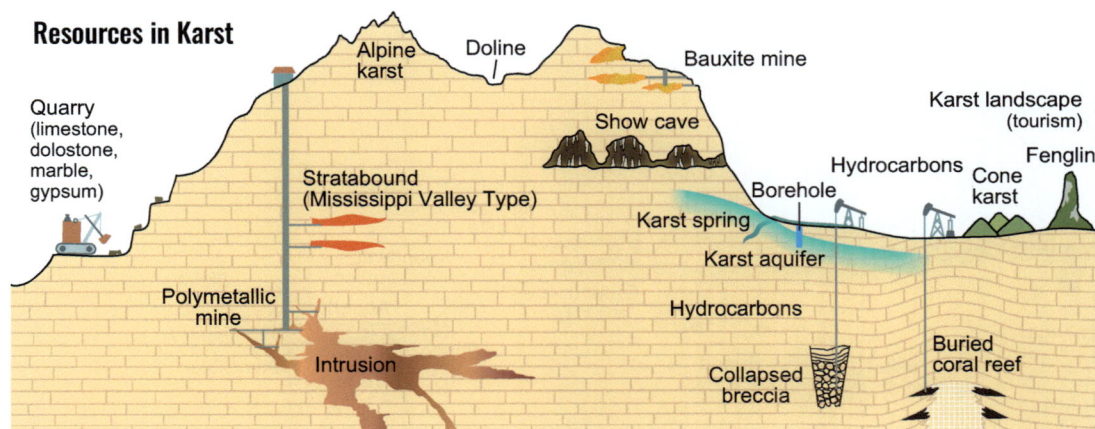

Figure 12.47 Schematic diagram showing the various resources in karst terrain.

make cement, a necessary ingredient of concrete, and lime, which is applied to acidic soils to raise the pH. Many karst rocks also host important reservoirs of crude oil and natural gas. Some of the most productive oil wells in the world are hosted in limestones, for example, in Iran and offshore Brazil. Many of these oil fields are actually set within buried paleokarst environments, such as ancient coral reefs, buried tropical karst, ancient hypogenic cave systems, or coastal karst.

Carbonate rocks are often also the host rock of important ore deposits. Karst depressions can act as traps for bauxite, a type of aluminum ore. The term bauxite is derived from the village of Les Baux, in southern France, where bauxite mining in buried dolines began over 150 years ago. Other placer deposits in karst depressions include cassiterite (a tin ore), various iron ores (siderite, goethite, limonite, and hematite), and even diamonds, sapphires, and rubies. The often-acidic weathering of minerals in a karstic setting can produce kaolin clays, used in the ceramics industry.

Karst landscapes are important recreational areas for both their natural and cultural value. Karst elements are parts of ≈50 UNESCO World Heritage Sites. Karst landscapes, such as the tower karst of southern China or the entrance dolines of Škocjanske Jame in Slovenia, have inspired artists and travelers for centuries. Show and tourist caves of all sizes and types are among the most popular recreational destinations in the world, visited by over 70 million people annually. Some of the world's most popular caves and karst features include the Puerto Princesa Underground River on Palawan in the Philippines, the Gunung Mulu National Park in Malaysia, Carlsbad Caverns in New Mexico and Mammoth Cave in Kentucky in the USA, the Aggtelek and Slovak Karst shared between Hungary and Slovakia, Škocjanske Jame in Slovenia, Pamukkale tufa buildups in Turkey, Desembarco del Granma National Park in Cuba, and the drowned Ha Long Bay and tropical Phong Nha Ke Bang National Park in Vietnam. Several caves are part of the UNESCO list for their extremely old (upper Paleolithic) paintings, including Altamira Cave in northern Spain, the many caves of the Vézère valley, including Lascaux, and the Chauvet Cave in Ardèche (**Fig. 12.40**), both in southern France. Being

delicate environments, these caves are often no longer accessible to the public, and as a result "replicate" caves have been artificially constructed. Despite the large amount of money invested in these projects, they remain economically profitable.

12.9 LIVING ON KARST LANDSCAPES

Karst environments are highly vulnerable systems in a number of different ways (**Fig. 12.48**). It was not until agricultural practices started that human impact on karst began to be important. Deforestation of karst landscapes, carried out to expand agricultural and residential land uses, causes rapid loss of soil. The surface then turns into a barren, rocky landscape, and because rock dissolution produces only limited amounts of insoluble material, soils are slow to recover. In temperate areas such as Slovenia, recovery of reforested areas, following appropriate management strategies, has taken as long as a century. These problems are especially important in densely populated karst areas such as southern China, the Philippines, and Java, where runoff is high and deforestation continues even today.

The porous bedrock of karst landscapes allows for rapid (and unfiltered) transfer of surface materials to underground locations (**Fig. 12.48**). Thus, the rapid transfer of pollutants to subsurface aquifers can and does occur, with minimal filtering and self-treatment. Surface pollutants can derive from agricultural practices (spreading of manure, fertilizers, and/or chemicals), urbanization (sewer systems and landfills), and industry (mine tailings and oil spills). Unfortunately, humans have long been using dolines as dumping areas for solid waste, and as locations into which waste fluids can be diverted. Many groundwater pollution issues have arisen because of this practice, which is now (thankfully) being intensively discouraged. Historically, even though caves have long been utilized in a variety of cultures, the human impact on caves has always been limited, being mainly confined to entrance areas.

Karst is also not without the occasional natural hazard. The most spectacular hazards in karst are collapse sinkholes (**Fig. 12.20**), but subsidence of any kind can damage infrastructure. The development of dolines is especially rapid in areas underlain by highly soluble evaporites. Large

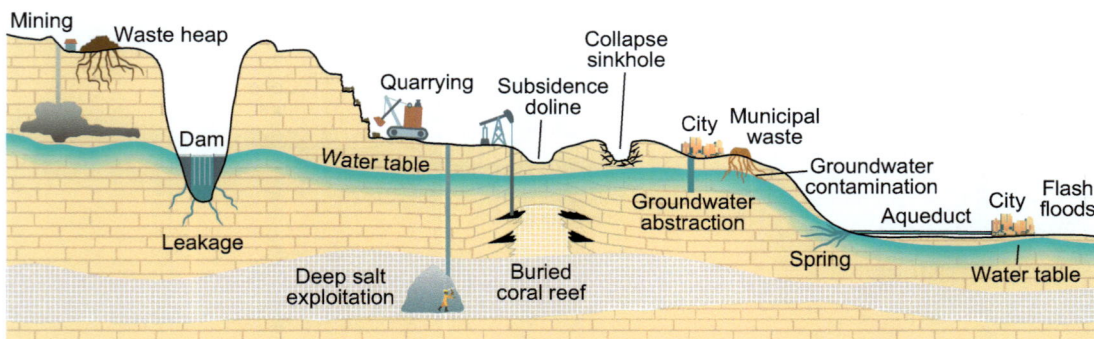

Figure 12.48 Schematic diagram showing some of the human-induced environmental impacts on karst areas.

underground voids can develop without signs of deformation at the surface, because an insoluble caprock forms a stable roof over the subsurface cavern. However, once a critical threshold is reached, and the roof becomes too wide and/or thin, a sudden collapse can propagate to the surface, forming a sinkhole (**Fig. 12.21**).

Human-induced sinkholes can be extremely hazardous due to their rapid formation. Where groundwater is abstracted for drinking water purposes, subsidence will be slow, but can nonetheless cause damage over time. Collapses are often triggered by lowering of the water table. Over-pumping has caused many sinkholes to develop in Florida (USA), which has shallow groundwater and is widely underlain by karst. Collapses here frequently result in loss of buildings, damage to roads and sometimes, loss of life. On the shores of the Dead Sea between Jordan and Israel, the progressively lowering of the lake level in recent years has led to the formation of thousands of sinkholes, parallel to the retreating shoreline. In coastal areas, saltwater intrusion, which occurs very rapidly in karst areas when pumping is carried out, can be a serious problem.

Another hazard, typical of karst areas, is that of flash floods. These events occur as rivers respond rapidly to exceptionally large storms. Flow rates in rivers on karst landscapes can vary by three orders of magnitude, increasing from a few liters to several cubic meters per second in only a few hours. Because so much of the hydrology in karst terrain is intimately connected with the underground, it is difficult to model and to predict such flash flood events.

REVIEW QUESTIONS

12.1 Provide a general definition of karst and its main geomorphological and hydrogeological characteristics.

12.2 What types of rock do karst features typically form on?

12.3 What is the chemistry behind the dissolution of carbonate rocks (limestone) and the formation of secondary calcite in caves?

12.4 Describe the mixing effect and why it so nicely explains limestone dissolution by acidic waters.

12.5 How does temperature affect limestone dissolution rates, and karst formation?

12.6 Define speleogenesis and speleothems.

12.7 Why are groundwater resources so vulnerable in karst areas?

12.8 State what the diagnostic landform of karst is and explain the different ways by which it forms. What other landforms are typical of karst from the mm- to the km-scale?

12.9 What is epikarst and what are some characteristic epikarst landforms/features?

12.10 Explain how karst features like swallow holes and rises (or karst springs) are related.

12.11 Describe the typical evolution of a karst landscape, incorporating the terms doline, uvala, and polje, among others. Place tower (cockpit) karst into this evolutionary sequence and then explain what must happen for a karst landscape to evolve to this last state.

12.12 What are the two main categories of caves and what are their characteristics and differences?

12.13 Explain the different ways by which caves form from tiny fissures to 10-m wide passages. How can we distinguish between the different modes of cave formation?

12.14 Explain the various types of speleothems, how each of them form, and why there are so many different shapes and colors of speleothems.

12.15 What are some of the different types of cave sediments? How can these sediments help us understand how a specific cave has evolved over time?

12.16 What can speleothems and cave sediment tell us about past environments? How is such information obtained?

12.17 What kind of resources can be obtained from karst landscapes?

12.18 What kind of natural and anthropogenic hazards can occur in karst areas?

FURTHER READING

De Waele, J. and Gutierrez, F. 2022. *Karst Hydrogeology, Geomorphology and Caves*. Wiley.

Ford, D. C. and Williams, P. W. 2007. *Karst Hydrogeology and Geomorphology*. Wiley.

Frumkin, A. (ed.). 2013. *Treatise on Geomorphology, vol. 6: Karst Geomorphology*. Elsevier.

Gunn, J. (ed.). 2004. *Encyclopedia of Caves and Karst Science*. Taylor and Francis.

Hill, C. A. and Forti, P. 1997. *Cave Minerals of the World*. National Speleological Society.

Klimchouk, A. B., Ford, D. C., Palmer, A. N., and Dreybrodt, W. (eds.). 2000. *Speleogenesis Evolution of Karst Aquifers*. National Speleological Society.

Palmer A. N. 2007. *Cave Geology*. Cave Books.

Patrick, K. J. 2004. *Pennsylvania Caves and Other Rocky Roadside Wonders*. Stackpole.

Waltham, T. 2008. *Great Caves of the World*. Firefly Books, Natural History Museum.

White, W. B. 1988. *Geomorphology and Hydrology of Karst Terrains*. Oxford University Press.

White, W. B., Culver, D. C., and Pipan, T. (eds.). 2019. *Encyclopedia of Caves*. Academic Press.

13 Soils and Paleosols

Randall Schaetzl

Soil means different things to different people. To a gardener, it is a medium for plant growth. To a civil engineer, it is a type of foundational material, or perhaps something to backfill around a house or in a septic drain field. To a hydrologist, soil functions as a source of water purification and supply. To some geologists, it is the overburden that buried all the rocks! But to geomorphologists and pedologists (**pedology** is the study of soils), **soil** comprises both organic and/or mineral materials, normally at the surface, that have been altered by biological, chemical, and/or physical processes. Another recent definition stresses the importance of biota in soil formation, defining soil as the "biologically excited layer" of Earth's crust. All that being said, most pedologists and soil geographers still use the definition that a soil is a natural, three-dimensional body that has formed at the surface, through the interactions of at least five soil-forming factors (climate, biota, relief, parent materials, and **time**).

Soils are truly the best "integrator" among the various Earth systems, because they reside at the atmosphere–biosphere–lithosphere interface. Soils take in and allow water to filter through them, so they are a key part of the hydrosphere. Almost all life depends – directly or indirectly – on soils. And from a geomorphologist's perspective, soils can provide excellent insight into how landscapes and landforms have evolved, because soils and landscapes develop concurrently. That is the theme of this chapter – how soils and landscapes co-evolve.

To best understand soils, they must be studied in space and time. Indeed, soils are four-dimensional systems, varying with depth (one dimension), across landscapes (two more dimensions), and over time (the fourth dimension). If we can understand how soils change over time, they become important windows into how landforms have similarly changed.

A geographic approach is often one of the most fruitful ways to examine soils. Like most of the components of Earth science, soils are inherently spatial, varying systematically across landscapes. To study "soil landscapes," we examine how and why soils are different on discrete parts of landscapes, for example, hilltops, side slopes, valley bottoms. And so, if geomorphology is the study of landforms and the evolution of Earth's surface, then soil geomorphology is the study of soils across landscapes. Soil geomorphology can help us understand landform evolution and age, surface stability and instability, surficial processes, and past climates. It is the study of the genetic relationships between soils and landforms – more specifically, the interrelationships of soils and slopes.

13.1 GENERAL "SOIL TRUISMS"

Perhaps a good way to begin this discussion is to state a few soil–landscape truisms, ideas, and connections:

- Soil complexity is more common than simplicity. And soils, as well as the soil pattern on the landscape, tend to get more complex with time.
- The principle of uniformitarianism applies to soils, that is, processes that formed old soils are still active today, in modern soils.
- Relatively few, really old soils (in a geological sense) exist. Most soils and surfaces are no older than the **Pleistocene Epoch**. Why? Over time, soils are eroded or buried, or they are modified so much that their "old" precursors are no longer recognized. In short, soils exist at a vulnerable location – the skin of the land.
- Soils are natural clay factories. Weathering in soils turns geologic materials to clay minerals and other, more soluble, byproducts (see Chapter 11). Globally, shales are often just soils that have been eroded and washed into the ocean basins, to become lithified at a later date.
- Humans can and do alter soils, inadvertently as well as intentionally. Not all soils "out there" are natural; humans have accelerated their erosion or led to their burial. Or humans have changed their chemistry. So beware, some soils may be telling us more about ourselves than about Earth.

13.2 THE STARTING POINT: SOIL PARENT MATERIALS

Soils form in unconsolidated geologic materials called regolith. They do not form in solid rock. All loose and unconsolidated materials above solid bedrock is called regolith, regardless of their origin. Rocks must first be weathered to some type of regolith before soil-forming processes can act on it.

Figure 13.0 This soil pit in a glacial kettle contains several buried soils, evidence of previous periods of landscape stability. Source: R. Schaetzl.

If one were to dig a hole to examine a soil, or inspect a deep roadcut, the **soil profile** would be just below the surface. But deeper, one would find material less altered (or completely unaltered) by soil processes (**Fig. 13.1**). Soils form in this **parent material** – what some might call the deep, "not-soil" material. Examples of soil parent materials include dune sand, glacial till, volcanic ash, or indeed, all manner of regolith. The wide variety of possible parent materials sets the stage for soil development; they set limits on what kinds of soil can form at a site. Gradually, over time, and driven by percolating water, generations of roots, and burrowing animals, the upper parts of the parent material develop into soil. The base of the soil gradually deepens, and in so doing, more of the parent material gets slowly converted into a soil. Knowing this, we can use thickness of the soil as a relative dating tool; older soils should be thicker (see Chapter 3).

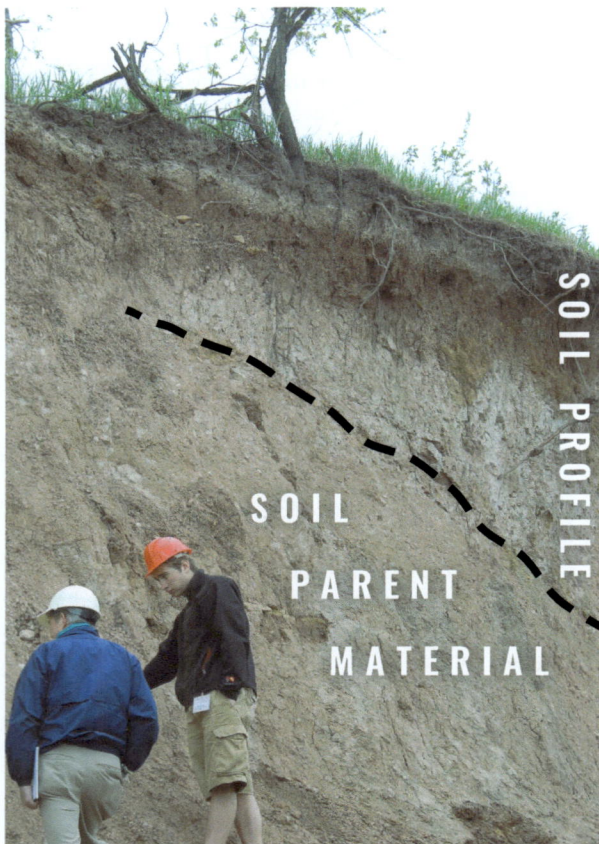

Figure 13.1 A road cut showing a soil profile, with its distinctive soil horizons, overlying the relatively unaltered soil parent material. Source: R. Schaetzl.

Figure 13.1 shows an example of a soil profile, which is a vertical section/cut through a soil, extending down, from the surface, into the parent material. We view a soil profile like we view a person's profile – from a two-dimensional "side-view." Usually, a soil profile also exhibits layers, or horizons. A **soil horizon** is a genetic soil layer – formed by

soil processes. Pedology is the study of soil development, and so, **pedogenesis** is the suite of processes that form soils, and their horizons. Soil horizons are approximately parallel to the land surface and differ from adjacent horizons in a number of ways, including physical, chemical, and biological characteristics such as color, structure, texture, consistency, and kinds and numbers of **organisms**. As we will see below, most soil horizons are formed directly or indirectly by water that percolates through the soil. This is confirmed whenever we see deep "tongues" in soil horizons (**Fig. 13.2**).

Figure 13.2 Deep tongues in this soil from northern Michigan (USA) indicate that percolating water, which often moves through soils along preferred pathways, is the driver by which most soil horizons are formed. Nonetheless, most soil horizons generally parallel the land surface. Scale in cm. Source: R. Schaetzl.

Two major types of parent materials occur on soil landscapes – **mineral** and **organic parent materials** (**Figs. 13.3, 13.4**). Most parent materials are of the mineral type; they are largely composed of sand, silt, clay, and coarser fragments like gravel and stones. Essentially, mineral parent materials were once rocks, but weathering and other geomorphic processes, for example, glaciation, broke the rock up into finer particles. Under just about every upland soil, one will find a mineral parent material (**Fig. 13.4C**).

Some parts of the soil landscape are underlain by organic materials, composed of plant materials in various stages of decay (**Fig. 13.4A, B**). At some depth below, mineral materials are present. Organic parent materials are mainly found in wet, low-landscape positions, where a high water table and colder soil temperatures inhibit decomposition of plant matter that accumulates there (**Fig. 13.4A**). Often, such sites have standing water at the surface. Any leaves, sticks, seeds, or other plant materials that fall into the water decay only slowly, due to anoxic conditions. Common names for such materials are **peat**, if only minimally decomposed (**Fig. 13.3B**), or **muck**, if decomposition is more complete. Most organic soils are found in cool–cold climate locations, and often on landscapes that were recently glaciated.

Another way to categorize parent materials is by their geologic history. Did the parent materials form *in situ*, or

Major types of soil parent materials

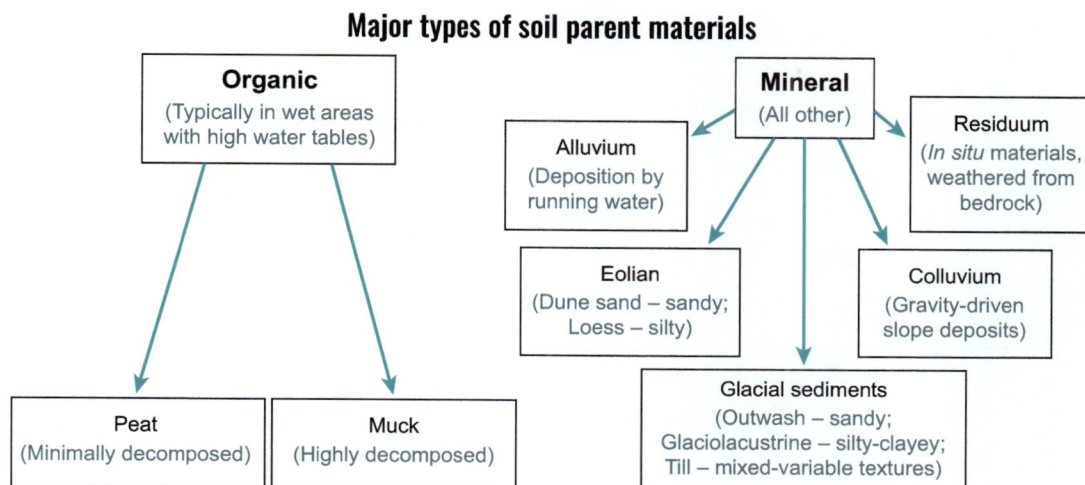

Figure 13.3 A flowchart listing the major types of soil parent materials.

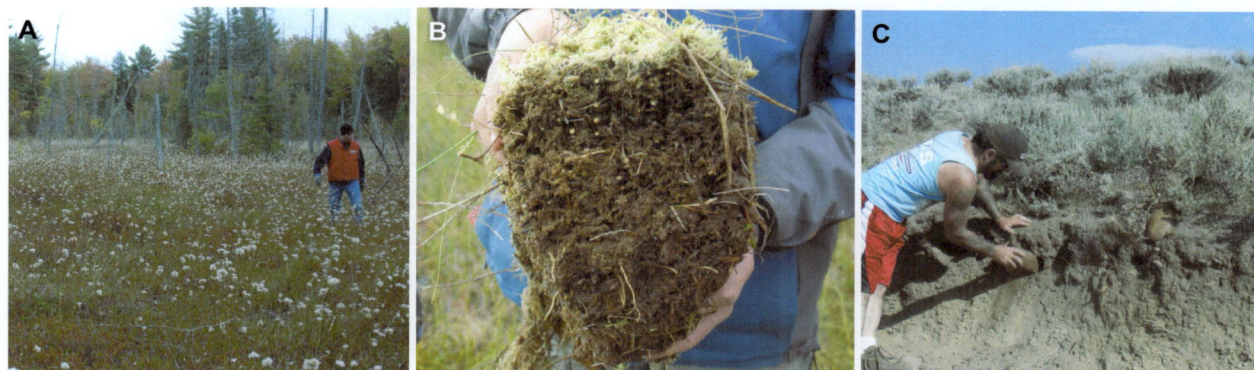

Figure 13.4 The two main types of soil parent materials. **A.** In this swamp in northern Michigan, the water table is near the surface. Because the water inhibits decomposition, thick accumulations of (**B**) peat – an organic parent material – can form. **C.** Most soils are formed in *mineral* parent materials like this one in Wyoming, from a mixture of sand, silt, and clay, with some rocks. Source: R. Schaetzl.

were they transported to their current location? *In situ* parent materials form when rocks weather in place; the resulting sediment is called **residuum**. Residual parent materials are common wherever bedrock is near to the surface (**Fig. 13.5**). Organic soil materials (**Fig. 13.4A**) are also a type of residual parent material. All other parent materials have been transported to their present location, usually by wind, running water, glaciers, or gravity.

13.3 SOIL PROCESSES AND SOIL HORIZONS

Soil processes, more correctly termed **pedogenic processes**, operate on parent material, over time, to form soils and soil horizons. They can be grouped into four main types: **additions**, **removals**, **translocations**, and **transformations**. Let's examine these processes and see how each forms the various horizons in soils.

Imagine a pile of raw parent material, perhaps beach sand or a landslide deposit, that lacks a soil cover. Some of the

first processes to affect this material, to start forming a soil, will be additions. Rain will add dissolved substances, dust will blow onto the soil, and plant materials (leaves, seeds, etc.) will accumulate on the surface. Accumulated plant materials are the most common and universal type of "addition." Where present, regardless of their state of decay, they form an **O horizon** (**Fig. 13.6**; **Table 13.1**). Most soils in their "natural" state have an O horizon.

Over time, the plant materials in O horizons decompose into more soluble organic substances; as a whole, we call decomposed organic materials **humus**. Let a banana (*sans* peel) ripen for a few weeks and you will have good humus! Microbial decomposition of raw organic material in the O horizon, to form humus, is a type of soil transformation.

Because humus is soluble, it is readily translocated in soil water into the mineral soil below – our first translocation process. **Translocations** are necessary to form most soil horizons. Humus translocated out of the O horizon, downward, by percolating water doesn't move far; it accumulates in the

Figure 13.5 An example of residuum. Quartz veins in the weathered granite remain visible at depth because quartz is a highly weathering-resistant mineral. Source: R. Schaetzl.

Table 13.1 The major types of soil horizons

Horizon	Characteristics
O	Layers dominated by organic material (litter and humus) in various stages of decomposition
A	Dark-colored horizons at the surface or below an O horizon, characterized by an accumulation of decomposed organic matter (humus) mixed into the mineral fraction
E	Light-colored horizons dominated by processes of eluviation, i.e., losses of weatherable minerals, clay, iron, aluminum, and/or humus, resulting in a concentration of mostly uncoated quartz grains and other resistant materials
B	Subsurface horizons usually dominated by illuviation, i.e., gains of clay, iron, aluminum, humus, carbonates. Some B horizons have formed by the residual concentrations of sesquioxides (Oxisols) and/or gleying (wet soils)
C	Mineral horizons, excluding bedrock, that have been minimally affected by pedogenic processes and whose characteristics largely reflect unaltered parent material
R	Hard bedrock that is so coherent as to make digging by hand impractical

The soil profile and soil horizons

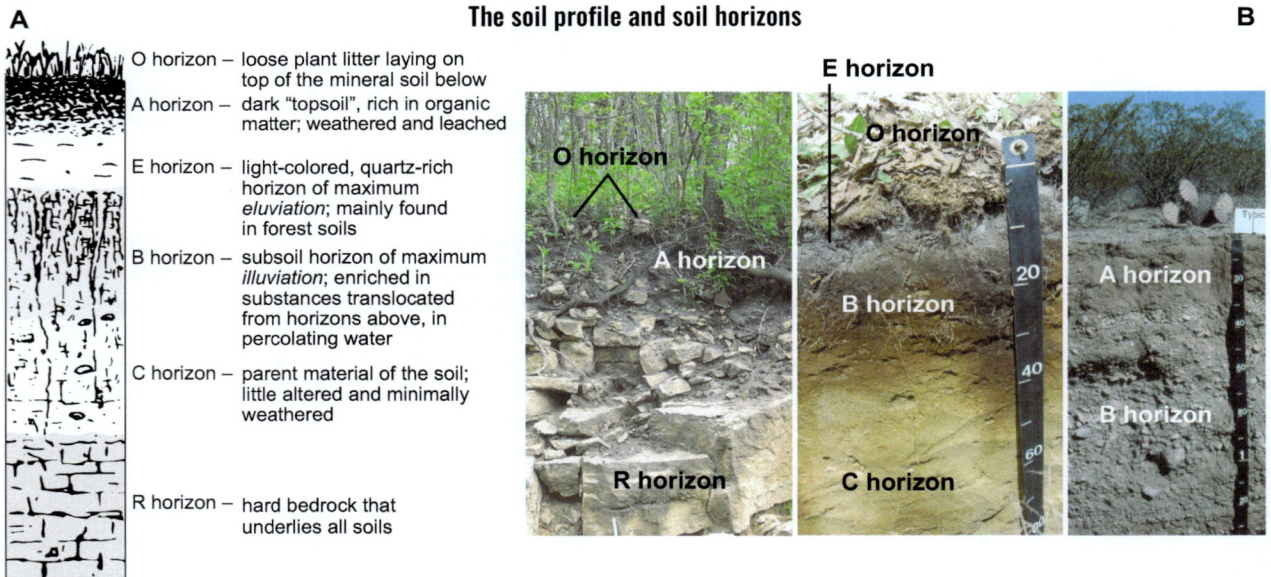

A.

O horizon – loose plant litter laying on top of the mineral soil below

A horizon – dark "topsoil", rich in organic matter; weathered and leached

E horizon – light-colored, quartz-rich horizon of maximum *eluviation*; mainly found in forest soils

B horizon – subsoil horizon of maximum *illuviation*; enriched in substances translocated from horizons above, in percolating water

C horizon – parent material of the soil; little altered and minimally weathered

R horizon – hard bedrock that underlies all soils

Figure 13.6 The major types of soil horizons. **A.** Shown graphically (see also Table 13.1). **B.** Examples of soil horizons, using images of actual soil profiles. The examples in B illustrate that not every soil contains all of the master soil horizons. Scale in cm. Source: R. Schaetzl.

upper part of the parent material, where it coats the sand, silt, and other particles. The dark humus coatings (you can think of them as soil "paint") form a dark-colored **A horizon**,

enriched in organic matter. A horizons are the uppermost mineral soil horizon, usually immediately below the O (**Figs. 13.6, 13.7**; **Table 13.1**). Most A horizons are thin, but in other soils, they can be thick and dark. To the non-specialist, the A horizon is "topsoil."

In humid climates, percolating water continues to translocate humus downward, thickening the A horizon. But humus is difficult to translocate very deeply. It's not *that* soluble, because it is geochemically attracted to soil particles, and many soil pores are simply too small. Most humus, therefore, gets "filtered out" in the A horizon. So, below the A horizon, the "clean" percolating water picks up other substances from the upper part of the soil profile and translocates them downward, even farther. These substances, commonly clay, iron, aluminum, and other soluble, mineral compounds, eventually get deposited in the **B horizon**. Most people would refer to the B horizon as the "subsoil." B horizons are usually enriched in clay, but in other soils, they become reddened from additions of iron oxide minerals (**Figs. 13.6**, **13.8**), and in soils forming in dry climates, the B horizon can accumulate additions of $CaCO_3$ (**Figs. 13.6**, **13.7**).

Translocation of substances *into* a soil horizon is called **illuviation**, whereas translocation of materials *out* of a soil horizon is called **eluviation** (**Fig. 13.6**). The B horizon is the zone of maximum illuviation, gaining materials from all of the overlying horizons. A horizons gain materials (humus) from the O horizon above, but lose materials to the B horizon below; they are simultaneously undergoing eluviation *and* illuviation.

Eluviation is optimized by ample water to percolate through the soil, by porous and permeable soils, and by acidic conditions which help to mobilize many soil constituents. In many soils, especially in wetter climates and below forest vegetation, eluviation is so strong in the upper part of the soil that a zone of extreme eluviation – an E horizon – forms immediately below the A horizon. The E horizon is typically light colored, because it takes on the color of uncoated quartz grains (**Figs. 13.6**, **13.8**; **Table 13.1**). In most E horizons, the coatings on these grains have been stripped and translocated to the B horizon, as have most other components such as clay, iron, and humus (**Table 13.1**). Many E horizons, therefore, appear almost snow-white in color (**Fig. 13.8**).

Figure 13.7 A soil from South Dakota, USA. In this dry climate, dust, rich in $CaCO_3$, is commonly in the air. Even though the climate is semi-arid, ample precipitation exists to translocate the light-colored carbonate minerals from the dust into the B horizon, where they accumulate. Note also the thick, dark A horizon in this soil. Scale in cm. Source: P. Schoeneberger.

Figure 13.8 A forest soil from northern Michigan, USA, with a well-expressed E horizon and a thin, dark B horizon below. The O horizon in this Spodosol is also noticeably thick. Source: R. Schaetzl.

The B horizon is the deepest horizon formed by pedogenesis. Most people consider the suite of O–A–E–B horizons as the complete soil profile. The horizons below the profile are more geologic in origin than they are pedogenic.

Geologic materials below the soil profile that are waiting to become soil include the **C horizon** (the parent material) and at some depth below that, bedrock (the **R horizon**) (**Fig. 13.6**; Table 13.1). Although we consider the C horizon to be unaltered parent material, in reality parts of the upper C horizon have likely been slightly altered by pedogenesis. Slowly, the soil profile thickens and encroaches down into the C horizon, converting geologic material to soil material. By comparing characteristics of the A, E, and B horizons to that of the C horizon below, we can estimate the amount of change that has occurred in the soil since time$_{zero}$ (when soil formation began). This information, along with overall soil profile thickness data, is useful, because soil development can usually be correlated to age (or time elapsed since time$_{zero}$; see Chapter 3). Soil characteristics vary with time – older soils generally have thicker profiles and are more weathered. But soils also vary considerably based on geomorphology; soils on some kinds of slopes or surfaces will form faster or differently than on other landscape positions.

13.4 SOIL TEXTURE

One of the most important soil characteristics is its **texture** – the term that describes its relative proportions of sand, silt, and clay. Texture can be thought of as how a soil "feels." It is determined only on the **fine-earth** fraction of the soil – particles < 2 mm in diameter (**Fig. 13.9**). Larger fragments, named **coarse fragments** and further described based on their sizes, include gravel, stones, and boulders. Soil analyses and comparisons, however, typically only utilize the

fine-earth fraction. Within that fraction, sand grains are the largest, clay the smallest. Sand grains are composed mainly of quartz and other primary minerals, derived from physical weathering of rocks. Most silts have similar mineralogy, but are simply smaller. But clay minerals have formed differently and have vastly different mineralogies, and as a result they behave very differently in soils. Clay minerals are derived by chemical weathering, and thus, are called secondary minerals. Common clay minerals include kaolinite, illite, smectite, vermiculite, and oxides of iron and aluminum. Clays impart "stickiness" to soils. Because they typically have negatively charged surfaces, clay minerals are capable of retaining nutrients (which plants can then take up). By virtue of their large surface area/mass ratio, clay minerals also have high water-holding capacities. Even a small amount of clay can change how a soil functions regarding plant growth and hydrology (water runoff vs run-in), and thus have a large influence on how geomorphically stable the soil is.

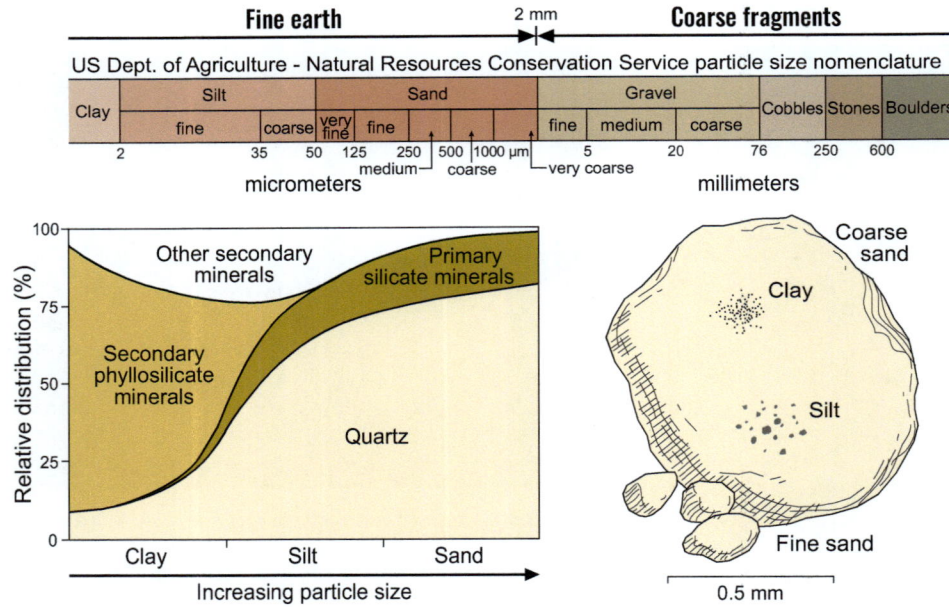

Figure 13.9 Particle size limits and names for the various mineral components of soils and sediments, as used in the US Department of Agriculture system.

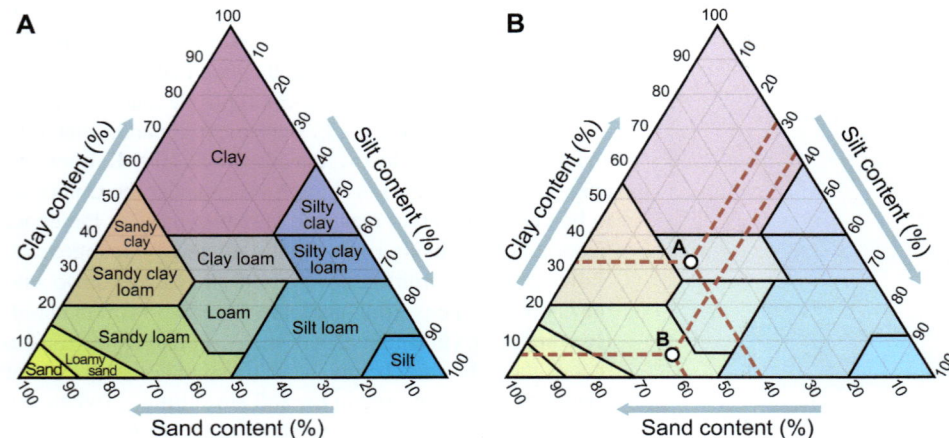

Figure 13.10 A. The standard USDA textural triangle. **B.** Examples are shown for soils with textures of (A) clay loam (28% silt, 40% sand, and 32% clay), and (B) sandy loam (37% silt, 57% sand, and 6% clay).

The relative proportions of sand, silt, and clay give a soil its distinctive "feel." A clayey soil feels sticky, a sandy soil feels gritty, and a silty soil feels smooth. Soils with a good mix of sand, silt, and clay have components of all three of these "feels" and are described as **loamy**. Soil scientists have developed a scheme to communicate the variety of soil textures, by using a ternary diagram known as the **textural triangle** (**Fig. 13.10A**). Loam textures are generally in the central part of the triangle. Soils heavily dominated by sand, silt, or clay are given that name as a texture. A soil dominated by silt but which has a fair amount of sand and clay would be a silt loam. Two examples are given in **Fig. 13.10B**, to illustrate how texture classes are determined by using the triangle. Because clay so dominates the "feel" of a soil, textures with "clay" modifiers occupy the largest proportion of the triangle, such that a soil with only 40% clay will still be classified as having a "clay" texture (**Fig. 13.10A**).

13.5 TYPES OF (MINERAL) SOIL PARENT MATERIALS

Soils form in unconsolidated parent materials, that is, regolith. A relatively young soil will closely resemble its parent material because pedogenesis has only minimally altered the parent material since the soil started forming. This is especially true for soils that date back only to the last major glaciation, between 25 and 10 ka. In these "postglacial" soils, it is usually possible, sometimes very easy, to determine the parent material. But in really old soils, especially in the tropics, there has been so much alteration by pedogenesis – over such long timespans – that often we cannot tell what the original parent material was.

That said, what are the major parent materials for soils? What is the best way to determine a soil's parent material? Answer #1: Dig. At some depth in most soils, maybe a meter or two, or perhaps three, parent material will be encountered. The first thing to notice is its texture – if it is sandy the parent material may be dune sand, beach sand, or glacial outwash. Lake sediment is usually rich in clay and silt. The siltiest soils have formed in loess – wind-blown silt (see Chapter 21). **Stratification** (layering) may suggest that the parent material is some type of alluvium. Many soils in hilly and mountainous regions have formed in bedrock residuum, which generally reflects the texture of the bedrock; sandstone residuum will be sandy, shale residuum will be clayey, etc. Granites and metamorphic rocks weather to coarsesandy soils because many of the minerals within are so large that they become sand in the regolith. On lower slopes of these mountainous areas, soils formed in colluvium and alluvium are likely; they often reflect the general texture and mineralogy of the rocks upslope. Soils formed in volcanic ash and tephra (see Chapter 7) are loamy; most are silt loam in texture.

The parent material list could go on even longer, but it is also important to consider answer #2: Look at the landform. Interpreting the landform is also very useful in determining the parent material for a soil. For example, one might expect gravelly alluvium on an alluvial fan, or silty-sandy playa sediment at the distal margin of the fan. In reality, this is a two-way street; the geomorphologist uses information from the soil (pit) to help better understand the landform type, and information on landform morphology to suggest the likely parent material for its soils. That is the nature of soil geomorphology, where everything "fits together."

13.5.1 Lithologic Discontinuities and Stone Lines

Although most soil profiles form in a single parent material, some have formed in two or more parent materials, stacked one on top of the other. In this case, a thin layer of one parent material (the younger of the two) lies atop the lower and older parent material. The contact between these two different parent materials is called a **lithologic discontinuity** (**Fig. 13.11**). Identification of a lithologic discontinuity can be very insightful to anyone interested in the geological history of a site. Lithologic discontinuities are common on aggrading surfaces, where parent materials accumulate by deposition, one on top of another, as seen in **Fig. 13.11**. Here, the discontinuity represents a clear break in type of past depositional systems. These types of discontinuities are typical for recently glaciated landscapes, or lowlands where materials such as alluvium or colluvium episodically accumulate on the land surface.

Lithologic discontinuities can also form on unstable, eroding surfaces. Here, the discontinuity represents a former erosion surface, which has subsequently been buried (by a second parent material) (**Fig. 13.12**). Often in such circumstances, the evidence of the erosion event is manifested as a layer of stones or gravel – a **stone line** – left behind because the erosional processes were unable to transport away these larger particles. Bear in mind that most lithologic discontinuities are *not* marked by a stone line.

Although a stone line – whether at the surface or in the subsurface – can indicate a former erosional surface, they more frequently form by bioturbation, or soil mixing by plants and animals (biota) (**Figs. 3.9, 13.13**). Many soil biota are constantly moving sediment within the soil, in search of food or for shelter.

Movement of soil by bioturbators takes three general forms. The most common of these involves upward transfers of fine sediment by insects and invertebrates, such as ants or termites, often forming distinct mounds (**Fig. 13.14**). This group of bioturbators is called **moundmakers**. As these upward sediment transfers occur, particles too large to be moved will slowly settle to the depth of burrowing, forming a subsurface stone line. Other biota burrow but do not make mounds; these are called **mixmasters**. They constantly churn the soils, but their actions are largely underground and unseen. But the end result is the same; large gravels and rocks pass through the burrows as the fauna tunnel, eventually settle to the depth of burrowing. Lastly, some soil animals do most of their work by digging, scratching, and burrowing at the surface; these are **cratermakers**. Most cratermakers are larger mammals like badgers, skunks, and aardvarks, but many birds scratch and dig away at the soil surface too, in search of food. This action, as well, gradually

Figure 13.11 Examples of soils with lithologic discontinuities – indicated by a dashed line. **A.** Within a fluvial terrace, the discontinuity in this soil indicates the change from rapidly flowing glacial meltwater (gravelly alluvium) to post-glacial alluvium, when the river was being supplied mainly by precipitation, and hence, it was no longer able to transport the larger gravels. **B.** This soil, on the bed of a former glacial lake, shows lake sediment at depth. After the lake dried up, sand dunes migrated onto the site, forming the lithologic discontinuity and marking a major change in depositional systems (and climate!). **C.** As the glacier was retreating from this site, it deposited thick sequences of sandy outwash. Later, a cover of wind-blown silt (loess) was deposited across the landscape, burying the outwash. Scales in cm; in part B the scale is in units of 10 cm. Source: R. Schaetzl.

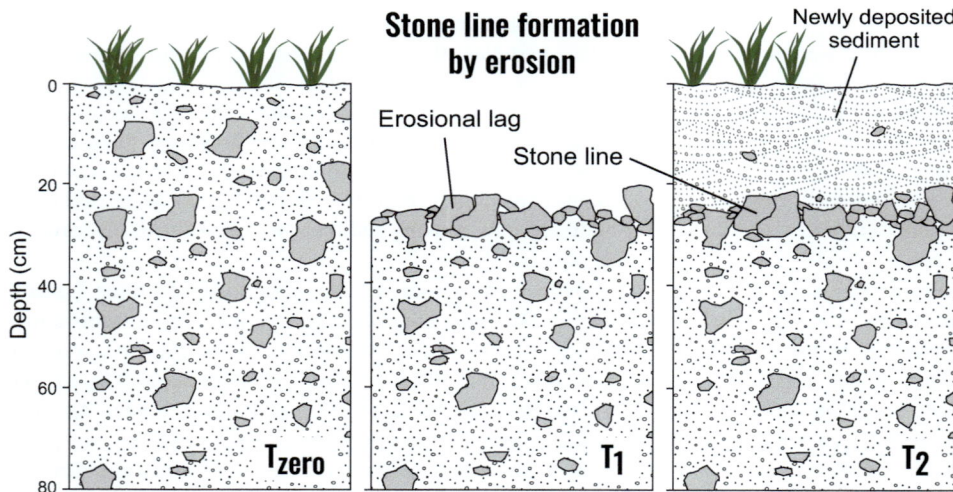

Figure 13.12 Diagram showing how a stone line can indicate a past erosional event. In this case, the stone line would also clearly mark a lithologic discontinuity.

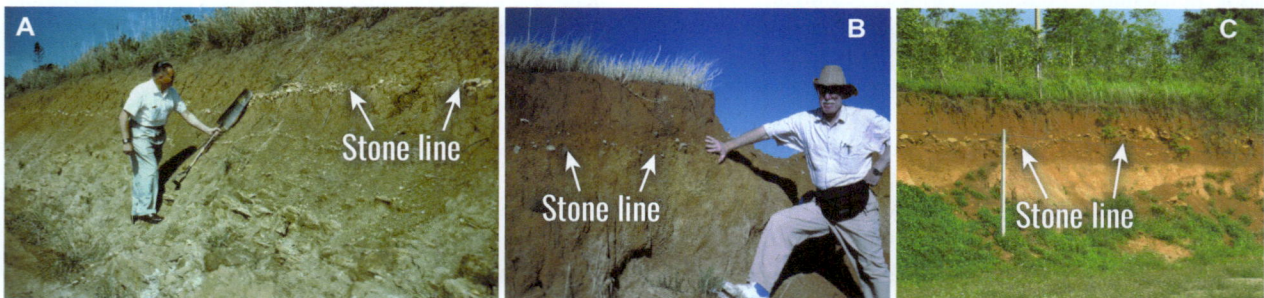

Figure 13.13 Buried stone lines formed by bioturbation, presumably by termite activity, in: (**A**) Kenya; (**B**) South Africa; and (**C**) Vietnam. Source: D. Johnson.

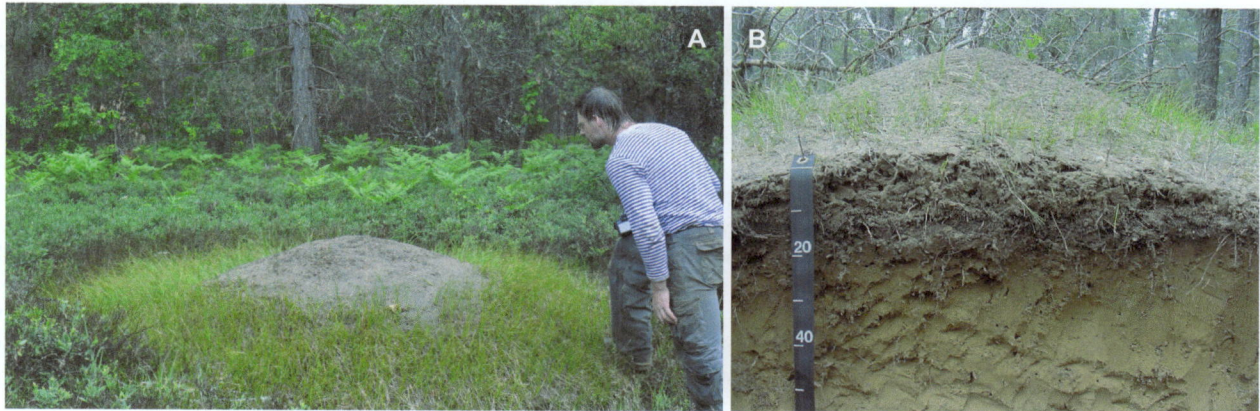

Figure 13.14 A mound formed by the ant *Formica insectoides*, in northern Michigan, USA, illustrating the process of bioturbation. Any particles too large for the ants to move, usually gravel-sized or larger, are slowly lowered in the soil to the depth of burrowing. Note the distinct A horizon in part **B**, indicative of the former soil surface. Source: R. Schaetzl.

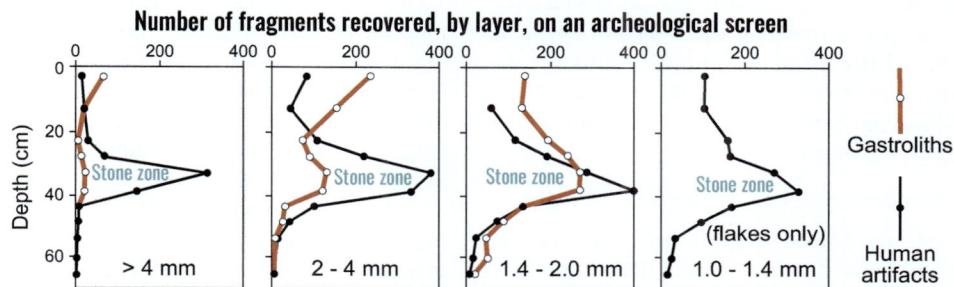

Figure 13.15 Depth distributions of human artifacts and gastroliths (stones deliberately ingested by birds to aid in digestion, and later either excreted or coughed up, landing on the soil surface) at an archeological site in Illinois. Note how all of these coarse fragments have been lowered to ≈35 cm depth by bioturbation (in this case, mainly earthworms), forming a stone line. Source: After Van Nest, J. (2002), reproduced with permission from Wiley & Sons.

builds up a layer of soil at the surface that is largely free of coarse fragments, and possibly with a stone line below.

These forms of bioturbation lead to the formation of a relatively stone-free layer – a **biomantle** – often above a subsurface stone line. The only instance where such activity would not form a stone line is in sediment that lacks coarse fragments entirely.

Geoarcheologists often utilize this type of knowledge. For example, at former sites of human habitation, many artifacts such as tools and projectile points no longer lie on the surface, but are discovered only by subsurface excavation (**Fig. 13.15**). After it became clear that bioturbation can not only "lower" surface clasts within the soil, and that bioturbation is more widespread than was formerly thought, geoarcheologists have become better able to interpret previous human habitation activities.

13.6 TYPICAL SOIL PROFILES AND PEDOGENIC PROCESSES

Perhaps the best way to understand soil processes is to examine some typical soil profiles and place them into their climate type, vegetation assemblage, or landscape position. In the discussion below, "typical" soils types are linked to

the processes that have formed them, and tied into a broader discussion of landscape evolution and geomorphology. The soils will be discussed using their taxonomic names. The US Soil Taxonomy system recognizes 12 different soil orders – the highest classification level.

As we have seen, soils acquire many characteristics because of inheritance from their parent material. Parent material is considered a **soil-forming factor** (not a process). The complete suite of soil-forming factors is traditionally given by the equation:

$$\text{Soil} = f(cl, o, r, p, t \ldots),$$

where the *soil* and its morphology are seen as functions of *climate*, *organisms* (plants mainly, but animals also affect soils), *relief* (or landscape position – hilltops, valley bottoms, etc.), and *parent material* (especially important in young soils), all operating over *time*. The "…" indicates that other, less important factors can sometimes come into play, such as fires, invasive species, and human actions such as cultivation or manuring. The five soil-forming factors all play a part in influencing which pedogenic processes are operative, that is, which actually *form* the soil. To reiterate: the factors "set the stage" for pedogenesis, whereas the processes "do the work." For example, in a soil landscape with a dry climate and covered with short grasses, certain soil-forming processes will be operative, whereas others will not. The climate and organisms factors set the "pedogenic stage," which facilitates certain processes and excludes others.

We have already discussed the four main types of soil-forming processes – additions, removals, translocations, and transformations. Within these groups are more focused types of processes, as discussed below.

In order to better determine the influence of each soil-forming factor, geomorphologists often examine a suite of soils, and within this suite we hold four of the five soil-forming factors constant. In this type of experimental setup, we know that most of the differences in those soils would then be due to that one factor which has been left to vary. For example, we may study five soils of different wetness/drainage characteristics. But all of them would have formed in the same parent material, under the same type of forest, in the same climate, and are all of the same age. In this case, the major differences among the soils would be due to the soil-forming factor *relief*. Let's first look at the soil-forming factor *organisms*.

13.6.1 Soils Influenced Strongly by Organisms

Mollisols are typically formed in grassland environments. Their defining characteristic is a thick, dark-colored A horizon, rich in humus (**Figs. 13.7, 13.16A**). The dark color, and the great depth to which it extends, derive not from humus that was translocated into the soil from an O horizon above (as discussed earlier), but from roots. Grassland ecosystems are root factories, producing thick mats of fine roots each year, many of which decay in place to form humus.

In essence, beneath grasses, the humus is "injected" directly into the soil by decaying roots, rather than having to be washed in from the O horizon above. Both root growth and root turnover are rapid and ongoing beneath grasses.

Contrast that to soils that have formed under forest, which have O horizons on the surface and thin A horizons below (**Figs. 13.6, 13.8, 13.16B**). Most of the organic matter additions to the soil come from leaf-fall, because woody-root turnover is slow. The comparatively thin A horizon in such soils has formed mainly due to humus that gets translocated into the mineral soil by percolating water. Tree roots are not large contributors of organic matter and humus, because they last for years or decades. Unlike the roots of grasses and forbs, they do not "turn over" annually.

13.6.2 Soils Influenced Strongly by Climate

Aridisols form in dry climates. Despite what you may think, percolating water remains an important process in Aridisols. To be sure, Aridisols form slowly, and for most of the time they remain dry, with little pedogenesis actually happening. That said, the influence of precipitation and percolating water is very important in these soils; it may happen infrequently during brief rainstorms, but when it does occur, it helps form B horizons enriched in soluble compounds such as carbonates, gypsum, and salts.

A horizons are thin in Aridisols; there is little vegetation to produce organic matter, and much of what falls to the surface decays and blows or washes away (**Fig. 13.17**). In many Aridisols, the A horizon is extremely porous, so much so that it is described as **vesicular** (full of vesicles, or small pores). The vesicular characteristic occurs as the soils grow upward by small additions of eolian (wind-deposited) dust. As this surface layer of dust wets up after a rain, small pores form and are retained after drying. Thicker and more vesicular A horizons are a hallmark of well-formed Aridisols on stable surfaces.

A typical Aridisol has a thin, vesicular A horizon that is low in organic matter. But their B horizons – the products of translocations/additions of soluble compounds – can sometimes be thick and even indurated (cemented) by these compounds. The little rain that falls on Aridisols must do its work immediately; there may be only one pulse of water into the soil every few months. Precipitation dissolves what soluble materials it can from the upper profile and translocates them to the subsoil. The soluble materials get deposited at roughly the mean depth of the wetting front. In slightly wetter climates, where wetting is typically deeper, this depth of translocation may be greater than in the more arid deserts. Hence, B horizons will be deeper, and perhaps thicker. B horizons simply accumulate these soluble compounds, because there is hardly ever enough rain to translocate the compounds out of the profile. In short, the B horizon continues to gain materials over time. Thus, as discussed in Chapter 3, the content of illuvial, soluble materials in Aridisols has proven to be a valuable soil chronometer.

How do the potentially soluble compounds come to be on the soil surface in the first place? Most of the soluble

Figure 13.16 A comparison of the typical morphology of (**A**) a grassland soil from Iowa, vs (**B**) a forest soil from Michigan, USA. Scales (in cm) are similar in both photos. Source: (A) J. Sandor; (B) R. Schaetzl.

substances that end up in Aridisol B horizons were weathered out of rocks or sediments elsewhere and blown onto the soil surface as dust. Thus, soils just downwind of sources of compounds like carbonates, salts, and gypsum will often have accumulated more of them in their B horizons. In areas downwind of carbonate rocks like limestone, carbonate-rich dust settles on soils, and their B horizons are mainly enriched in carbonates (**Figs. 13.7, 13.17**). Gypsum-rich soils are usually downwind of dry playa lakes, where the lake sediment is rich in gypsum and minerals containing sulfur. Soluble salts can even accumulate in soils where a desert abuts an ocean. As waves send droplets of sea spray into the air, the water evaporates. The tiny salt crystals that form in this manner are blown inland, and onto soils. In summary, desert soils are accumulators of what is blowing in the wind, and in deserts there seemingly is always something blowing around.

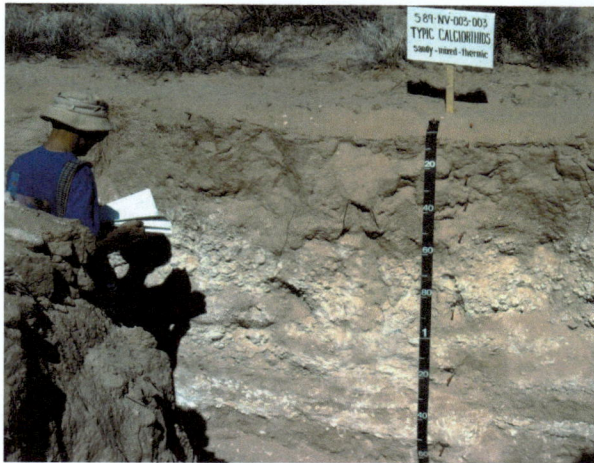

Figure 13.17 An Aridisol in Nevada, USA, showing the accumulation of carbonates (light colors) in the B horizon. Scale in cm. Source: R. Schaetzl.

On the opposite end of the climate spectrum from Aridisols are **Oxisols** – soils of the humid tropics. Here, soils are very old, and are continually leached as water percolates through them, to great depths. **Chemical weathering** is so strong in these environments (see Chapter 11) that most primary minerals have long been weathered away, leaving behind soils that are composed mainly of secondary clay minerals, mainly oxides like hematite, gibbsite, and kaolinite. Although rich in clay, these kinds of clays clump together to give the soils a sandy feel, which maintains their high porosity. Water continues to percolate through and through, leaching out nutrients; Oxisols are inherently infertile. The tell-tale feature of Oxisols is their red-orange color, indicating the dominance of oxide clays like hematite. A related soil order – **Ultisols** – are comparable in many ways, but form in the subtropics, and therefore are not as "extreme" in the characteristics discussed above (**Fig. 13.18**). But they are still red!

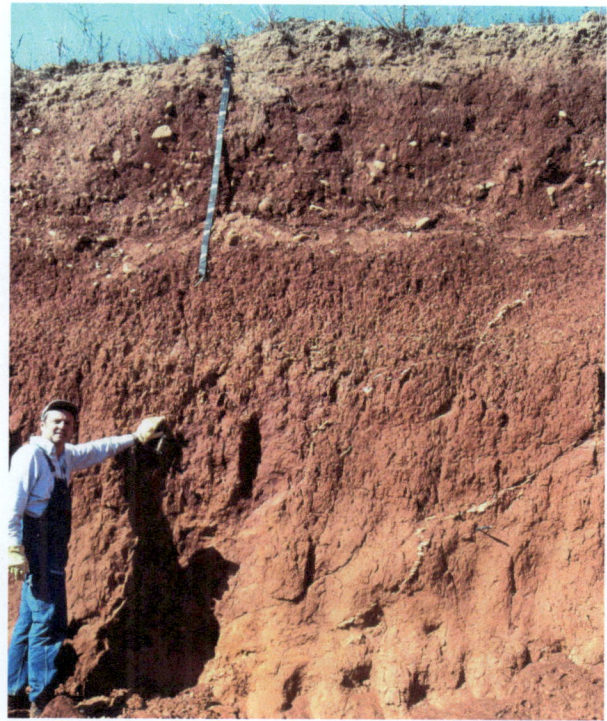

Figure 13.18 An Ultisol in North Carolina, USA. The red colors, typical of both Ultisols and Oxisols, are here indicative of hematite and kaolinite clay minerals. Source: R. Schaetzl.

In cool, humid climates, where coniferous forest is common, forest litter can accumulate to great thicknesses on the forest floor, due to (a) its acidic nature, (b) its high contents of waxes in the needles, and (c) the cool conditions, including long snowy winters. All of these factors slow decomposition of the litter. Snowmelt waters percolate deeply into the soils, as do fall rains. As a result, these soils – called **Spodosols** – are deeply leached and acidified. The acidic water helps to weather primary minerals and strip coatings off sand grains, as it translocates the byproducts to the B horizon. Red-brown, iron-rich, and wavy (reflective of the tortuous paths that water takes as it percolates) B horizons are the hallmark of Spodosols (**Figs. 13.2, 13.6, 13.8**). The best-developed Spodosols occur in sands, where translocations are unhindered, and in snowy climates.

Intense and frequent translocation processes also drive the formation of E horizons in Spodosols, some appearing snow-white because the quartz sands within have been stripped clean by the acidic soil water. (Sands are mostly quartz [**Fig. 13.9**].) Whatever *had* coated the sand grains in the E horizons (iron and humus, mainly) ends up accumulating in the B horizon. Spodosols epitomize the process couplet of removals-additions, leading to E–B horizon sequences that are very apparent in the field.

13.6.3 Soils Influenced Strongly by Relief/Topography

If **topography** represents the variation in elevation across a geomorphic surface, then relief is a more quantitative

Figure 13.19 Contrasts in color for a typical upland soil (**A**: North Carolina, USA) and a poorly drained, lowland soil (**B**: Michigan, USA). The soil in **B** normally has a water table that is even higher than shown here. The high water table and persistently wet conditions create gleyed conditions and gray colors. In upland soils, as in **A**, with a deep water table, oxygen can diffuse into the soil and maintain oxidizing conditions, leading to red-brown colors. Scales in cm. Source: R. Schaetzl.

measure of how variable that surface actually is. A landscape may have flat topography, or rolling topography, or highly incised (deep, narrow valleys) topography. Relief defines how rolling, flat, or incised the land surface is. Unlike topography, which is a qualitative descriptor, we can put a number on relief; it is the difference in elevation (in feet, meters, etc.) between the highest and lowest points on a landscape. However, it is important to note that, with regard to the five soil-forming factors, the term "relief" is actually better stated as topography.

Relief/topography affects soils by its impacts on (1) the depth from the soil surface to the water table, and (2) the steepness (gradient) of the land surface. Let's discuss these factors in turn.

If we exclude desert areas, then in most lowland or bottomland areas, the water table is close to the surface, or even at the surface. Conversely, on uplands, the water table is usually much deeper. When the water table is deep, translocation processes are able to operate more-or-less freely, driven by vertically percolating water. In such soils, oxygen is able to diffuse into the soil from the atmosphere (which is the main source of soil oxygen). Thus, the pores in freely aerated soils are rich in oxygen, which maintains oxidizing conditions in the soil. Iron-bearing minerals stay oxidized and soil colors take on shades of red, brown, orange, and yellow (**Fig. 13.19A**). Contrast that with wet soils, where much of the soil is below the water table. Saturated conditions exclude much of the oxygen that would normally diffuse into the soil, so that iron-bearing minerals quickly become chemically reduced. Soils in a reduced state are gray-colored, a condition we call **gleyed** (**Fig. 13.19B**). As you see, color alone can tell us a great deal about the hydrology of a soil, and these colors

persist for some time, even if that hydrology changes, or if the soil gets buried.

13.7 SOIL VARIATION ACROSS LANDSCAPES: SOIL GEOMORPHOLOGY

The previous discussion focused on soils that vary significantly due to the influence of climate, vegetation, and relief (which impacts wetness). To observe soils that vary in climate and vegetation, we may have to traverse great distances, from the tropics to the cold mid-latitudes, or from the prairies to the forests. Soil geomorphologists are mainly concerned with much smaller-scale variations, that is, across a field or up-and-down across a series of ridges. At this scale, the soil-forming factor of relief dominates how the soils develop, although variability in parent material is also important. Recall that, in this context, the term relief refers to more than just elevation differences; it includes all the ways that the land surface changes spatially.

Soils form on surfaces, and surfaces can be defined (a) genetically (as to origin), (b) geomorphically (from stable to unstable), and (c) morphometrically (as to shape and geometry). An example of a genetic classification would be a surface formed on a glacial moraine, or along an eroding bedrock escarpment. In both cases, the genetic origin of the surface is apparent in the description. Geomorphic stability and instability refer to the rates of erosion/deposition on the surface (**Figs. 3.3, 13.20**). Pedogenesis is allowed to operate unhindered on parent materials on stable surfaces. Thus, the potential exists for well-developed soils on stable surfaces. On unstable surfaces, soils either become buried or eroded, and as a result, the soils there will not be as well developed,

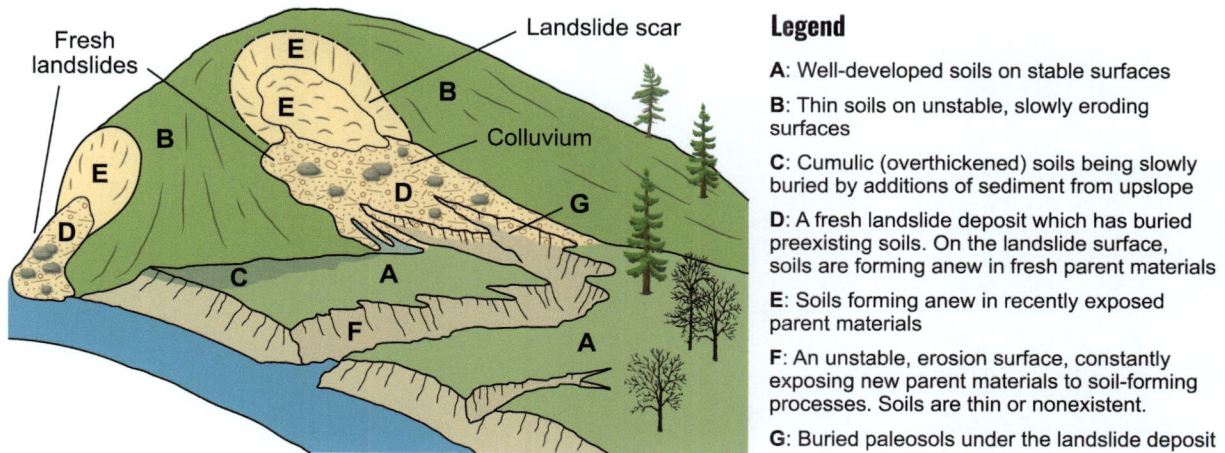

Legend

A: Well-developed soils on stable surfaces

B: Thin soils on unstable, slowly eroding surfaces

C: Cumulic (overthickened) soils being slowly buried by additions of sediment from upslope

D: A fresh landslide deposit which has buried preexisting soils. On the landslide surface, soils are forming anew in fresh parent materials

E: Soils forming anew in recently exposed parent materials

F: An unstable, erosion surface, constantly exposing new parent materials to soil-forming processes. Soils are thin or nonexistent.

G: Buried paleosols under the landslide deposit

Figure 13.20 A fictional landscape showing areas of surface stability and instability, illustrating how soils on those surfaces can help ascertain the degree to which surfaces are stable and unstable.

or they may be buried. Nonetheless, all soils can potentially inform us about the stability of non-stability of a surface, or landscape (**Fig. 13.20**).

Although we cannot always ascertain the genesis of a surface, soils can provide insight into its degree of stability or instability (**Fig. 13.20**). We can usually also gain some insight into stability/instability from the geometry or morphometry of a surface, that is, its overall "3-D shape." We do this by first partitioning the slopes in hilly areas into five major **slope elements**. From the top of the slope to its base, they are the summit, shoulder, backslope, footslope, and toeslope (**Figs. 13.21A, 13.22**). Each of these slope elements typically has an inferred amount of stability or instability.

Studying soils along a slope sequence is a simple yet elegant way to discern the spatial interrelationships between soils and topography/geomorphology. The complete sequence of soils from summit to toeslope is called a **catena** (Latin *catena*, "chain") (**Fig. 13.23**). In other words, a catena is a transect of soils from the summit, straight downslope, to the base of a hill. Soils along a catena can be thought of as links in a hanging chain, each one interrelated and in some ways, dependent on its immediate upslope and downslope neighbors. These linkages are primarily formed by lateral translocations of matter and energy, mainly in the downslope direction. Thus, any soil on a slope will occasionally receive matter and water from the soils upslope, and deliver some of the same to soils downslope. The soils are all "linked." But the linkages are not due only to downslope translocations. They are also due to association. Soils closer to each other simply have more in common with each other – due to the greater similarity of the soil-forming factors – than with soils farther away. For example, two adjacent soils are more likely to share similar parent materials and vegetation than are soils hundreds of meters apart along the same slope. In summary, the catena provides a time-tested, conceptual framework to help explain local-scale soil variability.

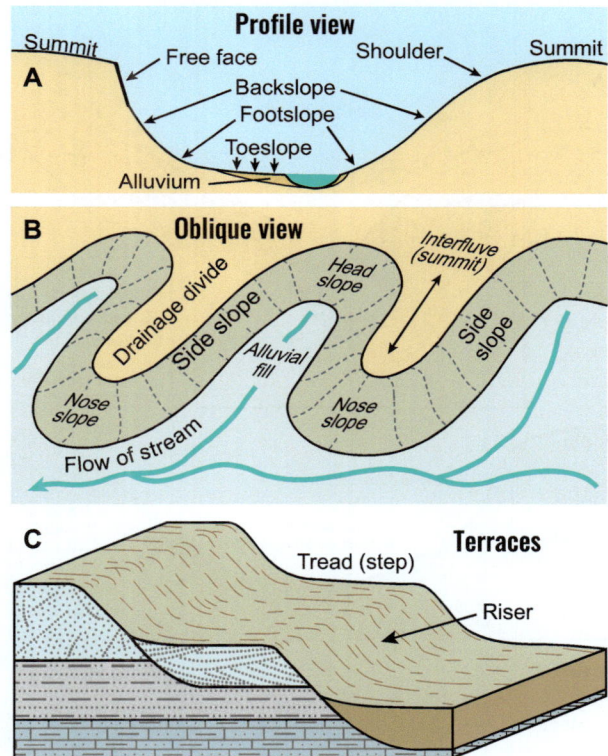

Figure 13.21 Various descriptors used to describe the elements of sloping and stepped surfaces.

Let's examine how and why soils vary along catenas, by using the five elements of slope outlined in **Fig. 13.21A** and illustrated for a "real" landscape in **Fig. 13.22**. **Summits** – the top part of the landscape – are usually either flat or convex. Broad, flat summits are generally quite stable surfaces, with minimal erosion or accretion of sediments. Here, soil development can proceed nearly uninterruptedly because little water is likely to run off, and therefore, most precipitation

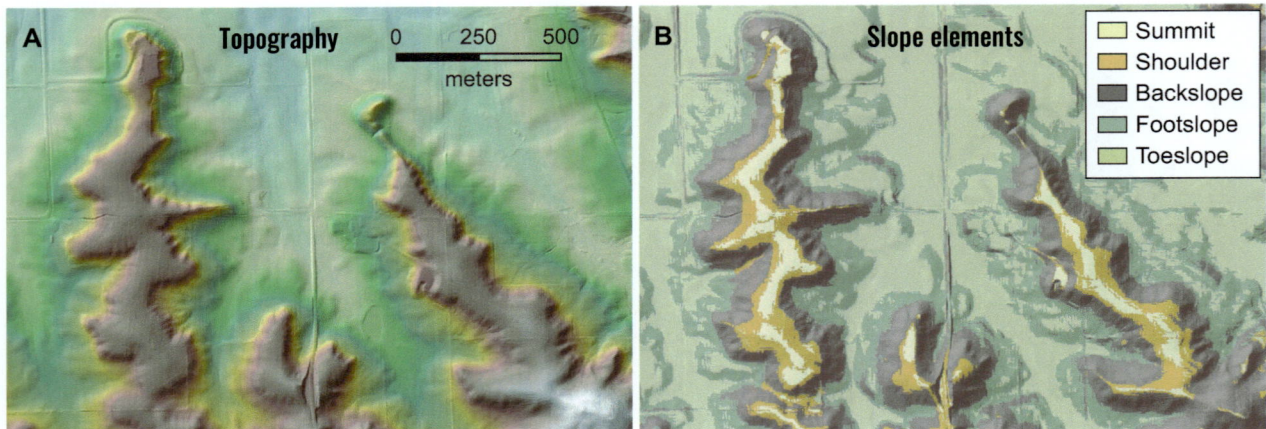

Figure 13.22 A landscape in western Wisconsin, USA, where sandstone hills rise above the surrounding lowlands. The topography, shown in **A**, has been digitally parsed into the five slope elements in **B**, using an algorithm derived by Miller and Schaetzl (2015) that determines slope elements based on slope gradient. Upland sites with <3.5% slope are summits, those with slope gradients between 3.5 and 7% are considered shoulders, and steeper slopes are classified as backslopes. In lowland areas, the same rationale is applied, with footslopes having gradients between 3.5 and 7%; toeslopes have gradients less than 3.5%. This example illustrates that all segments of a landscape are interconnected and that these interconnections are often predictable and repeatable.

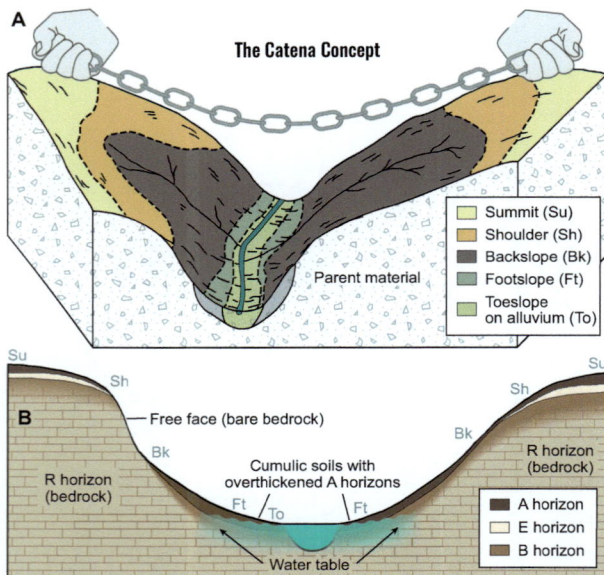

Figure 13.23 A. An illustration of the catena concept, along with the five slope elements. **B.** Typical soil thickness relationships along a catena in a humid climate where the water table is close to the surface in the toeslope position.

infiltrates. As a result, translocation processes are maximized and soils on summits are often the best developed of anywhere on the landscape. Soils on summits usually have more organic matter, thicker profiles, and better horizonation than soils on slope elements immediately downslope.

As summits become more convex, they (and their soils) grade into **shoulder slopes**. Convexity is the operative concept on shoulder slopes (**Fig. 13.21A**). The steepest shoulder slopes can be so nearly vertical as to be free faces (**Figs. 13.21A, 13.23**; see Chapter 8). Thus, water tends to run off

shoulder slopes, rather than infiltrate into the soil, leading to periods of erosion and, overall, less-developed and thinner soils. Shoulder slopes are the least geomorphically stable of the five slope elements. Runoff and erosion dominate to the point that erosion sometimes outstrips pedogenesis, resulting in thin soils or even exposed parent material, in other words no soil at all (**Fig. 13.24**). Probably the most erosion-prone shoulder slopes occur on **nose slopes** (**Fig. 13.21B**). And because much of the precipitation runs off shoulder slopes, soils here are also drier than at other locations along the catena, and with usually the deepest water tables.

Figure 13.24 A thin soil in Maine, USA, on a shoulder slope position. The bedrock is shallow here because erosion is accelerated due to runoff on the steep, convex slope. Source: R. Schaetzl.

Backslopes are the steepest and sometimes the driest parts of the landscape. They are slopes of transportation, lying between shoulder slopes above, dominated by erosion, and slope elements below, which often accumulate sediment. In a typical catena, backslopes are fairly straight, lacking significant curvature in profile view (**Figs. 13.21A, 13.25**). Debris and water move rapidly across backslopes, and therefore, much of the sediment is episodically "in transit." And their pathway of movement largely depends on the curvature of the slope (**Fig. 13.21B**). Mass movements such as creep, slump, and solifluction are common here. Needless to say, soils are thin on backslopes, and depending on local circumstances, sometimes even thinner and drier than on shoulders.

Figure 13.25 Soils are extremely thin, and bedrock crops out in many places, on this steep backslope in Wyoming's Wind River Mountains (USA). Source: R. Schaetzl.

Nearer the bottom of a catena, profile curvature becomes concave (**Figs. 13.21A, 13.22B**) and slope elements become sediment- and water-receiving positions. **Footslopes** are the most concave parts of the slope. Material carried downslope, in solution and in suspension, by throughflow and overland flow, begins to be deposited here, as the hillslope gradient lessens. Most of the sediment deposited on footslopes originated from eroding soils upslope, and what gets eroded first is the A horizon. That sediment then gets deposited on top of soils at the base of the slope. Therefore, many soils in the footslope (and toeslope) positions have overthickened A horizons and thicker overall profiles (**Fig. 13.23B**). The increasing wetness of the lower slope positions also promotes higher plant productivity, which in turn provides more litter to the soils on these sites, further enhancing A horizon development there. Also, the relatively cooler and wetter conditions at the base of the slope may slow decomposition of the litter. Together, these factors and processes combine to produce soils in the foot (and toeslope) positions that are high in organic matter, with thick A horizons. And now, for the first time, we begin to see indications of wetness in the soils (**Fig. 13.23B**), which gets even greater in the toeslope position.

Your toes are at the end of your foot; similarly, toeslopes are the outward extension of footslopes. **Toeslopes** are aggradational sites where both water and sediment accumulate – not only from upslope but also from streams that flow at the bottom of the slope and occasionally deposit overbank alluvium. The latter type of sediment accumulation was especially common in the United States after European settlers cleared and cultivated the forest and prairies. So much sediment washed onto toeslopes and into river channels there that the channels filled, forcing the rivers to flood more frequently. Each time they did, more sediment was deposited on the toeslopes. Because toeslopes can accumulate sediment from rivers in this way, they are often called **alluvial toeslopes**. In parts of the upper Midwest, USA, this post-settlement alluvium can approach a meter in thickness. The process whereby soils slowly grow upward by additions of sediment onto the top of the profile is called **cumulization**. Nowhere is cumulization better expressed than on toeslopes; soils with overthickened A horizons due to cumulization are referred to as **cumulic soils** (**Figs. 13.23B, 13.26**).

Figure 13.26 Scientists inspect a soil in a small stream valley on an alluvial toeslope. Note the dark, overthickened A horizon. Source: R. Schaetzl.

Sediments on toeslopes and footslopes tend to be finer-textured and better sorted than material upslope, because much of it has been carried downslope in slopewash. Water will transport the finest sediment farther downslope, and as a result, soils in downslope parts of catenas often have more clay and silt than soils in upslope areas, such as footslopes. For this reason, soils on shoulders and backslopes are often some of the sandiest on the catena. Soils on toeslopes are also naturally some of the wettest in a catena, usually due to high water tables (**Fig. 13.23B**).

Although sediment gradually accumulates on toeslopes, the process reaches its zenith in basins of closed (or internal) drainage. In these kinds of "**closed basins**," no process, except perhaps wind, can remove the sediment delivered to them, and thus, sediment accumulations may be particularly thick. In "open" toeslope areas, rivers may add sediment while in flood, but they also cut away at their banks and remove sediment, carrying it away, downriver (**Fig. 13.27**). Closed basins are, therefore, excellent long-term sediment traps; many of them preserve a long record of sedimentation.

Because the sediments usually contain buried wood and other plant remains which can be radiocarbon dated, the rate of sedimentation can be determined and tied to paleoclimate events, or to changes in land use.

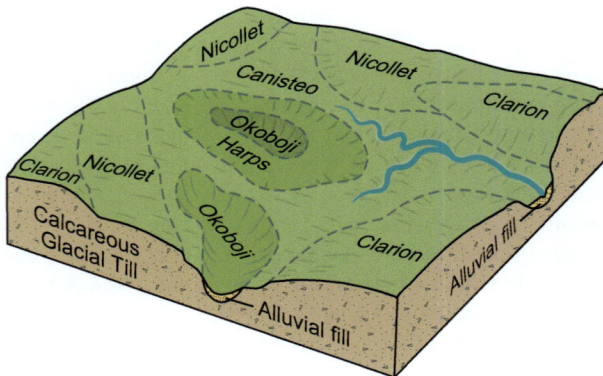

Figure 13.27 Block diagram of a soil landscape in Polk County, Iowa, USA, where closed depressions are common. The depressions here are kettles, formed where blocks of ice left behind by the last glacier melted out to create isolated, closed basins. The names on the map represent the typical soils found on each of the various landscape positions.

13.8 PALEOSOLS AND PALEOPEDOLOGY

We have seen the propensity for sediment accumulation at the bases of slopes, especially on toeslopes. If sediment accumulates slowly on surfaces like these, soil development can keep pace and the A horizon thickens, leading to a cumulic profile (**Figs. 13.23B, 13.26**). But if sediment accumulates more rapidly, or if a large amount of sediment is "dumped" onto a soil by, for example, a landslide, the soil would be effectively buried (**Fig. 13.20**). We call buried soils paleosols (*paleo*, old), and the study of paleosols is **paleopedology**.

The traditional definition of a **paleosol** is "a soil that formed on a landscape of the past." That definition would certainly apply to all buried soils; they formed during some past interval of time and are now buried and no longer "functioning" as surface soils do. But buried soils may be thought of as more than that; they are a unique type of sediment layer as well, and they preserve valuable information about that "past landscape" (**Fig. 13.28**).

Some soils in the tropics and subtropics exist on very old and stable landscapes. These soils certainly started forming in, and maybe even owe most of their morphology to, a past landscape and climate. Such soils are called **relict paleosols**, because they are relicts of a past soil-forming environment. How old does a soil have to be, to be considered relict? There is no one answer, but most geomorphologists would assume that a relict paleosol dates to at least the last interglacial period, ≈130 ka, or older.

Why do geomorphologists study paleosols? Well, if these soils formed in the past, then maybe they can inform us about that past world. Pollen and other plant macrofossils

Figure 13.28 A buried paleosol in Michigan, USA. This soil formed in gravelly, sandy, glacial outwash and was later buried by dune sand. Note that the soil has well-developed A, E, and B horizons. Source: R. Schaetzl.

preserved in paleosols can inform us about what the environment was like while the paleosol formed. The overall morphology of the soil can do the same. For example, a buried Mollisol would suggest that the soil formed in a past environment dominated by grasses and a subhumid/semi-arid climate. The weathered status of buried paleosols can also inform us about past climates, and other paleoenvironmental methods also exist to interpret buried soils. In short, by studying the buried soils and knowing the time interval within which it formed, we can learn a great deal about past environments. Certainly, for this application, buried paleosols are preferred.

Although a relict paleosol dates back to some past landscape or climate, it nonetheless remains at the surface today. Thus, it has been influenced by a variety of climates since it started to form, and those later climates may have changed the soil in countless ways. Relict paleosols are therefore referred to as being **polygenetic** – having multiple instances of "genesis." A modern soil that dates back only a few thousand years might have formed mainly under one climate and one type of vegetation; such a soil would be **monogenetic**.

Relict paleosols are sometimes also referred to as palimpsests. The word **palimpsest** (Greek *palimpsestos*, "scraped again") comes from ancient Egypt, where papyrus was a type of paper or parchment. Because papyrus was so valuable, it was erased (as best as possible, but not completely) and re-used, time and time again. Older writings remained partially visible on these palimpsests. Relict paleosols are like palimpsests because they contain all manner of "writings" from today and all the way back to time$_{zero}$. This type of mixed message can be difficult to interpret, but nonetheless we must always be aware of it.

Buried paleosols, on the other hand, have a definite time$_{zero}$ in the past and another endpoint – when they got buried (**Fig. 13.28**). Thus, they formed within, and their morphology represents, a defined time interval; they've often been "written on" only once. The period of time from beginning of soil formation (its time$_{zero}$) to its end (in this case, by burial, but in other soils, by erosion) is called a **soil-forming interval**.

The soil-forming interval of most modern soils (at the surface today) started at their time$_{zero}$ and continues today. That time$_{zero}$ often started shortly after a parent material was deposited, when the surface stabilized and soil formation could begin. For soils forming on slowly eroding surfaces or those developing in bedrock residuum, identifying an exact time$_{zero}$ is not only very difficult, but almost meaningless, as the soil and its morphology continually evolve along with the surface. As was mentioned at the start of this chapter, soils are excellent at integrating all aspects of Earth's environments, including of course, parent materials and surface stability/instability issues.

REVIEW QUESTIONS

13.1 What is meant by time$_{zero}$ of soil formation?

13.2 What are some common soil parent materials and how does each form?

13.3 In what kinds of environments do organic soil parent materials accumulate, and why there?

13.4 Define and differentiate between a soil profile and a soil horizon.

13.5 What are the six major types of soil horizons, and what are the characteristic of each?

13.6 Explain and differentiate eluviation from illuviation. What major types of soil horizons are dominated by each of these processes?

13.7 What is soil texture and what are the major components of soil texture? Can you effectively use a textural triangle to determine the texture class of a soil?

13.8 What is a lithologic discontinuity and what can it tell us about the geologic past of a site or surface?

13.9 What are the five elements of slope?

13.10 Explain the two ways that stone lines can form in soils.

13.11 How do soils vary along slopes, and why?

13.12 How do forest soils differ from those formed under grasses, and why?

13.13 Describe two types of horizons found only in Aridisols and explain how they form.

13.14 How can color help identify the wetness/water regime of a soil? Supply some examples.

13.15 What are the two main types of paleosols and in what types of settings are each found?

FURTHER READING

Birkeland, P. W. 1999. *Soils and Geomorphology*. 3rd ed. Oxford University Press.

Eash, N. S., Green, C. J., Razvi, A., and Bennett, W. F. 2008. *Soil Science Simplified*. 5th ed. Blackwell.

Hillel, D. 1991. *Out of the Earth*. University of California Press.

Logan, W. B. 2007. *Dirt: The Ecstatic Skin of the Earth*. Riverhead Books.

McRae, S. G. 1988. *Practical Pedology: Studying Soils in the Field*. Ellis Horwood Series in Soil Science. Ellis Horwood Ltd. Press, Halsted Press.

Paton, T. R., Humphreys, G. S., and Mitchell, P. B. 1995. *Soils: A New Global View*. Yale University Press.

Schaetzl, R. J. and Thompson, M. L. 2015. *Soils Genesis and Geomorphology*. 2nd ed. Cambridge University Press.

Steila, D. and Pond, T. E. 1989. *The Geography of Soils: Formation, Distribution, and Management*. 2nd ed. Rowman & Littlefield.

Warkentin, B. P. (ed.) 2006. *Footprints in the Soil: People and Ideas in Soil History*. Elsevier.

14 Mass Movements and Slope Processes

Markus Stoffel

Although a natural process, human actions and extreme climatic events can accentuate **slope instability**, leading to disastrous **slope failures** and loss of life, like the one that occurred in the Brazilian city of Petrópolis on February 17, 2022. Over 200 people died in the mudflows, caused by intense rainfall (258 mm in three hours) and the deforestation of upslope areas. Understanding how and why materials move downslope helps geomorphologists to predict where and when future **mass movement** events may occur.

Except for perhaps volcanic eruptions and earthquakes, the most impressive (and deadly) geomorphic "events" involve the downslope movement of rock, debris, and sediment – referred to as mass movements because the material moves *en masse*. In their simplest sense, mass movements represent the downslope transport of rock and soil materials. Examples range from massive, fast-moving landslides and debris flows, to the inexorably slow process of soil creep. In every case, material is on the move – pulled downslope by gravity and accommodated by circumstances that allow these materials to "move." Mass movements are commonly assisted by water, which acts both as a lubricant as well as providing added weight. The terms "mass movement" and "**mass wasting**" are equivalent; readers may see either or both terms in the literature.

Mass movements can occur virtually anywhere, even in the absence of extremely steep slopes and inhospitable terrain (**Fig. 14.1**). They occur both on land and under water, on cultivated lands, on barren slopes, and within forests. Many of the largest mass movements have gone down in history as major disasters, taking hundreds or even thousands of lives in a matter of minutes. This chapter focuses on slope processes, which are so often driven by mass movement.

14.1 SLOPE PROCESSES

Mass movements fulfill an important role within the geomorphic system. Rock and sediment on slopes everywhere are moving downward, toward a river valley or into a lowland. Rivers are capable of transporting that sediment away, and as a result, the landscape is lowered – a process called denudation. But rivers can only transport sediment that gets delivered to their channel. This "delivery" is mainly accomplished by two processes – mass movement and **slopewash** (**Fig. 14.2**). Without mass movement and slopewash, most rivers would be flowing in deep, narrow canyons. We find this situation on weathering-limited slopes in dry climates where weathering of bedrock is slow, such that rivers can readily carry away the little sediment delivered to them (**Fig. 14.2B1**). In areas with softer bedrock and in warmer, wetter climates, weathering processes operate much faster. On these transport-limited slopes, movement of material downslope cannot keep pace with its production by weathering and thus, a thick mantle of sediment covers the bedrock (**Fig. 14.2B2**). Regardless, the system operates similarly – sediment is moved downslope and eventually delivered to a river. Together, these connected slope processes facilitate landscape denudation.

In addition to recognizing two different end-members of slope *types* (transport-limited and weathering-limited), geomorphologists also recognize two different end-members of slope *processes*. On one end are the large, rapid, and dramatic mass movements such as rockfalls and landslides. The location and frequency of these processes is mainly affected by the underlying geology, or some sort of triggering mechanism. But on more stable slopes, especially where there exists a thick cover of sediment, mass movements are smaller and their efficacy is generally proportional to slope gradient, or steepness. This family of **diffusive slope processes**, which includes creep and slopewash, forms the other end-member of slope processes. The term *diffusive* implies that these

Figure 14.1 Deposits from a variety of mass movement events are shown here, in Bermejo Pass in the southern Andes Mountains of South America Source: M. Mergili.

Figure 14.0 A massive landslide (debris flow) in Nepal illustrates the impressive power of gravity on landscape development. Source: Nick Pedersen / Getty Images.

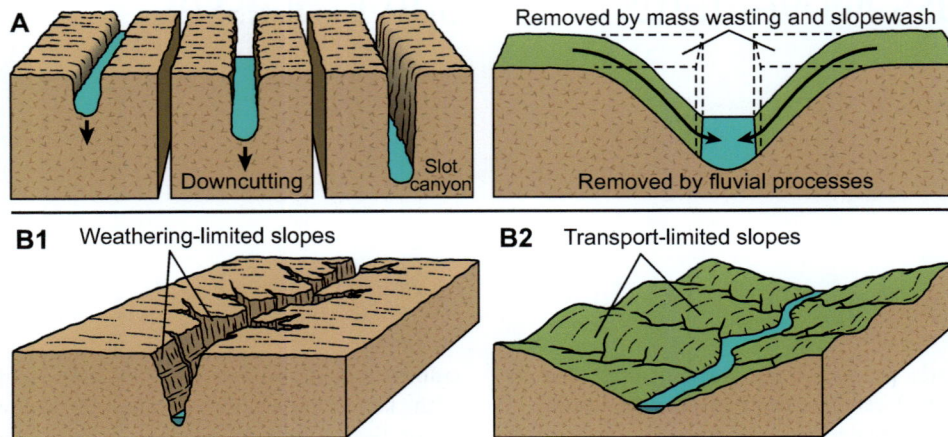

Figure 14.2 Schematic illustration of the general effects of mass wasting processes on landscape development. **A.** Most of the sediment that gets removed from valleys by streams is delivered to them by mass wasting and slopewash. Without mass movements, many rivers would flow in deep, narrow valleys, some in canyons! **B1.** In arid climates, where weathering rates are slow, many slopes are weathering-limited, and little sediment is made available for slope processes. Rivers here can carry away the sediment delivered to them, and use any excess power they have to cut deep canyons. **B2.** In areas of humid climate and on softer rocks, weathering operates faster. In this transport-limited landscape, a thick cover of sediment forms because slope processes are incapable of delivering all the sediment that is produced to the river, leading to wider valleys.

processes are not spatially focused, but are instead spread out across the entire slope, and operate without the assistance of wind, glacial ice, or channelized water. Diffusive slope processes operate slowly, and some are almost always ongoing. Even though diffusive slope processes often go unnoticed, over time they can transport massive amounts of sediment into lower slope positions.

14.2 SLOPE STABILITY

Because gravity is the main driver of slope processes, mass movements only occur on sloping surfaces. **Slope stability** refers to the ability of slopes to withstand mass movement and erosion. When viewed over short timespans, all slopes appear to be stable, that is, sediment on them does not appear to be moving. Stable slopes are amenable for human activities such as agriculture, construction, or recreation. Slopes where rock and sediment are moving (or have recently moved) downslope are experiencing *instability*. When this happens, we refer to the event as a **slope failure**.

Broadly speaking, all slopes can be considered along a continuum from stability to instability. Unstable slopes are failing or have recently failed. On stable slopes, where sediment is moving very slowly or not at all, soils are actively forming (see Chapter 13). Any stable slope may become unstable as conditions change. Geomorphologists strive to understand how and why slopes fail, so as to better manage Earth's physical systems.

A series of factors determine whether a slope is stable or unstable, including, among others, rock structure and lithology, hydrology, slope undercutting or loading, as well as shaking and tremors caused by earthquakes. An example of a major mass movement event, set up by favorable

rock structure, occurred near Yellowstone National Park, Wyoming, USA, in 1925 (**Fig. 14.3**). Here, the Gros Ventre River had cut a deep canyon into sedimentary rocks. Bedrock layers on one side of the valley were inclined parallel to the steep (20°), valley-side slope. A thick sandstone layer, which overlies less permeable shale across this slope, had been undercut by the river, leaving its front edge unsupported (an undercut slope). At this point, friction at the sandstone–shale contact is the main force holding the sandstone in place. Melt from a thick snowpack, followed by heavy rainfall, **saturated** the rock, added weight to the sandstone, and reduced the shear resistance at the contact between it and the underlying shale. The resulting Gros Ventre landslide, possibly triggered by small earthquake tremors, carried ≈40 million m³ of debris into the valley. Clearly, this was a case of an unstable slope, resulting in a massive slope failure.

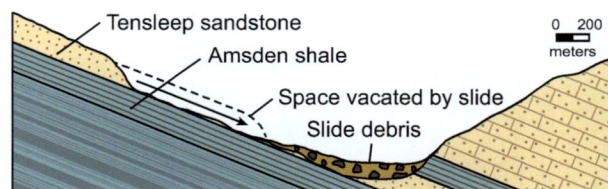

Figure 14.3 Cross-section of the Gros Ventre landslide. Note that the layering of bedrock on the side of the valley that failed was inclined parallel to the valley-side slope.

In the Gros Ventre landslide, slope failure occurred in bedrock. Often, however, slope failure occurs in the unconsolidated sediment that overlies the bedrock. Such sediment is termed **regolith**. Regolith has a more open and less cohesive character than bedrock, which can make it more prone to mass movement.

14.3 FORCES ACTING ON SLOPE SEDIMENTS

Let's first think of slope stability in theoretical terms. The main force driving mass movement is gravity, which pulls continuously downward on all materials. But other forces are also at work (**Table 14.1, Fig. 14.4**). On a flat surface, gravity – the **gravitational force** (*fg*) – will simply pull downward on a mass and it will not move. The surface will be stable (**Fig. 14.4A**). On inclined slopes, the gravitational force consists of two components – a component acting perpendicular to ("normal to") the slope *and* a component acting tangential, or parallel, to the slope. The perpendicular component of gravity is called **normal force** (*fn*) pushing objects *into* the slope and helping to hold them in place. Think of *fn* as friction, preventing or inhibiting movement. The tangential component of gravity is commonly referred

to as **shear force** (*fs*). Shear force pulls the mass *down* the slope, and increases as the slope steepens. The balance between the shear force and the normal force is called **shear strength** (*ss*). Where *fn* > *fs*, the mass will *not* move. On steeper slopes or when the regolith becomes less cohesive, *fs* exceeds *fn* and thus, slope failure may occur (**Fig. 14.4**).

14.4 SLOPE MATERIALS AND SLOPE STABILITY

To simplify the above discussion, consider slope stability in this way. Slopes fail whenever shear *stresses* on slope materials are high, and/or their shear *strength* is low. Geomorphologists often think of this relation by using a ratio called the **Safety Factor** (F_s); when F_s exceeds 1, slopes will be stable, otherwise they may fail:

F_s = resisting forces (shear strength) / driving forces (shear stress).

A number of factors may combine to increase the **shear stresses** on a slope and its materials, for example, removal of underlying or lateral support for the slope, additions of mass (overloading), or seismic vibrations, among others (**Fig. 14.5**). Most importantly, shear stresses are always increased as slopes steepen. As shear stresses increase, the likelihood of slope failure also increases. The shear strength of the slope and its materials work against these stresses to hold the slope material in place. Decreased shear strength can be caused by weathering and weakening of the sediments on the slope, planes of weakness parallel to the surface of the slope, loss of stabilizing roots due to devegetation, or additions of water, which can lubricate the sediments, among others (see below) (**Figs. 14.5, 14.6**). Other factors that promote slope failure by weakening shear strength and/or increasing shear stress include frequent freezing and thawing, high water contents in the regolith, and dry conditions followed by heavy rainfall. Slope stability is a balance (or imbalance) between these two opposing "forces" or factors.

Slope materials may consist of either bedrock or regolith. Regolith is also sometimes described as **debris** if it contains a large amount of coarse fragments such as rock and gravel, or as **earth** if composed primarily of sand or finer particles. Bedrock tends to be more durable than regolith, and hence, resists failure better, even if slope angles are

Table 14.1 The main components involved in slope stability

Component/ term	Nomenclature (see Fig. 14.4)	Definition/comments
Gravitational force	*fg*	Force pulling everything toward the center of the Earth. On a flat surface, normal and gravitational forces are equal.
Shear force	*fs*	A tangential component of gravity, acting parallel to the surface of a slope, increasing as the slope gets steeper. Slope failure will occur where shear forces exceed the shear strength of the slope materials.
Normal force	*fn*	Force acting perpendicular to a sloping surface on which the object rests, helping to hold it in place.

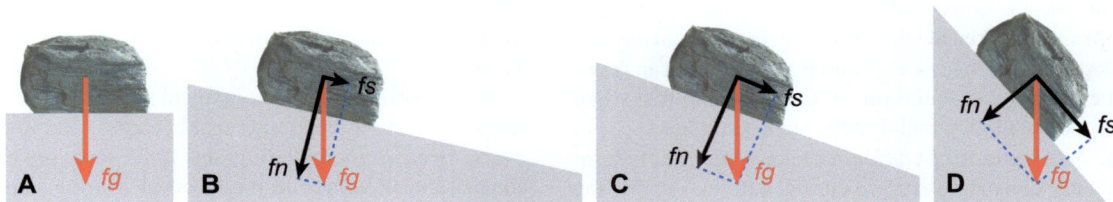

Figure 14.4 Theoretical diagram illustrating the relative strengths of gravitational force (*fg*), shear force (*fs*) and normal force (*fn*), as a function of slope steepness. Total gravitational force (*fg*) remains the same in all cases. **A.** On a flat surface, the only force acting on an object is gravity (*fg*). **B.** On gentle slopes, shear force (*fs*) is much smaller than normal force (*fn*), so the object is stable. **C.** As the slope steepens, shear force increases, and so the object is more likely to move. **D.** On steep slopes, shear force is so great that it overwhelms normal force, causing the block to move, i.e., the slope will fail.

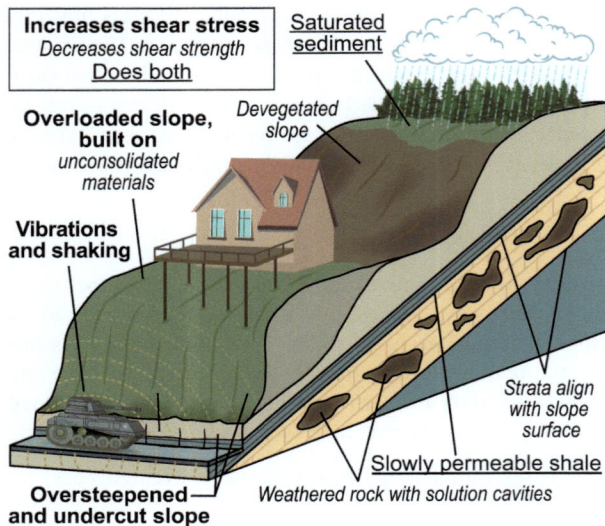

Figure 14.5 Schematic illustration showing several factors that may facilitate (individually or in combination) slope failure.

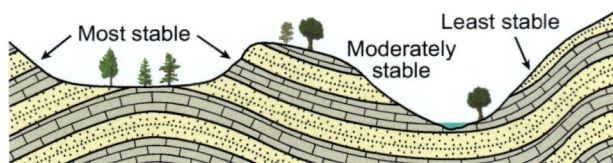

Figure 14.6 Relative stability of slopes as a function of the orientation of weaknesses (in this case bedding planes) relative to the slope orientations. Source: Graphic adapted from Earle (2019). Copyright © 2019 by Steven Earle, CC BY-SA 4.0.

steep. Crystalline rocks, for example, granite or basalt, are strong and preferentially form steep cliffs. In contrast, some metamorphic rocks, for example, schists, or soft sedimentary rocks like shale, are weaker and thus tend to be stable only on gentle slopes. Strength also varies among the various types of sedimentary rocks; although most limestones are strong, sandstones, shales, and mudstones are weaker. Variations in the composition or structure of rocks can also affect their strength.

Characteristics such as fracturing, stratification, and bedding can also impact bedrock strength. Where bedding planes crop out perpendicularly to the slope, mass movements are less common. The Gros Ventre landslide example illustrates how slopes are weakened when bedding planes align parallel to the slope (Figs. 14.3, 14.6). Similar weaknesses can occur when the layering, or foliation, in metamorphic rocks like schist happen to align parallel to the slope surface.

Joints are fractures or cracks in rocks. Many joints form due to the expansion induced during cooling, or by the release of pressure (unloading) after erosion of overlying rocks. Water or roots can enter joints, pry them farther apart, and potentially reduce the rock's strength. Where joints form parallel to the slope, a sliding surface can develop, along which larger blocks can become loosened and slide downslope.

Although regolith is much weaker than bedrock, it can still bind together in various ways. The condition that describes how strongly materials internally bind together is referred to as cohesion. A cohesive sediment, such as clay, binds strongly together – within itself and to other sediments. Alternatively, a sediment with poor cohesion, such as sand, will easily fall apart.

14.5 TRIGGERS THAT INITIATE SLOPE FAILURE

For most slope failures to occur, some sort of trigger is often required (Fig. 14.6). A triggering event starts the mass movement, that is, the trigger "gets it going." The most common triggers include rapid saturation by heavy rainfall or during snowmelt, seismic vibrations, volcanic activity, or even human activities such as road or building construction.

14.5.1 Water (Saturation)

Saturation is the single most important cause of mass movements; it can both change the shear strength of the sediment *and* act as a trigger. Saturation can occur due to intense rainfall, rapid snowmelt, and/or rising groundwater levels. Water dramatically affects the strength of the slope materials, especially in regolith but also in rocks. Water lubricates loose sediment, reducing its cohesion and lowering its shear strength. Water can also dissolve the cements that create the cohesion, particularly if the cement is calcite, gypsum, or halite. In all of these cases, high water contents will weaken the rock and promote downslope movement. Water can even reduce the strength of bedrock, especially where the rock is porous or has abundant fractures. Saturation is also important because it can increase the shear force by overloading, that is, adding excess weight to the slope.

Many earth materials, such as sand, are very porous (Fig. 14.7). In *dry* sand, the pore spaces are primarily filled with air. Dry sand is therefore held together only by the friction between grains, resulting in weak cohesion because of minimal grain-to-grain contacts. As the sand becomes wetter, the pores fill with water. Sand, like most regoliths, tends to be strongest when *moist*, when the water between individual grains holds them together by surface tension. Moist, unconsolidated material can therefore hold steeper slopes. Alternatively, in *saturated* sediment, the pores are filled with water, which decreases the frictional forces (cohesion) between the grains. Essentially, water in this context acts as a lubricant.

Saturation can affect fine-grained materials, like clay, differently. If fine-grained sediment becomes saturated or even *oversaturated*, individual grains will lose grain-to-grain contact and liquefaction might occur. During liquefaction, water-saturated sediments are transformed into a substance that acts like a liquid, flowing and deforming readily. Clay-rich sediments are particularly prone to liquefaction, because many types of clay minerals will readily adsorb water – as films around the grains. Thus, slopes on clayey sediments are more susceptible to slope failure when saturated than

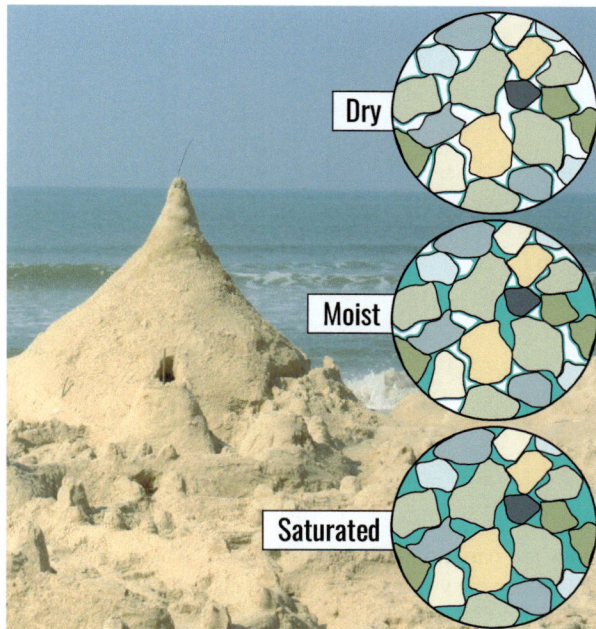

Figure 14.7 The perfect sandcastle. In completely dry sand, spaces will be filled with air, resulting in weak cohesion due to minimal grain-to-grain contact. Moist sand is the strongest, because water films hold grains together by surface tension. Saturated sand is weak due to its low cohesive strength. Source: Sreejith K via Wikimedia Commons.

are other regolith types. For example, during an episode of heavy rain associated with the passage of Hurricane Mitch on October 30, 1998, part of the flank of Casita Volcano in Nicaragua collapsed. Although the initial failure was small, the debris eroded older deposits from the volcano and incorporated additional water and wet sediment from along its path, increasing the volume of the mass movement by a factor of nine.

Sudden changes in groundwater levels can trigger slope instability. For example, if the water level in a lake or a river falls rapidly, groundwater levels in adjacent uplands cannot adjust rapidly enough, resulting in an artificially high water table on the slopes, inducing instability. This situation occurs often on steep riverbanks that fail – one way that mass movements can deliver sediment to rivers. Likewise, urbanization alters runoff. In urbanized areas, excessive lawn watering or leakage from buried water or sewer lines are other types of mass movement triggers.

Both the drawdown and the filling of reservoirs have triggered major mass movement disasters, the most dramatic being the Vajont (Italy) incident (**Fig. 14.8**). The Vajont River gorge that hosted the reservoir is steep sided and its banks had been undercut by the river. The limestone and claystone units forming its walls are interbedded with clay-rich, unstable layers that are inclined towards the axis of the gorge. In addition, caverns in the karstified limestone weakened it. On the evening of October 9, 1963, triggered by the drawdown of the reservoir and facilitated by recent heavy rains, a massive landslide of about 260 million m^3 rushed into the reservoir at an estimated velocity of 110 km/h, filling the reservoir behind the dam and displacing an estimated 500 million m^3 of water. The water formed an overtopping wave estimated at 250 m above the dam. The resulting wave and flood destroyed the Piave Valley downstream and claimed between 1,900 and 2,500 lives.

14.5.2 Oversteepening and/or Undercutting

Steepness is a key factor in conditioning a slope for failure, as steep slopes fail more often than gentle slopes due to their greater shear forces. Thus, when a stable slope becomes oversteepened, it is more likely to fail (**Fig. 14.9**). Oversteepening can occur either at the foot of the slope or on the upper portions of a slope.

Oversteepening develops at the foot of the slope by removal of sediment through erosion or excavation, a process geomorphologists call **undercutting**. Removal of sediment here will increase the overall slope gradient and reduce the slope's resistance to the gravitational force. Undercutting is one of the main ways that slopes become oversteepened (**Fig. 14.5**). It is particularly common where rivers have eroded their banks or where waves have eroded a bluff into a coastline. The Gros Ventre landslide is a classic example of how undercutting by a river can trigger a mass

Figure 14.8 The Vajont Dam (Italy) landslide disaster of October 9, 1963. **A.** The reservoir was built in a valley cut into very unstable limestones and claystones. The water of the reservoir saturated the nearby sediments, triggering (**B**) a massive landslide that rushed into the reservoir. **C.** The resulting flood destroyed several villages in the valley downstream. Source: (A) Keystone-France / Contributor / Getty Images Ltd; (B and C) Bettmann / Contributor / Getty Images.

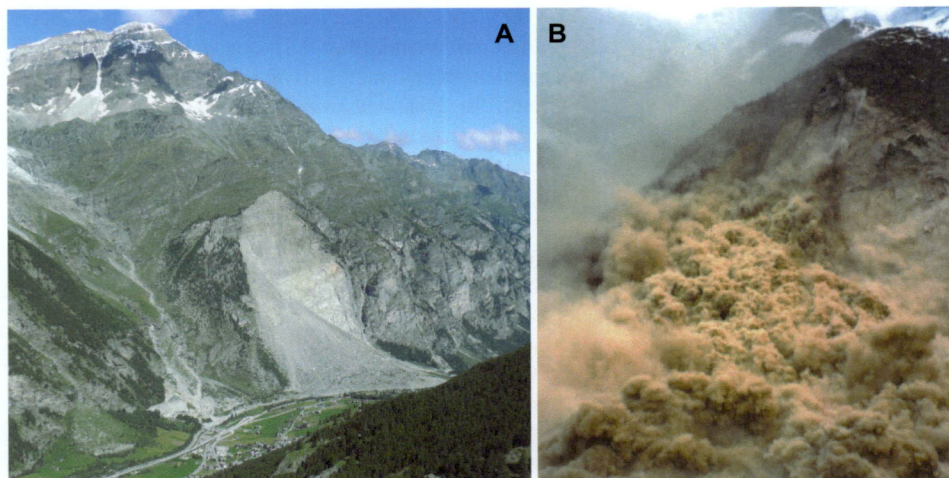

Figure 14.9 An oversteepened slope that led to rockfalls. **A.** Between April 18 and May 9, 1991, multiple rockfalls occurred at Grossgufer near Randa, in the Valais Alps, Switzerland. Note how the valley sides have been undercut and oversteepened by Pleistocene glaciers. Source: Wandervogel, CC BY-SA 3.0, via Wikimedia Commons. **B.** During these events, massive amounts of dust were generated, covering some nearby roofs with dust up to 30 cm thick. Source: Nationale Plattform Naturgefahren (PLANAT).

movement (**Fig. 14.3**). Undercutting is frequently caused by human actions, usually due to the construction of buildings and/or roads.

Oversteepening can also occur on the *upper* portions of a slope, typically when additional materials, such as "fill," are added due to construction activities. Oversteepening develops because a short (but steeper) slope segment is then formed at the lower end of the newly added sediment. These materials also may overload the slope by adding excess weight. Thus, adding sediment to the top of a slope can induce slope failure by oversteepening *and* by overloading.

14.5.3 Devegetation

The removal of vegetation commonly increases mass movement activity by reducing the shear strength of the slope sediment (**Fig. 14.5**). Devegetation removes many of the stabilizing root systems that add cohesion to rock and sediment. Removal of deep-rooted vegetation can also reduce the connection between the regolith and bedrock, which can induce mass movement at the interface. Plant roots also promote porosity, which in turn reduces runoff and saturation.

Devegetation can result from **clearcutting**, a logging method whereby all or most of the trees are removed from a site. Industrial logging is therefore often a key trigger of slope instability (**Fig. 14.10**). Clearcutting also indirectly triggers slope instability via the construction of logging roads, which are often made near the bases of slopes and thus, undercut the slope. Clearcutting also changes groundwater and soil moisture conditions. Saturation of regolith is more readily accomplished after the trees have been removed, increasing the likelihood of slope failure.

Conversion of land for agriculture, practiced since Neolithic times, is yet another way that slopes get devegetated. Conversion of land to agriculture leads to changes in vegetation cover, soil properties, and/or hydrology, and thus can contribute to slope instability. Recently, deforestation of widespread areas in Madagascar, the Amazon Basin, and the foothills of Nepal (for agriculture) has resulted in myriad slope failures and accelerated erosion. Wildfire-burned

slopes are particularly prone to instability because they not only reduce the vegetation cover but also change the chemistry of the soil, facilitating runoff (see Chapter 22).

Figure 14.10 Deforestation can change the slope hydrology, which in turn can trigger landslides, like in this example from Pe Ell, Washington, USA. Source: Courtesy of Washington State Department of Transportation.

14.5.4 Vibrations and Tremors

Shaking and vibrations can trigger mass movements by weakening rocks and regolith. The most obvious source of ground shaking is an earthquake. Earthquakes can destabilize slopes directly due to the seismic waves they generate, often inducing landslides, rockfalls, and rockslides. Indirectly, earthquakes weaken slope materials by forming cracks and fissures. Minor shocks induced by human-caused vibrations, such as the passage of heavy trucks or heavy machinery on roads, and mining explosions, can also be triggers (**Fig. 14.5**). Trees blowing in the wind cause underground vibrations as well – a natural trigger of smaller mass movements.

Mass movements associated with earthquakes are usually the result of seismic shaking and increases in pore water pressure. The shaking induces waves with complex and differing vertical and horizontal amplitudes and frequencies, all of which increase shear forces and reduce the shear strength of slope materials. In some materials, seismic waves can even induce liquefaction. Indeed, liquefaction failure is most often triggered by earthquakes. Alternatively, seismic shaking has also been shown to favor dilation of soil materials, facilitating rapid infiltration of water which can lead to saturation. Indeed, in many seismically active areas, more damage is caused by related mass movements than by the earthquake itself.

14.5.5 Volcanic Eruptions

Volcanic activity is another trigger of mass movement – one that has been responsible for some of the most devastating slope failures in human history. Mass movements can occur either in association with the eruption itself, or through the subsequent mobilization of the weak, often unconsolidated, deposits erupted from the volcano, such as ash. Should these deposits collapse, they can cause rockslides, landslides, and debris avalanches (see Chapter 7). These types of mass movements are sometimes called **flank failures**, to indicate that the mass movements occurred along the side of the volcano, during later periods of volcanic quiescence. Flank failures are usually triggered by seismic shocks or heavy rainfall. Likewise, when volcanic islands experience instability and periodic slope failures, massive submarine landslides and deadly tsunamis can often result.

Lava and pyroclastic materials from a volcanic eruption can mix with wet soil, ash, snow, ice, and mud on the slope of the volcano, producing rapidly moving mudflows or debris flows known by their Indonesian name – **lahars** (Fig. 7.41). Typically consisting of a mixture of rock, soil, ash, and water, lahars originate high up on a volcano, potentially travelling great distances and at great speeds (see Chapter 7). Lahars can also occur when the volcano is quiescent, as long as conditions exist for unconsolidated volcanic materials to mobilize. Floods caused by the sudden melting of a glacier by volcanic activity can also generate lahars; geomorphologists refer to these events as **jökulhlaups** (Figs. 14.11, 18.24).

Figure 14.11 A jökulhlaup in Skaftá, Iceland, on October 2, 2015. Source: T. Jóhannesson.

14.6 CATEGORIES OF MASS MOVEMENT

Mass movements are usually classified based on (1) the nature of the failure, that is, how the material moved, (2) the rate at which it moved, and (3) the type of material that failed (Figs. 14.12, 14.13). The nature of the movement is the most important defining characteristic of a slope failure. Four main types of motion exist. In **falls**, the material drops downward through the air, typically on steep slopes. **Topples** occur where a mass fails via rotation around an axis (or point). If instead the material is transported *en masse* along a sloping surface, but with minimal internal motion, we call it a **slide**. Lastly, when material moves downslope and exhibits considerable internal motion and mixing, we refer to the process as a **flow**. All mass movements have a **source area** from which the material originated, a **transit area** through which the mass moves, and a depositional (or **runout**) area where the material comes to rest.

One can also differentiate mass movements based on their speed. Some mass movements are very rapid and devastating, as in the case of falls and certain flows. Others can be almost imperceptibly slow, as in the case of slowly creeping sediments (Fig. 14.13).

14.6.1 Falls

A **fall** (or rockfall) is the airborne, downward motion of material from a steep cliff or slope (Figs. 14.12, 14.13). The material that falls may be rock (rockfall), earth (earthfalls), debris (debris falls), or various combinations. Falls can vary significantly in volume, from small grains to massive blocks. Falls are some of the most dramatic and rapid forms of mass movement.

Falling rocks and boulders typically detach from a steep cliff or escarpment along bedding planes, cracks, or joints, that is, where weathering has weakened the rock. This source area is usually a nearly vertical free face, from which rocks become dislodged (Fig. 14.14). Quite often the rocks are pried loose by frost action, as ice grows within joints. The size and shape of the falling rocks are influenced by the properties of rocks in the cliff, as well as the amount of disintegration while in transit. Once in motion, the material falls through the air as a **free fall** if the event occurs at a vertical or nearly vertical cliff (Fig. 14.15). Free fall speeds vary from between 20 and 150 km/h. It is not unusual for a falling mass to impact the slope while falling, often breaking apart upon impact. If the falling material strikes the slope, intermittent contacts with the slope may cause it to bounce (Fig. 14.15). The bouncing mass may then start to roll as the terrain flattens. At the same time, the bouncing or rolling mass may gather rotational momentum, potentially doing great damage far downslope (Fig. 14.16). Portions of a cliff that have seen recent rockfall activity can often be recognized by the freshness of the newly exposed face, sometimes referred to as **rockfall scars**.

Falling material, especially rocks, can be life-threatening. They can damage infrastructure, especially if large boulders

Bedrock   **Unconsolidated sediment (weathered rock, regolith, and soil materials)**

Falls

Rockfall

Debris fall
Earthfall

Topples

Rock topple

Debris topple
Earth topple

Slides

Rotational

Rotational slides (slumps) occur
in unconsolidated sediment,
not bedrock

Slump

Debris slump
Earth slump

Translational

Block slide

Rockslide

Debris slide
Earth slide

Flows

Rock avalanche

Flows occur in unconsolidated sediment

Debris flow
Mudflow

Quick clay flow

Permafrost

Solifluction

Creep

Debris
avalanche

Earthflow

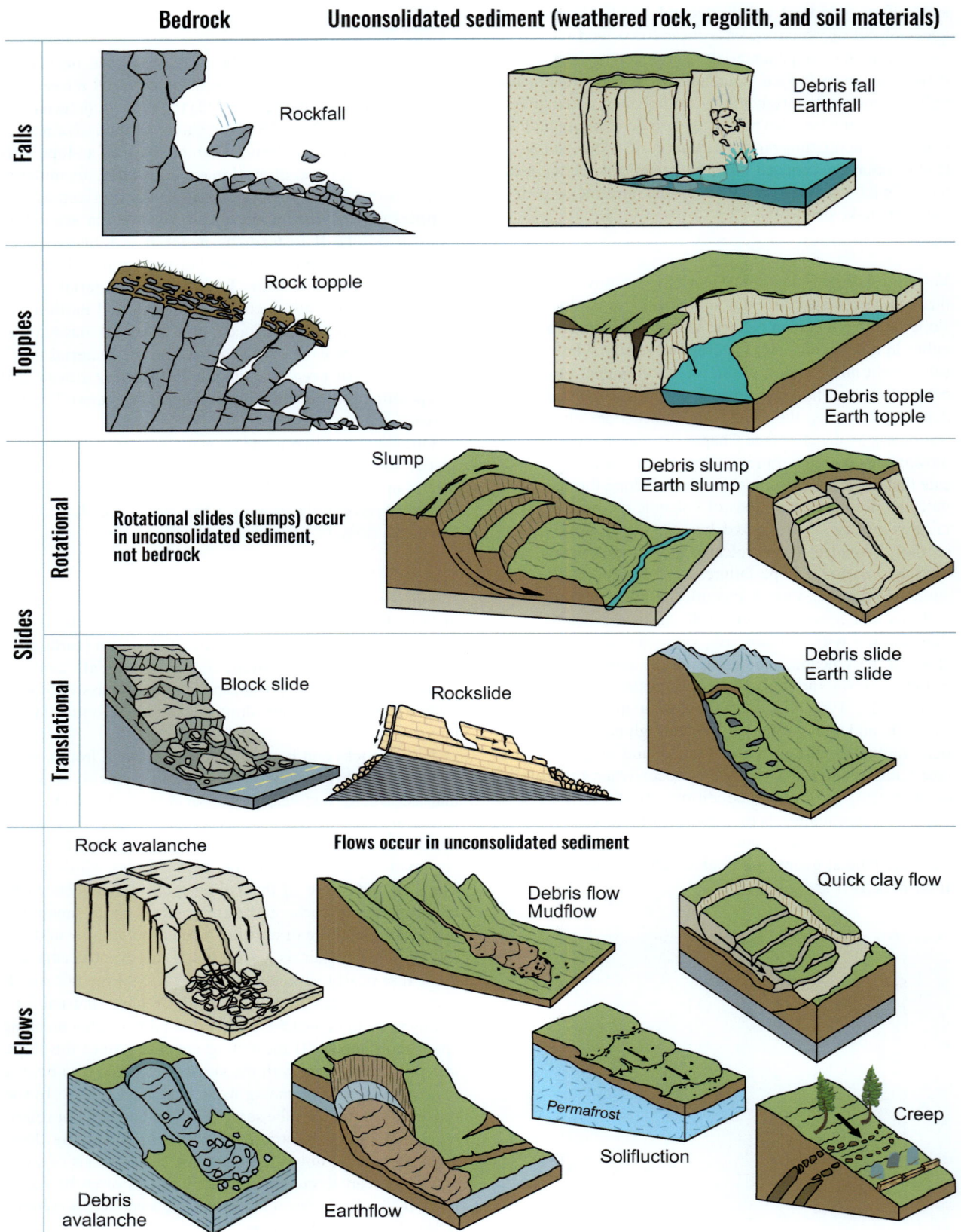

Figure 14.12 Mass movements classified according to the nature of the failure and the material involved.

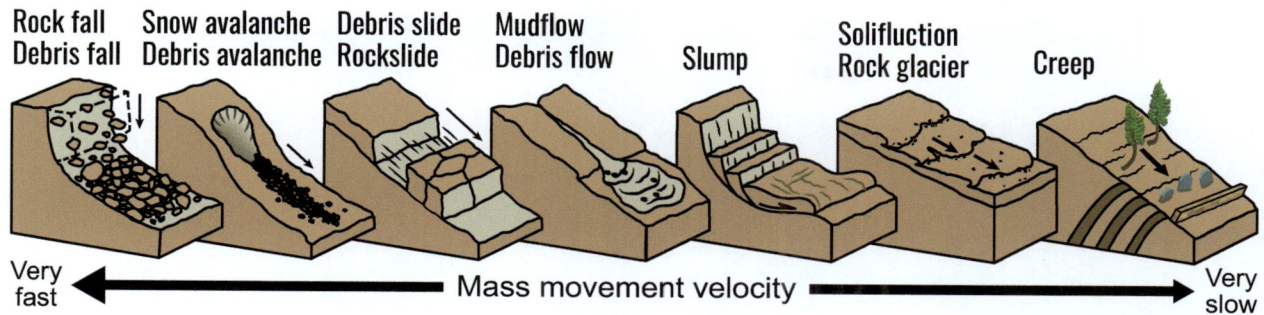

Figure 14.13 Illustration of the relative speeds of the various groups of mass movements.

Figure 14.14 On this slope in western Utah, USA, rocks being detached from the sandstone free face have fallen onto a layer of weathered shale below. Depending on the size and shape of the rock, its fall velocity, and the characteristics of the shale upon which it lands, the rocks may then slide or roll farther. The point – not all rockfall debris ends up as piles of talus at the bottom of a free face. Source: R. Schaetzl.

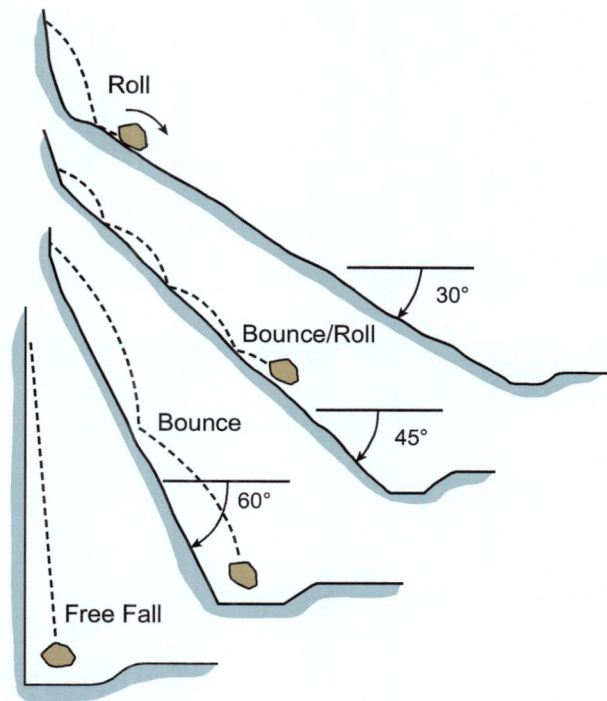

Figure 14.15 Effects of slope angle on the fate of rocks that are released from a free face. Source: Adapted from Ritchie, A. M. Evaluation of Rockfall and Its Control. Highway Research Record, No. 17, 1963, Figure 11, p. 25. Copyright, National Academy of Sciences. Reproduced with permission of the Transportation Research Board.

are involved. Such boulders can travel great distances, even when rolling (Figs. 14.16C, D, 14.17). Rocks rolling across hiking trails, roads, and railway lines have killed people in vehicles. For this reason, a broad range of mitigation measures have been designed to control or stop falling material and to mitigate damage in runout areas. Below the free face, rock curtains or similar slope covers, together with anchors (or rock bolts), can be installed to limit the release of rocks; these measures essentially stabilize the cliff. Where people or their assets are at risk, rockfall nets or retaining walls are commonly erected at the base of the free face to catch, and minimize damage from, falling and rolling rocks. Alternatively, explosive blasting of dangerous rock, debris, or earth materials can be used as a preventative measure to remove masses that are deemed likely to fall.

Rockfall debris accumulates at or near the base of the free face as **scree** or **talus** (Figs. 8.7, 14.18). Talus typically accumulates in piles of irregular and angular rocks. In areas of higher moisture, and where rockfall activity is less frequent, regolith, vegetation, and soil may obscure the talus. In cases where rockfall debris becomes channelized as it falls/rolls to the base of the cliff, talus cones are formed (Figs. 8.11, 14.18A). Talus typically accumulates at a nearly

uniform slope angle, referred to as the angle of repose (Fig. 8.8). The angle of repose is the steepest slope at which loose material remains stable without fragments rolling farther downslope (see Chapter 8). It is controlled by the frictional contact between grains, being about 30° for dry sands, but can be as steep as 45° for larger blocks and boulders. That said, 34°–36° is typical for most talus.

Most talus deposits exhibit some amount of sorting, with increasing clast sizes from the upper part to the lower end of the slope (Fig. 14.18B). This sorting occurs because larger boulders have higher momentum and thus tumble greater distances by rolling, sliding, and bouncing. The roughness

Figure 14.16 Rockfalls. **A.** A 2017 rockfall of ≈150 m³ at Gelmersee, Guttannen in the Bernese Alps, Switzerland, with (**B**) a close-up view of the free-falling rocks. **C.** A rockfall at Tramin, South Tyrol, Italy, on January 21, 2014, led to this huge dolomite block rolling downslope, before stopping in a vineyard. **D.** A second block stopped just upslope of a large building. This example illustrates that rockfalls can have runout areas that can extend far beyond the free face from which they originated. Source: (A, B) Kantonspolizei Bern; (C, D) C. Bonati.

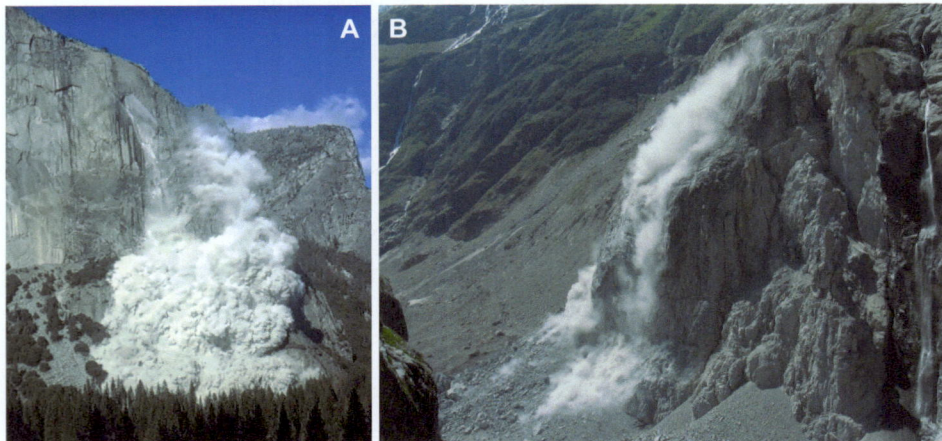

Figure 14.17 Major rockfalls. **A.** A rockfall at El Capitan, a sheer face of granite in Yosemite National Park, California, USA, on September 28, 2017. Source: P. Pawilkowski. **B.** A 2009 rockfall that developed from a heavily disintegrated rock mass on the Eiger east face, Grindelwald, in the Bernese Alps, Switzerland. This event may have been triggered by glacial undercutting of the slope, centuries earlier. Source: Hadi, Public domain, via Wikimedia Commons.

Figure 14.18 Talus. **A.** Talus here has developed due to intense frost weathering on this slope near Keremeos, British Columbia, Canada. Rocks at the site vary in their lithology, as reflected in the colors of the talus. Source: S. Earle. **B.** A talus slope showing good clast sorting, with larger blocks nearer the runout area. Source: R. Schaetzl.

and porosity of the talus surface also plays an important role in sorting. Although small pebbles can be stopped by irregularities on the slope surface or fall into holes between boulders, large boulders will mainly be stopped by hitting other, large obstacles, or by reaching the bottom of the deposit. As a consequence, the largest boulders tend to stop at the lower end of a talus slope, and materials of comparable size tend to end up in the same general sectors of the talus deposit (**Fig. 14.18B**).

Because talus deposits are generally coarse and porous, they tend to resist erosion, and runoff is rare. Thus, most talus deposits are stable, long-lasting features of mountain landscapes, even if rockfall activity has recently ceased. Indeed, in many mountainous areas, talus deposits date back to colder (glacial) climates, when freeze–thaw activity was rampant, prying rocks from free faces.

Rockfalls are affected by a number of variables, particularly geology. A free face formed in rocks that are favorable to rockfalls is often defined as **incompetent**. Competent cliffs are not as favorable to rockfalls, usually because the bedrock is better consolidated and less jointed. In addition to geology, a variety of environmental and climatic factors also affect the frequency of rockfalls by impacting the nature and degree of weathering. Any process that generates fractures, such as root growth and freeze–thaw activity, will promote rockfalls. The US National Park Service has studied the timing of more than 1,000 rockfalls in Yosemite National Park, California over the past 150 years (**Fig. 14.19**). These data have helped identify possible controls and triggers of rockfalls.

Attributing a rockfall to an individual cause is difficult, as a large degree of uncertainty persists about what exactly triggers an individual fall. Even at Yosemite National Park, where a very robust record of rockfalls exists (**Fig. 14.19**), correlating a given rockfall with a recognizable trigger is difficult. It is often better to assume that a suite of geological processes set the stage for rockfalls by slowly rendering the free face less competent. At Yosemite, these include oversteepening and undercutting of slopes due to glacial erosion, but also long-term weathering.

Physical and chemical weathering processes promote rockfalls by weakening the rock face. Physical weathering produces joints in the rock. Trees in sheer rock faces can then trigger a rockfall because their roots pry apart the rocks along joints. Water also seeps into these joints, where it may eventually freeze, widening the cracks and dislodging rocks through a process referred to as **frost wedging**. Successions of freeze–thaw cycles will slowly but incrementally pry rocks off from a free face. For this reason, most rockfalls in mountain environments occur in winter and early spring, that is during periods of subfreezing temperatures and/or during snowmelt. In cold alpine regions, rockfalls at previously stable sites may now be more likely due to recent changes to the climate, as more freeze–thaw cycles now occur there. Alternatively, at lower elevations where the climate is now warmer and wetter, rockfalls can become more frequent due to enhanced chemical weathering.

Water in joints in bedrock can also develop massive pressures, even if unfrozen. In the Randa rockfall in the Swiss Alps, which let loose ≈30 million m³ of debris, water that had penetrated into cracks and fissures ultimately caused the slope failure (**Fig. 14.9**). Substantial amounts of water were observed by eyewitnesses to spurt out horizontally from the rockwall as it failed.

14.6.2 Topples

In toppling, large blocks fall over by rotating outward over a pivot point (**Figs. 14.12, 14.20**). Parts of the original material may remain intact, yet reoriented, within the resulting debris pile. Toppling is a type of **overturning failure**, typically occurring in materials containing steeply dipping layers. Topples can be very destructive, especially in cases where the failure is sudden.

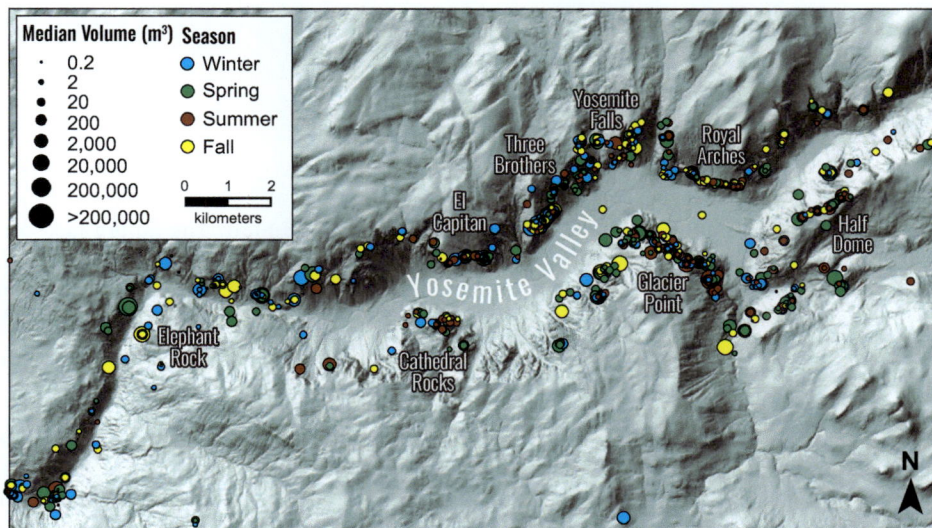

Figure 14.19 Rockfall events from 1857 to 2020 in Yosemite Valley, Yosemite National Park, California, USA. Shown here are only those events for which the location and date are known. Source: Map from Greg Stock and Autumn Helfrich of the US National Park Service.

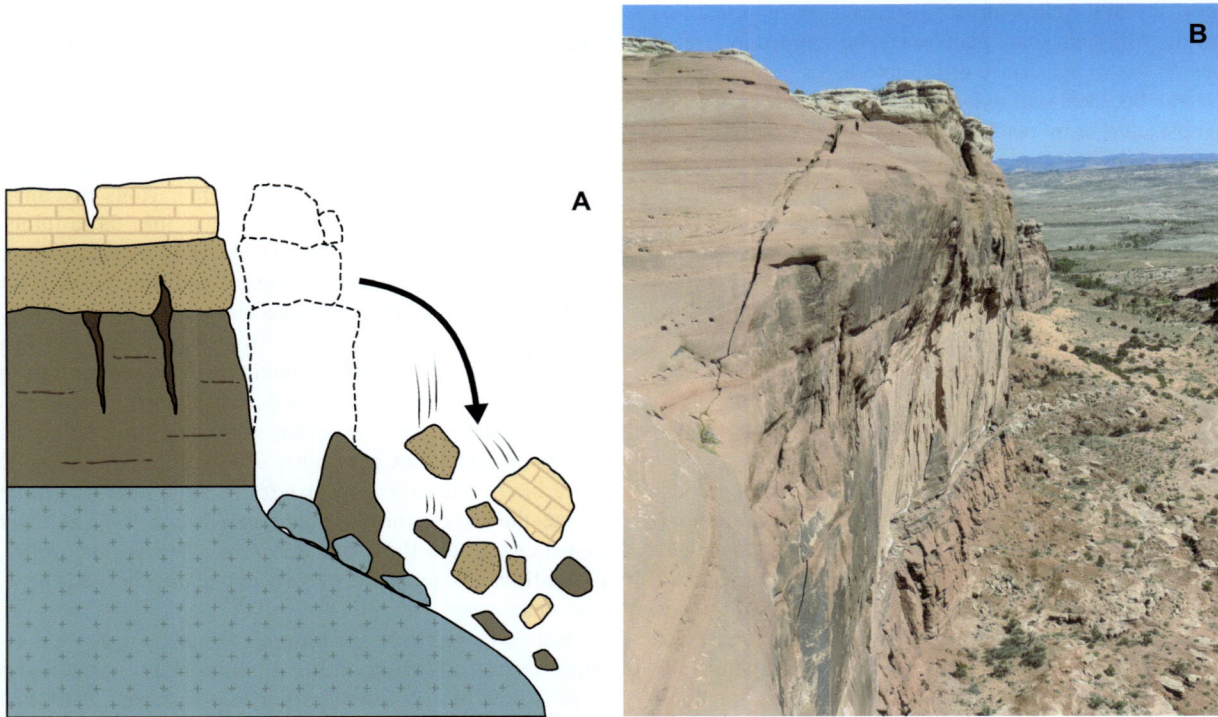

Figure 14.20 Topples. **A.** Schematic representation of a rock topple. **B.** A rock slab at the edge of a large sandstone mesa. The ≈450 m long and 40 cm wide crack at the edge of this sandstone mesa in Utah, USA indicates that the rock face may soon fail as a topple. Source: E. Bessette-Kirton.

Any kind of coherent material – rock, debris, or earth – can topple; the process is accordingly called a rock topple, debris topple, or earth topple (**Fig. 14.12**). Bedrock types that are prone to develop topples include massive beds of sandstone, limestone, igneous lithologies with steep joints, sedimentary strata pushed to vertical positions during folding, and metamorphic lithologies with nearly vertical foliation. Earth materials prone to toppling include jointed deposits of stiff clay or silt.

Topples can begin slowly, only to become extremely rapid as the collapse begins. They can be driven directly by gravity, pulling the mass down, or by mass that pushes from above. Many topples occur due to the prying actions of water or ice in cracks, or because of vibrations. Many toppled slopes were initially undercut, as is shown in **Fig. 14.20B**.

Depending on the nature of the failing material, two types of topples are distinguished. **Rock toppling** consists of relatively massive blocks of strong rock. Here, the mass relies on the rotational stability of its base and moves as a rigid mass (**Fig. 14.20B**). Rock topples usually lead to catastrophic failures. In contrast, **flexural toppling** occurs as thin, steep layers of relatively weak rock, particularly schist or phyllite, undergo slow bending or tilting, and eventually fall.

A recent, fatal, topple occurred on Grandview Surf Beach in Encinitas, north of San Diego, California, USA, where a ≈10 m long slab of the sandstone bluff fell (**Fig. 14.21**). Coastal erosion at the base of the cliff had undercut the slope

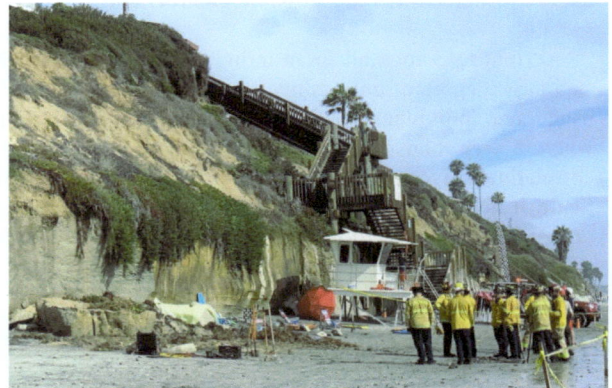

Figure 14.21 Scene after the rock topple at Grandview Surf Beach on August 2, 2019. Source: H. Palmour.

and destabilized the sandstone bluff. The failure face contained many roots, suggesting that root growth probably played a role in destabilizing the slope. The slab underwent toppling failure without any obvious warnings, tragically killing three people.

14.6.3 Slides

Geomorphologists define a **slide** as a mass (or block) that moves downslope as a unit (or series of units) along a defined sliding surface, usually called a **surface of rupture** or a **plane of failure**. Slides are common and occur in most environments. They can move in a matter of seconds or slide gradually and/or episodically over weeks and months. Commonly, the term **landslide** is applied to virtually any kind of large slope failure. Geomorphologists, however, only use the term landslide for mass movements that move along some kind of planar, sliding surface. Similar to the other mass movements discussed above, saturation, undercutting, and human activities such as devegetation, excavation, and vibrations are common triggers for slides. Slides can also be triggered by rapid drops in river levels after floods, or by rising water tables in reservoirs, streams, and lakes.

Because of their great variability in form, landslides are often subdivided, based on the depth to the plane of failure. If the sliding surface is located within the soil or the upper part of weathered bedrock, they are referred to as **shallow landslides**. Most shallow landslides do not involve failure of competent bedrock. Shallow landslides typically occur on slopes with highly permeable regolith overlying less permeable materials. Saturation of the upper, more permeable sediment overloads the slope and initiates the landslide. In contrast, **deep-seated landslides** usually have their sliding surface deeper than 10 m, that is, well below the root zone, and along a plane of weakness in the bedrock, such as faults or bedding planes. These slides involve soil, regolith, weathered rock, and even bedrock. Deep-seated landslides tend to be much larger than shallow landslides.

Landslides can also be subdivided based on their type of movement – translational landslides (Fig. 14.22), rockslides, and rotational landslides or slumps (Fig. 14.12). In **translational landslides**, the mass slides along a planar (or only slightly undulating) plane of failure, such as geological discontinuities at faults, joints, and bedding planes, or at the bedrock–regolith contact (Figs. 14.22, 14.23). They are particularly likely on slopes where these discontinuities parallel the land surface, and are one of the most common types of mass movements. When they occur in bedrock, translational landslides can also be called **block slides** (Figs. 14.12, 14.22C). On steep slopes, translational landslides may continue sliding for long distances, especially if the mass breaks up, which is typical for fast-moving and wet materials (Fig. 14.23A). Translational landslides can travel at a range of velocities, from those that are extremely slow to some that exceed 5 m/s.

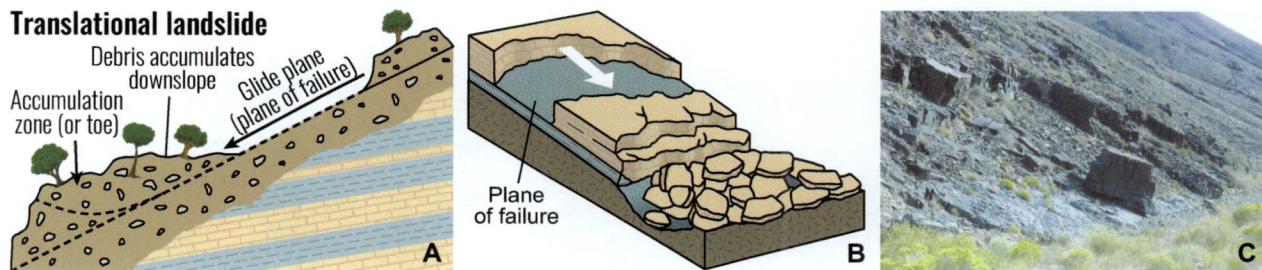

Figure 14.22 Schematic illustrations of translational landslides in (**A**) regolith and (**B**) bedrock. **C.** Photo of a very small translational landslide (or block slide) in bedrock. Source: Garry Hayes, from Geotrippers.com.

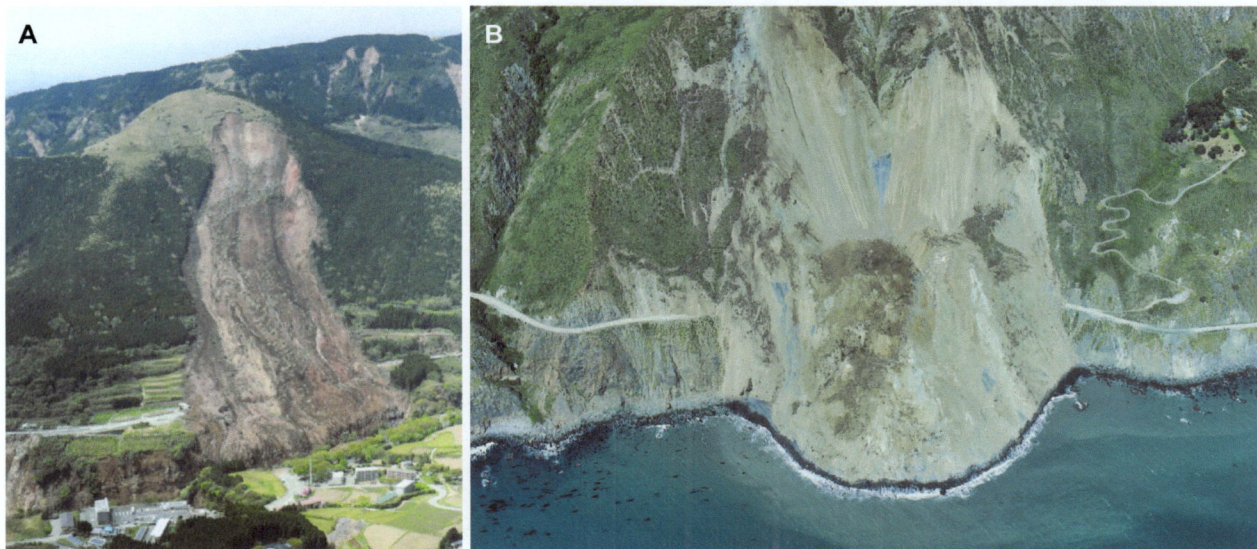

Figure 14.23 Translational landslides. **A.** The April 18, 2016 Minami-Aso translational landslide in Japan was triggered by seismic shaking. Source: Japanese Ministry of Defense; in public domain. **B.** The May 20, 2017 Mud Creek landslide near Big Sur, California, USA, destroyed a large section of US Highway 101. The repairs took almost two years to complete. Source: US Geological Survey; in public domain.

A **rockslide** is another type of translational landslide, again developing along a bedding plane, joint, or another plane of structural weakness in bedrock. Rock (or some other type of consolidated material) moves *en masse* for a short distance, then disintegrates (**Figs. 14.1, 14.22B, C**). Joint systems in the bedrock are critical for the development of a rockslide, because they not only weaken the bedrock but also, during failure, provide a plane along which the slide may travel. Indeed, as soon as the stress force exceeds the cohesive strength along any plane in a rock, the mass will fail. Rockslides are doubtless one of the most dangerous forms of mass wasting because they incorporate a sudden, incredibly rapid release of often massive amounts of bedrock (**Fig. 14.24**). Although rockslides usually occur on steep mountain fronts, they have also been observed to develop on slopes with gradients as low as 15°.

Rockslides can move very large volumes of rock – on the order of millions of metric tons. These large masses can then travel downslope at great speeds. The speed and potential devastation of a rockslide is often determined by the steepness of the slope.

Instead of a generally planar surface, **rotational landslides**, also called **slumps**, move along a curved, concave plane of failure that is parallel to the contour of the slope (**Figs. 14.12,**

14.25). Their "backward rotational" movement gives slumps their name. The slump block is characteristically circular or spoon-shaped and moves as a fairly coherent body, with little internal deformation. Nonetheless, even if the upper parts of the slump block retain their integrity, sediment in the **toe** of the block commonly breaks up and may even move as a flow. Rotational landslides range from extremely slow (< 0.3 m in 5 yrs) to moderately fast (1.5 m/month), and some can be very rapid. They occur preferentially on slopes with gradients between 20 and 40° and in homogeneous regolith such as glacial till or weathered volcanic sediment. They are also common in shale and clay-rich soils. Because homogeneous materials are not widespread in nature, rotational slides are less common than translational slides. Slumps are also frequently observed on slopes weakened and altered in some way by human activity, for example, by undercutting or devegetation.

In the field, the presence of the characteristic scarp (or scarps), the tilting of bedding or other surfaces, as well as the contorted surface of the toe, all point to previous slumping and hillslope instability (**Fig. 14.25**). Once identified, these features can help geomorphologists to identify slopes that could be prone to failure in the future (**Fig. 1.1**).

In a rotational landslide, or slump, the mass moves downward and outward. It *rotates*, such that the geological units

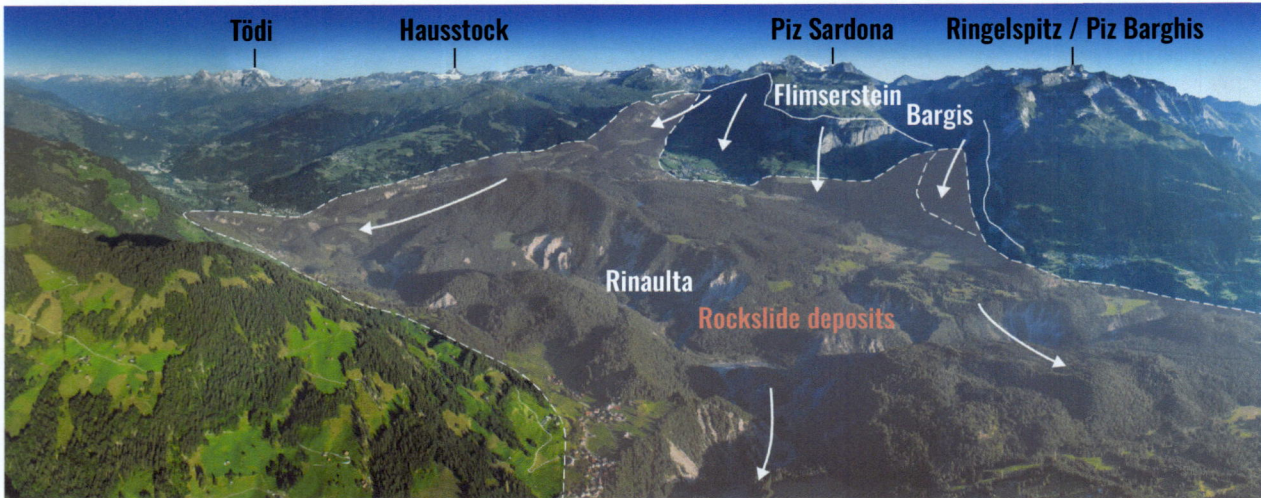

Figure 14.24 The landscape near Flims, Grisons, Switzerland, shows clear signs of a rockslide ≈9,450 years ago. That event filled the valley with ≈10 km³ of debris, forcing the river to find a new course across the deposits, into which it formed a prominent gorge. Source: UNESCO World Heritage Tectonic Arena Sardona; in public domain.

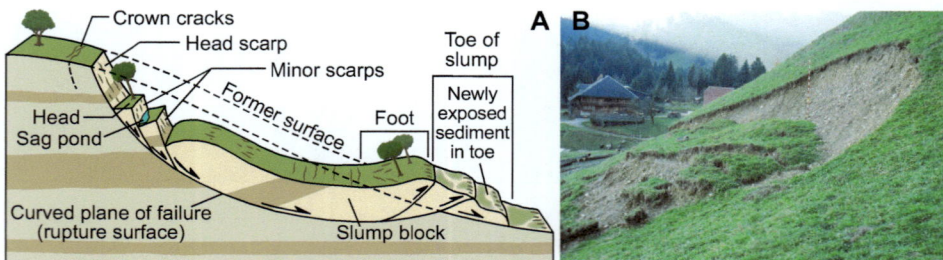

Figure 14.25 Rotational landslides, or slumps. **A.** Schematic representation of the various components of a rotational landslide. **B.** A small slump in Central Switzerland, formed in July 2002 after a heavy thunderstorm. Source: C. Rickli.

tilt backward (Fig. 14.25A). Because of this, rotational land-slides have very characteristic features that help to identify their rotational history. They often develop into a clear block or series of blocks, with each block having a steep, almost vertical, **scarp** and a backward-tilting, upper surface called a head. The uppermost scarp forms the landslide **crown**, that is, material that is still in place, immediately above the highest parts of the main (or head) scarp. In addition to the main plane of failure, several secondary sliding planes may develop in the slump block, breaking the larger block into a series of steps, or small terraces. The upper part of the slump block then develops into several compartments of stair-step-like surfaces called **minor scarps**.

Because of the rotation, the head (top) of the slump block comes to rest *below* the original ground surface. Depressional areas can then form between the base of the scarp and the back-tilted, upper slide surface (**Fig. 14.25A**). These areas can then trap water, often forming sag ponds (**Fig. 14.26**). The water in the pond may per-colate down and along the slide planes, lubricating them and setting off additional movement, sometimes days later. This process is another example of positive feedback in geomorphology – the slump generates sag ponds, which facilitate further movement, which in turn enlarges the catchment for the sag ponds, and so on.

The material at the base of a rotational landslide is referred to as the toe (**Fig. 14.25A**). Often, sediment within the toe will become saturated and move further downslope as a small flow. Removal of the toe, as is sometimes done by humans to "remediate" the sites, undercuts the slope and often sets the slump block back into motion.

Rotational landslides occur in a variety of sizes. Many landscapes have small slumps that are quickly remediated (**Fig. 14.25B**). But slumps can also be huge. In the Llusco (Peru) landslide, ≈30 ha of weathered, volcanic rocks were activated on February 23, 2018, destroying almost 100 homes (**Fig. 14.27A**). Activity here had been observed for months by the civil defense office of Peru, which considered the risk level to be so high that the area was evacuated.

Reactivation of the landslide occurred later, due to intense precipitation, additions of water by human activities, and stream erosion which undercut the slope. Another impres-sive rotational slide occurred at Holbeck, North Yorkshire, UK, in which a hotel was destroyed (**Fig. 14.27B**). The 1993 landslide displaced ≈1 million tons of glacial till to cut back the 60 m high coastal cliff by 70 m. Both of these landslides involved thick masses of homogeneous, weathered sediment – typical for rotational landslides.

14.6.4 Flows

A **flow** is a mixture of rocks, debris, and earth, with some water, that moves downslope, mixing internally as it moves (**Fig. 14.12**). Flows do not move along a distinct plane of failure, but instead they move at different speeds within, flow-ing most rapidly at the surface and slower at depth, facilitat-ing the mixing. At some depth, the rate of movement goes to zero. As they move downslope, most flows become less viscous, as the material loses coherence.

Flows are sometimes likened to toothpaste, honey, or ketchup. This type of flow behavior is typical of **non-Newtonian fluids**: they become more liquid-like when under pressure or shaken, and more cohesive when at rest. In other words, a flow will tend to become less viscous when in motion, and as it slows the material will become less prone to flow. In addition, once motion has started, and if the slope is steep enough, flows tend to continue downslope; the most difficult, and key, part is the trigger.

Flows can be differentiated based on composition (rock, debris, and earth) or water content (dry to wet), both of which dramatically influence flow velocities (**Figs. 14.12, 14.13**). Based on water content, flows can be separated into **granu-lar flows** and **slurry flows** (**Fig. 14.28**). Granular flows are the drier type, usually containing < 20% water. In these types of flows, the flow-like behavior occurs because sediment is mixing with air. In contrast, slurry flows are wetter, with 20–40% water. Both flow types can be further divided based on velocity. As water contents increase beyond ≈40%, slurry flows grade into stream flows.

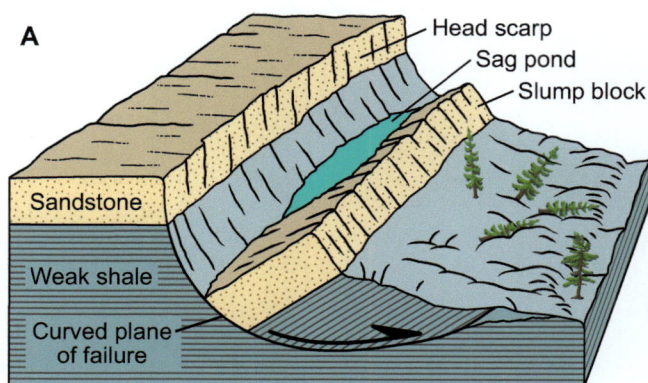

Figure 14.26 A. Sag ponds are water bodies occupying an enclosed depression (or **sag**) near the head of rotational landslides. Source: © A. N. Strahler (1992). Used by permission. **B.** A sag pond has formed at the head of the 2014 West Salt Creek (Colorado) rock avalanche. Source: J. Coe, via the US Geological Survey; in public domain.

Figure 14.27 Large rotational landslides. **A.** A rotational landslide at Llusco, near Cusco, Peru. Source: Galeria del Ministerio de Defensa del Perú, CC BY 2.0, via Wikimedia Commons. **B.** The Holbeck Hall landslide, in Scarborough North Yorkshire, England. This rotational landslide moved over the course of a few days, destroying the hotel at the top of the cliff. The rotated blocks can be seen as grass-covered "benches." Source: British Geological Survey – UK Research and Innovation; in public domain.

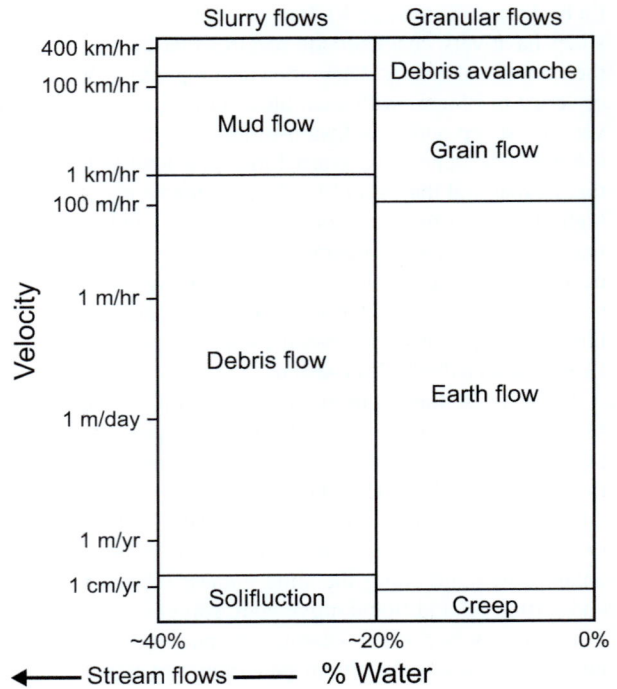

Figure 14.28 A classification of flows based on the relative amount of water involved and the nature of the material transported (rock, debris, and earth), which influence their velocity.

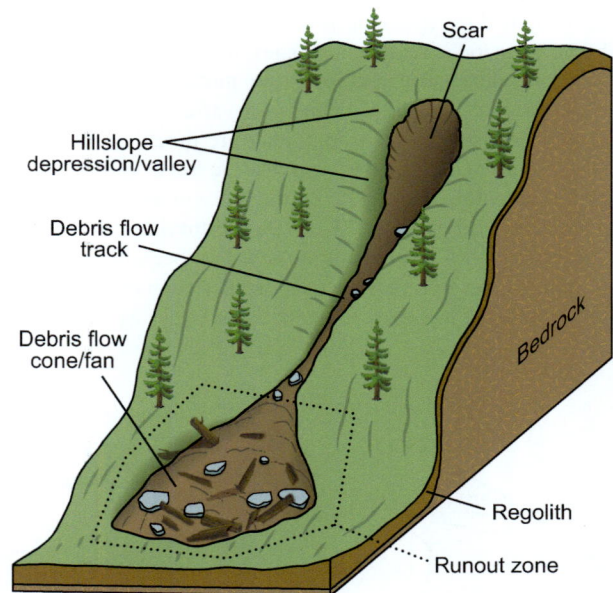

Figure 14.29 Debris flows often have distinct zones, such as the starting zone (now only a scar), the debris flow track, and the runout zone.

The most common type of flows – **debris flows** – are fast-moving flows that typically flow/tumble downslope within preexisting channels or tracks (**Figs. 14.12, 14.13, 14.29, 14.30A**). Most debris flows have starting (or initiation) zones on slopes or within steep gullies, in erodible, loose sediment. Debris flows come to a rest in runout zones, where they may form deposits of various sizes and shapes (**Fig. 14.29**).

Debris flows are often triggered by heavy, short-lived storms and/or by long-lasting, less intense, rainfall events. Snowmelt or the thawing of permafrost can also trigger them (see Chapter 20). In some cases, the bursting of natural dams and related glacial lake outburst floods have also triggered debris flows. Thus, most debris flows move as a saturated, or nearly saturated, mass. A debris flow at Kedarnath, in the Indian Himalayas, is an example of a lake outburst that led to a debris flow disaster (**Fig. 14.30A**). Here, unusually heavy precipitation falling on snow cover saturated the steep

Figure 14.30 Debris flows. **A.** The June 17, 2013 debris flow at Kedarnath in the state of Uttarakhand in the Indian Himalayas was triggered by intense rainfall on a snowpack. Source: V. Kaul. **B.** Boulders transported by a debris flow on January 9, 2018 in Montecito, California. The debris flow was initiated by a thunderstorm, on a slope recently devegetated by a wildfire. Source: US Geological Survey; in public domain.

slopes, leading to multiple, devastating debris flows and an outburst flood, killing at least 5,000 people. Debris flows can become more frequent or more intense when occurring on devegetated slopes.

The "debris" in debris flows consists of loose sediment and large organic materials such as brush and logs, all of which can combine with water to form a slurry. Sediment in the debris can be of very diverse shapes and sizes, commonly ranging up to boulder size. The fairly poor sorting of debris flow sediments distinguishes these deposits from most water-laid (fluvial) sediments. The consistency of debris flows can range from watery mud to thick, rocky flows carrying large boulders and trees (**Fig. 14.30B**). Debris flows therefore have very high bulk densities, approaching ≈ 2000 kg/m^3. Water acts as a lubricant, decreasing friction between grains, and also adds weight to the mass, because it replaces air in the voids. Generally, the wetter the flow, the greater its speed. Debris flows traveling down steep channels can attain speeds >10 m/s; some of the largest flows have been reported to exceed 50 km/h. Debris flows can be small and localized (a few 100 m^3) but much larger events have been reported in mountain regions worldwide, some with volumes >100,000 m^3.

Debris flows often develop distinctive components (**Fig. 14.31**). The bouldery front, or head, typically contains coarser debris such as boulders and logs, imparting a great deal of friction and force. Behind the head and the main debris flow body trails a liquefied mass of debris, with larger contents of sand, silt, and clay, and with fewer boulders.

The fine sediments here help maintain the high pore-fluid pressure, which in turn helps the debris flow to maintain its mobility. The last part of the debris flow – the runout – is a watery tail (**Fig. 14.31**). Debris flows often occur in a series of discrete pulses or surges; each pulse or surge may have its own distinctive bouldery head, body, and tail.

Having such distinctive parts while flowing, debris flows then leave behind recognizable types and sequences of deposits and landforms as they stabilize. Many have well-defined lobate forms, with boulder-rich snouts (**Figs. 14.29, 14.31**). The margins of (incised) debris flow paths are often characterized by **lateral levees**, which form as the liquefied, fine-grained body of the debris flow pushes aside coarser debris that accumulates in the front of the flow (**Fig. 14.32**). Once established, levee deposits and paths can influence the paths of ensuing debris flows. Debris flows usually terminate at the base of slopes, where they form fanlike **debris flow cones** or fans (**Fig. 14.29**).

The general public sometimes uses the term **mudflow** to describe all watery debris flows. Nonetheless, true mudflows consist mainly of grains that are sand-sized or smaller, with few large clasts. They are far less common than true debris flows. Mudflows often have the consistency of sloppy, wet concrete, splaying out across a broad runout zone (**Figs. 14.12, 14.33**). The significant proportion of mud and fine-grained materials in mudflows renders them more fluid than other types of debris flows. Water contents in mudflows can be up to one-third, allowing them to flow rapidly, often > 5 m/s and even across gentle slopes. The dense slurry

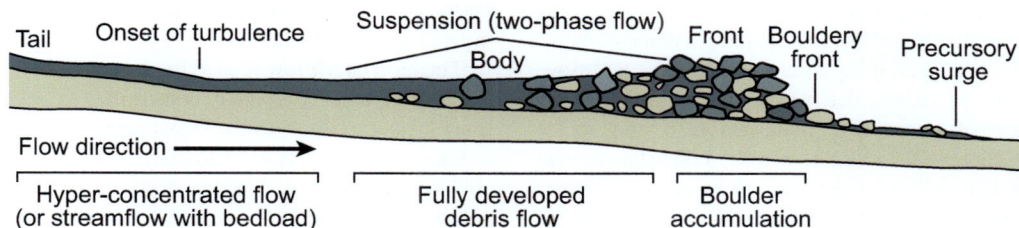

Figure 14.31 The various components of a debris flow, including the bouldery front, the main body, and the watery tail.

Figure 14.32 Incised channels, lateral levees, and lobate deposits of debris flows near Hühnerspiel, Zillertal Alps, Italy. Source: M. Mergili.

Figure 14.33 Mudflows **A.** A miniature mudflow with a characteristic lobate deposit next to a mud volcano of the Salton Sea in Southern California, USA. Source: T. McGuire. **B.** A mudflow at Chamoson (Valais, Switzerland), triggered by a thunderstorm on August 7, 2018 onto heavily weathered, clay-rich sediments. Note how the bridge had partially blocked the flow. Source: M. Schmid.

nonetheless can transport a few large boulders by "floating" them over large distances; some mudflows have even moved houses from their foundations. Like other types of debris flows, materials of different sizes become sorted across the runout zone.

Surprisingly, mudflows are not uncommon in arid and semi-arid regions, where they usually flow down gullies. Here, they almost invariably result from unusually heavy precipitation or a sudden snowmelt event.

Volcanic mudflows are known by their Javanese name **lahar**. Lahars are triggered by volcanic eruptions, and flow down the flanks of steep-sided volcanoes as fast-moving, watery slurries of pyroclastic materials (ash and rocky debris) (see Chapter 7). Flowing with the consistency, viscosity, and density of wet concrete, lahars are fluid when moving and solid at rest. Water – the key trigger of lahars – is made available from melting snow or ice on large volcanoes, or can originate from lakes in the volcano's summit area.

The deadliest lahar events have typically originated from eruptions which rapidly melted snow and ice on the volcano.

Lahars can be extremely devastating, destroying infrastructure and burying housing (**Fig. 7.41**). They can flow at velocities of tens of meters per second, and the largest known lahars have been reported to be ≈140 m thick, covering areas > 300 km² and with volumes exceeding 2 km³. In 1985, lahars from the Nevado del Ruiz volcano in Colombia buried the city of Armero under 5 m of mud and debris, killing ≈23,000 people.

The term **earthflow** is used for debris flows that are slower and less expansive, typically constrained to only a small section of a slope (**Fig. 14.12**). They contain mostly fine-grained sediment. Earthflows tend to start at a scarp or on a small cliff, and then flow downslope in confined, distinct channels. Many will then form "tongues" in runout zones. Like mudflows, the viscous flow associated with an earthflow develops due to saturation in the regolith. Flow velocities – ranging

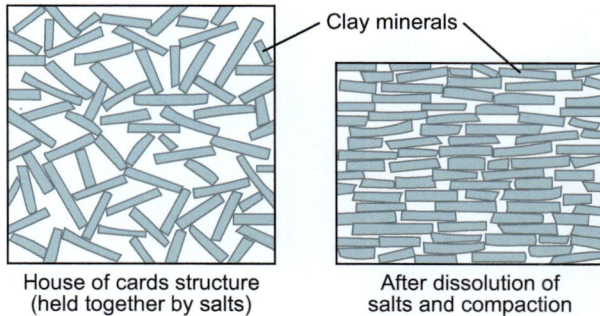

Figure 14.34 Quick clays. Their packing structure is lost as salts are lost from the clays, resulting in compaction, loss of cohesion, and slope failure.

from as little as 1 mm/yr to 20 km/h – are controlled by water content (the higher the water content, the faster they flow). Although in some respects earthflows appear similar to mudflows, earthflows tend to flow more slowly. The slowest earthflows – **creep** and **solifluction** – form a category of their own, and are discussed later in this chapter.

Because earthflows depend on saturation and liquefaction as preconditions, they are more common in humid climate regions, where periods of heavy rainfall or snowmelt will increase pore-water pressures and reduce the shear strength of the regolith.

Earthflows occur worldwide, but the most catastrophic, rapid earthflows have been reported in marine clays of the St. Lawrence Valley Lowlands of North America, along coastal Alaska and British Columbia, and in Scandinavia. These types of marine clays are often referred to as **quick clays**. When quick clays are dry, the clay minerals are arranged randomly, with considerable pore space (**Fig. 14.34**). In this type of structure, grains obtain their cohesion by salts, giving them strength. Once quick clays become wet and then disturbed, they fall apart and experience liquefaction, behaving almost like a fluid. This happens as water infiltrates into the pore spaces and becomes adsorbed onto the clay minerals. The water dissolves some of the salts, causing collapse and ultimately slope failure. Quick clay flows are retrogressive, usually starting at the contact with water bodies in lower slope positions, and slowly progressing inland and upslope. Nonetheless, steep slopes formed on quick clays may collapse much more rapidly, or in large chunks. On June 3, 2020, several buildings were swept into the sea by this type of earthflow in Kråknes, near Alta in northern Norway (**Fig. 14.35**). The disaster struck unexpectedly after a rainfall event. On the day preceding the event, a local resident noticed a tensile crack which may have favored the infiltration of water.

Avalanches are rapid flows of snow, ice, rock, and/or debris initiated along a slide plane (**Fig. 14.36**). Of these, **snow avalanches** are the best known. In a snow avalanche, a cohesive slab of snow lying on a weaker layer of snow (or ice) slides downslope, fracturing as it moves. Avalanches can also occur on glaciers if a large piece of ice detaches. Our focus is on **rock avalanches** and **debris avalanches**, as their occurrence changes landforms significantly, unlike snow and ice avalanches which do very little geomorphic work.

Rock avalanches originate in jointed and fragmented rock. By definition, they include mainly rock material. Debris avalanches, in contrast, involve large amounts of unconsolidated sediments. The main difference between a debris flow and a debris avalanche is channelization; debris flows are typically confined to some sort of channel, but debris avalanches are not (**Fig. 14.12**).

Rock and debris avalanches typically begin on mountainsides. As they move downslope, the rock breaks up. Small rock fragments can then act as the matrix, carrying larger particles with ease. In their source area, the sudden, catastrophic collapse leaves concave depressions in the bedrock.

Avalanches are often the fastest of all flows, with velocities estimated at up to 100 m/s. Their extreme velocities can be accentuated by cushions of steam and air, on which the fragmented masses are transported. Their great speeds promote extremely long runout distances, and some may even continue to flow further downslope as a debris flow or lahar. Across these runout areas, rock and debris avalanche deposits produce distinctive **hummocky topography**, with its many small hills, isolated depressions, and discontinuous drainage, as well as steep-sided distal and lateral edges. However, as a result of the large distances travelled, the

Figure 14.35 Damage caused by liquefaction of glaciomarine quick clays at Kråknes, near Alta, in northern Norway. The flow began at the base of the bluff and progressed inland for days after the initial failure. Source: J. E. Bakkeby and A. Bjordal, Norwegian Water Resources and Energy Directorate.

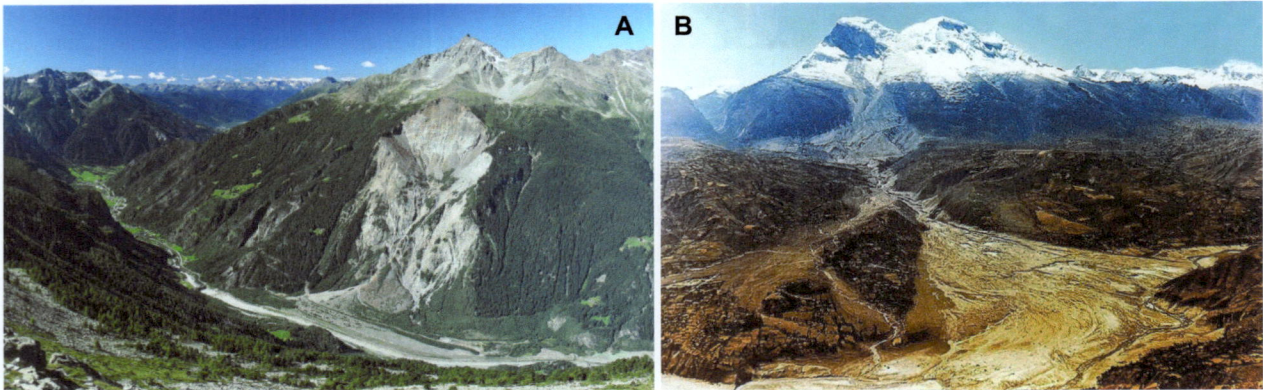

Figure 14.36 Major avalanches. **A.** The Monte Zandila rock avalanche in the Adda River valley of Val Pola, Italy claimed 27 lives. Source: M. Mergili. **B.** The Huascaran debris avalanche in Peru, on May 31, 1970. Over 30,000 people died from this event. Source: Courtesy Autoridad Nacional del Agua, Peru; in public domain.

overall thicknesses of these runout deposits can sometimes be fairly shallow.

Rock avalanches are also among the largest mass movements on Earth; some can involve volumes that exceed 100 million m³. Possibly the best-known rock avalanche, the Sentinel rock avalanche in Zion National Park, Utah, USA, was estimated at 300 million m³ in volume. Dimensions of debris avalanches differ greatly, based on the environments in which they occur.

Avalanches often occur at sites that are predisposed to mass movement, such as those that have been oversteepened, or on the inherently unstable slopes of steep volcanoes. Triggers for avalanches vary, but like so many other mass movements, often involve earthquakes or saturation.

The Val Pola avalanche (also known as Monte Zandila rock avalanche) in Valtellina in the Italian Alps occurred on July 28, 1987, after a period of exceptionally heavy rainfall (Fig. 14.36A). Prior to the event, the slope consisted of heavily fractured and deeply etched igneous and metamorphic rocks. Like most Alpine valleys, glaciation had created oversteepened side slopes. The catastrophic rock avalanche, with a volume of ≈40 million m³, flowed into the Adda River valley. The front of the avalanche ran 300 m up and onto the opposite slope before it split and diverted upstream and downstream, sending a wave of muddy water ≈3 km upstream. Avalanche deposits filled the valley to an estimated 30 to 60 m, with a maximum of 90 m.

The Huascaran debris avalanche was one of the largest historical events of its kind, and the deadliest rock-and-ice avalanche in recorded history (Fig. 14.36B). The event, triggered by a strong earthquake off the Peruvian coast, began on the steep, glacier-capped, west face of Nevado Huascarán, when seismic tremors dislodged a large (800 m wide) slab of rock. The slab fell almost 600 meters before landing on a glacier and sliding over 3 km down the glacier's surface, entraining ice, snow, and morainic material along its path. The mass movement then transformed into a very rapidly moving (estimated at 50–85 m/s) debris avalanche, sweeping away parts of the towns of Yungay and Ranrayrca and killing most of their inhabitants. The ensuing mudflow travelled so fast that it struck Yungay, 13 km from the avalanche origin, only a little less than two minutes after the earthquake first dislodged the rock. From its origin to its runout area, the event travelled over 16 km. The dislodged material not only filled the bed of the Rio Santa River, but was also remobilized in the form of a debris flow to travel an additional 160 km downstream, reaching the Pacific Ocean.

Solifluction (Latin *solum*, "soil," and *fluere*, "to flow") is the slow, downslope flow of saturated regolith in areas of permafrost (see Chapter 20). Solifluction is therefore most common in high-latitude (polar) or high-mountain regions where only the uppermost layer of the regolith thaws each summer – starting from the surface and extending downward less than a meter. The thawed upper part of the regolith is referred to as the **active layer**. Because permafrost is impermeable, the active layer is usually saturated during the summer, creating excessive pore pressures, which may then cause the saturated sediment in the active layer to flow slowly downhill. Flows are therefore most common in late spring and summer, when soils are saturated. They stop completely in winter when the active layer freezes.

Hillslopes affected by solifluction exhibit lobate or tongue-like forms of moving debris with steep fronts (or risers) and lower angled treads upslope, much like melted wax on a candle (Figs. 14.12, 14.37). Geomorphologists refer to these landforms as **solifluction lobes**. Although the bulk of the material in solifluction lobes is fine-textured, large blocks can also be carried along. Solifluction can occur on slopes as gentle as 1°, but is most common on slopes with gradients between 5° and 20°. Rates of movement typically range between 1 and a few cm/yr, and the thickness of solifluction deposits is normally only a few decimeters. Flow velocities and the depth of movement vary depending on slope gradient, local relief, thickness of the active layer, and water content.

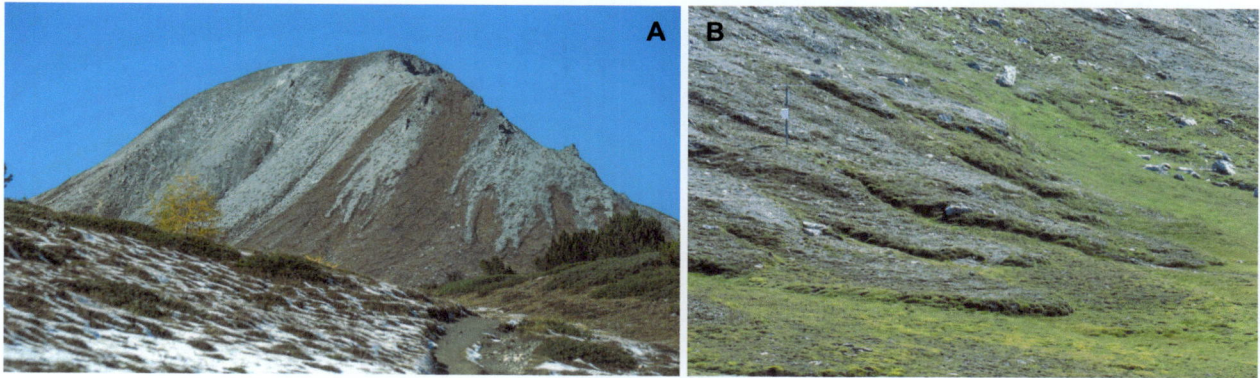

Figure 14.37 Solifluction lobes on Munt Chavagl, Swiss National Park, Switzerland. Source: H. Lozza.

The slowest form of mass movement, **creep**, is the almost imperceptible, downslope movement of loose debris and soil (**Figs. 14.12, 14.38**). In essence, creep is a type of continuous, slow earthflow. Creep, also sometimes called soil creep, is especially noticeable on steeper, but seemingly "stable," and vegetated slopes; here, the impression is that no downslope movement is occurring, because sediment is held in place by vegetation. Nonetheless, ongoing, slow movement *is* occurring. On moderate slopes (10–15°) in humid temperate environments, creep rates can be as slow as 1 to 2 mm/yr, whereas in semi-arid regions with cold winters, rates can be up to 5 to 10 mm/yr. Creep rates decrease with depth, such that at some depth, the sediment is stable and not moving. Rates of creep often increase immediately before a larger slope failure, thereby providing an indication of an imminent disaster. For example, prior to the Vajont landslide (**Fig. 14.8**), creep on nearby slopes had been measured at ≈1 cm/week. In September, one month before the landslide, creep rates had increased to ≈25 cm/day, until finally, the day before the landslide, creep rates exceeded one meter/day.

Creep is generally too slow to cause any obvious evidence of recognizable failure, such as open cracks or lobate deposits. Instead, signs of creep occur as bulges or low, wavelike swells on slopes. Exposed strata in road cuts and along stream banks may exhibit bent or steeply dipping layers in the downslope direction (**Fig. 14.38**). Blocks of rock may have been moved by creep, downslope from their outcrop. In short, we usually infer that creep is ongoing based on observations of features that are on top of, or shallowly emplaced into, the soil. Curved tree trunks, tilted power poles, and broken fences, are additional cultural "hints" that creep is active on a slope (**Fig. 14.38**). One has to be careful, however, as tilted gravestones and retaining walls, commonly taken as evidence for creep, can instead be due to faulty and unstable foundations.

Because gravity is always pulling materials downslope, all forms of disturbance on slopes – regardless of the processes involved – can result in a net downslope movement of sediment, and ascribed to creep. In general, two main processes are in play – one that is dominant in cool climates where freeze–thaw processes are common, and the other that is widespread in warmer climates.

In many mid-latitude locations, soils and regolith are repeatedly disturbed by near-surface processes such as the growth and melt of ice crystals during freeze–thaw events, the expansion and contraction caused by wetting and drying, and trampling and other types of disturbances by biota. Of these processes, creep is driven mainly by freezing and

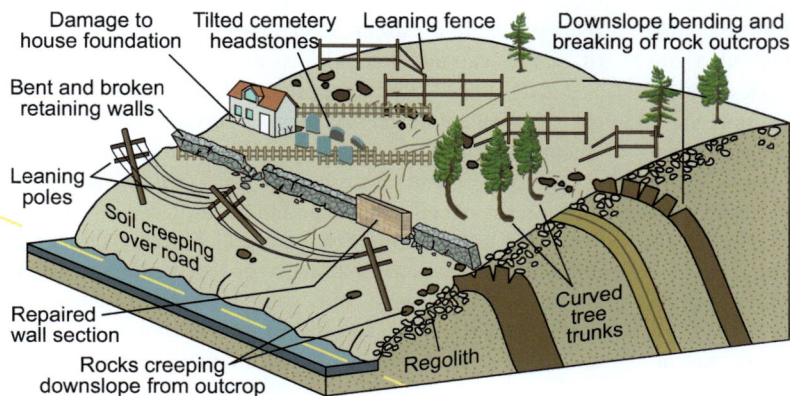

Figure 14.38 A schematic diagram showing the tell-tale effects of soil creep on both natural and cultural features that occur on slopes.

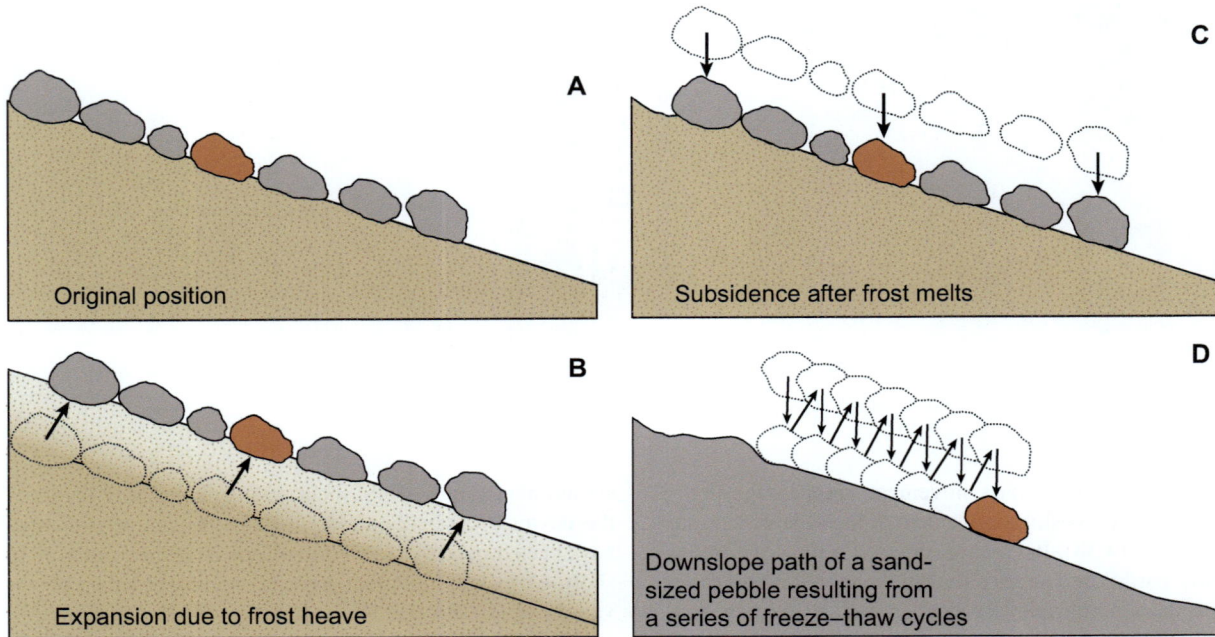

Figure 14.39 A schematic representation of the process of soil creep due to heave by the formation and melting of ice crystals. **A.** Particles on the bare soil surface are unprotected from the freeze–thaw cycles presented to them. **B.** As water in the soil freezes and expands, ice crystals lift soil particles upward, perpendicularly from the surface. **C.** As the ice melts, gravity pulls the particles down, vertically, thereby displacing them ever-so-slightly downhill. **D.** Repeated cycles result in the net downslope movement of regolith and soil – but faster for particles nearer the surface.

thawing. Freeze–thaw and wet–dry cycles lead to **heave**, that is, upward expansion as ice crystals form or as clays wet up. Heaving is most important in areas of frequent (often, daily) wet–dry or freeze–thaw cycles. During the wetting and/or freezing phase, rock and soil particles will expand and shift upward, perpendicularly to the slope (**Fig. 14.39**). Particularly important to this process is the formation of needle ice, which can force individual particles upward by as much as a few mm. Later, as the ice melts (or the soil dries), particles are allowed to settle back vertically, under the force of gravity. At the end of each cycle of expansion and contraction, rock and soil particles will move slightly downslope, over time transcribing an up-and-down, zig-zag path (**Fig. 14.39**). As would seem obvious, creep is therefore very pronounced in areas that undergo frequent freeze–thaw cycles, such as alpine areas and the high mid-latitudes. Wetting and drying cycles occur in environments where episodes of rainfall alternate with periods of desiccation. Surfaces with a thick cover of litter or organic materials are more insulated from these types of cycles, and as a result, have slower rates of creep. That is, creep is more prominent on barren slopes or on those with only a thin organic cover. Although heave only affects the upper few mm of sediment, underlying materials are dragged along as well, albeit at slower rates.

In the tropics, where soils are old and regolith is often very thick, the slow downslope movement of regolith is still ongoing, even though freeze–thaw processes are absent. Here, as in other areas, this movement is often driven by

bioturbation – soil mixing by plants and animals. Termites, ants, and myriad other soil organisms are particularly widespread in the tropics (**Fig. 22.5**). Almost incessantly, they bring sediment from the subsurface to the surface. Their nests and mounds are more porous and "open" than what is below, rendering the sediment susceptible to mobilization by wind and water. Any form of disturbance to the pedoturbated (mixed) sediment is likely to displace it slightly; in the long run, however, more sediment is moved downslope than upslope, because of gravity. Thus, the imperceptibly slow, downslope movement of sediment continues.

Bioturbation-driven mass movement is not just caused by mound-makers like ants or termites. Any form of disturbance will accomplish the same end result, although rates will vary. Birds and mammals scratch at the surface, and as they do, more sediment is displaced downslope than upslope. Even the slight disturbances caused by the hooves and paws of animals can facilitate this process.

Lastly, a particularly important driver of slow downslope mass movement in forested areas occurs as trees uproot (see Chapter 22, and **Fig. 22.4**). **Tree uprooting** tears up large masses of soil, and because more trees fall downslope than upslope, the net effect is for soil to be displaced downslope – one root mass at a time (**Fig. 14.40**). Unlike creep, which is a diffusive process that may be more-or-less continuous across a slope, uprooting is localized. Nonetheless, after thousands of years, most sites on a forested slope succumb to uprooting, and as they do, the effects of uprooting eventually get spread out, across the slope (**Fig. 14.40B**).

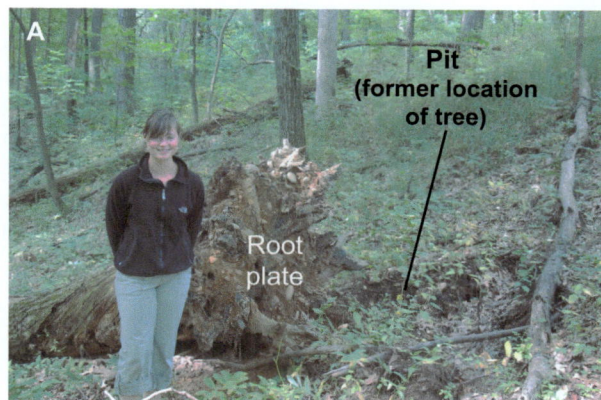

Figure 14.40 Mass movement by tree uprooting. **A.** A recently uprooted tree; note that, on this slope, most of the displaced soil will have been moved downslope, after the root plate decomposes. Source: R. Schaetzl. **B.** Volumes and distributions of mounds and pits formed by tree uprooting on a sloping, forested site in northern Michigan, USA. Arrows represent distance and direction from pit bottom to the corresponding mound, indicating the overall direction of the displaced regolith. Source: Used with permission of Elsevier, from Šamonil P., Valtera M., Schaetzl, R. J. et al. (2016); permission conveyed through Copyright Clearance Center, Inc.

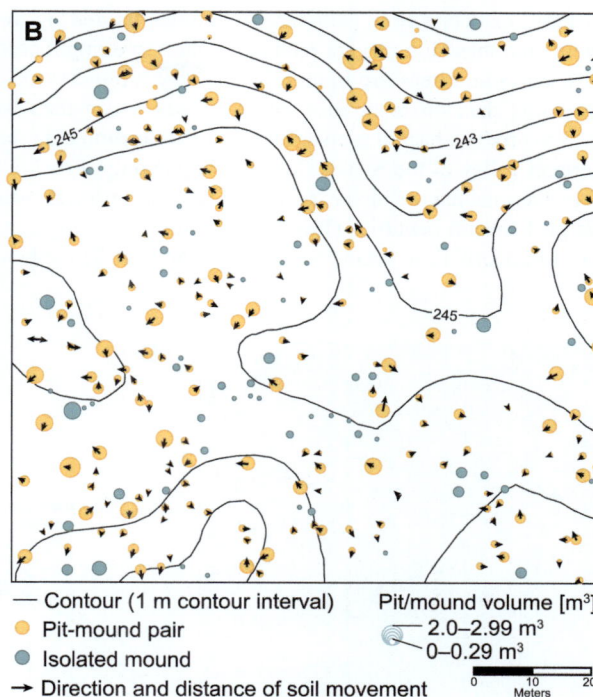

14.7 DATING MASS MOVEMENTS

Determining the age of a mass movement event, or a series of such events, is important to local inhabitants and authorities. Why? Knowing the frequencies of past and/or current mass movements helps assess the likelihood of future hazards. Such data also provide an indication of the general geomorphic stability of the region.

Geomorphologists may use one of several different methods to date a mass movement event. First, one can examine the freshness, or the degree of alteration, of sediment contained in the deposit. For example, in the hyperarid climate of Death Valley, California, USA, where debris flow deposits are not covered by vegetation, the deposits are dated based on the amount of rock varnish on the surface boulders (**Fig. 3.12**). Although this method cannot provide highly precise dates, it is very useful for determining the relative ages of a suite of deposits.

Dendrochronological (tree ring) data can, however, be used to provide precise estimates of mass movement activity. For example, trees buried in mass movement deposits tell of the exact age of the event. Where forests occur on avalanche or debris flow deposits, the trees can be directly dated to provide minimum-limiting ages for the event – the mass movement can be no younger than the oldest tree growing on the deposit (see Chapter 3). Lastly, trees growing close to avalanche,

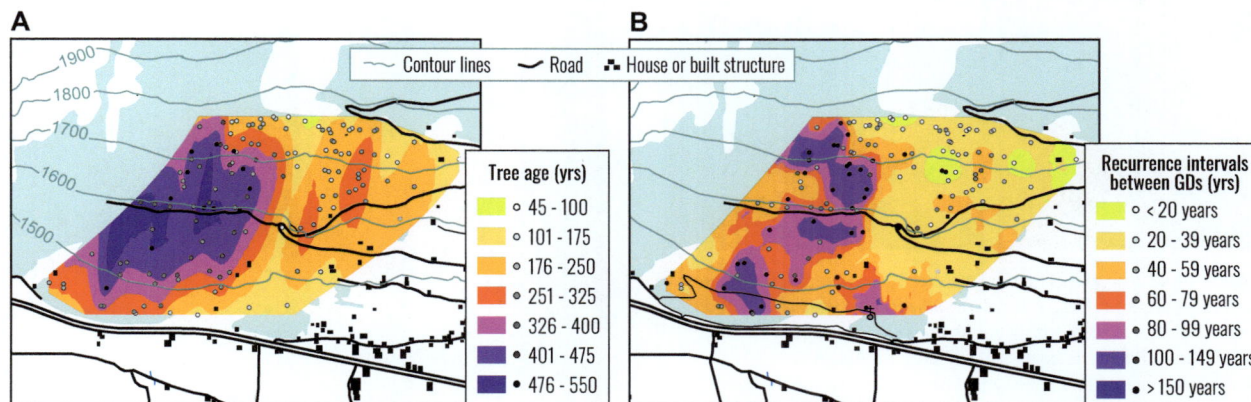

Figure 14.41 Tree age and rockfall frequencies across a talus slope near the village of Täsch, in the Valais Alps, Switzerland, as inferred from dendrochronological data. **A.** Trees are removed most frequently (and therefore, are the youngest) in the areas affected by the most frequent rockfall activity. Elevations in meters asl. **B.** Interpolated rockfall recurrence intervals, i.e., years between rockfall events, as indicated by tree growth disturbances. Source: Used with permission of Elsevier, from Stoffel M., Schneuwly D., Bollschweiler M. et al (2005); permission conveyed through Copyright Clearance Center, Inc.

rockfall, or debris flow paths are sometimes injured by the event, and these injuries are retained in their tree ring record as scars. Some trees are injured in this way multiple times, providing data on recurrence intervals, which can then be used to produce hazard probability maps. For example, in the Zermatt valley of the Swiss Alps, tree ring dating of rockfall scars has enabled the reconstruction of rockfall histories over the past several centuries (**Fig. 14.41**). Trees surviving century-old debris flow events were also mapped at Ritigraben,

in the Valais Alps, Switzerland, and used to reconstruct the patterns of past debris flow activity (**Fig. 14.42**). Here, the dating relied on scars and reaction wood in tilted trees, as well as on growth suppression in the trees as a result of partial burial.

In addition to data on past events, various tools have been developed to model debris flows of varying magnitude, so as to better understand and predict the reach of future events (**Fig. 14.42**). Validation of model outputs is normally based on knowledge of past events.

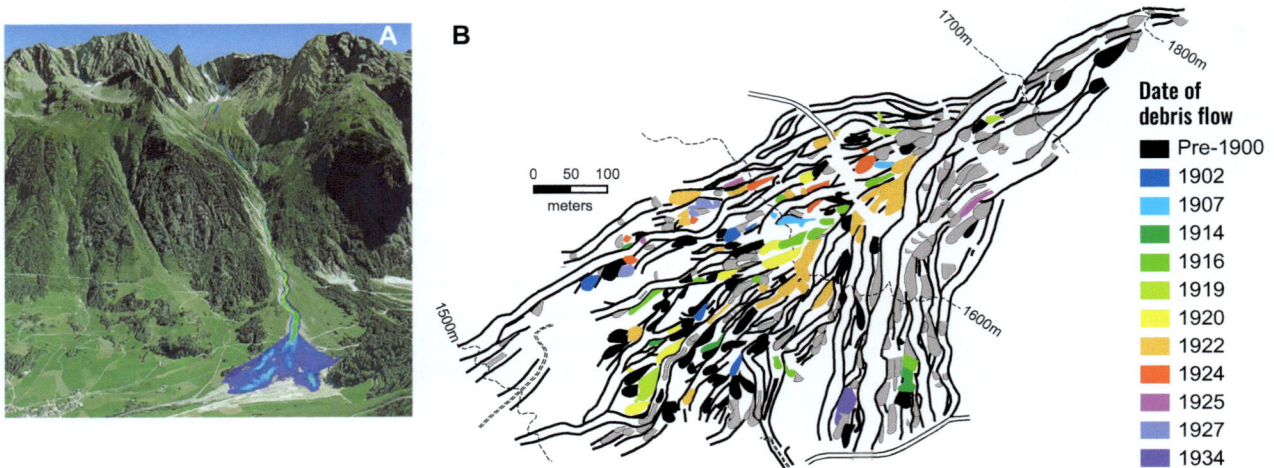

Figure 14.42 A. Modelling of the August 2005 debris flow at Rotlaui, Guttannen, in the Bernese Alps, Switzerland – one of the largest (≈0.5 million m³) debris flows in the Alps. Process models can improve our understanding of debris flows and assist in estimations of deposition and erosion depths, as shown by the different colors in this model. Source: Used with permission from Geotest Inc. **B.** Dating of past debris flow activity at Ritigraben, Valais, Switzerland, using dendrochronological data from larch, spruce, and pine trees buried in debris flow deposits and levees. Source: Used with permission of Elsevier, from Stoffel M., Conus D., Grichting M. A., et al. (2008); permission conveyed through Copyright Clearance Center, Inc.

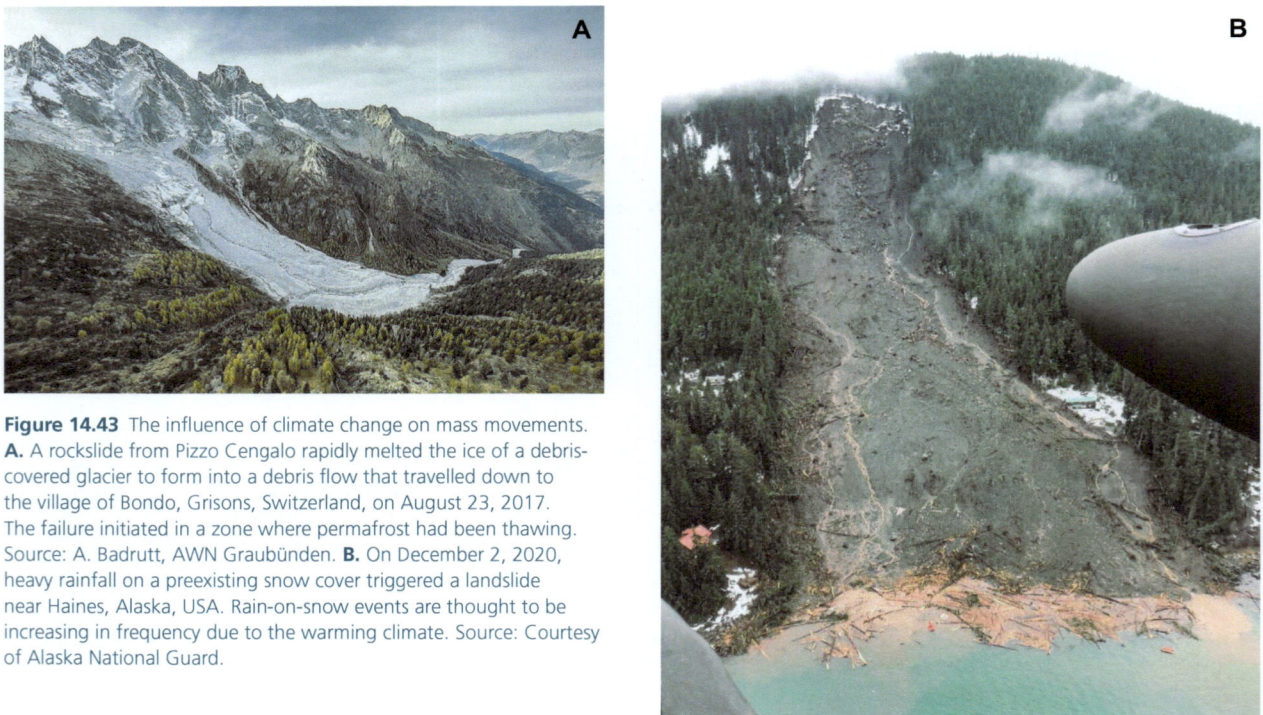

Figure 14.43 The influence of climate change on mass movements. **A.** A rockslide from Pizzo Cengalo rapidly melted the ice of a debris-covered glacier to form into a debris flow that travelled down to the village of Bondo, Grisons, Switzerland, on August 23, 2017. The failure initiated in a zone where permafrost had been thawing. Source: A. Badrutt, AWN Graubünden. **B.** On December 2, 2020, heavy rainfall on a preexisting snow cover triggered a landslide near Haines, Alaska, USA. Rain-on-snow events are thought to be increasing in frequency due to the warming climate. Source: Courtesy of Alaska National Guard.

14.8 IMPACTS OF CLIMATE CHANGE ON MASS MOVEMENTS

Mass movement activity is influenced by both water (as ice and as a liquid) and air temperatures. Meteorological data show clear warming trends in most regions of the globe, a tendency which is expected to continue and accelerate (see Chapter 4). These climate changes may lead to more frequent heavy precipitation (rainfall and snowfall) events which can, in turn, induce slope failures as water tables rise, as runoff is increased, and as regoliths become wetter. Warming climates can also intensify the cycling between wet and dry periods, which in turn may enhance the widening of cracks in rock and soil and thus lead to enhanced slope instability. Warmer temperatures can also enhance slope instability by promoting the thermal breakdown of rock and by decreasing the viscosity of groundwater. At altitudes and latitudes with permafrost, increasingly thick active layers are already accelerating slope instability (see Chapter 20). In summary, scientists widely agree that climatic changes will ultimately affect the frequency and/or magnitude of many different types of mass movements (Fig. 14.43).

REVIEW QUESTIONS

14.1   What are the major groups/categories of mass movement, from slowest to fastest? Identify a few key, differentiating characteristics of each.

14.2   What is the role of mass movement in the development of slopes and valleys, vis-à-vis rivers?

14.3   What are the main forces acting on all slopes that act to either set them in motion, or maintain their stability?

14.4   In general, what factors operate to cause slopes to fail, that is, to trigger their failure?

14.5   Why is a sandcastle made of moist sand less susceptible to failure than dry sand or saturated sand?

14.6   What factors determine or influence the fate of a rock that has fallen off a free face?

14.7   What factors act in concert to precondition a slope to topple?

14.8   How does a fall differ from a slide? What are the various types of slides and how do they differ?

14.9   Explain how positive feedback can operate on a slump block to "keep it moving."

14.10   What are the main types of flows and what are some criteria that can be used to differentiate among them? Which flow type is most common? Which is the fastest?

14.11   What are the various components of a typical debris flow?

14.12   What are quick clays and why are they susceptible to producing landslides?

14.13   What are the various kinds of avalanches and how do we differentiate between them?

14.14   Explain what factors must come together for solifluction to occur.

14.15   What features can be used to confirm that creep is occurring on a slope?

14.16   How does bioturbation induce the slow, downslope movement of regolith? Provide some examples.

14.17   What are some ways that large mass movements like debris flows or avalanches can be dated?

14.18   Why and how is climate change likely to influence the relevance of some triggers and thereby affect the frequency of mass movement activity?

FURTHER READING

Cruden, D. M and Varnes, D. J. 1996. Landslide types and processes. In: Turner, A. K. and Schuster, R. L. (eds.) *Landslides – Investigation and Mitigation*. Transportation Research Board Special Report 247. National Research Council, National Academy Press. pp. 36–75.

Highland, L. M. and P. Bobrowsky. 2008. *The Landslide Handbook – A Guide to Understanding Landslides*. US Geological Survey Circular 1325.

Selby, M. J. 1993. *Hillslope Materials and Processes*. 2nd ed. Oxford University Press.

15 Surface Water, Drainage Basins, and Sediment Transport

Douglas J. Faulkner and L. Allan James

Water is central to life. Geomorphologists know that running water also plays a key role in sculpting the land surface. This chapter covers **physical hydrology** – the science concerned with the occurrence, distribution, and movement of water – and the movement and storage of water-borne sediment within the various Earth systems. In this chapter, we focus on **streams** and how they transport sediment, from source to sink. The material presented here forms an important background for Chapter 16, which focuses on landforms developed by running water.

15.1 STREAMS AND THE HYDROLOGIC CYCLE

Geomorphologists consider a stream to be a body of water flowing in a channel. The definition of "stream" implies nothing about the depth and width of the water body. Streams can be small enough to jump or wade across; these are commonly called creeks, brooks, runs, branches, or a variety of other terms (**Fig. 15.1**). Larger streams, called rivers, can range in size, with some large enough to carry ocean-going ships (**Fig. 15.2**). There is no universally accepted size limit that separates rivers from creeks and other small streams.

The term "stream" implies nothing about the constancy of water flow within it. Streams that flow all the time, such as those in **Figs. 15.1B** and **15.2**, are called **perennial streams**. Some streams, such as **intermittent streams** and **ephemeral streams**, flow only occasionally. Intermittent streams flow continuously during part of the year. During the rest of the year, they flow only after heavy rain or snowmelt events. Ephemeral streams are *usually dry*, flowing only after heavy rain or snowmelt events (**Fig. 15.3**).

Almost every stream is a part of an extensive, branching *network*, consisting of a large primary stream, smaller streams that flow into it, and still smaller streams that flow into them (**Fig. 15.4**). The largest stream in a network is called the **trunk stream**, whereas the smaller streams are called **tributaries**. The streams at the upstream ends of a stream network (with no tributaries) are called **headwater streams** or **first-order streams** (see Chapter 16).

Figure 15.1 Small streams. **A.** Smail Creek in northwestern Lower Michigan. Source: R. Schaetzl. **B.** Shullsburg Branch in southwestern Wisconsin. Source: D. Faulkner.

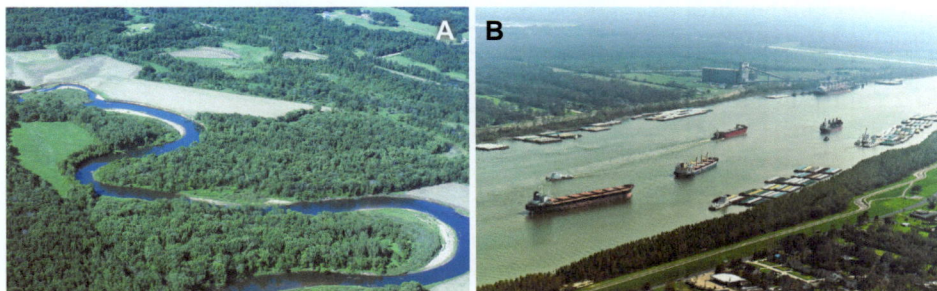

Figure 15.2 Two rivers of vastly different size. **A.** The Red Cedar River in western Wisconsin. Source: D. Faulkner. **B.** The Mississippi River near New Orleans, Louisiana. Source: USDA photo by Bob Nichols via CC BY 2.0; 05di1395-325.

Figure 15.0 Even small floods on the IJssel River in the Netherlands can be widespread on this low-relief landscape. Source: Frans Lemmens / Getty Images.

Figure 15.3 The bed of an unnamed ephemeral stream near Sedona, Arizona, (**A**) before and (**B**) immediately after an early spring thunderstorm. Source: D. Faulkner.

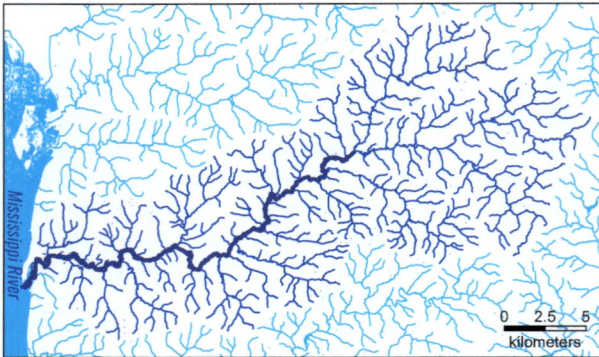

Figure 15.4 Map view of the Coon Creek stream network in southwestern Wisconsin. Neighboring stream networks are shown in lighter colors. Source: Data compiled from US Geological Survey 1:24000 topographic maps.

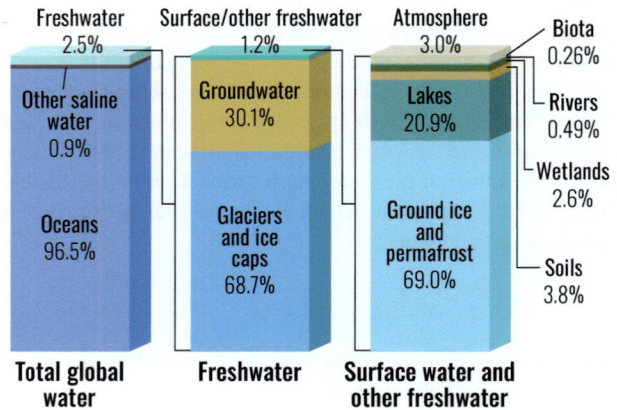

Figure 15.5 The major reservoirs of the hydrosphere, showing the amount of water in each. Source: From the US Geological Survey's Water Science School. https://www.usgs.gov/special-topics/water-science-school/science/where-earths-water.

Streams are a component of Earth's **hydrosphere**, which comprises all the water on, below, and above the surface. Water in the hydrosphere is stored in various reservoirs (**Fig. 15.5**). The largest reservoir is, by far, the oceans, which contain ≈96.5% of all the water on Earth. Freshwater streams, on the other hand, are one of the hydrosphere's smallest reservoirs. At any given time, the volume of water in all streams is 0.00015% of that in the entire hydrosphere. To put this amount into perspective, consider that all the Earth's streams and rivers contain less than half of the water currently in Lake Michigan (USA).

The movement of water between and through the various reservoirs of the hydrosphere is called the **hydrologic cycle**, which is commonly described in terms of reservoir volumes and flow rates (**Fig. 15.6**). For streams, flow rates are more important than the volume in storage at any one time. For example, the world's rivers flow into oceans at an average rate of ≈1.2 million m^3/second. At this rate, the total amount of water in rivers at any given moment (2,200 km^3) is emptied into the oceans every 20 days.

Like most cycles, the hydrologic cycle has no starting or ending points. Nonetheless, the oceans are a good place to start because they are the largest reservoir. Oceans constantly lose water to the atmosphere via **evaporation**, at an estimated rate of 413,000 km^3/year (**Fig. 15.6**). This water vapor

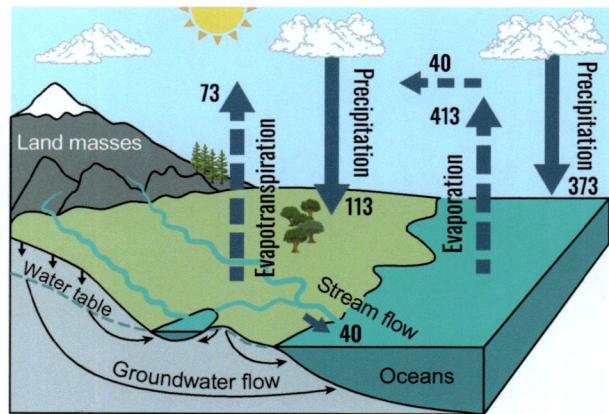

Figure 15.6 A generalized model of Earth's hydrologic cycle. Arrows indicate the volume of water that moves between reservoirs in thousands of km^3 per year. Source: From Trenberth, K. E., Smith, L., Qian, T., et al. (2007). Estimates of the global water budget and its annual cycle using observational and model data. *J. Hydrometeorology* 8:758–769. © American Meteorological Society. Used with permission.

eventually condenses and falls back to Earth as precipitation, either onto the land masses or back onto the oceans.

Water evaporates from land masses just as it does from the oceans. However, the movement of land-based water to the atmosphere involves not only evaporation from soil, lakes, and other open water bodies, but also **transpiration** from plants. Evaporation occurs from any source that has water to give, with evaporation rates strongly affected by temperature. On land, water taken up by plants passes through them and into the atmosphere through small pores called **stomata**. This process is called transpiration. The summed combination of evaporation and transpiration is known as **evapotranspiration**. Precipitation onto the land masses exceeds evapotranspiration from them by ≈40,000 km³/year. Of this surplus, a small amount (≈1,000 km³) ends up flowing to the oceans below the surface (in groundwater), leaving most of the excess to flow to the oceans in streams. Essentially, streams are constantly draining the surplus precipitation from the land masses to the oceans.

15.2 PATHWAYS OF WATER TO THE STREAM CHANNEL

Streams exist because more water falls onto land masses than leaves via evapotranspiration. But how this excess water becomes streamflow involves many possible pathways

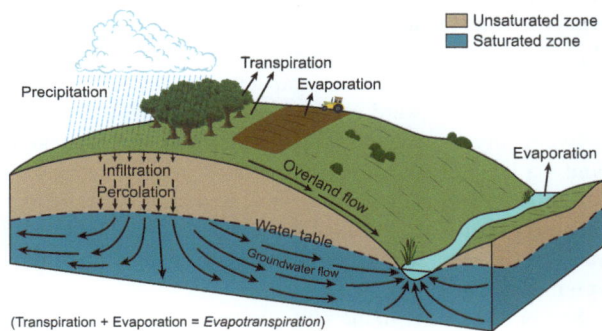

Figure 15.7 Possible fates of precipitation that falls onto land and its pathways to streamflow.

(**Fig. 15.7**). Because it starts with precipitation, that's where we will begin as well.

Precipitation that falls onto land can experience a variety of fates. One of these is **interception**, which occurs when precipitation lands on vegetation or organic litter lying on the ground. Intercepted water can then evaporate and cycle back to the atmosphere, or it may drip off or flow down plant stems and leaves, impacting the land surface below.

Precipitation that reaches the surface – either by falling directly on it or after being delayed by interception – can then move into the ground. This process is referred to as **infiltration**. Infiltrating water may initially become part of the **unsaturated zone**, in which only some of the pores are filled with water. Much of the water here occurs as thin films on mineral particles.

Whether rain infiltrates or not is largely determined by the **infiltration capacity** of the surface, defined as the maximum rate that water can enter it. Infiltration capacity is usually expressed in units such as millimeters/hour (mm/h). If the infiltration capacity exceeds the rainfall rate, all the water will infiltrate. For example, if the infiltration capacity is 20 mm/h, but the precipitation intensity is only 5 mm/h, all the water will infiltrate. Several factors determine the infiltration capacity of a surface, with one of the most important being the size of the void spaces (**pores**) between the mineral particles. In general, larger pores result in higher infiltration capacities, because water flows more easily into and through larger openings than smaller ones. In other words, soil containing larger pores is more permeable than one with mainly smaller pores. Because pores in coarse-textured (sandy) soils are almost always larger than pores in fine-textured (clayey) soils, sandy soils are more permeable and have high infiltration capacities. **Permeability** is also increased by organic materials and humus, which is a dark, organic substance produced by the decomposition of dead plants and animals. Humus helps aggregate (clump together) individual mineral grains (**Fig. 15.8A**). The large voids in humus-rich soils increase

Figure 15.8 Examples of how soil properties can affect its infiltration capacity. **A.** A humus-rich topsoil from North Dakota with well-developed soil aggregates and, hence, large pores. Source: J. Kelly, USDA-NRCS. **B.** A soil with numerous macropores formed by burrowing ants. Scale in cm. Soils like these encourage rapid infiltration because of their high infiltration capacities. Source: R. Schaetzl. **C.** A clay-rich soil with few macropores. Water infiltrates slowly into soils like this. Source: R. Schaetzl.

the soil's infiltration capacity. Large voids can also develop along **root channels** (open spaces formed by the decay of dead plant roots) and burrows of soil fauna such as worms and ants. Soils with these types of voids – called **macropores** – have very high infiltration capacities (**Fig. 15.8B**). In contrast, fine-grained soils that have few macropores also have low infiltration capacities (**Fig. 15.8C**).

In addition to the size of voids, infiltration capacity is also influenced by the soil's preexisting moisture content. A dry soil will take up water more rapidly than a wet soil, due to stronger **capillary forces** – the same forces that pull water into a sponge. Because capillary forces are stronger in drier soils, infiltration capacities are initially high, but then taper off as the soil becomes wetter, eventually falling to a constant value (**Fig. 15.9**). This latter value is a function of how fast water can move by **percolation** – the downward movement of water *already* in the soil. (Recall that infiltration refers only to water initially entering the soil.) After infiltration ceases, the soil water content will decrease due to evaporation and plant uptake. Then, if it rains again before the soil dries, infiltrating water will encounter a lower infiltration capacity than if the soil had been dry.

Vegetation cover dramatically affects infiltration capacity (**Fig. 15.9B**). When rain falls on bare soil surfaces, raindrop impacts can break up soil aggregates and form a thin crust (**Fig. 15.10A**). These crusts typically have low infiltration capacities. As a result, a bare soil that initially has a high infiltration capacity can quickly develop a low infiltration capacity. In addition, some soils contain clay minerals that shrink when dry and swell when wet. When dry, these soils shrink and crack, creating macropores that promote rapid infiltration (**Fig. 15.10B**). However, when the soils become wet and the clays within them expand, the cracks and other voids close, leading to a tremendous reduction in infiltration capacity.

Bedrock is typically much less permeable than soils and regolith, with lower infiltration capacities. Thus, when precipitation falls on exposed bedrock, infiltration typically occurs only into cracks. Likewise, materials common to urbanized areas, such as asphalt and concrete, are largely impermeable.

Water will continue to infiltrate as long as it is available (from precipitation) and the rate at which it reaches the ground surface is less than the infiltration capacity of the surface. Sometimes, so much water enters the soil that its **field capacity** is exceeded. Field capacity is the maximum amount of water that soil/sediment can hold against the force of gravity (**Fig. 15.11A**). When infiltration causes the water content of soil to exceed field capacity, gravity will pull the excess water downward via percolation (**Fig. 15.11B**).

Eventually, percolating water will enter the **saturated zone** – a layer of soil or bedrock where, even between precipitation events, all the voids are filled with water. Water in the saturated zone is called groundwater. The addition of water to the saturated zone in this way is called **groundwater recharge**.

Sometimes, water in the saturated zone may emerge back onto the surface as **groundwater discharge**. The most significant groundwater discharge sites occur where the water table (the top of the saturated zone) intersects the land surface (**Fig. 15.7**). Importantly for our purposes, at sites where the water table intersects a stream channel, groundwater discharge enters the channel as **baseflow**.

Precipitation falling onto the land surface may also end up as **overland flow**. Overland flow, also called **runoff**, refers to water flowing across the surface but not in a stream

Figure 15.9 Hypothetical infiltration capacity curves for four sites with initially dry soils. Each curve shows initially high infiltration capacities, which decrease over time as water infiltrates and the soil becomes wetter. **A.** Initially rapid decreases in infiltration capacity occur in both coarse- and fine-textured soils, although coarse-textured soils, such as the gravelly sand, maintain higher infiltration capacities because of their larger pores. **B.** Rapid decreases in infiltration capacity also occur on both vegetated and unvegetated surfaces, although vegetated areas maintain higher infiltration capacities due to their better aggregation and more macropores.

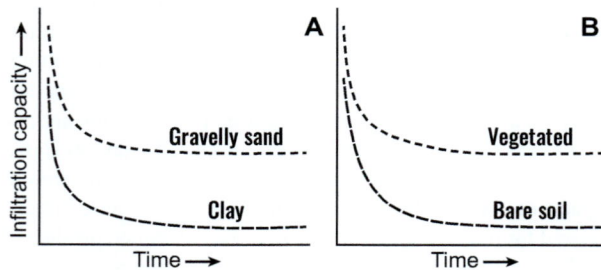

Figure 15.10 Soil conditions that can affect infiltration capacities. **A.** A soybean sprout breaking through a soil crust formed by raindrop impact and splash on exposed soil. Source: J. Rees. **B.** Desiccation cracks in a dry soil rich in expandable clay minerals, which shrink when dry and expand when wet. Source: D. Faulkner.

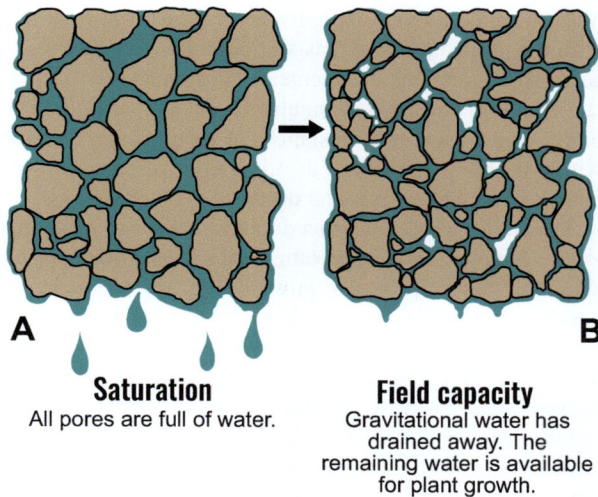

Saturation
All pores are full of water.

Field capacity
Gravitational water has drained away. The remaining water is available for plant growth.

Figure 15.11 Graphical illustrations of two different blocks of soil, in the state of (**A**) saturation and (**B**) field capacity.

channel. Runoff occurs when the rate of precipitation (or snowmelt) exceeds the infiltration capacity of the surface. This type of runoff, called **Hortonian overland flow**, is most common on surfaces with very low infiltration capacities, such as bare bedrock, paved surfaces, soils with surface crusts, and sparsely vegetated surfaces (**Fig. 15.12**). On surfaces with high infiltration capacities, such as those with thick vegetation and permeable soils, Hortonian overland flow is uncommon.

Rain falling on impermeable surfaces such as paved roads and parking lots immediately generates Hortonian overland flow (**Fig. 15.12A**). When rain falls on rough surfaces, typical of rural and natural landscapes, and doesn't entirely infiltrate, the excess won't immediately become overland flow. Some of it may collect in small depressions. If enough rain falls so that these depressions fill and overtop, the water will flow downhill as sheets and, more commonly, concentrated rivulets of overland flow (**Fig. 15.12B**).

Overland flow can even be generated on surfaces with high infiltration capacities. Water on these surfaces infiltrates into the soil on an upland, only to return to the surface *farther downslope* (**Fig. 15.13**). This happens when percolating water encounters a layer of less permeable material

at depth in the unsaturated zone. These **impeding layers** are often formed on clay-rich zones or layers of hard bedrock. The water may then pond (or perch) on top of the impeding layer and flow across it, through the more permeable material above, as **throughflow**. Farther downslope, this layer can become saturated with throughflow. When this happens, water will discharge onto the surface and become overland flow. Rain that falls on this saturated ground will become overland flow as well. Runoff generated in these ways is called **saturation-excess overland flow**. This is the most common type of overland flow on undisturbed landscapes with high infiltration capacities.

In short, water can reach the base of a slope in different ways and under different circumstances; it doesn't always happen via sheets and rivulets of water flowing downslope. Nonetheless, the various types of runoff will all flow toward low spots in the landscape, where they may merge to form a stream channel. Streamflow that results from runoff is called **stormflow**, because it often happens during a rainstorm. Sometimes a precipitation event produces so much runoff that stream channels become overfilled with stormflow, causing them to flood.

The surface characteristics that govern infiltration capacity vary greatly across the landscape. **Figure 15.13** presents an idealized dichotomy in which different sides of a valley have contrasting conditions that affect infiltration and the generation of overland flow. The left side has thick, coarse-textured, organic-rich soils and thick vegetation, leading to high infiltration rates, with little-to-no runoff generated when it rains. Conversely, the right side of the valley

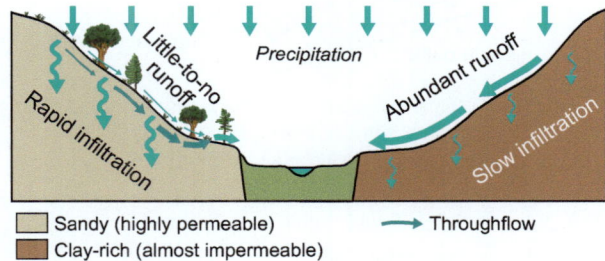

Sandy (highly permeable)
Clay-rich (almost impermeable)
Throughflow

Figure 15.13 An idealized landscape with valley sides of contrasting permeabilities that are being subjected to uniform rainfall. Runoff on the right side of the valley is predominantly Hortonian overland flow. Runoff on the left side is mostly saturation-excess overland flow.

Figure 15.12 Hortonian overland flow. **A.** Broad shallow sheets of runoff from a parking lot during a summer thunderstorm in Chantilly, Virginia. Source: Famartin, CC BY-SA 4.0, via Wikimedia Commons. **B.** Runoff from a bare agricultural field in Iowa after a spring thunderstorm. Source: L. Betts, USDA-NRCS.

has fine-textured soils, some impermeable surfaces due to urbanization, and minimal vegetation. Thus, infiltration is minimal, and runoff occurs easily and often.

15.3 DRAINAGE BASINS

A **drainage basin**, also called a **watershed**, is the land area from which water drains to a given point along a stream. In most instances, the water in a stream comes from precipitation that falls on its drainage basin, entering the stream via the pathways discussed above. Thus, streams are intimately connected to their drainage basins. Differences in the attributes of one stream vs another are largely due to differences in their drainage basins.

A drainage basin is defined for a specified point along a stream (the basin's outlet), and its spatial extent is determined by topography (**Fig. 15.14**). The outer boundary of a drainage basin, which runs along land that is topographically

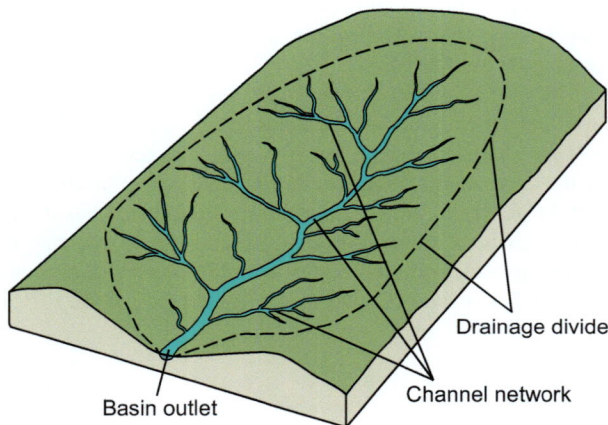

Figure 15.14 Schematic diagram of a drainage basin, showing the drainage divide, channel network, and outlet.

high, is called its **drainage divide**. Runoff from anywhere within a drainage basin has the potential to end up in a series of interconnected channels (a **channel network**) that coalesce into one main channel which conveys water to the outlet. In addition, most instances of groundwater recharge within a drainage basin will ultimately discharge into the basin's channel network and drain to its outlet as well.

Importantly, the outlet to a drainage basin can be defined at any location along a stream channel. Consider, for example, the drainage basins shown in **Fig. 15.15**. The largest drainage basin of Coon Creek is defined for its outlet on the Mississippi River. But also notice a smaller, upstream part of the same basin, defined for an outlet (the blue point) located about halfway up-basin from the ultimate end of Coon Creek. Next, consider the basin defined for an outlet on Timber Coulee Creek, a tributary to Coon Creek. Finally, look at the small basin defined for an outlet on Rullands Coulee Creek, which is a tributary to Timber Coulee. Notice that in each of these cases, the smaller basins are located farther upstream, essentially "nested" within larger basins.

Now consider the drainage basin of the Mississippi River (**Fig. 15.16**). At its outlet into the Gulf of Mexico, it is the largest drainage basin in North America. Every stream in the Mississippi's stream network, including Coon Creek, has a drainage basin that is smaller and nested within the basin of the Mississippi River. Indeed, for every drainage basin worldwide, as in the Coon Creek basin, an unlimited number of small, inset and nested basins exists, making it important that, for each basin, its outlet be clearly identified in order to be properly "defined."

A key aspect of every drainage basin is its size, which is the land area that lies inside its drainage divide. Drainage basin area is significant because it largely determines the amount of water that could potentially drain to the basin outlet. The volume of water actually *flowing* in a stream is known as its **discharge**. In the Coon Creek example

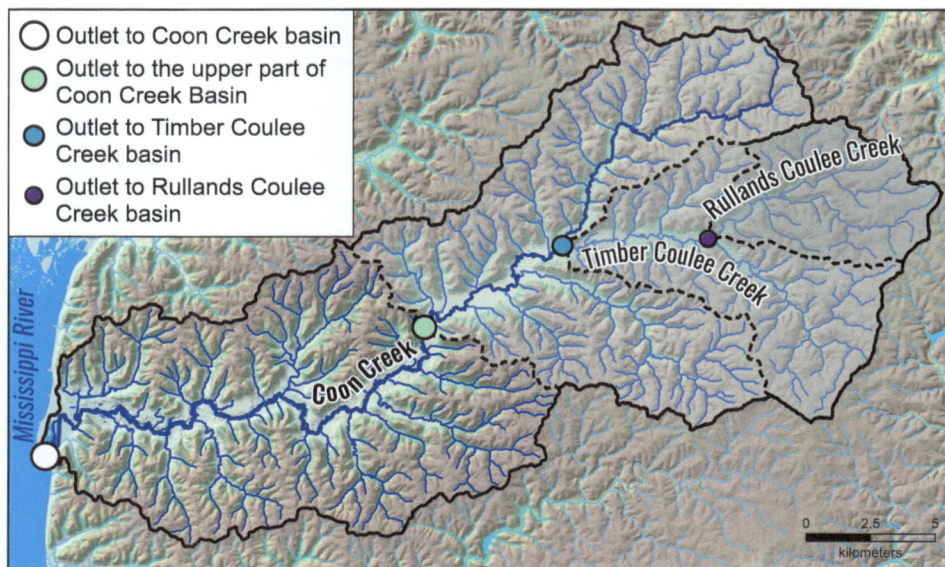

Figure 15.15 Shaded relief map of the Coon Creek drainage basin in southwestern Wisconsin, USA, with three smaller drainage basins nested within it.

Figure 15.16 The Mississippi River drainage basin. Nested within it are the drainage basins of its two largest tributaries (the Ohio and Missouri Rivers), as well as the relatively small Coon Creek basin.

Figure 15.17 A. Rullands Coulee Creek at the outlet of its drainage basin. **B.** Coon Creek near the outlet of its drainage basin. Source: D. Faulkner.

(**Fig. 15.15**), Rullands Coulee Creek has the smallest discharge because it has the smallest drainage area, whereas Coon Creek has the largest discharge because it drains the largest area (**Fig. 15.17**). Likewise, the Mississippi River, with the largest drainage basin in North America, has the largest average discharge of any stream on the continent (**Fig. 15.2B**).

15.4 GLOBAL STREAMFLOW

Discharge is a fundamental property of all streams. Formally, discharge (Q) is the volume of water that flows by a point along a stream over some unit of time. Typical units of Q are cubic meters per second (m³/s) or cubic feet per second (ft³/s, or cfs). Drainage area, as mentioned above, is an important factor that determines the discharge of a stream. It is not, however, the only factor. Climate is also important, because it determines how much precipitation falls onto the drainage basin and how much is lost to evapotranspiration.

Consider, for example, the ten largest rivers in the world (based on their discharge) (**Fig. 15.18**). The Amazon River in South America, which drains the largest drainage basin on Earth, is also (by far) the largest river on Earth. The second largest river is the Congo River in central Africa. However, its discharge is < 20% of the Amazon's, even though its drainage basin is ≈60% as large. The Amazon's much greater discharge is due to its wetter climate. The Mississippi River has the third largest drainage basin in the world yet ranks only sixth in terms of discharge, because much of its basin (especially its western half) has a semi-arid climate. Even more extreme is the Nile River, which has the fifth largest drainage basin on Earth but doesn't even come close to ranking in the top ten rivers with respect to discharge. It has a low discharge because much of its basin is in the hot, dry Sahara Desert. Finally, the relatively high average discharges of the Yenisey and Lena Rivers in eastern Russia rank among the world's largest, despite their relatively dry climates. Their unusually high discharge values are explained by low evapotranspiration rates in their cold climates. And so, both

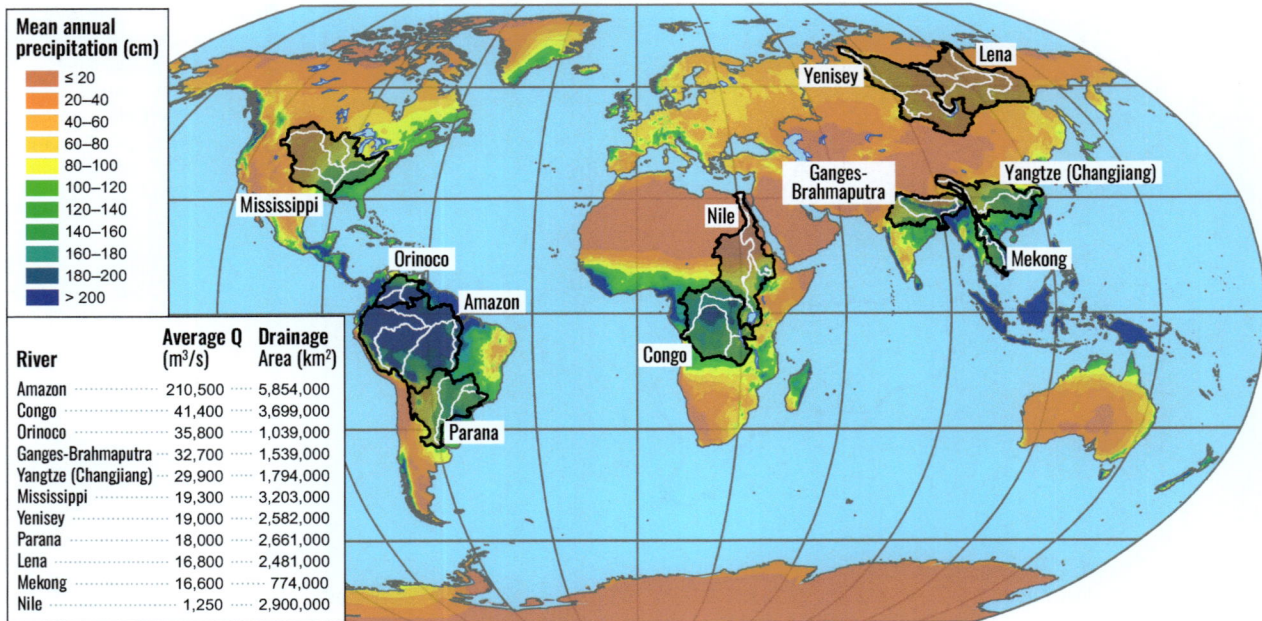

River	Average Q (m³/s)	Drainage Area (km²)
Amazon	210,500	5,854,000
Congo	41,400	3,699,000
Orinoco	35,800	1,039,000
Ganges-Brahmaputra	32,700	1,539,000
Yangtze (Changjiang)	29,900	1,794,000
Mississippi	19,300	3,203,000
Yenisey	19,000	2,582,000
Parana	18,000	2,661,000
Lena	16,800	2,481,000
Mekong	16,600	774,000
Nile	1,250	2,900,000

Figure 15.18 Global annual precipitation, and the world's ten largest river systems, ranked by average discharge (Q). Source: Data from Fick and Hijmans (2017); used with permission from Wiley & Sons/Royal Meteorological Society.

climate (precipitation AND temperature) and basin size are major players in determining the discharge of rivers.

15.5 FLOW IN STREAM CHANNELS

The flow of water in any given stream is variable, both temporally and spatially. Understanding how and why changes in streamflow occur is essential to understanding streams as agents of geomorphic work.

15.5.1 Hydrologic Classification of Streams

Although streams can be classified in several ways, a common classification scheme is based on streamflow regime. **Perennial streams**, also called **gaining streams**, flow continually. They gain water in the downstream direction due to discharge of groundwater into their channels. In humid regions, all but the smallest tributary streams are perennial streams. **Ephemeral streams**, which are common in arid regions, lack baseflow because the water table lies below their channels. They flow only when precipitation or snowmelt generates runoff. Ephemeral streams are also known as **losing streams**, because they lose water by infiltration into channel beds and banks. Thus, their discharge values often decrease in the downstream direction. As you might expect, some streams alternate between perennial and ephemeral in response to groundwater fluctuations. Called **intermittent streams**, they are gaining streams during wetter periods when the water table is high, but become losing streams during drier periods when the water table falls. Intermittent streams are common in climates with pronounced wet and dry seasons.

Even though ephemeral streams are typical of arid regions and intermittent streams are common in regions with wet/dry climates, both types can occur in humid regions as well. Small, headwater tributaries are commonly ephemeral because they are high in the landscape, well above the water table; they flow only during times of runoff generation. However, the downstream segments of these same streams, where they are lower in the landscape and therefore closer to the water table, may classify as intermittent. They flow constantly during seasons when the water table is high, but dry up when the water table falls. Further downstream, these same streams usually become perennial, as their channels are low enough in the landscape to be constantly fed with groundwater discharge.

15.5.2 Turbulence and Eddies

Streams flow downslope in response to gravity. Although the overall direction of streamflow is downhill, the water in a stream may flow in many directions locally, which can be expressed as flow lines. If a stream is very shallow and slow-moving, flow lines will generally be oriented downstream, parallel to one another (and to the channel bed and banks). Slow flow along parallel flow lines is called **laminar flow**, which is rare (**Fig. 15.19A**). Typically, chaotic motions are almost always present in rivers, even if not readily apparent. In this case, flow lines go in multiple directions, even though they are, on average, directed downstream. **Turbulence** is the term that describes the chaotic motion of flowlines (**Fig. 15.19**). It can occur across a range of scales. Highly turbulent streams with extremely chaotic flow lines may cause substantial erosion. Turbulence

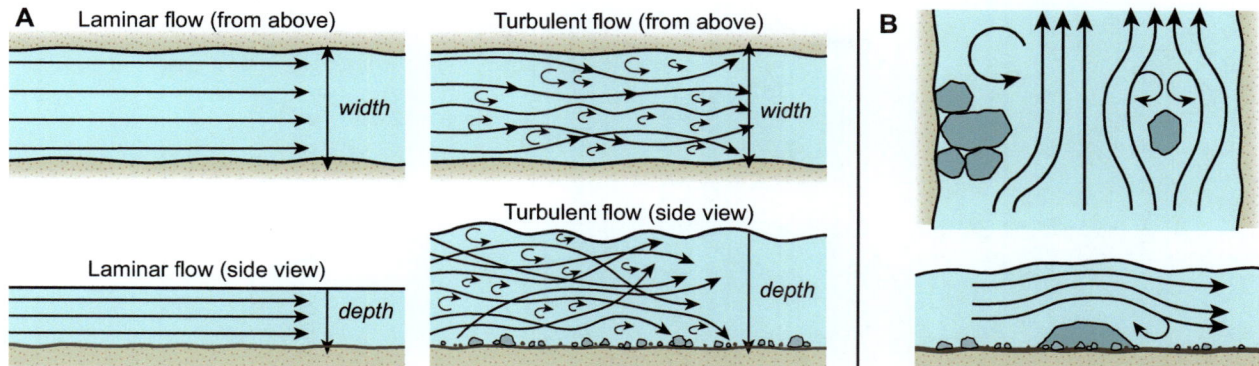

Figure 15.19 Types of flow, as shown by flowlines. **A.** Laminar and turbulent flow. **B.** Large-scale eddies in turbulent flow caused by streamflow obstructions (arrows indicate average flow direction) such as horizontal eddies behind a channel margin obstruction and a mid-channel obstruction, and a vertical eddy behind a channel-bottom obstruction.

involves swirling water called **eddies**. Small eddies can occur anywhere within overall turbulent flow. Large eddies typically develop on the downstream sides of rocks, islands, bridge piers, tight channel bends, or other protrusions along a channel bank (**Fig. 15.19B**).

15.5.3 Flow Velocity

Gravity drives the flow of water within stream channels. If unopposed by other forces, gravity would cause the flow in streams to accelerate downstream, flowing faster and faster. However, gravitational forces are opposed by friction exerted on the flowing water by the bed and banks. If examined over short distances downstream and through time, the flow in streams can either speed up (gravity > friction) or slow down (friction > gravity). Where flow velocities are constant, the driving force of gravity is being balanced by the resisting force of friction.

Flow velocities also vary *across* the channel. They are usually slowest next to the channel bed and banks, where frictional drag is greatest. As a result, rivers often flow fastest near the water surface, about midway between the banks, where there is generally the lowest amount of friction (**Fig. 15.20**). Many exceptions to this generalization exist, such as at channel bends, where momentum causes the fastest water to shift toward the outside bank.

Stream velocity is determined by many factors. Faster stream velocities typically occur when the channel bed and banks are smooth, and in deeper streams. Obviously, streams with steeper slopes also tend to flow faster, other things being equal. Nonetheless, the velocity of water flowing through a channel cross-section generally remains constant for a given discharge.

15.6 MEASURING STREAM DISCHARGE

As discussed previously, stream discharge (Q) is the volume of water that flows past a point over some unit of time, and thus is expressed in units of volume/time. Discharge is *calculated* as the cross-sectional area of a stream multiplied

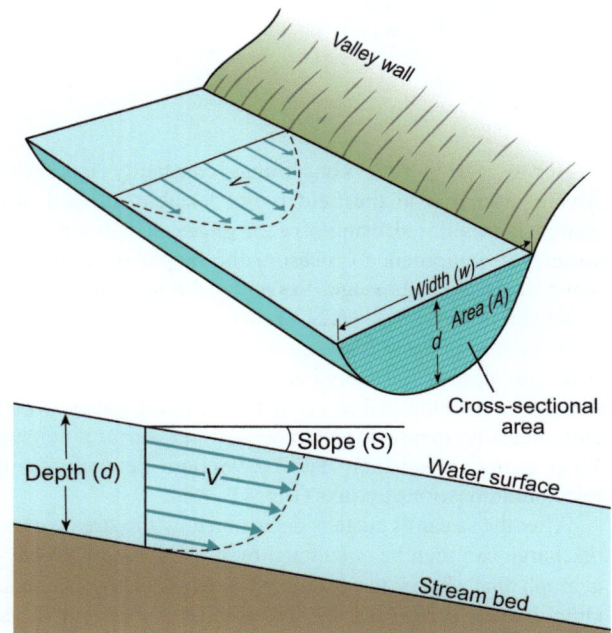

Figure 15.20 Schematic diagram illustrating flow velocities within a stream, as well as the various metrics used to calculate discharge. Source: © A. N. Strahler (1992). Used by permission.

by its average flow velocity (v). The cross-sectional area of a stream is calculated as its width at the top of the water surface (w) multiplied by its average depth (d). Thus, the equation for discharge is written as:

$$Q = w \times d \times v.$$

This equation is called the **continuity equation**. Using this equation, one need only know a stream's width, average depth, and average velocity to determine its discharge. That said, the field measurements needed to determine the discharge of a stream can take considerable time and effort (**Fig. 15.21**).

Figure 15.21 Time-lapse photo of stream discharge measurements. After measuring stream width, a hydrologic technician will measure the stream's depth and velocity at several points. To average out the effects of turbulence, velocity has to be measured at each point for ≈1 minute. Source: US Geological Survey, Idaho Water Science Center; in public domain.

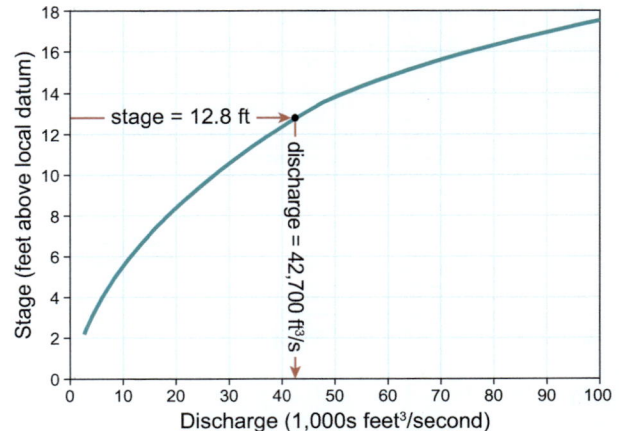

Figure 15.23 A stage–discharge rating curve for the Chippewa River at Durand, Wisconsin, using data from the US Geological Survey gage 05369500. The rating curve illustrates the empirical relationship between stream discharge and stage (height of the water surface).

Because it is time consuming, constantly measuring stream discharge in the field is not feasible. Instead, discharge is monitored using **stream gages**, which employ a variety of equipment to measure the height of a stream's water surface, or its stage. **Stage** can be measured with a simple staff gage (**Fig. 15.22A**), which is similar to a yardstick fixed in place in a stream. Continuous records of flow, which are needed for many purposes (including geomorphic analyses), are obtained at **recording stream gages**, which automatically measure and record water surface heights. Most gages upload their data by wireless networks using telecommunications uplinks (**Fig. 15.22B**).

After the stream's stage is determined by the stream gage, discharge can then be calculated by using a **rating curve** – a graph that shows the relationship of discharge to stage (**Fig. 15.23**). Rating curves are established by field measurements of discharge at various corresponding flow stages. Once a rating curve has been developed for a gaging station,

discharge can be readily obtained from stage height data (**Fig. 15.24**).

The discharge of a stream can be calculated for different time periods. Discharge for a point in time is called **instantaneous discharge**. Alternatively, discharge can be expressed as an average value, calculated from several instantaneous measurements, for example, average daily discharge or average annual discharge. Over a given period of time, the peak instantaneous discharge of a stream is always greater than its average discharge. For example, on May 22, 2019, the Chippewa River at Durand, Wisconsin, had a peak instantaneous discharge of 68,200 cfs (cubic feet/second), whereas the average for the day was 65,400 cfs. The average discharge of the river for the month was 26,800 cfs, and for the entire year of 2019 it was only 13,930 cfs.

Flood hazards and major geomorphic adjustments are usually associated with high discharges. Thus, bridges and other infrastructure built along streams are commonly designed to handle specific instantaneous discharges that might occur, such as floods of a given magnitude. Other

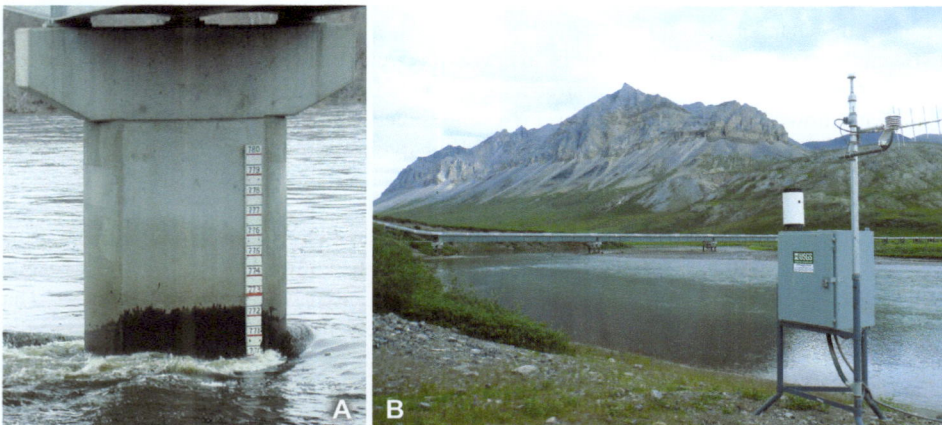

Figure 15.22 Stream gages. **A.** A staff gage (in feet) on the Chippewa River in western Wisconsin, shows the height of the water column during a spring flood. Source: D. Faulkner. **B.** A recording stream gage operated by the US Geological Survey on the Atigun River in Alaska. Note the satellite upload antenna. Source: J. Conaway. US Geological Survey; in public domain.

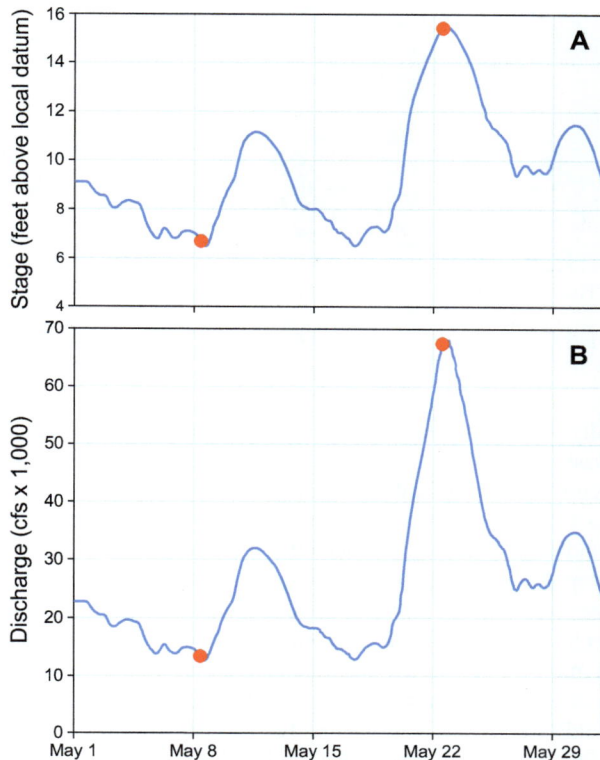

Figure 15.24 Graphs showing (**A**) variations in flow stage as measured at a recording stream gage on the Chippewa River at Durand, Wisconsin, for the month of May 2019, and (**B**) discharge, determined from the stage record, using the rating curve in **Fig. 15.23**. Dots indicate field measurements of stage and discharge; these are made periodically at stream gages to check the reliability of the rating curve.

land-use decisions, such as those related to activities like fishing, boating, and irrigation, may instead be based on longer-term (average) discharges.

15.6.1 Hydrographs

A **hydrograph** is a plot of stream discharge over time. Different types of hydrographs can be developed, depending on the duration of flow being examined. **Storm hydrographs**

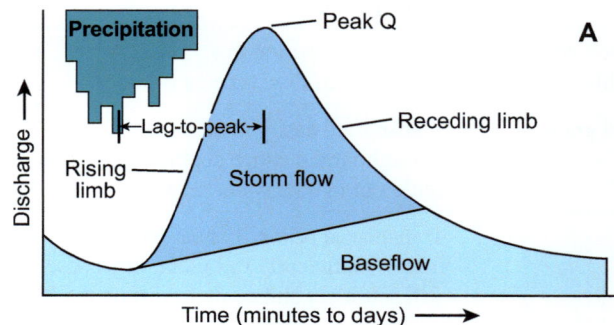

show the changes in discharge over the timespan of a single storm, that is, from minutes to days (**Fig. 15.25A**). Storm hydrographs record the response of a stream to a precipitation or meltwater event. A typical storm hydrograph consists of two sections. The section leading up to the peak discharge is known as the **rising limb**. The subsequent period of decreasing discharge is known as the **receding** or **falling limb**. Discharge in streams does not immediately increase as soon as the storm begins. Instead, discharge lags behind the onset of precipitation. This time delay (between the onset of rainfall and the onset of the rising limb) occurs because it takes time for runoff to arrive at the gage site where discharge is measured. This delay is typically measured as the time between the centroid of rainfall and the peak discharge, and is known as the **lag-to-peak time**. Storm hydrographs are useful indicators of a stream's response to precipitation in its drainage basin. They can also tell us about the effects of changing land use and landcover within the drainage basin.

In contrast to storm hydrographs, **annual hydrographs** depict changes in river discharge over the course of an entire year (**Fig. 15.25B**). They reveal the reliability of streamflow for a variety of uses (such as water supply and irrigation), and its seasonal variability.

15.7 FLOODS AND FLOOD FREQUENCY

Floods are of great importance to geomorphologists. Floods (flows that overtop the stream banks) occur most often in response to large storms or snowmelt events, but also to upstream dam breaks and other causes. In addition, flooding may be promoted by reductions in channel size, such as when the channel fills with sediment, causing the river to overflow its banks more frequently. Assessing the risk of floods is conventionally done by analyzing records of former peak flood discharges, in what is known as **flood frequency analysis**. One goal of such an analysis is to determine the **recurrence interval** (RI) of flows or floods of different sizes, which is important for risk assessments. The recurrence interval of a flood is the time, in years, that can be expected (on average) between events of that size (or larger). For example, a flood with a recurrence interval of 100 years

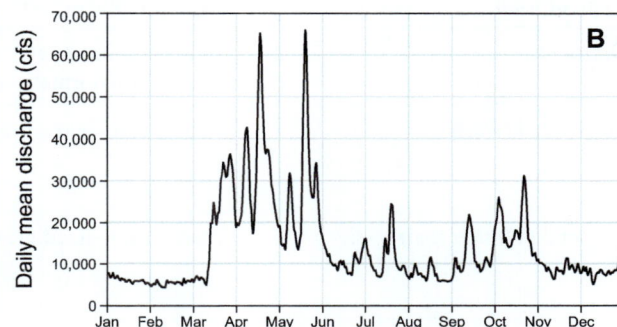

Figure 15.25 Hydrographs. **A.** An idealized storm hydrograph, showing changes in discharge at a site, from before, during and after a storm. **B.** The 2019 annual hydrograph of the Chippewa River at Durand, Wisconsin, showing discharge peaks derived from snowmelt and several large storms. Source: (B) US Geological Survey gage 05369500.

– the 100-year flood – is the flow that is likely to be equaled or exceeded – on *average* – only once every 100 years. Mathematically, RI is equal to the inverse of the probability (P) of an event occurring in a given year:

$$RI = 1/P \qquad \text{or} \qquad P = 1/RI .$$

For example, the 100-year flood has a one in a hundred (1%) chance of occurring in any given year. **Bankfull discharge** on many streams has a recurrence interval between one and five, implying that this level of discharge will typically occur once every one to five years. Larger floods have longer recurrence intervals; that is, they occur less frequently. Estimating the peak discharge and recurrence interval of large floods is important for designing bridges and other infrastructure. It is also important for regulating land use near rivers, so as to minimize the likelihood of flood damage.

15.8 SEDIMENT PRODUCTION AND TRANSPORT BY STREAMS

Streams are much more than just flowing water. They are also natural conveyor belts that transport sediment out of their drainage basins to its ultimate resting place, commonly the ocean or an internally drained basin, such as the Dead Sea in Israel. Along the way, sediment is passed downstream from one channel segment to another, and from tributaries to their trunk streams, many times over. In so doing, streams leave a profound imprint on the morphology of the land surface (see Chapter 16).

A stream does work by transporting sediment, which is called its **load**. The sediment load of a river can be classified based on its composition and on how it is transported (**Fig. 15.26**, **Table 15.1**). The **dissolved load** of a stream consists mainly of compounds that are held in solution by ionic bonds with water molecules. A stream can look "clean" and still contain much dissolved load – you simply cannot see it (**Fig. 15.27A**). The **suspended load** of a stream consists of particles that are small enough to be held in the water column simply by turbulence. The smallest particles – those most easily kept in suspension (generally clay and

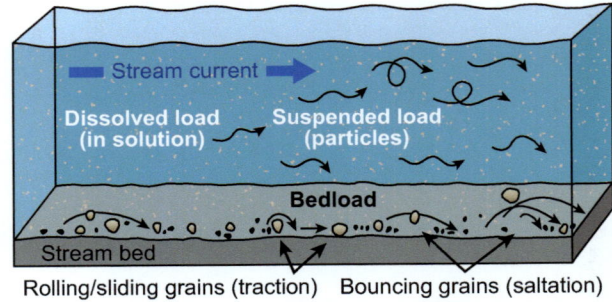

Figure 15.26 The three types of sediment load in streams, and their various modes of transport.

silt) – constitute a part of the suspended load called the wash load. **Wash load** is well mixed throughout the entire depth of flow and gives streams a muddy appearance (**Fig. 15.27B**). Suspended load can, however, also include larger sand-sized particles, provided that turbulence is sufficient to keep them in **suspension**. Larger, suspended load particles are typically only lifted off the channel bed when the flow is highly turbulent. Even then, this part of the suspended load is mainly concentrated near the bed. Because turbulence increases in faster-flowing streams, they can transport more sandy suspended load than more sluggish streams. Also, in most streams, overall suspended sediment load generally rises as discharge increases, because velocity and turbulence increase, along with discharge.

The largest sediment carried by rivers is called **bedload**. It consists of particles too big to be held in suspension. Instead, bedload moves by sliding or rolling along the bed, or by **saltation** (bouncing). Particles that slide or roll, which are generally the largest a stream can move, constitute the **traction** component of bedload. Saltating particles are small enough to get plucked or knocked off the bed, but too big to be held in suspension. As a result, they quickly fall back to the bed after bouncing along for a short distance. Saltating bedload can transition into suspended load if flow conditions become faster and more turbulent. Streams must expend energy to

Table 15.1 Characteristics of the various types of sediment load in streams

Type of load	Physical characteristics	Transport characteristics
Dissolved	Ionic compounds and dissolved substances, not normally visible	Can be transported in all rivers, regardless of flow type or velocity
Suspended (wash)	Mainly clay- and silt-sized grains	Transported easily in all streams throughout the entire depth of flow; may settle out in backwater areas
Suspended (other than wash)	Mainly sand-sized grains	Transported primarily near the bed when flow is sufficiently turbulent; otherwise, will settle out and come to rest on the bed
Bed	Sands and gravels	Transported in constant or intermittent contact with the bed, mainly in fast-flowing, turbulent streams

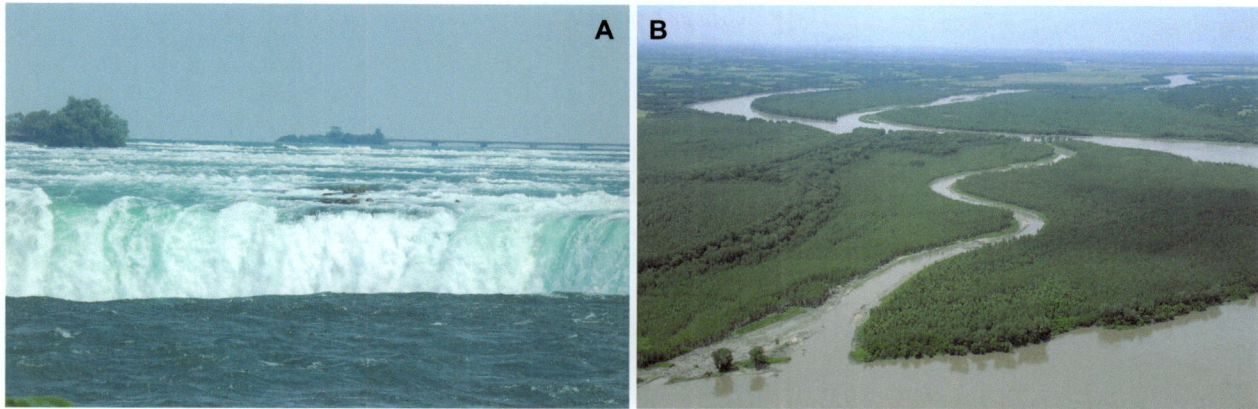

Figure 15.27 A. The Niagara River at Niagara Falls on the US–Canadian border. The water appears clear and clean but is rich with dissolved load. Over 139 tons of dissolved load typically pass over the falls every minute! Source: R. Schaetzl. **B.** Upstream view of the Missouri River in central Missouri, USA. The Missouri River's nickname is "Big Muddy," because of its muddy appearance, for which it can thank its high wash load content! Source: Hillebrand Steve, US Fish and Wildlife Service, Public domain, via Wikimedia Commons.

transport bedload and the larger components of suspended load. In contrast, dissolved and wash loads require little to no energy to carry.

Stream competence refers to the largest particle that a stream can move. Streams with greater competence (to move sediment) are typically deeper, faster, and more turbulent, which suggests that they have steeper water-surface slopes, necessary to create sufficient drag and lift forces (**Fig. 15.28**). **Drag forces** are generated when water flows around a bed particle, generating water pressures on the particle's upstream side that are greater than on its downstream side. The result is a net force *pushing it* in the downstream direction. **Lift forces** are generated when water accelerates as it flows up and over a bed particle, creating a zone of lower pressure on the top side of a particle, like an airplane wing (**Fig. 15.28**). These forces help *lift the particle* off the bed. Both drag and lift forces increase as in faster-flowing rivers. Thus, streams with faster near-bed velocities have greater competence than those with slower near-bed velocities.

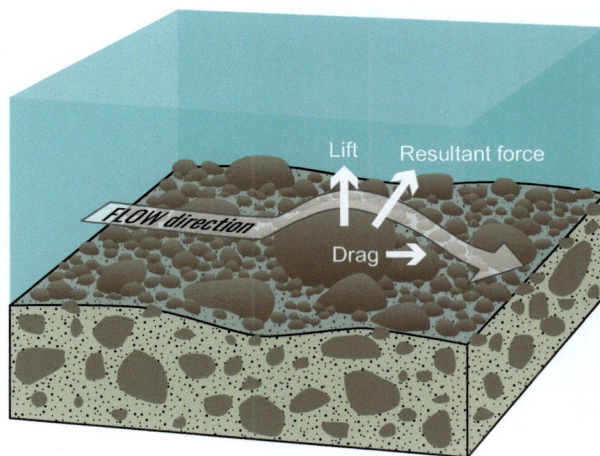

Figure 15.28 Schematic depiction of how lift and drag forces are generated on bed particles in streams.

Although competence varies from stream to stream due to differences in flow depth and velocity, it also varies spatially and temporally within a given stream. For example, as discharge increases during a flood, competence will also increase due to increasing depth and near-bed velocities, and because of increased turbulence. Later, stream competence will decrease as the flood recedes, causing sediment that was previously in transit to be deposited.

Stream capacity is the maximum amount of sediment a stream can transport over a period of time (assuming the sediment is available). Think of stream capacity as the potential for a stream to carry sediment. Just like stream competence, stream capacity increases as discharge and/or slope increase.

A key distinction is often made between streams in which the limiting factor to sediment movement is their capacity (**transport-limited streams**) vs sediment availability (**supply-limited streams**). In a transport-limited stream, the quantity of sediment moved is not limited by availability of sediment; rather, it is limited by the capacity of the stream to transport it. Thus, transport-limited streams tend to carry as much sediment as discharge and slope allow. Transport-limited streams typically have gravel or sandy beds, with lots of loose sediment that the stream will pick up and move – when it is able. In such streams, when discharge goes up, sediment loads go up. When discharge goes down, sediment loads go down. Alternatively, supply-limited streams have ample capacity to carry the available sediment, but have incomplete access to sediment that can be transported. Supply-limited streams commonly flow in channels with bedrock beds. Sediment transport in such streams is controlled more by deliveries of sediment to their channels than by fluctuations in their discharge.

15.8.1 Sources of Stream Sediment

The sediment load of a stream comes from an array of upstream sources (**Fig. 15.29**). A major source of sediment for streams is the hillslopes upstream in their drainage basins.

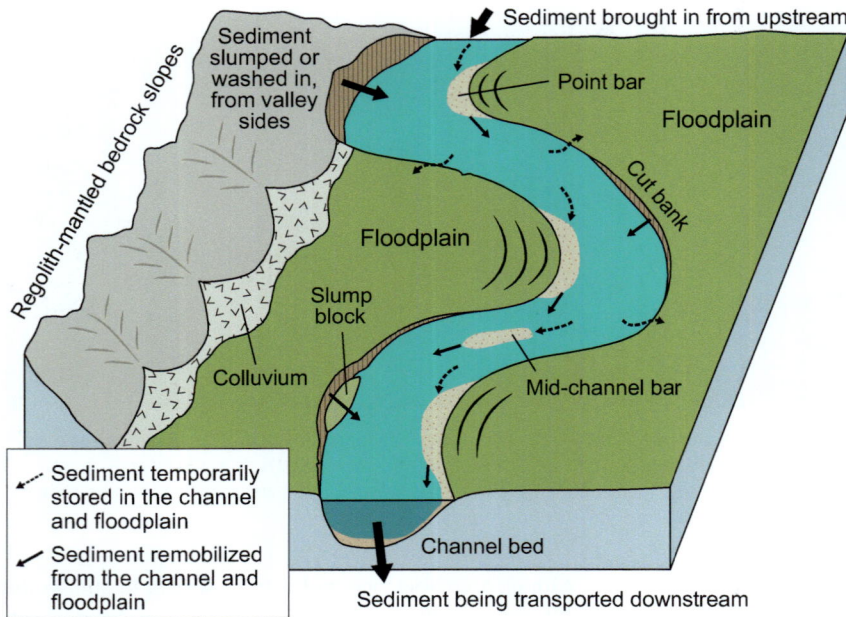

Figure 15.29 Sources of alluvium for a stream, including temporary storage sites within the channel (the bed and channel bars), as well as channel banks and valley-sides. Long-term storage sites, such as colluvium at the base of slopes, can also periodically provide sediment to channels. Source: Used with permission of Taylor & Francis, from *Fundamentals of Fluvial Geomorphology*, Charlton, Ro, 2008; permission conveyed through Copyright Clearance Center, Inc.

However, streams also mine sediment from their channel beds and banks. Material that is captured, transported, and deposited by streams is called **alluvium**.

Sediment is conveyed down hillslopes – to the channel – by runoff and various types of mass wasting processes (see Chapter 14). In some instances, this material is delivered directly into stream channels. In others, it is temporarily stored at the bottoms of slopes. Regardless, the sediment may eventually make its way into the channel, where it will become part of a stream's load.

Channel banks are an important source of a stream's sediment load. Bank erosion occurs when flow near the bank is rapid enough to scour and erode it. This type of erosion happens primarily on the outsides of channel bends, forming steep banks that are susceptible to slumping or toppling (**Fig. 15.30**). Thus, many streams derive much of their sediment from local sources, such as nearby hillslopes and channel banks.

15.8.2 Streams as Sediment Conveyor Belts

Sediment production for streams is especially pronounced where hillslopes feed sediment directly into channels, that is, from hillslopes that are strongly *coupled* (geomorphically linked) to stream channels (**Figs. 15.30, 15.31**). The region of a drainage basin where the slopes are relatively steep and coupled to stream channels is referred to as a **sediment production zone**.

Channels in the sediment production zone can be thought of as the start of a conveyor belt that moves sediment downstream toward its ultimate resting place (e.g., the oceans), but with long periods of inactivity and sediment storage along the way (**Fig. 15.32**). In other words, the conveyor belt is "jerky" because it is not always operating. For example,

Figure 15.30 Classical sources of alluvium. **A.** The Niobrara River in northern Nebraska, where coarse-textured sediment in the river banks is eroding and falling into the river. Source: D. Faulkner. **B.** Raccoon Creek in central Ohio is eroding fine-grained alluvium from its floodplain, along the outside of a channel bend. Source: James St. John, CC BY 2.0, via Wikimedia Commons.

Figure 15.31 The concept of a sediment production zone is illustrated well in this drainage basin in Afghanistan. Note that the slopes continue almost uninterruptedly down to the channels. Source: Mark Ray, Public domain, via Wikimedia Commons.

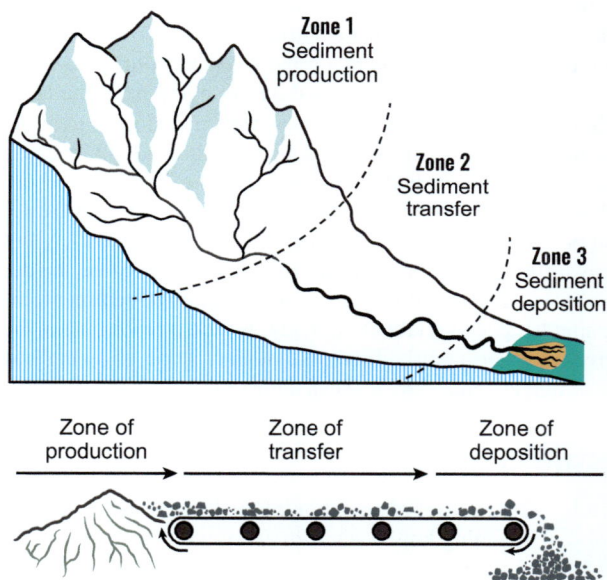

Figure 15.32 Depiction of the zones of sediment production, transfer, and deposition in a drainage basin. Sediment movement through these zones can be likened to a jerky conveyor belt, with periods of intermittent sediment storage and remobilization.

in the sediment production zone, sediment delivered from hillslopes toward stream channels is often temporarily stored at the bases of slopes (**Fig. 15.30A**) or in the channels themselves. Sediment storage at these locations can be brief, but it can also persist for millennia if the size of the material exceeds the stream's competence, or if the material is geomorphically "disconnected" from the channel. Eventually, however, weathering will reduce the size of large particles, rare large floods will occur, and connectivity will improve, enabling the stored sediment to be remobilized and transported downstream.

From the sediment production zone, the fluvial conveyor belt moves sediment through the **sediment transfer zone** (**Fig. 15.32**). In this zone, hillslopes and stream channels are generally not coupled. Rather, the hillslopes are separated from the river channel by a low-lying, flat valley floor, or **floodplain**. Sediment moves through the zone of transfer at various speeds and, of course, intermittently. It moves slowly during times of low flow (or not at all for parts of the year, in ephemeral streams) and rapidly during floods. The intermittent (jerky) nature of sediment transfer here is most notable for bedload, which typically travels only relatively short distances, and mainly during floods. Then, as floods wane, bedload transport slows and, eventually, stops. This "in-transit" characteristic of the sediment transfer zone is apparent where gravelly or sandy alluvium emerges above the stream surface during times of lower flows (**Fig. 15.33**). This type of bedload material – immobile during low flows – will remain so until a flood occurs that is large enough to remobilize and transport it downstream. In contrast to bedload, suspended load (and especially wash load) moves more continuously through the sediment transfer zone. However, even suspended load is subject to temporary storage. For example, overbank floods transport suspended sediment out of the channel, depositing it across the adjacent valley floor. These overbank flood deposits will remain in storage on the floodplain until the stream remobilizes them by eroding its banks or by scouring the floodplain surface during a flood.

Sediment *storage* and *remobilization* are important aspects of sediment movement in fluvial systems. Sediment storage on floodplains is particularly significant, enabling excess sediment that is transported during floods to

Figure 15.33 In the sediment transfer zone, sediment is often stored in the stream channel and floodplain until it can be remobilized and moved farther downstream (mainly during floods). In this photo, bars of coarse bed sediment emerge above the surface of the Chippewa River during low flow conditions. The wide floodplain of the Chippewa River contains vast amounts of stored alluvium. Over time, the river will erode laterally into this sediment, remobilizing and transporting it downstream. Source: D. Faulkner.

be temporarily stored on the wide, flat valley bottoms of larger rivers. Over time, however, the amount of sediment stored may be balanced (or nearly so) by the amount that gets remobilized, resulting in little or no net change in transfer-zone storage.

Eventually, sediment moving in the fluvial system will be deposited – typically at the end of the channel, in an ocean, a large lake, or an inland basin (**Fig. 15.32**). The key aspect of this **sediment deposition zone** is that the deposited sediment will experience little to no subsequent remobilization in the short term. Only when a major geomorphic change occurs, such as a drop in base level, might the sediment become remobilized.

15.9 THE GRADED STREAM CONCEPT

Streams are adaptable geomorphic systems. This is particularly true for transport-limited streams, which can readily *adjust their form* in order to transport the sediment load delivered to them. These adjustments can happen in a variety of ways, but transport-limited streams commonly adjust to varying sediment loads by altering their channel (water surface) *slope*. When these streams need to carry more load, they commonly steepen their gradient, and when they need to carry less, they often *lower* it. In this way, transport-limited streams tend toward a balance between the load delivered to them and their ability to carry it. This balance is known as **grade**. A river that has such a balance is said to be **graded**.

The concept of grade is often depicted graphically by a diagram referred to as **Lane's balance** (**Fig. 15.34**). The balance uses a "scale" to illustrate how transport-limited streams may degrade (erode) or aggrade (fill) their channels in response to changes in various factors such as discharge, slope, sediment load, and sediment size. Stream slope and discharge can be equated to **stream power** (see Chapter 16), which basically determines a stream's ability to transport sediment. When stream power goes up (due to increased discharges and/or steeper channel slopes), the balance is pulled down on the right side, causing the

indicator arrow to move to the left towards stream degradation; such streams can transport more sediment, and hence, they erode their beds. Conversely, when sediment load (or size) is increased, the balance shifts down on the left side, making the indicator arrow point towards aggradation, and sediment gets deposited. Deposition of sediment in the channel increases the overall slope of the channel, whereas degradation reduces it. Both tend to bring the system back to grade. (As described in Chapter 16, increases in channel sinuosity, or curvature, will also decrease the slope of the channel.) In summary, transport-limited streams have many ways to adjust, remaining graded as conditions change. They continually respond to changes in sediment load and size, as well as changes in discharge and slope. Lane's balance is a good way to visualize the adjustments streams can make to stay at grade.

Sometimes, streams are kicked out of grade by a disturbance. The damming of the North Fork Stillaguamish River in Washington by a landslide in 2014 provides an example of how a graded stream can be disturbed and then make adjustments to recover from that disturbance (**Fig. 1.1A**). The landslide instantaneously dammed the river, forming a lake that extended ≈4 km upstream. As the river overtopped the landslide dam, it encountered an over-steepened slope, but at the same time it was carrying a reduced sediment load, because so much sediment had been trapped in the lake. As a result, downstream from the dam, the river had excess power (greater slope and less sediment to carry). Thus, it immediately cut down (or **incised**) into the landslide deposit. Because the deposit was composed primarily of fine-textured sediment, incision happened relatively quickly. Within a few months, the river had re-established its pre-landslide slope, showing just how quickly some streams can adjust to stay at grade!

REVIEW QUESTIONS

15.1 Give a brief explanation of the hydrologic cycle, including how water moves from one reservoir to another through precipitation, evaporation, and transpiration.

15.2 Explain the differences between infiltration and percolation. What is the infiltration capacity of a soil and how is it related to the generation of surface runoff?

15.3 Explain and discuss what a drainage basin is, how it functions, and how basins can be "nested."

15.4 What are the differences between perennial, intermittent, and ephemeral streams? What are the various water components that contribute flow to each of these three types of streams?

15.5 What is the formula used to calculate stream discharge? Be sure to know the units, and exact meaning, of each term in the equation.

15.6 How is stream discharge determined at a stream gaging station? Include in your answer the concept of a rating curve.

15.7 The hydrograph in **Fig. 15.25A** is an example of a storm hydrograph for a perennial stream. Draw

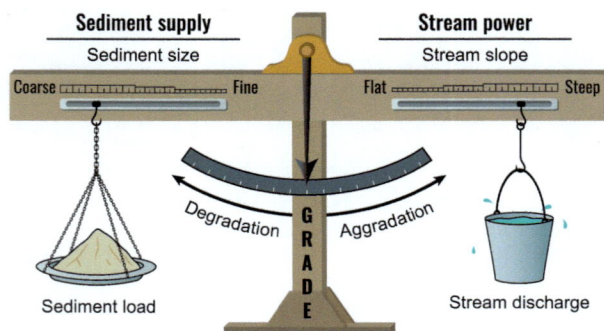

Figure 15.34 Lane's balance, a conceptual model of grade in streams, i.e., the balance between discharge, channel slope, and sediment load and size. Imbalanced conditions may lead to aggradation (channel filling) or degradation (channel-bed erosion).

and label a storm hydrograph for an ephemeral stream.

15.8   What is the probability that the 100-year flood on a given river will happen this year? What is the probability that it will happen next year? And the year after that?

15.9   Describe and contrast the three types of sediment load (dissolved load, suspended load, and bedload) and the means by which they are transported within streams.

15.10 Describe and contrast stream competence vs stream capacity, and what governs them.

15.11 If the sediment delivery to a stream increases without any change to the size of the sediment or to the slope or discharge of the stream, what will happen? Frame your discussion in terms of Lane's balance.

15.12 What is meant by a graded stream? Provide an example of how a change in one aspect of the fluvial system is compensated for, by the stream, so as to remain at grade.

15.13 Discuss the sediment transfer and production zones, and explain why the term "jerky" is apt when describing them.

FURTHER READING

Brooks, K. N., Ffolliott, P. F., and Magner, J. A. 2013. *Hydrology and the Management of Watersheds*. 4th ed. Wiley.

Davie, T. and Quinn, N. W. 2019. *Fundamentals of Hydrology*. Book 4 of 4: Routledge Fundamentals of Physical Geography. 3rd ed. Routledge.

Dunne, T. and Leopold, L. B. 1978. *Water in Environmental Planning*. W.H. Freeman.

Ward, A. D., Trimble, S. W., Burckhard, S. R., and Lyon, J. G. 2020. *Environmental Hydrology*. 3rd ed. CRC Press.

16 Landforms of Fluvial Erosion and Deposition

L. Allan James, Douglas J. Faulkner, and Randall Schaetzl

Rivers and their valleys have long been a source of contemplation and wonder. They are not only key geomorphic agents, but they are also economically important, acting as transportation arteries, sources of irrigation water and food, and as generators of hydropower. We also use rivers for drinking, waste disposal, and for a variety of recreational activities. Many geomorphologists consider running water to be the most dominant and important geomorphic process – shaping landscapes everywhere. Even in deserts, running water is often the most important and widespread geomorphic agent.

Most valleys have a stream or a river at their bottom. In ancient days, it was thought that water simply "found" its way into preexisting valleys, forming rivers there. Geomorphologists now know that most valleys were formed by the rivers currently within them, which moved sediment out and carved the valley over time. Geomorphologists have also shown that rivers can provide important insights into the evolutionary history of entire landscapes, thanks to their naturally inherent connections with other geomorphic systems. Rivers carry away the sediment delivered to them by slope processes (**Fig. 14.2**) and they provide sediment to eolian, coastal, and other systems.

Processes and landforms associated with running water are given the adjective **fluvial**. This chapter discusses fluvial landforms and landscapes at all scales and in all their forms – from the small, dry channels that only occasionally carry water, to the massive rivers whose names we all know. For each, similar and recognizable suites of landforms typically evolve, despite differences in climate, geologic materials, and the amount of time involved. Our emphasis is on **alluvial channels**, which are channels developed in river-deposited sediment (not in bedrock). This distinction is important because streams can more easily erode and reshape alluvium than bedrock, which results in fundamental differences between alluvial and bedrock channels. Rivers may pass from one type of channel to the other, and back again, as they flow. Nonetheless, many features, such as channel networks and longitudinal profiles, apply to both bedrock and alluvial channels.

16.1 CHANNELS AS DRAINAGE SYSTEMS

Fluvial processes operate as surface water follows topographic pathways from high to low elevations. In doing so, water initially forms small, headwater channels that join downstream into broader channel networks. These networks form drainage patterns with characteristic shapes that can be used to interpret geomorphic history and rock structure.

16.1.1 Initiation of Channelized Flow

The fluvial system begins as water runs off the landscape, often as **sheetflow**, which is shallow, overland flow on surfaces that lack channels. **Rills**, which are very small channels, may form where sheetflow concentrates downslope into faster flowing rivulets (**Fig. 16.1**). As rills merge and enlarge downstream or through time, they may form **gullies** that are larger, often steep-walled, channels. Like rills, gullies are often dry, usually holding water only during and shortly after precipitation events (**Fig. 16.1B, C**). Active rills and gullies sometimes represent serious erosional and sediment production concerns.

16.1.2 Channel Networks

Most streams form interconnected, branching patterns, referred to as channel networks. These patterns are of great interest to geomorphologists and hydrologists because they branch in distinctive, characteristic ways. Channel networks are also important because water in channels flows much faster and more efficiently than water in sheetflows. Thus, stream channels are more efficient at transporting sediment than are unchanneled systems. Similarly, the transport of sediment or pollutants is greater and faster in drainage basins with many well-connected channels than in basins that have few channels.

Channel networks can be examined using a hierarchical classification system of **stream orders**, first devised by Robert Horton and later modified by Arthur Strahler (**Fig. 16.2**). Stream ordering is used in many kinds of applications. The system categorizes stream segments according to their relative position within the overall channel network. In this method, unbranched (fingertip) channels that lack tributaries

Figure 16.0 The Wild River, in the Brooks Range of Alaska, USA, contains many of the classical fluvial landforms. Source: Patrick J. Endres / Getty Images.

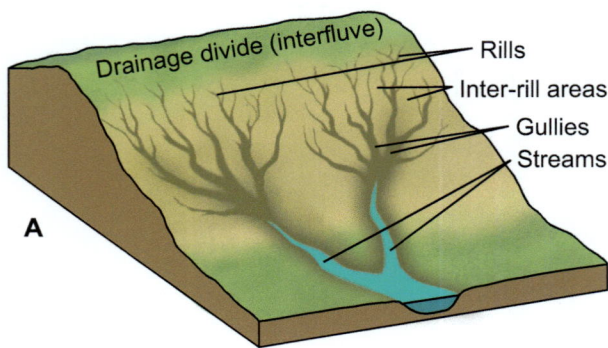

Figure 16.1 Rills and gullies. **A.** A schematic diagram showing the interconnections among inter-rill areas, rills, gullies, and larger forms of channelized flow. **B.** Rills and gullies on an unvegetated surface in Toadstool Geological Park, Nebraska, USA. Source: D. Faulkner. **C.** A deep gully in Khustayn Uul National Park, Mongolia. Source: M. Mergili.

Figure 16.2 The Strahler system of stream ordering for a fourth-order drainage basin.

Typically, stream orders are determined using large-scale topographic maps, or maps of similar scale, thereby allowing comparisons to be made.

An important feature of drainage basins that can be derived from maps of channel networks is their drainage density. **Drainage density** (D_d) is defined as:

$$D_d = \Sigma L / A_d,$$

where ΣL is the total length of streams in the drainage basin (remember, Σ means sum) and A_d is the area of the basin. Basins with higher D_d values have more channels per unit area (**Fig. 16.3A**), usually because more water runs off. The increased runoff may be due to a number of causes, such as precipitation intensity, slope of the land surface, or permeability of the underlying sediment. Drainage densities tend to be higher in semi-arid regions with sparse vegetation and in basins with low infiltration rates, because these areas tend to generate more runoff. Geology influences drainage density because coarse-grained rocks, such as granite, tend to weather to permeable regolith, which generates less runoff. As a result, fewer channels form on granite than on fine-grained rocks like shale (**Fig. 16.3B**). For most basins, drainage density increases with time, as new channels are continually formed and then lengthened by runoff.

High drainage densities facilitate rapid runoff from storms and snowmelt events, resulting in more variable flow regimes and more rapid erosion rates. Basins with higher D_d values are much more efficient at moving water and sediment (within channels) than are basins dominated by unchanneled, overland sheetflows.

Drainage density suffers from the same mapping limitations as stream orders. Use of small-scale maps that cover large areas will underestimate the lengths of channels and produce lower drainage densities than networks determined off large-scale maps. Although the "blue lines" on most topographic maps tend to underestimate channel lengths, alternate sources of data are often not available.

are designated as first-order streams. Fundamental to the stream ordering system is the notion that wherever streams *of similar order* join, the order of the downstream segment is increased by one (**Fig. 16.2**). Stream order does not change when streams of unequal order join, such as when a first-order stream enters a higher-order stream. One disadvantage of the method is that order numbers are highly dependent on the map scale used to derive the network. Thus, a network mapped from a small-scale map (one that covers a very large area) will have far fewer channels than when the same network is shown on a large-scale map, such as a 1:24,000 topographic map. This makes it difficult to compare stream orders constructed from maps of different scale.

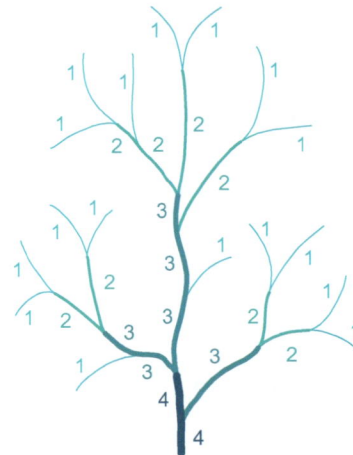

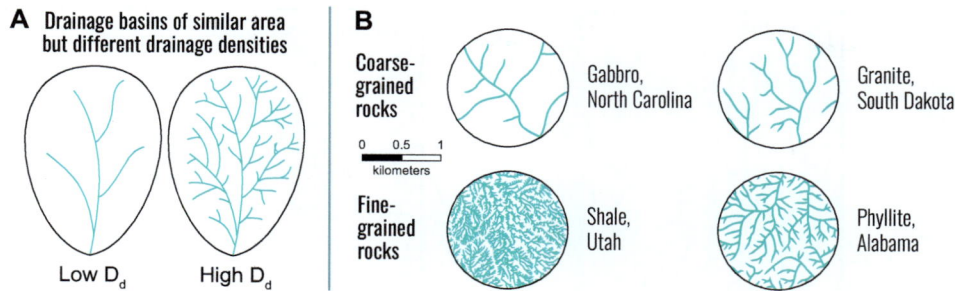

A Drainage basins of similar area but different drainage densities

Low D_d High D_d

B

Coarse-grained rocks

Gabbro, North Carolina

Granite, South Dakota

0   0.5   1
kilometers

Fine-grained rocks

Shale, Utah

Phyllite, Alabama

Figure 16.3 Drainage density. **A.** An illustration of different D_d values for two basins of the same size. **B.** Drainage densities on four different rock types. Areas underlain by coarse-grained rocks tend to have fewer channels than areas on fine-grained rocks. Source: Adapted from Ray, R. G. and W. A. Fischer (1960). Quantitative photography – a geologic research tool. *Photogramm. Engr.* 25:143–150.

16.1.3 Drainage Patterns

A great variety of channel network forms exist in nature, due to differences in geologic structures and materials, processes acting upon the system, and the stage of landscape development. Thus, over time, drainage basins develop distinct **drainage patterns**, that is, shapes of the channel networks as seen from above (**Fig. 16.4**). Examples of rock structural controls on erosion and topography were covered in Chapters 8, 9, and 10. Geomorphologists study drainage patterns to learn more about those large-scale processes and geologic structures (**Table 16.1**). Patterns are often examined from aerial photographs, maps, and other remotely sensed data.

The most fundamental drainage pattern – **dendritic** – is named for its tree-like branching pattern that develops randomly when the underlying landscape is geologically "uniform." In other words, the rocks impart little or no structural control on the stream pattern (**Fig. 16.5A**). Dendritic patterns are, for example, found where large swaths of the landscape are covered uniformly with thick, silt-rich loess

Table 16.1 Characteristics of the most common drainage patterns

Pattern	Causes and implications
Dendritic	Tree-like pattern associated with areas of uniform geology, fairly gentle slopes, and lack of structural control
Parallel	Similar to dendritic, but elongated due to steeper slopes, such that channels are quasi-parallel
Radial	Streams radiate outward from an isolated, structural dome or peak, such as a volcano
Trellis	Long, nearly linear, master streams with many, shorter tributaries joining nearly at right angles; common in areas of folded mountain belts, near cuestas, and along beach ridges
Annular	Circular-shaped pattern associated with cuestas that form around a dome (scarps face inward, as shown in **Fig. 8.30**) or basin (scarps face outward); master streams curve around the dome, and shorter tributaries join nearly at right angles
Deranged	Random and irregularly connected pattern of lakes, wetlands, and basins in karst, arid, or periglacial regions, or in recently deglaciated or volcanic terrain; streams appear to wander in seemingly odd and unexplained directions, flowing between many lakes and swamps; waterfalls and rapids are not uncommon
Centripetal	Streams converge into a central depression or bay
Rectangular	Strong joint control from underlying rocks causes streams to take sharp, right-angle turns, following the weakened rocks in fractured and highly jointed areas
Distributary	Channels branch, split (**avulse**), and diverge in the downstream direction, as on a delta or alluvial fan
Pinnate	"Featherlike" pattern with many small tributaries joining a single master channel or gully, typical of highly erodible materials, such as gullies eroded in loess
Contorted	Non-systematic pattern distorted by underlying metamorphic rocks
Compound	Two stream patterns that overlie each other, imparting characteristics of both to the resultant pattern

Figure 16.4 Twelve of the most common drainage patterns. See Table 16.1 for descriptions and implications of each. Black arrows are used to indicate overall flow direction (where not obvious). Compiled from numerous sources.

deposits. They also tend to form on flat-lying rocks without any structural orientation. Similarly, **parallel** patterns, which are elongated dendritic patterns, develop on regionally steepened surfaces, but where few other structural controls are present. As a result, the streams tend to flow more-or-less directly down the steep slope, taking on a quasi-parallel orientation (**Fig. 16.5B**).

Other drainage patterns reflect stronger controls by topography, such as **radial** and **centripetal** patterns, which may be influenced by steep regional slopes (**Fig. 16.4**). In radial patterns, rivers flow away from a topographically high point such as a volcano or structural dome, whereas in centripetal patterns, streams flow in towards a central low point, such as a structural basin or a lake.

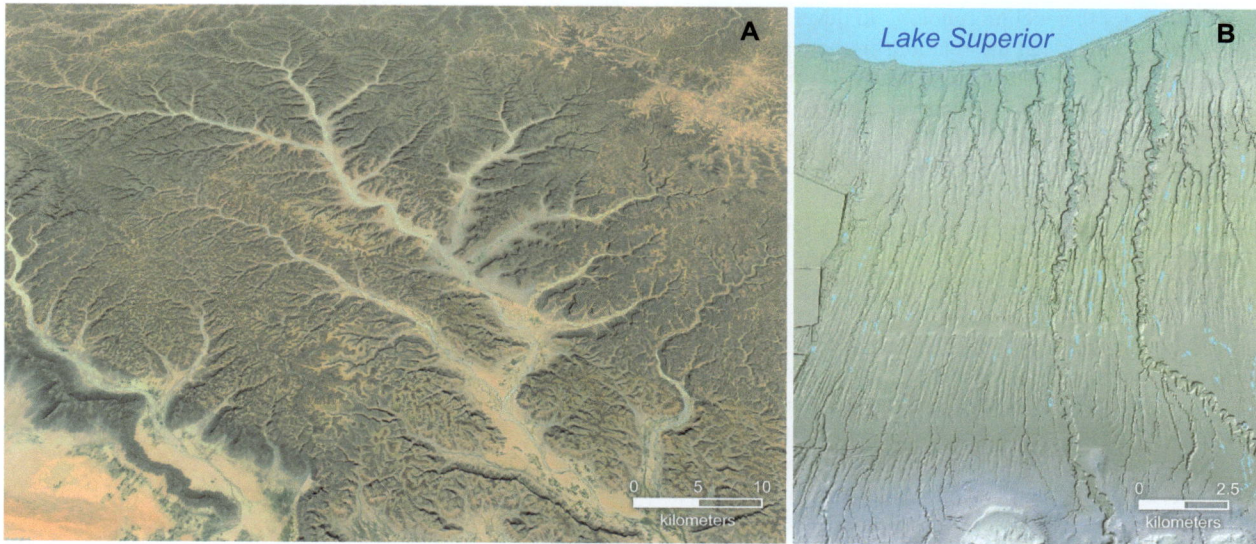

Figure 16.5 Aerial views of two common drainage patterns. **A.** A satellite oblique image of a dendritic drainage pattern in Yemen, where the lack of vegetation makes the channels very apparent. Source: NASA. **B.** A parallel drainage pattern with a high drainage density has developed on the regionally steep slopes that slope down into Lake Superior in Michigan's Upper Peninsula.

Two other patterns are related to long-term erosion of geologic structures on sedimentary rocks, where the rock layers have variable resistance to erosion. Weak rocks, such as shales, are easily eroded, but resistant rocks tend to form ridges (see Chapters 5, 8, and 9). In these landscapes, channels tend to be located on the weaker rocks, or on sloping surfaces that run along the sides of hard rock layers. Over time, these streams commonly exhibit a process known as **homoclinal shifting**, where the streams that are incising into gently dipping rocks shift position, laterally down the dip slope over time (**Fig. 9.23**). Thus, rivers that flow along strike valleys between **cuestas** (see Chapter 8) or within folded mountain belts (see Chapter 9) tend to shift down the slope of rock layers over time. Trellis and annular patterns both form on landscapes like this; they follow outcrops of more erodible rocks (**Fig. 16.4**). Where cuestas are linear as seen from above, trellis stream patterns develop, with short, parallel tributaries feeding into a long, master stream that cuts across the cuestas (**Fig. 9.27**). Circular or arcuate cuestas around domes and basins (**Figs. 8.27, 9.6**) typically develop an annular pattern with arcuate master channels and short tributaries.

In contrast to the patterns discussed above, which have distinct structural controls, some drainage patterns have poorly organized and weakly connected networks that do not reflect the underlying geology. For example, **deranged** patterns have seemingly randomly connected basins, lakes, or wetlands, with streams that flow in a variety of directions (**Fig. 16.4**). Deranged patterns are inefficient at draining water from the landscape and are typical of regions that were recently disturbed, such as by recent glaciation or burial by volcanic rocks, so that the drainage hasn't had time to become fully integrated. Many streams in these types of landscapes have randomly spaced **rapids** and small waterfalls, as the fluvial system has not yet had time to "smooth out" its longitudinal profile. Deranged drainage patterns also occur on karst landscapes and in periglacial or arid regions with playas, where fluvial processes are limited or overwhelmed by other processes. Another pattern that is not controlled by rock structures is the **distributary** network, which occurs where streams are not horizontally constrained and sedimentation rates are high. Channels in distributary networks often fill and shift position, such as on alluvial fans and deltas. Indeed, channels on deltas are often referred to as distributary systems (see Chapter 17). **Pinnate** drainage networks have numerous, small, side tributaries, formed in erodible materials (**Fig. 16.4**).

Finally, some drainage patterns reflect controls of underlying hard-rock geology but not necessarily on sedimentary rocks. For example, the high angularity of **rectangular** patterns is governed by joints or other features in the underlying bedrock (**Fig. 16.4**). **Contorted** patterns are distorted by underlying, structurally deformed rocks. **Compound drainage pattern**s describe two drainage patterns interlaced with each other. They reveal the impact of multiple processes or structures on the evolution of the drainage pattern and may or may not include controls by rock structures. **Figure 16.4** shows a compound example with annular and radial patterns.

16.2 CONCEPTS OF CHANNEL MORPHOLOGY

Many factors, such as drainage patterns, drainage densities, and slopes, affect how stream systems function. Even the shapes and sizes of the channels themselves influence the ability of the fluvial system to transport water and sediment. Channel shape and size, that is, **channel morphology**, affect the various fluvial geomorphic processes, heights of flood waters, and aquatic ecosystems. This section addresses (1) the tendency for channel morphology to develop into stable (equilibrium) forms, (2) the relationship between the magnitude and frequency of river flow and the resulting channel form, and (3) systematic changes in channel width, depth, and velocity that occur downstream, with changes in discharge. In essence, river channel shapes and forms are always evolving, and can tell us a great deal about how the fluvial system operates and how channels will likely respond to changes in climate or land use.

16.2.1 Dynamic Equilibrium and Effective Discharge

In fluvial geomorphology, the concept of **dynamic equilibrium** refers to how the size, shape, and pattern of a channel are in balance between (1) the forces applied to the channel boundary by the flowing water, and (2) the "resisting" forces exerted by channel materials. A channel is in equilibrium if, over time, the stream has just enough power to transport the sediment supplied to it. Often, such a channel or stream is referred to as being at "grade." A **graded river** has a slope that is perfectly adjusted to carry away just the amount of sediment delivered to it, but not more (see Chapter 15). When considered over time, a graded stream is neither progressively eroding nor depositing sediment, and is not showing systematic changes in shape or size. However, channel equilibrium ("at grade") conditions are highly dynamic, in that they can change over short periods of time. For example, a flood may enlarge the channel, but subsequent moderate-magnitude flows may then fill it back in, returning the channel to its previous condition.

It is important to recognize whether or not stream channels are in equilibrium, because false assumptions of equilibrium can result in overestimations of channel stability. Streams that are truly in equilibrium will tend to return to a stable condition after extreme events have come and gone. At any time, however, changes in stream power or sediment load may disrupt a channel, driving it out of equilibrium. Channels out of equilibrium may experience drastic changes, such as stream bank collapse, shifts in channel location, infilling of pools, or channel downcutting and erosion of bridge abutments.

Early in the twentieth century, it was often thought that channel morphology mainly reflected the work of extreme floods. Geomorphologists thought that, surely, the tremendous amounts of energy exerted by large floods must be responsible for the largest channel changes! Later, a theory was advanced that moderate-sized floods were more important in forming the prevailing size and shape of most channels (or at least, those in erodible materials, not bedrock). The logical argument made at the time was that, over a period of years, the work of the river carrying sediment was the product of the amount of sediment carried by a given flood, multiplied by the frequency of that size of flood (**Fig. 16.6**). This magnitude–frequency relationship is often described as the **Wolman–Miller Principle** after the two authors who proposed it. It was initially applied to channels dominated by suspended load (**Fig. 16.6B**). The peak in the

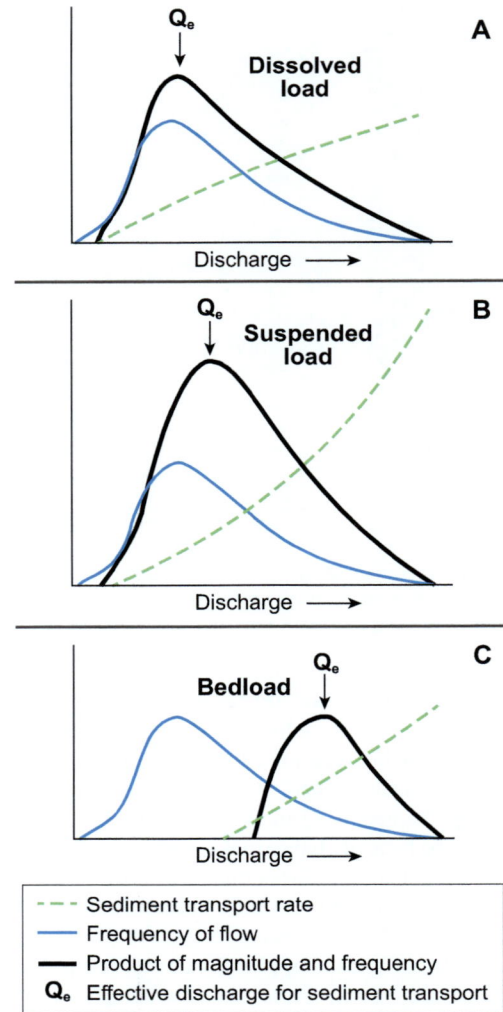

Figure 16.6 Magnitude–frequency diagrams for three types of fluvial sediment transport systems – (**A**) dissolved, (**B**) suspended, and (**C**) bedload channels. Each panel shows the theoretical sediment transport rates, which increase indefinitely with discharge, and flow frequencies, which are most frequent at moderate magnitudes. The mathematical product of transport rate and flow frequency yields the amount of sediment transported, which is maximal for moderate-magnitude events in dissolved and suspended load streams and is, therefore, the effective discharge for those streams. Source: After Ritter et al. Waveland Press, Inc., © 2011; all rights reserved.

magnitude–frequency product at modest discharges implies that the stream discharge that transports the most suspended sediment over time, often referred to as the **effective discharge**, was actually *not* an extreme flood. Rather, most of the "work" done by a stream was accomplished by more moderately sized discharges that occur, on average, every few years. Wolman and Miller concluded that moderate-magnitude flows (those that might be considered "high flows," but which nonetheless occurred relatively frequently) were primarily responsible for the channel morphology we see today. They summarized this point with this analogy:

A dwarf, a man, and a huge giant are having a woodcutting contest. Because of metabolic peculiarities, individual chopping rates are roughly inverse to their size. The dwarf works steadily and is rarely seen to rest. However, his progress is slow, for even little trees take a long time, and there are many big ones which he cannot dent with his axe. The man is a strong fellow and a hard worker, but he takes a day off now and then. His vigorous and persistent labors are highly effective, but there are some trees that defy his best efforts. The giant is tremendously strong, but he spends most of his time sleeping. Whenever he is on the job, his actions are frequently capricious. Sometimes he throws away his axe and dashes wildly into the woods, where he breaks the trees or pulls them up by the roots. On the rare occasions when he encounters a tree too big for him, he ominously mentions his family of brothers—all bigger, and stronger, and sleepier.
(Wolman and Miller, 1960: 73)

In other words, flows of moderate size, that is, those that repeatedly work and rework channels, are usually the most effective at shaping the channel. Extremely large floods may greatly alter channels and move a lot of sediment, but they occur rarely and the channel forms that they leave behind may not persist if moderate-magnitude flows are able to rework the materials. Conversely, small flows move only a limited amount of sediment and cannot substantially erode the channel bed or banks, so they are geomorphically less effective.

Since that seminal work, the Wolman–Miller principle for suspended sediment has been extended to systems dominated by dissolved load and bedload (**Fig. 16.6A, C**). Dissolved sediment can be transported by very low flows, so the peak of the mathematical product occurs at a discharge similar to the peak of the flow-frequency curve. In other words, high dissolved loads are carried by moderate-magnitude flows, in keeping with the Wolman–Miller principle. Coarse-grained materials (such as bedload) require larger and faster flows to transport, so the sediment-transport curve in **Fig. 16.6C** is shifted to right and the peak of the magnitude–frequency product occurs at higher discharges. Therefore, the morphology of channels with coarse bed materials tends to be formed by larger, less frequent flows (floods).

Other factors can also affect the magnitude–frequency relationship, thereby determining the effective discharge of a stream. First, moderate-magnitude flows, which are necessary to restore channel conditions following large floods, may be rare in some small drainage basins and arid-climate streams. In these basins, channel shape and size are more likely to reflect the work of floods. Second, rivers with hard-to-erode banks and beds may make it impossible for moderate flows to initiate erosion. For example, boulders, cemented alluvium, or dense root mats may protect channels from erosion, shifting the most effective discharge toward large, infrequent events (**Fig. 5.19**). In these rivers, only large, erosive events will be able to initiate substantial channel erosion and form persistent channel shapes and sizes.

16.2.2 Controls on Channel Morphology

Stream channels are often described as self-regulating systems where many variables are adjusted to discharge, and where changes in one aspect of the stream are balanced by changes in others. For example, the size and shape of channels (channel morphology) can change through time or change in the downstream direction in response to changes in water and sediment loadings. But how do these changes happen and how can they be measured?

The standard stream metrics of channel cross-section width, mean depth, and mean velocity (**Fig. 15.20**) all increase with discharge. In other words, a river is usually wider and deeper, and flows faster, when its discharge increases, either at a given site or in the downstream direction.

For consistency, geomorphologists often use the **bankfull channel** or bankfull stage as the basis for many analyses of channel size and shape, making it useful for channel restoration work. Field identification of bank tops is a science of its own! Geomorphologists define "bankfull" as the entire channel that extends from the channel bed up to the bank tops, beyond which (usually) is the floodplain. Usually, rivers flow below bankfull stage. By definition, flows above bankfull stage are in flood.

Discharge is usually a dominant factor in determining channel morphology. In almost all fluvial systems (except arid and karst systems), discharge increases downstream. Therefore, systematic increases in discharge downstream allow estimates to be made of channel width and depth (**Fig. 16.7**). The trend lines in **Fig. 16.7** indicate that channel width and depth increase relatively rapidly downstream, as discharge increases. Other features that increase systematically downstream include floodplain width and stream power. Conversely, channel slope *decreases* in the downstream direction. What may not be as intuitive is that mean flow velocity actually increases slightly downstream. Early studies of this increase in mean velocity downstream were surprising, because they showed that so-called "lazy" large rivers usually have slightly higher mean velocities than small mountain rivers, despite their turbulent whitewater.

The type of analysis of the fluvial system, shown in **Fig. 16.7**, where channel characteristics are statistically compared to discharge, is known as **hydraulic geometry**. Knowing how channel size and shape change in the downstream direction is a key element of channel design and management by river engineers and restorationists. Variations in hydraulic geometry data within or between rivers are influenced by several factors, including valley gradient, valley bottom width, resistance of the channel bed and banks to erosion, water and sediment deliveries from local tributaries, sediment texture, and antecedent floods. Resistance to channel bed and bank erosion is driven by factors such as cohesiveness of the bank and bed materials, exposures of bedrock, and vegetation along the banks, as well as any engineering structures that may exist.

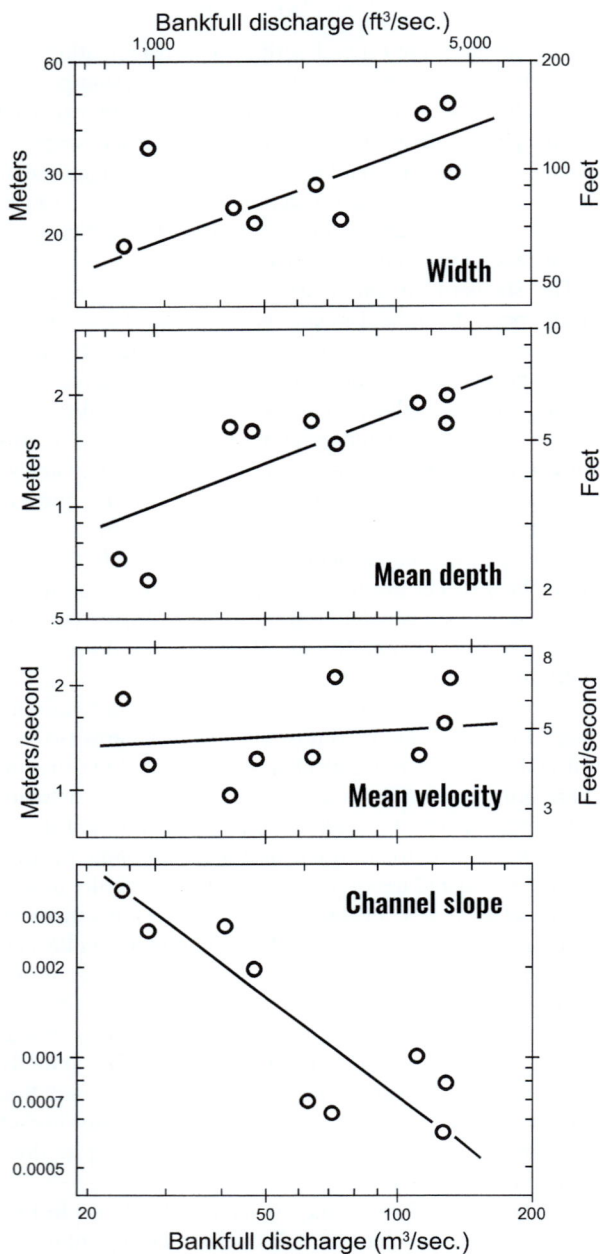

Figure 16.7 Changes in channel hydraulic geometry, as a function of discharge (which increases downstream) over a 40 km reach of Brandywine Creek, Pennsylvania, USA. Bankfull channel width, mean depth, and mean velocity increase with increases in discharge in the downstream direction, whereas channel slope decreases. Source: US Geological Survey, Department of the Interior/USGS. US Geological Survey Prof. Paper 271 by Wolman (1955); in public domain.

16.3 CHANNEL TYPES

Rivers are complex! No one rule or factor can be used to determine what kind of channel may develop at any one place. Channel morphology is governed by the materials within which the channel is developed, the amount and grain sizes of sediment carried by the river, the size and frequency of floods, channel-side vegetation, channel and valley-bottom gradients, and geomorphic history. This section reviews many channel features and processes, including the four main types of channels – straight, meandering, braided, and anastomosing.

16.3.1 Types of Floodplain Sedimentation

Rivers often deposit sediment along their margins, and much of this sediment is used to ultimately construct the river's floodplain. Three different fluvial sedimentation processes can generate floodplain deposits – vertical (overbank) accretion, lateral accretion, and island formation (**Fig. 16.8**). Diagnostic sedimentary structures associated with these processes allow sedimentologists to infer the processes and environmental conditions that were acting in the past.

Vertical accretion is the addition of sediment to floodplains that occurs during floods, as rivers overtop their banks and deposit sediment on the flooded areas. Thus, floodplain surfaces get progressively higher over time, as layers of horizontally bedded sediment are incrementally added. These sediments are typically fine sand, silt, and clay, but coarser sediment may even get deposited close to the channel edges.

Lateral accretion is the accumulation of sediment along channel margins, that is, on the sides of the channel. It may occur as a channel gets eroded on one side and coarse-grained sediment is deposited on the other side, where the current is slower. Landforms that develop from lateral accretion in these slower zones, on the inside of bends, are called **point bars**. Lateral accretion deposits are typically coarse grained, containing sand, gravel, or even cobbles. They lack horizontal bedding but often have angled layering known as cross-bedding.

Channels with multiple small islands or mid-channel bars are referred to as being **braided** (**Fig. 18.23B**). Mid-channel bar formation in braided channels is facilitated by highly variable, energetic flows and coarse-grained sediment such as sand and gravel. Almost all braided channels have sandy or gravelly beds and highly erodible banks. But when conditions change, some of the channels may fill with sediment, converting the braided system, in places, to a more contiguous floodplain. Why and how does this happen? It often starts when flow variability decreases, allowing mid-channel bars to stabilize and become vegetated. Sediment becomes trapped on these bars and in the channels between them by obstructions, such as gravelly deposits, log jams, or shrubs and trees. As additional sediment accumulates, many of the channels fill with sediment and are abandoned, transforming the braided stream into a broad, sandy floodplain.

16.3.2 Fluvial Landforms

A variety of landforms may develop in the different kinds of fluvial systems that occur in nature. Landforms associated with overbank deposition include natural levees and crevasse splays. **Natural levees** are low, ridge-like landforms that form along channel margins, as sandy sediment is deposited on the floodplain during floods (**Figs. 16.8, 16.9A, B**). Levees grow vertically over time, because overbank

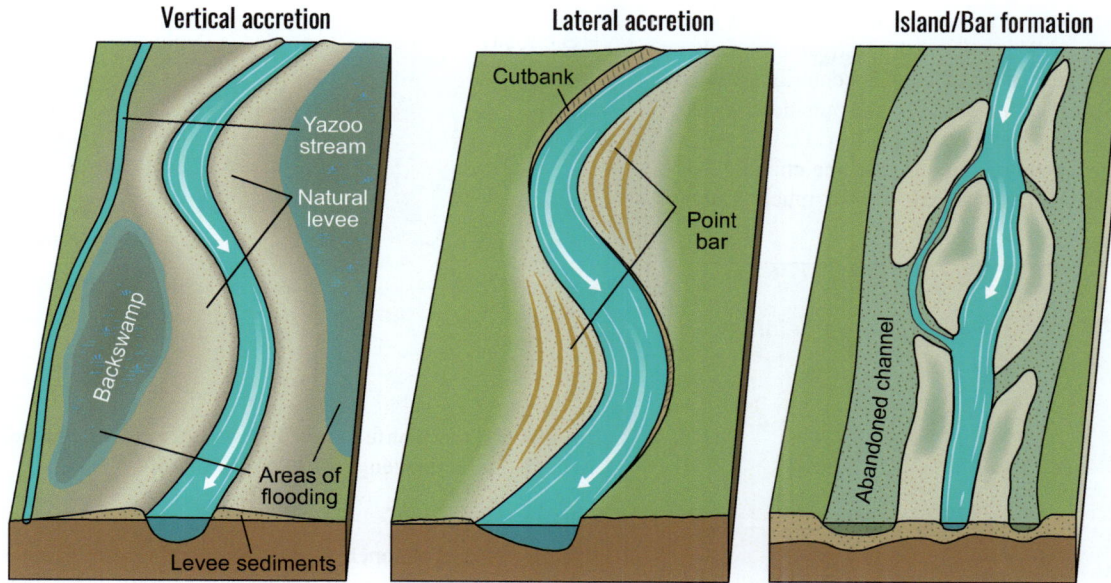

Figure 16.8 Three modes of floodplain sedimentation. Vertical accretion is dominated by deposition of suspended sediment on floodplain, during floods. Lateral accretion is dominated by construction of point bars as the channel migrates horizontally. Floodplain formation from channels with mid-channel bars is associated with filling and abandonment of channels.

deposition rates during floods are greatest near the channel. Here, the largest particles (sands) are deposited first, in areas near to the channel, while finer particles remain in suspension in the river or are deposited in low, **backswamp** areas of the floodplain farther from the channel (Fig. 16.9A, C). Thus, levee deposits are coarser-textured and thicker than overbank deposits farther from the channel. "Coarse near the source" is the rule of thumb for sediment deposition. Often, natural levees are the highest, driest parts of a floodplain.

Crevasse splays are floodplain overbank deposits of relatively coarse-textured material – typically sands – that are delivered through a narrow break in a natural levee. As waters pour out of these gaps in the levees, sandy sediment spreads out over the floodplain, in low, fan-like deposits. Unlike most forms of vertical accretion, crevasse splays are not dominated by suspended sediment but by sediment carried through a levee breach as bedload. Thus, crevasse splays are sandy, or even gravelly.

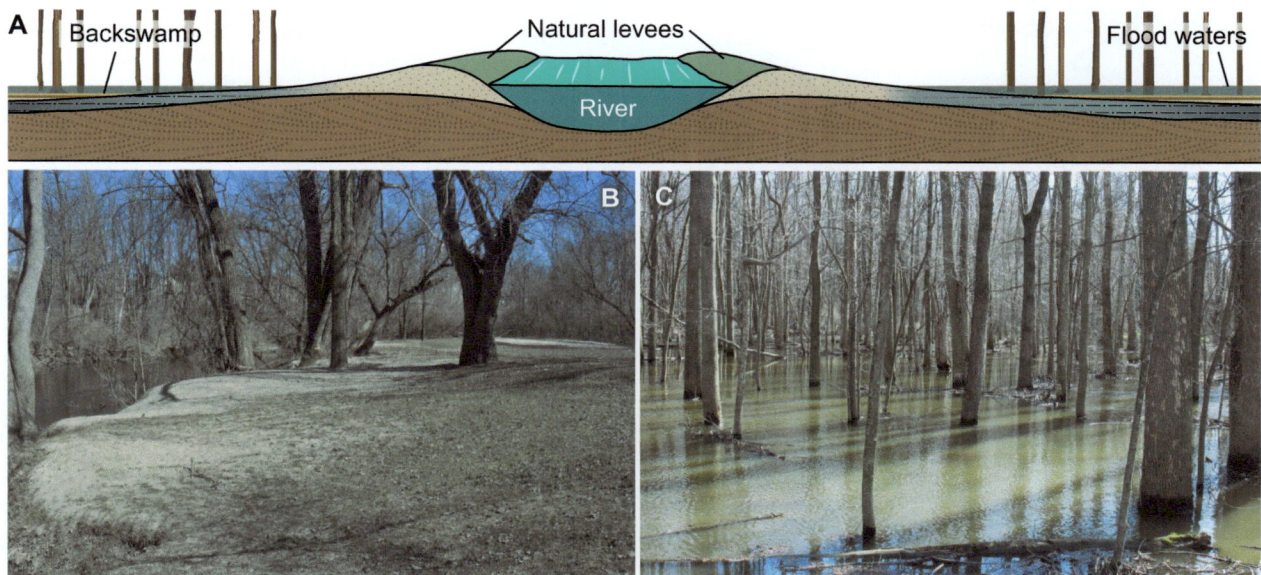

Figure 16.9 Floodplain landforms. **A.** Diagrammatic cross-section through a floodplain, showing natural levees adjacent to the channel and the low backswamp areas farther away. The height and thickness of the levees is exaggerated. **B.** Fresh, sandy sediments deposited adjacent to the Red Cedar River in Okemos, Michigan, USA, adding to the natural levee that already exists at this site. The river had flooded only a few days prior. Source: R. Schaetzl. **C.** Silt-rich waters stand in the backswamp of the flooded Muskegon River in central Michigan, USA. Source: R. Schaetzl.

The **thalweg** is defined as the line that follows the channel's maximum depth. When traced downstream, the thalweg is longer than the trace of the channel's centerline, because it typically stays close to the banks on the outside of channel bends and crosses over between bends (**Fig. 16.10A, B**). As a result, thalwegs are more sinuous than the channel itself. **Channel sinuosity** (S_i), a measure of how much a stream

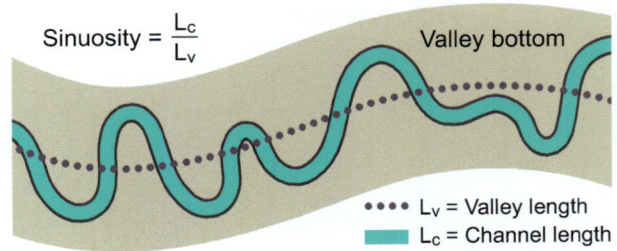

Figure 16.11 Channel sinuosity is the ratio of channel length to valley length. The dotted line traces the valley, which is shorter than the trace of the channel.

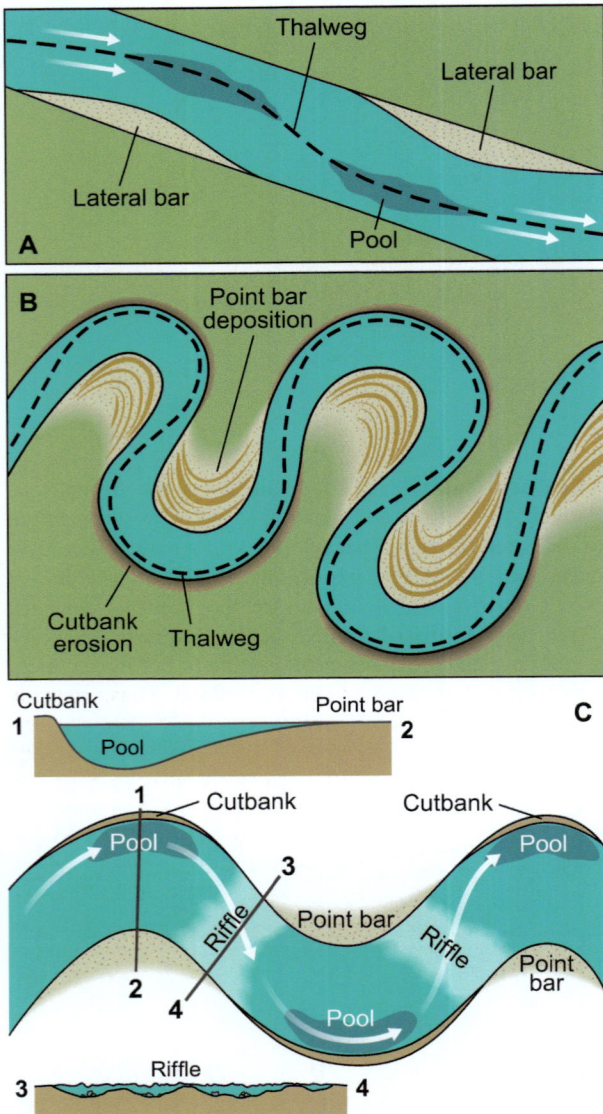

Figure 16.10 Channel patterns and features. **A.** Even straight channels can have a slightly sinuous thalweg, influenced by alternating bars. Channels like this can continue to meander more and more, eventually becoming highly sinuous, as shown in **B.** Here, a meandering channel has developed point bars on the insides of meander bends and cutbanks on the outside of bends. Note how the thalweg crosses over between meanders, such that the deep part of the channel always hugs the outsides of bends. **C.** Pool–riffle sequences often develop in meandering streams. Riffles typically occur between pools, where the thalweg crosses the center of the channel. Stream cross-sections commonly show a deep, asymmetrical pool (deepest near the outside bend) in meander bends, and a shallow riffle at the crossover point.

channel bends and turns, is defined as the ratio of channel length (L_C) to valley length (L_V) (**Fig. 16.11**):

$$S_i = L_C / L_V.$$

Perfectly straight channels have a sinuosity of one. Highly sinuous channels may have S_i values approaching three, four, or more. Sinuous (meandering) channels often develop pool and riffle sequences, as shown in **Fig. 16.10B** (and see below).

The slope (or gradient) of the channel, as often measured by the slope of the water surface, is also important because it governs (along with discharge) the power of a given river flow to move sediment. **Stream power** (Ω) is a measure of the ability of a river to pick up and carry sediment. It is proportional to the product of slope (S) and discharge (Q) (see Chapter 15):

$$\Omega = \gamma Q S,$$

where γ (often treated as a constant) is the specific weight of the fluid–sediment mixture. High discharges in a steeply sloping channel may have the power to move all but the largest cobbles and boulders, whereas the same discharge in a low gradient channel may not be able to move this same sediment. The slope of a channel (and hence, also its power) will decrease as sinuosity increases. Therefore, as with many variables in river systems, a change in one variable may be compensated for by changes in others.

16.3.3 Sediment Load, Bank Stability, and Channel Type

Channel type (straight, meandering, braided, or anastomosing) is strongly correlated to overall channel stability, which in turn is controlled by channel gradient and the type of sediment carried by the stream (**Fig. 16.12**). A good rule of thumb is that the main type of sediment that a stream has carried is reflected in the makeup of its banks.

Suspended load streams carry mostly sediment of silt and clay size, and tend to have stable, cohesive banks that are fairly resistant to erosion. These channels are associated with straight and meandering channels. Such channels are relatively narrow and deep, approaching a semi-circle in cross-section, which is the most efficient shape to convey suspended load. Thus, streams carrying mainly silty or clayey sediments in suspension often have relatively narrow, deep channels with low **width/depth ratios** (W/D) and higher flow velocities.

Type of sediment load

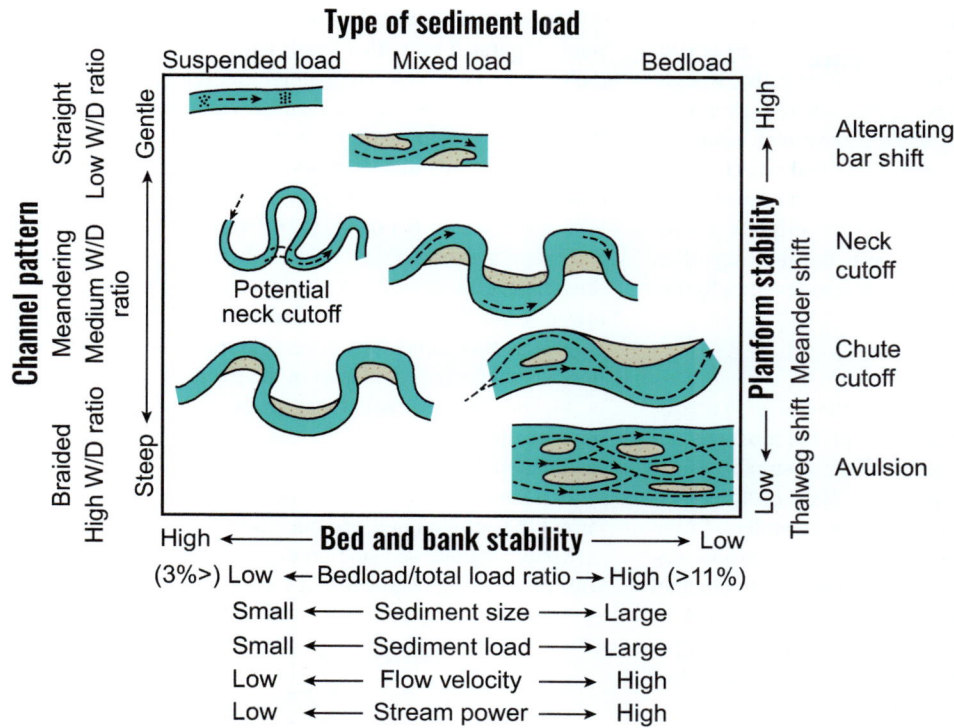

Figure 16.12 Straight, meandering, and braided river channels as a function of sediment load and channel stability. Source: After Schumm (1981), © SEPM Society for Sedimentary Geology.

Bedload streams, carrying a lot of coarse-grained sediment, usually have unstable, non-cohesive banks of sand or gravel. The erodible materials in these types of banks tend to collapse easily, adding coarse-textured sediment to the channel, forming braided channels with mid-channel bars. Channels carrying mostly bedload are relatively wide and shallow (high W/D ratios). In these types of channels, the maximum flow velocity occurs near the bed, which optimizes the transport of the coarse sediment. Mixed load streams have intermediate conditions, often with banks of interlayered material of different textures.

16.3.4 Straight Channels

By definition, **straight channels** have low or very low sinuosities (**Fig. 16.13**). They tend to form in areas of low valley gradients, where a straight thalweg maximizes the channel slope and helps the river transport the required sediment load. Straight channels are not common except where structural controls or

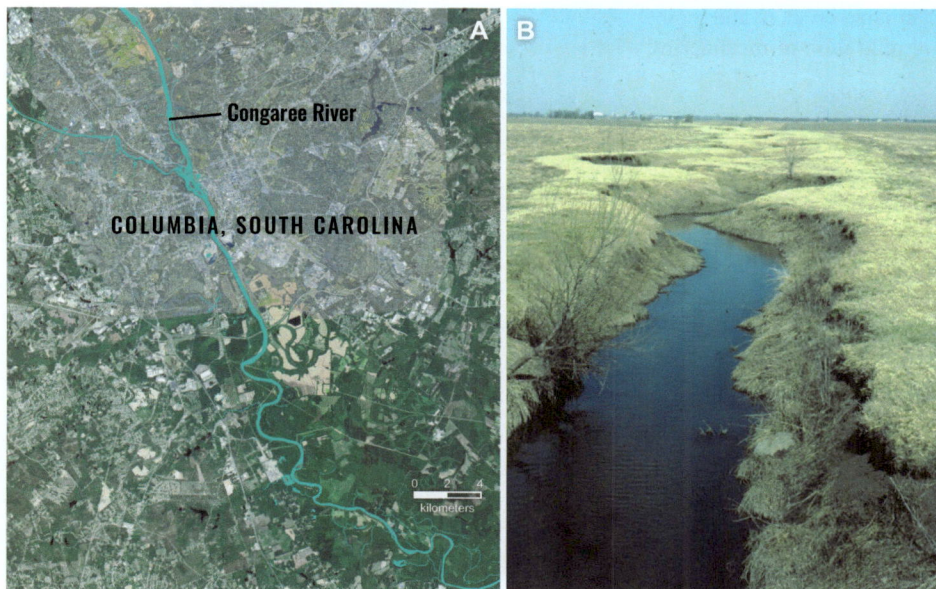

Figure 16.13 Straight channels. **A.** A satellite image of the Congaree River in South Carolina, USA, where it changes from a straight, bedrock-bedded channel (top half) to one that meanders within thick alluvium of the Atlantic Coastal Plain (bottom half). Source: NASA. **B.** A drainage ditch in central Illinois, USA, made straight originally, is reverting to a meandering pattern. Source: R. Schaetzl.

human disturbances dominate the system. For example, many drainage ditches in agricultural fields have been artificially straightened and are maintained that way (**Fig. 16.13B**). Straight channels can also form in bedrock valleys, where rock armors the bed and banks. Even if such streams have high values of stream power, they cannot disrupt the stable, bedrock channel.

Straight channels may look straight but many of them have submerged alternating bars that can migrate downstream. Thus, even in straight channels, the thalweg often meanders (**Fig. 16.10A**). Such channels may eventually become fully meandering over time.

16.3.5 Meandering Channels

Meandering channels have high sinuosities (**Figs. 16.10B, 16.11**). They are very common and, by far, the most studied of all the channel patterns. Meanders form as streams erode one side of the channel and deposit sediment on the other, which results in the formation of bends or meanders. In the fluvial system, channel meandering dissipates surplus energy and results in a more stable channel; a straight channel in the same setting would be steeper and more likely to erode. Recall that a meandering channel has a lower slope, and hence, lower stream power, than a straight channel of similar valley gradient and discharge.

In meandering channels, erosion is most pronounced on the outsides of the meander bends, where it forms steep **cutbanks**. Sediment deposition is focused on **point bars** on the inside of bends (**Figs. 16.10B, 16.14**). These processes are driven by centrifugal forces that push surface water toward the outside of the bend, where it then sinks and returns towards the bottom of the channel (**Fig. 16.14B**). This motion may result in a **helicoidal flow**, like a corkscrew, that pulls sediment from the stream bed upward and directs it onto the point bar. Helicoidal flow is often enhanced by eddies on the downstream side of point bars. As the thalweg crosses over from one bend to the next (**Fig. 16.10B**), the fastest flow velocities stay close to the cutbank, enhancing erosion there. Taken together, cutbank erosion and point bar deposition result in the lateral migration of meandering channels across the floodplain.

Meandering channels commonly have a sequence of **pools** separated by **riffles**. Pools are relatively deep, low-velocity features within the channel that tend to form near the outsides of meander bends, near the cutbanks (**Fig. 16.10C**). Riffles are shallow, higher-velocity zones underlain by coarser sediment that commonly occur where the thalweg crosses over from one side of the channel to the other, usually between meander bends (**Fig. 16.10C**). Many people would call these areas "rapids," because here the shallower water flows faster. In actuality, pools in pool–riffle sequences have lower mean flow velocities than riffles during low flows. However, this trend may reverse during high flows, when flow velocities may be faster in the pools.

Meandering channels range from (1) narrow, highly sinuous channels dominated by suspended load, to (2) mixed load channels that have wide, sandy point bars, to (3) those with islands, carrying more bedload (**Fig. 16.12**). These latter types of channels are transitional to braided channels.

Many distinctive landforms are commonly associated with meandering channels and their floodplains (**Fig. 16.15**).

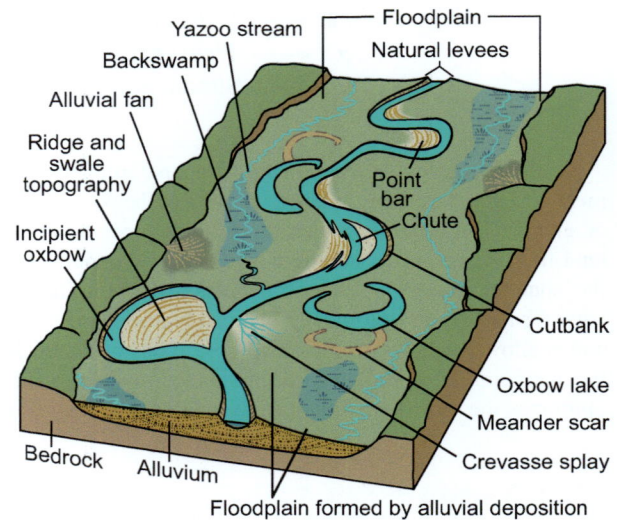

Figure 16.15 Typical landforms on the floodplain of a meandering river.

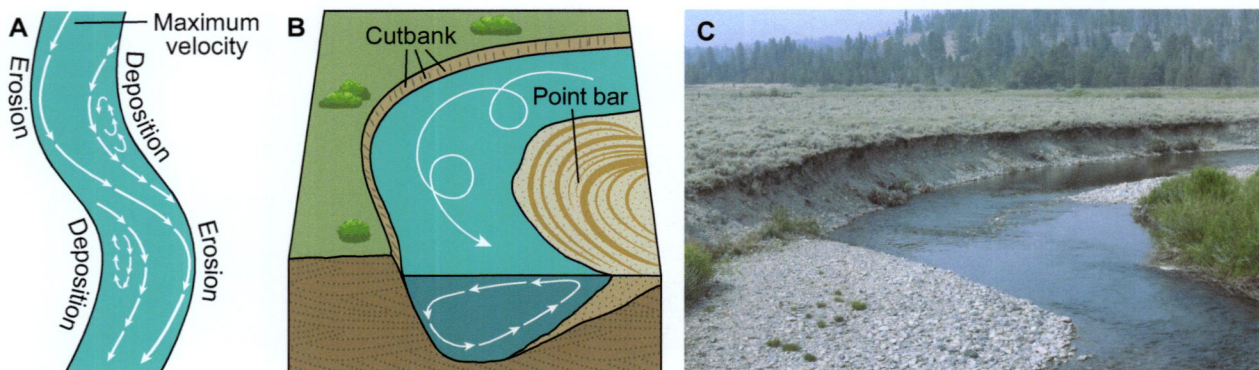

Figure 16.14 Features associated with meandering channels. **A.** Maximum flow velocities tend to occur on the outsides of bends. **B.** Currents at bends may develop helicoidal flows that concentrate erosion on the outsides of bends, and deposition on point bars on the insides of meander bends. **C.** View downstream of a meandering channel with a steep cutbank and alternating point bars. Source: A. James.

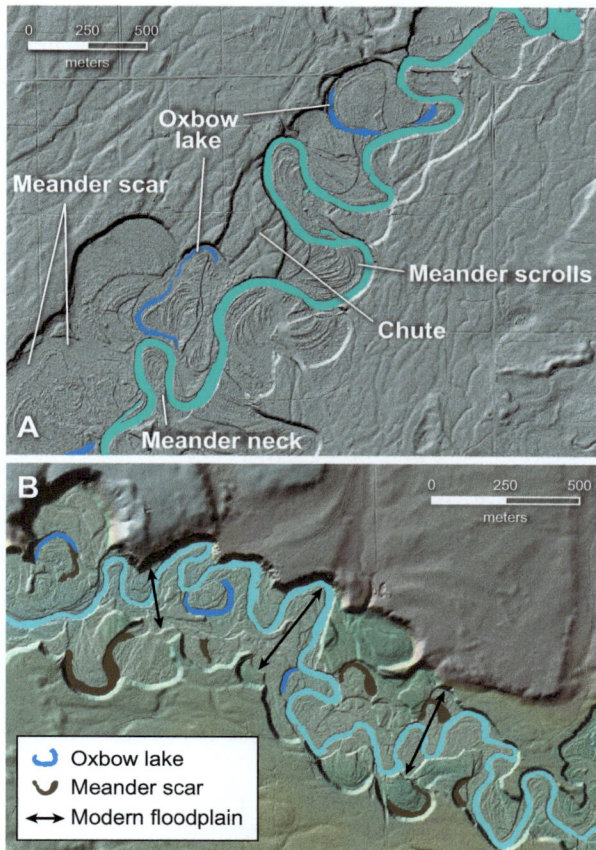

Figure 16.16 Typical features of the floodplains of meandering rivers, illustrated for the (A) Muskegon and (B) Pere Marquette Rivers in central Michigan, USA.

Obviously, cutbanks and point bars are widespread across such landscapes, but they have other landforms associated with them. For example, some point bars have arcuate ridges on their surface known as meander scrolls that represent individual episodes of point bar deposition (Fig. 16.16A). **Meander scrolls** are youngest immediately next to the channel, and become older farther away from it. Point bars with meander scrolls express **ridge and swale topography**. Point bar growth and development can be readily seen by careful inspection of this topography (Figs. 16.15, 16.16A). **Chutes** are secondary channels that cut across the surface of point bars and other low areas of the floodplain (Fig. 16.15). They are often dry during low flows but can carry large amounts of water during high flows. Floodwaters can enlarge chutes and convert the point bar that they cross into an island or a mid-channel bar. This process illustrates the young and ephemeral nature of many floodplain landforms – appearing to be stable, but potentially changing markedly during the next flood. **Yazoo streams** flow parallel to the main channel on the floodplain for some distance (Fig. 16.15). They are unable to cross the (higher) natural levees that flank the main channel. Only at some point downstream, usually where the main channel migrates over and intercepts the yazoo stream, can it enter the trunk stream. The name comes from the Yazoo River, in the US states of Mississippi and Louisiana. The Yazoo River flows parallel to the Mississippi River for 280 km, before entering the main channel.

Meander cutoffs are a common and important component of floodplain geomorphology. Lateral channel migration forms cutoff meanders by eroding into and through a thin **meander neck** (Fig. 16.17). After the neck is breached, the main flow is quickly diverted through the straight **reach** and abandons the meander loop, taking the steeper, straighter

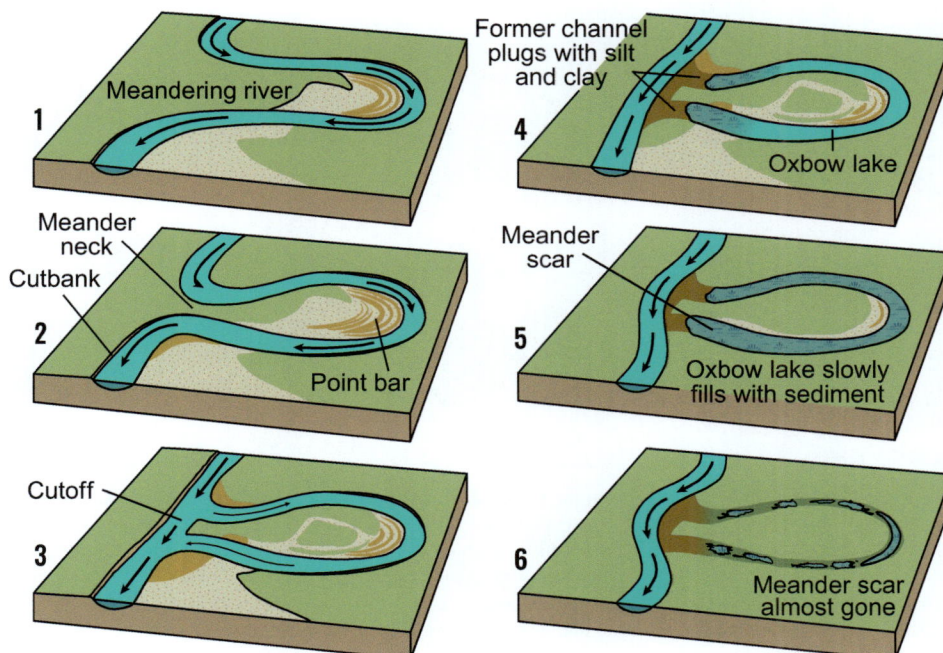

Figure 16.17 Diagrammatic sketch of the stages of a meander cutoff, leading to the formation of an oxbow lake. The lake eventually fills with sediment, forming a meander scar.

pathway down the valley. Sediment can then quickly block the former inlet and outlet to the meander bend. Isolated, cutoff meander loops formed in this way are called **oxbow lakes** (**Figs. 16.15**, **16.16**, **16.17**). These lakes eventually fill with sediment (primarily during floods) and organic matter (by ecological succession), to form arcuate, swampy depressions on floodplains known as **meander scars**. Oxbow lakes and their ever-shallowing successional phases provide important habitat diversity to the floodplain system. Meander cutoffs shorten flow paths and steepen the channel locally, which may accelerate erosion and deposition downstream until the system adjusts.

Historical maps, air photos, and detailed topographic data for floodplains often record the past positions of meandering river channels, revealing the complex nature of channel migration over time (**Fig. 16.18**). These maps can be important because political boundaries are often associated with rivers; the rivers migrate but the boundaries are permanent.

In some cases, channels migrate progressively across the floodplain, consuming all the floodplain sediments and landforms as they go. As seen in **Fig. 16.18A**, this situation occurred between 1765 and 1830 on the Mississippi River. In other cases, channels may jump from one position to another and leave intact areas behind, such as occurred at Moss Island between 1830 and 1881. This type of "sudden" change in channel position is known as a channel **avulsion**.

The geometry of meanders can be described using a variety of metrics, such as **amplitude**, **wavelength**, and **radius of curvature** (**Fig. 16.19**). These metrics may be highly variable for streams with irregularly shaped meanders, so multiple measurements are usually made. Then, the mean or median of the measurements is used to characterize the river. Meander geometry data like these are often correlated to bankfull discharge, which may allow estimates of past streamflows to be made from meander scars. For example, large paleomeanders of the Muskegon River in Michigan (USA), formed long ago, are still visible today (**Fig. 16.20**). This morphology confirms the vast size of the ancestral Muskegon River, when it carried glacial meltwater.

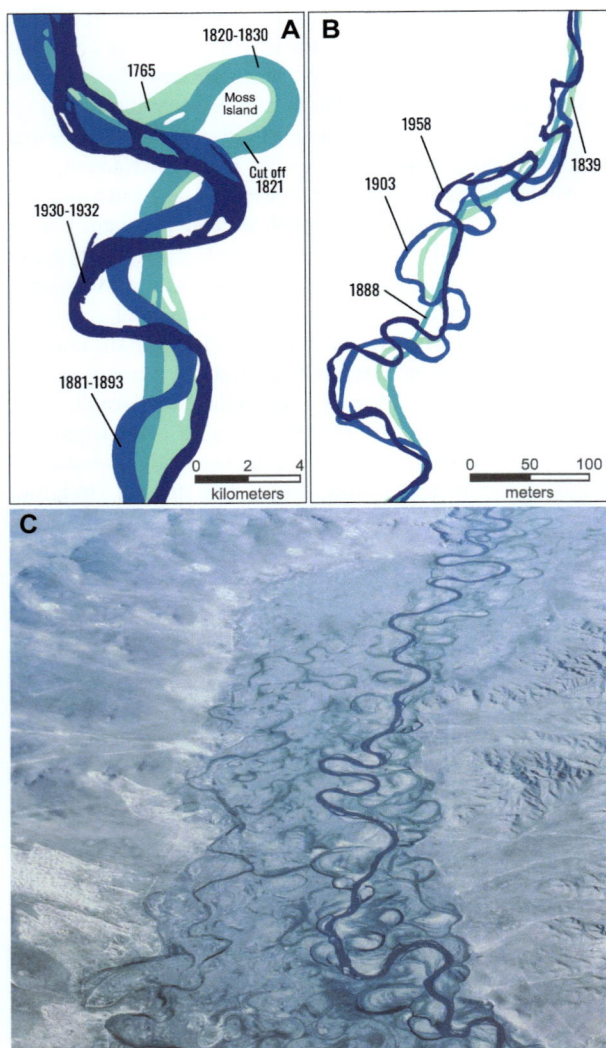

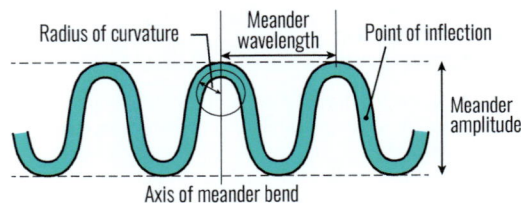

Figure 16.19 Terms associated with meander geometry, shown graphically.

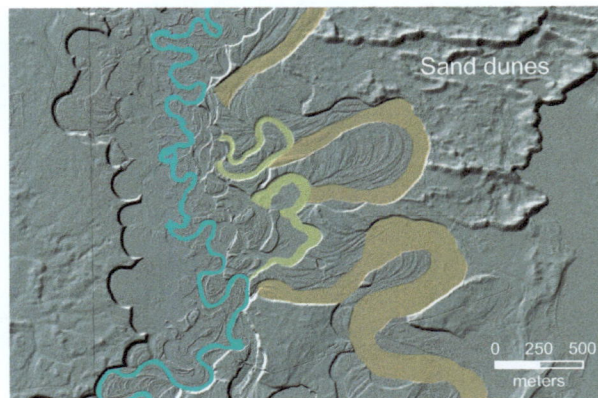

Figure 16.18 Meandering rivers change position through time. **A.** A map of the meanders of the Mississippi River in northern Tennessee from 1765 to 1932. Source: After Strahler, A. (1965), with permission from Wiley & Sons. **B.** A similar map of the River Sid in east Devon, UK from 1839–1958. Source: After Gregory, K. J. (1977), with permission from Wiley & Sons. **C.** Meanders and meander scars on the Tuul River, northern Mongolia, show the many positions the channel has occupied in the past. Source: M. Mergili.

Figure 16.20 Meanders on the Muskegon River floodplain in central Michigan, USA. Note the extremely large paleo-meanders (brown), which formed when the river had a much larger (glacially derived) discharge. The smaller meanders of the current river reflect its much lower discharge. This meander geometry suggests that the discharge of the river could have been 8× greater when the large meanders formed. Also note the intermediate suite of meanders (tan), which may reflect an intermediate stage in the fluvial system.

16.3.6 Braided Channels

The third type of channel (braided) is characterized by a network of small channels separated by sandy, mid-channel bars (**Fig. 16.21**). In most actively braiding systems, the bars are so ephemeral that they are largely free of vegetation. Braided channels form in relatively coarse sediment composed of sand and gravel, with wide, shallow cross-sections (high W/D ratios) and steep channel slopes. A hallmark of braided streams is that their discharges are usually highly variable; the term "flashy" is often used to describe these types of discharges. Together, these conditions are common in rivers that drain glaciers, in arid regions with variable flow regimes and coarse sediment, and in disturbed areas where large amounts of sediment fill channels. All of these situations involve deposition of large amounts of coarse sediment within the channel, leading to steeper overall gradients and increased stream power; both are necessary to transport the coarse bedload. Coarse-grained sediment is non-cohesive and does not usually form stable banks. Thus, frequent bank collapse delivers sediment to the channel, where it eventually gets stored as mid-channel bars.

Channels may transition from meandering to braided as the ratio of bedload to suspended load increases. During this transition, sandy point bars widen, chutes increase in number and size, and finally, numerous bars and islands form (**Fig. 16.12**). Ultimately, the entire channel becomes a series of shifting, coarse-grained, mid-channel bars known as **braid bars** that form a **braidplain** (**Fig. 16.21**).

The introduction of large amounts of coarse sediment to a meandering channel may cause it to change to a braided system, because the channel is receiving more sediment than it can currently transport. In this type of aggrading system, the excess sediment accumulates in the channel, and bars form. Conversely, if sediment loads decrease, or change such that the stream carries more suspended load and less bedload, the channel may revert back to a meandering system. This is a common sequence downstream of formerly glaciated regions, where valley bottoms were once filled with large amounts of coarse glacial outwash, and the systems were braided. After the glaciers melted and sediment loads declined, the channels converted to meandering systems (**Fig. 16.22**).

16.3.7 Anastomosing Channels

Anastomosing channels – the fourth major type of channel – are a broad series of interconnected channels (**Fig. 16.23**). In the past, anastomosing channels were grouped with braided channels, but we now recognize their important differences. Channel banks in an anastomosing system are usually more stable than in braided systems, and the islands are also not only more stable but also larger than bars in braided channels. Many anastomosing channels carry relatively high contents of silt and clay and are relatively narrow, whereas braided channels are usually wide, carrying mainly coarse-textured bedload. Anastomosing channels may be stabilized by vegetation on their banks, and are most common in low-gradient environments. Their channels are often formed by avulsions that shift the channel to an entirely new position. Although the term anastomosing describes the entire fluvial system, individual channels within it may be braided, meandering, or straight.

16.3.8 Distributary Fluvial Systems

Many fluvial systems have zones of sediment deposition, where channels frequently fill, and as a result, shift position over time. Alluvial fans form where this occurs on land (see Chapter 10) and deltas form where rivers flow into standing water (see Chapter 17). Alluvial fans are arcuate-shaped landforms, constructed dominantly by fluvial deposition (**Figs. 1.2, 1.3**). **Colluvial fans** are similar in form, but they accrue sediment mainly due to landslides, debris flows, and other hillslope processes. Many fans form by a combination of alluvial and colluvial episodes of sedimentation. All of these fans form where a channel exits a confined area (such as the mouth of a canyon) and flows out onto a flatter surface, where water can spread out, sink in, and slow down, resulting in channels that fill and shift through time (**Figs. 10.31, 10.32, 10.34, 10.35**).

Deltas develop where rivers flow into bodies of standing water. As the river meets the water body, channel cross-section areas increase substantially, and as a result, flow velocities and sediment transport capacities suddenly decrease, forcing deposition of sediment at the river mouth. Most deltas have an internal stratigraphy that contains three different types of layers, or beds (**Fig. 16.24**). This

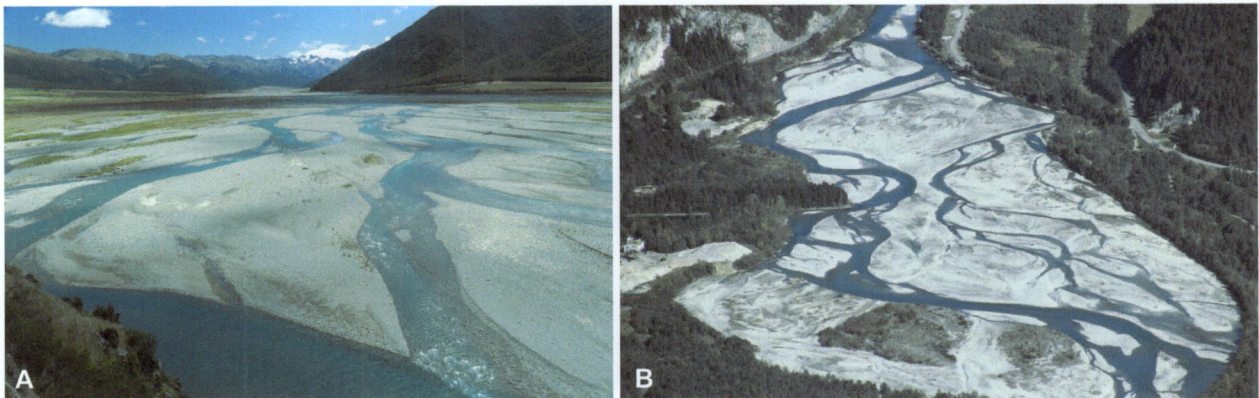

Figure 16.21 Braided streams and braidplains. **A.** A braided river in New Zealand that receives coarse sediment from a glacier upstream, and flows at a steep gradient. Source: M. Miller. **B.** Braided channels in the Lechtal Valley near Schartenberg, Austria. Source: M. Mergilli.

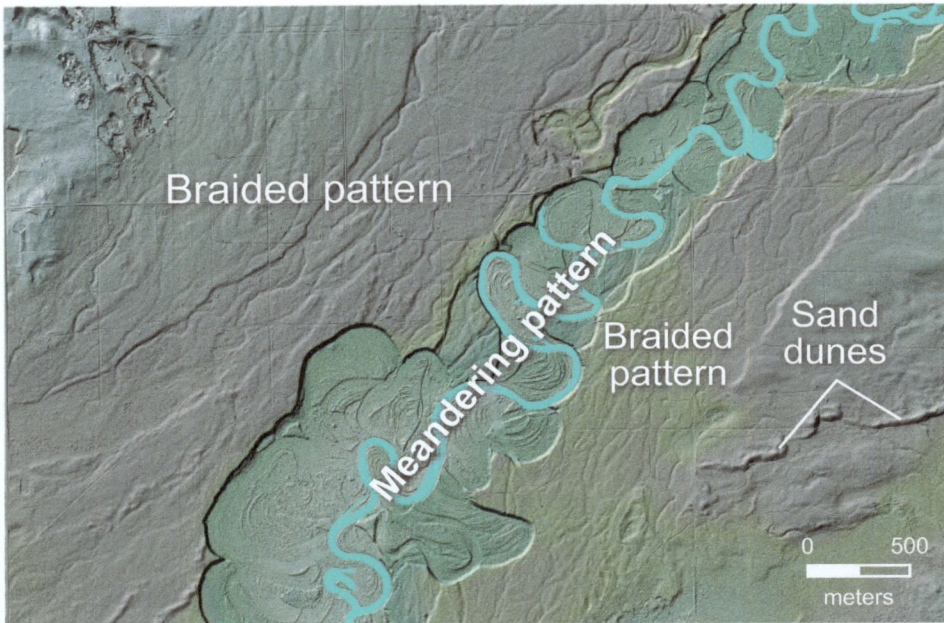

Figure 16.22 A shaded-relief map of a reach of the Muskegon River in central Michigan, USA. The high terraces of the river have a braided pattern that formed when the river carried large volumes of glacial meltwater and sandy glacial outwash. In the post-glacial period. Today, the river no longer has a meltwater source, and hence, carries much less water and mainly fine-textured sediment. Thus, a single channel, meandering pattern has formed.

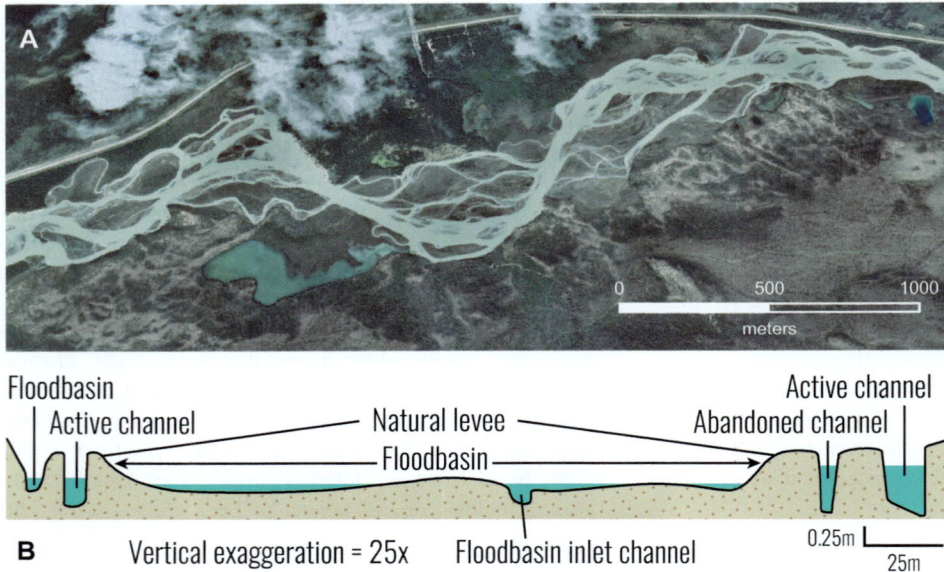

Figure 16.23 The anastomosing Alexandra River in British Columbia, Canada. **A.** A NASA image showing parts of the individual channels that alternate between braided and meandering conditions. Source: NASA. **B.** A diagrammatic cross-section of the Alexandra River, showing its both deep and shallow channels. Source: Used with permission of Elsevier, from Bart Makaske (2001); permission conveyed through Copyright Clearance Center, Inc.

conceptual model of delta sediments is known as a **Gilbert delta**; its three main components are listed below:

- **Bottomset beds** are composed of fine-grained suspended sediment deposited far out, in deep water. Bottomset beds later may be buried by the delta as it progrades into the water body, so they show up at the bottom of the delta's sedimentary sequence.
- **Foreset beds** are composed of dipping layers, formed as sediment spills over the steep delta front. Foreset beds define the expanding front of the delta.
- **Topset beds** are relatively flat-lying layers of coarse sediment, deposited on the top of the delta by distributary streams. Topset beds closely resemble the sediments in the bed of the upstream river.

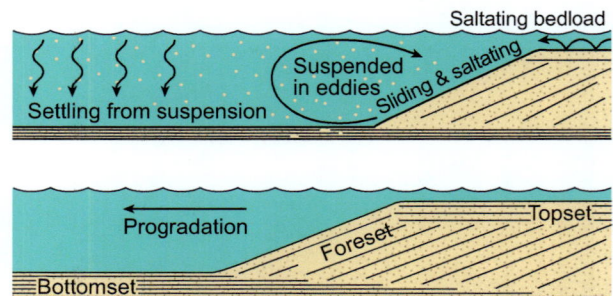

Figure 16.24 A schematic diagram of a Gilbert delta showing its growth over time, and its bottomset, foreset, and topset beds. See also **Fig. 17.38**.

16.4 REACH-SCALE FEATURES OF FLUVIAL CHANNELS

For practical reasons, geomorphologists often cannot study an entire river or drainage basin. Therefore, they may choose to study shorter segments called **channel reaches** – a local stretch or segment of a river channel with somewhat uniform characteristics. The length of a reach is typically between ≈10 to 40 channel widths. Landforms at the reach scale are important to an understanding of fluvial processes and aquatic habitats for stream assessments, restoration, or management.

16.4.1 Pool–Riffle Sequences

When examined on a reach-scale, river channels exhibit a variety of smaller features that often occur in a particular sequence (Table 16.2). As shown in Fig. 16.10C, pools are relatively deep parts of the channel, whereas riffles are relatively shallow-water areas. Water flows relatively slowly through pools during low-flow conditions. Pools form by a variety of processes, including scour by backwater eddies, lateral scour, plunge pools at the base of waterfalls, or as a consequence of damming by tributary fans, log jams, or other obstructions (Fig. 16.25A). **Step pools** are a special kind of

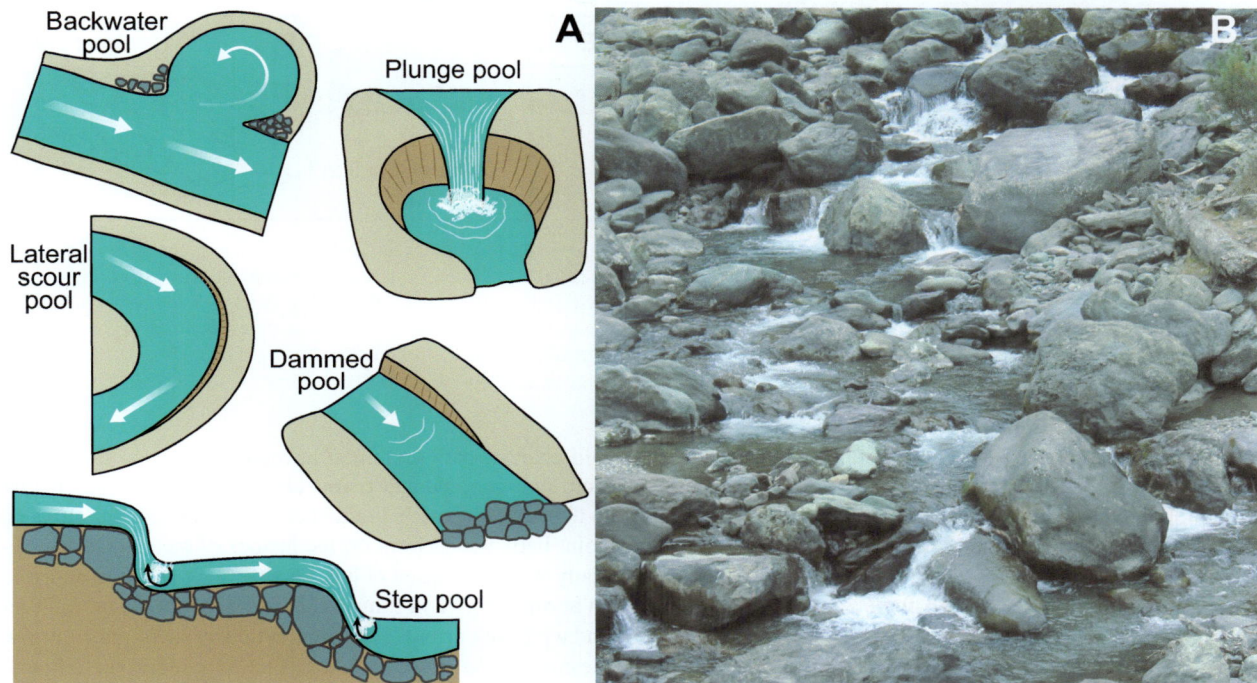

Figure 16.25 River or stream pools. **A.** A schematic diagram of the various types of pools in rivers. **B.** Step pools in a small mountain stream at Aru, Kashmir. Source: A. James.

Table 16.2 Some typical reach-scale features of streams

Feature	Characteristic
Pool	Deep portion of a river channel with a low water-surface slope
Glide	Downstream segment of a pool, where the flow accelerates as it transitions to a shallow riffle downstream
Riffle	A shallow, high-velocity feature that develops at thalweg crossovers on sand- or gravel-bed streams, and is characterized by small-scale turbulence
Run	Downstream part of a riffle, where the flow slows down as it transitions to the pool immediately downstream
Step pools	A sequence of small, evenly spaced falls and plunge pools in the channel
Rapids	Generic term for any zone of swift flow in a stream or river; in geomorphology, the term may apply to riffles or to any steep, rocky stretch in a canyon

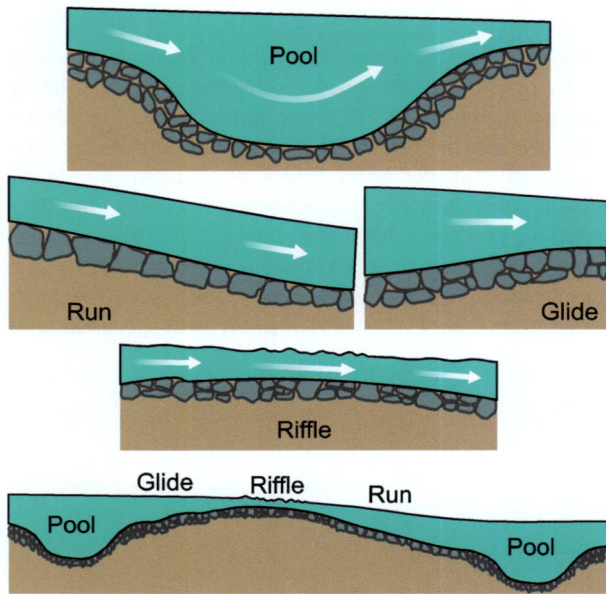

Figure 16.26 Pool–riffle sequences may form a sequence from pool, to glide, to riffle, to run, and to the next pool, as shown in this collage of channel cross-sections.

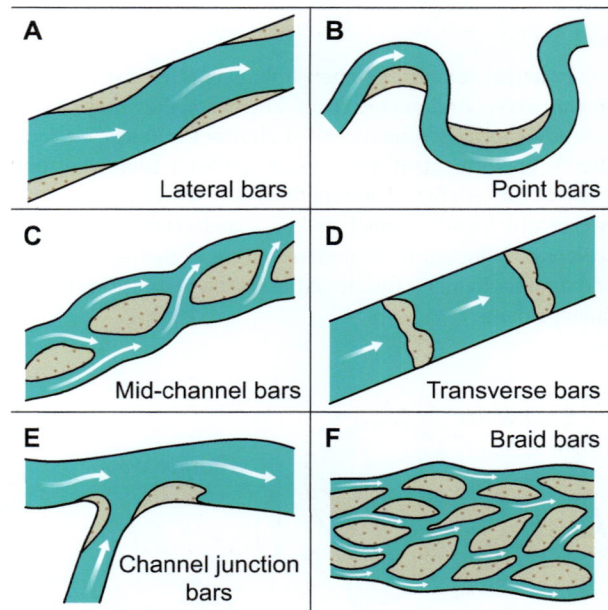

Figure 16.27 Types of sand and gravel bars in stream channels. **A.** Lateral bars along the sides of a channel. **B.** Point bars on the insides of meander bends. **C.** Mid-channel bars are like islands, but lower, not densely vegetated, and less stable. **D.** Transverse bars, formed perpendicular to the flow. **E.** Channel junction bars near a tributary confluence. **F.** Braid bars typical of braided channels.

pool. They represent sequences of evenly spaced undulations that alternate between steps (small waterfalls) and plunge pools, where waterfalls have deepened the channel bed (**Fig. 16.25B**). Step pools are defined by an underlying structure of large cobbles, boulders, or woody material. They form in small, relatively steep, coarse-grained streams with gradients > 2°, such as small mountain tributaries. Step pools tend to be evenly spaced, at intervals that are adjusted to stream power during high flows, and to the size of bed material. Coarser materials that can only be moved by high-energy flows generate higher step pools with longer wavelengths.

Gravel- and cobble-bed streams with pool-and-riffle sequences may also include runs and glides (**Fig. 16.26, Table 16.2**). **Glides** are zones of low turbulence in the downstream parts of pools, where the channel bed may even slope upstream. They often provide gravelly spawning habitat for fish. Looking down onto a glide, one may see a "V" in the pool tailwater, as the flow speeds up and approaches the turbulent downstream riffle. **Runs** are transitions between the turbulent riffles upstream and the placid pools downstream. Within runs, the flow slows down and tends to be well aerated by the riffle upstream.

16.4.2 Fluvial Bars and Bedforms

At an even finer scale within a channel reach, small topographic features may form on the bed, where they may impact local flows, habitats, and sediment storage. Sand and gravel bars are typical examples of such fine-scale features, constructed of non-cohesive sediment. Bars may take many forms, often falling into six general categories (**Fig. 16.27**). For example, **lateral bars** form along the channel margins,

mid-channel bars (common in braided streams) form away from the margins, and **transverse bars** form perpendicular to flows. Sand bars in braided channels are called braid bars. Point bars, which form on the insides of meanders, have already been discussed (**Figs. 16.14B, 16.17**). In response to an abrupt change in transport capacity or to eddies and secondary flow currents, **channel junction bars** may form at the confluence of two channels. They are important to fluvial geomorphology because they redirect flow directions, lead to habitat diversity, and help to store sediment that can be recruited by the stream during high, erosive flows. Bars can also be diagnostic of the amount and patterns of flow energy exerted by past floods.

Sand-bed streams may have small, fine-scale individual features on the bed, called **bedforms** (**Fig. 16.28**). For example, **ripples** are wave-like bedforms on a sandy stream bed. Ripples migrate downstream when sand eroded from the upstream side of a ripple saltates across the ripple and spills down the steep lee side, just as sand dunes do in the wind (**Fig. 21.21**). **Dunes** are similar wave-like forms on the stream bed that also migrate downstream, but are larger than ripples and move in deeper water. Both bedforms may be present in any given stream, with ripples migrating across the tops of dunes during low flows or in shallower parts of the channel. **Antidunes** are different in that they migrate *upstream*, as the water erodes the downstream side and deposits sand on the upstream side. Antidunes indicate high-energy flows (fast currents).

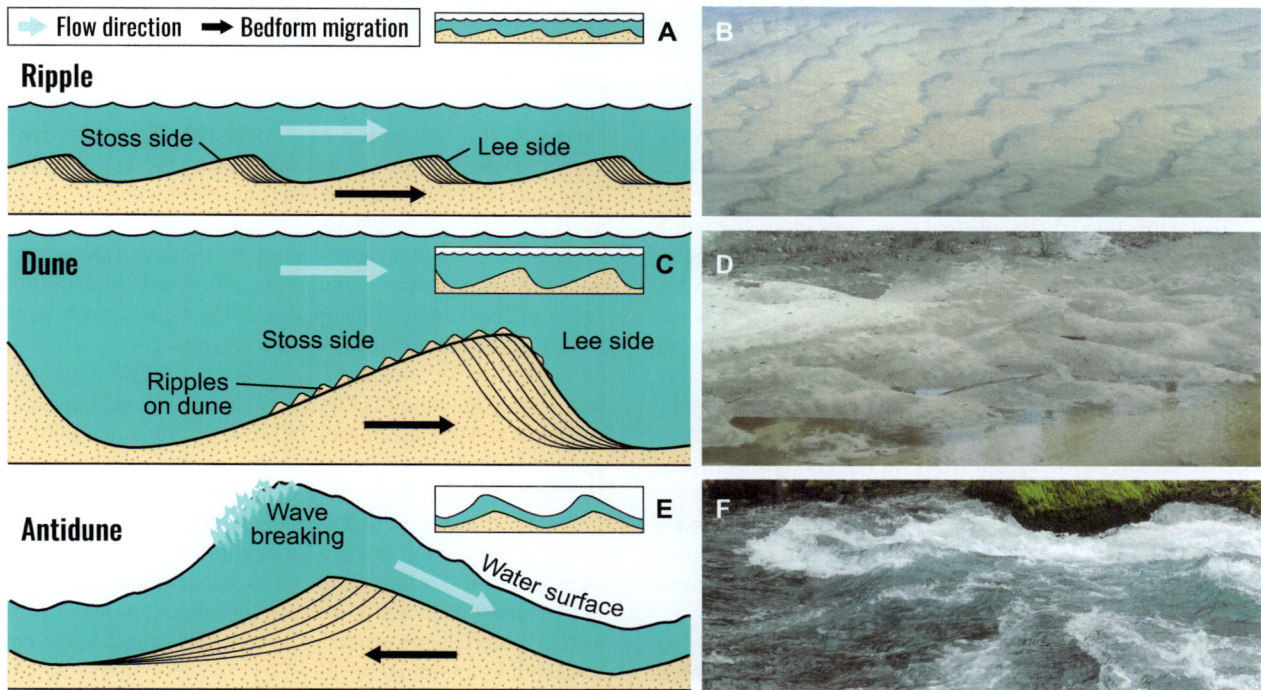

Figure 16.28 Fluvial bedforms. The flow is left to right in all examples. **A.** Ripples are small, crested forms that migrate downstream by growing on their lee side. **B.** Ripples in a sand-bed stream. Source: Michael C. Rygel, CC BY-SA 3.0, via Wikimedia Commons. **C.** Dunes are similar to ripples but larger. **D.** Dunes at a channel margin spaced ≈30–80 cm apart. Source: A. James. **E.** Antidunes form in high-velocity flows and may migrate upstream by growth on the upstream side. The water surface may even exhibit standing or breaking waves above the crest of each antidune. Note that, unlike ripples and dunes, the water surface over antidunes conforms approximately to the bedform shape. **F.** Breaking waves may form over bedrock, coarse boulders, or antidunes in sand or gravel. Source: A. James.

16.5 LONG-TERM VERTICAL ADJUSTMENTS IN FLUVIAL SYSTEMS

River channels are constantly changing. Recognizing such changes in channels is key to understanding how fluvial systems have evolved over geologic time. This information can then shed light on past environmental changes. This section focuses on the vertical adjustments that streams make, which often leave distinctive landforms as evidence.

16.5.1 Channel Gradients, Longitudinal Profiles, and Knickpoints

The **longitudinal profile** of a river is a graphical representation of its elevation, from its source to its mouth (end). Most channel gradients change from steep in headwater areas to more gradual in the downstream reaches, which leads to a concave-upward longitudinal profile (**Fig. 16.29**). Smaller streams (and headwater areas of larger rivers) have steeper gradients because they have relatively small discharges,

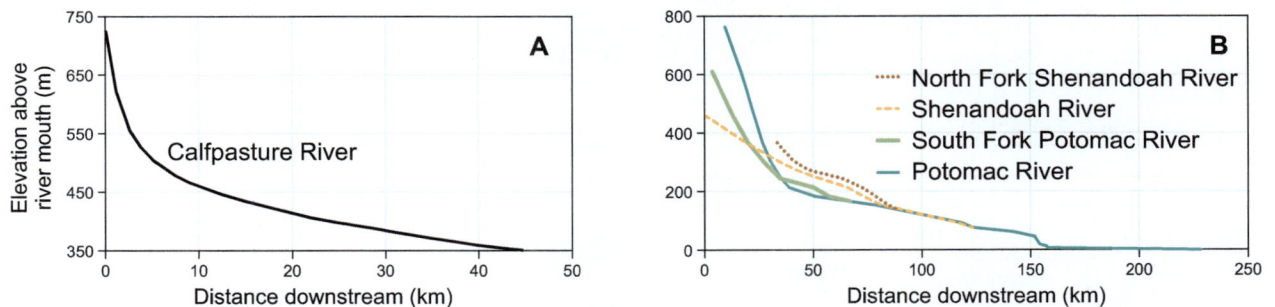

Figure 16.29 Typical longitudinal profiles. Note that, for all these rivers, the headwater reaches have much steeper gradients. **A.** The Calfpasture River shows a remarkably smooth, concave-upward profile. Source: After Hack (1957), via the US Geological Survey; in public domain. **B.** The longitudinal profiles of several other, large rivers. Note that the Potomac River (including its South Fork) has a distinct change in slope in its downstream reach, and a knickpoint nearer to its mouth. Source: After Gannett (1901), via the US Geological Society; in public domain.

so steep gradients are needed to generate enough power to carry their sediment load. Small channels also tend to carry coarser sediment, which also necessitates steep gradients. Farther downstream, stream loads become dominated by suspended sediment, and overall discharge values increase. Thus, the river has ample power, even at a lower slope, to carry its sediment, because of its increased discharge.

If streams flowed at a constant slope, stream power would increase downstream because discharge increases. Steep channels in the downstream reaches, with all that excess power, would be unstable. In such a case, the river would use that surplus power to erode its banks, producing a sinuous channel and, therefore, lowering its gradient. Therefore, steep gradients in large rivers are rare; they are not stable unless channel banks and beds are armored with bedrock, coarse sediment, or engineering works. So, to conclude, small streams carrying coarse sediment in upstream reaches need steeper slopes to carry their sediment load. Conversely, large rivers (with lower overall slopes) are efficient at transporting their sediment loads even at low gradients, because of their higher power values.

Abrupt topographic changes or breaks in a stream's longitudinal profile, such as a waterfall, are known as **knickpoints**. Knickpoints disrupt the smooth longitudinal profile of a river (**Figs. 16.29B, 16.30**). Streams will slowly eliminate these knickpoints, as they attempt to once again achieve

a smoother longitudinal profile. This adjustment does not begin everywhere at once. Instead, it starts at each knickpoint and propagates upstream (**Fig. 16.30C**). The time required to smooth out the profile can be very long, especially for rivers flowing on resistant substrates. Therefore, rivers can have one or more knickpoint "perturbations" along their longitudinal profile at any given time, all of which are slowly working their way upstream.

Knickpoints have many possible causes. Among them are faults, resistant rock outcrops, sediment inputs by tributaries or landslides, volcanic dams, log steps, beaver dams, armoring, or (commonly) a drop in base level (**Fig. 16.30A**). In layered rocks, a caprock (resistant top layer) may maintain a steep knickpoint (**Fig. 16.30C**). Slowly, the caprock is undercut and chunks of rock fall from it. But knickpoint migration is again slowed as a rock-armored plunge pool forms on the rocks that have broken off the retreating caprock. Only when abrasion and erosion remove the coarse rocks in the plunge pool can the knickpoint continue its migration upstream. Through time, knickpoints migrate upstream, get smaller/lower, and eventually become eliminated as resistant materials are eroded, allowing the stream to develop a smooth longitudinal profile.

Changes in channel **sinuosity** are directly related to channel gradient and also to a stream's longitudinal profile. For example, sinuous channels have lower gradients than

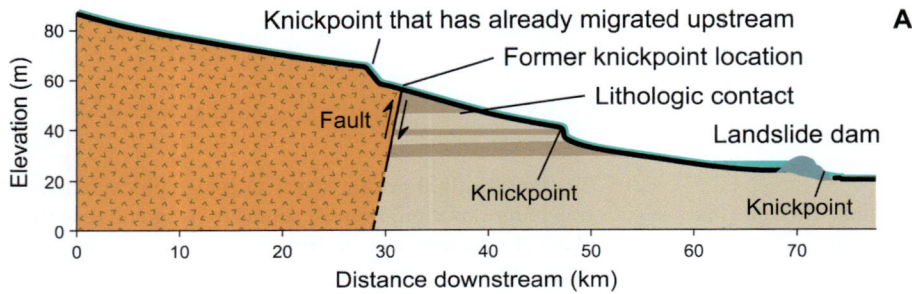

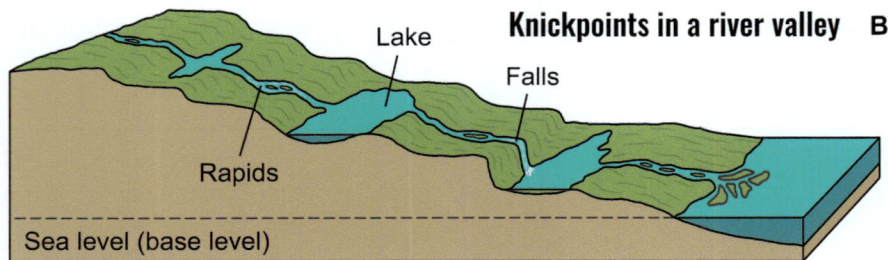

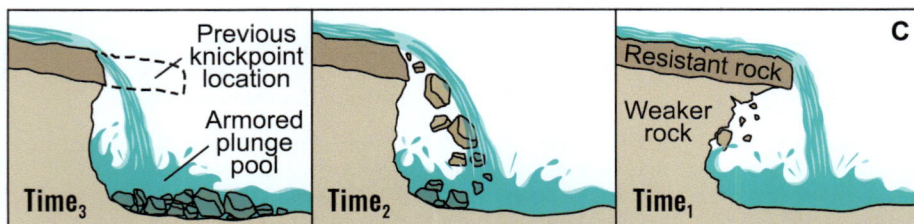

Figure 16.30 Longitudinal profiles and knickpoints. **A.** Knickpoints can be formed in different ways. Shown here are some typical causes: a fault (with upstream migration of that knickpoint), a resistant rock layer, and a landslide dam. Commonly, knickpoints also originate as the base level of a stream is quickly lowered. **B.** Knickpoints often exhibit waterfalls or rapids and may have lakes or marshes upstream. Source: © A. N. Strahler (1992). Used by permission. **C.** One method by which knickpoints retreat is the undercutting of a resistant caprock. Armoring the plunge pool with collapsed boulders may slow the rate of knickpoint retreat.

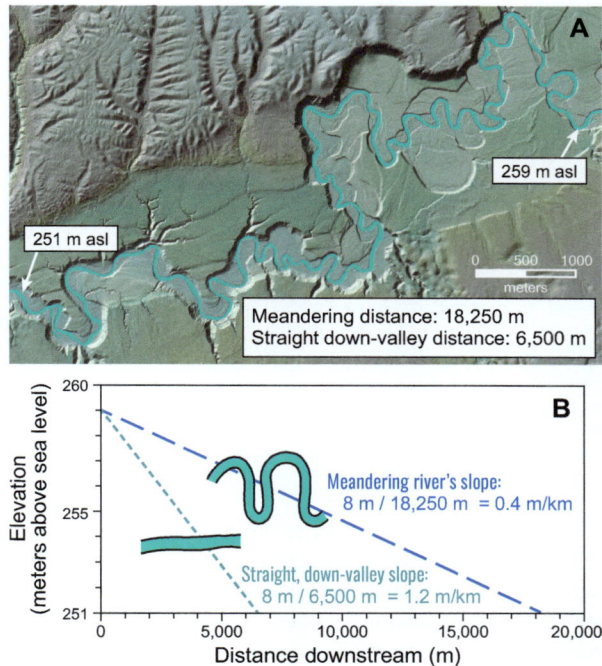

Figure 16.31 Illustrations showing that meandering rivers have lower gradients than do straighter rivers, other things being equal. **A.** A map view of the highly sinuous Manistee River in northern Michigan, USA. **B.** Elevation and distance data for the Manistee River along this reach illustrate how a straighter channel would flow over a shorter river distance and at a steeper gradient. Increases in sinuosity are a way that river channels decrease stream power and stabilize.

straighter channels in the same valley (**Fig. 16.31**). Changes in sinuosity over long stretches of a river can therefore result in substantial changes in the stream's longitudinal profile. In short, as rivers get more sinuous, their slopes are lowered, and so is their power.

Streams can and do adjust their sinuosity, as conditions of sediment load and stream power change. For example, a meander cutoff – formed either by natural or engineered processes – will shorten the path of the river and steepen its slope. When this happens, stream power increases through the steepened reach, facilitating channel erosion. Small knickpoints may even form at meander cutoffs and migrate slowly upstream. Streams that have excess power may use that power to meander more markedly by eroding their banks. The change to a more meandering pattern lowers their slope and diminishes their overall power. Such adjustments are far easier to accomplish in alluvial channels than in bedrock channels.

16.5.2 Terraces in River Systems

Channel aggradation and/or incision are natural parts of the history of many fluvial systems. Incision into alluvium often isolates the former alluvial floodplain above the incised, lower section (**Fig. 16.32**). These higher segments of the stream valley are called **alluvial terraces** (or **fill terraces**). They form when a river incises into its own floodplain (and

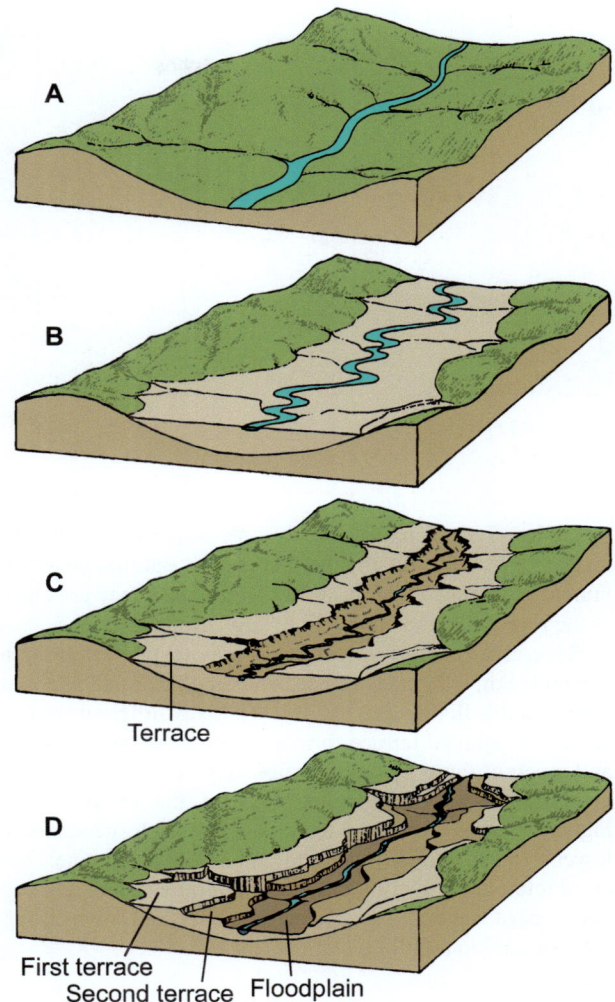

Figure 16.32 Formation of terraces may start (**A**) with a valley without much alluvium. The valley fills with alluvium (**B**), forming a broad floodplain. Due to some kind of change in the fluvial system, the river later incises into the alluvium to (**C**) form a terrace. Further incision (**D**) may form a second, lower terrace. Source: Adapted from Hamblin, W. K. (1978).

the alluvial fill it contains) and abandons it, forming a new floodplain surface at a lower elevation. Like all stepped surfaces, terraces consist of a **tread**, which is the relatively flat, former floodplain surface that forms the top of the terrace, and a scarp (or **riser**), which is the steep edge of the terrace that represents a former cutbank or channel margin (**Fig. 16.33**). Thus most terrace histories can be traced back to a fill event, which formed the tread, and a cut event, represented by the riser. For this reason, geomorphologists typically interpret terraces as evidence of past fill-and-cut events.

Alluvial terraces provide clear evidence of vertical channel cutting, or incision – either ongoing or at some time in the past. If extensive, terrace treads may allow for a reconstruction of the longitudinal profile of the former floodplain, which then can allow for insights into past channel

Figure 16.33 Alluvial terraces. **A.** Multiple terraces with high scarps and broad treads in the Altai Mountains, Russia. Source: Heljqfy Alexei Rudoy, CC BY-SA 3.0, via Wikimedia Commons. **B.** Terraces along the Zarafshan River in Tajikistan. Note the highest terrace at the arrow. Source: M. Mergilli.

dynamics. Terrace treads are not only important as indicators of environmental change, but also as optimal areas for human settlement and transportation, because they are close to the river but on relatively flat, "high ground."

Several terraces of varying ages may be present along a stream reach, with each terrace providing information about a prior, stable floodplain (**Fig. 16.33**). Many different combinations of alluvial terrace formation processes are possible, and much of these fluvial histories may be preserved in the geomorphology and sedimentology of the terraces. The highest terraces are always the oldest, and each successively lower tread is younger than the one above it. Channels incised into older alluvium may form multiple sets of terraces that record several subsequent episodes of downcutting without additional filling (**Fig. 16.34**).

Past climatic, tectonic, or anthropogenic events can be inferred from the longitudinal profiles preserved by terrace remnants. For example, the gradient of the terrace treads may be different than the modern channel, pointing to past environmental changes (**Fig. 16.35**). Terraces can be also buried by subsequent channel aggradation, so buried terraces may emerge from the modern floodplain upstream or downstream. This situation is common for rivers that enter the ocean. The substantial rise in global sea levels since the last glaciation has caused the gradients of modern streams approaching the coast to be less steep than their predecessors during the glacial period. Together, the lower modern channel gradients and rising sea levels led to burial of the lower reaches of these rivers and their floodplains. Away from the coast, these former floodplains may be represented

Alluvial terraces, paired

A1 One terrace, one fill.

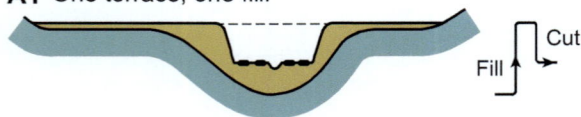

A2 Two terraces, three fills. Channel is in alluvium.

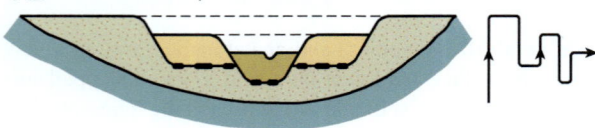

A3 Two terraces, three fills.

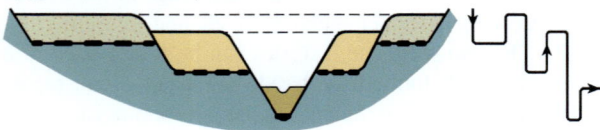

Alluvial fills: ▮ Young ▮ Medium ▮ Old
- - - Erosional surface ▮ Bedrock

B Alluvial terraces, unpaired

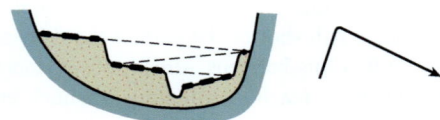

C Strath terraces, paired

Two strath terraces, no fills. Channel is in bedrock.

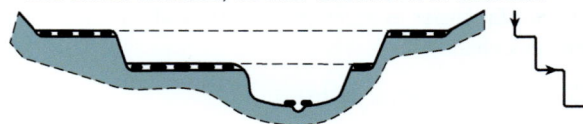

Figure 16.34 Examples of terrace types and their subsurface stratigraphy. Both the thin and thick dashed lines represent former floodplain surface elevations. Schematics of the timing of the sequences (lines at the right, with arrowheads) indicate the maximum and minimum levels of fill and cut episodes. The modern channel is shown as a very small notch. **A.** Three different scenarios for terrace fill and cut sequences. **B.** Unpaired terraces represent steady, persistent rates of channel incision, rather than episodic incision separated by periods of stability. **C.** Strath terraces are also formed by lateral channel erosion and incision, but unlike alluvial terraces, they are cut into bedrock. Source: Modified from Chorley et al. (1984) and Zonneveld, J. I. S. (1975).

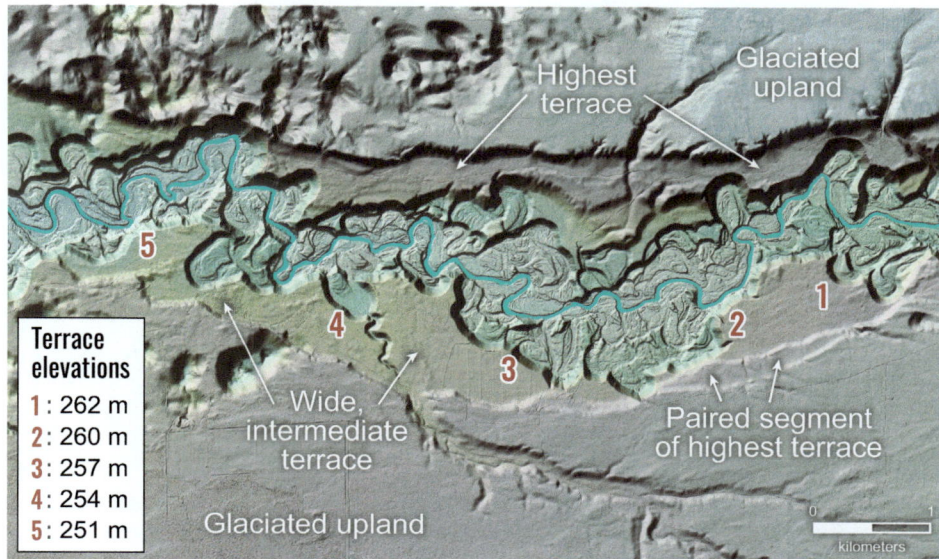

Figure 16.35 A complex suite of alluvial terraces on the Pine River in northern Michigan (USA). The river is incising into easily erodible, fine sands, deposited in a former glacial lake. Base level lowering since deglaciation drove the river to incise, forming the terraces. Note that the terrace elevations decrease downstream (right to left) – indicating the gradient of the former floodplain.

by terraces that can be traced downstream, as they dive into the subsurface.

Terraces may be paired or unpaired across the valley (**Fig. 16.34**). **Paired terraces** occur where terrace treads are at equal elevations on both sides of the valley. This situation indicates a fairly long period of floodplain stability, followed by an episode of downcutting. During the stable phase, the floodplain widened across the valley floor. During the cutting phase, the river incised into that floodplain, leaving behind a set of paired terraces. **Unpaired terraces** occur at different elevations on each valley side, usually indicating a more continuous degradation of the valley without periods of stability. In other words, the channel downcut rapidly even as it was migrating from side to side. Identification and interpretation of paired vs unpaired terraces can significantly help to understand the history of a fluvial system.

Strath terraces differ from alluvial terraces, which by definition are composed of alluvium. Strath terraces are cut into bedrock (**Fig. 16.36**). The treads of strath terraces formed during a long period of stability in the fluvial system, during which the river was neither cutting down nor aggrading. Instead, the channel was migrating laterally, cutting a broad "bench" or **strath** into the bedrock. Later, the river underwent a period of incision, abandoning the bedrock-floored floodplain and leaving the strath behind as a terrace.

16.5.3 Incised Meanders and Natural Bridges

Some rivers incise downward so rapidly that their channels stay generally fixed in one location, and if that channel initially had a meandering form, the result is a series of **incised meanders** (**Fig. 16.37**). This form is best preserved when the channel is cut into bedrock, which helps preserve the previous channel shape within deep, narrow, often sinuous, bedrock canyons. Incised meanders imply rapid vertical channel downcutting, often induced by tectonic uplift or lowered base level. Channels cannot normally form meanders in bedrock valleys, implying that the incised meanders had previously formed within alluvial channels at some higher elevation, above the currently exposed bedrock. Later, as the river incised downward into the bedrock, the meanders were superposed – or "let down."

Incised meanders come in two types (**Fig. 16.37A**). **Entrenched meanders** have steep, almost symmetrical, valley cross-sections. This morphology indicates that, during downcutting, lateral channel migration was negligible, and incision was almost entirely vertical. Downcutting must, therefore, have been quite rapid. In contrast, **ingrown meanders** have steep outer bend walls but relatively gentle slopes on the insides of bends, almost like steeply sloping point bars. This morphology indicates that lateral migration and an increase in meander amplitude occurred during the incision period. Entrenched meanders may have had steeper channel gradients and thus, may have incised more rapidly

Figure 16.36 Strath terraces of the Río Paine River, cut into bedrock, in the Torres del Paine National Park, Chile. Source: M. Mergili.

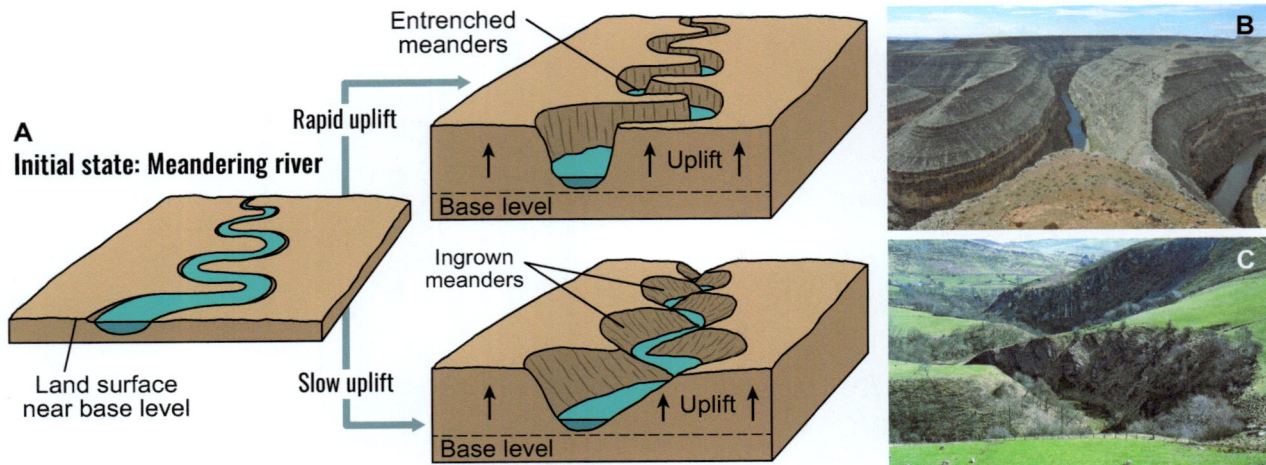

Figure 16.37 Incised meanders. **A.** Block diagrams illustrating a meandering channel before incision, and its two possible responses to subsequent downcutting. **B.** The famous goosenecks of the San Juan River in Utah, USA, are a classic example of entrenched meanders formed due to rapid tectonic uplift of the surrounding landscape. Note the steep, nearly vertical canyon walls. Photo from the Water Desk of the University of Colorado-Boulder. Source: Finetooth, CC BY-SA 3.0, via Wikimedia Commons **C.** Ingrown meanders on the Green River, near Moab, Utah. Note the gentler slopes on the insides of the meanders, formed as the river migrated laterally while it also (more slowly) cut down. Source: © Mitch Tobin/The Water Desk, with aerial support provided by LightHawk.

than ingrown meanders, but both types of incised meanders can sometimes occur in the same region.

In some settings, fluvial incision into bedrock can form **natural bridges**. These features are usually formed as rivers breach an incised meander neck that had formed in bedrock (**Fig. 16.38A**). Natural bridges should not be confused with **arches**, which are not formed by rivers but by weathering of bedrock (see Chapter 8) or by coastal erosion (see Chapter 17). For a natural bridge to form, some degree of channel lateral migration or widening must occur, to undercut the bedrock of a meander neck, forming the opening beneath the bridge. Natural bridges are relatively uncommon, because specific conditions must exist for them to form. Rainbow Bridge in Utah (USA) is the largest known natural bridge, at 84 m high and spanning 88 m long (**Fig. 16.38B**). It was formed at a meander neck cutoff through sandstone. Although the site had previously been difficult to reach, Rainbow Bridge is now easily accessible by boat from Lake Powell.

16.5.4 Transverse Valleys, Antecedence, and Superpositioning

As the name implies, **transverse valleys** occur where rivers cut across bedrock ridges or complex geologic structures. How some rivers do this and form transverse drainages has been a longstanding question for geomorphologists. Rivers would normally seek the least resistant path and flow *around* a geologic structure, on the surrounding softer rocks. Nonetheless, many large rivers flow in gorges that directly cut through bedrock uplands, such as anticlinal and synclinal ridges. Excellent examples occur in the Appalachian Mountains of the United States, as well as the Zagros Mountains of Iran.

A number of processes have been proposed to explain the formation of transverse valleys. Here, we describe two of them: antecedence and superposition. Interpretations of regional landscape evolution over geologic time differ, depending upon which of these two processes is invoked to explain it.

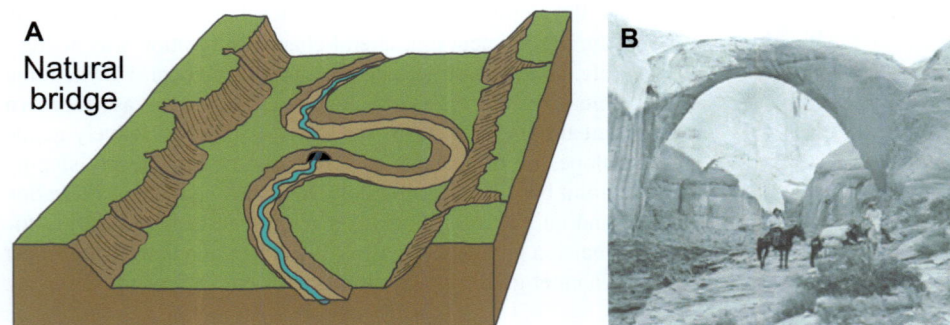

Figure 16.38 Natural bridges. **A.** An idealized view of the formation of a natural bridge, as a river meanders back against its own valley, cutting a window into bedrock. **B.** Rainbow Bridge in 1938. Source: B. W. James II.

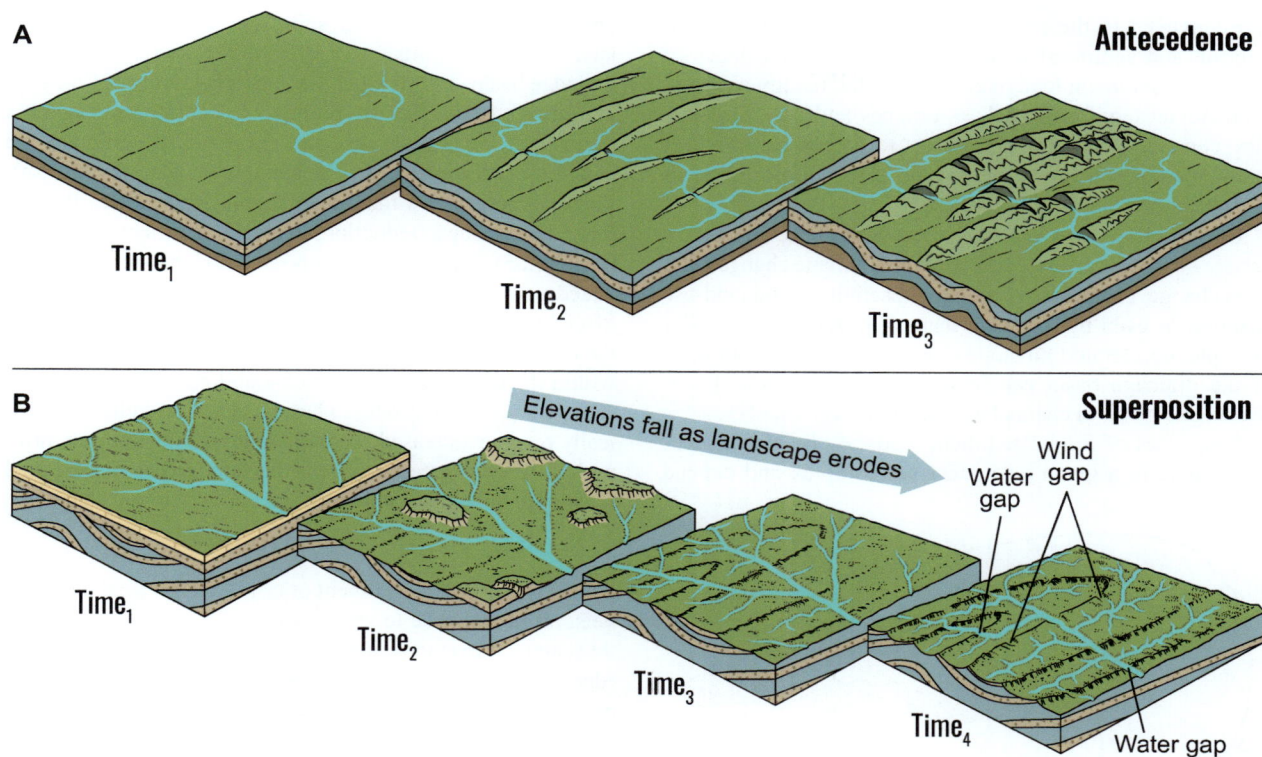

Figure 16.39 Antecedence vs superposition. **A.** The development of transverse drainage by antecedence. During Time$_1$, the stream flows in a seemingly random path before the geologic structures in the subsurface are exposed. At Time$_2$, the path of the river is maintained, even as the geologic structures become exposed and grow larger. By Time$_3$, the growing structures have formed large ridges. Note that the master channel hardly changed its position on the landscape, and is older than the structures. Source: After Selby (1985), who redrew it from Oberlander (1965). **B.** The development of drainage by superposition. During Time$_1$, the stream flows on relatively soft, flat-lying rocks that overlie buried geologic structures below, in harder rocks. Over time, the channel pattern is slowly impacted by the underlying harder rocks, as they become exposed by removal of the flat-lying rocks above. Ultimately, many of the river channels are "diverted" around some of the harder rocks, into (in this case) a trellis pattern. The large channels are able to maintain their positions but their smaller tributaries are unable to keep pace with the emergence of the structures from below, and become diverted or captured. Source: After Hamblin (1978).

1. An **antecedent river** is older than the bedrock structures that it cuts through, that is, the river existed before the structures formed. The river maintains its relative position during the slow events of folding and faulting (**Fig. 16.39A**).

2. A **superposed river** develops in overlying materials and then incises down onto older, preexisting bedrock structures below; that is, a superposed river is younger than the geologic structures that it cuts through (**Fig. 16.39B**). This process usually requires a long period of widespread erosion across large parts of the landscape. Typically, deeper bedrock structures were overlain by softer rocks in which the channel initially developed. The river then erodes down onto the older structures, but slowly enough that it is able to keep flowing in the same location through narrow canyons in the ridges known as **water gaps**. If regional lowering occurs rapidly, some channels may not be able to keep up and could get cut off. Abandoned water gaps that lost their rivers and were left high and dry are referred to as **wind gaps** (see Chapter 9). The largest channels of the superposed streams are able to maintain their positions, but their smaller tributaries are not, and so they become diverted and/or captured.

16.6 FLUVIAL RESPONSES TO CHANGES IN CLIMATE AND LAND USE

In addition to long-term vertical channel changes, rivers change in a variety of other ways, sometimes relatively rapidly, for example, rapid channel **aggradation–degradation episodes** (filling, then downcutting). Geomorphologists can use information from fluvial sediments and landforms to reconstruct these former environmental conditions and changes. For example, terraces may represent evidence of responses to climate change, or perhaps may be due to anthropogenic causes that drive land-use change. Distinguishing the effects of land-use change from those associated with climate change is a challenge.

16.6.1 Hillslope Erosion and Sediment Production

One of the primary ways that fluvial systems change over time involves variations in the amount of sediment generated and conveyed to the river as load, that is, **sediment production** and **sediment delivery**, respectively. Evidence of changes in sediment load over time is often readily apparent from fluvial landforms such as terraces and the sediments they contain. Geomorphologists try to determine which of many potential processes were responsible for past changes in this dynamic sediment "balance."

Changes in erosion and sedimentation rates are driven by a variety of processes, including tectonics, climate change, base level change, channel erosion, human activities, and land-use changes, or even by upstream dams that trap sediment. For example, accelerated hillslope erosion in the upstream parts of the drainage basin can increase sediment loads downstream and result in channel aggradation (**Fig. 16.40**).

Responses of fluvial sediment loads to climate or land-use change in a given catchment are complex and depend

on many factors, such as soil type and geologic materials, antecedent climate and vegetation, position within the drainage basin, and whether glaciation is involved. Channels in drainage basins that are prone to glaciation often aggrade during glacial periods, as large amounts of sediment are delivered to the system by the glaciers (as glacial outwash). These same systems then degrade and incise when the climate warms, due to reductions in glacial sediment production (**Fig. 16.35**). This stands in contrast with channels in unglaciated, semi-arid drainage basins, which tend to incise during glacial periods, when their climates turn cool and wet. Here, the fluvial response is governed by thicker upland vegetation within the drainage basin, which stabilizes soils and lowers sediment production rates. Thus, rivers in arid climates typically receive increased sediment loads during warmer, drier periods when vegetation cover thins out, leading to increased erosion on hillslopes and sediment delivery to the channel.

16.6.2 Channel Aggradation and Degradation

Rivers may fill with sediment or erode and incise their channels, all in response to changes in sediment load. Channels that cannot transport an increased sediment load will **aggrade** (deposit sediment), causing this excess sediment to be deposited on their floodplains or even within their channels. In cases of extreme aggradation, sediment can fill entire channels, cover floodplains, and even cause channels to become braided. Floodplain aggradation is accomplished during floods by overbank deposition (**Fig. 16.9**). Aggradation may occur in response to an increase in sediment load or size, or to a decrease in stream discharge or slope (stream power). These factors were described in Chapter 15 under the concept of grade and Lane's balance (**Fig. 15.34**). Evidence of channel aggradation includes shallow-but-wide channel cross-sections, or marshy floodplains with partially buried vegetation. Evidence for overbank sedimentation can often be observed in streambanks, particularly where the new sediment buries preexisting floodplain soils (**Fig. 16.41**).

Conversely, channels may **degrade** and cut down into their beds. Drivers of degradation include decreases in sediment load or grain size, or increases in discharge or slope

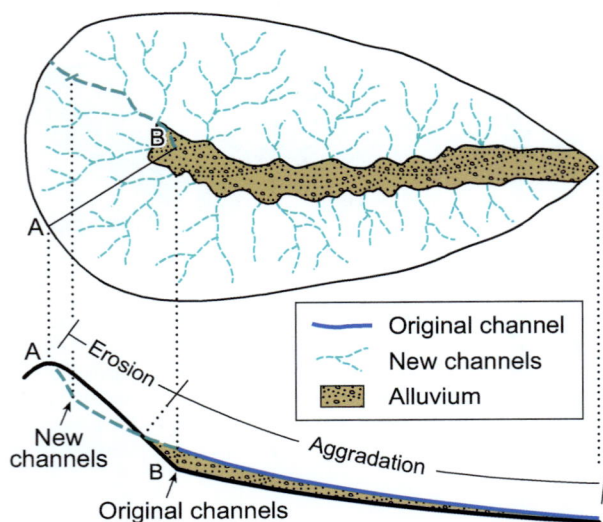

Figure 16.40 A small drainage basin showing channels extended headward by erosion associated with small tributaries. The sediment from these finger-tip channels is then transported downstream but largely stored there, causing the floodplain of the major trunk stream to aggrade.

Figure 16.41 Buried soils exposed in streambanks. **A.** A former floodplain soil buried by historical alluvium along the left valley margin of Hutchinson Creek, Buffalo County, Wisconsin, USA. The man's hand is touching the buried soil, 1.5 m below the modern floodplain. Note that the buried soil can be traced laterally to the surface soil on the valley side. **B.** A cutbank along Trout Creek, Buffalo County, Wisconsin, with an exposure of a well-developed floodplain soil buried by 1 m of historical alluvium. Source: D. Faulkner.

(stream power). Evidence of on-going degradation includes deep and narrow channels, high, steep channel banks, under-cut bridge pilings, and bedrock or coarse material in the bed (left behind as a lag deposit). Reductions in sediment load or increases in discharge may be driven by climate, tectonics, land-cover changes, or from dams upstream that trap sediment.

Because streams may aggrade or degrade for a number of different reasons, many fluvial systems have a complex history that involves both aggradation and degradation at various times in the past. Commonly, increased sediment production and channel aggradation are followed by a period of channel degradation, driven by decreased sediment production. This sequence forms an aggradation–degradation episode (**Fig. 16.42**). Such episodes often leave a distinct sedimentary record that geomorphologists can recognize.

Responses of the fluvial system to climatic or land-use perturbations may involve long **lag times** (delayed responses). Lag times get longer as the drainage basins increase in size, and for areas farther downstream (**Fig. 16.43**). In addition to lag times, the proportion of sediment that is transported downstream lessens, because much of the sediment is stored along the way. The complex linkages between climate, sediment production and transport, and channel responses are an area of climate change science where geomorphologists have much to contribute.

16.6.3 Human Interactions with Rivers

Humans are now a powerful and pervasive geomorphic agent. On a global scale, humans now move more sediment than is moved by natural processes. Thus, many drainage basins have a clear history of substantial land-use change from human activities such as mining, logging, or agriculture. Most of these land-use changes are accompanied by severe erosion, causing increasing sediment production and aggradation in the fluvial system. In some areas, land uses have changed multiple times, for example, eroded farm fields were abandoned, and reforestation followed. Then, in the fluvial system, sediment loads that initially were increased later get reduced. In parts of Europe, multiple such aggradation–degradation episodes occurred, as cultural activities have changed, historically, on the land surface.

Geomorphologists are increasingly acknowledging the geomorphic impacts of human activities on the fluvial system, which are often observed as **legacy sediment**, that is, human-induced sediment. Legacy sediment often contains toxic elements and chemicals, especially if generated from mining or agriculture. Thick deposits of legacy sediment are common along rivers in regions that were heavily farmed or mined. When stored on floodplains, legacy sediments can be identified where they overlie a buried soil from a former floodplain surface (**Fig. 16.41**). Subsequent channel incision may leave the aggraded surface behind, as an alluvial terrace, exposing both the legacy sediment and the buried soil in stream banks. Then, should this legacy sediment be

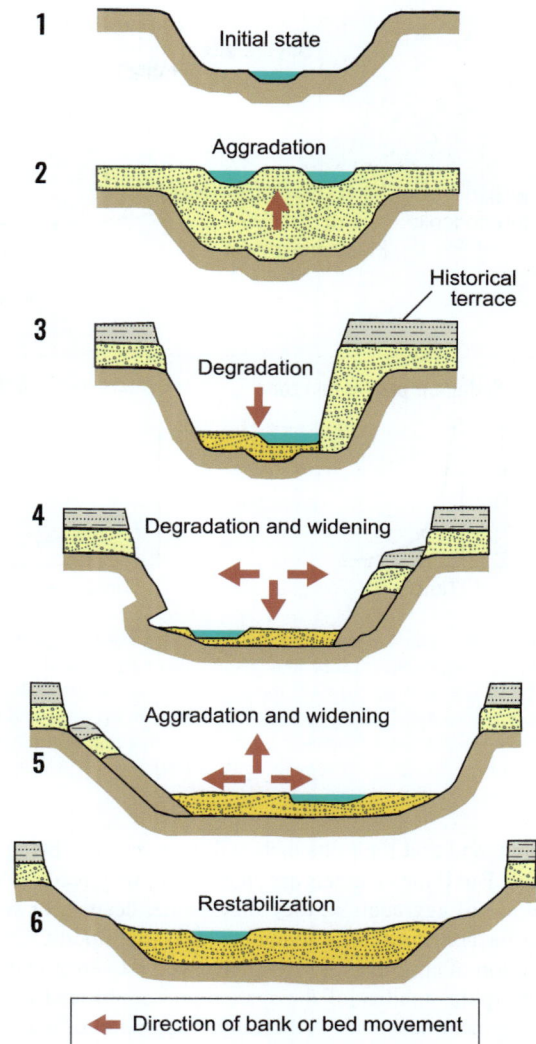

Figure 16.42 One possible model of channel changes through an aggradation–degradation episode. The initial channel (1) aggrades and fills with sediment (2), which may cause braiding if the sediment is coarse grained. In later phases of aggradation and during degradation, fine-grained overbank sediments may bury (3) the initial deposits. As sediment loads decrease, degradation begins as the channel downcuts (3) and then (4) widens. Because upstream areas are also incising, the sediment arriving from there may cause secondary aggradation (5) in areas downstream. Eventually, the system stabilizes (6). Source: Used with permission of Elsevier, from James (2018) and Simon and Rinaldi (2006); permission conveyed through Copyright Clearance Center, Inc.

eroded by the river, the enhanced sediment loads may have an effect on downstream channel morphology as well as water quality.

Recognition of legacy sediment is important for an understanding of river dynamics, bank stability, sediment loads, water quality, and the local history of the river. Where legacy sediment is widespread, thick, and young, the river may still be responding, as it attempts to reach a new equilibrium state. Many assume that river channels are in equilibrium, such that their size and shape are somewhat resilient to

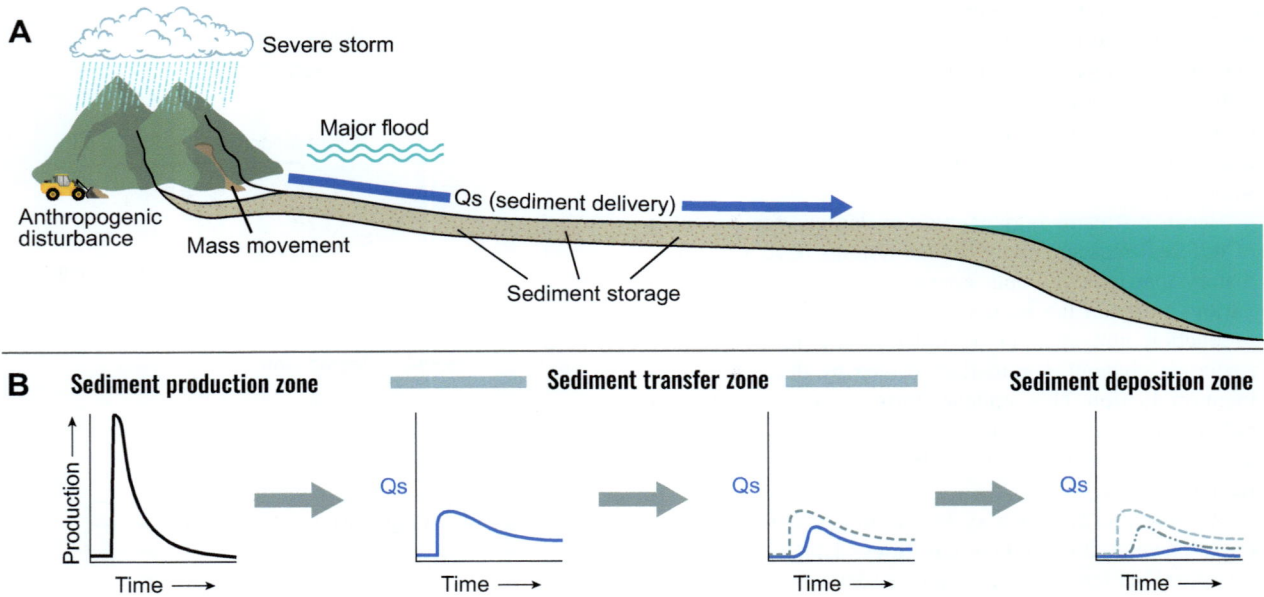

Figure 16.43 Sediment transfer within a drainage basin, illustrating the concept of lag times, following a significant erosional event which produces a large influx of sediment to the upper part of the drainage basin. **A.** A variety of events (triggers) can initiate a pulse of sediment to the fluvial system. That sediment then makes its way downstream. **B.** These graphs illustrate the timing of the down-valley propagation of that sediment pulse. Note the delays (lags) and reductions in sediment delivery (Qs) in the downstream direction, due to temporary sediment storage along the way, usually in floodplains. Notice also that not all of the sediment makes its way to the end of the fluvial system. Source: Used with permission of Elsevier, from Romans, B. W., Castelltort, S., Covault, J. A., et al. (2016); permission conveyed through Copyright Clearance Center, Inc.

change, and that their channels will recover after being disturbed. But if the channels are changing as they recover from a period of aggradation, they may erode, deepen, or widen even further in response to events like large floods. The assumption of equilibrium in this case would overestimate the geomorphic stability of the system and could lead to poor management decisions. A thorough integration of many fluvial features and processes, as well as the watershed history, is necessary to effectively manage rivers.

REVIEW QUESTIONS

16.1 What is the progression in channel size from non-channelized runoff (overland flow), through to fully channelized flow? What are the names given to the small- to intermediate-sized channels?

16.2 Be sure that you are able to put stream order numbers on a stream network, if given a map of the channels.

16.3 Explain the concept of drainage density and what factors (characteristics of the basin or the climate) affect it.

16.4 Describe the geologic structure, topography, process, or stage of development that is associated with each of the 12 drainage patterns in **Fig. 16.4**.

16.5 Do stream channel widths, depths, and flow velocities typically increase or decrease as one progresses downstream?

16.6 Describe the four main types of stream channels and the processes and sediments associated with them.

16.7 List and describe the main landforms of a meandering stream and its floodplain. Explain how oxbow lakes and meander scars form.

16.8 Describe and discuss how sediment load and bank stability vary among straight, meandering, and braided channels.

16.9 As you paddle downstream in a canoe through a pool–riffle–pool sequence, describe the local "in-stream" landforms in the order that you will encounter them, and the associated changes in flow depth, flow velocity, and bed material grain size.

16.10 What are the differences between ripples, dunes, and antidunes, and what can they tell us about the flow in the channel?

16.11 What are knickpoints and how do they form? Describe how streams "handle" knickpoints over time, relative to their longitudinal profile.

16.12 How do stream terraces form and what do they tell us about the geologic and climatic history of a drainage basin?

16.13 Explain the differences (in formation and sediment type) between an alluvial terrace and a strath terrace.

16.14 How do natural bridges form? How does their formation differ from arches?

16.15 Discuss the similarities and differences that occur in rivers and drainage basins that have experienced superposition vs antecedence.

16.16 A sediment exposure in a streambank has a dark horizontal layer ≈15 cm thick, buried by a meter of light-colored, horizontally layered sediment on top. What would be a good hypothesis for the origin of the dark layer? Of the overlying sediment?

FURTHER READING

Brierley, G. J. and Fryirs, K. A. 2012. *Geomorphic Analysis of River Systems: An Approach to Reading the Landscape*. Wiley.

Charlton, R. 2008. *Fundamentals of Fluvial Geomorphology*. Routledge.

Leopold, L. B. 2006. *A View of the River*. Harvard University Press.

Rhoads, B. L. 2020. *River Dynamics: Geomorphology to Support Management*. Cambridge University Press.

Wohl, E. E. 2014. *Rivers in the Landscape: Science and Management*. Wiley.

17 Coastal Processes and Landforms

Ethan J. Theuerkauf

Who doesn't love the beach? Beaches and coastlines are beautiful landscapes that provide a wealth of recreational, economic, and environmental benefits. In many locations, coastal areas are highly developed, which can make managing these dynamic landscapes challenging. The study of these landscapes is essential for developing land management practices that balance natural coastal processes with the challenges associated with coastal development.

Coasts are shaped by a variety of processes, such as waves, **tides**, and water level fluctuations. These processes operate on different timescales, ranging from short-lived storm events to sea level fluctuations that span millennia, and from local to global spatial scales. In some cases, coastal processes interact to enhance risk and vulnerability along the coast. For example, a hurricane that makes landfall at a spring (high) tide can be far more devastating than one that landfalls at a neap (low) tide. Add a variety of underlying geological configurations to the mix, and it becomes apparent that coasts are complex geomorphic systems. This chapter introduces the various coastal landforms found globally and the primary physical processes driving coastal evolution.

17.1 TYPES OF COASTS

Let's distinguish between a **coastline** (or **coast**), which refers to the entire zone where the mainland and water meet, and a **shoreline** (or **shore**), which is the specific (linear) interface between land and water. Often these terms are erroneously used interchangeably. The **beach** is the strip of land (often sandy, but can also be formed of rocks) that is just landward of the shoreline.

The location of the shoreline can be used as a metric for the general trajectory of a coastline over time. In other words, is the shoreline **regressing** (moving landward) or **prograding** (moving out, into the water)? In this vein, we can differentiate between two basic coastal trajectories – **erosional** and **depositional** (Fig. 17.1). Erosional (regressive) coasts are losing area over time, whereas depositional coasts are gaining area over time. Throughout this chapter we will discuss the specific processes and mechanisms of how and why erosion and deposition occur on coasts. For now, know that the geomorphic evolution of a coast is a function of the interaction between water level, waves, currents, underlying geology, and (where appropriate) human modifications.

Coasts can also be differentiated based on the dominant physical processes that formed them. **Wave-dominated coasts** are shaped by persistent waves, such as ocean swell, and generally have low (< 2 m) tidal ranges. Currents also play a prominent role in forming these types of coasts, and sandy beaches are common here as well, such as those along the Great Lakes of North America. **Tide-dominated coasts** typically only occur on the ocean, where sediment movement and landform development are primarily driven by daily tidal currents. Here, waves play only a minor role in moving sediment. In some places, the daily rise and fall of

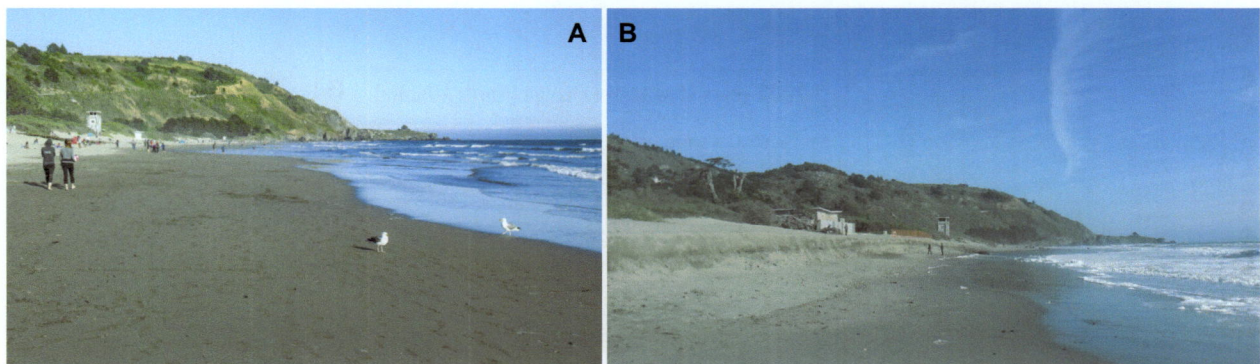

Figure 17.1 Beaches not only vary from place to place, but also through time. **A.** The wide, sandy beach is present during the summer at Golden Gate National Park in California (USA). **B.** In winter, however, it is narrower and has a sharp berm, caused by storm erosion. Source: M. Ngyen.

Figure 17.0 A summertime view of Blind Beach, near Jenner, California, USA, with its many stacks, and even a tombolo. Source: Moment / Getty Images.

the tide can often exceed four meters, as along the Bay of Fundy in Canada. **Mixed energy coasts** are influenced by both waves and tides, but often show a dominance towards one process.

17.2 WATER LEVELS AND COASTLINES

Water level exerts a first-order control on coastal evolution, as it dictates where physical processes such as waves and currents will have their greatest impact. Water levels in lakes and oceans vary widely over space and time. For example, **seiches** (short-term water level fluctuations) can raise water levels several decimeters in a matter of minutes (**Fig. 17.2**). Seiches occur when sustained onshore winds push water from one side of a body of water to the other, resulting in rapid changes in water level locally. In contrast, sea level fluctuations may occur over timespans of millennia, but affect coastlines globally. Daily tides are one of the most important water level fluctuations that marine and estuarine coasts experience, and as mentioned before, can even control the morphology of coastal landforms.

On ocean coasts, sea level is a major control on the evolution of coastal landforms. The geologic record provides abundant evidence that sea levels have fluctuated globally in response to glacial cycles. Global sea level is referred to as **eustatic sea level**. From about 120,000 to 20,000 years ago, large portions of Earth's land masses were glaciated, which resulted in drastically lower sea levels. Deglaciation began around 24,000 years ago, releasing water that was stored in

the glaciers back to the sea, and sea level has been rising ever since (**Fig. 17.3A**). However, the rate and magnitude of sea level rise has not been uniform over time. Brief periods of both rapid and slow sea level rise have been documented over the past 20,000 years and are largely correlated to rates of glacial melting. Generally, post-glacial sea levels attained stable positions about 6,000 to 7,000 years ago. Nonetheless, since the late 1800s, large portions of the world's coastlines have been experiencing accelerated rates of sea level rise (**Fig. 17.3B**). Recent sea level rise is widely thought to be related to greenhouse gas emissions from anthropogenic sources, causing atmospheric warming and accelerating the melting of glaciers worldwide (see Chapters 4 and 18). Sea levels are projected to continue to rise throughout the twenty-first century (**Fig. 17.4**).

Although global sea level is generally rising today, its actual rate of rise varies dramatically, depending on location. This condition is known as **relative sea level**, which accounts for both eustatic (global) sea level change as well as local factors such as rise and fall of the land mass due to tectonics.

During the Quaternary Period, the massive weight of continental glaciers depressed the crust (see Chapter 19). Then,

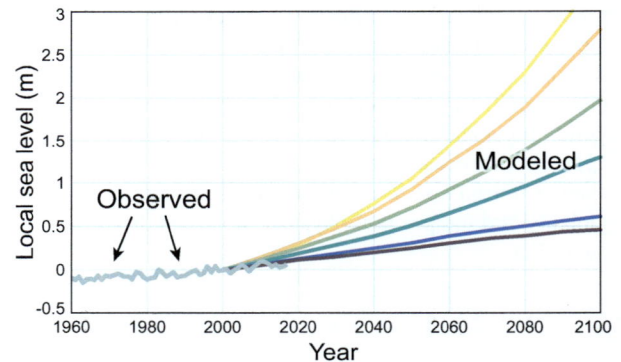

Figure 17.4 Sea level rise observed at a tide gauge in New York, NY, USA, coupled with various modeled projections of sea level rise for this location through 2100 AD. Source: Data from NOAA Tides and Currents; in public domain.

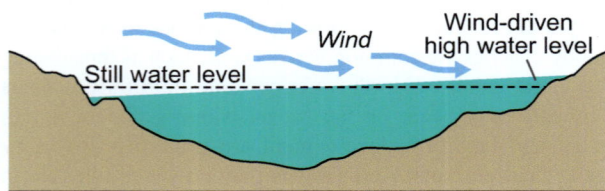

Figure 17.2 Diagram of a seiche, illustrating how sustained winds can cause water levels to rise quickly along one side of a basin.

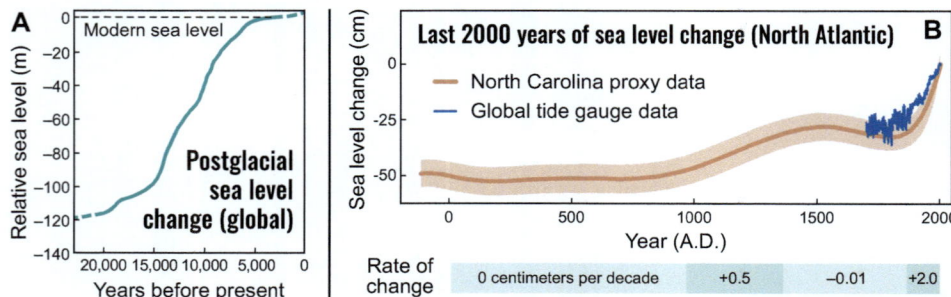

Figure 17.3 Sea level curves. **A.** Global sea level changes since the end of the last glaciation. Note that sea level rose quickly after deglaciation, but the rate of rise has slowed over the last few thousand years. Source: NASA graphic. **B.** High-resolution North Atlantic Sea level curve for the last 2,000 years, showing the accelerating rate of sea level rise since the late 1800s. Source: From the US National Climate Assessment; in public domain.

later, when the ice melted and the weight removed, the crust rebounded back – a process referred to as *isostatic rebound*. The rate of isostatic rebound varies by location, because the weight of the glaciers was not uniform, and the crust varies in its elasticity (**Fig. 19.42**). At some locations, such as along portions of the Alaskan coastline, uplift is occurring at such a rapid rate that relative sea level is actually falling. In contrast, some crustal areas are sinking, or subsiding, due to isostatic adjustment. This condition can be even further exacerbated on deltas, which grow in weight over time, as is occurring in the Mississippi River Delta region, near New Orleans, Louisiana, USA. Rates of relative sea level rise are three times higher along the Mississippi River Delta than globally, due to the combined effect of isostatic rebound, sediment loading on the delta, and human impacts such as oil and gas withdrawal that allow the delta sediments to compress. The high rates of relative sea level rise in this area threaten the natural environment, as well as human infrastructure and lives.

Other types of water level fluctuations on ocean coasts include nuisance flooding and storm surges (**Fig. 17.5**). As sea level rises but coastal land and infrastructure elevations remain the same, the normal rhythm of the tides causes flooding in areas that were not prone to flooding a century ago. This type of **nuisance flooding** – flooding that occurs at high tide during fair-weather – can cause road closures, overwhelm storm drains, and damage infrastructure (**Fig. 17.5B**). As sea levels continue to rise globally, nuisance flooding will only increase in frequency. **Storm surges** are rises in water level that occur in response to strong, onshore winds during a storm, such as a hurricane (**Fig. 17.5A**). Storm surges are a key factor in determining the amount of damage in such a storm. Flooding from storm surges can cause damage well inland from the coast, as evidenced by the catastrophic flooding in the New Orleans area during Hurricane Katrina in 2005.

17.3 TIDES

Along ocean and estuarine coasts, tides play a key role in shaping coastal landforms. Tides are the rise and fall of sea levels caused by the combined effects of the gravitational forces exerted by the Moon and Sun (but mainly the Moon). Depending on location, this can result in either a once- or twice-daily rise and fall of water level (**Fig. 17.6**). If the water level rises twice daily, the tide is referred to as **semidiurnal**, and if it only rises and falls once a day, it is referred to as **diurnal**. The magnitude of the rise and fall of the tide, referred to as **tidal range**, can be predicted based on lunar phases. Four times each month, the tide predictably deviates from its normal fluctuation. **Spring tides** occur during new and full moons when the Sun and the Moon are in alignment, and thus the gravitational pull of the Sun is added to the pull of the Moon, increasing the tidal range (**Fig. 17.6**). During a spring tide period, high tides are higher than normal, and low tides are lower than normal. **Neap tides** occur during the first and third quarter moons, when the Sun and the Moon are at right angles to each other. This alignment causes the

Figure 17.5 Storm surges and nuisance flooding. **A.** Diagrammatic representation showing how nuisance flooding and storm surges can interact to produce abnormally high water levels in coastal locations, resulting in enhanced damage. **B.** An example of nuisance flooding during a king tide at the Annapolis (Maryland, USA) dock. Note the boats moored in the slip, while water overlaps the street. Source: A. McGovern.

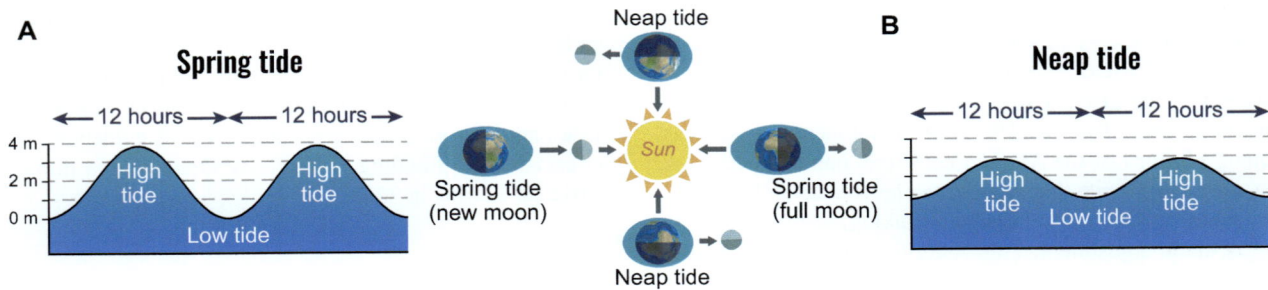

Figure 17.6 The relationship between Moon phases and tides. The graphs show the different tidal ranges during spring and neap tides. Note that a greater difference exists in water levels between high and low tide during a spring tide vs during a neap tide, when the tidal range is smaller.

Figure 17.7 Images of Halls Harbor on the Bay of Fundy, Newfoundland, Canada, at high and low spring tides.

Sun's gravitational pull to offset the pull of the Moon, resulting in slightly lower tidal ranges. If a hurricane makes landfall during a high tide that is also a spring tide, the damage from flooding and storm surges is typically greater than at other times.

Because tides are so important to coastal evolution, coasts can be classified based on their tidal range. These include **microtidal** coasts, where the tidal range is < 2 m, **meso-tidal** coasts with a range of 2–4 m, and **macrotidal** coasts with tidal ranges > 4 m. The Bay of Fundy in Newfoundland, Canada, with its > 11 m tidal range, is an excellent example of a macrotidal coast (**Fig. 17.7**). Here, boats commonly sit in mud during low tide!

17.4 WAVES AND CURRENTS

Waves – both their size as well as their orientation – are a very important force along coastlines (**Fig. 17.8**). As waves break along the shoreline, they cause water and sediment to move onshore (**swash**) and offshore (**backwash**), as well as parallel to the shore. The amount of erosion, sediment transport, and/or damage to coastal infrastructure is a direct function of wave characteristics, but particularly wave height.

The two primary components of a wave are the **crest**, the highest point of the wave, and the **trough**, its lowest point (**Fig. 17.8A**). The difference between the crest and the trough is the **wave height** (or **amplitude**) (**Fig. 17.8B**).

Because waves are driven mainly by wind, large wave heights are associated with stronger winds and storms. Large waves tend to generate more erosion, sediment transport, and coastal damage. Another important wave characteristic is the **wavelength** (L), which is the distance between wave crests (**Fig. 17.8A**). Wavelengths are long in deep water and shorten as waves move into shallow water.

In open water, as a wave passes a given spot, orbitals of water particles develop, from the surface of the water down to the **wave base**, which is generally considered to be equal to ½ the wavelength (**Fig. 17.8B**). Below the wave base, in theory, water is unaffected by the wave action above. In this "deep" water (depths greater than ½L), where the orbitals do not make contact with the bottom of the water body, waves also cannot transport sediment. Here, a wave is considered to be a **deep-water wave** (**Fig. 17.8C**). Waves in water shallower than ½L are considered **shallow-water waves**.

As a wave approaches the shore, it begins to encounter shallow water because its wave base is now deeper than the bed of the water body. As a result, the wave orbitals encounter friction, or drag, on the bed, slowing the forward motion of the waves. (The bed acts as a "drag brake" on the waves.) Sand transport then starts to occur along the bed. As the waves slow even further, in even shallower water, the top of the wave starts to travel faster than the bottom, causing it to steepen and fall over, as it becomes a **breaker** (**Fig. 17.9**).

After a wave breaks, water and sediment are transported up the **beach profile** as swash (**Fig. 17.10A, B**). After swash

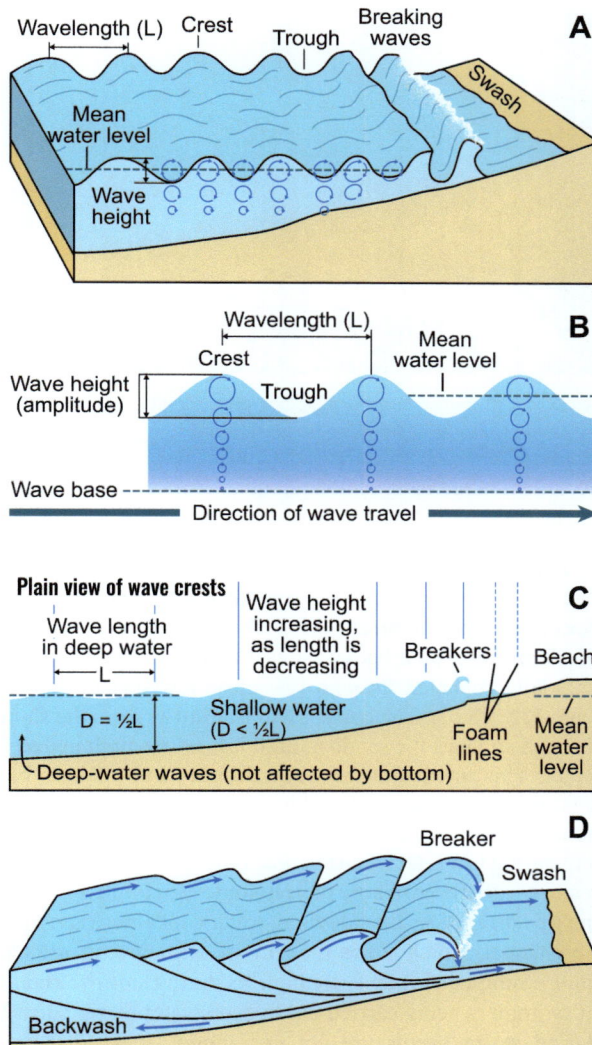

Figure 17.8 Characteristics of waves. **A.** Various components of waves. Note the wave orbitals that are generated as a wave passes over a location. Wave orbitals do not touch the bottom until the water depth shallows, i.e., until the water depth is shallower than the wave base. **B.** A detailed view of the various components of waves and their orbitals. **C.** The transition zone from deep-water waves to shallow-water waves. Once the wave transitions to a shallow-water wave, it will become oversteepened and break. **D.** After waves break, energy is transferred onto the beach, forming swash. Gravity then moves the water and sediment back towards the sea, as backwash.

Figure 17.9 A breaking wave along a steep shoreface at Sand Beach, Maine, USA. Source: James St. John, CC BY 2.0, via Wikimedia Commons.

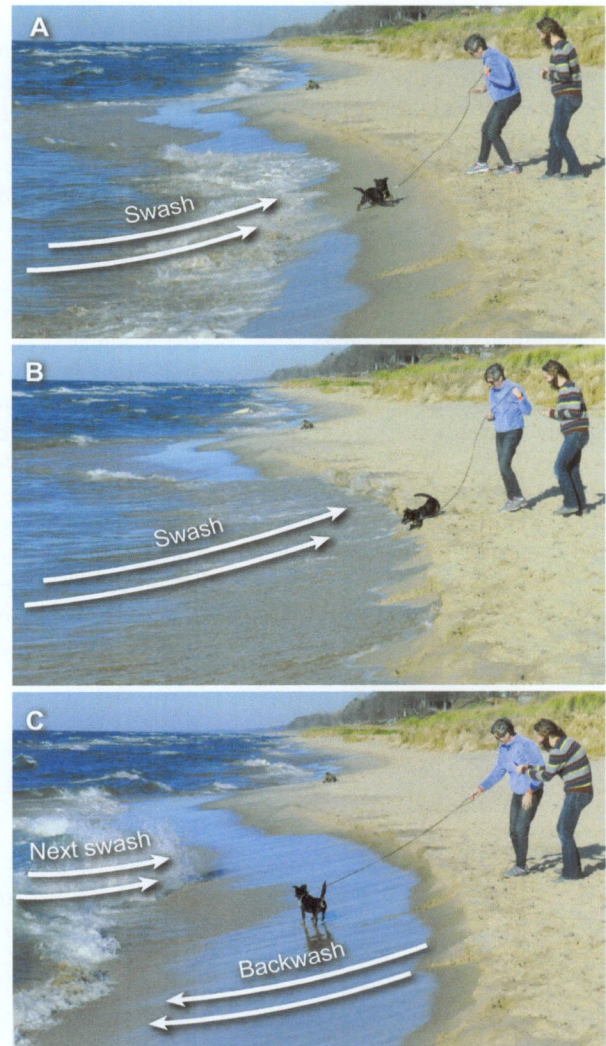

Figure 17.10 Wave action on the foreshore portion of a beach, illustrating the processes of swash and backwash. Source: R. Schaetzl.

has reached its maximum extent (which is controlled by the beach slope, grain size, and wave energy), it flows back to the body of water as backwash (**Fig. 17.10C**). Swash usually drives sand grains up the beach at an angle, whereas backwash typically allows these same grains to flow back, down the beach, in a direction more nearly perpendicular to the coastline (**Figs. 17.10C, 17.11**). Given that waves usually approach the shoreline at an angle, the motion of swash and backwash sets up a sand transport pattern along the shoreline known as **beach drift** (**Fig. 17.11**). In essence, beach drift

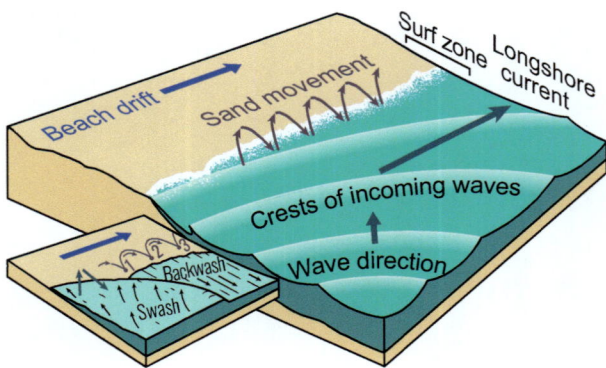

Figure 17.11 A conceptualization of how beach and longshore drift are generated by waves approaching the shoreline obliquely. Note the zig-zag motion of sand on the beach, which drives beach drift on sandy beaches. Longshore drift, or a longshore current, is focused in the shallow nearshore water. It, too, is generated as waves approach the shoreline at an angle.

Figure 17.12 Sediment accumulation on the updrift side of a groin in Ventura, California, USA, by longshore currents. Note the loss of beach sediment on the downdrift side of the jetty. Source: California Coastal Records Project; in public domain.

is the down-beach, zig-zag motion of sand grains, driven mainly by swash and backwash.

Water **currents** result from a variety of forces, including winds, tides, and water density differences. They can operate over a variety of temporal and spatial scales, ranging from oceanic-scale currents that are always present, such as the Gulf Stream, to rip currents that form for only a few minutes at a specific location. This chapter focuses on the most common currents found along coastal areas – longshore and rip currents.

Longshore currents form as waves approach the shore at an oblique angle and then reflect off the beachface (**Fig. 17.11**). This process sets up a current that flows just offshore, parallel to the shoreline, in the general direction that the waves approach the shore. Longshore currents are responsible for transporting sediment in shallow water, parallel to the coast, as the main part of a process called **longshore drift**. The impact of longshore drift is obvious along coasts that have engineering structures built perpendicularly to the shoreline, such as a **jetty** or a **groin**. Sand will accumulate on the updrift side of the structure while the beach will become starved of sand on the downdrift side, leading to coastal erosion (**Fig. 17.12**).

Rip currents are a type of coastal current. These fast-moving currents flow perpendicularly away from the coastline, through breaks in offshore sandbars. As waves break onshore, their backwash is directed offshore and preferentially funneled through low areas (swales) in offshore bars (**Fig. 17.13**). The current will flow away from the shore, through and past the sandbar, until it loses strength in deeper water. This interaction between currents and offshore bars sets up a rip current cell, which can persist for minutes to hours. In oceans and estuaries, rip currents are most prevalent during low tide, when the flow of water can interact more readily with the nearshore morphology, generating fast-moving rip currents. As the tide rises, water depth increases in the rip channel, which slows the current. Rip currents are also quite strong when waves are high, as during storms. Reports of swimmers being carried out to deep water by rip currents during storm events are not uncommon, making them a common cause of accidental death within the coastal zone.

17.5 OTHER COASTAL PROCESSES

Winds are important in the coastal zone, as they are the primary force behind waves and currents. They also directly move sediment through **eolian**, or wind-driven transport,

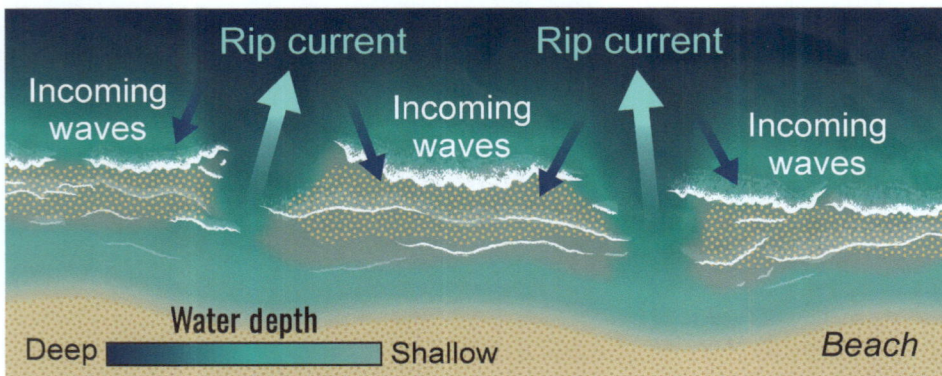

Figure 17.13 Schematic illustration of how rip currents develop along a coastline.

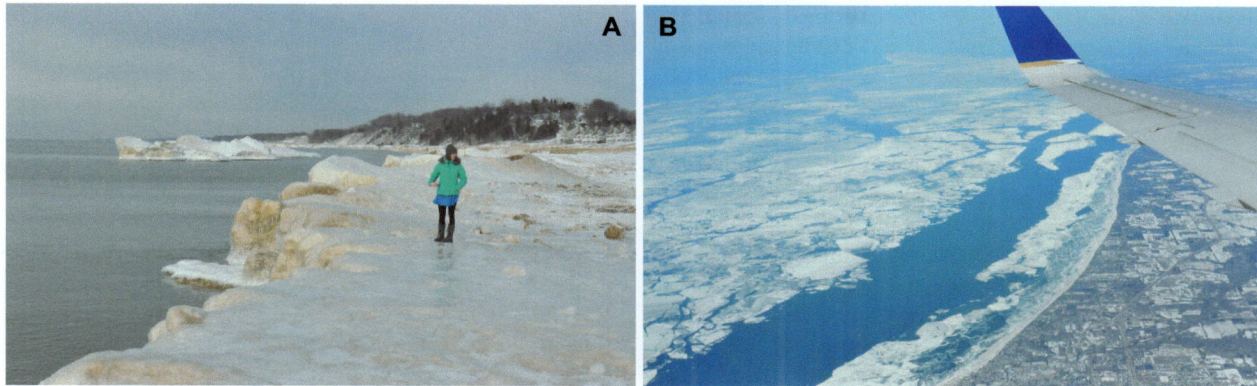

Figure 17.14 Winter shore ice. **A.** A sediment-rich ice ridge along the Lake Michigan (USA) shore. **B.** An aerial view of shore ice along that same Lake Michigan shore. Source: R. Schaetzl.

which is important for building coastal landforms such as dunes (see Chapter 21). Vegetated dunes can enhance coastal resilience in the face of storms and sea level rise.

Although water level fluctuations, winds, waves, and currents occur in some form along all coastlines, other processes are more regionally specific. An example of such a process is shore ice, which is an important control on coastal evolution in colder climates (**Fig. 17.14**). Ice along the shoreline can offer protection from wave erosion during strong winter storms. Why is this important? One potential outcome of warming global temperatures is shorter, warmer winters. Less ice in nearshore areas could facilitate enhanced erosion from winter storms. An argument counter to the "protection" of beaches by shore ice states that ice rich with sand can disconnect from the shore and move (float) out into the water, transporting sediment far offshore. For these reasons, geomorphologists view the role of shore ice in coastal systems as a complex one.

Various types of geomorphic events can dramatically impact coastal zones, some of which occur only in specific locations. For example, earthquakes can directly impact coastal areas by generating a **tsunami wave** – a fast-moving wave generated by the sudden tectonic displacement (faulting) at or below the sea floor (**Fig. 17.15**). Sometimes, the actual driver of the tsunami is a post-faulting event like a submarine landslide. Tsunami waves differ from storm-generated waves in that they are much larger and originate from a point source, causing waves to radiate outward in all directions, like a pebble tossed into a pond (**Fig. 17.15B**).

Tsunamis are actually classified as shallow-water waves even when they are in deep water, because of their extremely large wavelength (remember, a wave is considered a shallow-water wave if the water depth is less than half its wavelength). Tsunami waves travel quickly towards the shore, and even though the wavelength exceeds half the water depth, little energy is expended until the waves reach the coast. In the open ocean, tsunami waves have low (< 1 m) heights and long **wave periods** (the time required

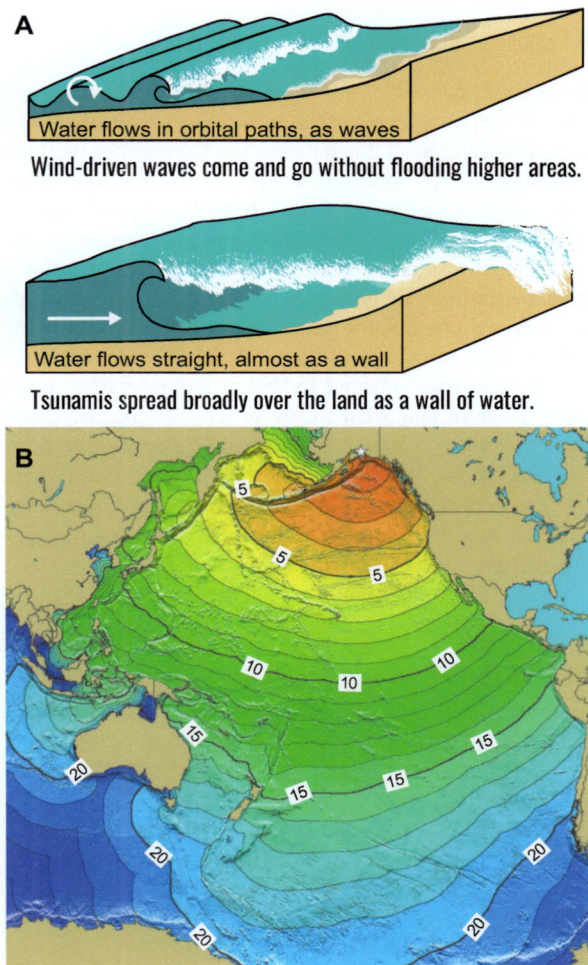

Wind-driven waves come and go without flooding higher areas.

Tsunamis spread broadly over the land as a wall of water.

Figure 17.15 Characteristics and outcomes of tsunami waves. **A.** A comparison of traditional, wind-driven waves vs tsunami waves. **B.** Tsunami travel times (hours) for the March 28, 1964 tsunami, which originated in Alaska, USA. Source: NGDC, CC BY-SA 3.0, via Wikimedia Commons.

for successive waves to pass by a given point) – sometimes as long as 10–30 minutes. Thus, on the open ocean, tsunami waves are barely detectable. But as they approach the shore and interact with the shoreface, they begin to slow down, which causes the wavelength to decrease and the height to increase, sometimes to 10 m or more (**Fig. 17.15A**). At this point, tremendous damage can occur in coastal areas.

Typically, the arrival of a tsunami is characterized by modest rising and lowering of the water level in the coastal area, followed by several massive waves. Tsunami waves can be extremely large, quickly inundating low-lying coastal areas, and even areas far inland, causing major damage and loss of life. Numerous examples exist in the history of tsunami waves that devastated coastal areas. The largest tsunami wave ever recorded, which was > 500 m high, occurred in Alaska's Lituya Bay in 1958, following a major earthquake and associated landslide.

17.6 COASTAL LANDFORMS

Coastal zones contain a mosaic of different landforms, each formed in response to environmental drivers such as storms, waves, and water level fluctuations. The interactions between coastal morphology and these drivers are referred to as **morphodynamics**. Changes that happen to the beach profile during summer and winter are excellent examples of morphodynamics. During the summer, with its smaller waves and fewer storms, beach-building generally occurs (**Fig. 17.1A**). At this time, waves preferentially direct sand *onshore* (towards the beach), leading to sand deposition. The beach widens. Also in summer, sand bars – offshore and below the water line – often migrate onshore and weld to the beach, helping the sandy beach grow even wider. All this deposition *steepens* the slope of the beach front, resulting in a **reflective beach** state, where more wave energy reflects off the beach and into the water. The beach profile during these conditions is referred to as a **fair-weather** or **summer profile** (**Fig. 17.1A**).

In contrast, during the fall and winter, when storms and larger waves are more common, the beach profile transitions to a **storm**, or **winter**, profile (**Fig. 17.1B**). The increased wave energy of winter storms enhances beach erosion. Sand eroded from the beach area is transported offshore and deposited in sandbars, which grow in size. As these bars grow, the beach profile is flattened, allowing it to dissipate the increased wave energy from the storms, because of the shallower wave base – a condition referred to as the **dissipative beach state**. Waves break offshore, on the bars, more than they would have during summer. As time progresses and wave conditions lessen during the subsequent summer, beach building will begin again, and the cycle will repeat.

It is important to note that not all beaches will follow this progression, nor does a given beach always follow the exact same progression. These are general states; the exact nature of a beach is a complex response to a range of conditions. For example, on the Great Lakes of North America, annual peaks in water level occur during the summer months, which alter the typical summer vs winter profile response.

17.6.1 Beach Morphology

Beaches are arguably the most identifiable coastal landform (**Figs. 17.1, 17.17**). Generally, most beaches share similar morphologic characteristics, although their composition and dimensions vary.

Beach morphology is often described by its profile (**Fig. 17.16**). The **beach profile** extends from any sand dunes on the landward side of the beach to the shallow water near the shore, where waves interact with the bed. Thus, the beach profile includes not only the sand that we think of as the "beach" but also some of the underwater areas. Let's begin this discussion by considering the three main *zones* common to most coasts (**Fig. 17.16**). The part of the underwater beach farthest from the shore – the **nearshore zone** – extends

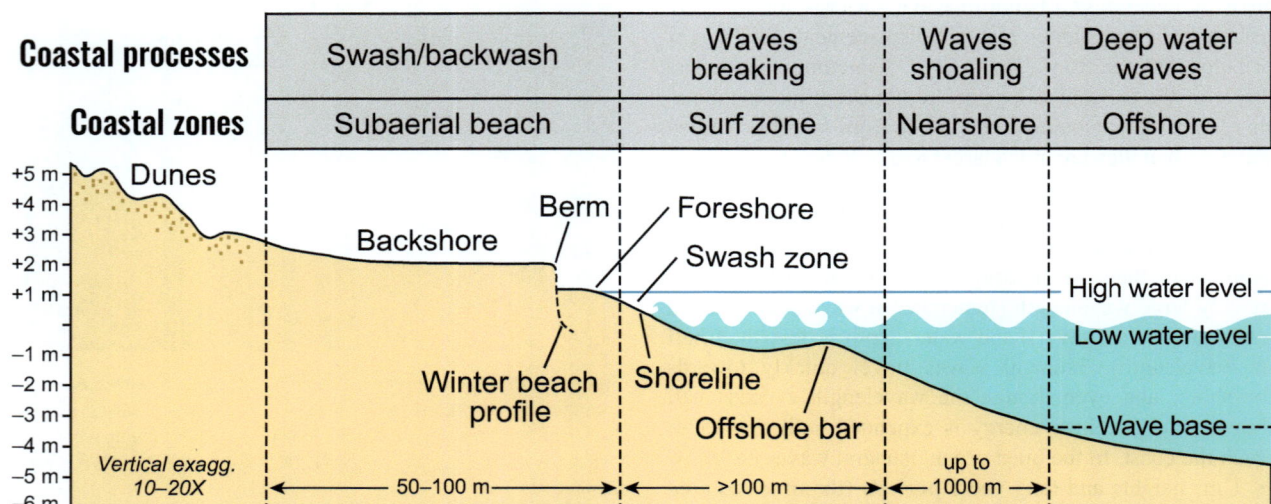

Figure 17.16 Schematic diagram of a typical beach and nearshore profile. Source: © A. N. Strahler (1992). Used by permission.

landward from where the bed first intersects the wave base to an offshore sand bar. In the shallow waters of the nearshore zone, waves begin to feel the bottom as they approach the beach. Immediately landward of the nearshore zone is the **surf zone** where those same waves break. This zone generally contains at least one offshore bar and trough, and extends landward to the **swash zone**. Here, waves run up and down (swash and backwash) on the **foreshore**, which is the portion of the subaerial (dry land) beach that slopes towards the water (**Figs. 17.10, 17.16**). The width of the swash zone varies in response to the steepness of the beach front. At the upper end of the swash zone is the actual shoreline. The shoreline is just that – a line that separates land (the sandy beach) from the water. The dry-land parts of the beach extends farther inland, from the swash zone to any uplands or dunes behind it. This area contains both the foreshore and the **backshore**. The backshore is the "back" part of the beach zone, extending from the limit of high water to dunes or other high ground farther inland. It is only affected by waves during exceptionally high tides or severe storms. Often, a short vertical step, or **berm**, divides the flatter backshore from the foreshore, which slopes towards the water. The berm may build basinward during the summer when quiet conditions prevail and may erode back a considerable distance during storms, as may happen in the winter (**Fig. 17.1**).

17.6.2 Spits, Baymouth Bars, and Tombolos

Many sandy coastal landforms are formed by the deposition of sand transported via longshore currents (**Fig. 17.17**).

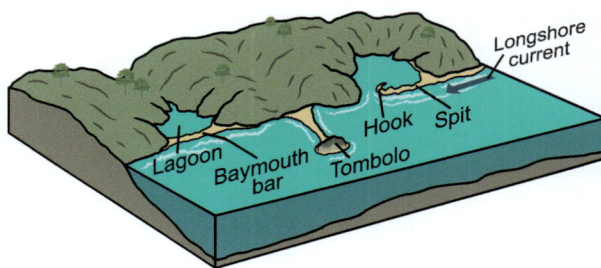

Figure 17.17 Some coastal landforms formed by transport of sediment via longshore currents.

For example, **spits** are elongated bodies of sand that are attached to the beach at one end (**Figs. 17.17, 17.18**). They grow downshore in the direction of the longshore current. Typically, they begin forming at, and thus, are anchored to, a river mouth or headland. Both are sources of sand for the growing spit. The river brings sand to the coastal zone, as does the eroding headland. The headland may also provide for an area of diminished current, or a "shadow" area, where sand can be deposited in the spit (**Fig. 17.17**). Spits will lengthen over time, making them excellent indicators of the direction of the dominant longshore current. Spits have some similarities to barrier islands (see below), except that spits are attached to the mainland at one end, whereas barrier islands are simply that – islands. Spits can, however, become barrier islands if they become detached from the mainland. Spits often bend or "recurve" at their distal (furthest downdrift) end, as the sand deposition patterns follow the currents. These recurved ends are called hooks (**Fig. 17.17**).

When a spit grows so long that it encloses a coastal bay or river mouth, it is referred to as a **baymouth bar** (**Figs. 17.17, 17.19**). Baymouth bars represent a balance between the longshore transport processes that bring sand across the mouth of the bay, and fluvial processes that want to push that sand out, into deeper water. Some baymouth bars are able to completely enclose the bay for long periods of time, because discharge from the river cannot maintain the opening, and thus, longshore transport of sand (which forms the bar) wins, and the baymouth bar is maintained (**Fig. 17.19**). This type of baymouth bar is common in drier climates. If large enough, rivers that empty into the bay can maintain an opening or gap in the bar. Even if this gap closes due to sand deposition, the bar can be breached during storms.

Another coastal landform generated by longshore drift is a **tombolo** – a body of sand that connects an offshore island to the mainland (**Fig. 17.17**). The sands accumulate through wave refraction and diffraction. As waves approach the island, they are refracted (bent) and generate converging longshore currents that deposit material behind the island – in a "shadow" area with its smaller waves. Eventually, these sediments accumulate to the point where they connect the mainland to the island. Once formed, the sand "bridge" is

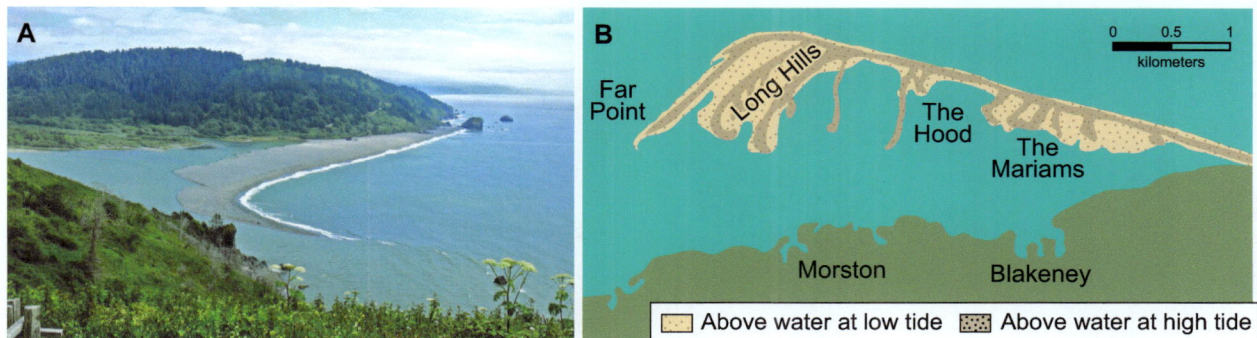

Figure 17.18 Spits. **A.** A large spit at the mouth of the Klamath River in California, USA. Source: L. Tanner via Flickr.com, licensed via CC BY 2.0. **B.** Drawing of the large spit that forms Blakeney Point, in Norfolk, England.

Figure 17.19 A baymouth bar completely encloses this small bay formed at the mouth of a small river, as it enters the Pacific Ocean, along the California (USA) coast. Source: California Coastal Records Project; in public domain.

Baymouth Bar

fairly stable, because the tombolo and its island will continue to modulate the wave and current patterns, promoting continued sand deposition in the "shadow" of the island, thereby maintaining the tombolo.

Spits, baymouth bars, and tombolos are all excellent examples of the dynamic coastal landscape. A coastal landform may start out in one form but evolve into something else as the currents and waves change over time, which are sometimes driven by the formation of the landform itself.

17.7 BARRIER ISLAND SYSTEMS

Barrier islands are sandy coastal features, elongated parallel to the coast (**Fig. 17.20**). Most are located along ocean coasts, but some can also occur in lakes and estuaries. The barrier island is only one part of a **barrier island complex**, which is a system of associated environments that evolve synchronously with the barrier island, in response to tides, sea level rise, storms, and human disturbance. Included in the barrier island complex may be beaches, dunes, tidal channels, tidal deltas, tidal flats, inlets (gaps) between the islands, saltmarshes, and backbarrier lagoons (**Figs. 17.20,**

17.21). Beaches and dunes on barrier islands behave similarly to those on mainland coasts. The primary difference is that erosion and accretion play perhaps more important roles in the overall evolution of a barrier island, as it responds to storms and changes in sea level.

During major storm events, barrier islands can be breached and form gaps called **tidal channels** (or **inlets**). These openings bisect the barrier islands and connect the water behind the barrier (the lagoon) to the open ocean (**Fig. 17.21**). The **lagoon** is the brackish (slightly salty) water body between the barrier island and the mainland. Water and sediment flow through tidal channels as the tide rises and falls, depositing sediment on both the lagoon and ocean sides of the tidal channel as **tidal deltas** (see below for more detail). Sediments, brought to the lagoon by rivers, can also be deposited at various locations throughout the lagoon. These are best seen during low tide, as **tidal flats**, particularly along the fringes of the mainland. Eventually, if enough sediment accumulates to raise tidal flats and deltas to an intertidal elevation, saltmarshes can colonize these areas.

Barrier islands are complex and interconnected systems. They result from the interplay between sea level and

Figure 17.20 Oblique aerial photo of Core Banks, North Carolina, a barrier island complex along the US Atlantic coast. Note the various components of the barrier island complex including the beach, dunes, and backbarrier marshes. Source: J. Himmelstein.

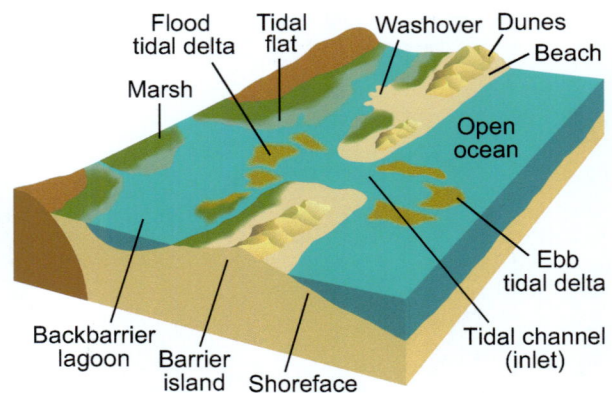

Figure 17.21 Schematic diagram of the components of a typical barrier island complex. Note that some of the features shown are coastal landforms, such as dunes and marshes, whereas others are clearly offshore.

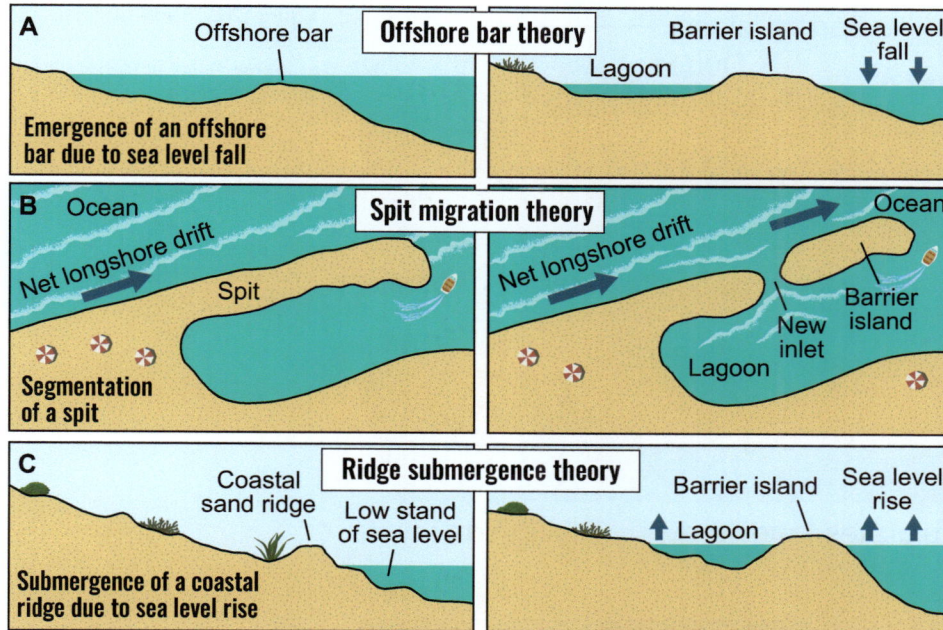

Figure 17.22 Schematic diagrams illustrating the three main theories of barrier island formation. **A.** An offshore bar becomes a barrier island due to sea level fall. **B.** Spit growth and eventual segmentation form the barrier island. **C.** An onshore ridge is submerged due to rising sea level during the Holocene, forming an offshore barrier island as sea level rises.

sediment availability and transport along the ocean coast. In order for a barrier island to form, there must first be enough sand to keep the island above sea level. But how does this sand initially get there? In that regard, three primary theories of barrier island formation are being debated (**Fig. 17.22**). The oldest theory on barrier island formation, the **offshore bar theory**, posits that barrier islands are/were essentially large, submerged sandbars that have subsequently thickened in response to wave deposition. Once the sandbar becomes subaerial (above and out of the water), eolian processes take over, sand accumulates (some as dunes), and the area stabilizes into a barrier island. The lack of evidence from field and laboratory investigations makes this theory challenging to validate. The second theory, the **spit migration theory**, suggests that barrier islands originate as spits connected to the mainland. The spits grow longer via longshore drift and, if breached during storm events, become separated from the mainland to form a barrier island. The final theory on barrier island formation, the **ridge submergence theory**, suggests that barrier islands originate from flooding of former beach and dune ridges. These ridges presumably formed when sea level was lower. At that time, sets of ridges formed and migrated outward as sea levels fell. Then, as sea levels rose during the Holocene (**Fig. 17.3A**), the ridges were breached, forming a lagoon and initiating the formation of a barrier island complex. From this point on, barrier island evolution continues in response to storms and sea level rise. Geomorphic evidence from ocean coasts supports both the ridge submergence and spit migration theories for barrier evolution.

Barrier island systems are currently being impacted by rapidly rising sea levels. The general response of a barrier island to sea level rise is to migrate landward via a process called

rollover. In this process, storm waves overtop the island, eroding sand from the ocean-side beach, transporting it *over* the barrier island, to the lagoon. The process of washing sand over the barrier island by strong waves and high water is called **overwash**. Erosion on the ocean side narrows and lowers the barrier island, making it vulnerable to any increase in water level during storm surges (**Fig. 17.23**). The largest storms may even breach the island entirely at its lowest and narrower parts, forming new inlets. Overwash sediments often get deposited on top of backbarrier marshes and within backbarrier flats, as **washover fans** – deposits of sand on the lagoon side of the barrier island (**Fig. 17.21**). Rollover allows the barrier islands to naturally migrate landward, presumably keeping pace with the rising sea level. Barrier islands that cannot naturally rollover because they are anchored by infrastructure are at risk of drowning or eroding away as sea level rises.

Like a river channel, inlets in barrier islands provide a pathway for sediment and water transport within the barrier island complex. Many inlets form during storms, when wave energies are high and the island is breached. Once formed, sands transported through the inlet by waves and tidal currents are deposited in the lagoon. As long as the inlet remains open, currents can continue to transport sediment through it and into the lagoon during the incoming tide. The landform that results from sediment transport *into* the lagoon is known as a **flood tidal delta** (**Fig. 17.21**). As the tide comes in, water and sediment flow into the lagoon. Alternatively, some sediments can be removed from the lagoon through the inlet during the falling tide and deposited on the ocean side of the barrier, as an **ebb tidal delta** at the mouth of the inlet. Once formed, flood tidal deltas can become colonized by saltmarsh vegetation, helping to stabilize the delta and the barrier island.

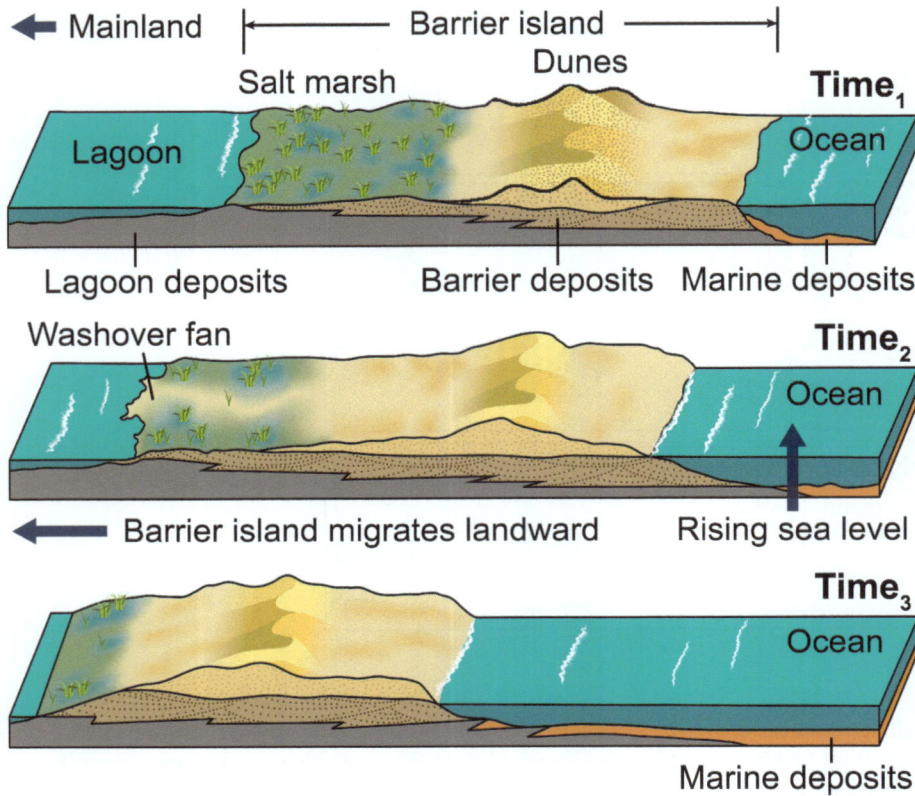

Figure 17.23 Barrier islands migrate landward through a process known as rollover. As sea level rises and storms impinge upon the island, sand is eroded from the ocean side of the island and washed over the island at its lowest points, where it may form washover fans. The rollover process allows barrier islands to move landward while still maintaining their elevation with respect to sea level.

As should be apparent, barrier islands must be able to migrate, landward or seaward, in order to keep pace with rising sea level. If these natural processes are not able to occur due to human disturbance, the island may drown or erode as sea level rises. This situation is currently occurring at many barrier islands throughout the world, because human development impedes overwash and inlet formation from occurring. Thus, many barrier islands are eroding, becoming much narrower and more vulnerable to flooding, and ultimately drowning in place. As a result, the infrastructure on barriers is often at risk.

17.8 COASTAL DUNES

Most beaches and coastal areas are sandy, and so **sand dunes** are common in coastal settings (**Fig. 17.24**). Although their morphology varies widely, ranging from small foredunes only a meter high, to towering sand dunes > 100 m high,

Figure 17.24 Coastal dunes. **A.** Coastal dunes along Lake Michigan at Indiana Dunes National Park in Indiana, USA. Source: A. Arbogast. **B.** Wave action and high lake levels can drive foredune erosion, as depicted in this image along the Lake Michigan coast, near Holland, Michigan (USA). Source: R. Schaetzl.

similar processes form and sustain all **coastal dunes**. Coastal dunes are the product of wind-blown, or eolian, sand transport. Their ability to form and evolve results from the interplay between wind patterns, sand availability, vegetation dynamics, and the geometry and geology of the area. Dunes are best developed along coastlines with high sand supplies, persistent onshore winds, and wide and gently sloping beaches, where vegetation does not stabilize the sandy beach surface. Dunes play an important role in protecting coastal infrastructure from storm damage, and are therefore a key aspect of a resilient coastal system. If dunes erode or are removed to make way for coastal development, the landscape behind them will become more vulnerable to erosion and (on barrier islands) overwash, as evidenced by portions of the Outer Banks of North Carolina, USA.

On many coasts, as one moves inland from the beach, the first dune encountered is often a young, or incipient, **foredune** (Fig. 17.24A). Coastal dunes, including foredunes, begin to form when dune grasses and other early successional vegetation begin to colonize parts of the backshore. **Wrack** – organic debris such as branches and seaweed that get deposited on the backshore by waves or the tide – often helps promote vegetation colonization at this stage. Colonizing vegetation and wrack deposits slow the wind and thus, trap wind-blown sand. As the incipient dune grows taller and can support even more vegetation, more sand is trapped, which then grows the foredune even more (Fig. 21.28). Foredunes eventually get covered with dune grasses and small, brushy plants. They are younger than the larger and more established dunes further landward (behind them).

Proceeding landward, away from the shoreline, are larger, often forested, dunes (Fig. 17.24A). These larger, "inland" dunes usually have a well-established vegetation cover. These more mature parts of a coastal dune system may also contain **blowouts** – portions of the dunes where vegetation and sand have been removed by wind erosion, often enhanced by human disturbance, such as walking. Once the vegetation is removed, onshore winds are funneled into these areas, opening them up even further and creating a positive feedback loop for continued dune regression (Fig. 21.16). For this reason, many coastal areas restrict walking or driving on dunes to preserve the dune vegetation and lower the potential for erosion.

An important reason to preserve and protect coastal dunes is that they provide an inherent source of material to nourish eroding beaches. Sand stored in foredunes, when released by coastal erosion, can help to build and stabilize beaches (Fig. 17.24B). If the erosion event is only temporary, such as during a storm, much of this sand will then get stored in foredunes. However, if prolonged beach and dune erosion occur, such as during periods of high Great Lakes water levels, even additional material sourced from dune erosion may not sustain the beach.

17.9 CLIFFED COASTS

Although most people think of coastal areas as broad, sandy beaches, a large portion of the world's shorelines are dominated by steep cliffs (Fig. 17.25). Many of these cliffs are composed of rock or other consolidated material, such as clay, rendering them resistant to erosion. Thus, cliffed coasts do not change as rapidly as sandy coastlines do. However, when they do change, it is usually sudden and of high magnitude.

Figure 17.25 Cliffed coasts are naturally beautiful and attractive as tourist destinations. **A.** The view from the top of the Manneporte on the white chalk cliffs of the Falaise d'Aval, France. Source: M. Mergili. **B.** Waves crash on the basalt cliffs of Hawaii Volcanoes National Park, USA. Source: M. Miller. **C.** Ireland's famous Cliffs of Moher. Source: Suicasmo, CC BY-SA 4.0, via Wikimedia Commons.

Figure 17.26 Erosion of cliffed coasts is problematic for structures built on top of them. **A.** An eroding cliff, along Lake Michigan, in the Great Lakes of the United States. The sandy sediments in this cliff are likely contributing to the existence of the wide beach at its base, which offers some protection from storms. Source: E. Theuerkauf. **B.** Erosion threatens an apartment complex along California's Pacific Coast. Note the rocks placed at the base of the cliff, to reduce erosion. Source: B. Inaglory via Wikimedia Commons.

Cliffed coasts often recede (or fail) due to undercutting at their base, where wave energy is focused (**Fig. 17.25B**). The material comprising the base of the cliff first becomes loosened by wave action. It can then be removed by waves or wind. Eventually, the face of the cliff becomes undercut, unstable, and collapses, sometimes catastrophically. After the cliff fails, recovery is not possible. In other words, cliffed coasts that erode or collapse cannot be rebuilt. Although clifftop views are highly sought-after as prime housing sites, it can be costly to ignore the geomorphic consequences. Failure of the cliff can result in the complete loss of the home or structure built on top of it (**Fig. 17.26**).

After a cliff fails, the material that falls to the beach will temporarily stabilize the cliff and help to protect it from further undercutting. However, long-term wave attack and longshore currents will eventually remove this material, likely resulting in another cycle of undercutting and slope failure. The good news – as waves remove this material from the base of the cliff, the beach system gets infused with this new sediment, which is then carried downdrift in the longshore current. In many coastal locations, this input of sediment is an important component of the coastal **sediment budget**, which is a method

of accounting for sand gains or losses along a stretch of coastline. Cliff erosion, along with rivers that flow into the lake or ocean, is a necessary input of sediment into the nearshore zone.

Cliffs erode and collapse in different ways (**Fig. 17.27**). Most commonly, waves undercut the cliff, leading to outright collapse or mass movement along the cliff face. However, erosion can also occur in response to subaerial erosion and mass movement on the *top* of the cliff. Runoff rates, which are a function of land cover type, slope steepness, and precipitation, greatly affect this type of erosion. As a general rule, cliffs built on softer materials are also more likely to erode than are cliffs of hard bedrock.

Along some coastlines, wave erosion can form a **wave-cut platform** or **bench** at the base of a cliff (**Fig. 17.28**). This platform forms at or just above the waterline, and if on the ocean, slightly below the high tide level. Formation of a wave-cut platform occurs as waves break on the shore, dragging particles back-and-forth, abrading the underlying surface. Thus, wave-cut benches can sometimes be used as geomorphic markers of past water levels. If tectonic uplift occurs or if sea level falls, a new, lower wave-cut platform will begin to form. Eventually, if uplift continues, the

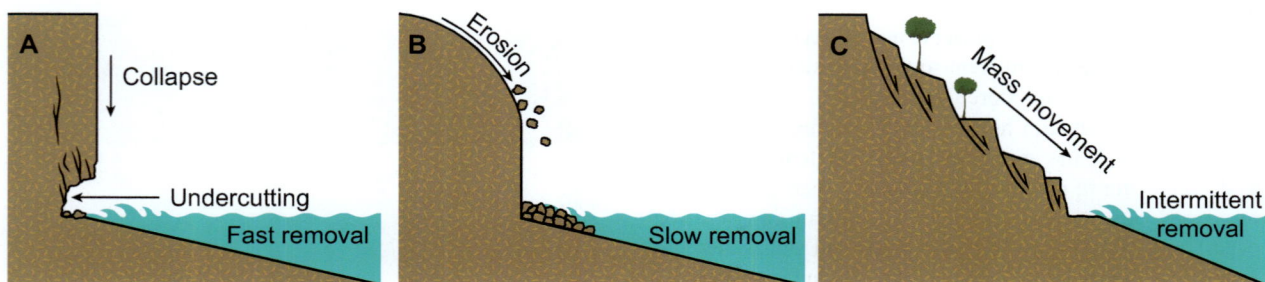

Figure 17.27 Mechanisms of cliff failure. **A.** Wave undercutting at the base of the cliff leads to instability and eventual failure. **B.** Runoff from the top of the cliff will lead to erosion from the upper parts of the cliff. **C.** Movement of groundwater through the cliff, combined with undercutting by waves, can lead to slumping and mass movements.

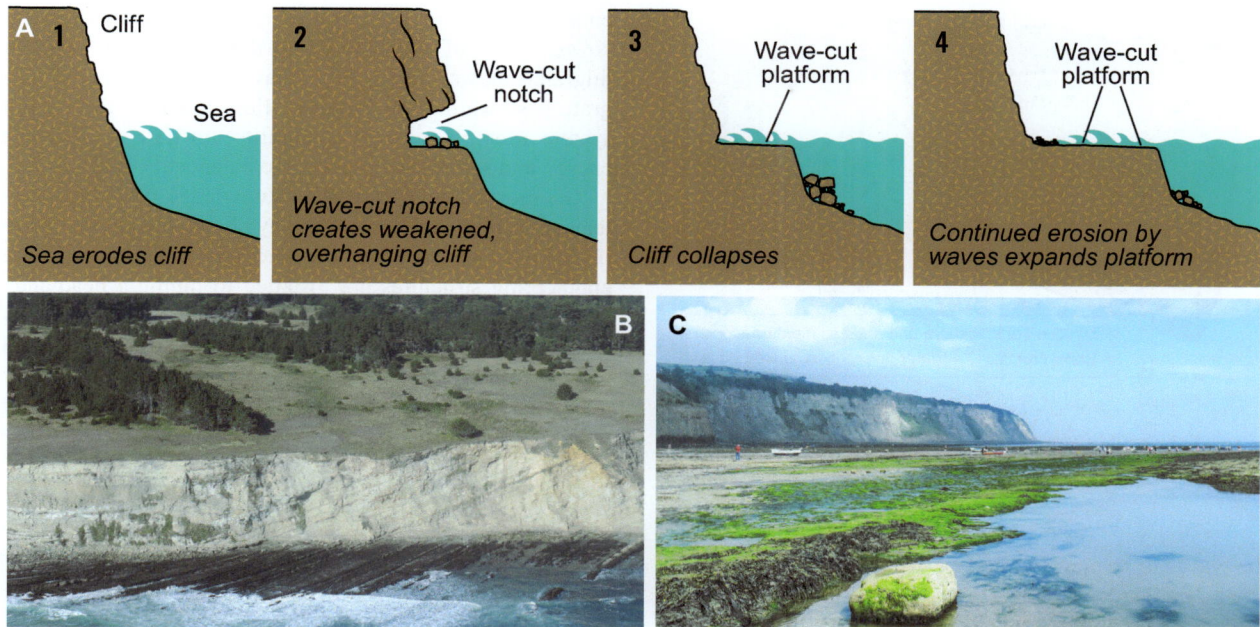

Figure 17.28 Wave-cut platforms. **A.** Evolution of a wave-cut platform follows a clear progression. Initially, a cliff is subjected to erosion (1) that leads to the formation of (2) a wave-cut notch. Eventually, this notch causes collapse of the overhanging bluff, allowing for the development of a wave-cut platform (3). Continued erosion removes the collapsed cliff material and further widens the platform (4). **B.** An example of a wave-cut platform forming at the water line along the California (USA) coast. Source: California Coastal Records Project; in public domain. **C.** A broad, wave-cut platform at low tide in Robin Hood's Bay, Great Britain. Source: A. Locking.

platform becomes isolated and exposed above the modern shoreline, as a **marine terrace** (Fig. 17.29). Application of surface exposure dating to this uplifted terrace can help develop a timetable for uplift along the coast, as has been done for parts of the California (USA) coast.

Another important geomorphic aspect of rocky and consolidated coasts is the evolution of headlands and embayments (Fig. 17.30). **Headlands** are areas that protrude out, into the water, whereas **embayments** (or bays) are sheltered

areas where the shoreline curves back towards the land. Think of headlands as land masses that protrude out, into the water, and embayments as the curved, often sandy, bays between headlands.

The curvature of a coastline, including its headlands and bays, impacts the coastal system in several ways. As waves approach the shore, their wave fronts bend and converge onto headlands by a process known as **wave refraction** (Fig. 17.30). Wave refraction causes the erosive power of

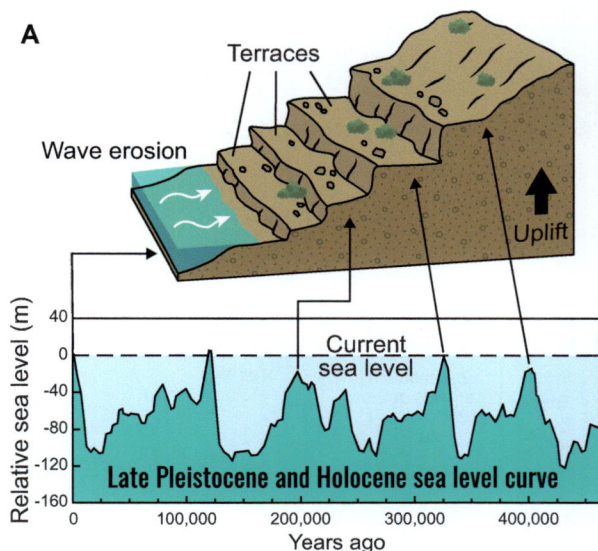

Figure 17.29 Marine terraces. **A.** Schematic diagram illustrating how uplifted marine terraces form – from the processes associated with coastal erosion that form the wave-cut platforms, as well as uplift of the land mass. Source: After the US Geological Survey Fact Sheet 2018–3002; in public domain. **B.** Uplifted marine terraces on San Clemente Island, off the California (USA) coast. The lowermost surface/terrace formed during a high sea level stand at ≈120,000 years ago. Source: D. Muhs.

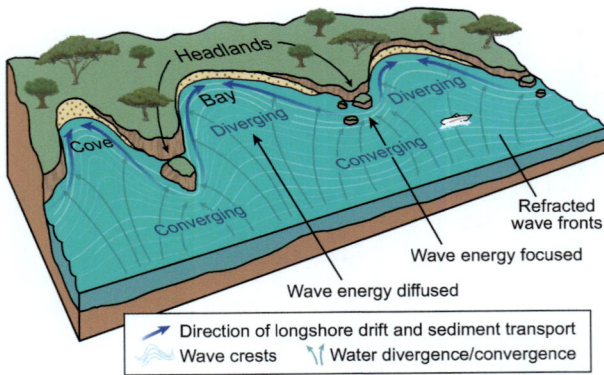

Figure 17.30 Wave processes and morphology of a coastline with headlands and bays. Wave energy converges on the headlands, eroding them. Waves diverge in bays, resulting in sand deposition there.

the waves to be concentrated on headlands. Rock is eroded from headlands and weathered into sand. Conversely, waves approaching embayments (bays) diverge due to refraction, dispersing their energy and forming smaller waves that gently wash up onto broad, sandy beaches (**Figs. 17.30, 17.31**). Over time, material is continually eroded from headlands and transported to embayments by longshore currents. As stated above, erosion of the headlands provides sediment for beaches – in this case, in the

embayments. Over time, these processes lead to a general *straightening* of the shoreline as the headlands regress, or migrate landward, and the embayments prograde, or migrate basinward. This example illustrates how the coastal system attempts to reach an equilibrium state where wave energy and sediment transport are balanced.

Geomorphologists use their knowledge of coastal systems to inform engineers who may choose to modify shorelines with various kinds of structures. Sometimes, these engineering structures are installed to mimic a headland–embayment configuration, forming wide beaches in embayments (where they otherwise might not occur). We will discuss these types of shoreline modifications later, but for now it is important to know that many of the geomorphic processes that occur naturally are also at work along "engineered" coasts.

Progressive erosion of rocky headlands often leads to a sequence of coastal landforms that are magnificent and unique (**Fig. 17.32**). As wave energy is concentrated on headlands, openings in the rock near the water line get exploited, widen, and may evolve into **sea caves**. Continued erosion sometimes enlarges the cave to form an **arch**. Sustained erosion will eventually lead to collapse of the arch, forming a **sea stack**. Finally, the sea stack will erode down to a **stump**, and if erosion continues, no trace of the former headland will remain above the water line. Spectacular examples of sea caves, arches, and stacks exist all over the world, but

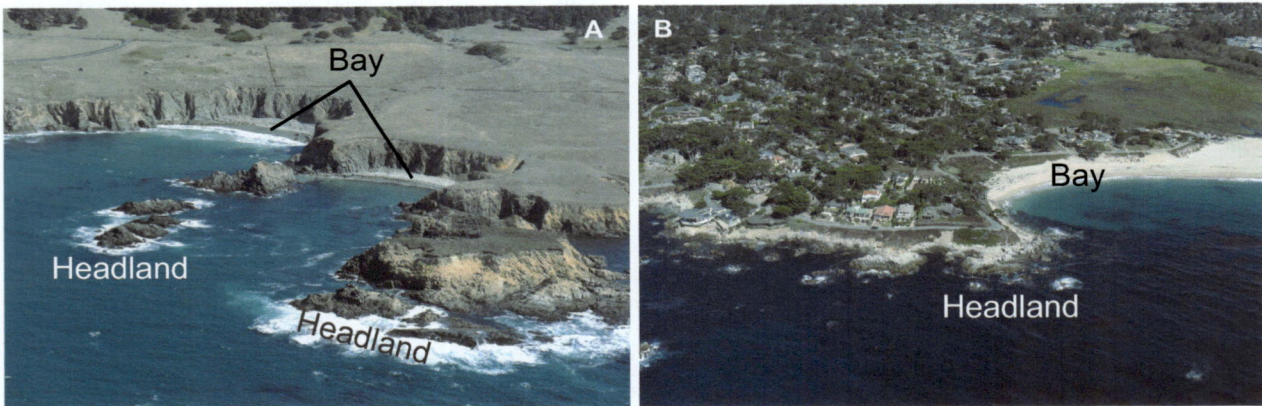

Figure 17.31 Oblique aerial views of classic headland-bay coastlines along the California (USA) coastline. **A.** A site ≈140 km north of San Francisco. **B.** A site on the Monterey Peninsula, near Carmel. Source: California Coastal Records Project; in public domain.

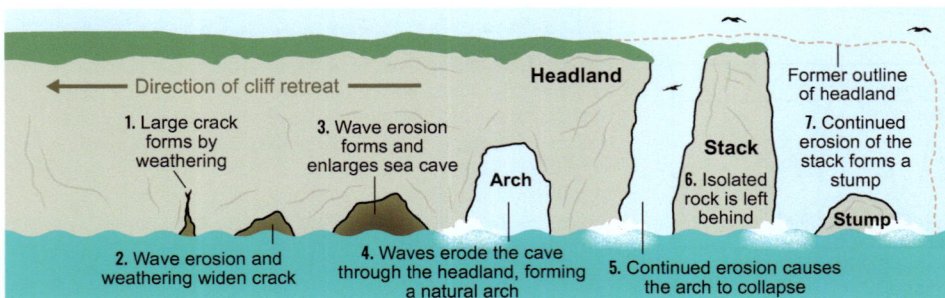

Figure 17.32 Illustration of the evolutionary sequence of headland erosion, as it forms sea caves, arches, stacks, and stumps.

Figure 17.33 Photos of various coastal landforms formed by headland erosion. **A.** Fingal's Cave, Isle of Staffa, Scotland. Source: Akela NDE, CC BY-SA 2.0, via Wikimedia Commons. **B.** Great Pollet Sea Arch in Ireland. Source: Pollet Great Arch by John M, CC BY-SA 2.0, via Wikimedia Commons. **C.** Haystack Rock, along the Oregon (USA) coast, an excellent example of a sea stack. Note the small stumps next to it. Source: Kirkouimet, CC BY-SA 4.0, via Wikimedia Commons.

some of the most famous include Fingal's Cave in Scotland, the Great Pollet Sea Arch in Ireland, the Twelve Apostles in Australia, and Haystack Rock in the United States (**Fig. 17.33**).

17.10 TROPICAL ISLAND COASTS: ATOLLS, BARRIER REEFS, AND FRINGING REEFS

Atolls, barrier reefs, and fringing coral reefs are coastal landforms found in tropical regions that often surround islands. They represent different evolutionary stages of the coastal system, as it responds to sea level rise and/or land subsidence. The famous naturalist Charles Darwin developed his **theory of atoll formation** in the mid-1800s, during his voyage on the *HMS Beagle* in the southern Pacific Ocean. It is still accepted today. Darwin suggested that atoll formation begins with a dormant or extinct volcanic island (or chain of islands). A **fringing coral reef** develops and grows along the edge of the island (**Fig. 17.34A1**). All the while, the island is slowly subsiding (or sea level is rising). As subsidence continues and/or sea level rises, the island's fringing reef grows upward and transitions into a **barrier reef**, separated from the island by a lagoon (**Fig. 17.34A2**). Finally, after the island becomes submerged and/or more eroded, the coral reef transitions into an **atoll**, in which coral, at or just above sea level, rings a lagoon (**Fig. 17.34A3, B**). Atolls are highly susceptible to future changes in sea level because their form is a consequence of the interplay between land subsidence

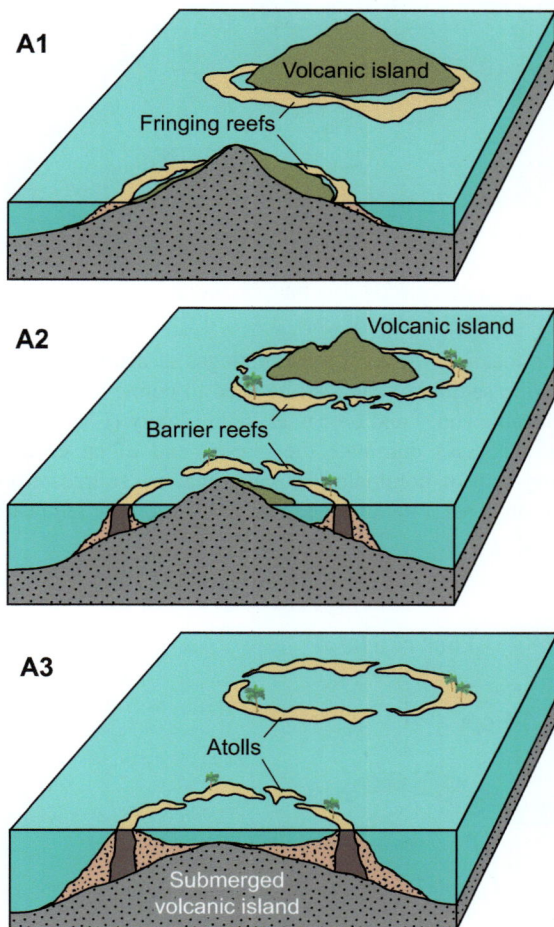

Figure 17.34 Atolls. **A1.** Initially, a fringing reef surrounds an inactive volcanic island. **A2.** As sea level rises and/or the island submerges, the fringing reef transitions into a barrier reef. **A3.** Over time, the island will become completely submerged by erosion, even as the barrier reef becomes an atoll. **B.** Atafu Atoll, in the central Pacific Ocean, a classic atoll, is shown here at low tide. Source: NASA.

and sea level. Many atolls support large human populations, such as in the Marshall Islands. These landscapes and the people living on them are becoming increasingly vulnerable to sea level rise.

17.11 ESTUARINE COASTS

At the interface between terrestrial and marine environments lie **estuaries** – flooded river valleys where freshwater from the river mixes with saltwater from the ocean. This mixing of fresh and saltwater plays an important role in the geomorphic evolution of estuaries, as their different densities can generate currents that erode, transport, and deposit sediments. Estuaries are sediment sinks, obtaining sediment from both the river and the ocean. During postglacial time, as sea level has risen, estuaries grew in size, making space for even more sediment. At times of sea level drop, rivers can incise into the freshly exposed sediment.

Estuaries evolve via changes in sea level, often driven by glacial cycles but also sometimes by tectonic activity. Tectonically derived estuaries are located along leading edge continental margins, such as the Alaskan coast of the United States. Here, the continental crust is sinking, relative to sea level. **Leading edge** estuaries tend to be elongated and narrow, and often occur along fault lines. The other, more common type of estuaries occur at drowned river mouths on the **trailing edges** of continents, such as the eastern seaboard of the United States. They result from postglacial sea level rise along an otherwise stable continental margin, where rivers enter the ocean along a coastal plain (**Fig. 17.35**). Other types of estuaries include **fjords**, which are glacial valleys that have been submerged by the rising postglacial seas (**Fig. 19.8**). In some instances, such as along the California coast, estuaries can form and evolve in response to both tectonic activity *and* sea level rise.

The fluvial and marine boundaries of estuaries are often defined by salinity (**Fig. 17.36**). Although this method seems straightforward, changes in freshwater input and circulation patterns can quickly alter the boundary between freshwater, sea water and the brackish water in-between.

Figure 17.35 A satellite image of Chesapeake Bay, the largest estuary in the United States and an excellent example of a drowned river valley estuary on the trailing edge of a continent. Source: NASA.

The locations of certain assemblages of sediment (such as marine sediment) can also be used to define the various geomorphic zones and processes in an estuary (**Fig. 17.36**). Wave processes dominate the marine end of an estuary, causing it to be dominated by sands that are sourced from the ocean coast. Fluvial processes, at work on the landward

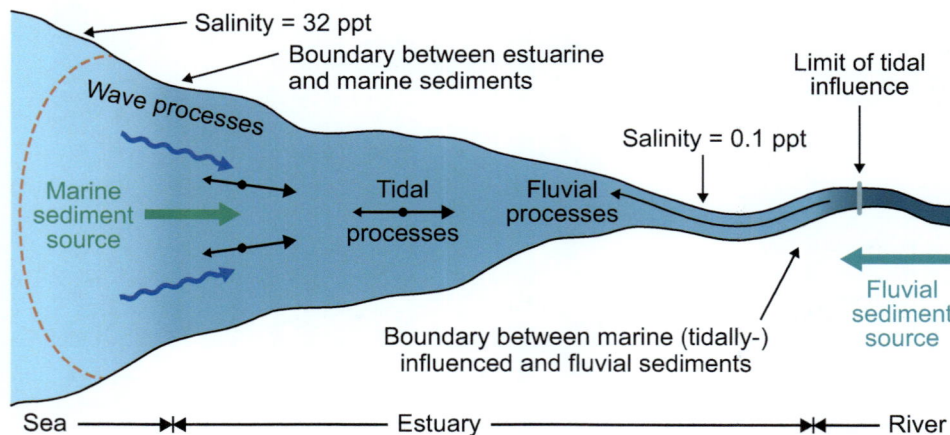

Figure 17.36 The various zones of a typical estuary, highlighting the limits of fluvial, tidal, and marine processes, as well as salinity variations. Source: After Dalrymple, R. W., Zaitlin, B. A., and Boyd, R. (1992), © Society for Sedimentary Geology.

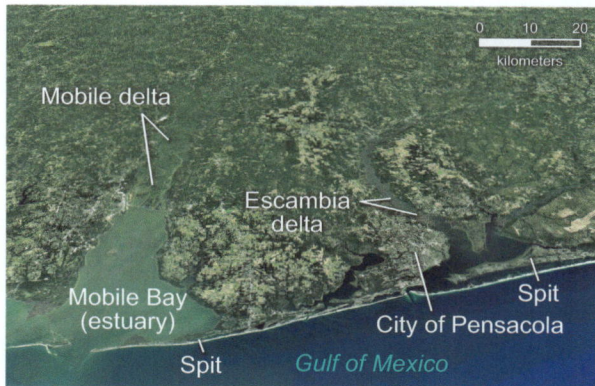

Figure 17.37 Satellite image of the Mobile Bay area of Alabama, USA, illustrating the various coastal landforms that have formed in this area. Mobile and Escambia deltas are both bayhead deltas. Source: NASA.

margin, mainly deposit silts and muds in the upper portions of the estuary. Tides in the estuary begin at the ocean side and end somewhere upstream. Therefore, the location of the transition between sands and muds is often determined by tidal influences. By identifying these sediment boundaries, researchers can then examine how and why the estuary has evolved in response to changing conditions such as storms and sea level rise.

The relative influence of fluvial vs marine processes has a strong control on the evolution of an estuary. If fluvial sediment supply exceeds the estuary's **accommodation space** (the space between the water surface and the bed that is available for sediment to accumulate), the upper end of the estuary will slowly begin to fill with alluvium. In this situation, a **bayhead delta** may form in the upper end of the estuary. Excellent examples of estuaries that contain bayhead deltas include Galveston Bay (Texas) and Mobile Bay (Alabama) in the United States (**Fig. 17.37**). Estuary evolution is also influenced by marine processes at its mouth, which supply marine sediments, but more importantly can deliver wave energy that redistributes sediments throughout the lower portion of the estuary. Over time, the relative influence of fluvial and marine processes may change as the estuary system evolves.

17.12 DELTAS

Deltas are named for their shape – the Greek letter delta (Δ). They develop where a river meets a large water body. For a delta to form and persist along a coast, a river must discharge more sediment than tidal and wave processes can remove and redistribute. Additionally, adequate accommodation must exist within the water body for these sediments to accumulate. **Passive plate margin** (or **trailing edge**) coasts with extensive river systems are best suited for delta formation because shallow water usually occurs just offshore. **Active plate margins**, or **leading edge** coasts, tend to have steeper dropoffs to deep water. Thus, it is more difficult for sediments to

accumulate, minimizing delta formation except in locations where sediment inputs from rivers is extremely high.

Deltas have always been important areas for human settlement due to their fertile soils and easy access to water. Deltaic locations have freshwater resources from the river, while also offering access to maritime commerce on the open ocean. Thus, humans have played an important role in the geomorphic evolution of many deltas. The largest influence humans have had on delta evolution involves the sediment delivery of the rivers that enter the deltas. Clearcutting land for agriculture and logging in upstream areas destabilizes the land surface and adds to the sediment load of rivers, helping to grow deltas downstream. Dams on rivers are a major barrier to sediment delivery to deltas, as they both restrict the discharge downstream and trap sediment, starving the delta. The Mississippi Delta is an excellent example of this type of human impact. Navigational locks and river diversions, such as those along the Nile River, also inhibit sediment discharge to deltas. The combination of postglacial sea level rise on the marine boundary of deltas and reduced sediment discharge on the terrestrial side has placed most deltas (and their associated ecosystems and human infrastructure) in peril.

Sea level plays an important role in delta formation and evolution. During the Pleistocene (when sea level was much lower than today), many rivers were flowing across the continental shelf and depositing sediments directly at the shelf edge. As the ice sheets started melting, sea level rose too rapidly for delta growth to keep pace. It was not until the rate of sea level rise slowed, ≈6,000 to 7,000 years ago (**Fig. 17.3A**), that most modern deltas began to form.

Deltas grow as sediment is deposited at the river mouth. How does this process develop over time? Initially, rivers form a series of **distributary** channels that branch and break off from the main river channel (**Fig. 17.38**). Distributary channels form when a river, entering a standing water body, is forced to drop much of its sediment in the slower currents of the standing water body. Much of this sediment is deposited immediately *in front of* the river mouth, forcing the stream to **bifurcate** (branch). Distributary channels continually migrate and split, or **avulse**, on the delta, as sediments accumulate at their mouths and push the channel in a new direction (**Fig. 17.38**). As this process continues, the delta grows and more channels form. Continued additions of sediment at the mouths of these distributary channels eventually generate the familiar deltaic form – provided that deposition of sediment exceeds the abilities of waves and tides to remove it. Some deltas have only a few, large distributary channels, while others are a maze of small distributaries that criss-cross the delta plain. As you can see, the dominant morphologic features of the **delta plain** are distributary river channels. These channels form via the processes of sediment deposition and avulsion, and as new channels are formed, older ones fill in with sediment during floods. The delta is a dynamic landscape indeed!

Most deltas can be subdivided into three zones. From land to water, they are the (1) **delta plain**, (2) **delta front**, and

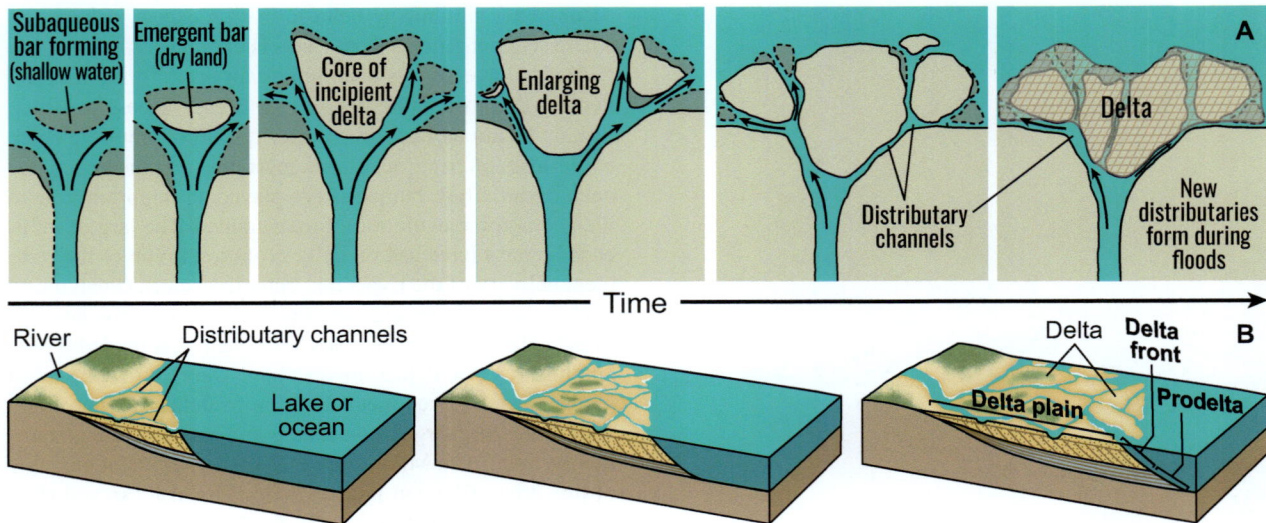

Figure 17.38 Schematic representations of the formation of deltas and their distributary channels, over time. Black arrows in **A** indicate flow in the channel. It is important to note that delta sediments not only deposit laterally as the delta plain and front evolve, but also get thicker by aggradation, as denoted in **B**.

(3) **prodelta** (**Fig. 17.38B**). The delta plain is the broad, low-lying, upper part of the delta, where fluvial processes play important roles. Rivers here meander lazily across the flat, swampy landscape. Marshes and tidal flats are common features of the delta plain. In contrast, marine processes are primarily at work on the delta front and prodelta. On the steeply sloping delta front, deltaic deposits grade downward to the sea floor. In the prodelta, the part of the delta furthest out to sea and forming in deep water, deposition is slow.

Let's examine the Mississippi River Delta, with its extensive areas of salt marshes. In order for the Mississippi's delta plain to continuously grow outward and negate the effects of land subsidence (common in deltas, because of the heavy load of sediments), sediment must continually replenish the delta plain. Loss of elevation relative to sea level rise, a major issue on the Mississippi Delta, makes this delta (like many others) highly vulnerable to coastal hazards such as hurricanes. Many parts of the City of New Orleans, on the delta, are at or below sea level due to a combination of delta subsidence and sea level rise. Hurricane Katrina, which struck the US Gulf Coast in 2005, brought a devastating storm surge to the Mississippi Delta and the New Orleans area. This type of vulnerability will only increase as sea levels continue to rise. The rising waters may completely inundate large portions of some deltas, making them unsuitable for human occupation.

The delta front lies beyond the delta plain, where it meets the marine environment. Waves, currents, and tides are at work here, redistributing sediment deposited at the mouth of the distributary channels. Bars commonly formed at the mouth of distributary channels also comprise part of the delta front (**Fig. 17.38B**). Even farther out, fine-grained material rich in silt and clay gets carried by waves and currents to the distal portions of the delta (the prodelta).

As we have discussed, deltas form and evolve due to the complex interplay between river, tides, and waves. The dominance of any one of these processes leads to distinct morphologies that form the basis for the standard delta classification scheme. That is, deltas may be (1) river dominated, (2) tide dominated, or (3) wave dominated (**Fig. 17.39**). Deltas that are dominated by a specific process are placed at the end members, but in reality, most deltas are hybrids of some type.

River-dominated deltas form where discharge of sediment from rivers is high and wave and tidal influences are low, enabling the river to build the delta rapidly out, into the water. Fluvial deposits of this type sometimes form long, finger-like extensions into the sea, giving the common name "bird's foot" to these types of deltas. Bird's foot deltas exhibit well-developed delta plains, characterized by long distributary channels (**Fig. 17.39B**). River-dominated deltas like the Mississippi Delta are typically found along gently sloping and broad continental shelves (on passive plate margins), where ample room is available for sediment to accumulate.

Tide-dominated deltas are found in locations where strong tidal currents overwhelm wave and river processes (**Fig. 17.39E**). As water rushes up (or down) the channels with the rise and fall of the tide, strong tidal currents form in the delta distributary channels. The daily ebb and flow of these currents redistributes sediments within the channels. Many of the distributaries on tide-dominated deltas become noticeably wider at their mouths due to the continual upstream-downstream, ebb-and-flow of tidal waters. These processes cause tide-dominated deltas to develop extensive tidal flats and associated marshes. The Ganges-Brahmaputra Delta in India is an example of a tide-dominated delta (**Fig. 17.39E**). Although large volumes of sediment are delivered to the delta during the monsoon season, they are eventually redistributed by tidal currents.

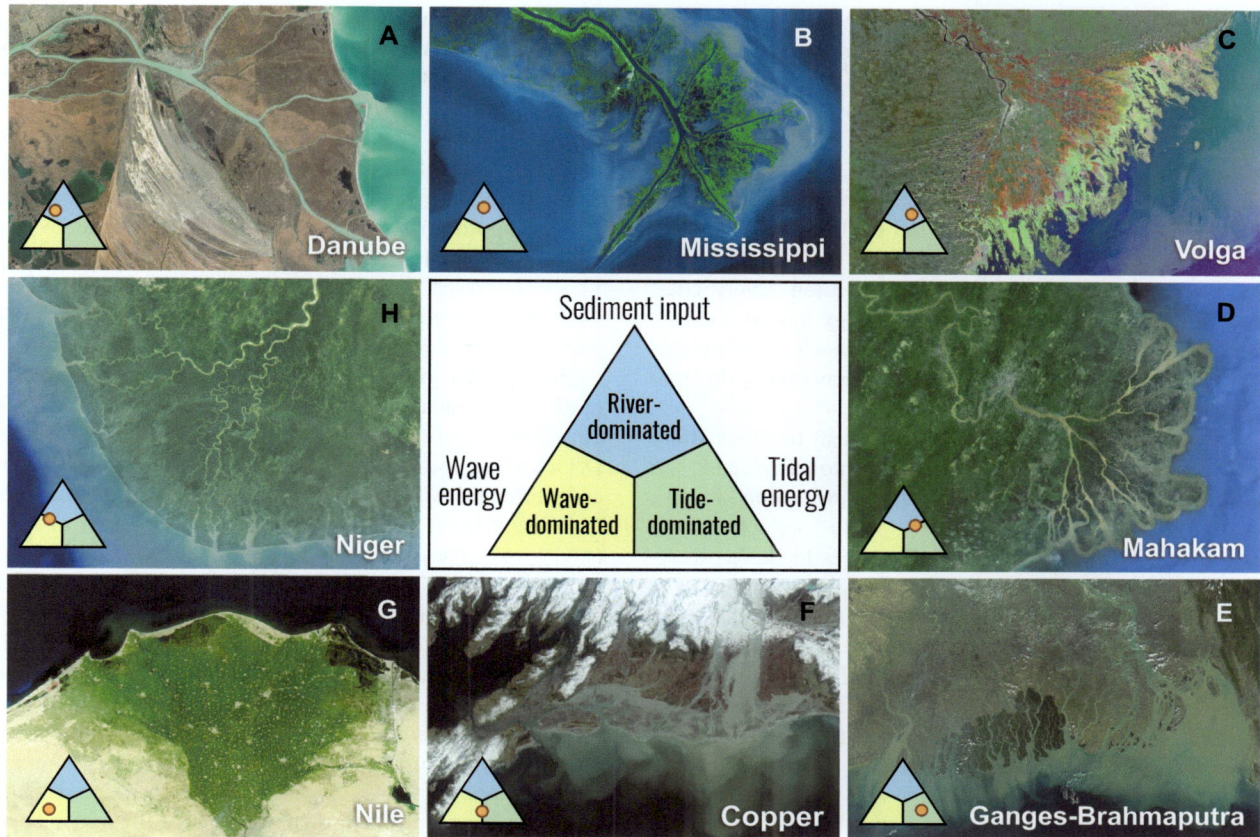

Figure 17.39 Ternary diagram of the various delta types and the dominant physical processes that shape their geomorphology. Examples of typifying deltas include (**A**) the Danube in Romania and Ukraine, a mixed wave- and river-dominated delta, (**B**) the Mississippi, a classic, river-dominated, "bird's foot" delta, (**C**) the Volga in Russia, a river-dominated delta with some tidal influence, (**D**) the Mahakam (Indonesia), a mixed river- and tide-dominated delta, (**E**) the Ganges-Brahmaputra, a tide-dominated delta, (**F**) the Copper in Alaska, a mixed wave- and tide-dominated delta, (**G**) the Nile in Egypt, a wave-dominated delta, and (**H**) the Niger in Nigeria, a mixed wave- and river-dominated delta. Source: (A, E) NASA Earth Observatory; (B, D, F, G, H) NASA; (C) European Space Agency.

Wave-dominated deltas are shaped by waves and longshore currents (**Fig. 17.39G**). These two processes work together to redistribute sediment brought to the coast, transporting it parallel to the coast, via longshore currents. As a result, wave-dominated deltas are wide and broad, rather than extending far into the sea (**Fig. 17.39G**). In locations where a strong directional control on longshore currents is lacking, the delta will develop a symmetrical arcuate shape, often with only a single distributary channel near its center. In such instances, the delta may develop long spits and barrier islands in the direction of the longshore currents. A classic example of a wave-dominated delta with spits and barrier islands along the delta front is the Niger Delta in Nigeria (**Fig. 17.39H**).

As mentioned earlier, most deltas are intermediate in form, shaped by both fluvial and marine processes. The Danube River Delta is an excellent example of such a delta (**Fig. 17.39A**). Sediment discharge is high enough that multiple distributary channels have formed, but waves in the Black Sea continue to smooth out the delta front.

17.13 COASTAL MANAGEMENT AND PROTECTION

People have lived in tandem with coastal systems for millennia. In some cases, we have learned to live in harmony with natural coastal processes, whereas at other times civilization and economic development have interfered with these processes. Therefore, it comes as no surprise that people have been modifying the coast for millennia. Nonetheless, "development" can have drawbacks, and is often at odds with natural coastal processes. For example, barrier island erosion and overwash during storms become hazards when immovable infrastructure, such as homes and roads, are situated on the island. To protect coastal infrastructure from erosion and/or flooding, people attempt to "manage" the coastal system. Broadly defined, **coastal management** refers to efforts to protect lives and property from coastal hazards. Coastal management can take a variety of forms, ranging from nourishing a beach with sand after a storm to installing concrete seawalls to hold the shoreline in place.

Historically, coastal management has been applied as a reaction to coastal erosion. However, new approaches to management that are more *proactive* are becoming popular. One such approach, applied in locations where erosion is rampant, is to simply retreat away from the shoreline and its associated hazards. New houses and structures are built farther from the coast, and existing structures are moved inland, thereby allowing the natural evolution of the shoreline to occur. Other strategies involve the installation of engineered structures that help slow erosion. Every coast is different and each management strategy has an intended purpose. However, some strategies can result in unintended consequences, especially for locations immediately down or up the coast.

A common type of coastal management involves **shore protection**, which can be divided into *hard* and *soft* approaches. As the name implies, shore protection involves ways to inhibit coastal erosion, that is, to "protect" it from **regression** (or flooding). Hard approaches to shore protection erect barriers along the coast, using materials such as steel, wood, and rock. Soft approaches utilize more natural materials such as sand and vegetation to slow erosion. Soft strategies typically aim to slow down, rather than completely stop, coastal erosion. Often, the type of strategy used relates to whether the goal is simply to protect the land behind the shoreline (and the structures that may be situated on it) or to maintain a sandy beach. Each approach requires a different strategy, and no one strategy can typically accomplish both goals.

The most common soft strategy of shore protection is **beach nourishment** – placing additional sand on a beach in an attempt to rebuild it. Although beach nourishment is very expensive, for some communities where beachfront property is highly valued, such as Miami Beach, Florida (USA), it can be justified. In a typical year, Miami Beach recently spent $40 million on beach nourishment for critically eroded sections of the coast, prompting many to question the economics of this approach. Sand can be dumped from trucks and graded into landforms such as berms and even dunes, to mimic natural systems. Sometimes, planting dune

grass is done in combination with nourishment to stimulate growth of foredunes. One advantage of beach nourishment is that it does not impede the natural processes of erosion and longshore drift. Nonetheless, it is not a permanent solution, as the same erosive processes that generated the need to nourish will likely remove the material again. Beaches that require nourishment often need it time and time again.

Hard approaches to shore protection can be subdivided into "shore-parallel" and "shore-perpendicular" types (**Fig. 17.40**). Shore-parallel strategies include **breakwaters**, seawalls, and revetments, and shore-perpendicular strategies include groins and jetties.

Perhaps the most sustainable hard approach to coastal erosion and management are **detached breakwaters**. They are essentially offshore walls, built parallel to the shore. Detached breakwaters are very effective at widening beaches and limiting coastal erosion because they slow (but do not technically interfere with) the longshore transport of sand (**Fig. 17.40**). They are meant to "break" the wave energy and protect the coastline behind them from erosion. They cause the beach to widen and grow, much like the formation of a tombolo. The negative aspect to detached breakwaters is their cost; installation of these features in deep water can be quite expensive. Breakwaters can also help to minimize shoaling inside harbors, which is why they are often placed just offshore from harbor entrances. Not all breakwaters are built offshore. Some are attached at one end to the coast, and some are emergent (with their tops above the water line), while others are submerged. However, in all instances they represent an expensive, but often effective, option for coastal management.

Another type of shore-parallel structure used for hard shore protection is a **seawall**. Metal, wood, or concrete seawalls are built along an eroding shoreline to "hold the line," in an attempt to prevent further erosion and flooding of the land behind them (**Figs. 17.40, 17.41**). Although seawalls can be an effective short-term strategy for minimizing erosion, they will eventually be undercut or broken by waves, causing them to collapse and fail. The process of seawall failure is straightforward. As waves impinge upon a seawall,

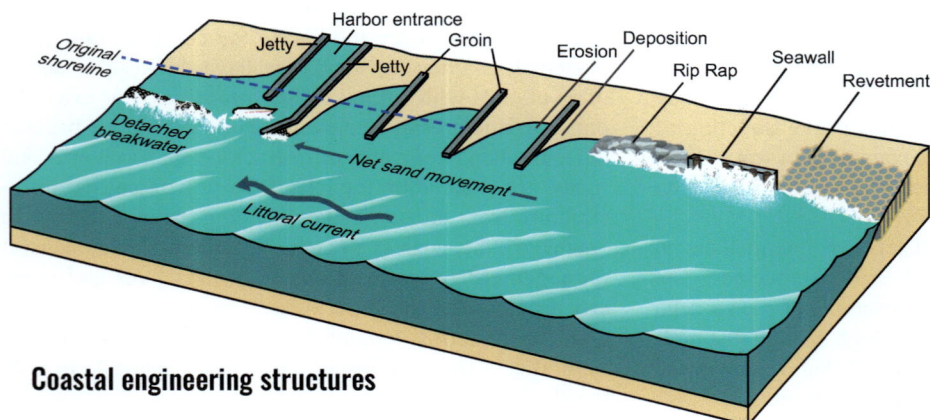

Coastal engineering structures

Figure 17.40 Many of the typical hard shore protection structures used to control coastal erosion and/or slow longshore sediment transport.

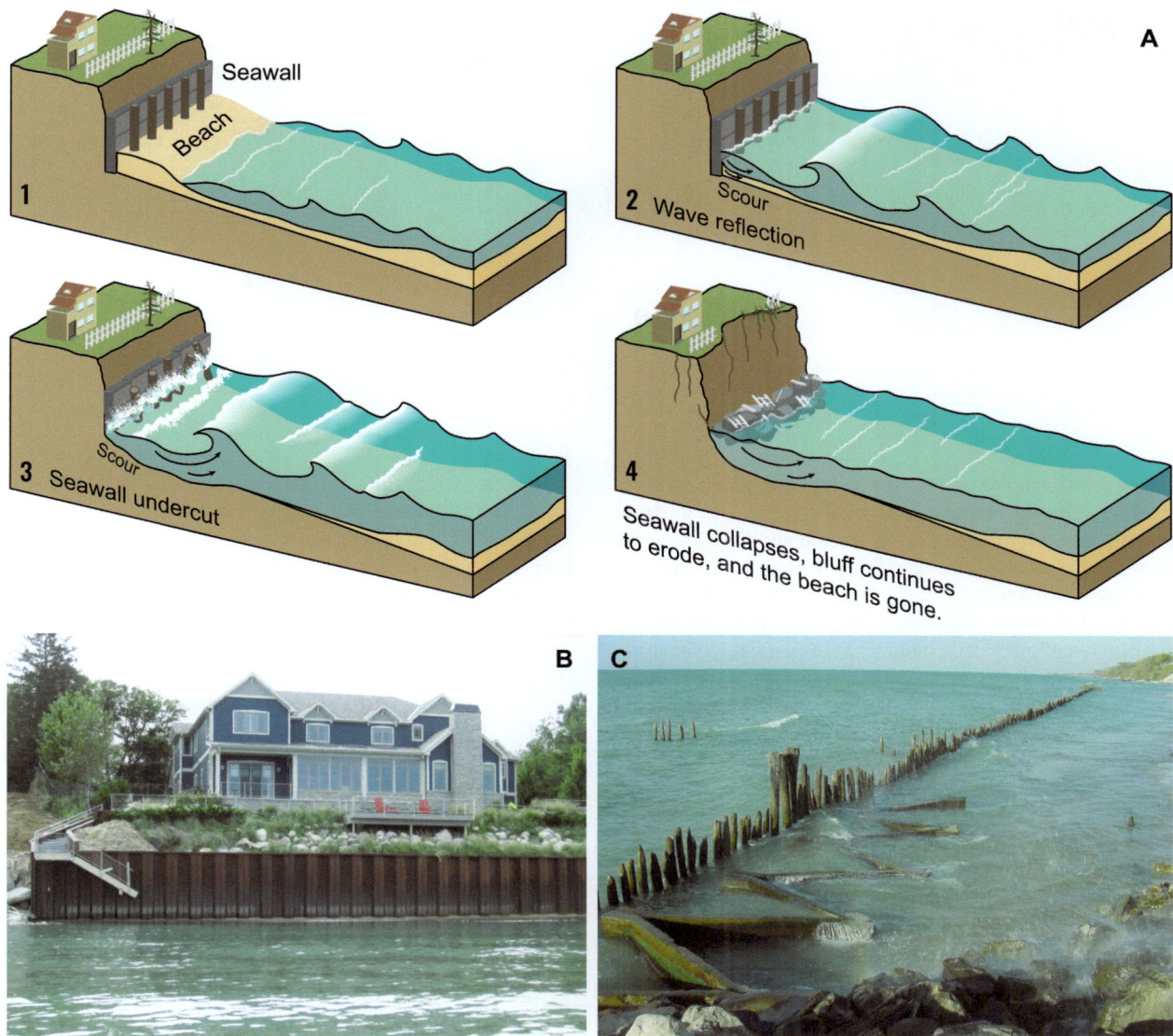

Figure 17.41 Seawalls. **A.** An illustration of how a seawall, initially intended to prevent coastal erosion, eventually fails. Source: After Marshak © 2019. Used by permission of W. W. Norton & Company, Inc. **B.** This steel seawall is protecting the house behind it, but note also that the beach in front of it is now gone, as evidenced by the staircase that ends well above the waterline. Source: G. Meadows. **C.** This landowner has seen the demise of three different seawalls along the Lake Michigan coast, in Michigan, USA. As a last resort, rip rap (in the lower right part of the image) has been laid down in the area behind the failing seawalls. Source: R. Schaetzl.

their energy is deflected downward, accelerating the erosion of the sediment at the base of the wall (**Fig. 17.41A**). Over time, this "reflective scour" will destabilize the wall and lead to failure, as well as loss of any beach sand that may have existed in front of the wall. Many coastal residents spend considerable time, effort, and money to install seawalls, only to find that, a few years later, their investment was for naught – their seawall has collapsed. Thus, it is important to note that seawalls erected behind (landward of) a beach almost always end up destroying the beach anyway. Clearly, the axiom that "seawalls destroy beaches" often rings true.

A particularly useful (yet still expensive) shore-parallel structure is known as a **revetment**. Revetments are similar to seawalls, but built to mimic a sloping surface rather than as a seawall, which is vertical. They are often made of large blocks of rock or concrete or stone, called **rip rap**, and are intended to reduce erosion by absorbing and dispersing the wave energy, instead of reflecting it back, as seawalls do (**Fig. 17.42**). Thus, an alternative name for these features is **rip rap revetment**. Revetments minimize (but do not eliminate) scour. Rip rap can also be placed along the shoreline as protection without being formed into a revetment (more like a short seawall). In all these shore-parallel designs, coastal erosion still occurs underwater, leading to the eventual failure of the structure and increased impacts from storm waves.

Figure 17.42 Rip rap. **A.** Rip rap (limestone blocks) has been placed randomly along this Lake Michigan coast in Wisconsin, USA. Source: R. Schaetzl. **B.** Heavy machinery is carefully placing rip rap along this coastline, forming a sloping revetment. Source: G. Meadows.

A second group of options for coastal protection involves structures erected perpendicular to the shore. These structures are generally meant to stop erosion and/or enhance deposition of sand along the shoreline by blocking or slowing the transport of sand moving in longshore and beach drift, allowing it to accumulate. Thus, the beach gets wider (**Fig. 17.40**). Two of the most common shore-perpendicular structures are jetties and groins. **Groins** are engineering structures consisting of a wall or a series of rocks, installed perpendicular to the shore. They minimize erosion *and* widen the beach by capturing sand moving in longshore drift. Groins (like jetties) work well for their intended purpose. They protect the shoreline on their updrift side, where they trap sand and widen the beach (**Figs. 17.12, 17.40**). However, sand captured by groins is often offset at downdrift locations, where the beach becomes starved of sand, causing the beach to narrow (**Figs. 17.12, 17.40**). This consequence is a common outcome of groins. Thus, landowners often feel compelled to install additional groins in the downdrift direction, to compensate for the effects of the one updrift. As the saying goes, "Groins beget more groins." One groin soon necessitates the installation of another, until the coast becomes a groin field (**Fig. 17.43**).

Jetties, like groins, mitigate coastal erosion by trapping sand. The difference stems from the locations where they are installed. Jetties are walls, usually installed in pairs at harbor mouths, where a river enters a larger body of water. One wall is installed on either side of the river channel, to minimize shoaling within the river mouth (**Fig. 17.12**). **Shoaling** (shallowing of the river by sediment deposition in the channel) would impact shipping and water transportation up and down the river. Like groins, jetties also trap sand on their updrift side. Their utility lies in the fact that they prevent sand from entering the harbor mouth, where it could accumulate and impede navigation. But just as groins do, jetties create sand starvation and beach erosion on their downdrift sides.

As development along coasts increases, and climate change results in rising sea levels and more extreme storms, the demand for coastal management actions will undoubtedly grow. To promote effective and sustainable coastal management, we must understand the intricacies of coastal processes. Fundamental coastal geomorphology concepts that you have learned in this chapter, such as longshore sediment transport and beach profile evolution, are the foundation for sustainable management strategies now and into the future.

Figure 17.43 A satellite image of a groin field along the Lake Michigan coast in Chicago, Illinois, USA. Note the sand accretion on the updrift side of each groin, resulting in sand starvation and erosion downdrift. Source: NASA.

REVIEW QUESTIONS

17.1   What are the various kinds of coasts, and what does each look like?

17.2   Describe how water levels fluctuate in coastal areas (e.g., sea level, tides, and storms). What physical processes cause these fluctuations?

17.3   Explain how sea level rise results in enhanced coastal flooding during both fair-weather conditions (as nuisance flooding) and during storms.

17.4   Describe the differences between spring and neap tides, and explain the reasons for these differences.

17.5   Describe the journey a wave takes from moving onshore as a deep-water wave, to eventually breaking and then running up the beach as swash. Explain when and why waves break, as well as how these wave processes relate to nearshore sediment transport.

17.6 Describe how longshore currents form. Next, discuss the role longshore currents play in coastal sediment transport and the formation of coastal landforms such as spits and tombolos. How do these differ from rip currents?

17.7 Sketch a beach profile from the dunes onshore to deep water. Describe how the features in this zone change from summer (fair-weather) to winter (storm) conditions. What are the geomorphic processes at work as the beach profile evolves from one condition to another? From one season to another?

17.8 What are tsunami waves and how do they typically form? How do they differ in form from traditional waves?

17.9 What is a spit, how does it form, and how are spits differentiated from baymouth bars?

17.10 What are the main parts of a barrier island complex? Where are barrier islands typically found? What geomorphic processes shape barrier islands on both long and short temporal scales, and how do these processes help barrier islands respond to rising sea levels?

17.11 Describe and discuss the three main theories of barrier island formation.

17.12 What are foredunes and what is their importance to the coastal system?

17.13 Describe the processes that lead to the formation of sea stacks. Start with a coastal headland that is exposed to wave erosion. How does this evolutionary process differ from both the formation of a wave-cut platform and a marine terrace?

17.14 Describe the process of wave refraction and how it causes coastlines to become "straighter" over time.

17.15 What is an estuary and how does it form? Describe the main parts/components of a typical estuary.

17.16 Describe how deltas form. Be sure to mention the process of avulsion and the role that distributary channels play in the formation of a delta.

17.17 What three main groups of processes or systems control the shapes of deltas? What is the typical delta morphology that develops within each of these?

17.18 How do atolls form and why are they particularly vulnerable to sea level rise?

17.19 Describe the differences between hard and soft approaches to coastal protection. Give an example of each.

17.20 Discuss some of the unintended impacts of shoreline protection and describe the coastal processes that are at work to generate these impacts.

17.21 Explain at least two ways that both global temperature rise and the resulting rises in sea level will affect coastal landforms and systems in the future.

FURTHER READING

Davidson-Arnott, R., Bauer, B., and Houser, C. 2019. *Introduction to Coastal Processes and Geomorphology*. 2nd ed. Cambridge University Press.

Davis, R. A., Jr. and Fitzgerald, D. M. 2020. *Beaches and Coasts*. 2nd ed. Wiley.

Jackson, D. W. T. and Short, A. D. 2020. *Sandy Beach Morphodynamics*. Elsevier.

Komar, P. D. 1998. *Beach Processes and Sedimentation*. 2nd ed. Prentice-Hall.

McCoy, K. and Bascom, W. 2020. *Waves and Beaches: The Powerful Dynamics of Sea and Coasts*. Patagonia Works.

Moore, L. J. and Murray, A. B. 2018. *Barrier Dynamics and Response to Changing Climate*. Springer.

Wolanski E., Day, J. W., Elliott, M., and Ranachandran, R. 2019. *Coasts and Estuaries: The Future*. Elsevier.

18 Glacial Systems

Patrick M. Colgan

Glaciers are perennial bodies of ice and snow whose movement is driven by gravity. They vary greatly in size and morphology; most glaciers cover small areas of a mountain slope, while the largest glaciers cover entire continents! Glaciers interact with the lithosphere as they erode their beds, depressing the land below them as they grow, and allowing the lithosphere to rebound as they shrink. Along the way, glaciers are effective agents of rock weathering, erosion, transport, and deposition, and important sources of water.

Glaciers add to the natural beauty of mountain and continental landscapes, both in currently glaciated landscapes and in relict landscapes formed during past ice ages. Nonetheless, their ice and water can also pose deadly hazards.

Glacial systems include the glacier and its adjacent lakes, streams, and landscapes – a system that is also closely linked to the atmosphere. The largest glaciers, **ice sheets**, are so massive that they influence Earth's gravitational field, global and local sea level, and even Earth's rotation. The study of glaciers is part of **glaciology** – the study of ice in all its forms. Ice occurs as intricate hexagonal crystals in the atmosphere, as baseball-sized crystals in glaciers, and as sheets on lakes and oceans. **Glaciologists** have diverse backgrounds, and include geologists, geographers, physicists, mathematicians, climatologists, computer scientists, and engineers. They share an interest in the beauty and physics of snow and ice. This chapter covers how glaciers are formed, how they are classified, how they advance and retreat, their key environments and processes, and their current decline in today's rapidly changing world.

18.1 IMPORTANCE OF GLACIAL SYSTEMS

Glaciers are vital to human livelihood. They supply water resources, aid in food supply, and are a focus of global tourism. Glaciers store water as ice in winter and then release meltwater in summer. Many regions of the world depend on glacial meltwater for drinking and irrigation, notably along glacier-fed rivers draining the Himalayas and the Tibetan Plateau. In North America, glaciers in the Cascades of Washington, the Rockies of Montana, and the Teton and Wind River Ranges of Wyoming are locally important to stream flow. The beauty of glaciers and glaciated landscapes also provide tangible economic benefits. Numerous National Parks around the world showcase glaciated mountains or formerly glaciated landscapes (**Fig. 18.1**).

Glaciers are important indicators of local, regional, and even global climate change. If the local climate changes, glaciers respond by fluctuating in length and/or thickness. Small glaciers are particularly sensitive to climate change and have a very short **response time** (in this case, the time needed to respond to a change in climate) of less than a decade. Large glaciers have longer response times. By using historical records of glacier retreat and advance, glaciologists can reconstruct past climates, which then helps us understand better how contemporary climates are changing.

Glaciers are also important controls on local climate; they provide cool and moist summer environments in mountains and polar regions, because their ice and associated lakes store high amounts of **latent heat**. Glacier ice absorbs energy from the atmosphere in the summer, which may raise the temperature of the snow and ice up to the melting point. Any excess heat goes into melting of the glacier's snow and ice, not into heating the air. Thus, the air stays cool. In winter, the glacier radiates stored heat back to the atmosphere. This ability to store and radiate heat moderates local temperatures in the same way that a large lake or sea will moderate local climates.

About 10% of Earth's land surface is currently covered by glaciers. Nonetheless, during the last glacial maximum ≈30% of Earth's surface was glaciated. Earth has experienced numerous **ice ages** or times when large glaciers called ice sheets were present. The most recent ice age began about 34 million years ago at the beginning of the Oligocene Epoch when an ice sheet first formed in East Antarctica. Mid-Miocene cooling led to permanent and larger Antarctic ice sheets and the establishment of ice sheets in the northern hemisphere. More recently, during the **Quaternary Period**, Earth has experienced ≈50 glacial and interglacial cycles (see Chapter 4). During **glaciations**, glaciers and ice sheets expand to cover 15–30% of the land. During **interglaciations**, glaciers and ice sheets melt back, typically covering ≈5–15% of Earth's land masses.

Figure 18.0 The Godwin Austen Glacier in the Karakorum Mountains of Pakistan illustrates the impressive degree to which glaciers can become crevassed. Source: Robert Boesch / Getty Images.

Figure 18.1 Examples of glaciers and formerly glaciated landscapes in National Parks. **A.** Grinell Glacier basin in Glacier National Park in northwestern Montana, USA. Source: T. Rains, US National Park Service; in public domain. **B.** View of the glacier Skaftafellsjökull in Vatnajökull National Park, southeastern Iceland. Source: Martin Falbisoner, CC BY-SA 4.0, via Wikimedia Commons. **C.** Franz Joseph Glacier in Westland Tai Poutini National Park, South Island, New Zealand. Source: P. Colgan. **D.** A formerly glaciated landscape in Cairngorms Mountains of Cairngorms National Park, Scotland. Source: P. Colgan.

18.2 CLASSIFICATION OF GLACIERS

A **glacier** is a perennial mass of ice and snow that moves due to gravity. Snow banks can be perennial, but they are not glaciers because they don't move. Sea ice moves, but is not considered a glacier because it forms from frozen sea water and is driven by currents and winds, not gravity. Glaciers are extremely variable in type and character, and as a result, strict classification schemes are almost always destined to leave some glaciers out.

Several systems of glacier classification are in use today. The most common glacier classification is based on size and form. **Mountain glaciers**, which form in high alpine areas, include **glacierets**, **cirque glaciers**, **valley glaciers**, **piedmont glaciers**, and **icefields**. Larger glaciers include ice caps and ice sheets. Ice caps and ice sheets are not contained within a valley, whereas most mountain glaciers are. The largest ice sheets can contain numerous features such as ice domes, ridges, **outlet glaciers**, **ice streams**, ice rises, and **ice shelves**.

Glacierets are small, isolated mountain glaciers in depressions or ravines on the side of a mountain (**Fig. 18.2A**). A cirque glacier is a mountain glacier that is confined to an amphitheater-shaped depression or basin (a **cirque**), which has been eroded by the glacier itself into the side of the mountain (**Fig. 18.2B**). A valley glacier is a mountain glacier confined within a former stream valley (**Fig. 18.2C**). Valley glaciers are up to >100 km long and >1,000 m thick. They usually get deeper and wider down-valley, as they erode the valley. All of these glaciers can vary enormously in their geography and physical properties.

Icefields are broad uplands of ice restricted to mountainous areas (**Fig. 18.2D**). Icefields are drained by valley and piedmont glaciers, which may even terminate in large lakes or the ocean. Glaciers that drain icefields or larger bodies of ice may spread out across a broad plain (**piedmont**) that slopes away from the mountain. This type of glacier is called a piedmont glacier. The Malaspina Glacier in southeastern Alaska is an excellent example of a piedmont glacier; it is the longest valley glacier in North America, at ≈190 km long (**Fig. 18.3A**).

Ice sheets and ice caps are the largest of all glaciers, with areal extents exceeding 50,000 km^2 and thicknesses of up to 4,000 m (**Fig. 18.3**). The Greenland and Antarctic ice sheets are the only ice sheets on Earth today (**Fig. 18.3C, D**). During the last glaciation, the **Laurentide ice sheet** (≈13

Figure 18.2 Glaciers, glacial features, and glacial landscapes. **A.** A glacieret near Aoraki/Mount Cook National Park in South Island, New Zealand. Source: P. Colgan. **B.** Glacial cirques and small cirque glaciers on Mt. Tjåmuhas in Abisko National Park, Sweden. Source: Dentren at English Wikipedia, CC BY-SA 3.0, via Wikimedia Commons. **C.** Skaftafellsjökull, an outlet (valley) glacier draining an ice cap in Vatnajökull National Park, Iceland. Source: H. Szydlowski. **D.** Icefields and valley glaciers leading to tidewater glaciers in Dobbin Bay, Nunavut, Canada. Source: NASA.

million km²) covered northern North America, and the Feno-Scandinavian ice sheet (≈6.6 million km²) covered northwestern Europe. In comparison, the Greenland ice sheet covers an area today of ≈1.7 million km² and the Antarctic ice sheets cover ≈14 million km². Currently, Antarctica is covered by two ice sheets – the East Antarctic and the West Antarctic ice sheet (**Fig. 18.4**). The East Antarctic ice sheet – the largest – is mainly grounded on land. It has several broad domes where ice flows outward and toward the coast, and through glaciers that flow through the Transantarctic Mountains and from there onto the Ross Ice Shelf (**Fig. 18.4**). The West Antarctic ice sheet is grounded below sea level (a marine-based ice sheet). It is drained by several large outlet glaciers terminating in the sea, and numerous ice streams that flow into the Ross and Ronne-Filchner Ice Shelves (**Fig. 18.4**). **Ice streams** are zones of rapidly flowing ice bordered by slower-moving ice.

If glaciers flow into the ocean or a deep lake, their outer margin may float, creating a **tidewater glacier**. Tidewater glaciers lose mass by **calving**, which is the process of ice

chunks breaking off from the glacier, forming icebergs. It is an especially important process by which ice shelves lose mass, for example, on the Greenland and Antarctic ice sheets. An **ice shelf** is a large body of floating glacial ice that develops on land then advances over water. Ice shelves can be hundreds to thousands of meters thick. Calving from ice shelves sometimes produces extremely large, tabular icebergs that may be hundreds of square kilometers in size and can take decades to melt (**Fig. 18.5**).

Glaciers can also be classified according to their temperature and climate. A **high polar glacier** contains ice that is continually below freezing. They are found in polar climates where mean annual air temperatures are well below 0 °C. Borings and ice cores drilled into the East Antarctic ice sheet show that ice at depth can be as cold as –50 °C to –60 °C. The geomorphic significance of high polar valley glaciers is that they generally lack liquid water at their base, which could (if present) enhance sliding and erosion. They also flow at lower velocities because ice is resistant to deformation at extremely low temperatures.

Figure 18.3 Some of the world's largest glaciers. **A.** The outlet glacier and piedmont lobe of the Malaspina Glacier in southeastern Alaska, USA. Source: NASA. **B.** The Vatnajökull ice cap in southeastern Iceland, which has an area of ≈8,100 km², making it the largest glacier in Europe. Source: European Space Agency. **C.** The Greenland ice sheet, as seen from the NASA Space Shuttle. Photo dated March 29, 1992. Source: NASA. **D.** A satellite image mosaic of Antarctica, which is covered by two major ice sheets and many smaller glaciers. Source: Image by David Pape using NASA data and Blue Marble software.

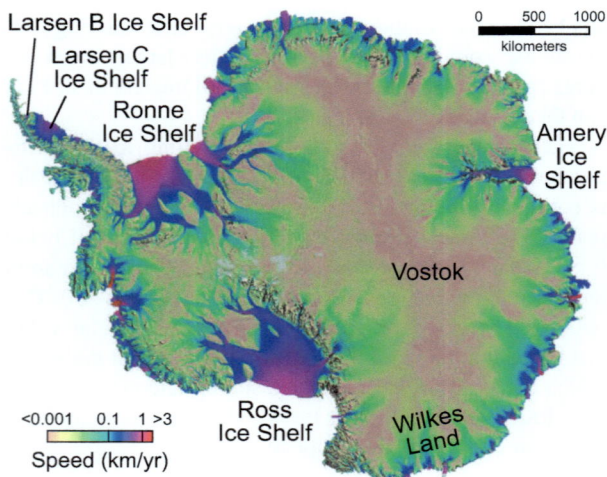

Figure 18.4 Estimated ice surface velocities (km/yr) of the Antarctic ice sheet derived from satellite radar interferometry. Source: Revised from a NASA image, originally from the US National Snow and Ice Data Center; in public domain.

Figure 18.5 A small tabular iceberg, which calved from the Antarctic ice shelf, floats by Clarence Island in the South Pacific. Source: David Stanley from Nanaimo, Canada, CC BY 2.0, via Wikimedia Commons.

In **subpolar glaciers**, the surface of the glacier melts in the summer. Meltwater then percolates down through the snow and ice and refreezes. In some cases, the water may pond at the surface of the glacier and then the water pressure may fracture the ice and force its way down to the bed of the glacier. Numerous Arctic and Antarctic glaciers, ice caps, and parts of ice sheets are currently subpolar as per their temperature regime. Subpolar glaciers can flow very rapidly when meltwater makes its way to the bed and facilitates bed sliding. If water pressure builds up, it can even cause the glacier to float and flow rapidly for days to months.

A **temperate glacier** is near or at its melting point throughout its thickness, except near the surface in the winter. In winter, the surface of the glacier cools down to below 0 °C. Then in summer this ice warms up and melts. Most glaciers today are temperate glaciers, especially near their margins. Temperate glaciers generally flow faster than high polar and subpolar glaciers, because warm ice deforms more readily than cold ice and water is available to enhance sliding.

Another way to classify glaciers is by their speed of flow. Most glaciers experience relatively steady flow, even though they may flow slightly faster in summer when more meltwater is present at the bed, and more slowly in the winter when meltwater is not present. Small, high polar and subpolar glaciers flow at rates that vary from as fast as a few meters per year to as great as tens of meters per year (mm/day). Larger subpolar and temperate outlet glaciers and ice streams may flow as rapidly as >10 km/yr (up to ≈30 m/day), which is still slow (a truly "glacial" pace) when observed directly, but extremely fast for a glacier. Ice streams and outlet glaciers have some of the highest recorded surface velocities. Ice

streams are common in Antarctica and drain both the West Antarctic and East Antarctic ice sheets (**Fig. 18.4**). **Outlet glaciers** flow from ice sheets and ice caps but are bordered by rock or sediment along their sides. Jacobshavns Isbrae in western Greenland is a large outlet glacier that drains a major part of the Greenland ice sheet and terminates in a large fiord, where it has a calving margin. Jacobshavns Isbrae glacier flowed at ≈5,000 to ≈7,000 m/yr during the 1980s and 1990s and then increased in velocity during the 2000s, when it experienced a maximum surface velocity of 13,000 m/yr during the 2002 and 2003 melt seasons. Much of the fast flow in the glacier appears to be related to surface meltwater that forms during summers, making its way to the bed of the ice. Velocity increases are also related to rapid calving and retreat of the floating tidewater margin of the glacier, which reduces the back pressure on the ice, allowing it to flow more rapidly.

Surging glaciers flow steadily for years but then experience a rapid velocity "surge" for a few months or perhaps even up to a year or two. Surging glaciers may increase in velocity more than 100 times during short periods of the **surge phase**, as compared to the **quiescence phase** – a time of normal, slow flow between surge events. Surging in the glaciers have been documented in Alaska (USA) in the Yukon and British Columbia in western North America, on Ellesmere and Axel Heiberg Islands in the Canadian Arctic, Svalbard, East Greenland, and Iceland in the North Atlantic, and in the Pamir, Caucasus, and Tien Shan Mountains of Asia.

Glaciologists are fascinated by surging glaciers. They recognize that glaciers surge primarily because of the buildup of subglacial water, which creates high pressures under the ice and causes parts of the glacier to float, allowing it to flow

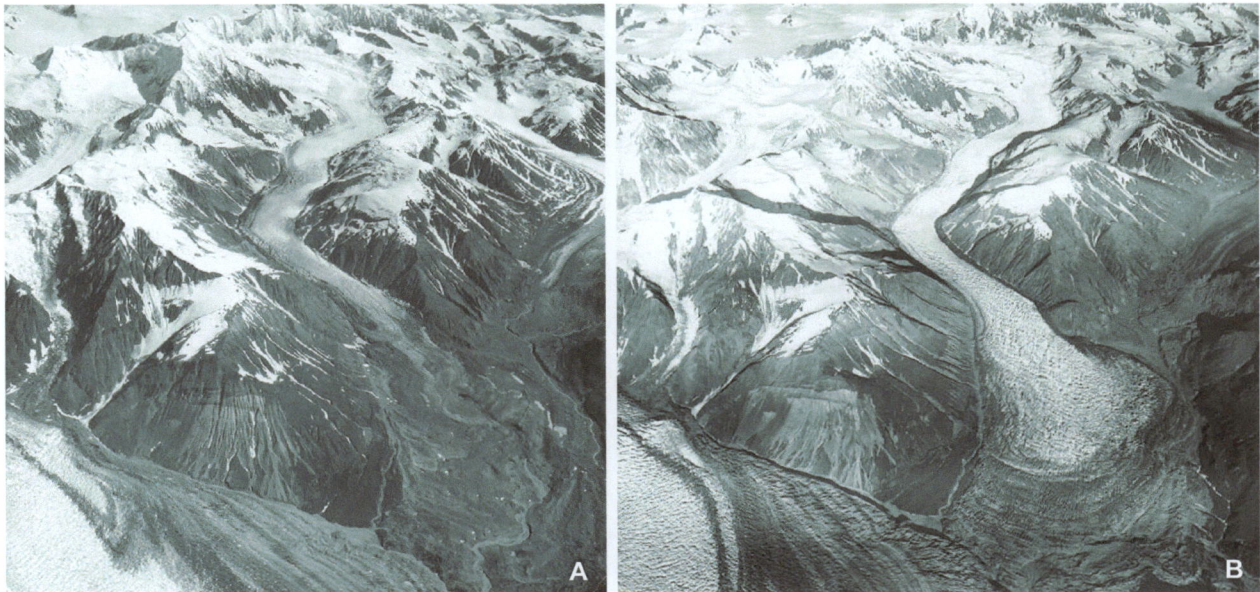

Figure 18.6 Oblique aerial photos of Variegated Glacier, Alaska (USA), before and after the 1964–1965 surge event. **A.** The pre-surge photo, taken in early fall of 1964. Note that the glacial margin is thin, stagnant, and covered in debris. **B.** The post-surge photo, taken in August 1965, after the surge had ended. Note how the ice margin is much thicker, with numerous compression ridges, and has advanced down-valley, compared to the pre-surge image. Source: A. Post from the US Geological Survey; in public domain.

rapidly. The water reduces resistance to flow along the bed of the glacier. During a surge, ice flows rapidly down glacier, thinning the upper parts of the glacier and thickening the ice nearer to its margin, causing the terminus to advance rapidly. The best-studied surging glacier is Variegated Glacier in southeastern Alaska (**Fig. 18.6**), which has surged numerous times in the last century. Surges have been documented by historical accounts and aerial imagery, occurring in 1905–1906, sometime before 1933, in 1946–1947, 1964/1965, 1982–1983, 1993, and again in 2003–2004, as well as more recently. These observations, along with records from other surging glaciers, indicate that surging glaciers experience quasi-periodic surge cycles on decadal scales.

18.3 TRANSFORMATION OF SNOW TO ICE

Glaciers form as snow is transformed into ice, and that ice begins to flow. In mountains and in polar climates, winters are snowy, and summers are cool and brief, such that snow can accumulate, thicken, and be transformed into glacial ice. In most of Antarctica, even very small accumulations of only a few millimeters of snow each year are preserved because air temperatures are always well below 0 °C. In temperate climates, snowfall can be up to tens of meters thick in the high alpine.

Over time, accumulated snow builds up, compacts, and becomes glacial ice. New, freshly fallen snow has low densities ($\approx$50 to $\approx$200 kg/m^3) (**Fig. 18.7**). **Firn** – transitional between snow and ice – is granular ice more than a year old. It has a density of $\approx$200 to $\approx$800 kg/m^3. Air and water can infiltrate into and through porous firn, until it becomes solid glacier ice. Glacial ice is denser because air bubbles within it have become closed and compressed. The density of glacial ice is normally between $\approx$800 and 917 kg/m^3, depending on the number and size of the air bubbles. Over time, compaction, melting, and refreezing processes cause the ice crystals in glacial ice to grow larger. These same processes cause smaller crystals to shrink and eventually disappear, as larger ice crystals grow. Ice crystals in thick ice sheets may grow to be the size of a baseball!

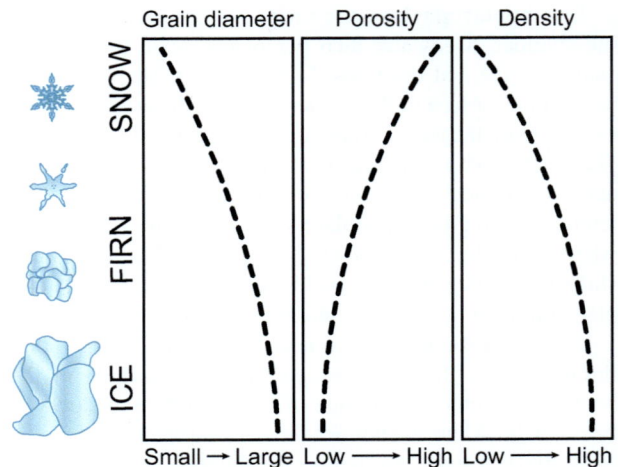

Figure 18.7 Snow is transformed into ice in a glacier along a series of steps. Snow is first transformed into firn, and then to ice. During this transformation, small ice crystals become larger, porosity decreases, and density increases.

Observers are often amazed by the beautiful blue color of glacial ice (**Fig. 18.8A**). Ice crystals are minerals, with the same hexagonal crystal form as quartz (**Fig. 18.8B**). Crystal form and bubbles in the ice cause it to absorb light in all the bands of the spectrum, except for the blue wavelengths, resulting in blue colors. That said, glacial ice can also appear as white, grey, or even black, depending on its content of debris and/or bubbles.

Field observations help glaciologists to generalize about how fast the snow-to-ice transformation proceeds (**Fig. 18.9**). The process occurs more slowly in colder climates than in warmer climates. It also occurs more rapidly in areas of high snowfall. Additionally, the process occurs faster in glaciers where melting occurs in summer, as the presence of water speeds up the process. Therefore, the time it takes snow to transform to ice depends mainly on climate. In temperate climates, snow changes into ice more rapidly than in polar climates, where there is less seasonal melting (**Fig. 18.9**).

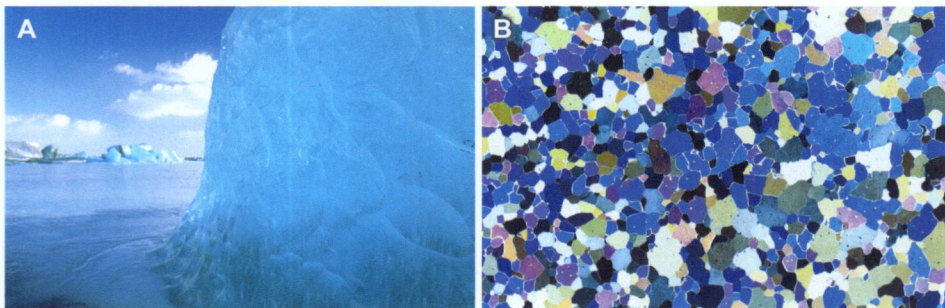

Figure 18.8 Glacial ice. **A.** A block of blue glacier ice in a proglacial lake in front of Breiðamerkurjökull, a large outlet glacier of the ice cap Vatnajökull, Iceland. Source: Andreas Tille, CC BY-SA 4.0, via Wikimedia Commons. **B.** Glacial ice in thin section, as seen under cross-polarized light, illustrating its crystalline structure. Source: US Geological Survey; in public domain.

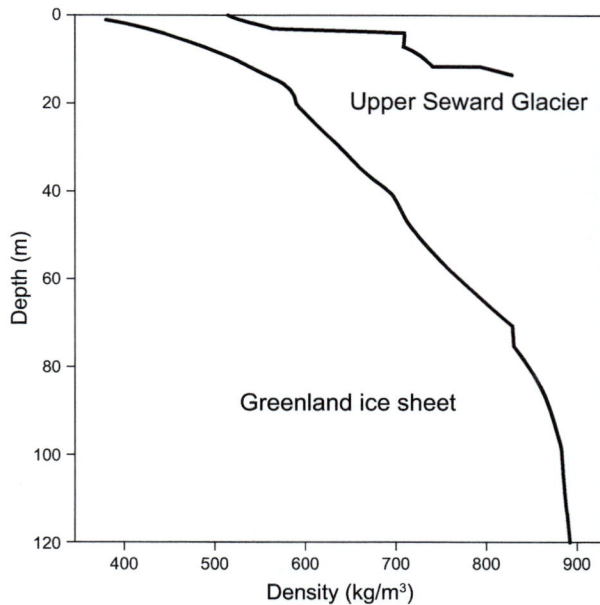

Figure 18.9 Plot of density (snow to firn to ice) versus depth in two different glaciers. The Upper Seward Glacier is a temperate valley glacier in the maritime (snowy but also rainy) climate of southeastern Alaska (USA). As a result, the transition of snow to ice occurs very quickly. In Greenland, this transition takes much longer, because there is so little available meltwater in the colder climate. Here, snow turns to ice mainly by compaction. Source: Used with permission of Elsevier, from Patterson (1994); permission conveyed through Copyright Clearance Center, Inc.

18.4 GLACIER MASS BALANCE

Glaciers are healthy if their mass is growing (or at least steady) over time. They are unhealthy and may ultimately perish if their mass is decreasing. Glaciologists monitor the health of glaciers by determining their **mass balance**,

defined as the difference between the accumulated ice mass minus the **ablated** (lost) ice mass, usually calculated over a one-year period.

Glaciers form and achieve net gains of ice in their **accumulation zones**, by processes including snowfall, rainfall, and wind drift and snow avalanche from higher slopes (**Fig. 18.10**). Snow and rain that fall onto a glacier's accumulation zone may become transformed into ice. Rain may fall onto a cold glacier and freeze, adding to the mass of the glacier. Snow may blow onto the glacier, especially downwind of a ridge. Avalanches may bring snow from higher slopes onto the glacier surface. All these processes add mass to the glacier and cause more ice to accumulate. The accumulation zone is the area where glacial ice is formed. Glaciers flow out and away from their accumulation zones. In mountain glaciers, the accumulation zone is at the highest altitudes, where snow is plentiful and temperatures are low (**Fig. 18.10**).

Data on glacial ice formation and accumulation are usually obtained by digging snow pits to determine how much snow fell during a year, and converting this thickness and snow density into millimeters of water equivalency (**Fig. 18.11**). The depth of new snow is determined by looking for the interface between new and old snow, which can often be identified by dusty or refrozen material at the top of the old snow (or firn) from the previous year.

Glacier **ablation** is the loss of ice mass. Ablation is maximal in the glacier's **ablation zone** (near the terminus), where it occurs by melting, wind erosion, evaporation, sublimation, avalanching, and iceberg calving (**Fig. 18.10**). Net ablation of ice exceeds net accumulation of ice in the ablation zone. Ablation is usually measured by using stakes as reference points to estimate the thickness of snow and ice melted during a year, and converting these data into water equivalency.

In valley glaciers, the ablation zone is at the lowest altitude, where higher temperatures increase the rates of melting and ablation (**Fig. 18.10**). In tidewater glaciers and in high

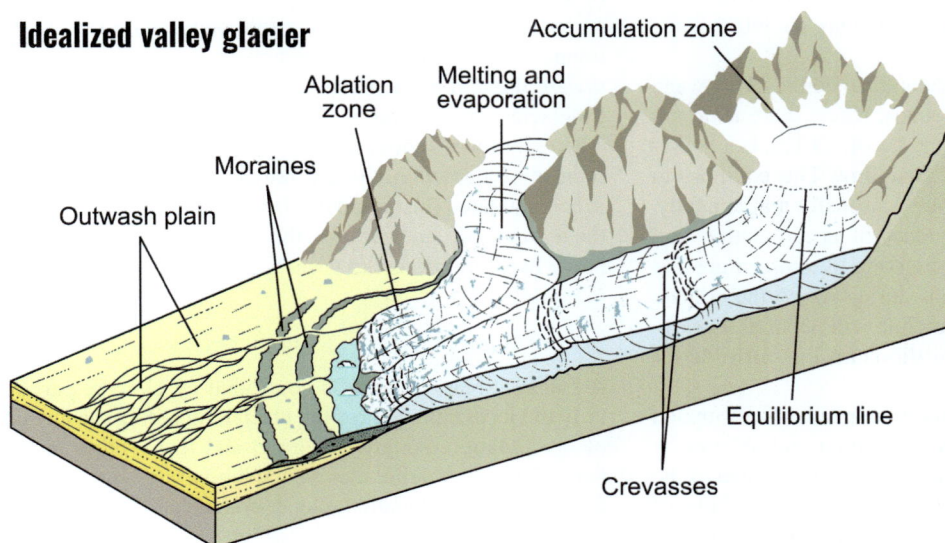

Idealized valley glacier

Figure 18.10 An idealized valley glacier system leading to a piedmont lobe on a nearby plain. Source: Adapted from Fig. 14.1 of E. H. Christiansen and W. K. Hamblin. 2015. *Dynamic Earth*. Jones & Bartlett Learning.

Figure 18.11 A glaciologist measuring the density and thickness of snow in a snow pit on the West Antarctic ice sheet. Source: NASA ICE, Public domain, via Wikimedia Commons.

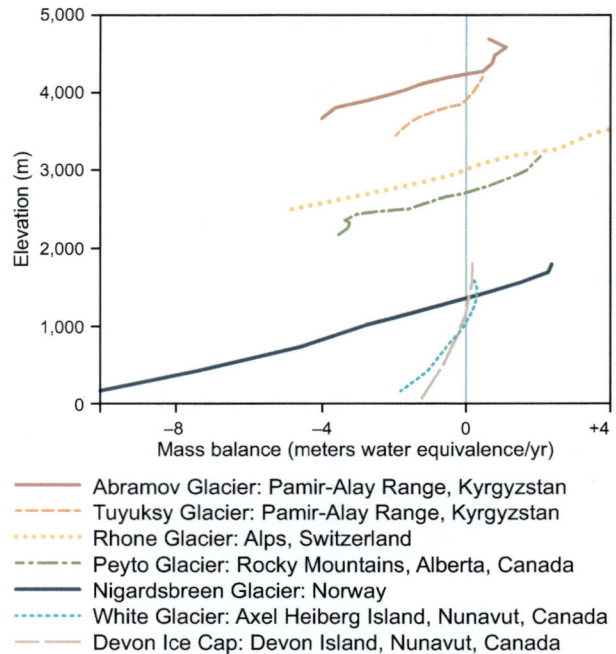

—— Abramov Glacier: Pamir-Alay Range, Kyrgyzstan
- - - - Tuyuksy Glacier: Pamir-Alay Range, Kyrgyzstan
· · · · · · Rhone Glacier: Alps, Switzerland
-·-·-· Peyto Glacier: Rocky Mountains, Alberta, Canada
—— Nigardsbreen Glacier: Norway
· · · · · · · White Glacier: Axel Heiberg Island, Nunavut, Canada
— — Devon Ice Cap: Devon Island, Nunavut, Canada

Figure 18.12 The measured, annual mass balance for seven glaciers from various climates. The vertical light gray line is the ELA where the glacier mass balance is zero. The glacier Nigardsbreen in Norway exists in a wet-cool, maritime climate. It loses up to 10 meters of water near its terminus at lower elevations, but gains almost 3 meters of ice in its accumulation zone at colder, higher elevations. Both the White and Devon Glaciers in the Canadian Arctic have low accumulation and ablation rates because of the cold, dry climate. Source: After Anderson and Anderson (2010), with permission from Cambridge University Press & Assessment.

polar ice shelves where little melting occurs, ablation may occur primarily by iceberg calving. Melting of ice near the margin will produce runoff into meltwater streams. Wind erosion removes snow from the glacier. Evaporation of water off the glacier surface is especially common in cold and dry climates. Glaciers also lose mass by **sublimation** – the direct loss of ice to water vapor, without an intervening (liquid) water stage. Sublimation is particularly common on cold, sunny days. Commonly, ablation is slower in partially shaded glaciers; in the northern hemisphere this occurs on north- and northeastern-facing slopes. Calving occurs as icebergs break off the glacier and float away in adjacent water bodies.

In healthy glaciers, the ablation zone is usually smaller than the accumulation zone, even though summer melting can remove large amounts of snow and ice near the terminus. For glaciers that are in equilibrium (with a steady terminus), the accumulation area is usually >60% of the total surface area of the glacier. In ice sheets or in glaciers with an active calving margin, the accumulation zone may be >90% of the total surface area.

The **equilibrium line** is the theoretical boundary between the accumulation and ablation zones, where there is no net gain or loss of ice over time (**Figs. 18.10, 18.12**). It is sometimes called the **snow line** or **firn line**. The equilibrium line on a glacier changes location seasonally and annually, depending on the relative amounts of accumulation and ablation, which are primarily a function of temperature and snowfall. Over long time periods during which the climate is fairly constant, the equilibrium line will occur at a nearly constant elevation, called the **equilibrium line altitude** (or ELA).

The net annual **mass balance** of a glacier is estimated by measuring the net annual accumulation and ablation at numerous areas on the glacier, multiplying by the area of the glacier, and then taking the difference between the volume of water accumulated vs water ablated. Mass balance changes

in a glacier usually lead to movement of the terminus (the end of the glacier), over periods of days to years. The movement of the glacier's terminus is called a **glacial advance** and **glacial retreat**. Glaciers that gain in mass over a period of years tend to advance, and glaciers that lose mass over a period of years usually retreat (although some simply get thinner). One key concept to remember: ice is always flowing toward the terminus, or else it is not moving. It never flows up-glacier, or backwards. The glacier's terminus advances when more ice is moving towards the terminus than can be removed by ablation. Glacial retreat occurs when more ice is being ablated (lost) than can be replaced by ice flowing to the terminus. At no time does the glacial ice turn around and move back up hill as the term "glacier retreat" might seem to imply!

18.4.1 Climate Controls on Glacier Mass Balance

A glacier's mass balance is controlled by climate, particularly the local climate near the glacier, including the amount of precipitation, summer melting, and many other factors that influence accumulation and ablation rates. Topographic factors, such as elevation, slope, and aspect also indirectly affect the local climate on the glacier.

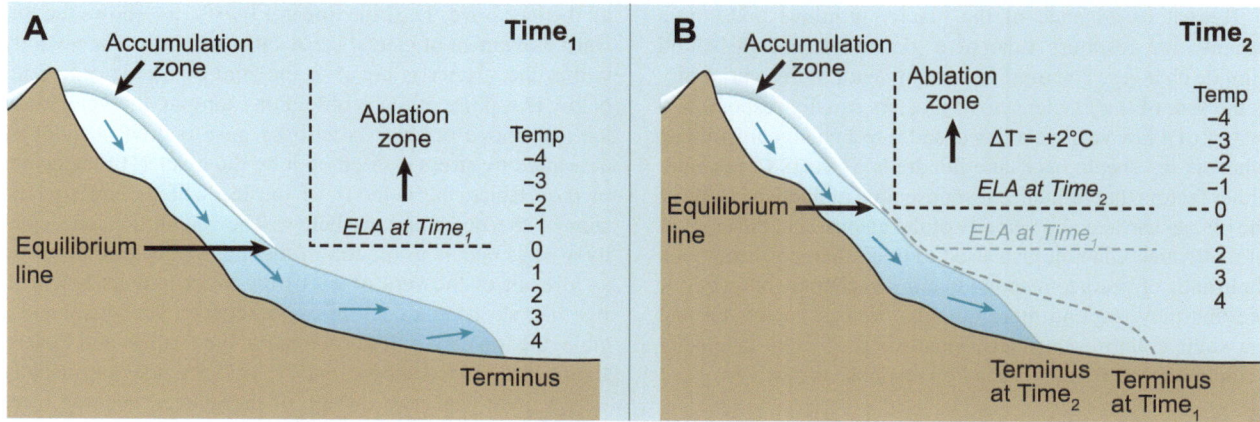

Figure 18.13 Glacier cross-sections that illustrate the concept of the equilibrium line. **A.** Cross-section and profile of a valley glacier in equilibrium at Time$_1$. **B.** Cross-section and profile of a glacier at Time$_2$, where the climate has warmed by 2 °C. The dashed vertical line is the boundary between the accumulation and ablation zones, i.e., the equilibrium line. The dashed horizontal line is the elevation of the equilibrium line, or the equilibrium line altitude (ELA).

In high-elevation areas on the glacier (above the equilibrium line), where the mean annual temperatures are below 0°C, most of the snow that falls will be retained through the summer (**Fig. 18.13**). On the other hand, in areas where the mean annual temperature is warmer than 0°C, below the **equilibrium line**, snowfall on the glacier surface has the potential to melt during the summer. **Figure 18.13** illustrates how climate change can alter the location of the ELA and the glacier terminus. In **Fig. 18.13A**, a cross-profile of a valley glacier in equilibrium, the terminus is neither advancing nor retreating, even though ice is flowing downhill (blue arrows). **Figure 18.13B** illustrates what would happen if the local climate warmed. Warmer temperatures would cause the ELA to rise, resulting in a larger ablation zone on the glacier. The terminus would respond by retreating, in order to balance the glacier and keep the ratio between accumulation zone and ablation zone ≈60% and ≈40% as in Time$_1$. Note that climate is the cause (or driver) of the behavior of the glacier (advance or retreat of the terminus). Alternatively, if the climate cooled or if snowfall increased, the size of the accumulation zone would increase, causing a decline in the ELA, driving an eventual advance of the glacier's terminus.

The systematic variation in ELA across the globe is directly related to climate (**Fig. 18.14**). Mean annual temperature and precipitation are the two most important factors in determining the ELA elevation. To confirm this, notice that glacier ELA values are lowest in the high latitudes and highest in the tropics (**Fig. 18.14**). Also, glaciers near the ocean where precipitation is high, such as in coastal Alaska (USA) and in the Coast Range of British Columbia (Canada) and Washington (USA), have lower ELAs than do glaciers in drier, more continental climates like the Brooks and Alaska Ranges of Alaska (**Fig. 18.14**). During the **Last Glacial Maximum**, ≈26,000 years ago, glacier ELAs around the world were ≈600–800 m lower than today, because global mean annual temperatures were ≈4–5 °C colder. In a warming world, we can expect that ELAs will rise. Glaciers

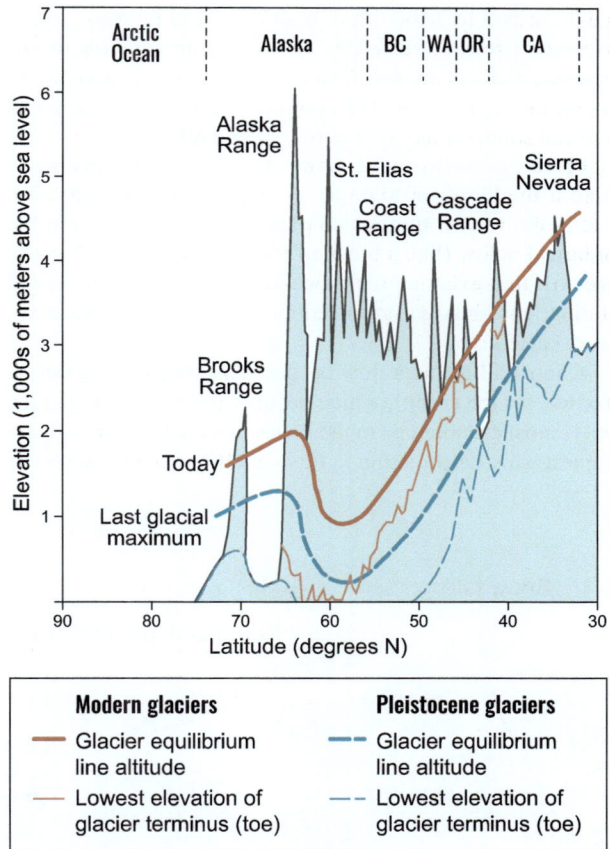

Figure 18.14 Variation in ELA of glaciers in time and space.
Source: After Anderson and Anderson (2010), with permission from Cambridge University Press & Assessment.

will disappear if the local ELA is higher than the highest mountain peak! This scenario has already occurred in many regions around the world, due to global warming in the twentieth and twenty-first centuries.

Retreat or advance of the glacier terminus takes time because the response time of a glacier often lags behind climate changes. The most important factor in this variability is the size of the glacier. Small glaciers usually respond in a matter of a few years to a few decades to a change in climate, whereas ice sheets may take hundreds of years to respond. Other factors that influence response time include the slope of the glacier surface, the velocity of the glacier, and the amount of meltwater. Glaciologists used to think that ice sheets took thousands of years to respond to climate change, but we now see that they respond more quickly, because meltwater and ice shelf collapse can cause positive feedbacks in glacier velocity that can greatly speed up response times.

18.5 GLACIER MOVEMENT AND FLOW

Glacier movement is driven by gravity, just like rivers and landslides. Valley glaciers are commonly described as rivers of ice and this is generally apt. Like rivers, glaciers have numerous tributaries that feed into a main glacier. Glacial ice is formed by snowfall in high mountain regions, just as tributary streams are fed by rainfall upstream. But unlike streams, glaciers are solid, so the analogy is not perfect. As we will see, glacial ice is a complex material that can act as a brittle solid, a fluid, or a plastic (**Fig. 18.15**).

For a glacier to move, it must either deform internally (like a mudflow) or slide along its bed (like a landslide). Along the way, it might also plow through and deform the sediment below (like a block of rock sliding over soft mud). We will first examine the processes of ice deformation and sliding, and then go on to sediment deformation processes (**Fig. 18.16**).

Almost all glaciers flow by the mechanism of **ice deformation** (or **ice creep**), which occurs throughout the glacier, but is most pronounced in the lower parts, where stresses are highest. Any deformation in the ice will cause movement of all the ice above. The blue line in **Fig. 18.16A** shows the forward movement of glacial ice at various depths. The velocity within the glacier is equal to the sum of the deformations below that point in the profile. For example, the velocity of ice just above bedrock is zero because bedrock cannot be deformed by stress exerted on it by the glacier. In the center of the glacier, the velocity of the ice will be equal to the sum of the deformations below. The maximum ice velocity always occurs at the top of the glacier because it moves as a result of the sum of *all* of the deformations below it. Just like shearing a deck of cards between your hand and a table, the top card will always move the farthest and fastest. **Figure 18.16** also shows that the velocity does not usually vary a great deal near the top of the glacier, because there isn't much deformation in the upper portion of the ice. On the other hand, there is usually a large change in velocity in the lower portion of the glacier, where most of the deformation is occurring.

Although ice is a crystalline solid, it can also contain water between the ice crystals. But this situation can only occur when ice is at the melting point, which varies with pressure. The melting point of ice is ≈0 °C at atmospheric pressure. The temperature at which glacial ice melts or freezes (the **pressure melting point**) decreases slightly with increased pressure. In thick glaciers, the weight (pressure) of the ice causes the melting point to decrease in the deeper ice. For example, in an ice sheet 3,000 meters thick, the melting point would be about –2 °C at its bed.

Ice deforms internally by processes such as **gliding** (slipping of the basal plane of the hexagonal crystals), **crystal movement** (where ice crystals move or rotate in relation to one another), or by **crystal growth or decay** (where water molecules move from one crystal to another without melting, a process similar to metamorphism in rocks). These mechanisms cause ice to be an unusual material.

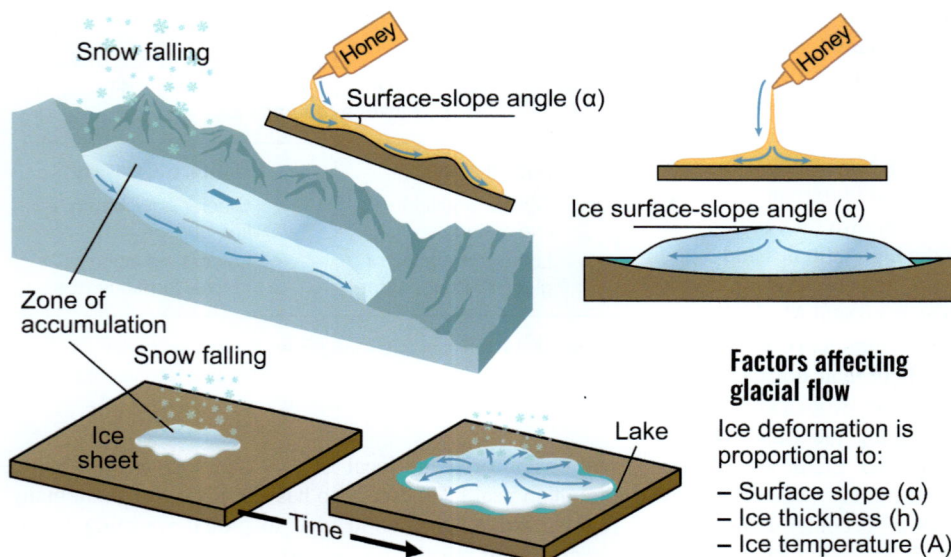

Figure 18.15 Important factors related to flow and movement in glaciers. A first approximation of glacier flow is a very viscous fluid like cold honey. The steeper the slope on the surface of the honey, the faster it will flow. Also, the thicker the honey, the faster it will flow, because this steepens the flow surface and provides a thicker layer to deform. Source: After Marshak © 2019. Used by permission of W. W. Norton & Company, Inc.

Snow falling

Honey

Surface-slope angle (α)

Honey

Ice surface-slope angle (α)

Zone of accumulation

Snow falling

Ice sheet

Lake

Time

Factors affecting glacial flow

Ice deformation is proportional to:

– Surface slope (α)
– Ice thickness (h)
– Ice temperature (A)

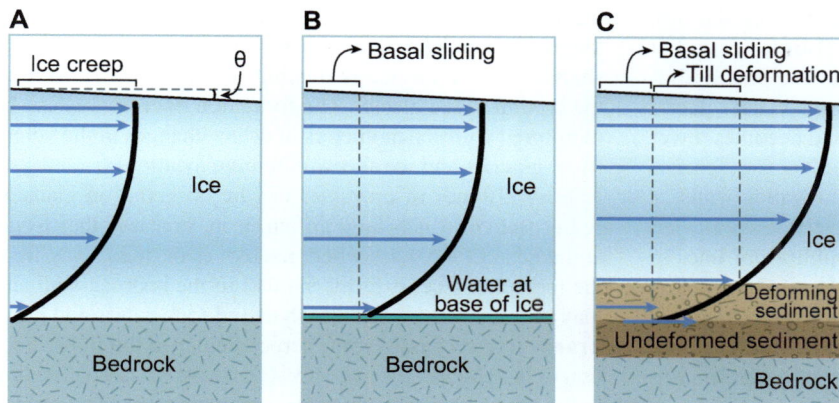

Figure 18.16 Vertical velocity profiles for three different glacier bed conditions. The black line shows the velocity with depth in the glacier. **A.** Ice on rigid bedrock where the glacier is frozen to its bed, as is common in many polar glaciers. Here, all of the movement occurs by ice deformation. **B.** Ice that experiences both ice deformation and sliding at the ice–bedrock interface, as is common in temperate or subpolar glaciers. Here, ice is near the pressure melting point, and water is present. **C.** Ice flowing across deformable sediment on top of bedrock. Here, three different processes can add to the surface velocity of the glacier – ice deformation, sliding, and sediment deformation.

Ice deformation occurs in all glaciers. But it dominates flow in cold glaciers where ice is below the pressure-melting point, and water for sliding on the bed is lacking, that is, the glacier is frozen to its bed. Internal deformation in cold glaciers occurs mostly near the bottom of the glacier, where the applied shearing stresses are highest (**Fig. 18.16A**). Deformation of ice can be estimated by **Glen's Law**, a simple empirical relationship ($\varepsilon = A\tau^n$), which predicts that the ice deformation rate (ε) is proportional to the applied shear stress (τ) cubed (n = 3), and A, which is a non-linear function of the temperature of the ice. Cold ice deforms less than does ice near its melting point. Shear stress (τ) is shearing force per applied area and can be estimated with another simple relationship ($\tau = \rho g h \sin \alpha$). Shear stress is directly proportional to ice density (ρ), the acceleration of gravity (g), the thickness of the ice (h), and the sine of the ice surface slope (α). Glen's Law predicts that ice at the bottom of a glacier will deform more than ice nearer the surface of the glacier, and that steeper parts of the glacier will deform more and flow faster than parts of the glacier where the ice surface slope is low. If the surface slope (α) is zero (perfectly horizontal), there will not be any ice deformation or flow.

Glaciers can also slide over their beds – like an ice skater slides across a frozen lake. Glacial sliding occurs when a thin layer of water is present at the bed of the glacier (**Fig. 18.16B**). This condition only occurs if the ice at the bed of the glacier is at its pressure-melting point. Up to ≈90% of the surface velocity of some glaciers may be due to sliding, for example, in ice streams or outlet glaciers. Sliding velocity increases as the water pressures along the bed increase. When water pressures are high enough to allow the glacier to float and slide very rapidly, the glacier may "surge" or form rapidly moving ice streams. Unfortunately, a general law for sliding glaciers (as in Glen's Law above) has not yet been developed. Understanding how sliding relates to the type of glacier bed (hard rock or soft sediment) and water pressure at the bed is one of the largest unsolved questions in glaciology.

Some of the surface velocity of a glacier may also result from the deformation of soft sediments below the ice

(**Fig. 18.16C**). In the early 1980s, it was discovered that soft sediment beneath Whillans Ice Stream in Antarctica was very weak and had little shear strength. This observation helped explain why ice streams were able to move so rapidly (thousands of meters per year), even though the driving stresses were very low and little deformation was occurring in the ice itself. In these ice streams, little movement occurred because of internal ice deformation. Instead, most movement was driven by bed sliding and deformation of weak subglacial (bed) sediments.

Velocity profiles through a valley glacier show that, like water in a stream channel, the highest velocities are at

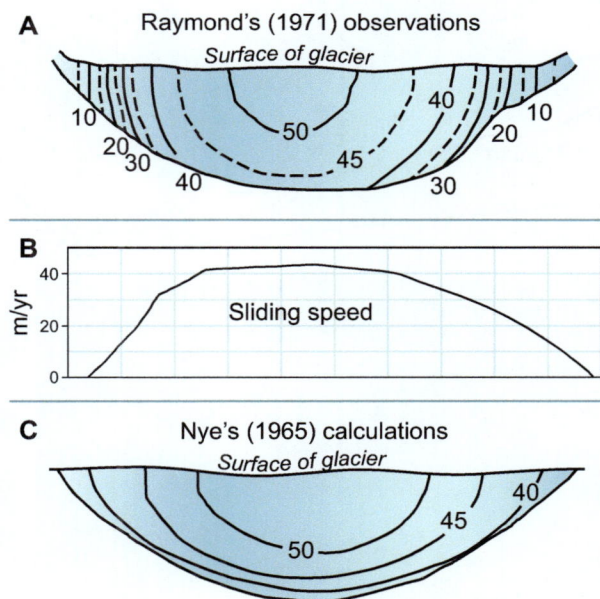

Figure 18.17 Rates of movement (meters/yr) of ice in the Athabasca Glacier in the Canadian Rockies. **A.** A cross-section of velocity within the glacier, in m/yr. **B.** Sliding speeds across the glacier (looking down). **C.** Calculated glacial flow velocities, using Glen's Law. The observations do not exactly match the calculations because sliding at the bed of the glacier is also important to the surface velocity of the glacier. Source: After Anderson and Anderson (2010), with permission from Cambridge University Press & Assessment.

the surface (**Fig. 18.17A**). Velocity decreases with depth, becoming zero at the bed in a glacier (if it lacks sliding). If bed sliding is occurring, velocity will also decrease, but will have some finite velocity in the basal ice due to sliding beneath. In a profile across the top of a valley glacier the highest velocity occurs in the middle and becomes lower toward the sides of the valley (**Fig. 18.17B, C**), because the sides are experiencing friction with the valley walls. A glacier's velocity will increase as one travels down-glacier, through the accumulation zone to the equilibrium line. In this zone, ice is extending, termed **extensional flow**. Below the equilibrium line, the velocity will begin decreasing and the ice will start to become compressed, as **compressional flow**. The maximum down-valley velocity of a glacier usually occurs near the equilibrium line.

18.6 DEFORMATION OF GLACIAL ICE

Crevasses are tension fractures in a glacier that indicate deformation and movement (**Fig. 18.18**). Crevasses form because, if deformed rapidly, snow and ice act as brittle solids. Most crevasses occur in the upper 30–50 m of the glacier; generally, below this depth, ice deforms and closes the crevasse faster than tensional stresses can pull it open.

Crevasses usually form where the ice velocity is increasing down-valley or where the ice margin spreads out in a tongue-shaped lobe, longitudinally. During surges and in ice shelves (where the ice is floating), crevasses may form at the bottom of the glacier. The formation of crevasses can be monitored with seismometers to detect changes in the velocity of glaciers and ice sheets in remote locations.

Several types of crevasses can be observed on glaciers. A **bergschrund** is a large, often deep, crevasse that forms at the head of a glacier where ice moves or pulls away from the mountain face. It forms parallel to the bedrock wall and may become filled with frost shattered rock debris and talus. **Transverse crevasses** form perpendicular to the flow direction of the glacier (**Fig. 18.18A–C**), where ice is accelerating in the accumulation zone. Transverse crevasses close as they pass below the equilibrium line and enter the ablation zone, where the surface velocity decreases. These crevasses may be covered by snow in the accumulation area, or over the entire glacier in the winter (**Fig. 18.18B, C**). **Longitudinal crevasses** form parallel to the flow direction of the glacier (**Fig. 18.18D**), because of the splaying of the glacial margin in the ablation zone. They usually form where a piedmont lobe or valley glacier spreads out at the bottom of a valley, and may even extend to the bed of the glacier near the terminus.

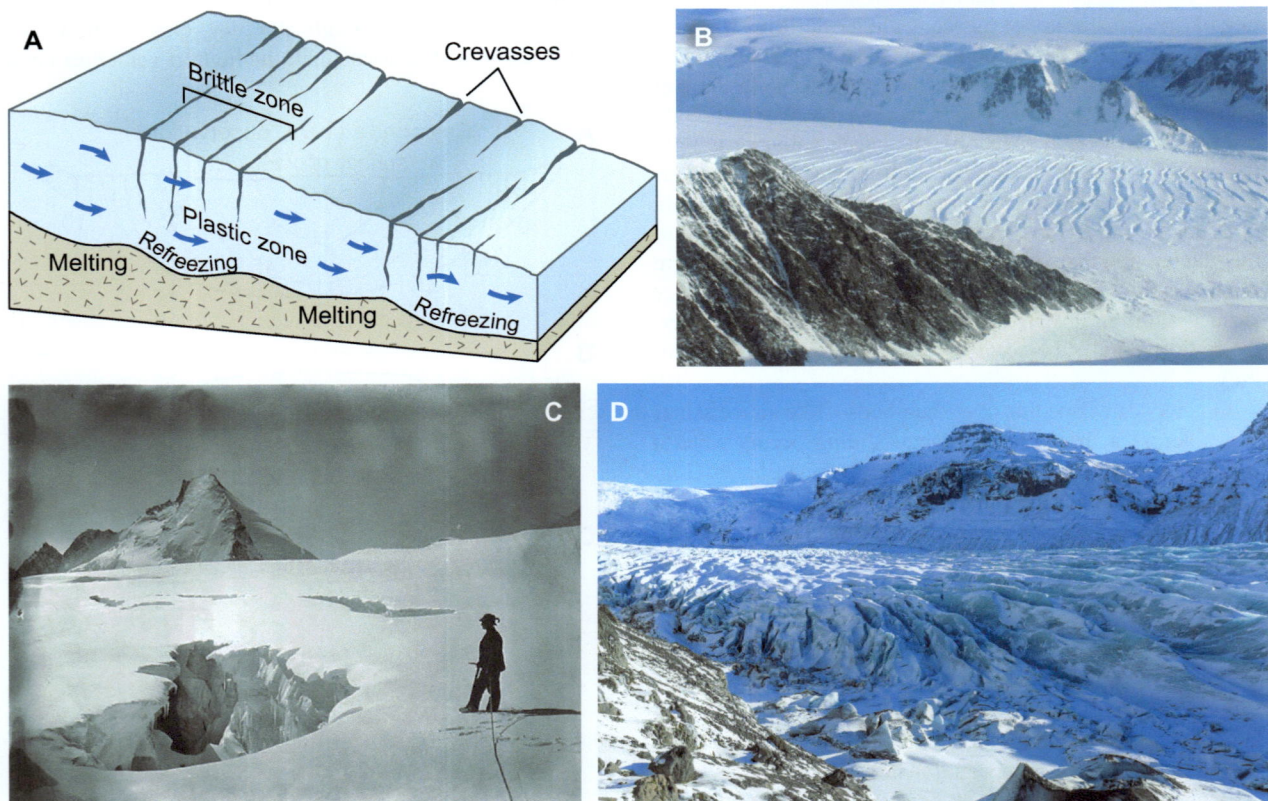

Figure 18.18 Glacial crevasses. **A.** Diagram showing how crevasses can form in glaciers as ice is pulled apart. **B.** Transverse crevasses on Crane Glacier, Antarctica. Source: US National Snow and Ice Data Center; in public domain. **C.** A partially covered transverse crevasse on Stockje Glacier in the Pennine Alps. Source: H. Fielding Reid in 1894, from the Glacier Photograph Collection, Boulder, Colorado, USA. **D.** Longitudinal crevasses, which run parallel to glacier flow, in Svínafellsjökull, Iceland. Source: David Phan from Toulouse, France, CC BY 2.0, via Wikimedia Commons.

18.7 DISTRIBUTION OF GLACIERS AND ICE SHEETS

Glaciers are present today on every continent except Australia, and cover more than 15 million km² (10% of Earth's land area) (**Fig. 18.19**). This estimate includes all glaciers, ice caps, and ice sheets, but does not include ice shelves, sea ice, or seasonal snow and ice cover. Glaciers store ≈70% to 75% of the world's freshwater, but most of this freshwater is not readily available, because it is contained in the Greenland and Antarctic ice sheets. If all land-based glaciers melted, sea level would rise by ≈65–70 meters. Mountain glaciers and ice caps are found near sea level at high latitudes, and even in the tropics, for example, Mt. Kilimanjaro, Kenya, in Mexico, in the northern Andes, and in New Guinea, where they are restricted to the highest elevations.

More than 90% of all glacial ice is located on the continent of Antarctica, where glaciers cover ≈98% of the land area, and ice shelves and seasonal sea ice cover large areas of the adjoining continental shelf (**Fig. 18.20A**). The thickest ice occurs in the East Antarctic ice sheet, where it is up to ≈4,700 m thick! The West Antarctic ice sheet is grounded in many of the deep marine basins, even though they are well below sea level. For example, the ice sheet is >3,000 m thick and grounded ≈2,555 m below sea level in the Bentley Subglacial Trench in Marie Byrd Land (**Fig. 18.20A**).

The Greenland ice sheet covers an area of ≈1.71 million km² or ≈79% of Greenland's land area (**Fig. 18.20B**). The thickest ice here is >3,200 m thick, with a maximum elevation of ≈3,000 m above sea level. Much of the ice sheet flows to the west or east of the linear dome of ice, feeding large outlet glaciers that drain through fringing mountain ranges and into the sea as tidewater glaciers. It is likely that the iceberg that sank the *Titanic* had calved from a tidewater glacier draining the Greenland ice sheet. Greenland also has large areas of valley glaciers, ice fields, and ice caps that are separated from the main ice sheet.

18.8 GLACIAL ENVIRONMENTS

Glaciers produce a variety of depositional environments, especially near their margins, where water is often present. These environments produce a diversity of landforms and sediments. Glaciologists recognize four environments: supraglacial, englacial, subglacial, and proglacial (**Fig. 18.21A**).

The **supraglacial environment** includes the top of the glacier, and also contains several sub-environments (**Fig. 18.21A**). The **dry snow zone** is cold enough that there is no melting; it is only present in polar glaciers. The **wet snow zone** experiences melting during summer; meltwater from this zone flows downward and freezes in the ice and colder snow below. Water and slush formed here can flow downslope and collect in melt ponds. Near the terminus, in the ablation zone, all the snow on top of the glacier melts in summer, exposing old ice that has flowed there. In summer, the supraglacial environment contains melt ponds. Meandering channels of meltwater form here as well, melted down into the ice by water flowing along the top of the glacier (**Fig. 18.21B**). Some of these streams drain into **moulins** – vertical shafts melted down into the ice. In all these zones, sediment may fall on top of the glacier as eolian dust or volcanic ash, or from landslides and avalanches.

The **englacial environment** occurs within the glacier, and is mainly composed of ice, but also contains debris. The englacial environment is cut by numerous crevasses, fractures, and tunnels. Supraglacial meltwater drains into and through it, toward the glacier bed and the terminus (**Fig. 18.21B**). Water may be stored in winter in englacial cavities. Debris that enters the glacier is transported and modified in the englacial environment and can become abraded in this environment through clast to clast abrasion and crushing.

The **subglacial environment** is under or within the basal ice of the glacier (**Fig. 18.21A**). Much of the sediment load of the glacier is picked up, transported, and deposited within

Figure 18.19 Distribution of glaciers on Earth today. The Antarctic and Greenland ice sheets are the two largest bodies of ice on Earth.

Glaciers and ice shelves on the modern landscape

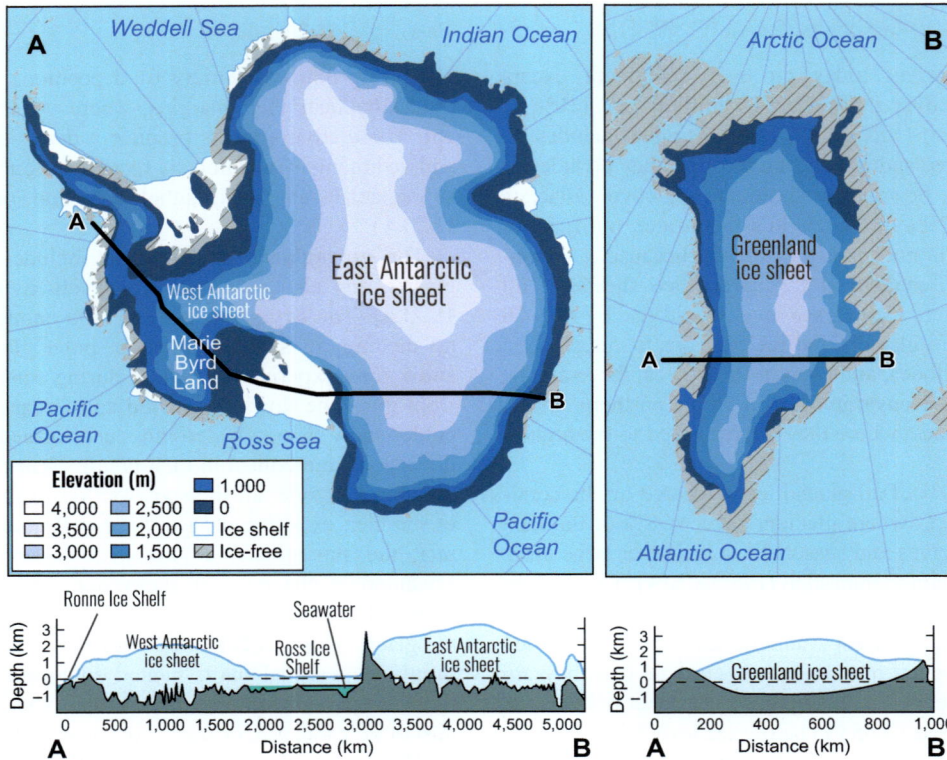

Figure 18.20 Maps of Earth's remaining ice sheets. **A.** Surface elevations of the East and West Antarctic ice sheets, with cross-sections shown below. Note that the bed of the East Antarctic ice sheet is mostly above sea level, whereas most of the bed of the West Antarctic ice sheet is below sea level. **B.** Surface elevations of the Greenland ice sheet, with a cross-section shown below.

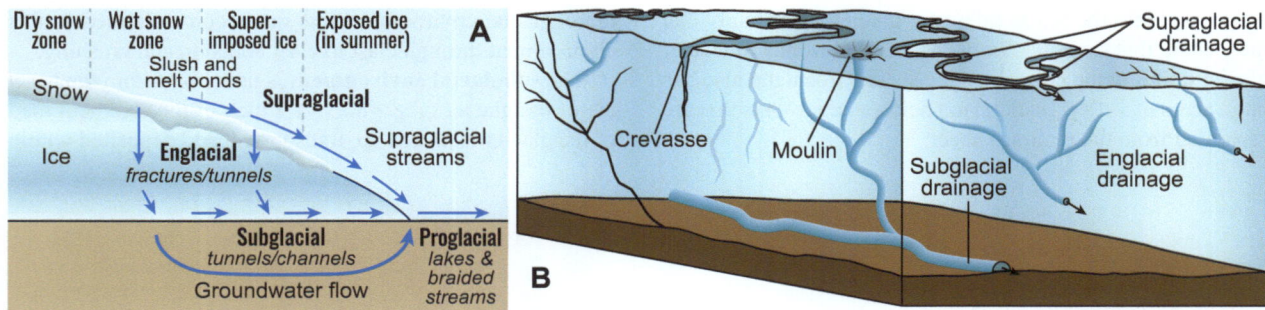

Figure 18.21 Different views of the various environments and meltwater flow paths (arrows) in a glacier.

this zone. If the temperature of the subglacial environment is at the pressure-melting point, it will contain water in thin sheets or channels. If the temperature is below the pressure-melting point, any water there will freeze onto the glacier. Water may also be stored near the bed in subglacial lakes, cavities, and channels (**Fig. 18.21B**). Although the subglacial environment is difficult to study, it has been accessed through boreholes, tunnels, and using seismic reflections or ice-penetrating radar.

Lastly, the **proglacial environment** refers to areas in front of the glacier, transitional between glacial and non-glacial environments (**Fig. 18.21A**). Proglacial environments accumulate large volumes of meltwater transported via streams, and even commonly stored in proglacial lakes (**Fig. 18.22**). Here, deposition occurs mainly by meltwater and in glacial lakes. Sediment deposited in the proglacial environment can be overridden by the glacier during an advance or surge.

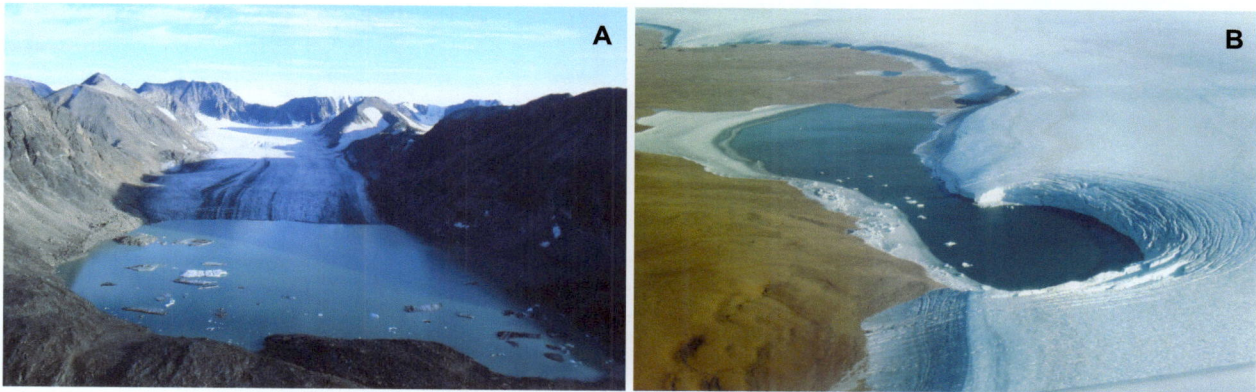

Figure 18.22 Proglacial lakes. **A.** A small proglacial lake in front of a valley glacier east of Southwind Fiord on Baffin Island, Nunavut, Canada. Source: Mike Beauregard from Nunavut, Canada, CC BY 2.0, via Wikimedia Commons. **B.** A proglacial lake along the margin of Humbodlt Glacier, which drains the Greenland ice sheet in northwestern Greenland. Source: NASA photograph by John Sonntag/Operation IceBridge, Public domain, via Wikimedia Commons.

18.9 WATER AND GLACIERS

Glaciers are vast storehouses of water, which gets released upon melting of the ice, as **glacial meltwater**. Meltwater flowing on, in, and under ice has been observed in the marginal areas of many glaciers. Melt ponds that form every summer around the periphery of, for example, the Greenland ice sheet often drain in less than a few hours, as the weight of the water forms cracks in the thick, cold ice, facilitating drainage to the bed. Melting ice in the ablation zone can form many interesting surface melt features, all of which contradict the common misconception that glaciers are homogeneous blocks of blue ice and white snow!

Melt features form in the supraglacial environment because of the complex interplay between solar radiation, air temperature, debris on the glacier, and the meltwater which forms on the glacier surface. Streams carry surface meltwater to moulins, which transport the water to the bed of the glacier. **Debris bands** and **ice foliation** form due to longitudinal shearing of ice and debris, forming lineations of debris and ice crystals that parallel the ice flow direction, like foliation in metamorphic rocks like schist or gneiss. A walk on a glacier reveals a plethora of unique melt features that affirm the interesting and varied history of the ice!

Proglacial lakes are common in most glacial systems, especially in temperate glaciers (**Fig. 18.22**). They are especially common today because most glaciers in the world are melting and retreating. Proglacial lakes occur in situations where the land surface slopes *toward* the ice margin. Glaciers, ice caps, and ice sheets commonly erode deep basins called **over-deepenings** in the ablation zone, behind the terminus, or near their calving margins. As a glacier begins to retreat, over-deepenings form natural basins for proglacial lakes. Glaciers also commonly deposit **end moraines** that can act as a natural dam for proglacial lakes, trapping water between them and the terminus. Normally the moraine is lowest in elevation along the mid-line axis of the glacier, where it can become a natural spillway for yet another proglacial lake in front of another end moraine.

Proglacial streams normally form down-valley from proglacial lakes, or from the glacier terminus itself (**Fig. 18.23**). They are fed by meltwater flowing off the top of the ice or from subglacial tunnels. These tunnels may be spaced regularly along the terminus in ice caps or ice sheets, or merge into a single tunnel in narrow, valley glaciers. Proglacial streams usually merge down-valley into one large trunk stream, often with a distinctly braided pattern because of the abundance of sand and gravel that these streams carry (**Figs. 18.23B, 19.19**).

Jökulhlaups (Icelandic for "glacier run"), also known as glacial lake outburst floods, are large releases of stored meltwater. In Iceland, jökulhlaups mainly occur when volcanic activity beneath the glacier results in rapid melting of ice. Much of this meltwater is stored in the subglacial or supraglacial environment and then suddenly released as a large flood.

In 1996, a major outburst flood occurred at Vatnajökull Ice Cap in Iceland (**Fig. 18.24**). First, subglacial volcanic eruptions began, eventually breaking through the ice cap. Meltwater from the event filled a subglacial lake, which then overflowed and cut a channel under the ice to the terminus of the ice cap via an outlet glacier. Water then burst out at the ice margin and flowed for about two days. The maximum flood discharge was ≈50,000 m³/s, or about ten times the mean discharge of the Mississippi River and slightly less than some of the larger Mississippi floods. During the jökulhlaup, large ice blocks were carried by the floodwaters, later forming kettles in outwash sediments. Numerous channels were cut into the outwash plain and massive amounts of sediment were deposited on this plain or carried into the Atlantic Ocean.

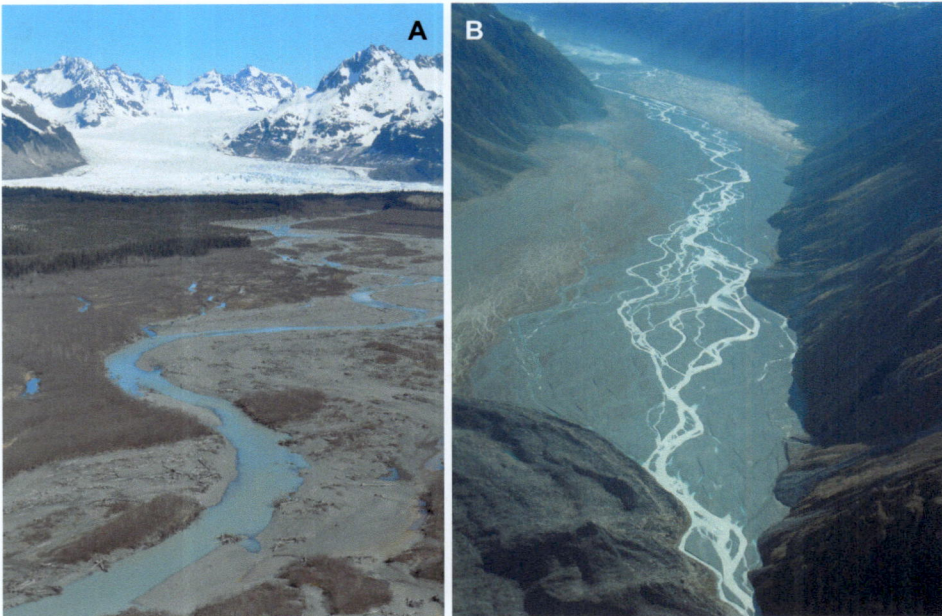

Figure 18.23 Meltwater streams. **A.** A proglacial stream, carrying gravel, sand, and fine silt as glacial outwash, in front of Sheridan Glacier, Alaska. Source: J. Crusius, US Geological Survey; in public domain. **B.** The braided Murchison River, fed by the Murchison Glacier, in Aoraki/ Mount Cook National Park, New Zealand. Source: Avenue, CC BY-SA 3.0, via Wikimedia Commons.

Figure 18.24 Image of the November 6, 1996 jökulhlaup in Iceland, as the flood poured out of a subglacial tunnel in the ice margin of Skeiðarárjökull, an outlet glacier of the ice cap Vatnajökull. Water, icebergs, and sediment are moving along the ice margin and then out through several floodways toward the Atlantic Ocean. Source: O. Sigurðsson, Iceland Meteorological Office; in public domain.

18.10 EROSION, TRANSPORT, AND DEPOSITION OF SEDIMENT BY GLACIERS

Glaciers are powerful forces in forming and modifying landscapes (Fig. 18.25). They can erode even the hardest bedrock because the "tools" the ice carries are just as hard as rock. Glaciers can transport large slabs of rock and massive amounts of sediment for hundreds of kilometers, seemingly with ease. The main processes of glacial erosion are those related to the glacier itself, such as abrasion and quarrying, or by the water associated with glaciers (see Chapter 19). Glacial erosion can occur at both small scales, as in abrasion and quarrying, and at large scales, in processes that push or move very large slabs of Earth materials up to several km^2 in area.

Abrasion is grain-by-grain erosion, or wearing down, of bedrock and sediments. Abrasion occurs in glacial, fluvial, coastal, and eolian systems, essentially wherever clastic particles are made to forcibly impact each other. Abrasion occurs on both Earth and Mars! Indeed, wind erosion via abrasion appears to be a very important process on Mars (see Chapter 23). Abrasion by glaciers is especially efficient because they not only produce many grain-to-grain collisions, but also because these collisions occur with high contact forces due to the weight and shear stresses that develop under the ice.

One useful way to think of glacial abrasion and the factors that control its efficiency is the sandpaper analogy. The effectiveness of sanding wood is related to several variables. First, you need a hard grit (or tool) to sand the wood. This is why the mineral corundum (hardness of 9 on the **Mohs scale**) is commonly used in sandpaper. Corundum is harder than most of the material you would use it on, such as wood, plastic, or soft metal. Diamond (hardness of 10) is used in some especially difficult sanding and cutting operations, such as for granite or metals. A second important factor in abrasion involves how fast you move the sandpaper, as energy per unit time is important in doing the work of abrasion. Third, the downward stress applied on the sandpaper is also important; if you don't push down, you won't produce enough force to make the sanding efficient. On the other hand, too much force will stop the sanding process completely, as the paper gets "locked" on its substrate. Finally, you must also clear away the products of abrasion (dust) or the sandpaper will clog up and the grit will no longer be in contact with the underlying surface. In glaciers, subglacial water is critical in transporting the abraded material away from the glacier bed. Also, in glaciers just as in sanding, both the surface is abraded as well as the tools. Over time, glacial "tools" can

1. Bergshrund

2. Supraglacial debris, including a debris flow (2A) from the surrounding upland, on the top of glacier

3. Lateral moraine formed by rockfall onto the ice surface; debris is then transported in the supraglacial path

4. Sediment in the englacial transport path

5. Sediment in the subglacial transport path, some of which has been plucked from the bed

6. Sediment being transported along the bed of the glacier, in the subglacial transport path

7. Debris septa formed along the bed as subglacial sediment moves within the glacier, down-ice of a bedrock divide between two tributary glaciers

8. Subglacial sediment brought into the englacial zone by shearing and thrusting within the ice

9. Medial moraine formed down-ice of a junction between two lateral moraines

10. Weakly expressed medial moraine

11. Supraglacial sediment exposed in a debris septa in the ablation zone

12. Grooves and striations eroded into bedrock

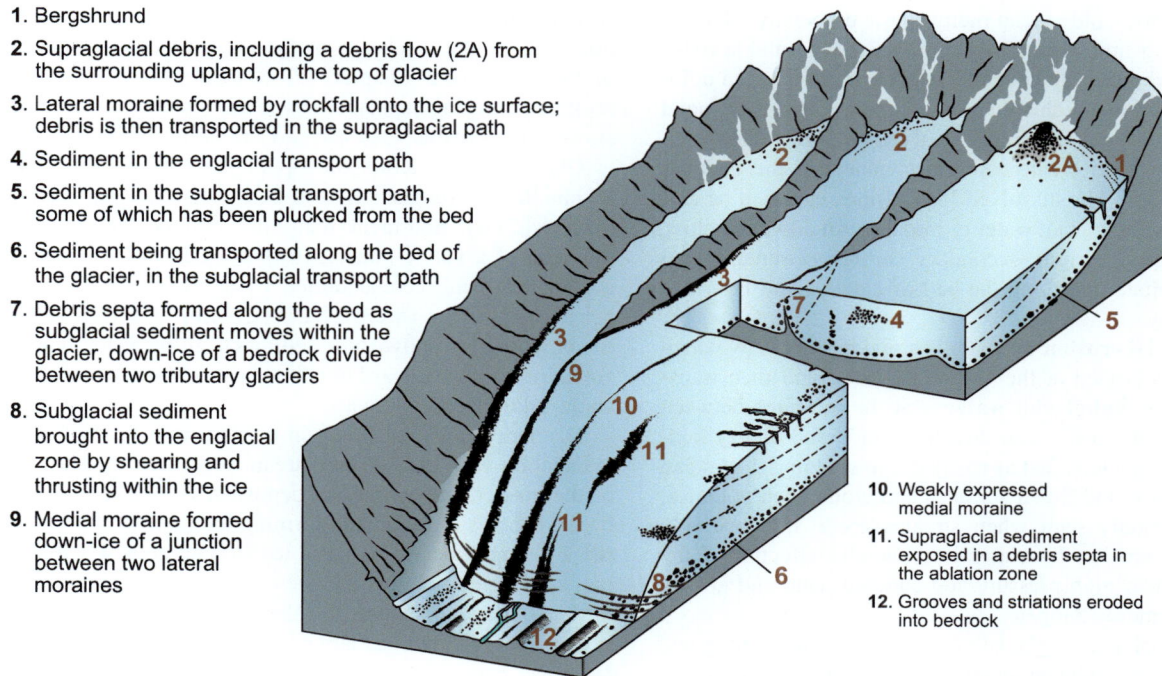

Figure 18.25 Transport paths, sediments, and landforms in a typical valley glacier. Source: Used with permission of Elsevier, from Benn and Evans (2010); permission conveyed through Copyright Clearance Center, Inc.

be abraded away, just as in sandpaper; even diamond cutting tools eventually wear out!

Together, glacial abrasion and crushing of grains produce large amounts of material of various sizes, but particularly notable are **silt**-sized particles (0.002–0.05 mm, or 2–50 μm in diameter). Glacial abrasion is a primary process in silt production, and as a result, glacial sediment is a major source of silt-rich loess, worldwide (see Chapter 21 and **Fig. 21.32**).

Quarrying (or **plucking**) is the process by which boulders and larger-sized rocks are pulled from the bedrock beneath the glacier. Quarrying can occur as ice intrudes into joints and fractures and then, by freezing on, pulls out fragments and incorporates them into the glacier. This happens where subglacial water is freezing onto the bed. Quarrying commonly occurs below the bergschrund and down-ice from icefalls and other bedrock obstructions.

Sediment may be entrained into the ice by processes of **regelation** at small (mm to cm) scales and by quarrying and freeze-on of slabs of rock/sediment at a larger (meter to km) scales (**Fig. 18.26**). Regelation is the process of melting due to lowering the melting point, for example, to –1 °C, by the higher pressures that form at the bed of the glacier, followed by refreezing (the "regelation") when the pressure is lower in down-ice areas. Pressure doesn't cause melting in itself; only heat causes melting. This is a common misconception of the term "pressure-melting point"; pressure only lowers the melting point so that melting can occur with the amount of heat that is present. If there isn't ample available heat, no melting will occur. The heat at the bed of a glacier comes from geothermal flux and friction. As shown in **Fig. 18.26**,

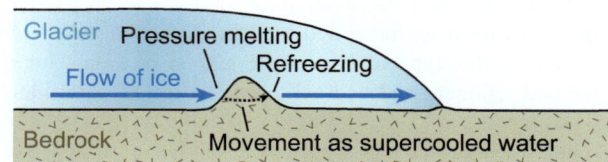

Figure 18.26 Diagrammatic sketch of the process of regelation (pressure melting and refreezing).

regelation occurs at obstructions in the bed of the glacier; ice melts on the up-ice side as pressures increase, lowering the melting point. The melted ice (water) flows around the obstruction and refreezes on the down-ice side, where the pressures are lower. Refreezing there can cause plucking and erosion. In valley glacier systems, sediment is also added to the glacier surface via avalanches and landslides from the surrounding uplands.

The largest clasts carried by glaciers are called **erratics**; some erratics can weigh >100 tons. The name "erratic" comes from the fact that their lithology is usually different than the local bedrock, because they have been transported to their resting place from an area of different bedrock geology.

Solution is another process by which glaciers erode landscapes. In this process, chemical weathering causes minerals to dissolve and become ions in solution. Glacier environments in the past have been thought to be areas where chemical weathering occurs only slowly, but this misconception overlooks the presence and importance of water in those environments. Cold glacial meltwater is especially effective at dissolving limestone, dolostone, and other soluble rocks because carbonate minerals like calcite, aragonite, and dolomite usually have high solubilities in cold water. Subglacial water may dissolve minerals directly at the bed and has even been observed to create solution channels in the bedrock below glaciers. The dissolved load of ions can then be carried away in subglacial water or into the groundwater below. If conditions are right, secondary carbonate minerals can also precipitate out along the bed of glaciers and be transported away.

Subglacial erosion by water is a major process by which sediment is eroded at the bed of a glacier and then transported away. **Subglacial water** – at the interface between the ice and the bed – can directly abrade and quarry bedrock below glaciers, just as rivers do. In so doing, water can form polished and fluted surfaces and erode deep potholes, as is commonly seen when streams erode into bedrock. Subglacial erosion by water is especially effective if the water is under high pressure and contains sand and gravel, which commonly happens.

Subglacial and proglacial deformation of preexisting bed materials can also be an important part of the suite of glacial erosion processes. Glaciers push and deform sediments (and even bedrock) in front of them, like a bulldozer. Masses of sediment and bedrock also freeze onto the bed of the glacier, facilitating the erosion and transportation of "rafts" and chunks of frozen sediment and boulders by the ice. We see evidence of this type of large-scale glacial erosion where deformed sediment and bedrock occur on top of older glacial sediments.

Once eroded, glaciers are capable of transporting large masses of sediment. **Competence** is a term used to describe the largest-sized clast a stream or glacier can transport. Most large rivers have a competence of boulders that are meters in diameter; in glaciers, boulders larger than a house can easily be moved! **Capacity** describes the total mass of sediment carried by a river or glacier. For rivers, capacity is largely governed by the discharge of the river and the turbulence of the water (see Chapter 15). Glaciers have very high sediment-carrying capacities. Even though they move much slower than streams, they can carry far more sediment.

Glaciers carry debris on top of the ice (supraglacial), within the ice (englacial), and below the ice (subglacial) (**Fig. 18.21A**). Sediment that falls into bergschrunds at the head of a valley glacier, or that which falls onto it by avalanches, can become buried by snow, making the ice debris-rich (**Figs. 18.25, 18.27**). Continued accumulation of snow and ice above such sediment, in the accumulation zone, forces it downward into the glacier. There, it enters the englacial transport pathway.

Glaciers deposit sediment in myriad ways, and often in association with gravity and stream processes. At the base of the glacier, debris may be deposited one clast at a time by the process of lodgment, forming **lodgment till** or **basal till**. Lodgment occurs as basal ice slowly melts while sliding, causing particles of all sizes to come into contact with the bed. As friction causes the clast to stop moving, the clast is deposited. Sometimes larger clasts, partially contained in basal ice, may plow up subglacial sediments. Entire bodies of debris-rich ice may also stop moving and slowly melt out. Deforming subglacial debris can stop moving as well, and be deposited. Lodgment till or basal till is usually well mixed, massive, and can be very dense due to the weight of the ice above it when it was deposited.

Englacial and supraglacial sediment may be deposited by processes of mass wasting, as debris in the ice is released by melting and then moved downslope by gravity as a debris flow. Some of this debris may also be transported by supraglacial streams on top of, and near the margins of, the ice (**Fig. 19.29**). Supraglacial debris may fall or wash into crevasses and moulins and be deposited there.

Glacial sediments directly deposited by glaciers are usually poorly sorted and unstratified, matrix-supported

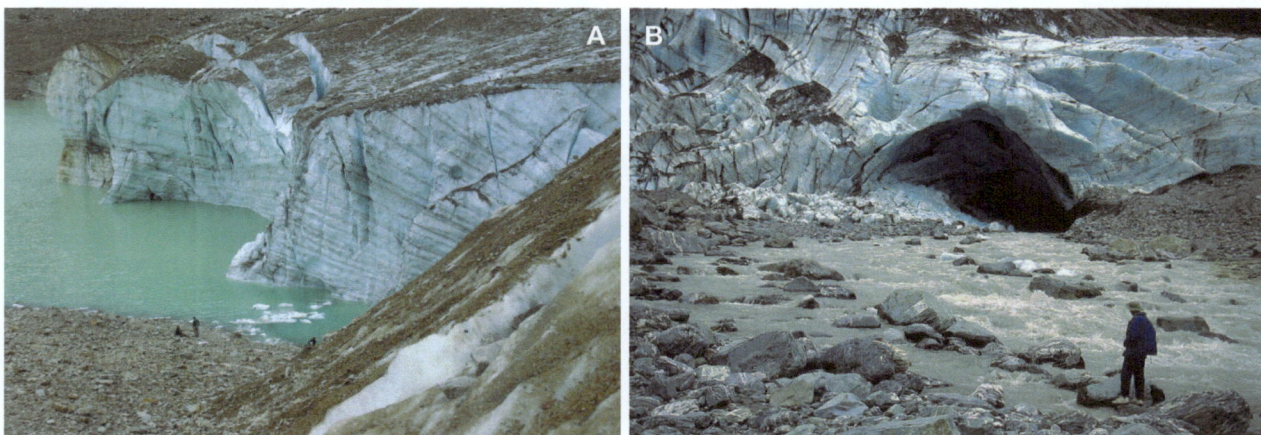

Figure 18.27 Examples of debris-rich glaciers. **A.** Supraglacial debris, as well as debris bands and ice foliation (dark bands), in the Cavell Glacier, Jasper National Park, Alberta, Canada. Source: Wing-Chi Poon, CC BY-SA 2.5, via Wikimedia Commons. **B.** A subglacial tunnel exiting at the debris-rich terminus of Fox Glacier, New Zealand. Source: Michael Hambrey, www.glaciers-online.net.

diamicton called till. **Diamicton** is a type of matrix-supported sediment where larger clasts such as gravels are set within sand, silt, and clay particles. Diamicton is a purely descriptive textural term that does not imply genesis or origin. **Glacial till** is diamicton that is directly deposited by glacier ice (see Chapter 19). Sediments transported by meltwater are better sorted and are usually **stratified** (deposited in layers). Glacial till can also form by deposition of sediment on top of, and on the sides of, the glacier; in this case it is usually less dense and can contain sorted layers. This type of glacial sediment is often called **meltout till** or **supraglacial till**. Unfortunately, it is commonly difficult to interpret the exact process by which till was deposited. Thus, these genetic terms are usually difficult to confirm in the field.

18.11 GLACIER CHANGES TODAY

All over the globe, glaciers are retreating, primarily because of human-induced climate change, bringing warmer temperatures and/or changes in snowfall and cloudiness (**Figs. 18.28, 18.29**). All these climate factors cause a rise in the glacier ELAs and lead to negative mass balance conditions. The World Glacier Monitoring System, initiated in 1894 as part of the International Glacier Commission at the 6th International Geological Congress in Zurich, Switzerland, compiles mass balance data from numerous mountain glaciers from around the world. Since 1950, the change in mean glacier mass balance has been negative for every year except five (**Fig. 18.30A**). Although every glacier in the world did

Figure 18.28 Changes in the Muir (foreground) and Riggs (background) Glaciers over a 63-year timespan. Both glaciers are in Glacier Bay National Park, Alaska, USA. **A.** Lower portion of Muir Glacier, August 1, 1941. **B.** The same location in August 1950. Source: (A and B) William O. Field of the US Geological Survey; in public domain. **C.** In this August 31, 2004 image, Muir Glacier has melted back and is no longer visible from this vantage point. Riggs Glacier is in the distance. Source: B. F. Molnia, US Geological Survey; in public domain.

Figure 18.29 Changes in Grinell Glacier, Glacier National Park, Montana, USA. The upper part of the glacier in (**A**) 1938 and (**B**) 2016. The lower margin of the glacier in (**C**) 1887 and (**D**) 2013. Source: US National Park Service; in public domain.

Figure 18.29 (continued)

not retreat during this time, the data indicate that the average mass balance of all the glaciers monitored in nine regions was negative (**Fig. 18.30B**).

Worldwide, valley glaciers have lost, on average, the equivalent of ≈20 m of water equivalence over their surface during the 1950–2019 period (**Fig. 18.30B**). Although the warming climate has resulted in the thinning and retreat of glaciers in most regions, the greatest mass losses have occurred in glaciers in Western Canada and the United States, Central Europe (Alps), Central Asia, and Alaska (**Fig. 18.30C**). Glaciers that have experienced lower mass losses are located in the Canadian Arctic, Scandinavia, the Caucasus and Middle Eastern region, Svalbard and Jan Mayen Islands, and in the southern Andes (**Fig. 18.30C**). Even in these regions, however, glacier retreat has been significant, and locally, some glaciers have completely melted away.

Mass balance changes for the Greenland and Antarctic ice sheets are difficult to estimate because of their sheer size and complexity. Current estimates are based on satellite and aircraft-based measurements of laser-derived surface elevation changes, or total gravity change measurements. Gravity data collected by the GRACE missions (NASA and ESA Gravity and Climate Experiment) show that the Greenland and Antarctic ice sheets have been losing mass since the late 1990s (**Fig. 18.31**). More detailed satellite and aircraft-borne laser altimetry data from NASA ICESat and ICEBridge missions demonstrate that, although ice in the central regions of the Greenland ice sheet is thickening, most of the ice sheet near the margins is rapidly thinning. The same data in Antarctica show that parts of the West Antarctic ice sheet are thinning faster than the East Antarctic ice sheet. The most rapid changes in the West Antarctic ice sheet are being driven by warm sea water moving under the calving margins of tidewater outlets and melting them from below.

18.12 HAZARDS ASSOCIATED WITH GLACIERS

Natural hazards related to runoff from glaciers usually involve sudden outbursts of meltwater due to moraine-dam failures of proglacial lakes, or sudden drainage of supraglacial or subglacial lakes. These events, called **glacial lake outburst floods** (GLOFs), cause loss of life and property down-valley in mountain regions, as well as near ice caps and ice sheets. Perhaps the best-studied glacial hazards are the Icelandic GLOFs known locally as jökulhlaups, which have been well-documented there for the past 1,000 years (**Fig. 18.24**).

Over the last 50 years, GLOFs have become more common in the North American Cordillera, the European Alps, the Andes, and especially in Central Asia and the Himalayas. In October of 1994 in Bhutan, a moraine dam failed and flooded the Pho Chu valley, killing 21 people and damaging homes, farms, and a culturally significant monastery. More recently, in 2021, a glacier outburst flood in the northern Indian State of Uttarakhand led to the deaths of over 200 people and destroyed two hydroelectric dams. Many of the casualties were workers at the hydroelectric plants. Himalayan glaciers have been losing mass each year since 2000 – double the amount of melting that occurred between 1975 and 2000.

Avalanches of snow, ice, and rock are also common hazards in glacial landscapes, causing large numbers of deaths. In a warming world, glacier hazards from GLOFs and snow and ice avalanches are likely to increase in frequency.

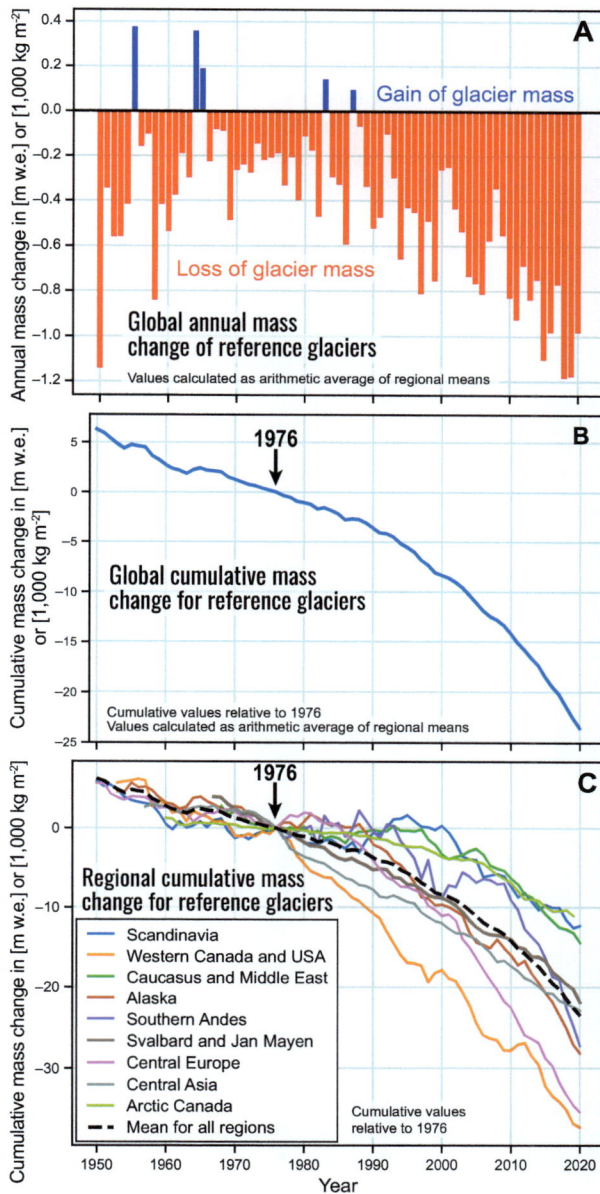

Figure 18.30 Changes in selected small glaciers worldwide, where detailed mass balance measurements have been collected continuously from 1950–2019. **A.** Changes in mean annual net mass balance for 30 reference mountain glaciers in nine regions. These glaciers only had positive mean mass balances in 1955, 1964–65, 1983, and 1987. **B.** Cumulative mass change for 30 reference mountain glaciers in nine regions around the world. **C.** Mean cumulative net mass balance for reference mountain glaciers, by region. Source: After Zemp, M., Gärtner-Roer, I., Nussbaumer, S. U., Bannwart, J. et al. (Eds.) (2020), © World Glacier Monitoring Service.

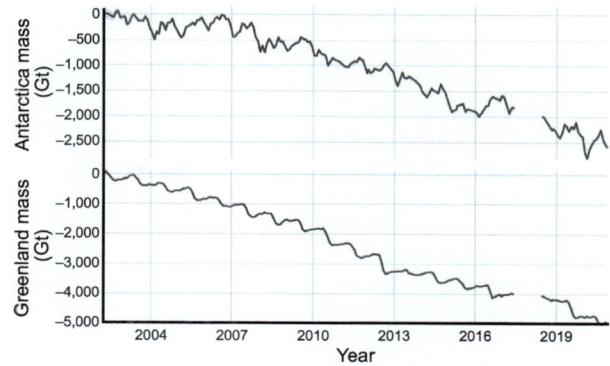

Figure 18.31 Gravity measurements from NASA's GRACE mission, showing the estimated loss of mass in the Greenland and Antarctica ice sheets from 2002 to 2020. GRACE uses two satellites to measure the changes in Earth's gravity over time. Most of the changes in gravity over ice sheets are due to annual mass changes in ice. Note how the ice masses change seasonally. Longer-term changes are a function of ice gain or loss, in this case, loss. These data provide clear evidence that these ice sheets are currently melting.

REVIEW QUESTIONS

18.1 Why are glaciers important to humans and our economy?

18.2 What is the definition of a glacier and what are the main types of glaciers on Earth today?

18.3 How do surging glaciers differ from other kinds of glaciers?

18.4 Explain the transformation from snow to ice. What is the intermediate step along this pathway?

18.5 What is meant by glacier mass balance? What are the components of this type of mass balance equation?

18.6 What are the two main zones in almost every glacier, and what are the main processes that occur in these zones? What is the name given to the boundary between these two zones?

18.7 What are the main ways that glacial ablation is accomplished?

18.8 Explain how a change in local climate can cause a change in the location of the glacier terminus.

18.9 By what two main mechanisms do glaciers flow?

18.10 Where are most of Earth's glaciers today?

18.11 What are the four main "environments" of glaciers?

18.12 Provide some examples of processes that add mass to a glacier, and processes that cause a glacier to lose mass. How do these contribute to the mass balance of the glacier?

18.13 How do glaciers erode bedrock and sediment? How do they transport and deposit sediment?

18.14 What is meant by a proglacial stream or lake?

18.15 What is a jökulhlaup and how does it form?

18.16 Describe how the process of regelation operates, and under what specific bed conditions it is promoted.

18.17 What are some of the major types of sediment that glaciers carry, and where are they located/ transported within the ice?

18.18 What evidence is there for glacier retreat from around the world?

18.19 What are some of the more common types of glacial hazards?

FURTHER READING

Benn, D. I. and Evans, J. A. 2010. *Glaciers and Glaciation.* 2nd ed. Hodder Education.

Florentine, C. 2019. *Glacier Retreat in Glacier National Park, Montana.* US Geological Survey.

Patterson, W. S. B. 1994. *The Physics of Glaciers.* 3rd ed. Pergamon.

Sugden, D. E. and John, B. S. 1976. *Glaciers and Landscape.* Edward Arnold.

19 Glacial Sediments and Landforms

Randall Schaetzl and Patrick M. Colgan

Ice sheets have dramatically shaped the landscape across the northern regions of North America and Europe. Ice sheets are so vast that they are sometimes referred to as **continental glaciers**. Their deposits have directly influenced human history by rerouting river systems and by providing nutrient-rich parent materials for soils. Abundant lakes and rivers, many of which were newly formed by the ice, became early transportation arteries and supplied aquatic resources to early cultures. Indirectly, glacial sediments were transported by wind to form thick and extensive blankets of **loess** – home to many of the world's best soils. Ice sheets reduced the overall relief of the landscape, as valleys were widened and filled, providing for ease of transportation, growth of agriculture, and the rise of civilizations.

Despite being restricted to mountains, valley, or **alpine**, glaciers have also impacted our world. These glaciers, still present today in diminished extent, shaped many mountainous landscapes, forming majestic, craggy peaks, deep blue lakes, and beautiful alpine valleys. Many societies rely on glacial meltwater for irrigation and drinking water, even today.

For most midlatitude and polar landscapes, glaciation has been a key geomorphic process. Glaciers have done more to sculpt and modify these landscapes than wind, water, or gravity. How did glaciers accomplish all this work? What do the outcomes of glacial processes look like? Answering these questions is the main focus of this chapter. We look forward to having you join us on this glacial expedition.

19.1 GLACIATION AND CLIMATE HISTORY

In Chapter 4, many aspects of Earth's past climates were explained. To summarize, Earth remains in a period of cold temperatures, which began millions of years ago (Fig. 4.9). However, human-caused emissions of greenhouse gases are causing many of Earth's glaciers to thin and retreat.

Geologists define the start of the most recent part of the Late Cenozoic Ice Age – the **Quaternary Period** – at ≈2.58 million years ago, and it continues today. During the Quaternary, land masses in the high latitudes (and mountains) were subject to repeated glacial advances (termed "glaciations") and retreats (termed "interglaciations"), generally coinciding with intervals of cold and warm climate, respectively (Fig. 4.10). Geologists concerned with Earth history subdivide the Quaternary into the **Pleistocene** and **Holocene Epochs**. During the Pleistocene, glaciers and ice sheets were very active, periodically advancing and retreating dozens of times.

Earth is currently in an interglacial called the Holocene Epoch. The Holocene is simply the latest of a series of interglacial periods. Although the retreat of the last ice sheets took thousands of years to come to completion, geologists currently define the Pleistocene–Holocene boundary at ≈11,700 years ago. The Pleistocene–Holocene boundary is marked by major changes in climate, as recorded by oxygen isotopes in Greenland ice cores. Throughout the Holocene, Earth's climate has been relatively warm and (with a few small exceptions) stable, and large ice sheets have generally been restricted to Greenland and Antarctica (see Chapter 4).

When the glacial history of the Quaternary Period is examined, it becomes clear that ice sheets have been extremely active, shaping and reshaping vast swaths of Earth's landscapes. Landscapes have been eroded, flattened, gouged out, and/or scraped away, whereas others have been buried by many meters of glacial sediment. Across much of Earth's formerly glaciated terrain, nearly all lakes, hills, and valleys owe their existence to past glaciers; very little of the pre-Quaternary landscape remains.

19.2 INTERPRETING GLACIAL SEDIMENTS AND LANDFORMS

Most geomorphologists learn about past **glaciers** by studying modern glacial environments as analogs. By observing and measuring what modern glaciers do, we hypothesize about past glaciers and glacial deposits. In short, we apply the principle of **uniformitarianism** – which states that past geological actions behave the same as present geological actions. In essence, the laws of nature do not change with time. But we must always remember that large-scale events have also occurred in Earth's history for which we have no direct experience, for example, planet-killing meteorite impacts, or giant glacial megafloods. Even so, physical processes and natural laws have generally operated similarly in the past, as they do today.

This principle works well for valley glaciers, which are still plentiful on Earth. Using modern analogs to learn about

Figure 19.0 The Trift Glacier flows down the Urner Alps of Switzerland. Source: Federica Grassi / Getty Images.

Pleistocene ice sheets is more difficult, though, as ice sheets do not currently occur in mid-latitude locations. Some ice caps in Iceland and in Arctic Canada may provide small-scale analogs. Nonetheless, the large scales over which ice sheets reshaped the land in North America and Europe have no modern counterparts. However, decades of research into glaciers and their dynamics has helped us to understand the many landforms they left behind. We begin with a discussion of glacial erosion.

19.3 PROCESSES AND LANDFORMS OF GLACIAL EROSION

By virtue of their mass and seemingly continuous motion, glaciers are capable of tremendous amounts of erosion. Like the tale of the tortoise and the hare, slow and steady wins the (glacial erosion) race!

Glacial erosion implies that the ice is removing material from its bed, whether that bed is composed of bedrock or unconsolidated sediment. If the bed is composed of unconsolidated material, erosion lowers the bed but leaves only a few distinct landforms that we might be able to identify. If the glacier has been flowing across bedrock, erosional features are more easily preserved, and these become tell-tale indicators of glacial erosion. Often, these features also tell

us exactly which direction the ice was flowing, because they align with the former ice flow.

Abrasion and plucking/quarrying – the two main forms of glacial erosion – wear down even the hardest bedrock (see Chapter 18). Ice near its melting point is soft, but the tools it carries – rocks, gravel, sand, etc. – are more than capable of wearing down the rock under the bed of the glacier. Just as paper is softer than wood, ice is softer than rock. But sandpaper, paper with grit embedded within, can easily wear down wood. In comparison to a glacier acting as sandpaper, the rock fragments (the "grit") in a glacier can wear down bedrock, especially if the fragments are carried at the bed of the ice.

19.3.1 Small-Scale Landforms of Glacial Erosion

Landforms of glacial erosion range across a variety of scales, from the smallest scratches to deep valleys. This discussion assumes that an eroding glacier has already removed most of the unconsolidated material from its bed, and so we now focus on erosion of the bedrock.

As ice carrying the tools of erosion (rocks, gravel, sand, and silt) slides across bedrock, it can make scratches called **striations** (or **striae**), in the rock (**Fig. 19.1A, C**). Striations always parallel the direction of glacial flow. They are formed by abrasion, whereby rocks within the ice, under pressure

Figure 19.1 Features of glacial erosion. **A.** A schematic diagram of the various, small-scale, erosional features produced by glaciers. Source: © A. N. Strahler (1992). Used by permission. **B.** A groove formed in bedrock, by glacial abrasion. Source: R. Schaetzl. **C.** Striations. Source: (left) R. Schaetzl; (right) Walter Siegmund, CC BY-SA 3.0, via Wikimedia Commons. **D.** Chatter marks and crescentic gouges. Arrows indicate the direction of ice flow. Source: R. Schaetzl.

due to the weight of the glacier, are dragged across bedrock, like the sandpaper analogy used above. Striations can only form when glaciers are sliding.

Factors that affect how much abrasion a glacier can produce include (1) the concentration of debris in the lowest part of the glacier, (2) how much water exists at the base of the glacier, (3) the sliding velocity of the ice, and (4) the hardness of the rocks (see Chapter 18). Water is important because it removes the sediment formed by the scratching, keeping the rock surface exposed for more abrasion. High polar glaciers, with no water at the bed, are less effective at erosion, and produce few striations. Softer rocks like limestone are easily striated while hard rocks like quartzite are less easily striated.

Sometimes, debris being carried in the ice gets dragged with such force that it causes the bedrock to fracture, forming half-moon shaped **crescentic gouges** or **chatter marks** (**Fig. 19.1A, C**). Some geologists believe that chatter marks form as exceptionally hard and sharp-edged rocks are dragged over a bedrock surface, episodically sticking in place but then suddenly fracturing the bedrock, releasing stress, and then moving again.

Glaciers often carry basal debris (near the bed of the ice). When dragged across the bed, this debris can produce large, deep **glacial grooves**, even in hard bedrock (**Fig. 19.1A, B**). On the other hand, if ice is carrying smaller tools (sand and silt), it can produce **glacial polish** as fine as any kitchen countertop, especially on granite and other hard rocks. Glacial polish is simply a mass of minute striations, visible only with magnification. Rocks with this kind of polish can even reflect sunlight (**Fig. 19.2**).

Striations, glacial polish, and other erosional features we see today indicate not only what was initially produced, but also how well they survived post-glacial weathering processes. Erosional features formed in hard, fine-grained bedrock lithologies persist longer, that is, the rock weathers slowly and therefore, are better preserved. In soft rocks, the rock surface quickly weathers, taking the features with it.

Quarrying (or **plucking**) is a larger-scale process of glacial erosion than abrasion. It occurs when meltwater is present at the glacial bed. Meltwater refreezes in fractures and joints in the bedrock, and also onto the ice itself. Then, when the glacier moves, it can pull/pluck pieces of the rock out of the bedrock and incorporate them into the glacier. Quarrying can erode material from pebble size up to gigantic boulders. Meltwater at the bed of the glacier is essential for quarrying to occur. In general, meltwater can be produced in three different ways and locations: (1) at the glacial surface, where it may then flow down to the glacial bed in moulins and crevasses and freeze, (2) at the bed, where frictional heat melts some of the basal ice, or (3) by a process called **regelation** (see Chapter 18).

A common landform of glacial erosion – one that is produced by abrasion, plucking, *and* fracturing of bedrock – is a **roche moutonnée** (**Fig. 19.3**). These features are very common in glacially eroded, bedrock-dominated landscapes, and come in a wide range of sizes. The term comes to us from the French (*roche* rock, *moutonnée* sheep), perhaps because of the rounded top of the feature. Roches moutonnées are abraded, smoothed, rounded, and usually striated on their stoss (up-ice) side, and quarried, fractured, and jagged on their lee (down-ice) side. A quick inspection of a roche moutonnée will provide a clear indication of the past ice flow direction.

Roches moutonnées form best in bedrock that is usually already jointed, prior to glaciation. Quarrying (breaking

Figure 19.2 Glacial polish. **A.** A glacially polished rock outcrop on a city park in Helsinki, Finland. Source: Raymond M. Coveney, CC BY-SA 3.0, via Wikimedia Commons. **B.** Glaciers can even polish coarse-grained rocks like this granite outcrop. Source: R. Schaetzl.

rocks out) occurs as water penetrates and freezes in these joints. Air- and water-filled cavities can also form down-ice from bedrock obstructions, which can lead to large pressure differences that can fracture the down-ice side of a roche moutonnée. In bedrock that is less jointed and hence less prone to quarrying, cavity formation, and fracturing, a smoother but still streamlined bedrock landform known as a **whaleback** or **rock drumlin** may form (**Fig. 19.4**). Whalebacks are formed mainly by abrasion, as the ice slides across bedrock (**Figs. 19.4, 19.5A, B**). In summary, landforms dominated by abrasion tend to become smoothed and streamlined, whereas quarrying and fracturing cause the bedrock to become jagged and rough.

Before we leave the discussion of smaller-scale, erosional landforms, let's look at a common *hybrid* landform, formed partly by glacial erosion and partly by glacial deposition. **Crag-and-tail** landforms are formed where a hard, resistant rock crops out under the glacier (**Fig. 19.4**). Unable to erode it completely, the ice abrades and even may quarry around the rock, leaving a jagged bedrock high. Down-ice, uneroded rock remains, or glacial sediment is deposited in the lee of the crag. This tail marks the direction of glacial

Roche moutonnée

Figure 19.3 Roches moutonnées. **A.** Schematic diagram of a roche moutonnée, showing the erosional processes associated with its stoss and lee sides, and how the ice fractures as it passes over the bedrock obstruction. **B.** A small roche moutonnée, showing striations on the stoss side and a fractured lee side; coins for scale. Source: R. Schaetzl. **C.** Lembert Dome, a large roche moutonnée in Yosemite National Park, California, USA. Even though this feature is formed in coarse-grained granite bedrock, parts of it are nicely polished. Source: R. Schaetzl.

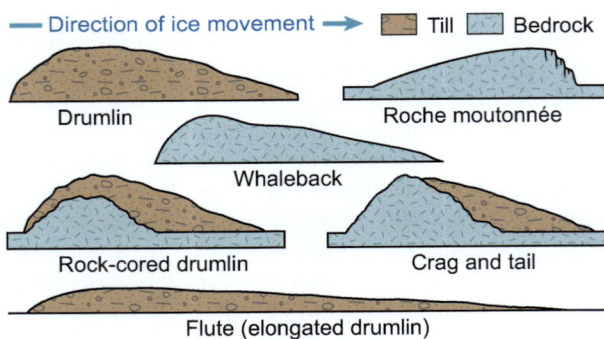

Figure 19.4 The various types of streamlined glacial landforms, each of which can help indicate the direction of ice flow. Landforms are not drawn to scale, and each one can vary greatly in size.

flow. Crag-and-tail features are unequivocal indicators of glacial flow direction because the crag is always up-glacier, and the tail is always down-glacier. Such features demonstrate that sliding glacial ice can form landforms by either erosion and/or deposition, even in the same area.

19.3.2 Large-Scale Landforms of Glacial Erosion

Landforms of glacial erosion occur at a variety of scales. Abrasion can produce everything from microscopic striations to deep bedrock troughs. Plucking can pull small fragments out of bedrock but can also rip off the entire back half of a mountain. There is no question: glacial erosion is a powerful force that typically operates uninterruptedly for millennia at a time.

Widespread erosion by ice sheets is epitomized by the landscapes of central Canada. Here, the core of the North American continent – the Canadian Shield – has been levelled by wave after wave of glaciation associated with the **Laurentide ice sheet** (**Fig. 19.5**). Repeated episodes of glacial erosion have made this landscape flatter, with more smoothed and rounded forms than it had previously. Many of the low bedrock uplands are streamlined in the direction of glacial flow, roches moutonnées and whalebacks of all sizes are common, and pre-glacial regolith is thin or absent across much of the landscape, especially on uplands. With bedrock so close to the surface, the Canadian Shield is often rife with lakes and swamps. In summary, erosion by ice sheets tends to preferentially wear down uplands and transport the sediments down-ice, often filling in preexisting valleys, resulting in a low-relief, bedrock-dominated landscape.

Erosion in *alpine settings*, however, produces a very different kind of landscape. Valley glaciers are capable of eroding deeply into bedrock, leading to accentuated relief. The most striking landforms of glacial erosion occur in these settings.

Glacial erosion in mountains produces deep, wide, **U-shaped valleys** (**Fig. 19.6**), many of which are surrounded by higher valleys, jagged peaks, and ridges called cirques, horns, and arêtes respectively. Most deep valleys in mountainous regions formed initially by running water and mass wasting, that is, they are fluvial valleys. These types of valleys are usually V-shaped in cross-section (**Fig. 19.7A**). As glaciers occupy these valleys, they modify the valley shape, mainly by erosion. Valley glaciers, especially the larger

Figure 19.5 Images of the Canadian Shield in northern Canada, where repeated erosion by ice sheets has generally smoothed and sculpted the bedrock landscape. Source: R. C. Paulen.

Figure 19.6 Images of U-shaped valleys formed by alpine glaciation. **A.** The upper Green River Valley, Wind River Mountains, Wyoming, USA. Source: R. Schaetzl. **B.** The St. Mary Valley in Glacier National Park, Montana, USA. Source: R. Schaetzl. **C.** Parabolic Valley in Norway. Source: Thatcher Clay, CC BY 2.0, via Wikimedia Commons.

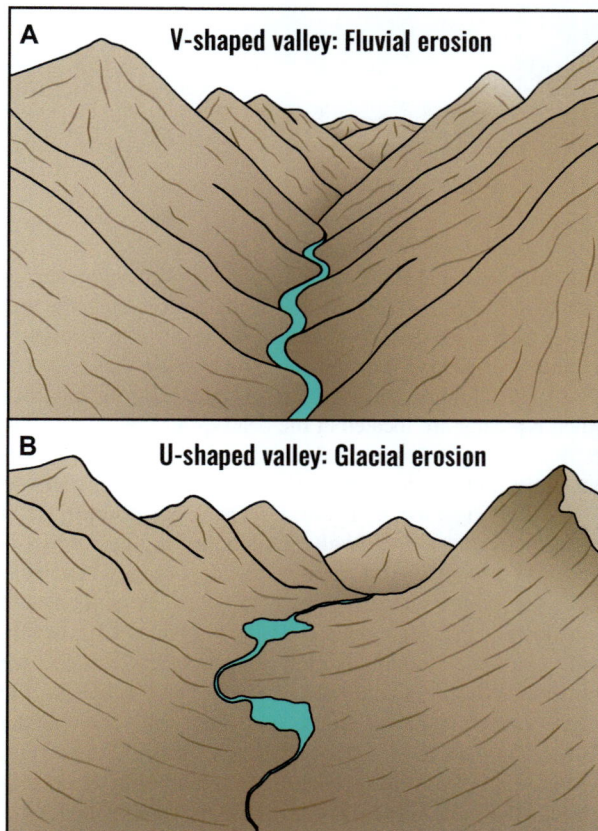

Figure 19.7 Idealized valley cross-sections, as formed by (**A**) running water (V-shaped) and (**B**) glaciers (U-shaped).

ones, are highly effective at widening the valley as much as deepening it, and hence the valley becomes more U-shaped in cross-section (**Fig. 19.7B**). The valleys also generally become slightly straighter than the preexisting river valley. U-shaped valley forms are indicative of extensive glaciation at some time in the past. At places along the valley, the ice margin can produce an over-deepening, where some of the highest velocities and sites of erosion occur; many of these are at or near the former **equilibrium line** (see Chapter 18). End moraines commonly dam up these glacial troughs in various locations, producing deep lakes. Commonly, the flat

bottoms of these types of glaciated valleys results from infilling of sediments after the ice leaves.

Glacial valleys drowned by the sea are called **fjords**, or **fiords** (Old Norse, *fjörör*, "inlet" or "estuary"). In Scandinavia, home to some of the world's longest and deepest fjords, many fjords are tens of kilometers long (**Fig. 19.8**) and have bottoms well below sea level. Glacial ice is capable of eroding below sea level, because it will usually start to float only when about 90% of its thickness is under water. But the main reason that fjords extend so far below sea level, however, is because sea level was much lower during the peak glacial periods (see Chapter 4).

Part of the reason that valley glaciers are so effective at erosion stems from the fact that they are rife with debris, which are used as tools to deepen and widen the valley – via subglacial water erosion, quarrying, and abrasion. Ice sheets commonly derive most of their load from the bed, but in valley glacier systems, debris can enter the glacier from the bed, or as rocks fall onto the ice surface from the neighboring valley walls. The latter process is more robust, because freeze–thaw activity (necessary to pry rocks from the bedrock walls) is common in cold, glacial uplands. Thus, valley glaciers typically have a plentiful supply of "tools" to erode the landscape.

A good way to envision how valley glaciers form their impressive array of erosional landforms is to consult **Fig. 19.9**, which shows conceptually the before, during, and after stages of mountain glaciation. Prior to glaciation, weathering and mass wasting are at work to lower the side slopes of the valleys, delivering that sediment to the rivers below. Mountains like this, which have not been affected by glaciation, have rounded summits and V-shaped valleys (**Figs. 19.7A, 19.9A**). Although thicknesses vary, most slopes have a cover of regolith and soil, implying that most parts of the slope (below treeline) are vegetated.

As the climate cools, snow begins to accumulate in the higher elevations, but the thickest snow is not on the ridge crests. Instead, the deepest snow accumulates on the upper slopes, just off the crest, usually in the headwaters of small rivers, where the slope gradients are lower and where snowdrifts can accumulate. Eventually, snow thicknesses reach a point where the base becomes, under pressure, glacial ice (see Chapter 18). The onset of glaciation then begins; small glaciers

Figure 19.8 Fjords, as seen from the (**A**) air (Norway) and (**B**) water (New Zealand). Source: Marli Miller.

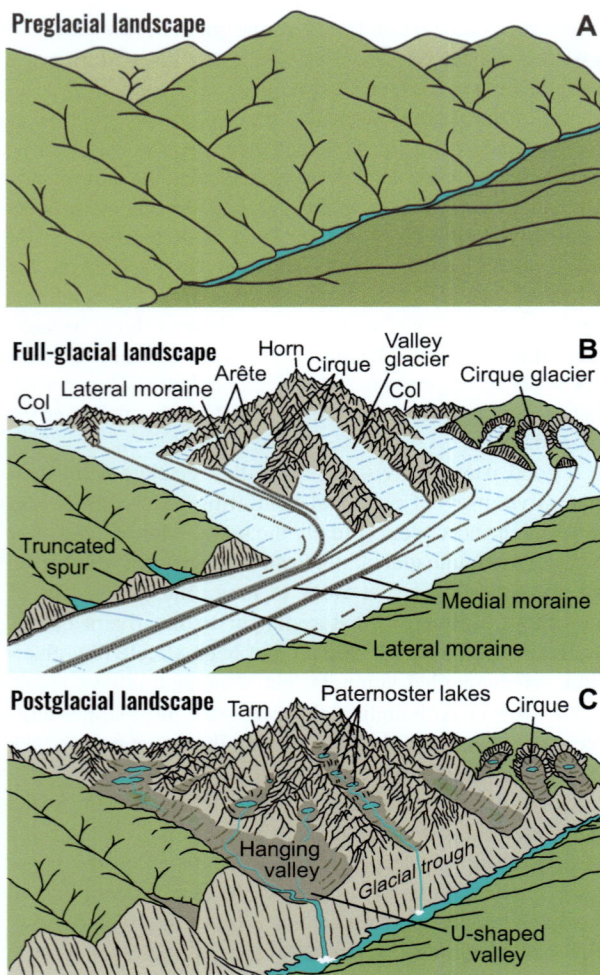

Figure 19.9 The suite of landforms typical of a highly glaciated alpine landscape. **A.** Prior to glaciation, the landscape has rounded summits and V-shaped valleys formed by fluvial incision. **B.** The same landscape at full glaciation, showing some of the distinctly glacial landforms that have developed. **C.** After the ice has melted back, the full suite of alpine glacial (erosional) landforms, in this case all formed on hard bedrock, is apparent. Source: © A. N. Strahler (1992). Used by permission.

form, thicken, and start to flow. Glacial erosion, coupled with frost shattering of the rock walls surrounding the ice, deepen and widen the upper parts of the valley into a bowl- or amphitheater-shaped **cirque**, or cirque basin (**Figs. 19.9, 19.10**). By definition, cirque glaciers originate in cirque basins, which typically are located at the head of a river valley (**Fig. 19.9B**). As they enlarge and flow down-valley, cirque glaciers extend beyond the cirque proper, becoming **valley glaciers**.

On their upslope end, cirques terminate abruptly in a steep, rocky cirque **headwall**. Here, quarrying of bedrock is especially pronounced, forming this steep, craggy headwall. Often the cirque glacier is not completely in contact with its headwall; instead, a deep crevasse known as a **bergschrund** (German, "mountain cleft") occurs between the ice and the headwall. As snow melts, water may seep into the bergschrund and (later) freeze onto both the rock and the glacier. Then, as the ice flows, the rock face is quarried and the headwall of the cirque is eroded further back. Rocks that fall into the bergschrund also provide the glacier with additional tools for abrasion.

Glaciers that flow out of the cirque and down-valley as valley glaciers scour, deepen, and widen their valleys. Just as rivers join at accordant tributary junctions (**Fig. 19.9A**), valley glaciers also join at similar junctions. However, in the

Figure 19.10 Cirques. **A.** Three small cirques in the Wind River Mountains of Wyoming, USA. Source: R. Schaetzl. **B.** Lower Curtis Glacier, a cirque glacier in the North Cascades Range of Washington State, USA. The photo, taken in 2003, shows the retreat of the ice margin since 1985 (red line). Source: US Forest Service; in public domain. **C.** A classically shaped cirque in Canada's Yukon Territory. Note the well-formed arêtes on both sides of the cirque. Source: Marli Miller.

case of glaciers, the accordance is at the *top* of the ice, not the base (**Figs. 19.9B, 19.11**). In other words, the rock floors among glacial valleys are commonly discordant, or stepped. This discordance develops because larger glaciers are able to erode their valleys (sometimes called **glacial troughs**) much deeper than can small, side-valley tributary glaciers. As a result, where side glaciers meet the main trunk glaciers, the bedrock floor falls precipitously, forming a **hanging valley** (**Fig. 19.9C**). Waterfalls, sometimes spectacular ones, are a common feature at hanging valley junctions (**Fig. 19.12**). Over time, these streams may begin to cut V-shaped notches in the trough of the hanging valley. Unquestionably, no other landform better epitomizes alpine glaciation more than hanging valleys; no other process can so readily form them.

Over time, valley glaciers continue to deepen and widen their valleys. Indeed, most glacial valleys have experienced several episodes of glaciation during the Pleistocene, each one scouring out more bedrock. Eventually, the bedrock walls between the valleys become narrower, as the valleys between them widen. These uplands become increasingly narrow **arêtes**, eroded not only by ice but by intense frost

Figure 19.11 The Marjorie Glacier in Alaska's Glacier Bay National Park and Preserve (USA), showing the accordance of the ice surfaces between the main trunk glacier and its tributary. Note the many distinctive glacial horns in the background. Source: US National Park Service; in public domain.

Figure 19.12 Hanging valleys in Yosemite National Park, California, USA. **A.** Bridalveil Falls has formed at a classic hanging valley junction with the deep, main Yosemite Valley (center). **B.** Yosemite Falls is a double waterfall, again at the junction of a tributary valley with the main valley. Source: R. Schaetzl.

weathering on their steep sidewalls (see below). Rocks broken off the walls by ice and freeze–thaw fall onto the glacier and are carried away. Rocks that fall off valley walls during an interglacial period will be transported down-valley by flowing water, or maybe by the next glacier! In the end, the valleys remain deep and generally free of sediment.

Knife-like, jagged arêtes are the hallmark of heavily glaciated mountains (**Figs. 19.9C**, **19.10C**). Arêtes may form at the intersection of two valley walls or where the headwalls of two cirques intersect. **Cols** are saddle-shaped lows through arêtes (or similarly shaped ridges), formed as glacial erosion makes the arêtes ever thinner and hence, lower; they are often used as mountain passes. Where three or more arêtes intersect, a sharp, high peak called a **horn** is often present. Horns are often three- or four-sided, with each edge descending steeply into a cirque headwall. The term "horn" originated from the Matterhorn, in the Swiss-Italian Alps, perhaps the world's most famous glacial horn (**Fig. 19.13**). The angular shape of horns is more a function of mass wasting, preexisting bedrock joints, and fractures than it is the direct result of glacial erosion. Because horns are often the highest parts of a glaciated alpine landscape, they are also the most likely to stand up above the snow and ice fields during glaciation. Horns that stand above widespread fields of snow and ice are called **nunataks** (Inuit *nunataq*).

Figure 19.13 The 4,478 m-high Matterhorn, a classic glacial horn. Source: Liridon, CC BY-SA 4.0, via Wikimedia Commons.

Arêtes and horns are residual features of the bedrock landscape – left behind and standing above valley glaciers. Nonetheless, distinctive erosional landforms also form beneath the ice, in the glacial trough. Valley glaciers are notorious for eroding more deeply in some areas than in others. For example, cirques are usually over-deepened, resulting in a basin and a rock step that define their down-valley limit. Continuing down-valley even farther, rock steps become a common feature of the glacial trough. The over-deepened parts of the glacial trough are sometimes located near areas of valley constriction or where a tributary glacier enters the main flow. Regardless, these steps, and the scoured rock basins between, lead to an uneven down-valley profile. Subsequent to glaciation, lakes often exist in many of these over-deepened basins. If a lake exists in the cirque, it is referred to as a **tarn**, or **tarn lake**. If a sequence of lakes continues down-valley, beyond the tarn, a chain of descending lakes, often called **paternoster lakes**, may form (**Figs. 19.9C**, **19.14**, **19.15**). The term paternoster is from the Latin (*pater* "father," *noster* "our"), a name coined because the string of lakes reminded early explorers of strings on a rosary, one prayer of which is the Lord's Prayer, also known as the Our Father.

19.4 GLACIAL SEDIMENTS AND DEPOSITIONAL LANDFORMS

As we have seen, glaciers produce a distinctive suite of erosional landforms. The debris that the glaciers erode eventually forms depositional landforms. In Chapter 18, some discussion was devoted to how and where glaciers entrain (pick up), transport, and then deposit sediment. Our focus in this chapter will be on sediments deposited by glaciers, and the resultant landforms. **Glacial drift** is an outdated term that is sometimes still used in glacial literature; it refers to any sediment of glacial origin. We mention it here only for completeness.

Traditionally, geomorphologists have recognized three broad types of glacial sediment: till, glaciolacustrine sediment, and glaciofluvial sediment. **Glacial till** is generally unsorted and unstratified sediment directly deposited by glaciers, sometimes reworked by gravity and meltwater after it is released from the ice (**Fig. 19.16A**). Some tills that form

Figure 19.14 Lakes in glacial valleys. **A** and **B**. A glaciated valley in North Cascades National Park, Idaho (USA), showing Upper, Middle, and Lower Thornton Lakes. Upper Thornton Lake (a tarn) joins the other down-valley lakes to form a series of paternoster lakes. Source: (A) Walter Siegmund, CC BY-SA 3.0, via Wikimedia Commons; (B) NASA. **C.** A classic sequence of paternoster lakes in a glacial valley in the Colorado Rockies (USA). Source: Marli Miller.

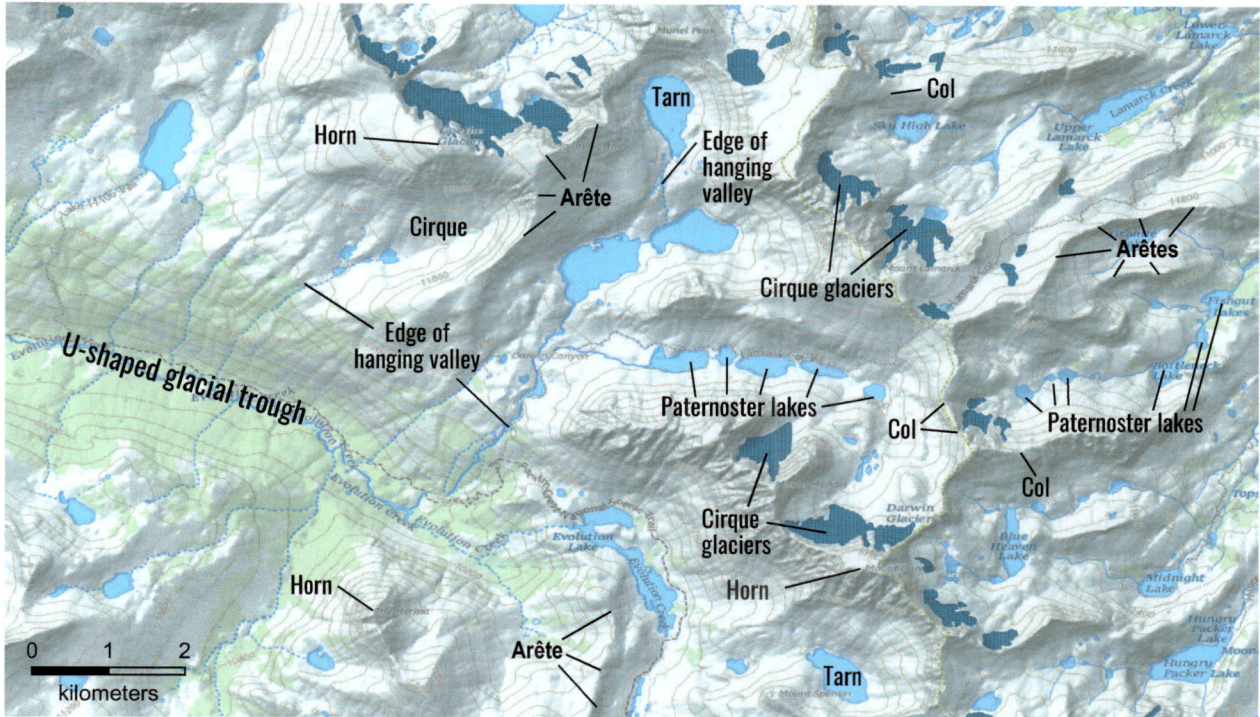

Figure 19.15 A section of a classic, (formerly) heavily glaciated, high mountain landscape, in the Sierra Nevada Mountains of California, USA, ≈100 km ENE of Fresno, showing many classical alpine glacial landforms.

Figure 19.16 The three main types of glacial sediment. **A.** Glacial till, which is usually unsorted and unstratified. **B.** Glaciofluvial sediment (outwash), which is usually coarse-textured and stratified, deposited by glacial meltwater. **C.** Glaciolacustrine sediment, which can be clay- and silt-rich (left), or a mixture of sands with finer sediments (right), is deposited in glacial lakes. Source: R. Schaetzl.

on top of the ice, or as debris melts out at the base of the ice, can exhibit crude stratification or layering. Gravity and some water may rework this material slightly to produce stratification, manifested as interbedded lenses of (usually) sandy sediment.

Unlike rivers or wind, glaciers are capable of transporting large boulders as easily as sand or clay. Then, when the ice melts, these materials can be deposited as a heterogeneous, mixed, sediment called glacial till. Till is often comprised of a mixture of sediment sizes, that is, it is texturally unsorted. Till is seldom transported far from where it was originally deposited. Instead, as the ice melts, it is simply "dumped" in place. Therefore, till can be deposited within the supraglacial, englacial, or subglacial environment, but never in a proglacial setting (in front of the glacier).

Early geologists were often puzzled by the rocks they found in regolith that were completely unlike the bedrock below; how could they have gotten there? These rocks were located "erratically" across the landscape, with no discernable local origin. The term has stuck; **erratics** are glacially transported rocks, different in lithology from the local bedrock. Although many people use the term erratic only for large boulders, it strictly applies to any rock that, because of past glacial activity, is now seemingly "out of place" on the modern landscape.

Till is deposited when glacial ice melts, which can occur at any location within the glacier. Ice melts fastest at the top of the glacier, causing sediment trapped in the ice to melt out

as **supraglacial till**. This form of till is typically loose and open, with low densities, as it was never compressed under the weight of the glacier. Till can also form at the base of the glacier. **Subglacial till**, or basal till, is often dense and "tight" due to compaction from the glacier. Often, it has fabric (a preferred orientation), meaning that the long-axes of elongated rocks within it are aligned parallel (or sometimes perpendicular) to the direction of flow. Geologists can use fabric data within basal till to provide insight into the former direction of ice flow.

Actively flowing glaciers transport sediment like a giant conveyor belt (**Fig. 19.17**). In an **active glacier** (to differentiate from a stagnant glacier, see below), the ice is constantly moving forward, or at least its "still-stands" are short-lived. This concept is important; ice *never* flows backward, up-valley, or toward its accumulation zone (see Chapter 18). It can only flow forward or melt in place, just as pancake batter would never flow in, toward the center of the pancake. Sediment within the ice (englacial), below it (subglacial), or on the ice surface (supraglacial) all move forward with the ice, toward its terminus – the end of the conveyor belt. Sediment is added to the conveyor belt (picked up by the ice) at any location along the way, as the glacier entrains debris from its bed, or in the case of valley glaciers, as debris falls from the valley walls. Sometimes, shear planes develop near the margin of the ice, where the glacier is frozen to its bed (**Fig. 18.25**). As the upper parts of the ice are pushed from behind, they shear across it, dragging sediment from

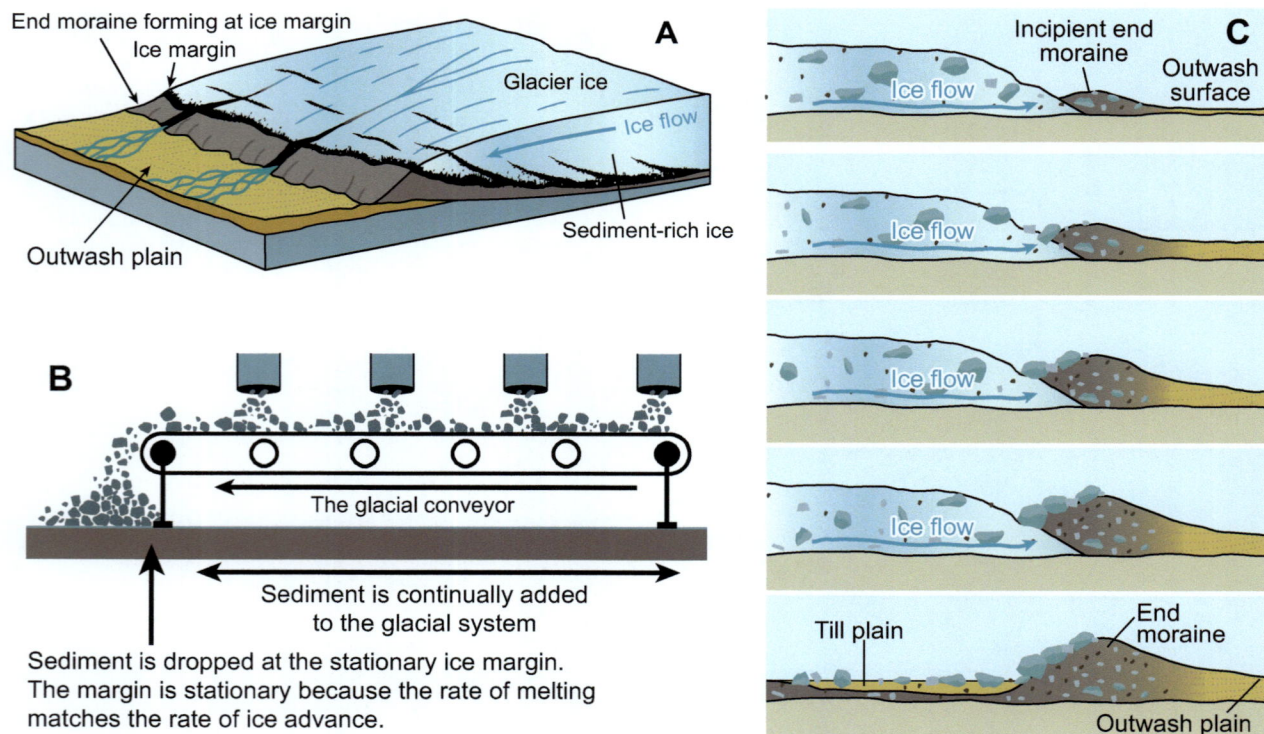

Figure 19.17 Models showing how active ice moves sediment forward like a large conveyor belt. Many glacial end moraines form in this way. Source: (B) After Marshak © 2019. Used by permission of W. W. Norton & Company, Inc.

Figure 19.18 Coarse-textured, gravelly outwash, deposited near the glacial margin (**A**) contrasts with sandy outwash, with almost no gravel, which was deposited farther from the same ice margin (**B**). Notice how well stratified (layered) these outwash deposits are. Source: R. Schaetzl.

the lower part of the glacier (where the ice is sediment-rich) to the supraglacial environment, where it gets deposited. More importantly, melting of ice in the ablation zone leads to upward flow of the ice, which can bring sediment frozen in the bed of the glacier to the surface, as well as concentrate any supraglacial debris. Ultimately, much sediment is often transported to, and released at and near, the glacier's terminus.

Sediment that is released from ice can also be secondarily picked up and transported by glacial meltwater. Indeed, much of it is, given that meltwater flows abundantly from glacial margins. When these deposits are eventually left behind by the meltwater streams, we refer to them as **glaciofluvial sediment**, or **outwash** (**Figs. 19.16B, 19.18**). Meltwater leaving the ice front is not capable of carrying the largest rocks and boulders, and so these are left behind. However, many meltwater streams do occasionally flow so fast and with enough power that they can readily move large and medium gravel and sand away from the glacier. Then, at some distance from the ice, these streams slow and are forced to deposit their sediment as outwash. Thus, outwash becomes finer textured farther from the ice margin. Geologists looking for potential gravel mines know this, and search for it near former ice margins. Farther away, outwash deposits primarily contain sands, which have less economic value (**Fig. 19.18B**). Most outwash deposits are sandy, with distinct layers (**stratification**) that correspond to periods of higher or lower discharge (**Figs. 19.16B, 19.18**). Outwash is also typically well *sorted*, with each layer being composed of particles that fall within a fairly narrow range of sizes. Water (in this case, meltwater), like wind, is an effective sorting agent. Glacial ice is *not*, which explains why glacial till is typically unsorted.

19.4.1 Proglacial Sediments and Landforms

Perhaps the most underestimated aspect of glaciation is the vast amount and power of meltwater that is produced by glaciers. This would have been especially true for ice sheets; water was everywhere, and therefore, much of the sediment left behind has some sort of glaciofluvial history. Vast quantities of water would have water poured off ice sheets each summer. Much of this meltwater flowed off the glacial surface, but much of it also gushed out of tunnels within and

beneath the ice. As long as the **proglacial** (Greek *pro*, "before") land surface sloped away from the ice, meltwater was free to flow outward, away from the ice margin, and carry sediment with it. Laden with sand and gravel as bedload, these meltwater streams would have formed wide braidplains, with multi-thread braided channels, countless sand bars, and constantly shifting channel patterns (**Figs. 19.19, 19.20**). Milky-colored water, rich in silt, flowed readily away from the ice and far downstream. This silt would either end up in ocean basins or be picked up by the wind from dried river valleys and blown away as **loess** (see below). But the coarser sediments, particularly sand and gravel, were deposited close to the ice; most of it was not carried very far at all. Deposition of layer-upon-layer of sand and gravel by these streams would have buried the preexisting land surface, replacing it with a flat **outwash plain**, or **sandur** (Icelandic *sandur*, sand) (**Figs. 19.19, 19.20, 19.21**). Outwash plains are typically sandy, flat landscapes. Occasionally, former channels (now dry) are still evident on the surface. As mentioned above, nearer the former ice margin, outwash plains can be a very good source of gravel.

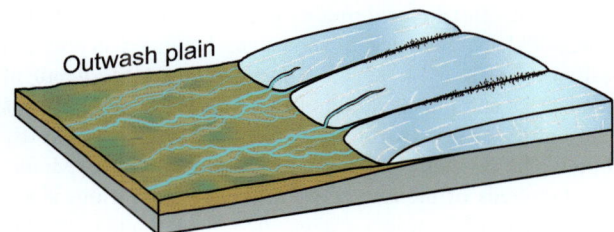

Figure 19.19 Schematic diagram showing the typical morphology of an outwash plain, and how it relates to the ice front.

Often, as the glacial terminus retreats, outwash streams traverse landscapes that had only recently been uncovered by the ice. Parts of this terrain may then become buried in sand and gravel from meltwater streams. As a result, low areas are filled in, and the overall landscape becomes much lower in relief. Ice blocks often remain scattered across these landscapes, having been left behind by the ice when the terminus was more advanced. These blocks eventually become partially or completely buried by the outwash. This

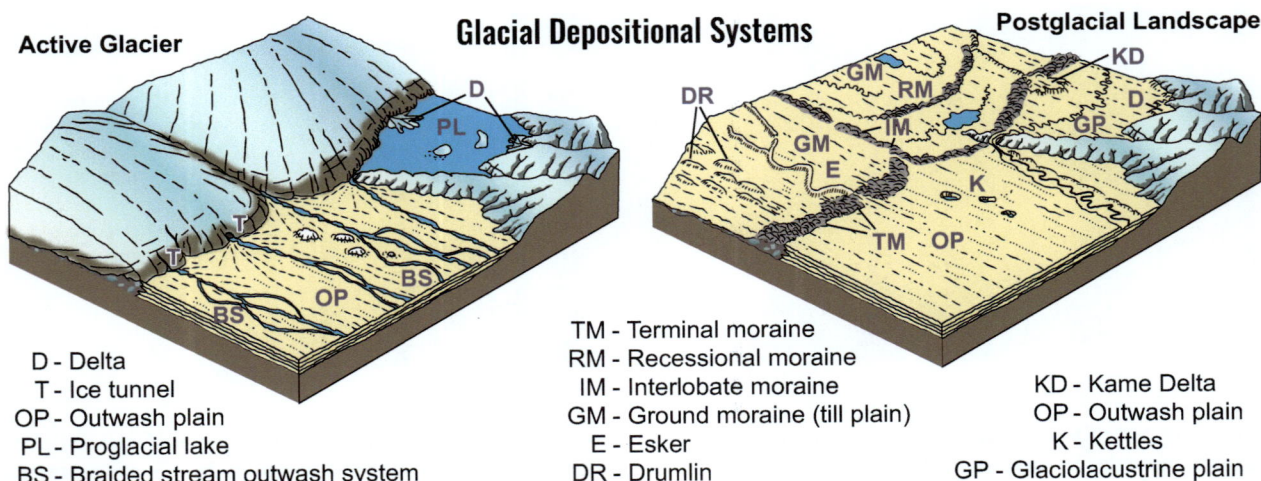

Figure 19.20 The major landforms associated with the margin of an active ice sheet. Source: © A. N. Strahler (1992). Used by permission.

Figure 19.21 The road stretches out onto this vast, flat outwash plain in northern Michigan, USA. Source: R. Schaetzl.

buried ice can take centuries or even millennia to melt, but when it does, each ice block leaves behind a depression called a **kettle** (**Fig. 19.22A**). Depending on the depth of the kettles and the elevation of the local water table, kettles can be dry or filled with water. The latter are called **kettle lakes**. The northern United States, from Montana to Maine, has hundreds of thousands of kettle lakes of various sizes. Almost all of Minnesota's lakes, the famed "Land of 10,000 Lakes," are kettle lakes.

Outwash plains are variously "kettled," depending on how many ice blocks were present when the outwash buried them. Some outwash plains have no, or few, kettles. If a considerable number of kettles exist, the landscape is commonly called a **pitted outwash plain**. And if there seems to be more kettles than outwash plain surface, the landscape is referred to as a **collapsed outwash plain**; much of this type of landscape owes its origin to collapse after the ice blocks melted. **Figure 19.22B** illustrates how, sometimes, these different types of landscapes can commonly exist adjacent to each other.

Ice-contact stratified sediment (ICSS) is a type of outwash deposit that is worth noting. As the name implies, this type of sediment formed *in contact with* the ice, but because it is stratified, we know that it was deposited there by meltwater. It is not a type of till. Meltwater streams that carry sand and gravel far out, beyond the ice margin, deposit sand and gravel on these landscapes as gravelly or sandy outwash. On the other hand, ICSS is deposited in *close* contact with ice, typically either in crevasses, moulins, or tunnels. **Moulins** are near-vertical shafts that often span the glacier from its surface to base, trapping ICSS within, as meltwater and finer sediment find various pathways through and out of the glacier (**Fig. 19.23**). Because the meltwater is "trapped" within the ice tunnel, and because flow paths are often downward in moulins and tunnels, the water can flow very fast and carry larger sediment than meltwater streams beyond the ice margin. As a result, ICSS is typically much more gravel-rich than traditional outwash in an outwash plain, because some of the finer sands are washed out of the moulin, leaving the gravels behind.

As we will see later, ICSS landforms typically form beneath **stagnant** glaciers, that is, those that are not moving, but instead are melting away in place. These types of glaciers are rife with tunnels and meltwater conduits, where ICSS-rich landforms form and can be preserved. Had the ice been active, that is, flowing or sliding, such features would have been destroyed by subsequent ice flow.

The last type of sediment directly associated with glaciers is deposited within proglacial lakes – those lakes ponded in front of the glacier. **Proglacial lakes** have a shoreline *on* the glacier itself (**Figs. 19.20, 19.24**). This type of environment is typical when the land surface slopes *toward* the ice margin, trapping water in front of the ice and forming vast but often shallow lakes. Paleoshorelines of these lakes tell us of their extent and indicate how dynamic they were, with lake levels rising and falling as outlets became uncovered or closed, due to fluctuations of the ice terminus. The deep

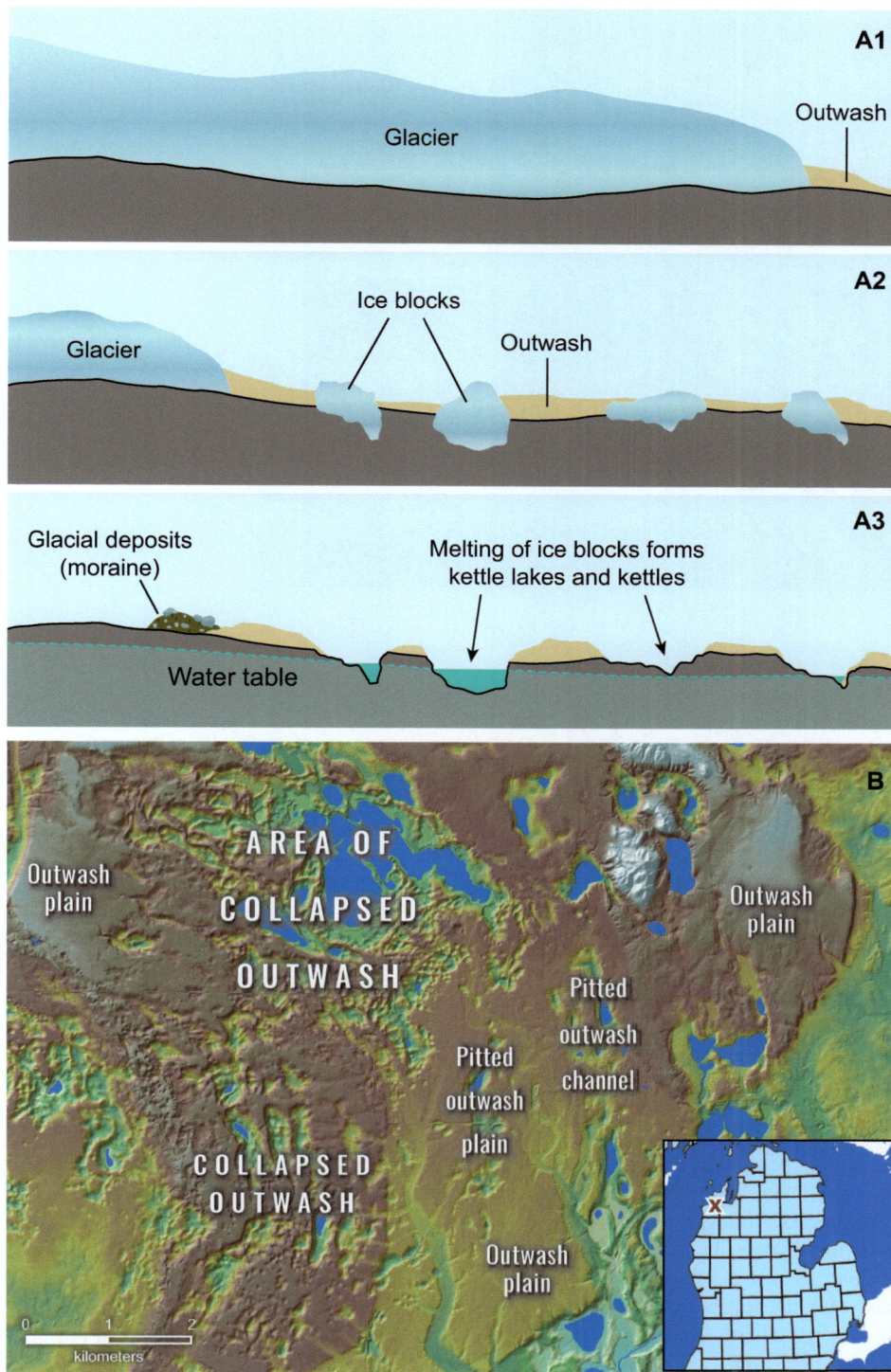

Figure 19.22 Kettles. **A.** Schematic illustration of how kettles form. Ice blocks from a more expansive glacial advance get dropped, and before they can melt, are buried (partially or completely) with outwash originating from the retreating ice. Upon melting, the ice blocks form kettles or kettle lakes, depending on depth to the water table. **B.** A highly kettled landscape in the northern Lower Peninsula of Michigan, USA (the "X" in the inset map denotes the location of this landscape). Note that only the deeper kettles contain water. Notice also that many kettles on the collapsed outwash plain have steep rims, implying that the preexisting landscape consisted mostly of ice blocks; the outwash simply filled in the cracks between them.

parts of proglacial lakes slowly fill with layer upon layer of fine clay, silt, and sand, referred to as **glaciolacustrine sediment** (**Fig. 19.16C**). Sometimes, this sediment is stratified, typically with (1) a coarse-textured layer that forms in the summer, when the lake is open, waves and currents are active, and hence, the water is turbid, keeping the finer sediment in suspension, and (2) a finer-textured layer that forms in winter, when the lake is frozen and the water beneath the ice is still. In this quiet setting, even the finest clays can settle out in lake-bottom deposits. Thus, some glaciolacustrine sediments contain annual alternations of fine and coarse layers called **varves** (Swedish *varv*, "turn" or "in

Figure 19.23 A large moulin on the Snowbird Glacier, Alaska, USA. Source: Paxson Woelber, CC BY-SA 4.0, via Wikimedia Commons.

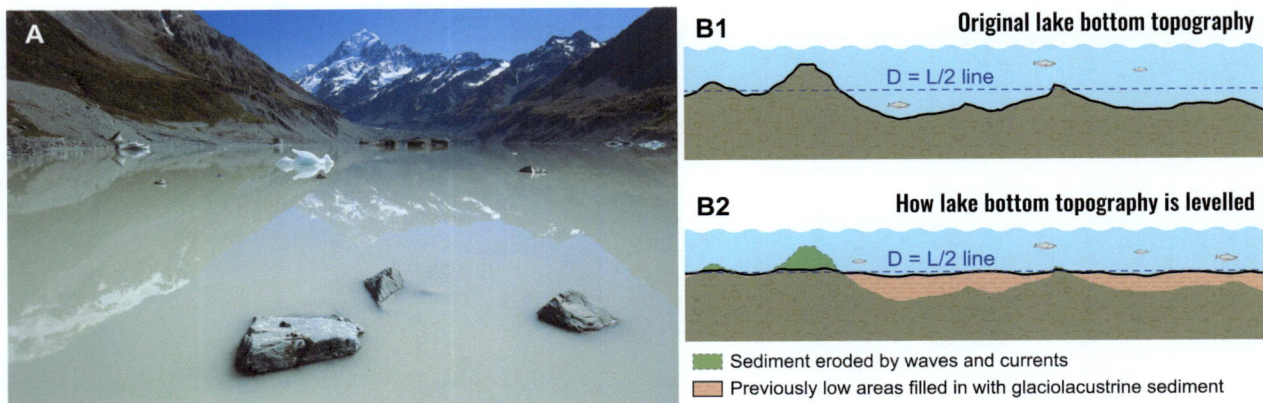

Sediment eroded by waves and currents
Previously low areas filled in with glaciolacustrine sediment

Figure 19.24 Proglacial lakes. **A.** An iceberg floats in the cloudy, silt-rich proglacial lake of the Hooker Glacier, New Zealand. Source: Marli Miller. **B.** Schematic diagram showing how waves and currents in proglacial and other types of lakes can lead to the levelling of the lake bottom over time.

layers"). Each varve comprises an *annual couplet* of layers. Geomorphologists can use varves to estimate the length of time that a proglacial lake may have existed. Although many glaciolacustrine sediments are rich in silt and clay, thick, massive layers of glaciolacustrine sand are also common.

Just as the action of meltwater streams levels out the proglacial landscape by filling in low areas with sand and gravel, proglacial lakes do the same. Low areas get filled with fine-textured glaciolacustrine sediment, while high areas may become eroded by the action of waves (**Fig. 19.24B**). Recall that waves impact and variously affect the bottom of a water body, down to a depth that is half their wavelength, that is, the **wave base** (see Chapter 17). Many proglacial lake basins were quite shallow, and often had very large waves due to strong winds off the glacier. Thus, their waves impacted the lake bed to great depths. This wave

action gradually eroded high areas on the lake bottom, with much of this sediment then deposited in nearby lows. The deep wave base and shallow lake bottom led to ideal circumstances for this type of glaciolacustrine "planation" or erosion. As a result, **glaciolacustrine plains** are exceptionally flat landscapes (**Fig. 19.25**). Thus, on many lake plains, the surficial sediment is not lake *clay*, but instead is simply whatever was present on the lake floor prior to its erosion. Even today, on some lake plains in Michigan, USA, deep, linear grooves are still visible, formed as the fierce glacial winds drove icebergs across the lake, scouring the bottom as they moved through the shallow lakes.

Finally, we mention the last sediment with (sometimes) a glacial origin – loess. **Loess** is a wind-blown sediment, rich in silts and very fine sands (**Fig. 21.32**). Although loess can form in deserts from abrasion (sand grains impacting each

Figure 19.25 The monotonously flat surfaces of glacial lake plains – these images are from the Glacial Lake Saginaw plain in southern Michigan, USA. Source: R. Schaetzl.

other and becoming silt-sized as a result), most loess forms due to glacial processes. Glaciers are massive "silt factories," grinding vast quantities of sediment into sand, and then into silt. No other surficial system produces as much silt as glaciers do. Most of this silt is washed away from the ice in meltwater. In the meltwater, the largest and heaviest sediments (gravel and sand) are deposited as outwash at sites nearer the ice margin, as meltwater velocities slow. However, silt-and clay-sized particles are easily carried much farther. This silt-rich meltwater coming off the ice makes its way into large river valleys, which flow to the sea, sometimes with stops in glacial lakes along the way. These river valleys flood during the summer, as meltwater volumes explode, but largely dry up in winter. Strong winds – common during glacial cycles – lift the silt from the dry floodplains and deposit it downwind as loess, covering the landscape with a blanket of silt. Much of the clay that moves through this glacier-to-sea conveyor belt ends up in the ocean basins; loess contains only small amounts of clay.

Many of the world's largest rivers were prodigious loess sources, for example, the Mississippi, Missouri, Wabash, Huang He (or Yellow), and Danube, among many others. The only criterion necessary is that their headwaters were at or very near former glacial margins. Loess deposits blanket the landscape for tens of kilometers on both sides of these large meltwater valleys (see Chapter 21).

19.4.2 Landforms Associated with Deposition by Active Ice

Until now, we have been focused on landforms of glacial erosion and glacial depositional landforms that form in proglacial environments, that is, in front of the ice. A large family of landforms also form *in direct contact* with this ice – at the ice margin or at the base of the glacier. Because of the nature of this depositional environment, most of these ice-contact landforms are composed of glacial till. The most common and widely discussed of these types of features are moraines.

Moraines are depositional landforms that form at the margin of a glacier (valley glacier or ice sheet), either at its terminus, or at its margins/sides. The most easily conceptualized type of moraine forms at the very end of the glacier – an end moraine (**Figs. 19.17A, 19.20**). These types of moraines typically form when the ice sheet is *active and advancing forward*. Active ice moves sediment forward toward the margin (**Fig. 19.26A**), like the conveyor belt analogy discussed above (**Fig. 19.17B, C**). While the ice is bringing sediment forward, it has only three options, depending on its mass balance (advance rate of the ice vs retreat rate of the margin), as depicted graphically in **Fig. 19.26**:

Option B1. As the terminus advances, sediment brought to the margin and deposited is quickly buried by, dragged along, and spread out under the ice. No end moraine is formed, or if one forms, it is over-ridden. The sediment deposited under the ice forms a low-relief, rolling landscape called a **till plain** or **ground moraine** (**Fig. 19.20**).

Option B2. The terminus retreats faster than the advancing ice can move forward. Recall that, based on the balance between the rate of ice advance and the rate at which the margin ablates, an ice margin can retreat even if the ice itself is moving forward. In this scenario, the glacier deposits sediment quickly and spreads it out along its path of retreat. As a result, a distinct end moraine will not form. Again, much of the landscape so "uncovered" may be a till plain, or if meltwater processes dominate, outwash plains may form in some locations. This type of scenario often results in detached ice blocks such that many kettles may form on this "uncovered" landscape (**Fig. 19.22B**).

Option B3. The ice margin is stable, staying in one general location for a period of time. This scenario depicts how most large end moraines form (**Fig. 19.17C**). The glacial "conveyor belt" brings sediment forward and deposits it at the same location, along the ice front, for years, decades, or centuries at a time. A large end moraine, mainly composed of till, then forms at the ice margin (**Fig. 19.17C**).

In reality, glacial margins always fluctuate position. Some end moraines are several kilometers across, reflecting the annual advance and retreat of the ice margin that may have continued for decades or longer, but with the margin staying in one general location. These types of subtle advances and retreats only serve to make the end moraine larger and wider; subtle advances "bulldoze" material into the preexisting moraine and as a result, it grows even larger. Very small

An active ice sheet bringing sediment forward, toward the margin.

A

The ice margin advances, burying and dragging out any sediment that had formerly accumulated at the margin. No end moraine is formed.

B1

The ice margin retreats, but because the margin is never in one location for very long, no end moraine is formed.

B2

The ice margin is stable. As sediment continues to accumulate at the margin, a conspicuous end moraine is formed.

B3   Outwash

Figure 19.26 Three possible scenarios for an *actively advancing* ice sheet, showing how an end moraine can only form when the ice margin is stable.

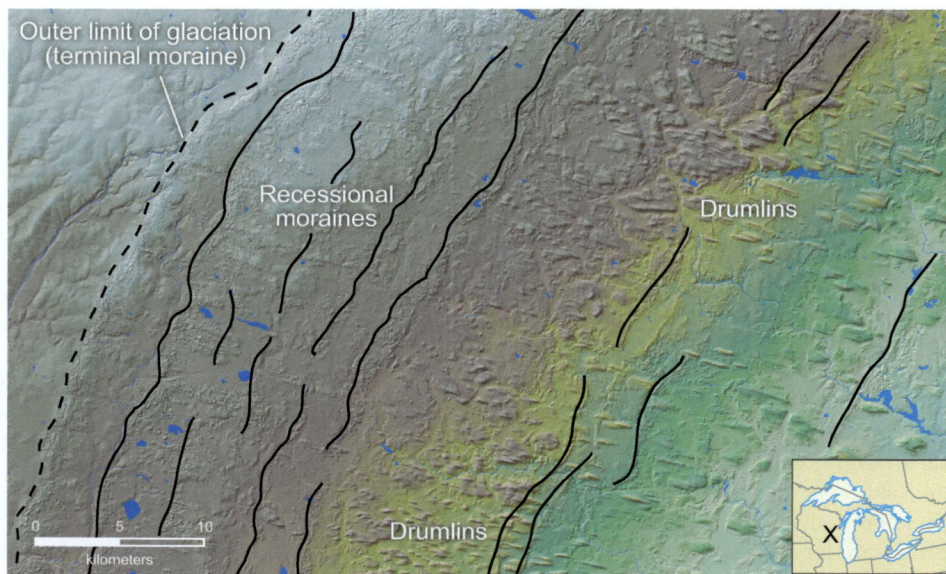

moraines may even form annually as the terminus is retreating. This occurs because a small winter readvance may push up sediment into a small moraine, but then the ice margin retreats slightly in the ensuing summer due to ablation. If this occurs for decades, a series of small annual moraines may be superimposed on a till plain.

End moraines are rolling, sometimes hummocky, ridges that mark the location of a former ice margin, or they may be a single, small ridge (**Fig. 19.27**). The term "hummocky" refers to landscapes that have many small hills and lowlands, often isolated and not "connected" by streams. Patterns of looping end moraines across the landscape help geomorphologists to understand the general retreat pattern of the glacier. End moraines are usually composed of till, but often contain layers of outwash or other types of sediment, either deposited by the ice or bulldozed into the moraine during advance. Kettles are also common in end moraines, as ice blocks are readily detached from glacial margins and quickly buried (completely or partially) there. Gaps in end moraines may occur where a meltwater channel cuts through the ridge.

Geomorphologists categorize end moraines based on their formation and location. The outermost end moraine, formed as the glacier advanced to its farthest extent, is its **terminal moraine** (**Figs. 19.20, 19.27**). Each glacial period (or major advance) can have only *one* terminal moraine. Terminal moraines do not have to be the largest of all the moraines formed by a glacier. Some glacial advances did not even form a terminal moraine, or formed only a small one.

As the ice margin retreats from its terminal position, it occasionally stabilizes, forming another end moraine. We distinguish these end moraines from the terminal moraine by calling them **recessional moraines** (**Figs. 19.20, 19.27**). Any one glacial advance can have many recessional moraines, which we usually distinguish by their local name, for example, Johnstown moraine, Marquette moraine. Some recessional moraines form as the ice margin stabilizes at a

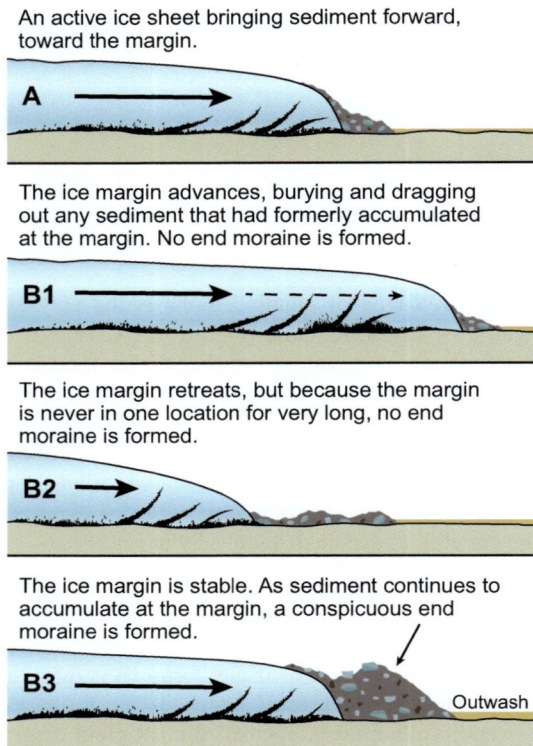

Figure 19.27 Relief model of a part of eastern Wisconsin, USA, showing the various glacial landforms associated with the retreat of the Green Bay Lobe of the Laurentide ice sheet. On this map, any areas not classified as moraines or drumlins would be till plains. Compare this sequence of landforms to the conceptual model shown in **Fig. 19.20**. Note: the moraine positions are shown by lines drawn at their outermost margins.

location (**Figs. 19.17C**, **19.26B3**), but they also commonly form as the retreating ice margin suddenly readvances to a new position (as in a surge). Often, it is difficult to determine which mode of formation is responsible for a given recessional moraine.

Ice caps and ice sheets usually flow as discrete lobes that follow preexisting valleys, deepening, filling, and generally modifying them with each successive advance (**Fig. 19.28**). During each advance, most glacial lobes will build their own terminal moraine and, very likely, a series of recessional

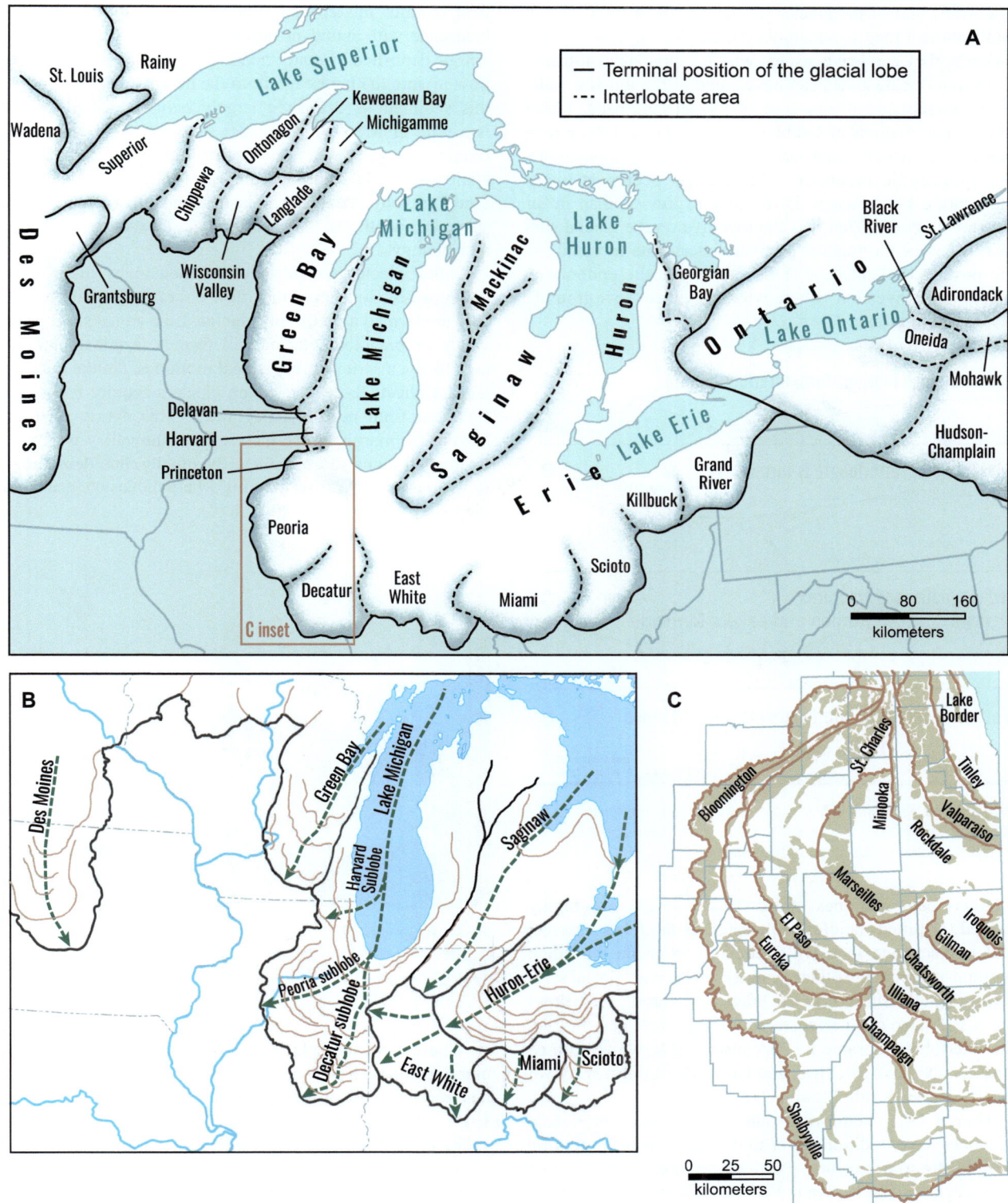

Figure 19.28 The glacial lobes and sublobes of the Laurentide ice sheet for the Great Lakes region (USA) during the last major glacial phase. **A.** The major glacial lobes and interlobate areas. **B.** The major glacial lobes, glacial flowlines, and recessional moraine positions. **C.** The recessional moraines in Illinois have been mapped in great detail by the Illinois State Geological Survey. Source: (C) After Hansel and Johnson (1996), © Illinois State Geological Survey.

moraines. The area of (usually) side-to-side contact between two glacial lobes is called an **interlobate zone** (**Figs. 19.20, 19.28**). Sediment is brought to the interlobate zone from the lobes on either side, both as till being carried by the ice, and as outwash, washed into the interlobate suture or "crease" by meltwater streams (**Fig. 19.29**). Interlobate areas tend to be the lowest part of the glacier surface, and as a result, water and sediment tend to accumulate there, having few options whereby they can leave, except within supraglacial streams. Thus, interlobate zones can have some of the thickest sediment assemblages of any type of glacial landscape. These thick accumulations of sediment are called **interlobate moraines** (**Fig. 19.20**). Interlobate zones are classic examples of **topographic inversion**. This process occurs in many landscapes, but nowhere is it better expressed than in an interlobate zone. What was the low area on the glacial surface becomes the highest part (the interlobate moraine) of the postglacial landscape. This example is the epitome of topographic inversion, where what was once "low ground" is now "high ground."

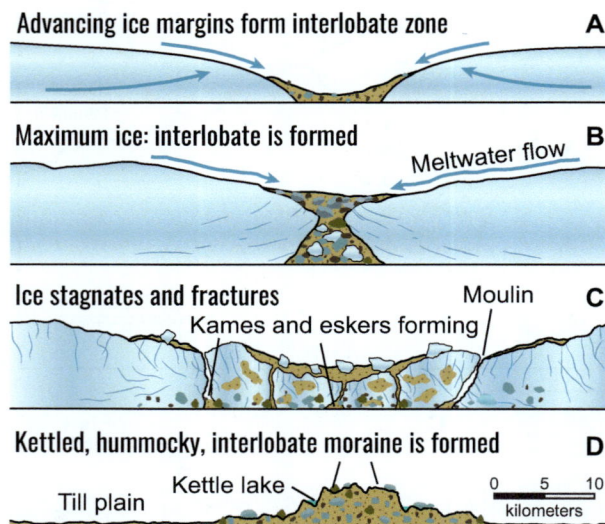

Figure 19.29 Diagrammatic sketch of how a broad, interlobate zone, or moraine, forms. Ablation in and near the ice margin brings subglacial sediment to the top of the glacier.

Interlobate moraines (**Figs. 19.20, 19.29**) are highly kettled (or pitted) landforms, because as the lobes intersect and interact, one lobe will frequently override the other, burying and breaking off large masses of ice with glacial sediment. High densities of **kettle lakes** (and kames; see below) are a hallmark of interlobate zones and moraines. Indeed, the interlobate moraine in Wisconsin, USA, which formed between the Lake Michigan and Green Bay Lobes, is known locally as the "Kettle Moraine" (**Fig. 19.28A**).

Hummocky, kettle-rich landscapes like interlobate moraines have traditionally been described as having **kame and kettle topography**. Portions of the highly kettled outwash landscape shown in **Fig. 19.22B** would also classify

as kame and kettle topography. The lakes and the low areas – the kettles – formed as buried ice blocks melted. The high, irregularly shaped hills that comprise much of the rest of the landscape are the kames. A **kame** is a constructional glacial landform of indeterminate origin that is usually formed of ICSS. The various types of kames will be discussed in later parts of this chapter. But for now, consider that in a landscape with abundant buried ice, much of the sediment will be deposited by water running off melting ice blocks, or flowing into the landscape from the nearby glacier. Because this sediment is deposited largely *between* ice blocks, it is a type of ICSS. And because it is deposited mainly by running water, ICSS is often sandy or gravelly. Thus, in a kame and kettle landscape, many of the hills (kames) are sandy/gravelly features. And just as the kettles can be of any shape and depth, the kames can be of any shape and surface morphology.

Valley glaciers are usually hemmed in by high, steep valley walls. But should they flow out onto the plain below, they are termed a **piedmont glacier**. Like an ice sheet, each valley or piedmont glacier will form a terminal moraine and, likely, a series of recessional moraines. Unlike ice caps and ice sheets, however, valley glaciers acquire rocks and sediment from nearby valley walls. This debris accumulates at the margin of the glacier (near the valley wall) and is dragged along, down-valley. Eventually, this debris may form a **lateral moraine**, marking where the ice once abutted the valley wall (**Figs. 19.9B, 19.30**).

Figure 19.30 Looking down-valley, notice the two large lateral moraines that have been deposited on either side of the retreating Glacier de Moiry in the Swiss Alps. Source: Ebe.wiki, CC BY-SA 3.0, via Wikimedia Commons.

Where two valley glaciers merge down-valley, the lateral moraines of each will coalesce to form a **medial moraine** that continues down the center of the glacier (**Figs. 19.9B, 19.31**). Medial moraines are distinctive features of large valley glacial systems, each one traceable up-ice to its position as a lateral moraine. Most lateral moraines are simply supraglacial debris riding along on the ice surface, although

Figure 19.31 Medial moraines. **A.** A very distinctive medial moraine on the Bighorn Glacier in Yukon Territory, Canada (1961). Source: US Geological Survey; in public domain. **B.** Medial moraines in Wrangell-St. Elias National Park, Alaska, USA. Source: US National Park Service; in public domain. **C.** Medial moraines in the Columbia Icefield, in the Chugach Mountains of southern Alaska (1972). Source: US Geological Survey; in public domain.

they can continue at depth. Because medial moraines can be readily followed back to their source, they have been used to locate ore bodies. After an ore is discovered in a medial moraine, finding the rock outcrop from which it came is relatively straightforward – follow it up-valley!

Previously, we have been discussing landforms that develop at or beyond the glacier terminus (proglacial), such as end moraines and outwash plains. Now we will turn to the landforms and landscapes of the subglacial environment, that is, those formed *under* the ice. **Till plains**, for example, form beneath the ice (**Fig. 19.20**). Many other glacial landforms also share a subglacial origin.

An easily recognizable landform of the subglacial environment, formed by actively and often rapidly flowing ice, is a **drumlin** (**Fig. 19.20**). Drumlins are streamlined hills, typically 0.1–5 km in length, a few tens of meters in height, and up to a few hundreds of meters across (**Figs. 19.32, 19.33**). Drumlins form beneath actively flowing and sliding ice sheets. They are less common in valley glacial settings, where erosion, rather than deposition, is typical of the subglacial system. Drumlins typically occur in groups of dozens to thousands called **drumlin fields** or swarms. Because drumlins align with the direction of ice flow, they are of great help in determining the direction of past ice advances. Most drumlins are ovate- or spoon-shaped, but they can become extremely elongated. Elongated drumlins are named **flutes**; they are long and narrow with a length to width ratio greater than 25 (**Fig. 19.4**). Flutes are often constructionally and sedimentologically the same as drumlins, but longer and narrower. The length/width ratio of both drumlins and flutes has been interpreted by some as an indication for the former ice velocity, as elongated landforms have recently been observed under and in front of fast-flowing ice streams or outlet glaciers in Antarctica. Because drumlins are so common, worldwide, you may encounter the term "drumlinoid" for other types of streamlined glacial landforms.

Typically (but not always), elongate, or "inverted spoon-shaped" drumlins have a steep "end" on their up-ice (stoss) side, and a long, gently sloping "tail" that extends

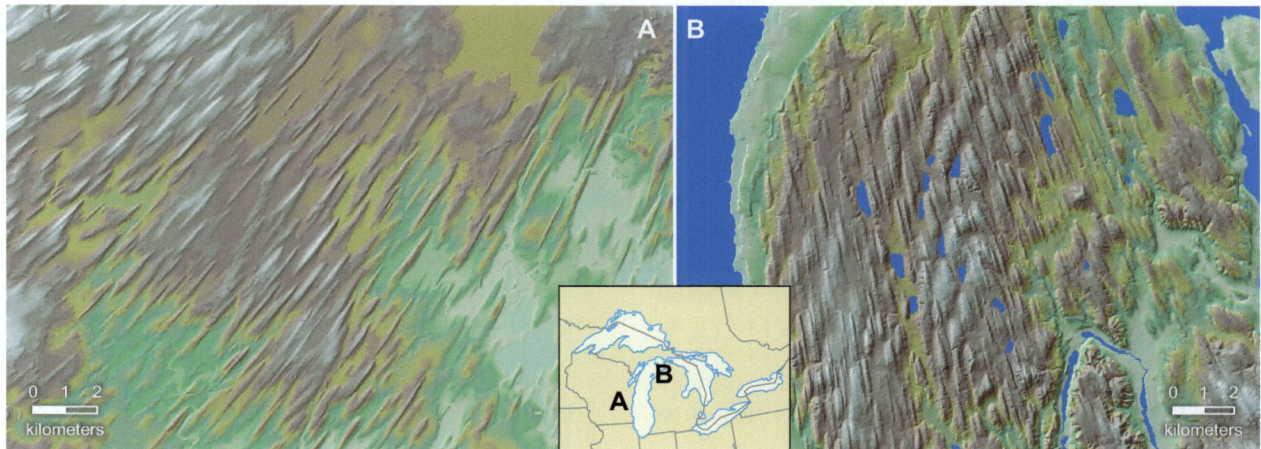

Figure 19.32 Drumlin landscapes. **A.** Part of the drumlin field that covers much of southeastern Wisconsin, USA. **B.** Part of the Antrim-Charlevoix drumlin field of northwestern Lower Michigan, USA. Excellent drumlin fields are also present in upstate New York, Ireland, and parts of Ontario, Canada.

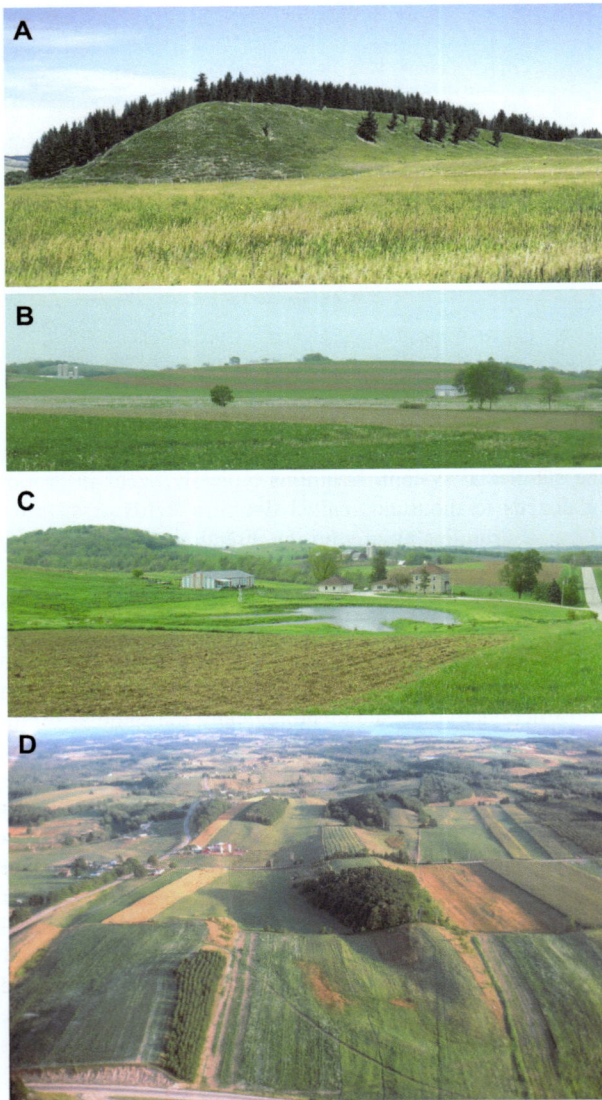

Figure 19.33 Drumlins and drumlin landscapes. A and B. Individual drumlins. C and D. The drumlinized landscapes portrayed in Fig. 19.32A and B, respectively. Source: (A) Courtesy of the University of Toronto-Scarborough; (B–D) R. Schaetzl.

drumlins are formed in till, some are composed entirely of outwash-like material. Geomorphologists are still debating exactly how drumlins form, but one thing is clear – they form only beneath ice that is advancing. Beneath such a rapidly sliding ice sheet, sediment can be eroded and deformed as irregularities in the glacial bed interact with deforming ice. The continuum of streamlined features – from short and stubby to long and narrow – reflects the varying geologic and hydrologic conditions of the bed, the velocity of ice flow, and ice thickness.

19.4.3 Landforms Associated with Deposition by Stagnant Ice

Geomorphologists are increasingly recognizing that many glacial landforms form in association with *stagnant* ice. Until now, our discussion has focused on landforms associated with *active* ice, that is, ice that is flowing forward, even if the terminus is stationary (Figs. 19.17C, 19.26). But as often happens, glaciers will advance – often rapidly – and then stop moving, only to stagnate and melt in place. This type of scenario is most common in the ending stages of a glacial period. As stagnant ice melts, most of the sediment released from the melting ice is picked up and deposited by meltwater or in debris flows. As a result, most "stagnant ice" landforms are more texturally sorted (sandy) and may contain considerable amounts of ICSS. Basal or subglacial till is uncommon in such settings because it is commonly buried, but poorly sorted supraglacial sediments are common.

Recall that outwash plains form beyond the ice margin, as braided, meltwater streams deposit sand and gravel across the proglacial landscape (Fig. 19.19). Usually these outwash plains merge gradually (upstream) into an end moraine, even if it is a muted version of one. However, if the ice margin is stagnant, a variant of an outwash plain – a **head of outwash** – will often form in the same general (proglacial) position (Fig. 19.34). A classical head of outwash can be likened to thicker and "more sloping" outwash plain. Although both heads of outwash and outwash plains slope away from the former ice margin, the meltwater streams that formed the head of outwash carried so much water (due to the rapidly melting ice) and sediment that large amounts of sediment were rapidly deposited near the ice. Hence, heads of outwash have a steeper slope, and near the former ice margin they tend to be thicker and very gravel-rich.

Stagnant ice margins do not produce end moraines, but instead are marked by the steep, ice-contact slope at the head of outwash. This slope forms as (1) meltwater deposits sand and gravel in the head of outwash, burying some of the stagnant ice, which then later (2) melts, leading to (3) collapse of the upper end of the head of outwash, forming a steep inner margin. Often this collapsed zone is marked by an abundance of kettles, for example, a pitted or collapsed outwash plain.

Large or even massive quantities of sediment are often deposited at stagnant ice margins, because (1) the ice margin remains in one location for long periods of time, and (2) the

down-ice (Fig. 19.33A). This morphology is the opposite of the other major type of streamlined glacial landform – a **roche moutonnée** (Figs. 19.3, 19.4). Roches moutonnées are steeply sloping on their *down-ice* side, the opposite of drumlins, and are formed in bedrock (Fig. 19.4). **Crag-and-tail** features, mentioned earlier, are a third type of streamlined glacial landform, developed as ice deposits sediment in the lee of a bedrock obstruction. If such an obstruction is buried by glacial sediment, it is usually called a **rock-cored drumlin** (Fig. 19.4).

Geomorphologists view flutes and drumlins as both erosional and constructional features, formed in whatever material the base of the glacier was able to erode, deform, and redeposit into a streamlined form. Although most

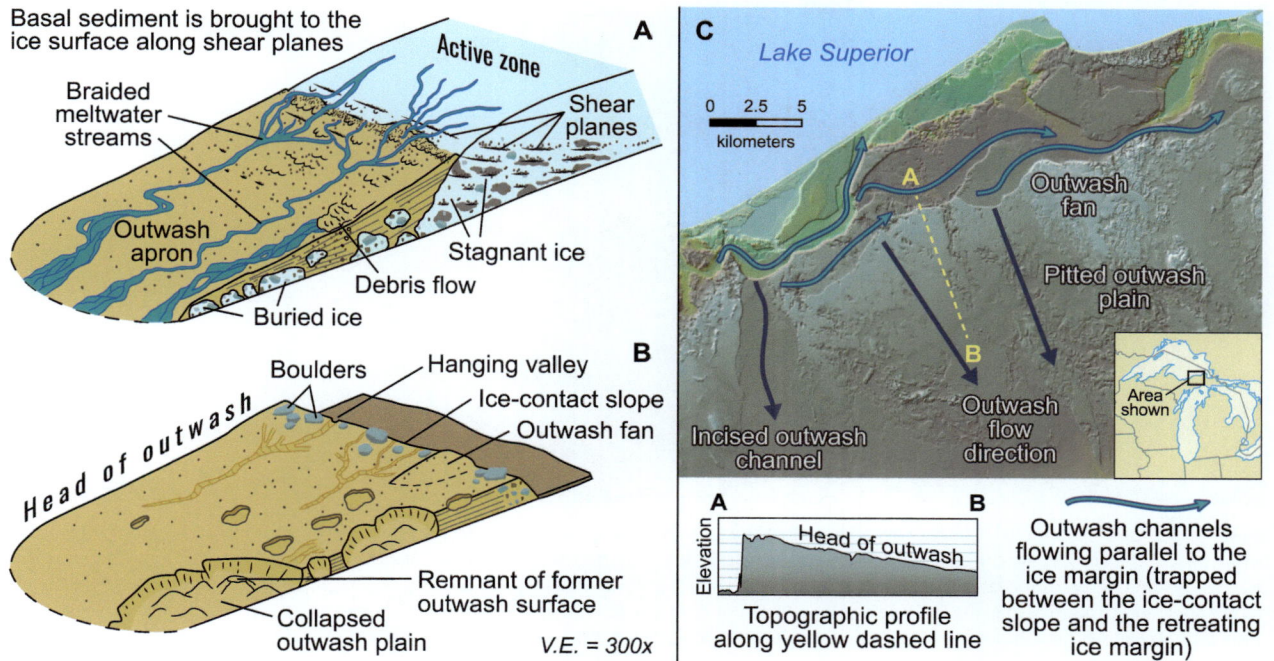

Figure 19.34 Diagrammatic illustration of the formation of a head of outwash (after Blewett et al., 2009) **A.** As it is forming. **B.** After the ice has melted. **C.** A landscape example of a head of outwash.

ice is typically melting more rapidly than an active ice front would melt. Stagnant ice may also imply that the climate is warm or warming, leading to rapid melting.

Sediment is brought to a stagnant ice margin by supraglacial meltwater streams, many of which are flowing within crevasse networks, as well as by water streaming out of subglacial tunnels (**Fig. 18.21**). These meltwater pathways and the impacts they leave on the landscape often remain unmodified, because the stagnant ice melts away in place. If the ice were to become reactivated and surge forward, these tunnels and the deposits they contain could be destroyed, buried, or "levelled" by the advancing ice mass.

The vast quantities of meltwater flowing toward stagnant ice margins can be highly erosive if the water is sediment-poor and/or under pressure, or if it flows through subglacial tunnels. Highly pressurized meltwater will readily erode into the underlying sediment, as well as up, into the ice. A common landform thought to have been eroded by subglacial meltwater is a **tunnel channel**. These features likely formed as pressurized meltwater flowed subglacially, eroding into the ice, as well as down into the glacial bed. Tunnel channels, also called tunnel valleys, are often flat-floored and may be tens of kilometers long and as much as two kilometers wide (**Fig. 19.35**). Many are eroded into bedrock, pointing to the great pressures on the flowing water inside the tunnel. Most tunnel channels coalesce and eventually exit at the glacier terminus, where the sand and gravel they are carrying is deposited, producing large, fan-shaped landforms. In the northwestern part of the Lower Peninsula of Michigan, USA, the dense network of tunnel channels even extends into the underlying

bedrock; East Grand Traverse Bay is eroded into bedrock and its floor is below sea level (**Fig. 19.35A**). Note that, like many instances of tunnel channel formation, water in these tunnel channels was flowing *upslope*; this situation could only occur if the tunnel channel below the glacier was sealed and pressurized with meltwater. In some tunnel channels, the roof of ice above them could have been well over a kilometer across, but still thick enough to withstand the pressure from below.

As water that forms the tunnel channel wanes in intensity, and as the glacier continues to melt, the roof of ice above it will collapse. Chunks, blocks, and various pieces of ice then riddle the floor of the tunnel channel, as it goes from a subglacial to a subaerial environment. Then, should the glacier continue to melt, the (now, open) channel may fill with outwash, surrounding and burying the ice blocks. Later, as the ice blocks melt, ridges of sand and gravel will be exposed, with kettles on either side. These sinuous ridges of sand and gravel, snaking between and among the ice blocks, are **eskers** (**Figs. 19.20, 19.36, 19.37**). Eskers and kettles are common features of tunnel channels and help identify them as such. "Open" meltwater channels do not form eskers.

Eskers are sometimes found in tunnel valleys, but in most instances, they form in smaller tunnels that link moulins and crevasses with tunnels that exit at the glacier terminus. As meltwater from a stagnant glacier seeks an exit route to the terminus, small tunnels can form, both subglacially and englacially (**Figs. 18.21, 19.20**). Many of these tunnels exit at the glacier margin, where they may feed proglacial lakes or enter an outwash plain. Unlike tunnel channels, water within tunnels that form eskers is not pressurized, and is

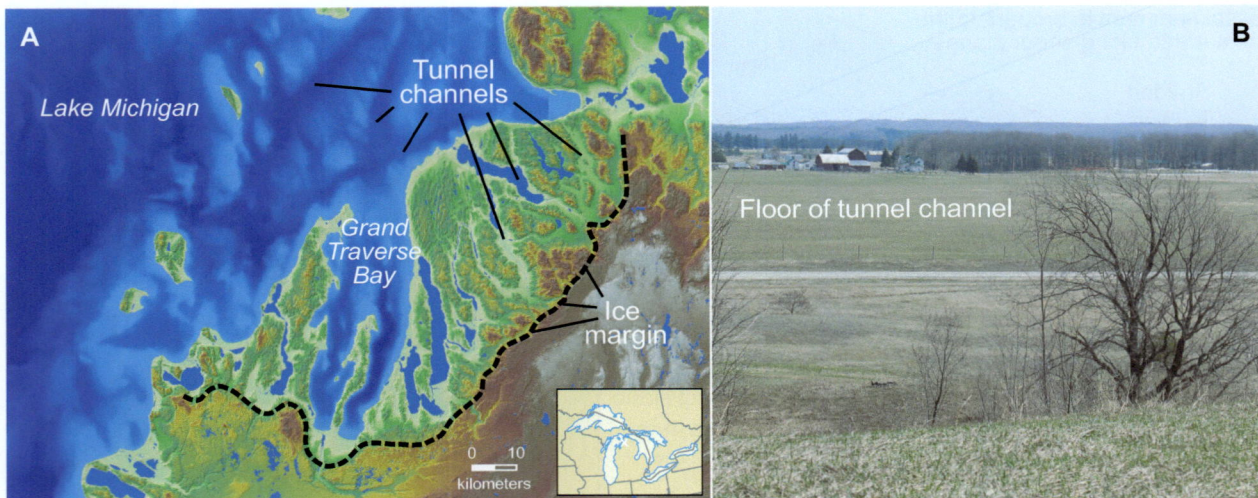

Figure 19.35 Valleys interpreted to have formed as tunnel channels. **A.** A terrain map of northwestern Lower Michigan (USA) and the bathymetry below Lake Michigan. This area has many tunnel channels, most of which started deep within the Lake Michigan basin and connect to tunnel channels on land. Many of the latter are now flooded and occur as long, narrow lakes. Most of the tunnel channels end at what was assumed to be a stagnant ice margin, marking the inside of a vast head of outwash. **B.** Photograph of the floor of one of the tunnel channels shown in A. These tunnel channels have abnormally flat floors because, after deglaciation, they were flooded by a glacial lake. Source: R. Schaetzl.

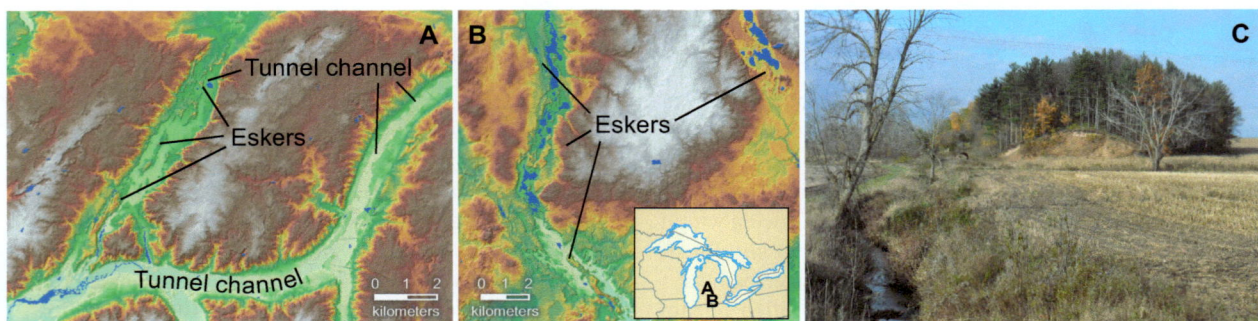

Figure 19.36 Tunnel channels in southern Michigan, USA. As shown in **A** and **B**, eskers are often common within the tunnel channels, although as can be seen in the figure, some tunnel channels lack them entirely. **C.** An esker in a typically flat-floored tunnel channel, being mined for sand and gravel. Note the pine trees on the esker – a giveaway sign that this is a sandy site. Source: R. Schaetzl.

Figure 19.37 Classical landforms of stagnant ice systems. **A.** The landscape of the kettle (interlobate) moraine in southeastern Wisconsin has many glacial landforms associated with stagnant ice, such as eskers, kames, kettles, and meltwater channels. **B.** The classic meandering shape of the Parnell Esker is shown here, as well as in A. **C.** A moulin kame, one of several also shown in A. Source: Image and photos from the Wisconsin Dept. of Natural Resources; in public domain.

flowing more slowly. Thus, esker tunnels can quickly fill with sand and gravel because subglacial streams are usually choked with sediment. Thus, they reflect a depositional setting, and almost all of the deposition occurs as the glacier is melting. After the glacier has completely melted away, all that remains are the glaciofluvial sands and gravels from the former stream bed in the tunnel, that is, the esker.

Eskers are commonly composed of sandy, gravelly ICSS. However, they can also contain masses of unsorted sediment, or even basal till. Aggregate producers know that eskers are excellent sources of gravel aggregate, and as a result, many have been mined and are no longer fully present on the landscape (**Fig. 19.36C**). Just as subaerial streams meander, water in a glacial tunnel also often takes on a meandering form. Thus, many eskers have a tell-tale meandering morphology (**Figs. 19.20, 19.37B**). Like drumlins, eskers occur in groups, defining the area where stagnation was widespread. Most eskers are not continuous and probably formed episodically (and time-transgressively) as the ice variously wasted away. Some eskers are more continuous (up to tens of kms) and have tributaries, which suggest that they formed as a system of subglacial streams under a large area of stagnant ice.

Other landforms associated with stagnant ice are grouped under the general term kames. By definition, **kames** are composed of ICSS, deposited in contact with (usually) stagnant ice. As discussed earlier, kames are often associated with kettles; kame-and-kettle topography is a giveaway for an area that experienced widespread ice stagnation (**Fig. 19.22B**).

Kames form in crevasses, tunnels, and various types of openings in a stagnant ice mass, and can even form in association with supraglacial streams (**Fig. 19.38**). Eskers could

be viewed as a special type of kame in which the water was flowing mainly horizontally within a subglacial tunnel. In contrast, many kames form in nearly vertical, often cone-shaped, moulins (**Fig. 19.23**). The resultant landform is a **moulin kame**, which resemble gumdrops or inverted cones (**Fig. 19.37A, C**). The cone shape is formed as ICSS is deposited in a vertical tunnel. The sediment in the tunnel then collapses as the ice melts. **Delta kames** (or kame deltas) are another type of landform composed of ICSS, this time forming as water flows off the ice and into a proglacial lake (**Figs. 19.20, 19.38**). **Kame terraces** form where the ice margin abuts an upland, allowing meltwater streams to deposit sediment in contact with both the ice and the upland (**Figs. 19.20, 19.38**). Although most common in alpine glaciers, kame terraces can also form when a glacier retreats from its end moraine, allowing outwash to become trapped between the retreating ice front and the upland (moraine) (**Fig. 19.39**).

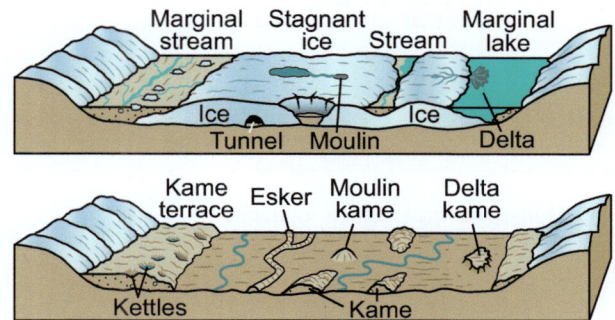

Figure 19.38 The various landforms found in association with stagnant ice. Source: © A. N. Strahler (1992). Used by permission.

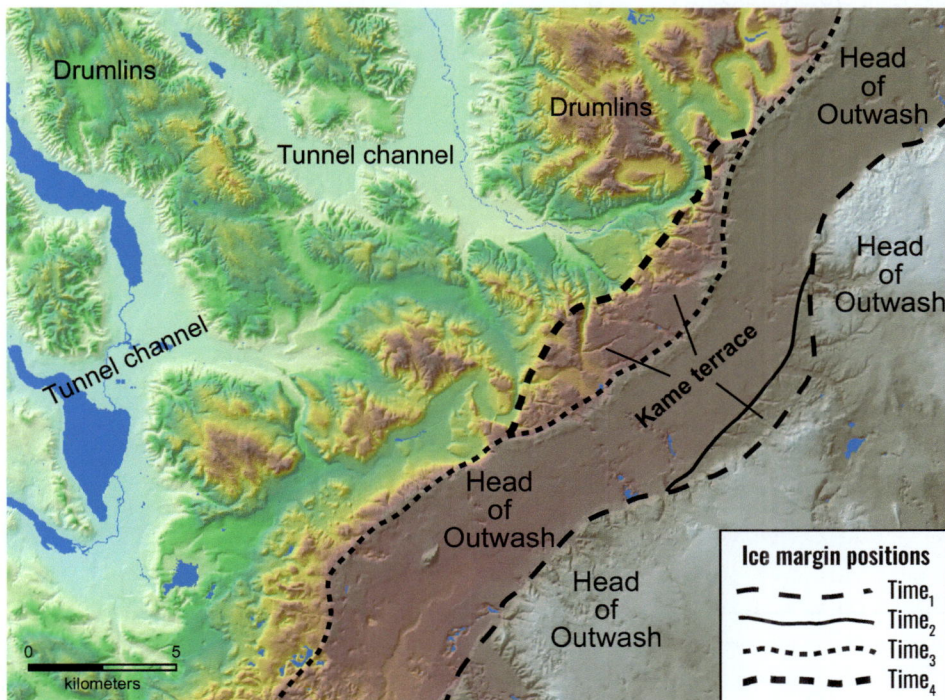

Figure 19.39 A landscape in northwestern Lower Michigan, USA, that has excellent kame terraces. This map is a larger-scale version of the landscape shown **Fig. 19.35A**. Deglaciation here was dominated by stagnant ice. The stagnant ice margin at Time$_1$ formed a broad head of outwash that continues off to the east. By Time$_2$, the ice margin had retreated slightly, trapping meltwater between the margin and the proximal side of the head of outwash, forming a kame terrace. Following retreat to the position shown at Time$_3$, part of the ice margin again retreated off the head of outwash to form a second kame terrace.

A

Debris

Shear plane

Active ice

Stagnant ice

B Slush ponds and ice-walled lakes start to form

Ablation retarded by
thick cover of debris,
and permafrost
on landscape

More rapid
ablation on
clean but
stagnant
ice

C Sediment begins to accumulate in
the bottom of ice-walled lake

D Melting ice contributes meltwater and
outwash to surrounding landscapes

E Slow melting due to cold temperatures extends
the life of the lakes, sometimes for millenia

F Lakes get larger, deeper, and more numerous,
as they slowly melt the surrounding ice

G After the ice has melted, ice-walled
lake plains become prominent uplands
on the ice-stagnation landscape

Rim ridge

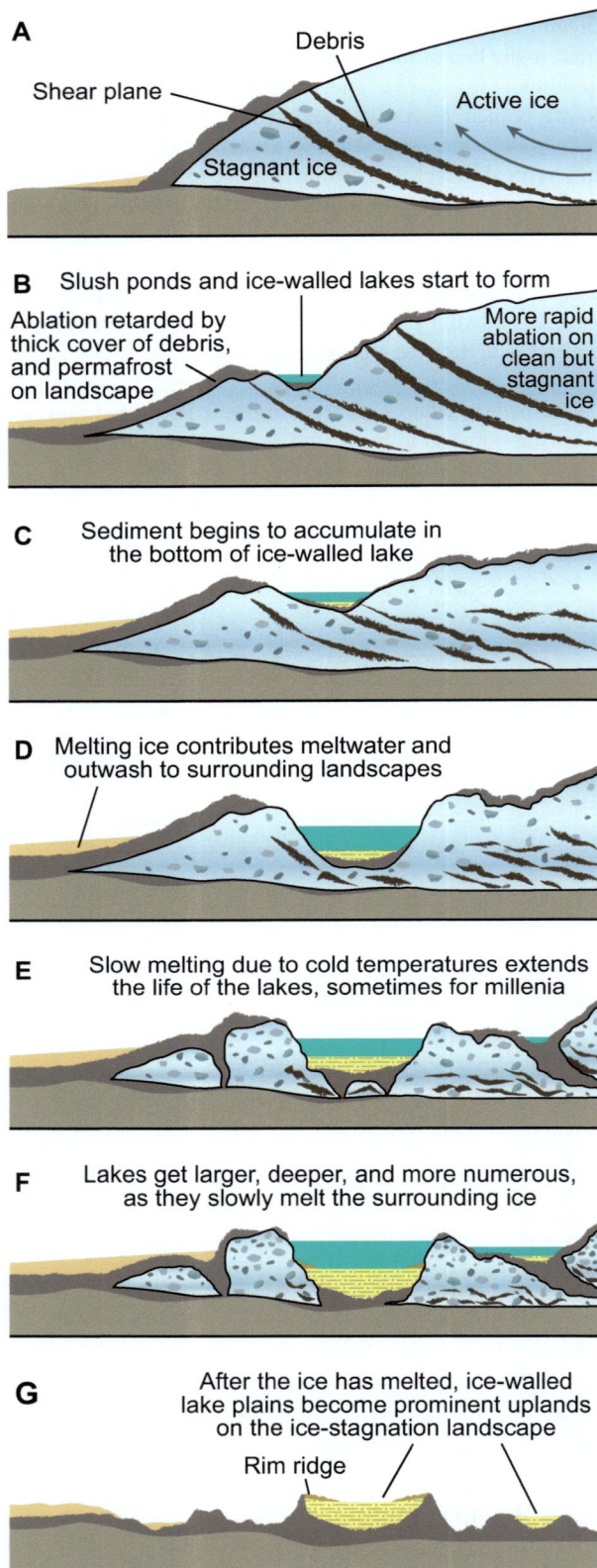

Figure 19.40 Schematic diagram of the formation of ice-walled lake plains near the margin of a stagnant ice sheet.

A unique landform formed in stagnant ice settings is an **ice-walled lake plain**. These features have only recently been recognized. As a result, geomorphologists are only now beginning to realize their ubiquity and importance. Most ice-walled lake plains form where widespread stagnant ice was covered with supraglacial sediment (as is typical). Ice blocks and heaps of sediment chaotically dot this type of ice-cored landscape, forming depressions in the ablating ice that, in places, can retain meltwater lakes (**Fig. 19.40**). Much of the ice is buried by supraglacial sediment, and the landscape may be stabilized by permafrost, slowing its melting and extending the lakes' lifespan. Over time, silty-sandy lake sediment fills in the ice-walled lakes, carried in from the surrounding ice. This sediment can be many meters thick, and often preserves pollen and other plant fragments, which are not only dateable but provide a clear picture of the paleoenvironment. In parts of the Great Lakes region of the United States, the sediment in ice-walled lake plains records thousands of years of paleoecological information.

Ice-walled lake plains are readily identified as circular to oval-shaped plateaus, often with a slightly elevated, sandy "rim ridge" (**Figs. 19.40G, 19.41**). Because many of these lake plains are subtle and have low relief, they have gone

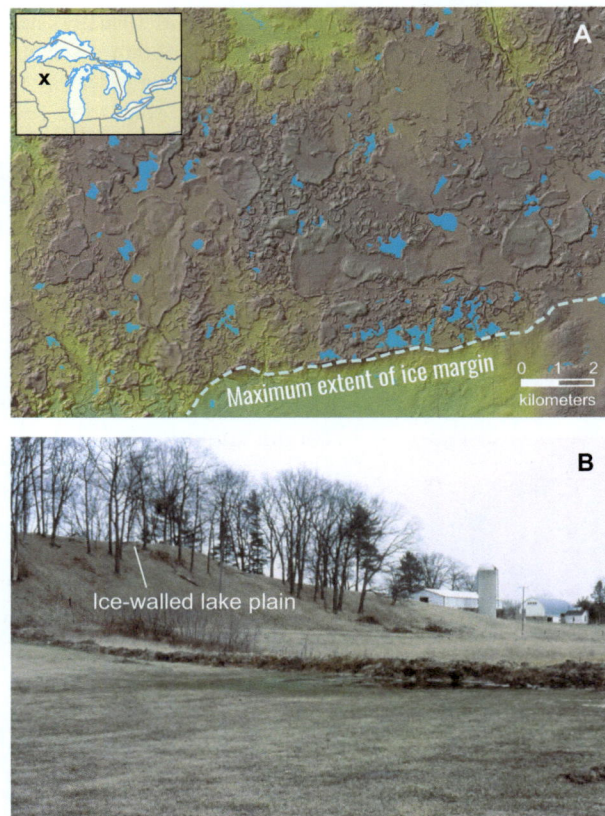

Maximum extent of ice margin

0 1 2
kilometers

B

Ice-walled lake plain

Figure 19.41 Ice-walled lake plains. **A**. This end moraine in central Wisconsin (USA) is rife with ice-walled lake plains. **B**. The thickness of the sediment in this ice-walled lake plain is apparent, based on its elevation above the surrounding lowland. Source: (B) K. Syverson.

completely unnoticed until recently, when detailed topographic data became available.

19.5 THE POSTGLACIAL PERIOD

As the last ice sheets receded from the landscapes of North America and Eurasia, change was everywhere. The climate was rapidly warming, although it was sometimes briefly interrupted with cold intervals (see Chapter 4). Plants and animals had migrated south as the ice sheets were advancing, taking refuge in warmer locations known as **refugia**. Only as the climate warmed and ice melted, leaving behind its barren landscape of rocks, sediment, and water, could these plants migrate back north. Here, they often found the bare soil to be a receptive substrate, even though permafrost was present in many locations, delaying the initial colonization for some.

Kettles and kettle lakes formed as myriad ice blocks melted, allowing supraglacial sediment to be reworked by water and gravity. Many kettle lakes would partially drain or fill to become bogs and wetlands. Streams incised the landscape, attempting to become graded (see Chapter 15). In doing so, suites of fluvial terraces were developed for many of the larger rivers. Sea level rose, flooding many coastal locations that had for millennia been dry land. It was a time of rapid change.

One legacy of continental glaciation that not only affects most of the deglacial landscape, but which is ongoing even today, is **glacioisostasy**. This term is used to indicate that the crust is either being depressed or rebounding back from the immense weight of glaciers, and especially from an ice sheet. Burdened with this mass, the lithosphere during each glacial advance was warped downward, especially where the ice was thickest (**Fig. 19.42A**). Areas immediately outside of the ice sheet actually bulged upward because the lithosphere is slightly elastic at regional scales, forming a **glacial forebulge**. After the ice melted, the crust in a previously depressed area rose, first rapidly and then slowing with time, a process known as **glacioisostatic rebound**.

Rebound rates and locations can be reconstructed by examining landforms that were once horizontal, such as lake shorelines or fluvial terraces, but which are now tilted. If we know the age of these features, the rates and direction of postglacial uplift can then be calculated (**Fig. 19.42B**). Today, highly precise global positioning systems can measure uplift across glaciated regions, confirming the geomorphic evidence on the ground.

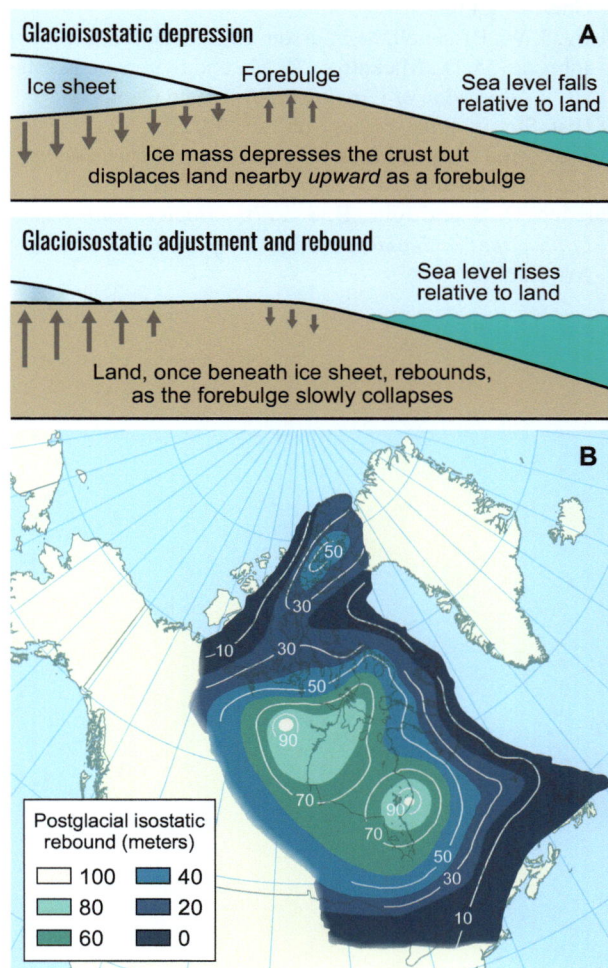

Figure 19.42 Glacioisostasy. **A.** Schematic diagram of crustal depression and rebound associated with a large ice cap. **B.** Map of isostatic rebound of Earth's crust in North America, following the last glaciation.

REVIEW QUESTIONS

19.1 Discuss and distinguish between the concepts of glacials and interglacials, with respect to the Ice Age of the Quaternary Period.

19.2 State the two main ways in which glaciers erode, and describe how each functions.

19.3 Describe and discuss each of the several bedrock landforms formed by glacial erosion.

19.4 Compare and contrast the morphology, composition, and formation of roches moutonnées, crag-and-tails, drumlins, and flutes.

19.5 What are the main bedrock (erosional) landforms associated with alpine glaciation, and how does each form?

19.6 What is the characteristic shape of a glaciated valley? If drowned by the sea, what would this valley be called?

19.7 How do hanging valleys form and what are their characteristics?

19.8 Describe the sedimentological differences between glacial till, glacial outwash, and glaciolacustrine sediment. How and where does each sediment form or be deposited? State if each deposit is stratified and/or sorted.

19.9 How do classical end moraines form? Compare and contrast them with another ice-marginal landform, a head of outwash.

19.10 What are the typical landforms you might observe on a till plain of an ice sheet? Discuss how each is formed.

19.11 With respect to outwash plains, what factor determines how many kettles they have? What are the names given to the different types of outwash plains, depending on kettle density?

19.12 How can glaciolacustrine sediment tell us the length of time that a glacial lake may have existed?

19.13 What conditions must occur for proglacial lakes to form? Why are glacial lake plains so flat?

19.14 What is loess and how do its origins relate to glaciation?

19.15 Describe and discuss the three sediment-depositional scenarios associated with an active ice margin, and what types of ice-marginal landforms will develop under each scenario.

19.16 What are the three main types of end moraines formed by ice sheets, and where are these features with respect to each other, on the glaciated landscape? What sediments dominate each type of moraine?

19.17 Why are interlobate moraines so kettled?

19.18 Differentiate between medial and lateral moraines, as to their mode of formation, and location within the glacial valley.

19.19 What is the characteristic shape of a drumlin, and how do they form? Which "end" of a drumlin indicates formation in the "down-ice" direction?

19.20 Under what kind of conditions does a head of outwash form, and what types of sediment is a head of outwash composed of?

19.21 What are tunnel channels, how do they form, and why do so many contain small eskers within them?

19.22 What are the similarities and differences between how interlobate and medial moraines form?

19.23 What shape does a moulin kame take on, and how does it form?

19.24 How is ICSS different from "normal" outwash, in regard to composition and depositional systems?

19.25 How do ice-walled lake plains form? Are they associated with stagnant or active ice margins?

19.26 Discuss isostatic rebound – what is it, and why does it happen?

FURTHER READING

Anderson, B. G. and Borns, H. W. 1994. *The Ice Age World: An Introduction to Quaternary History and Research with Emphasis on North America and Northern Europe During the Last 2.5 Million Years*. Oxford University Press.

Attig, J. W., Bricknell, M., Carson, E. C., Clayton, L., Johnson, M. D., Mickelson, D. M., and Syverson, K. M. 2011. *Glaciation of Wisconsin*. Wisconsin Geol. Nat. Hist. Survey Educ. Series Publ. 36. 4th ed.

Lowe, J. and Walker, M. 2015. *Quaternary Environments*. 3rd ed. Routledge.

Sugden, D. E. and John, B. S. 1976. *Glaciers and Landscape: A Geomorphological Approach*. Edward Arnold.

20 Periglacial Processes and Landforms

Stephen A. Wolfe

The term periglacial describes areas subject to repeated freezing and thawing and the processes associated with the growth of ice within soil and rock. Although originally referring to processes and climates adjacent to glaciers, "periglacial" now applies more broadly to cold-climate processes where frost action predominates. Earth's cold, periglacial landscapes span both polar regions and many high elevation and mountainous areas. These landscapes are unlike any others, with ice-formed landforms such as pingos (**Fig. 20.0**) ice-wedge polygons, sorted circles, and rock glaciers found only in these cold landscapes.

20.1 FUNDAMENTALS OF THE PERIGLACIAL ENVIRONMENT

Periglacial landscapes may seem to be in a state of suspended animation due to the sub-freezing ground temperatures. Although desolate and frozen on the surface, they are always changing below, driven by the movements of water and ice. An entire suite of periglacial processes results from this intensive seasonal ground freezing and thawing, known as **frost action**. In both appearance and genesis, periglacial landforms are truly unique.

Ice – but not glacial ice – drives the formation of landforms in periglacial landscapes. Many periglacial landscapes were never actually glaciated, being too removed from glacier sources and too dry and windswept to accumulate glacial ice. In other areas, former glaciers have melted and permafrost still preserves relict portions of glacial ice beneath frozen layers of sediment. Landforms on cold, periglacial landscapes derive their morphology from ice formation, as well as thermal expansion and contraction, and the melting of ice. All of these processes are mediated by temperature changes.

One last point to remember – ice *melts* but permafrost *thaws*.

20.1.1 The Ground Thermal Regime

In periglacial environments, temperatures in the upper ground are < 0 °C in winter and > 0 °C in summer. Permafrost occurs below most periglacial landscapes where ground (rock or regolith) remains below 0 °C for at least two consecutive years. The depth at which temperatures remain < 0 °C throughout the year defines the top position of the

permafrost table. Although temperatures change within the upper part of the permafrost throughout the year, they remain < 0 °C, and water within the soil or rock typically remains frozen. Permafrost can range in thickness from just a few meters near its warmer limit, to many hundreds of meters in the coldest regions. The thickness and extent of permafrost depends upon the seasonal air temperatures, combined with local conditions of vegetation cover type and thickness, sediment type, bedrock exposure, and aspect, all of which affect how the ground retains and loses heat.

It is important to note that the upper part of the ground on periglacial landscapes *does* undergo seasonal freezing and thawing (**Fig. 20.1**). In permafrost settings, this active layer typically ranges in thickness from < 20 cm in very cold regions to > 1.5 m in sub-Arctic regions. The active layer is generally thinnest where the air and permafrost temperatures are coldest, such that summer thawing is minimal. However, surface cover has a large impact on active-layer thickness, with bedrock and exposed soil having thicker active layers than areas with more insulating organic covers.

At some greater depth, the range of seasonal temperatures decreases, such that it remains constant throughout the year – the **depth of zero annual amplitude** (**Fig. 20.1**). Permafrost temperatures at this depth can range from near 0 °C to below

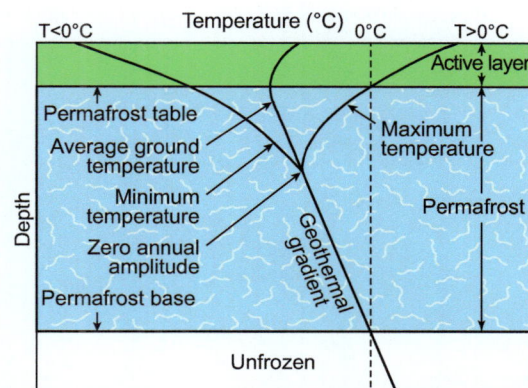

Figure 20.1 The "trumpet curve" diagram shows the annual maximum, minimum, and average temperatures in the active layer and permafrost. Source: After Burn (2012), with permission from Wiley & Sons.

Figure 20.0 These two pingos in Canada's Northwest Territories formed due to periglacial processes. Source: Tawna Brown / Alamy Stock Photo.

−15 °C. The depth to the zero annual amplitude relates to several factors including, importantly, the ability of the ground to conduct and dissipate heat, as well as the amount of water in the soil. In "cold" permafrost, where heat is readily conducted through the ground, the depth of zero annual amplitude can exceed 15 meters, but in warmer areas this depth is commonly only between 2 and 5 meters, due to unfrozen water which suppresses temperature changes near freezing. Under equilibrium conditions, temperatures below the depth of zero annual amplitude typically increase with depth at a rate of about 1 °C for every 30–40 m, due to the **geothermal gradient,** a result of heat originating deep within the Earth.

20.1.2 Permafrost Distribution

Continuous permafrost occurs within very cold landscapes, where it underlies 90–100% of the ground surface (**Fig. 20.2**). Ground temperatures are usually coldest in the continuous permafrost zone, where mean annual air temperatures are typically below −7 °C. Permafrost here is commonly several hundred meters thick, and even exceeds 1,200 m northeast of Siberia in Russia, where it has aggraded over millennia in unglaciated areas. In continuous permafrost environments, summer air temperatures

are typically too cold to support the growth of trees. The maximum limit of where trees are capable of surviving defines the **treeline,** beyond which is **tundra,** where the ground is covered with herbs, mosses, lichens, and low shrubs (**Fig. 20.3A**).

Mostly barren landscapes called **polar deserts** occur poleward of the tundra (**Fig. 20.3B**). Temperatures during winter months in polar deserts typically average < −30 °C and precipitation is typically < 25 cm/year, occurring mostly as snow. Tundra and polar deserts also experience extremes in solar radiation because the sun remains continuously below the horizon in winter and above the horizon in summer.

Landscapes underlain by **discontinuous permafrost** (50–90% coverage) and **sporadic permafrost** (10–50% coverage) occur in warmer regions (**Figs. 20.2, 20.4**). Nonetheless, in the discontinuous permafrost zone, permafrost may still be many tens of meters thick and mean annual air temperatures are typically between −4 °C and −7 °C. Equator-facing slopes that receive more sunlight may not contain any permafrost, whereas more shaded, polar-facing slopes may have thick permafrost (**Fig. 20.5**). In the sporadic discontinuous zone, whether permafrost is present depends on localized conditions of exposure, vegetative cover, and

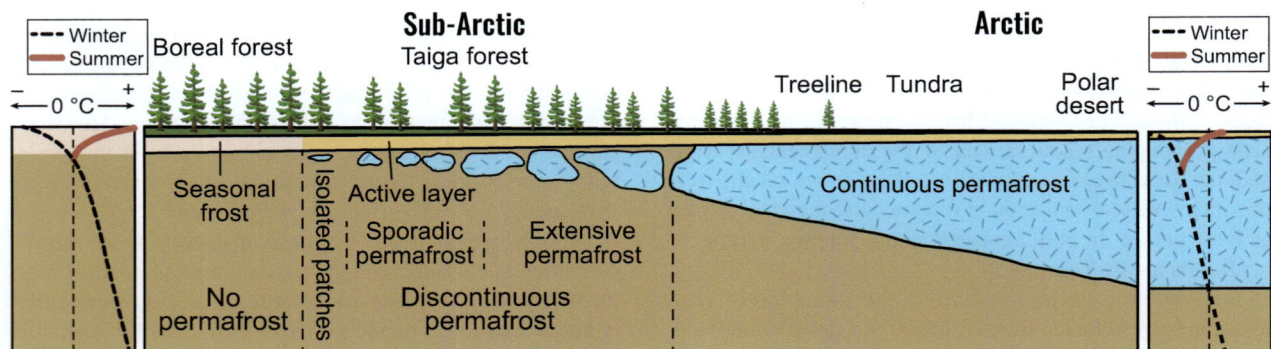

Figure 20.2 Permafrost distribution across a latitudinal gradient in the northern hemisphere. Also shown are the ground temperature profiles in non-permafrost terrain with seasonally frozen ground (left) and in permafrost (right) environments, indicating how the permafrost and active layer relate to ground temperatures that are < 0 °C throughout the year. Source: After R. J. E. Brown, © University of Toronto Press 1970.

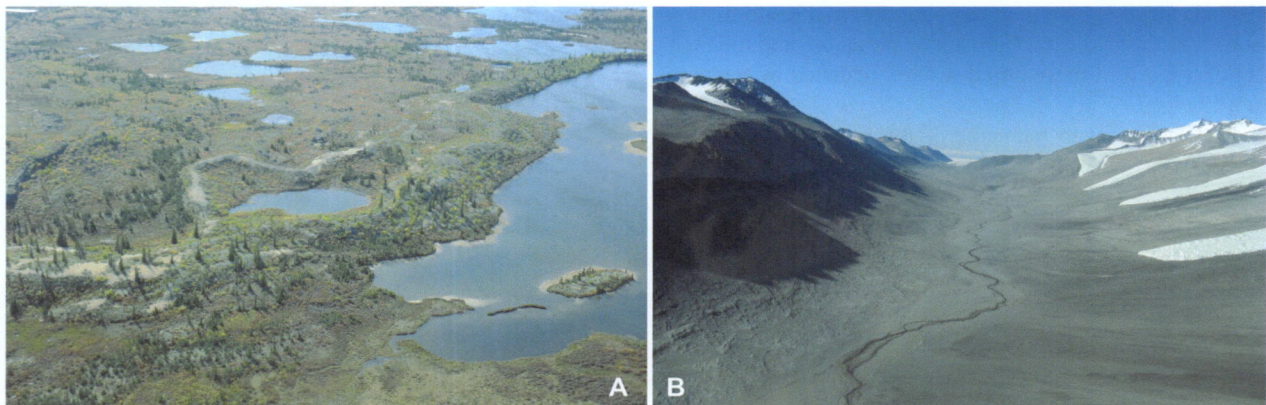

Figure 20.3 Sub-Arctic and Arctic permafrost landscapes. **A.** The treeline–tundra boundary in the Northwest Territories of Canada. Source: S. Wolfe (NRCan Photo ID 2021-237). **B.** A polar desert in the McMurdo Dry Valleys, Antarctica. Source: J. A. Gillies.

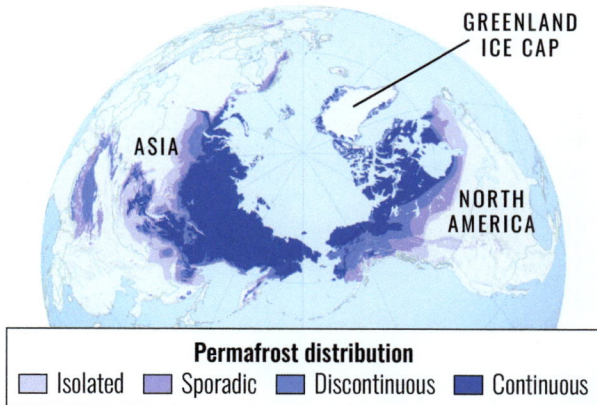

Figure 20.4 The distribution of permafrost in the northern hemisphere. Compiled from various sources.

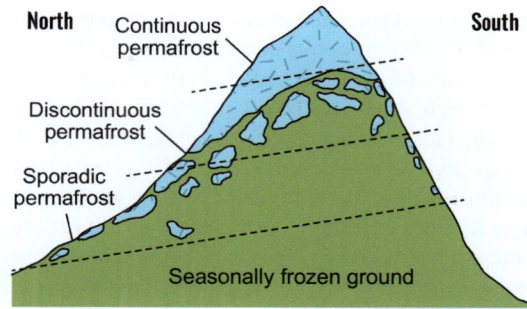

Figure 20.5 Permafrost distribution with aspect and elevation for a mountain in the sub-Arctic northern hemisphere. Permafrost is more abundant on poleward (north)-facing slopes than equatorial-facing slopes, due to differential receipts of solar radiation. The extent and thickness of alpine permafrost also increases with altitude. Source: After US Geological Survey; in public domain.

other factors that influence local ground temperatures. Bedrock exposures that are not well-insulated by vegetation cover transmit heat into the ground effectively, and consequently may not contain permafrost. In contrast, thick surface organic layers act as insulation, enabling the ground below to support permafrost. Coniferous forests are the most common type of vegetation in these regions.

Near the warmer limits of discontinuous permafrost, it may only be present as isolated patches (< 10% of the surface) (**Fig. 20.2**). In these areas, permafrost exists outside of thermal equilibrium, being either in a state of decay or as **relict permafrost**, persisting only because of past (colder) climate conditions. In the northern hemisphere, these areas typically occur within peatlands in the **boreal forest**. South of this region, which still includes vast areas of boreal forest, permafrost is absent, and the landscape includes an upper layer of **seasonally frozen ground** (**Fig. 20.2**). In these areas, frost penetrates the upper soil each winter, but no permafrost occurs below.

Permafrost regions occupy between 20% and 25% of Earth's land surface, mostly in the northern hemisphere (**Fig. 20.4**). About half of both Canada and Russia are within the

permafrost zones, and about 80% of Alaska. In the southern hemisphere, continuous permafrost underlies most of the exposed ground in Antarctica. Permafrost has formed at these high-latitude locations where the seasonality of incoming solar radiation is high and winter air temperatures are very low. Elevation also affects permafrost, as air temperatures are generally cooler at higher altitudes. As a result, permafrost is sometimes common at high elevations, such as the European Alps, the South American Cordillera, and the Rocky Mountains, as well as on the high-elevation plateaus of China, Mongolia, and Tibet (**Fig. 20.5**).

20.1.3 Taliks

Even within permafrost areas, some ground remains unfrozen. **Taliks** are bodies or layers of perennially unfrozen ground within permafrost. Different types of taliks can form, depending on the terrain conditions (**Fig. 20.6**). For example, lakes and rivers that do not freeze to the bottom are commonly underlain by closed or open taliks due to the heating effect of the water (**Fig. 20.6**). **Closed taliks** are thawed areas surrounded (and underlain) by permafrost, whereas **open taliks** are unfrozen from the ground surface all the way to

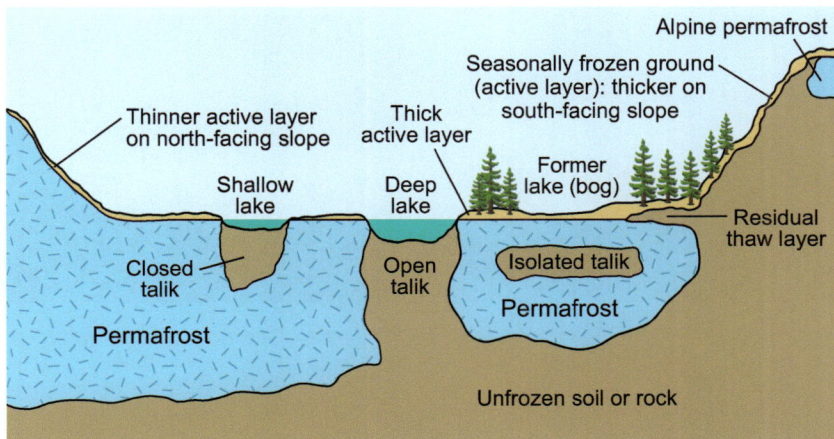

Figure 20.6 Cross-section through a region of discontinuous permafrost, showing the distribution of permafrost, taliks, active layers, and seasonally frozen ground. Note the effects of aspect, elevation, and former lakes.

great depths. In contrast, **isolated taliks** are completely surrounded by permafrost – above and below and on all sides. Most isolated taliks form because the overlying ground was once thawed, such as where a lake formerly existed but then drained or filled in, causing permafrost to later form near the surface (**Fig. 20.6**). An additional layer of thawed ground known as the **residual thaw layer** can occur between the active layer and the underlying permafrost. This layer may form where surface warming causes the underlying permafrost to thaw to a greater depth. The deep thaw may have been driven by wildfire or some other disturbance that removed the vegetation cover, allowing the permafrost below to thaw from the top-downward, to a "greater-than-normal" depth.

Taliks are common in the discontinuous permafrost zone (**Figs. 20.2, 20.5**). As taliks increase in area over time, the thickness and extent of permafrost typically declines. Surface water and groundwater flowing through taliks can introduce additional heat into the nearby permafrost, which causes further warming and thawing. Changes in climate, including increased air temperatures, alterations in rainfall and snowfall, and the accompanying changes in vegetation can also have a large effect on the formation and distribution of taliks.

Permafrost stores large amounts of carbon (as organic matter) in frozen peatlands and where it is particularly thick. Thawing of this permafrost promotes the decay of organic matter, releasing CO_2 and methane to the atmosphere. This results in a positive feedback loop – climatic warming caused by anthropogenic emissions of greenhouse gases adds additional greenhouse gases to the atmosphere, from thawing peatlands and other permafrost areas, causing further atmospheric warming. From there, a domino effect continues to amplify climatic warming. Many scientists who study global environmental change are very concerned about this potentially damaging, positive feedback loop.

20.2 PERIGLACIAL PROCESSES

Frost action, perhaps the most important periglacial process, describes the processes stemming from repeated freezing and

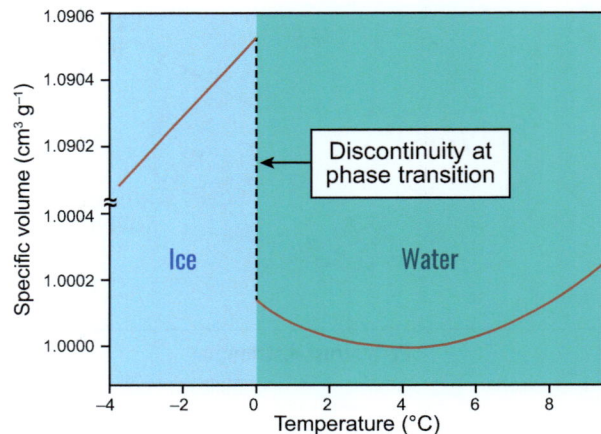

Figure 20.7 The relationships between temperature and specific volume. A 9% volumetric expansion occurs at the phase transition between water and ice. Water is most dense at about 4 °C, and ice shrinks as it gets colder. Together, all these processes can lead to thermal contraction cracking and the formation of ice wedges.

thawing of soil, sediment, and rock. It occurs within the active layer in permafrost landscapes. In more temperate areas, frost action occurs within seasonally frozen ground. Frost action involves the formation of ice during ground freezing, and the melting of ice as the ground thaws. As water changes into ice in the ground, it results in (1) volume increases, (2) water migration and ice segregation to form ice lenses, and (3) expansion and contraction of ice in the frozen sediment.

The change from water to ice has a profound effect on the properties of geologic materials. As water freezes to ice it expands by ≈9% (**Fig. 20.7**), forcing apart sediments and rock fractures. Expanding ice can exert extreme pressures (up to 2,000 atmospheres) on the surrounding soil and rock. This process can cause **frost jacking** as water freezes within voids, forcing large rocks to be uplifted (**Fig. 20.8A**). Freezing and thawing of bedrock can also result in **frost shattering**, as water within the rock fractures expands during freezing, causing fractures in rocks to widen (**Fig. 20.8B**; see Chapter 11).

A key process associated with frost action is **ice segregation**, which results as water in soil and rock migrates toward

Figure 20.8 Examples of frost action. **A.** Frost jacking of bedrock in the Northwest Territories, Canada. **B.** Frost shattering of rocks on Ellesmere Island, Nunavut, Canada. Source: (A and B) S. Wolfe (NRCan Photo IDs 2021-238, 239). **C.** Frost heave in a soil. Source: D. Riseborough.

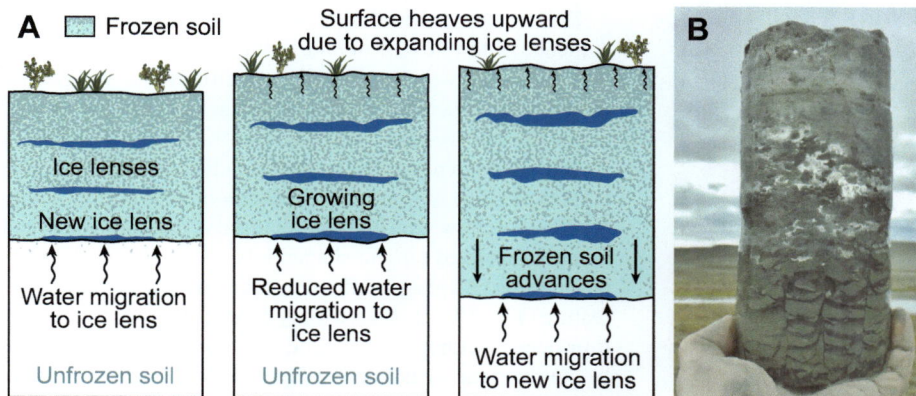

Figure 20.9 Ice segregation. **A.** Schematic diagrams of the process of ice segregation, showing how lenses can form at the freezing front in soils, as they freeze. Note how the soil surface has heaved upward due to the excess ice. When water availability in the soil is reduced, the freezing front advances downward, to form a new ice lens. In this way, multiple layers of ice lenses are formed. **B.** A core sample, taken from near the top layer of permafrost, showing silty sediments with segregated ice lenses. Source: M. Paquette.

areas of lower temperature, forming distinct ice lenses that can grow over time (**Fig. 20.8C**). This process is responsible for the uplift of soil particles and much of the **frost heave** we see of the ground surface in winter, including of roadways and sidewalks in cold climates. Ice segregation is driven by **cryosuction**, which is the pressure reduction that develops as ice forms in pores; unfrozen porewater is then drawn towards the colder ground (**Fig. 20.9A**). This process forms additional ice and frost heave, beyond the expansion caused by the formation of pore ice alone. Ice segregation is most effective in silts, which allow significant unfrozen water to exist at subzero temperatures due to its small particle size. Thus, water in silty sediments can travel through the unfrozen zone, along a cryosuction gradient, driven by a nearby, growing ice lens. Cryosuction is ineffective in sands and gravels, which are too coarse to produce ice segregation, and is less effective in clays due to their lower permeability, which slows the movement of water toward the growing ice lenses. Segregated ice lenses can range in thickness from a few millimeters to > 20 cm, and can take on a variety of forms, from blocky structures to continuous lenses several meters in length (**Fig. 20.9B**). Ultimately, ice segregation can lead to **excess ice**, a phenomenon in which the ice volume exceeds the volume of pores originally in the sediment. Thus, the ice pushes the soil upward and outward (**Fig. 20.9A**). Excess ice causes permafrost and soils to be highly thaw-sensitive, resulting in **thermokarst** features, as the ice lenses melt (see below).

As soils freeze, excess ice may heave the surface upward (**Fig. 20.9A**). Later, as the ice melts, **thaw settlement** may occur, causing a loss of volume and reduction in soil strength. This process of cyclical heave and settlement occurs primarily within the active layer (above the permafrost)

but may also occur within seasonally frozen ground, that is, without permafrost, particularly in mountainous terrain. This repeated upward and downward movement is seldom uniform across the surface, and instead results in the differential movement and displacement of sediment. Thus, thaw settlement forms uneven and irregular surface topography and is responsible for many sorted **patterned ground** features in periglacial environments, as well as the heaving and buckling of roads.

Frost heave and thaw settlement are widespread in periglacial regions and are responsible for a variety of periglacial landforms. The cyclical heaving and settlement of soil materials can cause mixing of soils and soil horizons, a process called **cryoturbation**. This process can also result in the formation of patterned ground on the surface, as well as unusual sedimentary structures within the ground (**Fig. 20.10**). Although a variety of processes may act to form cryoturbation structures and patterned ground, differential frost heave and settlement is usually a part of the mix.

In soils that undergo freezing and thawing, stones can slowly make their way to the surface. Farmers in cool climates know this – every year there are more stones on the soil surface, even though they were removed the year before! The uplift of stones in cryoturbated soils is the result of the combined effects of the processes of frost heave in winter, followed by thaw settlement in summer. Two processes, known respectively as **frost-push** and **frost-pull**, are particularly effective at lifting stones up. The processes are driven by the growth of ice lenses (**Fig. 20.11**). With frost-push, an ice lens forms below the stone as the ground freezes, pushing the stone upward. As the ground thaws in spring, the ice lens melts, creating a void below the stone. Because a small

Figure 20.10 Cryoturbation in a soil from the western Arctic in Canada. The patchy, dark brown layers of peat overlie gray-brown silty sediments with ice lenses. Note the contorted, mixed character of these sediments. Source: S. Wolfe (NRCan Photo ID 2021-240).

Sloping ground in the treeless tundra commonly shows evidence of the slow, flow-like movement of the active layer – a process known more broadly as **solifluction** – the general term for the processes that drive masses of soil downslope, caused by freeze–thaw activity. Solifluction is a hallmark slope process in periglacial regions. **Solifluction lobes** move at rates of a few mm/yr to up to 12 cm/yr, and on slopes with gradients as low as 10°. They can take the form of broad sheets or isolated tongue-shaped lobes, or as bench-like terraces; in some areas, solifluction lobes are a dominant slope feature (**Fig. 20.12**). A subsidiary process known as **gelifluction** occurs within waterlogged soils overlying permafrost. In gelifluction, as the active layer thaws in summer it becomes saturated, because water cannot percolate into the permafrost below. The saturated soil flows slowly downslope on top of the underlying impermeable permafrost. Materials moving downslope by gelifluction also typically take the form of lobes, creating step-like landforms on hillsides (see Chapter 14).

amount of soil often fills this space, the stone ends up slightly above its original position. With frost-pull, freezing around the stone causes ice lenses to form, lifting and heaving the soil and any rocks within. A void space, rather than an ice lens, may develop beneath the stone, which may later become filled with soil, again causing the stone to settle above its original position.

Frost heave and thaw settlement on hillslopes can result in the net downslope movement of soils and sediments. This **frost creep** (see Chapter 14) occurs as the soil heaves upward, perpendicularly to the slope of the ground surface, during freezing, and then re-settles during thaw in a nearly vertical direction (**Fig. 14.39**). Repeated heaving and settling results in a net downslope movement of soil materials. Frost creep (or simply, creep) is a slow process, but leaves a legacy on the land surface (**Fig. 14.38**).

Figure 20.12 The front of a solifluction lobe in the Yukon of Canada. Source: A. Lewkowicz.

Figure 20.11 Schematic diagram showing the frost heave processes that lift up stones in soil. **A.** Frost-push, where an expanding ice lens below pushes the stone upward. **B.** Frost-pull, where the heaving ground around the stone pulls it upward. Source: After Murton, J. in Ballantyne, C.K. (2018); reproduced with permission from Wiley.

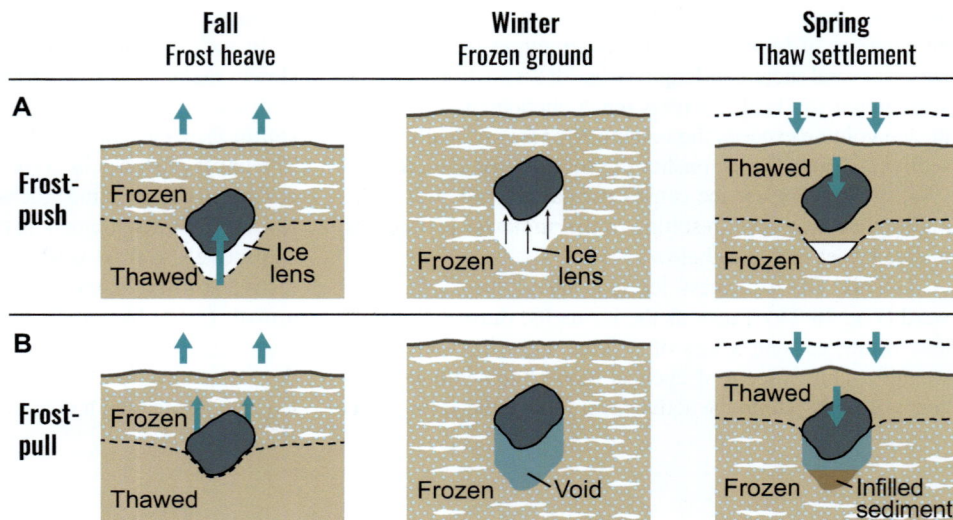

20.3 GROUND ICE AND ASSOCIATED PROCESSES

Various periglacial processes cause ice to form within permafrost, often producing large volumes of excess ice. Ice within frozen soil or rock is known as **ground ice**. Several different types of ground ice exist, but the two most common forms are **pore ice** and **segregated ice**. Both form as water freezes within sediments. Pore ice forms as pore water freezes *in-situ* within sediment. If the sediment is at or near saturation (i.e., most or all of the pores are filled with water) the sediment will expand when the water freezes. Pore ice is very common in sandy and coarse-grained sediments. Segregated ice forms as pore water migrates towards the freezing surface to form lenses of nearly pure ice (**Figs. 20.9, 20.13**). The formation of segregated ice lenses is one way in which excess ice can form in soils. Silty soils are particularly vulnerable to this process, as explained above.

Intrusive ice forms when water within soil or rock is forced under pressure into an unfrozen layer and then freezes (**Fig. 20.14**). This freezing can occur very rapidly.

Figure 20.14 Seasonal frost mounds (aka frost blisters) containing intrusive ice, Yukon. Source: A. Lewkowicz.

For example, water injected through a fracture in permafrost may freeze almost instantly. However, large water lenses may freeze over many years if they contain dissolved solutes or are under pressure, which lowers the freezing point. Intrusive ice is very important within periglacial geomorphology, and helps to form various types of landforms, including **pingos** and **seasonal frost mounds** (see below).

Another frost action process, quite common in periglacial landscapes, is **thermal contraction cracking** which forms **wedge ice**. This process occurs in permafrost areas where the soil is frozen year-round. Like many solids, ice expands and contracts with changes in temperature (**Fig. 20.7**), illustrating how dynamic ice can be on these types of landscapes. A rapid temperature drop of more than 15 °C in permafrost soils containing ice can cause enough contraction to force the ground to crack open. These cracks, averaging ≈1 cm wide and up to 5 m deep, extend through the frozen active layer and into the permafrost below (**Fig. 20.15A**). Water from melting snow can then fill the crack and freeze, forming an incipient, V-shaped "wedge" of ice. As the wedge ice forms within the crack it will then expand. Across successive winters, cracks tend to re-occur within the ice infill, which has a lower tensile strength than the surrounding frozen soil. The infilling of cracks and accumulation of ice within the permafrost occurring over many years typically results in wide, wedge-shaped, downward-tapering bodies of ice with vertical bands or foliations (**Fig. 20.15B**). As this **ice wedge** grows, sediments surrounding it are commonly deformed and pushed upwards to accommodate the added volume, resulting in raised ridges adjacent to the ice wedge.

Soils containing wedge ice undergo thermal contraction cracking as temperatures drop. The ice and the soil then contract even further. The stresses formed in this way are relieved by contraction and cracking. The optimal form of stress release in frozen ground is a polygonal pattern, typically forming **ice-wedge polygons** (**Fig. 20.16A**). Ice-wedge polygons occur today mainly in the tundra, where colder

Figure 20.13 Segregated ice lenses (blue) in permafrost, within silty clay sediments. Source: S. Wolfe.

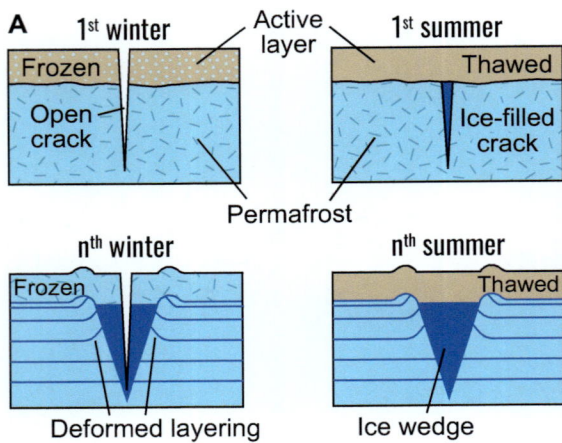

Figure 20.15 Wedge ice. **A.** Schematic diagram showing the formation of an ice wedge. Source: After Lachenbruch (1962) © Cambridge University Press, reproduced with permission. **B.** An exceptionally large ice wedge on Northern Richards Island, NW Territories, Canada. Note the person with their arms extended above their head, for scale. Source: S. Wolfe (NRCan Photo ID 2021-242).

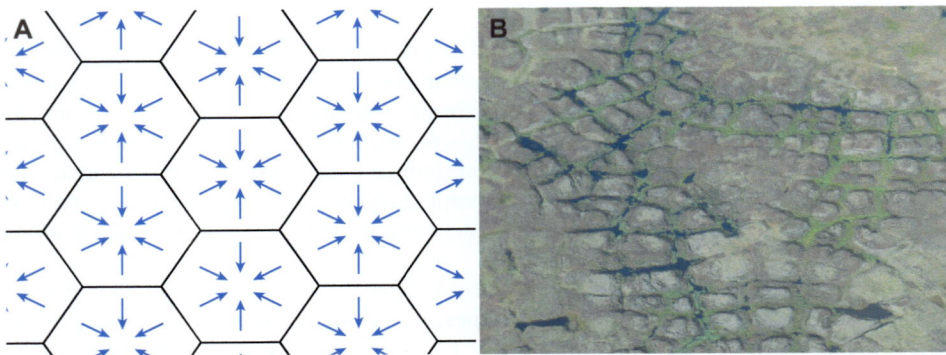

Figure 20.16 Polygonal cracking due to thermal contraction. **A.** An idealized ground thermal contraction pattern formed by stress release, as frozen ground cools. **B.** Ice-wedge polygons in northern Yukon, Canada. Source: S. Wolfe (NRCan Photo ID 2021-243).

ground temperatures promote thermal contraction cracking and ice-wedge growth. Ice-wedge polygons commonly vary from three-sided to five-sided and typically range between 15 and 40 m in diameter (**Fig. 20.16B**). The polygonal pattern commonly evolves over time to form networks of smaller, interconnected polygons (**Figs. 20.16, 20.17**). Ice wedges and ice-wedge polygons are some of the most common features of periglacial landscapes. They develop in peaty terrain, within glacial till or on outwash plains, and in areas that share similar soil, drainage, and thermal conditions. On the ground, ice-wedge polygons are described as being **low-centered** or **high-centered**. Low-centered polygons have low, flat centers, and shallow troughs bounded by

ridges. These polygons typically develop in flat areas and may contain ponded water in their centers (**Fig. 20.17A**). High-centered polygons have deep troughs at their edges, where ice wedges occur, and the polygons commonly lack ridges (**Fig. 20.17B**). They are commonly associated with thermokarst terrain, caused by the melting of the ice wedges within the polygonal network.

The growth of ice wedges provides another example of positive feedback in geomorphology. Initially, ice-wedge cracks develop within the contracting permafrost, but as they continue to grow, more and more ice forms inside them. Thermal contraction occurs more effectively in ice than in the nearby frozen soil, causing thermal contraction

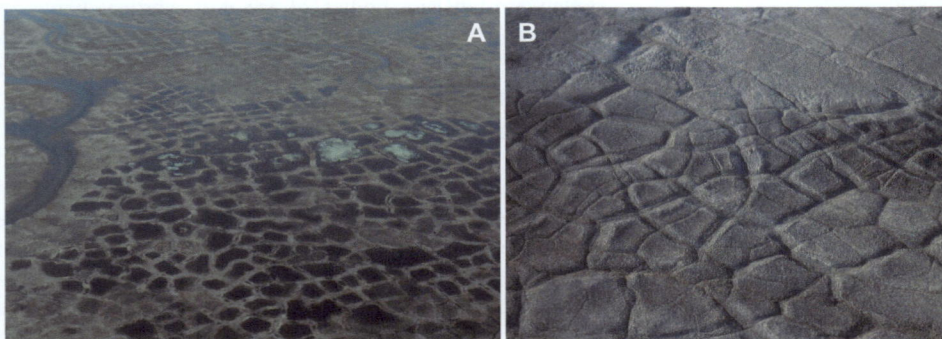

Figure 20.17 Polygonal networks on periglacial landscapes. Photos of (**A**) low-centered and (**B**) high-centered polygons, Ellesmere Island, Nunavut, Canada. Source: A. Lewkowicz.

cracks to continue to develop *within* the wedge ice. They become wider and deeper as the ice wedges continue to grow, sometimes resulting in very large ice wedges (**Fig. 20.15B**). This positive feedback may be countered by a process whereby the primary trough (crack) gets deep enough to trap sufficient snow to insulate the ground from cooling, thus causing secondary cracks to occur at the expense of the primary one.

20.4 ICE-WEDGE PSEUDOMORPHS AND SAND WEDGES

As permafrost thaws, the soil surrounding an ice wedge collapses into the space formerly occupied by the ice, creating an **ice-wedge pseudomorph** or **ice-wedge cast** (**Fig. 20.18A**). These features typically fill in the areas where ice wedges or contraction cracks once occurred and are therefore a clear indication of past permafrost conditions. In periglacial environments with limited snow cover, contraction cracks may not be filled with ice, leaving them open for periods of time. In such cases, they can fill with eolian sand, blown in by the wind, forming a **sand wedge** instead of an ice wedge (**Fig. 20.18B**). Unlike ice wedges, which melt and collapse, sand wedges are retained even after the permafrost thaws, leaving the sediment surrounding the sand wedge intact (**Fig. 20.18C**). Sand wedges provide evidence of past environments that were cold, dry, and windy enough that sand was blowing across the land surface, filling the open wedges with sand.

The polygonal form of ice-wedge and sand-wedge polygons can persist as relict periglacial features long after the permafrost has thawed. Relict polygons on modern (but warmer) landscapes indicate that they were clearly once

Figure 20.19 Relict thermal contraction polygons near New Sharon, New Jersey, USA, formed on unglaciated terrain south of the Laurentide ice sheet. Troughs between polygons are darker because they are wetter and have more organic material. This September 2020 image, taken when crops were mature but before harvesting and plowing, enhances the differences in soil moisture. Source: NASA.

subject to permafrost and periglacial processes. Such polygons are common in the eastern and midwestern United States, where they formed within a few tens of kilometers of the margin of the Pleistocene glaciers. Their presence hints at the harsh, wind-swept landscapes that existed at that time (**Fig. 20.19**).

20.5 PERIGLACIAL FEATURES AND LANDFORMS

A wide variety of features form and then thaw or melt on the ever-changing periglacial landscape. Let's examine some of the more important ones.

Figure 20.18 Ice-wedge pseudomorphs and sand wedges. **A.** An ice-wedge pseudomorph resulting from the melting of an ice wedge developed in floodplain gravels, England. Source: J. Murton. **B.** A sand wedge actively forming in the Northwest Territories, Canada. **C.** A relict sand wedge in Wisconsin, USA, dated by luminescence to ≈19,000 years ago. Note the upturned strata immediately adjacent to the infilled sand. (See also **Fig. 3.0**.) Source: (B and C) S. Wolfe (NRCan Photo ID 2021-244).

20.5.1 Patterned Ground

Patterned ground of various kinds is a common characteristic of periglacial landscapes, from high-latitude tundra and polar desert environments to high-elevation (alpine) locations. This term describes the semi-regular patterns and sorted features commonly observed on the ground surface. Broadly speaking, patterned ground features form due to the repeated freezing and thawing of moist sediment in the active layer. Ice-wedge polygons (**Fig. 20.17**) are one of the most obvious forms of patterned ground, but many others occur in periglacial settings, due to the constant movement of the ground. Patterned ground features may occur as sorted and non-sorted features. The difference lies in the degree of sorting of sediment particles, commonly observed on the surface. Non-sorted features include patterned ground, earth hummocks, water tracks, and frost boils which may show little or no evidence of surface particle sorting. In comparison, sorted features include circles, polygons, networks, and stripes, where finer particles are commonly noted near the centers and coarser materials found towards the edges, due to the continual movement of the soil.

Of all the **non-sorted patterned ground** features, the most common are **earth hummocks**. These vegetated mounds are typically 10–60 cm high and 0.5–2.0 m wide, separated by narrow troughs (**Fig. 20.20**). Earth hummocks make hiking across the tundra extremely awkward! Multiple mechanisms have been proposed for the formation of earth hummocks. All of these proposed mechanisms involve the movement of fine-textured soils and water within the active layer due to annual freezing and thawing, including ice lens formation and the migration of pore water. The net effect of these processes is the uplift of the hummock center.

Frost boils are another type of non-sorted patterned ground feature formed by differential frost heave and settlement. These circular areas of bare soil occur within the tundra (**Fig. 20.21A**). Frost boils range from 0.5 to 3.0 m in diameter, although some are smaller. They are a common ground surface feature on soils that contain a high proportion of fine-grained sediment (**Fig. 20.21B, C**). Commonly, frost

Figure 20.21 Frost boils. **A.** A schematic diagram of the internal structure of a frost boil during summer. **B.** A frost boil formed in a glacial till in the Kivilliq region of mainland Nunavut, Canada. **C.** Frost boils forming as a non-sorted net in fine-grained till. Source: (B and C) I. McMartin (NRCan Photo IDs 2021-246, 247).

boils have wet, saturated mud on their upper surface. This mud may have been ejected from the subsurface to form the boil. In essence, pressure exerted on the underlying soil during thaw consolidation causes liquefied mud to be forced upward, under pressure, to the surface (**Fig. 20.21A**).

As in the examples above, patterned ground develops best on unvegetated or sparsely vegetated landscapes. The patterns become particularly evident in soils that contain gravel and coarse stone clasts set within finer sediment. Here, the gravel and rocks can become "sorted" and separated from the finer sediment. On level ground, the sorted patterns are usually circular. **Sorted circles** tend to have well-defined centers of fine material

Figure 20.20 Earth hummocks in the Richardson Mountains, northern Yukon, Canada. Source: S. Wolfe (NRCan Photo ID 2021-245).

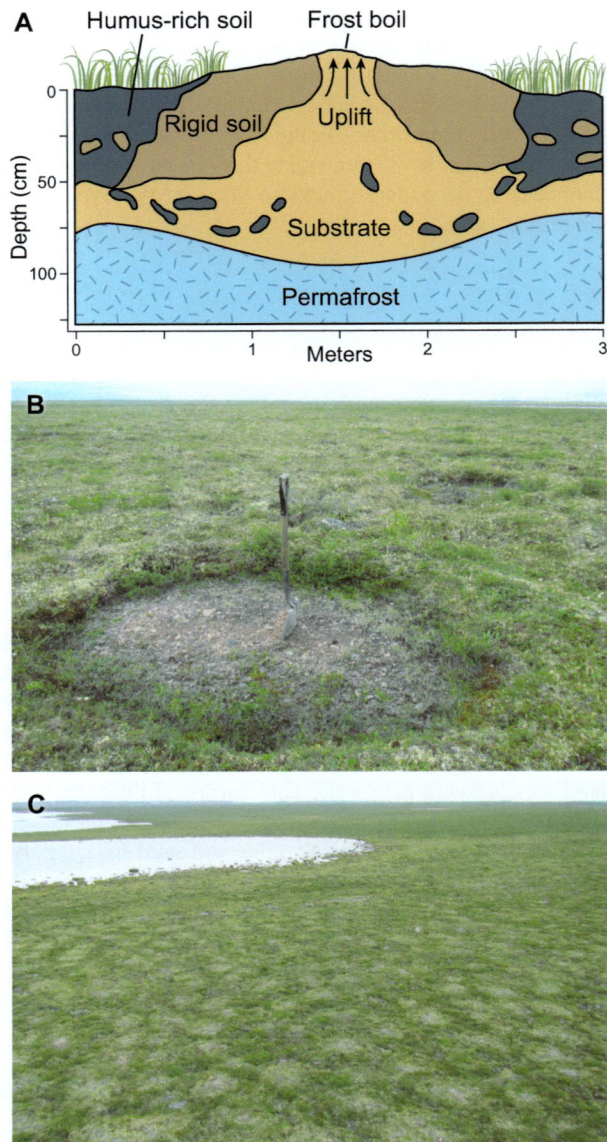

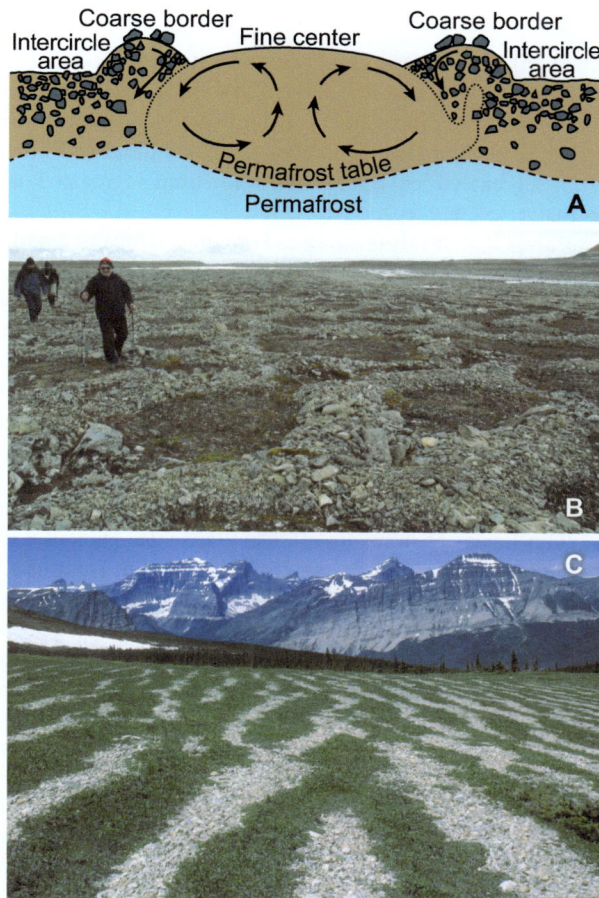

Figure 20.22 Patterned ground features. **A.** Schematic diagram of net sediment movement within sorted circles. Source: Used with permission of Elsevier, from Hallet and Prestrud (1986); permission conveyed through Copyright Clearance Center, Inc. **B.** Sorted circles in Spitzbergen, Svalbard, Norway. Source: P. Owczarek. **C.** Relict stone stripes in Montana, USA. Source: M. Miller.

enclosed by a perimeter of stones and coarse clasts (**Fig. 20.22A, C**). A variant on this form – **sorted nets** – may exhibit more irregular shapes, with thinner perimeters of coarse clasts. It is common for the coarse clasts to be embedded in fine sediment along the perimeter of the circles and nets, and even to be vertically inclined. Circles and nets may vary in size from 0.5 to 3.0 m in diameter (**Fig. 20.22B**). As the ground surface becomes more sloping, these features become elongated in the downslope direction, and may eventually develop into **sorted stripes** (**Fig. 20.22C**). Most models proposed for the formation of patterned ground envision a circulatory movement of material upwards and outwards from the feature's center, causing sediment to migrate to the outer edges, and with downward and returning movement occurring at the perimeters (**Fig. 20.22A**). In summary, sorted circles and stripes are the products of a variety of mixing processes (cryoturbation), involving a seasonally changing interplay of soil, ice, and water.

Hummocks, frost boils, and other patterned ground features occur in periglacial landscapes with a soil cover. Features like blockfields, also known as **felsenmeer**, will form in areas of barren rock and stone (**Fig. 20.23**). **Blockfields** are accumulations of angular rocks on hillslopes, mountainsides, and plains. Periglacial processes responsible for the weathering of bedrock to form blockfields include the frost shattering of rock along joints, as well as the differential thermal contraction of minerals that may have cracked the rocks in the first place. Blockfields typically appear as collections of bare, loose boulders covering the surface. The rocks usually lack any sediment between them. Instead, the angular boulders typically overlie bedrock to a depth of a few meters. Blockfields are common in present-day Arctic and alpine areas, especially on slopes (**Fig. 20.23A**). Intriguingly, relict blockfields also occur in areas of former periglacial environments,

Figure 20.23 Blockfields. **A.** Contemporary blockfield formed in a modern periglacial environment – the Richardson Mountains of northern Yukon, Canada. Source: S. Wolfe (NRCan Photo ID 2021-248). **B.** A relict blockfield in the Appalachian Mountains of Pennsylvania, USA. Source: K. Patrick.

where they cover uplands and low mountains. Examples include mid-latitude areas such as central Europe, the United Kingdom, and the Appalachian Mountains of the United States (**Fig. 20.23B**). Whereas the main hypothesis for the formation of blockfields is frost weathering in periglacial environments, other hypotheses suggest that they may be relict from much older, warmer, and more humid, deeply weathered surfaces.

Another feature observed on slopes in periglacial landscapes, particularly in alpine areas, are **rock glaciers**. These intriguing features are lobate or tongue-shaped bodies of frozen rocky debris containing interstitial ice and ice-rich sediment (**Fig. 20.24**). Active rock glaciers move slowly downslope through the deformation of the ice between and below the rocks. The rocks are essentially riding along, above, and within the slowly deforming ice mass. Nearer the edges of the rock glacier, the rocks may be more often stationary, because the interstitial ice may have melted.

Rock glaciers occur as glacigenic and talus glaciers. **Glacigenic rock glaciers** are located near the termini of "traditional" glaciers and contain within them a core of buried glacial ice. The rocks are simply riding along on top of the glacial ice, having arrived at that location by being incorporated into, or falling onto, a preexisting glacier. **Talus rock glaciers** occur below, and develop within, talus (see Chapters 8 and 14). They may have mostly segregated ice within frost-susceptible sediment. Many have a permafrost core. These rock glaciers form bench-like lobes at their lower end (**Fig. 20.24A**). In contrast, in some cases, hybrid rock glaciers may form.

As rock glaciers slowly flow downslope, arcuate ridges and furrows form due to buckling of the rocky, debris-rich material over the ice-rich sediments (**Fig. 20.24B**). Active rock glaciers have steep fronts and range in length from a few tens of meters to > 2,000 m, and vary from 50 to 1000 m in width.

20.5.2 Frost Mounds

Many distinctive periglacial landforms develop from the accumulation of ground ice, derived from the migration of soil and groundwater in sediment. These features, known collectively as **frost mounds**, include palsas and lithalsas formed from segregated ice, pingos that form from segregated and intrusive ice, and frost blisters.

Palsas and **lithalsas** are raised mounds, typically circular or oval in shape. They develop as permafrost aggrades into the ground, producing segregated ice lenses as the ground freezes. These features are common within the discontinuous permafrost zone and near the limits of permafrost. Although palsas and lithalsas are similar, they differ in that palsas are found within peatlands whereas lithalsas are common on former lake basins or submerged marine areas that lack a thick peat layer at the surface.

Palsas are peat-covered permafrost mounds that may also contain layers of mineral soil. They are inter-layered with segregated ice lenses. Palsas commonly range from about 1–5 m high and < 100 m across (**Fig. 20.25B**). They are commonly found individually or in clusters in the discontinuous permafrost zone and near the limits of permafrost, on wet, peaty landscapes (**Fig. 20.25C**). The growth of ice lenses within the ground heaves the surface upward, forming the palsa (**Fig. 20.25A**). This process begins by the formation of ice below ground, enhanced at the palsa site due to thin snow cover or the modification of local drainage conditions. To initiate palsa growth, the freezing front must penetrate sufficiently in winter such that ice and permafrost form locally, which creates a small mound of up to several centimeters high, surrounded by unfrozen ground. The slightly elevated mound cools more than the surrounding ground because less snow accumulates on the exposed, windblown surface. Because the interior temperature of the growing palsa is consistently lower than that of adjacent ground, ice lenses continue to grow within it by drawing in water from surrounding unfrozen areas. The growing ice lenses push

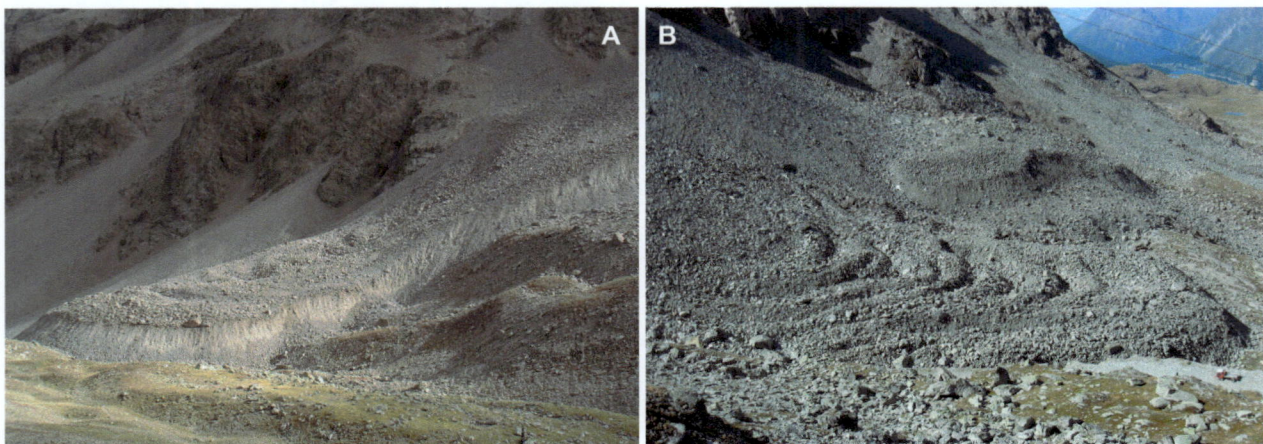

Figure 20.24 Rock glaciers. **A.** The Laurichard talus rock glacier in the French Alps. Note the talus slopes that occur at the head of the glacier, and feed rocks into it. The Laurichard glacier flows at speeds of ≈1.5 m/yr. Source: Xbodin, CC BY-SA 3.0, via Wikimedia Commons. **B.** The Murtèl rock glacier, Switzerland. Note the arcuate ridges that form from buckling over frozen materials. Source: S. Gruber.

Figure 20.25 Palsas. **A.** Schematic diagram of the formation of a palsa, as ice lenses develop during permafrost aggradation. Winter-time conditions are shown. Source: Used with permission of Elsevier, from Seppälä (2011); permission conveyed through Copyright Clearance Center, Inc. **B**. A palsa in MacMillan Pass, Northwest Territories, Canada. Source: A. Lewkowicz. **C.** Palsas on the Tavvavuoma mire in Sweden. Source: S. Backe.

the surface up even higher, further limiting snow accumulation and enhancing winter cooling and ice lens growth. This is yet another example of positive feedback in geomorphic systems: palsa growth leads to thinner snow cover, which in turn enhances ice formation and continued palsa growth. The contrasting thermal properties of the peat overlying the palsa further enhance palsa growth – in winter, the frozen peat effectively transfers heat out of the ground to cool it, whereas in summer the dry peat provides increased thermal insulation from the warm air. Thus, the palsa continues to grow in all seasons.

Just like in their initial growth phase, the decay of palsas is tied to the thermal conditions of the site. When fully formed, palsas are high spots in an otherwise wet, peaty landscape (**Fig. 20.25B, C**). Once the palsa reaches the point

where it has a high and distinctly convex upper surface, the peat cover on the palsa begins to dry out, crack, and slide off. Wind then erodes what is left of the peaty cover, initiating melting of the ice lenses inside. At this time, the process of ice melt and palsa degradation is set in motion – another positive feedback.

Lithalsas are similar to palsas, but they lack the thick surface peat cover. Instead, they have only a thin organic cover. Lithalsas grow in areas of fine-grained, frost-susceptible sediment with an available groundwater supply, and extensive, but warm, discontinuous permafrost which promotes slow freezing and ground ice formation. Contemporary lithalsas are found near water bodies including stream valleys, shorelines, large lake basins, and ponds (**Fig. 20.26A**). Their irregular shape and size is

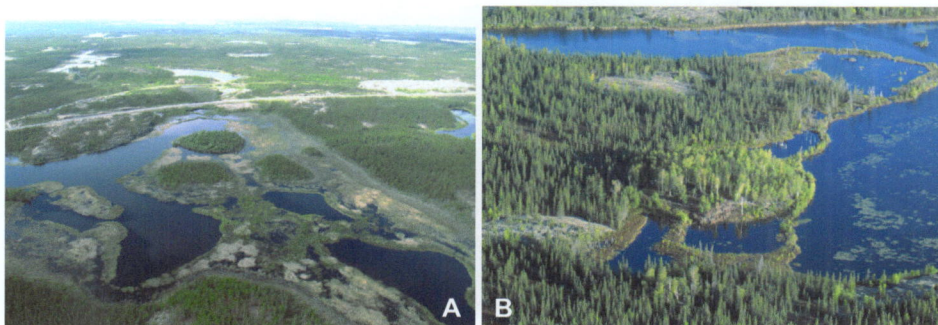

Figure 20.26 Lithalsas. **A.** Lithalsas near Yellowknife, Northwest Territories, Canada. **B.** Degraded lithalsas with thermokarst ponds. Source: S. Wolfe (NRCan Photo IDs 2021-249, 250).

typically related to the pattern of permafrost aggradation, the thickness of frost-susceptible sediments, and the distribution of the water source.

When the permafrost within them thaws, palsas and lithalsas will collapse, forming depressions called **thermokarst ponds**. Commonly, the ponds will have large circular rings of raised earth around their perimeters, marking their former edges (**Fig. 20.26B**).

Pingos (Inuit, "conical hill") are large, ice-cored hills that range from ≈30 to 600 m in diameter and up to 50 m in height, with a circular or ovoid base (**Figs. 20.0, 20.27**).

Figure 20.27 At 49 meters in height, Ibyuk Pingo, located near Tuktoyaktuk in the Northwest Territories, is the largest pingo in Canada and the second largest on Earth. Source: C. Tarnocai.

Some estimates place the number of pingos on Earth at over 11,000. A surface layer of frozen soil typically covers the massive core of ice within the pingo. The soil and vegetation cover on the top of a pingo may be fissured and cracked, due to rupturing by the progressive growth of the ice core below. Just as can happen with palsas and lithalsas, if the ice within the pingo melts, it may collapse, leaving behind a pond and a distinctive circular rim, almost like a small crater.

Pingos occur in both the continuous and discontinuous permafrost zones. They are differentiated as open- or closed-system pingos, based on how they form. To preview, closed-system pingos form as permafrost grows within drained lake basins surrounded by continuous permafrost. Alternatively, open-system pingos typically form as ice grows within groundwater "springs" at the base of a slope.

Closed-system pingos, also called hydrostatic pingos, are common in the continuous permafrost zone. The formation of closed-system pingos typically begins with the draining of a lake underlain by sandy sediments. Once drained and in contact with cold atmospheric conditions, permafrost begins to form in the barren, saturated sands (**Fig. 20.28A**). As permafrost grows downward and inward across the former lake bottom, porewater in the sands freezes and expands, pressurizing the remaining unfrozen water and forcing it towards the ground surface. This water then freezes and pushes the ground surface upward. Water continues to be expelled toward the growing ice mass, forming a large body of ground ice below a conical hill, or pingo. The ground ice

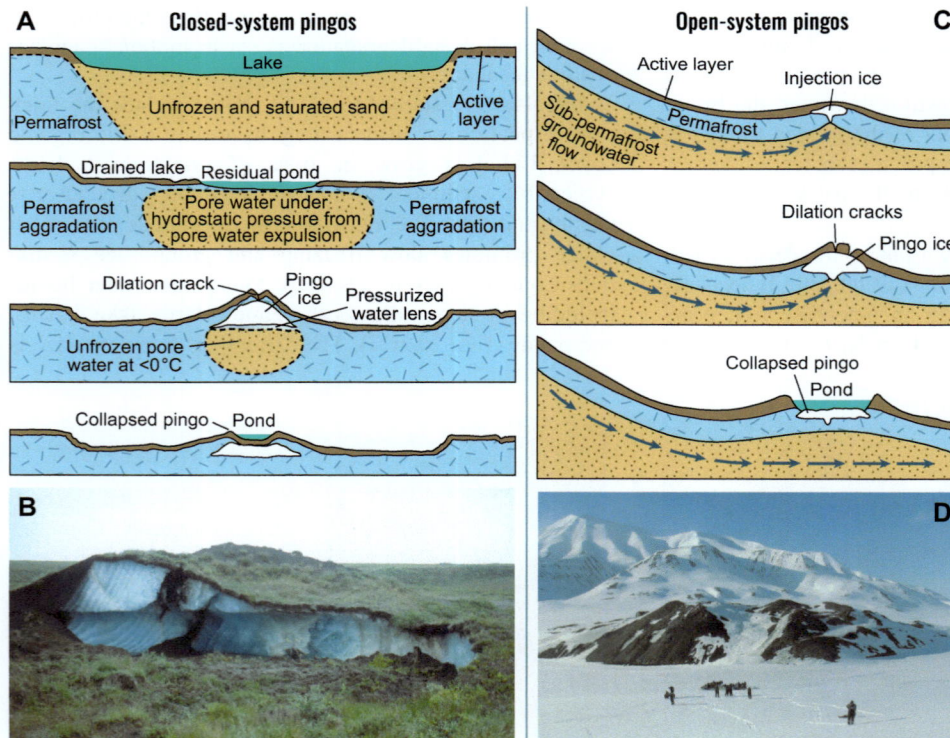

Figure 20.28 Pingos. **A.** Schematic diagram showing the formation of a closed-system pingo. **B.** A collapsed, closed-system pingo in the western Canadian Arctic, showing the massive ground ice in the core. Source: J. Ross Mackay. **C.** Schematic diagram showing the formation of open-system pingos. **D.** An open-system pingo in Svalbard. Source: A. Lewkowicz.

within closed-system pingos is commonly a combination of intrusive ice and segregation ice (**Fig. 20.28B**). A pocket or lens of unfrozen water beneath the pingo is commonly present, remaining under pressure from the continued pore-water expulsion. Seasonal changes in pressure can cause pingos to rise and fall in height and, if the overlying cap of permafrost is penetrated, water can flow under artesian pressure to the surface.

Open-system pingos, also known as hydraulic pingos, are typically found within areas of discontinuous permafrost on slopes or valley bottoms where groundwater flows from uplands to lowlands and is then forced upward, emerging at the ground surface (**Fig. 20.28C**). The groundwater flowing toward the surface forms an ice lens that continues to grow in size, building the pingo. Open-system pingos occur in isolation or in small groups, in areas with the proper combinations of topographic relief and water, such as on hillslopes and in valley bottoms (**Fig. 20.28D**). Open-system pingos are typically smaller than closed-system pingos. **Frost blisters**, also known as seasonal frost mounds (**Fig. 20.14**) are similar to open-systems pingos, but form annually. The ice within frost blisters melts – partially or completely – each summer.

20.6 THERMOKARST FEATURES

The issue of thawing permafrost resulting from global warming is a major concern. Both the thermal and physical states of permafrost are closely tied to climatic conditions. As the climate warms, so does the permafrost. The term **thermokarst** describes the subsidence, erosion, and instability of the ground surface resulting from the thawing of ice-rich permafrost and the melting of ground ice. Subsidiary processes include the accumulation of surface water bodies that form as the ground subsides. Thus, a periglacial landscape dominated by thermokarst has irregular surface topography with ponds, marshy hollows, and small hummocks. The name thermokarst comes from the fact that the thawing landscape resembles karst, with its many depressions caused by dissolution of limestone (see Chapter 12).

The primary cause of thermokarst is a disruption of the ground thermal regime, driven by warming air temperatures or disturbance at the ground surface, such as by construction, fire, or deforestation. These changes are occurring over much of the Arctic today. As the ground thermal regime warms, the active layer thickens and the permafrost gets thinner (**Fig. 20.29**). This process, known as **permafrost degradation**, can also lead to the formation and expansion of taliks and residual thaw layers. Other causes of thermokarst include erosion of the surface, which exposes the ice-rich permafrost or massive ice to the atmosphere. Atmospheric warming can cause water in ponds and lakes to also warm, driving the expansion of taliks. Several landforms in periglacial landscapes are the direct result of thermokarst.

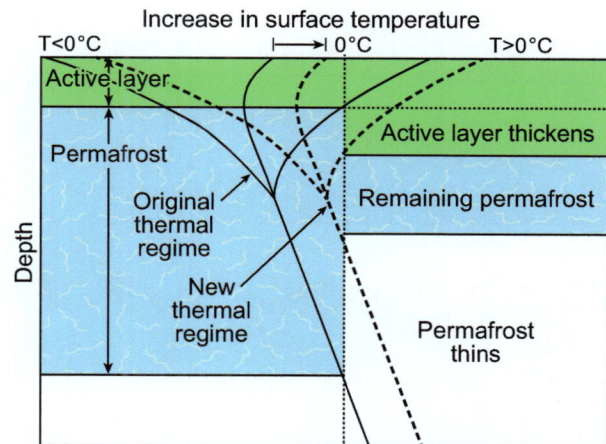

Figure 20.29 Graph showing the effects of increased surface temperatures on the ground thermal regime, particularly an increase in active-layer thickness and a decrease in permafrost thickness. Source: After French, H. M. (2007), with permission from JoÚ Wiley and Sons.

20.6.1 Thermokarst Ponds

Ground surface subsidence forms depressions on the land that may fill with water. These **thermokarst ponds** are common in areas of ice wedges and palsas (**Fig. 20.30A**). They form as the saturated soils subside below the surface of the water table (which is often quite shallow). Larger-scale subsidence can form thermokarst lakes and vast wetlands (**Fig. 20.30B**). In sub-Arctic permafrost regions, uneven subsidence of the surface due to thermokarst often causes trees to tilt severely, a phenomenon known as a drunken forest.

Thermokarst is common within wetlands with thick accumulations of peat that overlie ice-rich, fine-grained sediments. The formation of thermokarst ponds and taliks in peatlands may exacerbate CO_2 and methane emissions by enhancing decay of the peat, further contributing to the emissions of greenhouses gases.

20.6.2 Active-Layer Detachment Failures and Slumps

As thermokarst processes become more pronounced, thawing of permafrost can cause **active-layer detachment failures** on sloping surfaces. Active-layer failures involve the downslope movement of the active layer and the surface vegetation on it, while the undisturbed permafrost below remains solidly frozen. Although both are forms of mass wasting, active-layer detachments are much faster than the slower motion that occurs with solifluction. The detachment is driven by melting of ground ice in the near surface, which causes increased porewater pressures in the thawed sediment. The thawed surface materials can then slump, flow, and slide across the permafrost. In tundra environments, detachments may be initiated by warm air temperatures or by periods of intensive precipitation that saturate the active layer.

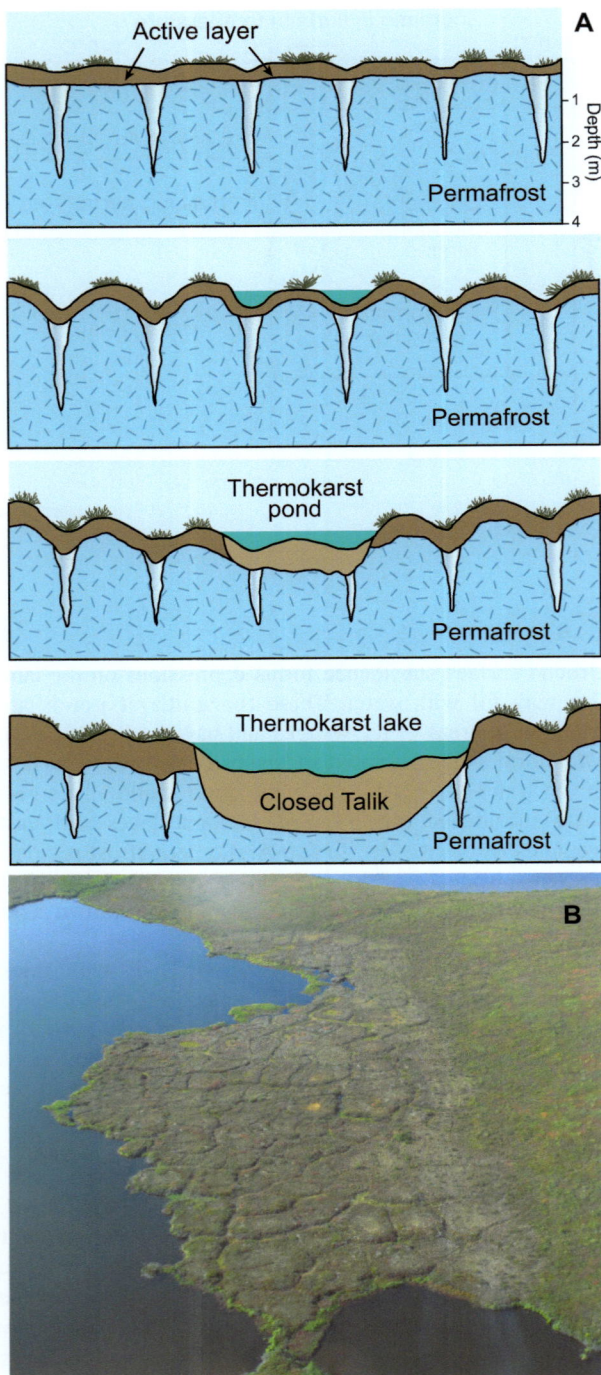

Figure 20.30 Thermokarst. **A.** Schematic diagrams illustrating the thawing of permafrost terrain containing ice wedges. Thermokarst leads to thickening of the active layer, melting of the ice wedges, ground subsidence, and the progressive formation of thermokarst wetlands, ponds, and lakes. **B.** High-center polygons on the perimeter of a thermokarst lake on the Tuktoyaktuk Coastlands of the Northwest Territories, Canada. Source: S. Wolfe (NRCan Photo ID 2021-251).

Active-layer failures commonly range from < 30 to 100 m wide, and vary in length considerably, up to 700 m long (**Fig. 20.31A**). They are most common on low slopes (4°–7°) to

moderately steep (< 20°) valley sides (**Fig. 20.31B**). In forested environments they can also be caused by fire burn scars, by deforesting by logging, or by avalanche activity, which changes the thermal regime of the site.

Active-layer failures may develop into larger mass wasting features called **thaw slumps**, if ice-rich material within the permafrost thaws. Recall that in active-layer detachments, the underlying permafrost remains solidly frozen. In thaw slumps, permafrost thaw is the primary means of soil loss and erosion. Thaw slumps are bowl-shaped, with vertical headwalls, exposing the permafrost below. The headwalls retreat upslope as ice-rich sediments continue to thaw (**Fig. 20.32A**). Thaw slumps are typically deeper than active-layer detachment failures because they develop down, into the permafrost. Active-layer detachments move along the top of existing, stable permafrost. Most thaw slumps create basins that are 1–5 hectares in area, outlined by their bowl-shaped headwall. The term **mega slump** is used to describe large slumps with basin areas > 5 hectares, and sometimes as large as 50 hectares. Headwall heights on thaw slumps typically range from 1 to 10 m, but mega slumps can have headwalls ranging up to 100 m high (**Fig. 20.32B**). The continued thawing of permafrost causes the headwalls in thaw slumps to retreat upslope, sometimes as fast as several meters/year. Thaw slumps commonly have watery and muddy debris tongues on their downslope margins, which continue to flow downslope. Debris tongues flowing from active thaw slumps can raise the suspended sediment loads in downstream lakes and streams by several orders of magnitude. Because these types of flows typically follow stream and river channels, they can temporarily dam the flow of water in rivers and streams.

Although warming air temperatures set periglacial landscapes up for this type of slope failure, the initiation of the slumps is usually due to localized erosion. For example, along rivers or at coastlines, erosion can expose ice-rich permafrost, starting the slump. Thaw slumps can also begin near the crest of a slope. In this case, warm summer temperatures, rainfall, or fire may initiate active-layer detachments that can then grow into thaw slumps. Thaw slumps eventually stabilize when materials accumulate and cover (insulate) the ice-rich headwall.

20.6.3 Beaded Streams and Thermokarst Gullies

Beaded streams are unusual features found only in periglacial landscapes. They are characterized by a series of ponds connected by narrow channel segments. They develop as networks of ice-wedge polygons in permafrost decay by thermal erosion. The beaded component of these streams typically forms along the connecting intersections of ice-wedge polygons (**Fig. 20.33A**).

Thermokarst gullies form similarly to beaded streams, that is, from the thermal erosion of permafrost. Thermokarst gullies are commonly associated with ice-wedge polygon networks. As the ground thermal regime gets warmer, ice in the troughs melts and fills them with water. The troughs provide natural flow paths for the water, and any ice in the

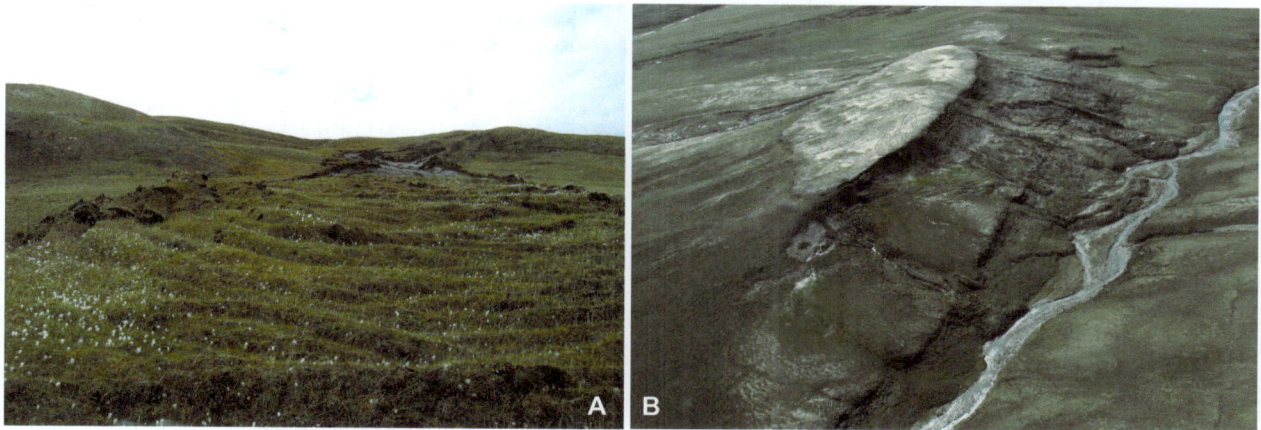

Figure 20.31 Active-layer detachment failures on Ellesmere Island in Nunavut, Canada. **A.** Multiple failures have occurred on this slope. **B.** Widespread surface instability, caused by active-layer detachment failures. Source: A. Lewkowicz.

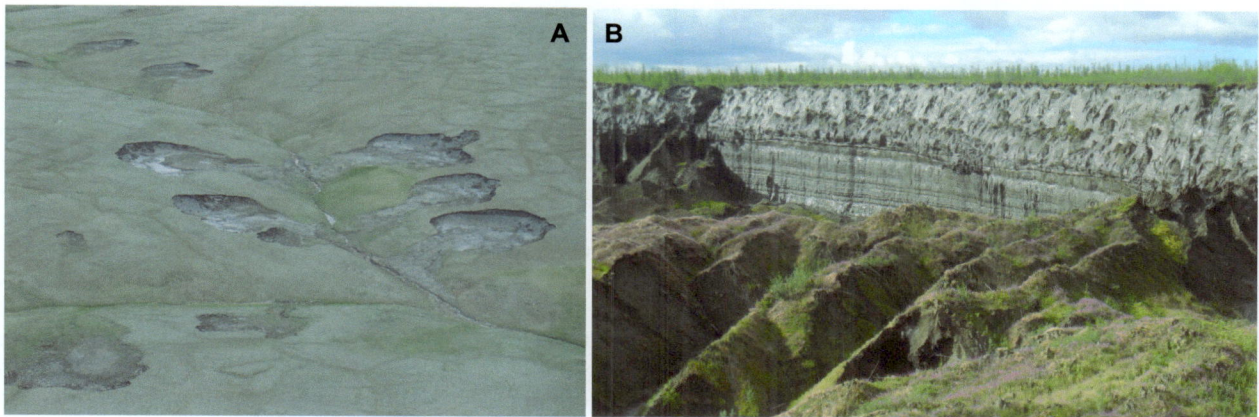

Figure 20.32 Thaw slumps. **A.** Multiple slumps on Banks Island, Northwest Territories, Canada. Source: R. Smith (NRCan Photo ID 2021-252). **B.** The headwall (≈100 m high) of the Batagay mega thaw slump in Siberia. Source: J. Murton.

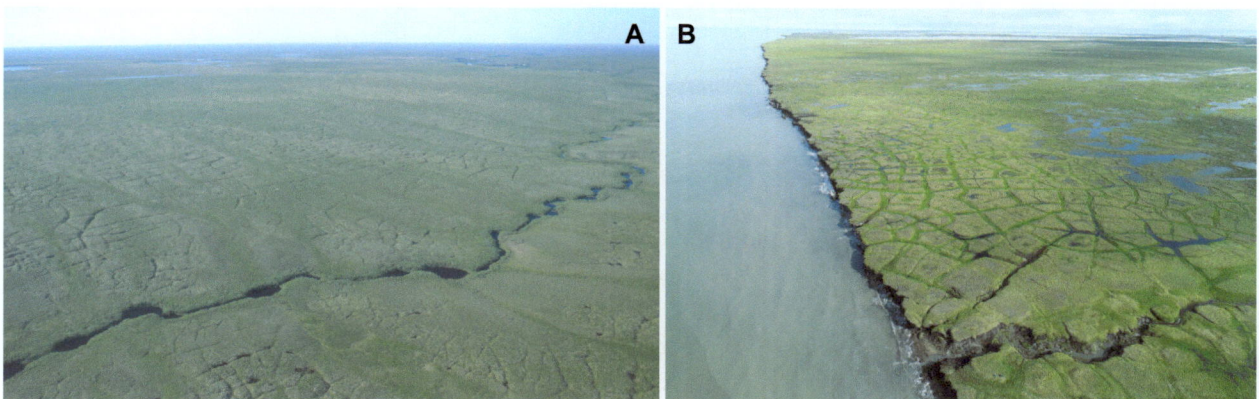

Figure 20.33 Thermokarst landforms and features. **A.** Thawing ice wedges and a beaded stream in landscape of high-center ice-wedge polygons on the Alaskan Arctic Coastal Lowlands. **B.** Thermokarst gullies in ice-wedge troughs along the Alaskan Coast, formed as ice in the wedges melts, and runoff is focused into these low areas. Source: B. Jones.

former wedge is then easily melted. Thermokarst gullies are commonly incised 1–5 m deep and may extend for hundreds of meters (**Fig. 20.33B**). They provide effective exit routes for water in ponds and lakes, often leading to rapid drainage.

20.7 EFFECTS OF CLIMATE CHANGE ON PERIGLACIAL ENVIRONMENTS

Climate is ever-changing, but perhaps nowhere are the effects of modern climate change and warming more apparent than on periglacial landscapes.

20.7.1 Polar Deserts

Few places on Earth are more geographically or climatically extreme than polar deserts. These landscapes of barren areas of rock, soil, and ice are associated with the Antarctic and Greenland ice sheets, and the High Arctic Island

Figure 20.34 The barren polar desert of Taylor Valley and Commonwealth Glacier (77° 39′ S) in Antarctica. Source: J. A. Gillies.

archipelagos of the Franz Yosef Land (Russia), Svalbard (Norway), and the Queen Elizabeth Islands of Canada that are poleward of 75° N. Excluding the vast areas covered by glacial ice, polar deserts represent ≈3% of Earth's total land area. Located poleward of the Arctic and Antarctic Circles, these areas receive only low-intensity sunlight in summer, and experience extended periods of darkness during winter. In short, polar deserts are areas of climate extremes – in daylight and in temperature (**Fig. 20.34**).

Polar deserts are defined as areas where the warmest average monthly air temperature is < 10 °C and the mean annual precipitation is < 250 mm. Nonetheless, many polar deserts are considerably colder and even drier. For example, the average annual temperature at Alert on Ellesmere Island, Canada (the most northerly terrestrial meteorological station on Earth) is –18 °C, with the warmest month averaging only 3 °C. Here, the mean annual precipitation is only ≈150 mm, with 90% falling as snow. In the McMurdo Dry Valleys of Antarctica, the average annual temperature is about –16 °C, and annual snowfall is only ≈110 mm.

Permafrost is continuous across polar deserts, where it may even exceed 600 m in thickness. Average ground temperatures are also very cold. For example, the average ground temperature at Alert is –15 °C (**Fig. 20.35B**). Much of the polar desert surface has little or no vegetation cover due to the scarcity of water. In areas of thicker snow accumulation, snowmelt water can accumulate in low-lying areas and run off through shallow channels on top of the permafrost, producing non-sorted stripes called **water tracks** (**Fig. 20.35A**). In these areas, soil moisture in summer is adequate for some surface vegetation to grow. The active layer is typically thinner beneath these "oases," than beneath the surrounding polar desert surfaces, due to the insulation provided by plants.

Given the extremely cold and dry conditions of the polar deserts and knowing that many areas have only thin layers of soil over bedrock, most geomorphic processes are

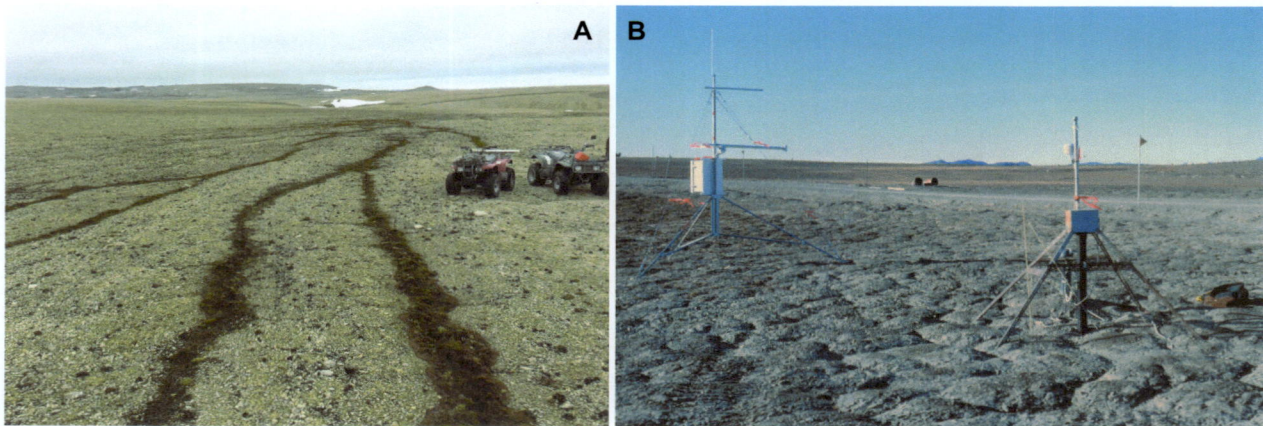

Figure 20.35 Polar desert landscapes. **A.** Water tracks within the polar desert landscape of Ward Hunt Island, Nunavut, Canada. Snowmelt-derived water flowing within the active layer provides moisture for plant growth in this otherwise dry landscape. Source: M. Paquette. **B.** The intensely cryoturbated ground surface at Alert, northern Ellesmere Island. Source: S. Smith (NRCan Photo ID 2021-253).

restricted to the active layer. Cryoturbation within the active layer can produce sorted and non-sorted circles, as well as frost mounds (**Fig. 20.35B**). Ice wedges are present, but because of the limited moisture and the abundance of weathered bedrock, they may not be very wide or deep. Overall, ice contents in the near surface tend to be low but may be enriched with segregation ice beneath vegetation patches.

Even the polar deserts are not immune to the effects of climatic warming. Instead, these areas are experiencing **polar amplification**, as air temperatures near the poles increase at higher rates than in the mid-latitudes. Polar amplification is being driven largely by a decrease in surface albedo (solar reflectance) as sea ice coverage decreases, leading to increases in open ocean and energy absorption. Polar amplification is very pronounced in the Arctic, where recent increases in annual air temperatures have been more than three times the global average (**Fig. 20.36**). Ground temperatures in polar desert regions are in turn highly responsive to changes in air temperature due to the lack of insulating vegetation cover. As a result, ground temperatures are also rising, causing active layers to thicken, as the permafrost table deepens. Increases in precipitation, both as snow and as rain, make active-layer soils even wetter. Together, these changes can induce active-layer failures and thaw slumps (**Fig. 20.37A**). Ice wedges can also melt, leading to surface subsidence (**Fig. 20.37B**).

20.7.2 Arctic Coastal Lowlands

As seen earlier in **Fig. 20.4**, over 90% of the land surface across a broad belt of northern North America, Europe, and Asia has continuous permafrost. Most contemporary periglacial landscapes reside within these regions, bounded to the north by the Arctic Ocean. Thick, ice-rich sediment and abundant ground ice underlie most of the Arctic coastal lowlands. Typical periglacial features include segregated ice, ice wedges, pingos, and buried glacial ice.

As with the polar deserts, Arctic coastal lowlands are experiencing rapid warming due to polar amplification, including warmer winter and summer air temperatures, and increased summer rainfall. They are also experiencing accelerated rates of coastal erosion from rising sea levels and the extended open-water season. Unconsolidated, ice-rich sediments exposed at the coastline can erode rapidly by wave action, forming **wave-cut notches** (**Fig. 20.38A**). These notches are geomorphically unstable, such that large blocks of frozen sediment can collapse into the sea and erode further as the ice-rich permafrost thaws (**Fig. 20.38B**). With rising sea levels, ice-wedge polygons near the coast are being eroded, and some are being inundated by sea water (**Fig. 20.38C**). Overall, rising air temperatures and increased precipitation are accelerating the thawing of permafrost and slumping of the ground surface, exposing ground ice with permafrost to the atmosphere. These areas are then particularly vulnerable to coastal erosion (**Fig. 20.38D**).

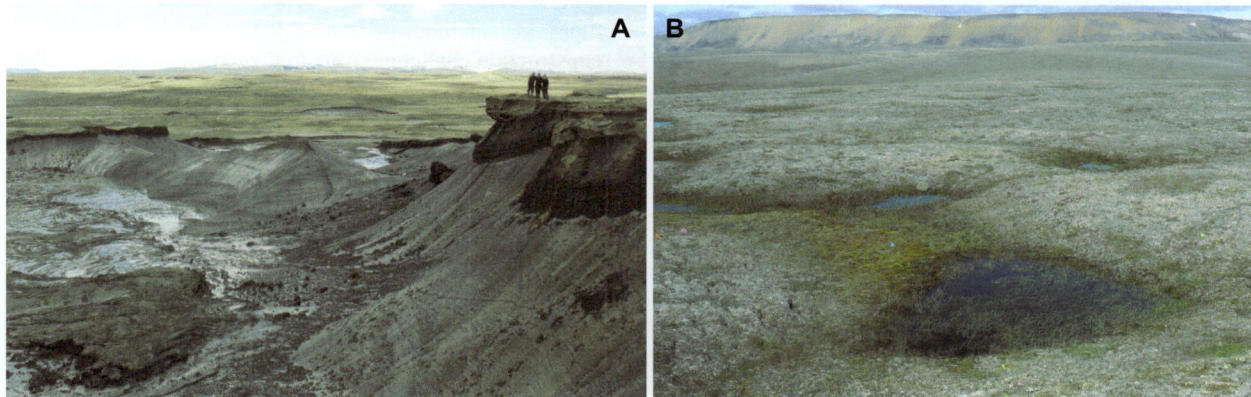

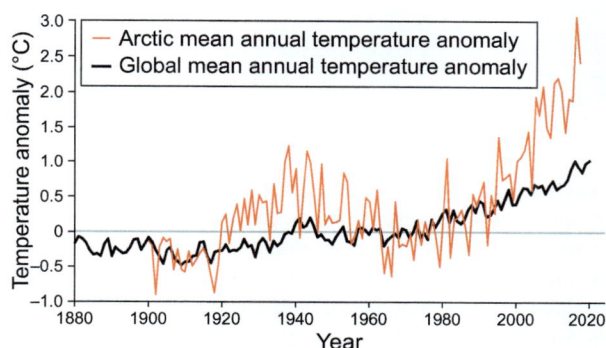

Figure 20.36 Changes (anomalies) in global air temperatures, indicating a warming to 2020 of ≈0.8 °C, relative to the 1951–1980 global average. In comparison, temperatures in the Arctic rose by ≈2.5°. Source: Compiled from various sources, including NASA.

Figure 20.37 Geomorphic phenomena caused by warming temperatures in polar deserts. **A.** Massive, segregated ice is exposed in a thaw slump, caused by thawing and loss of the active layer, Fosheim Peninsula, Ellesmere Island. Source: A. Rudy. **B.** Ground subsidence and ponding caused by melting of the top portions of ice wedges, Mould Bay, Prince Patrick Island, Nunavut, Canada. Source: L. Farquharson.

Figure 20.38 Thermokarst features representative of Arctic coastal lowlands. **A.** A wave-cut notch, exposing ice wedges along the Yukon (Canada) coast. Source: S. Wolfe (NRCan Photo ID 2021-254). **B.** Collapse of ice-wedge polygons along the Alaskan coast, caused by wave undercutting. Source: B. Jones. **C.** Flooding of low-center ice-wedge polygons along the Arctic Coastal Plain, Alaska, USA. Source: B. Jones. **D.** Thaw slumps exposing buried ice along the Yukon coast, Canada. Source: S. Wolfe (NRCan Photo ID 2021-255).

20.8 ENGINEERING CHALLENGES IN PERIGLACIAL ENVIRONMENTS

Engineering in periglacial environments faces the unique challenges of designing and constructing infrastructure on seasonally and perennially frozen ground. Roads and airfields can suffer from differential heave and settlement, and buildings anchored on permafrost can subside if the underlying permafrost thaws (**Fig. 20.39**). Construction on permafrost must take into consideration how the new infrastructure will affect the thermal and physical state of the ground. Where permafrost occurs, construction designs are typically implemented to either minimally impact and/or preserve the permafrost or, if possible, cool the ground and enhance permafrost conditions. A common practice is to undertake construction in winter to avoid thawing of the ground surface. Another

way to overcome these challenges involves the installation of a gravel layer beneath roads and airstrips to preserve the permafrost below. However, drainage systems must then be installed to prevent saturated conditions within the gravel, which can cause thawing of the permafrost. These issues are becoming even more important, given the changes that polar climates are currently undergoing.

Heat from buildings can also thaw the underlying permafrost. To avoid this problem, buildings are commonly constructed on stilts, or **piles**, secured into the frozen ground. The piles both support the building and prevent its heat from being conducted into the ground. Another approach involves insulating the foundation and/or keeping the underlying ground cool by circulating cold air through underground pipes. Mountainous periglacial landscapes present particular construction challenges, due to their steep slopes and frozen

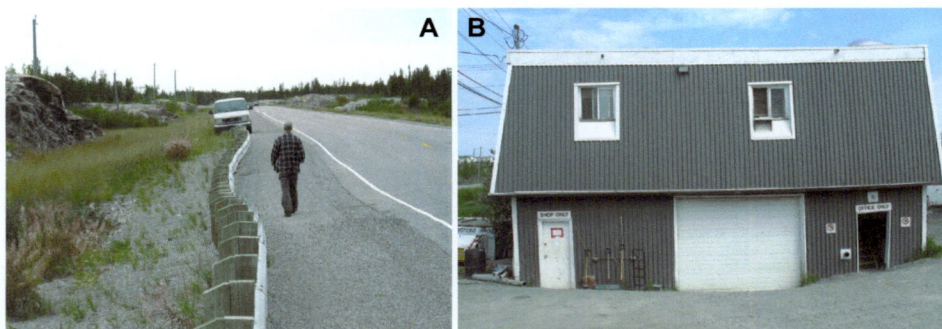

Figure 20.39 Examples of damage to infrastructure caused by thawing of permafrost. **A**. Lateral slumping of the road embankment and the shifting of guard rails caused by ground subsidence. **B**. Subsidence of a building caused by thawing of permafrost beneath its foundation, Yellowknife, Northwest Territories, Canada. Source: S. Wolfe (NRCan Photo IDs 2021-256, 257, 258).

Figure 20.40 Examples of engineering approaches used to accommodate the challenges of permafrost. **A.** Structures called thermopiles have been imbedded into permafrost to help support a building in Inuvik, Northwest Territories, Canada. Source: E. Hoeve. **B.** An airport terminal under construction in Iqaluit, Nunavut, Canada, showing insulation installed beneath the foundation footings, as well as thermosyphons (white pipes), which cool the ground. Source: E. Hoeve. **C.** A section of the above-ground Trans-Alaska pipeline. Notice the pipes (thermopiles) used to maintain frozen conditions in the ground below. Source: P. Morse (NRCan Photo ID 2021-259).

bedrock. Thus, piles in such setting must be installed deep into the bedrock.

Pipelines in periglacial landscapes also require a variety of unique engineering approaches (**Fig. 20.40C**). On the Trans-Alaska (crude oil) pipeline, for example, some sections are buried in coarse and well-drained soils that contain little ground ice. Other sections are built above ground in areas that contain more ground ice, and the supporting pilings are refrigerated.

REVIEW QUESTIONS

20.1 What is the difference between periglacial and permafrost environments?

20.2 What is the active layer and how does it change throughout the year?

20.3 How does the distribution, thickness, and continuity of permafrost change across low-relief, polar landscapes and within mountains?

20.4 What are the different types of taliks and how do they differ in their formation?

20.5 What are the different types of processes driven by frost action?

20.6 What are some of the various forms of ice discussed in this chapter?

20.7 How do freezing and frost heave act to move stones upward in soils?

20.8 What is solifluction and how does it operate?

20.9 What is patterned ground and what hypotheses have been put forward to explain its formation?

20.10 What are ice wedges and how do they form? How do they differ from sand wedges?

20.11 How are high-centered polygons formed?

20.12 How do palsas and pingos form? Describe how they differ.

20.13 Describe the formation of, and explain the similarities and differences between, blockfields and rock glaciers.

20.14 What is thermokarst, and what are the different processes and resulting landforms involved?

20.15 How is global climatic change affecting the periglacial landscape, and what is meant by polar amplification?

20.16 Describe the positive feedback relationship between global climatic warming and the thawing of permafrost.

20.17 What are some of the challenges involved with constructing infrastructure on permafrost? Explain some of the engineering designs that are used to accommodate these challenges.

FURTHER READING

Ballantyne, C. K. 2018. *Periglacial Geomorphology*. Wiley.

French, H. M. 2017. *The Periglacial Environment*. 4th ed. Wiley.

Jones, A., Stolbovoy, V., Tarnocai, C., Broll, G., Spaargaren, O., and Montanarella, L. (eds). 2010. *Soil Atlas of the Northern Circumpolar Region*. European Commission, Publications Office of the European Union.

Price, L. W. 1972. *The Periglacial Environment, Permafrost, and Man*. Association of American Geographers.

Smith, M. W and William, P. J. 1990. *The Frozen Earth: Fundamentals of Geocryology*. Cambridge University Press.

Washburn, A. L. 1980. *Geocryology: A Survey of Periglacial Processes and Environments*. Wiley.

21 Eolian Processes, Deposits, Landforms, and Systems

Stephen A. Wolfe and Randall Schaetzl

Eolian, (or **aeolian**) simply stated, refers to the wind. Eolian processes and landforms involve the erosion, transport, and deposition of sediment by wind. Of the major geomorphic agents (wind, water, ice, and gravity), wind is perhaps the one that is most readily observed, and the one that is often in play across Earth's surface. Wind is everywhere, and its effects are easy to find. That said, wind as a geomorphic agent is mainly felt on landscapes where vegetation cover is minimal and where sediment is exposed to the power of the wind. Even landscapes that are currently vegetated have beneath them a cover of sediments with ties to eolian systems – but from a different, usually drier and windier past. In this chapter, we will discuss the variety and importance of eolian processes, sediments, and landforms, in all manner of different places!

21.1 IMPORTANCE OF EOLIAN SYSTEMS

Wind is everywhere, on our planet and on many others (see Chapter 23). It **entrains** (picks up) and transports many different kinds and sizes of sediment, if not adequately protected by vegetation. Any activity that disturbs the surface, such as military maneuvers, road construction, recreation, deforestation, or agriculture, can change a stable landscape to one that is vulnerable to the wind.

Throughout history, eolian processes have affected the lives of people everywhere. Some winds are so predictable and commonplace, and affect society so much, that they have been given names. For example, **Harmattan winds** pick up and transport dust from the Sahara Desert, out onto

the Atlantic Ocean, carrying it as far as the Caribbean Sea. **Santa Ana winds** bring hot, dry air from the interior of the deserts of the southwestern United States, across California, to the Pacific Coast. These strong winds can fuel devastating, fast-moving brush fires. **Chinook**, or "snow-eater," winds blow downslope out of the Canadian Rocky Mountains onto the prairies, warming and drying as they descend, rapidly melting the winter snowpack.

Winds are, of course, a dominant force in geomorphology. The persistent effects of wind can form dramatic sand dunes and sand seas by transporting enormous amounts of sand, and by sculpting soil and rock into spectacular wind-abraded features. Many world regions, and particularly North America, contain stabilized dune fields and sand sheets, which are today vegetated. Droughts and disturbance by vehicles, and/or poor land-use practices, can destabilize these landscapes and allow this sediment to again be transported by the wind. These effects were clearly seen during the 1930s' **Dust Bowl** era in the western United States (**Fig. 21.1**). Much was learned from the Dust Bowl about adapting agricultural practices to suit semi-arid environments. However, it remains important to remember the lessons and practices that were learned through these mistakes, particularly in areas where aridity and drought are predicted to increase with climatic change.

Examples abound of sediment – mostly sandy sediment – that was once transported and deposited by the wind. Even today, strong winds can erode sediments, forming huge dust clouds and obscuring visibility (**Fig. 21.2**). Eolian sand can pile up and bury buildings, or even entire villages.

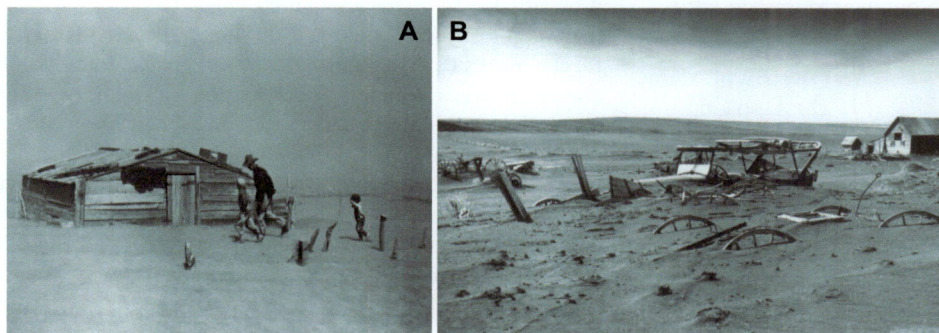

Figure 21.1 Classic images of the 1930s' Dust Bowl in the western United States. **A.** A farmer and his sons walking into a dust storm in Oklahoma. Source: Arthur Rothstein, Public domain, via Wikimedia Commons. **B.** Farm machinery buried by dust in South Dakota. Source: Sloan, Public domain, via Wikimedia Commons.

Figure 21.0 The sand dunes in White Sands National Monument, New Mexico, USA, are white because they are composed of gypsum sand. Source: Zack Frank / 500px / Getty Images.

Figure 21.2 A dust storm / dust cloud, sometimes called a **haboob**, approaches a US military base in Iraq on April 27, 2005. Source: US Marine Corps photo by Cpl. Alicia M. Garcia, Public domain, via Wikimedia Commons.

Sandstorms are a frequent occurrence in many desert regions, where there is little one can do to mediate the hazard; the most common strategy is to simply wait them out.

And yet, eolian sediment is not simply about sand. Smaller airborne particles are also carried by the wind. They can, at times, pose serious hazards to human health, depending on their size. Generally, smaller particles are more problematic. Scientists call such substances "particulate matter." Because size is important, airborne particles are categorized into groups (**Fig. 21.3**), using the designation PM (particulate matter). The largest particles that are worrisome are 2.5 to 10 μm (microns) in diameter, labelled as PM_{10}. Humans are often able to cough or sneeze out some of these coarser particles. Fine particles are < 2.5 μm in diameter ($PM_{2.5}$), whereas ultrafine particles are < 0.1 μm in diameter. The worry is that fine, and especially ultrafine, particles can become trapped in the lungs. The smallest of these are so minute that they can pass through the lungs and enter the bloodstream. However, no matter the size, particles that enter your lungs can harm your health. Although wind is the vector by which these particles are transported, the ultimate sources of much airborne particulate matter are automobiles, mining, and agricultural activities, among others.

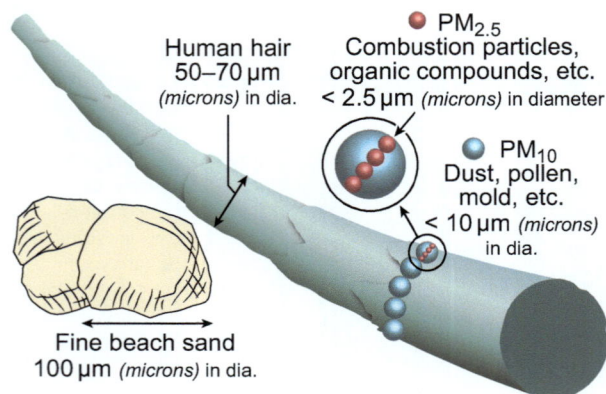

Figure 21.3 Illustration of the sizes of various airborne particles, as compared to a human hair.

21.2 MEASURING THE WIND

Wind develops due to differences in atmospheric pressure, as air flows from areas of high pressure to low pressure. Pressure differences commonly form between cool, dense air and warm, buoyant air. Wind is the result of the air moving across these pressure gradients to reduce the pressure differences. Where pressure gradients are greater, winds tend to blow faster.

Air, like water, is a fluid, and typically moves turbulently. The terms **wind speed** and **wind velocity** each refer to wind movement. Wind speed simply refers to how fast the air is moving, but not to its direction. Traditionally, cup **anemometers** are used to measure wind speed (**Fig. 21.4A**). Eolian transport can occur when wind speeds near the surface exceed some threshold of sediment transport. On a bare, dry, sandy surface, this threshold is about 6 m/s at a height of 10 m above the ground. **Wind vanes** are used to measure the wind direction. Wind velocity, however, refers to the simultaneous wind speed and direction. Modern anemometers are capable of measuring the wind velocity in three dimensions (**Fig. 21.4B**).

Wind speeds typically increase with height. Slower wind speeds near the ground are caused by resistance to flow from vegetation and the generally "rough" land surface. Because wind is always moving turbulently, its flow is often complex on slopes and across obstacles.

21.3 ENTRAINMENT AND TRANSPORT OF SEDIMENT BY WIND

Wind transports sediment by **saltation** (bouncing) of sand grains along the ground, and by **suspension** of smaller particles higher up in the atmosphere. These sediments can later accumulate as distinctly eolian landforms of varying shapes, sizes, and compositions – it's not just about sand dunes! True, eolian sediment is commonly sandy and piled high into sand dunes, but much eolian sediment is also silt-sized, blanketing vast areas of the landscape. Indeed, many landscapes today have a cover of eolian material, or were recently influenced by wind erosion and transport.

The **entrainment** (to "pick up" from the ground surface) and transport of sediment by wind is known as **deflation** when it occurs due to forces applied by wind. To move particles, the weight of the particle and the **interparticle attractive forces** must be overcome by the wind. These interparticle attractive forces, which include van der Waals forces, water adsorption forces, and electrostatic forces, cause particles to resist being pulled apart. For example, sands on a beach are held in place by both gravity and the forces acting upon them by interparticle attractive forces. When wet, the adhesive and cohesive forces of the water between the grains help to keep the sand particles together, all in one place. In order to lift and move this particle away from its spot on the beach, the force of the wind must be greater than the interparticle attractive forces, which hold the sand particles together.

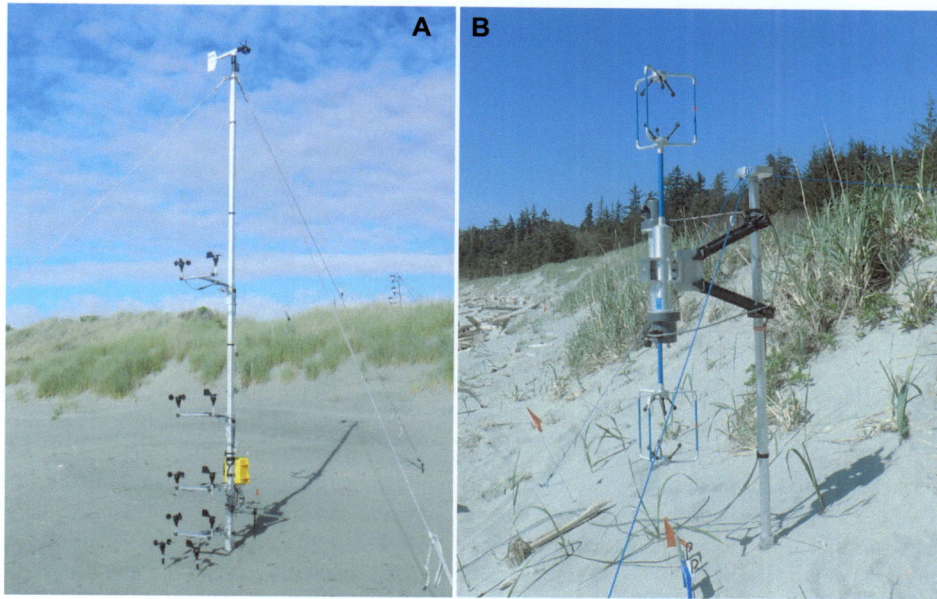

Figure 21.4 Anemometers as part of weather stations. **A.** An anemometer and wind vane tower array used to measure wind speeds and directions across different heights near Arcata, California, USA. **B.** An array of two 3-D sonic anemometers set up on a dune on Calvert Island, British Columbia, Canada. Source: I. Walker.

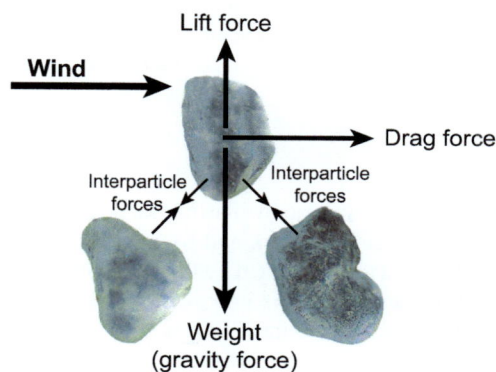

Figure 21.5 The forces acting on a sand particle as it is exposed to the wind.

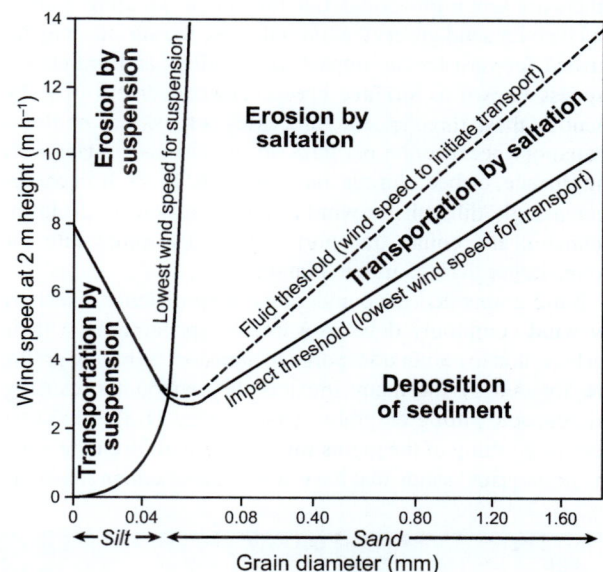

Figure 21.6 Threshold wind speeds required for the initiation of sediment transport by saltation and suspension, and for sustained transport in saltation. Source: After Chepil (1945), with permission from Wolters Kluwer Health, Inc.

What wind-generated forces are needed to get grains in transport? Winds apply both a **drag force** and a **lift force** to individual particles (**Fig. 21.5**). Similar to an airplane wing, faster winds create lower pressures across the top of a grain than it may have at the bottom. This differential pressure creates a lift force. With sufficient wind speed, the drag force "pushes" grains across the surface while the lift force causes them to rise off the surface. The wind speed required to lift sand grains is known as the **fluid threshold** (wind is a fluid). The fluid threshold is the direct result of grain movement caused by the force of the air only (**Fig. 21.6**).

As the grains are lifted into the air, they can also be moved by stronger winds higher above the ground (**Fig. 21.4A**). When the once-airborne grains fall and re-impact the surface, they impart additional energy to other grains, causing them to move or bounce off the surface. In this way, a mass of jumping and bouncing, or saltating, sand grains can be sustained, even at wind speeds that are less than the original speed needed to initially move them. This speed is known as the **impact threshold**, owing to the additional force of

the impacting sand grains (**Fig. 21.6**). When the wind speed drops below this threshold, all of the particles that were being carried by the wind fall to the ground; deposition occurs when sufficient wind strength to sustain transport no longer exists. So, in summary, when wind speeds increase above the fluid threshold, the force of the wind is sufficient to cause net removal of sediment and hence, erosion.

Saltation – the repeated jumping and bouncing motion of grains across the ground surface – is the most common form of eolian sand transport. About 75–80% of wind transport occurs by saltation. By definition, sand grains range from

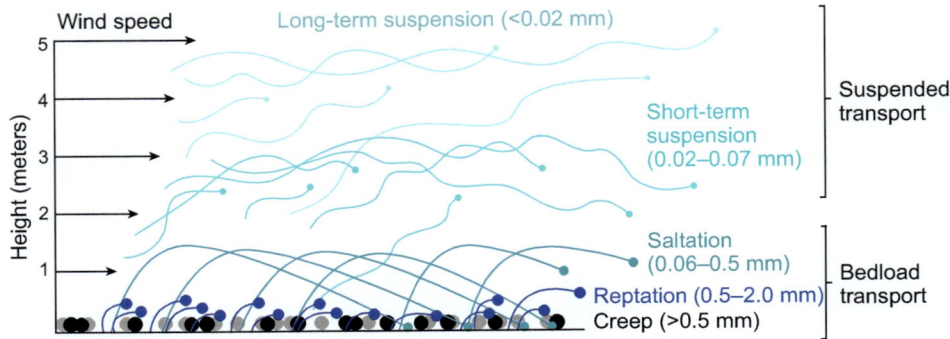

Figure 21.7 The mechanisms of sediment transport by wind involving creep, reptation, saltation, and suspension, reaching higher into the wind speed profile. Source: Used with permission of Elsevier, from Pye (1987); permission conveyed through Copyright Clearance Center, Inc.

0.05 to 2.0 mm in diameter; most eolian sands range between 0.07 and 0.5 mm in diameter. The impact of saltating sand grains on the surface transfers energy to surrounding grains on the surface, causing them to be ejected upward and transported downwind (**Fig. 21.7**). Several particles can be sent into transport from a single grain impact, producing a rapid cascade of sediments – a "saltation cloud." Some sand grains make only single short hops or **reptations**, whereas others are sent high enough that they begin saltating. Larger and heavier sand grains that do not leave the surface may be pushed forward by the impact of saltating sand grains – a process known as **surface creep**. The movement of sediment by these three specific processes is known as **bedload transport** because the particles are in various contact with the surface, or bed, during transport (**Fig. 21.7**). It becomes increasingly difficult for wind to move particles > 2 mm in diameter, and eolian transport is typically impossible for grains larger than ≈5 mm in diameter.

Sand grains that have undergone long-distance transport by wind commonly develop a frosted appearance on their surface, due to eolian transport and abrasion. Saltating grains are abraded by the many impacts they encounter, causing microscopic pitting on grain surfaces (**Fig. 21.8**). The characteristic pitting of the grains caused by saltation can be used to "fingerprint" sands that have some sort of eolian transport

in their history, so as to distinguish them from sands that have weathered out of sandstone, or which have been produced by glacial outwash, for example.

In addition to bedload transport, smaller particles like very fine sands and silts can be entrained by wind via **suspension** (**Figs. 21.2, 21.9**). Suspended particles can be carried for long distances if the winds are strong, and especially if they are very turbulent. Silt and clay particles are so small that they can be transported in long-term suspension for hundreds of kilometers through the air. Even sand grains can be carried in short-term suspension (or **modified saltation**) for several hundred meters.

21.4 CONTROLS ON EOLIAN TRANSPORT

Wind can erode; it does so by blasting sand grains into sediment, by lifting sediment up, and by transporting it downwind. Eolian erosion is affected by factors related to both the wind and the sediment. It is important to consider the specific conditions of each of these factors when considering the potential for eolian erosion and transport, and the resulting eolian landforms.

Factors related to the wind (its **regime**) include speed, duration, and direction. As we have seen, wind must first be of sufficient speed to move sediment particles. Most

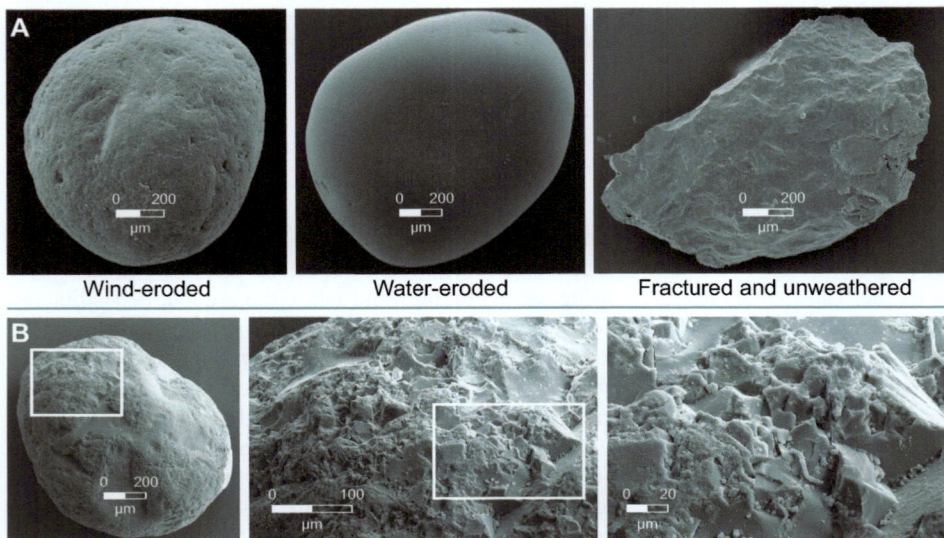

Figure 21.8 Magnified images of sand grains. **A.** Scanning electron microscope (SEM) images of sand grains that have been transported by wind and water, as compared to an unweathered grain that has not been transported. **B.** Progressively more enlarged SEM images of sand grains with microscopic pitting due to impacts, presumably formed during saltation. Source: B. Woronko.

Figure 21.9 The Sahara Desert is a major global dust source. Shown here is a Terra-MODIS image of plumes of silt-rich dust coming off the Sahara Desert on June 24, 2009. Source: NASA Earth Observatory.

locations on Earth encounter wind speeds sufficient for sediment transport during at least a portion of the year. Locations where the wind speeds are usually low, or do not occur often, are considered to be **transport limited**. In other words, transportable sediment is available but the means of transport, in this case, the wind, is limiting.

Sediment eroded by wind is deposited downwind, in areas where the wind slows. In addition, shifts in climate over time cause some areas that were erodable in the past to be less so today. Wind direction is also an important component of the **wind regime**. For example, seasonal and annual variations in wind directions and speed play an important role in determining the shape and size of sand dunes (see below).

Other factors associated with eolian landscapes may be more closely related to the sediment regime than to the characteristics of the wind. For eolian processes to really affect the landscape, they require sediment of appropriate size and amount, such that it is capable of being transported by wind. This factor is referred to as *sediment supply* – of appropriate size such that it can be moved by wind. Sand, silt, and clay particles can all be transported by wind and are common almost everywhere on Earth. River deposits, beaches, and deserts commonly have abundant supplies of sand. Recall that wind cannot transport gravel- or larger-sized particles, and wind transport is ineffective in terrain without finer materials, for example, a landscape of bare bedrock. Such landscapes are referred to as **supply limited**. On other landscapes, wind and sediment may be sufficient for transport but the sediment cannot be picked up by the wind – the surface is too wet, cemented together by chemical agents or clay, or protected by vegetation. As a result, the sediment – even though it is of appropriate size – cannot be deflated and moved by the wind. Most supply limited landscapes like these are covered with vegetation, for example, forests, tundra, and grasslands. Others may be rocky or wet, or have salt crusts or a cover of snow and ice. As you can see, **sediment availability** revolves around whether the sediment is either freely available to the wind or protected from it. Hot dry deserts, like the Sahara, lack vegetation, making most of the sediment on the landscape available for transport. Dune fields that are today covered with vegetation (due, usually, to a climate change) exemplify supply limited landscapes.

Vegetation cover is the most common control on sediment availability for transport by wind, whether it occurs naturally or it is deliberately planted to reduce sediment loss by wind, as in a "windbreak." Vegetation plays several important roles here (**Fig. 21.10**). First, it covers and protects the surface, that is, roots and fallen vegetation "hold onto" sediment that would otherwise be exposed to the wind. Second, by protruding upward into the wind profile, vegetation reduces wind speeds near the surface, forming a **boundary layer** of slower wind speeds. Third, vegetation traps particles in transport, causing them to be deposited, either within the vegetation cover or immediately downwind. In deserts and other sandy regions, the surface cover of rocks performs many of the same roles as vegetation (**Fig. 21.11**). Soil moisture and various types of surface crusts also decrease sediment availability by increasing interparticle adhesive forces, making them more difficult to be entrained. For this reason, shading by vegetation cover not only lowers wind speeds, but also helps to retain surface moisture, preventing soil particles from

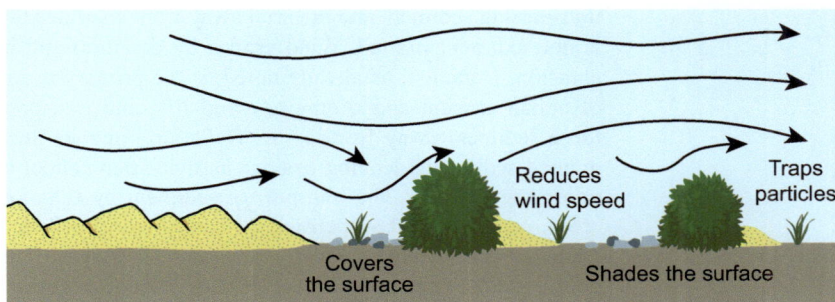

Figure 21.10 Diagram showing how vegetation and lag surfaces can control eolian erosion. Source: Adapted from Stephen A. Wolfe, William G. Nickling, Progress in Physical Geography: An International Review of Geographical Work in the Natural and Environmental Sciences. Copyright © 1993 by (Sage Publications). Reprinted by Permission of Sage Publications.

Figure 21.11 Rocks, surface crusts, and vegetation can all act to reduce the availability of surface sediment for deflation. **A.** A rocky desert soil developing a surface gravel lag. Tape scale in cm. **B.** Surface salt crusts and sparse vegetation in a semi-arid landscape (Wyoming, USA). Source: R. Schaetzl.

becoming loose and dry, and hence, more easily entrained. In fact, the ground surface does not require a thick cover of vegetation to be protected from wind erosion. Sparse vegetation, including desert shrubs, dune grasses, and even dead standing stubble on farm fields, can act as an effective means of controlling wind erosion.

21.5 EOLIAN PROCESSES FROM SOURCE TO SINK

Eolian features and landforms develop from the processes of erosion, transport, and deposition (**Fig. 21.12**). That is, eolian landforms can have both erosional and depositional histories. Source areas for eolian sediment are those where erosion occurs, that is, where these sediments are available for deflation. "Sinks" are areas where eolian sediments are deposited, ending their transportation pathway.

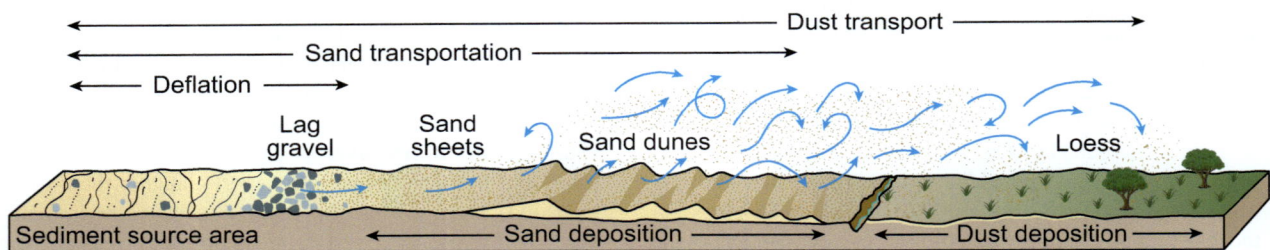

In source areas, sediment is freely available for transport. Eolian sources may include landscapes with particularly windy regimes, or sandy-silty regions in deserts with little vegetation cover. Dry lake beds, beaches, sandy river valleys, cleared lands, or areas undergoing drought can all act as source areas for wind erosion. Sediments removed from these areas are transported by either saltation or in suspension, until they are deposited in areas where the wind can no longer transport them. In some cases, the **transport pathway**, which is the route between the erosional source area and depositional sink, can be very short. For example, wide, sandy beaches are sources for nearby coastal dunes (**Fig. 21.13**). In other cases, the transport pathway can be very long, as in some deserts and sand seas, which are hundreds of kilometers long. Some winds, like the **Harmattan**, also lift dust from the Sahara Desert into suspension high

Figure 21.12 The eolian system, showing the sediment source area and the various transport pathways and depositional areas for sand and dust.

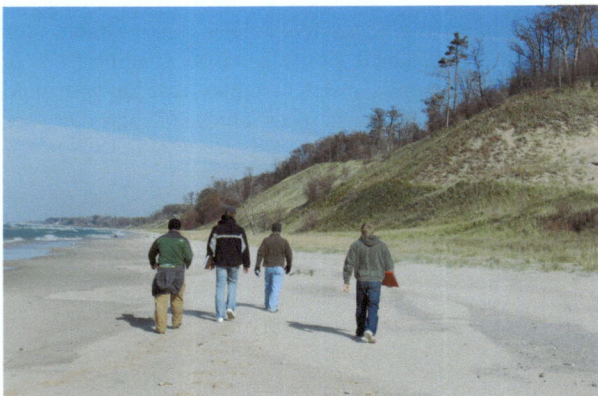

Figure 21.13 Beach sands along the shore of Lake Michigan (USA) provide an immediate source area for nearby coastal dunes. Source: R. Schaetzl.

into the atmosphere, carrying it across continents and oceans (**Fig. 21.9**).

21.6 EOLIAN EROSION

Eolian erosion occurs where wind strengths are capable of transporting sediment and material away from a surface that is not sediment limited. Wind erodes by deflation and by abrasion. Deflation, as already noted, is the primary means of eolian erosion and occurs as wind lifts and transports loose sediment away from a site. Deflation removes finer-grained sediments, leaving heavier particles that cannot be transported behind. The transport of sediment by saltation, creep, reptation, and suspension all contribute to deflation. Abrasion is a secondary effect of deflation, caused as transported grains impact the surface or other particles, breaking the material into small pieces or flaking off parts of the grains, many of which are then deflated and enter the transportation pathway.

Some areas may lose smaller grains to wind erosion (deflation) until an **erosional lag** of larger particles develops on the surface (**Fig. 21.14A**). As an erosional lag forms, the surface becomes supply limited (**Fig. 21.14B**). Typically, lag deposits are only one or two stones thick, but that is enough to protect the underlying sediment from further deflation. However, it is now believed that wind erosion is greatly diminished even when only about half of the surface is covered with rocks. Despite the remaining bare surface being "unprotected" from the wind, the lag of coarse materials effectively protects the intervening finer sediments from further erosion. And so, like a scattered vegetation cover, a sparse cover of pebbles and stones may be quite effective in protecting a surface from wind erosion.

A **desert pavement** is a dense cover of rocks on the surface, usually in a desert and often containing a layer of fine-grained sediment underneath (**Fig. 21.15A**). Desert pavements often resemble lag deposits formed by deflation, such as the one shown in **Fig. 21.14B**. It had long been thought that, because deserts are so windy and often have only sparse vegetation, desert pavements formed mainly by deflation. But this thinking has recently changed. Let's examine the evidence.

Many desert pavements are so well formed that the stones almost appear to interlock, like a jigsaw puzzle (**Fig. 21.15A**). Little bare soil is exposed at the surface, and like erosional lag deposits, the pavement is only one stone layer thick. But unlike lag deposits formed by deflation, beneath many desert pavements lies a stone-poor soil layer that typically has a

Figure 21.14 Eolian surface lag deposits. **A.** Schematic diagram showing how a surface lag deposit can be formed by deflation. **B.** A surface lag of pebbles and stones, formed by deflation and protecting the bare sand from further erosion, near Great Slave Lake, Northwest Territories, Canada. Some stones have become faceted by abrasion from saltating sand grains to form ventifacts. Source: S. Wolfe.

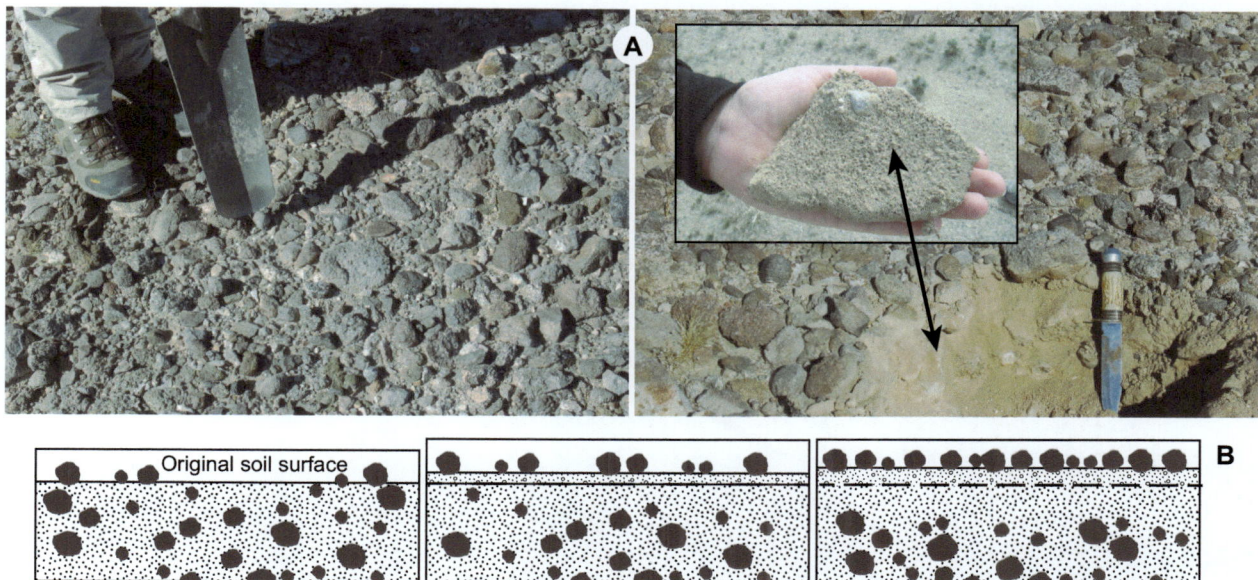

Figure 21.15 Desert pavement. **A.** Typical desert pavements in the Anza-Borrego Desert of southern California, USA. Note that the stones are typically only one layer thick. **B.** Conceptual model showing how eolian dust, added to a stony surface, eventually works its way beneath stones to form a stone-poor, vesicular horizon, below a desert pavement. Source: R. Schaetzl.

vesicular fabric with many bubble-like pores. When walking on such surfaces, the gravel that comprises the desert pavement sinks a few millimeters into the soft sediment below. The key to the formation of desert pavements is the vesicular layer – if desert pavements had formed by deflation alone, stones should be abundant immediately below, as well as on top of, the surface (Fig. 21.14A). In desert pavements, however, this is not the case (Fig. 21.15A).

Geomorphologists now believe that desert pavements form by (1) the accumulation of rock clasts on surfaces, along with (2) the detachment and slow uplifting of those clasts as fine, usually eolian, sediments infiltrate between and below them (Fig. 21.15B). For this process to work, a source of small rocks as well as inputs of eolian dust must both be present. Rough surface topography and large interstices between rocks facilitate the trapping and accumulation of eolian dust. Rains wash the trapped dust into cracks that occur between and below the surface rocks. Slowly then, a stone-free, vesicular horizon forms below a layer of rocks at the surface. In essence, the rocks on the surface are actually *lifted* by the dust that has settled into the fractures below them, assisted by clays in the dust that expand when wet. Dust that lands on the surface but which is not immediately incorporated below the rocks is susceptible to deflation. Once formed, the desert pavement protects the underlying, accumulated dust from further erosion, while at the same time the rocks can continue to trap more dust.

Vesicular pores and tubules dominate the stone-poor sediment below the desert pavement. They form due to the repeated wetting and drying of the soil. How does this process work? During and after a rainfall event, the surface of the soil seals by puddling, and then, as the wetting front advances downward, gas pressures are elevated within the sediment, forming vesicles or bubbles. In short, desert

pavements "are born and maintained at the surface." They need not evolve from rocks scattered below – a key component of any erosional model such as in Fig. 21.14A.

In the deserts of the Middle East, where desert pavements are common and much of the landscape is wind-swept and rocky, the term **reg** is used to describe a surface of closely packed, interlocking rocks. Contrast this with the term **erg**, which is a "sand sea," or an area of widespread sand and dunes (see below). Regs can form either by eolian erosion or by the pavement-forming process described above.

Deflation is also common in sandy locations where the vegetation cover is sparse, or where disturbance has reduced this cover in localized areas. Vegetated coastal dunes, for example, are prone to localized or extensive deflation, forming areas of open-sand called **blowouts** (Fig. 21.16). Blowouts can form due to excessive trampling or off-road vehicle traffic, which disturbs the vegetation that would normally stabilize the surface. Winds coming off the nearby water body are funneled into and through the blowout, further enlarging it and maintaining it as an area of localized strong winds and erosion. For this reason, boardwalks are commonly placed across coastal dunes to prevent erosion and, hence, slow the initiation of new blowouts.

Rocks, as well as soils cemented by various substances, cannot be easily eroded by deflation. Nonetheless, they are still susceptible to eolian abrasion by saltating sand grains. Quartz sand grains, in particular, are very effective at abrading other surfaces because they are quite hard and difficult to break by small impacts alone. Impacts from saltating sands are an effective way of removing small amounts of material from rock surfaces, producing chip marks and even grooves on both the rocks and sand grains (Fig. 21.8). Fine-grained silt and clay particles may actually smooth and polish rock surfaces, but are not considered to be as effective at abrasion

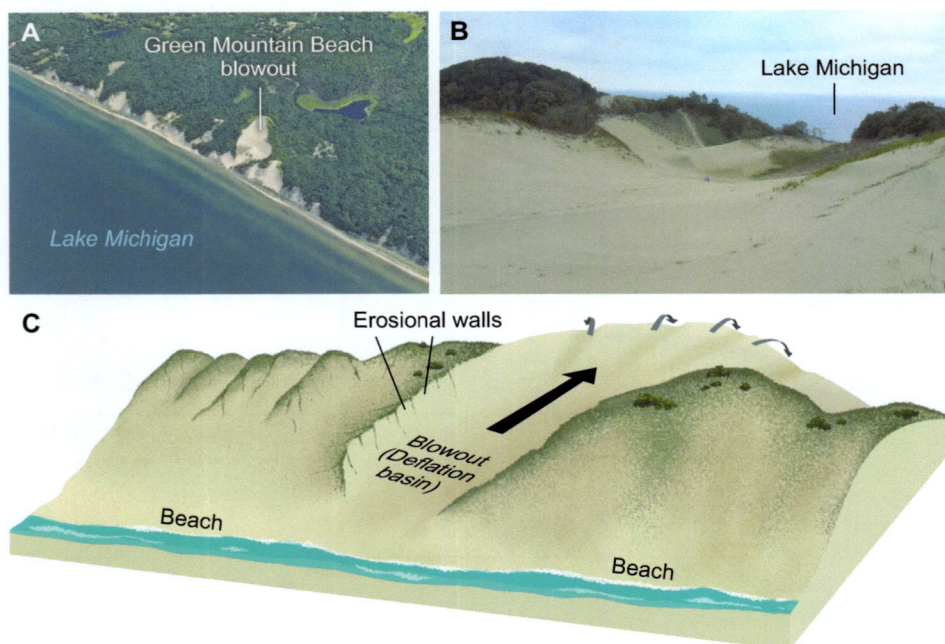

Figure 21.16 Blowouts. **A.** An oblique (NASA) view of the forested dunes and bare-sand blowout areas on the coastal dunes of Lake Michigan, Michigan, USA. Source: NASA. **B.** A view of the Green Mountain beach blowout, looking upwind, toward Lake Michigan. Source: R. Schaetzl. **C.** Schematic illustration of the formation of a blowout within a suite of coastal dunes. Source: Used with permission of Elsevier, from Hesp (2002); permission conveyed through Copyright Clearance Center, Inc.

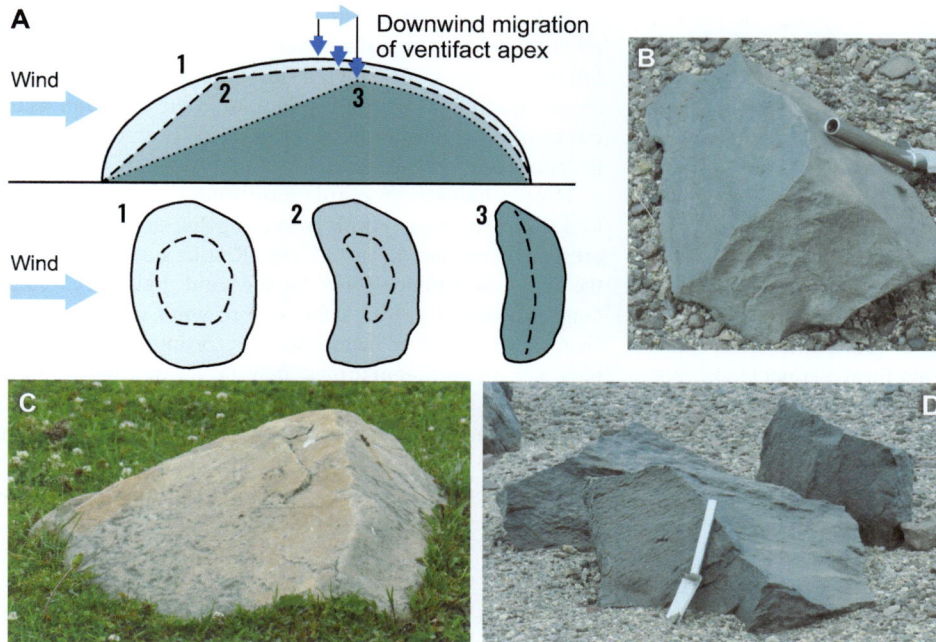

Figure 21.17 Ventifacts. **A.** Schematic diagram showing how ventifacts may form over time, in cross-section and plan view. Note how the planar surface forms on the windward side of the ventifact, and how the apex of this surface migrates downwind over time. Source: After Knight (2019), with permission from John Wiley and Sons. **B., C.,** and **D.** Examples of ventifacts; note the well-faceted edges. Source: R. Schaetzl.

Figure 21.18 Yardangs and other similar wind-eroded and sculpted bedrock forms. **A.** This wind-sculpted rock in Death Valley National Park, California (USA) illustrates how wind erosion by saltating sand is focused in the lowest meter of the atmosphere. Source: R. Schaetzl. **B.** and **C.** The wind-sculpted landscapes of the Lut Desert in Iran illustrate a variety of yardangs and similar landforms. Source: (B) Ninara from Helsinki, Finland, CC BY 2.0, via Wikimedia Commons; (C) J. Radebaugh.

as are sands. Intricate grooves, known as flutes, are formed in rocks by intensive eolian abrasion, and can indicate the dominant direction of strong winds.

Ventifacts are rocks that have been continually impacted by saltating sand grains, causing the abraded surface to become smooth and faceted (Fig. 21.17). Ventifacts commonly have one or more faceted faces, the orientation of which is sometimes an indication of the prevailing wind direction. They can range in size from small faceted grains to rocks > 3 m in height, and are most common in areas of high sand supply with little or no vegetation cover. Ventifacts are indicative of strong winds, either currently or in the past.

Yardangs (Turkish *yar*, "steep bank") are another type of wind-eroded landform, again caused by abrasion of bedrock or fine-grained consolidated sediments such as silts and clays. They are usually eroded into soft bedrock, implying that one of their formative processes is the deflation of

surrounding surface materials (Fig. 21.18). This is in contrast with ventifacts, which are typically rocks and boulders that are simply laying on the surface. Most yardangs have a streamlined form, paralleling the dominant wind direction. Some have the form of an inverted boat hull, being more blunt-shaped on the windward face and having a tapered, elongated lee side, or tail. They may range from about 1 to 200 m in height and can be several kilometers long. As is obvious in Fig. 21.18, they often form in clusters.

21.7 EOLIAN DEPOSITION

Taken en masse, eolian deposits consist of sediments that range in size from wind-blown sand to dust. Wind-blown sand typically forms landforms of various types and sizes, called **sand dunes**. Because dunes are formed by transport of sand over distances of meters to kilometers, the wind has usually had ample opportunity to texturally *sort* the sediment. Sorted sediment has a narrow range of particle sizes; dune sands are typically well-sorted and range from 0.07 to 0.5 mm in diameter. Indeed, the well-sorted nature of sands is a giveaway to sedimentologists and geomorphologists that what they are seeing has an eolian origin, whether they are examining sandstones or dunes.

Sometimes, sand deposits simply (thinly) cover the landscape; these **sheet sands** and **cover sands** do not form distinct landforms. These deposits are often poorly sorted, meaning that they are not just sand, but also contain silts or maybe even small gravels. Silt and dust that has been transported and redeposited is known as **loess**. Like sheet sands and cover sands, loess also does not form distinct landforms.

21.7.1 Ripples

Eolian sand landforms can take on a variety of shapes and sizes. Ripples – perhaps the smallest eolian feature – are also the most characteristic feature formed in wind-transported sands (**Fig. 21.19**). The transport processes of saltation, reptation, and creep are very active on rippled sand surfaces. Ripples form as small crests and troughs in asymmetric wave-like patterns that are aligned perpendicularly to the direction of sand transport. The windward face of a ripple has a slope of about 10° toward the wind; saltating sand grains heavily bombard this side of the ripple. In contrast, the leeward sides of ripples slope at an angle of about 30°

away from the wind and are thus more sheltered. Sand grains on the windward side that reach the crest of the ripple may fall or slide down the leeward side, where they will rest. The crests and troughs of ripples migrate in the direction of the transporting grains, as the windward side is eroded and sand is deposited on the lee side.

Ripple heights, or **ripple amplitudes**, are typically 0.5 to 1.0 cm, whereas **ripple wavelengths**, or distance from crest to crest, are ≈5 to 15 cm. Ripple size is related to the saltation distance (how far the sand grains are "bouncing"), which is in turn a function of wind speed and sand size, as well as other factors such as sorting. Very large ripples, known as **megaripples**, may be formed where a distinctive mix of coarse sand and fine sand is present. In these instances, the finer sands are transported primarily by saltation and the coarser sands by creep. Megaripples typically have wavelengths of up to 25 cm, often with a lag of coarse sand on the ripple crest (**Fig. 21.19B**).

21.7.2 Sand Dunes: The Prototypical Eolian Landform

Sand dunes form as wind-blown sands accumulate into distinct landforms. Dunes are classified based on their shape and mode of formation, as well as their current state of stability – active, stabilized, or partially stabilized. Although they are viewed as deposits, active sand dunes can also be thought of as part of a sediment transport pathway

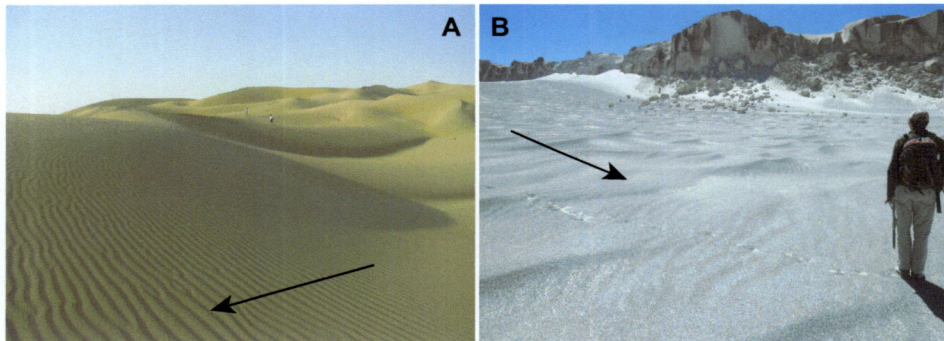

Figure 21.19 Sand ripples, with the wind direction that formed them shown by arrows. **A.** Ripples on large sand dunes near the Liwa Oasis, UAE. Source: Peter Dowley from Dubai, United Arab Emirates, CC BY 2.0, via Wikimedia Commons. **B.** Megaripples in Abra Pomez, Argentina. These megaripples are up to 2 m in height, and grains of up to several cm in diameter are found on the crests. Note the sinuous crestlines. Source: T. Gough.

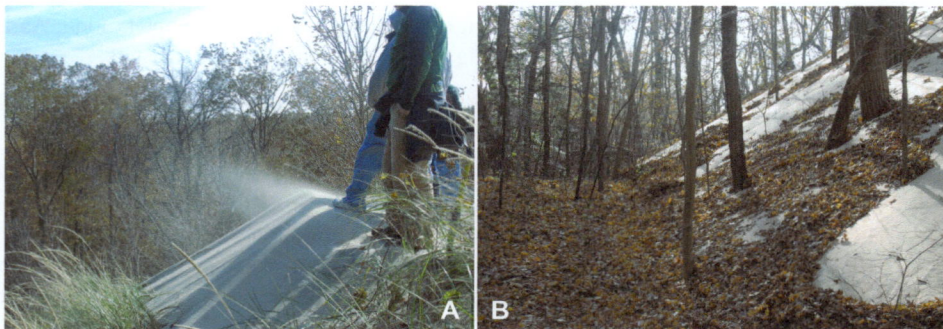

Figure 21.20 Sands in eolian transport. **A.** Sands typically blow and saltate over the crests of dunes, and as they do, they slide and roll down the slip face, (**B**) burying anything in their path. Source: R. Schaetzl.

(Fig. 21.20A) and, like ripples, sand is transported across them, allowing the dunes to migrate over time. Migrating dunes can bury obstacles that are in their way. Buildings, roadways, forests, and agricultural fields that are in the transport pathway of dunes may get buried (Fig. 21.20B). Even dunes that are partially stabilized or appear stationary may continue to have sand transported across their surfaces.

Stable dunes are currently inactive. They may have become inactive because of increased vegetation cover (usually due to climate change toward wetter conditions) and/or lower wind speeds. Thus, stable dunes provide indications about past conditions that were different than today.

Like ripples, most sand dunes have characteristic profiles in cross-section, with a windward slope, crest, and lee slope (or **slip face**) (Fig. 21.21A, B). Wind accelerates across the windward side of the dune, allowing sand grains to be transported upslope, mainly by saltation, toward the crest. The saltating sand may form ripples on this (windward) slope, which is commonly an angle of about 10° to 15°. Wind speeds typically reach their maximum values near the highest point (crest) of the dune, beyond which it decelerates, allowing the sands to be deposited. Most of the saltating sand falls just beyond the dune crest and then slides and avalanches down the slip face, which typically maintains an angle of between 30° and 35° (Figs. 21.20, 21.21B), also known as the **angle of repose** (see Chapter 8), which is the maximum angle that can be maintained by the dry, loose sand grains.

When wind speeds are strong, wind accelerating up the windward slope can, beyond the crest, separate into two components – a process known as **flow separation** (Fig. 21.21A). Winds considerably above the dune may continue moving rapidly forward, potentially taking with them some sand that may travel in a type of suspension well beyond the dune (Fig. 21.20A). But simultaneously, winds nearer the dune descend beyond the leeward slope (slip face), swirling and even reversing direction. Under these circumstances, saltating sand grains entering the fast-flowing current are sent long distances beyond the dune, whereas sand grains nearer

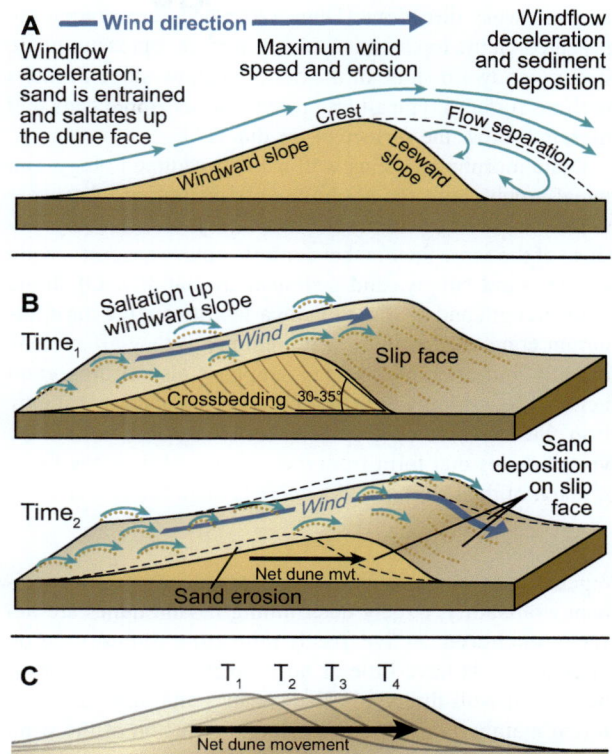

Figure 21.21 Wind and sediment transport characteristics of typical sand dunes, showing how they form and migrate over time, as they develop their classical stratigraphy.

the dune surface fall and slide down the slip face. Together, these processes cause the dune to migrate downwind, while all-the-time maintaining a similar form (Fig. 21.21B, C).

As a dune migrates by sediment being deposited on its slip face, it usually acquires a sedimentology that includes layering known as **cross-bedding** (Figs. 21.21B, 21.22). Cross-bedding is a hallmark indicator of sand dunes and persists into the bedrock record. The steeply sloping deposits that represent former slip faces also provide indications of

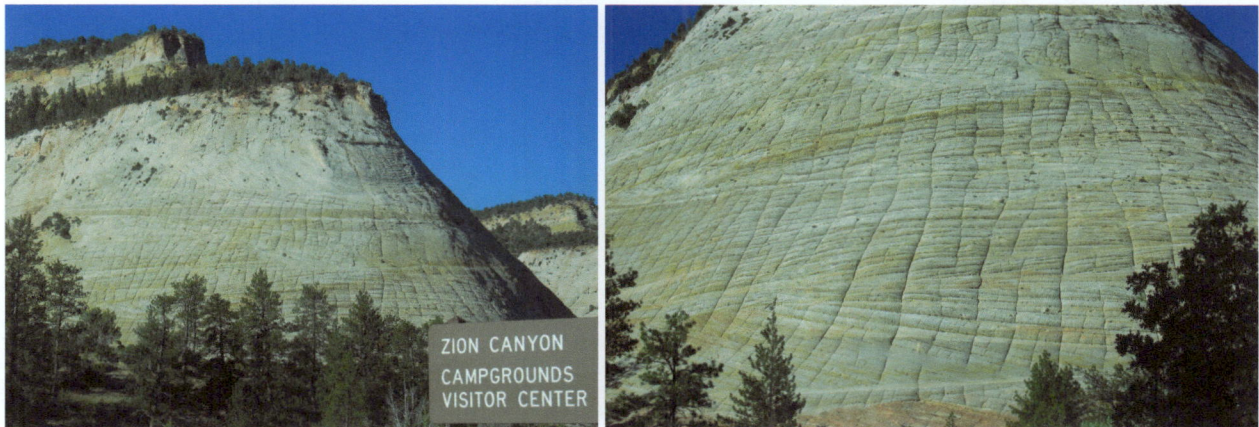

Figure 21.22 Cross-bedded sandstones in Zion National Park, Utah, indicate that these rocks formed as sand dunes in the geologic past. Source: R. Schaetzl.

the past wind directions. Dune cross-beds are commonly truncated on the top by an erosional surface representing the former windward slope of the dune. Shallow-slope deposits on these surfaces typically represent former ripple beds that migrated across the surface of the dunes.

Dune morphologies, as well as their sedimentology, can reveal much about the environmental conditions that were present when they formed. Their morphologies vary because of the same factors that control eolian transport – wind regime, sand supply, and sediment availability. Of these, wind strength and its directional variability are the most important components of the wind regime, as they affect dune mobility and orientation. Sand supply is also an important factor in determining both the size and type of dune that may form. Various models have been developed to illustrate the type of dunes that form from these factors; this is the focus of much of the text that follows.

21.7.3 Freely Migrating (Active) Dunes

Vegetation is the most important component affecting sediment availability, largely determining if sand dunes are **active** or **anchored**. Active, **freely migrating dunes** occur on landscapes that have little or no vegetation cover and thus, they move with the wind. Their shapes and sizes are controlled mainly by (1) the strength and direction of the wind and (2) the sand supply (**Fig. 21.23, Table 21.1**). Anchored dunes are no longer moving.

Several types of freely migrating dunes are described by geomorphologists, such as **barchan** and **transverse dunes**, as well as **reversing, linear,** and **star dunes** (**Fig. 21.24**). Because these dunes are actively migrating, studies of these types of dunes provide key insights into contemporary eolian processes. Then, what we learn about these types of dunes can be translated to areas of anchored or stabilized dunes, to help us better understand past environments.

Most active dunes are simply bare sand. As the amount of vegetation on the landscape increases, however, dune forms change and the dunes themselves become more stable and less prone to downwind movement.

Small, individual dunes may form in areas of low sand supply, whereas larger, interconnected dunes tend to form as the sand supply increases. Dunes that form with winds from a generally uniform direction, such as barchan and transverse dunes, usually have only one prominent slip face. Wind directional variability adds another component to the dune form by causing greater complexity in sediment transport directions. As a result, these dunes can have two or more prominent slip faces, and more complex morphologies.

The simplest (and fastest-moving) of the freely migrating dunes is the **barchan dune** (**Fig. 21.25**). These dunes develop in wind regimes with a single (unimodal) wind direction, or only a narrow range of wind directions, and where the sand supply is low compared to the ability of the wind to transport sediment (**Table 21.1**). Typically, barchans develop where vegetation cover is minimal and therefore, only minimally affects the local wind regime, which explains why they are so common on Mars (**Fig. 21.25A, B**). Because of the limited sand supply where they form, barchans often migrate in isolation (**Fig. 21.25**).

Barchan dunes have characteristic crescentic shapes, with their "horns" or arms pointing *downwind*. Ripples are common on their upwind slopes. The slip face of the dune – on the inside of the crescent – is concave, meaning that it curves inward towards the dune center. Most barchan dunes range in height between 3 and 10 m, but can be as tall as 25 m (**Table 21.1**). Very large barchans, known as **mega barchans**, can exceed 50 m in height and can even have small dunes superimposed on them. Barchans migrate freely, typically between 1 to 10 m/ yr (**Fig. 21.25C**), with smaller dunes moving much faster than larger dunes. Barchan dunes are found in many desert areas, on

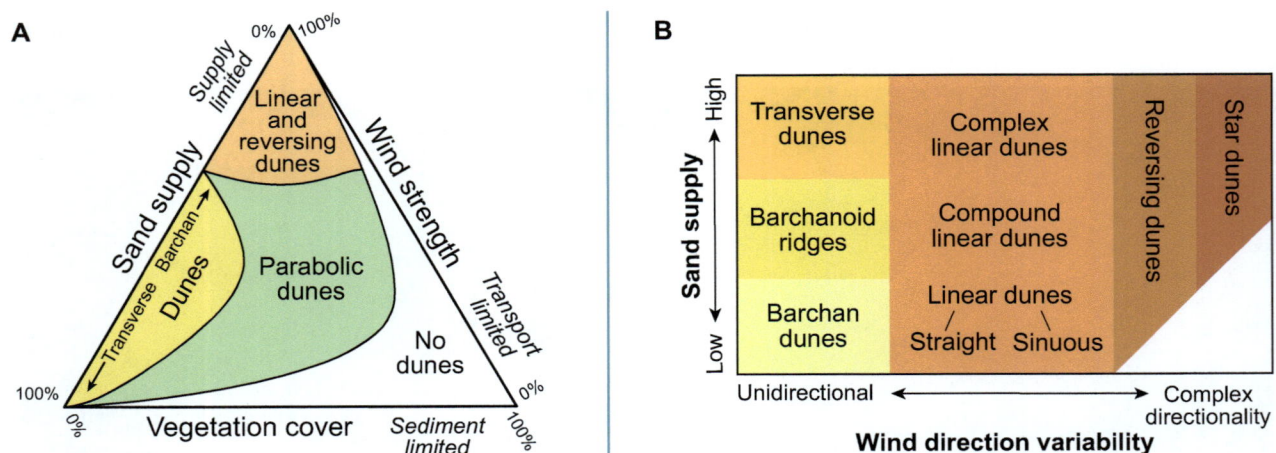

Figure 21.23 Illustrations showing how the morphology of sand dunes is controlled by factors such as (**A**) wind regime, sand supply, and sediment availability, and (**B**) sand supply and wind direction variability. Source: (A) After Hack (1941), reprinted by permission of the publisher (Taylor & Francis Ltd, http://www.tandfonline.com); (B) After Lancaster (1995); reprinted by permission of the publisher (Taylor & Francis Ltd, http://www.tandfonline.com).

Table 21.1 Characteristics, wind directionality, and sediment regimes of the major types of freely migrating dunes

Dune type	Height (m)	Wind directionality	Mode of change	Dune shape	Conditions
Barchan	3–25	Uniform	Migration	Crescentic	Low sand supply relative to wind strength
Transverse	8–30	Uniform	Migration	Straight to sinuous	High sand supply relative to wind strength
Linear	5–100	Bi-modal seasonal winds, 90° to <180° apart	Extending and low lateral migration	Straight to sinuous	Low sand supply relative to wind strength
Reversing	2–20	Bi-modal but in opposite directions	Semi-stationary	Straight to sinuous	Moderate sand supply relative to wind strength
Star	50–400	Several directions, based on season	Stationary	Pyramidal or star	Very high sand supply

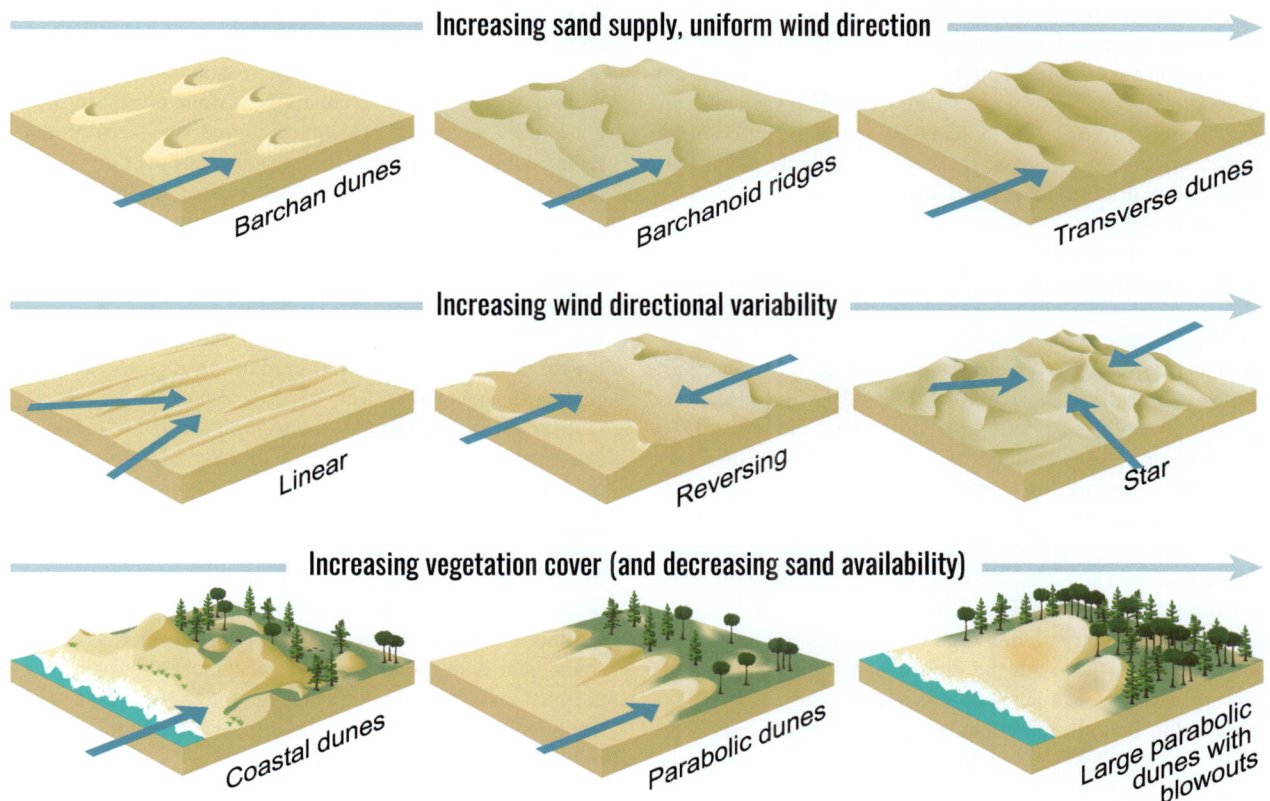

Figure 21.24 Diagrams illustrating how the various types of dune morphologies vary as a function of sediment supply, wind directional variability, and vegetation cover.

the margins of sand seas, in areas with diminished sand supply, and within sediment transport pathways between source areas and depositional sinks. As the sand supply increases, the sides of the dunes can merge together to form single, larger dunes known as **barchanoid ridges** (**Figs. 21.23B, 21.24, 21.25B**).

As the name implies, **transverse dunes** have crest-lines that are perpendicular (or transverse) to the dominant transporting winds, and thus they have a single, prominent slip face (**Figs. 21.23, 21.24**). They represent a continuation of the sand supply gradient shown in **Fig. 21.23** from barchans (minimal sand) through to barchanoid ridges (more sand) to transverse dunes (most sand). In contrast to barchans, transverse dunes occur where the sand supply is high relative to the wind strength. In these kinds of areas

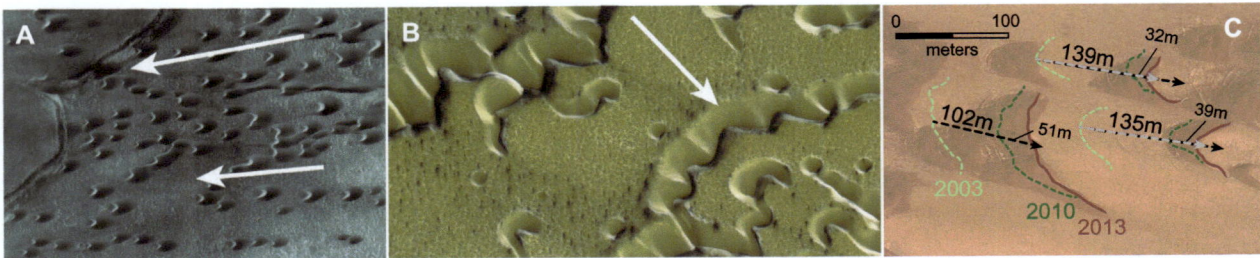

Figure 21.25 Barchan dunes (**A**) and barchanoid ridges (**B**) on Mars. Arrows indicate dominant wind direction. Source: Courtesy of NASA/ JPL-Caltech/Univ. of Arizona. **C.** Barchan dunes in China, showing their direction and distance of movement. Source: Used with permission of Elsevier, from Hu, F., Yang, X., Li H. (2019); permission conveyed through Copyright Clearance Center, Inc.

(where the sediment supply is increasing), it is common to observe barchan dunes linking together to form barchanoid ridges and then transverse dunes. All of these dune types, however, form under a generally unimodal wind regime (**Table 21.1**).

Transverse dunes are sometimes sinuous, with long, wave-like forms. They vary considerably in size. Crest lengths can vary from < 100 m to > 3 km, with dune heights from 8 to 30 m. Crest-to-crest spacings are also variable, ranging from ≈150 m to 2 km. Transverse dunes are common in sand-rich deserts, where they can completely cover very large areas, referred to as **sand seas**.

Just as transverse dunes are long and generally linear, but aligned perpendicular to the dominant wind direction, **linear dunes** are sinuous-to-straight dunes that are generally oriented parallel to the dominant wind direction (**Fig. 21.26**). In some areas, and where they are small, this type of dune is referred to as a **sand stringer**. Most agree that these dunes tend to form in areas with bi-modal wind directions, commonly with seasonally distinctive wind directions (**Fig. 21.26A**) that are separated by 90° or more (**Table 21.1**). Nonetheless, some linear dunes may also form parallel to a dominant unimodal wind direction. They are the longest dune form, ranging from 20 to ≈200 km in length. Linear dunes tend to vary from ≈5 to 100 m in height, with crest-to-crest spacings of 200–500 m. Thus, they are narrower and more symmetrical than transverse dunes. Because linear dunes form in line with the dominant winds, sands saltate *along* the dunes, not across them. Hence, they do not have a clear slip face. In many landscapes with linear dunes, the inter-dune areas are rocky, gravelly, and swept clean of sand. These dunes tend to form where they are either partially anchored by topographic obstacles or are partially stabilized by vegetation.

Seif dunes (Arabic *saif*, sword) are a subtype of linear dune. Seifs are sinuous, with peaks and saddles, and are common in many deserts. They generally lack vegetation and are associated with winds of seasonally opposed directions that are at least 90° apart. Seif dune crests, in profile, commonly have a succession of curved slip faces produced by infrequent cross winds. As the angles of the two opposing (cross) wind directions increase, seif dunes may become even more sinuous. Then, as the angle approaches 180°, they change form to become **reversing dunes**, wherein sands are transported in opposite directions at different times of the year (**Fig. 21.24**, **Table 21.1**). Both linear dunes and reversing dunes have two opposing slip faces along each side. Linear dunes are very common in desert regions, with some sand seas occupied entirely by these dunes.

Star dunes are large – to extremely large – pyramid- and star-shaped dunes with three or four radiating arms and an equal number of slip faces (**Figs. 21.24, 21.27**). They form the tallest of the dune types, commonly reaching 400 m in height (**Table 21.1**). Star dunes occur in clusters in areas with high (seemingly unlimited) sand supplies, and form on winds that blow from several, seasonally opposing directions. Just as other dunes may grow laterally or migrate across the landscape, star dunes tend to stay in place, changing shape and, at times, even growing upward. Although they are a type of freely migrating dune, they can appear to be semi-stationary

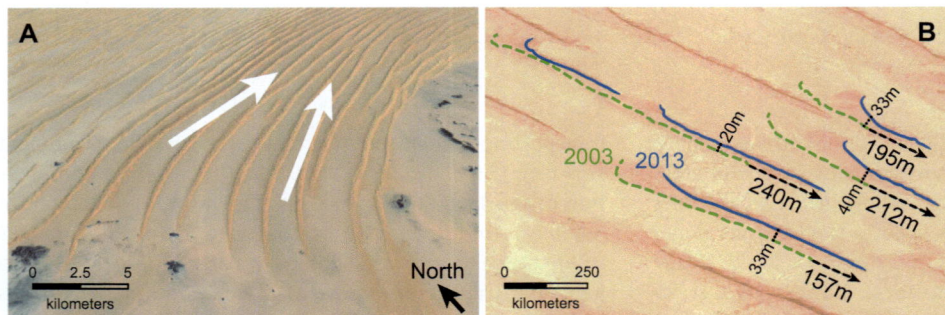

Figure 21.26 Linear dunes. **A.** A NASA image of linear dunes in the Ar Rub' al Khali Sand Sea, Saudia Arabia. White arrows indicate the dominant wind directions. Source: NASA. **B.** Seif dunes in China, showing their direction and speed of movement. Source: Used with permission of Elsevier, from Hu, F., Yang, X., Li, H. (2019); permission conveyed through Copyright Clearance Center, Inc.

as the sand accumulates towards their centers. The Arabic word for a landscape of seemingly unlimited sand supply – where star dunes are common – is *erg*, translated as "sand sea." Strictly speaking, an **erg** is defined as a desert area that contains at least 125 km^2 of wind-blown sand, and where sand covers most of the surface. Star dunes are often the most common type of dunes in the vast ergs of the Sahara Desert (**Fig. 21.27**).

Figure 21.27 Star dunes in the Issaouene Erg, in the Sahara Desert of Algeria. Source: NASA Johnson Space Center; in public domain.

21.7.4 Anchored (Stabilized) Dunes

Anchored dunes, also known as **fixed** or **stable dunes**, are partially or completely held in place by vegetation (**Fig. 21.28**). Plant cover plays a major role in controlling dune form and in dune stabilization by affecting sediment availability (**Fig. 21.10**). Vegetation reduces or eliminates sand transport by modifying and slowing air flow (**Fig. 21.10**). Plant roots also help to hold the sand in place. The morphologies of anchored dunes are relict from a time in the past. Thus, establishing the age of anchored dunes can be helpful to our understanding of past environments, when these dunes may have been freely migrating.

Some anchored dunes, such as **coppice dunes**, are immobile, while others, like **parabolic dunes** may occasionally migrate, but retain anchored arms due to the stabilizing effect of vegetation (**Fig. 21.24**). Parabolic dunes

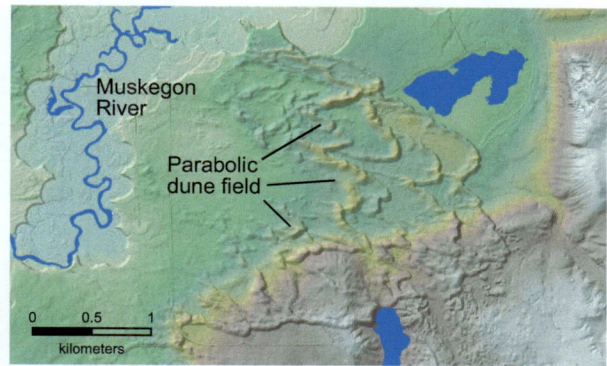

Figure 21.28 This landscape in central Michigan, USA, illustrates the sand source (the broad Muskegon River valley) and the dunes that have formed as sand was transported out of the valley and onto the low upland nearby. The parabolic dunes, which are currently stable (or anchored), had been migrating to the east–southeast. They were last active about 10,000 years ago, at which time the climate became less favorable for their migration, and they stabilized.

are crescentic-shaped landforms, but unlike barchans, they have fixed arms pointing *upwind* and a blunt-nosed, convex, dune head that moves downwind (**Figs. 21.24, 21.29**). As viewed from above, parabolic dunes have essentially the opposite form of barchan dunes, relative to wind direction. While they are forming, the dominant winds blow *into* the open end of the parabola, such that the slip face of parabolic dunes is on the *outside* of the feature. Parabolic dunes range considerably in size, from 5 to 70 m high, with arms ranging from 20 m to > 30 km long.

Parabolic dunes form in semi-arid, cold-climate, and coastal environments, where moisture and vegetation both act to reduce the availability of sand and to modify the dune form. In semi-arid environments, barchan and parabolic dunes can be found in close proximity; barchans are not anchored by vegetation whereas parabolic dunes are. Often, the "arms" of parabolic dunes are anchored by patches of vegetation, or even a single shrub, allowing the dune arms to "stretch" and lengthen over time.

Active barchan dunes can change form and "convert" into more stabilized forms of parabolic dunes if environmental

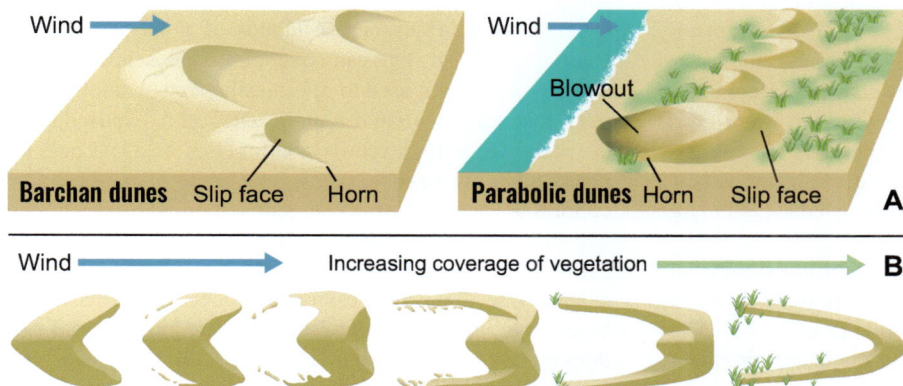

Figure 21.29 Diagrams showing the differences between barchan and parabolic dunes. **A.** Diagram contrasting the morphologies of barchan and parabolic dunes. **B.** Evolution of a barchan dune into a parabolic dune, associated with a decrease in aridity and an invasion of patches of vegetation.

Figure 21.30 Foredunes along the (**A**) Lake Michigan coast and (**B**) Lake Superior coast, USA. Source: R. Schaetzl.

conditions become less arid and the dune arms become vegetated/fixed (**Fig. 21.29B**). With increased aridity and loss of anchoring vegetation, barchan dunes can then re-form. However, if the vegetation cover continues to increase, the area of active sand may be further reduced to the point where only small areas of active sand remain within the more "active" arms.

The areas of active deflation between the arms of parabolic dunes are typically known as **blowouts** (**Figs. 21.16B, 21.29A**). Blowouts are common in coastal dunes, forming the interiors of large parabolic complexes. The blowouts are areas of preferred sand transport, as winds coming off the water are funneled into, up, and through the walling barrier of coastal dunes.

Coastal dunes and **foredunes** occur inland, along lake and marine shorelines (**Figs. 21.16, 21.29A, 21.30**). They develop parallel to the shoreline, with beach sand being their primary sand source. Winds blowing onshore or along the shoreline drive saltating sands from the beach onto the dunes, which are commonly held in place by invading dune grasses. These dunes can vary in height from 2 m to more than 75 m high. Most are parabolic in form. Winds accelerate up the windward side of the foredune, transporting sand in saltation up to and over the top of the dune, where it descends down the slip face (**Fig. 21.16C**). Flow separation (**Fig. 21.21A**) is common (under strong winds) on the lee sides of these dunes.

Coastal foredunes are smaller and nearer to the shoreline than do the larger, coastal dunes (**Fig. 21.30**). They typically do not have slip faces and generally resemble haphazard piles of sand, generally parallel to the coastline. Being close to the shore, they are prone to erosion by high water levels and large waves. But they are also important, because they help protect shorelines from excessive erosion, and act as a source of sediment supply to re-nourish the beach after large storms.

Coppice dunes, also known as **nebkha dunes**, are formed by the accumulation of sand among and around scattered shrubs or low-lying vegetation (**Fig. 21.31**). They are typically less than 1 to 2 m high and are circular to tear-drop shaped, with the dune tip elongated downwind,

anchored at a plant. The vegetation slows the wind, locally, allowing the sand to be deposited under, around, and near the plant, commonly a woody shrub. This process is not unlike what happens in parabolic dunes, except that for parabolic dunes, both arms are anchored. Most of the sand in a coppice dune is immediately downwind of the anchoring plant, and usually partially buries it. As the coppice dune grows, the plant roots lengthen to keep the plant above the surface. In effect, these dunes form in a wind shadow.

Figure 21.31 Coppice dunes in the Chihuahuan Desert, west of El Paso, Texas, USA. Source: S. Hall.

21.8 LOESS

Loess is one of the most widespread eolian sediments, composed mainly of silt-sized quartz particles. Most loess is transported in suspension clouds of great areal extent, leading to deposits that can cover entire landscapes. **Dust** is also a popular term used to refer to silt-sized or smaller sediment that has an eolian origin. When freshly exposed, loess is usually tan or light brown in color (**Fig. 21.32**). Loess deposits make for excellent soils, and some of the best agricultural areas in the world are situated on widespread loess deposits, such as the Pampas of Argentina, the central Great Plains of the United States, and the Steppes of central Asia (**Figs. 21.33, 21.34**). Loess deposits occur as belts, sometimes covering many thousands of square kilometers. Loess does not form any distinctive landforms, but typically blankets the preexisting landscape.

Figure 21.32 An exposure of thick loess deposits in southern Russia, near the Caspian Sea. The darker bands within the loess are *paleosols* – buried soils. They represent periods of time when loess deposition was slowed or stopped, allowing soil to form on the former surface. Subsequently, another deposit of loess buried that surface, and soil. Source: R. Schaetzl.

Loess is derived from silt-rich source areas. The best "silt-factories" are glaciers, which are very efficient at grinding rock and sediment into silt-sized particles, commonly called **glacial flour** because it has a flour-like consistence when dry. Meltwater from modern glaciers typically contains large amounts of glacial flour, which gives the water an intense, light blue color (**Fig. 21.35**). Much of the loess around the world has a glacial origin, because when glaciers covered much of North America and Europe (**Figs. 21.33, 21.34**), meltwater rich in glacial flour filled river valleys and spread across outwash plains. When the meltwaters declined each winter, strong winds deflated silt particles from these dry, unvegetated source areas, assisted by saltating sands. The silty loess was then transported in suspension and deposited downwind across large parts of the landscape, most of it on westerly and northwesterly winds. Today, the thickest loess deposits occur just east of the major meltwater rivers such as the Mississippi and Missouri in the USA, and the Danube in Europe. These deposits get thinner farther downwind. Thus, many of the world's major loess deposits lie just beyond the margin of the vast continental glaciers.

A second mode of origin for loess occurs in deserts. "Desert loess" forms from a variety of processes, including the shattering of sand grains by the growth of salt and ice crystals, crushing and grinding of sands and gravels in river valleys, and the breakdown of sands by saltation impacts. Desert loess is particularly common just downwind of dune fields and sand sheets, confirming the genetic link between saltating sand (bedload transport) and suspended silt mobilization.

Loess deposits in parts of China are the thickest and perhaps oldest on Earth, commonly 150 m thick, and in some locations > 330 m thick. The loess is so thick that people even build their homes directly into the deposits (**Fig. 21.36**). Most of the Chinese loess lies downwind from desert basins, suggesting that for the past 2.5 million years, silts have been transported from these deserts by the winds of the winter

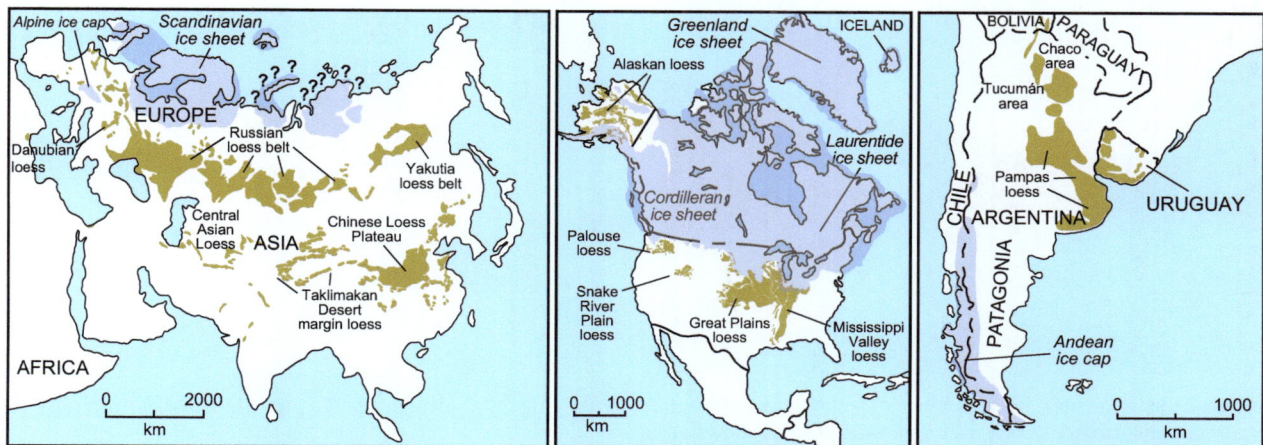

Figure 21.33 Maps of Earth's major loess deposits, with the extent of the last major ice sheets also shown. Source: After Muhs, D. R., Prins, M. A., and B. Machalett (2014). Used with permission.

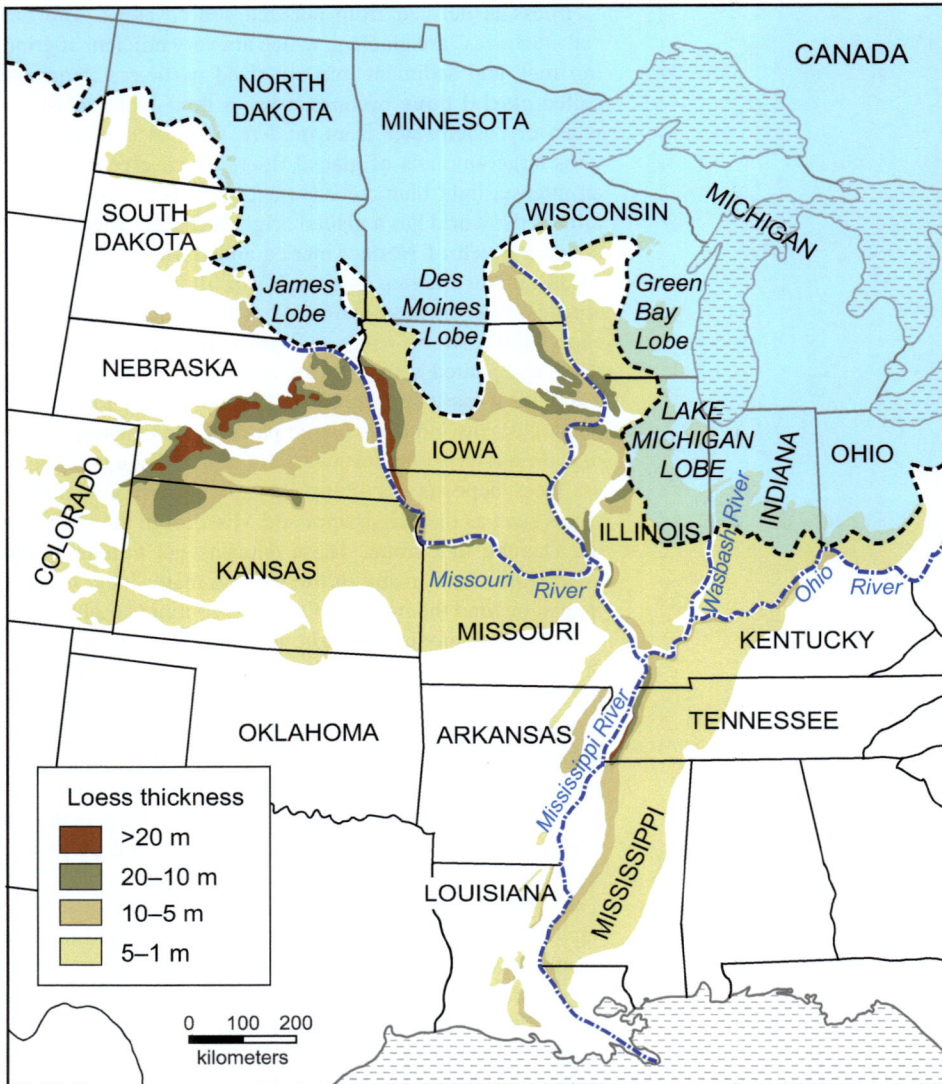

Figure 21.34 A map of the loess deposits of the Midwestern United States, with the maximum extent of the MIS 2 Laurentide ice sheet also shown. Source: After Muhs, D. R., Bettis III, E. A., Roberts, H. M. et al. (2013). © Cambridge University Press, reproduced with permission.

Figure 21.35 Lake Louise, in Banff National Park, Alberta, Canada, gets its beautiful blue-turquoise color from suspended silt, known as glacial flour. Note the glaciers in the background – the source of the silt. Source: Needpix.com; in public domain.

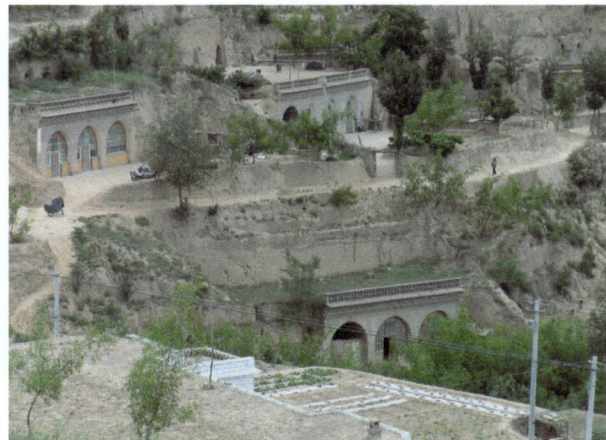

Figure 21.36 Homes built directly into thick deposits of loess, near Yulin City, Shaanxi Province, China. Raw loess is heated to create the hard bricks used in making the exterior walls of these homes. Source: X. Miao.

Asian monsoon. Therefore, the Chinese loess is – at least partially – of desert origin. Nonetheless, many of the silt- and sand-rich sediments in those deserts likely were transported there by meltwater rivers originating from nearby glaciated mountains, and so maybe the ultimate source of the Chinese loess is also glacial. The production of desert and glacial loess can still be observed today in some settings (**Fig. 21.37**).

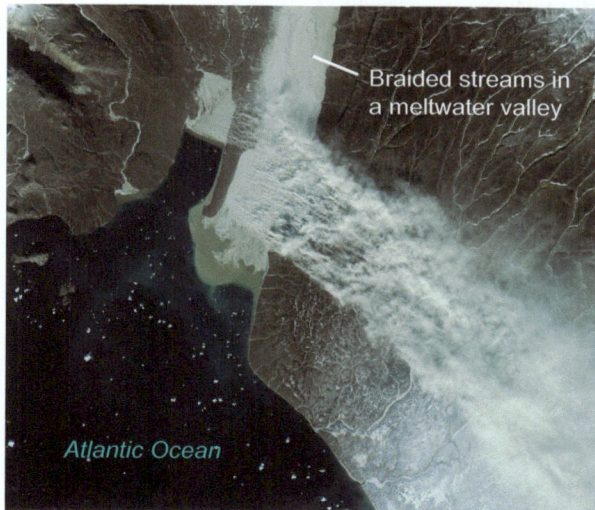

Figure 21.37 A Landsat 8 and Sentinel-2 image of silts being blown off a glacial meltwater valley near the southern coast of Greenland on September 29, 2018. Source: NASA Earth Observatory.

Dust clouds can cross oceans. For example, satellite images show dust transport from the Sahara Desert, west, across the Atlantic Ocean (**Fig. 21.9**). Some of this dust makes it all the way to North America (**Fig. 21.38**). Thus, the global impacts of long-distance transport of silt and smaller-sized particles are becoming increasingly recognized. Dust is also routinely transported from the deserts of the American Southwest to the offshore Channel Islands. Terrestrial loess deposits, as well as cores from ocean basins and ice sheets, confirm that dust transport is truly global in scale and that it has been ongoing for millennia.

Atmospheric dust is becoming a critical cog in our understanding of climate change. Climatologists have long known that ice crystals and water droplets in clouds need to nucleate around **aerosols** – solid or liquid particles such as dust or pollen. Recent research has shown the ice crystals that comprise thin, wispy cirrus clouds most commonly nucleate around mineral dust particles. As humankind adds more dust to the atmosphere via activities like burning, deforestation, agriculture, and industry, the densities of such aerosols in the upper atmosphere is increasing. If, in fact, increased aerosol contents can produce more cirrus clouds, climate may be affected globally. Will more clouds cool the climate by reflecting sunlight, or will they trap more outgoing heat? The jury is still out! But dust clearly plays a role.

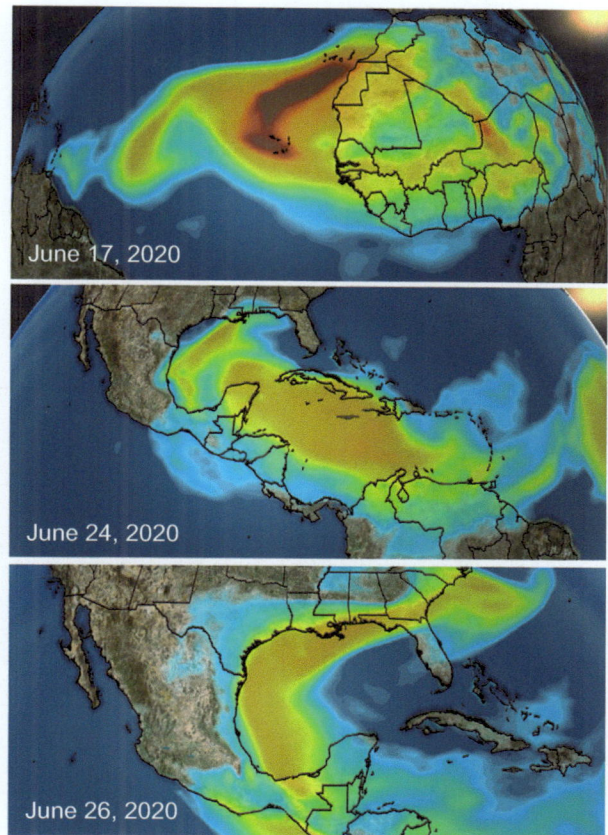

Figure 21.38 Modeled flow of a June 2020 dust storm that began in the Sahara Desert of North Africa and, over the course of several days, traversed the Atlantic Ocean.

21.9 EOLIAN SYSTEMS AND LANDSCAPES

So much of Earth's surface is comprised of eolian landforms and sediments that entire landscapes owe their character to this one geomorphic process. Large sand seas span thousands of square kilometers in the Saharan and Arabian Deserts, and even where sand is not present, the erosive effects of past winds define the landscape. Because so many desert landscapes owe their geomorphic signatures to eolian processes, they are often divided into three types, all of which have direct eolian histories:

(a) Sandy deserts, called **ergs** or sand seas. Dunes are widespread on these landscapes.
(b) Gravelly deserts, called **regs**. These are landscapes of eolian and fluvial erosion, where desert pavements are common.
(c) Rocky deserts. The word **hammada** (Arabic *hamāda*, rocky desert) is often used to describe desert landscapes that consist mainly of rocky plateaus and uplands, with almost no eolian sand. This type of desert is quite common; little sand is available for transport.

21.9.1 Sediment-State Theory

In this section on eolian systems, our focus is on sandy environments such as sand seas, dune fields, and coastal dunes. We examine how these landscapes evolve over time and what factors are at work in their evolution. In the preceding sections, we emphasized the factors of wind regime, sediment supply, and sediment availability. Remembering that these factors work together to drive different eolian erosional and depositional processes, we can now consider a "systems approach" to understanding eolian landscapes.

Although several factors and processes are *internal* to the eolian system, here we focus on *external variables*, or drivers, that lead to changes in the eolian system. One is obviously climate, and another is time.

Figure 21.39 shows an example of a sediment-state approach to eolian processes. The model considers climate and time as the two primary external factors (**Fig. 21.39A**). Other external variables, such as mountain building, sea level changes, or glaciation are also important over long timescales, but for simplicity we omit them here. In the

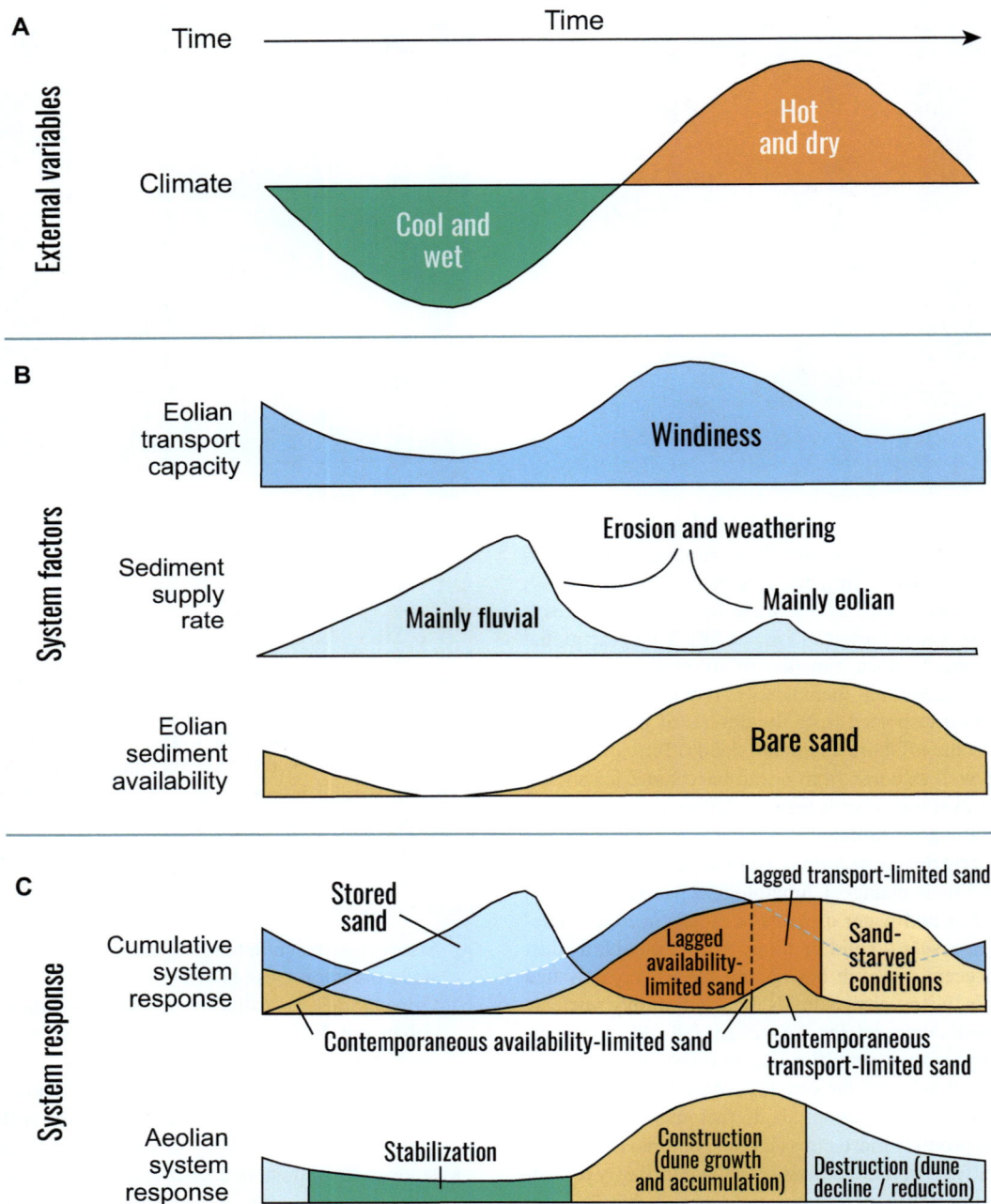

Figure 21.39 Example of the eolian system sediment-state theory, applied to desert sand dunes and sand seas. Source: After Thomas, D. S. G. (2011), with permission from Wiley & Sons.

model, the climate alternates between cool–wet and hot–dry intervals, over thousands of years. In turn, each system factor responds uniquely. In our example, eolian transport capacity, as represented by overall windiness (**Fig. 21.39B**), cycles *with* the climate but is offset. Peak eolian transport capacity is reached at a time when the climate is *becoming* arid, because wind strengths are related to major shifts in air pressure patterns, which are most unstable during times of transition from cooler to warmer conditions. The next system factor is the sediment supply (**Fig. 21.39B**); it shows fluvial erosion operating during the cool–wet period, producing the sandy and silty sediment, and a much smaller secondary peak due to eolian impact and abrasion (saltating sands) during hot–dry conditions. The final system factor indicates how much sediment is actually available for transport (**Fig. 21.39B**). In our example, sediment availability cycles similarly to climate, being less available when the climate is cool and wet, and the landscape is fully vegetated. Eolian sediment availability is highest when it is hot and dry, and the vegetation cover is low and soil moisture is minimal.

Eolian systems respond to these changing factors, as shown in **Fig. 21.39C**. Plotting the factors and the system responses together helps resolve the sediment-state of the eolian system. In total, nine possible sediment states can result from these interactions, but for simplicity we will highlight and discuss just a few.

During the cool–wet period, sediment supplied by fluvial processes is not available for eolian processes due to vegetation cover and higher soil moisture conditions, and because the transport capacity of the wind is also low. Therefore, sands get stored, and dunes stabilize (**Fig. 21.39C**). As the climate gets more arid, eolian transport capacity increases and the "stored" sediment becomes increasingly available for transport. The input of sediment to the eolian system initially lags behind the ability of the wind to transport it. Nonetheless, sediment becomes increasingly available as the climate gets drier (**Fig. 21.39C**). Eventually, the system is so arid that it is no longer limited by sediment availability, but by the transport capacity of the wind. As a result, sand dunes form and grow. But the sediment supply is not enough to sustain these dunes, so the system begins to re-work the sand that is already there. At this point, the system becomes sediment-starved (**Fig. 21.39C**), and the dune fields may transform into a rocky or gravelly desert with only scattered dunes. Of course the sand must go somewhere, and may be piled into large star dunes, or simply blown into lakes, river valleys, or out and onto the continental edge. With declining aridity and reduced wind strength, the remaining sand dunes again become stabilized.

This eolian sediment-state model fits well to desert sand seas such as the Saharan Desert of North Africa and the Mojave Desert of the southwestern United States (**Fig. 21.40**). However, it can also be applied to other eolian landscapes, by considering the different ways in which a changing climate may affect eolian systems in different environments and, where necessary, by introducing additional variables.

Figure 21.40 The Kelso Sand Dunes, in the Mojave Desert of California, USA. Source: M. Sweeney.

21.9.2 Postglacial Dunes on the Canadian Prairies

For many sand seas and dune fields around the world, determining when, and how long, eolian systems have functioned is difficult. The dune fields of the southern prairies in western Canada formed from sandy sediments deposited as the last glaciers melted. And so, the history of these dune fields began at the time of deglaciation, ≈15,000 years ago. By understanding the past climate and vegetation history, and considering how the wind regime, sediment supply, and sediment availability may have responded to these changes, we can imagine how this landscape responded to these changes. Geomorphologists studying the dunes on the Canadian Prairies have developed a model of how the eolian system responded to these postglacial climatic changes (**Fig. 21.41A**).

During deglaciation, sand dunes first formed in response to cold-dry **katabatic winds** that flowed off the nearby ice sheet. The sands in these dunes were derived from glacial outwash sediments; sediment supply was not a problem at this time and some of this sediment was stored for future dunes. Nonetheless, glacial outwash was the only sand available on this landscape, meaning that the eolian system would remain supply limited throughout its history, with little if any new sand becoming available later on. Dunes formed at this time may have been unvegetated transverse dunes and, where limited in size by the outwash sand deposits, barchans.

As the glaciers retreated further, ≈13,500 years ago (**Fig. 21.41A**), katabatic winds were replaced by warmer and moister winds from the Pacific Ocean. As a result, the dune fields stabilized, leaving preserved parabolic dunes on the landscape. Between about 9,000 and 5,000 years ago, long periods of dry climate reactivated the dune fields. Winds reworked the existing eolian sediment. In many cases, former stabilized sand dunes were completely destroyed and formed into new dunes during this Mid-Holocene dry period (**Fig. 21.41A**).

During the last 4,500 years, cycles of drought have led to repeated re-activation of the dune fields on the Canadian Prairies (**Fig. 21.41B**). During times of drought, small barchanoid dunes would form, but would then transform into parabolic dunes as they stabilized (**Fig. 21.29B**). During more humid times, such as at present, only small blowouts

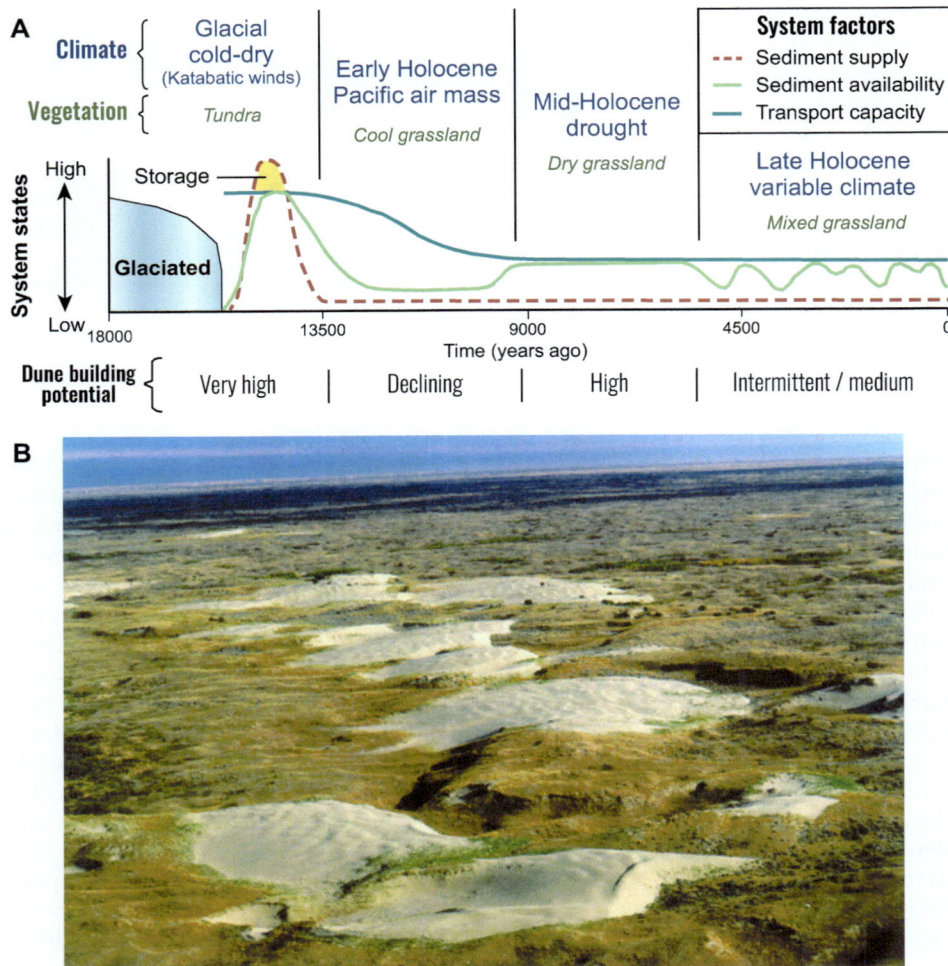

Figure 21.41 The eolian system of the southern Canadian Prairies. **A.** Conceptual illustration of eolian system sediment-state theory applied to the dune fields of the Canadian Prairies. Source: Used with permission of Elsevier, from F. Halfen, N. Lancaster, S. Wolfe (2016); permission conveyed through Copyright Clearance Center, Inc. **B.** Isolated and partially active dunes and blowouts on the stabilized dune field of the Great Sand Hills, southern Saskatchewan, Canada. Source: S. Wolfe.

and pockets of active dunes occur on the landscape. Even so, it would commonly take decades or centuries for these dune fields to completely stabilize, as the wind regime is strong and maintains a high sediment transport capacity.

21.9.3 Perched Dunes of the Great Lakes (USA) Region

Not all dunes are in dryland places. Wherever an abundance of unvegetated, sandy surfaces exist, dunes can form. Beaches and sandy bluffs are excellent examples of these types of sand sources, giving way to suites of coastal dunes. For most coastal dunes, the beach is their main source of sand (**Fig. 21.13**). But along parts of the Great Lakes' coastline in the central United States, **perched dunes** occur on bluffs high above the water line, well removed from the beach proper. Their name reflects the fact that they sit sometimes as much as 90 m above the water. Two excellent areas of perched dunes are in Sleeping Bear Dunes National Lakeshore and the Grand Sable Dunes, both in Michigan, USA (**Fig. 21.42A**). These dune fields occur in isolated fields that rest on high bluffs composed of sandy glacial sediments.

It is not the beach sands, but the sandy glacial sediments in the bluffs that are the main source of the sand for these perched dunes. Strong storms with onshore winds transport the sands up the bluff face, forming dunes far above the lake. Contemporary research at these sites, however, has indicated that the evolution of these dunes is complex; they are tied directly to lake level fluctuations that, in turn, govern the amount of sand that is made available from the bluffs.

Sand can only be deflated from the bluffs when they are unvegetated, and this mainly occurs when lake levels are high – and have been high for prolonged periods of time. During high-water periods, waves crash against the base of the bluff, causing it to destabilize. Vegetation on the bluff slides and tumbles into the lake, creating large areas of bare sand (**Fig. 21.42C1**). Onshore winds then transport sand up and over the bluff, to form dunes (**Fig. 21.42C1**). In short, the dunes grow when lake levels are high.

Alternately, when lake levels drop, as they inevitably do, a wide beach forms at the base of the bluff, and vegetation colonizes the stable bluff. As vegetation expands across the bluff face, it seals off the source of sand for the perched dunes above. Because sand is no longer supplied to the dunes, the cover of vegetation in the dune field also expands, further stabilizing the dunes and resulting in the

A **Grand Sable Perched Dune Field**

Dunes

Glacial outwash

Lake Superior

B

Dunes

Buried soil (paleosol)

C **The Perched Dune System**

① Waves destablize bluffs and strong winds blow sand on to plateau

Eolian sand (dunes)

Onshore winds

Eroding bluff

Lake Superior

Glacial sediments

High lake phase

② Limited wave erosion stabilizes bluff

Soils form in new dunes

Eolian sand (dunes)

Stable bluff

Lake Superior

Glacial sediments

Low lake phase

③ Waves destablize bluffs and strong winds again blow sand on to plateau

Eolian sand (dunes)

Onshore winds

Eroding bluff

Soil gets buried

Glacial sediments

High lake phase

④ Limited wave erosion and stable bluff

Soils form in new dunes

Eolian sand (dunes)

Stable bluff

Glacial sediments

Low lake phase   Buried soil

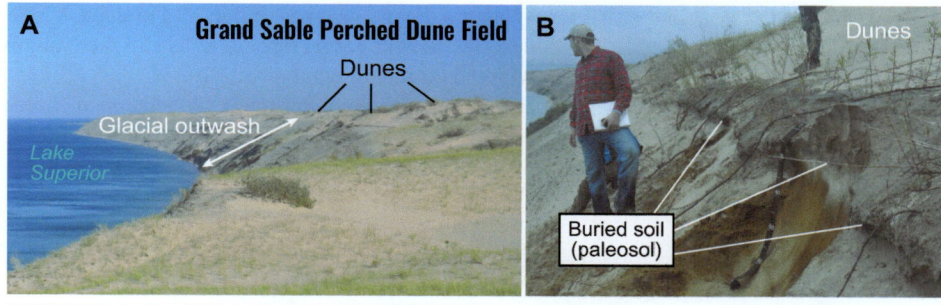

Figure 21.42 The perched dune system. **A.** The Grand Sable Dune field on the Lake Superior shoreline (USA). Source: R. Schaetzl. **B.** The buried soil that lies below the dune field, and immediately atop the outwash that comprises the bluff. Source: R. Schaetzl. **C.** The model that explains how dune growth is tied to lake level fluctuations. Source: Used with permission of Elsevier, from Anderton J. B., Loope W. L. (1995); permission conveyed through Copyright Clearance Center, Inc.

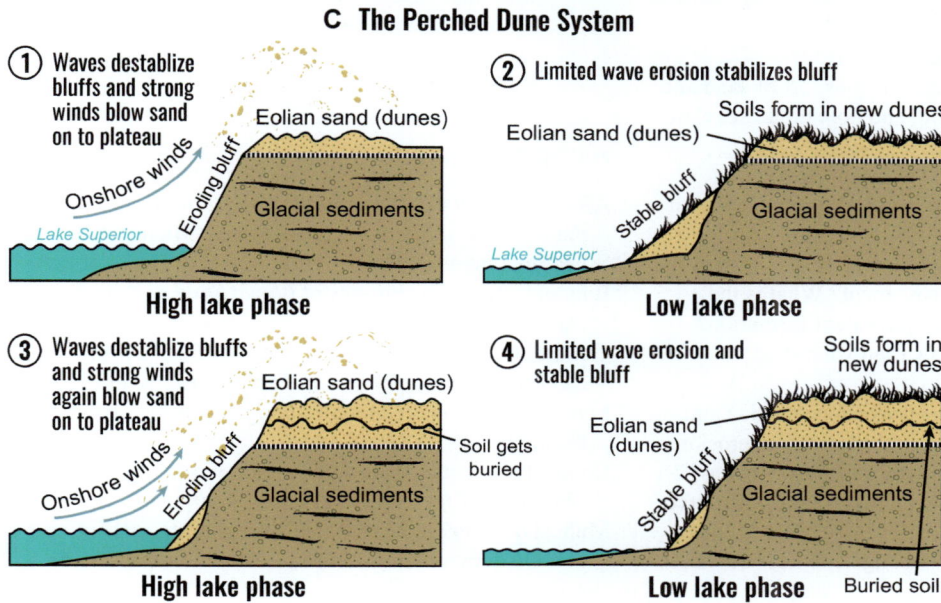

formation of an extensive soil cover across the dune field (**Fig. 21.42C2**).

The cycle of dune growth begins anew when lake levels rise again (**Fig. 21.42C3, C4**). Soils that had developed on the dunes while they were stable (when lake levels were low) get buried by fresh deposits of eolian sand. These buried paleosols are present throughout the dune field, even today (**Fig. 21.42B**). The age of wood or organic materials that may have gotten buried along with the soils can be established by radiocarbon dating, providing a date for the rise in lake level. Because lake levels have risen and fallen throughout the postglacial period, the perched dunes preserve within them a record of lake level fluctuations.

Instead of considering climate as the external factor in dune formation, we can consider the cycle of changing lake levels as the driver of the eolian system. For simplicity, we will consider the wind regime to remain constant with time. When lake levels are low, the sediment supply may be high,

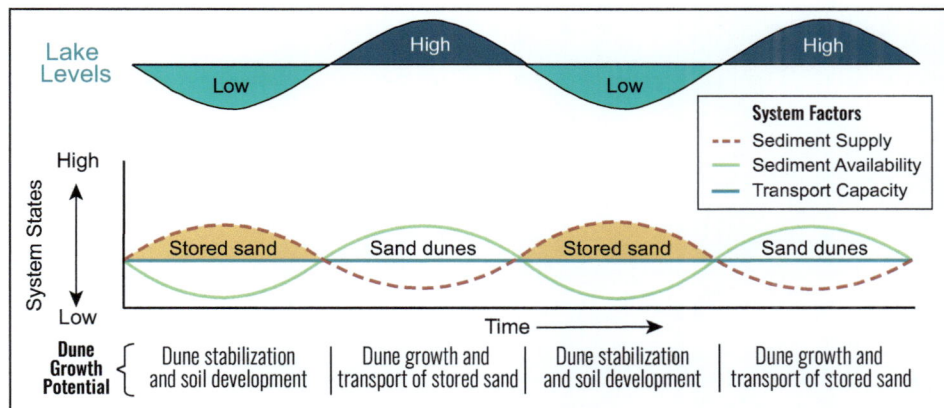

Lake Levels

High   High

Low   Low

High

System States

Low

Stored sand   Sand dunes   Stored sand   Sand dunes

Time →

System Factors
- - - Sediment Supply
—— Sediment Availability
—— Transport Capacity

Dune Growth Potential

Dune stabilization and soil development | Dune growth and transport of stored sand | Dune stabilization and soil development | Dune growth and transport of stored sand

Figure 21.43 Conceptual illustration of sediment-state theory applied to a perched dune system along a lake shoreline.

but it is stored within the stable coastal bluff and thus, sediment availability is low due to vegetation cover on the bluff. This is the time of dune stabilization and soil development. When lake levels are high, the stored sediment is released by wave erosion and bluff destabilization, and transported by the wind to form the perched dunes. The perched dune system provides an excellent example of how dunes can help unravel the geomorphic history of a place.

REVIEW QUESTIONS

21.1 What factors determine whether sediment can be picked up and transported by the wind?

21.2 What are the various modes of transport of sand and silt-sized particles by wind, and how do these relate to the sediment grain size?

21.3 Describe and explain what is meant by transport-limited and sediment-limited systems. Give an example of each.

21.4 How does wind produce lag deposits of surface gravels, such as a desert pavement? What is the alternative model of desert pavement formation?

21.5 What is a blowout and how (and where) do blowouts form?

21.6 What are the various ways in which vegetation and coarse fragments, like gravel, control eolian erosion?

21.7 What is meant by a "sediment transport pathway"? Provide an example of such a pathway, from source area to depositional area.

21.8 What are ventifacts and yardangs and how do they form?

21.9 What are two primary means of wind erosion and describe how they differ.

21.10 How does sand move up and over dunes, and what kinds of subsurface sedimentology result from this kind of transport and deposition?

21.11 What three (or four) major factors control the type of dunes that form in an area of blowing sand?

21.12 What are the major types of dunes? For each, describe how its form (especially its slip face) aligns with the dominant wind direction.

21.13 By what means do dunes typically become anchored or stabilized? Provide an example of a stabilized dune type.

21.14 What is loess and what are the two main types of course areas?

21.15 What are the three primary factors of eolian systems? Describe how each contributes to the sediment state of dune fields.

21.16 Describe the theory behind the perched dune model. What can it tell us about past environments along the Great Lakes' shore zone?

FURTHER READING

Albert, D. A. 2000. *Borne of the Wind: An Introduction to the Ecology of Michigan Sand Dunes*. Michigan Natural Features Inventory.

Bagnold, R. A. *The Physics of Blown Sand and Desert Dunes*. Methuen.

Goudie, A. S., Livingstone, I., and Stokes, S. 1999. *Aeolian Environments, Sediments and Landforms*. Wiley.

Lancaster, N. 1995. *Geomorphology of Desert Dunes*. Routledge.

Livingstone, I. and Warren, A. 1996. *Aeolian Geomorphology: An Introduction*. Longman.

Muhs, D. R., Prins, M. A., and Machalett, B. 2014. Loess as a Quaternary paleoenvironmental indicator. *Pages Magazine* 22:84–85.

Thomas, D. S. G. (ed.). 2011. *Arid Zone Geomorphology: Process, Form and Change in Drylands*. 3rd ed. Wiley-Blackwell.

Trenhaile, A. S. 1998. *Geomorphology: A Canadian Perspective*. Oxford University Press.

Warren, A. 2013. *Dunes: Dynamics, Morphology, History*. Wiley.

22 Biogeomorphology

David R. Butler

Plants and animals are, unquestionably, important geomorphic agents. Nonetheless, their key roles in the geomorphic system have only recently been properly appreciated and studied. In fact, the term **biogeomorphology** was only introduced in 1988, by Professor Heather Viles, as an approach to geomorphology that explicitly considers the role of organisms.

Biogeomorphology focuses on the influence of plants, animals, and microorganisms on landforms and geomorphic processes, and vice versa. This chapter examines how the field of biogeomorphology has expanded since its formal definition in 1988. We will discuss the role of plants in geomorphology, usually simply referred to as **phytogeomorphology**, as well as the role of animals, whose role in landscape evolution is captured by the term **zoogeomorphology**. Despite the emphasis that researchers have placed on the role of macroorganisms in geomorphology, some more recent, pioneering work has also shown that microorganisms are also important. Finally, we will cover **ecosystem engineering**, which examines how living organisms build and/or manage their environment. From nest-building by birds to beaver dams, anything constructed of rocks, mud, and/or pebbles and pellets, or of organic matter mixtures – all are considered zoogeomorphic bioconstructions.

22.1 CONCEPTS IN BIOGEOMORPHOLOGY AND ZOOGEOMORPHOLOGY

Biogeomorphic processes can be grouped into three categories: **bioerosion**, **bioprotection**, and **bioconstruction**. Bioerosion refers to the weathering and/or erosion of the land surface by organic means, whether by plants, animals, or microorganisms (**Fig. 22.1**). Bioerosion may be the most studied of all aspects of biogeomorphology.

Bioprotection involves the direct (active) or indirect (passive) roles of organisms in reducing (or completely preventing) the action of other Earth surface processes. Examples range in scale from microorganisms producing biofilms on rock surfaces, to much larger examples such as the role of vegetation in stabilizing sediment in rivers, and controlling the development of mid-channel bars (**Fig. 22.2**). Animals can also act as agents of bioprotection. By damming streams, beavers create ponds that reduce the stream velocity and limit channel incision. This in turn

Figure 22.1 Bioerosion of a streambank in Yellowstone National Park, Wyoming, USA, by a bison using its hooves to dig. Source: D. R. Butler.

Figure 22.2 Woody material helps stabilize and protect the gravel bar of the Middle Fork of the Flathead River (Montana, USA) from erosion. Small trees and shrubs can then become established, further stabilizing the bar. Source: D. R. Butler.

induces sedimentation in the ponds, which further protects the landscape from erosion (**Fig. 22.3**).

Bioconstructions include all materials that biota build, directly or indirectly, such as crusts, films, mounds, dams, or reefs of earth materials. Coral reefs are one of the most widely studied forms of bioconstruction. Australia's Great

Figure 22.0 A beaver dam can block even large streams and impact the local geomorphology in many different ways. Source: DeniseBush / Getty Images.

Figure 22.3 Beaver dams in Michigan (USA). **A.** This large beaver dam has recently blocked a small creek, forming a large, deep pond. **B.** The view upstream from a much older dam, showing the sedimentation that has taken place in the pond. Source: R. Schaetzl.

Barrier Reef is so large that it can be seen from space. Other bioconstructions, such as beaver dams and some bird nests, are an active area of research in zoogeomorphology (**Fig. 22.3**). Nonetheless, little work has been done by geomorphologists on nests and micro-bioconstructions such as biofilms and biocrusts.

Some processes in biogeomorphology are difficult to pigeonhole into one of the categories described above. For example, let's examine **tree uprooting**. When a tree falls over or is ripped out of the ground by wind or a snow avalanche, it often leaves a pit on the landscape, marking the former location of the roots (**Fig. 22.4**). Soil that slumps off the root "plate" often forms an adjacent mound. Pit formation, in this case, exemplifies the process of bio*erosion*. However, the adjacent mound is a bio*construction*. Pit-and-mound topography formed in this manner characterizes many forest landscapes, and illustrates that a single process (uprooting)

transcends simple categorization. Broadly stated, though, tree uprooting is an example of **bioturbation** (or **biopedoturbation**), the mixing of soil by plants and animals. This example is specifically termed **floralturbation**, whereas the mixing of soils by animals is called **faunalturbation**.

Bioturbation is a globally significant source of fresh surface soil material and organic matter. It accomplishes this by mixing organic material throughout the upper sections of a soil and by bringing sediment to the surface. In the process, soil porosity and permeability are increased. As an example of the tremendous global impact of bioturbation, it has been estimated that average global rates of soil turnover by some species of ants are in the range of 5,000 kilograms of soil per hectare per year!

Let's examine another example from zoogeomorphology. Termites excavate vast quantities of sediment while producing their underground tunnels and nests. During this

Figure 22.4 Effects of tree uprooting on geomorphology. **A.** This Red Pine tree in Pennsylvania, USA, was uprooted by heavy snowfall. Note the distinct pit that now exists at the former site of the roots. **B.** A massive Douglas Fir tree has been uprooted here. Uprooting is one of the few geomorphic processes that can tear up bedrock, and as a result, dramatically increase the rate of bedrock weathering. **C.** Photo showing how the soil torn up by an uprooted tree can slump back into the pit. Because not all of the soil will end up back in the pit, a mound–pit pair will eventually be formed at this site. **D.** Pits and mound microtopography, formed by persistent tree uprooting in a northern Wisconsin landscape. Source: R. Schaetzl.

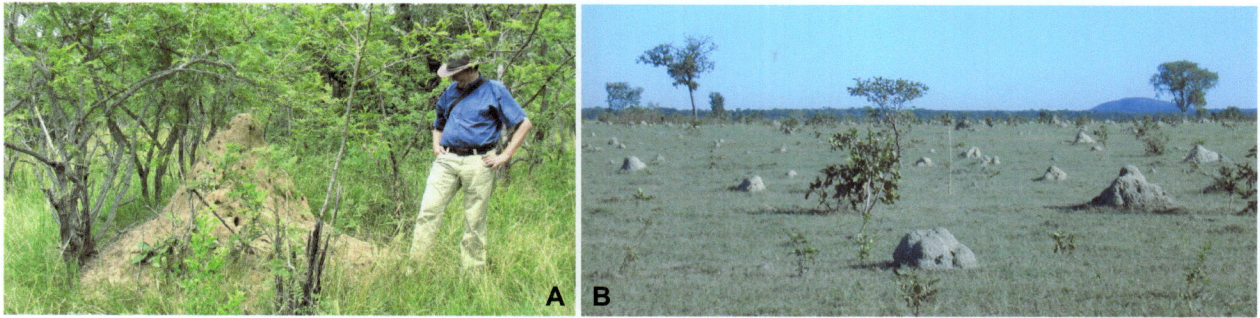

Figure 22.5 Termite mounds. **A.** A termite mound in Kruger National Park, South Africa. Source: D. R. Butler. **B.** Myriad termite mounds on a plain in Brazil. Source: A. Arbogast.

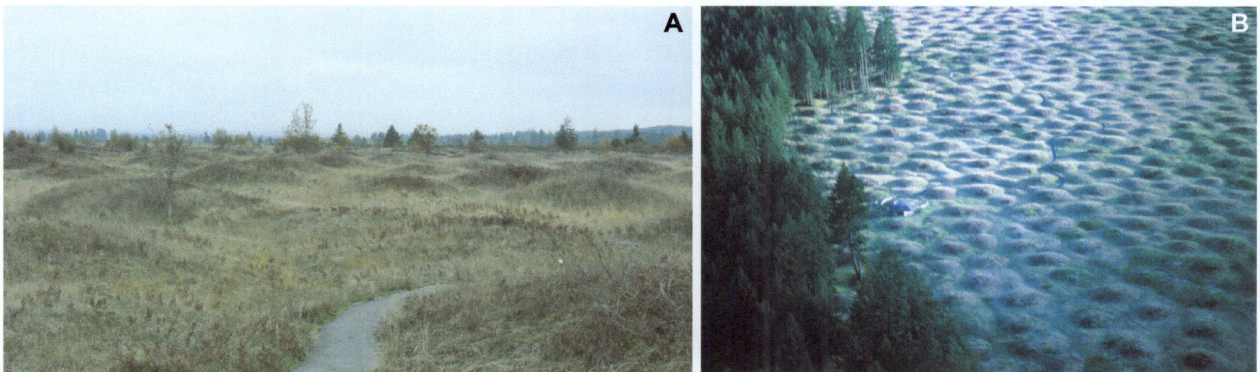

Figure 22.6 Mima Mounds on the Mima Prairie, Washington, USA. Source: (A) R. Schaetzl; (B) D. Johnson.

process, much of the excavated sediment is deposited on the surface and cemented together with organic material (such as excrement), forming large, long-lived termite mounds that characterize many tropical and subtropical landscapes (**Fig. 22.5**). Is the termite mound a bioerosion or a bioconstruction feature? Such quandaries are common in the field of biogeomorphology!

One last impressive example of the impact of bioturbation shows how the continued, albeit slow, actions of fauna can "remake" entire landscapes. Across many parts of the western United States, burrowing by pocket gophers has led to uniquely mounded, almost pimple-like, landscapes. The mammals burrow for food and form mounds for protection from predators. Being territorial by nature, they space their mounds uniformly across the landscape. The best-known example of these mounds is on Mima Prairie in Washington, USA (**Fig. 22.6**). The term "mima mounds" has commonly been applied to mounds formed by these types of gophers.

22.2 GEOMORPHIC INFLUENCE ON THE DEVELOPMENT AND DISTRIBUTIONS OF BIOTA

Little work has been done on the role of landforms and geomorphic processes on the development or distribution of animals or microorganisms. Because plants do not move from place to place as fauna do, most work has examined how geomorphology and geology can affect plant distributions and development. Much of this work has been carried out by **ecologists**, scientists who study ecology and the environment, rather than by geomorphologists. Ecologists have shown how certain plant species will only grow on specific rock types or soils. In California, for example, some plant species will only grow on soils or rocky surfaces containing the mineral serpentine (**Fig. 5.18**). In the White Mountains of California, ancient bristlecone pine trees at upper treeline (the upper altitudinal limit at which trees can grow) only grow on slopes with soils developed on dolomite bedrock; they will not grow on the sandstone soils adjacent to dolomite outcrops. This relationship between the bristlecone pines and dolomite bedrock makes for some striking spatial patterns in the field (**Fig. 22.7**).

A related field of ecology examines the effects of **slope aspect** (compass orientation) on the distribution of plants and animals. In the northern hemisphere, north-facing slopes are typically cooler and moister than south-facing slopes. Specifically, northeast-facing slopes are the coolest and most humid, whereas the southwest-facing slopes are the hottest and driest. These microclimatic differences can lead to distinctly different plant communities on slopes of varying aspect, especially where the slopes are steep (**Fig. 22.8**).

22.2.1 Banded Vegetation and Ribbon Forests

The effects of geomorphic processes and landforms on plant distributions span a variety of forms, across a range of scales. These effects frequently produce various spatial patterns on the landscape.

Figure 22.7 Bristlecone Pines flourish only on soils developed on dolomite (light-colored areas), but not on the reddish soils developed on sandstone. Source: D. R. Butler.

Figure 22.8 A valley in Glacier National Park, Montana, USA. The slope and vegetation to the right of the valley face northward; the slope and vegetation on the left face southward. The cooler, moister, north-facing slope is forested because the trees here require more moisture to grow than do grasses; hence, grasslands dominate the warmer, drier south-facing slope. Source: D. R. Butler.

Figure 22.9 Treeless avalanche chutes are surrounded by darker, coniferous forest in Glacier National Park, Montana, USA. Source: (A) D. R. Butler; (B) NASA.

One common landscape pattern formed by geomorphic processes involves stripes, or "bands." Within many mountain ranges, treeless "stripes" are found within surrounding forests (**Fig. 22.9**). Often, these stripes are produced by snow avalanches, which break and uproot trees as snow and ice move downslope. Small, flexible, deciduous shrubs and low-growing herbaceous vegetation are the only flora that can withstand the battering pressures of frequent avalanches. These treeless stripes are referred to as **avalanche chutes** or paths, and illustrate the preferred pathways of these dangerous mass wasting events.

Roughly linear stripes of vegetation at smaller spatial scales also occur in other landscapes. In many cases, these stripes are attributable to localized geomorphic processes. For example, "tiger stripes" of vegetation are found in some arid and semi-arid landscapes, associated with linear sand dunes. Another example occurs in the alpine tundra, where the concentration of moisture and fine sediments at the leading edge of solifluction terraces/lobes impacts how tundra vegetation establishes itself, often producing distinct lateral bands (**Fig. 22.10A**; see Chapter 20). Some striped patterns may not reflect current geomorphic processes at all, but rather are a reflection of past geomorphic processes. **Ribbon forests** are roughly linear stripes of trees (**Fig. 22.10B, C**). In Glacier National Park, Montana, erosion by Pleistocene glaciers is responsible for these patterns. Glaciers scoured most effectively along the strike of the bedrock and parallel to bedding planes. The resulting vegetation pattern is one of forested ribbons on the higher, drier, bedrock ridges, with thin soils, grass, and herbaceous vegetation in the intervening meadows (**Fig. 22.10B, C**). This landscape pattern would not exist if not for the glacial scouring – a geomorphic process.

Figure 22.10 Linear vegetation patterns. **A.** Small stripes of vegetation mark the lateral fronts of solifluction terraces. **B** and **C.** Linear "ribbons" of forest growing on slightly elevated bedrock ridges of Flattop Mountain, Montana, separated by wetter, lower glades with thick, fine-textured, meadow soils. **C.** NASA image of Flattop Mountain, showing ribbons of forest. All images are from Glacier National Park, Montana, USA. Source: (A and B) D. R. Butler; (C) NASA.

22.2.2 Hillslope Stability and Vegetation Patterns

A variety of processes can impact vegetation patterns on hillslopes. These processes can expose fresh soil materials and create unstable surfaces, on which vegetation cannot easily become established. Alternatively, the fresh surfaces may act as a preferred substrate for some plants, as long as the surface does not continue to be disturbed. Typically, vegetation on a slope acts to stabilize the surface and reduce erosion from overland flow and fluvial incision. Litter, such as leaves and other fresh plant matter on the soil surface (see Chapter 13), intercepts rain and dissipates its erosive energy. Vegetation reduces the variability in soil moisture by slowing losses via evaporation. Vegetation also adds organic matter to the soil, which assists in binding soil particles together, helping to stabilize slopes and reduce erosion. Plant roots also bind soil together, further stabilizing slopes. Additionally, aboveground biomass (especially tree trunks and foliage) creates microtopography that slows overland flow and increases hydraulic roughness.

Alternatively, removal of vegetation often destabilizes slopes and leads to increased erosion. The various modes of vegetative disturbance that affect geomorphology include wildfires, animal grazing, vegetation conversion to rangeland, agriculture and cropland abandonment, deforestation, insect outbreaks, military maneuvers and their corresponding impacts, and climate change. As you can see, the interactions of vegetation with hillslopes are highly complex!

A variety of mass movement processes, such as rockfalls and debris flows, produce highly unstable surfaces that may not be vegetated for decades or even centuries afterward (**Fig. 22.11**). Geomorphic disturbances on more localized scales can, as we have seen, uproot trees (**Fig. 22.4**). Animals can also uproot trees, such as elephants, who commonly push over trees on the African savanna. Beavers build dams that create ponds that saturate soils upstream, which in turn makes nearby trees more susceptible to uprooting (**Fig. 22.3B**). Meteorological events such as tornadoes, hurricanes, and localized straight-line winds can also uproot trees.

A good rule of thumb is that the more vegetation on a slope, the more stable it will be. That said, vegetative cover is not an ironclad guarantee that a slope is stable! If

Figure 22.11 A series of large debris flows in Glacier National Park, Montana, USA, blocked the stream at the base of the slope, producing two small lakes. Because the debris is very rocky, little vegetation has colonized the surfaces in the decades-to-centuries since. The lake on the left was produced by a 1946 event, whereas the lake on the right was impounded by a 1910 event. Being older, it is now more vegetated. The ages of the mass wasting events were determined by dendrochronology (see Chapter 3). Source: D. R. Butler.

the sediments on a slope are saturated, and/or experience a tectonic (earthquake) trigger, even well-vegetated slopes can become unstable and initiate sediment displacement downslope (see Chapter 14).

22.3 WOODY MATERIAL IN STREAM SYSTEMS

Much of the recent work in biogeomorphology has focused on the role of woody material in streams and rivers. In small, steep, low-order streams, the focus has been on **log steps** in streams and how they can stabilize the landscape and inhibit fluvial erosion. In high-order, low-gradient rivers, research has focused on the role of large wood in impacting fluvial erosion or deposition. How can it stabilize mid-channel bars and lead to vegetation succession, and what kinds of hazards are created by the presence of large woody material in streams?

Woody material can enter a stream system in various ways. Meteorological phenomena such as tornadoes, hurricanes, and strong straight-line winds may uproot trees and deliver wood into streams. Forest fires can kill trees that subsequently decay in place or topple onto the surface; these trees can eventually make their way into stream channels. In mountainous landscapes, avalanches uproot trees and deliver large quantities of woody material to stream channels. Other forms of mass movement such as debris flows can also deliver wood into stream channels. Failed beaver dams introduce large quantities of woody material downstream, as can wood from beaver lodges and food caches. Woody materials can also enter stream systems via lateral streambank erosion (at cutbanks) during times of high flow.

22.3.1 Log Steps

Log steps in streams develop when woody material extends across a stream channel, acting like a small dam. The step then changes the water surface elevation, as flowing water spills over the log (Fig. 22.12). A series of log steps across the channel produces waterfalls that both reduce the amount of energy available for downcutting and erosion, while also forming pools upstream where sediment can be deposited. The reduction in stream energy and velocity in the pools fosters sediment storage, rather than allowing the river to flush the sediment downstream. Studies have shown that the greatest number of log steps occur in third-order streams, but have the greatest effect in dissipating potential stream energy in (larger) fourth-order streams (see Chapter 16 for a discussion of stream orders).

22.3.2 Large Woody Material

As noted above, large wood can be delivered into rivers of all sizes, in a variety of ways (Fig. 22.13). Increasingly, geomorphologists and ecologists are viewing large wood as a vital component of fluvial ecosystems. Anglers know

Figure 22.12 Log steps in a low-order stream in a snow avalanche path in Glacier National Park, Montana, USA. Streamflow is low in this late-summer image. Source: D. R. Butler.

the importance of woody material and log jams in rivers, because they provide key habitat for fish and their prey.

Historically, many nations have developed programs to remove large wood from rivers, so as to reduce hazards to stream commerce, to minimize the likelihood of log jams that imperil bridges and riverbank structures, or to divert the river into a new channel. For these reasons, the reintroduction of large wood into streams and rivers may seem counterintuitive to many. Nonetheless, it has recently become an

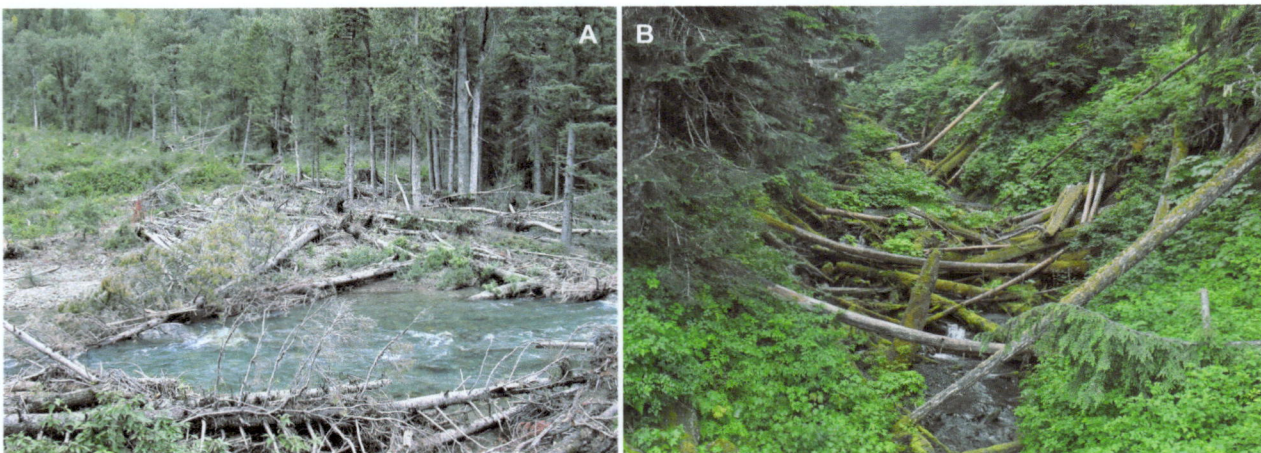

Figure 22.13 Large woody material in stream channels. **A.** In the floodplain of McDonald Creek in Glacier National Park, northwestern Montana, USA, large stacks of woody material were deposited during floods in 1964 and 1975, and by recent snow avalanches. The wood has protected and stabilized the gravel bars and allowed them to become further stabilized by fostering new vegetation growth. Source: D. R. Butler. **B.** Woody material in a steep, lower-order stream channel on the flanks of Mt. Baker, Washington State, USA. Source: R. Schaetzl.

important component of river ecosystem management, as an attempt to restore rivers to a more natural state.

The effects of large woody material in and along rivers can vary by its position relative to the channel. For example, wood may lodge on the inside or the outside of a **cutbank**. Large wood along a cutbank may "armor" the bank and reduce cutbank incision and meander migration, whereas wood on **point bars** may have very different impacts on stream flow. At higher stream discharges, large wood will be less effective at armoring streambanks and re-directing flow than it might be at lower discharges.

Large wood in the channel also maintains and/or modifies the hydraulic conditions of channel flow through its influence on bed roughness, which in turn provides flow resistance. In short, wood in the channel slows flow velocities, which can induce sedimentation and bar formation. As a result, vegetation growth near and in the channel, and on mid-channel bars, is promoted. As woody material decreases the stream's velocity, it also deflects water into backwaters and forms eddies behind wood dams.

22.4 SALT MARSHES AND TIDAL FLATS

Log steps and large woody material illustrate how vegetation can reduce stream velocity and inhibit channel erosion. The same general principle (vegetation reduces flow velocity and stabilizes the landscape) can be seen in coastal environments such as in salt marshes and on tidal flats (**Fig. 22.14**). There, the twice-daily tidal effects of high and low tide have powerful influences on sediment movement and stabilization. Tidal flats are nearly horizontal topographic surfaces that are flooded and drained by the rise and fall of tides. Biofilms (recall their role as a bioprotector of underlying surfaces) are important for sediment stabilization on tidal flats. Salt marshes occupy a narrow range of elevations above tidal flats, between mean sea level and mean high water points. When tidal flooding occurs, the presence of salt-marsh vegetation induces inorganic sediment deposition by reducing the velocity of the moving water, especially near rivers that flow into the marsh.

22.5 FIRE

Although the effects of fire as a biogeomorphic agent may not be apparent at first glance, fire is an important active (direct) and passive (indirect) geomorphic agent. Fire not only impacts the vegetation cover, but also the soils in which the vegetation grows, and even bedrock and boulders.

22.5.1 Fire as a Direct Geomorphic Agent

Have you ever sat around a campfire and heard a loud "crack!" as the rocks around it popped apart from the heat of the campfire? The same thing happens in nature during forest fires. Ground fires that burn very hot can fracture rocks with their intense heat (**Fig. 22.15**). These "popping rocks" have been reported from a variety of locations, including forested mountain sites in Arizona, Montana, Wyoming, and California. The resultant rock fragments and flakes both armor the surface, slowing erosion, and eventually entering the slope debris cascade.

22.5.2 Fire as an Indirect Geomorphic Agent

The indirect effects of fire are considerably more widespread and much more significant from a geomorphic perspective than the fire-induced spalling of boulders. Fire does much more than burn off surface vegetation. It also burns the **litter** layer (organic materials) on the forest floor, and in doing so can bake the underlying soil surface to create **hydrophobic (water-repellent) soils**. Soil hydrophobicity can last from a few months to several years before it slowly dissipates. At the site illustrated in **Fig. 22.15**, the soil surrounding the fire-spalled boulders became hydrophobic, as noted by the inability of ground vegetation to re-colonize the soil. Water falling onto the surface ran off, rather than infiltrating. Hydrophobicity was still present here, six years after the fire. The removal of the litter layer by fire, even in locations where the soil does not become hydrophobic, exposes the soil surface and makes it even more susceptible to rainsplash, erosion, and runoff.

Soil hydrophobicity accelerates surface runoff, which can lead to rilling and gullying, generating large amounts of sediment downslope. The subsequent additions of sediment to

Figure 22.14 Salt marsh landscapes. **A.** A salt marsh along a stream inlet into Mobile Bay, Alabama, USA. **B.** A salt marsh (the vegetated area at center-right) and tidal flat at Mont St. Michel, France. Source: D. R. Butler.

Figure 22.15 In 2006, a fire at the alpine treeline in Glacier National Park burned through stands of gnarled, low-growing conifers. The fire burned so hot that it caused rocks to shatter and fragments of rock to spall off underlying boulders. The fire led to the formation of hydrophobic soils that had yet to recover six years later when these photos were taken. Source: D. R. Butler.

streams may then alter stream channel morphology, especially in areas of steep slopes. Rainfall on hydrophobic soils is the cause of many of the **mudflows** in Southern California landscapes, most of which happen shortly after fires scar the landscape. Where fires have occurred in popular recreation areas, signs warning the general public to avoid certain trails or roads during rainstorms are necessary because of the increased likelihood of debris flows and/or flash flooding.

22.6 ZOOGEOMORPHOLOGY

The study of animals as geomorphic agents has a rich tradition, extending back to Charles Darwin. In his final years, Darwin studied bioturbation (soil mixing by fauna and flora; see Chapter 13) by various soil fauna, particularly earthworms, and the subsequent micro-landforms they produced, in his home garden in England. Darwin conducted a series of earthworm experiments that would span 40 years. Through experimentation and observation, he noted that the actions of earthworms and other soil fauna gradually lowered objects like rocks, coins, and other artifacts, originally on the surface, into the soil. This process occurs as sediment below them is brought to the surface by fauna. Darwin even noted that the great monuments at Stonehenge had sunk into the ground several centimeters, assumedly by the same process. All forms of burrowing animals do this kind of geomorphic work (**Fig. 22.16**).

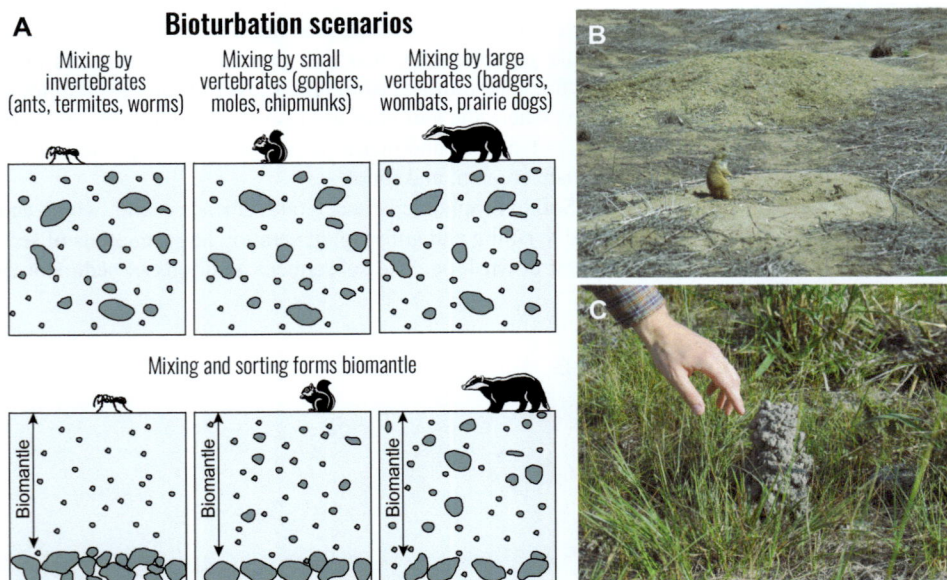

Figure 22.16 Effects of bioturbation on soils, particularly the formation of biomantles. **A.** Each of the various species of soil fauna have a maximum size particle that they can move upward and outward while burrowing. Particles larger than this size will eventually settle to the depth of burrowing, forming a comparatively stone-poor biomantle in the sediment above. **B.** Prairie dogs can move small stones to the surface. Source: R. Schaetzl. **C.** This "chimney" of stone-poor mud was formed by a crayfish. These crustaceans burrow into wet soils, down to the water table, but maintain an open tunnel to the surface. Over time, they bring large amounts of sediment to the surface, burying stones and other large objects that they cannot move. Source: R. Schaetzl.

Darwin noted that coins, formerly on the surface, had sunk several centimeters due to earthworm bioturbation. Eventually, a **biomantle** develops above the sunken coins or objects. The biomantle is a layer of mixed sediment that overlies a layer of rocks and coarse fragments too large for the bioturbators to move or bring to the surface (see Chapter 13). Each group of soil fauna has their own "limits" as to what size objects can be moved upward (**Fig. 22.16**).

Following Darwin's work, a 1939 textbook by Arwin Lobeck included photographs and comments on the geomorphic effects of beavers, termites, and coral. Nonetheless, the field of zoogeomorphology did not develop its theoretical foundations until the late twentieth century. Quantitative measurements of the amount of geomorphic work conducted by animals, and the landforms they create, are distinctly late twentieth- and early twenty-first-century phenomena.

The zoogeomorphic impacts of animals can be highly interrelated and complex, and equally intertwined with the effects of vegetation. In a well-known study from Yellowstone National Park, it was shown that the early twentieth-century removal of wolves via trapping resulted in a roughly 70-year period in which few predators existed to control the park's elk population. The growing population of elk removed much of the riparian vegetation along low-gradient, meandering streams in northern drainage basins in the park. This loss of riparian vegetation negatively impacted the beaver population there by removing their food sources. The combined effects of elk overgrazing and beaver population decline led to stream incision in the northern parts of the park. After wolves were reintroduced into the park in 1995–1996, and beaver trapping ceased, stream browsing of riparian vegetation by elk was strongly curtailed, fostering vegetation re-establishment along stream channels. As a result, beaver populations began to rebound. However, because the streams in some basins had become deeply incised during the low-wolf period, beavers have had more difficulty in damming those streams and reestablishing the preexisting meandering stream patterns. This example of population dynamics illustrates how complex and intertwined biota are with the natural landscape. Animals of all kinds, from the smallest ectothermic (cold-blooded) invertebrates and vertebrates (fish, amphibians, and reptiles) to endothermic (warm-blooded) vertebrate birds and mammals, are involved in zoogeomorphic activity, as shown in the flowchart in **Fig. 22.17**.

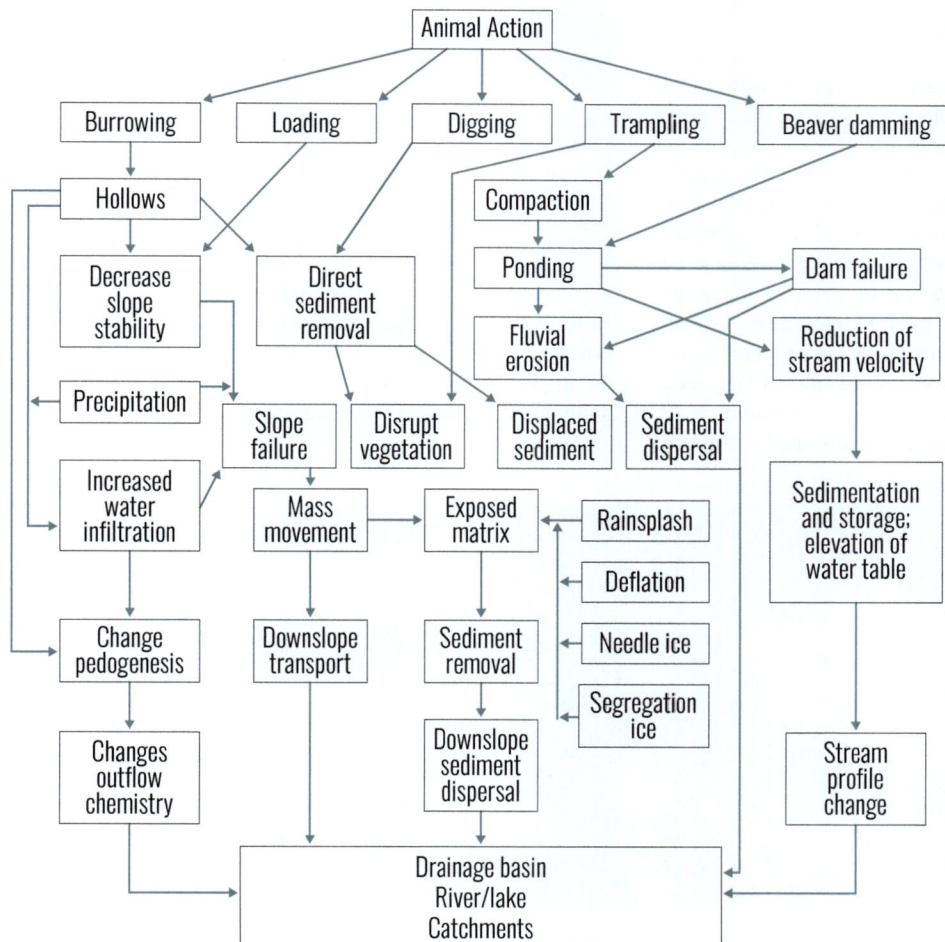

Figure 22.17 Flowchart illustrating some of the more spatially and volumetrically significant forms of zoogeomorphic activity on terrestrial landscapes. Oceanic activity such as coral construction is not included here. Source: Used with permission of Elsevier, from K. Hall, N. Lamont (2003); permission conveyed through Copyright Clearance Center, Inc.

22.6.1 Burrows and Burrowing

Burrowing as a category encompasses the underground excavations of animals. These excavations often include a variety of specialized "rooms," including birthing/natal rooms, toilet rooms, what are in effect "living rooms," as well as a series of connecting tunnels. Surface deposition of excavated sediment goes hand-in-hand with burrowing. That said, some excavated sediment always stays within the tunnels and is simply compacted and trampled down into the floor, as the tunnels are abandoned. Deposition of sediment on the soil surface may take the form of randomly deposited spoil mounds (**Fig. 22.18**), either as roughly circular bio-concentrations such as ant mounds, or as irregularly shaped mounds like those produced by termites (**Fig. 22.5**).

Burrowing fauna also create other small but distinctive surface depressions. These include **day beds**, shallow depressions excavated by an animal in which to rest during the heat of the day (**Fig. 22.19**), and **wallows** (**Fig. 22.20**), excavations used for the purpose of rolling in mud or dust. Coating themselves with dust and/or mud offers protection from biting insects and may also reduce sunburn for animals in the tropics with sensitive skin, such as elephants.

Animals of many different sizes excavate day beds, ranging from ground squirrels up to large mammals such as rhinoceroses, grizzly bears, and bighorn sheep (**Fig. 22.19A**). Domesticated animals such as your family dog may even excavate day beds to lie in during the heat of the day. Day bed excavations produce surface irregularities that serve as sources for sediment dispersal, because they "open up" the surface by removing vegetation.

Wallows tend to be larger than day beds, forming elliptical or roughly circular-shaped depressions sufficiently deep for a large animal such as a bison, Cape Buffalo, or elephant to roll around in and coat themselves with dust and/or mud (**Fig. 22.20**). Wallows may range up to 10 m in diameter, and can be tens of centimeters to over a meter deep. As wallows become compacted over time, they act as small water catchments after rainstorms, in some areas offering the only surface water sources for great distances. The amount of excavated sediment associated with an individual wallow is

Figure 22.18 Excavated animal mounds and burrows. **A.** The burrow and spoil mound of an Olympic Marmot in Olympic National Park, Washington, USA. Source: D. R. Butler. **B.** An American Badger at its burrow entrance (and associated spoil mound) in the Laramie Basin of Wyoming, USA. Source: D. R. Butler. **C.** Mounds made by moles in silty loess, Serbia. Source: R. Schaetzl.

impressive in and of itself. When you consider that the North American Bison is believed to have created over 100 million wallows on the Great Plains prior to European contact, the importance of animal wallows to that region's sediment budget and hydrology becomes clear.

Figure 22.19 Day beds, of (**A**) a desert bighorn sheep, and (**B**) a white rhinoceros. Source: D. R. Butler.

Figure 22.20 Typical animal wallows. **A.** A Cape Buffalo in its wallow (Kruger National Park, South Africa). Source: D. R. Butler. **B.** An elephant wallow. Note the deep elephant footprints within the dried mud (Kruger National Park, South Africa). Source: D. R. Butler. **C.** A North American Bison wallow on the shortgrass prairie in western Kansas, USA. Source: J. Gill.

22.6.2 Digging

Digging, as contrasted with burrowing, is primarily carried out by animals in search of food, either through excavating at the surface or by tunneling underground and eating items found there. Many examples exist of this type of biogeomorphic activity. Rodents burrow underground and eat items such as roots and tubers. The grubbing effects of snow geese, which use both their beaks and talons to excavate food sources in the Arctic tundra, have been well studied by ecologists, but little work has been carried out by geomorphologists. Grizzly Bear excavations have been studied in Alaska and in Glacier National Park, Montana. Grizzlies excavate for a variety of food sources such as roots and tubers, insects, and burrowing mammals (**Fig. 22.21A**). The amount of sediment excavated can be spatially extensive as well as volumetrically impressive. Even whales get in on the fun – gray whales feed on invertebrates that occupy the sea floor in shallow areas of the Pacific Ocean off the coasts of Canada and the United States. The whales "slurp" enormous amounts of sediment off the sea floor, leaving behind large, irregular bioerosion pits on the ocean floor. They filter this sediment through their baleen plates, retaining the food and expelling the sediment in plumes (visible from airplanes!) across the seafloor.

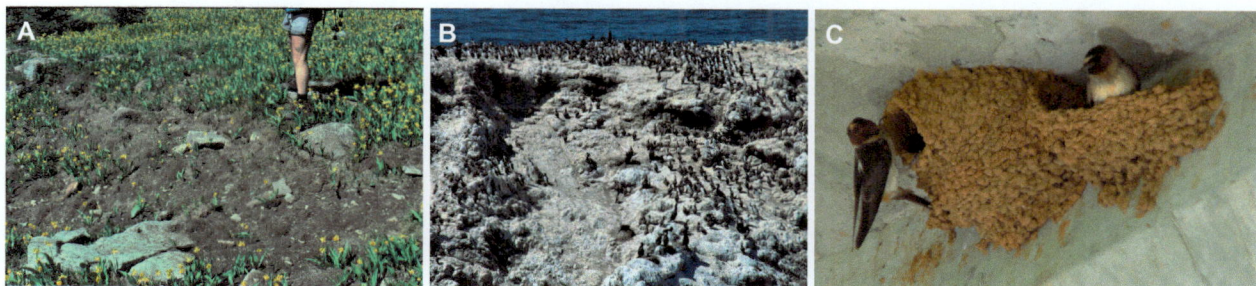

Figure 22.21 Direct and indirect effects of digging by animals. **A.** A grizzly bear has excavated this area in search of glacier lily bulbs. Source: D. R. Butler. **B.** Widespread incision caused by bird talons and bird excrement-induced chemical weathering in a sea stack comprised of basalt, Yaquina Head, Oregon. Source: D. R. Butler. **C.** A pair of cliff swallows in their mud nests. Source: Ken Thomas, Public domain, via Wikimedia Commons.

Digging is also carried out by animals when they excavate and consume soil directly, a phenomenon called **geophagy**, literally meaning "earth eating." Geophagy is accomplished by intentionally eating soil, which they excavate with claws, beaks, talons, or tusks, depending on the animal. A variety of mammals and birds practice geophagy. The practice is believed to be done in order to provide nutritional supplements to the animal's diet – essentially nature's equivalent to taking a vitamin – or in some cases to act as a coagulant to offset diarrhea. Excavation and consumption of rocks is known as **lithophagy** (literally, "stone (or rock) eating"), as is done by a variety of reptiles and birds. The stones are used as grinders to break down food in the gizzards or stomachs of the animals.

Digging may also be carried out by birds, in order to excavate materials used in constructing pebble nests (by penguins, for example) or mud nests (by numerous species of swallows and starlings). Material is excavated by both talons and beaks, and then carried in their beaks to the nest site. Individual swallow nests are typically comprised of up to 1,600+ individual mud pellets. Each nest can weigh as much as 1.3 kg (**Fig. 22.21C**). Large, flightless birds such as the megapodes of Australia and greater Australasia use their talons to rake up soil and surface material to construct large, egg-incubating mounds of earth material, with diameters of well over two meters and up to a meter high.

An inadvertent form of digging occurs when animals with sharp claws, hooves, or talons persistently travel over exposed bedrock or soil surfaces. Soil surfaces and streambank edges may become deeply chiseled and incised by this activity (**Fig. 22.1**) (Also see the discussion of trampling, below.) Exposed bedrock surfaces in Antarctica have become heavily incised and polished by the 10,000+ daily trips of penguins passing over bedrock in coastal areas. The concentration of numerous bird talons, working in concert with chemical weathering from bird excrement, can deeply incise the underlying bedrock, even resistant rock types such as basalt (**Fig. 22.21B**).

22.6.3 Loading

Loading is a zoogeomorphic process attributable to the weight of animals on the soil surface. Loading can induce soil creep on hillslopes. Persistent loading, especially by hooved grazing animals, whether wild or domesticated, can lead to the development of mounds and fine-scaled terracettes (**Fig. 22.22**). **Terracettes** are stairstep-like landforms formed by the downslope pressure of animal hooves on steep hillslopes. Heavier animals obviously impose greater forces on the hillslope. Terracettes are reported from locations around the world, in a variety of climates. The common denominator in almost all cases is the presence of hooved animals, be they wild animals such as elk, deer, reindeer, bison, or mountain goats; or domesticated grazing cattle, sheep, goats, yaks, or camels. Studies show that terracettes can develop in a few weeks, under intensive grazing.

22.6.4 Trampling

Animal **trampling** has both direct (active) and indirect (passive) geomorphic effects. Trampling can actively break up soils, especially along streambanks where animals cross from one side of a stream to another (**Fig. 22.1**). Trampling reduces or removes vegetative cover, which in turn can lead to erosion and gullying. Surface compaction is an obvious consequence of trampling, reducing the infiltration capacity of the underlying surface. Reduced infiltration capacity can lead to ponding of water, as well as accelerated runoff and erosion. Such actions are virtually unquantified on a worldwide basis, but are nonetheless important in many areas.

Imagine the impact that the massive bison herds must have had on the soils and landscapes of the American Great Plains during the precontact period. Millions of hooves, traversing thousands of kilometers each year, making billions of imprints in the soil! Add to that the myriad wallows that the animals made to dust themselves, to ward off flies (**Fig. 22.20C**). The dust clouds alone would have been impressive (**Fig. 22.23**). Geomorphologists have determined that large areas of the Great Plains are covered with

Figure 22.22 Terracettes formed on steep hillslopes by hooved animals. **A.** Terracettes produced by elk in Glacier National Park, Montana, USA. Source: D. R. Butler. **B.** Terracettes produced by cattle on a hillslope in Nebraska, USA. Source: R. Schaetzl.

Figure 22.23 Small dust plumes rise from the trampling of a small bison herd in Custer State Park, South Dakota, USA. The dust plumes, and damage due to trampling, from native herds a thousand times as large must have done extensive geomorphic work to the Great Plains landscape. Source: R. Schaetzl.

fields of sand dunes, the largest of which – the Sand Hills of Nebraska – is >250 km across. These dune fields must surely have been densely trampled as the buffalo herds crossed them, destroying the grass vegetation and opening the dunes to the wind and (likely) reactivating the movement of the sand. Each hoof print may be small, but as is typical in zoogeomorphology, the overall impact of the actions of so many animals can be unimaginably large.

22.6.5 Beavers and Beaver Dams

We could think of beavers as a **keystone species** – one on which other species largely depend, an organism that supports the wider biological community. In theory, if a keystone species is removed from the ecosystem, many aspects of the ecosystem change drastically. Keystone species help hold the ecosystem together. Beavers are considered by ecologists as a classic keystone species because of their function as ecosystem engineers, particularly the dams they build (**Figs. 22.3, 22.24**; see also **Fig. 22.0**). In this case, if beavers are removed, so are their dams, and many aspects of the ecosystem change.

Beaver dams, one of the most widely studied forms of bioconstruction, have significant zoogeomorphic impacts. Two species of modern beavers exist, the North American beaver (*Castor canadensis*) and the Eurasian beaver (*Castor fiber*). Although both species were hunted to near extinction for their fur, both have recovered to some extent. Nobody knows precisely how many beavers or beaver ponds existed in North America prior to European contact, but estimates range from 60–400 million beavers living in 15–250 million ponds. As we shall see below, not all beavers live in dams on ponds; some excavate burrows in the banks of streams and ponds.

Beavers build dams to pond water, which provides them with a safe habitat. They are semi-aquatic rodents that are much more at home in water than they are on land, where their movement is slow and clumsy. Building lodges in the midst of the ponds, or burrowing into the banks of a pond, protects beavers from predators. The pond also allows the expert-swimming animals access to food sources (trees and shrubs) that surround the pond, without needing to be on land. Beaver ponds also act as storage sites for food caches, with twigs and sticks stored underwater for consumption at a later date, including over the course of the winter in environments where the surface of the pond freezes.

Geomorphologists study beavers and their dams because of their profound zoogeomorphic impacts. Dams drastically reduce stream velocity, which lowers the stream's ability to downcut and erode. Sediment is deposited on the floor of the beaver pond (**Fig. 22.3B**). Sedimentation continues throughout the life of a beaver pond, until the water becomes too shallow for the continued presence of beavers, at which point the pond is abandoned and the beavers relocate. However, as long as the dam remains intact, sedimentation continues. Eventually, the pond disappears after it is completely filled with sediment, at which point a **beaver meadow** occupies the valley floor where once there was a pond (**Fig. 22.24**). That sediment may eventually be downcut as well, when a new stream crosses the meadow, and the process is renewed.

Beaver dams frequently occur in sequences along the course of a stream, such that the stream develops a

Figure 22.24 Formation of a beaver meadow from a beaver pond. **A.** A beaver pond in Glacier National Park, Montana, USA in 1992, with open water and a canal running from the pond toward the trees at upper right. **B.** By 2002, much of the pond had begun to fill with sediment and was covered in shallow water-tolerant vegetation. **C.** By 2012, the pond had completed the transition to a beaver meadow. A forest fire had burned through the area in 2006, accelerating surface runoff and adding additional sediment into the former pond, completing the infilling process. Source: D. R. Butler.

"stair-step" profile of flat pond surfaces separated by short, intervening areas of less altered streamflow. These pond step sequences reduce stream velocity and induce in-pond sedimentation of any sediment that may have passed through dams upstream. Studies have shown that as many as 20 dams/kilometer may be emplaced by beavers in especially favorable stream valleys.

How much sediment is stored in beaver ponds in North America? One study estimated that the amount stored in beaver ponds prior to European contact at up to 125 billion cubic meters! Even in modern times, present-day beaver ponds in North America are estimated to store up to 3.85 billion cubic meters of sediment. Imagine, then, how much clearer streams must have looked prior to European contact, when rivers flowed with an enormously reduced suspended sediment load because of the presence of so many beaver dams!

Beaver dams are, in and of themselves, impressive bioconstructions, comprised of woody sticks and logs as well as other debris, held together with mud on the upstream side of the dam. When viewed from downstream (Fig. 22.25), a beaver dam looks like an incredibly well-stacked jumble of logs and sticks because you don't see the mud on the other side. In fact, you only see that mud if the beaver dam fails and the pond drains.

Beaver dams can last for decades. At one location in Glacier National Park, Montana, dams built after a major flood in 1964 have withstood the test of time; they are still in existence in 2020. An enormous beaver dam in Wood Buffalo National Park in northern Alberta, Canada, currently known to be the largest such dam in the world at over 850 m long, has existed for at least three decades and may be at least half a century old.

Nonetheless, although many beaver dams can last for decades, they also eventually fail. Heavy rainstorms, sometimes exacerbated in colder regions by rapid snowmelt, can lead to dam failure and rapid-to-catastrophic pond drainage. When dams break and ponds partially or fully drain, the large amounts of sediment trapped behind the dam is revealed (Fig. 22.26). The floor of the former beaver pond often reveals a muddy surface littered with woody material (drowned logs, former beaver food caches, and material carried in by the flood that caused the dam to burst) (Fig. 22.27).

When the dam shown in Fig. 22.25 failed, water rushed into the pond downstream and caused that pond's dam to fail as well. These events then drained the lower pond and revealed the pond floor there – a cascading effect (Fig. 22.27). Unearthed by the drainage of the lower pond were bioerosion features that had been excavated by beavers beneath the pond surface. Several beaver bank burrows had been excavated into the edge of the drained pond, indicating that the beaver population there had exceeded the capacity of the lodge (not shown) in that pond. A beaver canal was also revealed by the pond drainage. Beavers excavate canals at the base of ponds to produce deeper areas where they can swim to access food caches and bank burrows around the edge of ponds. Beavers also excavate

Figure 22.25 Looking upstream at a beaver dam in Glacier National Park, Montana, USA in 1991 (A) and 2002 (B). Note how the beavers used two (now dead) slightly leaning conifers, one tree on either side of the dam, to help stabilize it. Source: D. R. Butler.

Figure 22.26 Upstream images of the same beaver dam shown in Fig. 22.25, before and after dam failure. A. In summer 1994, with a pond at the front of the dam and another pond downstream. B. One year later, in the summer of 1995. A "freak thunderstorm" in June caused the dam to burst, with the gap on the left side of the dam. Note the exposed mud in front of the broken dam and on the floor of the former pond. C. By the summer of 2002, vegetation had mostly covered the sediments on the floor of the former pond, and had completely covered the surface of the dam. Source: D. R. Butler.

Figure 22.27 Bioerosion features excavated by beavers, revealed after the pond drained, following dam failure. **A.** An excavated canal in the center of the photo leads to a bank burrow at the former pond shoreline. Note the thickness of sediment stored on the former pond floor. Food caches of wood are visible at upper right as well as at lower left. **B.** A partially drained pond reveals several bank burrows and an underwater entrance to a beaver lodge. Source: D. R. Butler.

canals into surrounding forest areas to extend their access to food sources, and sometimes dig canals to connect adjacent ponds in areas of low-lying terrain.

One particularly interesting recent case of introducing beavers onto a landscape has gone horribly wrong from an ecological viewpoint. North American beavers were introduced into southern Patagonia, Argentina, by the Argentine government in 1946, with the goal of boosting the economy of the region. The beavers represent furbearing animals that can be trapped and sold by locals. However, with no natural enemies in the region, the population expanded without control to the point where beavers have essentially overrun many areas of Tierra del Fuego and the adjacent mainland tip of Patagonia. The rapid increases in the beaver population completely overwhelmed the number of trees available for dam building. The beavers adapted by excavating thousands of cubic meters of peat and mineral sediment which they used for dam construction.

22.7 CORAL REEFS

Coral reefs are the largest zoogeomorphic bioconstructions on the planet. Corals are marine invertebrates that live in compact colonies of many, genetically identical, individual coral polyps, all of which have $CaCO_3$ skeletons. Basically, the corals extract solutional calcium from seawater and use it to construct a solid structure within which live the coral polyps (**Fig. 22.28A**).

Most coral reefs occur in shallow, clear, tropical or subtropical, seas. Reefs grow very slowly, with growth rates estimated to add about one centimeter in height each year. Large coral reefs such as Australia's Great Barrier Reef are, therefore, believed to be quite old, up to around 10,000–20,000 years (**Fig. 22.28B**). Coral reefs are among Earth's most diverse ecosystems, providing homes for at least 25% of all marine species.

A variety of classification schemes exist for coral reefs, but common to most of these are three primary forms of

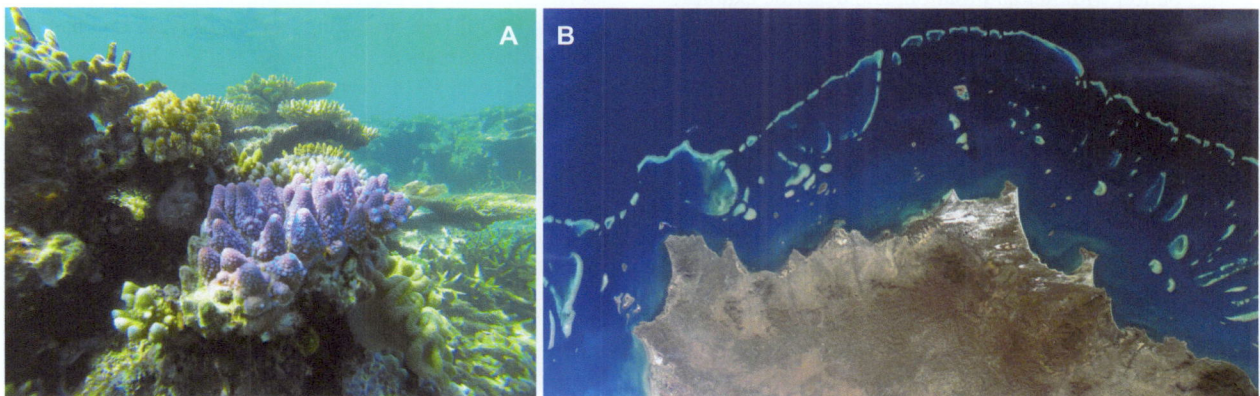

Figure 22.28 Coral is a living organism that lives in clear, warm, sea water. Each coral polyp excretes an exoskeleton of $CaCO_3$, which over many generations can grow into massive colonies of coral. **A.** Coral is beautiful and colorful. Source: Dmitry Brant, CC BY-SA 4.0, via Wikimedia Commons. **B.** The largest bioconstruction on Earth – the Great Barrier Reef, off the Australian coast – is visible from space. Source: NASA Earth Observatory.

reef, first classified by Charles Darwin and later reiterated by geomorphologist William Morris Davis: fringing reefs, barrier reefs, and atoll reefs (see Chapter 17).

Fringing reefs are also called shore reefs. They are typically attached directly to, or very near, the shore, with only a narrow channel or lagoon separating them from the shore. Fringing reefs are the most common form of coral reefs. They follow coastlines and may extend for many kilometers. Fringing reefs are usually < 100 m across, but some may be hundreds of meters wide. They form on shores at low water level and expand seaward as they slowly grow in size, sometimes taking thousands of years to form. Some of the best fringing reefs in the world are those that surround most of the Red Sea.

Barrier reefs resemble the later stages of fringing reefs with their lagoons, but barrier reefs are typically much larger and their offshore outer edge is formed in the open ocean rather than next to a shoreline. Australia's Great Barrier Reef is the world's largest barrier reef. It is roughly 300–1,000 m from the shore and stretches for over 2,000 km (**Fig. 22.28B**). Another large example of a barrier reef is the Belize Barrier Reef off the coast of Belize, in Central America.

Atoll reefs are roughly circular coral reefs that form a nearly continuous ring of coral encircling a lagoon, without a central island (**Fig. 17.34**). It is believed that they originate as fringing reefs surrounding a volcanic island. Over time, the central volcano is eroded, leaving behind the coral ring/reef. Sinking of the seabed or rising sea level can also have a similar result – a coral atoll. **Atolls** are quite common in the South Pacific. The Indian Ocean also contains several examples including the Seychelles and the Maldive Island chains.

REVIEW QUESTIONS

22.1   Define biogeomorphology and contrast it with zoogeomorphology. How do these terms relate to or compare with the term ecosystem engineering?

22.2   What is bioturbation? Define and contrast floralturbation with faunalturbation, and describe examples of each.

22.3   In what ways can geomorphology affect the distribution and patterns of vegetation?

22.4   What are some ways that linear or striped vegetation patterns form?

22.5   Describe some of the interactions between the presence or absence of vegetation on a hillslope that is prone to mass movements like avalanches or

rockfalls. How do these interactions impact slope stability/instability?

22.6   From a geomorphic perspective, describe why woody material in streams and large rivers is important.

22.7   What are some of the geomorphic impacts of salt marshes?

22.8   How is fire an active/direct geomorphic agent? What is soil hydrophobicity? In what other passive/indirect ways does fire have geomorphic impacts?

22.9   What are the primary processes by which animals sculpt the Earth's surface? Provide examples of landforms formed by bioerosion and by bioconstruction. Describe how beavers are agents of both bioconstruction and bioerosion.

22.10   Describe and discuss the importance of beaver dams as a geomorphic agent.

22.11   What are the various types of coral reefs and why do they fall under the rubric of zoogeomorphology?

FURTHER READING

Butler, D. R. 1995. *Zoogeomorphology: Animals as Geomorphic Agents*. Cambridge University Press.

Butler, D. R. and Hupp, C. R. (eds.). 2013. *Treatise on Geomorphology*, vol. 12: *Ecogeomorphology*. Academic Press.

Guilcher, A. 1988. *Coral Reef Geomorphology*. Wiley.

Kruckeberg, A. R. 2002. *Geology and Plant Life: The Effects of Landforms and Rock Types on Plants*. University of Washington Press.

Malanson, G. P. 1993. *Riparian Landscapes*. Cambridge University Press.

Naylor, L. A. 2005. The contributions of biogeomorphology to the emerging field of geobiology. *Palaeogeog. Palaeoclimatol. Palaeoecol.* 219:35–51.

Ruiz-Villanueva, V., Piégay, H., Gurnell, A. M., Marston, R. A., and Stoffel, M. 2016. Recent advances quantifying the large wood dynamics in river basins: new methods and remaining challenges. *Rev. Geophys.* 54:611–652.

Schaetzl, R. J., Burns, S. F., Small, T. W., and Johnson, D. L. 1990. Tree uprooting: review of types and patterns of soil disturbance. *Phys. Geog.* 11:277–291.

Viles, H. (ed.) 1988. *Biogeomorphology*. Basil Blackwell.

Viles, H. 2020. Biogeomorphology: past, present and future. *Geomorphology* 366: 106809.

Wohl, E. 2019. *Saving the Dammed: Why We Need Beaver-Modified Ecosystems*. Oxford University Press.

23 Planetary Geomorphology

Devon M. Burr

Exploration of planetary bodies beyond Earth is occurring at an ever-increasing rate. What used to be points of light in the night sky are now amazing, complicated, and intriguing objects of geologic study. For extraterrestrial bodies with solid surfaces – such as rocky planets, asteroids, and icy bodies – the study of planetary bodies as geologic objects includes careful scrutiny of their surfaces. Planetary exploration is an examination of geomorphology, as our interpretations of other planetary surfaces are largely guided by geomorphic studies done on Earth. At the same time, planetary landforms developed in different geologic conditions than on Earth – such as under different gravities, in different materials (like ice instead of rock), and beneath different atmospheric pressures or compositions.

This chapter illustrates that various geomorphic processes observed on Earth occur on other planets as well, and also how the resultant landforms contrast with those found on Earth. We start this discussion with an introduction to the sub-discipline of planetary geomorphology and the approaches used by planetary geologists to explore extraterrestrial landscapes.

23.1 THE SOLAR SYSTEM

Planetary geomorphology, which focuses on the geomorphology of planetary bodies in outer space, is the story of going from astronomy to geology. Planetary bodies were first explored by ancient astronomers, such as the Greeks, Chinese, and Native Americans, who all made insightful observations of the sky. Modern astronomers continue this Earth-based exploration of planetary bodies using sophisticated telescopes; some like the Hubble and James Webb telescopes are mounted in space. With the advent of the space age, humankind has sent spacecraft missions to fly by, orbit, and even land on these bodies. This new information has provided an opportunity for these planetary objects to become objects of interest for geologists.

One of the first geologic aspects of a planet to be observed is its surface. As we know, surface features are the playground of geomorphologists. Thus, we can pursue the study of planetary geomorphology on any body with a solid surface. The four rocky planets in the inner Solar System that are closest to the Sun (Mercury, Venus, Earth, and Mars) all have solid surfaces, as do the millions of much smaller bodies in the asteroid belt (Fig. 23.1). Other geomorphic bodies include dwarf planets such as Pluto, Ceres, and Makemake (pronounced "mah-kay mah-kay"), and Kuiper Belt Objects. The gas giants Jupiter and Saturn, and the ice giants Uranus and Neptune, of the outer Solar System lack solid surfaces. However, their natural satellites – their many moons – are icy bodies with solid surfaces, like comets. So, there is a plethora of opportunities to study geomorphology beyond Earth (Fig. 23.2).

23.2 TERRESTRIAL AND EXTRATERRESTRIAL LANDFORMS

Across the Solar System, planetary landscapes are shaped by processes similar to those on Earth, but which occur in different materials and under varying conditions, and so exhibit diverse suites of landforms. The single most common landform in the Solar System is the **impact crater**, formed by the supersonic contact of a space rock with a planet's surface (Fig. 23.3). Impact craters are less common on Earth than on other planetary bodies for a variety of reasons, including erasure by plate tectonics, the effects of biota, and processes such as weathering, erosion, and mass wasting. Nonetheless, a few dozen impact craters have been identified on Earth, including the ≈1.2 km-wide Meteor Crater in northern Arizona (USA) and the immense Vredefort Crater in South Africa (Fig. 23.4). Impact craters on Earth are also found on the ocean floor, such as the Chicxulub Crater off the Yucatán Peninsula of Mexico that contributed to the extinction of the dinosaurs.

Impacts by **meteorites** – rocks from space – are hugely energetic events that melt and vaporize rock. The only anthropogenic activity that can even approach this energy release is a nuclear explosion. The landforms that result from such impacts vary in size (width and depth) and shape, depending on impact energy (Fig. 23.5). Because kinetic energy is the product of mass times velocity squared ($KE = mv^2$), the impact energy of a meteorite is largely a function of impact velocity. Crater *shapes* are also a function of the strength of the surface material. Impacts that occur either with low kinetic energies

Figure 23.0 Impact craters, like these in the eastern Hellas region of Mars, are the most common of all the planetary landforms. Source: Stocktrek Images / Getty Images.

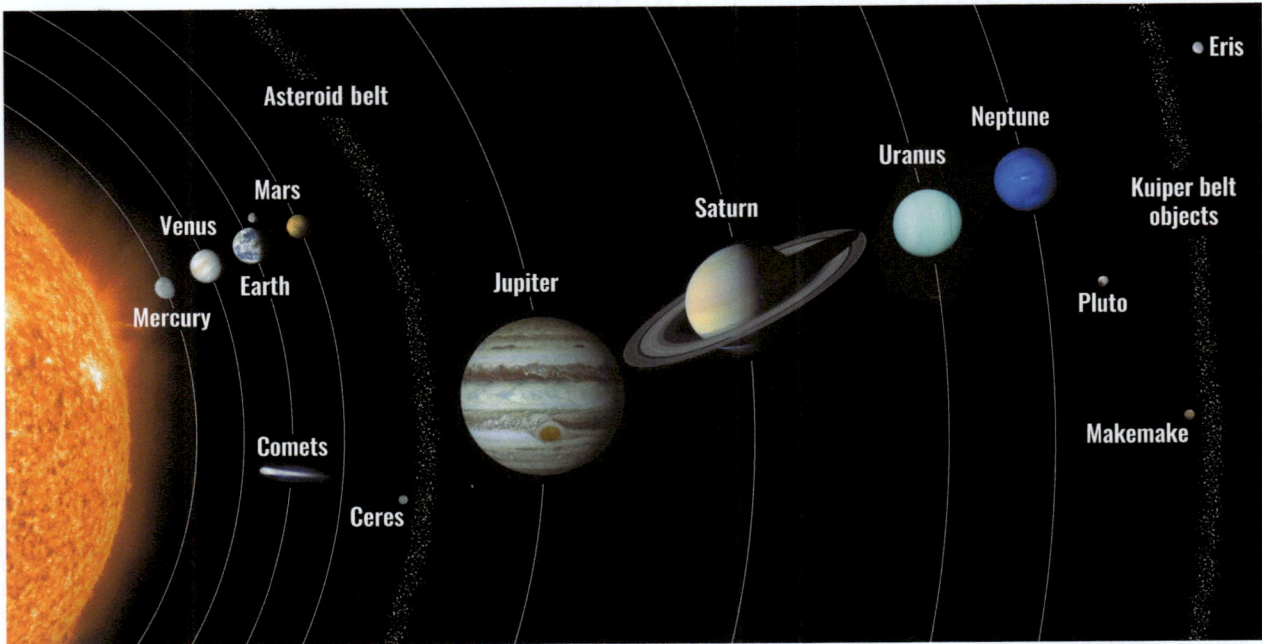

Figure 23.1 Our Solar System consists of eight planets and many moons/satellites. This diagram shows the major planetary bodies in order from the Sun, although neither the distances nor the sizes of the objects are shown exactly to scale. The only moon shown is that of Earth.

Figure 23.2 A visual representation of the rocky and icy solid-surface planetary objects in our Solar System, by relative area. The Earth and Titan (the largest moon of Saturn) even have some liquid surfaces, although the lakes on Titan are rather other-worldly, being filled with liquid hydrocarbons and nitrogen!

or into "weak" surfaces, such as ice, form simple bowl-shaped craters (**Fig. 23.5A**).

In addition to the formation of molten materials and vapor, meteorite impacts also fracture rocks. The broken up rock that is thrown out from the impact site, often inter-mixed with molten material, is called **ejecta**. Ejecta from craters can extend across the land surface for up to several times the crater radius, forming extensive **ray** systems in which narrow swaths of ejecta extend beyond a more con-tinuous ejecta blanket (**Fig. 23.6**). Impacts that occur with higher kinetic energies or which collide into harder surfaces

(such as rock) can form complex crater shapes with central peaks or central rings. For these larger craters, much of the excavated crustal material collapses in on itself, pushing up a mound that forms a central peak or complex (**Fig. 23.3**). At the same time, the rock beneath the crater rebounds, or bounces back up, to add to the central structure. The highest energy impacts may form multi-ring basins with extensive ejecta and ray systems.

The rate of impact cratering on planetary bodies has been *decreasing* since the formation of the Solar System. As aster-oids in the Solar System have been destroyed over time by

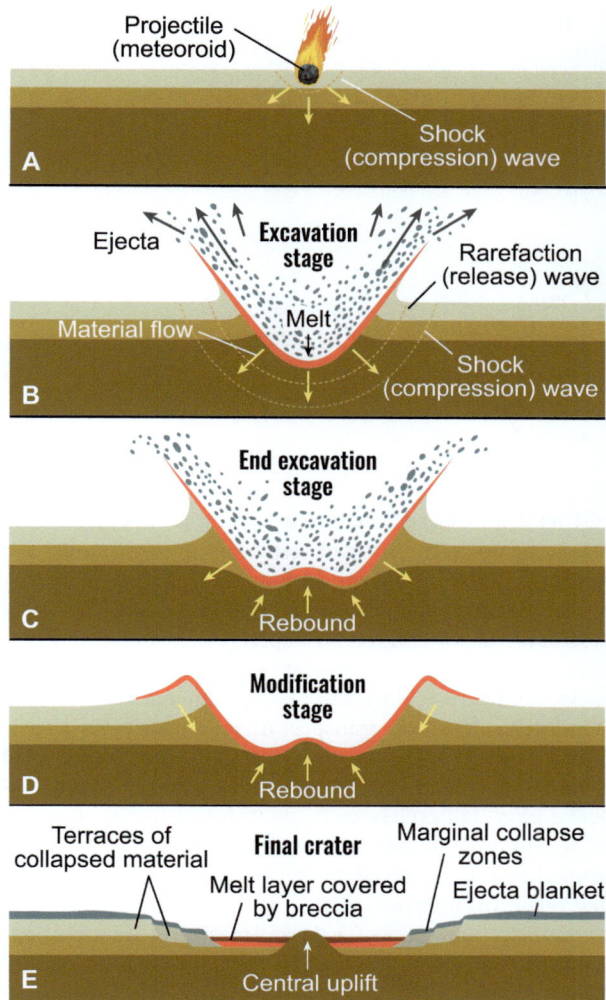

Figure 23.3 Diagram showing the multistage process that may form a complex impact crater with a central peak. As a meteorite impacts the surface (**A**), it throws out ejecta (**B**) and creates a shock wave that compresses the material. This shock wave causes material to flow downward and outward, resulting in a depression. **C** and **D**. The shock wave is followed by a rarefaction wave that releases the compression. The resultant rebound of the material in the crater center, along with the collapse of excavated material, forms a central peak. **E**. The crater is then modified by slumping and block faulting of wall material. The final form exhibits evidence for all these processes. Source: Adapted from a classroom illustration by Bevan M. French/David A. Kring/Lunar Planetary Institute/University of Arizona.

Figure 23.4 Impact craters on Earth. **A.** Meteor Crater, >1200 m across and 167 m deep, is located in northern Arizona, USA. Because it was formed only ≈50,000 years ago, it is a particularly good example of a simple crater. Source: Shane.torgerson, CC BY 3.0, via Wikimedia Commons. **B.** Vredefort Crater, located in South Africa, may have originally been 180–300 km wide. More than two billion years of erosion has made its initial size difficult to determine. Source: NASA Earth Observatory.

impacts with other planetary bodies, meteorite impacts on planetary bodies have decreased in frequency. Additionally, the asteroids that remain tend to be smaller and form smaller impact craters.

Both impact cratering from outside the planetary body and tectonic forces generated within the body can produce tectonic landforms. Impacts can produce *extensional* stresses on planetary bodies by uplifting the surrounding land surface during impact. Similar stresses on planetary bodies, as on Earth, come from intrusions of molten material, for example, magma, pushing up the overlying land surface. Commonly, these stresses form fissures and rifts, where the surface has been pulled apart (**Fig. 7.7**). Such fissures are found on all the inner planets and almost all of the icy moons of outer Solar System planets.

Compressional tectonic forces can cause the surface to push up into distinct ridges, which are seen on all the inner Solar System planets. These features, sometimes called **wrinkle ridges**, appear different than extensional landforms like fissures because they are in positive relief (**Fig. 23.7**). Landforms produced by transform tectonic motion often look like fissures but show some offset of features on either side of the fissure, like a strike-slip fault (**Fig. 23.8**). Long, narrow trenches on planetary bodies are referred to as **fossae** (Latin *fossa*, "ditch, trench") (singular: fossa).

Volcanic landforms on extraterrestrial bodies have some similarities with those on Earth. Shield volcanoes, like those that make up the Hawaiʻian Islands, are found on both Venus and Mars, where the lower gravity and lack of plate tectonics enables them to achieve large

Simple crater   **Complex crater**   **Peak-ring basin**   **Multi-ring basin**

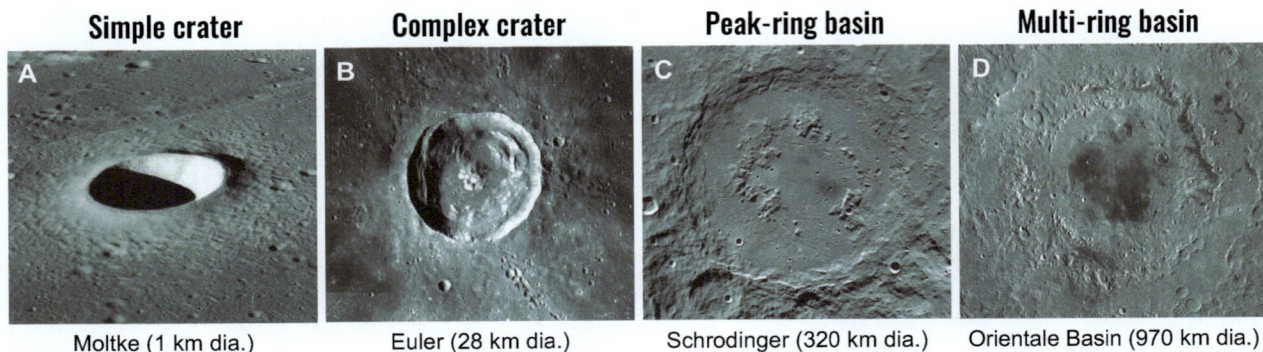

Moltke (1 km dia.)   Euler (28 km dia.)   Schrodinger (320 km dia.)   Orientale Basin (970 km dia.)

Figure 23.5 Impact craters vary in morphology and extent depending on the energy of the impact and the strength of the surface. These lunar impact craters show changes in crater shape and size with increasing impact energy, from (**A**) a small, simple crater, to (**B**) a more complex crater, (**C**) a large, peak-ring impact basin, and (**D**) a massive, multi-ring impact basin. See also **Fig. 23.36**. Source: NASA.

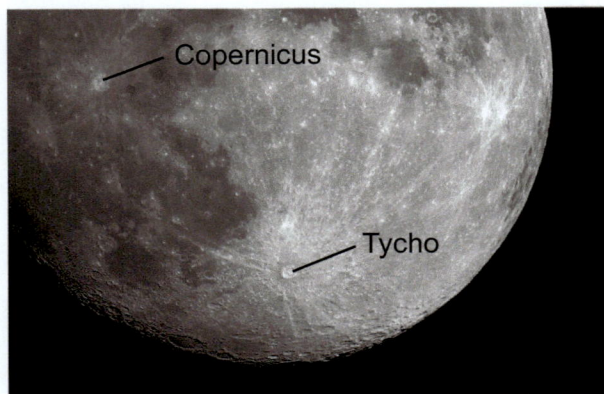

Figure 23.6 Tycho and Copernicus – lunar craters with extensive ray systems. Tycho (85 km dia.) is the youngest large crater on the Moon's nearside and exhibits a very extensive ray system of debris blasted out by the crater-forming impact about 108 Ma. Copernicus (93 km dia.) is surrounded by dark mare (lava plains) which contrast nicely with its shorter bright rays. Source: NASA.

sizes. For example, the **shield volcano** Olympus Mons on Mars (**Fig. 23.9**) is almost as large as France! All the inner planets show areas of vast lava flows, some as large as the continental United States. Some of the moons of the outer planets also exhibit **cryovolcanic flows**, that is, a lava-analog slush of ice and water that is produced by extrusion of cryomagma (i.e., the water) from below, or from heating of the crust by impacts.

Many surficial geomorphic processes on Earth also occur on extraterrestrial bodies. For example, mechanical rounding of cobbles occurred in ancient rivers on Mars, as it does today on Earth (**Fig. 23.10**). Lacustrine landforms – formed in lake basins – have been observed in ancient lake basins on Mars, as well as in present-day lakes and lake basins on Titan. Some processes, however, are very different, such as the chemical reactions that break down rock on asteroids, or ice in the crust of outer Solar System moons.

On all planets, sediment is moved by suites of processes that wear down the surface, generally grouped under the terms **gradation** or **denudation**. Because all planetary bodies have gravity, the most common gradational process is **mass wasting** (see Chapter 14), which often produces landslides. Long runout landslides are found on rocky bodies like Mars and also on Iapetus (**Fig. 23.11**). Wind-driven (**eolian**) transport of sediment is most common on bodies with atmospheres, like Mars, Venus, and Saturn's largest moon Titan (**Fig. 23.12**). However, evidence of eolian processes is being increasingly observed even on airless bodies like comets, where the "wind" is formed by out-jetting of gases (**Fig. 23.12E**). Landscapes that result from fluvial, glacial, and ground-ice processes, which are common on Earth, are most prevalent on bodies with atmospheres. Nonetheless, landforms formed by running water or flowing ice have also been interpreted from images of asteroids and icy moons.

23.3 EXPLORATION OF PLANETARY SURFACES

Traditionally, geomorphologists begin by working on the ground, and then follow up with remotely sensed data to expand on, or test, these findings. Increasingly, however, initial reconnaissance of terrestrial geomorphology is made remotely – from space. Only after extensive remote data collection and analyses are surface assets deployed to collect land-based data.

Remote sensing technologies gather information (data) from a surface from afar. Examination of planetary surfaces commonly begins with telescopes, and then progresses to orbiting satellites. To see landscapes up close, spacecraft missions are required. Although these missions are expensive, advances in miniaturization are making these endeavors more affordable. Remotely sensed data from planetary landscapes commonly include images taken with sophisticated cameras. This kind of imaging is often easiest for us to understand because it usually corresponds to what we see with our eyes or in photographs. However, other wavelengths

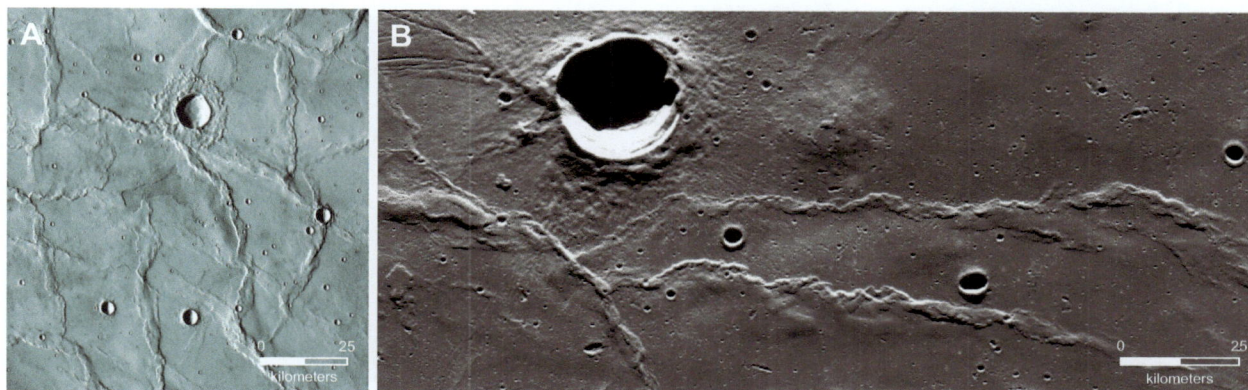

Figure 23.7 Wrinkle ridges on (**A**) Hesperia Planum on Mars, and (**B**) Mare Tranquilitatis on Earth's Moon. Source: NASA.

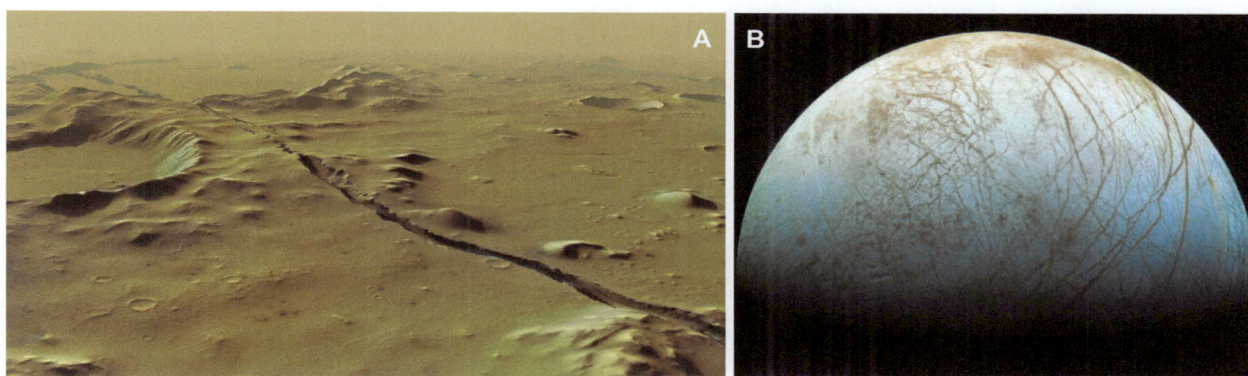

Figure 23.8 Fissures on other planetary bodies. **A.** This small section of the Cerberus Fossae on Mars has fissures with widths of ≈250 m. Source: ESA. **B.** Europa, a moon of Jupiter. Many of the fissures shown here are a few km wide and > 200 m deep. Europa is thought to have a subsurface ocean, and so is a foremost astrobiology target. Source: NASA.

within the electromagnetic spectrum can also provide important information. For example, using wavelengths slightly longer than those of visible light provides information on mineralogy, whereas even longer wavelengths can be used to gather information on surface roughness.

The moon has been an exception to the general process of planetary exploration. The moon was not well-explored remotely before the Apollo astronauts landed there, starting in 1969. Human exploration was made possible because the moon is so close to Earth. The Apollo missions were less about science and more about beating the Soviets to the moon. However, the last Apollo mission to the moon, Apollo 17, included a geologist who recovered important rock samples that provided evidence for a lunar magnetic field, and other samples that suggested the presence of lunar volcanism (**Fig. 23.13**). Recent analyses of these samples have even revealed trace amounts of water.

For all other planetary bodies (besides the moon), surface missions are only done after extensive remote sensing, using flyby or orbiting spacecraft. Surface missions might include a stationary lander, such as the early landers on Mars, or the series of Venera spacecraft that the Soviets landed on Venus. More advanced surface assets include rovers of various types, most of which have gone to Mars.

Some planetary geomorphologists perform their work by exploring landscapes on Earth that might be analogous to those on extraterrestrial bodies. Even though there are many differences between the two, this method allows geomorphologists to collect "ground truth" data on how the planetary landscapes *might* have formed. After all, the laws of physics and chemistry are universal, even though they might produce different results under dissimilar conditions.

Physical and chemical experiments can also provide data for understanding landscape formation. For example, wind tunnel experiments can be used to understand more about eolian landscapes on Mars or Titan, and cryogenic experiments can be used to learn more about how chemistry shapes landscapes on icy moons. With these experimental or terrestrial data, set within the relevant physical and chemical laws, planetary geomorphologists can construct computer models to better understand planetary landscapes.

Figure 23.9 Olympus Mons, the largest volcano in the Solar System. It is 624 km in diameter, 25 km high, and rimmed by a 6 km-high scarp, with an 80 km-wide caldera. The volcanoes that comprise the large island of Hawai'i are about the size of just its central caldera. Source: Viking orbiter image from NASA.

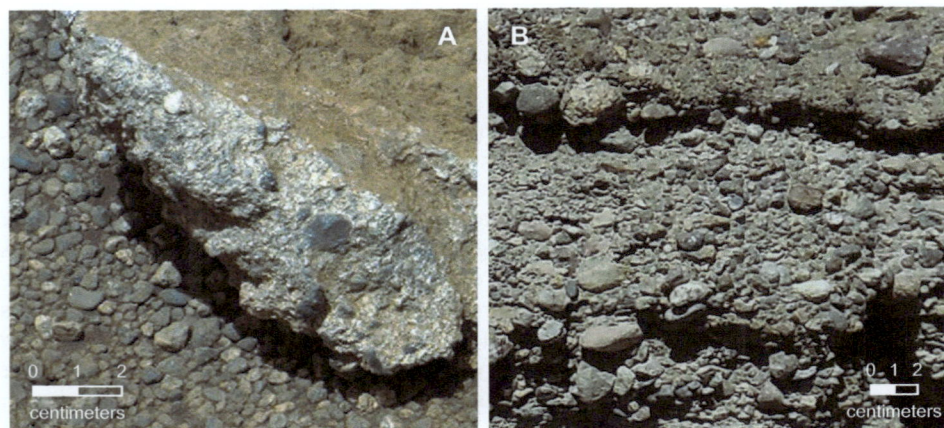

Figure 23.10 Rounded cobbles near the Mars Science Laboratory landing site in Gale Crater (**A**) compare favorably in size and shape with rounded rocks in alluvial fan deposits in the Atacama Desert, Chile (**B**). The Martian cobbles were deposited by running water a few billion years ago, whereas the Atacama Desert cobbles are only a few years to decades old. Source: NASA/JPL-Caltech/MSSS and PSI.

23.4 EARTH'S MOON

The surface of Earth's Moon is dominated by impact craters, which range in size from a fraction of a millimeter in diameter to giant impact basins (**Figs. 23.5, 23.6, 23.14**). The near side of the Moon, that is, the side facing Earth, has many large impact basins (**Fig. 23.15**). However, their surfaces are mainly largely smooth, dark **maria** (Latin *maria*, "sea"; also the root of the English word, "marine"). The name maria derived from early astronomers in the seventeenth century, who mistook them for seas. The Moon's maria are actually vast lava flows, formed when meteorite impacts formed giant impact basins that then filled with magma from the lunar interior. The rounded outlines of many of these basins point to their formation by impact. The Imbrium Basin is a good example of a multi-ring maria basin (**Fig. 23.15**). Its outermost ring is formed on the Apennine Mountains, its intermediate ring (850 km dia.) forming smaller mountain chains, and its innermost ring (570 km dia.) is defined by arc-like ridges where lava buried the ring structure.

Copernicus Crater, visible on the Moon's nearside, is a relatively fresh crater, believed to have formed less than a

Figure 23.11 Image of a long, runout landslide on Iapetus, an icy moon of Saturn. The length of the deposit is about 10 times the height of the cliff that it came from. The white arrowhead indicates the direction of solar illumination. Source: After Singer, K., McKinnon, W., Schenk, P. et al. (2012), with permission from Springer Nature.

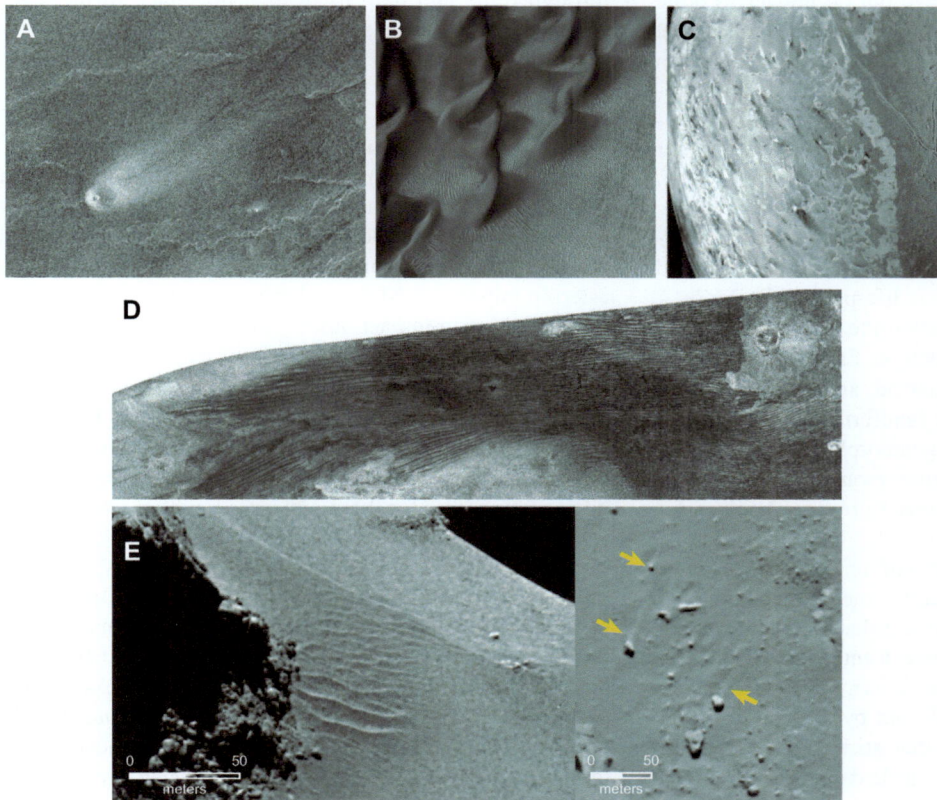

Figure 23.12 Features formed by wind used to be thought to form only on bodies with atmospheres, as shown in **A–D**. However, they are also observed on bodies without continuous atmospheres, like comets (**E**). Shown here are (**A**) a wind streak on Venus, formed in the lee of a volcanic cone, (**B**) transverse sand dunes on Mars and (**C**) linear dunes on Triton, an icy moon of Neptune, (**D**) wind streaks on Triton, as material vented from the subsurface is transported downwind, and (**E**) dune-like deposits (left) and wind-streak like deposits (right, at arrows) on Comet 67P. Source: NASA.

Figure 23.13 Geologist-astronaut Harrison "Jack" Schmitt uses an adjustable sampling scoop to retrieve lunar samples at the Taurus-Littrow landing site. The "dirty" appearance of the astronaut's suit illustrates how dusty the lunar surface is. Source: NASA.

billion years ago (**Figs. 23.6, 23.15**). Its system of bright rays, composed of material thrown out on impact, is quite prominent at full Moon. The rest of the near side, and almost all the far side, of the Moon are composed of highlands with numerous impact craters (**Fig. 23.14H**). Most of these simple, bowl-shaped craters are smaller than ≈15 km in diameter. Tycho is another relatively fresh rayed crater, similar to Copernicus, located on the lunar highlands (**Figs. 23.6, 23.14F, 23.15**).

Impact craters and their ejecta dominate lunar geomorphology. Dust from these impacts – the result of continued impacts over the ≈4.5-billion-year lifespan of the Moon – blankets the lunar surface, and even the astronaut's spacesuits (**Fig. 23.13**). Unlike craters on Earth, lunar craters have not been modified by rain, wind, and other terrestrial processes. Volcanic and tectonic landforms associated with impacts also characterize lunar geomorphology. For example, crater rims can form circular mountain chains, and the fractures generated by the impact provide openings for magma to erupt onto the surface.

Although the most obvious form of lunar volcanism is the effusive basaltic lava flows that fill the maria, the lunar landscape also has about 100 dark deposits that mantle the underlying relief. These landscapes have formed from explosive eruptions. **Pyroclastic beads** (Greek *pur*, "fire" and *klastós*, "broken") found in samples collected by the Apollo astronauts contain water, indicating that such explosions were powered by water. The dark pyroclastic deposits on the lunar surface provide clear evidence of explosive lunar volcanism, likely with assistance by water in some cases.

In addition to the tectonic activity associated with the > 3 Ga maria basins, tectonic activity on the Moon is likely

also occurring today. Wrinkle ridges that formed by contraction (**Fig. 23.7B**), fresh blocks of rock on the lunar surface, and fissures with sharp edges all point to contemporary tectonism. Seismometers placed on the lunar surface by Apollo astronauts revealed weak moonquakes deep below the surface that have been attributed to gravitational interactions with Earth, as well as shallower earthquakes within the lunar crust. The latter are thought to be due to the shrinking of the Moon as its interior cools. Recent observations of tectonic morphologies on the lunar surface indicate that the deep and/or shallow moonquakes are strong enough to fracture the lunar crust. In summary, the Moon has been a convenient training ground for geomorphologists studying extraterrestrial volcanism and tectonics.

In addition to providing evidence of water in the moon, samples collected by the Apollo astronauts have provided critical age information, confirming that most of the lunar basins formed > 3.8 Ga. The lunar highlands are even older. Scientists have been able to develop correlations between the ages of these lunar samples and the density of impact craters in the areas where the samples were collected. These correlations allow planetary geomorphologist to estimate the ages of lunar (and even other) landscapes, and demonstrate that older surfaces have more impact craters (**Fig. 23.16**). Thus, lunar impact craters provide important clues about impact cratering processes through time, and help us to date planetary surfaces for which we do not have samples.

23.5 MERCURY

Like the Moon, Mercury is a small body dominated by impact craters. Mercury is the closest planet to the Sun (**Fig. 23.1**), whose massive gravity pulls in **asteroids** (rocks out in space). Sometimes, instead of getting pulled into the Sun, the asteroids strike Mercury. As on the moon, asteroid impacts on Mercury form small impact craters and large, multi-ring impact basins several hundred kilometers across, as well as ejecta blankets and cratered plains. The largest impact basin on Mercury – the Caloris Basin – contains many small impact craters (**Fig. 23.17**). The many superposed craters indicate that the Caloris Basin was most likely formed early in Mercury's history, that is, there has been enough time for these craters to develop *after* Caloris formed.

On the opposite side of Mercury from Caloris, the landscape is hilly with linear valleys. Planetary geomorphologists infer that the asteroid that impacted Mercury to form the Caloris Basin created a compression wave that passed through the planet, deforming the surface rocks on the opposite side! The impact basin itself shows fissures or rifts oriented radially to the basin (**Fig. 23.17A**), indicating that they were formed by crustal extension, when it rebounded up during the impact event.

Some of the impact craters and basins on Mercury, which were originally depressions, have slowly rebounded over time, so that the crater floors are now flat. Other craters and

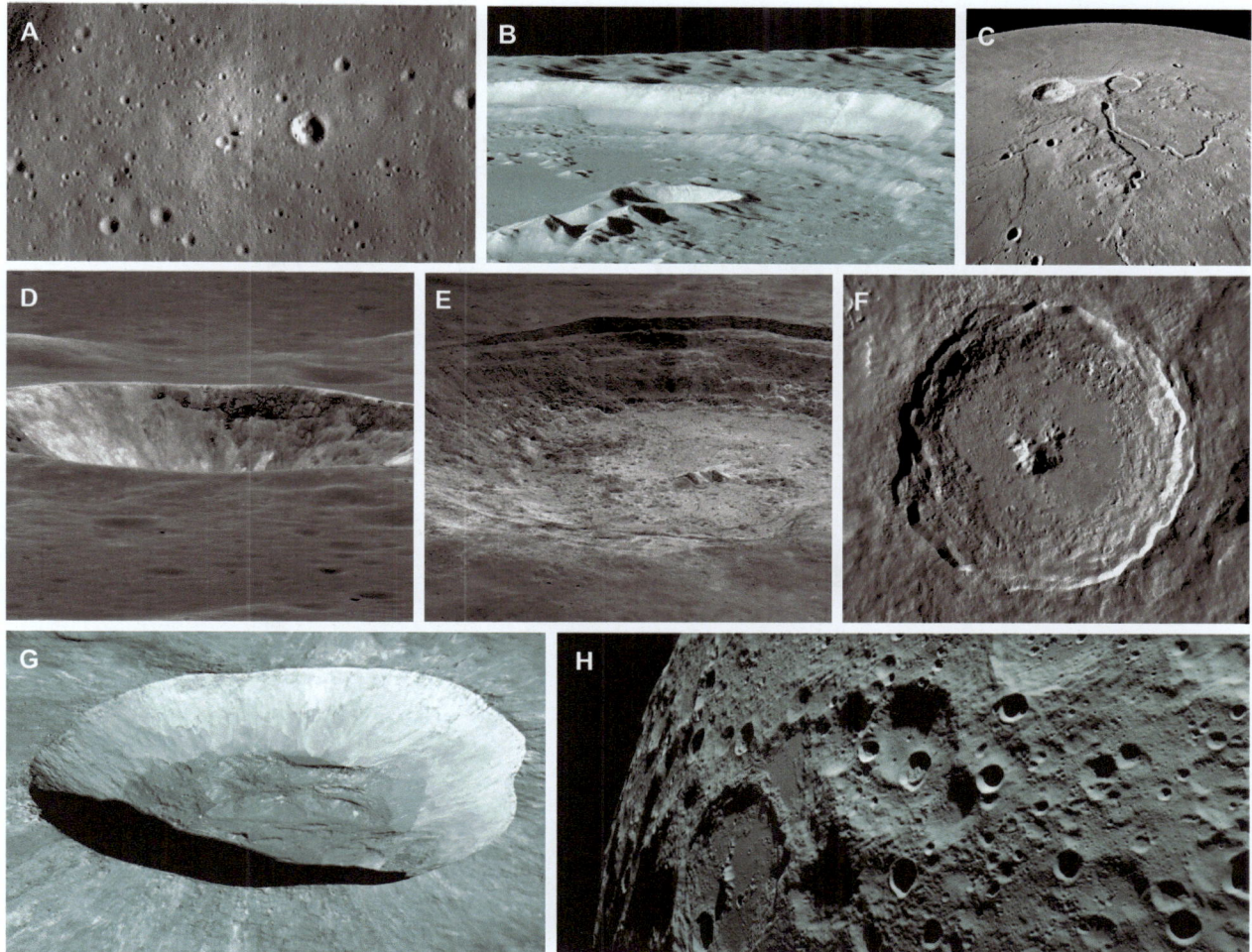

Figure 23.14 NASA images of the Moon's surface. **A.** The comparatively low-relief landscape of the first manned lunar landing site, in July 1969 – Mare Tranquilitatis (The Sea of Tranquility). The bright white base of the lunar module and its shadow are visible in the center of the image. **B.** This image of only a portion of Antoniadi crater, on the Moon's dark side, illustrates its vast size. The crater wall in the background rises 4 km above the floor, and the bottom of the small foreground crater is the lowest point on the surface of the Moon, ≈7.6 km deep. **C.** This Apollo 15 image shows the "Cobra head" volcanic vent, out of which poured lava that flowed down the Aristarchus Plateau, before spilling out onto the lava plains of Oceanus Procellarum. To the left of the vent is Aristarchus Crater. **D.** The 5.7 km diameter Wallach Crater, as viewed from the Lunar Reconnaissance Orbiter. This site is formed in a thin layer of dark, basaltic lava (observed right below the crater rim) overlying brighter-colored anorthosite rock. The impact ejected the anorthositic rocks, which can be seen in the crater walls and on the surrounding landscape. Many of the deepest and oldest rocks are now on top of the shallowest and youngest rocks – a case of inverted stratigraphy. **E.** Aristarchus Crater, ≈40 km in diameter and 3.5 km deep, is one of the brightest features visible from Earth. The crater is bright because of its relatively young age (450 Ma). **F.** Tycho Crater, only 108 million years old, as viewed from the Lunar Reconnaissance Orbiter (see also **Fig. 23.15**). Its central peaks rise 1.6 km above the crater floor, and the rays of ejected material span out almost 1,500 km from the crater. **G.** The 22 km-wide Giordano Bruno crater on the far side of the Moon. The height and sharpness of the rim are impressive, as are the rolling hills and rugged nature of the crater floor. Giordano Bruno crater is one of the youngest craters on the Moon – estimated to be only 4 million years old. **H.** The highly cratered far side of the Moon. Source: NASA.

basins have flat floors because they have been filled by lava, like the lunar maria. Outside of impact basins, lava flows on Mercury have formed smooth plains elsewhere. However, some of these plains might have been formed by ejecta from impacts. More ejecta falls back onto Mercury than it does on the Moon, because of that planet's larger iron–nickel core, which makes Mercury's gravity (3.7 m/s²) stronger than that of the Moon (1.6 m/s²). Thus, ejecta from Mercury are less likely to escape into space.

Mercury's core has influenced its geomorphology in another important way – by producing tectonic features.

Iron and nickel shrink when they cool. Mercury's iron–nickel core has cooled and shrunk over the ≈4.5 Ga history of the Solar System, resulting in tectonic contraction at the surface. Planetary geologists have observed many large thrust faults on Mercury, formed as one part of the crust is thrust up over another (**Fig. 23.18**). Based on the size of these faults, planetary scientists can calculate how much Mercury has shrunk, as well as the size and composition of Mercury's core. Thus, important information about the interior of Mercury has been gleaned from the study of its geomorphology.

Figure 23.15 The near side of the Moon shows several adjacent impact basins with diameters of up to 1,200 km, filled with dark lava, forming maria. The maria, such as Mare Imbrium, show a few bright impact craters with ejecta, such as Copernicus. To the north and south of these maria are brighter lunar highlights, which are comprised of numerous impact craters on the lunar surface. Like Copernicus, Tycho Crater is a young, rayed crater. Source: NASA.

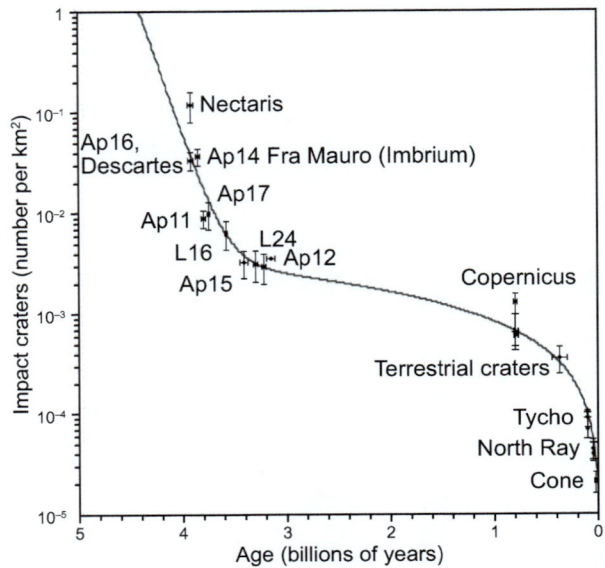

Figure 23.16 The correlation between the density of craters on the lunar surface vs surface age. Source: After G. H. Heiken, D. T. Vaniman, B. M. French, ©1991, Cambridge University Press.

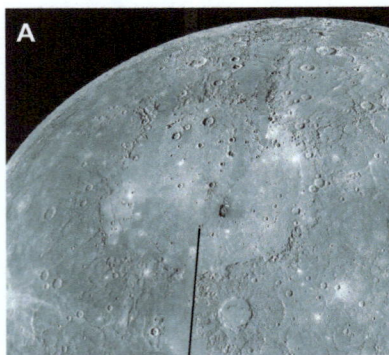

Caloris Basin

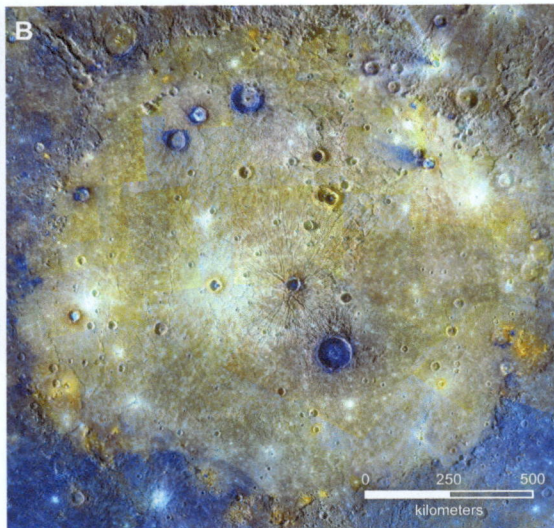

Figure 23.17 Images of Mercury's surface. **A.** The 1,525 km-wide Caloris Basin, the largest feature on the surface of Mercury, was formed by the impact of a large meteorite in the early Solar System. Source: Smithsonian Air and Space Museum. **B.** This color mosaic of the Caloris Basin allows the correlation of geologic features. Lava that flooded the basin interior appears orange. After the flooding, craters excavated material from beneath the surface. The larger of these craters exposed darker material (blue in this image) from beneath the surface lavas, likely giving a glimpse of the original material of the basin floor. Source: NASA.

23.6 VENUS

Venus is sometimes called "Earth's evil twin." Although it is similar in size and density, Venus' atmosphere, surface conditions, and geomorphology are all very different from that of Earth. The hot, dense atmosphere of Venus is comprised of 95% CO_2, and rains sulfuric acid. The surface temperature is ≈ 900 °F – hot enough to melt lead. The Venusian geomorphology reflects these extreme conditions. The dense atmosphere of Venus cannot be penetrated by visible light, so spacecraft cannot take visible light photographs of the surface, but we can image the surface using synthetic-aperture radar (SAR), which can provide information about surface roughness.

Conditions on Venus make impact craters appear very different than they do on the Moon or Mercury. Near-crater ejecta are bright in SAR images, indicating that the ejecta are rough, rocky, and blocky, as on the Moon and Mercury. Additionally, Venusian craters often exhibit long flows of material that once was fluid (**Fig. 23.19**). The high temperatures at the Venusian surface may cause impact ejecta to become fluidized, forming lava, as the rocks on the hot Venusian surface melt at the impact site.

As on Mercury, where the number of craters on the Caloris Basin was used to infer that it had formed early in Mercury's history, the numbers of craters on Venus can be used to learn about its history as well. Venus has surprisingly few craters. Part of the reason for the low crater densities is that many asteroids burn up in (or are significantly slowed down by) the dense atmosphere. Thus, fewer hypervelocity impacts occur on the surface than if Venus had a thinner atmosphere. Scientists also think that Venus has been **resurfaced**, a process by which a planetary surface is covered over with new material, or eroded, so that many of the early impact craters are no longer apparent. Based on its somewhat sparse crater

Figure 23.18 Enterprise Rupes (at the white arrows) is tens of kilometers wide and has > 3 km of relief. It is the surface expression of a thrust fault, and the largest lobate scarp on Mercury. Source: NASA.

population, planetary geomorphologists have estimated that resurfacing on Venus has occurred recently, ≈500 Ma. Although 500 million years seems like a long time, it's only the last 10% of the history of the Solar System.

What would have caused this resurfacing? On Earth, resurfacing is driven in large part by plate tectonics. However, in the Solar System, only Earth shows evidence of tectonic activity (with the possible exception of Europa, an icy moon of Jupiter). The best idea so far as to what caused this resurfacing on Venus is that, because Venus does *not* have plate tectonics, the heat in the interior of the planet builds up until it causes the surface to completely overturn. However, this hypothesis cannot be adequately tested without better data.

Venus does show evidence for some tectonic activity. Some high-elevation parts of Venus appear complexly textured and deformed. These areas exhibit troughs and grabens, indicating crustal extension. Areas with ridges and folds indicate crustal compression (**Fig. 23.20**). These high-standing terrains, or **tesserae** (singular, tessera), might be very old areas that have been deforming over a long period of time. Alternatively, recent sinking, or downwelling, of material inside Venus might have led to compression at the surface, above the region of downwelling and associated extension.

Whereas tesserae are found in the Venusian uplands, lowlands exhibit pervasive volcanism and lava flows. These **flood lavas** were produced by massive outpourings of molten rock. Some of these lava plains have channels, formed where the margins of the lava have hardened while the lava in the middle of the channel continued to flow. One of these channels, the longest known channel in the Solar System, is Baltis Vallis, ≈2 km wide and ≈7,000 km in discernable length (**Fig. 23.21**) – longer than the Nile River in Africa.

Wind is also an active geomorphic agent on Venus. Wind streaks (**Fig. 23.12A**), which may be either dark or bright, often point to variable wind directions. A few dune fields have been identified on Venus from SAR imagery. However,

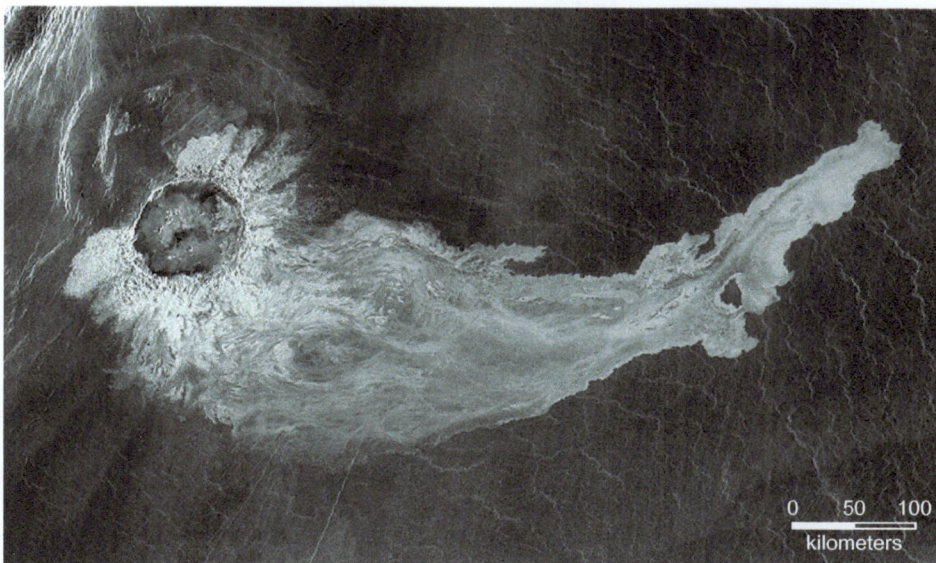

Figure 23.19 Addams Crater (87 km dia.) is an example of a Venusian impact crater with long (> 600 km), flowing ejecta. Source: NASA.

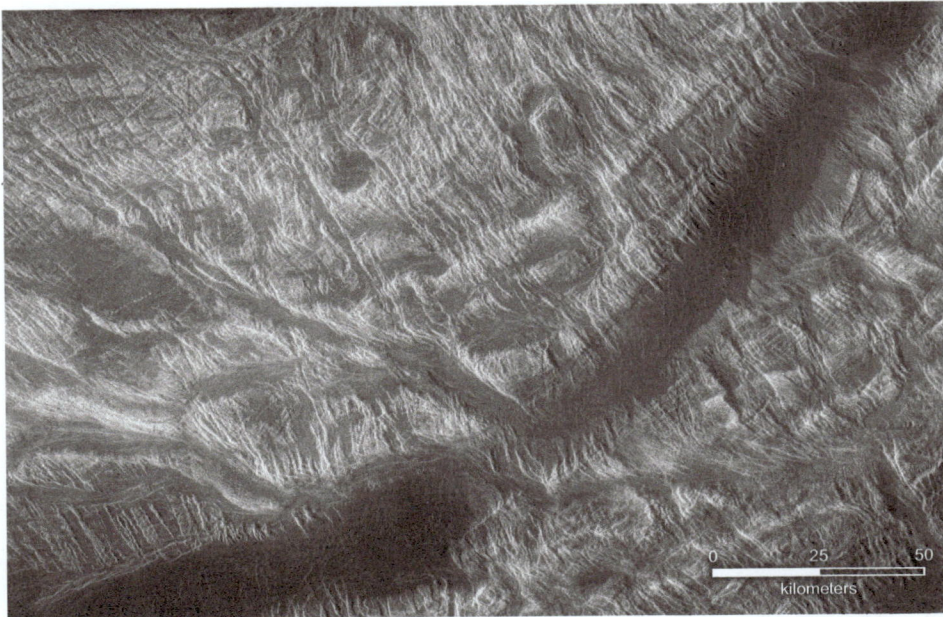

Figure 23.20 Portion of the Ovda Regio, a large highland terrain or tessera on Venus whose complex fabric attests to a long history of tectonic deformation. The underlying fabric of the linear ridges and valleys trends NE–SW, whereas a later (cross-cutting) set of features trends NW–SE. The large valley in the image was likely caused by crustal extension and later became filled with lava. Source: NASA/JPL.

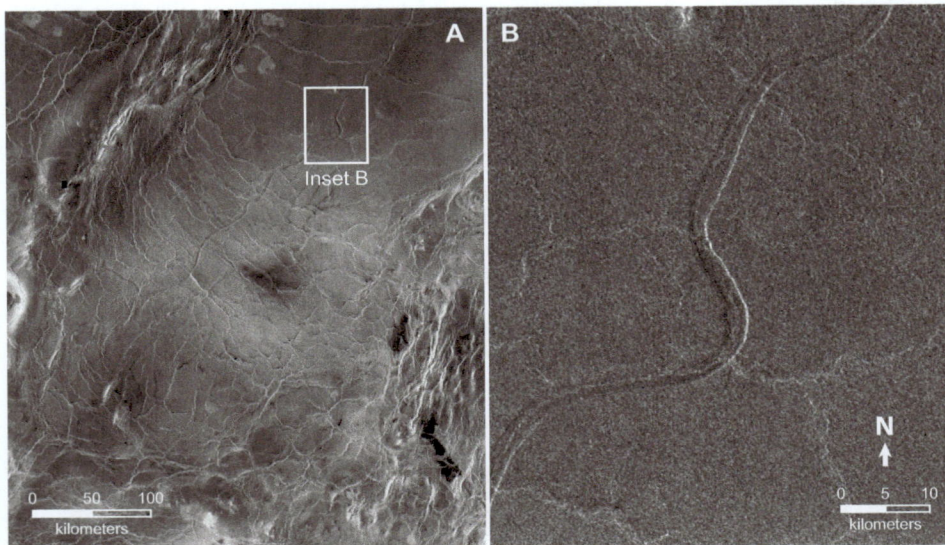

Figure 23.21 SAR images of the Venusian surface, showing a 200 km-long segment of Baltis Vallis, the longest lava channel discovered on Venus to date. Because both ends of the channel are obscured, its original length is unknown. Source: NASA.

there are likely many more dunes that are not visible at the resolution of the spacecraft systems.

23.7 MARS

Mars is a wonderland of planetary geomorphology. In addition to the extraterrestrial landscapes we have already discussed, Mars exhibits a variety of other landforms, often closely analogous to those on Earth. Unlike Earth, Mars does not exhibit plate tectonics, nor did it undergo the global resurfacing that apparently occurred for Venus. As a result, Mars provides a longer record of early terrestrial processes than does either of those two planets. In addition, we have far better data for Mars than for any other extraterrestrial body, enabling both global and highly local investigations.

Like almost all other planetary bodies, Mars hosts a variety of impact craters. Indeed, the vast northern lowlands, making up almost half of Mars, might be one giant impact basin (**Fig. 23.22**)!

The northern Martian lowlands are lightly cratered, implying that they have been resurfaced. In contrast, the southern highlands of Mars are heavily cratered, showing that they are ancient, exposed to impacts from space for > 4 billion years (**Fig. 23.22**). Like some craters on Mercury and the Moon, many of the older Martian craters have been filled with lava, flattening out their floors (**Fig. 23.23**). Other Martian craters with eroded rims are flat-floored and infilled due to weathering and transport of material into the crater. Although Mars is a desert today, this infilling points to a wetter past climate in which water flowed into craters, carrying

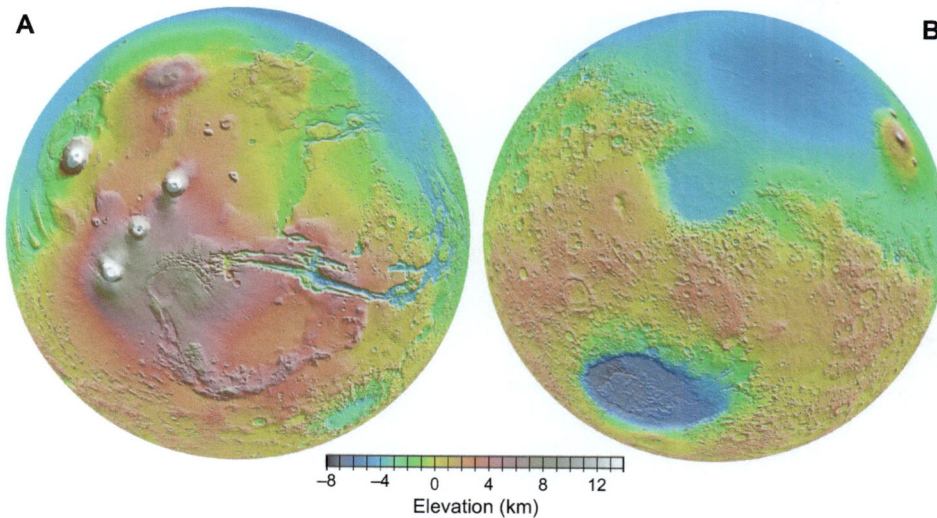

Figure 23.22 False-color, topographic images of Mars. **A.** The Tharsis region, a high-elevation volcanic area with four large volcanoes. Some of the northern lowlands (in blue) are visible around the northern edges. **B.** The northern lowlands (in blue) and the Hellas impact basin in the south (purple). Source: NASA composite images.

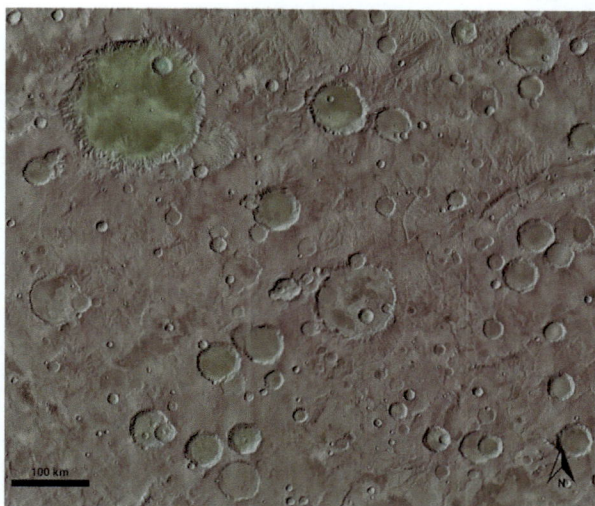

Figure 23.23 A computer-generated image of the Martian southern highlands, showing numerous flat-floored craters, infilled with alluvium. Source: Image by Devon Burr, using data from NASA/JPL-Caltech/Arizona State University.

Figure 23.24 This small, fresh, impact crater on Mars is surrounded by a large, rayed, blast zone of ejecta. Because the terrain where the crater formed is dusty, removal of the reddish dust by the blast makes the blast zone appear blue in this enhanced color image. Source: NASA HiRISE.

and depositing sediment (alluvium). These infilled craters are striking evidence for dramatic climate changes on Mars.

As illustrated earlier (**Fig. 23.16**), estimates of the age of planetary landscapes are often generated based on the number of craters per area and the current impact cratering rate. High-resolution data of the Martian surface indicates that small impacts are still occurring (**Fig. 23.24**), helping to date the youngest Martian surfaces. We have similarly young impacts on Earth, like the meteor that impacted near Chelyabinsk, Russia in 2013.

Like the Moon, Mercury, and Venus, Mars has many volcanic landforms. The same style of fissure-fed volcanism that occurs on those bodies also occurs on Mars, producing extensive lava plains. These plains are often fed from tectonic fissures hundreds of kilometers long, where magma has made its way from the interior, up and onto the surface. Small vents and/or cinder cones also feed the lava onto these plains. The lava-covered Cerberus Plains of Mars, which are larger than the continental United States, were formed by fissure eruptions (**Fig. 23.25**). The Cerberus Plains are the youngest region of Mars, dated by crater counting to only a few tens of millions of years. The Hesperia Planum region is another, older (> 3 Ga) example of a lava plain. The age gap between the Cerberus Plains and Hesperia Planum indicate that lava flows have occurred on Mars throughout almost its entire history.

Other volcanic features on Mars, in addition to vast lava plains, include gigantic volcanoes. Although Martian volcanoes are not as explosive as some stratovolcanoes on Earth, they are much taller and larger. The Tharsis volcanic province on Mars hosts several extremely large volcanoes. The largest of these volcanoes, Olympus Mons (**Fig. 23.9**), is more than twice as tall as Mt. Everest and has 100 times the volume of Mauna Loa on the island of Hawaiʻi, the largest volcano on

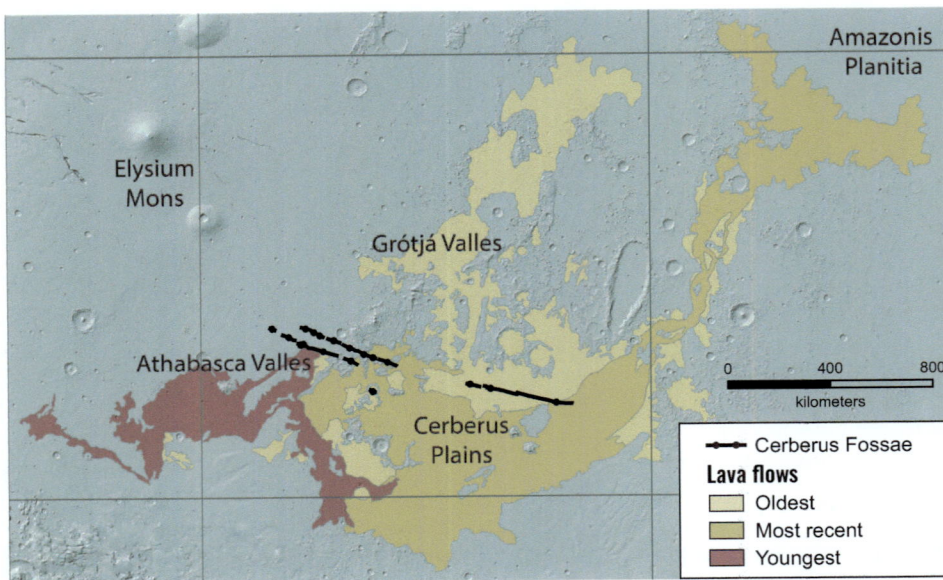

Figure 23.25 Map of some of the youngest known lava flows on Mars. The light-toned areas, including Athabasca Valles, the Cerberus Plains, and Grótjá Valles, are areas of extensive flood lavas. The black fossae indicate tectonic fissures, inferred to be the source of the Cerberus lavas. Source: Used with permission of Elsevier, from J. R. C. Voigt, C. W. Hamilton (2018); permission conveyed through Copyright Clearance Center, Inc.

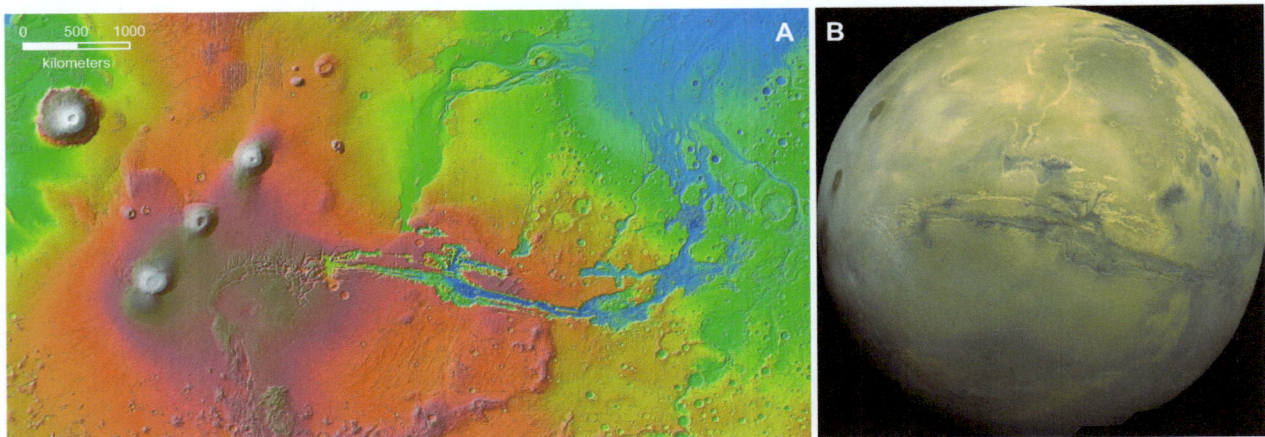

Figure 23.26 Tectonic and volcanic features on Mars. **A.** A color-enhanced image of the topography of the Valles Marineris canyon system. To the west of Valles Marineris is the Tharsis region with its impressive shield volcanoes, including Olympus Mons in the northwestern corner of the image. **B.** Valles Marineris is so large that it is even visible at planetary scales. This view, a photomosaic of 102 Viking Orbiter images, is similar to that which one would see from a spacecraft at 2,500 km above the surface. The center of the scene shows the entire Valles Marineris, extending from Noctis Labyrinthus, the arcuate system of grabens to the west, to the chaotic terrain to the east. Many large, ancient river channels begin in the chaotic terrain and in the north-central canyons, and flow north. Many of these channels flowed into a basin called Acidalia Planitia, the dark area in the extreme north of the picture. The three Tharsis volcanoes (dark red spots), each about 25 km high, are variously visible to the west. Source: NASA.

Earth. In fact, the entire chain of Hawai'ian Islands (from Kauai to Hawai'i) would fit inside Olympus Mons! The Elysium Mons shield volcano shown in **Fig. 23.25** reaches to > 14 km in height!

Martian volcanoes grow to be so massive in part because Mars lacks plate tectonics. On Earth, the movement of tectonic plates can eventually separate volcanos from their magma source below. This separation does not happen on Mars. Instead, the stationary Martian crust allows the subsurface magma chamber to continuously feed molten materials to the same location on the surface for billions of years. In addition, the lower gravity on Mars allows the lava to build the volcano to greater heights.

The fissures (fossae) that feed magma to the volcanic plains on Mars can form in swarms, as they do on Earth (**Fig. 23.25**). Individually, each fissure may only be a few kilometers in width. However, fissure swarms can collectively be hundreds of kilometers wide and long. The Cerberus Fossae – fissures that once fed magma to the Cerberus Plains – are over 5,000 km long (**Figs. 23.8A, 23.25**).

The Valles Marineris is perhaps the largest single tectonic feature on Mars (**Fig. 23.26**). Named after the Mariner 9 spacecraft that discovered this feature in 1972, Valles Marineris is a series of interconnected chasms and troughs that can reach depths of 7 km. Combined, these features extend for 4,000 km, or ≈20% of the circumference of Mars!

The Grand Canyon in Arizona, for comparison, is ≈800 km long and 1.6 km deep. Most of the width of Valles Marineris is a result of extensional tectonic forces, pulling apart the crust. However, some channels and landslides indicate that the Valles Marineris was also widened by erosional processes.

Both the ancient channels observed around Valles Marineris (Fig. 23.26) and the flat-floored craters, interpreted to be filled with alluvium (Fig. 23.23), provide clear evidence that in the past, the climate on Mars was warm and wet enough to produce snowfall. Even more pervasive evidence of surface water is found in the extensive valley networks that cover the southern highlands of Mars (Fig. 23.27). Many of these networks branch, like fluvial networks on Earth. Most of these networks are located on the southern highlands, and have been dated to the early part of Martian history. Geomorphologists believe that these dense, branching networks were formed by running water in the past. Some networks, however, are not so well developed and may have been formed by groundwater processes. Both explanations require rainfall, either to develop runoff or to provide groundwater. Thus, both types of valley networks point to a wetter paleoclimate on Mars. Nonetheless, the mechanism(s) that led to this warmer, wetter paleoclimate is (are) not yet known.

Alluvial fans provide another piece of evidence for a warmer and wetter Martian paleoclimate. On Earth, alluvial fans form where rivers enter a flat area, losing their capacity to carry sediment and deposit it in fan-shaped landforms (Fig. 1.3, and Chapter 10). Alluvial fans on Earth often develop at the bases of steep mountain fronts. On Mars, large alluvial fans have been observed on the insides of impact crater walls (Fig. 23.28), and many smaller examples occur in a variety of locations.

Rivers form valleys – on Mars and elsewhere. But their deposits can sometimes be *inverted* to become positive landforms. This **inversion of topography** happens when the surrounding terrain is less resistant to erosion than the fluvial deposits themselves. Erosion removes the softer sediments of the surrounding landscape, leaving the fluvial deposits standing up in positive (or inverted) relief (Fig. 23.29). On Mars, the erosion is commonly driven by wind. Inverted fluvial deposits are less common on Mars than are valley networks, but they do occur (Fig. 23.29B, C), such as in the

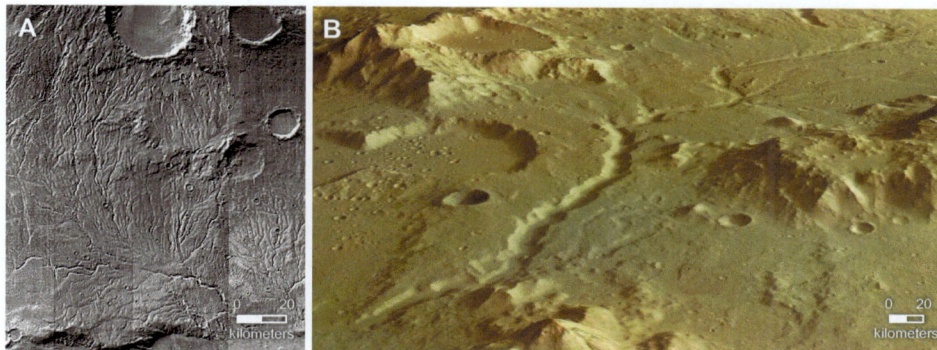

Figure 23.27 Two contrasting valley networks on the Martian southern highlands. **A.** Warrego Valles, an area with dense networks of fluvial channels. Source: NASA. **B.** A dry river channel in the Libya Montes region, near the equator on Mars. The valley, in places as wide as a few kilometers, snakes between hummocky mountain terrain and is fed by numerous tributaries. Source: ESA/DLR/FU Berlin via Astrobiology Magazine, July 18, 2017.

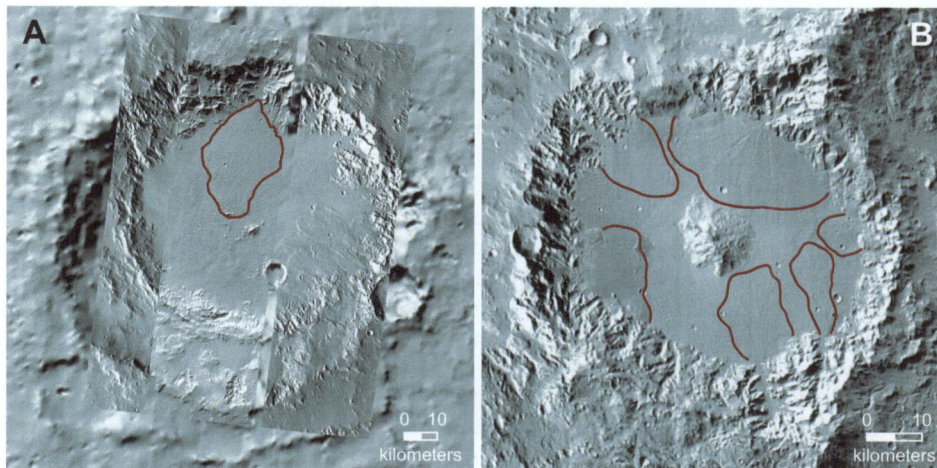

Figure 23.28 Alluvial fans (outlined in red) on Mars, within flat-floored impact craters. The fan material was derived from the crater rim by running water, likely via snowmelt. Source: After Moore, J. M., and A. D. Howard (2005), with permission from John Wiley and Sons.

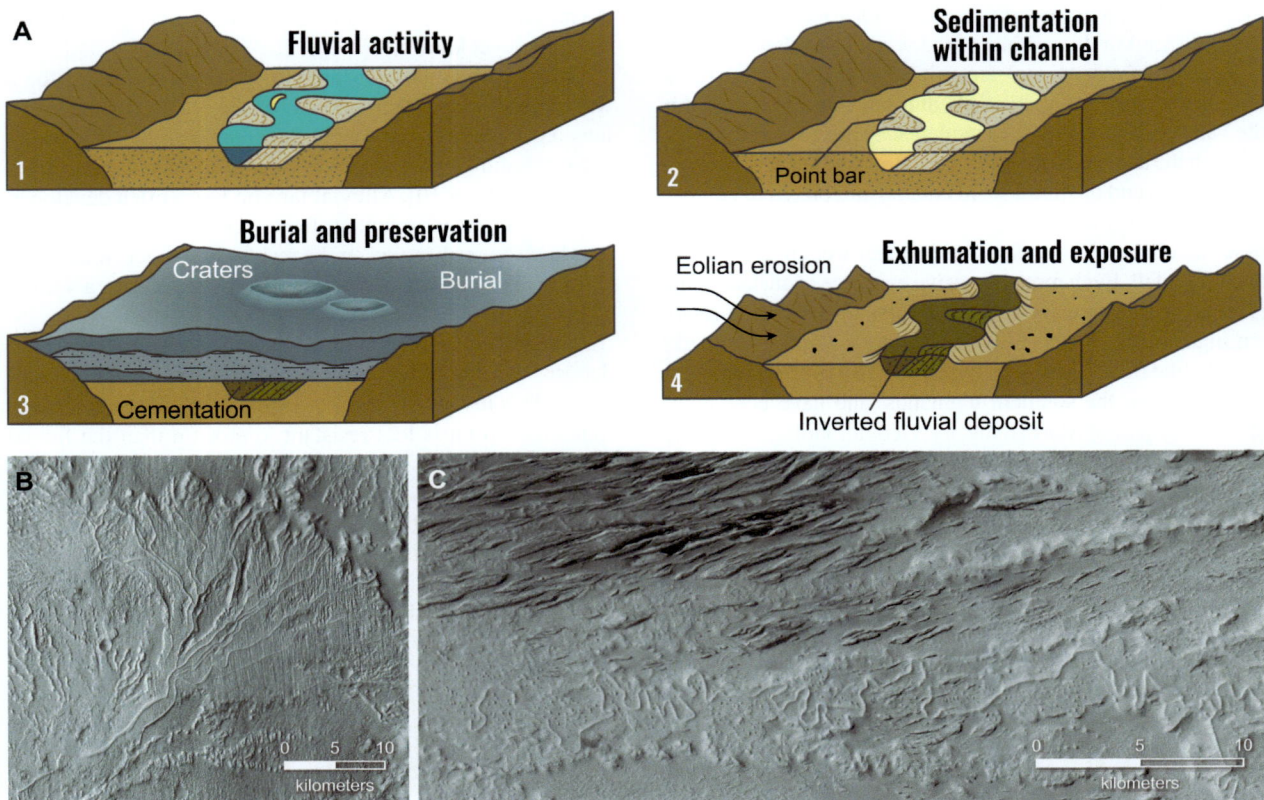

Figure 23.29 Inverted fluvial deposits on Mars. **A.** Diagram showing the multiple agents involved in the formation of inverted fluvial deposits in the Aeolis Dorsa region. 1. Rivers meander across the landscape. 2. Sediments fill the river channel. 3. Burial preserves the landscape while geochemical cementation and induration hardens the deposits. 4. Erosion of the surrounding sediment exposes the fluvial deposits in positive (inverted) relief. Source: Adapted from Williams et al. (2013) with permission from Elsevier, and from Jacobesn and Burr (2017) with permission from the Geological Society of America. **B.** Clearly shown here are inverted meandering forms. To the north, the overlying sediments are being eroded back to form yardangs (see below). Source: NASA. **C.** Inverted fluvial forms traverse an alluvial fan. Source: NASA.

Aeolis Dorsa region and in Valles Marineris. The greatest number of inverted river deposits is found near the boundary between the southern highlands and the northern lowlands, where they have been preserved through an extensive overlying blanket of sediment.

We have seen that alluvial fans can form as water flows into the flat floors of craters, usually through inlet channels (**Fig. 23.28**). Inflowing water (or groundwater) might also form a lake in the crater or even a delta. On Mars, evidence is clear that ancient rivers entered lakes that existed in craters, forming deltas. In one Martian crater (Gale Crater), the Mars Science Laboratory "Curiosity" detected layers of sand and mud, with pebbles and larger sediment nearby, as evidence for a series of ancient lakes fed by streams. In addition to forming in crater basins, lakes likely also formed on Mars in sedimentary basins and in the tectonic rift zone of Valles Marineris (**Fig. 23.26**). As they do on Earth, Martian deltas might have preserved organic sediments from the landscape drained by the rivers. NASA chose Jezero Crater as the landing site for its 2020 Perseverance rover mission because it hosts a classic deltaic deposit, with potential for the preservation of organic materials (**Fig. 23.30**).

Snowfall, which might have melted to produce rivers, deltas, and alluvial fans on Mars, can also – instead of melting – become compacted into ice. With continued snowfall and compaction, ice can form glaciers. Debris-covered glaciers almost certainly occur on Mars today. Planetary scientists conservatively use the non-genetic term "viscous flow features" or "glacial-like landforms" for features on Mars that resemble debris-covered glaciers, because any ice that these features may contain cannot be observed directly from satellites. Nonetheless, planetary geologists believe that glacial ice exists beneath the debris because these features resemble classic glacial landforms such as **moraines** (**Fig. 23.31**) and **eskers** (**Fig. 23.32**) (see Chapter 19). Thus, glaciers provide clear evidence for snowfall on Mars. Eskers also suggest that localized melting has occurred recently beneath some of these glaciers.

Ice can also form in the subsurface on Mars. As ground ice on Earth freezes and thaws, either annually or less often, it can form distinctive patterns on the surface (see Chapter 20). Similar patterns on Mars suggest that ground ice is common in its mid- to high latitudes. However, ground ice on Mars likely sublimates after it thaws, rather than melts, because of the low atmospheric pressure. As a

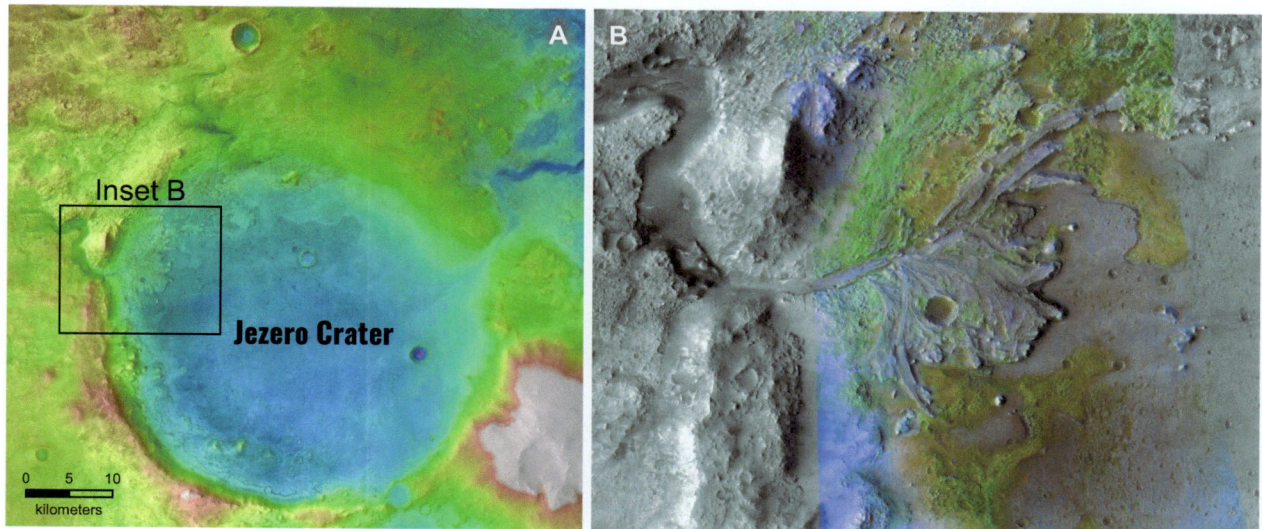

Figure 23.30 Jezero Crater on Mars. **A.** A color elevation map of Jezero Crater, which likely contained a paleolake and was the landing site for NASA's Mars 2020 Perseverance rover mission to look for past life. Ancient rivers carved the inlets on the upper left sides of the crater. The canyon carved by outlet/overflow flooding is visible in the upper right side of the crater. The inset box in **A** shows the location of the delta shown in **B**. Source: NASA.

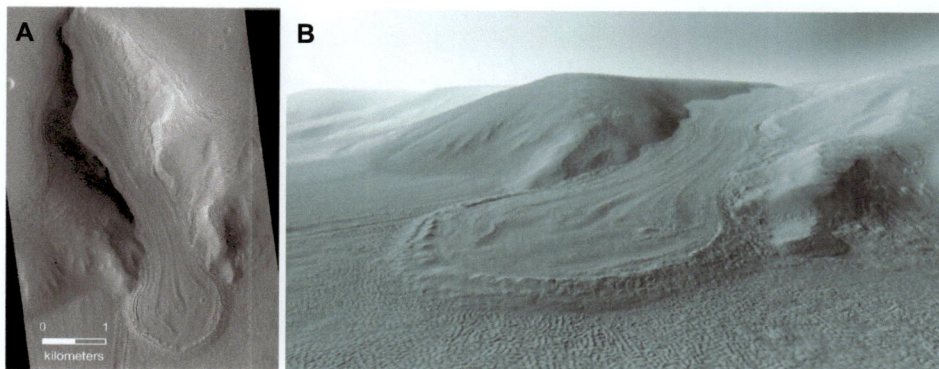

Figure 23.31 A possible debris-covered glacier on Mars. **A.** The glacier flows down this valley. The top of the image would be the zone of accumulation of snow and ice, where debris has also accumulated from the valley walls. The glacier and debris then flowed down-valley and out onto a neighboring plain. **B.** An oblique view of the same landform; note the conspicuous end moraine. Source: NASA HiRISE.

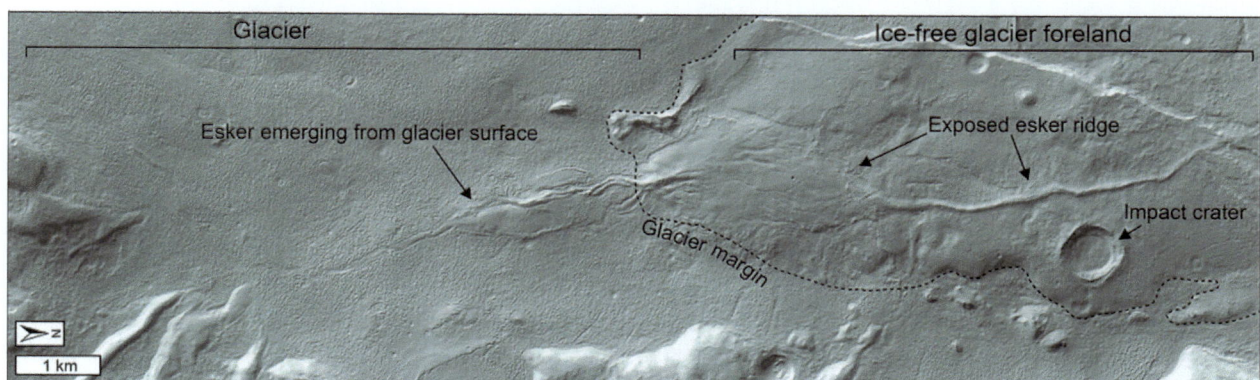

Figure 23.32 A possible esker emerging from a debris-covered glacier in Tempe Terra, Mars. Source: Frances Butcher at NASA, JPL-Caltech, MSSS.

result, the patterns on the land surface look almost "other worldly" (**Fig. 23.33**). The types of patterns in the mid-latitudes of Mars have led planetary geologists to argue for repeated episodes of climate change on Mars, in which ice advances from the poles to the mid-latitudes and retreats again, driven by variations in the tilt of the Martian axis. On the cold dry deserts of the Martian surface, ground

ice processes remain an active geomorphic agent, even today.

In addition to cratering by small impacts and deformation by ground ice, another contemporary and highly active geomorphic agent on Mars is sediment transport by wind (see Chapter 21). Eolian landforms are found on Mars and on a surprising number of other planetary bodies (**Fig.**

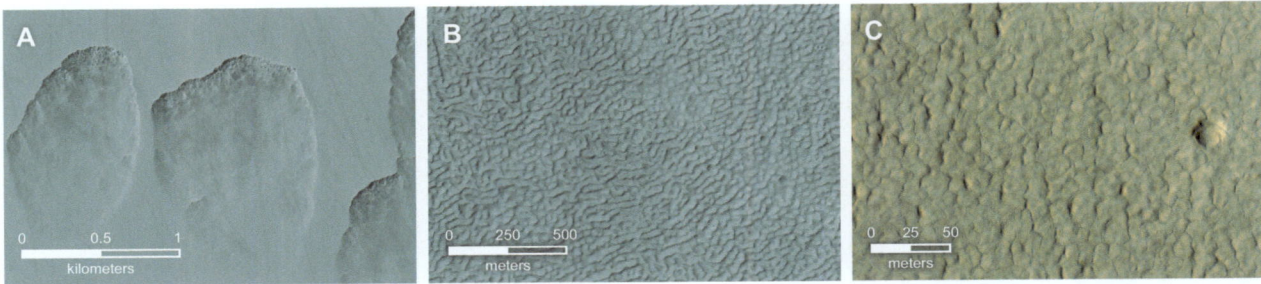

Figure 23.33 Images of possible ice-sublimation landforms on Mars. **A.** Scalloped terrain, showing kilometer-wide pits with pole-facing, backwasting escarpments thought to be due to ice ablation. Similar features are occurring today in the Arctic, driven by thawing permafrost (**Fig. 20.32A**). **B.** "Brain coral" terrain with its riffle-like curvilinear ridge-and-trough textures is usually found at the margins of glacial flow features. **C.** Dissected terrain of hummocky inter-plateau deposits, believed to result from loss of ice cement and subsequent surface subsidence. In all images, north is at the top. Source: HiRISE Images. After Douglas and Mellon (2019). Images from NASA/JPL/Univ. of Arizona.

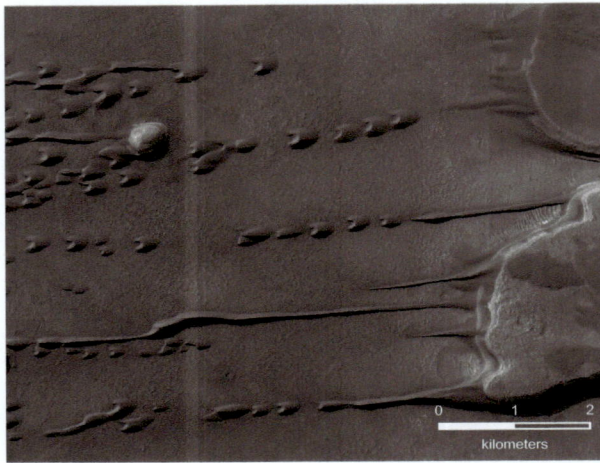

Figure 23.34 Image of mesa-like landforms upwind of longitudinal dunes, which break up into individual barchan dunes farther downwind. The orientation of the dunes suggest that the transport direction is from the sand-rich mesas on the right, toward the left. Source: NASA HiRISE.

23.12). Sand dunes are quite common on Mars, and many are exceptionally well-formed. As discussed in Chapter 21, dune morphologies provide information about sand availability, as well as wind strength and variability. For example, barchan dunes form where there is a limited availability of sand, with horns pointing in the downwind direction (**Fig. 21.25**). Although the ultimate source of sand on Mars is still unknown, the occurrence of barchan dunes adjacent to layered mesas and knobs suggests that, at least in those instances, the sand has been derived from eolian erosion of the bedrock (**Fig. 23.34**).

Blowing sand is also an effective agent of erosion. Wind-blown sand can "sand-blast" rock, forming **ventifacts** (**Fig. 21.17**) and **yardangs** (**Figs. 21.18, 23.35**). Because the Martian atmosphere is ≈100 times less dense than Earth's atmosphere, wind speeds likely have to be very high to transport sand. At these high wind speeds, sand has a lot of force when it impacts rock, causing it to be very erosive. As a result, Mars has a lot of yardangs – wind-eroded, elongated landforms, typically elongated parallel to the dominant wind

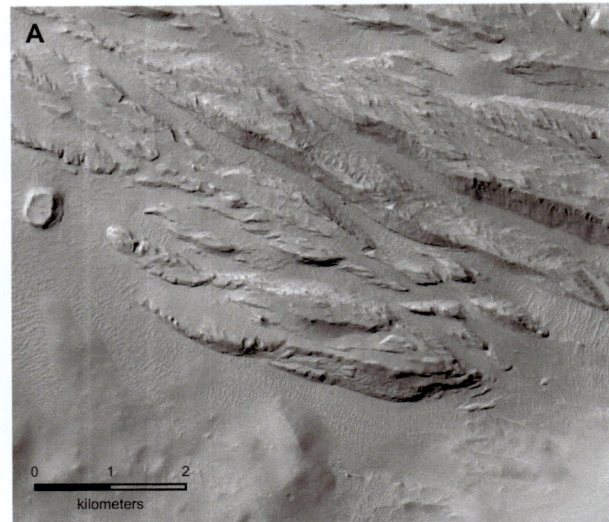

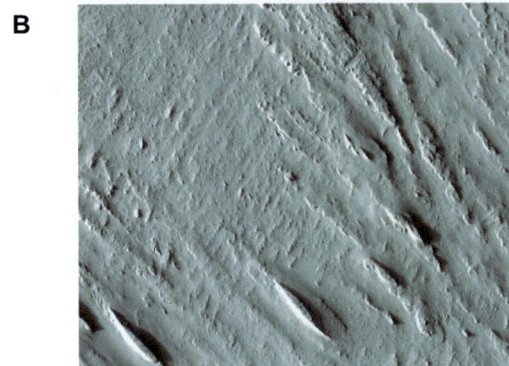

Figure 23.35 Yardangs on Mars. **A.** Well-developed yardangs that have formed from winds (blowing from the lower right). **B.** In this image, the prominent yardang orientation is from upper left to lower right. However, a second, more subtle, orientation occurs almost at right angles (upper right to lower left), suggestive of variable wind directions. Source: NASA.

direction (**Fig. 23.35**). Sometimes, yardang orientations criss-cross, providing evidence for variable wind directions (**Fig. 23.35B**). Indeed, eolian landforms provide important information about both the Martian surface and the atmosphere.

23.8 ICY GEOMORPHOLOGY IN THE OUTER SOLAR SYSTEM

Unlike the rocky planets of the inner Solar System, the planets of the outer Solar System have gaseous outer layers which do not form distinct geomorphic features. However, the many natural satellites or moons in the outer Solar System, along with Kuiper Belt Objects like Pluto, do have solid surfaces. These icy bodies show many of the same geomorphic landforms as their rocky counterparts in the inner Solar System.

As the single most dominant geomorphic process in the Solar System, impact cratering on the outer Solar System bodies generally results in the same features as on the rocky inner planets. However, because of the different response of ice (than rock), impact morphologies may be more extensive on icy bodies. For example, in the inner Solar System, complex impact craters form central ring complexes, for example, Orientale Crater on the Moon has three rings (Fig. 23.5D). However, on icy bodies, such basins might have many more rings. Valhalla, the largest impact structure in the Solar System, has up to 10 times the number of rings as the Orientale basin (Fig. 23.36).

Some impact cratering processes, such as **isostatic rebound**, may actually happen more quickly or thoroughly in ice than in rock. Recall that flat-floored craters formed on Mercury through slow uplift/rebound of the rock. Craters rebound more rapidly and completely on icy bodies, even at the very cold temperatures in the outer Solar System. As a result, flat-floored craters are common on icy worlds (Fig. 23.37).

As on the rocky planets, tectonic features are also prevalent on the icy bodies of the outer Solar System. For example, the icy crust of Europa shows extensive tectonic rifting,

as a result of deformation as it orbits Jupiter (Fig. 23.8B). Europa's deformation/flexing produces friction inside the moon, heating it and likely resulting in melting of its icy crust and possibly even forming a subsurface ocean. For this reason and others, Europa is a prime target for NASA's Europa Clipper mission, which launched in 2024.

Many other icy satellites also show tectonic fissures. One of the most spectacular is Enceladus, a moon of Saturn. It

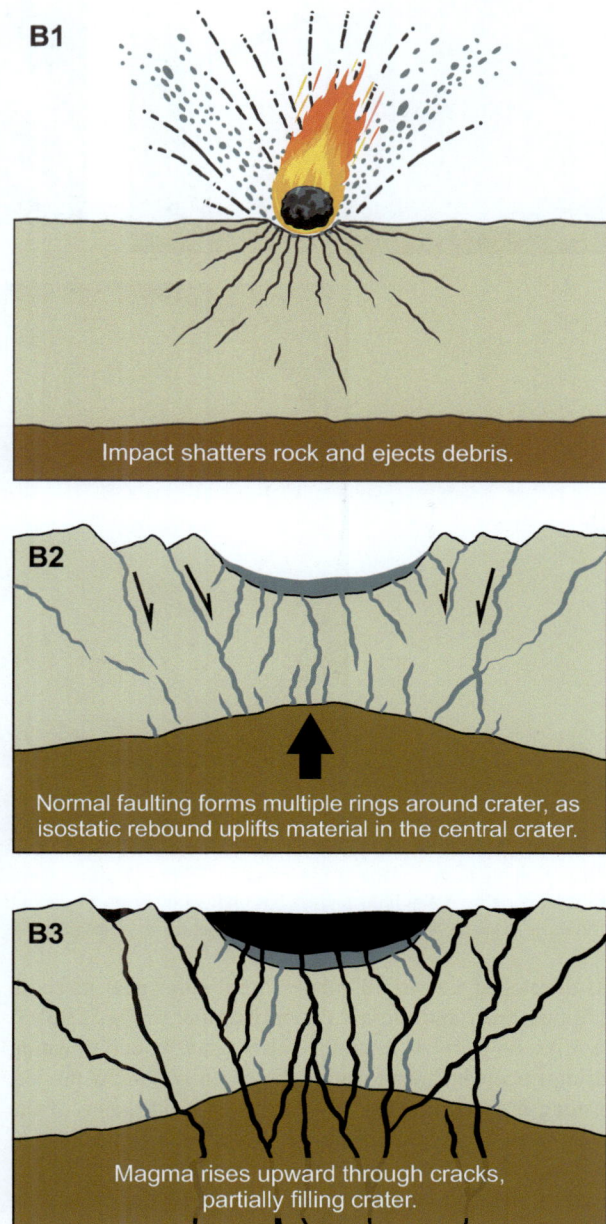

Figure 23.36 A. This mosaic of Voyager 1 images of Jupiter's icy moon, Callisto, shows the Valhalla multi-ring impact basin, consisting of a bright inner region about 600 km across and outer rings that extend up to 4,000 km in diameter, making it the largest impact feature in the Solar System. The outer rings are interpreted as fractures in the crust, formed due to impact. Source: NASA/JPL. **B.** Following the impact, "cleaner" ice or liquid water from beneath the surface flowed up through the impact-generated faults and infilled the central depression, creating the flat, bright, central plains, followed by, over time, isostatic rebound of the crater. Although this diagram depicts the situation for an impact on a rocky satellite, some commonalities likely exist for Callisto as well. Source: After Hamblin (1978).

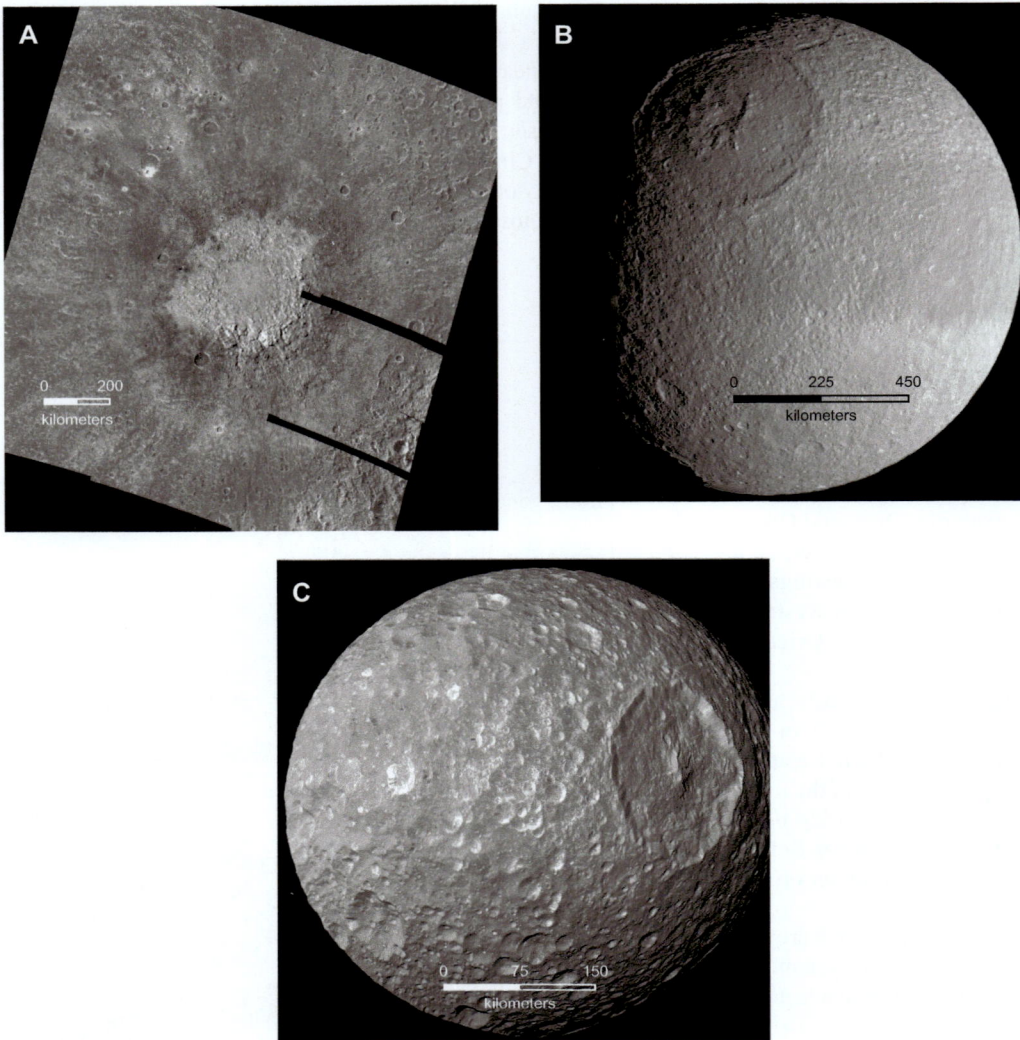

Figure 23.37 Impact craters on some of the Solar System's icy planetary bodies. **A.** An image mosaic of the Lofn impact crater on Callisto, an icy satellite of Jupiter. The impact crater's bright, flat, central area is surrounded by darker ejecta and ejecta rays. The low density of impact craters on the ejecta blanket, as compared to more distant terrain, together with the brightness of the central area, provide evidence that the crater is relatively young. **B.** This image of Tethys, an icy satellite of Saturn, shows the massive Odysseus crater, with its flat floor and central ring complex. Like the floor, the central ring complex has undergone isostatic rebound, making it smaller than when it originally formed. **C.** This image of Mimas, another icy satellite of Saturn, shows the Herschel crater, also with a flat floor and muted central ring complex. Source: NASA/JPL.

has developed a series of subparallel fissures near its south pole that open and close as it orbits Saturn (**Fig. 23.38A, B**). Like on Europa, frictional (also called "tidal") heating during Enceladus' orbit warms the moon and melts the ice, forming a subsurface ocean. During its orbit, some of the subsurface water freezes, putting pressure on the rest of the subsurface water. As a result, the overlying ice cracks, ejecting water and other molecules into space, generating periodic plumes (**Fig. 23.38C**).

As in the inner Solar System, volcanism also occurs in the outer Solar System. However, at the low – in some cases cryogenic (below ≈93 °K) – temperatures of the outer Solar System, and with the icy composition of outer Solar System moons, the volcanic material is not molten rock, but water!

Indeed, plumes of water and other materials from the tectonic fissures on Enceladus (**Fig. 23.38C**) are examples of this type of explosive, low-temperature volcanism. This phenomenon, termed **cryovolcanism**, involves the eruption of water and other liquid- or vapor-phase volatiles onto the cold surfaces of icy satellites. Cryovolcanism requires the presence of liquid water or other compounds inside an icy moon, commonly accomplished by tidal heating and the movement of liquids to the surface. These conditions might exist on a few icy satellites besides Enceladus, although the supporting evidence is uncertain.

Triton, a satellite of Neptune, has perhaps the best evidence for cryovolcanism. The evidence for *effusive* cryovolcanism is in the form of smooth plains (**Fig. 23.39A**).

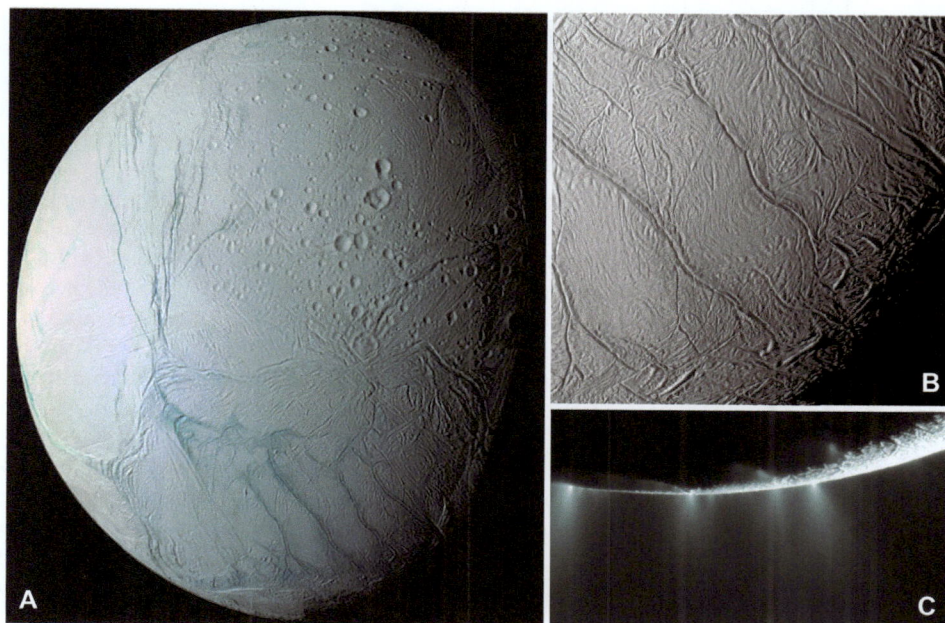

Figure 23.38 Enceladus, a moon of Saturn. **A.** This global view of Enceladus shows its icy crust with a series of subparallel tectonic fissures (bluish in this false color mosaic) near its south pole (bottom of the image). **B.** A close-up view of Enceladus shows the distinctive pattern of these continuous, ridged, slightly curved, and rough fissures. **C.** Plumes of water and other molecules spray out into space from fissures on Enceladus. Source: NASA.

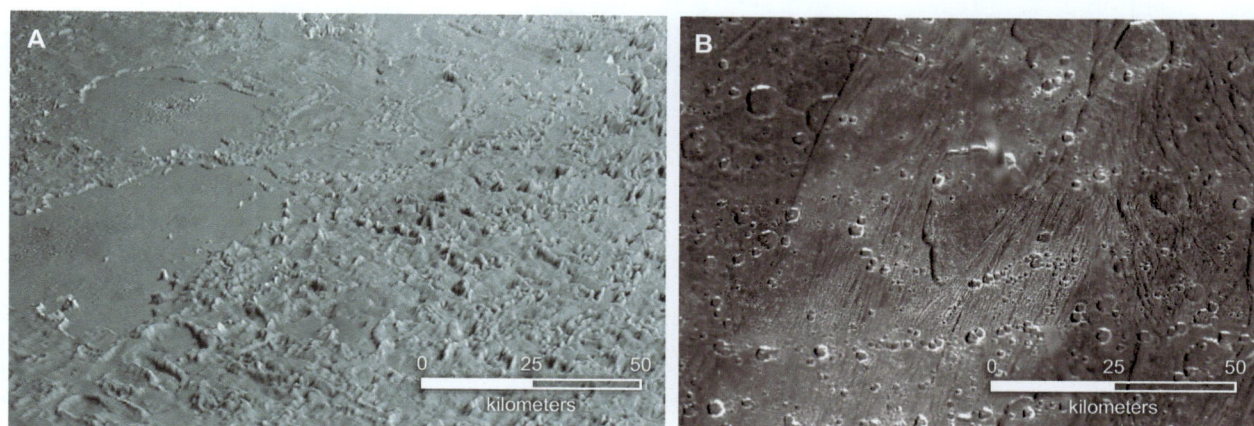

Figure 23.39 Cryovolcanic features on outer Solar System moons. **A.** Smooth plains on Neptune's largest moon, Triton, occur within caldera-like rims. The plains are interpreted as effusive, cryovolcanic flows. The rugged surface outside of these smooth plains is evidence for rising cryomagmas of water or ice. **B.** In this image of Jupiter's icy satellite, Ganymede, the central depression, with its scalloped cliffs, resembles a collapsed caldera, formed after material had erupted onto the surface. Source: NASA/JPL.

Explosive cryovolcanism on Triton occurs as geysers/plumes of nitrogen gas and entrained dust. How does this process work? Triton is so cold (−390 °F) that nitrogen – which occurs as a gas in Earth's atmosphere – freezes, forming layers of ice on the surface. Triton's nitrogen geysers are located at its subsolar point. The theory for their formation is that the sunlight at this hottest point on Triton penetrates the translucent nitrogen ice and heats the darker surface underneath. This heating vaporizes the lower layers of nitrogen ice until the pressure is sufficient for it to erupt through the overlying ice. Nitrogen gas geysers are then formed as gas bursts through this ice crust.

Ganymede, the largest icy satellite of Jupiter (and the largest in the Solar System), has scalloped, lobate-shaped depressions, sometimes hosting surfaces with ridges (**Fig. 23.39B**). The ridges and their lobate shapes resemble lava flows on Earth. Other images show closed scalloped depressions that appear more caldera-like and might serve as sources for bright, cryovolcanic flows of water and icy slush.

Cryovolcanism likely also occurs on **Kuiper Belt Objects** – small icy bodies in the distant outer Solar System (**Fig. 23.1**). Pluto, now considered a dwarf planet, is a well-known Kuiper Belt Object and was the focus of the New Horizons Mission during a 2015 flyby. Data from that flyby revealed large fissures on the surface, with distinct shapes and colors (**Fig. 23.40**). Because Pluto is not in orbit around a giant planet, as Europa and Enceladus are, it does not undergo the same level of tidal heating.

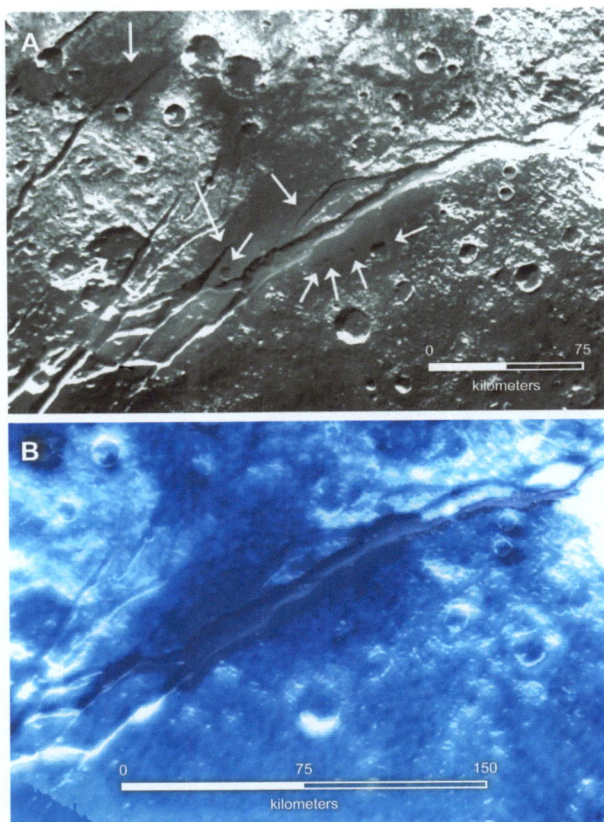

Figure 23.40 Evidence of cryovolcanism on Pluto. **A.** Examples of fissures mantled with water ice (at the arrows). **B.** Enlargement of the center of A, showing a location with a particularly high concentration of water ice in the darker blue. Source: Used with permission of Elsevier, from D. P. Cruikshank, O. M. Umurhan, R. A. Beyer, et al. (2019); permission conveyed through Copyright Clearance Center, Inc.

The crust of Titan is water ice. So, by analogy to sand on Earth – which is derived from erosion of the rocky crust – sands on Titan would be expected to be composed of grains of water ice. Instead, spectroscopy shows that the sand in the dunes on Titan is carbon-rich material. One hypothesis for the source of this organic sand is aerosols in Titan's atmosphere. However, it is still not clear how these very tiny, carbon-rich aerosol particles could get transformed on the surface into grains of sand, which are roughly one million times more massive than the aerosols they might have formed from. The Dragonfly mission to Titan, a lander scheduled for launch in 2028, will provide important information about the organic chemistry on Titan and help answer the geomorphological mystery of Titan's sands.

Planetary bodies have atmospheres of varying compositions and densities. In fact, Titan's atmosphere is denser than that of Earth. However, dunes can form in even the most tenuous of atmospheres (as on Mars), and even on comets due to outgassing (**Fig. 23.12E**). Images from the New Horizons Mission show that dunes have even formed on Pluto, which has an atmospheric pressure $\approx 1/100{,}000$ that of Earth (**Fig. 23.42**). As on Titan, the dunes on Pluto are organic in composition. Wind in atmospheres with such low-density conditions might not seem sufficient to lift sand grains. However, the process of sublimation – the transition of ice straight to a gas – might help with the lifting of the grains, which can then be transported by wind from mountainous uplands to the nearby plains.

An area of flat plains on Pluto – named Sputnik Planitia – exhibits large (> 10 km dia.) polygonal cells, outlined by darker lines (**Fig. 23.42**). Compositional information shows that Sputnik Planitia is composed of nitrogen ice, found elsewhere on Pluto but particularly concentrated there. Modeling indicates that this nitrogen ice should be forming convection cells, with warmer nitrogen ice rising up in the centers of the cells and sinking along the margins. The darker edges of these cells may correspond to the areas that simply trap darker sediment.

Sputnik Planitia could be considered one giant glacier, spanning more than 1,000 kilometers – about the size of Oklahoma and Texas (USA) combined. Pluto also has small, nitrogen-ice glaciers that flow down mountain valleys, like **valley glaciers** on Earth. However, Pluto's glaciers act differently than those on Earth. Liquid water is denser than ice, so glacial meltwater on Earth (or Mars) drains to the bottom of the glacier and helps transport sediment. Conversely, liquid nitrogen is less dense than nitrogen ice. As a result, liquid nitrogen rises *upward* within Pluto's glaciers, and as it emerges on top of the glacier, it may erupt as jets or geysers.

Interestingly, some of Pluto's surface is also composed of water ice, which is slightly less dense than nitrogen ice. As Pluto's nitrogen ice glaciers move down-valley, some of those water-ice "rocks" will "float" up through the nitrogen-ice glacier. A strange place indeed!

However, the deposits around the fissures show evidence not only of water, but also of ammonia. Ammonia (NH_3) acts as an antifreeze, allowing water to be liquid at very cold temperatures. Thus, the data suggest that, even in the especially frigid temperatures of Pluto, **cryomagma** – in this case, liquid water with some ammonia – has erupted onto the surface, through fissures.

As mentioned above, eolian landforms occur throughout the Solar System. In the outer Solar System, Saturn's moon Titan hosts vast dune fields in its tropical regions (**Fig. 23.41B**). The dunes on Titan resemble longitudinal dunes on Earth, forming roughly parallel to the predominant wind direction (see Chapter 21; **Fig. 23.41**), although smaller areas of **transverse dunes** are also discernable. Nearly 20% of Titan is covered with linear dunes. Dunes were unexpected here because the source for sand was (and still is) unknown. The atmosphere of Titan is rich in carbon compounds (also called "organic" compounds, although they have not been derived from biological processes), and the lakes and seas on Titan likewise have carbon-rich liquids, such as methane (CH_4) and ethane (C_2H_6).

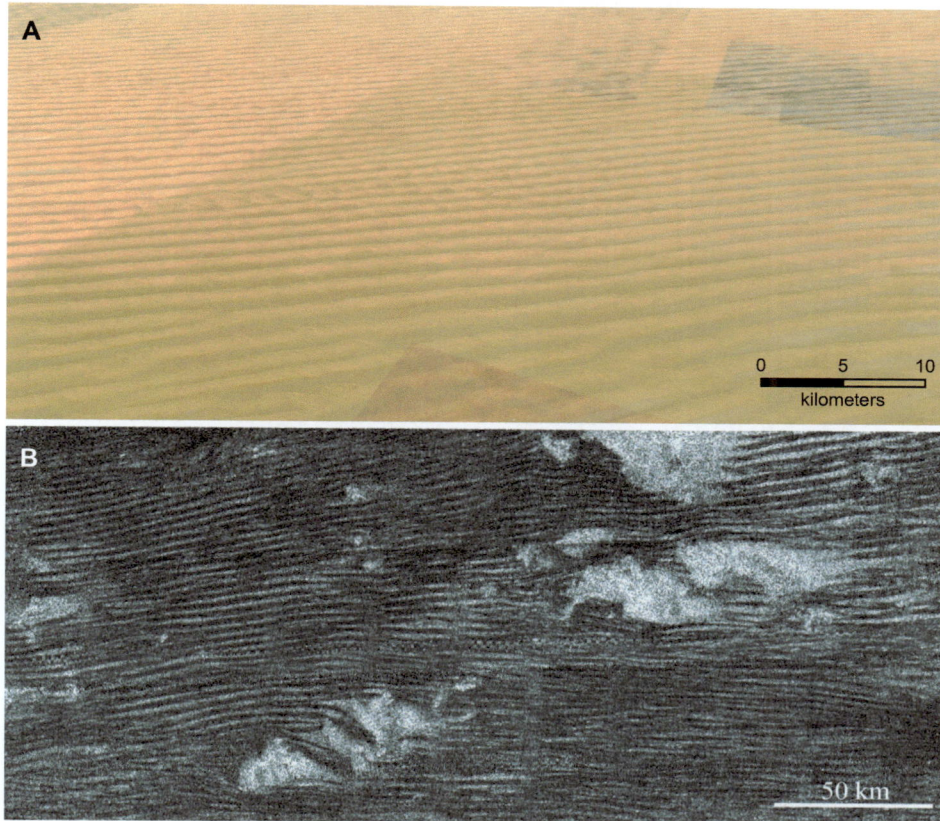

Figure 23.41 Dunes on Earth and Titan. **A.** Satellite image of longitudinal sand dunes in Saudia Arabia. Source: NASA. **B.** SAR image of longitudinal dunes in an equatorial sand sea on Titan. The dunes appear as brighter lines where the radar energy has reflected off the sides of the dunes. The dunes stream around the brighter (rougher and taller) uplands. Source: J. Radebaugh.

Figure 23.42 A New Horizons image of icy mountains encircling Pluto's Sputnik Planitia plain. In the lower left (and bottom) of the image, within the bright polygons bounded by dark lines, the smaller, subtle, subparallel lineations on the plain are interpreted as dunes, composed of carbon-rich material transported from the mountains. Source: NASA/JPL.

REVIEW QUESTIONS

23.1 What are the major planets of the Solar System? Which planetary bodies have rocky surfaces? Which have icy surfaces?

23.2 What is the most common extraterrestrial landform in the Solar System? Why do we not see lots of these landforms on Earth? Where *do* we see a lot of landscapes formed by this process?

23.3 What one tectonic feature is seen on both rocky and icy planetary bodies? Does this feature form the same way on all of these bodies?

23.4 What are the main stages that most impact craters undergo, and what kinds of landforms typically result from a meteorite impact?

23.5 Explain the formation of the maria on Earth's Moon.

23.6 How do rays form near to impact craters?

23.7 What is the general relationship between the density of impact craters on a surface, and age of the surface?

23.8 Much of the surface of Venus has been resurfaced. What does this term mean and what does resurfacing entail?

23.9 What are the two types of volcanism observed in the Solar System? Where in the Solar System does each type of volcanism occur?

23.10 Explain why volcanoes grow so much larger on Mars than on Earth.

23.11 What two explanations are typically offered for the flat floors on Martian impact craters? On Mercury's impact craters?

23.12 List four bodies that have wind-driven (eolian) landforms, and the types of landforms observed on each body. Can eolian landforms be formed on bodies without an atmosphere? If so, how?

23.13 What landforms have been used as evidence that lakes and rivers were once present on Mars?

23.14 Inverted fluvial deposits have been studied on Mars. What are these features and how do they form?

23.15 Describe the geomorphic evidence for glaciation on two planetary bodies besides Earth. Does the ice on these bodies have the same composition? Why or why not?

23.16 Explain how ice in the subsurface can form geomorphic landscapes visible at the surface. Where do such ground ice landscapes form?

23.17 What types of eolian features are commonly observed on Mars?

23.18 What are some unique geomorphic features of the icy bodies of the outer Solar System?

FURTHER READING

Greeley, R. 2013. *Introduction to Planetary Geomorphology*. Cambridge University Press.

Hargita, H. and Kereszturi, Á. (eds.). 2015. *Encyclopedia of Planetary Landforms*. Springer.

McEwen, A. S., Hansen-Koharcheck, C. and Espinoza, A. 2017. *Mars: The Pristine Beauty of the Red Planet*. University of Arizona Press.

McSween, H. Y., Moersch, J. E., Burr, D. M., Dunne, W. M., Emery, J. P., Kah, L. C., and McCanta, M. C. 2019. *Planetary Geoscience*. Cambridge University Press.

Planetary Geomorphology Image of the Month. https://planetarygeomorphology.wordpress.com/

Soare, R. J., Conway, S. J., and Clifford, S. M. (eds). 2018. *Dynamic Mars: Recent and Current Landscape Evolution of the Red Planet*. Elsevier.

Glossary

1st order of relief Relief associated with continental and oceanic tectonic plates, i.e., the difference in elevation between the ocean basins and the continents (Chapter 10)

2nd order of relief The relief associated with entire mountain ranges and large valley systems, typically formed in association with tectonic plate collision or divergence (Chapter 10)

3rd order of relief The relief associated with most landforms, as affected by erosional and depositional processes, as opposed to the movement of tectonic plates, i.e., typically what is seen on the ground by an observer (Chapter 10)

A horizon The uppermost mineral soil horizon, commonly called topsoil, that is usually dark in color due to additions of humus (decomposed organic matter) (Chapter 13)

'A'ā A type of solidified lava that has a sharp, angular, and jagged surface (Chapter 7)

Ablate, Ablation The loss of ice mass on a glacier due to melting, sublimation, calving, wind erosion, evaporation, and avalanching (Chapter 18)

Ablation zone The zone on a glacier, below the equilibrium line altitude (ELA), where ablation exceeds accumulation (Chapter 18)

Abrasion The grain-by-grain erosion of bedrock and sediment by wind and water, caused as transported grains impact the surface of other particles, breaking the material into small pieces or flaking off parts of the grains. Also a process by which rocks within ice are dragged across bedrock by a glacier, making scratches on the rock's surface (Chapters 18, 19, 21)

Accommodation space The volume or space in a landform, basin, or system that can accept new sediment, typically used in reference to alluvial fans, or coastal infilling as in deltas (Chapters 10, 17)

Accretionary wedge A geological structure at a tectonic subduction zone, formed by the scraping and compression of sediments from the subducting plate, as well as material from the overriding plate. These sediments are compressed and folded into a complex structure comprised mainly of metamorphic rocks (Chapter 6)

Accumulation zone The zone on a glacier, above the equilibrium line altitude (ELA), where accumulation exceeds ablation (Chapter 18)

Accuracy With regard to numerical dating, the property that describes how close the estimated value (date) is to the actual value of the material being dated (Chapter 3)

Active dune A sand dune that is currently freely migrating (Chapter 21)

Active glacier A glacier that is moving at a velocity greater than a few meters per year. Contrast with a stagnant glacier, where the ice is not flowing or flowing at a very low rate (Chapter 19)

Active layer In an area where permafrost occurs at depth, the active layer occurs near the surface and undergoes seasonal freezing and thawing, which may extend to 20 cm to 1.5 m below the surface (Chapters 14, 20)

Active plate margin The "leading edge" margin where a continental plate and an oceanic plate meet and is experiencing subduction. Where they meet the ocean, active margins typically have a trench offshore, and thus quickly transition into deep water offshore (Chapters 6, 17)

Active volcano A category of volcano that has erupted since the retreat of the last continental glaciers (Chapter 7)

Active-layer detachment failure The downslope movement of only the active layer and the surface vegetation cover, caused by warming of the ground thermal regime, while the permafrost below remains solidly frozen (Chapter 20)

Additions A group of soil processes that add substances or mass to the surface of the soil (Chapter 13)

Adsorb, adsorption Taking water onto the exterior of a substance, as opposed to absorption (taking water into the substance) (Chapter 14)

Aeolian An alternate spelling of the word eolian (Chapter 21)

Aerosol Any solid particle in the air, such as volcanic ash, dust, or pollen (Chapters 4, 21)

Age An interval of time measured back from the present, e.g., 289 years ago (Chapter 3)

Aggradation–degradation episode An interval of time when increased sediment production and channel aggradation is followed by a period of channel degradation, driven by decreased sediment production (Chapter 16)

Aggrade, aggradation To deposit sediment so as to add to the elevation of a surface. In fluvial geomorphology, to raise a channel bed and/or floodplain surface (Chapters 10, 16)

Aggregate Sand- and gravel-rich materials that are commonly used for infrastructure development, e.g., concrete (Chapter 1)

Albedo The reflectivity of a surface (Chapters 4, 11, 20)

Alcove *See* apse (Chapter 11)

Alluvial channel A river channel developed in river-deposited sediment, i.e., in alluvium, not in bedrock (Chapter 16)

Alluvial fan A fan-shaped landform, formed as sediment accumulates where a steep-gradient stream discharges onto an unconfined plain or basin (Chapters 1, 8, 10, 16, 23)

Alluvial terrace A relict floodplain, underlain by alluvium, that is isolated above the current floodplain due to stream incision and/or degradation (Chapter 16)

Alluvial toeslope The bottom part of a slope – a toeslope – but in this case, one that occasionally receives sediment from a nearby river, as it floods (Chapter 13)

Alluvium Material that is picked up, transported, and deposited by streams (Chapter 15)

Alpine glacier Another name for a valley or mountain glacier (Chapter 19)

Alteration feature A change to a rock surface produced by weathering, such as a weathering rind, varnish coating, crust, and/or a case-hardened layer (Chapter 11)

Alveoli Closely spaced (cm-scale) cavities on a rock surface, commonly found on sandstone, granite, and limestone surfaces, and formed by weathering (Chapter 11)

Amplitude In coastal geomorphology, this term refers to wave height, from crest to its trough. In fluvial geomorphology, the term refers to the distance between the centerlines of two successive meander bends in a river, or how far the meanders extend outward from the down-valley axis line (Chapters 16, 17)

Anastomosing channel   A type of river channel with broad, interconnected channels and banks that are more stable and with larger islands than in braided systems (Chapter 16)

Anchored dune   A sand dune that is partially or completely held in place by vegetation, i.e., not actively moving. Contrast with freely migrating dune (Chapter 21)

Andesite   A fine-grained, volcanic, extrusive, igneous rock that is similar in composition to diorite. It is intermediate in composition between silica-poor basalt and silica-rich rhyolite (Chapters 5, 7)

Anemometer   An instrument used to measure wind speed (Chapter 21)

Angle of repose   The steepest (but still stable) slope that loose material maintains, without clasts rolling or bouncing farther downslope (Chapters 8, 10, 14, 21)

Anisotropic   An adjective describing rocks or sediment that vary in one or more characteristic across different planes, i.e., it is not homogeneous. Thus, it has unequal resistance to stresses from different directions (Chapter 5)

Annual hydrograph   A graph that shows the changes in a river's discharge over the course of an entire year (Chapter 15)

Annular drainage pattern   A ringing or generally circular stream pattern associated with cuestas that form around a dome or basin. Master streams curve around the dome while the tributaries are shorter and tend to join nearly at right angles (Chapters 8, 16)

Anomalous fading   In luminescence dating, the effect that occurs due to loss of dose (stored beta particles) in grain traps over time while the sample is buried (Chapter 3)

Antecedent river   A river that is older than the geologic structure that it cuts through, having maintained its relative position/location during the slow events of geologic folding and faulting (Chapter 16)

Anthracite   A harder type of coal that has been metamorphosed (Chapter 5)

Anthropocene   A newly proposed but informal distinction in Earth's geologic timescale, referring to the (current) phase of time when the world has been modified by human action. If formally adopted, the Anthropocene Epoch would follow the Holocene Epoch, and would be defined by the many significant human impacts on the planet (Chapters 1, 4)

Anticlinal ridge   A ridge formed along the axis of an anticline, usually because the rocks there are more resistant to erosion (Chapter 9)

Anticlinal valley   A valley formed along the axis of an anticline, due to erosion of less-resistant rocks there (Chapter 9)

Anticline   A geologic structure in which the rocks at its core have been folded upward (Chapter 9)

Anticlinorium   A large, broad anticline whose overall structure contains many smaller folded structures (Chapter 9)

Antidune   A sandy, wave-like form on the bed of a river that migrates upstream, against the fast-flowing current (Chapter 16)

Antigravitative canyon   A furrow, channel, or small (inverted) canyon in a cave ceiling, formed by upward erosion by running water, as the floor of the cave gradually fills with sediment. The sediment on the cave floor reduces the space that the water can occupy and shields the floor from

dissolution and erosion, forcing the water to erode into the ceiling (Chapter 12)

Aphanitic   A type of igneous rock with fine-grained textures, formed by rapid cooling of lava on or near the surface (Chapter 7)

Aphelion   The date when a planet is farthest from the Sun in its orbit (Chapter 4)

Apse   A hole or gap that forms by erosion in a vertical slab of bedrock. As an apse forms, large slabs of rock become detached along curved fractures, opening a large recess near the base of the rock (Chapters 8, 11)

Aquifer   A layer of naturally occurring, permeable rock or sediment in the subsurface that can readily provide potable groundwater for human or other uses (Chapter 12)

Arch   When referring to geologic structure, an arch is a large, elongated dome formed as rocks are folded upward. When referring to an erosional feature, an arch is a landform that develops as weathering and wind and water erosion work together to form a gap through a resistant, vertical slab of rock (Chapters 8, 9, 11, 16, 17)

Archean Eon   A formal subdivision of geologic time from ≈4 to 2.5 Ga (Chapter 4)

Arête   A long, high, knife-edged bedrock ridge between two deeply eroded glacial valleys (Chapter 19)

Aridisol   A soil formed in a dry climate, typically with a thin A horizon and often with an accumulation of soluble materials such as $CaCO_3$, gypsum, or salts in the B horizon (Chapter 13)

Arkose   A type of sandstone that includes many feldspar grains (Chapter 5)

Ash fingerprinting   The technique used to determine the volcanic source of an ashfall bed, and correlate it to others of known origin, by examining its geochemistry or the shape and size of the ash shards (Chapter 3)

Asteroid   A rocky or metallic body orbiting the Sun in the inner Solar System (Chapter 23)

Asthenosphere   A ductile layer within the Earth's upper mantle, located beneath the crust (lithosphere), known for its high temperatures and semi-solid nature, allowing it to flow slowly under pressure (Chapters 6, 7)

Asymmetrical fold   A geologic structure in which rocks are folded and the axial plane of the folded rocks is tilted from the vertical (Chapter 9)

Asymmetrical hogback   A hogback with a steeper dip slope than is typical for hogbacks (Chapter 8)

Atoll, Atoll reef   A roughly circular, often partially submerged, coral reef that forms a nearly continuous ring, encircling a lagoon without a central island (Chapters 17, 22)

Avalanche   A type of mass movement that is dominated by snow, ice, rock and/or debris (Chapters 14, 22)

Avalanche chute   A treeless stripe or band on the side of a mountain, produced by frequent avalanches that damage and/or destroy the woody plants that grow there (Chapter 22)

Avulse, avulsion   The process whereby stream channels suddenly migrate to a new position and/or split, typical of deltas and alluvial fans (Chapters 16, 17)

Axial plane   The plane or surface that marks the center of a folded rock structure, dividing the fold as symmetrically as possible (Chapter 9)

B horizon   The soil horizon, commonly called the subsoil, that represents the zone of maximum illuviation, gaining materials from the horizons above via percolating water (Chapter 13)

Back-arc basin   A basin formed in an area where oceanic crust is being subducted, but above the crust that is NOT being subducted. They form due to crustal extension and thinning, and are typically associated with an island arc (Chapter 6)

Backshore   The "back" (landward) part of the shore zone, extending from the limit of high water to the sand dunes or other high ground farther inland. It is only affected by waves during exceptionally high tides or severe storms (Chapter 17)

Backslope   The steepest and sometimes driest part of a slope, midway down the slope, between the shoulder slope and the footslope (Chapter 13)

Backswamp   The low area on a floodplain, far from the channel, where finer particles settle out of floodwaters (Chapter 16)

Backwash   Water and sediment moving offshore as waves break on the beach (Chapter 17)

Backwasting   The parallel retreat of an eroding slope, while maintaining the same local relief, gradient, and inclination. Typical of dryland areas (Chapter 2)

Badlands topography   A landscape formed on erodible rocks where erosion is rapid, forming transport-limited slopes and intricate mazes of rills and gullies on bare bedrock (Chapter 8)

Bajada   A series of coalesced alluvial fans (Chapter 10)

Balanced rock   A bedrock landform with a hard caprock that develops as a pinnacle erodes. Before it topples and while the caprock remains, the caprock becomes seemingly "balanced" on top of an increasingly narrow neck of weaker rock (Chapter 8)

Ballena   A long ridge or a series of ridges formed as an intervening upland between incised valleys on an alluvial fan erodes and rounds-off with time (Chapter 10)

Bankfull discharge   The discharge of a river when streamflow is completely filling its channel. Flooding occurs when the discharge is greater than bankfull discharge (Chapter 15)

Bankfull stage   The stage of a river when its entire channel, extending from the channel bed up to the bank tops, is full of water (Chapter 16)

Barchan dune   A freely migrating sand dune that develops in wind regimes with one dominant wind direction, where the sand supply is low compared to the ability of the wind to transport it, and where vegetation cover is minimal. Arms of barchan dunes point downwind (Chapter 21)

Barchanoid ridge   A ridge of wind-blown sand formed as the edges and sides of multiple barchan dunes merge together (Chapter 21)

Barrier island   A long sandy island, elongated parallel to an ocean coastline (Chapter 17)

Barrier island complex   A system of coastal environments associated with barrier islands, including landforms, water bodies, and environments that evolve synchronously through time in response to sea level rise, storms, and human disturbance (Chapter 17)

Barrier reef   Much larger than a fringing reef, a barrier reef is a coral reef with an offshore outer edge formed in the open ocean, separated from the mainland or from a nearby island (Chapters 17, 22)

Basal till   Glacial sediment with a texture of diamicton that is deposited at or near the base, or bed, of a glacier or ice sheet. Basal till is usually massive and dense and commonly has a clast fabric due to the alignment of clasts either parallel or perpendicular to the former glacier flow direction (Chapters 18, 19)

Basalt   A dark-colored, fine-grained, extrusive igneous rock that forms from the rapid cooling of low-viscosity lava, rich in magnesium and iron (Chapters 5, 7)

Basaltic lava   A type of lava that is low in viscosity and gas content, and often hotter than other types of lava. Thus, it flows readily and is usually associated with effusive eruptions (Chapter 7)

Base level   The level, or elevation, below which a land surface cannot be eroded, or below which a stream cannot incise. The ultimate base level is sea level (Chapter 2)

Base level polje   A large plain in karst landscapes that floods during wet periods because its bottom is near the water table (Chapter 12)

Baseflow   Streamflow that is due to groundwater discharge into the stream channel (Chapter 15)

Basin   A geologic structure in which the rocks have been gently down-warped over broad areas, usually with a circular or nearly circular shape as seen from above (Chapters 8, 9)

Batholith   A very large to massive, irregularly shaped pluton, formed where magma has cooled and hardened (typically into granite) below the surface (Chapters 6, 7)

Bayhead delta   A delta that forms at the upper end of an estuary, as the fluvial system deposits sediment there (Chapter 17)

Baymouth bar   A spit or sandbar that elongates to the point where it can partially or completely enclose a coastal bay, estuary, or river mouth (Chapter 17)

Beach   The part of the coast that extends from the shoreline (land–water interface) landward, typically to a dune or a bluff. Most beaches are composed of unconsolidated sediments such as sands and gravels (Chapter 17)

Beach drift   Water and sediment moving generally parallel to and along the shoreline, driven by swash and backwash processes (Chapter 17)

Beach nourishment   A type of "soft" shore protection strategy that involves placing additional sand on a beach, to rebuild and restore it (Chapter 17)

Beach profile   The side-looking profile of the coastal zone, extending from any sand dunes on the landward side of the beach to the shallow water near the shore (Chapter 17)

Beaded stream   A stream with a series of ponds, connected by narrow channel segments, which develops as areas of ice-wedge polygon networks decay by thermal erosion (Chapter 20)

Beaver meadow   The former site of a beaver pond, now filled with sediment and covered in meadow vegetation on the valley floor (Chapter 22)

Bedform   A small, fine-scale, individual geomorphic feature, such as a small ripple, on the sandy bed of a stream (Chapter 16)

Bedload   The part of the sediment load of a stream that consists of sediment that is too large to be held in suspension, and so it typically gets transported along the channel bed by bouncing, rolling, and sliding processes (Chapter 15)

Bedload transport   The movement of sediment by saltation, traction, and surface creep – in streams or in the air
(Chapters 15, 21)

Bedrock   The solid rock that is either at the land surface
or which occurs at depth, immediately below the regolith
(Chapters 5, 8)

Bedrock-controlled landscape   A landscape where the overall topography is largely determined by the resistance of the
rocks to erosion, and most of the landscape owes its form to
landforms developed on bedrock (Chapter 8)

Benthic foraminifera   Foraminifera that live on the sea floor
(Chapter 4)

Bergschrund   A large, often deep, crevasse that forms at
the head of a valley glacier, where an ice mass has moved
or pulled away from the mountain face or headwall
(Chapters 18, 19)

Berm   A short, vertical step, cliff, or mound on a beach, just
landward of the foreshore, formed either by wave erosion or
deposition (Chapter 17)

Bifurcate   A term meaning to split, or branch (Chapter 17)

Biochemical weathering   A type of chemical weathering that
is either driven by or strongly influenced by biota such as
plants and microbes (Chapter 11)

Bioconstruction   A crust, film, mound, dam, reef, or
similar structure, built of earthen materials by biota
(Chapter 22)

Biocrust   A crust dominated by microbes (such as
cyanobacteria and fungi) and lower plants (such as lichens)
on a rock or the soil surface. *See also* biological soil crust
(Chapter 11)

Bioerosion   The weathering and/or erosion of the land
surface by organic means, whether by plants, animals, or
microorganisms (Chapter 22)

Biogeomorphology   The approach to geomorphology that
explicitly considers the role of organisms as geomorphic
agents (Chapter 22)

Biological soil crust   A crust that forms on the soil surface by
a suite of symbiotic organisms, commonly cyanobacteria,
fungi, and lichens (Chapter 5)

Biomantle   A relatively stone- and gravel-free layer of soil
and sediment at the soil surface, formed by the actions
of soil biota, which bring finer sediment to the surface,
allowing coarse materials to settle to the maximum depth of
burrowing (Chapters 3, 13, 22)

Biopedoturbation   Another name for bioturbation
(Chapter 22)

Biophysical weathering   A type of physical weathering that
is either driven by, or strongly influenced by, biota such as
plants and animals (Chapter 11)

Bioprotection   The direct (active) or indirect (passive) role
of organisms in reducing (or completely preventing) the
actions of other Earth surface processes (Chapter 22)

Biotite   A dark-colored, sheet-silicate, primary mineral with
the common name "mica" (Chapter 5)

Bioturbation   Soil mixing by plants and/or animals (Chapters
3, 13, 14, 22)

Bird's foot delta   A type of delta dominated by fluvial deposition, causing it to develop long, finger-like channels at its
outer margin (Chapter 17)

Blind valley   A river valley in a karst landscape that ends
abruptly against a steep rock slope, where the river usually
disappears into a swallow hole (Chapter 12)

Block   A category of pyroclastic debris, > 64 mm in diameter
and having rough and jagged edges (Chapter 7)

Block faulting   The faulting process by which crustal
extension causes large bodies of rock, typically hundreds of
kilometers in extent, to shift in position (up, down, and/or
rotate) relative to adjacent blocks (Chapter 10)

Block slide   A type of translational landslide that occurs in
bedrock (Chapter 14)

Block stream   Large accumulations of loose boulders
covering a surface, assumed to have been formed by
periglacial activity. Also called a felsenmeer or stone run
(Chapter 11)

Block-by-block disintegration   The physical weathering of
rocks primarily along bedding planes, producing rectangular-shaped blocks (Chapter 11)

Blockfield   An accumulation of angular rocks on hillslopes,
mountainsides, and plains. Also called a block stream or
felsenmeer (Chapters 5, 11, 20)

Blowout   An area of active deflation in a sand dune or along
a dune ridge, often forming between the arms of a parabolic
dune (Chapters 17, 21)

Bølling–Allerød   An abrupt, warm and moist, interstadial
period from ≈14.7 to 12.9 ka, during the final stages of the
last glacial period (Chapter 4)

Bomb   A category of pyroclastic debris, > 64 mm in diameter
and with elongated and fluidal and/or spiral shapes (Chapter 7)

Border polje   A large plain in karst landscapes in which
streams have eroded laterally into the bedrock on the flanks
of the depression (Chapter 12)

Boreal forest   A type of coniferous forest found in cold climates, typically grading into tundra on its colder margin
(Chapter 20)

Bornhardt   A monolithic, dome-shaped, bedrock hill, often
isolated within a large plain (Chapter 11)

Bottomset bed   Part of a delta's depositional sequence, a
bottomset bed is composed of fine-grained suspended sediment deposited in nearly horizontal layers, far out, in deep
water (Chapter 16)

Boulder   A rock clast that is >25 cm in diameter (Chapter 11)

Boundary layer   A slower layer in wind or flowing water,
caused by friction of the wind or water with the underlying
surface (Chapter 21)

Bowen's Reaction Series   The relationship between molten
rock material that is cooling/melting to the sequencing of
crystallization of minerals within it. As minerals crystallize,
they leave behind a liquid of slightly different composition
(Chapter 5)

Boxwork   A type of cave formation dominated by calcite
veins, formed in former cracks in the bedrock but which
now protrude out from the walls due to differential erosion
(Chapter 12)

Braid bar   A sandy, mid-channel bar in a braided fluvial
channel (Chapter 16)

Braided channel   A type of river channel that is wide and
shallow, with a sand and gravel-bed, and with multiple
small islands and mid-channel bars (Chapter 16)

Braidplain   The broad plain formed by a braided river system and its associated mid-channel bars (Chapter 16)

Breached fold   An anticline or syncline that contains distinct
gaps, formed by fluvial erosion (Chapter 9)

Breakdown   Sediment and rock that have broken off of
the roof and walls of a cave and fallen to the cave floor
(Chapter 12)

Breaker   A type of wave where the top part travels faster
than the bottom, causing it to break by steepening and
falling over on itself (Chapter 17)

Breakthrough time   The time required to develop an underground conduit large enough (usually 5–10 mm in diameter) to support turbulent flow (Chapter 12)

Breakwater   A coastal protection structure that is like a wall, typically built offshore and parallel to the coastline, to slow the longshore transport of sand and to encourage beach widening (Chapter 17)

Brittle   An adjective used to describe materials that do not easily deform under stress, but instead snap and break (Chapters 10, 18)

Buried surface   A land surface that has been buried by younger rock or sediment (Chapter 3)

Burrowing   The underground excavation of soil and rock by fauna (Chapter 22)

Butte   An isolated, flat-topped landform, smaller than a mesa but larger than a pinnacle and formed through escarpment retreat in (typically) horizontally bedded sedimentary rock (Chapter 8)

C horizon   The soil horizon that occurs so deep below the surface that it is composed of unaltered or nearly unaltered parent material (Chapter 13)

Calcareous tufa   Deposits of secondary calcite in caves, which usually take the form of speleothems. Also called travertine in external depositional settings (Chapter 12)

Calcrete   Soil materials that have become cemented by accumulations of illuvial calcium carbonate ($CaCO_3$) (Chapter 11)

Caldera   A depression in the summit of a volcano, formed as (or after) the volcano erupts and collapses. Larger than a summit crater (Chapter 7)

Calving   The process of ice chunks breaking off from a glacier and floating away in water (Chapter 18)

Capacity   The maximum amount of sediment that can be carried by a river or a glacier, but is most commonly used for streams, as in "stream sediment-transport capacity" (Chapters 15, 18)

Capillary force   The phenomenon whereby liquid is drawn into pores and can thereby ascend through these pores, primarily due to adhesive forces with the pore walls and adhesive forces within the liquid (Chapter 15)

Caprock   A hard rock layer that holds up a landform or an escarpment, and which is usually underlain by softer and more easily eroded rocks (Chapter 8)

Carbonate rocks   Sedimentary rocks primarily composed of carbonate minerals, such as calcite ($CaCO_3$) and dolomite ($CaMg(CO_3)_2$). The two major types of carbonate rocks are limestone and dolostone or dolomite (Chapters 5, 12)

Carbonation   A type of solution (chemical) weathering that affects limestone and carbonate rocks (Chapters 11, 12)

Case hardening   The process whereby a hard, protective, surface layer develops within and on top of bedrock due to mineral (often carbonate) precipitation (Chapter 12)

Castle koppie   A term for a tor or kopje, often used in an African context (Chapter 11)

Catastrophism   Historically, this concept refers to the discarded notion that much of Earth's surface was shaped by short-lived, violent, and possibly supernatural events. Today, the concept refers to the possibility that short-lived, violent, natural events can have a long-lasting effect on the shape of some landscapes (Chapter 2)

Catena   The sequence of soils along a slope, from summit to toeslope (Chapter 13)

Cave   A natural void in the ground or in bedrock, usually large enough for humans to enter. Many caves form due to dissolution of bedrock (Chapter 12)

Cave pearl   A small, free, spherical deposit of secondary calcite that forms in cave pools (Chapter 12)

Cavern   A cave for which no evidence exists of former human occupation (Chapter 12)

Cenozoic Era   A formal subdivision of geologic time from ≈66 Ma until the present (Chapter 19)

Centripetal drainage pattern   A type of drainage pattern where streams converge into a central depression or bay (Chapter 16)

Channel junction bar   A depositional bar that forms at the confluence of two river channels (Chapter 16)

Channel morphology   The general shape and size of stream channels (Chapter 16)

Channel network   A branching series of stream channels characterized by smaller channels merging into/with larger channels in the downstream direction (Chapters 15, 16)

Channel reach   A local stretch or segment of a river channel with somewhat uniform characteristics (Chapter 16)

Channel sinuosity   *See* sinuosity (Chapter 16)

Chatter mark   Another name for a crescentic gouge (Chapter 19)

Chelation   A chemical weathering process whereby a chemical complex or ligand removes a metal ion from a mineral (Chapter 11)

Chemical sediment   With respect to sedimentary rocks, sediment that has usually precipitated out of an aqueous solution, primarily in oceans or lakes (Chapter 5)

Chemical weathering   A general class of weathering reactions, sometimes called decomposition, in which the chemical and mineralogical makeup of rocks and minerals is changed due to a variety of chemical reactions. The results of chemical weathering include a different suite of (secondary) minerals, chemical elements, and ions (variously available to be transported in solution), and the release of unweathered minerals through granular disintegration (Chapters 5, 11)

Chert   A type of sedimentary rock composed of fine-grained silica as microcrystalline quartz (Chapter 5)

Chinook   Local name for a wind, also known as a "snow-eater," that blows downslope out of the Canadian Rocky Mountains onto the prairies below, warming and drying as it descends and rapidly melting any snow that may exist there (Chapter 21)

Chonolith   A pluton that is like a stock, but with a flatter and definable base (Chapter 7)

Chronofunction   *See* soil chronofunction (Chapter 3)

Chronology building   In dendrochronology, the method of obtaining cores from dead (but standing) trees and wooden beams inside old buildings, and matching them together, to extend a tree ring chronology back in time (Chapter 3)

Chute   A secondary channel cut into and across the surface of a point bar or floodplain (Chapter 16)

Cinder cone   A small, short-lived, conically shaped volcano, that is usually formed by a single eruption of mostly pyroclastic debris (Chapter 7)

Cirque   An erosional, ampitheater- or bowl-shaped depression or basin formed by glacial erosion on the upper slopes of a mountain (Chapter 19)

Cirque glacier   A small glacier that is confined to a cirque (Chapters 18, 19)

Clast   Any type of rock fragment (Chapter 5)

Clastic   An adjective used to describe sediment that is composed mainly of parts and pieces of other rocks and minerals, such as sand, silt, and clay (Chapter 5)

Clay   Particles that are < 0.002 mm in diameter (USDA classification). Geologists commonly use < 0.004 mm in diameter (Udden–Wentworth classification). The term "clay minerals" refers to a group of sheet silicate minerals such as kaolinite and illite, and does not have a specific size connotation (Chapter 13)

Clearcutting   A logging method whereby all or most of the trees are removed from a site (Chapter 14)

Cliff-and-bench topography   A term to describe the topography of a (usually arid) landscape formed on horizontally bedded sedimentary rock, which has many mesas and buttes (Chapter 8)

Climate   The average of weather conditions at a given location in the atmosphere, as viewed over timescales equal to or greater than 30 years. Meteorologists define climate as a 30-year mean of all weather-variables, e.g., temperature, precipitation, etc., which is updated every decade. In soil studies, climate is one of the five soil-forming factors (Chapters 4, 13, 18)

Climatic geomorphology   The study of landforms as they are affected by various climate types, in an attempt to explain the development and characteristics of landscapes under different climatic conditions (Chapter 4)

Climatic Optimum   The warmest part of the Holocene, peaking at around 8–7 ka. Also known as the Holocene Thermal Maximum (Chapter 4)

Clint   A small upland area between the open fractures in a barren limestone surface. Contrast with grike (Chapter 12)

Closed basin   A drainage basin where water and sediment flow in but rarely ever flow out. See also endorheic basin (Chapters 10, 13)

Closed talik   A body of unfrozen ground that is surrounded by permafrost below and on all sides (Chapter 20)

Closed-system pingo   A periglacial landform that develops when porewater freezes and expands, pressurizing the remaining unfrozen water and forcing it toward the ground surface. Also called a "hydrostatic pingo" (Chapter 20)

Coal   A combustible, black or brownish-black, sedimentary rock, formed when dead plant matter decays into peat and is converted into coal by the heat and pressure of deep burial (Chapter 5)

Coarse fragments   In soils, all fragments >2 mm in diameter, including gravel, stones, and boulders (Chapter 13)

Coast, Coastline   The zone where land and water meet (Chapter 17)

Coastal dune   A sand dune that forms on beaches and along lake and marine shorelines (Chapter 21)

Coastal foredune   A low, stabilized sand dune that forms on beaches and which parallels to the coastline of lakes and oceans. They generally do not have a slip face, but instead resemble haphazard piles of sand (Chapter 21)

Coastal karst   A coastal landscape where fresh- and saltwater mixing corrosion forms caverns and drives dissolution close to the water table (Chapter 12)

Coastal management   Efforts to protect lives and property from coastal hazards such as bluff erosion, silting-in of harbors, and/or storm damage (Chapter 17)

Cockpit karst   A type of karst landscape with broad valleys separated by tall, cone-shaped residual hills that act as topographic divides (Chapter 12)

Coefficient of thermal expansion   The rate at which a material expands with increases in temperature (Chapter 11)

Cohesion   The condition whereby a substance binds together internally, i.e., within itself (Chapter 14)

Col   A saddle-shaped low, or "pass," through an arête, formed as glacial erosion makes the arête thinner, lower, and smoother (Chapter 19)

Collapse   The brittle deformation (breakage) of bedrock into an underlying void, usually a cave or cavern (Chapter 12)

Collapsed outwash plain   A pitted outwash plain with many more kettles than in a typical low-relief, pitted outwash plain (Chapter 19)

Colluvial fan   Similar in form to an alluvial fan, it accrues sediment mainly due to landslides, debris flows, and other hillslope processes (Chapters 10, 16)

Colluvium   Unconsolidated sediment moved downslope under the influence of gravity (Chapters 9, 13)

Column   A type of speleothem formed when stalactites and stalagmites grow together and connect (Chapter 12)

Comminuted   To make something smaller, usually implying some sort of physical breakdown (Chapter 10)

Competence   The largest particle a stream or glacier can transport (Chapters 10, 15, 18)

Competent   A free face, rock layer, or surface with a geology that is generally stable and unfavorable to rockfalls, usually because the bedrock is well consolidated and minimally jointed (Chapter 8, 9, 14)

Composite cone   *See* stratovolcano (Chapter 7)

Compound drainage pattern   The drainage pattern that forms where two different types of stream patterns merge/overlap, imparting characteristics of both to the resultant pattern (Chapter 16)

Compressional flow   A phenomenon of glacial ice flow in which ice is being compressed, usually below the equilibrium line of the glacier where the glacier velocity is decreasing towards the terminus (Chapter 18)

Cone karst   Another name for cockpit karst (Chapter 12)

Cone-type geyser   A geyser that erupts steam and water in a tall eruption (Chapter 7)

Conglomerate   A clastic sedimentary rock made up of a mixture of detrital particles of various sizes, but this mix must include some rounded gravel (Chapter 5)

Constructional surface   A geomorphic surface formed by the accumulation of sediment (Chapter 3)

Contact metamorphism   Localized metamorphism that occurs where hot magma or geothermal fluids make contact with adjacent rock (Chapters 5, 7)

Continental crust   The crust that primarily comprises the continents – generally thick, less dense, and composed of granitic and felsic rocks (Chapters 6, 7)

Continental glacier   An outdated term for an ice sheet (Chapter 19)

Continental shield   A part of a land mass where ancient (typically Precambrian in age) igneous and metamorphic rocks form the core, and oldest part, of a continent (Chapter 5)

Continuity equation   The equation for discharge (Q) in rivers, written as $Q = w \times d \times v$, where w is stream width, d is mean flow depth, and v is mean flow velocity (Chapter 15)

Continuous permafrost   The condition whereby permafrost underlies 90–100% of the ground surface (Chapter 20)

Continuous series   A branch of Bowen's Reaction Series where plagioclase minerals crystallize out of the melt, starting with calcium-rich plagioclase and progressing to sodium-rich plagioclase (Chapter 5)

Contorted drainage pattern   A type of non-systematic drainage pattern distorted by the structure of the underlying metamorphic bedrock (Chapter 16)

Convergent plate boundary   The boundary between two lithospheric plates that are moving together and colliding (Chapter 6)

Coppice dune   An immobile sand dune, anchored by the accumulation of sand among and around scattered shrubs, grasses, or creeping vegetation (Chapter 21)

Coquina   A type of limestone formed from layers of shells and coral (Chapter 5)

Coralloid   A type of cave deposit that resembles small balls or gravels, hence its common name "cave popcorn" (Chapter 12)

Core   The innermost Fe-rich, metallic part of Earth that has two regions – an inner and an outer core (Chapter 6)

Corestone   A generally rounded rock fragment within an otherwise more weathered layer (Chapter 11)

Correlated age dating   A method of dating that is able to provide information about time intervals between events, but requires numerical dates on similar features or landforms, for geologic correlation (Chapter 3)

Cosmogenic isotope   An isotope produced in conjunction with solar radiation, many types of which are useful for dating (Chapter 3)

Country rock   The "background" rock which surrounds a landscape feature (Chapter 7)

Cover sand   An eolian sand deposit that covers broad swaths of the landscape. Similar to sheet sands, but cover sands are typically more poorly sorted (Chapter 21)

Crag-and-tail   A streamlined glacial landform from cm-scale to km-scale in size, that develops where a hard, resistant rock crops out under the glacier and is not eroded completely, leaving a jagged bedrock high, with glacial sediment deposited in the lee of the crag to form a tail (Chapter 19)

Crater rays   Lines of ejecta that extend beyond and outward from an impact crater (Chapter 23)

Cratermaker   A type of fauna that digs, scratches, and burrows into and on the soil surface, making craters and pits (Chapter 13)

Craton   Another name for a continental shield, i.e., the old, geologic core of a continent (Chapter 5)

Creep   The almost imperceptible downslope movement of loose debris and soil; the slowest form of mass movement (Chapters 14, 20)

Crescentic gouge   A half-moon shaped gouge in the bedrock, formed as debris in glacial ice is dragged across the bedrock surface, causing shearing and brittle failure of bedrock (Chapter 19)

Crest   In reference to coastal processes, the highest point of a wave (Chapter 17)

Cretaceous Period   The last period of the Mesozoic Era, between 145.5 and 65.5 Ma (Chapter 4)

Crevasse   A deep, tensional fracture in a glacier or ice sheet (Chapters 18, 19)

Crevasse splay   A deposit of relatively coarse material on a floodplain that forms as sediment is delivered through a narrow break in a natural levee, typically at the beginning stages of a flood (Chapter 16)

Critical zone   Earth's permeable, near-surface depositional layers, containing the soil, where the atmosphere, biota, and the lithosphere meet and interact (Chapter 11)

Cross-bedding   A sediment structure with inclined depositional layers, formed on former sand dune slip faces, that can even persist into the bedrock record (Chapters 16, 21)

Crossdating   The technique of matching patterns of rings from wood recovered from different locations (Chapter 3)

Crown   The part of a rotational landslide that consists of the slope segment that is still in place, immediately above the highest parts of the main scarp (Chapter 14)

Crust   In soils and surficial geology, a crust is an accumulation of indurated and/or cemented material within (but usually towards the top of) a soil profile. In geology, the crust is the hard, rigid, outer layer of the Earth, comprised of solid rock (Chapters 5, 7, 11)

Cryo-   A prefix referring to frozen conditions and/or ice (Chapters 20, 23)

Cryomagma   Liquid water with some ammonia that has erupted onto the surface of icy bodies in the outer Solar System (Chapter 23)

Cryosuction   The pressure reduction caused as ice forms in pores, drawing unfrozen porewater towards the colder, freezing ground (Chapter 20)

Cryoturbation   The physical mixing of soils by ice growth and thaw settlement (Chapter 20)

Cryovolcanic flow   A lava analog found on some planetary bodies that is a slush of ice and liquid water produced by extrusion of cryomagma from below (Chapter 23)

Cryovolcanism   The eruption of water and other liquid- or vapor-phase volatiles onto the cold surfaces of icy bodies in the outer Solar System (Chapter 23)

Crystal growth and decay   The process in glaciers (a part of ice deformation) whereby water molecules move from one crystal to another, with or without melting (Chapter 18)

Crystal movement   The movement or rotation of ice crystals in relation to one another, common to glaciers, and is a part of ice deformation (Chapter 18)

Crystalline rocks   A broad category of rocks composed of interlocking crystals, typically equated to igneous and metamorphic rocks (Chapters 5, 8)

Cuesta   A bedrock-influenced landform, developed on slightly dipping sedimentary rock, with a steep escarpment on one side and a long, gentle dip slope on the other (Chapters 8, 9, 16)

Cuestaform plain   A series of cuestas that comprise a landscape, and sometimes also known as a "belted plain" (Chapter 8)

Cumulic soil   A soil with an overthickened A horizon due to cumulization, typically found on toeslopes (Chapter 13)

Cumulization   The process whereby soils slowly grow upward by additions of sediment onto the top of the profile, followed by incorporation of that sediment into the soil itself (Chapter 13)

Cupola   A deep, hemispherical concavity on a cave roof, typically found in thermal caves by condensation–corrosion processes (Chapter 12)

Current   The flow or movement in water that results from a variety of forces, including gravity, wind, tides, and/or density differences (Chapter 17)

Curtain of fire   A wall-like structure of erupting basaltic lava, spewing upwards (Chapter 7)

Cutbank   The steep, erosional bank formed on the outside of a bend in a meandering stream channel (Chapters 16, 22)

Cycle of erosion   Often called the Geographic Cycle, a conceptualization of how landscapes evolve, formulated by geologist/geographer William Morris Davis in the early twentieth century (Chapter 2)

Dansgaard–Oeschger cycles   Climate oscillations involving rapid, short-term warming events spanning years to decades, followed by cooling events spanning hundreds of years (Chapter 4)

Date   A specific point in time, for example, June 17, 2018 (Chapter 3)

Day bed   A shallow surface depression excavated by an animal for the purpose of resting during the heat of the day (Chapter 22)

Debris   A term used to describe regolith that contains a large amount of coarse clasts, such as gravel and rock fragments (Chapter 14)

Debris avalanche   A fast-moving avalanche that involves large amounts of unconsolidated sediments and is not confined to a channel (Chapter 14)

Debris band   A layer of sediment and debris within glacial ice (Chapter 18)

Debris flow   A fast-moving mix of debris and water that typically flows and/or tumbles downslope within preexisting channels or tracks. Many people refer to a debris flow as a landslide (Chapters 7, 14)

Debris flow cone   A cone-shaped deposit formed at the distal end of a large debris flow (Chapter 14)

Debris flow fan   A fan-shaped landform that is a hybrid between a talus cone and an alluvial fan, formed at the distal end of a debris flow (Chapters 10, 14)

Declination   The difference in degrees between magnetic north and true (geographic) north (Chapter 3)

Decomposition   Another name for the suite of processes that comprise chemical weathering, whereby the chemical and mineralogical makeup of rocks and minerals is changed to other forms (Chapter 11)

Deep-seated landslide   A landslide where the sliding surface is >10 m deep, and typically involving soil, regolith, weathered rock, and bedrock (Chapter 14)

Deep-seated weathering   Weathering that is operating far below the land surface, typically dominated by chemical processes (Chapter 11)

Deep-water wave   Waves that occur in water bodies where the bottom of the water body is deeper than the wave base, and therefore cannot transport sediment along the bed of the water body (Chapter 17)

Deflation   The entrainment and transport of sediment by wind (Chapter 21)

Degrade, degradation   To erode or cut down (Chapter 16)

Delamination   The breakage, detachment, and sinking of a portion of the lower continental crust and mantle lithosphere, separating it from the upper continental crust (Chapter 6)

Delta   A landform that develops where a river enters into a large water body, driving the deposition of sediment into the body of water (Chapters 16, 17)

Delta front   The steeply sloping part of a delta, just beyond the delta plain, that lies where the delta meets the ocean water and where the deltaic deposits slope down to the sea floor (Chapter 17)

Delta kame   A delta that forms as water flows off a glacier and into a proglacial lake. Also called a kame delta (Chapter 19)

Delta plain   The broad, low-lying, topmost part of a delta, where fluvial processes play important roles and rivers meanders lazily across a flat, swampy landscape (Chapter 17)

Dendritic drainage pattern   A branching, "tree-like" stream pattern associated with areas of uniform geology, gentle or moderate slopes, and lack of structural control (Chapter 16)

Dendrochronology   The science that utilizes tree ring data to establish the ages of trees, surfaces, events and landforms (Chapters 3, 14)

Dendrogeomorphology   The science that uses tree rings to date geomorphic events (Chapter 3)

Denudation   The removal of mass by erosion from a landscape that either gradually decreases elevations over time and reduces its relief, and/or counteracts the effects of uplift (Chapters 1, 2, 5, 6, 8, 11, 14, 23)

Depositional coast   A coast that increases in area over time, due to net deposition of sediment (Chapter 17)

Depositional environment   A place where sediment gets deposited (Chapters 4, 5)

Depositional surface   Another name for a constructional surface (Chapter 3)

Depth of zero annual amplitude   The depth in the soil, regolith, or bedrock below which temperatures remain constant throughout the year (Chapter 20)

Deranged drainage pattern   A type of drainage pattern where rivers follow seemingly odd, random, and unexplained directions, and in which many channels are connected to lakes, wetlands, and basins. Deranged drainage occurs in karst, arid, or periglacial regions, or on recently deglaciated or volcanic terrain (Chapter 16)

Desert pavement   A dense cover of small closely spaced rocks on the surface, usually in a desert and often overlying a layer of porous, fine-grained sediment (Chapters 3, 10, 21)

Detached breakwater   A structure, usually constructed of stone or concrete, in a water body to reduce coastal erosion by forcing waves to break offshore (on the structure). Most detached breakwaters appear as "walls" that parallel the coastline, in the shallow nearshore zone (Chapter 17)

Detachment fault   A deep, low-angle normal fault associated with large-scale extensional tectonics and block faulting (Chapter 10)

Detrital rocks   A broad category of rock composed of particles that have been weathered out of other, preexisting rocks (Chapter 5)

Detrital sediment   Sediment composed of particles or clasts of former rocks and minerals (Chapter 5)

Devonian Period   A period of the Paleozoic Era, between 419.2 and 358.9 Ma, that is often known as the Age of Fishes (Chapter 4)

Diamict, Diamicton   A purely descriptive term for a matrix-supported sediment where gravel clasts are embedded within a matrix of sand-, silt-, and clay-sized particles. The term does not imply genesis or origin, because it can be formed by a variety of processes, including various mass movements or glacial deposition (Chapter 18)

Diastrophism   The large-scale deformation of Earth's crust, involving bending and/or folding stresses and processes (Chapters 1, 5, 7, 8, 10)

Differential erosion   Erosion that occurs at varying rates on different kinds of rocks, such that harder, more resistant rocks erode slower than weaker, less resistant rocks (Chapters 5, 9)

Differential weathering   Weathering processes that preferentially remove material from some locations or rocks, based on their hardness or resistance (Chapters 5, 11)

Diffusive slope processes   Slow processes including creep, rain splash, and sheetwash that are spread across the entire slope and are largely governed by slope gradient (Chapter 14)

Dike   A tabular-shaped pluton formed as magma invades and unconformably cuts across the strata in sedimentary or other rock types (Chapter 7)

Diorite   An intrusive igneous rock that is a darker version of granite, but with less quartz and more plagioclase feldspar (Chapter 5)

Dip   The angle of inclination of a rock layer, measured downward from the horizontal (Chapter 8)

Dip slope   The land surface that generally follows the dip (inclination) of the underlying rock (Chapters 8, 9)

Dirt cracking   A physical weathering process whereby dust enters existing rock fractures and, by swelling and precipitation of carbonate, breaks apart the rock (Chapters 10, 11)

Discharge   The volume of water flowing past a specified point in a stream, over a given amount of time, and which is measured in units of volume/time (Chapters 15, 16)

Discontinuous permafrost   The condition whereby permafrost underlies 50–90% of the ground surface (Chapter 20)

Discontinuous series   A branch of Bowen's Reaction Series in which each of several minerals crystallizes within a limited temperature range (Chapter 5)

Dissipative beach state   The situation on a beach whereby erosion has formed a flatter beach profile, dissipating the wave energy offshore, such that waves break offshore and are smaller when they reach the beach (Chapter 17)

Dissolution   Another name for solution weathering, wherein all or part of a rock gets dissolved in water and its associated weak acids (Chapters 11, 12)

Dissolved load   The part of the sediment load of a stream that consists of compounds that are held in solution by ionic bonds with water molecules (Chapter 15)

Distributary   A river channel on a delta or alluvial fan that is forced to drop much of its sediment, causing it to branch, split, and diverge in the downstream direction (Chapters 16, 17)

Distributary drainage network   A type of drainage network found on alluvial fans and deltas, where the channels branch out (become more numerous) in the downstream direction (Chapter 16)

Diurnal tide   A tide that rises and falls once a day (Chapter 17)

Divergent plate boundary   The boundary between two lithospheric plates that are moving apart, allowing molten material to well up from below to form new oceanic crust (Chapters 6, 7)

Doline   An isolated, closed, and often bowl-shaped depression in karst landscapes, commonly also called a sinkhole (Chapter 12)

Dolomite   A harder, more magnesium-rich version of limestone with the chemical makeup $CaMg(CO_3)_2$ (Chapter 5)

Dome   A geologic structure in which the rocks have been gently up-warped over broad areas, usually with a circular or nearly circular shape as seen from above (Chapters 8, 9)

Dormant volcano   A category of volcano that has not erupted in the past 10,000 years but is expected to erupt again (Chapter 7)

Dose rate   In luminescence dating, the rate at which beta particles (electrons) are produced by radioactive decay and released into the surrounding sediment (Chapter 3)

Downwasting   The term used to describe how erosion results in flattening of the landscape through a gradual decrease in relief over time, and accompanied by a decrease in the steepness of hillslopes (Chapters 2, 18)

Drag force   The force generated when wind or water flow around a particle, creating more pressure on the upstream side than on its downstream side, thereby pushing the particle downstream (Chapters 15, 21)

Drainage basin   The land area from which water drains downhill to a given point along a stream, called the drainage basin outlet (Chapter 15)

Drainage density   A measure of the density of channels within a drainage basin, calculated as the total length of channels in a basin (L) divided by the basin area (A) (Chapter 16)

Drainage divide   The outer boundary of a drainage basin that runs along land that is topographically high, separating it from neighboring basins (Chapter 15)

Drainage pattern   The shape of a network of river channels as seen from above (Chapter 16)

Drapery   A type of speleothem, often called "cave bacon," formed by droplets of water flowing down along an inclined part of the cave roof, depositing calcite along their trail. The presence of impurities like iron oxide often give draperies a striped, bacon-like appearance (Chapter 12)

Driving force   A force that acts on rock and sediment to form landforms by either building up or tearing down the landscape (Chapter 1)

Drumlin   A streamlined hill of glacial sediment that commonly forms beneath actively flowing glaciers, ice caps, and ice sheets (Chapter 19)

Drumlin field, swarm   A group of drumlins that can sometimes number from dozens to thousands (Chapter 19)

Dry snow zone   The area, only present on the highest elevations of polar glaciers and ice sheets, where it is cold enough that there is no melting at any time of the year (Chapter 18)

Ductile   An adjective used to describe materials that easily bend (Chapters 6, 10)

Dune   A constructional landform of wind-blown sand or a similar feature on the sandy bed of a river. If formed by wind, it is commonly called a sand dune (Chapters 16, 21)

Dust   A term used to refer to silt-size (or smaller) sediment that has an eolian origin (Chapters 4, 21)

Dust Bowl   The period of severe dust storms in the 1930s, primarily affecting the Great Plains region of the United States, caused by a combination of drought, poor land management practices, and strong winds. Dust Bowl storms caused widespread ecological and economic damage (Chapter 21)

Dynamic equilibrium   Term for the behavior of a geomorphic system in which average form or shape changes only gradually over time (Chapters 2, 16)

E horizon   A light-colored soil horizon immediately below the A horizon, where eluvial processes are maximally expressed (Chapter 13)

Earth   A term used to describe regolith that is primarily composed of sand or finer-sized particles. Also the name of our home planet (Chapter 14)

Earth hummock   In periglacial landscapes, a non-sorted patterned ground feature where vegetated mounds are separated by narrow troughs (Chapter 20)

Earthflow   A type of debris flow that is a slow, plastic mass movement of mostly fine-grained sediment (Chapter 14)

Earthquake   A geologic event, occurring along a fault, caused by the rupture and sudden movement of rocks that have been strained beyond their elastic limits, crerating vibrations (seismic waves) in the crust (Chapters 10, 14, 17)

Ebb tidal delta   A landform that develops as sediment is transported through inlets in barrier islands, out of the lagoon and into the ocean, best expressed as the tide is rising (Chapter 17)

Eccentricity   Cyclical changes in the shape of Earth's orbit from a more circular to a more elliptical shape (Chapter 4)

Ecologist   A scientist who studies ecology and the environment (Chapter 22)

Ecosystem engineering   The science that examines how living organisms build and/or manage their environment (Chapter 22)

Eddy   A motion in a river involving swirling, turbulent water (Chapter 15)

Effective discharge   The stream discharge that transports the most suspended sediment over time (Chapter 16)

Effusive eruption   A volcanic eruption involving magma that is not overly viscous, and thus flows quite readily and is not explosive (Chapter 7)

Ejecta   Broken up rock thrown out of a crater during a meteorite impact, often intermixed with molten material (Chapter 23)

Elevation   The height above a given datum, such as sea level (Chapter 6)

Eluviation   The translocation of materials out of a soil horizon and into a lower one, usually by percolating water (Chapter 13)

Embayment   A sheltered area or bay along a coast where the shoreline curves back inward, toward the land (Chapter 17)

End moraine   A ridge of glacial sediment formed at the outer margin of a glacier, marking a former position of the terminus (Chapters 18, 19)

Endogenic   An adjective pertaining to the isostatic, tectonic, and/or volcanic (internal) processes that originate within the Earth system, such as doming, folding, and faulting. Sometimes also spelled "endogenetic" (Chapters 1, 5, 6, 7)

Endorheic basin   A drainage basin that has no outflow to an external body of water such as a river or an ocean, and so must lose water through evaporation and seepage into the ground. Sometimes called a closed basin (Chapter 10)

Englacial environment   The environment that occurs within the glacier itself, containing mainly ice, but also debris, and which is cut by numerous crevasses, fractures, and tunnels (Chapter 18)

Entrain, entrainment   To "pick up" from the ground surface, usually implying by wind, water, or ice (Chapter 21)

Entrenched meander   An incised meander that has steep, almost symmetrical, valley cross-sections, indicating that, during downcutting, lateral channel migration was negligible, and incision was almost entirely vertical (Chapter 16)

Eocene Epoch   A formal subdivision of geologic time, within the Cenozoic Era, that lasted from ≈56 to 34 Ma (Chapter 4)

Eolian   An adjective that refers to wind or wind-related processes. Also spelled aeolian (Chapters 17, 21, 23)

Eon   The longest unit of time on the geologic timescale (Chapter 4)

Ephemeral stream   A stream that is usually dry, flowing only after heavy rains or snowmelt events (Chapters 10, 15)

Epicenter   The location on the land surface directly above the focus point of an earthquake in the subsurface (Chapter 10)

Epigenic cave   A cave formed by descending (downward-percolating) fluids that derive their aggressiveness from the surface environment (Chapter 12)

Epikarst   Karst features formed in the highly permeable and weathered upper part of bedrock, i.e., near the surface (Chapter 12)

Epoch   A unit of geologic time that is shorter than a period, and which may be further subdivided into stages (Chapter 4)

Equifinality   In geomorphology, the condition whereby different processes can eventually form generally similar features (Chapter 11)

Equilibrium line   The boundary between the accumulation and ablation zones on a glacier, where no net gain or loss of ice mass occurs (Chapters 18, 19)

Equilibrium line altitude   The elevation of the equilibrium line on a glacier (Chapter 18)

Equivalent dose   A measure of the number of electrons or beta particles stored or trapped in a sample that may later be analyzed for luminescence dating (Chapter 3)

Era   A unit of geologic time that is shorter than an eon and which is typically subdivided into periods (Chapter 4)

Erg   An area of widespread sand and dunes, or a "sand sea," in a desert, where wind-blown sand covers most of the surface (Chapter 21)

Erosion   The suite of processes whereby rock debris and sediment are moved from one location to another, usually lower, location (Chapters 8, 11)

Erosion surface   A land surface formed by an erosion event (Chapter 3)

Erosional coast   In coastal geomorphology, a coast that loses area, or has lost area, over time (Chapter 17)

Erosional lag   Stones, rocks, and gravel left behind on a surface after erosion has removed many of the finer sediments (Chapter 21)

Erratic   A clast of non-local rock that has been eroded, transported, and later deposited by a glacier. The name comes from the fact that their lithology is different than the local bedrock because they have been transported to their current location from an area of different bedrock geology. Erratics can be pebbles to large boulders in size and can be transported up to thousands of kilometers from their place of origin (Chapters 18, 19)

Escarpment   A steep slope, also sometimes called a cliff, wall, or bluff, usually formed by erosion (Chapters 6, 8, 9)

Esker   A sinuous or linear ridge of sand and gravel formed as ice-contact stratified sediment is deposited within an englacial or subglacial tunnel. Eskers can also form as part of a supraglacial stream system (Chapters 19, 23)

Estuary   A long, often narrow, semi-enclosed basin, where freshwater from a river mixes with saltwater from the ocean, typically formed by the drowning of river valleys or glacial troughs by postglacial sea level rise (Chapter 17)

Eustatic sea level   Another name for global sea level (Chapter 17)

Evaporation   The passive process whereby water changes form to water vapor (Chapter 15)

Evaporite, Evaporite deposit   A natural salt or mineral deposit left after the evaporation of a body of water, or as forms in a body of water due to excess evaporation (Chapters 5, 7)

Evapotranspiration   The combination of the processes of evaporation and transpiration (Chapter 15)

Excess ice   The condition in which the volume of ice in a sediment is greater than the total pore volume of the previously unfrozen sediment (Chapter 20)

Exfoliation   The physical breakup and loss of large sheets of rock from areas of exposed bedrock (Chapter 11)

Exfoliation dome   A large, rounded, bedrock hill that owes its shape to exfoliation – the formation and loss of large sheets of bedrock due to expansion of the rock by unloading processes (Chapter 11)

Exogenic   An adjective pertaining to processes that originate exterior to (on, or above) the land surface, and which tend to wear down the crust, including weathering, erosion, and mass wasting processes. Sometimes also spelled "exogenetic" (Chapters 1, 5, 6)

Exorheic   A drainage basin whose rivers flow out and connect to other basins or bodies of water that ultimately drain into the ocean (Chapter 10)

Exotic stream   A river that originates in a wetter and more humid location but then flows into and through an arid region (Chapter 8)

Explosive eruption   A volcanic eruption that occurs when viscous, silica-rich, molten rock traps gases, causing pressures to build up in the subsurface, until a violent eruption occurs (Chapter 7)

Extensional flow   A phenomenon of glacial ice flow in which the ice is extending, usually occurring above the equilibrium line, where ice velocity is increasing down-glacier (Chapter 18)

Extinct volcano   A category of volcano that is not erupting and is not expected to erupt again (Chapter 7)

Extraterrestrial processes   Processes that originate outside Earth's system (Chapter 1)

Extrusive igneous rock   An igneous rock that forms at the surface as lava quickly cools (Chapter 5)

Facies   The characteristics of a rock across a geographic area, including its chemical, physical, and biological features, which distinguishes it from adjacent rocks (Chapter 8)

Fair-weather beach   A beach that typically forms in summer under low-wave energy conditions, and which is characterized by a well-developed berm that separates the flat backshore from the foreshore (Chapter 17)

Fall   A type of mass movement where the material drops downward, through the air, typically associated with steep slopes (Chapter 14)

Fall line   An area on a landscape where there is a noticeable drop in elevation, typically from an upland to a coastal plain, associated with a change in geology, leading to waterfalls or cataracts as rivers that cross this topographic "edge" (Chapter 5)

Falling limb   Another name for the receding limb of a storm hydrograph (Chapter 15)

Fanglomerate   A type of gravelly, stratified, alluvium commonly found in alluvial fans (Chapter 10)

Fan-head incision   The process whereby rivers flowing across alluvial fans start to incise their channels, beginning at the very top of the fan – its head – and working farther down-fan over time (Chapter 10)

Fault   A planar fracture or discontinuity in a rock across which there has been significant displacement (Chapters 3, 5, 8, 10)

Fault line   The trace of a fault on the land surface (Chapter 10)

Fault plane   The zones or plane along which rocks slide or move, during faulting (Chapter 9)

Fault scarp   The face or slope formed by the upward or downward movement of rock along a fault plane (Chapters 6, 10)

Fault trace   Another name for a fault line (Chapter 10)

Fault zone   Wide corridors of variously fractured rock, usually associated with several parallel faults (Chapters 5, 10)

Faulting   The process whereby rocks are broken and moved along a rupture surface, due to internal stresses (Chapter 10)

Faunalturbation   The mixing of soil and sediment by animals (fauna) (Chapter 22)

Feedback   Term for outputs of a system that are routed back as inputs to the same system, as part of a chain of cause-and-effect that forms a circuit or loop (Chapter 4)

Feldspar   A group of silicate, primary minerals, the two most common forms of which are plagioclase feldspar and potassium feldspar (Chapter 5)

Felsenmeer   An area of large, angular, dense, cobbles and boulders on a flat or gradually sloping surface, assumedly formed under a colder climate when freeze–thaw activity was more intense. Also called a stone run or blockfield (Chapters 5, 11, 20)

Felsic   A category of rock-forming minerals that are rich in silica and low in iron (Chapter 5)

Felsic lava   *See* rhyolitic lava (Chapter 7)

Felsic rock   A type (group) of rock dominated by the mineral feldspar and various other minerals rich in silicon, causing them to be lighter colored and have comparatively low specific gravities (Chapter 5)

Fenglin   The local name given to tower karst in China, consisting of towers rising from a karst plain, making it the most extreme form of karst (Chapter 12)

Fengcong   The local name given to cone karst in China (Chapter 12)

Ferricrete   A soil horizon that has become cemented by various iron oxide compounds (Chapter 11)

Field capacity   The maximum amount of water that soil/sediment can hold against the force of gravity, after it was first saturated and then allowed to freely drain (Chapter 15)

Fill terrace   Another name for an alluvial terrace in a fluvial system (Chapter 16)

Fin   A vertically oriented slab of bedrock formed by erosion along parallel and nearly vertical joints (Chapter 8)

Fine-earth   In soils, all particles < 2 mm in diameter, including sand, silt, and clay (Chapter 13)

Fingerprinting   A method of using unique geo-signatures of size, shape, elemental composition, mineralogy, hydration, or weathering of a mass of sediment to match it to another of similar origin (Chapter 3)

Fiord   An alternate spelling of fjord (Chapter 19)

Firn   The transitional state, usually more than a year old, between snow and glacial ice. Firn is granular ice with a density between ≈200 and ≈800 kg/m^3 (Chapter 18)

Firn line   *See* equilibrium line (Chapter 18)

First-order stream   An unbranched (fingertip) stream, i.e., a stream channel that lacks tributaries (Chapters 15, 16)

Fissure eruption   A volcanic eruption of primarily basaltic lava that occurs when magma flows up through cracks in the surface and pours out effusively (Chapter 7)

Fixed dune   An adjective describing a sand dune that is anchored in place, usually by vegetation. Contrast with freely migrating dune (Chapter 21)

Fjord   A glacial valley that has become submerged by rising post-glacial sea levels (Chapters 17, 19)

Flank eruption   A volcanic eruption that pours out of a vent on the side of a volcano, rather than from a main, central vent (Chapter 7)

Flank failure   A type of mass movement that occurs along the side of a volcano during volcanic quiescence (Chapter 14)

Flank margin cave   A type of cave that forms from the mixing of fresh and salt waters in a coastal setting (Chapter 12)

Flared slope   A shallow basal concavity found on many granite outcrops, formed by weathering and erosion (Chapter 11)

Flatiron   A hogback that has such a steep dip that distinctive, often V-shaped, gaps are eroded into it, and as a result, the uplifted bedrock layers between those gaps develop a jagged or pointed appearance (Chapters 8, 9)

Flexural isostatic uplift   Warping of the crust caused by differential loading and unloading (Chapter 6)

Flexural topple   A type of mass movement where thin, steeply inclined layers of relatively weak but solid rock slowly bend, tilt, and fall over in a toppling motion (Chapter 14)

Flood   The state of flow in a river when it has overtopped its banks (Chapter 15)

Flood basalt   A massive eruption of lava that covers broad regions, producing expansive areas of horizontally bedded layers of basalt (Chapter 7)

Flood frequency analysis   The statistical examination of past river flow data, as derived from streamflow gages, to determine the likelihood of a flood of a given size, in a given year (Chapter 15)

Flood lava   An outpouring of low-viscosity lava, expansive enough to cover a large region (Chapter 23)

Flood tidal delta   A landform that develops as sediment is transported into a lagoon, through inlets in a barrier island – a process best expressed as the tide is falling (Chapter 17)

Floodplain   A low-relief land area in a river valley, carved out by the river as it erodes into the valley walls, and which is subject to flooding during periods of high water (Chapters 15, 16)

Floralturbation   The mixing of soil by flora (plants) (Chapter 22)

Flow   A type of mass movement where material travels downslope with considerable internal motion and mixing (Chapter 14)

Flow separation   The process whereby wind that is accelerating up the windward slope of a sand dune then separates into two components beyond (downwind of) the dune crest (Chapter 21)

Flowstone   A calcite deposit formed when water flows as a sheet over the walls and floor of a cave (Chapter 12)

Fluid threshold   The speed required for a fluid to lift grains off the bed, or surface (Chapter 21)

Flute   A small, linear feature worn into a rock surface by weathering and erosion processes, or the name for a narrow and highly elongated drumlin (Chapters 11, 19, 21)

Fluvial   An adjective referring to processes and landforms associated with running water (Chapter 16)

Fluviokarst   A landscape where erosion by running water dominates over traditional karst processes such as dissolution (Chapter 12)

Focus point   The initial slippage point from which an earthquake originates (Chapter 10)

Fold and thrust belt   Mountainous topography formed due to folding and thrust faulting of bedrock, resulting in parallel ridges of resistant rocks separated by valleys underlain by comparatively weaker rocks (Chapters 6, 9)

Fold axis   The line (along the rock strike) that marks the top or bottom of a folded rock structure, and which connects points of maximum curvature (Chapter 9)

Fold belt   Part of a fold and thrust belt where thrust faults are relatively inconspicuous at the surface (Chapter 9)

Fold limbs   The sides or flanks of a folded rock structure (Chapter 9)

Fold nose   The tip, or end, of a folded and plunging rock structure as seen from above, indicating a change in the strike of the rock (Chapter 9)

Folding   The bending of rocks due to stresses within the crust (Chapters 5, 10)

Folia   Calcite linings in (often thermal) caves, resembling bracket fungi (mushrooms) (Chapter 12)

Foliated, foliation   Visible banding, plates, or sheets in rocks, as is common in some metamorphic rocks (Chapters 5, 14)

Foot cave   An indentation in bedrock, formed from the local intensification of weathering at the intersection of a rock outcrop and the rock surface that forms its base, or at the water table level. Also called a rock shelter (Chapter 11)

Footslope   The slope element that is on the most concave part of the slope, occurring just below the backslope and grading into a toeslope in the downhill direction (Chapter 13)

Footwall   The part of the rock in a reverse or normal fault that, if one were to stand on the fault plane, would be below your feet (Chapter 10)

Foraminifera, Foram   A phylum of single-celled organisms (plankton) that form calcite-rich shells in seawater, which when recovered from seafloor sediments have proven useful as a global climate archive (Chapter 4)

Forcing   A variable that controls or has an impact on the global climate system (Chapter 4)

Foredune   The first, often small and ephemeral, dune encountered, moving landward from the beach (away from the water) (Chapters 17, 21)

Foreland basin   A basin that develops adjacent and parallel to a mountain belt and which is often filled with debris eroded from the advancing crust. The basin is formed by crustal flexure, driven by tectonic collision (Chapter 6)

Foreset bed   Part of a delta's depositional sequence, composed of sediment that deposited in (often steeply) dipping layers, as it spills over the steep deltaic front (Chapter 16)

Foreshore   The subaerial (dry land) part of a beach that slopes down into the water (Chapter 17)

Fossa   A long, narrow depression or trough on the surface of an extraterrestrial body (plural: fossae) (Chapter 23)

Fountain geyser   A geyser whose eruption involves mainly bubbling and splashing water, but not much steam (Chapter 7)

Free face   A slope so steep anything not physically attached to it is free to fall off due to gravity alone (Chapters 8, 14)

Free fall   Material freely falling downward through the air (Chapter 14)

Freely migrating dune   A dune that occurs on landscapes that have little or no vegetation cover, and thus is free to move with the wind. Contrast with fixed dune (Chapter 21).

Fringing coral reef   A coral reef that is typically attached directly to the shore or is very near the shore of an island, with only a narrow channel or lagoon separating it from the shore (Chapters 17, 22)

Frost action   Processes associated with intensive, seasonal freezing and thawing (Chapter 20)

Frost blister   A feature similar to an open-system pingo, but forming annually and experiencing ice melt (partially or completely) each summer (Chapter 20)

Frost boils   A patterned ground feature that resembles circular areas of bare soil, typically covered with tundra vegetation, and which form where the soil contains a high proportion of fine-grained sediment (Chapter 20)

Frost creep   A more descriptive name for creep (Chapter 20)

Frost heave   *See* heave (Chapter 20)

Frost jacking   The process whereby water freezes within sediment voids, forcing large rocks to be uplifted (jacked) (Chapter 20)

Frost mound   Any number of surface features that form in periglacial landscapes due to the migration of water and its subsequent freezing in soils and sediments (Chapter 20)

Frost shattering   The process whereby water within fractures of rock expands during freezing, causing fractures to widen and the rock to fracture (Chapter 20)

Frost weathering   A type of physical weathering involving the growth of ice crystals and ice lenses in rocks (Chapter 11)

Frost wedging   The process whereby water seeps into joints and freezes, widening the cracks and dislodging rocks (Chapter 14)

Frost-pull   The process whereby ice lenses form within sediment, lifting and heaving the soil and any rocks within. Any void spaces that may develop beneath stones may later become filled with soil as the ice lenses melt, causing the stone to settle above its original position (Chapter 20)

Frost-push   The process whereby stones are pushed upward in soil by the formation of ice lenses below them. When the ice lenses melt, a void forms below the stone, leaving it slightly above its original position (Chapter 20)

Fumarole   A hydrothermal feature that continually emits mixtures of steam and other gases (Chapter 7)

Ga   The abbreviation for "billions of years ago" (Chapter 3)

Gabbro   An intrusive igneous rock at the dark, mafic end of the igneous rock spectrum, rich in mafic minerals (Chapter 5)

Gaining stream   Another name for a perennial stream, because it continuously gains baseflow water from groundwater (Chapter 15)

Gap   In geology, a gap is a short, narrow, stream-carved valley that cuts through bedrock ridges (Chapter 9)

Gastrolith   A stone deliberately ingested by a bird to aid in digestion (Chapter 13)

Gelifluction   A slope process that occurs within waterlogged soils overlying permafrost, where the active layer first thaws in summer and becomes saturated, allowing the saturated soil to flow slowly downslope on top of the underlying frozen permafrost (Chapter 20)

Genetic   Pertaining to genesis, or known origin (Chapter 2)

Geomagnetic excursion   A short change or reversal in Earth's magnetic field. During an excursion, Earth's magnetic pole moves sharply toward the equator and then returns to a more stable position near a pole (Chapter 3)

Geomorphic surface   A surface on a landform that can be delineated, defined, and mapped (Chapter 3)

Geomorphic threshold   The condition of a geomorphic system that, once exceeded, triggers adjustments that transform the system into a new, more stable configuration (Chapter 2)

Geomorphology   The study of landforms and the evolution of Earth's surface (Chapters 1, 13)

Geophagy   Literally "earth eating," geophagy is when animals or humans excavate and deliberately consume soil and sediment (Chapter 22)

Geothermal gradient   The change in temperature, with depth, within the upper part of Earth's crust (Chapter 20)

Geyser   A hydrothermal feature with a specific type of underground geometry and rock structure that permits spasmodic bursts of hot water and/or steam to erupt through a vent in the land surface (Chapter 7)

Geyserite   The name given to sinter deposits near geysers (Chapter 7)

Gilbert delta   A conceptual model of delta sedimentology, involving topset, foreset, and bottomset beds (Chapter 16)

Glacial, Glaciation   An event and time period within a longer Ice Age when glaciers and ice sheets expand to cover much of the land surface (Chapters 4, 18)

Glacial advance   The apparent forward movement of a glacier's terminus that occurs when ice accumulation is greater than ablation, but may also occur because of a glacier surge (Chapter 18)

Glacial drift   An outdated term that refers to any sediment of glacial origin. Sometimes simply referred to as "drift" (Chapter 19)

Glacial flour   Rock and sediment that has been ground into silt-sized particles by a glacier and which has a flour-like consistency when dry. Glacial flour causes inland lakes and other water bodies to have a soft bluish-green appearance in the sunlight (Chapter 21)

Glacial forebulge   The upward bulge in the lithosphere, immediately in front of the ice sheet (Chapter 19)

Glacial groove   A large, deep groove formed by glacial erosion of bedrock (Chapter 19)

Glacial lake outburst flood (GLOF)   The sudden drainage of a supraglacial or subglacial lake (Chapter 18)

Glacial meltwater   Water that is released upon melting of glacial ice (Chapter 18)

Glacial polish   A type of glacial erosion that leads to a smooth, polished bedrock surface with an almost mirror-like finish (Chapter 19)

Glacial retreat   The apparent backward movement of a glacial terminus that occurs when ablation losses exceed ice accumulation (or its forward flow) (Chapter 18)

Glacial system   The system that includes the glacier and its adjacent lakes, streams, and related landscapes (Chapter 18)

Glacial terminus   The end or outer margin of a glacier (Chapter 18)

Glacial till   *See* till (Chapters 18, 19)

Glacial trough   Another name for a deeply eroded glacial valley (Chapter 19)

Glacier   A perennial body of ice and snow whose movement is driven by gravity (Chapters 18, 19)

Glacier mass balance   The mass difference between the accumulated ice of a glacier minus the ablated (lost) ice mass, usually calculated over a one-year period (Chapter 18)

Glacier net annual mass balance   An estimate of the annual gain or loss of ice by a glacier, calculated by measuring the net annual accumulation and ablation rates at numerous areas on the glacier, multiplying those data by the area of the glacier, and then taking the difference between the volume of water accumulated (gained) vs the volume ablated (lost) (Chapter 18)

Glacieret   A small, isolated glacier that fills a depression or ravine on the side of a mountain (Chapter 18)

Glacigenic rock glacier   A rock glacier that contains a core of buried glacial ice, usually located downslope of a "traditional" glacier (Chapter 20)

Glaciofluvial sediment   Sediment released from ice that is then picked up and transported by glacial meltwater, where it is eventually deposited by meltwater streams on a plain (outwash) or is confined to a valley (valley train) (Chapter 19)

Glacioisostasy, Glacioisostatic rebound   The condition whereby Earth's lithosphere, which was depressed by the immense weight of ice sheets, rebounds back and gains elevation (Chapter 19)

Glaciokarst   A karst landscape where glacial erosion dominates over traditional karst processes such as dissolution (Chapter 12)

Glaciolacustrine plain   An exceptionally low-relief landscape, formed by wave erosion and sediment infilling within former glacial lakes. Also called a "glacial lake plain" (Chapter 19)

Glaciolacustrine sediment   Sediment deposited in proglacial lakes, composed of thin, finely bedded strata of clay, silt, and sand (Chapter 19)

Glaciologist   Someone who studies glaciology, including but not limited to geologists, geographers, physicists, mathematicians, climatologists, computer scientists, and engineers (Chapter 18)

Glaciology   The science of ice in all its forms (Chapter 18)

Glen's Law   An empirical relationship, developed by glaciologist John Glen in the 1950s, that predicts that the ice deformation rate is proportional to the cube of the applied shear stress (Chapter 18)

Gleyed   The condition that forms in soils that lack free oxygen, usually due to saturation, causing them to develop gray colors (Chapter 13)

Glide   The downstream segment of a pool in a river channel, and a zone of low turbulence in the lower part of the pool, where the flow accelerates as it transitions to a shallow riffle zone downstream (Chapter 16)

Gliding   One way that ice deforms and moves, involving the slipping of the basal plane of the hexagonal ice crystals (Chapter 18)

Gneiss   A coarse-grained, highly foliated metamorphic rock, indicative of high-grade metamorphism (Chapter 5)

Graben   A down-dropped block of crust, usually formed due to normal faulting, which may then evolve into a lowland or trough (Chapters 10, 23)

Gradation   Another name for denudation (Chapter 23)

Grade   The way in which rivers, over time, adjust so that they have sufficient energy to transport the sediment supplied to them. Or, the balance between the sediment load delivered to the stream and the capacity of the stream to transport it (Chapters 2, 15)

Graded river   A river that has achieved mutual adjustment between flow and form, so as to transport the amount of sediment delivered to it (Chapters 2, 15, 16)

Grain-by-grain disintegration   The breaking apart of rocks along mineral crystal edges. Also known as granular disintegration (Chapter 11)

Granite   A felsic, intrusive, igneous rock that is 20–60% quartz, with most of the remaining minerals being feldspar and a small percentage of other, dark, accessory minerals (Chapters 5, 7)

Granular disintegration   *See* grain-by-grain disintegration (Chapter 11)

Granular flow   A type of mass movement that flows but contains <20% water (Chapter 14)

Gravitational force   The force pulling downward on everything, toward the center of the Earth (Chapter 14)

Great Basin   A physiographic region in the western United States where rivers drain internally, rather than to the sea (Chapter 10)

Greenhouse climate   A period of warmer climate in Earth history (Chapter 4)

Greywacke   A type of sandstone that exhibits poor sorting and contains > 15% clay-sized particles (Chapter 5)

Grike   The low, enlarged fractures and/or fissures in a barren limestone surface. Contrast with clint (Chapter 12)

Groin   An engineering structure consisting of a wall or a series of rocks installed perpendicularly to the shore to minimize erosion by capturing sand moving in the longshore drift (Chapter 17)

Groove   A small-scale, linear-shaped feature worn into a rock surface, formed as a result of weathering and erosion processes (Chapter 11)

Ground ice   Ice within frozen soil or rock (Chapter 20)

Ground moraine   Another name for a low-relief till plain with a cover of glacial till (Chapter 19)

Groundwater   Water that completely fills the pore spaces in soils and bedrock (Chapters 12, 15)

Groundwater discharge   Water in the saturated zone, emerging back onto the surface (Chapter 15)

Groundwater recharge   The addition of water to the saturated zone by percolation through the layers above (Chapter 15)

Grus   A gravelly residual sediment formed by physical weathering of rocks, typically of granite, and dominated by grains of minerals such as quartz and feldspar (Chapters 5, 11)

Guano   Excrement produced by bats or birds, sometimes found in caves (Chapter 12)

Gully   A channel, slightly larger than a rill, that is formed by running water (Chapter 16)

Gypcrete   A soil horizon that has become cemented by gypsum (Chapter 11)

Haboob   A type of intense dust storm, often manifested as a "wall of dust" as high as 1500 m along the leading edge of a weather front or thunderstorm (Chapters 10, 21)

Hadean Eon   A formal subdivision of geologic time from the beginning of Earth history until ≈4 Ga (Chapter 4)

Half-blind valley   A stream channel downstream of a swallow hole within which water only flows when the swallow hole becomes incapable of taking the entire discharge of the river (Chapter 12)

Half-graben   A block bounded by a fault (or parallel faults) only on one side that typically rotates as faulting takes place (Chapter 10)

Hammada   A term, literally meaning "rocky desert," for a desert landscape that consists mainly of rocky plateaus and uplands with almost no eolian sand (Chapter 21)

Hanging valley A valley eroded by a valley glacier that drops precipitously into a larger glacial valley, often leading to the formation of a waterfall at the junction (Chapter 19)

Hanging wall The part of the rock in a reverse or normal fault that, if one were to stand on the fault plane, would be "hanging" above your head (Chapter 10)

Harmattan wind The local name for a wind that transports dust from the Sahara Desert to the Atlantic Ocean, sometimes carrying it as far as the Caribbean Sea (Chapter 21)

Head The term given to case-hardened rims standing between tafoni (or niches) on complex, cliff-side topography. In mass wasting, the uppermost surface of a rotational landslide (slump) (Chapters 11, 14)

Head of outwash A variant of an outwash plain, a sloping landform that is formed in front of a stagnant ice margin by deposition of glaciofluvial sediment (Chapter 19)

Head slope A type of hillslope that, in plan view, has a concave form, and which is commonly the headwater of first-order streams. Contrast with nose slope (Chapter 13)

Headland An area along a coastline that protrudes outwards, into the water (Chapter 17)

Headwall The steep bedrock wall at the upper end of cirque, formed as the glacier erodes into the mountainside (Chapter 19)

Headwater stream A first-order stream at the upper reaches of a stream network (Chapter 15)

Heave The process where, during a wetting and/or freezing phase, rock and soil particles get shifted upward perpendicularly to the surface of the slope, and then upon thawing or drying, are let straight down. Sometimes called frost heave (Chapters 14, 20)

Heinrich event A climatic event that is assumed to have been driven by huge releases of ice (as icebergs) from ice sheets (Chapter 4)

Helicoidal flow A corkscrew-like flow within a stream channel that pulls sediment from the stream bed upward and directs it onto a point bar (Chapter 16)

Helictite A small, slender speleothem that grows in all directions, not just downward, driven by capillary flow (Chapter 12)

High polar glacier A glacier that contains ice that is continually below 0° C (Chapter 18)

High-centered ice-wedge polygon A type of ice-wedge polygon that has deep troughs at its edges, within which ice wedges occur. These types of polygons commonly lack ridges (Chapter 20)

Hinge line Another name for the fold axis in folded sedimentary rocks (Chapter 9)

Hogback A bedrock landform that is similar to a cuesta, but formed on much more steeply dipping rocks (Chapters 8, 9)

Holocene Epoch The current warm (interglacial) phase of Earth's climate, from ≈11.7 ka to the present, making it the last part of the Quaternary Period. The only geological Epoch defined by climate change (rapid warming), as recorded in the NGRIP core from the Greenland ice cap. (Chapters 4, 19)

Holocene Thermal Maximum *See* Climatic Optimum (Chapter 4)

Holokarst True karst, i.e., a landscape where dissolution is the primary, active, land-forming process (Chapter 12)

Homoclinal ridge A ridge that has formed on resistant, dipping beds on the limb of folded rock (Chapter 9)

Homoclinal shifting The lateral (sideways) shifting of the streams and ridges on a body of rock, typically on fold limbs, as erosion exposes area of softer rock. The streams on the gently dipping, layered rocks shift laterally down the dip slope over time (Chapters 9, 16)

Homocline A geologic structure where the rock layers (either sedimentary or igneous) dip uniformly in the same direction, meaning they have the same general inclination in terms of both direction and angle (Chapter 9)

Honeycomb weathering A type of weathered rock morphology that involves alveoli and small cavities in rock surfaces (Chapter 11)

Hoodoo A tall, narrow bedrock spire, often pointed at its summit, that forms by weathering and erosion of weaker (often jointed) rocks that surround it (Chapter 8)

Hook The curved or recurved end of a spit (Chapter 17)

Horn A high, jagged peak surrounding glaciated mountain valleys, usually occurring where one or more arêtes meet (Chapter 19)

Hornfels A dark-colored, non-foliated, metamorphic rock formed by contact metamorphism without pressure stress (Chapter 5)

Horst A up-faulted block of crust, or a block of crust that has not moved but others near it have been down-faulted (Chapter 10)

Hortonian overland flow Runoff of water across the land surface that occurs when the rate of precipitation or snowmelt exceeds the infiltration capacity of the surface (Chapter 15)

Hot spring An open pool of hot water being heated by a subterranean heat source (Chapter 7)

Hotspot The surface expression of a plume of rising magma within the mantle, typically associated with volcanism (Chapters 6, 7)

Hummocky topography A type of surface topography with many small, irregularly shaped hills and depressions (Chapter 14)

Humus All manner of dark, amorphous, organic substances produced mainly by the decomposition of plant materials and which impart black or dark brown coloration to soils and sediments (Chapters 11, 15)

Hydration A type of chemical weathering involving the incorporation of water molecules into a mineral lattice (Chapter 11)

Hydraulic geometry A type of analysis of the fluvial system where channel characteristics are statistically compared to discharge. These analyses can be performed at a site, over time, or in the downstream direction for the same period of time (Chapter 16)

Hydrograph A graphical plot of stream discharge over time (Chapter 15)

Hydrographic apex The location at the top of an alluvial fan where the stream exits the mountain front and from which the distributary channels spread out (Chapter 10)

Hydrologic cycle A description of the movement of water between and through the various reservoirs of the hydrosphere (Chapter 15)

Hydrolysis A type of chemical weathering that involves the decomposition of minerals through reactions that "split" water molecules into H^+ and OH^- components (Chapter 11)

Hydrophobic soils Soils that are water repellent (Chapter 22)

Hydrosphere   All the water on, below, and above the surface of the Earth (Chapter 15)

Hydrothermal metamorphism   A localized type of metamorphism that occurs as hot, volatile solutions percolate into and react with the protolith (surrounding country rock) (Chapter 5)

Hydrothermal processes/activity   Geological processes associated with eruptions of hot water and/or steam (Chapters 6, 7)

Hypogenic cave   A cave formed by ascending fluids that acquire their solutional aggressiveness from sources deeper underground (Chapter 12)

Ice age   Colloquial name for an event and period of time when large ice sheets are present on Earth. Geologically, the Quaternary Period (which we are still in) is an ice age, although ice ages have occurred in the Late Cenozoic, Late Paleozoic, Ordovician, and during the Precambrian (Chapters 4, 18)

Ice creep   A synonym for ice deformation (Chapter 18)

Ice deformation   The internal deformation of ice. Also called "ice creep" (Chapter 18)

Ice foliation   The characteristic of glacier ice that has bands of distinctive crystal sizes and/or bubbles and debris, usually caused by shearing stress and deformation that a glacier experiences as it flows over its bed (Chapter 18)

Ice segregation   A process that occurs as liquid water in soil and rock migrate toward areas of lower temperature, forming distinct ice lenses that can grow over time (Chapter 20)

Ice sheet   The largest of all glaciers, with an extent of >50,000 km² and with thicknesses of up to 4,000 m. Sometimes called a continental glacier (Chapters 18, 19)

Ice shelf   A large body of floating glacial ice that develops on land but which extends out, over water (Chapter 18)

Ice stream   A zone of rapidly flowing ice bordered by slower-moving ice (Chapter 18)

Ice-contact stratified sediment   A type of outwash deposited by meltwater but in contact with glacial ice (Chapter 19)

Icefield   A broad upland of ice in high, mountainous areas, from which emerge valley and piedmont glaciers (Chapter 18)

Icehouse climate   Periods of colder climate in Earth history, many of which could be considered Ice Ages (Chapter 4)

Ice-walled lake plain   A landform that develops where widespread areas of stagnant ice become covered with supraglacial sediment, allowing ice blocks to form depressions in the ablating ice that retain meltwater lakes, which then fill with sediment (Chapter 19)

Ice wedge   A V-shaped body of ice that forms in the active zone of a permafrost landscape, and which may extend downward, into the permafrost (Chapter 20)

Ice-wedge cast   The infilled material that has the V-shape of what was once an ice-wedge (Chapter 20)

Ice-wedge polygon   Part of a broader pattern of polygonal, often three- to five-sided, forms, developed and driven mainly by thermal contraction cracking in periglacial regions (Chapter 20)

Ice-wedge pseudomorph   Another name for an ice-wedge cast (Chapter 20)

Igneous rock   A major category of rock that forms as molten geologic material (lava or magma) cools and crystallizes (Chapters 5, 7, 8)

Illuviation   The translocation of substances into a soil horizon from above, usually driven by percolating water (Chapter 13)

Impact crater   A depression formed by the hypervelocity contact of an asteroid or comet with the surface of a planetary body (Chapter 23)

Impact threshold   The speed required to move grains, but less than the speed required to initially displace them (Chapter 21)

Impeding layer   A less-permeable layer of material in the subsurface, with a lower infiltration capacity than the layer above, which causes percolating water to perch and possibly flow laterally along the top of it (Chapter 15)

Incise, incision   To erode or cut down, as is commonly done by river systems that have excess stream power (Chapter 15)

Incised meander   A meander in a stream channel, formed as a stream that is initially meandering across a low-relief surface incises downward so rapidly that the channel stays generally fixed in that location, forming a deep valley with a meandering form (Chapter 16)

Incompetent   Term referring to rocks or rock layers that are relatively weak and easily deformed under stress, often deforming more plastically than competent rocks, or which when forming a free face, often release rocks via mass wasting processes (Chapter 14)

Infauna   Soil-dwelling animals (Chapter 11)

Infiltrate, Infiltration   Water that moves from the air or land surface, into the ground below (Chapter 15)

Infiltration capacity   The maximum rate at which water can enter soil or sediment, vertically (Chapter 15)

Ingrown meander   A type of incised meander that has steep outer bend walls but relatively gentle slopes on the insides of bends, shaped like steep point bars, indicating that lateral migration of the meanders and an increase in meander amplitude occurred during the incision period (Chapter 16)

Inlet   Another name for a tidal channel (gap) in a barrier island complex that connects the lagoon to the open ocean (Chapter 17)

Inlier   The situation that occurs on a landscape where isolated outcrops of older rock occur at the surface, even though across most of the landscape, younger rock overlies older rock. In an inlier, older rock is exposed at the surface, isolated among younger rocks that surround it (Chapters 8, 10)

Inner core   Part of Earth's geologic core, at the very center of the Earth. It is a solid sphere that is responsible for Earth's magnetic field (Chapter 6)

Inselberg   An isolated bedrock hill that rises abruptly above a surrounding bedrock plain or pediment (Chapters 5, 10, 11)

Insolation   A synonym for solar radiation (Chapter 11)

Insolation cycle   The process of solar heating during the day and cooling at night (Chapter 11)

Insolation weathering   Another name for thermal weathering, driven by sunshine and the heat associated with it (Chapter 11)

Instantaneous discharge   The discharge of a stream at a given point in time (Chapter 15)

Interception   The capture of precipitation as it lands on vegetation or organic litter lying on the land surface, before reaching the soil below (Chapter 15)

Interglacial, Interglaciation   A time period within a longer Ice Age when glaciers and ice sheets generally melt back or have melted back (Chapters 4, 18)

Interlobate moraine   A glacial end moraine formed at the intersection of two glacial lobes (Chapter 19)

Interlobate zone   The area of side-to-side contact between two glacial lobes (Chapter 19)

Intermittent stream   A stream that flows continuously during part of the year, driven by inputs of baseflow, but which during the rest of the year flows only after heavy rains or snowmelt events (Chapter 15)

Internal drainage   The condition where rivers in an area do not drain to the sea, but instead drain internally, into an endorheic basin (Chapter 10)

Interparticle attractive forces   Forces that include van der Waals forces, water adsorption forces, and electrostatic forces, all of which cause particles to resist being pulled apart (Chapter 21)

Interstadial   A time period, not as long as an interglacial period (or interglaciation), when an ice sheet briefly retreats (Chapter 4)

Intrusive ice   Ice that forms when water within soil or rock is forced under pressure into an unfrozen layer, in which it then freezes (Chapter 20)

Intrusive igneous rock   A type of igneous rock that cools relatively slowly from magma, deep within the crust. Granite is the prototypical intrusive igneous rock (Chapter 5)

Inversion of topography (topographic inversion)   A phenomenon whereby a low-lying rocks, more resistant to erosion than the surrounding rocks, erode more slowly and eventually become an upland (Chapter 5)

Island arc   A chain of volcanic islands formed where two oceanic plates are converging and one gets subducted (Chapters 6, 7)

Isochronous   The property of being the same age everywhere (Chapter 3)

Isolated talik   An isolated body of unfrozen ground that is surrounded by permafrost on all sides, as well as above and below (Chapter 20)

Isostasy   Vertical adjustments of the land surface in response to changes in the loading of mass onto it (Chapters 2, 6, 10)

Isostatic rebound, adjustment   The rebound, or uplift, of parts of Earth's crust due to mass being removed from it, typically in response to the melting of an ice sheet or simply due to long-term erosion (Chapters 6, 9, 17, 23)

Isotope   For a given chemical element, an isotope is a variant, with the same number of protons but different numbers of neutrons, and different masses (Chapter 4)

Jetty   An engineering structure installed at harbor mouths, usually as two walls along each shoreline of a river that exits at a harbor, to minimize shoaling within the river mouth and harbor (Chapter 17)

Joint   A fracture or crack in a rock that forms due to the expansion induced during cooling, or other types of stresses (Chapters 5, 8, 14)

Jökulhlaup   A type of flash flood triggered by the sudden emergence of a heat source, e.g., a volcanic eruption, beneath a glacier (Chapters 7, 14)

ka   The abbreviation for "thousands of years ago" (Chapter 3)

Kame   A hill, composed of ice-contact stratified sediment, that has formed under or in contact with a glacier (Chapter 19)

Kame and kettle topography   A type of extremely hummocky glacial topography where many ice blocks were left behind and buried, forming kettles. The hills between the kettles are considered kames or kame-like features, formed where sediment filled holes in or next to the ice (Chapter 19)

Kame terrace   A landform that develops where the margin of a glacier abuts an upland, allowing meltwater streams to deposit sediment in contact with both the ice and the upland (Chapter 19)

Karren   Elongated grooves and fissures weathered and worn into a limestone surface, typically separated by sharp ridges that have been formed by a combination of erosion and dissolution (Chapters 11, 12)

Karst   A type of land system dominated by the dissolution of rocks and which form a distinctive terrain (Chapter 12)

Karst spring   Location where the water from a subterranean river rises to the surface and begins to flow again as a stream on the land surface (Chapter 12)

Karst valley   Another name for a blind valley formed on a karst landscape (Chapter 12)

Karstified   The condition whereby a soluble rock unit has developed a complex series of conduits, fractures, and large openings, due to dissolution (Chapter 12)

Katabatic wind   A cool and dry wind, driven by density differences, that flows off a nearby ice sheet or glacier (Chapter 21)

Kegelkarst   Another name for cockpit karst (Chapter 12)

Kettle   A depression on the land surface left behind after an ice block melts within an area of glacial sediment (Chapter 19)

Kettle lake   A kettle that is filled with water (Chapter 19)

Keystone species   A species on which other species largely depend, or which has importance in ecosystem functionality, such as an organism that supports the entire biological community (Chapter 22)

Kimberlite   An intrusive igneous rock and a rare variant of peridotite, most commonly known to be a host matrix for diamonds (Chapter 5)

Klippe   A mountain formed in older but resistant rocks that occurs on top of younger, more erodible rock, usually having been emplaced there by thrust faulting (Chapter 10)

Knickpoint   An abrupt topographic change or break, such as a waterfall, in a stream's longitudinal profile (Chapter 16)

Kopje   Another name for a tor or castle koppie (Chapter 11)

Kuiper Belt   A region in the Solar System, beyond Neptune's orbit, composed of icy bodies and dwarf planets (Chapter 23)

Laccolith   A pluton, like a sill, but which forms when intruding magma forces the overlying rocks to bulge and bend upward, taking on a broad, mushroom-shape, with a flat base and a domed upper surface, and usually with a feeder pipe below (Chapter 7)

Lacustrine   Adjective referring to lakes (Chapter 5)

Lag, Lag time   A delayed response to a stimulus (Chapter 16)

Lagoon   The brackish water body between a barrier island and the mainland (Chapter 17)

Lag-to-peak time   The time of delay in a stream system, between the centroid of rainfall and the peak discharge, as depicted on a storm hydrograph (Chapter 15)

Lahar   A type of mudflow that consists of a mixture of rock, soil, volcanic ash, and water, formed as hot volcanic materials mix with water from streams and/or melted snow and ice (Chapters 7, 14)

Laminar flow   A type of slow flow in a river, wherein the water moves in parallel flow lines (Chapter 15)

Landform   A natural feature on a landscape, formed in rock or sediment, that can be observed, described, mapped, and classified (Chapter 1)

Landscape   An organized assemblage of genetically interconnected and interrelated landforms (Chapter 1)

Landslide   A mass of rock, sediment, and soil, sometimes called a debris flow, that travels rapidly and destructively downslope (Chapters 7, 14)

Lane's balance   A conceptual and graphical depiction of the factors and forces that affect the concept of grade in a river system, as shown in the form of a balance. It is useful for understanding channel equilibrium (grade) conditions in relation to water and sediment inputs (Chapter 15)

Lapilli   A category of pyroclastic debris, composed of particles that are mostly the size of small gravel or peas (Chapter 7)

Last Glacial Maximum   Often abbreviated LGM, the time of maximum ice extent during the last glaciation, at around 26–19 ka, depending on location (Chapters 4, 18)

Late Cenozoic Ice Age   A geologic event during which cold temperatures and ice sheets were a major part of Earth history, starting ≈34 Ma during the early Oligocene Epoch, when extensive glaciers began forming in Antarctica (Chapters 4, 19)

Latent heat   Heat stored in water that gets released as it turns to ice, or as water vapor condenses into water (Chapter 18)

Lateral accretion   The accumulation of sediment along the margins and sides of a stream channel, as on point bars (Chapter 16)

Lateral bar   A bar that forms along the edge or margin of a river channel (Chapter 16)

Lateral levee   Debris that accumulates at the edges or margins of a debris flow deposit, often resembling small ridges (Chapter 14)

Lateral moraine   A moraine of a valley glacier that marks where the edge of the ice once abutted the valley wall (Chapter 19)

Laterite   Soil material that has become cemented by iron oxides (Chapter 11)

Laurentide ice sheet   The name of the ice sheet that formed over Canada, the northern United States, and other parts of northern North America during the Quaternary Period (Chapters 4, 18, 19)

Lava   Molten rock materials that have erupted onto the land surface (Chapters 5, 7)

Lava dome   A hemispherical to irregularly shaped mound of volcanic rock, formed by one or more eruptions of silicic lava above a vent (Chapter 7)

Lava tube   *See* pyroduct (Chapter 7)

Leading edge   See Active plate margin(Chapters 6, 17)

Left-lateral   A type of strike-slip fault in which the movement, as seen across the fault line, is to the left (Chapter 10)

Legacy sediment   Historically, human-produced, i.e., legacy, sediment carried and deposited by rivers, often containing toxic materials and chemicals (Chapter 16)

Lichenometry   A dating method used to assess the age of surfaces, based on the size and coverage of crustose lichens on rocks at the surface (Chapter 3)

Lift force   The force generated when wind or water accelerates as it flows up and over a particle, forming a zone of lower pressure on the top side of the particle, enabling it to be pulled upward, off the surface (Chapters 15, 21)

Lignite   A soft, brown form of coal derived from naturally compressed peat (Chapter 5)

Limestone   A type of sedimentary rock formed from precipitated calcium carbonate, with the chemical makeup $CaCO_3$ (Chapters 5, 8)

Limestone pavement   An extensive, bare rock plain, disrupted by open fractures which have become enlarged by the dissolution of limestone (Chapter 12)

Limestone sink   Another name for a doline or a sinkhole (Chapter 12)

Linear dune   A sinuous-to-straight sand dune that is generally oriented parallel to the dominant wind direction or directions (Chapter 21)

Linear ridge   A low ridge formed in association with a strike-slip fault, where rocks, crushed during a faulting event, have been pushed slightly upward (Chapter 10)

Liquefaction   The process by which unconsolidated, water-saturated sediments are transformed into a substance that acts like a liquid, flowing and deforming readily (Chapter 14)

Listric fault   A fault that dives into the crust at shallow angles, characterized by a decreasing angle of dip with depth, giving it a curved, concave-upward fault plane (Chapters 6, 10)

Lithalsa   A frost mound comprised of segregated ice, typically circular or ovoid in shape and found on former lake basins or submerged marine areas that lack a thick peat layer at the surface (Chapter 20)

Lithified   To turn into rock, usually by compaction, heating, and/or additions of cementing agents (Chapter 5)

Lithologic discontinuity   The contact zone between two different parent materials, stacked one upon the other, in a soil or sedimentary sequence (Chapter 13)

Lithology   The physical characteristics of a rock, including color, texture, grain size, and composition (Chapters 5, 8)

Lithophagy   Literally "stone (or rock) eating," the practice of animals deliberately excavating and consuming rocks (Chapter 22)

Lithosphere, lithospheric plate   Geologically, the crust and the part of the mantle above the asthenosphere, composed of a coherent, semi-rigid layer that is broken up into a number of sections, or plates (Chapters 6, 7)

Litter   The layer of organic materials on the top of the soil surface that, in soil science, is called an O horizon (Chapters 13, 22)

Little Ice Age   A cool climatic interval in the late Holocene, between ≈1500 and 1800 AD, during which glaciers re-advanced across much of the world (Chapter 4)

Load   Sediment transported by streams (Chapter 15)

Loading   A zoogeomorphic process attributable to the weight of animals pushing down on the soil surface (Chapter 22)

Loamy   An adjective used to describe soils with a generally uniform mix of sand, silt, and clay, such that the "feel" of any one does not dominate (Chapter 13)

Local relief   The difference in elevation between the local uplands and the nearby lowlands (Chapter 2)

Lodgment till   One type of basal till, deposited beneath a glacier as it is forced, or "lodged," into the bed (Chapter 18)

Loess   Wind-blown, dominantly silt-sized, sediment (Chapters 4, 19, 21)

Log step   Woody material that extends across a stream channel, which then elevates the water surface, often forming a small waterfall (Chapter 22)

Longitudinal crevasses   Crevasses that form parallel to the flow direction of the glacier, developing because of the splaying of the glacial margin in the ablation zone (Chapter 18)

Longitudinal profile   A graphical representation of the elevation of a river, a river valley, or a landform, from its source to its mouth, i.e., down-valley (Chapters 1, 5, 16)

Longshore current   A shore-parallel current that forms in waters just offshore, driven as waves approach the shore at an oblique angle and then reflect off the beach face (Chapter 17)

Longshore drift   The process whereby sediment is transported in shallow water, parallel to the coast, driven by longshore currents (Chapter 17)

Lopolith   A type of pluton resembling an "inverted" laccolith, i.e., a large, lenticular, igneous intrusion with a depressed central region (Chapter 7)

Losing stream   Another name for an ephemeral stream that loses water in the downstream direction because the water sinks into the bed of the channel (Chapters 10, 15)

Low-centered ice-wedge polygon   A type of ice-wedge polygon that has low, flat polygon centers, and shallow troughs bounded by ridges, typically forming in flat areas and which may contain ponded water within the polygon centers (Chapter 20)

Luminescence dating   A numerical dating method that determines the last time a sediment was exposed to sunlight (and then, presumably, buried). This definition refers to *optically stimulated* luminescence dating (Chapter 3)

Ma   The abbreviation for "millions of years ago" (Chapter 3)

Maar   A shallow volcanic crater with steep sides composed of tephra, formed by one or more explosions that develop as hot magma comes into contact with shallow groundwater (Chapter 7)

Macrofossil   A plant or animal fossil large enough to be visible to the naked eye (Chapter 4)

Macropore   A large void in a soil or sediment (Chapter 15)

Macroporosity   The amount of large void spaces, such as between fractures and bedding planes in a rock (Chapter 5)

Macrotidal   A type of coast where the tidal range is > 4 m (Chapter 17)

Mafic   A category of rock-forming minerals that are low in silica and rich in iron and magnesium (Chapter 5)

Mafic lava   *See* basaltic lava (Chapter 7)

Mafic rock   A type of dark and comparatively "heavy" rock that is dominated by minerals with large amounts of magnesium and iron in their crystalline structures (Chapter 5)

Magma   Molten rock beneath the land surface (Chapters 5, 7)

Magma chamber   A large pool or reservoir of molten rock (magma) beneath the Earth's surface, which typically feeds the volcano above (Chapter 7)

Magmatic   Processes and features associated with molten rock beneath the land surface (magma) (Chapter 7)

Magnetic susceptibility   The response of a soil or sediment sample to an applied magnetic field (Chapter 4)

Magnetostratigraphy   A numerical dating technique used to correlate and date rock and sediment sequences by analyzing their magnetic properties, particularly the direction and intensity of their natural remanent magnetization (NRM). This technique is based on the knowledge that Earth's magnetic field has reversed polarity throughout geological time, thereby providing a global stratigraphic framework (Chapter 3)

Mantle   The largest (by volume), solid layer of the Earth, located between the asthenosphere and the core, composed of dense ultramafic magnesium- and iron-rich materials. Mantle material is soft and flows plastically, under great depths and pressures (Chapter 6)

Mantle convection   The slow currents in Earth's mantle that drive the movement of the crustal plates above (Chapter 7)

Marble   The metamorphic equivalent of limestone (Chapter 5)

Mare   Vast, dark-colored, low-elevation, nearly smooth plain on the Moon, formed by lava flows in the past (plural: maria) (Chapter 23)

Marine   Having to do with the ocean (Chapter 5)

Marine oxygen isotope stages   Periods in the paleoclimate record recovered from, and based on, the deep-sea record in foram shells (Chapter 4)

Marine terrace   A wave-cut platform that is exposed above the modern shoreline due to coastal erosion, followed by episodic, tectonic uplift (Chapter 17)

Mass balance   In glaciology, the mathematical difference between the accumulated ice mass minus the ablated (lost) ice mass. A positive mass balance implies that the glacier has gained ice, and vice versa (Chapter 18)

Mass movement   The downslope transport of rock, soil, and ice materials under the influence of gravity (Chapter 14)

Mass wasting   Another name for the process of mass movement (Chapters 5, 14, 23)

Massive   The trait wherein a rock lacks clear layers, instead having a generally homogeneous composition (Chapter 5)

Maturity   The second stage of W. M. Davis' cycle of erosion, in which streams develop floodplains, and during which the relief on the landscape is maximal. In mineralogy, how evolved or old a sediment might be (Chapters 2, 5)

Maunder Minimum   A period of minimum solar activity, when sunspot numbers are exceptionally low (Chapter 4)

Maximum-limiting date   A numerical age on a feature or surface that establishes that it can be no older than the age reported (Chapter 3)

Maze cave   A cave composed of a complex and dense grid of intersecting passages (Chapter 12)

Meander amplitude   A measure of the width of a belt of river meanders on a floodplain (Chapter 16)

Meander cutoff   The situation that forms when a cutbank on the side of a river channel erodes into and through a thin meander neck, leaving the former channel of the river (the meander) "cut off" from the main channel (Chapter 16)

Meander neck   The narrow piece of land between two meanders of a river (Chapter 16)

Meander scar   A low, arcuate area on a floodplain formed when an oxbow lake fills (or partially fills) with sediment and organic materials (Chapter 16)

Meander scroll   An arcuate ridge on the surface of point bars that represents an individual episode of deposition (Chapter 16)

Meander wavelength   The downstream distance, within a belt of meanders on a floodplain, from one meander bend to the other (Chapter 16)

Meandering channel   A type of fluvial system where the river flows in one main channel that meanders and winds across its floodplain (Chapter 16)

Mechanical weathering   *See* physical weathering (Chapter 11)

Medial moraine   A moraine formed where two lateral moraines of a valley glacier meet and coalesce (Chapter 19)

Medieval Warm Period   A period of warmer global climate during the Holocene, between ≈900 and 1250 AD, when it was so warm that Norse settlers (Vikings) were able to colonize Greenland (Chapter 4)

Mega barchan   A large barchan dune, often exceeding 50 m in height (Chapter 21)

Mega slump   A large thaw slump with a basin area exceeding 5 hectares (Chapter 20)

Megacusp   A feature found in some thermal caves, as the ceiling and walls acquire wavy morphologies due to condensation corrosion (Chapter 12)

Megaripple   A very large ripple on a sand dune, often formed where a distinctive mix of coarse sand and fine sand is present (Chapter 21)

Meltout till   A type of basal till deposited via the melting of glacial ice. As basal melting occurs, sediment and clasts are slowly deposited below the glacier (Chapter 18)

Mesa   An isolated, flat-topped landform that is larger than a butte and formed through escarpment retreat in (typically) horizontally bedded sedimentary rock (Chapter 8)

Mesa-and-butte landscape   A (typically arid) landscape with many mesas and buttes, and usually underlain by horizontally bedded sedimentary rocks (Chapter 8)

Mesotidal   A type of coast where the tidal range is 2–4 m (Chapter 17)

Metamorphic core complex   A deep-seated rock mass that forms where shearing occurs along a detachment fault, such that the adjoining rocks are metamorphosed (Chapter 10)

Metamorphic rock   A major category of rock that forms as preexisting rocks are changed by high pressures and/or temperatures deep in the crust (Chapter 5)

Meteorite   A solid piece of debris from an object, such as a comet, asteroid, or meteoroid, that originates in outer space and survives its passage through an atmosphere to reach the surface of a planet or planetary body (Chapter 23)

Microporosity   The amount of small void spaces in a rock or sediment, between or within grains and/or mineral crystals (Chapter 5)

Microtidal   A type of coast where the tidal range is < 2 m (Chapter 17)

Mid-channel bar   A bar, typically sandy, that forms in the middle of a channel and which is not connected to the bank. They are like islands, but lower, not densely vegetated, and less stable (Chapter 16)

Midden   A heap of dung, urine, and general refuse in a cave (Chapter 12)

Mid-oceanic ridge   The ridge formed at a tectonic spreading center, where new oceanic crust is being formed, usually in the center of an ocean basin (Chapters 6, 7)

Mineral   An inorganic, naturally occurring, crystalline solid with specific physical properties and chemical composition (Chapter 5)

Mineral parent material   Geologic material composed of sand, silt, clay, and coarser fragments like gravel and stones, within which soils may develop. Contrast with organic parent material (Chapter 13)

Minimum-limiting date   A numerical age on a feature or surface that establishes that it can be no younger than the age reported (Chapter 3)

Minor scarp   A small, steep face on the upper part of a slump block, formed as parts of the main block break up and slide further downslope (Chapter 14)

Miocene Epoch   A formal subdivision of geologic time, within the Cenozoic Era, from ≈23 to 5.3 Ma (Chapter 4)

Mixed energy coast   A coast that is influenced by both waves and tides (Chapter 17)

Mixing corrosion   The process that explains why groundwater becomes aggressive and is able to dissolve more soluble rock, as occurs when it mixes with water percolating downward, through the regolith above (Chapter 12)

Mixmaster   A type of fauna that burrows in the soil but instead of making mounds on the surface, they mainly churn the soil (Chapter 13)

Modified saltation   Another name for short-term suspension (Chapter 21)

Mogote   An upland "tower" of bedrock in a tower karst landscape (Chapter 12)

Mohorovičić discontinuity   The boundary, also known as the Moho, between the mantle and the crust (Chapter 6)

Mohs scale   Developed by the German mineralogist Friedrich Mohs, the scale ranks the scratch resistance of minerals, ranging from 1 (softest) to 10 (hardest) (Chapter 18)

Mollisol   A type of soil with a thick, dark-colored A horizon rich in humus, typically formed in grasslands (Chapter 13)

Monadnock   An isolated, relict upland found on old, low landscapes called peneplains in W. M. Davis' cycle of erosion. They typically form on resistant bedrock, or are located far from the major streams (Chapter 2)

Monocline   A mass of folded rock with only one dipping limb (Chapter 9)

Monogenetic   The characteristic of having only one instance of "genesis" or formation. Contrast with polygenetic (Chapter 13)

Monsoon   An atmospheric circulation typified by changing winds and weather twice annually. The South Asian Monsoon is the classic example – with warm, rainy summers and dry, cool winters (Chapters 4, 6)

Moraine   A glacial depositional landform that forms at the margin of the ice, either at its terminus or at its side (Chapters 19, 23)

Morphoclimatic zone   A zone in which most of the landforms correlate to a certain type of climate (Chapter 4)

Morphodynamics   The interactions between coastal morphology and the various hydrodynamic drivers such as waves, currents, and water levels (Chapter 17)

Moulin   An open, vertical shaft that has been cut and/or melted into the ice (Chapters 18, 19)

Moulin kame   A gumdrop-shaped kame that formed inside a moulin and is usually composed of ice-contact stratified sediment (Chapter 19)

Moundmaker   A type of biota that burrows in the soil and moves sediment to the surface, where it accumulates in mounds (Chapter 13)

Mountain glacier   A general term for a glacier in alpine (high mountainous) regions (Chapter 18)

Muck   A type of highly decomposed, organic, soil parent material, typical of bogs and wetlands (Chapter 13)

Mud playa   A type of playa surface where the dominant surface sediment is silt and clay (Chapter 10)

Mud pot   A hydrothermal feature within which warm mud, formed where steam and acidic fluids have weathered the surrounding rocks into silt and clay, continuously boils and gurgles at the surface (Chapter 7)

Mud volcano   A mud pot forming in particularly viscous mud, so that it bubbles with very large, intact blobs of mud, forming a small mound of sediment at the opening (Chapter 7)

Mudflow A watery type of debris flow that consists mainly of grains that are sand-sized or smaller, and with a very small proportion of large clasts (Chapters 14, 22)

Mudstone A type of sedimentary rock made up of silt and clay (Chapter 5)

Natural bridge A bedrock feature formed as a river breaches an incised meander neck, leaving behind a rock span (bridge) across the river (Chapter 16)

Natural levee A low, ridge-like landform that develops as sediment is deposited on the floodplain near the channel, while the stream is in flood (Chapter 16)

Neap tide A tide that occurs during the first and third quarter moons, when the Sun and Moon are at right angles to each other, leading to lower-than-normal tidal ranges (Chapter 17)

Nearshore zone The part of the coastal zone farthest from the shoreline, between the beach and the various offshore zones (Chapter 17)

Nebkha dune Another name for a coppice dune (Chapter 21)

Negative feedback A feedback in a system that eventually causes the system to change direction, to be reduced, or to stop entirely (Chapter 4)

Niche The term given to tafoni separated by case-hardened rims (or heads) on complex cliff-side topography (called niches and heads) (Chapter 11)

Non-foliated Rocks that lack sheets or bands and thus appear "granular" (Chapter 5)

Non-Newtonian fluid A substance that becomes more liquid-like (less viscous) when under pressure or shaken, and more solid and cohesive when at rest (Chapter 14)

Non-sorted patterned ground A term for various features of the periglacial landscape, including earth hummocks, water tracks, and frost boils (Chapter 20)

Normal fault, faulting A type of faulting produced by extension of the crust, causing one or more blocks to collapse downward (Chapter 10)

Normal force A force operating on a mass on a slope, pushing it into the slope and holding it there (Chapter 14)

Normal polarity The time (like today) when Earth's magnetic North Pole is near the current North Pole. Contrast with reverse polarity (Chapter 3)

Nose slope A type of hillslope that, in plan view, has a convex form, which facilitates more runoff than on other types of slopes. Contrast with head slope (Chapter 13)

Nuée ardente A hot, pyroclastic flow of volcanic gases, ash, and rock fragments, sometimes called a glowing gas cloud, that glows red in the dark (Chapter 7)

Nuisance flooding Flooding of normally dry coastal areas that occurs at high tide during fair weather (Chapter 17)

Numerical (absolute) dating A method used to determine an estimate of the numerical age of a feature, surface, or sediment, using some type of chronometer that is naturally built into the system (Chapter 3)

Nunatak A rock summit or ridge that is completely surrounded by glacial ice, typically in ice fields, ice caps, and ice sheets (Chapter 19)

O horizon A soil layer, usually at the top of the soil, that is dominated by organic material that has fallen onto the surface from the plants growing there (Chapter 13)

Obducted, Obduction A geological process whereby denser oceanic crust (and even upper mantle material) is scraped off a subducting oceanic tectonic plate at a convergent plate boundary and thrust on top of an adjacent tectonic plate (Chapter 6)

Obliquity Cyclical changes in Earth's orbit related to the degree of tilt of the planet with respect to the orbital plane (Chapter 4)

Obsidian An dark, black, extrusive igneous rock that is essentially volcanic glass, and thus is amorphous (lacking mineral crystals) (Chapters 5, 7)

Oceanic crust, plate The comparatively thin, dense crust that underlies most of Earth's oceans and which is primarily composed of basalt and gabbro (Chapters 6, 7)

Offset steam A stream channel that is offset, i.e., takes a sharp bend, because it flows across a strike-slip fault (Chapter 10)

Offshore bar theory A geomorphic theory that suggests that barrier islands are large, submerged sandbars that have grown vertically in response to wave deposition (Chapter 17)

Oil shale A type of shale rich in petroleum products (Chapter 5)

Old age The last stage of W. M. Davis' cycle of erosion, in which the landscape has been worn down to a nearly flat plain (a peneplain) near sea level (Chapter 2)

Open talik A body of unfrozen ground that extends from the land surface down to great depth in the subsurface, and which is surrounded by permafrost on all sides (Chapter 20)

Open-system pingo A periglacial feature, smaller than a closed-system pingo and occurring in isolation (or in small groups), formed when groundwater flows toward the surface and freezes into an ice lens that continues to grow in size. Also called "hydraulic pingo" (Chapter 20)

Ophiolite A remnant of oceanic crust that has been thrust (obducted) onto continents (Chapter 6)

Organic parent material Plant materials in various stages of decay, mainly found in wet, low landscape positions that have high water tables, within which soils may develop. Contrast with mineral parent material (Chapter 13)

Organisms In soil studies, organisms are one of the five soil-forming factors (Chapter 13)

Orogenic belt A mountain belt formed as a result of an orogeny (Chapter 6)

Orogenic forces Forces associated with orogenies (mountain-building events) (Chapter 6)

Orogeny Mountain building events (Chapters 6, 10)

Outer core Part of Earth's geologic core but unlike the solid inner core, the outer core is composed of molten material (Chapter 6)

Outlet glacier A glacier that flows out of an ice sheet or ice cap, and is bordered by rock or sediment (Chapters 18, 19)

Outlier An isolated upland that lies ahead of or outside of the main body of the cuesta or hogback but has the same geology (Chapter 8)

Outwash Fluvial sediment that is usually composed of well-sorted and stratified sand and gravel, released from ice and picked up and transported by glacial meltwater, to eventually be deposited by meltwater streams. Also called glaciofluvial sediment (Chapter 19)

Outwash plain A low-relief, gently sloping plain formed as meltwater (and usually braided) streams deposit sand and gravel (outwash) in front of a melting glacier (Chapter 19)

Over-deepening A deeply eroded area behind the terminus of a glacier, often behind an end moraine or a calving margin, where the glacial valley is deeper than along other places in the valley (Chapters 18, 19)

Overland flow   Water flowing across the land surface, but not in a stream channel. Commonly referred to as runoff (Chapter 15)

Overloading   Adding excess weight to a surface (Chapter 14)

Overthrust   The product of thrust faulting where one block is thrust over the surface/top of another block (Chapter 10)

Overturned fold   A fold with such a severely tilted axial plane that the rocks on one fold limb are overturned (Chapter 9)

Overturning failure   Another name for the mass movement called toppling (Chapter 14)

Overwash   The process of storms on the open ocean washing sand onto and over a barrier island, to the lagoon side (Chapter 17)

Oxbow lake   An isolated, arcuate segment of a river channel that has been cut off from the main channel but remains filled or partially filled with water (Chapter 16)

Oxide clay   A type of clay mineral dominated by oxides of iron and aluminum, and which forms due to long-term weathering under warm, wet climates (Chapter 4)

Oxisol   A type of soil that is reddish in color and which typically forms on old surfaces in the humid tropics, where extreme weathering has left them to be dominated by oxide clays (Chapter 13)

Oxygen isotope record   The record of past climates as recorded in the isotopic changes in foram shells in deep ocean sediments (Chapter 4)

Pacific Ring of Fire   A series of mountain ranges and volcanoes that generally circle the Pacific Ocean, and are associated with the subduction of oceanic plates (Chapter 6)

Pāhoehoe   A type of hardened lava surface texture that exhibits a stringy, rope-like pattern (Chapter 7)

Paint pot   A mud pot that is particularly colorful, with red and pink muds, instead of the typical gray mud (Chapter 7)

Paired terraces   Terraces that have treads at roughly equal elevations on both sides of the river valley (Chapter 16)

Paleocene–Eocene Thermal Maximum   A warm climatic event widely referenced as a potential analogue for future global warming on Earth (Chapter 4)

Paleoclimate   A climate of the geologic past (Chapter 4)

Paleogene Period   A formal subdivision of geologic time within the Cenozoic Era that lasted from ≈66 to 56 Ma (Chapter 4)

Paleolake   A former lake that no longer exists. The preferred term for what were formerly called pluvial lakes (Chapter 10)

Paleomagnetism   The magnetic signature of Earth that is preserved in rocks and sediments (Chapter 3)

Paleopedology   The study of paleosols and the environments associated with them (Chapter 13)

Paleosol   A soil that formed on a landscape of the past. Although most paleosols are buried, the term also includes extremely old soils that may never have been buried, and which remain at the land surface (Chapters 3, 13, 21)

Palimpsest   A soil or other natural feature that retains evidence of different development episodes in the past; some of the features formed during the past interval have not yet been completely erased by contemporary environmental conditions, and thus, provide information about that older time period (Chapter 13)

Palsa   A raised mound or small hill, typically oval in shape, found within peatlands in periglacial landscapes (Chapter 20)

Palynology   The study of past climates using pollen data recovered from archives like peat bogs and lake sediment (Chapter 4)

Pangaea   A supercontinent that existed during the late Paleozoic and early Mesozoic Eras (Carboniferous Period, ≈335 Ma), and which consisted of the following, older continental units: Gondwana, Euramerica, and Siberia (Chapter 9)

Parabolic dune   A type of crescent-shaped, anchored sand dune that may occasionally migrate, with fixed arms pointing upwind and a blunt-nosed, convex, dune head that, when active, migrates downwind (Chapter 21)

Parallel drainage pattern   A type of drainage pattern, similar to dendritic but where the channel patterns are more elongated and quasi-parallel due to steeper slopes (Chapter 16)

Parent material   The initial sediment in which soils form, often equated to the C horizon. In soil studies, parent material is one of the five soil-forming factors (Chapter 13)

Partial bleaching   In luminescence dating, the effect that occurs when the traps in the sediment grains have only been partially emptied (or re-zeroed) prior to burial, implying that the sample inherited some amount of dose (Chapter 3)

Passive plate margin   The edge of a continental plate, where oceanic and continental crusts meet but both are moving in the same direction, as one unit. *See also* Trailing edge (Chapters 6, 17)

Paternoster lake   A lake in a chain of lakes that are often linked together (by streams) down glaciated valleys, beyond the tarn lake (the uppermost lake) (Chapter 19)

Patterned ground   The semi-regular patterns and sorted features commonly observed on the ground surface in areas underlain by permafrost (Chapter 20)

Peat   A type of raw and/or only minimally decomposed, organic parent material, common to bogs and wetlands (Chapters 4, 13)

Pediment   An erosion surface cut on bedrock, often covered by a thin veneer of sediment that is slowly but intermittently being transported downslope (Chapters 2, 10)

Pediment pass   The low "pass" on an up-faulted mountain range, sometime occurring where the pediments on each side intersect (Chapter 10)

Pediplain   In Lester King's landscape evolution model, a pediplain is a widespread area of coalesced pediments, typical of dryland areas that have undergone long-term erosion and denudation (Chapter 2)

Pediplanation   Lester King's view of how dryland areas are worn down and undergo denudation but retain steep slopes in upland areas that are undergoing backwasting (Chapter 2)

Pedogenesis   The suite of processes that, collectively, form soils and their horizons (Chapter 13)

Pedogenic processes   All those processes that are involved in the formation of soils (Chapter 13)

Pedology   The science of how soils form and are distributed on the landscape (Chapter 13)

Pedoturbation   Soil mixing (Chapters 3, 13)

Peneplain   A low-relief plain formed by long-term erosion, formed in the last stage of W. M. Davis' cycle of erosion (Chapters 2, 4)

Perched dune   A sand dune that occurs on bluffs high above a water body, well removed from the beach proper (Chapter 21)

Percolate, Percolation   The downward movement of water that is already in soil or sediment (Chapter 15)

Perennial stream   A stream that flows constantly due to baseflow contributions from groundwater (Chapter 15)

Peridotite   An intrusive igneous rock at the dark, mafic end of the igneous rock spectrum, rich in mafic minerals like olivine and pyroxene (Chapter 5)

Periglacial   Adjective describing the condition of repeated freezing and thawing, and the processes associated with the growth of ice within soil and rock (Chapters 4, 20)

Perihelion   The position at which (or date when) a planet is closest to the Sun in its orbit (Chapter 4)

Period   A unit of geologic time that is shorter than an era. Periods are typically subdivided into epochs (Chapter 4)

Permafrost   Conditions in the ground where soil or rock remain below 0° C for at least two consecutive years, due to the cold climate (Chapters 4, 20)

Permafrost degradation   The process whereby permafrost becomes thinner and possibly less continuous, due to a changing (warming) ground thermal regime (Chapter 20)

Permafrost table   The depth at which temperatures remain at < 0° C throughout the year, or in other words, the top of the permafrost layer (Chapter 20)

Permeable   The condition of a material that allows liquids or gases to pass through it (Chapter 15)

Permian Period   The last period of the Paleozoic Era, between 298.9 and 252.2 Ma (Chapter 4)

Petrocalcic horizon   A soil horizon that has become cemented by secondary $CaCO_3$ (Chapter 11)

Phacolith   A lens-shaped pluton that forms within folded sedimentary rock (Chapter 7)

Phaneritic   A descriptive term for igneous rocks that form as magma cools in the subsurface, developing large mineral crystals that are visible to the naked eye (Chapter 7)

Phanerozoic Eon   A formal subdivision of Earth's geologic timescale, from ≈541 Ma until the present (Chapter 4)

Phenology   The study of changes in the seasonal timing of biological events and life cycles, typically undertaken to provide information about climate and environmental change (Chapter 4)

Phreatic eruption   A type of steam-driven volcanic eruption, formed as magma or hot rock heats groundwater, causing it to boil and flash to steam (Chapter 7)

Phreatic zone   The saturated zone below the ground surface, where all of the pores are filled with water (Chapter 12)

Physical hydrology   The science concerned with the occurrence, distribution, and movement of water (Chapter 15)

Physical weathering   The suite of processes that lead to the disintegration, i.e., physical breakup, of rocks and minerals. Also called mechanical weathering (Chapter 11)

Phytogeomorphology   The science that studies the role of plants in geomorphology (Chapter 22)

Piedmont   A gently sloping surface that leads away from the mountain front in an area of uplift, typically either depositional (formed by alluvial fans) or erosional (formed by pediments) (Chapters 10, 18); Formal name for a physiographic region in the eastern United States, located between the Appalachian Mountains and the Atlantic Coastal Plain, and characterized by relatively low, rolling hills (Chapters 6, 9)

Piedmont angle   The angle that depicts the change in slope between a pediment and a retreating escarpment on the nearby upland (inselberg) or small mountain range (Chapter 10)

Piedmont glacier   A valley glacier that has flowed down and out of an upland and then spreads out across a broad plain, forming a broad, spoon-shaped lobe of ice (Chapters 18, 19)

Pile   A stilt- or pole-like feature used to elevate buildings above the ground in areas of permafrost (Chapter 20)

Pillar   A tall, isolated, erosional landform, formed in bedrock and usually capped by harder and more resistant rock. Also called a rock pinnacle (Chapters 8, 11)

Pillow lava   Basaltic lava that is erupted underwater, and hence cools rapidly into blob-shaped structures that resemble pillows (Chapter 7)

Pingo   A large, ice-cored hill with a surface layer of frozen soil, typically covering the massive core of ice (Chapter 20)

Pinnacle   A narrow but tall landform with a hard caprock, formed as buttes and mesas slowly erode away. Also known as a spire, column, or pillar (Chapter 8)

Pinnate drainage pattern   A type of "featherlike" drainage pattern with a dense network of small tributaries joining a single master channel or gully, typical of highly erodible materials, such as loess (Chapter 16)

Pipe   A cylindrical conduit that feeds a volcanic vent at the surface or in the subsurface (Chapter 7)

Pisolith   In a surface karst terrain, a pea-sized concretion dominated by bauxite and clay minerals (Chapter 12)

Pitted outwash plain   An outwash plain with a considerable number of kettles, formed due to the melting of ice blocks that had been buried in the outwash. Collapsed outwash plains have even more kettles per unit area (Chapter 19)

Plane of failure   The underground surface or plane along which a mass movement like a slide will travel (Chapter 14)

Planetary geomorphology   The science that focuses on the geomorphology of planetary bodies in outer space (Chapter 23)

Plate   In the context of tectonics, a plate is a rigid part of Earth's lithosphere which moves slowly on top of more malleable material below (Chapter 7)

Plate tectonics   The endogenic forces associated with the movement of Earth's crustal plates. Also, the theory that Earth's rigid outer layer, the lithosphere, is broken into several large, plates. These plates interact with each other at their boundaries, producing faults, earthquakes, volcanoes, and driving mountain building (Chapter 7)

Plateau   A large, usually flat, upland that is typically capped by a resistant rock layer or caprock (Chapter 8)

Playa   A lake or dry lake floor in the center of an endorheic basin, where water, sediment, and soluble substances can accumulate but not exit. Typically, playas form in areas of dry climate (Chapter 10)

Playa lake   An ephemeral lake on a playa surface, formed typically after a rainfall event (Chapter 10)

Playfair's Law   The concept or theory that each stream carves its own valley, and that each valley is proportional in size to the stream that carves it (Chapter 2)

Pleistocene Epoch   A formal subdivision of geologic time, within the Cenozoic Era, that lasted from ≈2.58 Ma to ≈11.7 ka. The Pleistocene is the first part of the Quaternary Period, during which time glaciers and ice sheets were very active, followed by the Holocene Epoch (Chapters 4, 13, 19)

Pliocene Epoch   A formal subdivision of geologic time, within the Cenozoic Era, that lasted from ≈5.3 to 2.6 Ma (Chapter 4)

Plucking   Another name for the quarrying of bedrock by glacier ice (Chapter 19)

Plunge   The angle that indicates the inclination (relative to the horizontal) of a mass of folded rock, measured along the axis of the fold (Chapter 9)

Plunging fold   A fold in bedrock that is tilted downwards in direction parallel to the fold axis. Plunging folds have inclined fold axes (Chapter 9)

Pluton   A body of igneous rock that develops as magma cools underground (Chapters 5, 6, 7)

Pluvial lake   A lake that formed in the past, when the climate was cooler and wetter, but today this area is warm and dry, and thus, has no (or few) lakes. The preferred term today is paleolake (Chapter 10)

Point bar   A (typically sandy) landform that forms from lateral accretion in zones of slower flow in a river channel, on the insides of meander bends (Chapters 16, 22)

Point-source dissolution   The process of concentrated infiltration of surface waters in locations where the bedrock is particularly vulnerable and permeable, leading to accelerated dissolution in those locations (Chapter 12)

Polar amplification   The condition whereby air temperatures near the poles are increasing at a higher rate than in the mid-latitudes, being driven by changes in albedo (as sea ice and glaciers decrease in area) and the overall contemporary trend of climatic warming (Chapter 20)

Polar desert   A mostly barren landscape that occurs poleward of the tundra, where precipitation is < 25 cm/year (Chapter 20)

Polje   A large, flattish plain in a karst landscape, bordered by steeply rising marginal slopes (Chapter 12)

Pollen   Small grains produced by most types of flowers of seed plants for the purpose of sexual reproduction (Chapter 4)

Polygenetic   The characteristic of having multiple instances of "genesis" or episodes of formation. Contrast with monogenetic (Chapter 13)

Polygonal cracking   Weathering features on rock surfaces, sometimes called elephant skin weathering, that appear as convex polygonal plates separated by shallow cracks, and are sometimes also associated with a weathering rind or crust (Chapter 11)

Pool   A relatively deep portion of a river channel with a low water-surface slope and, typically, slow current (Chapter 16)

Pore   A void in a solid, through which gases, liquids, or microscopic particles can pass (Chapter 15)

Pore ice   Ice that forms as the pore water freezes *in situ* within soils (Chapter 20)

Porosity   A measure of the amount of void space in rock or sediment (Chapter 5)

Positive feedback   A feedback in a system that eventually causes that system to increasingly do the same thing over and over again, or to become enhanced. It occurs when a change in a variable triggers a response, which then drives additional change in the same direction (Chapter 4)

Post-depositional modification   A physical or chemical change to a land surface or the materials on it, that can be correlated to age (Chapter 3)

Precambrian   The earliest interval of geologic time, from the beginning of Earth history until about 540 million years ago, when hard-shelled life forms then became abundant. The Precambrian, an informal unit of geologic time, accounts for about 88% of Earth's history (Chapter 4)

Precession   Cyclical changes in Earth's axial tilt relative to the Sun, at different points in the orbit (Chapter 4)

Precipitate   A category of rock that forms via chemical precipitation of minerals out of a watery solution (Chapter 5)

Precision   In most geomorphic applications, the statistical uncertainty associated with an age estimate of a sediment, landform, or geomorphic surface (Chapter 3)

Pressure melting point   The temperature at which glacial ice melts (or freezes), which is a function of both temperature and pressure. As pressure increases, the pressure melting point is lowered (Chapter 18)

Primary mineral   A mineral that forms with the rock, out of molten material or by precipitation from fluids (Chapters 5, 13)

Principle of ascendancy and descendancy   The geologic principle that states that an erosion surface is younger than the youngest deposit or surface that it cuts across, or truncates. Also, the principle that a surface is older than any valleys cut into it, as well as any deposits that lie within those inset valleys (Chapter 3)

Principle of cross-cutting relationships   The geologic principle that states that when a rock unit (or fault) cuts across a body of rock or sediment, it must be younger than the host rock (Chapter 3)

Principle of inclusions   The geologic principle that states that inclusions of rock or sediment within a host rock must be older than the formation that contains (surrounds) them (Chapter 3)

Principle of original horizontality   The geologic principle that states that layers of sediment are generally deposited in a horizontal position originally, before any subsequent changes occur to their geology (Chapter 3)

Principle of superposition   The geologic principle that states that in an undeformed sequence of sedimentary rocks or other similar sediments, each bed is older than the one above and younger than the one below (Chapter 3)

Principle of uniformitarianism   The geologic principle that has traditionally been interpreted to mean that the present is the key to the past. Natural laws are used to explain natural processes, and these laws do not change with time (actualism). Actualism is commonly used today instead to avoid confusion with earlier tenets of uniformitarianism as held by Charles Lyell and many nineteenth-century scientists. Some of the earlier tenets of uniformitarianism as stated by Lyell are no longer accepted as true, e.g., gradualism of all processes, and stasis of Earth's state (Chapters 2, 19)

Process geomorphology   An approach to geomorphology that focuses on the processes that produce landforms, more than just their shape, distribution, or configuration (Chapter 2)

Prodelta   The most distal, deep-water portion of a delta (Chapter 17)

Proglacial environment   The environment that includes areas in front of a glacier – transitional between glacial and non-glacial environments (Chapters 18, 19)

Proglacial lake   A lake ponded in front of a glacier, with a shoreline on the glacier itself (Chapters 18, 19)

Proglacial stream   A meltwater stream that forms in front of a glacial terminus and which, by deriving water and sediment from it, may form an outwash plain or a valley train (Chapter 18)

Prograding In coastal geomorphology, a prograding shoreline is gaining sand and moving outwards, into the water body (Chapter 17)

Proterozoic Eon A formal subdivision of geologic time from ≈2,500 to ≈541 Ma (Chapter 4)

Protolith The original rock from which a metamorphic rock has formed (Chapter 5)

Proxy A type of substitute or stand-in for something, which in geology is used to infer past environmental conditions, especially climate, when direct measurements are not available (Chapter 4)

Pumice A light, porous type of pyroclastic igneous rock, formed during explosive volcanic eruptions when lava is ejected into water or air as a froth, i.e. a foam-like solution containing masses of gas bubbles (Chapter 5)

Pyroclastic beads Glass-like particles discovered on Earth's Moon by the Apollo astronauts, indicative of explosive volcanic activity in the past (Chapter 23)

Pyroclastic cone Another name for a cinder cone volcano (Chapter 7)

Pyroclastic debris Material explosively ejected from a volcano (Chapters 7, 23)

Pyroclastic flow A mixture of rock, lava, ash, and volcanic gas that moves down the sides of a volcano at speeds greater than 80 km/h (Chapter 7)

Pyroclastic surge A volcanic hazard similar to a pyroclastic flow, but with much higher proportions of gas to rock (Chapter 7)

Pyroduct A natural conduit or tube, formed as low-viscosity lava flows away from the site of an eruption and cools, forming a surface crust, allowing the molten lava below to continue to flow. When emptied, the conduit resembles a tube or elongated cave. Also called a lava tube (Chapter 7)

Quarrying The process, also called plucking, by which joint- and fracture-bounded boulders and large rocks are eroded from bedrock beneath a glacier, as meltwater refreezes in fractures and joints in the bedrock, allowing the glacier to pull blocks out of the bedrock and incorporate them into the ice (Chapters 18, 19)

Quartz A primary silicate mineral with the formula SiO_4 (Chapter 5)

Quartzite The metamorphic equivalent of sandstone and, sometimes, conglomerate (Chapter 5)

Quaternary Period A formal geologic term for the time period from 2.58 Ma to the present, including the last and most intense glaciations of the Late Cenozoic Ice Age (Chapters 4, 18, 19)

Quick clay A deposit of marine clays that liquefies when disturbed and which may then form earthflows (Chapter 14)

Quiescence phase In surging glaciers, a time of steady, normal flow between times of more rapid flow events (Chapter 18)

R horizon In soil horizon terminology, the hard bedrock that underlies all soils (Chapter 13)

Radial drainage pattern A type of drainage pattern where streams flow outward from an isolated, structural dome or peak, such as a volcano (Chapter 16)

Radiating dike A dike formed in a (now extinct) volcano that radiates outward from its central area, seen most easily after much of the volcano has been eroded away, leaving the dike to stand up as a ridge of resistant rock (Chapter 7)

Radiocarbon An unstable isotope of carbon (^{14}C), that decays at a known rate (Chapter 3)

Radiocarbon dating Also known as carbon-14 dating, it is a numerical dating method used to determine the age of organic materials up to around 50,000 to 60,000 years old. The method establishes the amount of carbon-14 (a radioactive isotope of carbon) remaining in a sample after the organism has died. Because the rate of decay of the carbon-14 isotope is known, it allows scientists to estimate the time elapsed since the organism's death (Chapter 3)

Radius of curvature The radius of a meander bend of a river (Chapter 16)

Raft cone A stalagmite-like deposit formed by sunken calcite rafts beneath dripping points in, often, thermal lakes (Chapter 12)

Rain shadow The relatively drier area on the lee (downwind) side of a mountain range (Chapter 6)

Rapids Generic term for any zone of swift (and often shallow) flow in a stream or river. May apply to riffles or to any steep, rocky stretch of a river (Chapters 5, 16)

Rating curve A graph that shows the relationship of stream discharge to stage (water-surface height) (Chapter 15)

Rays *See* crater rays (Chapter 23)

Razorback An erosional bedrock landform formed on nearly vertically dipping strata (Chapter 8)

Reach A short stretch of a river channel with relatively uniform characteristics (Chapter 16)

Reaction wood Wood (tree rings) that forms and varies in width due to some sort of stress (Chapter 3)

Receding limb A section of a storm hydrograph that begins at peak discharge and continues through the period of decreasing discharge (Chapter 15)

Recessional moraine A glacial moraine formed as the ice margin retreats from its terminal position and stabilizes at a new location during a period of deglaciation (Chapter 19)

Recording stream gage A gage that acquires continuous records of stream flow, using equipment that automatically measures and records water-surface height (Chapter 15)

Rectangular drainage pattern A type of drainage pattern found where the strong joint control in the underlying rocks causes streams to take sharp, right-angle turns, following the weakened rocks in fractured and highly jointed areas (Chapter 16)

Recurrence interval The time, in years, that can be expected (on average) between events of that size or larger. With regard to river flow data, the recurrence interval is the expected time interval/period between flows (typically, floods) of a given size, or larger (Chapter 15)

Redox A shortened name for reduction-oxidation reactions (Chapter 11)

Reflective beach A beach condition where most of the wave energy is reflected off the beach and back into shallow water offshore (Chapter 17)

Refugia Unglaciated, and commonly warmer and wetter, locations where plants and animals were able to survive during cold periods of advancing glaciers and ice sheets (Chapter 19)

Reg A type of desert landscape characterized by a surface covered with closely packed, interlocking pebbles and cobbles (Chapter 21)

Regelation The process of melting and refreezing of ice that occurs at the bed of a glacier, due to lowering the pressure melting point in areas of higher pressure, e.g., up-ice of a bed obstacle, followed by refreezing when the pressure is lower, e.g., down-ice of a bed obstacle (Chapters 18, 19)

Regime   The overall characteristics of a system (Chapter 21)

Regolith   Any unconsolidated material at the land surface, regardless of origin, overlying bedrock at depth (Chapters 5, 8, 9, 11, 13, 14)

Regressive, Regressing, Regression   In coastal geomorphology, a regressive shoreline is eroding and moving landward (Chapter 17)

Relative dating   A dating method that establishes the age of a sediment, surface, or landform in relative terms, such as older or younger, i.e., no numerical age estimates are determined (Chapter 3)

Relative sea level   The local rise in sea level, or how the height of the ocean rises and/or falls relative to the land at a particular location (Chapter 17)

Relict paleosol   An extremely old soil, still forming at the land surface but which (due to its age) has acquired many of its characteristics from a past soil-forming environment (Chapter 13)

Relict permafrost   Permafrost that is in a state of disequilibrium, persisting today but which formed in a past, colder climate (Chapter 20)

Relief   The relative elevation difference between high and low points on a landscape. In soil studies, relief is one of the five soil-forming factors (Chapters 2, 6, 10, 13)

Removals   As a group of soil processes, removals include all processes that remove substances from the surface of the soil by erosion, or which translocate substances out of the bottom of the soil profile in percolating water (Chapter 13)

Reptation   Single, short hops made by particles being transported in air or water (Chapter 21)

Residual   An adjective referring to what is left behind after other material has been removed (Chapter 11)

Residual regolith   *See* residuum (Chapter 11)

Residual thaw layer   A layer of thawed ground that occurs between the active layer and the underlying permafrost (Chapter 20)

Residuum   Unconsolidated material that has accumulated *in situ* by the weathering of rock (Chapters 11, 13)

Resisting framework   A general classification of the part(s) of landforms that generally resist erosion, including rock type, stratigraphy, the underlying geologic structure, and biota (Chapters 1, 7)

Response time   The time required for a natural system to respond to a perturbation, often initiated by a change in climate, such as a glacier response time, or other driving force, e.g., tectonics (Chapter 18)

Resurfacing   The process by which a planetary surface is either covered over with new material (typically lava), or eroded so much that the original surface has been lost (Chapter 23)

Reverse fault, faulting   A type of faulting produced by crustal compression, causing one block to move upward relative to the other (Chapter 10)

Reversed polarity   As has happened in the past, the circumstance where Earth's magnetic North Pole is near the current South Pole. Contrast with normal polarity (Chapter 3)

Reversing dune   A type of sand dune that moves in opposite directions at different times, driven by seasonally changing wind directions (Chapter 21)

Revetment   A shore protection method, constructed parallel to the coast and intended to reduce wave impact on an eroding coast. Revetments can be built out of concrete or stone, and generally match the existing land slope, i.e., they are not vertical like seawalls. Thus they tend to generate less scour than seawalls (Chapter 17)

Rhyolite   An extrusive (typically pink) igneous rock, similar in composition to granite but with much smaller mineral crystals, which is high in silica and low in iron (Chapter 5)

Rhyolitic lava   A type of highly viscous, silica-rich lava from which gases cannot easily escape (Chapter 7)

Ribbon forest   A roughly linear strip of trees, usually attributed to localized geomorphic processes (Chapter 22)

Ridge and swale topography   A type of topography that is typical of point bars with meander scrolls (Chapter 16)

Ridge submergence theory   A theory suggesting that barrier islands originate from flooding of old beach and dune ridges (Chapter 17)

Riffle   A shallow, high velocity, turbulent feature of a stream channel. In a meandering, sand- or gravel-bed stream, riffles often develop at thalweg crossovers (Chapter 16)

Rift, Rift zone   The break, or crack, in a lithospheric plate (crust), formed by two crustal plates that are moving apart, or diverging (Chapter 7)

Rift valley   A linear lowland that develops at a divergent, tectonic plate boundary, formed as blocks of rock drop down due to normal faulting, and which is often bounded by steep walls (Chapter 7)

Right-lateral   A type of strike-slip fault in which the movement, as seen across the fault line, is to the right (Chapter 10)

Rill   A very small channel formed by running water and which is usually dry (Chapter 16)

Rimstone dam   A feature of secondary calcite that forms as water flows across the floor of a cave, preferentially depositing calcite (to form the dam) at locations where it has more turbulence, such as when it flows down a steep incline (Chapter 12)

Rip current   A fast-moving, dangerous, coastal current that flows perpendicularly away from the coastline through breaks in offshore sandbars (Chapter 17)

Rip rap   Construction material used for erosion control and shoreline protection, typically consisting of large, angular rocks or boulders, often made of limestone, recycled concrete, or granite. Rip rap is used to armor shorelines, stream banks, and other structures against the erosive forces of water, waves, and ice (Chapter 17)

Rip rap revetment   A structure built of large blocks and concrete chunks, placed on the coast for the purpose of reducing coastal erosion (Chapter 17)

Ripple   A small, wave-like crest and trough on the surface of sand dunes and on sandy river beds (Chapters 16, 21)

Ripple amplitude   A measure of the height of ripples on sand dunes, typically 0.5 to 1.0 cm (Chapter 21)

Ripple height   *See* ripple amplitude (Chapter 21)

Ripple wavelength   The distance from ripple crest to crest, typically 5 to 15 cm for ripples on sand dunes (Chapter 21)

Rise   *See* karst spring (Chapter 12)

Riser   Another name for the scarp (edge, escarpment) on a fluvial terrace (Chapter 16)

Rising limb   The section of a storm hydrograph that represents the increase in river discharge over time, typically in response to a rainfall event. Normally, the rising limb spans the time between the initiation of runoff and peak discharge (Chapter 15)

River-dominated delta   A type of delta, commonly with the long, narrow channels of a bird's foot-type delta, formed in settings where sediment inputs from rivers dominate the delta system (Chapter 17)

Roche moutonnée   A streamlined glacial landform formed by erosion of bedrock. The bedrock is abraded, smoothed, rounded, and usually striated on its stoss (facing up-ice) side, and quarried, fractured, and jagged on its lee (facing down-ice) side (Chapter 19)

Rock   A solid, naturally occurring mass that is either an aggregation of minerals, or aggregations of particles of other rocks, categorized by the minerals it includes, its chemical composition, and how it formed (Chapter 5)

Rock avalanche   A fast-moving avalanche that originates in jointed and fragmented rock and includes mainly rock material material (Chapter 14)

Rock city   Another name for a ruiniform landscape (Chapter 11)

Rock cycle   The cycle that involves the various geological processes associated with the formation of rocks, such as melting, cooling, erosion, compaction, and metamorphism. The cycle illustrates the ways that rock-forming materials are interconnected and recycled (Chapter 5)

Rock drumlin   Another name for a whaleback or glacially eroded and smoothed bedrock form (Chapter 19)

Rock fabric   Term describing the arrangement of minerals within rocks (Chapter 5)

Rock glacier   A lobate or tongue-shaped body of frozen, rocky debris containing interstitial ice and ice-rich sediment that slowly flows downslope (Chapters 8, 20)

Rock pinnacle   A tall, isolated bedrock feature, usually capped by harder and more resistant rock (Chapter 11)

Rock salt   A type of sedimentary rock formed of halite (NaCl, or rock salt) (Chapter 5)

Rock shelter   A type of indentation formed from the local intensification of weathering at the intersection of a rock outcrop and the rock surface (or the water table level) that forms its base. Also called a foot cave (Chapter 11)

Rock topple   A type of mass movement where relatively massive blocks of strong rock fall over and topple, usually leading to brittle, catastrophic failures (Chapter 14)

Rock varnish   A dark, almost black, manganese-rich coating that forms on rocks, particularly those in warm, dry climates. Coatings often get progressively thicker and darker over time. In older literature, the term desert varnish had been popular (Chapters 3, 5, 10, 11)

Rock-cored drumlin   A streamlined glacial landform that resembles a classical drumlin in shape, but which has a core of bedrock, with glacial sediment draping that bedrock core (Chapter 19)

Rockfall   The movement of rocks as they break off a free face and fall freely downward, often onto a deposit of talus below (Chapters 8, 14)

Rockfall scar   A name given to a newly exposed rock face formed after a rockfall event (Chapter 14)

Rockslide   A type of translational landslide that occurs along a bedding plane, joint, or any other plane of structural weakness in bedrock, where rock (or, sometimes, other consolidated material) moves en masse for a short distance, then disintegrates (Chapter 14)

Rollover   The process that drives barrier islands to move landward. Rollover occurs as storm waves overtop the barrier island, eroding sand from the ocean-side beach and transporting the sand over the island and into the lagoon (Chapter 17)

Root channel   The void formed in soils as a root grows and then decays (Chapter 15)

Rotational landslide   A type of landslide, commonly called a slump, that slides along a curved, concave plane of failure, typically parallel to the contour of the slope (Chapter 14)

Ruiniform landscape   A type of bedrock landscape formed by weathering, as vertically oriented joints widen across a large area, producing a complex network of slots and corridors, interspersed with pinnacles and towers (Chapter 11)

Run   A segment of a stream channel, transitional between turbulent riffles upstream and placid pools downstream. In the run, the flow slows down and the water tends to be well aerated (from the riffle upstream) (Chapter 16)

Runnel   A small-scale, linear-shaped feature worn into a rock surface by weathering and erosion (Chapter 11)

Runoff   Another name for overland flow – water flowing across the land surface but not in a channel (Chapter 15)

Runout area (zone)   The location where sediment that has moved downslope in a mass movement finally comes to rest (Chapter 14)

Sabkha   An Arabic name for a salt playa (Chapter 10)

Safety Factor equation   The equation used to explain slope stability: F_s = resisting forces (shear strength) / driving forces (shear stress). See text for more detail (Chapter 14)

Sag   An enclosed depression on a geomorphic surface (Chapter 14)

Sag pond   A small water body occupying an enclosed depression near the head of a rotational landslide (slump). In an area of strike-slip faulting, a sag pond may form within/ above the fault zone, where rivers are able to preferentially erode into the surface (Chapters 10, 14)

Sagging   The ductile deformation (slow bending) of bedrock caused by the lack of basal support (Chapter 12)

Salar   The Hispanic term for a salt playa (Chapter 10)

Salt playa   A type of playa where the dominant surface is formed on precipitated salts of various types, formed by evaporation of salty water in the playa lake (Chapter 10)

Salt weathering   A type of physical weathering involving the stresses associated with the repeated crystallization and growth of salts within porous rocks (Chapter 11)

Saltation   The repeated jumping and bouncing motion of grains across the ground surface, driven by wind, or along the bed of a river channel, driven by flowing water (Chapters 8, 15, 21)

Sand   Particles that are between 0.05 and 2.0 mm in diameter, i.e., larger than silt or clay but smaller than gravel (Chapter 13)

Sand dune   A landform composed of wind-blown sand (Chapters 17, 21)

Sand sea   An area in a desert where eolian sand completely covers the surface across an area of great extent (Chapter 21)

Sand stringer   A small linear sand dune (Chapter 21)

Sand wedge   A V-shaped feature on a former periglacial surface that forms as a contraction crack fills with eolian sand instead of ice (Chapter 20)

Sandstone   A clastic sedimentary rock made up of mostly sand-sized particles (Chapters 5, 8)

Sandur   An Icelandic term for an outwash plain (Chapter 19)

Santa Ana   Local name for a wind that brings hot, dry air from the interior of the deserts of the southwestern United States, across California, and to the Pacific coast (Chapter 21)

Saprock   Rock that has started to weather, but less than a third of its primary minerals have been altered (Chapter 11)

Saprolite   Weathered rock which still retains its original rock fabric (Chapter 11)

Saturated zone   A layer of soil or bedrock in which all the voids are filled with water (groundwater) (Chapter 15)

Saturated, Saturation   The state whereby all the pores in a rock or sediment are filled with water (Chapter 14)

Saturation curve   Lines on a mixing corrosion curve that represent situations where water is saturated with CO_2 (Chapter 12)

Saturation-excess overland flow   A type of overland flow that occurs as rain intercepts areas where the water table is near the surface. It also includes return flow – a term for groundwater that discharges onto the surface (Chapter 15)

Scallop   An asymmetrical, shell-like hollow on the wall of a cave, formed due to erosion and dissolution processes from turbulently flowing water (Chapter 12)

Scarp   A steep slope. Abbreviation of "escarpment" (Chapters 14, 16)

Schist   A foliated metamorphic rock but with finer-scale foliation than gneiss (Chapter 5)

Scoria   A dark-colored, porous, volcanic rock that forms when molten lava containing vesicles (gas bubbles) cools and solidifies (Chapter 7)

Scoria cone   Another name for a cinder cone volcano (Chapter 7)

Scree   See talus (Chapters 11, 14)

Sea cave   A cave formed by wave erosion in a cliff along a coast (Chapter 17)

Sea stack   A tall, isolated, residual rock formation, standing above the water line, that has formed due to the collapse of an arch or erosion of bedrock headlands (Chapter 17)

Seamount   A shield volcano that has been so eroded that it is now underwater (Chapter 7)

Seasonal frost mound   See frost blister (Chapter 20)

Seasonality   The overall differences and variations in weather and climate between winter, spring, summer, and autumn, for a given location (Chapter 4)

Seasonally frozen ground   The condition where it is cold enough for frost to penetrate the upper soil each winter, but there is no permafrost at depth (Chapter 20)

Seawall   A metal, wood, or concrete structure that is built along and parallel to an eroding shoreline, to "hold the line" by preventing erosion and flooding of the land behind it (Chapter 17)

Secondary carbonate   Carbonate minerals that have been translocated into the subsurface horizons of soils, where they have accumulated, as is common in dryland soils. Also, carbonate minerals that have been precipitated out of solution, as in a cave (Chapter 3)

Secondary mineral   A mineral (typically a clay mineral) that has been altered by weathering into a form that is chemically or mineralogically different from its preexisting counterpart (Chapters 5, 11, 13)

Sediment availability   The condition that indicates whether or not sediment is freely available for transport, vs protected from transport (Chapter 21)

Sediment budget   A method of accounting for sediment gains and losses in a geomorphic system (Chapter 17)

Sediment delivery   The amount of sediment conveyed to a river from the surrounding hillslopes (Chapter 16)

Sediment deposition zone   The lower (or end) part of the fluvial transport system where sediment is deposited and experiences little to no short-term, subsequent remobilization, e.g., a delta (Chapter 15)

Sediment maturity   A term used to describe how weathered and "evolved" a sediment is (Chapter 5)

Sediment production   The amount of sediment generated by nearby hillslopes which has the potential to make its way into a river channel, where it may then travel farther down the fluvial system (Chapter 16)

Sediment production zone   The part of a stream's drainage basin where the slopes are relatively steep and "coupled" to the channel (Chapter 15)

Sediment transfer zone   The part of the fluvial transport system, downstream of the sediment production zone, through which sediment moves (typically episodically), potentially ending up in the sediment deposition zone (Chapter 15)

Sedimentary rock   A major category of rock that forms as particles of previously weathered rocks are transported to a depositional basin, where they eventually become lithified. Some sedimentary rocks form as evaporite minerals precipitate and settle out of a supersaturated solution (Chapters 5, 8, 9)

Segregated ice   Nearly pure ice lenses that form in rock and sediment as pore water migrates toward the freezing surface (Chapters 11, 20)

Seiche   A short-term water level fluctuation, formed when strong winds push water from one side of a large water body or elongated lake, to the other (Chapter 17)

Seif dune   A type of linear dune, common in deserts, that is more sinuous than a typical linear dune, with distinct peaks and saddles (Chapter 21)

Seismic wave   Another name for the shock wave formed in rocks when they break or move during an earthquake (Chapter 10)

Seismology   The study of earthquake processes and seismic waves (Chapter 6)

Semi-diurnal tide   A tide that rises and falls twice daily (Chapter 17)

Serpentinite   A metamorphic rock formed of ultramafic minerals (such as serpentine), that typically forms in oceanic subduction zones (Chapter 5)

Shaft   In karst landscapes, a deep, narrow sinkhole or doline (Chapter 12)

Shale   A clastic sedimentary rock made up of mostly clay-sized particles (Chapter 5, 8)

Shallow landslide   A landslide where the sliding surface is located within regolith or the upper part of weathered bedrock (Chapter 14)

Shallow-water wave   Waves that occur in areas where the bottom of the water body is shallower than the wave base, causing the waves to drag on the bed, and transport sediment along the bed of the water body (Chapter 17)

Shattering   The physical breakup of rocks along seemingly random planes of weakness (Chapter 11)

Shear force   The tangential component of gravity that pulls a mass down a slope, and which increases as the slope steepens (Chapter 14)

Shear strength   The relationship between shear force and normal force, for masses of sediment on slopes. Materials with greater shear strength will resist movement and deformation better than materials with low shear strength (Chapter 14)

Shear stress   The stresses put on materials and sediment that tend to make them prone to movement and deformation (Chapter 14)

Sheet sand   A sandy, typically eolian, deposit that covers large areas of a landscape (Chapter 21)

Sheetflow   The shallow, overland flow of water, across surfaces that lack channels (Chapter 16)

Sheeting   The physical breakup and loss of thin sheets from the surface of a rock (Chapter 11)

Shield volcano   The largest type of volcano in terms of volume, diameter, and height, and formed mainly in basaltic lava (Chapter 7, 23)

Shoaling   Sediment deposition in a channel or shallow water body (Chapter 17)

Shore, Shoreline   The (linear) interface or contact line between land and water (Chapter 17)

Shore protection   A group of usually engineering strategies designed to protect the coast from erosion and/or flooding (Chapter 17)

Shoulder slope   The least stable of the five slope elements and dominated by runoff and erosion, the shoulder slope is immediately downslope from the summit (Chapter 13)

Shutter ridge   In an area of strike-slip faulting, a shutter ridge forms where a small hill has been moved in front of an offset river channel, potentially blocking it (Chapter 10)

Sieve flow   A mass wasting event and deposit that is mainly rocks and gravels, with almost no "matrix" material like sand, silt, or clay (Chapter 10)

Silcrete   A soil horizon that has become cemented by silicate minerals, but mainly quartz and various amorphous, silica-rich substances (Chapter 11)

Silicate   A type of rock-forming mineral that is comprised of various groups of SiO_2 compounds and structures, with or without accessory ions. The most common silicate mineral is quartz (Chapter 5)

Sill   A tabular-shaped pluton that forms as magma invades sedimentary rock, moving within that rock conformably, i.e., usually along and between bedding planes (Chapter 7)

Silt   Particles that are 0.002–0.05 mm in diameter (USDA), i.e., larger than clay but smaller than sand. Geologists commonly use 0.004–0.0625 mm for silt (Udden–Wentworth scale) (Chapters 13, 18)

Siltstone   A clastic sedimentary rock made up of mostly silt-sized particles (Chapter 5)

Sinkhole   Another name for a doline, a sinkhole is a closed depression on a karst landscape, formed directly or indirectly by dissolution of the underlying bedrock (Chapter 12)

Sinter   A silica-rich precipitate formed from the water of previous geyser eruptions (Chapter 7)

Sinuosity, channel sinuosity   A measure of how much a stream channel bends and turns, as defined as the ratio channel length (L_c)/valley length (L_v) (Chapter 16)

Slaking   A type of physical weathering, driven by wetting and drying, which often causes very thin sheets to peel off a rock surface (Chapter 11)

Slate   The low metamorphic equivalent of shale (Chapter 5)

Slide   A type of mass movement where the material is transported en masse along a sloping surface but with minimal internal motion (Chapter 14)

Slip face   The steep slope on the lee side of a sand dune, approximating the angle of repose for sand, down which sand grains slide, fall, and roll. Sand grains typically saltate up the windward face of the dune, before they slide, fall, and roll down the slip face (Chapters 21, 23)

Slope aspect   The compass orientation of a slope, such as north-facing or southwest-facing (Chapter 22)

Slope decline   The gradual lowering of the angle of a slope by erosion (Chapter 10)

Slope element   One of the five major parts of a slope, based on position and steepness: summit, shoulder, backslope, footslope, and toeslope (Chapter 13)

Slope failure   The condition wherein a slope has become unstable and sediment has moved (or is moving) downslope (Chapter 14)

Slope instability   The condition whereby slopes are experiencing mass movement and accelerated erosion (Chapter 14)

Slope stability   The ability of a slope to withstand mass movement and erosion (Chapter 14)

Slopewash   The movement of sediment downslope, assisted by running water, but not in a channel (Chapter 14)

Slump   Another name for a rotational landslide (Chapter 14)

Slurry flow   A wetter type of (mass wasting) flow that contains 20–40% water (Chapter 14)

Snow avalanche   An avalanche dominated by snow and ice (Chapter 14)

Snow line   The boundary between exposed firn and snow on a glacier, which changes with season and commonly migrates higher during summer. The snow line at the end of summer/early fall approximates the ELA for that year (Chapter 18)

Snowball Earth   A term for periods of comparatively cold climatic conditions and glaciation in Earth history (Chapter 4)

Soda straw   A thin, initially hollow, type of stalactite that forms on the roofs of caves, as calcite precipitates out of water dripping from a small hole in the roof of a cave or cavern (Chapter 12)

Soil   A natural, three-dimensional body that has formed at the surface, through the interactions of at least five soil-forming factors (climate, biota, relief, parent materials, and time). Most soils have naturally formed layers called horizons (Chapters 3, 13)

Soil chronofunction   The statistical relationship between a soil property and time (soil age) (Chapter 3)

Soil geomorphology   The study of soils across landscapes, using soil attributes and properties to better understand the age of geomorphic surfaces, soil formation rates, and surface stability over time (Chapter 3, 13)

Soil horizon   Layers in a soil formed by soil-forming processes (Chapter 13)

Soil profile   A vertical cross-section or cut through a soil, extending from the surface, down into the parent material (Chapter 13)

Soil-forming factor   A factor that establishes the parameters within which a soil will form. The five soil-forming factors (climate, organisms, relief, parent material, and time) influence which pedogenic processes are operative, how the soil forms, and what morphology may develop (Chapter 13)

Soil-forming interval   The period of time from the beginning of soil formation to its end (typically, the present) (Chapter 13)

Solifluction   The slow downslope flow of saturated regolith in the active zone, in areas of permafrost (Chapters 14, 20)

Solifluction lobe   A landform formed by the slow downslope flow (solifluction) of saturated regolith above permafrost, with lobate or tongue-like shapes and with steep fronts and lower angled treads upslope (Chapters 14, 20)

Solstice   The date when a given hemisphere (Northern or Southern) is angled most directly toward, or directly away from, the Sun (Chapter 4)

Solution   The process of chemical weathering whereby minerals dissolve and become ions in solution, as in the chemical process of dissolution (Chapters 11, 18)

Solution weathering   A type of chemical weathering affecting primarily carbonate rocks and rocks composed of soluble substances, which occurs as the minerals in these rocks are dissolved in water or weak acids. Also called dissolution (Chapter 11)

Solutional aggressivity   The ability of water and acids to dissolve rock (Chapter 12)

Sorted, sorting   A measure of the degree of uniformity in grain size for a sediment. A well-sorted sediment has particles that are restricted to a fairly narrow range of sizes (Chapters 5, 19)

Sorted circle   A patterned ground feature with a well-defined center of fine material enclosed by a wide perimeter zone of coarse clasts (Chapter 20)

Sorted net   A patterned ground feature with a well-defined center of fine material enclosed by a perimeter of coarse clasts, but which is more irregularly shaped and with narrower perimeters of coarse clasts, than are sorted circles (Chapter 20)

Sorted stripe   A long line of stones that trends downslope, formed as sorted features like nets and circles become elongated downslope (Chapter 20)

Source area   The location from which the material in a mass movement has originated (Chapter 14)

Spalling   Another name for sheeting (Chapter 11)

Spatial   Pertaining to location, or space (Chapter 2)

Speleogenesis   The natural processes of void production and enlargement in soluble rocks of karst terrains (Chapter 12)

Speleothem   A feature composed of secondary mineral deposits, formed in flowing, dripping, ponded, or seeping water in caves (Chapters 4, 12)

Spheroidal weathering   A process of weathering in bedrock, whereby concentric layers are eroded off, like the skin of an onion, causing the rocks to take on a more rounded form (Chapter 11)

Spit   An elongated body of sand that is attached to a beach at one end, growing longer down the coast and formed by sediment moving in the longshore current (Chapter 17)

Spit migration theory   A theory that suggests that barrier islands originate as spits, connected to the mainland, and then grow by longshore drift, only to eventually become separated from the mainland (Chapter 17)

Spodosol   A deeply leached and acidic soil formed in cool, humid climates, distinguished by thick, whitish E horizons and reddish brown, wavy B horizons enriched in humus, aluminum, and iron (Chapter 13)

Sporadic permafrost   The condition whereby permafrost underlies 10–50% of the ground surface (Chapter 20)

Spreading center   The location where two tectonic plates diverge, or pull apart. Also called a rift zone (Chapter 7)

Spring mound   A mounded deposit of secondary carbonate that has formed around a spring or in a lake (Chapter 11)

Spring tide   A particularly high tide that occurs during new and full moons, when the Sun and Moon are in alignment (Chapter 17)

Stable dune   Another name for an anchored dune or a dune that is not currently moving (Chapter 21)

Stack   A tall, isolated bedrock feature that forms in coastal settings due to erosion by waves (Chapter 11)

Stadial   A period in time when an ice sheet briefly advances, but not as long as a glacial episode or glaciation (Chapter 4)

Staff gage   Similar to a yardstick fixed in place, a staff gage is used to measure a stream's stage (water-surface height) (Chapter 15)

Stage   In fluvial geomorphology, stage is the height of a stream's water-surface. In geologic time, a stage is the shortest formal unit of geologic time (Chapters 4, 15)

Stagnant, Stagnant glacier, Stagnant ice   Glacial ice that is barely moving (< 1 m/yr) or not flowing, i.e., is melting away in place (Chapter 19)

Stalactite   A type of speleothem formed in caves, typically conical-shaped and hanging from the cave roof (Chapter 12)

Stalagmite   A type of speleothem that forms on the cave floor and grows upward over time, formed as calcite precipitates in water that has dripped off the ceiling and hit the floor (Chapter 12)

Star dune   A large to extremely large sand dune with a pyramidal/star shape, usually with three or four radiating arms and an equal number of slip faces (Chapter 21)

Steady-state equilibrium   The behavior of a geomorphic system in which the average form or shape does not change over time, despite short-term fluctuations in form or shape (Chapter 2)

Steam vent   A hydrothermal feature consisting of a vent in the bedrock that emits mainly steam. Also called a funarole (Chapter 7)

Step pool   A comparatively deep part of a stream channel, formed between steps (waterfalls or rapids), where the water is shallower. Step pools often form as a sequence of small, evenly spaced falls and plunge pools in the channel (Chapter 16)

Stepped structural surface   The name given to the platform surfaces on an erosional, bedrock-controlled landscape that forms on horizontally bedded sedimentary rock. Here, erosion has formed several distinct flat "steps" (Chapter 8)

Stock   A large, irregularly shaped pluton, but smaller than a batholith, formed as magma cools underground (Chapter 7)

Stomata   Small pores on the surfaces of leaves and other aerial parts of most plants, responsible for regulating gas (and, importantly, water vapor) exchange between the plant and the atmosphere (Chapter 15)

Stone line   A generally horizontal or slightly wavy layer of stones or gravel in a soil (Chapter 13)

Stone run   Another name for a block stream or felsenmeer (Chapter 11)

Stoping   A succession of upward propagation of collapses in bedrock (Chapter 12)

Storm beach profile   A beach profile characterized by a flattened profile, caused by enhanced beach erosion (Chapter 17)

Storm hydrograph   A plot of discharge for a stream that spans the period of time before, during, and after a single, runoff-generating storm (Chapter 15)

Storm surge   The rise in water level that occurs in response to pressure changes and onshore winds during a storm, typically a hurricane, as winds drive water and waves toward coastal areas (Chapter 17)

Stormflow   Enhanced streamflow that occurs due to additions of runoff (Chapter 15)

Straight channel   A stream channel that has a very low sinuosity, i.e., is almost straight, typically forming in areas of low valley gradient (Chapter 16)

Strain   The deformation of a material under the action of applied forces (stresses) (Chapter 10)

Strata   A layer, as in "rock strata" (Chapter 5)

Strath   A rock-floored floodplain formed as a stream cuts laterally into and across bedrock (Chapter 16)

Strath terrace   A strath that has been isolated above the current floodplain, due to incision, leaving the strath terrace as a relict bedrock surface (Chapter 16)

Stratified, Stratification   The characteristic of having distinct layers (Chapters 13, 18, 19)

Stratovolcano   A large, tall and often isolated volcano that is characterized by violent eruptions of viscous, silica-rich lava and pyroclastic materials that may eventually harden into andesite, dacite, scoria, and tuff. Also called a composite cone (Chapter 7)

Stratum   An individual layer of rock or sediment (plural: strata) (Chapters 5, 8, 9)

Stream   A body of water that flows in a channel (Chapter 15)

Stream capacity   *See* capacity (Chapter 15)

Stream competence   *See* competence (Chapter 15)

Stream gage   A device used to measure the height of a stream's water-surface, which is then used to determine and monitor its discharge (Chapter 15)

Stream ordering, Stream order   A way of categorizing stream segments according to their relative position within their overall channel network (Chapter 16)

Stream piracy   The natural process whereby a stream is captured and then diverted from its own channel, into a neighboring (pirating) stream channel (Chapter 9)

Stream power   The measure of the ability of a river to pick up and carry sediment, defined generally as the product of slope (S) and discharge (Q) (Chapters 15, 16)

Stress   The force applied to solid materials (Chapter 10)

Striation, Striae   A linear scratch in a rock surface, caused by glaciers dragging rocks across it (plural: striae) (Chapter 19)

Strike   The compass direction of a line formed by the intersection of a rock layer and a horizontal plane (Chapters 8, 9)

Strike ridge   An elongated ridge that parallels the strike of the bedrock (Chapter 9)

Strike valley   An elongated lowland or valley that is aligned with the strike of the bedrock, typically occurring between strike ridges or escarpments on a cuestaform plain (Chapters 8, 9)

Strike-slip fault, faulting   A type of faulting produced by transform (side-to-side) motion, with little or no vertical motion (Chapter 10)

Strombolian-type eruption   A type of volcanic eruption that looks like a fireworks display, typical of eruptions of ash and other pyroclastic materials at cinder cones (Chapter 7)

Structural polje   A type of polje that develops edges along structural elements in the bedrock, with floors flattened by erosion and alluviation of the underlying, less permeable or less soluble rock (Chapter 12)

Structure   The three-dimensional characteristics of a rock unit or body, typically including its deformational history (Chapters 5, 8)

Stump   A sea stack eroded to a smaller form by wave action (Chapter 17)

Subaerial surface   A surface that is on dry land, i.e., under the air (Chapter 3)

Subaqueous surface   A surface that is overlain by water (Chapter 3)

Subcritical cracking   A physical weathering process, common in deserts, which results from a suite of chemo-physical mechanisms that ultimately break rocks apart. Also called dirt cracking (Chapter 11)

Subglacial environment   The environment that is under or within the basal ice of a glacier (Chapters 18, 19)

Subglacial erosion   The process by which sediment is eroded at the bed of a glacier by ice and/or water, and then transported away (Chapter 18)

Subglacial till   A type of glacial sediment that forms under the glacier, making it dense and "tight," due to compaction from the weight of the ice. Another name for basal till (Chapter 19)

Subglacial water   Water at the interface between the ice and the bed of a glacier (Chapter 18)

Sublimation   The direct loss of ice to vapor, without an intervening liquid (water) stage (Chapters 18, 23)

Subpolar glacier   A glacier whose ice (at depth) is below the pressure-melting point but whose surface warms and melts in the summer. The meltwater percolates down through the snow and ice, and typically then refreezes (Chapter 18)

Suffosion   The downward migration of unconsolidated material through open fissures and pipes in the bedrock (Chapter 12)

Summer beach profile   Another name for a wide, sandy, fair-weather beach profile (Chapter 17)

Summit   The top part of a landscape or hill, and one of the five major slope elements (Chapter 13)

Summit crater   A circular depression or central vent, less than 1.6 km in diameter, near the summit of a volcano, similar to but smaller than a caldera (Chapter 7)

Superposed, Superposition   To be "let down," as in rivers that once flowed on rocks that existed at a higher elevation. Owing to long periods of erosion, the rivers have been let down (superposed) onto the current landscape, such that their pattern and location often bear little relation to the current geologic structure (Chapters 9, 16)

Superposed river   A river that is younger than the geologic structures that it cuts through, developing first in overlying materials and then incising down onto the older, preexisting bedrock structures below by superposition (Chapter 16)

Supply-limited stream   A stream whose sediment load is limited by the amount of sediment that is made available to it. Such streams could transport more sediment if it were available (Chapter 15)

Supply-limited system   A system or landscape in which the limiting factor to sediment transport is the strength and/or ability of the weathering system that produces the sediment. Typical of dry climates and bedrock-influenced landscapes (Chapters 15, 21)

Supraglacial environment   The environment on top of a glacier, i.e., on the glacier's surface (Chapter 18)

Supraglacial till   Glacial sediment that forms and is deposited near the surface of the ice, as the ice melts and releases it (Chapter 19)

Surf zone   The area on a beach immediately landward of the nearshore zone, where the waves often break (Chapter 17)

Surface creep   The process by which grains on a bed are pushed forward by the impact of saltating sand grains (Chapter 21)

Surface exposure dating   A group of dating methods designed to estimate when a particular geomorphic surface last stabilized (Chapter 3)

Surface of rupture   The plane along which a mass movement like a slide will travel (Chapter 14)

Surge phase   A period of increased velocity in surging glaciers that lasts from a few months to a year (Chapter 18)

Surging glacier   A glacier that may have generally been flowing steadily but which is now experiencing a rapid "surge" in velocity, which may last for a few months or more (Chapter 18)

Suspended load   The sediment load of a stream that consists of particles small enough to be held in the water column by turbulence alone (Chapter 15)

Suspension   A type of transport typical for smaller particles, where the particles remain suspended in a fluid (air or water) for long periods of time (Chapters 15, 21)

Swallow hole   A surface opening in the bedrock in karst terrain, into which a river may flow, diverting its water into underground caves and passageways (Chapter 12)

Swarm   Another name for a drumlin field with a large number of drumlins (Chapter 19)

Swash   Water and sediment moving onshore as waves break on the beach (Chapter 17)

Swash zone   The zone on a beach where swash and backwash of waves is ongoing and commonplace (Chapter 17)

Symmetrical fold   A fold with a vertical axial plane, such that both fold limbs appear as near mirror-images, structurally (Chapter 9)

Symmetrical hogback   A bedrock landform developed where the rock has a very steep, or nearly vertical, dip, such that both the dip slope side and the escarpment side have almost the same slope gradient, and are so steep that they may even be exposed as free faces (Chapter 8)

Synclinal ridge   A ridge formed along the axis of a syncline, usually because the rocks there are more resistant to erosion (Chapter 9)

Synclinal valley   A valley formed along the axis of a syncline, developed there by erosion into weaker rocks (Chapter 9)

Syncline   A geologic structure in which rocks at its core have been folded downward (Chapters 1, 9)

Synclinorium   A large syncline whose overall structure contains many smaller folded structures (Chapter 9)

Syntaxis   In geology, the convergence of geologic structures towards a single point (Chapter 6)

Tafone, Tafoni   A cavernous, usually elliptical, weathering feature, with concave inner walls, and which often forms on rocks at the bases of outcrops (plural: tafoni) (Chapter 11)

Talik   A body or layer of perennially unfrozen ground in a region of permafrost (Chapter 20)

Talus   Rock debris that has fallen from a free face or steep, rock slope (Chapters 8, 11, 14)

Talus cone   A deposit of talus whose shape is cone-like, made from falling material from an eroding free face that has valleys cut into it, such that many of the rocks are directed to fall into an area at the base of a valley, forming the head of the cone (Chapters 8, 14)

Talus rock glacier   A rock glacier that develops and forms below, and within, boulder-strewn talus slopes (Chapter 20)

Talus slope   A deposit of talus below a retreating escarpment that develops a shape that is long and continuous along the escarpment front (Chapter 8)

Tarn, Tarn lake   A lake that exists in the bottom of a glacial cirque (Chapter 19)

Tectonic aneurysm   A model of mountain building whereby erosion by deeply incising rivers and glaciers, plus the mass of the mountain, weakens the crust, allowing for an enhanced upward flow of crustal material, leading to accelerated uplift and mountain building (Chapter 6)

Tectonic plate   *See* lithospheric plate (Chapter 7)

Tectonic processes, forces, Tectonism   Processes that deform Earth's crust. Generally speaking, the science of tectonics is concerned with Earth's large-scale structural, geological features (Chapters 1, 6)

Temperate glacier   A glacier that is near or at its pressure-melting point throughout its thickness, except near the surface in the winter (Chapter 18)

Temporal   Pertaining to age, or time (Chapter 2)

Tephra   *See* pyroclastic debris (Chapters 3, 5, 7)

Tephrochronology   A numerical dating technique that uses tephra deposits to determine the chronologies of stratigraphic sediment sequences (Chapter 3)

Terminal moraine   The outermost end moraine formed by a glacier (Chapter 19)

Terminations   The comparatively short, rapid periods of ice sheet retreat and collapse during the Quaternary Period (Chapter 4)

Terra rossa   A reddish-colored, aluminum- and iron-rich, soil and residual material, formed by the dissolution of impure limestone, which leaves behind these residual materials (Chapter 12)

Terracette   A stairstep-like landform formed on steep hillslopes by animals (especially hooved ones) walking across the sloping surface (Chapter 22)

Terrestrial   An adjective referring to locations on land (Chapter 5)

Tessera   Areas of high-standing terrain on Venus, hypothesized to be indicative of crustal deformation and/or geologic folding (Chapter 23)

Textural triangle   A graphical, ternary diagram, with axes for sand, silt, and clay, used to classify sediments into texture classes (Chapter 13)

Texture   A name that describes a soil's relative proportions of sand, silt, and clay (Chapter 13)

Thalweg   The imaginary line that follows the line of maximum depth in a river channel (Chapter 16)

Thaw settlement   The process that occurs as ice within soil melts, allowing the surface to settle and slump down (Chapter 20)

Thaw slump   A type of mass wasting feature in periglacial landscapes, with bowl-shaped vertical headwalls and which also exposes permafrost. The slump headwall retreats as more ice-rich sediments thaw (Chapter 20)

Theory of atoll formation   A theory first proposed by Charles Darwin that suggests atoll formation begins with a dormant or extinct volcanic island (Chapter 17)

Thermal contraction cracking   The contraction, or shrinking in volume, of a substance as its temperature falls, typically below 0° C, forming cracks (Chapter 20)

Thermal stress fatigue   The condition whereby rocks get gradually weaker due to repeated cycles of heating and cooling (Chapter 11)

Thermal weathering   A physical weathering process driven by expansion and contraction of rocks and regolith by solar heating or fire (Chapter 11)

Thermokarst   The subsidence, erosion, and instability of the ground surface resulting from the thawing of ice-rich permafrost and the melting of ground ice (Chapter 20)

Thermokarst gully   A low, water-filled, channel-like feature, commonly incised 1–5 m deep, in areas of thawing permafrost with ice-wedge networks. The gullies may extend for hundreds of meters, providing exit routes for water in ponds and lakes formed by thermal erosion (Chapter 20)

Thermokarst pond   A water-filled area of surface subsidence in periglacial regions, caused by the thawing of permafrost, forming depressions that fill with water (Chapter 20)

Threshold   *See* geomorphic threshold (Chapter 2)

Throughflow   Percolating water in soil, regolith or bedrock that ponds on top of an impeding (impermable) or slowly permeable layer and then flows across it, laterally, through the more permeable material above (Chapter 15)

Thrust fault   A type of reverse formed as one layer of rock is thrust up and over another, by compression, along a low-angled fault plane (Chapters 6, 9, 23)

Thrust faulting   A type of reverse faulting produced by crustal compression, where one rock slides over another along low thrust angles (Chapter 10)

Thrust ramp   A steeply dipping, upwardly off-shooting, fault, connected to deeper, more horizontal thrust faults (Chapter 9)

Thrust sheet   A layer of rocks displaced by thrust faulting, such that overlapping strata become stacked upon each other (Chapter 9)

Tidal channel   A water-filled channel that separates two barrier islands, connecting the backbarrier lagoon to the open ocean. Also called an inlet or tidal inlet (Chapter 17)

Tidal delta   A landform developed as sediment gets deposited by tidal currents flowing through gaps (inlets) in barrier islands, either on the lagoon side (as a flood tidal delta) or the ocean side (as an ebb tidal delta) of the inlet (Chapter 17)

Tidal flat   A low-lying area in a lagoon, behind a barrier island and often exposed at low tide, formed by sediment deposition either by rivers that enter the lagoon or via sediment influx through tidal channels (Chapter 17)

Tidal range   The magnitude of the rise and fall of the tide (Chapter 17)

Tide   The rise and fall of the sea caused by the combined effects of the gravitational forces exerted by the Moon and Sun (but mainly by the Moon) (Chapter 17)

Tide-dominated coast   A coast that typically only occurs on the ocean, at locations where sediment movement and landform development are primarily driven by daily tidal currents (Chapter 17)

Tide-dominated delta   A type of delta on the ocean coast, dominated by tidal flats and tidal marshes, strongly influenced by currents rushing up and down distributary channels, making them noticeably wider at their mouths than they are upstream (Chapter 17)

Tidewater glacier   A glacier that flows into the ocean or a lake, typically causing the outer end to float (Chapter 18)

Till   Poorly sorted, usually unstratified, matrix-supported diamicton deposited by a glacier (Chapters 18, 19)

Till plain   A low-relief, rolling landscape formed on sediment deposited beneath a glacier. Another name for ground moraine (Chapter 19)

Tilted block faulting   A type of normal faulting where blocks of rock rotate and fall back, against each other (Chapter 10)

Time   In soil studies, time (or the age of the soil) is one of the five soil-forming factors (Chapter 13)

Time-transgressive   Term used to imply that something varies in age across its extent (Chapter 3)

Time$_{zero}$ The time when soil formation began at a site, or when a geomorphic surface last stabilized (Chapters 3, 13)

Toe   The end part, or base, of a rotational landslide, or slump (Chapter 14)

Toeslope   The slope element at the base of a slope, where sediments accumulate and wetness is typically greatest (Chapter 13)

Tombolo   A narrow body of sand that connects an offshore island to the mainland (Chapter 17)

Topographic apex   Topographically, the top, or head, of a landform, such as an alluvial fan (Chapter 10)

Topographic inversion   A process where what was once "low ground" becomes "high ground" due to preferential erosion of some areas more than others (Chapters 3, 19)

Topographic relief   *See* relief (Chapter 10)

Topography   The variation in elevation across a geomorphic surface (Chapter 13)

Topple   A type of mass movement where an earthen or rock mass moves and falls via rotation around an axis or point (Chapter 14)

Topset bed   Part of a delta's depositional sequence, a topset bed is a relatively flat-lying layer of sediment, deposited on the top of the delta by distributary systems of the river entering the delta (Chapter 16)

Tor   A large, free-standing, residual mass of isolated rock, displaying rounded forms (Chapters 4, 11)

Tower   A tall, isolated bedrock remnant, usually capped by a layer of harder and more resistant rock (Chapters 8, 11)

Tower karst   Another name for cockpit karst (Chapter 12)

Traction   The part of the bedload of a river that slides and rolls along the channel bed as the water flows downstream (Chapter 15)

Trailing edge   The margin where a continental plate meets an oceanic plate and where that margin is tectonically inactive and thus is not experiencing subduction. Trailing margins typically have shallow water offshore. *See also* Passive plate margin (Chapters 6, 17)

Trampling   The act of repeatedly walking on a surface by animals (Chapter 22)

Transform fault, faulting   Side-by-side motion in rocks and regolith. Another name for strike-slip faulting (Chapters 6, 10)

Transform plate boundary   A boundary between tectonic plates where the plates are sliding past each other laterally, and no subduction is occurring (Chapters 6, 7)

Transformations   As a group of soil processes, transformations include all processes that transform mineral or organic matter into different forms, usually by weathering (Chapter 13)

Transit area   The place through which a mass movement (usually a slide or flow) moves (Chapter 14)

Translational landslide   A landslide where the mass slides along a planar or only slightly undulating plane of failure (Chapter 14)

Translocations   As a group of soil processes, translocations include all processes that relocate substances in the soil, usually by transferring it from one soil horizon to another (Chapter 13)

Transpiration   The active process whereby water, taken up by plants, is then passed through small pores (stomata), into the atmosphere (Chapter 15)

Transport pathway   The route between an erosional source area and a depositional sink (Chapter 21)

Transported regolith   A form of regolith that has been transported to the site from somewhere else by wind, water, ice, or gravity (Chapter 11)

Transport-limited slope   A slope and/or surface where weathering proceeds rapidly enough and erosion is comparatively slower, such that most of the slope has a cover of regolith and soil (Chapters 8, 11, 14)

Transport-limited stream   A stream whose sediment load is limited by its ability or capacity to carry sediment. Essentially, sediment is available for the stream to carry, but it is unable to pick up and transport all of the sediment. Such streams tend toward a graded state (Chapter 15)

Transport-limited system   A system/landscape in which the limiting factor to sediment movement is the strength of the transport vector, e.g., wind or water. Sediment is available for transport but not all of it can be moved, or is being moved (Chapter 21)

Transpression   The combination of compression and shearing in rocks, often associated with crustal compression along strike-slip faults (Chapters 6, 10)

Transtension   Crustal extension that occurs along strike-slip faults (Chapter 6)

Transverse bar   A bar that forms within a river channel, and which is oriented perpendicularly to the downstream flow direction (Chapter 16)

Transverse crevasse   A crevasse or crack in a glacier that forms perpendicularly to the flow direction, usually where the ice is accelerating in the accumulation zone (Chapter 18)

Transverse dune   A dune whose crestline is perpendicular (or transverse) to the dominant wind direction, and which usually has a single, prominent slipface (Chapters 21, 23)

Transverse valley   A river valley that cuts across a geologic structure or structures (Chapter 16)

Trap   A landscape built of multiple layers of lava flows (Chapter 7)

Travertine   A deposit of secondary calcite found in caves (Chapter 7)

Tread   The relatively flat, former floodplain surface that forms the top of an alluvial or strath terrace (Chapter 16)

Tree uprooting   The uprooting of trees by the wind or by overloading of ice and snow, causing the tree to topple and its roots to be torn from the ground (Chapters 14, 22)

Treeline   The upper altitudinal (or polar) limit at which trees can grow, due to increasingly cold temperatures and/or windy conditions (Chapters 20, 22)

Trellis drainage pattern   A type of drainage pattern with long, nearly linear, master streams that have many, shorter tributaries joining at nearly right angles. Common in areas of folded mountain belts, near cuestas, and along beach ridges (Chapters 9, 16)

Triangular facet   The eroded remnants of a fault scarp, set between V-shaped valleys that have dissected the scarp (Chapter 10)

Triassic   A period of the Mesozoic Era, between 251.9 and 201.4 Ma (Chapter 4)

Tributary   A stream in a drainage network that flows into another, usually larger, stream (Chapter 15)

Trough   The lowest point of a wave or series of waves (Chapter 17)

Trunk stream   The largest stream in a drainage network (Chapter 15)

Tsunami, Tunami wave   A fast-moving wave generated by a sudden tectonic displacement of the crust, at or below the sea floor (Chapter 7, 17)

Tufa   A secondary carbonate deposit, like a porous limestone, formed when CO_2 is removed from water saturated with $CaCO_3$, leading to precipitation of the $CaCO_3$ (Chapter 11)

Tufa curtain   A curtain-like feature formed where water saturated with $CaCO_3$ flows out of a rock opening, allowing CO_2 to be removed, leading to precipitation of the $CaCO_3$ (Chapter 11)

Tufa tower   Tall deposits of secondary carbonate that precipitate under the water surface of lakes, typically as a result of biological or chemical processes, or both (Chapter 10)

Tuff   A low density, light-colored, porous rock formed as volcanic ash cools and lithifies (Chapter 7)

Tundra   A plant community in a climate too cold for trees to grow, and so is dominated by low-growing herbs, mosses, lichens, and low shrubs (Chapter 20)

Tunnel channel   A long, narrow valley formed as pressurized meltwater flows subglacially, eroding into the ice and down into the bed below the glacier. Also called a tunnel valley (Chapter 19)

Turbulence   The chaotic motion of flow lines in a stream or river (Chapter 15)

Ultisol   A type of reddish, acidic, and highly weathered soil, common to the humid subtropics and typically rich in kaolinite clays (Chapter 13)

Ultramafic   A category of rock-forming minerals that have very little silica and are very rich in iron and magnesium (Chapter 5)

Undercutting   The process where material is removed from the base or foot of a slope, usually through erosion or excavation (Chapter 14)

Undersaturated water   In karst, undersaturation refers to water that is capable of dissolving more CO_2, if it is available (Chapter 12)

Uniformitarianism   The geologic principle that states that the same natural laws and processes that operate today have operated in the past, or put another way, the present is the key to the past (Chapter 2, 13, 14, 19)

Unloading The release of weight from rocks and surfaces as weathering and erosion strip away material, driving the formation of expansion fractures or cracks in the rock (Chapters 6, 11, 14)

Unpaired terraces Fluvial terraces that are often of the same age but which occur at unequal elevations on each side of a river valley (Chapter 16)

Unsaturated zone A zone in soil or bedrock, usually just below the surface, where only some of the pores are filled with water and the rest are filled with air. Also called the vadose zone (Chapter 15)

U-shaped valley An alpine valley that is U-shaped in cross-section, due to extensive glaciation at some time in the past (Chapter 19)

Uvala In a karst landscape, a lowland formed as two or more dolines coalesce and deepen, intersecting each other in complex morphologies (Chapter 12)

Vadose zone *See* unsaturated zone (Chapter 12)

Valley glacier A mountain glacier confined within a former stream valley (Chapters 18, 19, 23)

Varve A couplet of thin layers of sediment that have formed in a glacial lake, typically one pair per year. One layer forms in the summer, and one in winter (Chapter 19)

Vein A small, sometime millimeters thick, feature formed as magma or mineralized fluids invade very small conduits, fractures, and cracks in rock (Chapters 5, 7)

Vent An opening in the surface from which lava and pyroclastic material are erupted (Chapter 7)

Ventifact A rock eroded by abrasion by blowing sand grains, typically with several smooth, faceted surfaces (Chapters 8, 21, 23)

Vertical accretion The process by which sediment is added to floodplains during floods, as rivers overtop their banks and deposit sediment on the flooded areas (Chapter 16)

Vesicular The condition of being full of vesicles (small pores), typical of the A horizons of many desert soils (Chapter 13, 21)

Viscosity A measure of a fluid's resistance to flow (Chapter 7)

Volcanic arc A string of volcanic islands formed where two oceanic plates are converging, such that one is subducted below the other (Chapter 7)

Volcanic ash A type of pyroclastic debris composed of tiny shards of volcanic glass and ground-up rock, often erupted along with sulfur-rich gases (Chapters 3, 5, 7)

Volcanic tremor A long-duration, more or less continuous, volcanic vibration, associated with activities such as the movement of underground magma, oscillations in magma reservoirs, and eruptions of volcanic gases (Chapter 7)

Volcanism The process by which molten rock and associated gases erupt onto the surface, forming volcanoes and related features. Sometimes spelled "vulcanism" (Chapters 5, 7)

Volcano A hill or mountain composed wholly or in part of hardened lava and/or pyroclastic debris that has been forcefully ejected from the subsurface (Chapter 7)

Wallow A surface depression excavated by an animal for the purpose of rolling in mud or dust (Chapter 22)

Wash load The smallest particles of the suspended load of a stream, generally clay- and silt-sized sediment. Wash load grains have such a small settling velocity that they can be kept in suspension by flow turbulence alone (Chapter 15)

Washover fan A low, sandy landform that develops on the lagoon side of a barrier island due to washover of sand by waves during large storms (Chapter 17)

Water gap A gap or valley through a bedrock ridge that contains an active, through-flowing river or stream (Chapters 9, 16)

Water table The top of the saturated zone in the subsurface (Chapters 12, 15)

Water track A narrow channel formed as water flows across the land surface in a polar desert, above the permafrost (Chapter 20)

Watershed *See* drainage basin (Chapter 15)

Wave amplitude *See* wave height (Chapter 17)

Wave base The depth in a water body that is generally considered to be equal to half the wavelength of the waves above. Below this point, i.e., in water deeper than the wave base, waves do not typically interact with the bed and therefore, do not transport sediment (Chapter 17, 19)

Wave height The difference in height (elevation) between the top of a wave crest and its adjacent trough (Chapter 17)

Wave period The time required for successive waves to pass by a given point (Chapter 17)

Wave refraction The process whereby the wave fronts in a lake or water body bend and converge onto headlands as they approach the coast (Chapter 17)

Wave-cut bench *See* wave-cut platform (Chapter 17)

Wave-cut notch An undercut slope on a beach, formed by erosion of waves at the bottom part of the slope (Chapter 20)

Wave-cut platform A terrace-like surface that forms at or just above the waterline, as waves break on the shore, dragging particles back-and-forth, abrading and forming the surface (Chapter 17)

Wave-dominated coast A coastal landscape whose geomorphology has been formed mainly by the action of persistent waves and which has low tidal ranges (Chapter 17)

Wave-dominated delta A type of delta shaped by strong waves and currents, which redistribute sediment along the delta front, causing the delta to develop a wide, broad shape, or a symmetrical arcuate shape (Chapter 17)

Wavelength In coastal geomorphology, the horizontal distance between wave crests, typically applied to water waves. In fluvial geomorphology, the term refers to the length (down-valley) from one meander bend to the next, along the same side of the river (Chapters 16, 17)

Weathering The *in-situ* breakdown of rocks and minerals, chemically and/or physically (Chapter 5, 8, 11)

Weathering basin Another name for a (slightly larger) weathering pit (Chapter 11)

Weathering front The lowest detectable limit of weathering activity in bedrock and regolith at the land surface (Chapter 11)

Weathering pit A small, closed depression formed by differential weathering and erosion on horizontal or gently sloping rock surfaces, commonly affecting granite, sandstone, and limestone (Chapter 11)

Weathering profile All the various zones and layers within the regolith, from the land surface down to the weathering front, that have been affected by physical, chemical, and biological weathering processes (Chapter 11)

Weathering rind A discolored ring or crust that develops on rocks due to weathering, which commonly correlates to surface exposure of the surface on which the rocks rest (Chapters 3, 11)

Weathering-limited slope   A slope or surface on which the agents of erosion and denudation outpace the agents of weathering, and thus the slope tends to be dominated by bare bedrock, or which may have only a thin cover of rego-lith (Chapters 8, 11, 14)

Wedge ice   A wedge-shaped, downward-tapering body of ice, formed just below the surface often containing vertical bands or foliations, and is typically found in periglacial landscapes (Chapter 20)

Wet snow zone   The area on top of a glacier or ice sheet where snow experiences melting during the summer (Chapter 18)

Wetting and drying   A type of physical weathering, driven by periods of wetting and drying, leading to slaking, granular disintegration, cracking, and flaking of rock (Chapter 11)

Whaleback   A smooth, streamlined, bedrock landform, shaped mainly by abrasion as glacial ice slides across bedrock (Chapter 19)

Width/depth ratio   The ratio of a stream's width (at the top of the water surface) to its mean depth (Chapter 16)

Wind gap   A gap or valley through a bedrock ridge that lacks an active, through-flowing river or stream. Most wind gaps are old water gaps that lost their stream, and have thus been left high and dry (Chapters 9, 16)

Wind regime   Characteristics of the wind in an area, including its speed, duration, and direction (Chapter 21)

Wind speed   Refers to the velocity, not the direction, of the wind (Chapter 21)

Wind vane   A tool used to measure wind direction, usually only in the horizontal dimension, relative to north (Chapter 21)

Wind velocity   Refers to the simultaneous wind velocity and direction (Chapter 21)

Window   A gap or opening in a fin (Chapter 8)

Wineglass valley (canyon)   A narrow, slot-like canyon that forms at the mouths of uplifted range fronts by fluvial incision of the range (Chapter 10)

Winter beach profile   Another name for a storm beach profile (Chapter 17)

Wolman–Miller Principle   The magnitude–frequency relationship that posits that, over a period of years, the work of a river is the product of the amount of sediment transported by a given flood (or flow), multiplied by the frequency of that size of flood (or flow) (Chapter 16)

Wrack   Organic debris, like wood, seaweed, and leaves, deposited on the backshore of a beach by waves or the tide (Chapter 17)

Wrinkle ridge   A ridge on the surface of a planetary body, formed by compressional tectonic forces (Chapter 23)

Xenolith   A fragment of a rock trapped within another rock (Chapter 7)

Yardang   A streamlined ridge, aligned parallel to the dominant wind direction and formed by eolian abrasion of bedrock or fine-grained, weakly consolidated sediment (Chapters 21, 23)

Yazoo stream   A stream that flows on a floodplain, paralleling the main channel for some distance because it is unable to cross the natural levees that flank the main channel (Chapter 16)

Yedoma   Organic-rich, frozen, loess-like sediments (Chapter 4)

Younger Dryas   A climatic event that represented a brief return to glacial conditions over the northern hemisphere during the general warming that had otherwise characterized the Holocene (Chapter 4)

Youth   The first stage of W. M. Davis' cycle of erosion, in which a low, flat landscape is uplifted, enabling streams to erode deep, widely spaced canyons, as they attempt to reach base level (Chapter 2)

Zoogeomorphology   The science that studies the role of animals in geomorphology (Chapter 22)

References

Acworth, R. I. 1987. The development of crystalline basement aquifers in a tropical environment. *Quart. J. Engineer. Geol.* 20:265–272.

Anderson, R. S. and Anderson, S. P. 2010. *Geomorphology: The Mechanics and Chemistry of Landscapes.* Cambridge University Press.

Anderton, J. B. and Loope, W. L. 1995. Buried soils in a perched dunefield as indicators of Late Holocene lake-level change in the Lake Superior basin. *Quat. Res.* 44:190–199.

Ballantyne, C. K. 2018. *Periglacial Geomorphology.* Wiley.

Benn, D. I. and Evans, D. J. A. 2010. *Glaciers and Glaciation.* 2nd ed. Hodder.

Beschel, R. E. 1950. Flechten als Altersmasstab rezenter moränen. *Gletscherkd. Glazialgeol.* 1:152–161.

Blewett, W., Lusch, D., and Schaetzl, R. J. 2009. The physical landscape. In: Schaetzl, R. J., Darden, J. T., and Brandt, D. (eds.) *Michigan Geography and Geology.* Pearson Custom Publishing. pp. 249–273.

Bloom, A. L. 1969. *The Surface of the Earth.* Foundations in Earth Science Series. Prentice-Hall.

Bloom, A. L. 1998. *Geomorphology A Systematic Analysis of Late Cenozoic Landforms.* 3rd ed. Prentice-Hall.

Brown, R. J. E. 1970. *Permafrost in Canada: Its Influence on Northern Development.* University of Toronto Press.

Bruthans, J., Soukup, J., Schweigstillová, J., Vaculiková, J., Smutek, D., Mayo, A. L., and Falteisek, L. 2013. Origin of "rock cities", pillars and cleft-conduits in kaolinite-bonded sandstone: New insight from study in sandstone quarry where landforms recently evolve. *Proceedings, 2013 Intl. Congress of Speleology.* Brno, Czech Republic. pp. 247–252.

Burn, C. R. 2012. Permafrost distribution and stability. In: French, H. M. and Slaymaker, O. (eds.) *Changing Cold Environments: A Canadian Perspective.* Wiley. pp. 126–146.

Chakraborty, T., Taral, S., More, S., and Bera, S. 2020. Cenozoic Himalayan Foreland Basin: An overview and regional perspective of the evolving sedimentary succession. In: Gupta, N. and Tandon, S. (eds.) *Geodynamics of the Indian Plate.* Springer Geology. Springer. https://doi.org/10.1007/978-3-030-15989-4_11

Chepil, W. S. 1945. Dynamics of wind erosion. III: The transport capacity of the wind. *Soil Sci.* 60:475–480.

Chinn, T. J. H. 1981. Use of rock weathering-rind thickness for Holocene absolute age-dating in New Zealand. *Arctic Alpine Res.* 13:33–45.

Chorley, R. J. and Beckinsale, R. P. 1980. G.K. Gilbert's geomorphology. *Geol. Soc. Am. Spec. Paper* 183:129–142.

Chorley, R. J. and Kennedy, B. A. 1971. *Physical Geography: A Systems Approach.* Prentice-Hall.

Chorley, R. J., Schumm, S. A., and Sugden, D. E. 1984. *Geomorphology.* Cambridge University Press.

Christiansen, E. H. and Hamblin, W.K. 2015. *Dynamic Earth.* Jones & Bartlett Learning.

Cílek, V., Williams, R., Osborne, A., Migoń, P., and Mikuláš, R. 2007. The origin and development of sandstone landforms. In: Härtel, H., Cílek, V., Herben, T., Jackson, A., and Williams, R. (eds.), *Sandstone Landscapes.* Academia Press. pp. 34–43.

Clayton, L., Attig, J. W., Ham, N. R., Johnson, M. D., Jennings, C. E., and Syverson, K. M. 2008. Ice walled lake plains: Implications for the origin of hummocky glacial topography in middle North America. *Geomorphology* 97:237–248.

Cruikshank, D. P., Umurhan, O. M., Beyer, R. A. et al. 2019. Recent cryovolcanism in Virgil Fossae on Pluto. *Icarus* 330:155–168.

Dalrymple, R. W., Zaitlin, B. A., and Boyd, R. 1992. Estuarine facies models: Conceptual basis and stratigraphic implications. *J. Sediment. Res.* 62:1130–1146.

Davis, W. M. 1932. Piedmont benchlands and Primärrümpfe. *Geol. Soc. Am. Bull.* 43:399–440. https://doi.org/10.1130/GSAB-43-399

Delcourt, P. A ., Nester, P. L., Delcourt, H. R., Mora, C. I., and Orvis, K. H. 2002. Holocene lake-effect precipitation in northern Michigan. *Quat. Res.* 57:225–233. https://doi.org/10.1006/qres.2001.2308

Dorronsoro, C. and Alonso, P. 1994. Chronosequence in Almar River fluvial-terrace soil. *Soil Sci. Soc. Am. J.* 58:910–925.

Douglas, T. A. and Mellon, M. T. 2019. Sublimation of terrestrial permafrost and the implications for ice-loss processes on Mars. *Nature Comm.* 10. Article #1716.

Earle, S. 2019. *Physical Geology.* 2nd ed. Open Textbook under a Creative Commons Attribution 4.0 Intl. License. https://opentextbc.ca/physicalgeology2ed/

Epstein, J. B. 1966. Structural control of wind gaps and water gaps and of stream capture in the Stroudsburg area, Pennsylvania and New Jersey. US Geol. Surv. Prof. Paper 550A.

Fick, S. E. and Hijmans, R. J. 2017. WorldClim 2: New 1-km spatial resolution climate surfaces for global land areas. *Intl. J. Climatol.* 37:4302–4315.

Fossen, H. 2010. *Structural Geology.* Cambridge University Press.

Frankel, K. L. and Owen, L. A. 2013. Transform plate margins and strike-slip fault systems. In: Owen, L. A. (ed.) *Treatise in Geomorphology, vol. 5: Tectonic Geomorphology.* Academic Press. pp. 37–70.

French, H. M. 2007. *The Periglacial Environment.* 3rd ed. Wiley.

Frisch, W., Meschede, M., and Blakey, R. C. 2011. *Plate Tectonics: Continental Drift and Mountain Building.* Springer.

Gannett, H. 1901. Profiles of rivers in the United States. US Geol. Surv. Water Supply Paper 44.

Gilbert, G. K. 1877. *Geology of the Henry Mountains.* US Government Printing Office.

Goudie, A. S. and Viles, H. A. 2010. *Landscapes and Geomorphology: A Very Short Introduction.* Oxford University Press.

Graly, J. A., Bierman, P. R., Reusser, L. J., and Pavich, M. J. 2010. Meteoric ^{10}Be in soil profiles – a global meta-analysis. *Geochim. et Cosmochim. Acta* 74:6814–6829. https://doi.org/10.1016/j.gca.2010.08.036

Gregory, K. J. (ed.). 1977. *River Channel Changes.* Wiley.

Gupta, S., Collier, J. S., Garcia-Moreno, D., Oggioni, F., Trentesaux, A., Vanneste, K., De Batist, M., Camelbeeck, T., Potter, G., Van Vliet-Lanoë, B., and Arthur, J. C. R. 2017. Two-stage opening of the Dover Strait and the origin of island Britain. *Nature Comm.* 8. https://doi.org/10.1038/ncomms15101

Hack, J. T. 1941. Dunes of the western Navajo Country. *Geog. Rev.* 31:240–263.

Hack, J. T. 1957. Studies of longitudinal stream profiles in Virginia and Maryland. US Geol. Surv. Prof. Paper 294B.

Halfen, A. F., Lancaster, N., and Wolfe, S. 2016. Interpretations and common challenges of aeolian records from North American dune fields. *Quat. Intl.* 410:75–95.

Hall, K. and Lamont, N. 2003. Zoogeomorphology in the Alpine: some observations on abiotic–biotic interactions. *Geomorphology* 55:219–234.

Hallet, B. and Prestrud, S. 1986. Dynamics of periglacial sorted circles in western Spitsbergen. *Quat. Res.* 26:81–99.

Hamblin, W. K. 1978. *The Earth's Dynamic Systems*. Burgess.

Hammond, W. C., Blewitt, G., and Kreemer, C. 2016. GPS Imaging of vertical land motion in California and Nevada: implications for Sierra Nevada uplift. *J. Geophys. Res. Solid Earth* 121:7681–7703.

Hansel, A. K. and Johnson, W. H. 1996. Wedron and Mason Groups: lithostratigraphic reclassification of deposits of the Wisconsin Episode, Lake Michigan Lobe area. *Ill. State Geol. Surv. Bull.* 104.

Heiken, G. H., Vaniman, D. T., and French, B. M. (eds.) 1991. *Lunar Sourcebook: A User's Guide to the Moon*. Cambridge University Press.

Hesp, P. 2002. Foredunes and blowouts: initiation, geomorphology and dynamics. *Geomorphology* 48:245–268.

Horton, R. E. 1945. Erosional development of streams and their drainage basins: hydrophysical approach to quantitative morphology. *Geol. Soc. Am. Bull.* 56:275–370. https://doi.org/10.1130/0016-7606(1945)56[275:EDOSAT]2.0.CO;2

Hu, F., Yang, X., and Li, H. 2019. Origin and morphology of barchan and linear clay dunes in the Shuhongtu Basin, Alashan Plateau, China. *Geomorphology* 339:114–126.

Huff, W. and Owen, L. A. 2013. Volcanic landforms and hazards. In: Owen, L. A. (ed.) *Treatise in Geomorphology, vol. 5: Tectonic Geomorphology*. Academic Press. pp. 148–192.

IPCC, 2021. Summary for policymakers. In: *Climate Change 2021: The Physical Science Basis*. Contribution of Working Group I to the Sixth Assessment Report of the Intergovernmental Panel on Climate Change [Masson-Delmotte, V., P. Zhai, A. Pirani, S. L. Connors, C. Péan, S. Berger, N. Caud, Y. Chen, L. Goldfarb, M. I. Gomis, M. Huang, K. Leitzell, E. Lonnoy, J. B. R. Matthews, T. K. Maycock, T. Waterfield, O. Yelekçi, R. Yu, and B. Zhou (eds.)]. Cambridge University Press, pp. 3–32. https://doi.org/10.1017/9781009157896.001

Jackl, M. 2013. *Bedrock Geology Map of Delaware Water Gap*. US National Park Service Geologic Resource Inventory.

Jacobsen, R. E. and Burr, D. M. 2017. Dichotomies in the fluvial and alluvial fan deposits of the Aeolis Dorsa, Mars: implications for weathered sediment and paleoclimate. *Geosphere* 13:2154–2168.

James, L.A. 2018. Ten conceptual models of large-scale legacy sedimentation – a review. *Geomorphology* 317:199–217.

Kamp, U. and Owen, L. A. 2013. Polygenetic landscapes. In: Owen, L. A. (ed.) *Treatise in Geomorphology, vol. 5: Tectonic Geomorphology*. Academic Press. pp. 370–393.

Knight, J. 2019. Wind erosion. In: Livingstone, I. and Warren, A. (eds.) *Aeolian Geomorphology: A New Introduction*. Wiley. pp. 61–80.

Lachenbruch, A. H. 1962. Mechanics of thermal contraction cracks and ice-wedge polygons in permafrost. *Geol. Soc. Am. Spec. Publ.* 70. 69pp.

Lancaster, N. 1995. *Geomorphology of Desert Dunes*. Routledge.

Lebold, J. G. and Wilkinson, C. 2018. *Roadside Geology of West Virginia*. Mountain Press.

Lichter, J. 1998. Rates of weathering and chemical depletion in soils across a chronosequence of Lake Michigan sand dunes. *Geoderma* 85:255–282.

Linton, D. L. 1955. The problem of tors. *Geog. J.* 121:470–486.

Lisiecki, L. E. and Raymo, M. E. 2005. A Pliocene-Pleistocene stack of 57 globally distributed benthic δ18O records. *Paleooceanography and Paleoclimatology*. https://doi.org/10.1029/2004PA001071

Lobeck, A. K. 1939. *Geomorphology: An Introduction to the Study of Landscapes*. McGraw-Hill.

Mabbutt, J. A. 1977. *Desert Landforms*. MIT Press.

Maher, B. A. 2016. Palaeoclimatic records of the loess/palaeosol sequences of the Chinese Loess Plateau. *Quat. Sci. Revs.* 154:23–84. https://doi.org/10.1016/j.quascirev.2016.08.004

Makaske, B. 2001. Anastomosing rivers: a review of their classification, origin and sedimentary products. *Earth-Science Revs.* 53:149–196.

Marshak, S. 2019. *Earth Portrait of a Planet*. 6th ed. Norton.

Mejía, A. I. and Niemann, J. D. 2008. Identification and characterization of dendritic, parallel, pinnate, rectangular, and trellis networks based on deviations from planform self-similarity. *J. Geophys. Res.* 113, F02015. https://doi.org/10.1029/2007JF000781

Mellor, A. 1985. Soil chronosequences on Neoglacial moraine ridges, Jostedalsbreen and Jotunheimen, southern Norway: a quantitative pedogenic approach. In: Richards, K. S., Arnett, R. R., and Ellis, S. (eds.) *Geomorphology and Soils*. Allen and Unwin. pp. 289–308.

Meng, A. A. III and Harsh, J. F. 1988. Hydrogeologic framework of the Virginia Coastal Plain. US Geol. Surv. Prof. Paper 1404-C.

Migoń, P., Duszyński, F., and Goudie, A. 2017. Rock cities and ruiniform relief: Forms–processes–terminology. *Earth-Science Revs.* 171:78–104.

Miller, B. A. and Schaetzl, R. J. 2015. Digital classification of hillslope position. *Soil Sci. Soc. Am. J.* 79:132–145.

Moberg, A., Sonechkin, D. M., Holmgren, K., Datsenko, N. M., and Karlén, W. 2005. Highly variable northern hemisphere temperatures reconstructed from low- and high-resolution proxy data. *Nature* 433:613–617. https://doi.org/10.1038/nature03265

Moore, J. M. and Howard, A. D. 2005. Large alluvial fans on Mars. *J. Geophys. Res.* 110. E04005.

Muhs, D. R., Bettis, E.A. III, Roberts, H. M., Harlan, S. S., Paces, J. B., and Reynolds, R. L. 2013. Chronology and provenance of last-glacial (Peoria) loess in western Iowa and paleoclimatic implications. *Quat. Res.* 80:468–481.

Muhs, D. R., Prins, M. A., and Machalett, B. 2014. Loess as a Quaternary paleoenvironmental indicator. *Pages Magazine* 22:84–85.

Murck, B. W. and Skinner, B. J. 1999. *Geology Today Understanding our Planet*. Wiley.

Nye, J. F. 1965. The flow of a glacier in a channel of rectangular, elliptic or parabolic cross-section. *J. Glaciol.* 5:661–690.

Oberlander, T. 1965. *The Zagros Streams*. Syracuse Geographical Series No. 1. Syracuse University Press.

Oldroyd, D. R. 2013. Geomorphology in the first half of the twentieth century. In: Orme, A. R. and Sack, D. (eds.) *Treatise in Geomorphology, vol. 1: The Foundations of Geomorphology*. Academic Press. pp. 64–85.

Owen, L. A. 2004. Cenozoic evolution of global mountain systems. In: Owens, P. N. and Slaymaker, O. (eds.) *Mountain Geomorphology*. Edward Arnold. pp. 33–58.

Owens, P. N. and Slaymaker, O. (eds.) 2004. *Mountain Geomorphology*. Edward Arnold.

Palmer, J. 2009. Cave exploration as a guide to geologic research in the Appalachians. *J. Cave and Karst Studies* 71:180–192.

Palmer, J. and Neilson, R. A. 1962. The origin of granite tors on Dartmoor, Devonshire. *Proc. Yorkshire Geol. Soc.* 33:315–340.

Patterson, W. S. B. 1994. *The Physics of Glaciers*. 3rd ed. Pergamon.

Peterson, F. F. 1981. Landforms of the Basin and Range Province. Nevada Agric. Exp. Station Tech. Bull. 28.

Porter, S. C. 2001. Chinese loess record of monsoon climate during the last glacial-interglacial cycle. *Earth-Sci. Revs.* 54:115–128. https://doi.org/10.1016/S0012-8252(01)00043-5

Powell, J. W. 1875. *Exploration of the Colorado River of the West and its tributaries: Explored in 1869, 1870, 1871, and 1872, under the direction of the Secretary of the Smithsonian Institution*. Geological Survey unnumbered monograph. US Government Printing Office.

Pye, K. (1987) *Aeolian Dust and Dust Deposits*. Academic Press.

Ray, R. G. and Fischer, W. A. 1960. Quantitative photography – a geologic research tool. *Photogramm. Engr.* 25:143–150.

Raymond, C. F. 1971. Flow in a transverse section of Athabasca Glacier, Alberta, Canada. *J. Glaciol.* 10:55–84.

Ritchie, A. M. 1963. Evaluation of rockfall and its control. In: *Stability of Rock Slopes*. Highway Research Record 17:13–28.

Ritter, D. F., Kochel, R. C., and Miller, J. R. 2011. *Process Geomorphology*. 5th ed. Waveland.

Romans, B. W., Castelltort, S., Covault, J. A., and Walsh, J. P. 2016. Environmental signal propagation in sedimentary systems across timescales. *Earth-Sci. Rev.* 153:7–29. http://dx.doi.org/10.1016/j.earscirev.2015.07.01

Ruhe, R. V. 1956. Geomorphic surfaces and the nature of soils. *Soil* 82:441–455.

Šamonil, P., Valtera, M., Schaetzl, R.J., Dusan, A., Vašíčková, I., Daněk, P., Janík, D., and Tejnecký, V. 2016. Impacts of old, comparatively stable, treethrow microtopography on soils and forest dynamics in the northern hardwoods of Michigan, USA. *Catena* 140:55–65.

Sauer, D., Schellmann, G., and Stahr, K. 2007. A soil chronosequence in the semi-arid environment of Patagonia (Argentina). *Catena* 71:382–393.

Schaetzl, R. J. 1994. Changes in O horizon mass,thickness and carbon content following fire in northern hardwood stands. *Vegetatio* 115:41–50.

Schaetzl, R. J. and Thompson, M. L. 2015. *Soils: Genesis and Geomorphology*. 2nd ed. Cambridge University Press.

Schumm, S. A. 1960. The shape of alluvial channels in relation to sediment type. *USGS Prof. Paper* 352B: 17–30.

Schumm, S. A. 1977. *The Fluvial System*. Wiley.

Schumm, S. A. 1981. Evolution and response of the fluvial system, sedimentologic implications. In: Ethridge, F. G. and Flores, R. M. (eds.) *Recent and Ancient Nonmarine Depositional Environments: Models for Exploration*. SEPM Spec. Publ. 31:19–29.

Schumm, S. A. and Lichty, R. W. 1965. Time, space and causality in geomorphology. *Am. J. Sci.* 263:110–119.

Selby, M. J. 1985. *Earth's Changing Surface*. Clarendon.

Seppälä, M. 2011. Synthesis of studies of palsa formation underling the importance of local environmental and physical characteristics. *Quat. Res.* 75:366–370.

Seton, M., Miller, R. D., Zahirovic, S., Williams, S., Wright, N., Cannon, J., Whittaker, J., Matthews, K., and McGirr, R. 2020. A global dataset of present-day oceanic crustal age and seafloor spreading parameters. *Geochem. Geophys. Geosyst.* https://doi.org/10.1029/2020GC009214

Shiklomanov, I. 1993. World fresh water resources. In: Gleck, P. (ed.) *Water in Crisis: A Guide to the World's Fresh Water Resources*. US Geological Survey. Available at: www.usgs.gov/special-topic/water-science-school

Simon, A. and Rinaldi, M. 2006. Disturbance, stream incision and channel evolution: the roles of excess transport capacity and boundary materials in controlling channel response. *Geomorphology* 79:361–383.

Singer, K. N., McKinnon, W. B., Schenk, P. M., and Moore, J. M. 2012. Massive ice avalanches on Iapetus mobilized by friction reducing during flash heating. *Nature Geosci.* 5:574–578.

Stoffel, M., Conus, D., Grichting, M. A., Lièvre, I., and Maître, G. 2008. Unraveling the patterns of late Holocene debris-flow activity on a cone in the Swiss Alps: chronology, environment and implications for the future. *Global Planet. Change* 60:222–234.

Stoffel, M., Schneuwly, D., Bollschweiler, M., Lièvre, I., Delaloye, R., Myint, M., and Monbaron, M. 2005. Analyzing rockfall activity (1600–2002) in a protection forest – a case study using dendrogeomorphology. *Geomorphology* 68:224–241.

Strahler, A. N. 1965. *Introduction to Physical Geography*. Wiley.

Strahler, A. H. and Strahler, A. N. 1992. *Modern Physical Geography*. 4th ed. Von Hoffman.

Summerfield, M. 1991. *Global Geomorphology*. Routledge.

Thomas, D. S. G. 2011. Aeolian landscapes and bedforms. In: Thomas, D. S. G. (ed.) *Arid Zone Geomorphology: Process, Form and Change in Drylands*. 3rd ed. Wiley. pp. 427–453.

Trenberth, K. E., Smith, L., Qian, T., Dai, A., and Fasullo, J. 2007. Estimates of the global water budget and its annual cycle using observational and model data. *J. Hydrometeorol.* 8:758–769.

Trewartha, G. T., Robinson, A. H., and Hammond, E. H. 1967. *Physical Elements of Geography*. 5th ed. McGraw-Hill.

Tucker, G. E. and Slingerland, R. L. 1994. Erosional dynamics, flexural isostasy, and long-lived escarpments: a numerical model. *J. Geophys. Res.* 99(B6):12,229–12,243.

Valvasor, J. W. 1689. *Die Ehre des Hertzogthums Crain*. 4 Vols, Endter, Ljubljana.

van Andel, T. H. and Murphy, J. B. 2024. Plate tectonics. In *Encyclopedia Britannica*, 5 December, www.britannica.com/science/plate-tectonics.

Van Nest, J. 2002. The good earthworm: How natural processes preserve upland Archaic archaeological sites of western Illinois, U.S.A. *Geoarchaeol.* 17:53–90.

Vincent, K. R., Bull, W. B., and Chadwick, O. A. 1994. Construction of a soil chronosequence using the thickness of pedogenic carbonate coatings. *J. Geol. Ed.* 42(4): 316–324.

Voigt, J. R. C. and Hamilton, C. W. 2018. Investigating the volcanic versus aqueous origin of the surficial deposits in Eastern Elysium Planitia, Mars. *Icarus* 309:389–410.

Wallace, R. E. 1978. Geometry and rates of change of fault-related fronts, north-central Nevada. *J. Res. US Geol. Surv.* 6:637–649.

Washington Division of Geology and Earth Resources. 2008. Cascadia Deep Earthquakes. Open File Rept. 2008-1.

Wesson, R., Helley, E., LaJoie, K., and Wentworth, C. 1975. Faults and future earthquakes. In: *Studies for Seismic Zonation of the San Francisco Bay Region.* US Geol. Surv. Prof. Paper 941:A5–A30.

Westerhold, T., Marwan, N., Drury, A. J., Liebrand, D., Agnini, C., Anagnostou, E., Barnet, J. S. K., Bohaty, S. M., DeVleeschouwer, D., Florindo, F., Frederichs, T., Hodell, D. A., Holbourne, A. E., Kroon, D., Lauretano, V., Littler, K., Lourens, L. J., Lyle, M., Pälike, H., Röhl, U., Tian, J., Wilkens, R. H., Wilson, P.A., and Zachos, J. C. 2020. An astronomically dated record of Earth's climate and its predictability over the last 66 million years. *Science* 369:1383–1387.

Whitlock, C., Bartlein, P. J., and Watts, W. A. 1993. Vegetation history at Elk Lake. In: Bradbury, J. P. and Dean, W. E. (eds.) *Elk Lake, Minnesota: Evidence for Rapid Climatic Change in the North-Central United States.* Geol. Soc. Am. Spec. Paper 276:251–274. https://doi.org/10.1130/SPE276

Wolfe, S. A. and Nickling, W. G. 1993. The protective role of sparse vegetation in wind erosion. *Prog. Phys. Geog.* 17:50–68.

Wolman, M. G. 1955. The natural channel of Brandywine Creek, Pennsylvania. US Geol. Surv. Prof. Paper 271.

Wolman, M. G. and Miller, J. P. 1960. Magnitude and frequency of forces in geomorphic processes. *J. Geol.* 68:54–74.

Wyllie, P. J. 1976. *The Way the Earth Works: An Introduction to the New Global Geology and its Revolutionary Development.* Wiley.

Young, A. 1972. *Slopes.* Oliver and Boyd.

Zeitler, P. K., Meltzer, A. S., Koons, P. O., Craw, D., Hallet, B., Chamberlain, C.P., Kidd, W. F., Park, S.K., Seeber, L., Bishop, M., and Shroder, J. 2001. Erosion, Himalayan geodynamics and the geomorphology of metamorphism. *Geol. Soc. Am. Today* 11:4–8.

Zemp, M., Gärtner-Roer, I., Nussbaumer, S. U., Bannwart, J., Rastner, P., Paul, F., and Hoelzle, M. (eds.). 2020. *Global Glacier Change Bulletin No. 3 (2016–2017).* ISC(WDS)/IUGG(IACS)/UNEP/UNESCO/WMO, World Glacier Monitoring Service.

Zonneveld, J. I. S. 1975. River terraces and Quaternary chronology in the Netherlands. *Geologie en Mijn.* 19:277–285.

Index